L. F.

THE TOTAL SYNTHESIS
OF NATURAL PRODUCTS

The Total Synthesis
of Natural Products

VOLUME 2

Edited by
John ApSimon

Department of Chemistry
Carleton University,
Ottawa

A WILEY-INTERSCIENCE PUBLICATION

JOHN WILEY & SONS, New York · London · Sydney · Toronto

Library of Congress Cataloging in Publication Data:

ApSimon, John.
  The total synthesis of natural products.
  Includes bibliographical references.

  1. Chemistry, Organic—Synthesis. I. Title.

QD262.A68   547′.2   72-4075
ISBN 0-471-03252-2 (V.2)

Printed in the United States of America

10 9 8 7 6 5 4 3 2 1

Contributors
to Volume 2

J. W. ApSimon, Carleton University, Ottawa, Canada
C. H. Heathcock, University of California at Berkeley,
                Berkeley, California
J. W. Hooper, Bristol Laboratories of Canada, Candiac, P.Q., Canada
D. Taub, Merck, Sharp and Dohme, Rahway, New Jersey
A. F. Thomas, Firmenich SA, Geneva, Switzerland

Preface

Throughout the history of organic chemistry we find that the study of natural products frequently has often provided the impetus for great advances. This is certainly true in total synthesis, where the desire to construct intricate and complex molecules has led to the demonstration of the organic chemist's utmost ingenuity in the design of routes using established reactions or in the production of new methods in order to achieve a specific transformation.

These volumes draw together the reported total syntheses of various groups of natural products and commentary on the strategy involved with particular emphasis on any stereochemical control. No such compilation exists at present and we hope that these books will act as a definitive source book of the successful synthetic approaches reported to date. As such it will find use not only with the synthetic organic chemist but also perhaps with the organic chemist in general and the biochemist in his specific area of interest.

One of the most promising areas for the future development of organic chemistry is synthesis. The lessons learned from the synthetic challenges presented by various natural products can serve as a basis for this ever-developing area. It is hoped that these books will act as an inspiration for future challenges and outline the development of thought and concept in the area of organic synthesis.

The project started modestly with an experiment in literature searching by a group of graduate students about six years ago. Each student prepared a summary in equation form of the reported total syntheses of various groups of natural products. It was my intention to collate this material and possibly publish it. During a sabbatical leave in Strasbourg in the year 1968-1969, I attempted to prepare a manuscript, but it soon became apparent that if I was to also enjoy other benefits of a sabbatical leave, the task would take many years. Several colleagues suggested that the value of such a collection would be enhanced by commentary. The only way to encompass the amount of data collected and the inclusion of some words was to persuade experts in the various areas to contribute.

Volume 1 presented six chapters describing the total syntheses of a wide variety of natural products. The subject matter of Volume 2 is somewhat more related, being a description of some terpenoid and steroid syntheses. These areas appear to have been the most studied from a synthetic viewpoint and as such have added more to our overall knowledge of the synthetic process.

A third volume in this series will consider diterpenes and various alkaloids, and suggestions for other areas of coverage are welcome.

I am grateful to all the authors for their efforts in producing stimulating and definitive accounts of the total syntheses described to date in their particular areas. I would like to thank those students who enthusiastically accepted my suggestion several years ago and produced valuable collections of reported syntheses. They are Dr. Bill Court, Dr. Ferial Haque, Dr. Norman Hunter, Dr. Russ King, Dr. Jack Rosenfeld, Dr. Bill Wilson, Mr. D. Heggart, Mr. G. W. Holland, Mr. D. Lake, and Mr. Don Todd. I also thank Professor G. Ourisson for his hospitality during the seminal phases of this venture. I particularly thank Dr. S. F. Hall, Dr. R. Pike, and Dr. V. Srinivasan, who prepared the indexes of Volumes 1 and 2.

JOHN APSIMON

Ottawa, Canada
May 1973

Contents

The Synthesis of Monoterpenes   1

    A. F. THOMAS

The Total Synthesis of Sesquiterpenes   197

    C. H. HEATHCOCK

The Synthesis of Triterpenes   559

    J. W. APSIMON AND J. W. HOOPER

Naturally Occurring Aromatic Steroids   641

    D. TAUB

Compound Index   727

Reaction Index   733

THE TOTAL SYNTHESIS
OF NATURAL PRODUCTS

THE SYNTHESIS OF MONOTERPENES

A. F. Thomas

Firmenich SA,
Geneva, Switzerland

1. Introduction   2
2. The Telomerization of Isoprene   3
3. 6-Methylhept-5-en-2-one   4
4. 2,6-Dimethyloctane Derivatives   8
    A. Hydrocarbons   8
    B. Alcohols   14
    C. Aldehydes and Ketones   26
5. Substances Derived from Chrysanthemic Acid   34
    A. The Santolinyl Skeleton   36
    B. The Artemisyl Skeleton   40
    C. The Lavandulyl Skeleton   43
    D. Chrysanthemic Acids   49
6. Cyclobutane Monoterpenes   58
7. Cyclopentane Monoterpenes   59
    A. Plinol   61
    B. Cyclopentanopyrans   62
    C. 1-Acetyl-4-isopropenyl-1-cyclopentene   87
    D. Campholenic Aldehyde   88
8. The p-Menthanes   88
    A. Hydrocarbons   88
    B. Oxygenated Derivatives of p-Menthane   93
9. The m-Menthanes   137
10. 1,1,2,3-Tetramethylcyclohexanes   138
    A. Safranal   138
    B. Karahana Ether   139
11. The o-Menthanes   139
12. Cycloheptanes   140
    A. Thujic Acid, Shonanic Acid, Eucarvone, and Kara-
        hanaenone   140
    B. Nezukone and the Thujaplicins   143

1

13. Bicyclo [3,2,0] Heptanes   144
     A. Filifolone   144
14. Bicyclo [3.1.0] Hexanes   145
15. Bicyclo [2.2.1] Heptanes   149
16. Bicyclo [3.3.1] Heptanes   154
17. Bicyclo [4.1.0] Heptanes   157
18. Furan Monoterpenes   159
     A. 3-Methyl-2-Substituted, and 3-Substituted Furans   159
     B. 2,5,5-Substituted Tetrahydrofurans   165
19. Oxetones   166
20. Tetrahydropyrans   167
21. Hexahydrobenzofuran-2-ones   169

1. INTRODUCTION

The total synthesis of monoterpenes is not a subject that has
attracted a great deal of attention. To be sure, each time
some terpenoid curiosity is isolated, there is a certain amount
of effort expended to synthesize it, generally as part of a
structural "proof," but few chemists have spent much time on
attempting to synthesize, say, pinene. One of the reasons for
this lack of interest is certainly the vast natural resources
of the more complex ring systems (especially the pinane, bor-
nane, and carane systems), so that industry has had little
need of total synthesis of these structures and has confined
itself to partial syntheses within the systems. These partial
syntheses, moreover, are themselves particularly interesting
in view of the lability of many of the systems, and have, indeed, often provided the examples for reaction mechanism and
stereochemical studies. Supplies of raw materials, however,
are not always as easily accessible as they once were and
total synthesis from cheaper materials cannot afford to be
completely ignored.
     On the whole, there are three reasons for synthesis. The
first of these has just been mentioned--the preparation of
"unusual" monoterpenes that are not readily available from
natural sources, with a view to checking the correctness of
their structures, examining their properties, and so on. The
second type of synthesis is the so-called biogenetic type.
These purport to imitate a route that approximates to what is
believed to happen in the plant. There are not too many of
them, but although they are designed from strictly theoretical
viewpoints, they could be extremely important, especially
since there is, so far, no good synthetic route to even quite
simple monoterpenes that are in large supply in nature. Finally, there are the "industrial" syntheses, about which a further point must be made. The legislation in some countries

is becoming increasingly concerned with whether a compound is
known to be naturally occurring or not. Many of the uses of
monoterpenes by industry are in perfumes, cosmetics, flavors,
and so on, and the idea of the legislators in these fields is
that whatever occurs naturally in plant or animal products has
been in existence, and possibly in use, for a long time, and
so is less likely to be harmful than a new untried substance.
Without discussing the merits of this position, it must be
remembered that naturally occurring materials are, on the
whole, asymmetric, so if an industry wishes to use synthetic
compounds that are going to be placed on or in human beings,
it may be under some pressure to synthesize not only the racemate, but the correct optical antipode, since we do not know,
a priori, what the physiological differences between the anti-
podes may be. This factor mitigates, to some extent, against
the use of purely synthetic starting materials, since total
synthesis of optically active substances implies resolution at
some stage. Legislation is, unfortunately, not consistent,
since patent law does not allow (at least in the United States
and Germany) the protection of natural products, no matter
what efforts were made to isolate them, assign structures to
them, and synthesize them.

There is a fourth type of synthesis that appears from
time to time, and which might be called the "building block"
type, where the desired molecule is put together by joining
small parts of it. Apart from its use as an intellectual ex-
ercise (such syntheses are rarely of any industrial use), this
approach does have an advantage where labeled molecules are
required, since it frequently allows the placing of a particular atom in the molecule in a clear-cut way. For this reason,
such syntheses have been included in this chapter. The literature is largely complete up to 1970; more recent work will
be found in the "Specialist Periodical Report on Terpenoids
and Steroids" (published annually by the Chemical Society,
London).

2. THE TELOMERIZATION OF ISOPRENE

The ready availability of isoprene makes it an attractive
starting point for the synthesis of monoterpenes, and several
routes involving the addition of halogen acids (see, for example, the next section on methylheptenone) are described elsewhere in this chapter, in addition to the historical dimerization to dipentene. It would naturally be much more useful to
have methods available for the direct dimerization and simultaneous hydroxylation of isoprene, and considerable effort has
been put into this aspect, particularly in recent times in

Estonia and Japan. The mixtures obtained are of considerable
complexity, and unfortunately much of the work is in journals
that are difficult to obtain, including a review of the telomerization using hydrogen chloride (i.e., via the hydrogen
chloride adduct).[1] Under these conditions, the C_{10} fraction
can contain as much as 45% of geranyl chloride.*[2] Phosphoric
acid telomerization of isoprene gives α-terpinene and allo-
ocimene as the main C_{10} hydrocarbons, together with geraniol
and terpineol.[3,4] In the presence of acetic acid, the phosphoric acid telomerization reaction leads to the acetates of
geraniol, lavandulol, and other compounds, besides a complex
mixture of monoterpene hydrocarbons.[5] There are other reactions of isoprene that lead to mixtures containing terpenoids,
for example, the hydrocarbon will react with magnesium in the
presence of Lewis acids, and the complex thus obtained gives
adducts with aldehydes but again only as mixtures.[6] Isoprene
is also dimerized by lithium naphthalene in tetrahydrofuran to
linear monoterpene homologs,[7] passing oxygen through the mixture giving then 30 to 40% of C_{10} alcohols and 30% of C_{10}
glycols. Although the alcohols include 10% each of nerol and
geraniol, most of the remainder are not natural products.[8]

3. 6-METHYLHEPT-5-EN-2-ONE

Although not strictly speaking a monoterpene, 6-methylhetp-5-
en-2-one (2) is a common constituent of essential oils, particularly of the *Cymbopogon* (lemongrass) species. Some important
terpene syntheses start from it, and it is also the chief product from the retro-aldol reaction and certain oxidations of
citral and its derivatives. In view of its key position, it
has been given a separate section on its synthesis.

     Any synthesis of methylheptenone (2) must take into ac-
count the fact that it is sensitive to acid,[9,10] and can undergo cyclizations to hydrogenated xylenes and tetrahydropyrans,
for example, during the decomposition of its semicarbazone by
acid.[9]

     Only the more recent syntheses will be given, and it is
interesting that these are all based on some form of electrocyclic reactions. One of the earliest of this type is that of
Teisseire, involving the transesterification of ethyl acetoacetate with 2-methylbut-3-en-2-ol (3).[11] The allyl ester (1)

*Formulas of these substances will be found in the sections
devoted to more clearly defined syntheses described later.
One example of telomerization is also discussed in greater detail in the section devoted to linalool, nerol, and geraniol
(p. 17).

obtained undergoes the Carrol reaction,[12] resulting in a β-
ketoacid through a reaction akin to the Claisen rearrangement,[13] and this ketoacid then loses carbon dioxide under the
reaction conditions to yield the product (2). An earlier technique for this type of reaction consisted in mixing the alcohol
with diketene;[13] when 3 and diketene is added to hot paraffin
containing a trace of pyridine, the methylheptenone (2) can be
distilled from the mixture.[14]

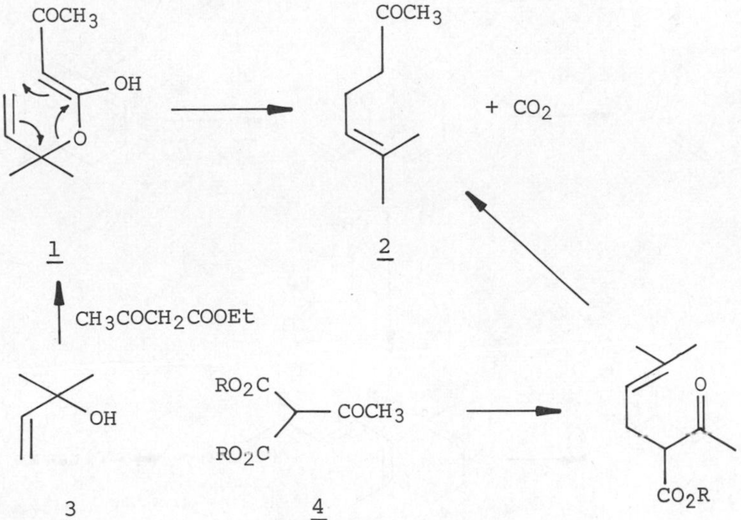

A further variant uses condensation of the allyl alcohol (3)
with an acylmalonic ester (4) at 130-200°, when alcohol and
carbon dioxide are lost, giving this time a β-ketoester (5)
that is convertible to methylheptenone by ketone hydrolysis.[15]
This type of procedure forms the basis of one of the best
known commercial preparations of methylheptenone (2), required
as a vital intermediate for syntheses of linalool (6), the
ionones (7), and vitamin A, and which is illustrated in Scheme
1.[16]

Scheme 1

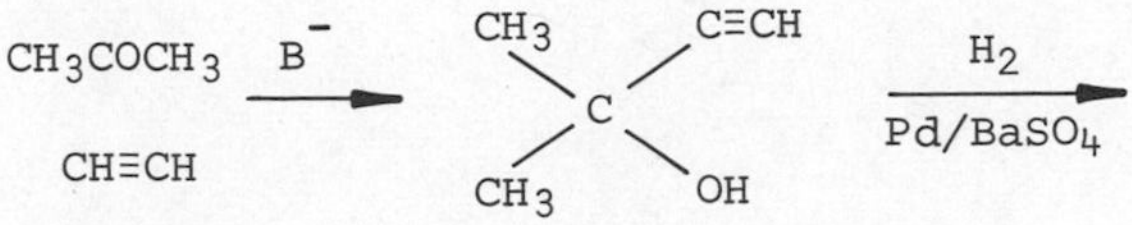

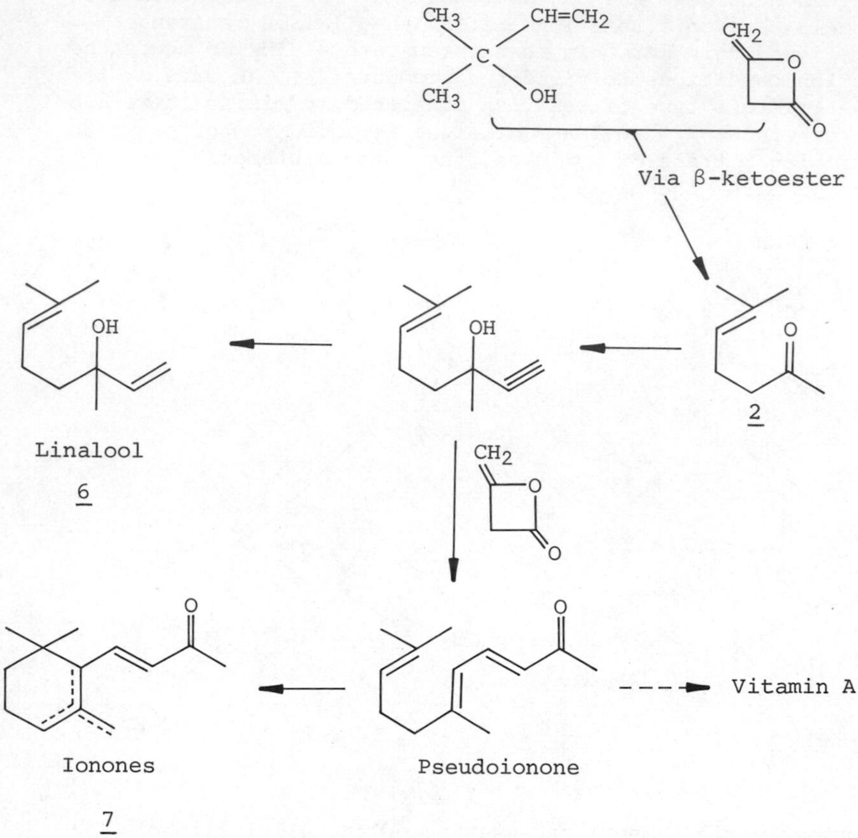
CH3
CH=CH2
C
CH3
OH
CH2
O
O
Via β-ketoester
OH
OH
O
2
Linalool
6
CH2
O
O
O
O
Vitamin A
Ionones
7
Pseudoionone

The thermal rearrangement of allyl ethers was described by Julia et al. in 1962,[17] and based on this idea, Saucy and Marbet synthesized methylheptenone from 2-methylbut-3-en-2-ol (3) and ethyl isopropenyl ether (8):[18]

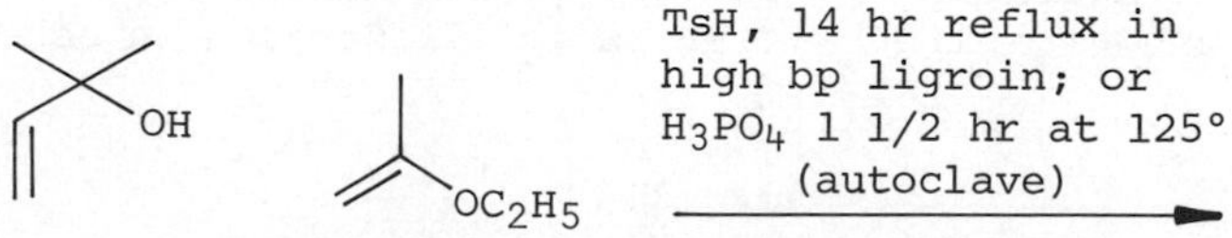
OH
OC2H5
TsH, 14 hr reflux in
high bp ligroin; or
H3PO4 1 1/2 hr at 125°
(autoclave)

3 8

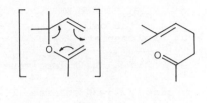

2

A somewhat more classical approach, namely, building up the molecule by a standard ketone synthesis from a functionalized isoprene is mentioned here particularly in view of the importance of the particular C_5 unit involved. The addition of halogen acids to isoprene (9) occurs initially by 1,2-addition, leading to 2-chloro-2-methylbut-3-ene (10) (or its bromo analog when hydrogen bromide is employed). This compound can even be isolated in a relatively pure state provided the addition is not carried through to completion.[19] Under the normal conditions of addition, however, using an excess of acid, the main product is the primary chloride (11). For example, one mole of isoprene and two to three moles of concentrated hydrochloric acid at 0-40° for 1 to 4 hr gives 63% of the primary chloride (11) and 8% of the tertiary chloride (10).[20] Formation of the primary chloride is also favored by elevated temperature and the presence of moisture.[21] Treatment of the sodium derivative of ethyl acetoacetate with the primary chloride (11) thus leads, after conventional hydrolysis and decarboxylation of the β-ketoester (5), to methylheptenone (2).[22] Direct condensation of acetone with the primary chloride (11) is also reported in a Russian patent.[23]

$$\text{9} \xrightarrow{\text{HCl}} \text{Cl} \text{10} \longrightarrow$$

9

10

$$CH_3CO\overset{-}{C}HCO_2C_2H_5$$

CH_2Cl

11

5 $\longrightarrow$ 2

4. 2,6-DIMETHYLOCTANE DERIVATIVES

A. Hydrocarbons

Myrcene, Ocimene, and Alloocimene

Both myrcene (12) and ocimene (13a, *cis* and 13b, *trans*) are common constituents of essential oils. In addition, myrcene is made on the industrial scale by pyrolysis of β-pinene (14),[24],[25] a reaction that also gives rise to a small amount of α-myrcene (15),[26] not yet reported as a natural product. One of the problems associated with *cis*-ocimene (13a) is its ready transformation to the non-naturally occurring allo-ocimene (16) [the only reported occurrence of the latter in a plant oil has been attributed to rearrangement of *cis*-ocimene (13a) during workup[28]]. Consequently, it is not surprising that pyrolysis of (+)-α-pinene (17) in liquid form over a nichrome wire at 600° gives, in addition to 48% of ocimene, 16.5% of alloocimene (16), together with racemized α-pinene and dipentene (18).[29]*

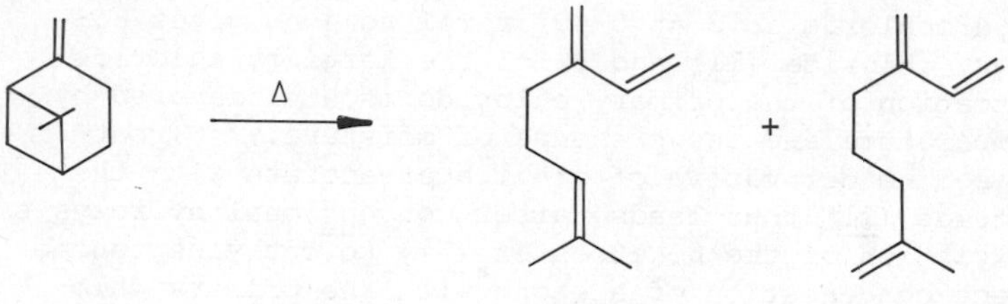

14	12	15
β-Pinene	Myrcene	

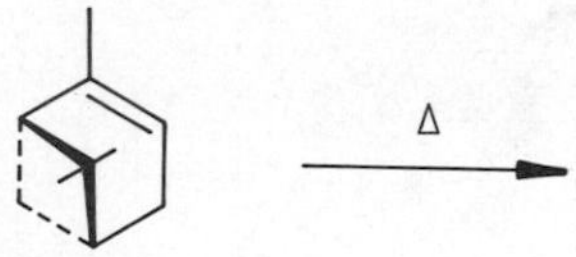

17
(+)-α-Pinene

13a ⇌ 13b

*"Dipentene" will be used in this chapter to denote the race-mate, "limonene" being reserved for the optically active iso-mers.

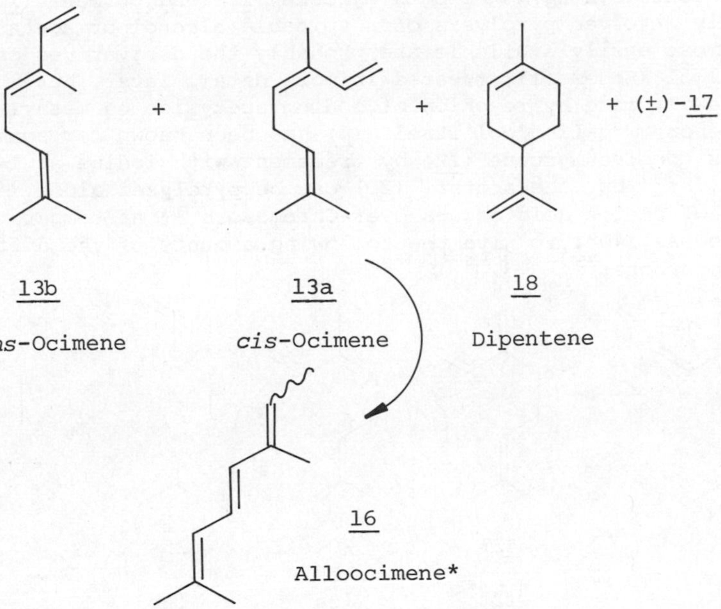

13b	13a	18	
trans-Ocimene	*cis*-Ocimene	Dipentene	

16

Alloocimene*

Isomerization of α-pinene (17) to ocimene can also occur photochemically,[30],[31] and Kropp has investigated this and related reactions.[32]   Direct irradiation of α-pinene in low yield was already known to give a product contaminated by dipentene (18) and other products.[33]   Sensitized irradiation of α-pinene in xylene or xylene-methanol was now found to give only *cis*-ocimene (13a) after short reaction times, but with longer periods of irradiation an equilibrium between *cis*- and *trans*-ocimene is set up, although the product is never seriously contaminated with dipentene (18),[32] as is the case with the pyrolytic and γ-ray radiolytic conversions of α-pinene to ocimene.[34]   This photochemical reaction of α-pinene is somewhat unexpected in giving no β-pinene, unlike the similar reaction of dipentene (18), that gives a 13% yield of p-mentha-1(7),8-diene (19) on sensitized irradiation.[32]

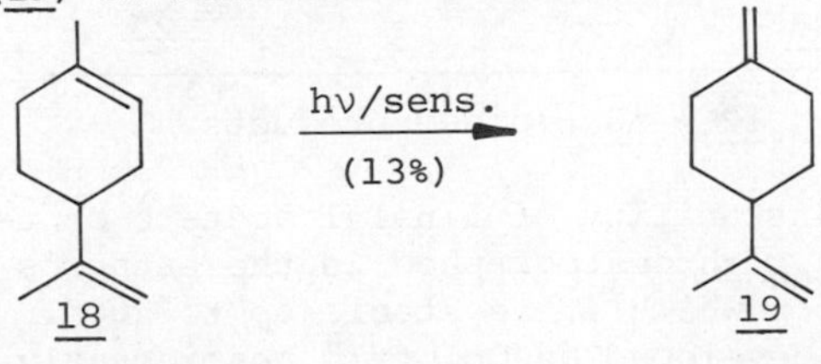

hν/sens.

(13%)

18   19

*For a detailed discussion of the various stereoisomeric alloocimenes, see Crowley.[27]

Total synthesis of both myrcene (12) and ocimene (13) generally involves pyrolysis of a suitable alcohol or acetate. The most easily available are probably the derivatives of linalool (Scheme 1), treated in more detail later, but which is easily made by reaction of sodium acetylide on methylheptenone.[35]   Linalool itself (6) has been known for some years to give myrcene (12) by treatment with iodine at 140-150°,[36,37] but the acetate (20) can be pyrolyzed alone,[37] while loss of acetic acid occurs over Chromosorb P* at temperatures as low as 140°, to give the following amounts of the different hydrocarbons:[39]

20	43%	20%	35%	2%
	12	13a	13b	18

Pyrolysis of various other acetates to give ocimenes and myrcene has also been described.[26,38]   Dehydration of geraniol (21) or nerol (22) (related to linalool allylically) in the presence of potassium hydroxide at 200° for 10 minutes† also gives 60% of myrcene and other products, including allo-ocimene.[41]

Geraniol	Nerol
21	22

12 + 16 + other products

*In view of the instability of linalyl acetate on certain supports, it is always chromatographed in the author's laboratory on Chromosorb-W, on which it is stable up to 200°.
†This technique, developed by Ohloff[40] has recently been re-examined by Bhati both for alcohol dehydration and isomerization (see below, under menthone).

A recent synthesis of myrcene on the "building block" principle (i.e., putting one unit together with another in logical sequence until the desired molecule is reached) has been described by Vig et al.[42] It starts with ethyl 2-carbethoxy-5-formyl butanoate (23) and follows Scheme 2.

Scheme 2

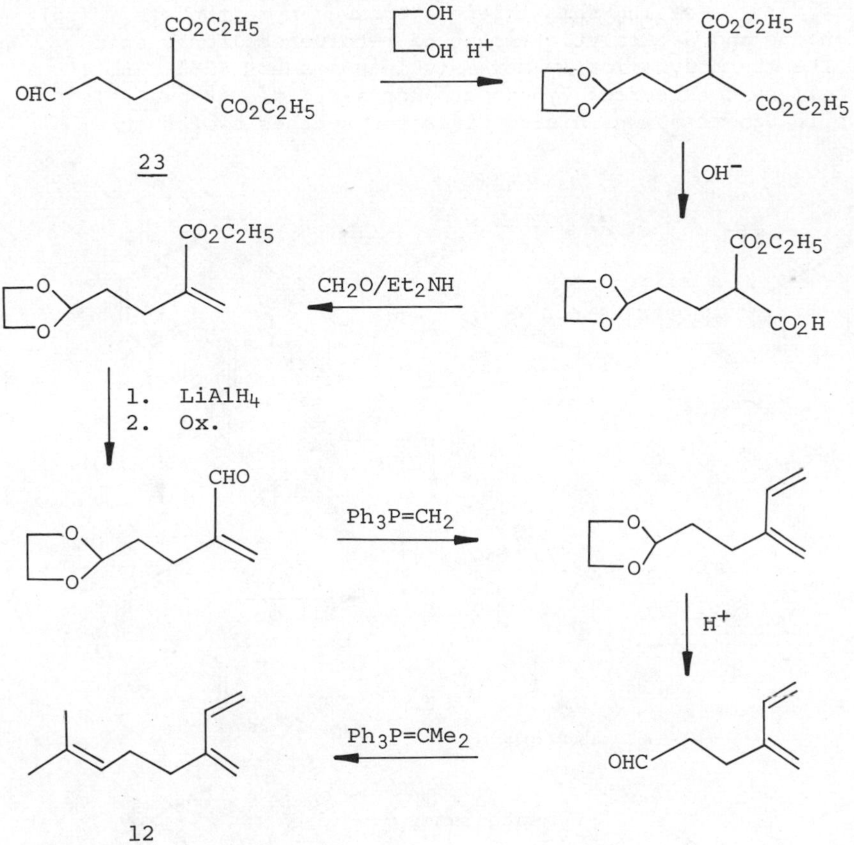

Hymentherene and Achillene

2,6-Dimethylocta-2,4,7-triene (26) was originally reported as a natural product, and given the name hymentherene;[43] later, however, it turned out that the substance isolated was a mixture of two known monoterpenes.[44] The name has been retained in this chapter for convenience, and also because it is still possible that the substance will be found in nature. The

related *cis*-2,6-dimethylocta-1,4,7-triene (27a) has been isolated from *Achilla filipendulina* by Dembitskii et al.[45,46]
    Early syntheses of "B-hymentherene" were unsuccessful[47,48]
and the first time the substance was certainly isolated is by
Schulte-Elte,[49] who started from the known[47] (6S)-(+)-2,6-
dimethylocta-1,3,7-triene (25), the (6S)-configuration of the
positively rotating isomer of this substance having been
established previously by correlation with (-)-*trans*-
pinane.[50,51] Heating this triene ("*trans*-α-hymentherene," (25)
in benzene and a catalytic amount of *p*-toluenesulfonic acid
gave the thermodynamically more stable β-isomers (26a, 26b).
The full synthesis from 2,6-dimethylocta-2,7-diene[50] (24) is
shown in Scheme 3, which also gives the synthesis of natural

Scheme 3

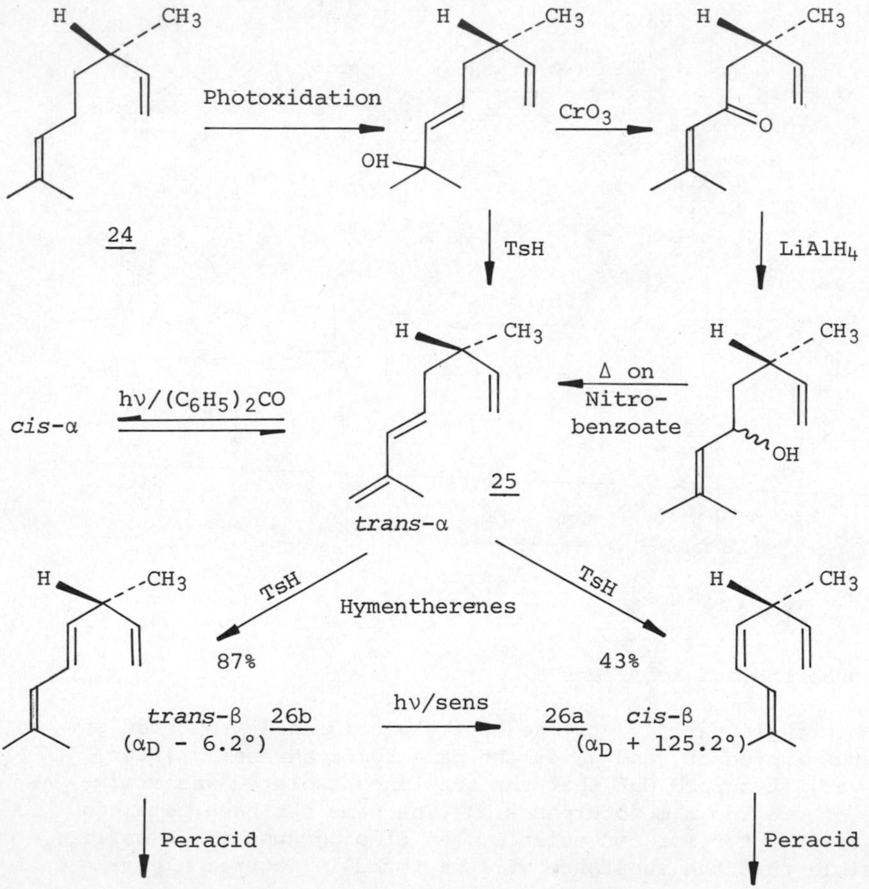

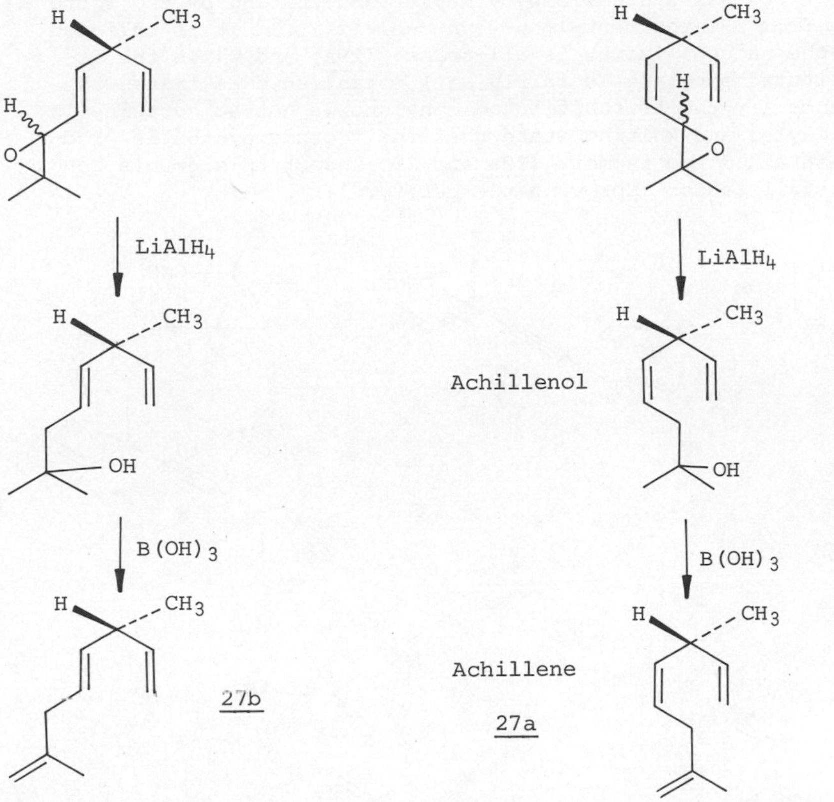

trans (α_D - 3°)   cis (α_D + 54°)

achillene (27a) that Schulte-Elte achieved from cis-β-hymen-
therene (26a).[49]   Although the reaction of p-toluenesulfonic
acid on trans-α-hymentherene (25) leads to only 3% of the
natural cis-configuration about the C-4 double bond, irradiation of the trans-compound (26b) in the presence of a sensitizer leads to a mixture containing 43% of the cis compound
(26a).[49] Achillenol [the alcohol obtained immediately before
achillene (27a) in the synthesis] has also been reported recently as a natural product,[52] but the rotation does not agree
with Schulte-Elte's synthetic material.[49]

Cosmene

The only natural monoterpene tetraene is cosmene, 2,6-dimethyl-
octa-1,3,5,7-tetraene (29a), isolated from *Cosmos bipinnatus*,
Cav., and other compositae by Sörensen and Sörensen.[53]   The

hydrocarbon was synthesized by Nayler and Whiting by the route shown from 3-methylpent-1-en-4-yn-3-ol (28).[54] It is believed that the natural isomer is all-*trans*, (29a) and while the di-substituted ethylene is fairly well established as *trans*, the evidence for the trisubstituted ethylene is not so certain, in fact Nayler and Whiting state that their crude synthetic prod-uct contained two isomers (29a and 29b) about this double bond, recrystallization improving the purity.[54]

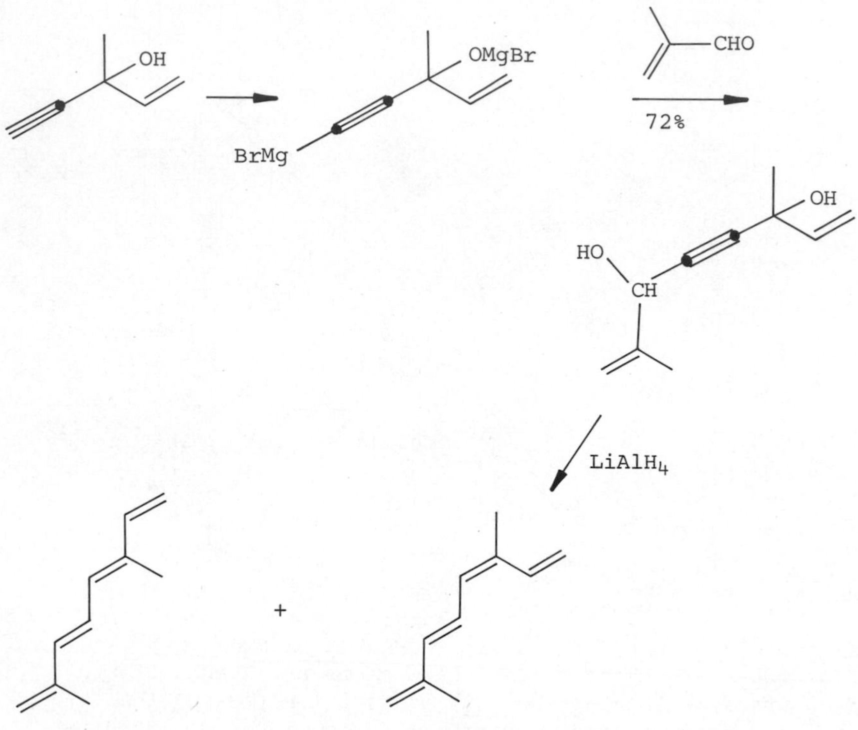

Cosmene   29b

   29a

B.   Alcohols

Citronellol

3,7-Dimethyloct-6-en-1-ol is one of the most widely distributed monoterpene alcohols both as the alcohol and the corresponding acetate. It occurs naturally in both (+)- and (-)-forms, al-most always as the β-form (i.e., isopropylidene, e.g., 30),

although isolated reports of the α-form (31) occurring in
nature do exist. In this connection, it is worth mentioning
the confusion in the literature about the terms "rhodinol" and
"rhodinal." The older literature refers to a mixture, but
later it has been held that citronellol and "rhodinol" are
identical.[55,56] Eschinazi[57] and Naves and Frey[58] consider
"rhodinol" to be (+)-α-citronellol (31), and "rhodinal" to be
the corresponding α-aldehyde. On the other hand, Chemical
Abstracts refers to "rhodinol" as 3,7-dimethyloct-6-en-1-ol,
but "rhodinal" as 3,7-dimethyloct-7-en-1-ol.[59] It is the
author's opinion that the name should no longer be used at all,
and all references be made to citronellol, α-citronellol
and citronellal.

30

(+)-Citronellol

31

(+)-α-Citronellol

Syntheses of citronellol are commercially important because supplies of the natural material are insufficient, and
since geraniol is available from myrcene (see below), there
are many syntheses described in the literature from geraniol
or geranic acid by reduction (see Ref. 44 for a list), and
these largely conventional syntheses will not be mentioned
here. Somewhat more interesting is the synthesis from 2,6-
dimethylocta-2,7-diene (32) using tri-isobutylaluminum then
oxidizing[51] (the presence of zinc isopropoxide being bene-
ficial[60]), this reaction being the equivalent of a hydroboration.

+ (i-C$_4$H$_9$)$_3$Al $\longrightarrow$ $\xrightarrow{\text{O}_2/\text{Zn}(\text{i-PrO})_2}$ CH$_2$OH

32

Another synthesis is important as an illustration of the

use of N-lithioethylenediamine to displace double bonds. This
reagent is mentioned below in similar connections, and, in the
synthesis of citronellol, it was found that dihydrogeraniol
(34), made from 6-methylheptan-2-one (33) via a Reformatsky
reaction[61] is converted in 70% yield to citronellol.[62]

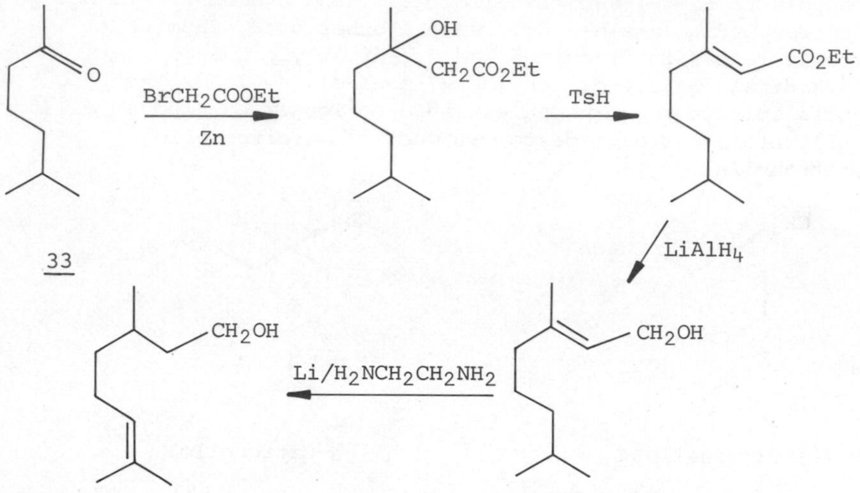

All the syntheses of citronellol from geraniol result, of
course, in the racemate, while the most important natural diastereomer is the (-)-form, and synthetic approaches to the
chiral form are rare.

Linalool, Nerol, and Geraniol*

Together with citronellol, these alcohols and their acetates
constitute the most **widely** distributed monoterpene alcohols in
nature. Geraniol is not only of some value in itself, but
also as an intermediate to citronellol which is more useful to

*A detailed discussion of the allyl rearrangements occurring
between linalool and the other two alcohols, recognized to
occur in the last century[63] will not be given since it belongs
properly to the domain of physical organic chemistry.

the perfumery industry. The natural supplies of linalool and
its acetate are insufficient to meet requirements, and these
facts, together with the central position these alcohols occupy in the monoterpene field make a knowledge of their synthesis desirable. The syntheses described, particularly in
the patent literature, are very numerous, and the reader is
referred to Ref. 55, 64, and 65 for lists of the earlier methods. The most useful syntheses of linalool (6) have as their
basis either acetone and acetylene (i.e., via methylheptenone,
see Scheme 1 above[15]), or pinene, one of the most readily
available hydrocarbons, from most types of natural turpentine.
One of the latter methods involves the hydrochlorination of a
β-pinene (14) pyrolysate (i.e., myrcene (12) see above), which,
in the presence of cuprous halides gives a mixture of geranyl
(37) and neryl (36) halides together with linalyl (35) and
α-terpinyl (38) halides (Scheme 3).[66] These halides can be
converted under various conditions to geraniol, nerol, or
linalool derivatives (see Ref. 55, p. 530), or neryl halides
to α-terpineol (39), water hydrolysis at elevated temperatures
favoring the formation of linalool (6) (Scheme 4).

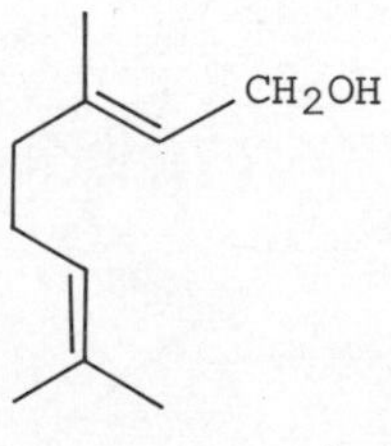

Geraniol

21

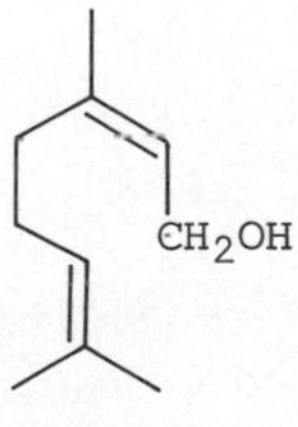

Nerol

22

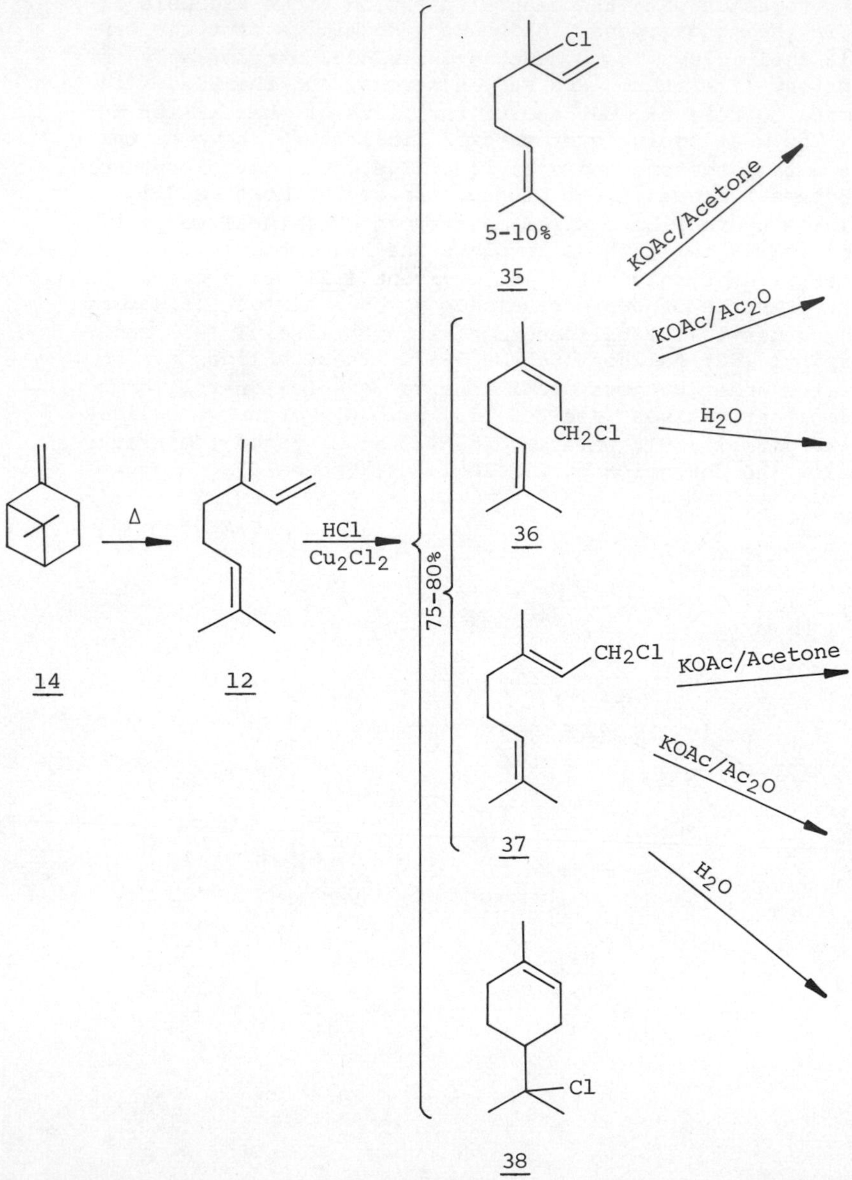

14 12 35 36 37 38

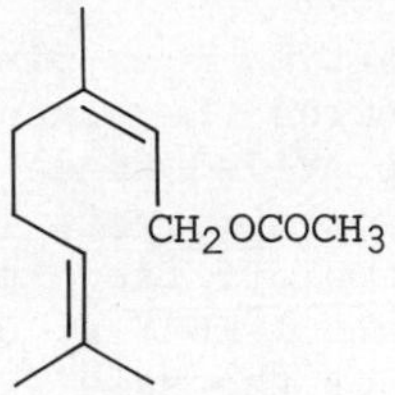

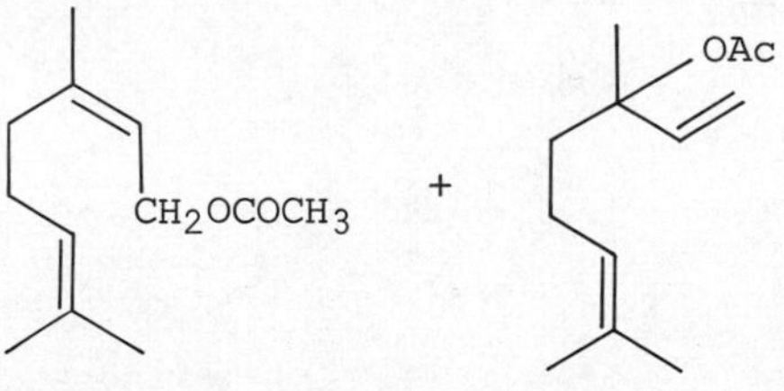

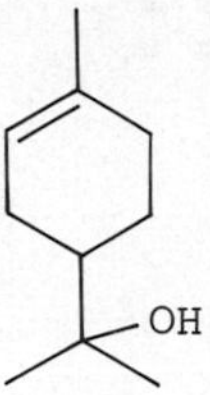

α-Terpineol

$\underline{39}$

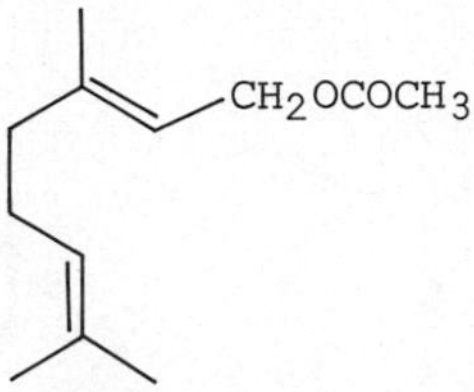

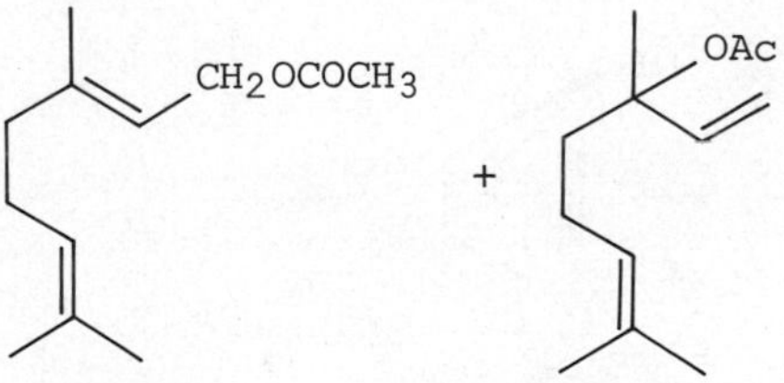

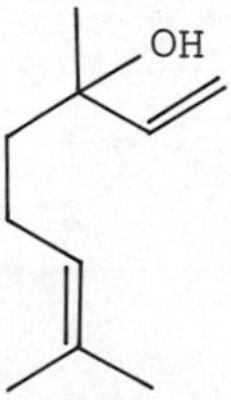

Linalool

$\underline{6}$

A much more elegant synthesis starting from pinene is
that of Ohloff and Klein.[67] This requires the pinanols (40a,
40b, 40c, and 40d, see below), then, provided these are chiral,
optically active linalool is obtained on pyrolysis (Scheme 5).
The *cis*-isomers react more easily and Ohloff found that (+)-
cis-pinanol (40a) gave 86% of (-)-linalool (6a) at 600°, while
(-)-*trans*-pinanol (40d) only gave 62% of the same linalool.

Scheme 5

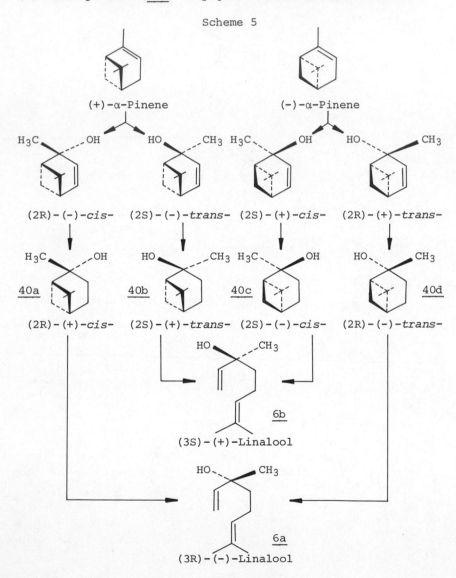

(+)-α-Pinene   (-)-α-Pinene

(2R)-(-)-*cis*-   (2S)-(-)-*trans*-   (2S)-(+)-*cis*-   (2R)-(+)-*trans*-

40a   40b   40c   40d

(2R)-(+)-*cis*-   (2S)-(+)-*trans*-   (2S)-(-)-*cis*-   (2R)-(-)-*trans*-

6b

(3S)-(+)-Linalool

6a

(3R)-(-)-Linalool

In certain regions, notably India, the supply of citral
(41) is sufficient (from lemongrass) to make up the lack of
linalool from this source.   Thus epoxidation of citral with
alkaline hydrogen peroxide followed by treatment of the 1,2-
epoxide (42) with hydrazine hydrate (the Wharton reaction[68])
in methanol at 0° gives a 30 to 35% yield of linalool (6).[69,70]

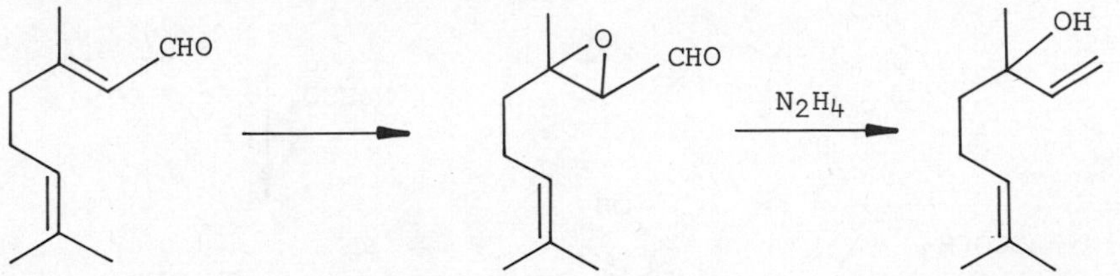

Citral 41   42   6

A related method uses the epoxide (44) of geranyl iodide
(43), which, on treatment with p-toluenesulfonylhydrazine and
calcium carbonate in dimethylsulfoxide, gives a quantitative
yield of linalool.[71]

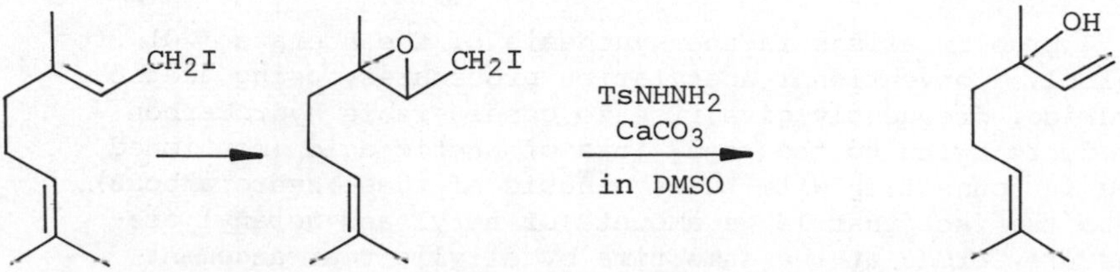

43   44

A synthesis that is somewhat different from those so far
described begins with α-acetyl-γ-butyrolactone (45) which is
converted to the chloroketone (46) with concentrated hydrochloric acid, then methylated with a Grignard reaction to the
tertiary alcohol (47).   Dehydration of the latter with sodium
bisulfate gives mainly the desired 1-chloro-4-methylpent-3-ene
(48), that can be converted into linalool (6) via the lithium
derivative with methyl vinyl ketone.[72]

45 → c. HCl / H₂O → CH₃COCH₂CH₂CH₂Cl (46) → CH₃MgI → 47

47 → NaHSO₄ → 6 (via 1. Li, 2. CH₂=CH-COCH₃) and 48

20 ← 6

A problem arises in the synthesis of the acetate (20) from linalool. Conventional acetylation procedures, using acetic anhydride, frequently give rise to considerable hydrocarbon byproducts owing to the ready loss of acetic acid (mentioned above in connection with the synthesis of these hydrocarbons), and to the fact that large amounts of neryl and geranyl acetates are formed at the same time by allylic rearrangement. These mixtures are tedious to separate, since their boiling points are fairly close to one another. The difficulty has been overcome by Jirát and Vonášek[73] by preparing the sodium derivative of linalool with sodium hydride in an inert solvent and treating this with diethyleneglycol diacetate. Kogami et al. have given an account of the problem[74] and suggested that the best method is by using ketene in the presence of acid esterification catalysts.[75,76]

Acid reagents convert linalool into geraniol and nerol[77] [thionyl chloride, for instance, gives only geranyl chloride (37)[78]] so all the syntheses mentioned for linalool constitute syntheses also for these two alcohols, but in view of the easier separation of the corresponding acid methyl esters (49 and 50), syntheses requiring the pure geometrical isomers most frequently pass through this stage. Many methods are available for the conversion of methylheptenone (2) to a mixture of methyl geranate (49) and methyl nerolate (50), from

the Reformatsky reaction[79] to the use of dimethoxycarbonyl-
methyl phosphonate,[80] but all these are standard synthetic
techniques, the detail of which need not be described here.   A
particularly interesting account of this type of approach is
that of Weedon's laboratory, since they confirmed the geometry
that was accepted traditionally by NMR spectrometry.[81]   An
earlier stereospecific synthesis of nerol involved addition
of carbon dioxide and hydrogen across a triple bond using
nickel carbonyl.[81a]

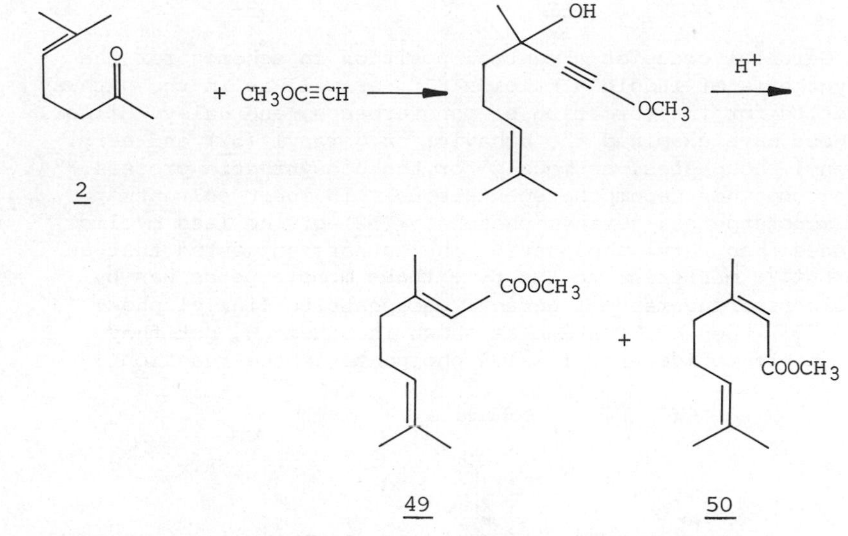

At the beginning of this chapter, it was briefly mentioned
that methods are available for converting isoprene directly
into mixtures of monoterpenoids.   One such method involves the
telomerization below 20° in 85% formic acid in the presence of
10% perchloric acid.   After 5 hr, then hydrolysis, dipentene
(18), and terpinolene (51) are obtained in addition to the
alcohols linalool (6), geraniol (21), nerol (22), and α-
terpineol (39).[82]

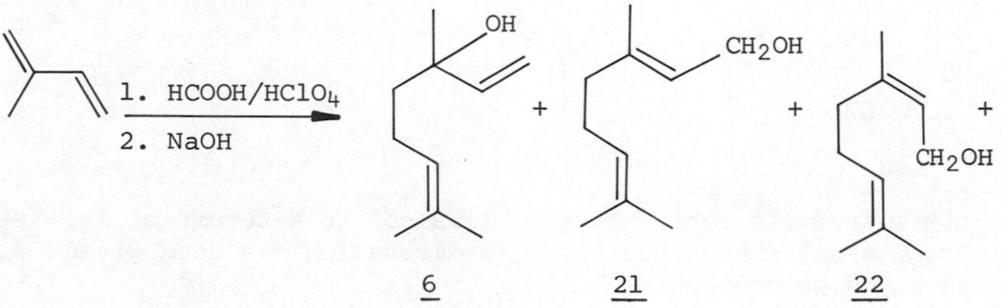

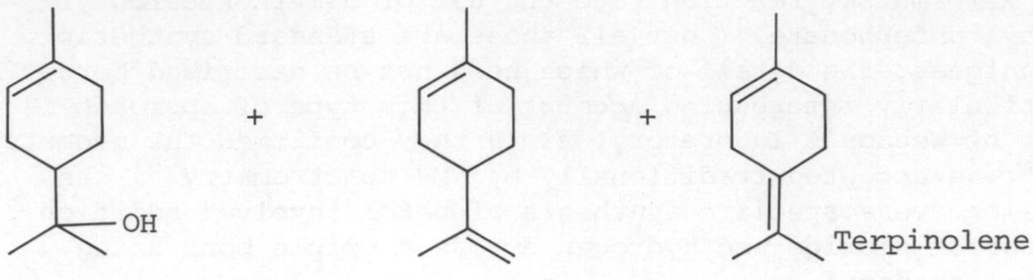

39   18   51

Geraniol occupies a central position in schemes for the biosynthesis of indole alkaloids[83-85] as well as in the pathway suggested for the formation of monoterpenes, and Haley, Miller, and Wood have examined the behavior of geranyl (52) and neryl diphenyl phosphates, as models for the biosynthetic process.[86]* These compounds decompose spontaneously in inert solvents to give monoterpenes--geranyl phosphate (52) giving less cyclic terpenes than neryl phosphate. The authors suggested that an alternative mechanism to the *p*-menthane monoterpenes was by allylic rearrangement of geranyl phosphate to linalyl phosphate (53) then cyclization as shown in Scheme 6, but they were unable to identify linalyl phosphate in the reaction mixture.

Scheme 6

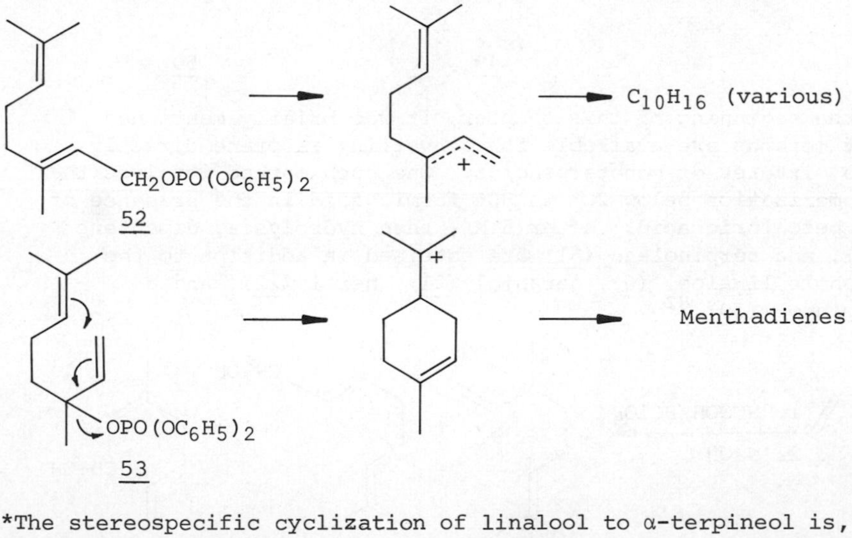

*The stereospecific cyclization of linalool to α-terpineol is, of course, a well-known reaction. A discussion has been given by Prelog and Watanabe.[87]

Hotrienol

3,7-Dimethyl-1,5,7-octatrien-3-ol (55) occurs as the 3S-(+)-
enantiomorph in Ho leaf oil,[88] while the R-enantiomorph has
been isolated from tea, both black[89] and green.[90] Before it
had been reported as a natural product, Matsuura and Butsugan
had prepared it from one of the photooxidation products of
(-)-linalool (6a). The glycol (54), on treatment with sulfuric
acid in acetone, yields R-hotrienol (55).[91] The route followed by Nakatani et al. also starts from R-linalyl acetate
(20a), reaction of N-bromosuccinimide on which gives a mixture of three bromacetates 56a, 56b, and 56c, from which hydrogen bromide can be removed by diethylaniline, leading to the
acetate (57) of the R-isomer, from which the alcohol (55) can
be obtained by conventional hydrolysis.[89]

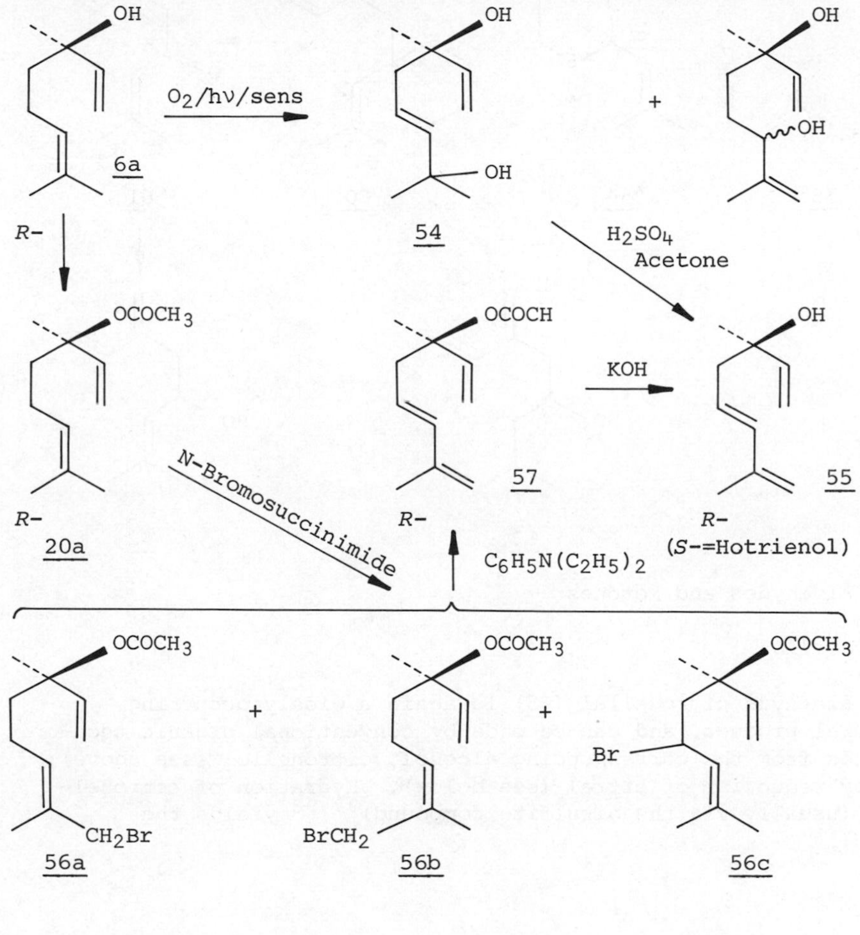

2-Methyl-6-methyleneocta-2,7-diene-4-ol (62) and 2-Methyl-6-methylene-oct-7-en-4-ol (63)

These two aocohols were isolated from the sex attractant of the bark beetle, *Ips confusus*,[92] and have been synthesized by Reece et al.[93] The initial reaction followed a procedure suggested by Corey and Seebach,[94,95] and consisted in coupling bromomethylbutadiene (58)[96] with the thioketal anion (59, or its dihydro-compound). The thioketal (60) thereby obtained was converted to the ketone (61) with silver nitrate, or, in the case of the dihydro-compound, with mercuric chloride-cadmium carbonate. The ketones were then reduced to the corresponding alcohols with sodium borohydride.

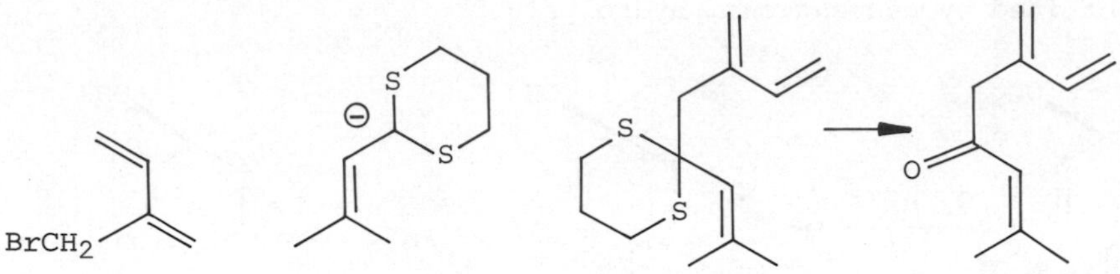

C.   Aldehydes and Ketones

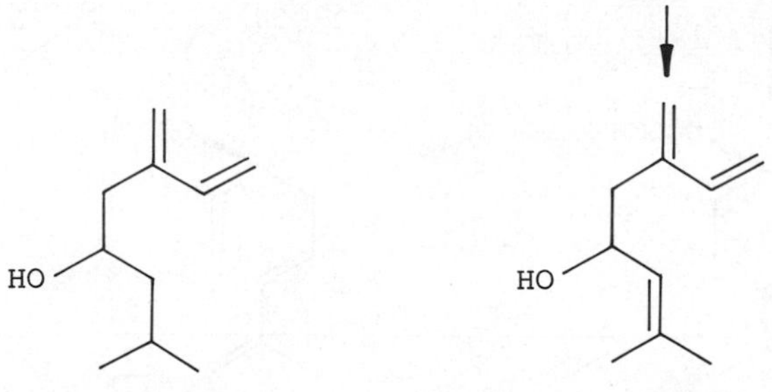

Citronellal

The aldehyde citronellal (65) is again a widely occurring natural product, and can be made by conventional organic techniques from the corresponding alcohol, citronellol (see above), or by reduction of citral (see below). Hydration of citronellal (usually via the bisulfite compound)[97,98] yields the

valuable hydroxycitronellal* (66) that is not, however, reported as being naturally occurring. A study of the dehydration of hydroxycitronellal to a mixture of α- (67) and β-
citronellals has been published by Eschinazi.[57] Since citronellal can be made by pyrolysis of isopulegol (64),[99,100] any
synthesis of this substance constitutes a synthesis of
citronellal.

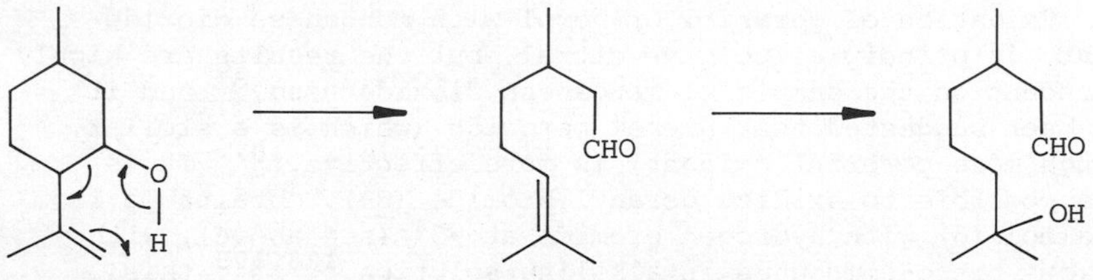

64   65   66

Isopulegol   Citronellal

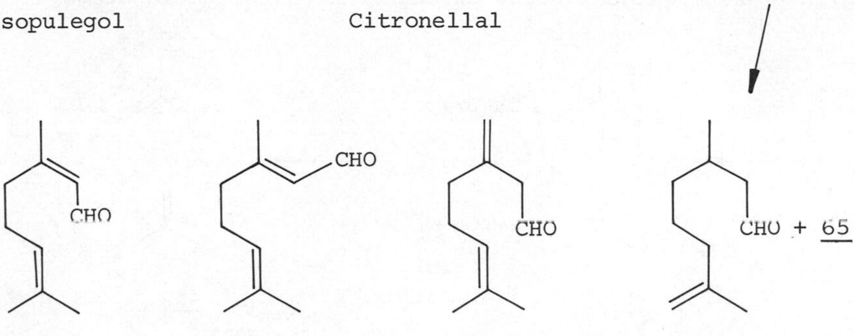

41b   41a   68   67

Citral b   Citral a   Isocitral   α-Citronellal

Citral

Both natural and synthetic citral is a mixture of the two possible isomeric forms, citral a (=geranial, 41a) and citral b
(=neral, 41b). The two isomers have been well characterized
by NMR analysis,[101,102] and in the commonly encountered mixtures the *trans-* form (geranial, 41a) is generally the predominating isomer. In addition, heating the mixture of citral
a and citral b to over 130° causes isomerization to a third
isomer, isocitral (68), to occur[99] (in addition to cyclization
reactions that occur at higher temperatures and under the

*The conditions necessary for high conversion to the hydrate
have been examined by several authors (see, e.g., Ref. 98).

influence of acid catalysts*). Citral can be made from
geraniol (21) by many varieties of oxidation procedure (see
Ref. 103, for a list, together with many other syntheses), but
particularly potassium dichromate in acetic acid, which, after
refluxing for 1 hr, then steam distilling, is reported[104] to
give almost quantitative yields. Various catalytic methods
are also described,[105,106] the use of citronellal as a hydrogen
acceptor being advised in some cases.[106]

Oxidation of geraniol or nerol with manganese dioxide
ought, in principle, to give citral, but the results are highly
dependent on the sample of manganese dioxide used,[81] and it
has been suggested that nickel peroxide (which is a similar,
though more powerful oxidant) is more effective.[107] It is
also possible to oxidize geranyl bromide (69), obtainable from
linalool (6) with hydrogen bromide at -5° (see above), with
various nitro-compounds in alkaline solution,[108,109] this
method representing one of the ways of obtaining citral from
methylheptenone without going through the geranic acid stage.

$$6 \longrightarrow \quad \underline{69} \quad \longrightarrow \quad \underline{41}$$

There are many other ways of synthesizing citral from
methylheptenone, and only a brief survey will be given here
to illustrate certain facets of typical terpene syntheses.
One type involves the addition of a two-carbon unit, as a
vinyl ether, to an acetal (e.g., of methylheptenone) in the
presence of a Lewis acid; a review of this reaction has appeared.[110] Loss of two molecules of ethanol from the acetal
thus obtained (70)† leads to citral.[111] Another method of
adding a two-carbon unit directly to **methylheptenone** represents a branch along the linalool synthesis described previously. The product of the reaction between methylheptenone
and acetylene (dehydrolinalool, 71) can be acetylated, and the

*For examples of citral cyclizations (see, e.g., p-isopropenyl-
methylbenzene, isopiperitenol, etc.).
†Methods for this elimination are mentioned in the review
quoted,[110] see also some more recent results using this type
of synthesis.[112]

70

71 R = H
72 R = COCH3

73

41

74

75

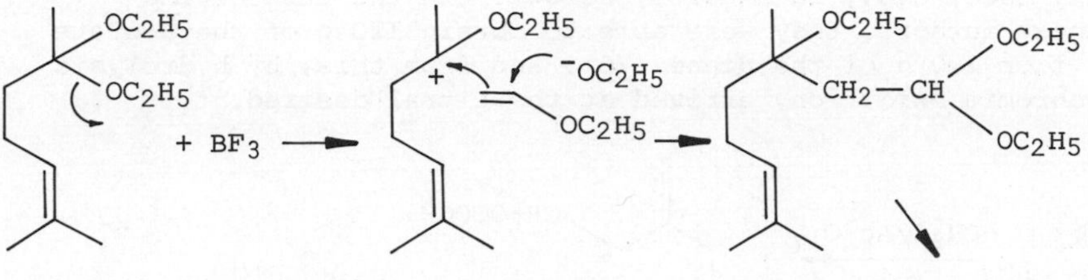

resulting acetate (72) pyrolyzed. This gives the intermediate allene acetate (73), which on hydrolysis yields citral.[113,114]

By comparison, an example of a synthesis involving successive additions of one carbon, first by formation of the hydroxymethylene compound (74), then Grignard addition of the protected aldehydoketone (75) and final dehydration,[115] is very cumbersome. If one-carbon units must be added, then it would appear better to start from a C_9 compound, as in the example of the use of the Prins reaction by Suga and Watanabe, when they obtained 7-methyl-3-methyleneoct-6-enyl acetate (77) from the reaction of formaldehyde in acetate anhydride with 2,6-dimethylhepta-1,5-diene (76).[116] Generally speaking, the Prins reaction in the case of these open chain olefins leads to complex mixtures that are not always easy to separate and

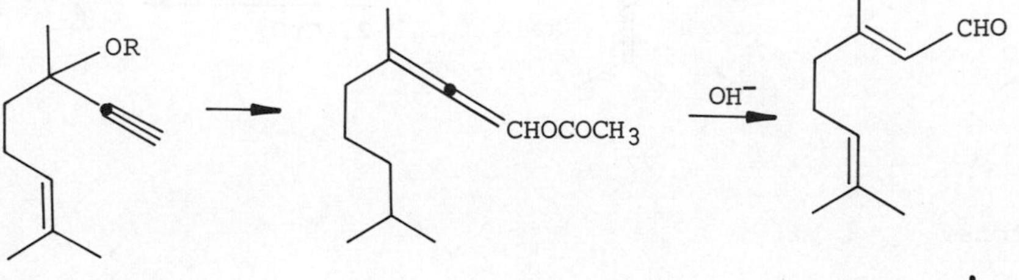

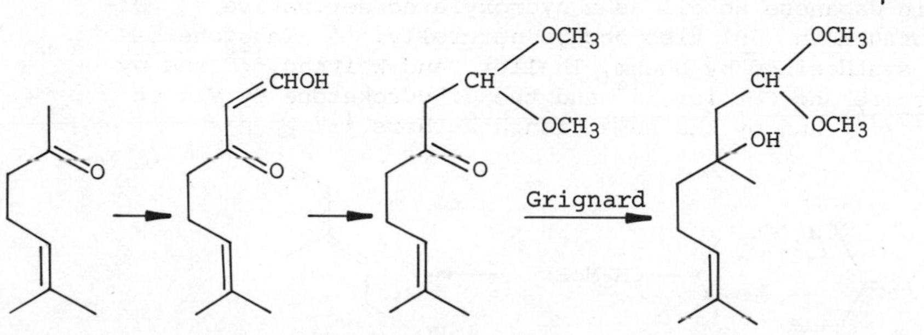

purify (see, e.g., lavandulol, below). In the case of the
Japanese authors, they were able to obtain 120 g of the acetate
(77) from 200 g of the diene (76), and from this, by hydrolysis
and chromic oxidation, arrived at the citral desired.[116]

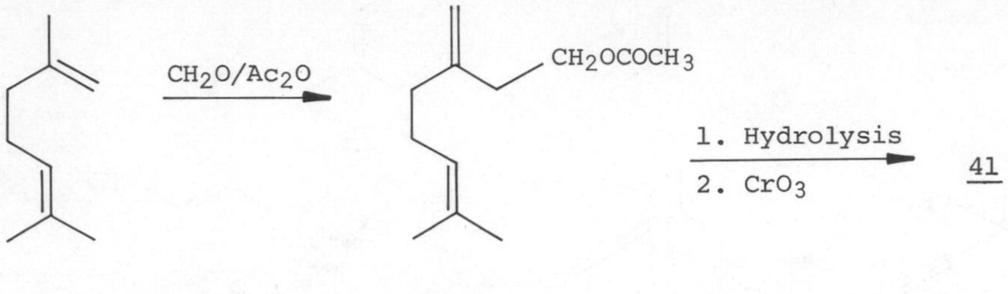

76   77

Tagetones

Tagetone, 3,7-di-methylocta-1,3-dien-5-one (78), occurs in
Tagetes glandulifera[117] and other *Tagetes* spp. (compositae),
and in Japanese Ho oil as a hydroxylated derivative.[88] Dihydrotagetone (79) also occurs naturally.[117] Tagetone has
been synthesized by Boehm, Thaller, and Whiting,[118] and by
Teisseire and Corbier,[119] and the dihydroketone by Vig et
al.[120,121] and by the same French authors.[119]

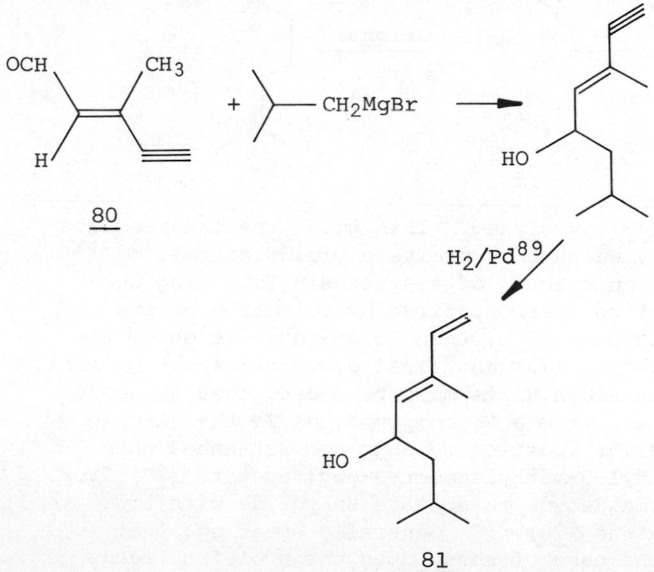

80

81

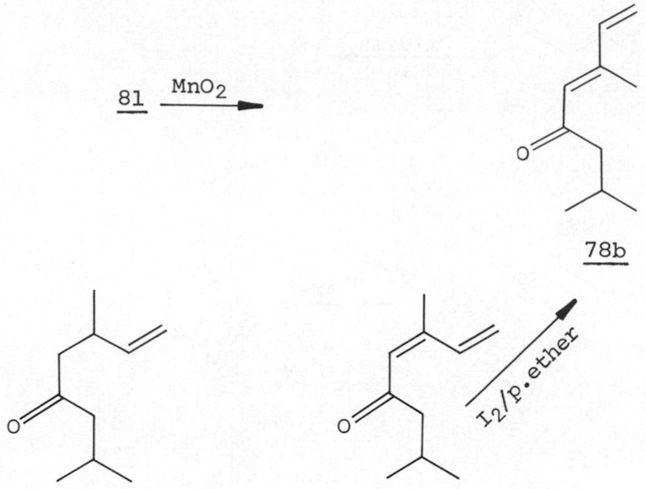

81 $\xrightarrow{\text{MnO}_2}$

78b

79   78a

The main substance in *Tagetes glandulifera* is *cis*-
tagetone (78a), although small amounts of the *trans*-isomer
(78b) occur, and it is the latter that Boehm et al. synthesized, at the same time proving the structure of the *cis*-isomer
by its conversion to the synthetic material with iodine in
petroleum ether.[118] Grignard reaction of isobutyl bromide
with the *trans*-enynal (80) gave the complete carbon skeleton
of tagetone, all that was then necessary was to reduce the
triple bond to a double bond by means of the Lindlar catalyst,[122] then oxidize the allyl alcohol (81) to a ketone with
manganese dioxide. The last step did not give very good
yields, but was adequate for the purpose of structural proof.

The synthesis of dihydrotagetone (79) by Vig et al. consists in again making the carbon skeleton by an organometallic
reaction on a 3-methylhexyl oxygenated compound. They prepared 3-methylpent-4-enal (82) by a Claisen rearrangement of
crotyl alcohol vinyl ether, and after converting this aldehyde
to the corresponding acid chloride (83), obtained the ketone
(79) directly by reaction of the organocadmium reagent from
isobutyl bromide.[120] Vig's later syntheses of dihydrotagetone[121] are shown in Scheme 7.

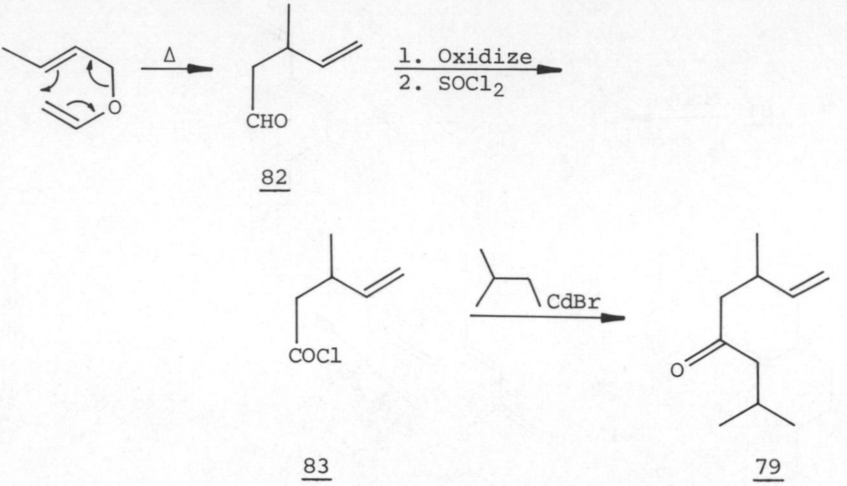

82

83   79

Scheme 7

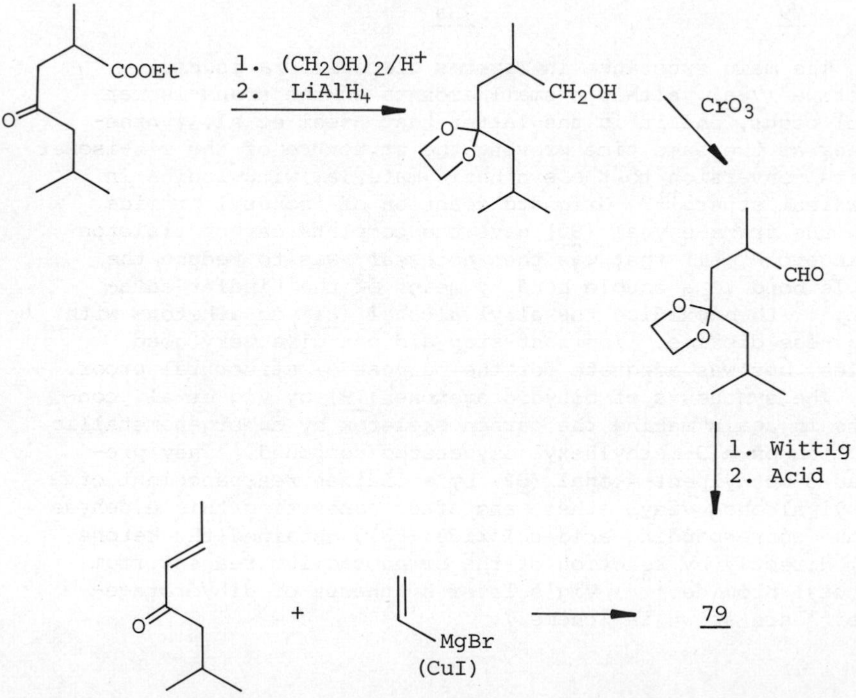

The French synthesis (Scheme 8) starts from methyl isobutyl ketone (84), which is first converted to the enol ether (85a),

Scheme 8

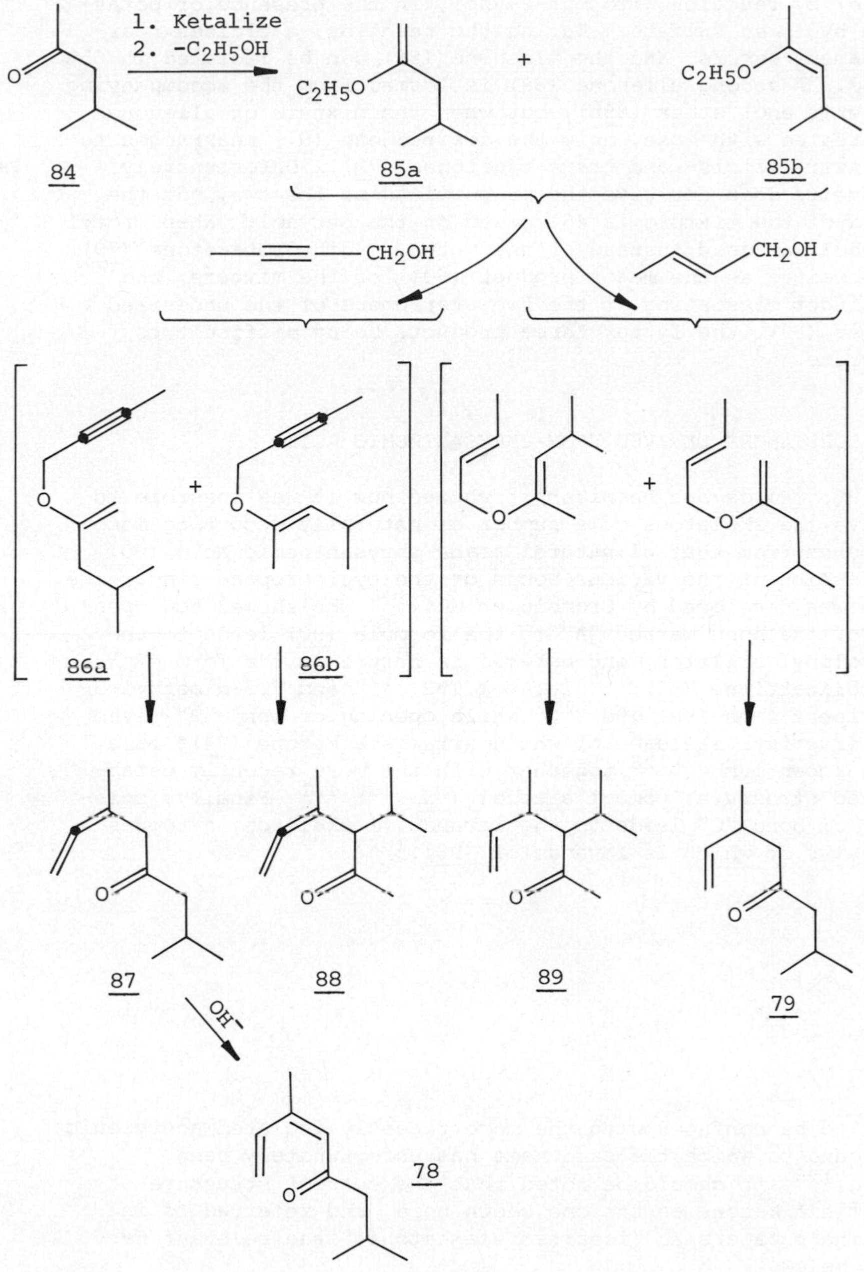

1. Ketalize
2. $-C_2H_5OH$
84
C_2H_5O
85a
+
C_2H_5O
85b
CH_2OH
CH_2OH
+
86a
86b
+
87
88
89
79
OH^-
78

from which the acetylenic ether (86) is made (but not isolated) by reaction with but-2-ynol, in the presence of potassium hydrogen sulfate. During the reaction, a Claisen rearrangement occurs, and the allenone (87) can be isolated in 65%
yield. A second allenone (88) is formed from the accompanying
isomeric enol ether (85b), but when the mixture of allenones
is treated with base, only the desired one (87) rearranges to
a mixture of cis- and trans-tagetones (78). Unfortunately,
Teisseire does not give the proportions of the two, but the
yield of the mixture is 45% based on the butynol. When crotyl
alcohol was used instead of the butynol, dihydrotagetone (79)
is obtained as the major product (60%) of the mixture, the
remainder consisting of the two stereomers of the undesired
isomer (89), the latter three products being difficult to
separate.[119]

5.   SUBSTANCES DERIVED FROM CHRYSANTHEMIC ACID

In 1965, Bates and Paknikar[123] showed how it was possible to
derive the skeletons of a number of naturally occurring monoterpenes from that of natural trans-chrysanthemic acid (90)
by fission of the various bonds of the cyclopropane ring. The
idea was developed by Crombie et al.,[124] who showed how opening of the bond marked "A" in the formula (90) leads to the
santolinyl skeleton, encountered in nature in the form of
santolinatriene (91),[125] lyratol (92),[126] and 2,5-dimethyl-3-
vinylpent-4-en-2-ol (93),[127] while opening of bond "B" gives
the artemisyl skeleton of which artemisia ketone (94)* is a
long-known member,[128] together with the more recently established structure, yomogi alcohol (95).[130,131] Finally, scission of bond "C" leads to the lavandulyl skeleton, a representative of which is lavandulol (96).[132]

*Not to be confused with the more recently isolated acetylenic
compound to which the same name has unfortunately been
given.[129] It should be noted that the correct structure of
artemisia ketone is the one shown here, and referred to in
Asahina's papers as "isoartemisia ketone," see relevant section below.

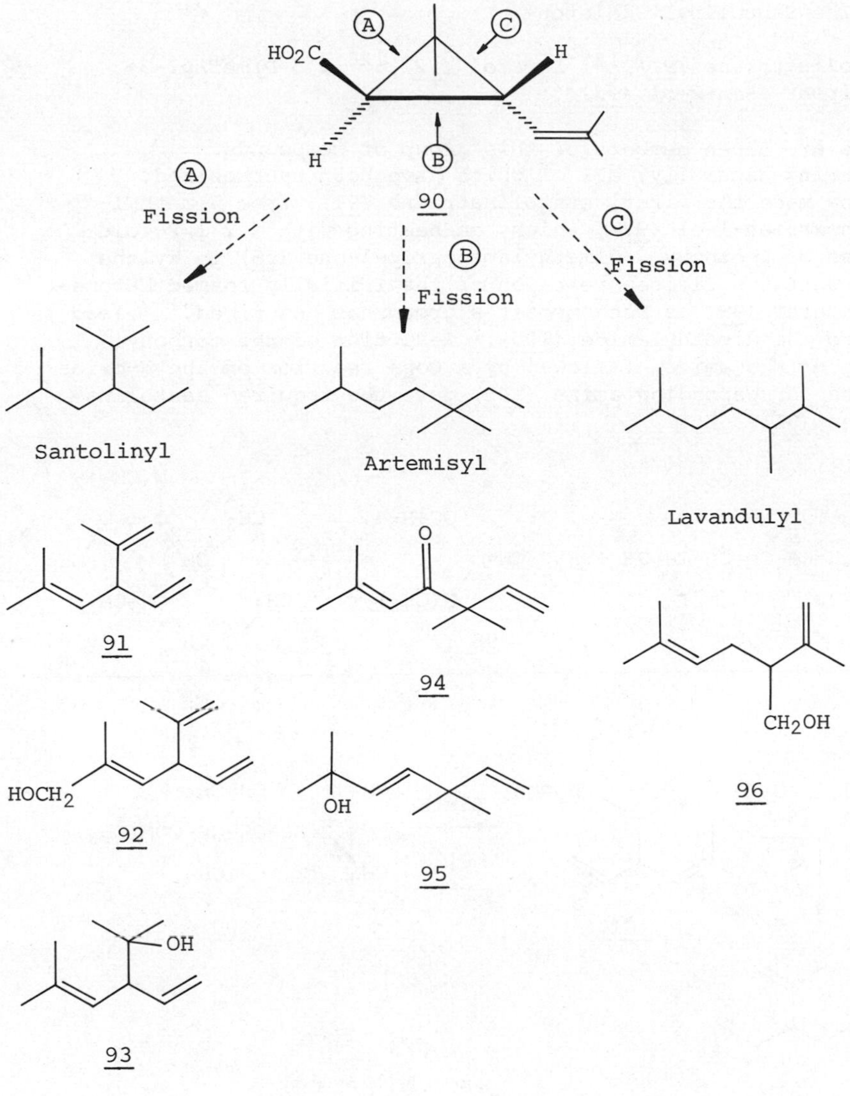

HO₂C

A

C

H

A
Fission

H

B

90

B
Fission

C
Fission

Santolinyl

Artemisyl

Lavandulyl

91

94

92

HOCH₂

95

OH

96

CH₂OH

93

OH

A. The Santolinyl Skeleton

Santolinatriene (91),[125] *Lyratol (92),*[126] *2,5-Dimethyl-3-vinylpent-4-en-2-ol (93)*[127]

There are three members of this group of compounds occurring naturally, all of which have been synthesized. Sucrow made the first, santolinatriene (91), from 5-methyl-2,4-hexadien-1-ol (97), which, on heating with a 2 1/2-fold excess of l-ethoxy-1-dimethylaminoprop-1-ene (98) in xylene, underwent the Claisen reaction of the initially formed ketene-O,N-acetal (99) as Eschenmoser's group has described,[133] leading to the dimethylamide (100). Reduction of the carbonyl group of the amide, followed by a Cope reaction of the N-oxide of the corresponding amine (101) gave the required santolinatriene (91).[134]

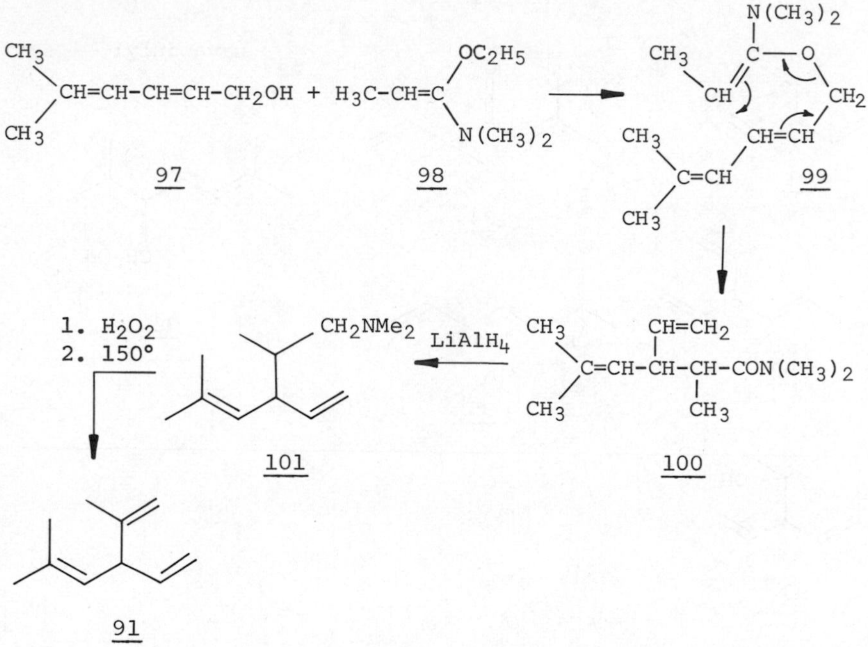

Trost and LaRochelle have described the possibility of a "biogenetic pathway" for the construction of an artemisyl skeleton by attachment of an **enzyme** to the sulfur atom in a sulfur ylid (102). After the rearrangements of this ylid (to 103), a carbonium ion can be formed (104) that can give rise to either the unchanged artemisyl skeleton--the santolinyl

skeleton (by migration of the vinyl group) or the chrysanthemic acid structure (by ring closure) as shown in Scheme
9.[135]

Scheme 9

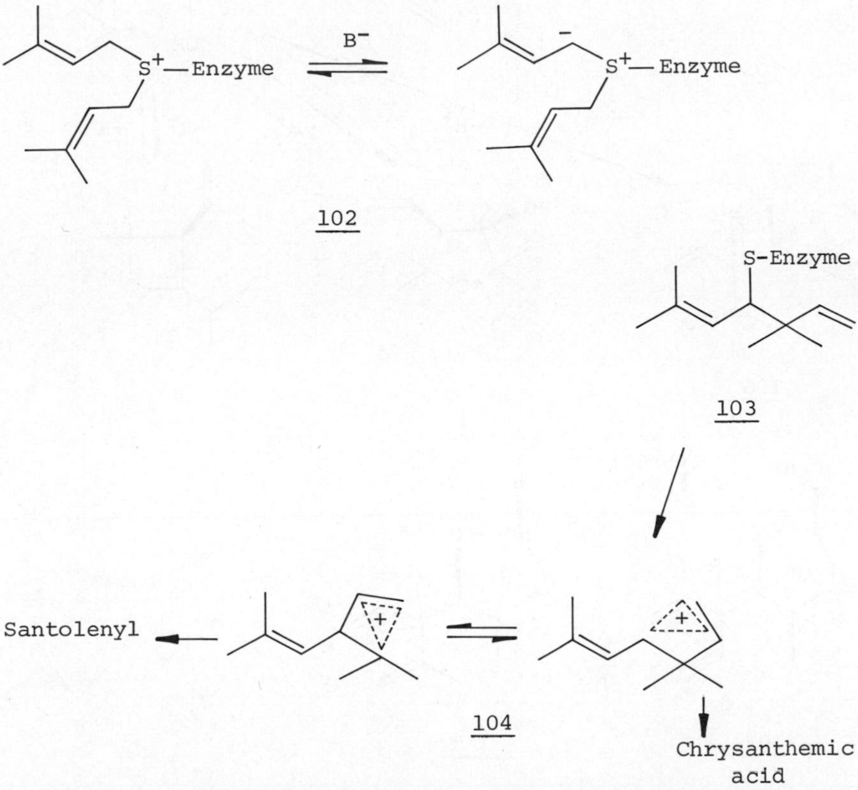

102

S-Enzyme

103

Santolenyl

104

Chrysanthemic
acid

The first step of this "biogenesis" as far as the artemisyl skeleton has been realized in the laboratory (see below),
and it has been shown that treatment of yomogi alcohol epoxide
(105, below) with p-toluenesulfonic acid yields the santolinyl
skeleton, either as cyclized substances (see Scheme 10) or as
a diol (106).[136] If the acid-catalyzed rearrangement of
yomogi alcohol epoxide (105) is carried out in the presence of
an excess of benzaldehyde, the two acetals (107a, 107b) are
the main products from the reaction, and treatment of the mixture with butyllithium in hexane[137] leads to santolinatriene
(91), although the yield is very low, competing reactions predominating (Scheme 10).[138]

Scheme 10

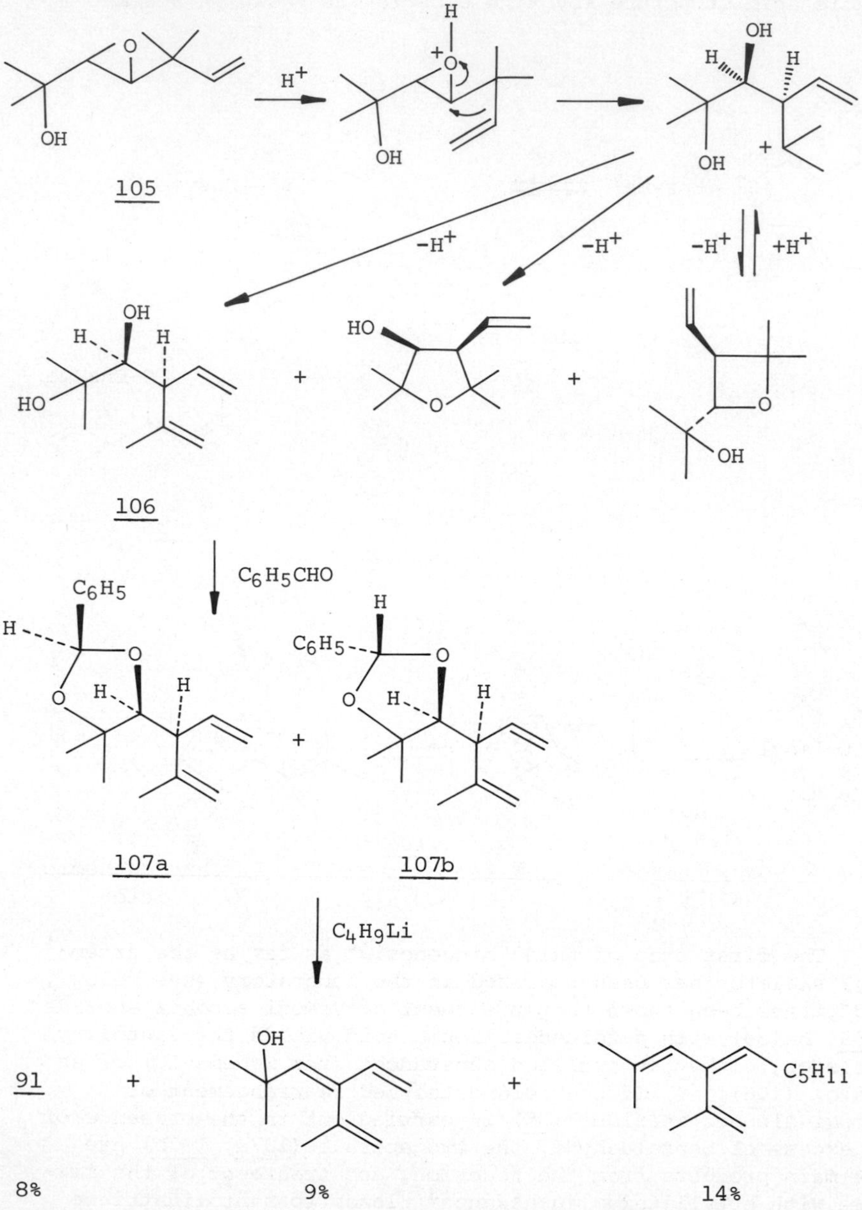

Sucrow and Richter's synthesis of the Chrétien-Bessière monoterpene (93)[127] is illustrated in Scheme 11,[139] and Sucrow's

Scheme 11

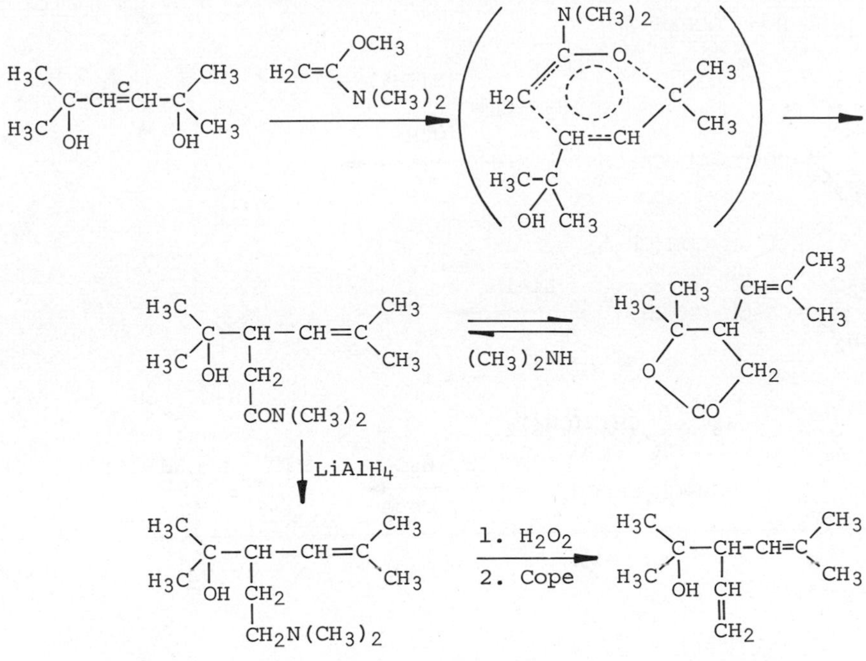

lyratol (92)[126] synthesis is in Scheme 12.[140] The principal
of the two syntheses is practically the same as for the same
author's earlier santolinatriene synthesis (see above).

Scheme 12*

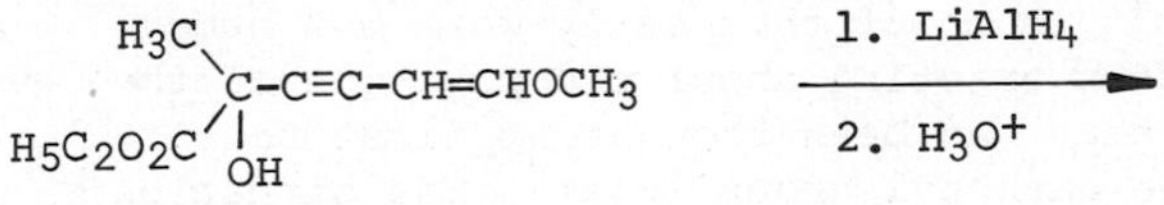

*In Scheme 12, the alcohol (108a) is acetylated (to 108b)
before proceeding with the lithium t-butoxyaluminum hydride
reduction.

$$H_3C \diagdown C=CH-CH=CH-CHO \xrightarrow{\text{LiAlH(t-BuO)}_3}$$
$$ROCH_2 \diagup$$

108a R = H

108b R = COCH$_3$

$$H_3C-CH=C \diagup^{N(CH_3)_2}_{\diagdown OCH_3}$$

$$H_3C \diagdown C=CH-CH=CH-CH_2OH \longrightarrow$$
$$AcOCH_2 \diagup$$

$$H_3C \diagdown \quad \diagup CON(CH_3)_2$$
$$CH$$
$$H_3C \diagdown C=CH-CH-CH=CH_2 \xrightarrow{\text{LiAlH}_4}$$
$$AcOCH_2 \diagup$$

$$H_3C \diagdown \quad \diagup CH_2N(CH_3)_2$$
$$CH$$
$$H_3C \diagdown C=CH-CH-CH=CH_2 \quad \begin{array}{l} 1.\ H_2O_2 \\ \hline 2.\ Cope \end{array}$$
$$HOCH_2 \diagup$$

$$H_3C \diagdown C \diagup CH_2$$
$$\diagdown CH-CH=CH_2$$
$$H_3C \diagdown C=C \diagup$$
$$HOCH_2 \diagup \quad \diagdown H$$

92

B.   The Artemisyl Skeleton

Artemisia Ketone, Alcohol, and Acetate, and Yomogi Alcohol

The first authentic synthesis of artemisia ketone (94) was
that of Colonge and Dumont in 1944,[141] although the carbon
skeleton in its reduced form had been made much earlier by
Ruzicka and Reichstein.[142] Much of this early work was ham-
pered by the confusion then existing about the position of the
conjugated double bond, that had been thought at first to be
in the βγ-position to the carbonyl group (94a). The situation
was resolved by Takemoto and Nakajima in 1957,[143] who demonstrated that the earlier "isoartemisia ketone" was actually
the correct structure for the natural product (94) and suggested that the unconjugated isomer (94a) did not exist. Actually, Colonge's synthesis gives initially the unconjugated

isomer* in only small amounts by reaction of 2,2-dimethylbut-3-enoyl chloride (109) with isobutene in the presence of stannic chloride,[141] the intermediate carbonium ion (110) cyclizing to give the six-membered ring carbonium ion (111) that stabilizes itself by ring contraction to the undesired major product (112).[145]

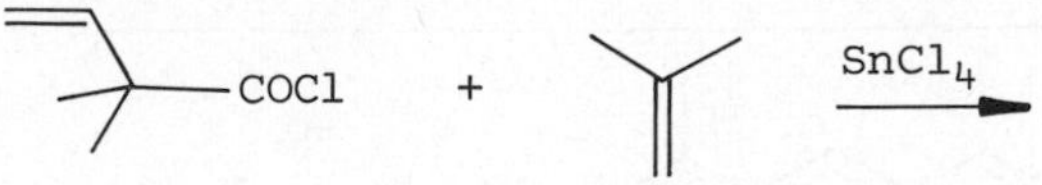

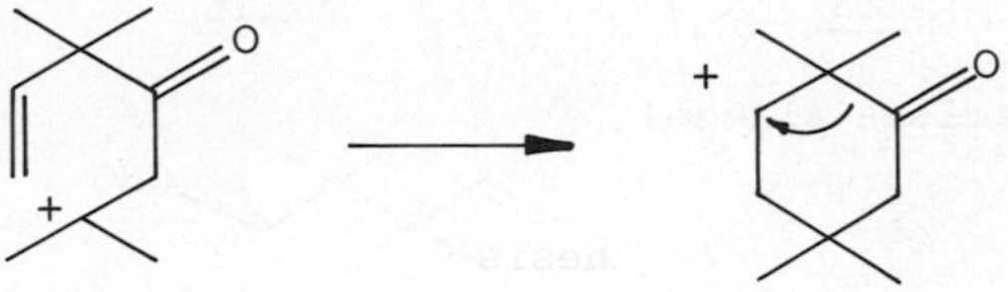

Based on the rearrangement of allyl sulfonium ylides,[146,147] Rautenstrauch,[148] and independently, Baldwin et al.,[149] developed a synthesis of artemisia alcohol, from which the acetate and the ketone are accessible. When diprenyl ether (113) is treated with butyl lithium at -25°, the main product (67%) is artemisia alcohol (114), by a [2,3]-sigmatropic change, together with three byproducts (115 to 117).

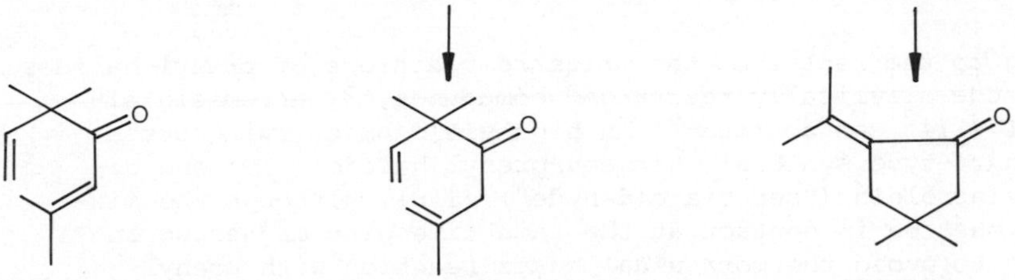

*This reaction has more recently been checked by Battersby and Rowlands, and the unconjugated ketone (94a) found to rearrange very readily to artemisia ketone (94).[144]

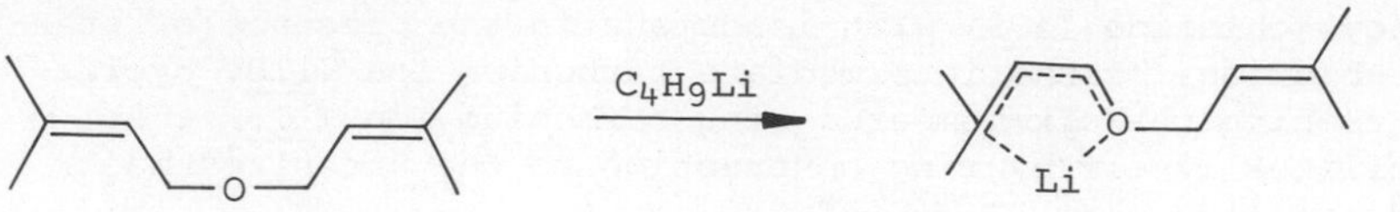

113

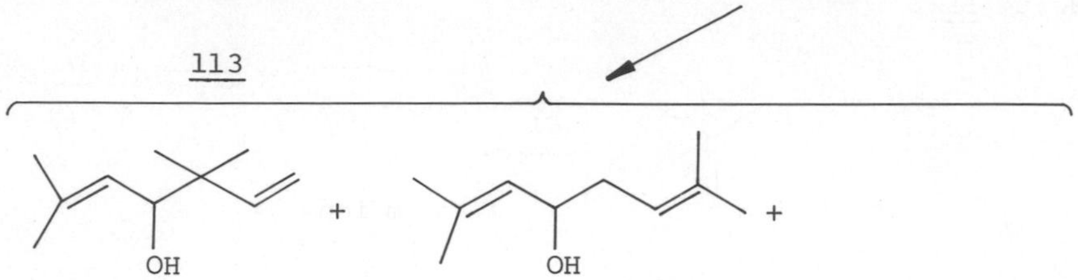

114   115

Artemisia Alcohol

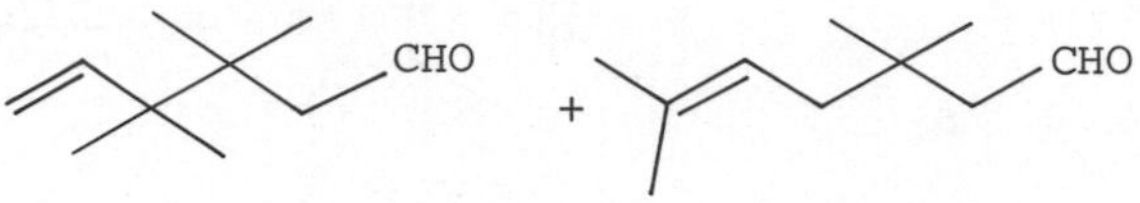

116   117

Owing to the fact that the Grignard reactions of prenyl halides
give the allylically rearranged compounds,[150] artemisia alcohol (114) can be made[151] in high yield by carrying out a
Grignard-type synthesis between prenyl halide (119) and di-
methylacrolein ("senecia aldehyde") (118), although the mixture must be in contact at the same time with magnesium in
order to avoid the more usual Wurtz reaction with prenyl
halides.[150,151]

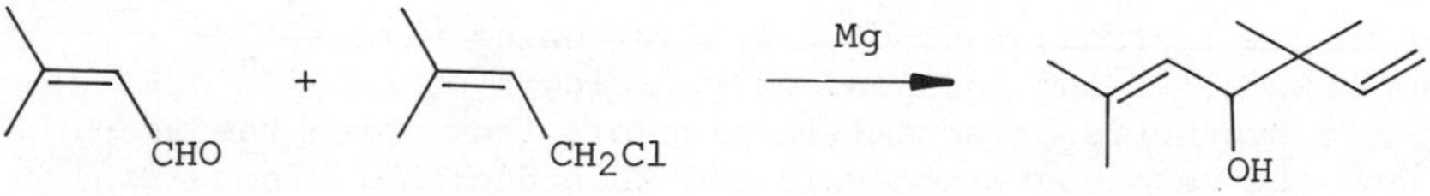

118   119   114

Yomogi alcohol (95) has also been synthesized from prenyl
chloride (119), the first step this time being the Wurz coupling with magnesium,[150] this giving ready access to the diene
(120) with all the carbon atoms in the correct position.
Photooxidation in the presence of a sensitizer yields two
photoperoxides, that can be converted with sodium sulfite to
the two alcohols, yomogi alcohol (95) and (121), in the ratio
of about 5:1.[130] Although conventional oxidation procedures

applied to yomogi alcohol yield traces of artemisia ketone, the main product is always the alcohol epoxide (105) that can be induced to rearrange to the santolinyl skeleton, but which does not readily give artemisia alcohol or ketone.[136]

119   120

95   121

Another synthesis of yomogi alcohol by Sucrow from the known[152] 2,2-dimethylbut-3-enal uses a Wittig reaction.[153,154]

$$H_2C=CH-\underset{\underset{CH_3}{|}}{\overset{\overset{CH_3}{|}}{C}}-CHO \quad + \quad (C_2H_5O)_2P\overset{-}{O}CHCOOC_2H_5 \longrightarrow$$

$$H_2C=CH-\underset{\underset{CH_3}{|}}{\overset{\overset{CH_3}{|}}{C}}-CH=CHCOOC_2H_5 \xrightarrow{CH_3Li} H_2C=CH-\underset{\underset{CH_3}{|}}{\overset{\overset{CH_3}{|}}{C}}-CH=CH-\underset{\underset{CH_3}{|}}{\overset{\overset{CH_3}{|}}{C}}-OH$$

C. The Lavandulyl Skeleton

Lavandulol, β-Cyclolavandulal

Schinz and co-workers started a series of studies on the synthesis of lavandulol shortly after its discovery in lavendar,[132] most of the early versions of which consisted in building up a suitable hydroxylated derivative (such as 122[155] or 123[156]) and then dehydrating them by, for example, pyrolysis of the acetates. A disadvantage of this type of synthesis is that the dehydration gives rise to other double bond isomers, such as the not naturally occurring isolavandulol (124) or tetrahydropyran types, such as 125, a type of cyclization that had been recognized since 1932.[157]

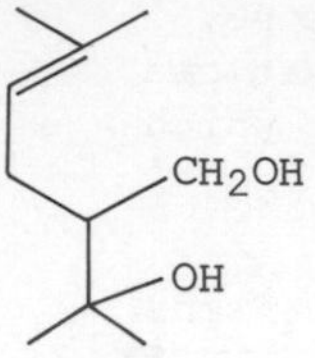

122 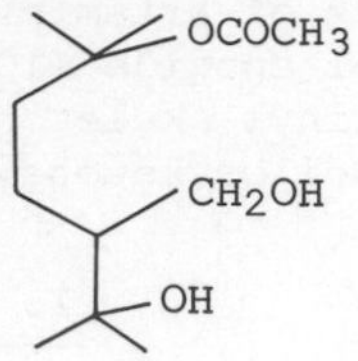   123 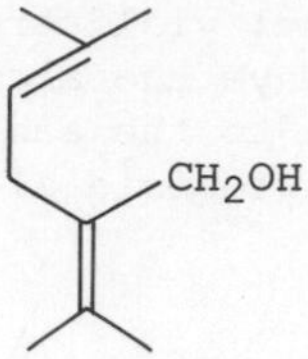   124 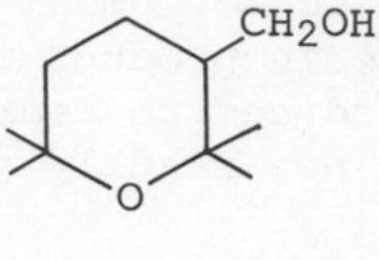   125

This disadvantage was overcome in the synthesis of Brack and Schinz[158] by starting with 4,4-dimethylhex-5-en-2-one (126). By means of ethoxyacetylene, this could be converted into the acetylenic ether (127), that gave the unsaturated ester (128) in the presence of acid. This ester was rearranged thermally in a Cope rearrangement to lead exclusively to the ethyl ester (129) corresponding to the alcohol lavandulol (96), which could then be obtained conventionally.

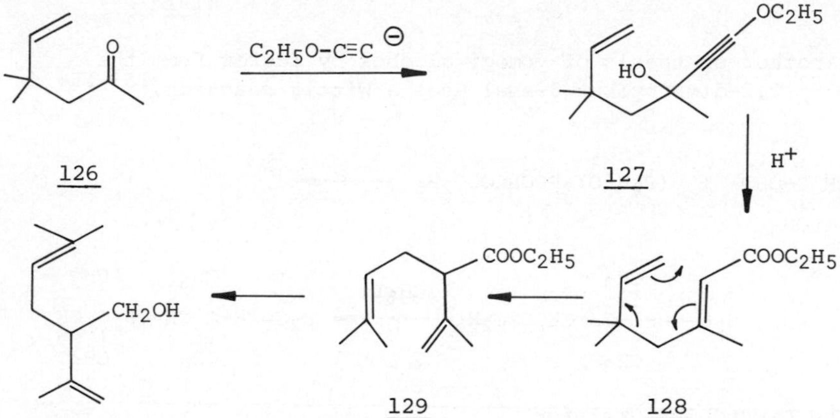

126   127

129   128

96

Another synthesis that depends primarily on an electrocyclic rearrangement is that of Matsui and Stalla-Bourdillon,[159] and is based on the fact that the enolates of allyl esters rearrange in a Claisen-like reaction.[160] Accordingly, the ester from 1,1-dimethylallyl alcohol and 3-methylbut-2-enoic acid (i.e., 130), when treated for 2 hr with sodium hydride in toluene at 110° gives a 7:3 ratio of lavandulate (131) to isolavandulate (132).

$$\underline{130} \xrightarrow{\text{NaH}} \qquad \longrightarrow$$

$$\underline{131} \qquad + \qquad \underline{132}$$

A synthesis of lavandulol by Teisseire and Rinaldi[161] consists in adding the dimethylallyl group to ethyl aceto-acetate to obtain the β-ketoester (133). The carbonyl group of this compound can be converted smoothly to a methylene group by means of the Wittig reaction, leading to the ethyl ester (129) of **lavandulic acid.**

$$CH_3COCH_2COOC_2H_5 \quad \xrightarrow{(NaNH_2)} \quad CH_3CO-CH-COOC_2H_5 \quad \xrightarrow{\text{Wittig}}$$

$$BrCH_2CH=C \begin{smallmatrix} CH_3 \\ \\ CH_3 \end{smallmatrix}$$

$$CH_2CH=C \begin{smallmatrix} CH_3 \\ \\ CH_3 \end{smallmatrix}$$

$$\underline{133}$$

$$\begin{array}{c} CH_2 \\ \parallel \\ CH_3C-CH-COOC_2H_5 \\ \mid \\ CH_2CH=C \begin{smallmatrix} CH_3 \\ \\ CH_3 \end{smallmatrix} \end{array}$$

$$\underline{129}$$

Prenyl bromide can, in fact, be coupled to give the lavandulyl skeleton directly; zinc chloride in carbon tetra-chloride yields the dibromide (134) that loses one molecule of hydrogen bromide in the presence of potassium acetate in acetic acid (to 135), the second bromine being replaced by hydroxyl via the corresponding Grignard reagent. The lavan-dulol thereby obtained (96) is accompanied by the coupling product from the competing Wurtz reaction (136). The appar-ently simpler route from the monobromide (135) leads to elimi-nation of hydrogen bromide resulting in the hydrocarbon (137).[162] This same hydrocarbon results when the allyl rearrangement product (138) from isolavandulol* (124) is dehydrated thermally.[163]

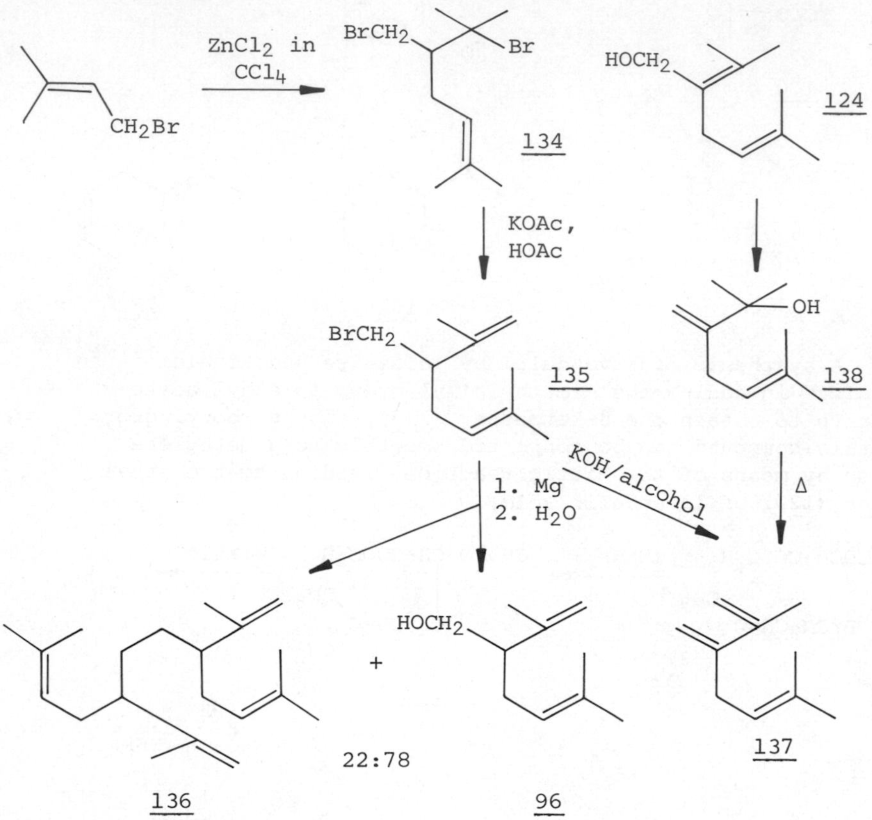

ZnCl$_2$ in CCl$_4$

BrCH$_2$ ... Br   134

HOCH$_2$ 124

KOAc, HOAc

BrCH$_2$ 135

OH   138

KOH/alcohol

1. Mg
2. H$_2$O

Δ

HOCH$_2$

137

+

22:78

136   96

One of the problems associated with the use of the Prins reaction for the preparation of compounds in the lavandulyl series, for instance by reaction of formaldehyde with the acetate (139),[156] is the presence of a large number of byproducts, only 33% of lavandulyl acetate (140) being obtained by this route.[156,163] Another attempt by Baba et al.[164] starts from the same alcohol (141), but converts this first to a mixture of hydrocarbons (76)[†] and (142) that is separated by fractional distillation. Treatment of the symmetrical hydrocarbon

*This rearrangement can occur when isolavandulol is distilled slowly over columns containing metal packing.[163]
†The Prins reaction of this hydrocarbon leading to isogeranyl acetate has been discussed above.[116]

with formaldehyde in acetic acid gives 15-20% of lavandulol.[164]

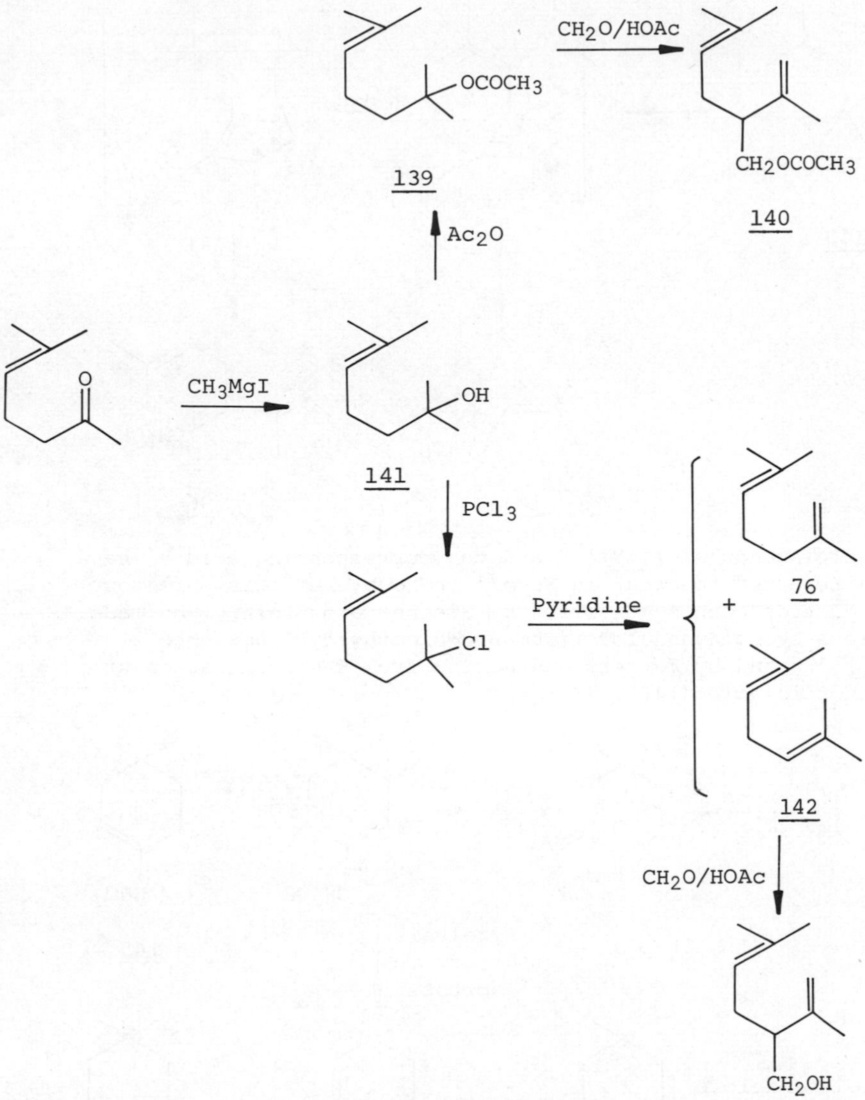

Lavandulol (<u>96</u>) has also been obtained by irradiation of chrysanthemol [<u>143</u>, obtained by the reduction of chrysanthemic acid (see below)] with a high-pressure mercury lamp. The main product apart from lavandulol is 3-methylbut-2-enol (<u>144</u>), both substances being obtained in about 20% yield:[165]

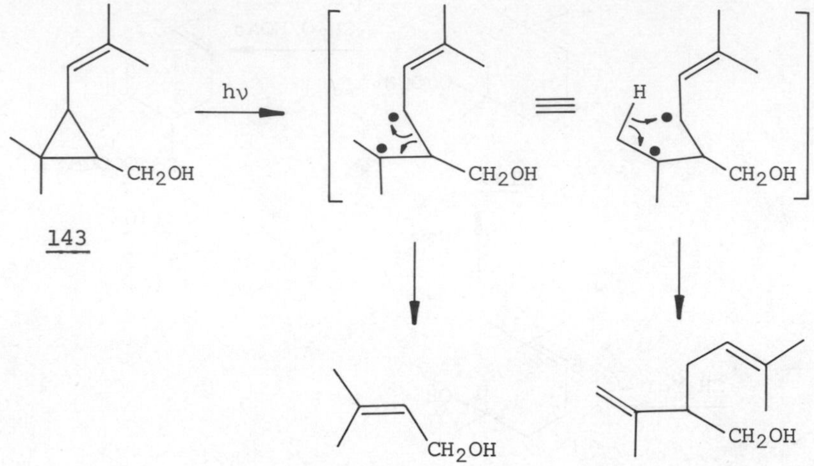
hν
H
CH2OH
CH2OH
CH2OH
CH2OH
CH2OH

143

144

96

β-cyclolavandulal (145)[166] and the corresponding acid[167] have been reported to occur in *Seseli indicum*, but this skeleton had already been synthesized by Steiner and Schinz, who made various cyclolavandulols from 3,3-dimethylcyclohexanone (146),[168] and by Ferrero and Schinz from the cyclization of lavandulic acid (147).[169]

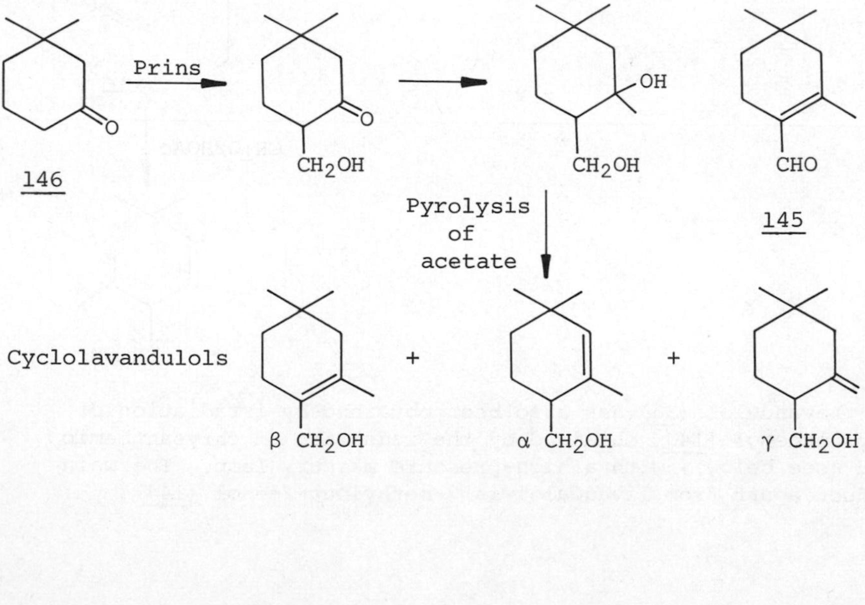
Prins
O
O
CH2OH
OH
CH2OH
CHO
Pyrolysis
of
acetate
Cyclolavandulols
+
+
β CH2OH
α CH2OH
γ CH2OH

146

145

147

D. Chrysanthemic Acids

Dalmatian pyrethrum flowers, *Chrysanthemum cinerariifolium*,
contain the insecticidal pyrethrum* that was shown in 1924 to
be an ester of *trans*-chrysanthemic acid (90b, R = H) by
Staudinger and Ruzicka.[171] A somewhat different structure had
been developed independently by Yamamoto[172] but was later
abandoned. In view of the insecticidal properties of many
chrysanthemic acid esters, the synthesis of compounds having
this carbon skeleton has received more attention than most
monoterpenes, and indeed the first synthesis dates from the
same time as the structural proof.[173] The doubtful purity of
the starting material used by Staudinger et al. made the
purification of the products extremely difficult at the time,
however, and it was not until 20 years later that Campbell and
Harper were able to repeat the earlier synthesis and show that
the natural chrysanthemic acid is actually the (+)-*trans*-isomer
(90b, R = H).[174] This synthesis consists in the addition of
ethyl diazo-acetate to 2,5-dimethylhexa-2,4-diene (148),†
giving a mixture of the *cis*- (90a, R = C_2H_5) and *trans*- (90b,
R = C_2H_5) racemic esters, that were hydrolyzed to the acids.
The latter were separated by crystallization, and the *trans*
acid resolved by crystallization of the quinine salt. In much
the same way, Crombie et al. were able to make the dicarboxylic acid [pyrethric acid (149, R = H) also a constituent of
pyrethrum] from 2,5-dimethylhexa-2,4-dienoic acid (150).[176]

148   90a   90b

*For a review of the earlier work in this field, see the de-
scription of the chemistry of the pyrethrins by Crombie and
Elliott.[170]
†A similar synthesis using the t-butyl ester has also been
described.[175]

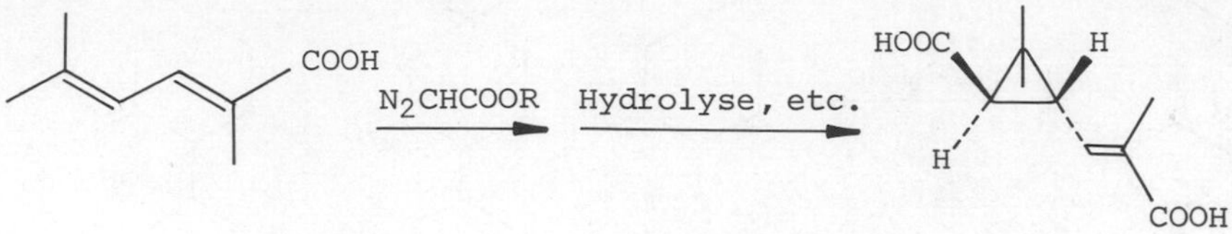

150   149

It was found by Julia et al. that bicyclo[3.1.0]hexan-
2-ones can be readily converted into chrysanthemic acids.[177]
Thus, the oxime of tetramethylbicyclo[3.1.0]hexan-2-one
(151), prepared from the diazoketone (152), undergoes the
Beckmann reaction with phosphorus pentachloride in ether,
yielding mainly the nitriles of *cis*-chrysanthemic (153 and
cis-isochrysanthemic (154) acids (Scheme 13). The potassium

Scheme 13

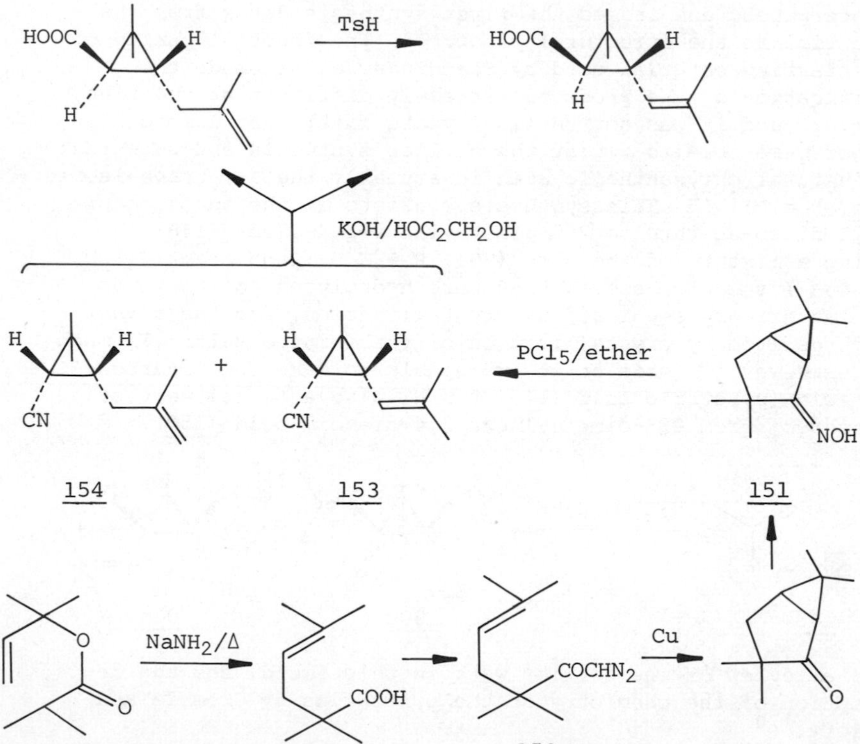

154   153   151

152

hydroxide in ethylene glycol required to hydrolyse the nitriles causes simultaneous isomerization* to the *trans*-acids;[179] double bond isomerism is not, however, complete under these conditions, but can be effected by *p*-toluenesulfonic acid in xylene.[180]

A somewhat different application of a bicyclo[3.1.0]-hexan-2-one is the synthesis of *cis*-isopyrethric acid (155) by Julia and Linstrumelle from 6,6-dimethylbicyclo[3.1.0]hexan-2-one (156),[181,182] a compound that was correlated with the known[183] *cis*-homocaronic acid (157). The bicyclic ketone was made either from the diazoketone (158), or from pinonic acid (159) obtainable from pinene (17) (Scheme 14),[184] but the

Scheme 14

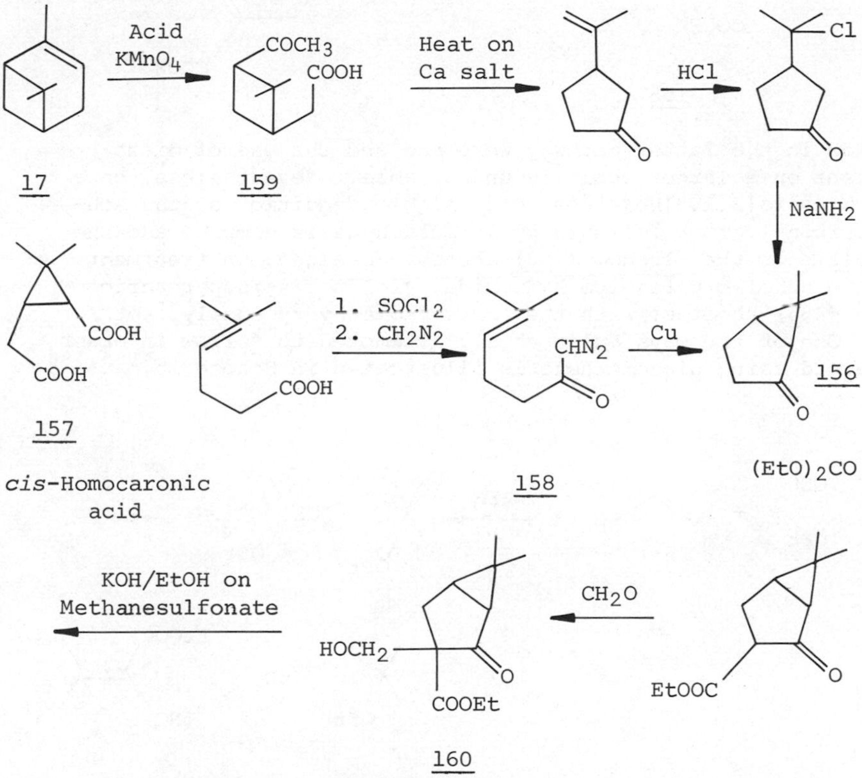

17   159

157

cis-Homocaronic
acid   158   (EtO)$_2$CO

KOH/EtOH on
Methanesulfonate   CH$_2$O

160

*Actually, the easiest way of isomerizing *cis*-chrysanthemate esters to the *trans* compounds is simply by pyrolysis of the *cis*-ester at 240°-260°, when the *trans*-ester is obtained in good yield.[178]

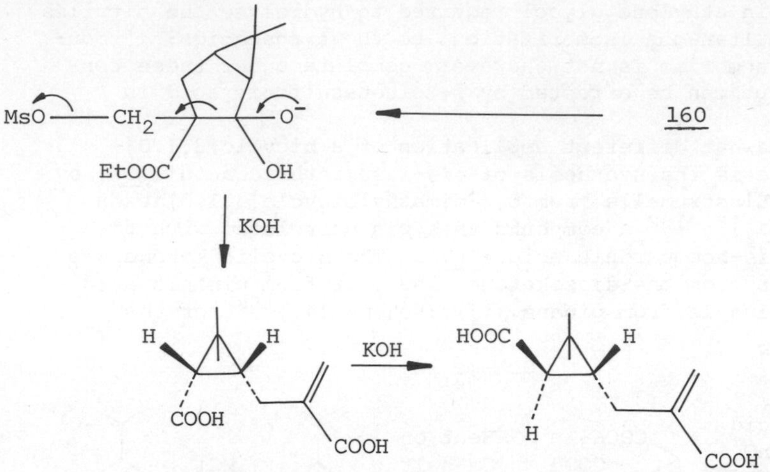

yields in the latter pathway were bad and the use of diazomethane on a larger scale is undesirable. Nevertheless, once
the bicyclo[3.1.0]hexanone is available, addition of the eth-
oxycarbonyl group followed by formaldehyde is simple, and the
mesylate of the alcohol (160) thereby obtained, on treatment
with alcoholic potassium hydroxide, yields *cis*-isopyrethric
acid (155) together with the *trans*-isomer very easily.

One of the ways Julia et al. attempted to follow in order
to avoid using diazomethane is illustrated in Scheme 15,[185,186]

Scheme 15

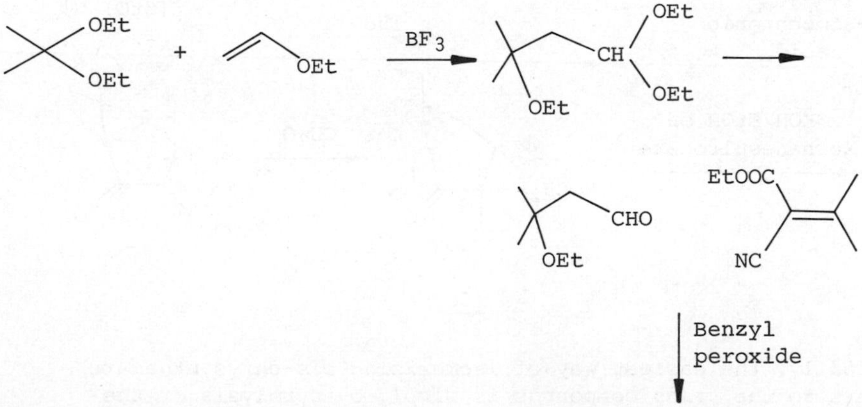

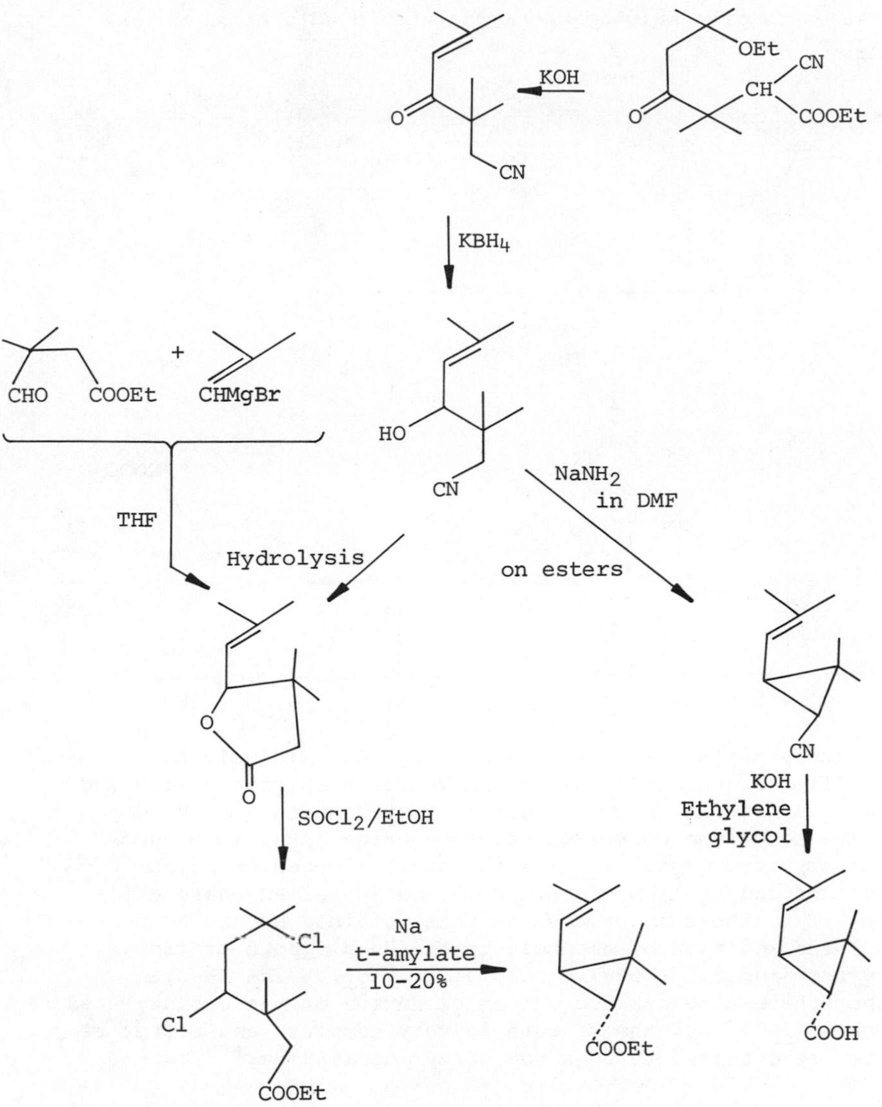

KOH

OEt
CN
CH
COOEt

CN

KBH₄

CHO   COOEt   +   CHMgBr

HO

CN

NaNH₂
in DMF

THF

Hydrolysis

on esters

CN

KOH
Ethylene
glycol

SOCl₂/EtOH

Cl

Cl

COOEt

Na
t-amylate
10-20%

COOEt

COOH

but this too involves a rather large number of steps, and a more rapid method for the synthesis of chrysanthemic esters is that of Martel and Huynh, that consists in allowing an ester of 3-methylbut-2-enoic acid to react with dimethylallyl sulfone (<u>161</u>), together with two equivalents of potassium t-butoxide in tetrahydrofuran, when the intermediate ion (<u>162</u>)

cyclizes to give only *trans*-chrysanthemic acid ethyl ester (163).[187]

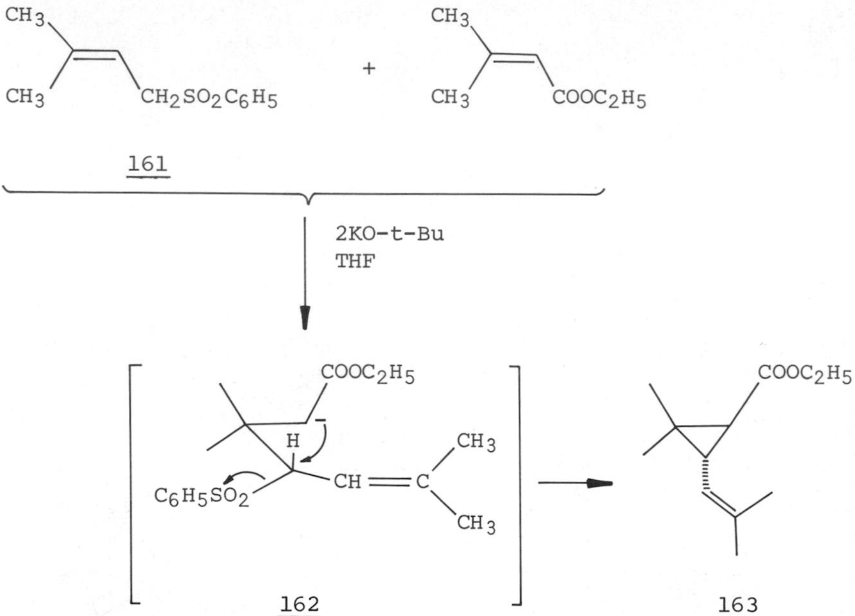

161

2KO-t-Bu
THF

162   163

There are a few syntheses of chrysanthemic acid that start from Δ^3-carene (164), readily available from Polish and especially Indian (*Pinus longifolla*) turpentine.[188] Volkov and Khachatur'yan converted carene epoxide (165) into the monobenzylated glycol (166) with benzyl alcohol in sulfuric acid, and dehydration of this followed by permanganate oxidation led to the same product as that obtained by the acid-catalyzed addition of alcohols to the double bond of *cis*-chrysanthemum dicarboxylic acid (i.e., 167). The Russian authors have also carried out an oxidation of the debenzylated alcohol (168), but the mixture is very complex, and better results are obtained through the benzyl derivative.[189]

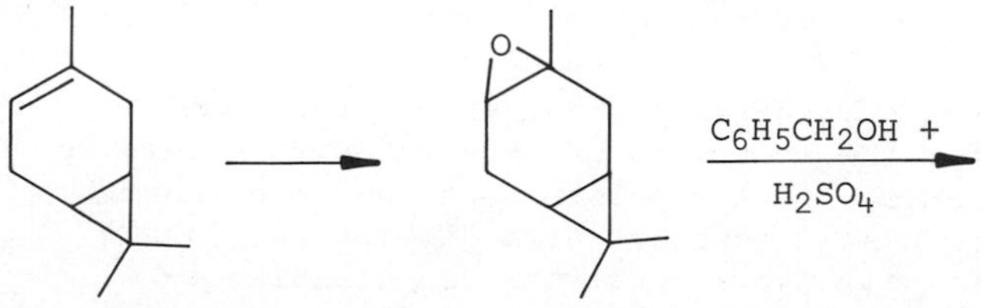

Δ^3-Carene   164   165

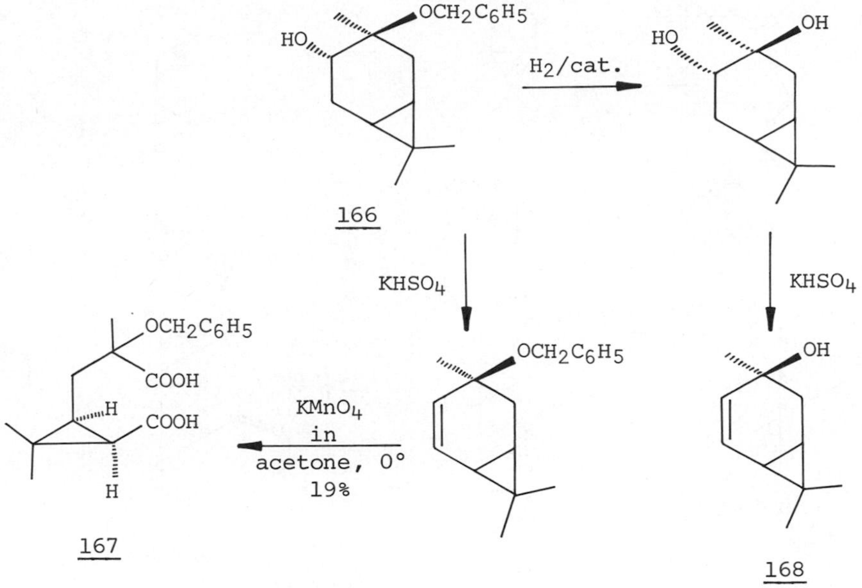

166

167

168

A synthesis of Yoshioka et al.[190] also starts from Δ^3-carene, the first step being ozonolysis to the ketoaldehyde (169).[191] The latter can be converted into either (-)-*trans*- or (+)-*trans*-chrysanthemic acids (Scheme 16).

Scheme 16

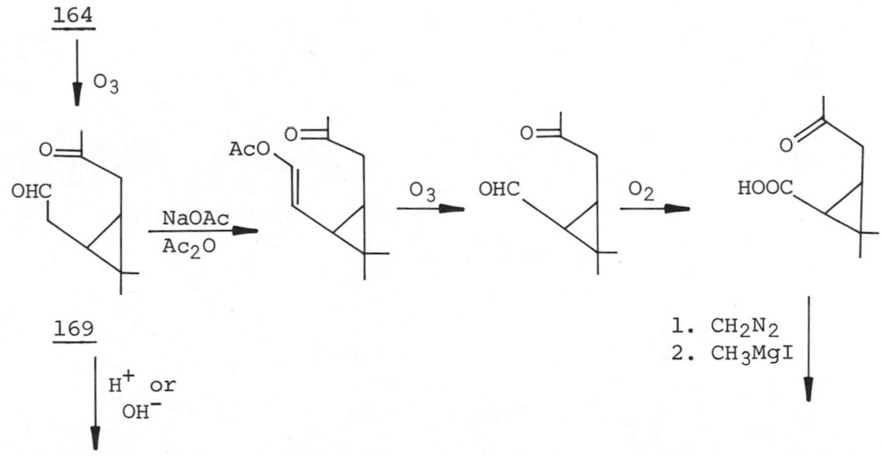

164

169

Chrysanthemic Acids

The synthesis of pyrethric acid (or its esters) from chrysanthemic acid (or its esters) has been carried out in several places, always following a very similar route (Scheme 17).

Scheme 17

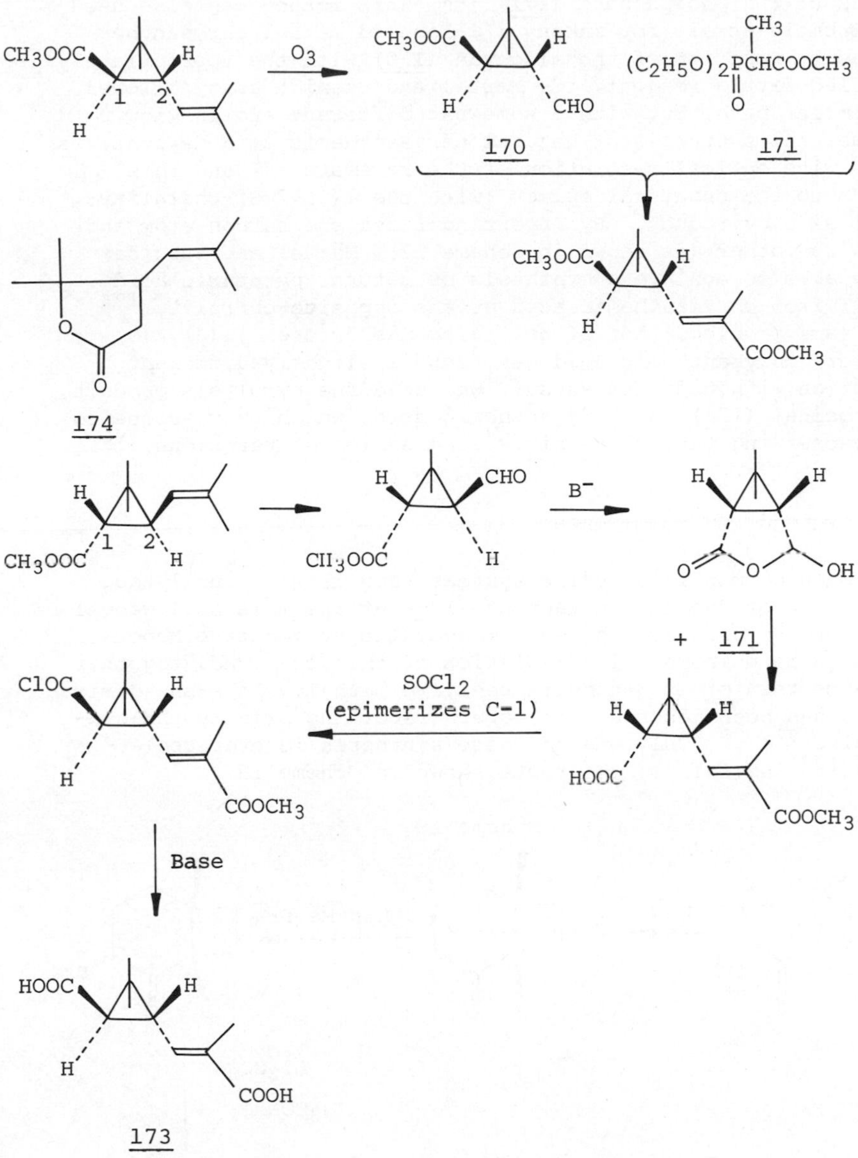

Ueda and Matsui made all four geometrical isomers of (±)-
pyrethric acid by converting the appropriate (*cis*- or *trans*-)
methyl ester of chrysanthemic acid to the aldehyde with ozone
(170, for example, represents the aldehyde obtained from the
trans- ester), and putting on the appropriate side chain by
means of the phosphonate (171).[192] This method was also used
by Crombie et al. for making ^{14}C-labeled methyl chrysanthe-
mates by reaction of the aldehyde (170) with the appropriately
labeled Wittig reagent.[193] Martel and Buenida also followed
a similar path, but with a somewhat different end in view.
Commercial synthesis of natural chrysanthemic acid derivatives
ends with optical resolution of the racemate,[194] and this
leads to the unnatural epimer [with the 1(S),2(S)-chirality,
172] as a byproduct. By inverting first one carbon atom and
then the other (as shown in Scheme 17), Martel and Buenida
were able to achieve a synthesis of natural pyrethric acid
(173) from chrysanthemic acid of the opposite chirality.[195]
The same problem, that of utilizing the "wrong" [1(S),2(S)-
epimer] chrysanthemic acid has found a slightly different
solution with Ueda and Matsui, who used the pyrolysis product,
"pyrocine" (174), of chrysanthemic acid, which they succeeded
in racemizing through a fairly long series of reactions.[196]

6. CYCLOBUTANE MONOTERPENES

Although common in bicyclic systems (see later), until the
isolation of the sex attractant (175) of the male boll weevil
(*Anthonomus grandis* Boheman), monocyclic cyclobutane monoterpenes were unknown. The isolation of this compound, together
with certain other 3-ethylidene-1,1-dimethylcyclohexane derivatives has been described in papers reporting briefly the synthesis.[197,198] Only one of these syntheses is stereoselective,[198] and follows the route shown in Scheme 18.

Scheme 18

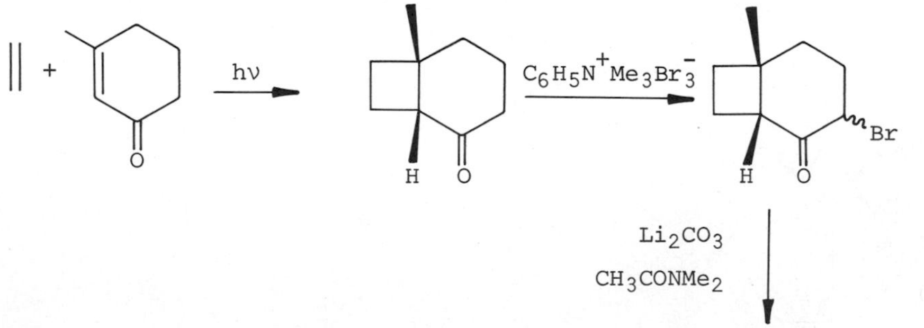

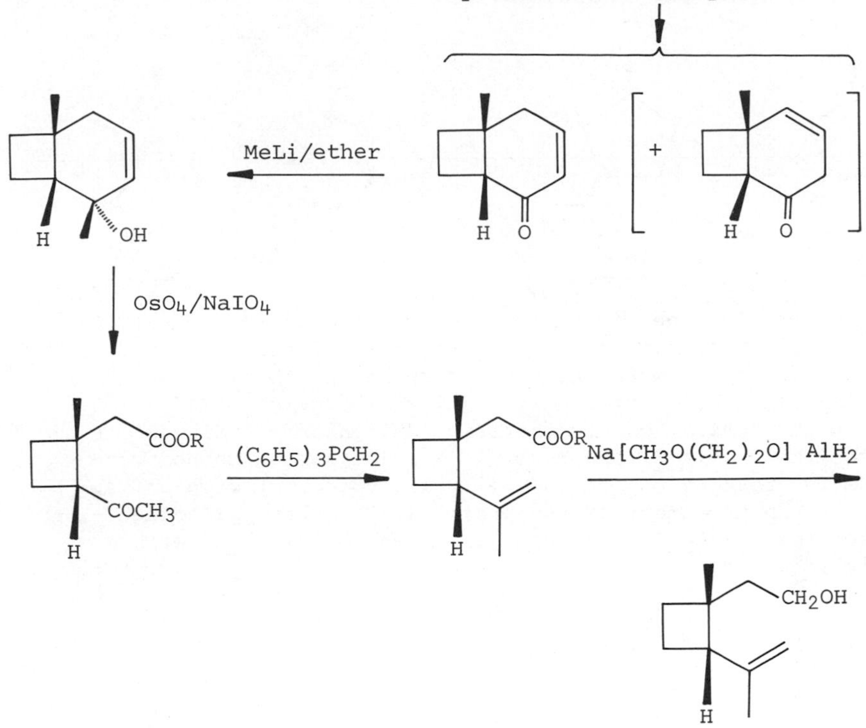

7. CYCLOPENTANE MONOTERPENES

Although the only two monoterpene cyclopentane hydrocarbons
("chamene"[199] and "osmane"[200]) to be claimed as naturally occurring were both later shown to be spurious (the former being
probably a mixture of other monoterpenes[201] and the latter
arising from the solvent used in the extractions[202]), the
basic hydrocarbon of the cyclopentane monoterpenes, 2,3-
dimethylisopropyl cyclopentane (-), for which the name "iridane" has been suggested,[203,204] is of some importance since
it was suggested as the unit from which the indole alkaloids[205,206] and the quinuclidine ring of quinine[207] might
arise biogenetically. All the possible stereoisomers of the
structure (176a to 176d) have been synthesized by Sisido et
al.[202] and by Crowley.[27]

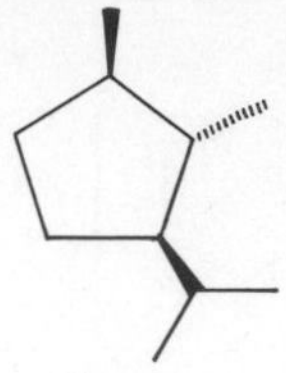

176a 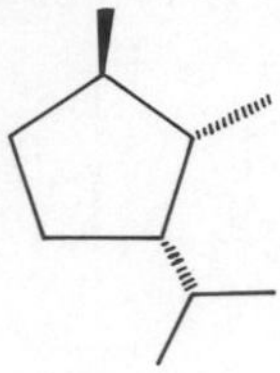   176b 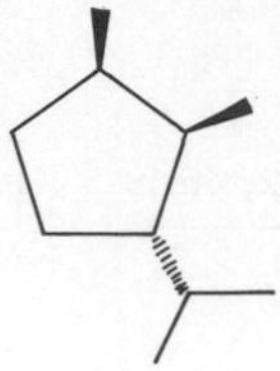   176c 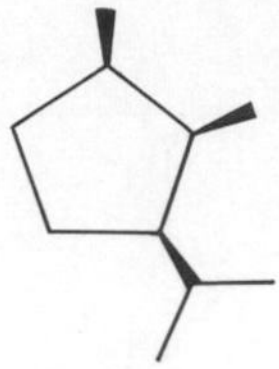   176d

Generally, one of the two methods is used for access to
the cyclopentane monoterpenes. The first of these is by
cyclization of an open-chain monoterpene, examples of which
are the dehydration of linalool (6) to the plinols (177)[208]
(see next section) or the photocyclization of citral (41) to
photocitrals A (178) and B (179).[209] The second method consists in breaking a bond in a bicyclic compound, such as happens in the treatment of thujone[210] (180) or thujyl toluene-
sulfonates[211] (181) with acid, or pinene oxide (182) with a
Lewis acid.[212]

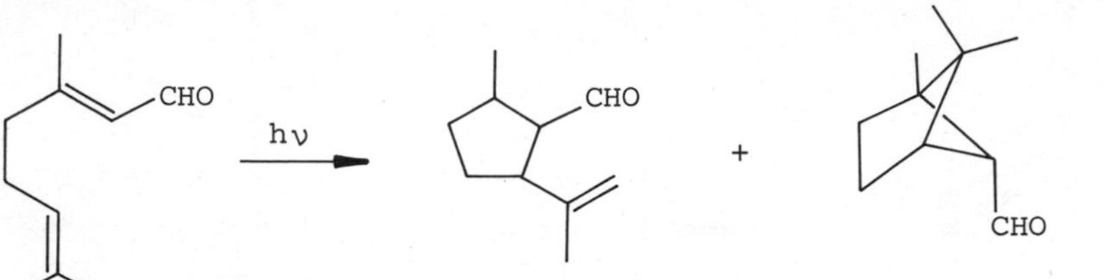

178
Photocitral A
(Major product)

179

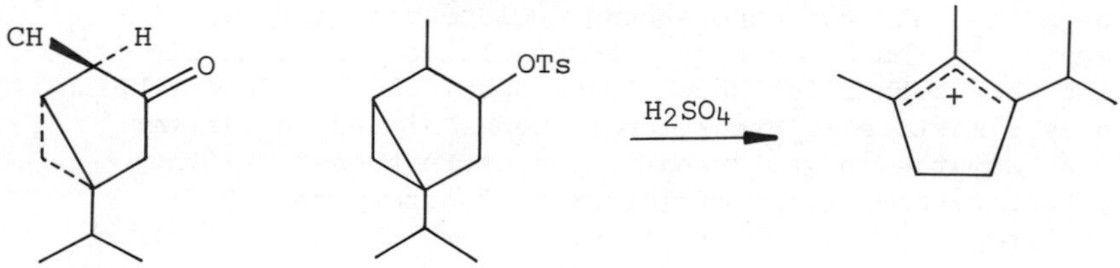

41

180   181, all
Thujone   isomers

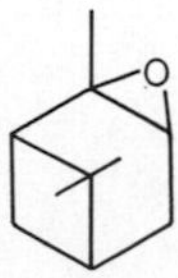

182

Many of the cyclopentane monoterpenes are lactones, but there is one less oxidized substance occurring naturally, which is dealt with first.

A. Plinol

A monoterpene alcohol, plinol, occurs in the higher boiling fraction of camphor oil,[213] the total structure of which, including the chirality (*R*) at C-3, was established by Sebe and Naito[214] (177d). Synthesis of plinol was first reported by Ikeda and Wakatzuki,[213] but the pyrolysis of linalool they report is actually more complex, and the structure attributed to another of their products[215] was incorrect, and the whole situation was reviewed in some detail by Strickler, Ohloff, and Kováts.[208] All four possible plinols (177a to 177d) are formed in the reaction in the yields shown. They can be

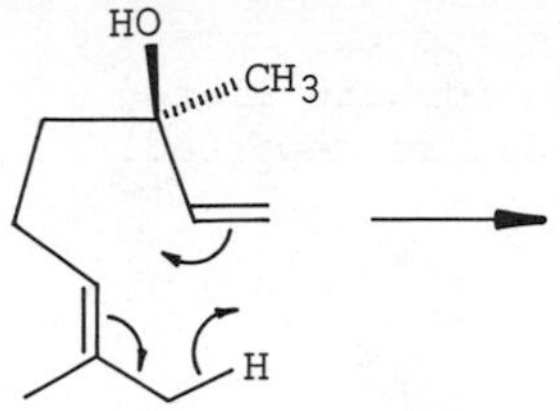

6

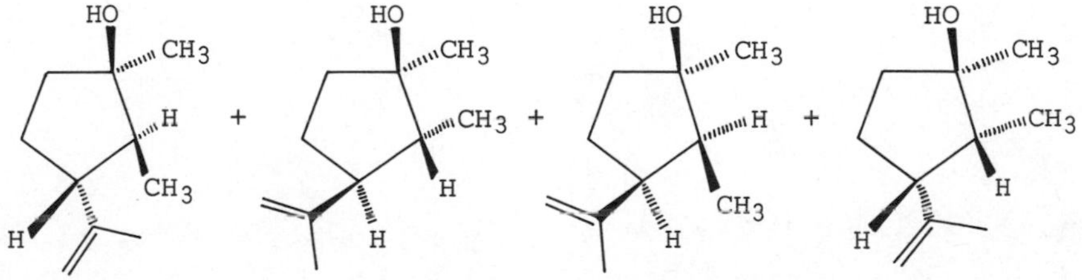

| 8.7% | 4.8% | 28.2% | 10.7% |
| a | b | c | d |

Plinols (177)

separated gas chromatographically,[216] and it was shown that the natural product corresponded to plinol-d (177d).

B.   Cyclopentanopyrans

Although these substances are not very widely distributed in
nature, their effect in the surroundings where they do occur
is so dramatic that a great deal of work has gone into an
examination of their isolation, structure and synthesis. The
compounds isolated from various species of *Iridomyrmex* and
other ants are used by them as agents of defense against pred-
atory insects,[217] and are, in fact, potent insecticides,[218]
while many of the compounds isolated from catnip oil (*Nepeta
cataria*) and other vegetable sources are highly attractive to
cats and other animals of the *Felidae* and *Chrysopidae*.

　　Table 1 gives a list of some of these compounds, with
their sources, and historical data necessary to locate the
first syntheses. The latter are by no means the only syntheses and some selection has been necessary in what is presented in greater detail below. There are some acids and
esters not listed, but which are closely related to the compounds shown. Some more recently isolated iridoids are listed
in Ref. 219, but no syntheses of these compounds are reported.

TABLE 1.

Structure	Name and Formula Number in Text	Source (Literature)	Structure	
			Literature	Synthesis
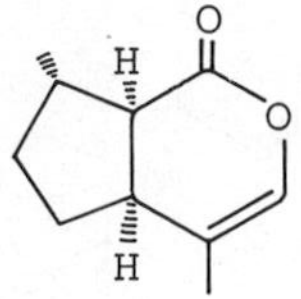	Nepetalactone, (cis,trans-) 183b	Nepeta cataria 220	221, 222	223
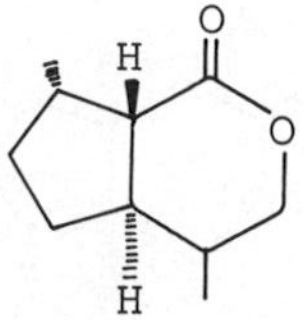	Iso-nepetalactone (trans, cis-) 183a	Nepeta cataria	222	223
	Neorepetalactone 190	Actinidia polygama 224	224	224
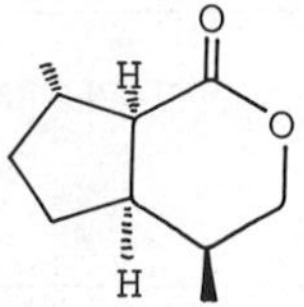	Dihydronepetalactone	Actinidia polygama 224	224	--

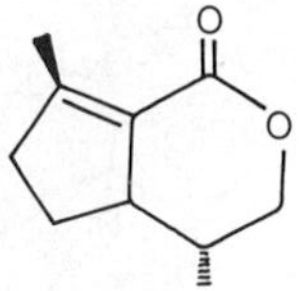

63

TABLE 1. Continued

Structure	Name and Formula Number in Text	Source (Literature)	Structure Literature	Synthesis
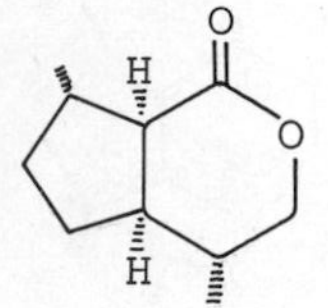	Isodihydronepetalactone	*Actinidia polygama* 224	224	--
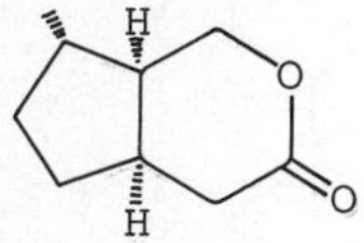	Iridomyrmecin	*Iridomyrmex humilis* 225 *Actinidia polygama* 224	226	227, 228
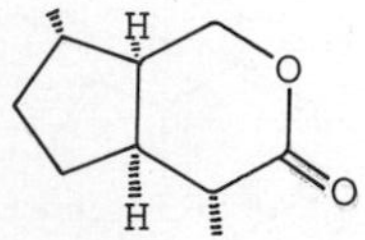	Isoiridomyrmecin 194	*Iridomyrmex nitidus* 229	229	227, 228
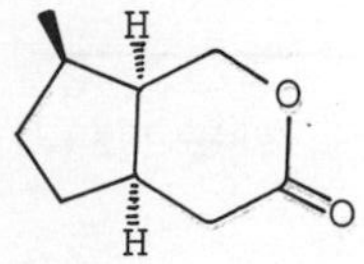	Boschnialactone 199	*Boschniaka rossica* 230	230	230

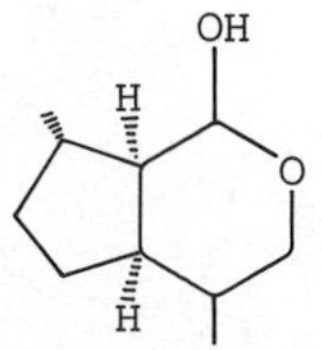 Neomatatabiol 200a	*Actinidia polygama* 231	231	231
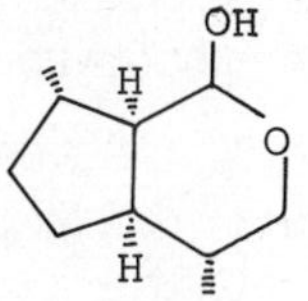 Isoneomatatabiol 200b	*Actinidia polygama*	231	231
Iridcdial 189a	*Iridomyrmex conifer* *Iridomyrmex defectus* 229	229	228
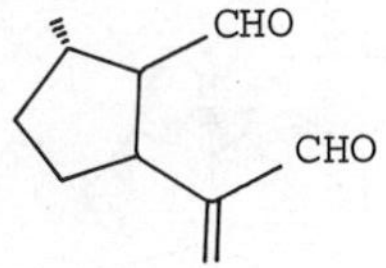 Dolichodial[233] *Anisomorpha*[234] (latter of unknown stereochemistry)	*Dolichoderus* 232 and *Iridomyrmex* 233 spp *Anisomorpha buprestoides*	233 234	cf. 235
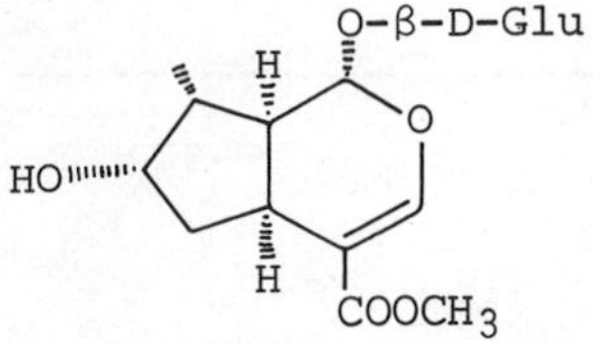 Loganin (pentaacetate, 210)	*Strychnos nux Vomica*[236] and other plants	237	238

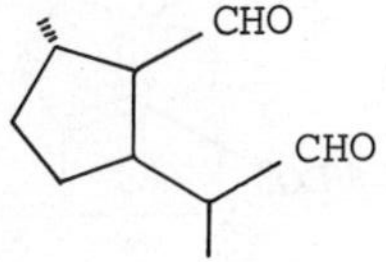

TABLE 1. Continued

Structure	Name and Formula Number in Text	Source (Literature)	Structure Literature	Synthesis
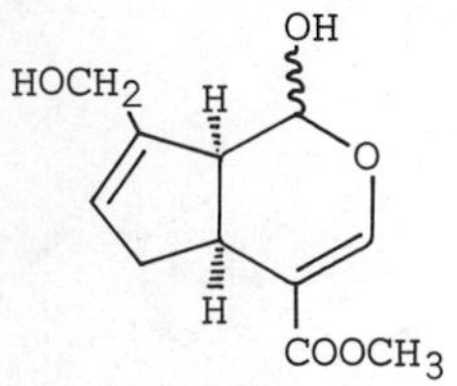	Genipin 217	*Genipa americana* 239	240	241
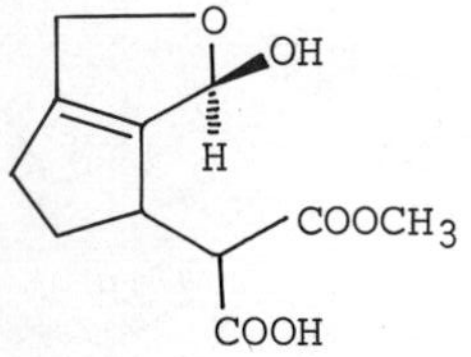	Genepinic acid 220	*Genipa americana* 242		
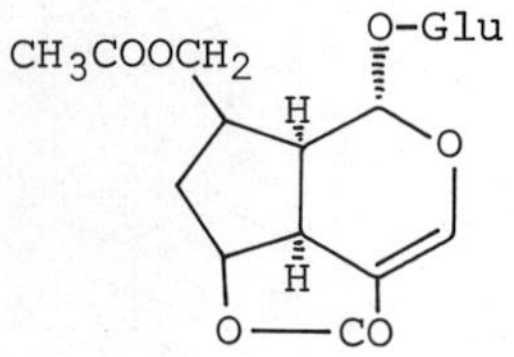	Asperuloside	*Coprosma spp.*	243	

Structure	Name	Species	Ref	Ref
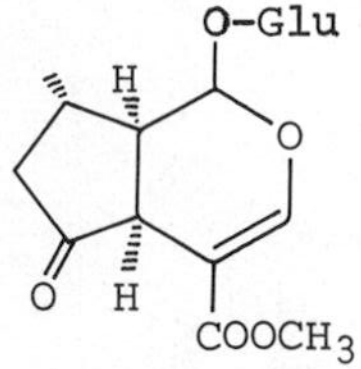	Verbenalin	*Verbena officinalis* 244	244	245 (aglucone, 220)
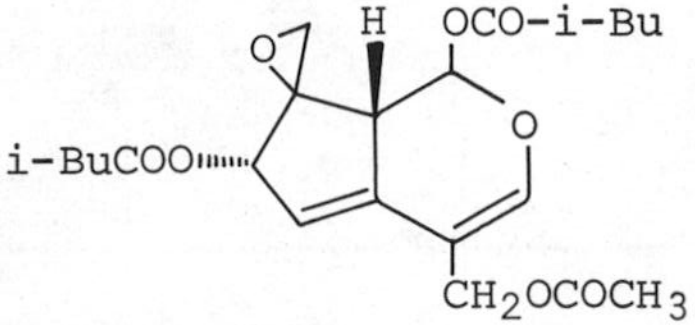	Valtratum	*Valeriana officinalis* *Kentranthus ruber* and other spp 246	246	--
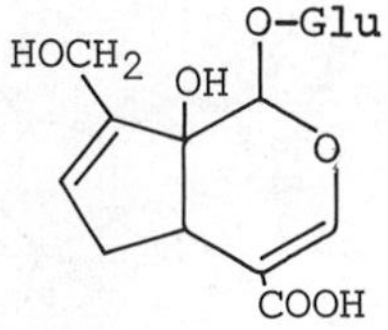	Monotropein	*Monotropa hypopithys* 247	248	--
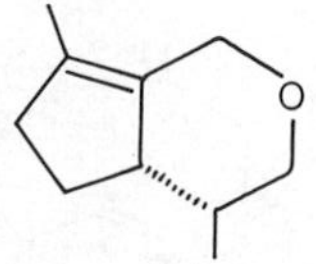	205	*Actinidia polygama* 249	249	249
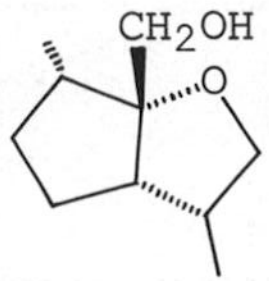	206	*Actinidia polygama*	249	249

67

TABLE 1. Continued

Structure	Name and Formula Number in Text	Source (Literature)	Structure Literature	Synthesis
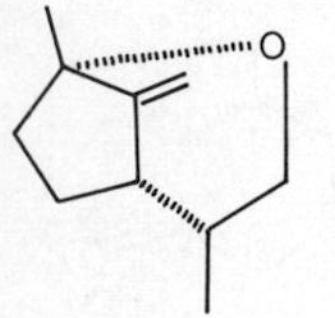	Matatabiether 207	*Actinidia polygama* 224	249	250
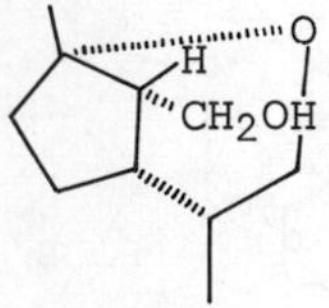	208	*Actinidia polygama*		240

Lactones

Nepetalactones. The total synthesis of nepetalactone (183) by
Sakan et al.[223] (Scheme 19) is long and does not take stereo-

Scheme 19

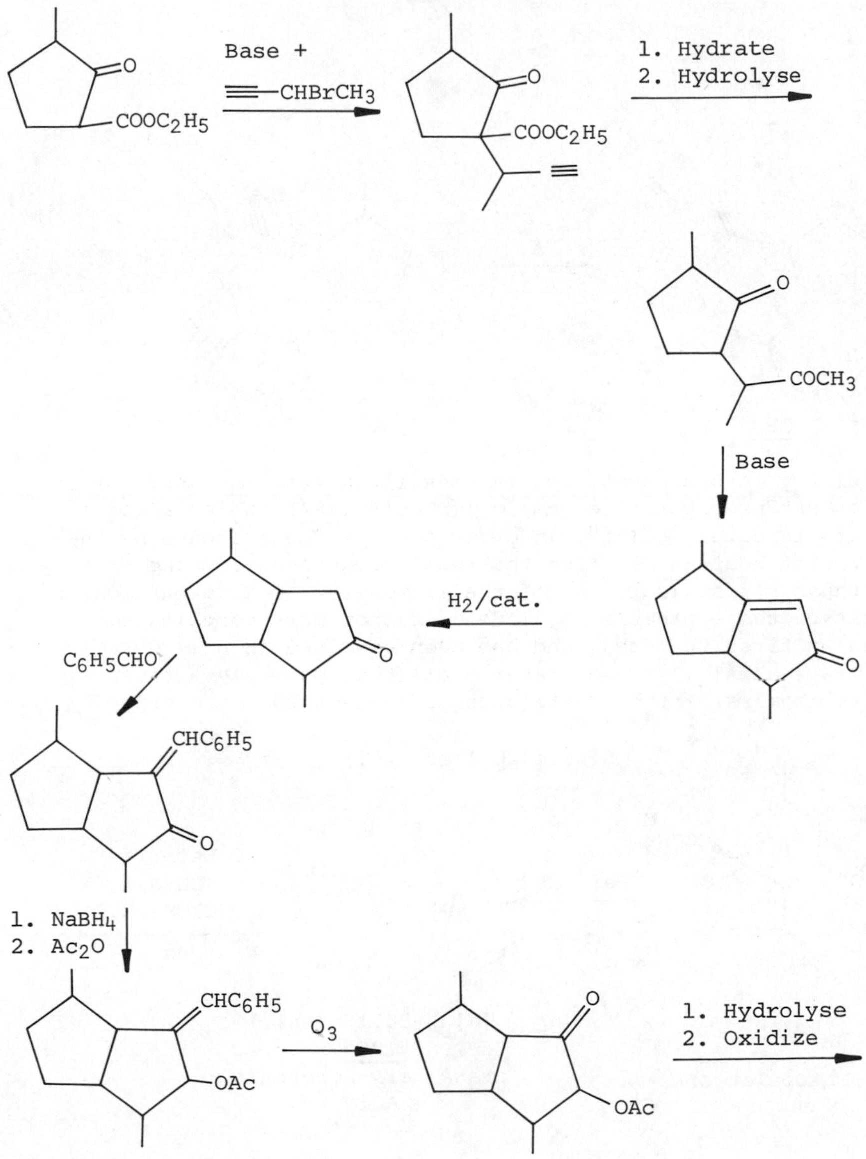

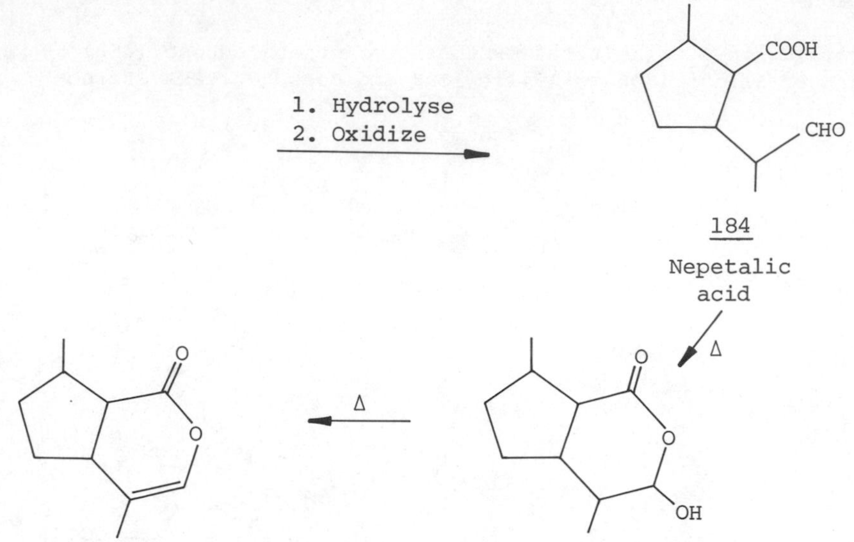
1. Hydrolyse
2. Oxidize
COOH
CHO
O
O
OH
Δ
Δ
O
O

184
Nepetalic acid

183

chemistry into account, but it does illustrate the need for an efficient preparation of nepetalic acid (184) (which also occurs in catnip oil[220]) in order to make the lactones by the action of heat--a reaction that was recognized from the days of the earliest isolation of the lactones.[220] This particular facet of the synthesis is, indeed, rather more complicated than at first realized, and has been examined in some detail by Trave et al.,[251] who prepared all the optically active stereoisomers of the nepetalactones (Scheme 20). In order to

Scheme 20 (Ref. 251)

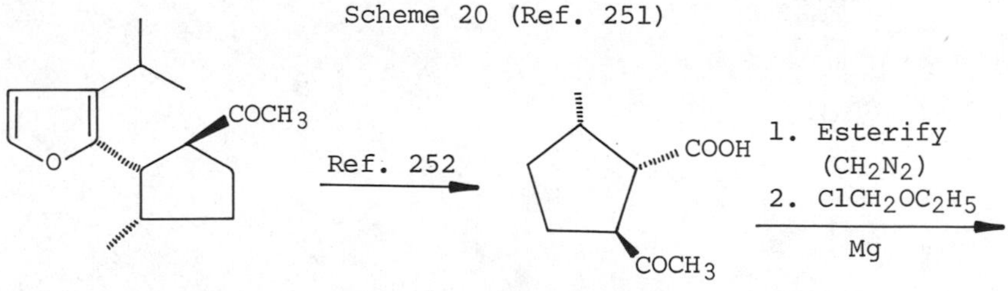
O
$COCH_3$
Ref. 252
COOH
$COCH_3$
1. Esterify
(CH_2N_2)
2. $ClCH_2OC_2H_5$
Mg

185

Furopelargone

$[\alpha]_D$ -56.1° (chl.)
186a
trans, cis-Nepetonic Acid

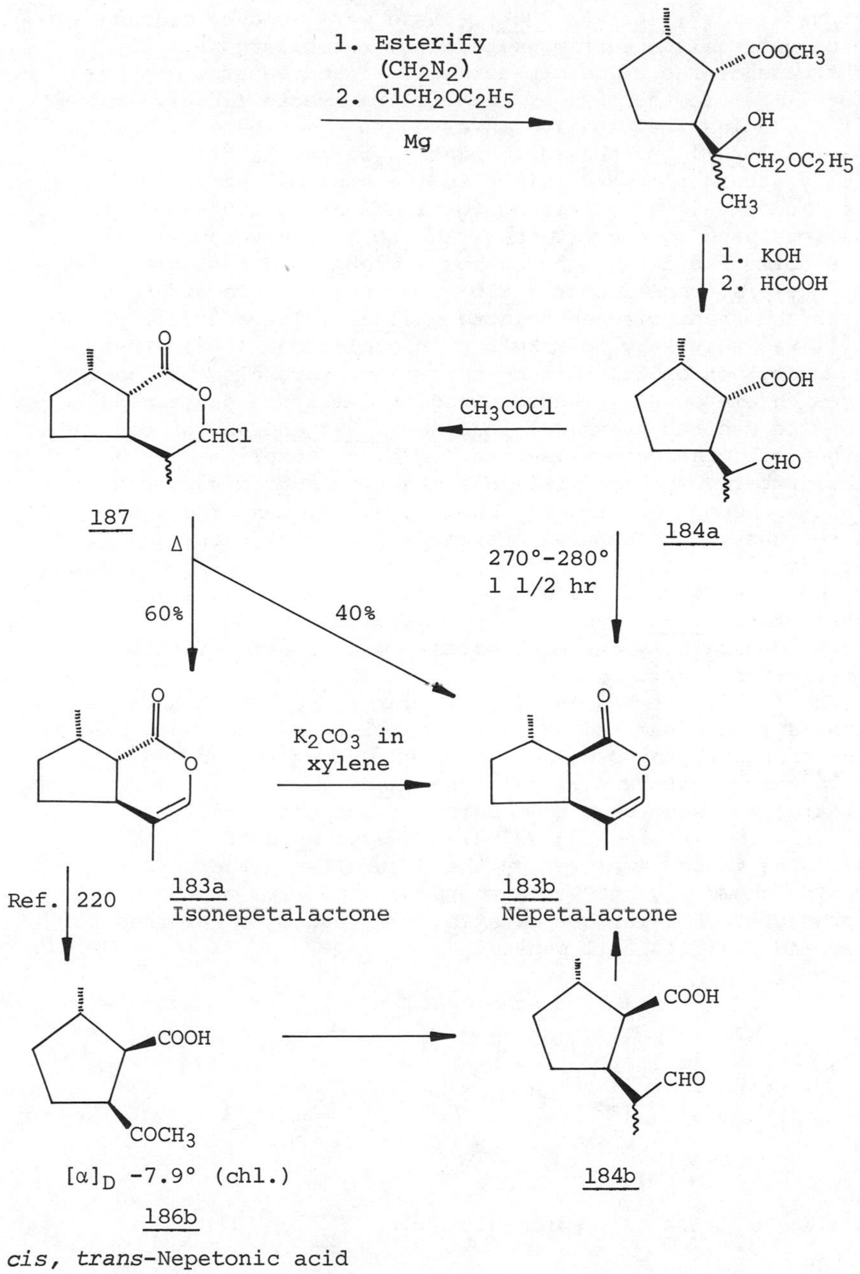
1. Esterify
(CH_2N_2)
2. $ClCH_2OC_2H_5$
Mg
$COOCH_3$
OH
$CH_2OC_2H_5$
CH_3
1. KOH
2. HCOOH
O
O
CHCl
187
CH_3COCl
COOH
CHO
184a
Δ
60%
40%
270°-280°
1 1/2 hr
O
O
K_2CO_3 in
xylene
O
O
183a
Isonepetalactone
183b
Nepetalactone
Ref. 220
COOH
$COCH_3$
$[\alpha]_D$ -7.9° (chl.)
186b
cis, trans-Nepetonic acid
COOH
CHO
184b

be certain of the chirality of their starting materials, the
nepetonic acids (186a and 186b), these were made by degradation of naturally occurring materials, furopelargone (185)[252]
in the case of the *trans,cis*-acid (186a) and natural nepetalactone (183) in the case of the *cis,trans*-acid (186b). The
nepetonic acids are readily converted to the nepetalic acids
(184a and 184b), but the lactonization of the latter is not
entirely straightforward, since in the case of *trans,cis*-
nepetalic acid (184a), heating for 1 1/2 hr at 270°-280° (the
conditions used to convert the acids to nepetalactones) involves epimerization of the carboxyl group, and a single lactone, the *cis,trans* isomer (183b) is obtained. In order to
prepare the *trans,cis*-nepetalactone (183a), Trave et al. found
that it was necessary to make the chlorolactone (187) first,
using acetyl chloride, then by thermal dehydrohalogenation in
the gas chromatograph, they arrived at a mixture containing
60% of the desired *trans,cis*-lactone (183a) and 40% of the rearranged *cis trans*-nepetalactone (183b). Inexplicably, the
optical rotation of the Italian synthetic *trans,cis*-lactone
(183a), although the same in absolute value, was of opposite
sign to that of the natural lactone isolated by Bates and
Sigel.[222]

The conversion of isonepetalactone (*trans,cis,* 183a) into
nepetalactone (183b) appears to be base catalyzed, since heating the former in xylene with potassium carbonate effects the
isomerization.[224]

The Italian work does not, of course, represent a total
synthesis in the same sense as the earlier Japanese work does,
and neither does the synthesis carried out by Achmad and
Cavill[253] from natural (+)-pulegone (188) (except in so far as
the latter has also been synthesized); but the latter synthesis is also interesting as being a synthesis of iridodial
(189). The method employed by the Australian authors is
shown in Scheme 21, but leads to the enantiomers of the natural products. For further details, the reader is referred to
Achmad and Cavill's full paper.[254]

Scheme 21

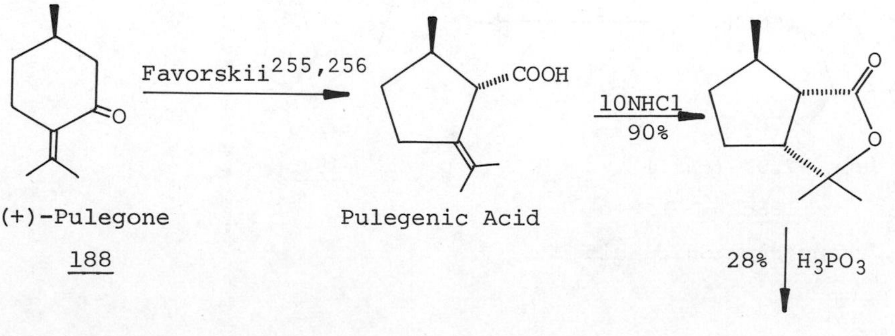

(+)-Pulegone   Pulegenic Acid

188

Favorskii[255,256]

10NHCl
90%

28% | H3PO4

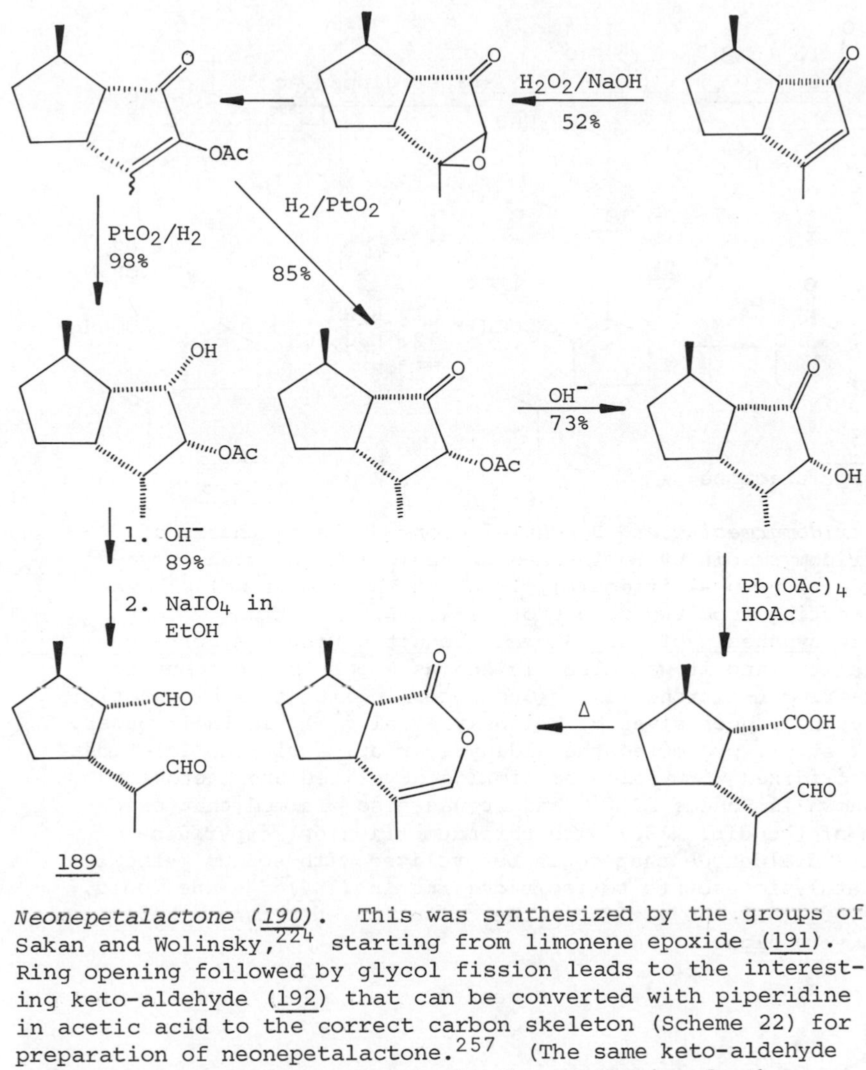

189

Neonepetalactone (190). This was synthesized by the groups of
Sakan and Wolinsky,[224] starting from limonene epoxide (191).
Ring opening followed by glycol fission leads to the interesting keto-aldehyde (192) that can be converted with piperidine
in acetic acid to the correct carbon skeleton (Scheme 22) for
preparation of neonepetalactone.[257] (The same keto-aldehyde
(192) has been used by Wolinsky in the synthesis of other
cyclopentanes, using different reagents--aqueous alkali, for
example--to effect different ring closures.) The remainder of
the synthesis consists in oxidizing the aldehyde function and
placing an oxygen atom in the isopropenyl side chain, and
finally ring closure by heating.

Scheme 22

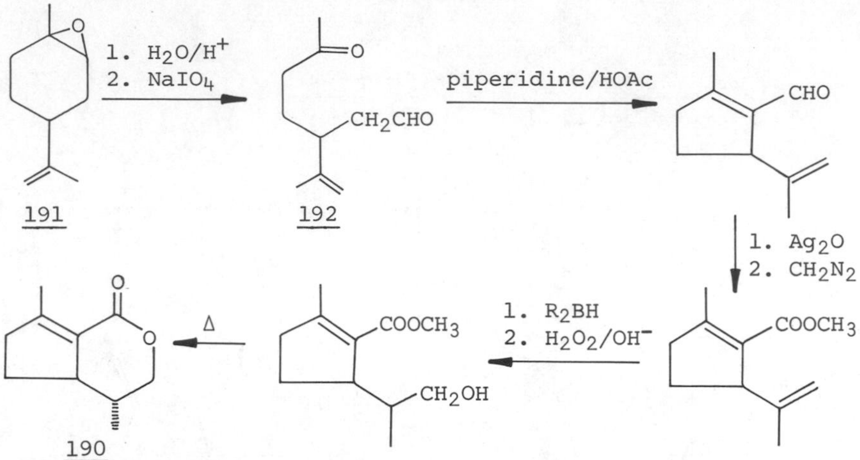

191

192

piperidine/HOAc

1. Ag$_2$O
2. CH$_2$N$_2$

1. R$_2$BH
2. H$_2$O$_2$/OH$^-$

Δ

190
Neonepetalactone

The Iridomyrmecins and Boschnialactone.   The synthesis of
isoiridomyrmecin by Robinson's group (which included a synthesis of natural iridodial, 189a, at the same time),[228] was
interesting from two points of view.   First, it was a very
simple synthesis of the natural product having the correct
chirality, and it was also claimed as a possible biogenetic
route from L-citronellal.   (Other possibilities of biogenetic
routes have been given by Wolinsky et al.[258])   In their paper,
Clark et al. protected the aldehyde group of citronellal first,
then oxidized with selenium dioxide, cyclized and lactonized
as shown in Scheme 23,[228] and it was also claimed that oxidation of the diol (193) with chromium trioxide in pyridine
gave a dialdehyde that could be cyclized with sodium methoxide
in catalytic amounts to isoiridomyrmecin (194)[259]--one feels,
however, that the yields in the oxidation step cannot have
been very high.

Scheme 23

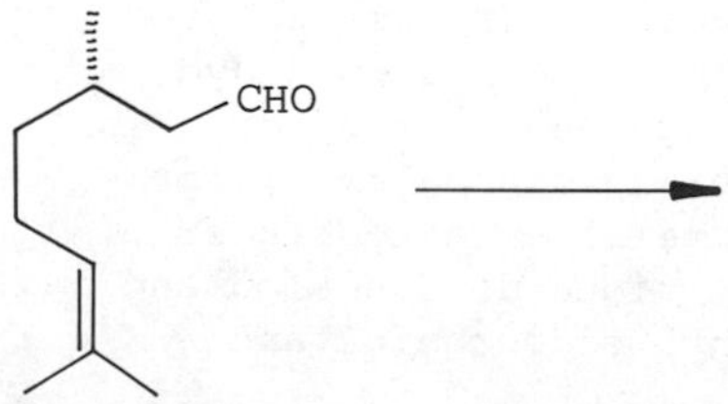

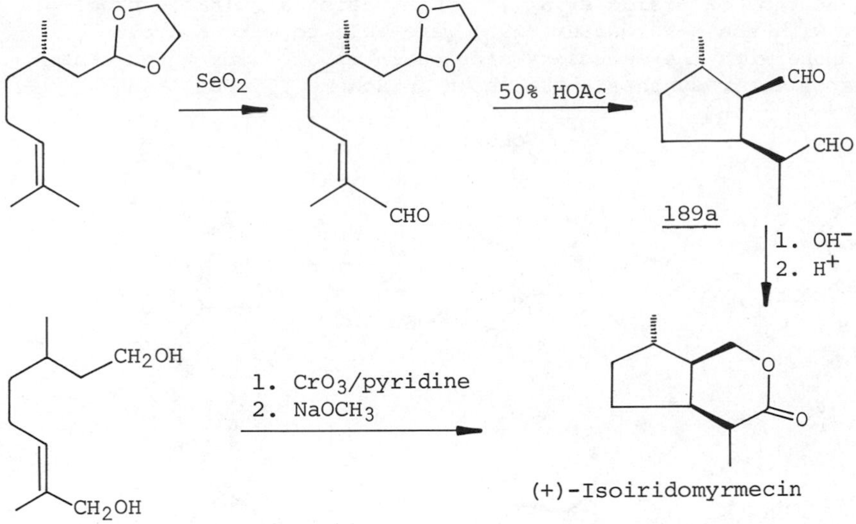

189a

193

(+)-Isoiridomyrmecin

194

Another synthesis of (+)-isoiridomyrmecin (the enantiomer of the natural product) from citronellal has been described by Cavill and Whitfield.[260]

Korte's first iridomyrmecin synthesis[227] involves a Prins reaction, which, as we have seen in earlier cases (notably lavandulol), is not generally a very good reaction for the synthesis of monoterpenes, but Korte developed a second synthesis which not only avoids this, but also has the virtue of simplicity,[261] and is shown in Scheme 24.

Scheme 24

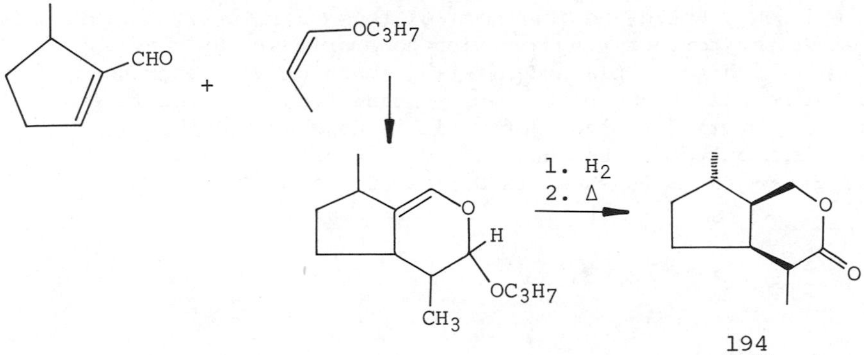

194

Another synthesis starting with a preformed cyclopentane ring is that of Sisido et al.,[262] who, using a suitable bromo-ester with the β-ketoester (195) were able to make a cyclo-pentanone with the necessary side chain (196). The subsequent steps of their synthesis are shown in Scheme 25, but the

Scheme 25

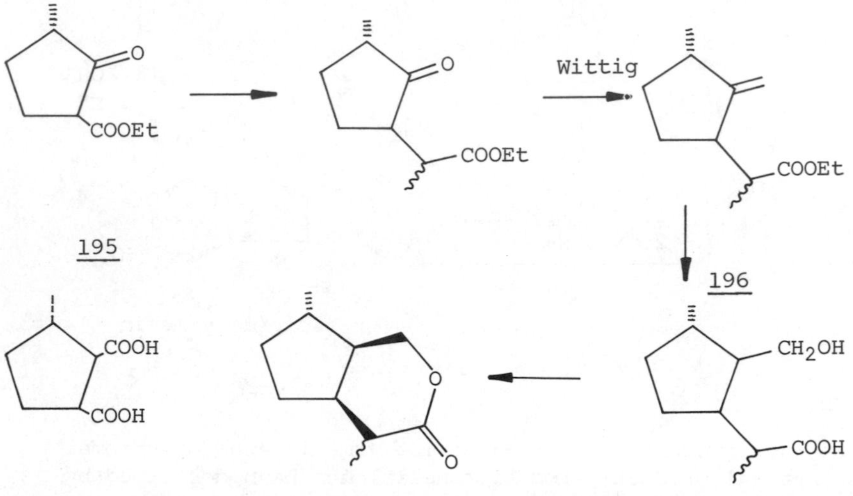

195

196

197   194

synthesis is somewhat long. Like many of the synthetic irido-myrmecins, this product was shown to have the stereochemistry desired by permanganate oxidation to the corresponding (±)-nepetalinic acids (197); Sisido et al. have, in fact, pub-lished a detailed discussion of the stereochemistry and con-formation of the various iridolactones (including boschnialac-tone).[263]

Wolinsky, believing that many of these earlier syntheses had given the correct ring junction more or less "by chance," set out to achieve a planned *cis*-ring junction with a *trans*-situated methyl group in the cyclopentane ring.[257] The first stages are similar to those of Cavill's nepetalactone syn-thesis from pulegone, but once the γ-lactone (198) has been made, the route is as shown in Scheme 26.

Scheme 26

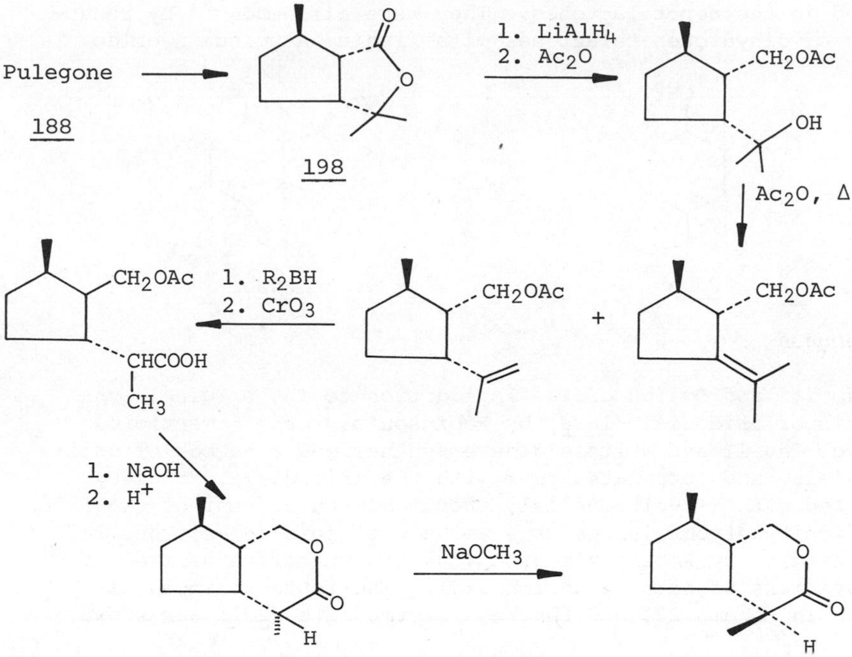

The noriridomyrmecin, boschnia lactone (199) has been synthesized by Sakan et al.;[230] they made the *cis-* and *trans-* fused lactones, that were not, of course, optically active, but racemates (Scheme 27).

Scheme 27

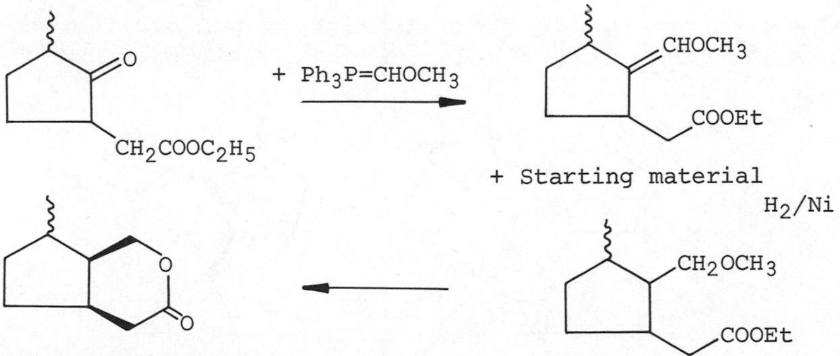

(±)-Boschnia Lactone

199

The Neomatatabiols (200a and 200b). These compounds, isolated
from *Actinidia polygama* are chrysope attractants, and are related to the nepetalactones. They were also made[231] by reduction of dihydronepetalactones with lithium aluminum hydride.

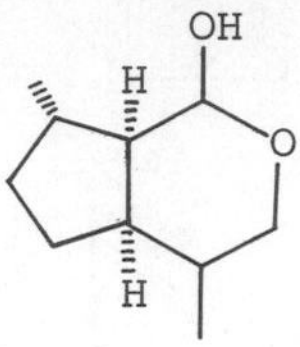

200a 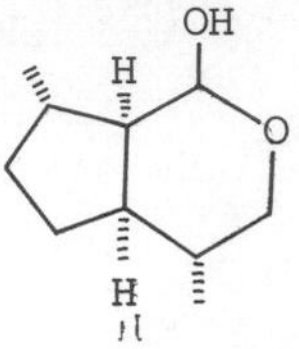   200b

Aldehydes

Iridodial and Dolichodial. In addition to the original synthesis of iridodial (189a) by Robinson's group[228] mentioned
above, Cavill and Whitfield have synthesized a number of doli-
chodials, and correlated them with the iridodials.[235]   They
started with (+)-citronellal, presumably on account of the
difficulty of obtaining large amounts of pure (-)-citronellal
referred to by Robinson's group,[228] and therefore arrived at
enantiomers of natural dolichodial. The route followed is
shown in Scheme 28, and the key intermediate (201) was shown

Scheme 28

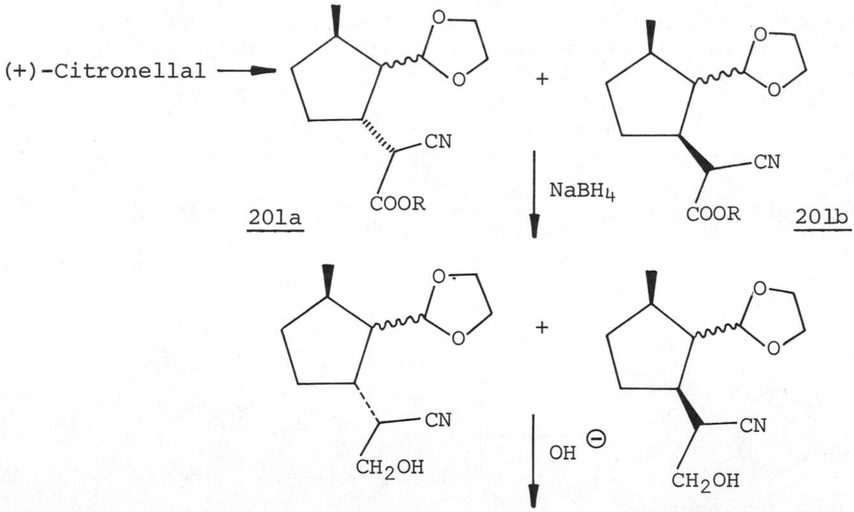

(+)-Citronellal ⟶

201a   +   201b

NaBH₄

+

OH⊖

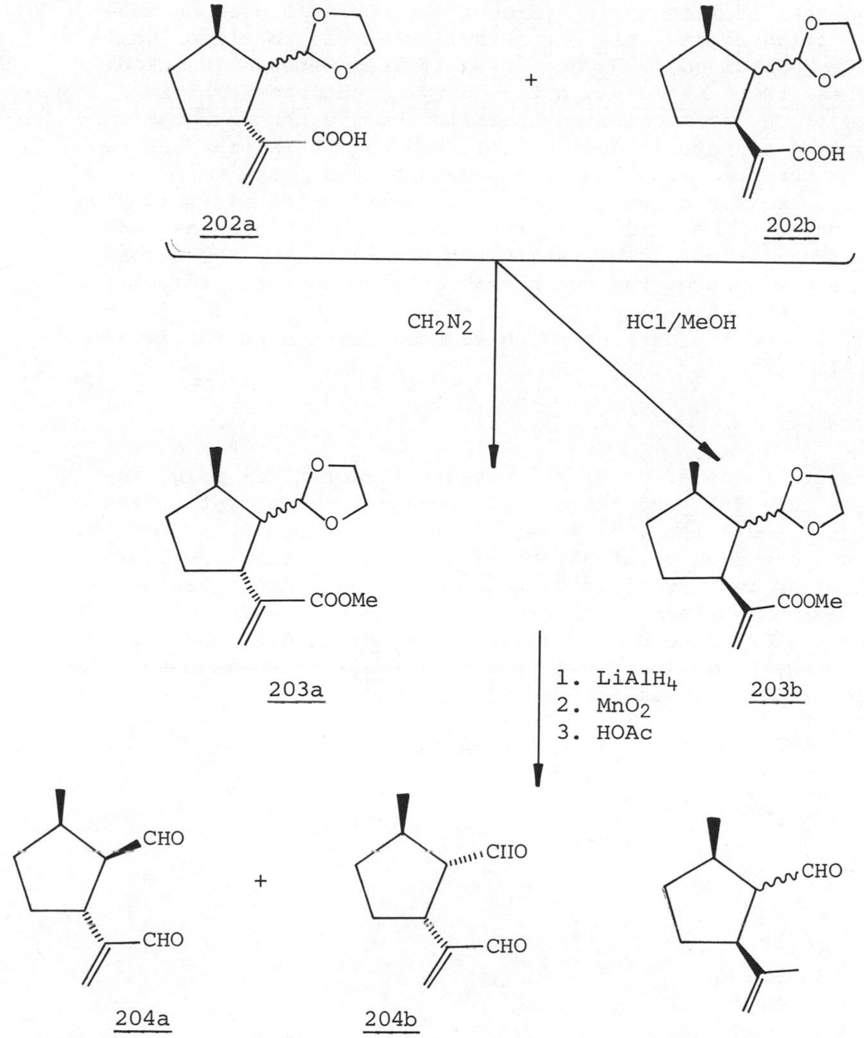

CH₂N₂ gives 203a; HCl/MeOH gives 203b

202a + 202b

203a   203b

1. LiAlH₄
2. MnO₂
3. HOAc

204a + 204b

Enantiomers of natural dolichodial

Iridodial

to be present as two isomers, the two structures shown being ascribed to them. At the stage of the methylation from the acid (202) to the corresponding methyl ester (203), it was

found that if methanolic hydrogen chloride was used as the
methylating agent, only one methyl ester was obtained, while
two were obtained (203a and 203b) if diazomethane was used.
Of these two substances, after carrying out the remaining
steps of the synthesis, one peak in the gas chromatogram corresponded to natural dolichodial, and when this peak was re-
chromatographed at a lower temperature, two peaks were observed, as is the case with natural dolichodial and iridodial
but not with the iridodial prepared by the Robinson method
(189a, see above). This splitting of the peaks is ascribed
to the presence of the two possible isomers of the formyl
group in the natural product (i.e., 204a and 204b for dolichodial, and the corresponding reduced substances for iridodial).[235]

Ethers

There are a number of ethers present in *Actinidia polygama*
(e.g., 205-208), and several of these have been synthesized by
Wolinsky and Nelson.[249] Their routes are shown in Scheme 29.
The synthesis of matatabiether (207) follows the same route
as that of Isoe et al.,[250] and in all these papers, as well as
the earlier one from both groups,[224] the conversion of matatabiether (207) into neonepetalactone (190) is described [ring
opening with formic acid to the diol (209), then manganese dioxide oxidation, see Scheme 29].

Scheme 29

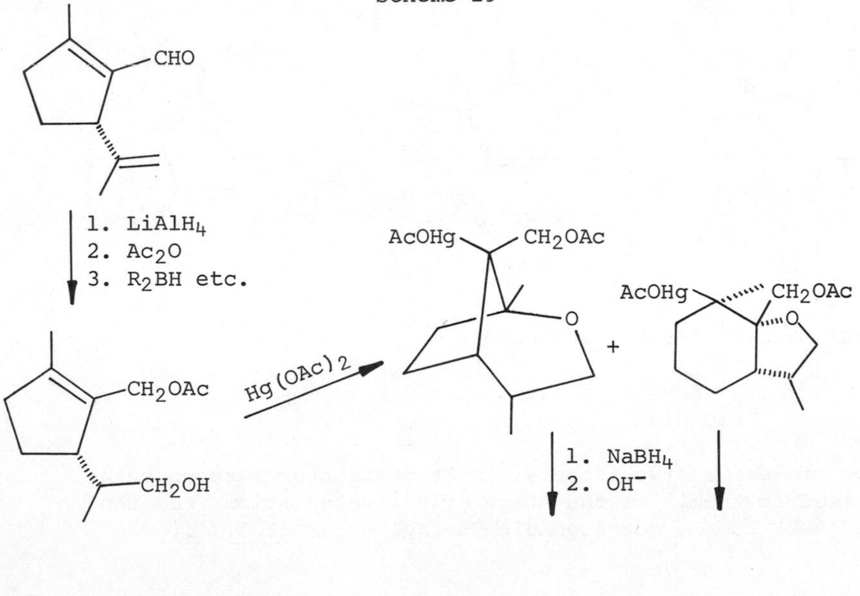

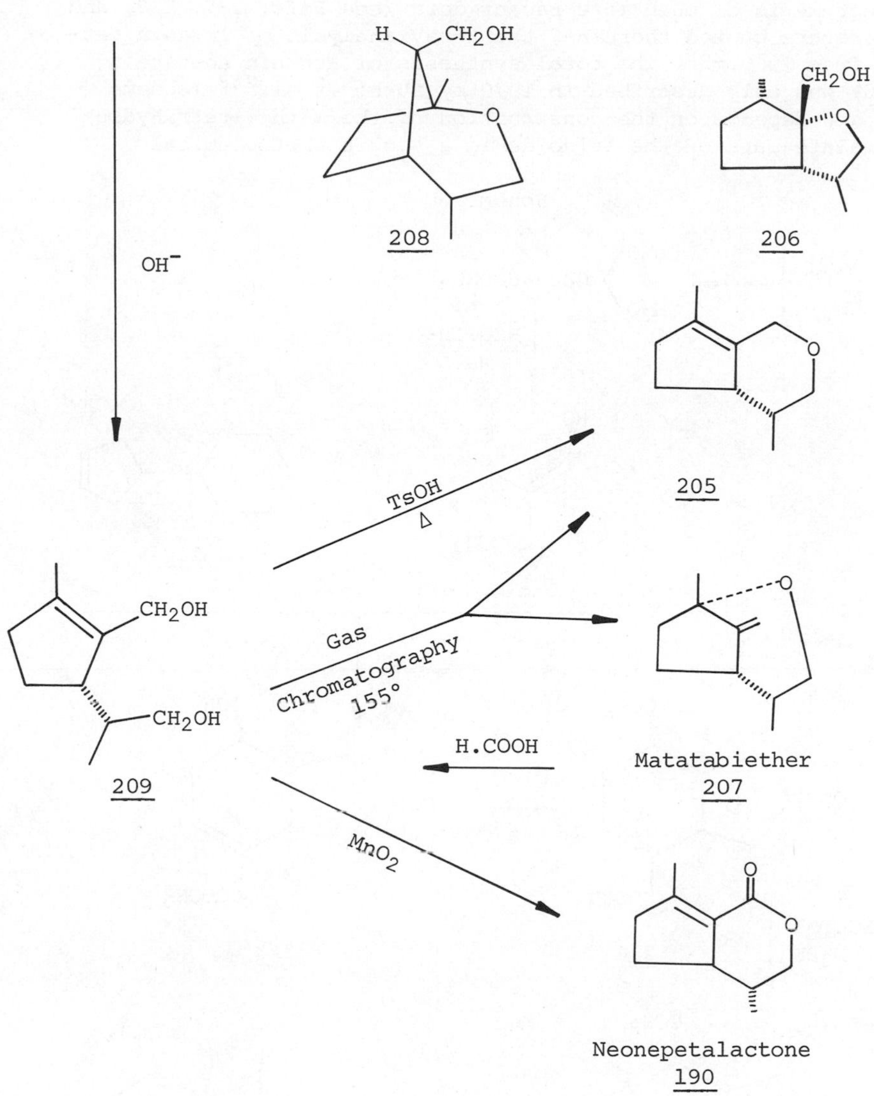

208

206

205

209

Matatabiether
207

Neonepetalactone
190

Esters, Glucosides, etc.

Loganin. Loganin has a position of special importance, since
it has been shown to be a key intermediate in the biosynthetic pathway to the *Corynanthe, Aspidosperma, Iboga,*[264-266]
and *Ipecacuanha*[267] indole alkaloids. Although discovered in
the last century in *Strychnos nux vomica,*[236] certainty of

structure is of much more recent date (see Refs. 237, 238, and literature quoted therein), the x-ray analysis of loganin dating from 1969.[268] The total synthesis of loganin acetate (210) was only described in 1970 by Büchi et al.[238] (Scheme 30) and depends on the construction of the methyl tetrahydrocoumalate part of the iridoids in a single photochemical

Scheme 30

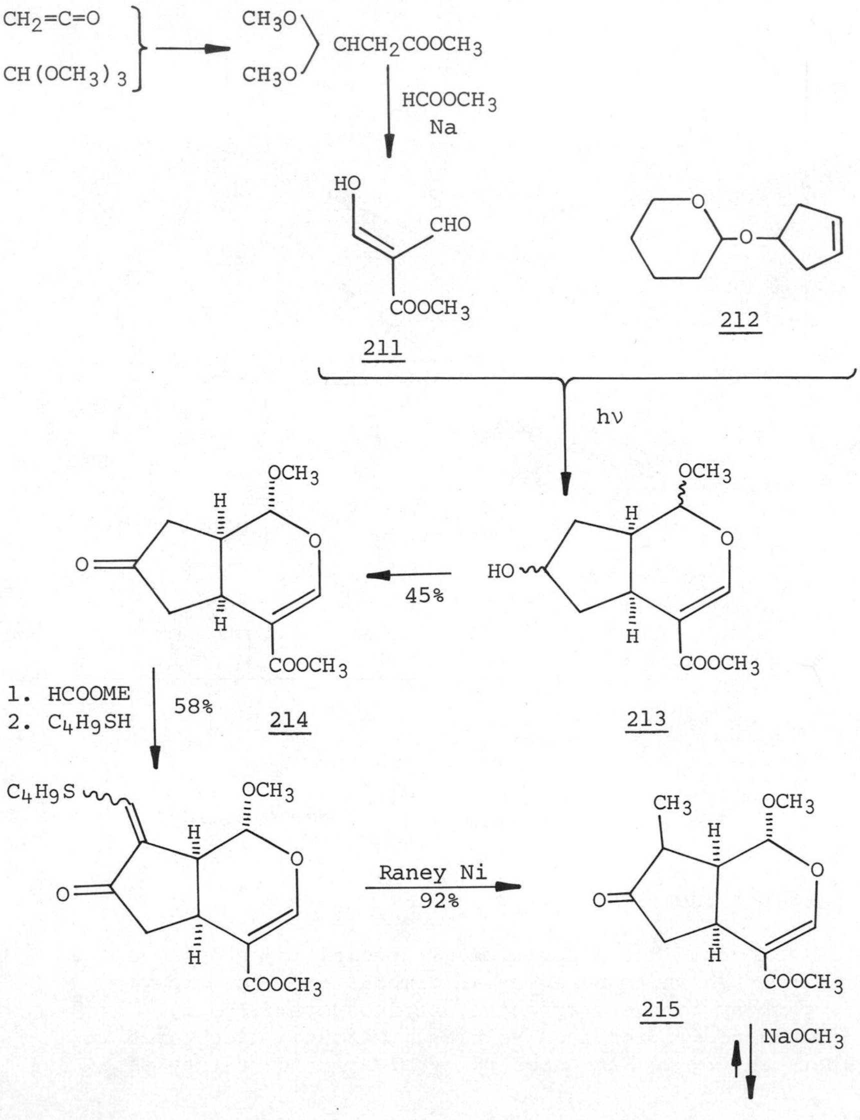

1. HCOOME
2. C_4H_9SH 58%   214   213

$\xrightarrow{\text{Raney Ni}}$ 92%   215

NaOCH$_3$

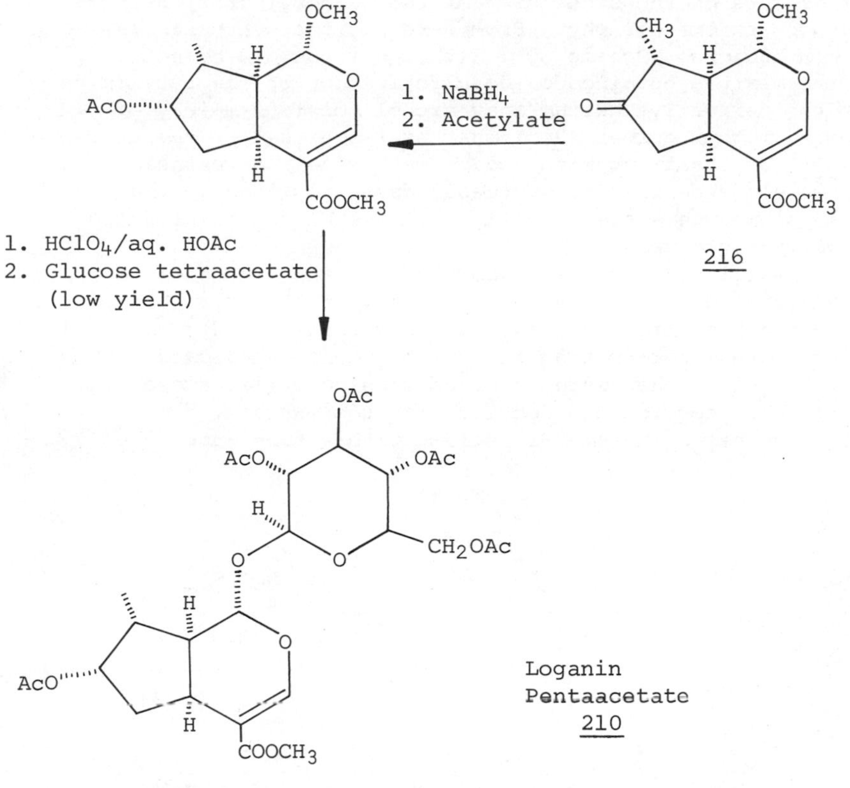

1. NaBH$_4$
2. Acetylate

1. HClO$_4$/aq. HOAc
2. Glucose tetraacetate
   (low yield)

216

Loganin
Pentaacetate
210

operation, an extension of de Mayo's method for the synthesis
of δ-diketones by photochemical cycloaddition of enolized
β-diketones to olefins.[269] 2-Formylmalonaldehydic acid methyl
ester (211) is prepared in two steps from trimethylortho-
formate and ketene followed by condensation of the acetal thus
obtained with methyl formate, then ultraviolet irradiation of
this tricarbonyl compound (211) in 3-cyclo-pentenyl tetrahydropyranyl ether (212) followed by treatment of a methanol
solution of the crude photoproducts with Amberlite IR-120
cation exchange resin, gave a mixture of the liquid hydroxy-
acetals (213) that were oxidized to the corresponding ketones,
the major (desired) isomer being isolated from the mixture by
crystallization, and whose configuration represents the most
stable isomer with cis-fused rings and an axially oriented
methoxy group (214). This ketone is converted to a mixture of
hydroxymethylene derivatives and then without purification, to
the butylthiomethylene derivatives. It is fortunate that the
main reaction is on the side of the carbonyl where it is desired to introduce the methyl group, the small amounts of the

other isomers on the other side of the carbonyl group being
removed by chromatography. From here on, the synthesis is
more conventional (Scheme 30), although it should be noted
that the mixture obtained on desulfurization of the butylthiomethylene derivatives with Raney nickel consists mainly of
the wrong methyl epimer (215) that is isomerized to the correct
one (216) by treatment with sodium methoxide. Nevertheless,
the real elegance of this synthesis certainly lies in the
initial steps where the natural iridane skeleton is produced
in a very direct way.

The earliest of the more complex iridane syntheses is
probably that of genepin by Büchi et al. in 1967.[241] Genepin
(217) has a long history, since it is responsible for the dark
blue persistent colors that the plant (*Genipa americana*) forms
with aminoacids, that were observed some centuries ago on the
skin of those coming into contact with the extract.[239]
Büchi's synthesis (Scheme 31) starts with a substance (218)

Scheme 31

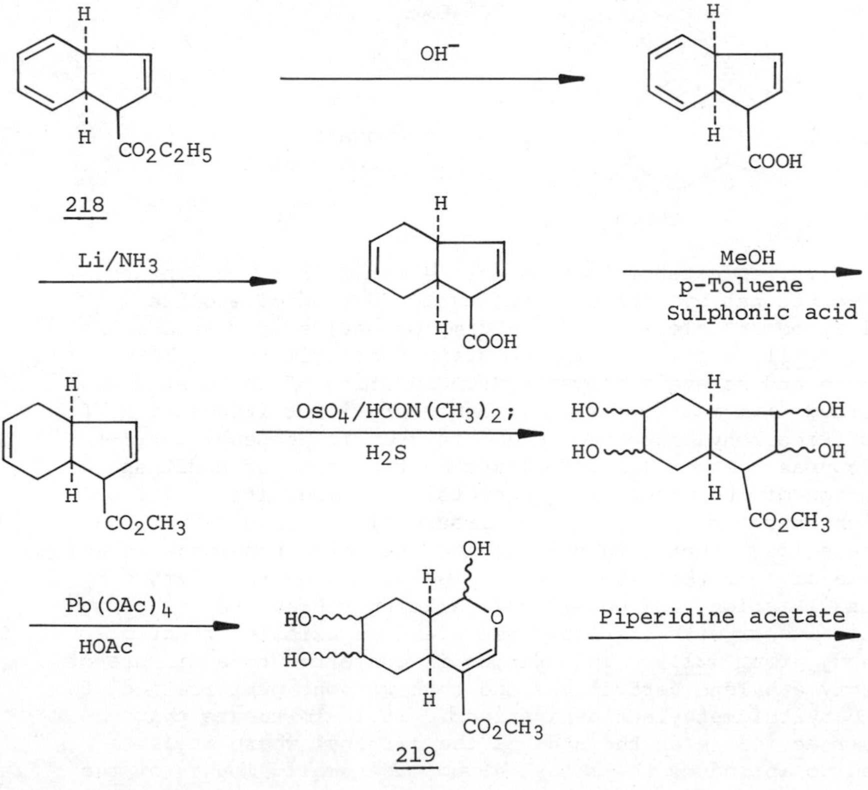

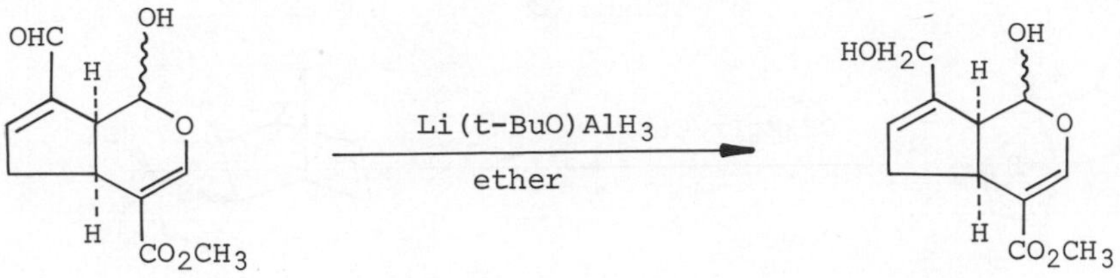

OHC, OH structure → Li(t-BuO)AlH₃ / ether → HOH₂C, OH structure

$$\text{Li(t-BuO)AlH}_3$$
ether

217

(±)-Genipin

having the correct *cis*-ring junction, and while the route passes through many intermediates that are isomeric mixtures, or, at any rate isomerically undefined, the final product has only one more asymmetric center than the starting material, so that it was not necessary to elucidate the stereochemistry of the intermediates. The vital step that assured the success of this synthesis was the lead tetraacetate glycol fission (to 219), where use was made of the fact that cyclopentanediols cleave more rapidly than cyclohexane diols, enabling the former to be cleaved in the presence of the latter.

Other iridane derivatives from *Genipa americana* (genipinic acid (220) and a norterpene, genipic acid,[242] remain to be synthesized.

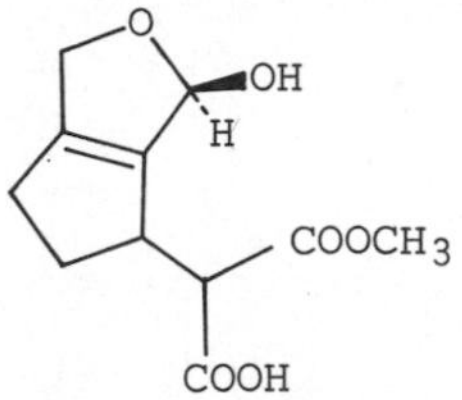

220

The aglucone verbenalol (221) from the *Verbena* glycoside, verbenalin, has been synthesized by Sakan and Abe by the route shown in Scheme 32.[245]

Scheme 32

CH₃MgBr/CuBr →

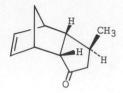

1. Reduce
   (reagent?)
2. Ac₂O/pyr.

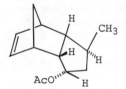

KMnO₄
Ref. 234
←

Ref. 271

Pyr. reflux
60%

OsO₄
100%

86

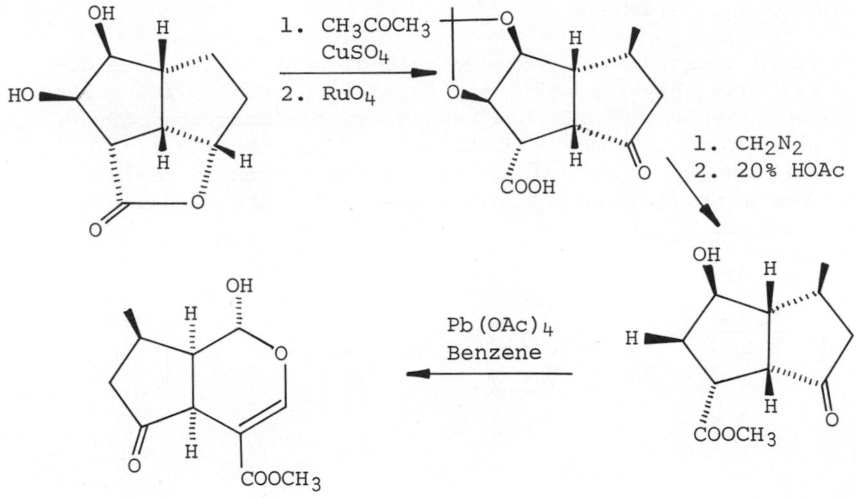

221

(±)-Verbenalol

C. 1-Acetyl-4-isopropenyl-1-cyclopentene

In 1947, H. Schmidt isolated a compound from *Eucalyptus globulus*[272] to which an incorrect structure was at first attributed, but which turned out to be the title compound, as Wolinsky and Barker found accidentally.[273]   In the course of work on limonene epoxide (191) they treated the glycol obtained from it by acid-catalyzed opening of the epoxide ring with sodium periodate, when the keto-aldehyde (192) was obtained, which has already been referred to in another connection.   As stated above, this ketoaldehyde undergoes ring closure with piperidine acetate to give a cyclopentenealdehyde used in the neonepetalactone synthesis, but potassium hydroxide catalyzes a different ring closure, to 1-acetyl-4-isopropenyl-1-cyclopentene (222), that turned out to be identical with Schmidt's natural product.

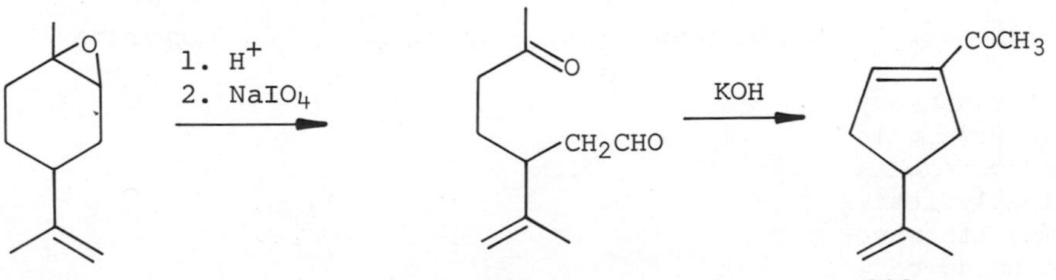

191   192   222

D. Campholenic Aldehyde

Campholenic aldehyde (2,2,3-trimethylcyclopent-3-ene acetaldehyde, 223) has only recently been found in nature in the oil of
Juniperus communis,[274] but has been known for more than 60
years as a product from the photolysis of camphor (224).[275]
It is most conveniently synthesized by the action of Lewis
acids (boron trifluoride,[276] zinc bromide,[277] etc.) on α-
pinene epoxide (182).

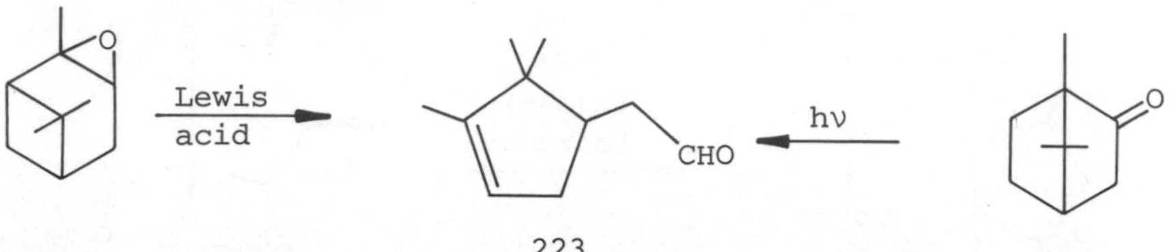

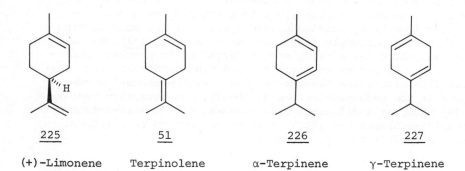

182

223

Campholenic Aldehyde

224

Camphor

8.   THE *p*-MENTHANES (EXCLUDING BICYCLIC SYSTEMS)

A. Hydrocarbons

*Limonene, Terpinolene, α-terpinene, γ-terpinene, and the
Phellandrenes*

225

(+)-Limonene

51

Terpinolene

226

α-Terpinene

227

γ-Terpinene

It is customary
to reserve this
name for the
optically active
forms, the race-
mate being called
dipentene

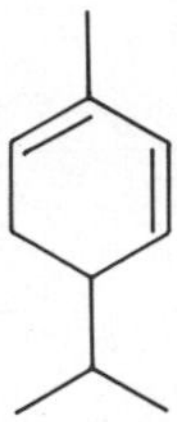

228	229
α-Phellandrene	β-Phellandrene

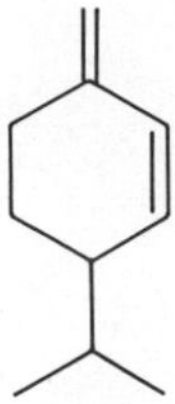

The *p*-menthadienes, particularly those illustrated above are
very common in natural oils, and very large amounts of limonene, for example, are available from citrus oils. In addition,
the racemate can be obtained (as every organic chemistry text-
book informs us) by the dimerization of isoprene in a Diels-
Alder synthesis. This is not the only Diels-Alder synthesis
that has been used to give limonene, a 5-carbon to 4-carbon
coupling having been used by Vig et al.,[278] followed by addition of the methylene group by a Wittig reaction to the ketone
(230):

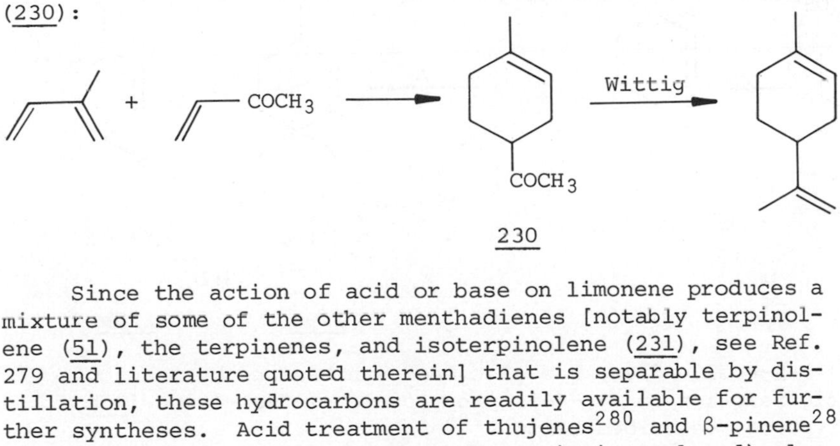

230

     Since the action of acid or base on limonene produces a
mixture of some of the other menthadienes [notably terpinolene (51), the terpinenes, and isoterpinolene (231), see Ref.
279 and literature quoted therein] that is separable by distillation, these hydrocarbons are readily available for further syntheses. Acid treatment of thujenes[280] and β-pinene[281]
(in these papers a cationic exchange resin is employed) also
leads to menthadienes. Although the phellandrenes are not
readily available by methods of this nature, β-phellandrene
(229) and β-terpinene (232) are the main products of the high
temperature (600°) pyrolysis of sabinene (233),[282] which,
since sabinene has itself been synthesized (see below), con-
stituted a synthesis of β-phellandrene. Since β-phellandrene
is converted into α-phellandrene (228) by acid (abietic acid
for 2 hr at 180° being reported to give a very pure

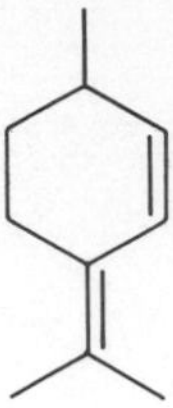

231	232	233
Isoterpinolene	β-Terpinene	Sabinene

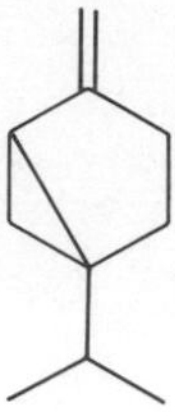

product[283]), any synthesis of β-phellandrene is a synthesis of α-phellandrene.

A synthesis of the phellandrenes that should be mentioned is that of Kuczyński and Zabża.[284] They made 1-hydroxymenth-2-ene (234) by a rather long method (it is itself a natural product, and is described below), and then pyrolyzed its phenylurethane, when a 3:1 mixture of (+)-β-phellandrene (229a) and (+)-α-phellandrene (228a) was obtained.

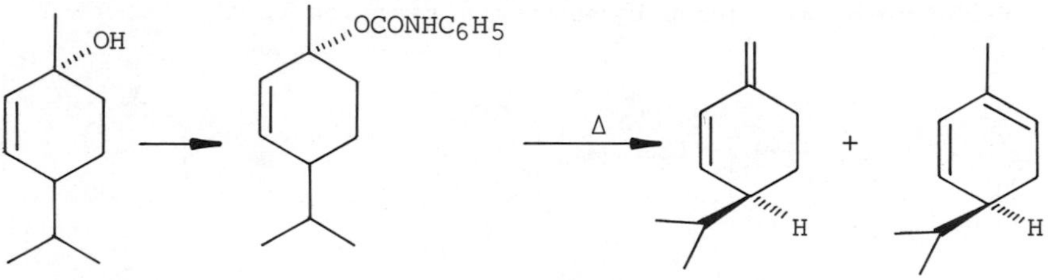

234		(+)-β	(+)-α
			3:1
		229a	228a

(+)-β-Phellandrene has also been synthesized by means of a Wittig reaction on (+)-4-isopropylcyclohex-2-en-1-one [235, (+)-cryptone, described below].[285]

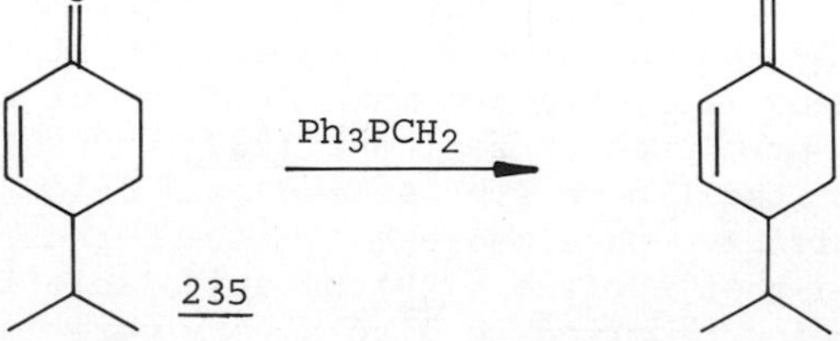

235   Ph₃PCH₂ →   229

1,3,8-Menthatriene (236)

Garnero et al. have isolated this compound from parsley (*Petro-selenium sativum*) in which it is reported to be responsible for the typical odor of the plant,[286] and a synthesis has been described by Birch and Subba Rao, although the product they obtained was not completely characterized, and was reported to be too unstable to allow further purification.[287]   Neverthe-less, the ultraviolet and nuclear magnetic resonance spectra reported by them are in agreement with the natural product. Their synthesis involves making the acetylcyclohexadienes by the route shown (Scheme 33), and carrying out a Wittig reac-tion.   The main problem is in the gas chromatographic separa-tion of the products.   The author has also achieved a syn-thesis of this menthatriene (and its isomer 237, also described by Birch and Subba Rao[287]) by the route shown in the lower part of Scheme 33.[288]   Although there was some *p*-cymene (238) formed, the impurity found to hinder purification was rather *p*,α-dimethylstyrene (239), and it was necessary to chromato-graph the products on both polar (Carbowax) and nonpolar (silicone oil) columns in the gas chromatograph in order to obtain the menthatrienes completely pure.   This was, however, achieved, and they were both sufficiently stable for all spec-tral and combustion analyses to be made.   In solution they are stable at room temperatures for several weeks, at any rate.[288]

Scheme 33

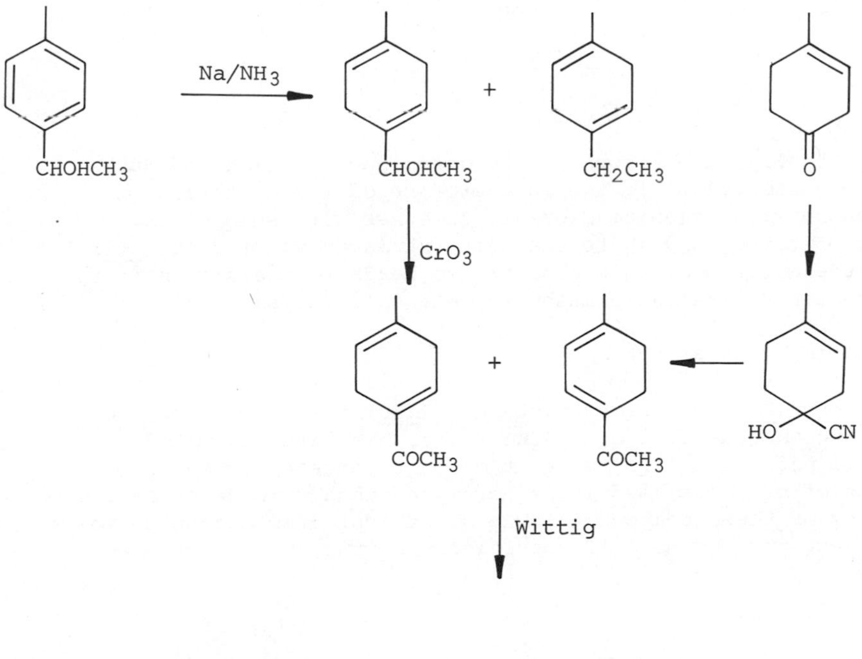

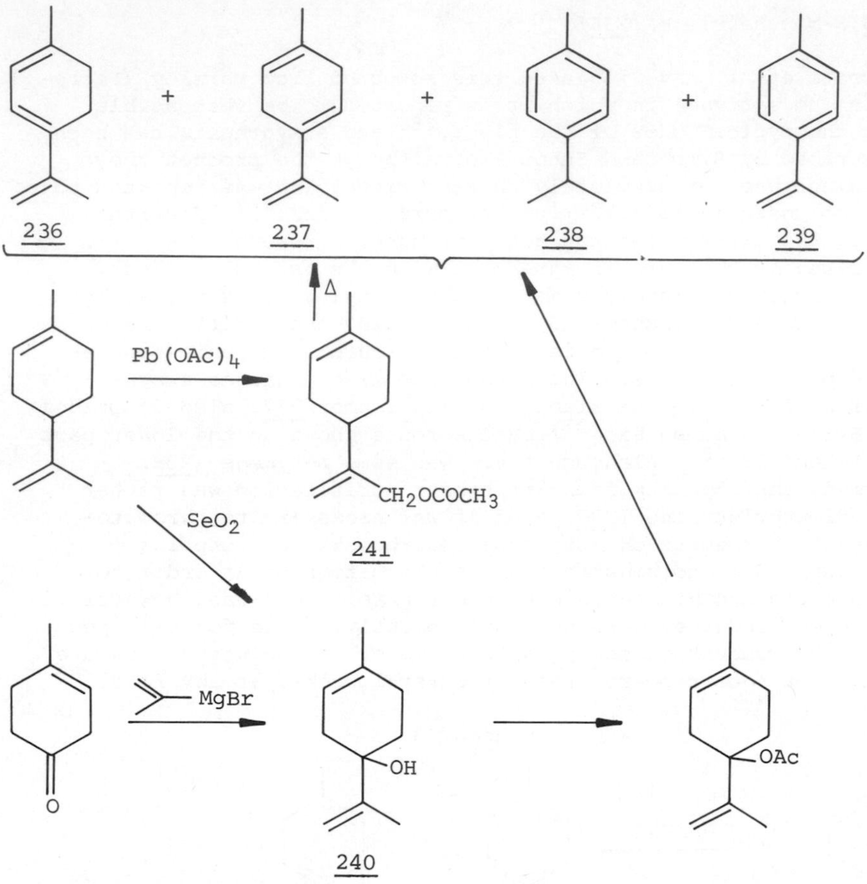

236   +   237   +   238   +   239

Finally, the alcohol (240) described by Birch and Subba Rao[287] (see below) is the main product of the oxidation of limonene with selenium dioxide, together with some of the alcohol corresponding to the lead tetraacetate product (241), and the acetate of this alcohol too leads to the same mixture of menthatrienes and dimethylstyrene on pyrolysis.[288]

Aromatic Hydrocarbons

p-Cymene (238) and dimethylstyrene (239) are very widely occurring in essential oils (the author has never examined a single natural oil so far that did not contain at least a trace of *p*-cymene) and it is hardly necessary to describe syntheses of these compounds. One might just draw attention to the fact that under acid conditions, citral is converted by

steam distillation into dimethylstyrene (239) and other hydro-
carbons--indeed, the presence of natural citric acid in an oil
is enough to effect some conversion,[289] making one wonder what
percentage of the total dimethylstyrene reported in various
oils is in reality an artefact!

H$^+$, distil.

239

B. Oxygenated Derivatives of *p*-Menthane

Of the vast literature about the many different *p*-menthane
alcohols, aldehydes, ketones, and acids, a certain selection
has been made. Very little of the older work has been included, since this can be found in the work that is quoted.
Many more of the syntheses concern products that are not
naturally occurring and that fall outside the scope of this
book. Finally, there is much work, particularly in the patent
literature, that is merely of a repetetive nature, and only
work of a wider interest has been included. A departure from
the earlier classification into alcohols, then ketones, then
other functional groups has been made, since it is more convenient to treat the oxygenated *p*-menthanes according to the
position of the oxygen atom or atoms, and relate the alcohols
to corresponding oxidized compounds.

The Tertiary Alcohols

Menthan-8-ol. The only saturated menthanol having an oxygen
at C-1, C-4, or C-8 to be reported naturally with any degree
of certainty is menthan-8-ol (*trans*) (242) that was isolated
from American pine oil residues by Zeitschel and Schmidt,[290]
and reported in flotation oil from *Pinus silvestris* by
Bardyshev and Livshits.[291] Even in these cases, however, one
is tempted to consider it as an artifact, since the German
work was carried out on oil that had previously been purified
by the "terpin hydrate" method, that is, it had been in contact with acid, and the antecedents of the Russian oil,

although not known with any precision, were certainly not such as would remove all doubt of a disproportionation reaction that could be responsible in both cases for the presence of the substance (arising, of course, as a minor by-product from α-terpineol). The geometry of *trans*-menthan-8-ol (<u>242</u>) was established by the discoverers,[290] confirmed by Keats[292] and by the synthesis from trans-4-methylcyclohexanecarboxylic acid (<u>243</u>) using a Grignard reaction by van Bekkum et al.[293] The Dutch authors also made the *cis*-alcohol (of higher melting point, 46.5°-47.5°) from the *cis*-acid.[293]

mp 35°-36°

<u>243</u>   <u>242</u>

1-Oxygenated p-Menthanes. *Menth-2-en-1-ol (234), β-Terpineol (244), Menth-3-en-1-ol (245), and γ-Terpineol (246)*. Trans-menth-2-en-1-ol (<u>234</u>) has been isolated from the oil of *Chamaecyparis obtusa* (Hinoki)[201] and raspberries,[294] and trans-β-terpineol (<u>244</u>) has long been known as a constituent of turpentine.[295,296] Their synthesis is most easily accomplished by the dye-sensitized photooxidation of menth-1-ene (<u>247</u>) in the first case, and limonene (<u>225</u>) in the second, from which the *trans*-alcohols are the main products. (A thorough review of the earlier literature has been given by Schroeter[297] and Schenck et al.[298]). For β-terpineol, the dienol (<u>248</u>) must then be partially reduced. The photochemical addition of hydroxylic compounds has been examined extensively by Kropp, and irradiation of limonene in aqueous solution with a trace of xylene as sensitizer leads to a 1.2:1 mixture of *cis*- and *trans*-β-terpineols.[299] Other chemical methods (e.g., the mercuric acetate oxidation of these hydrocarbons[296]) may also give primarily the *trans* product, but the Grignard reaction of cryptone (<u>235</u>, see below) yields a mixture of both isomers. The absolute configuration of natural menth-2-en-1-ol is not reported, but that of the material obtained by ring-opening of optically active piperitone epoxide* (<u>249</u>) has been

*This substance is also naturally occurring; see Ref. 300 for sources.

firmly established.[300]

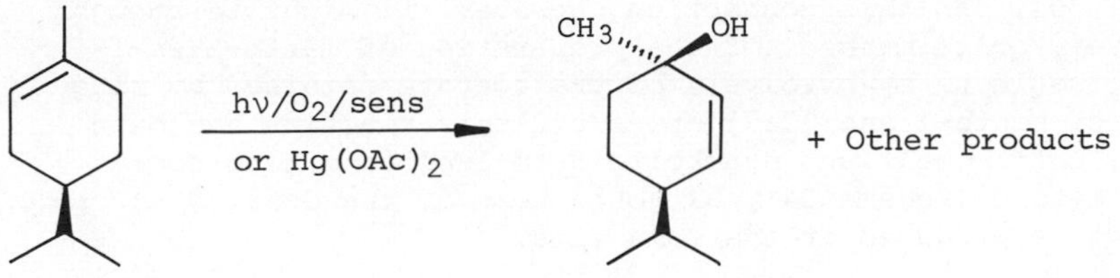

$h\nu/O_2$/sens
or $Hg(OAc)_2$
CH_3
OH
+ Other products

247 (±) 234

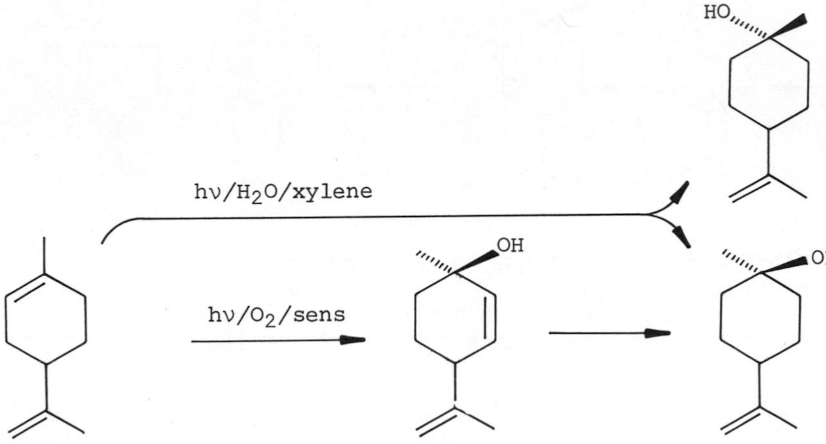

$h\nu/H_2O$/xylene
$h\nu/O_2$/sens
HO
OH

225 248 244

β-Terpineol

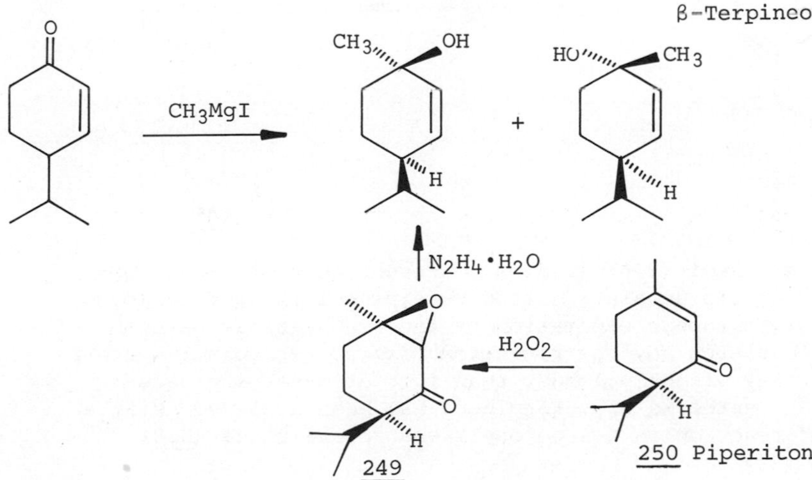

O
CH_3MgI
CH_3
OH
H
+
HO
CH_3
$N_2H_4 \cdot H_2O$
H_2O_2

249 250 Piperitone

β-Terpineol may also be made from limonene by ring opening of the epoxides; for a review of the relevant literature
see Ref. 301. In this connection, however, it might be thought
that a very quick method for the preparation of menth-2-en-1-
ol (234) would be by pyrolysis of the acetate obtained by ring
opening of menth-1-ene (247) epoxide. This reaction has been
found by Leffingwell and Shackelford to lead to a ring-con-
tracted ketone (Scheme 34), although some of the desired alcohol (234) is produced at the same time.[302]

Scheme 34

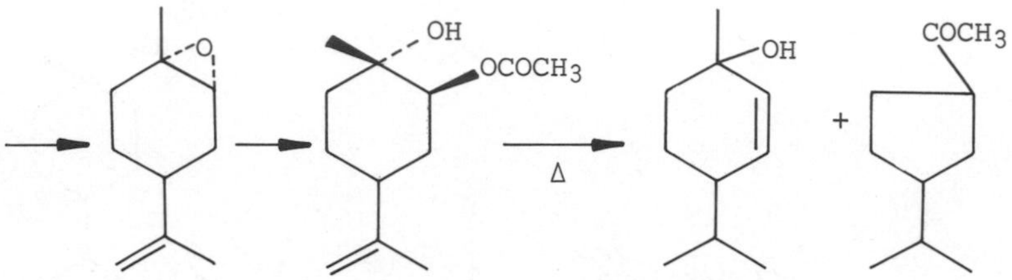

234

Menth-3-en-1-ol (245) may be a natural product (in
Origanum vulgare, for example[303]), and was made by Wallach and
Heyer by a Grignard reaction on 4-isopropylcyclohex-3-enone
(251),[304] a compound discussed below in the section on cryptone.

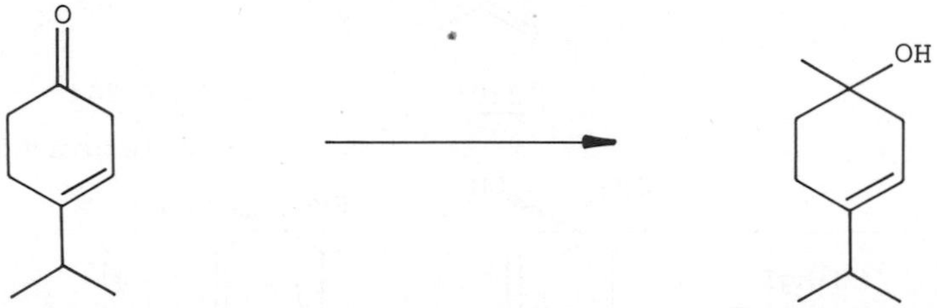

251   245

γ-Terpineol (246) is always present in commercial terpineol, but its separation from α-terpineol is very tedious.
Gas chromatographic separation of the acetates has been employed, but even so, the retention time of γ-terpinyl acetate
is only very slightly longer than that of α-terpinyl acetate.[305] Synthesis of γ-terpineol has been achieved using a
Grignard reaction on 4-isopropylidenecyclohexanone (252).[306]

The latter can be made by the route shown in Scheme 35 from γ-acetyl-γ-isopropenylpimelic acid (253),[306] but this is somewhat tedious, and if γ-terpineol is required in the laboratory, it is preferable to prepare it by the original method of Bayer.[307,308] Limonene (225) is dihydrobrominated with hydrogen bromide,[309] then the dibromide is further brominated to the tribromide (254) with bromine in the presence of light. This tribromide (254), on treatment with zinc in acetic acid gives γ-terpinyl acetate (255) in fairly good yield.[308]

Scheme 35

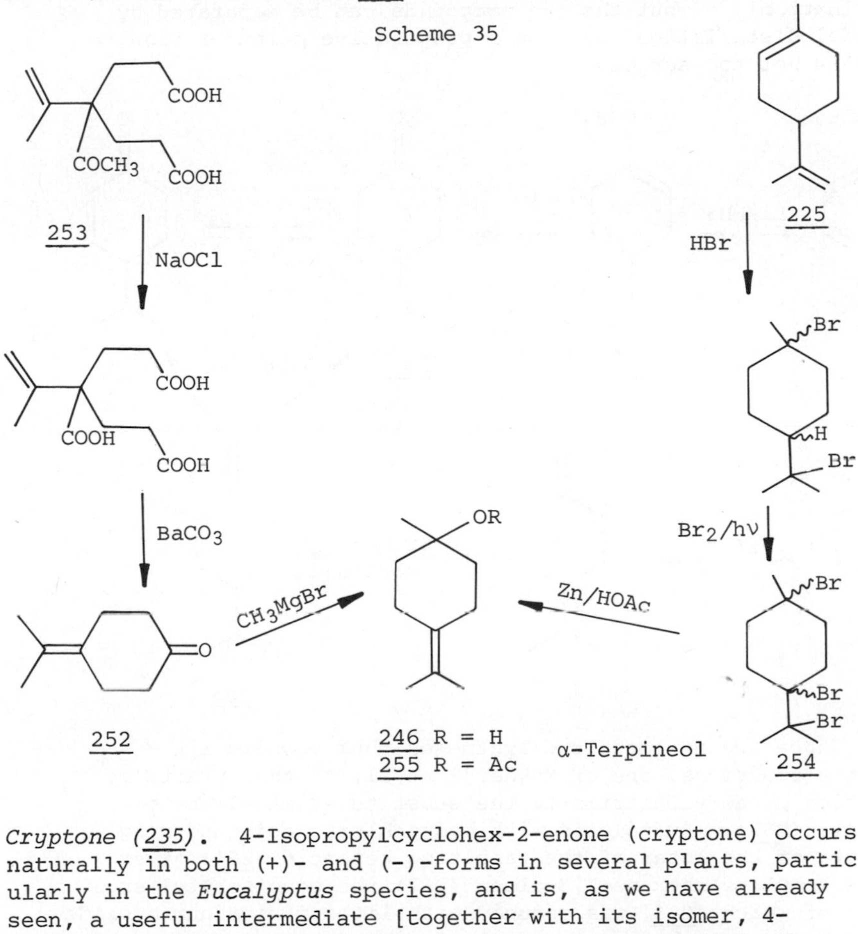

Cryptone (235). 4-Isopropylcyclohex-2-enone (cryptone) occurs naturally in both (+)- and (-)-forms in several plants, particularly in the *Eucalyptus* species, and is, as we have already seen, a useful intermediate [together with its isomer, 4-isopropylcyclohex-3-enone (251)] in the preparation of other substances. It can be synthesized, by analogy of the sodium in liquid ammonia reduction of substituted anisoles,[310] by the reduction of 4-isopropylanisole (256) with lithium in liquid ammonia, when the enol ether (257) thus obtained, is converted directly to (±)-cryptone semicarbazone with semicarbazide hydrochloride[311,312] (Soffer showed that lithium was distinctly better than sodium for this reduction.[311]) The problem that is here encountered, is that there is an equilibrium set up between cryptone (235) and its isomer, 4-isopropylcyclohex-3-enone (251) in acid conditions, so that recovery of cryptone from the semicarbazone must be done very carefully.[313] This equilibrium interferes even in the dehydrobromination of 4-isopropyl-2-bromocyclohexanone (258) from the reduction of 4-isopropylphenol (259) and subsequent bromination),[311] but the two compounds can be separated by careful distillation, so from a preparative point of view, this is not too serious.

There are at least two syntheses that involve ring-closure reactions, one of Mukherji et al.[314] that starts by addition of acrylonitrile to the substituted ethyl acetoacetate (260). Hydrolysis with hydrochloric acid and ring closure of the ester of the resulting ketoacid (261) gives 4-isopropylcyclohexane-1,3-dioen (262) the monoisobutyl enol ether of which (263) was reduced with lithium aluminum hydride to cryptone. Perhaps more simple is the synthesis of Stork et al.[315] involving addition of but-1-en-3-one to the piperidine enamine (264) of isovaleraldehyde. Without isolating the resulting dicarbonyl compound, it is cyclized with hydrochloric acid to yield (according to Stork et al.) 74% of cryptone. Actually, this synthesis too leads to a mixture of the two isomers (235 and 251); indeed, it is an excellent method for making either of them, since the undesired one can always be equilibrated afterwards and redistilled. We have found that cryptone is always the major product from the Stork synthesis, although one might have expected the equilibrium mixture (reportedly only 40% of the conjugated ketone in this case[316]) to be obtained.[317]

260 261 235 263 262

264 235 251

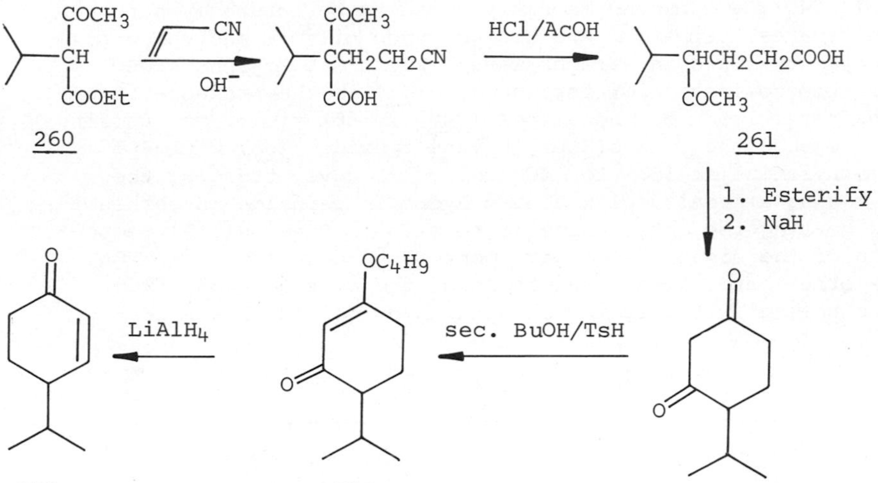

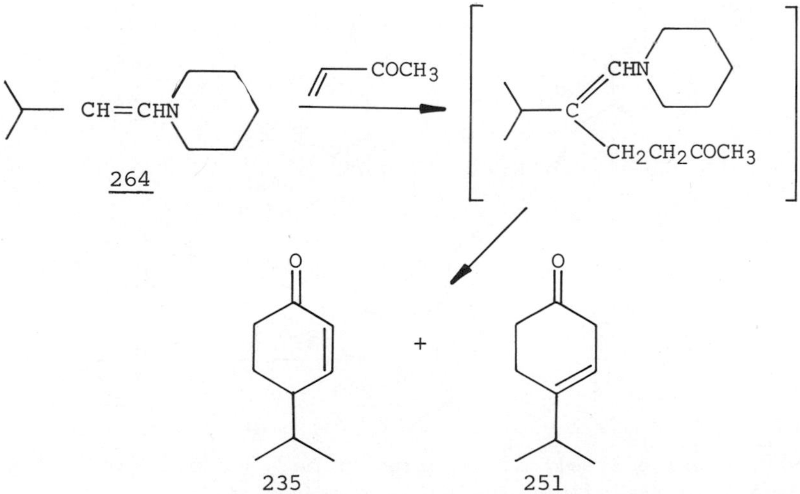

Since cryptone has been resolved by Soffer and Günay as the *p*-carboxyphenylhydrazone quinine salt,[318] these syntheses all constitute total syntheses of the natural product.

Menth-1-en-4-ol, Menth-1(7)-en-4-ol (265 and 266). The first of these terpene alcohols (265) is very widely distributed in nature and was synthesized in various ways by Wallach in the last century, for example, by the action of oxalic acid on 1,4-terpin (267).[319] It is also available from mentha-1,8-dien-4-ol (240), a compound recently found to be naturally occurring (in Japanese pepper[320] and *Citrus junes* oil[321]) and which has been synthesized in various ways. Klein and Rojahn showed that photooxidation of terpinolene (51) in the presence of a sensitizer leads to the dienol (240) in 56% yield.[322] Leffing-well opened the ring of terpinolene epoxide (268) with 40% dimethylamine at 130° to 140° for 2 1/2 days, treating the resulting aminoalcohol with 50% hydrogen peroxide in methanol and heating the amine oxide (Scheme 36). Raney nickel reduction of the dienol (240) gave menth-1-en-4-ol (265).[323] On the other hand, direct reduction of the epoxide (268) from terpinolene* with lithium aluminum hydride also leads to

Scheme 36

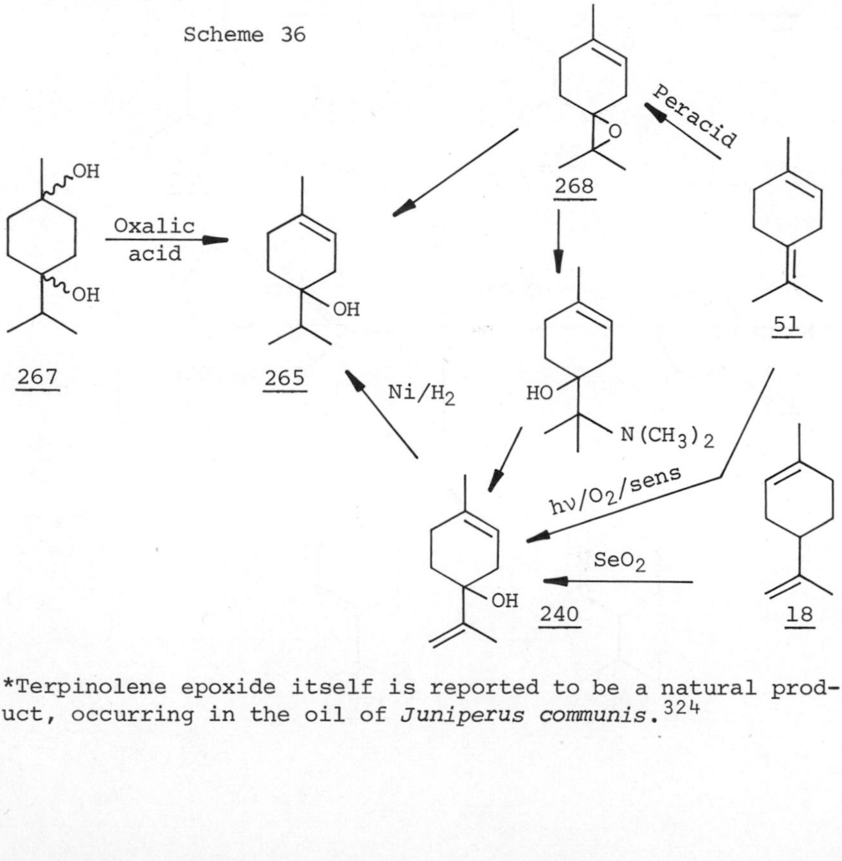

*Terpinolene epoxide itself is reported to be a natural product, occurring in the oil of *Juniperus communis*.[324]

menth-1-en-4-ol (265).[325] Mentha-1,8-dien-4-ol (240) is also
available from dipentene with selenium dioxide.[288,326,327]

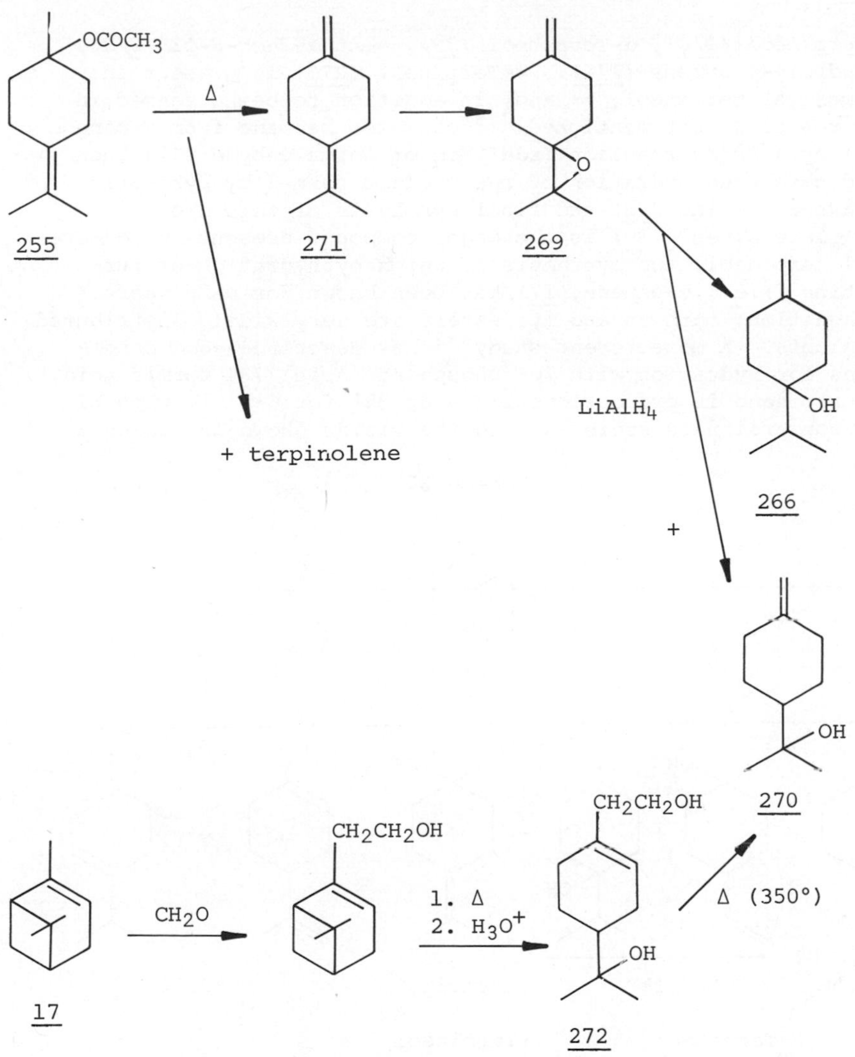

The corresponding alcohol with the exocyclic double bond
(266) has only been reported to occur in commercial terpineol,[328] and it was synthesized by lithium aluminum hydride ring
opening of the appropriate epoxide (269), δ-terpineol (270)
being also obtained during this reaction (see below).[328]

Mentha-1(7),4(8)-diene (271), required to prepare the epoxide (269), was made by the pyrolysis of γ-terpenyl acetate (255).[305]

δ-Terpineol (270), α-Terpineol (39), Mentha-2-en-8-ol (273), 8-Hydroxy-p-cymene (274). δ-*Terpineol* (270) is present in commercial terpineol,[329] and, in addition to being formed in the reaction just mentioned,[328] can also be made from α-pinene (17) by a Prins reaction (addition of formaldehyde[330]), then acid catalyzed hydration of the product formed by pyrolytic cleavage (giving 272) and final pyrolysis of this diol.[329]

α-terpineol (39) is the major compound present in commercial terpineol, the synthesis of which by hydration of turpentine (i.e., α-pinene, 17) has been known for many years.* In addition, both it and its esters are very widely distributed in plants. A more recent study[332] has described good conditions for hydration with 70% phosphoric acid, 70% formic acid, and α-pinene in proportions 1:3:5 at 35° for 6-8 hr, when 61-63% conversion is achieved with the yields shown in Scheme 37.

Scheme 37

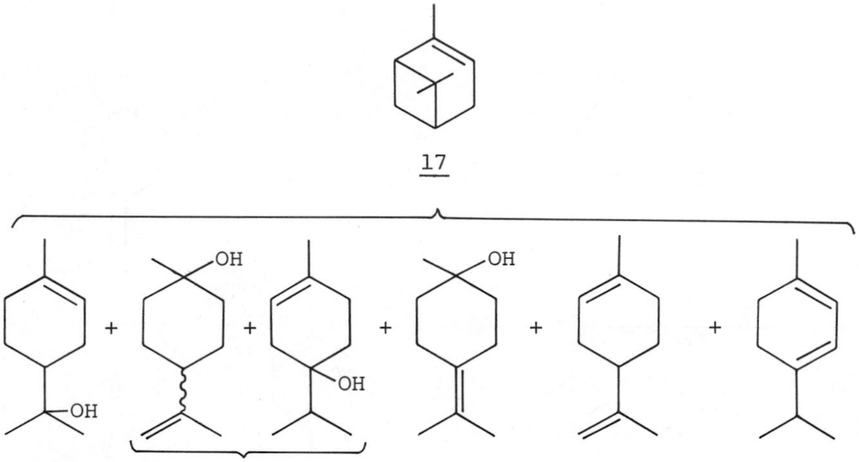

17

68.5%   3.1%   11.5%   10.7%

β-Terpineol   γ-Terpineol
        Terpin-1-en-4-ol

*For a list of the various techniques that have been used, see Ref. 331.

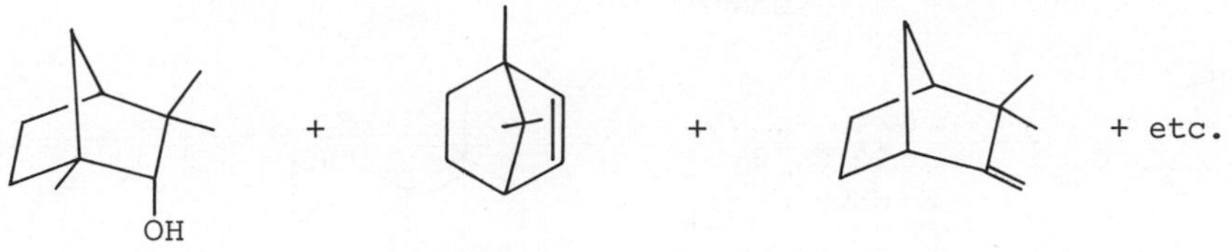

2.0%   0.2%   4%

α-Fenchol   Bornylene   Camphene

The conventional syntheses of this compound by various Grignard reactions on 4-substituted methylcyclopentenes belong more to the realm of chemical history than to a summary of this nature, and are given in all the standard textbooks of terpene chemistry [see, e.g., Ref. 331].

Menth-2-en-8-ol (*trans* 273) has been made by Chabudziñsky and Kudik,[333] but has not yet been positively identified as a natural product.   8-hydroxy-*p*-cymene (274) also occurs natural-ly and can be made by a variety of oxidative methods from *p*-cymene (238) cf., e.g., the photo-oxidation[334]).

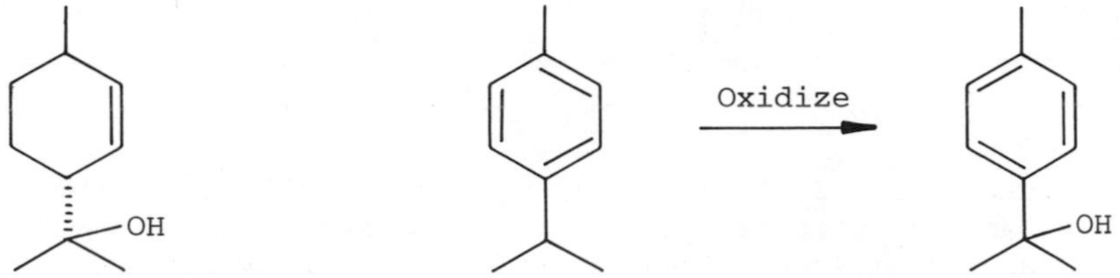

273   238   274

The 2-Oxygenated Menthanes

Phenols and Their Ethers.   The most oxidized 2-oxymenthane derivative found so far in nature is 2-methyl-5-isopropenyl-anisole (275), occurring in the oil of *Chamaecyparis ob-tusa*.[201]   This has been synthesized[335] from the known methyl 3-nitro-4-methylbenzoate (276)[336] by the route shown (Scheme 38), direct oxidation of the methyl ether of carvacrol (277) with a variety of reagents having led to oxidation of the 1-methyl group.[335]

Scheme 38

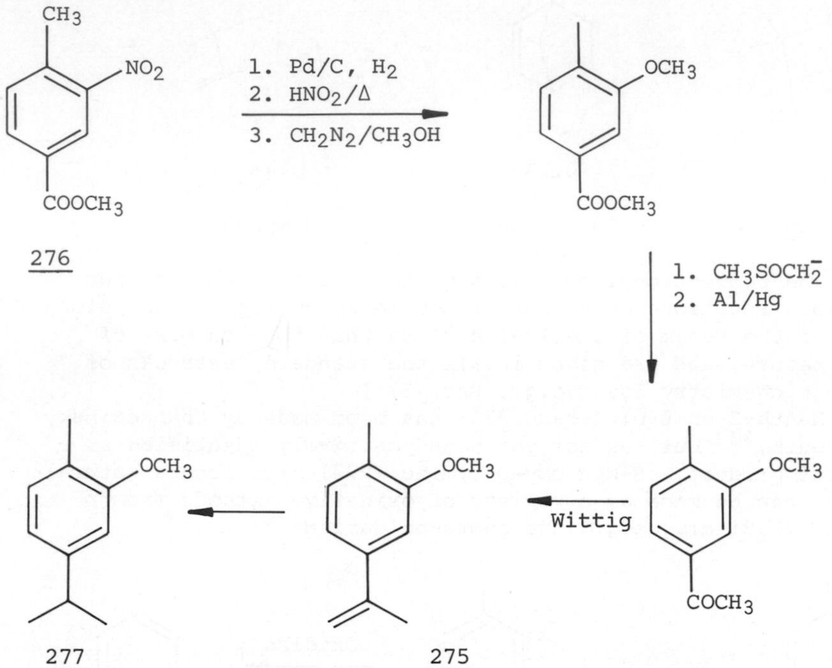
CH3
NO2
1. Pd/C, H2
2. HNO2/Δ
3. CH2N2/CH3OH
OCH3
COOCH3
COOCH3
276
1. CH3SOCH2−
2. Al/Hg
OCH3
OCH3
Wittig
OCH3
COCH3
277
275

Both carvacrol (278) and its methyl ether (277) are found naturally, and while it is not such an important product commercially as thymol (see below), it can be made from cymene (238, above) by sulfonation and treatment with base.[337] Most methods of this nature that start from *p*-cymene (238) lead, however, to a product that is usually contaminated by thymol, and a method leading to carvacrol of high purity in 39% overall yield has been described by Taylor et al. This involves treatment of *p*-cymene (238) with thallium trifluoroacetate in trifluoroacetic acid. The resulting thallium organic compound (279) is substituted in only the 2 position, and is converted to carvacryl trifluoroacetate (280) by treatment first, with lead tetraacetate in trifluoroacetic acid, then triphenylphosphine.[338]

238 → [structure 279] + Tl(OCOCF₃) →
1. Pb(OAc)₄
2. (C₆H₅)₃P
→ [structure 280] OCOCF₃ → NaOH → 278

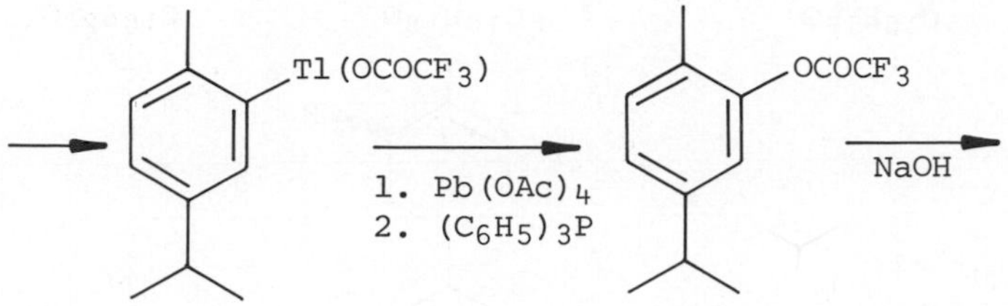

279   280

   Of course the carvone-derived compounds (see below) can
all be converted into the aromatic carvacrol by suitable oxidation techniques, but only carvone itself (281) is already of
the correct oxidation stage, and treatment of its oxime with
5N hydrochloric acid results in a fairly high yield (74-77%)
of carvacrol.[339]   The rearrangement can also be carried out
with lithium in ethylenediamine.[340]

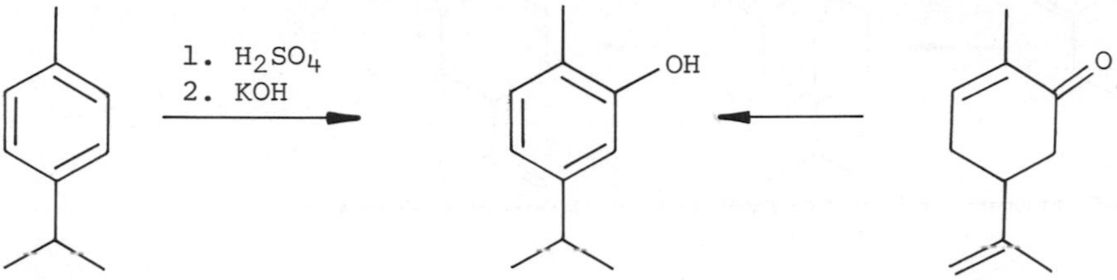

1. H₂SO₄
2. KOH

226   278   281
                    Carvacrol   Carvone

Carvone and Its Reduction Products.   A list of these compounds
is given (Scheme 39); many, indeed most of them occur naturally, and they can generally be prepared by selective reduction of carvone, choosing suitable reagents.   A full discussion
of the earlier work, together with the stereochemistry of the
isomeric compounds obtained, has been given by Schroeter and
Eliel.[341]

Scheme 39

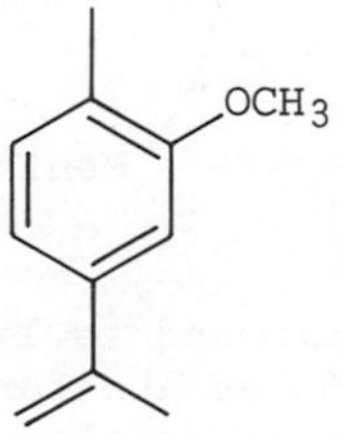

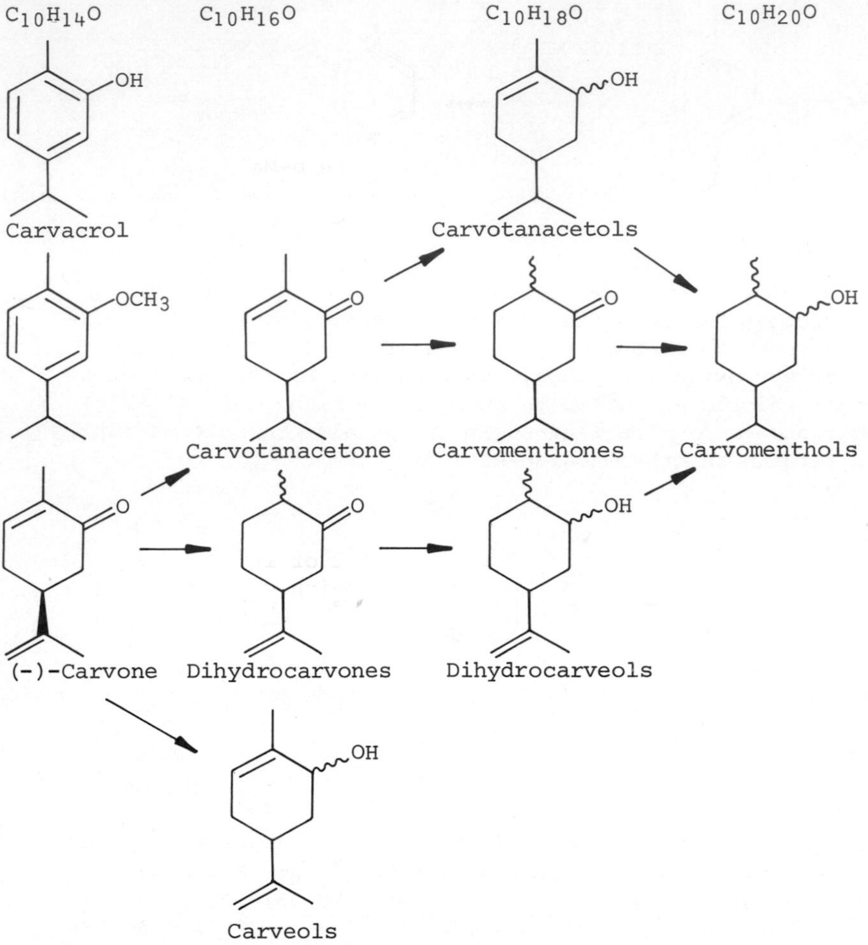

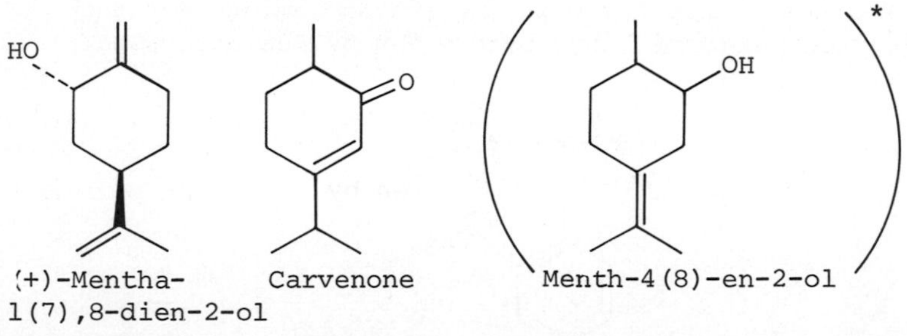

$C_{10}H_{14}O$ $C_{10}H_{16}O$ $C_{10}H_{18}O$ $C_{10}H_{20}O$

Carvacrol   Carvotanacetols

Carvotanacetone   Carvomenthones   Carvomenthols

(-)-Carvone   Dihydrocarvones   Dihydrocarveols

Carveols

(+)-Mentha-   Carvenone   Menth-4(8)-en-2-ol
1(7),8-dien-2-ol

*The propionate of this compound is listed in the Chemical
Abstracts Index for Vol. 66 as a natural product, but the
original publication makes no mention of it.[342]

Arrows represent routes that can be followed with suitable reducing agents (see Schroeter and Eliel[341]).

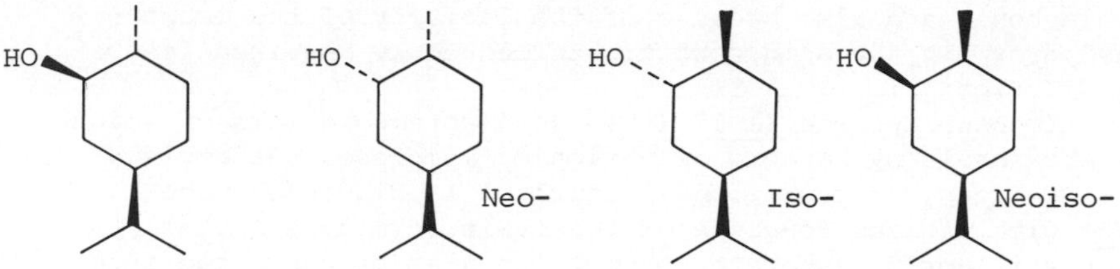

Carvomenthols: (The appropriate prefixes are used also with
other compounds of the series to indicate stereochemistry.)

Of all these compounds, carvone (281) is probably the
most widely distributed, and commercially the most important,
in view of its "spearmint" odor.   It occurs in both optically
active isomers, and these have been interconverted[343]
following Scheme 40.

Scheme 40

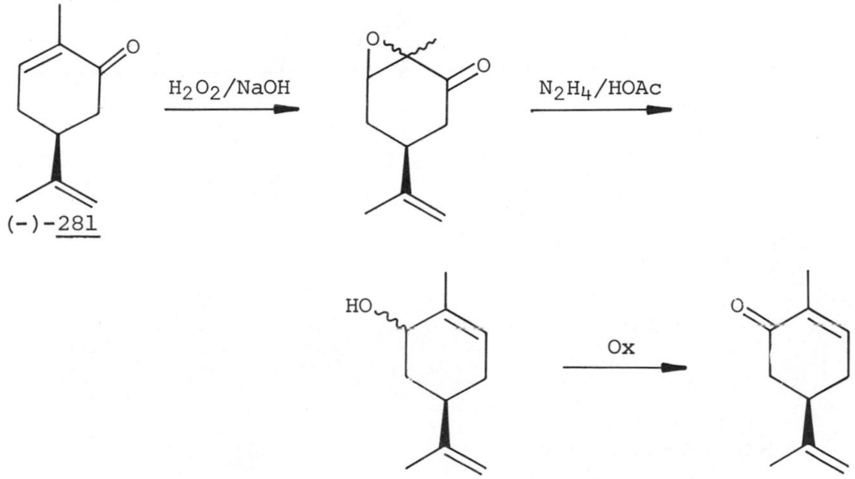

Carvone itself is available, usually in poor yield and
mixed with other substances, from limonene (225) by various
oxidation techniques, the best of which is probably chromium
trioxide-pyridine complex in methylene chloride, which leads
to a mixture of 36% carvone (281) and 33% piperitenone

(282).[344] In these allylic oxidation reactions, it is frequently found that menth-1-ene gives a cleaner reaction than
limonene, owing to competing reactions involving the other
double bond, and also because of the lability of the isopropenyl group in the acid conditions frequently required (see,
e.g., Treibs[345]).

Limonene epoxide (191) (the two diastereoisomers of which
are obtainable by peracid oxidation of limonene, see review,
Ref. 300) can, of course, be isomerized to dihydrocarvone
(283) with various reagents of the Lewis acid type, notably
boron trifluoride etherate,[346] but can also be converted into
mixtures containing carvone (up to 20%) by heating for prolonged periods (3 days) with aluminum oxide.[347] The yields
of the various 2-oxygenated menthane derivatives using this
method are shown in Scheme 41, which also shows the yields of

Scheme 41

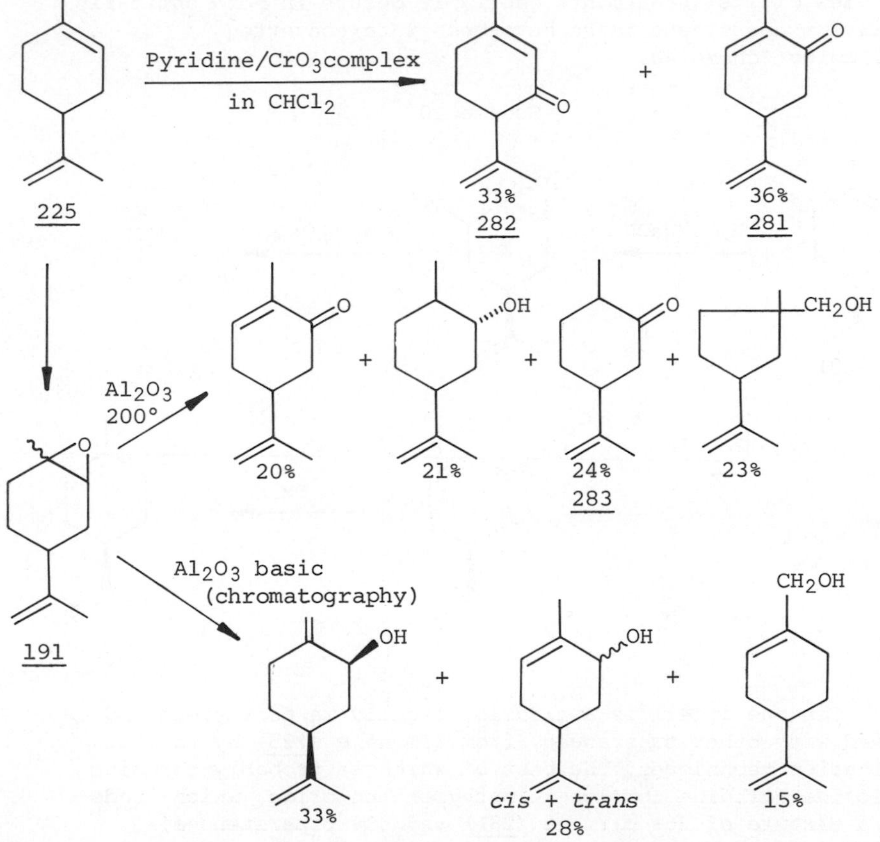

the products obtained by simple chromatography on basic alumina,[348] a method known to catalyze ring opening of epoxides.[349]

Preparations of carvone from limonene do not involve the assymetry at C-4, provided a symmetrical ion over positions 1, 2, and 6 is not formed. Thus, the sequence of steps from (+)- limonene epoxide (191) ring opening to the 1,2-diol, pyrolysis of the diacetate (284), hydrolysis to carveol then oxidation gives (-)-carvone [(-)-281].[350]

(+)-Limonene ⟶

1. H_3O^+
2. Ac_2O

(+)-191

1. Δ
2. OH^-
3. Oxidize

284   (-)-281

Oxidation with selenium dioxide is somewhat less specific and depends on the solvent. (+)-Menth-1-ene can be oxidized to *trans*-carvotanacetol (285) in ethanol with 44-55% retention of stereochemistry,[351] and the mechanism has been discussed by Schaefer et al.[352] The reaction has also been effected on (+)-limonene.[288,326,327] Attack of selenium dioxide on a double bond in a cyclohexane system occurs so that oxygen is inserted initially into the axial position, and the axial allylproton is abstracted (leading to 286). Bimolecular displacement by the water present in the mixture now occurs through an S_N2' path leading mainly to the axial *trans*-carveol (287) having 80% retention of configuration in the case of (+)-limonene, although here this is not the main product, which is **racemic** mentha-1,8-dien-4-ol (288).[288,326,327]

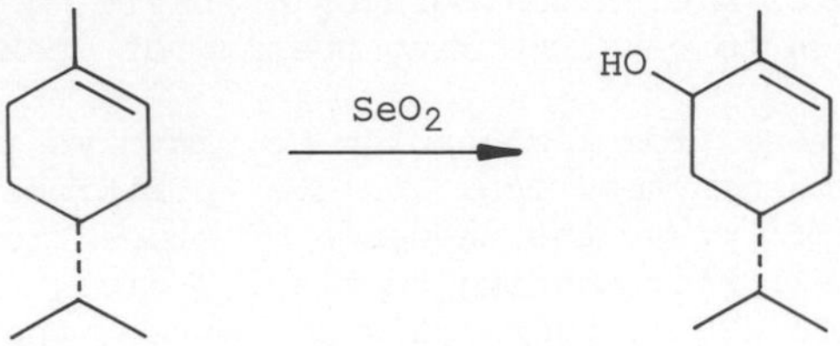

$$\underline{285}$$
(+) in Ethanol
(±) in CH_3COOH

These oxidations take a different course in acetic anhydride owing to the formation of ion pairs by the acetic acid present,[352] and in this type of solvent, both (+)-menth-1-ene (290)[351] and (+)-limonene (225)[288] give racemic carveyl acetate (Scheme 42), but optically active (+)-trans-mentha-1-(7),8-dien-2-yl acetate (289),[288] the acetate of an alcohol

Scheme 42

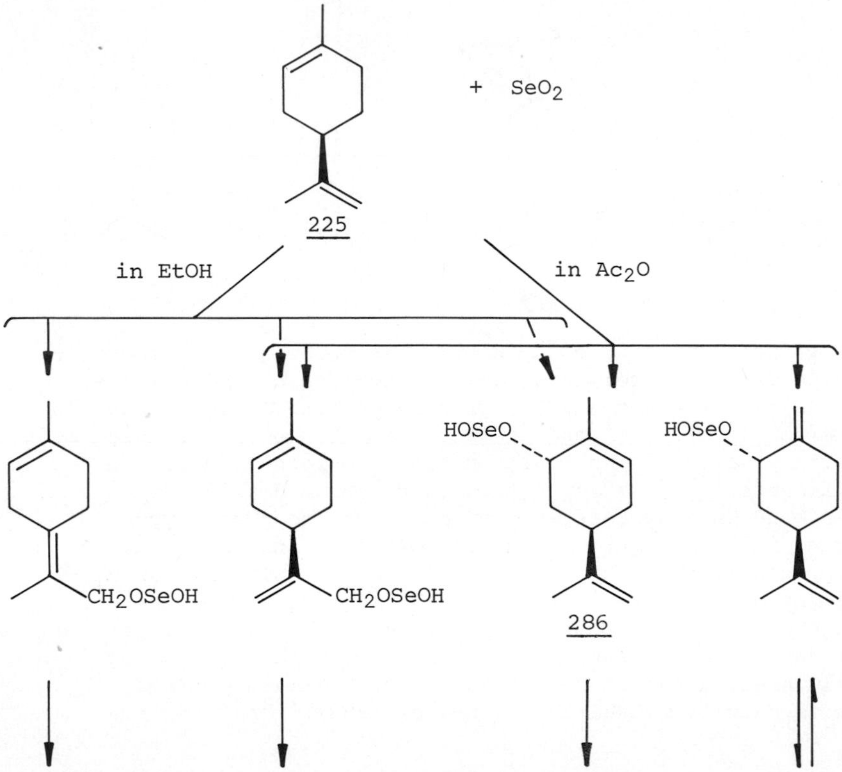

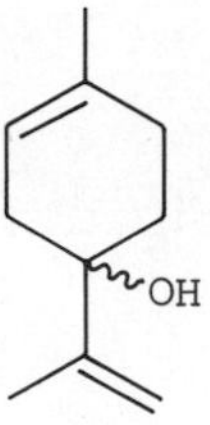

288

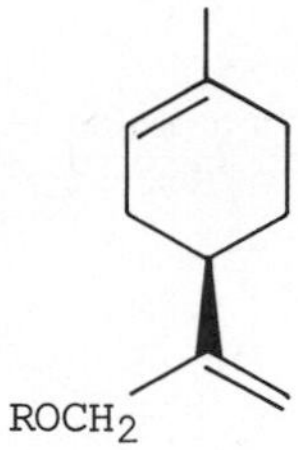

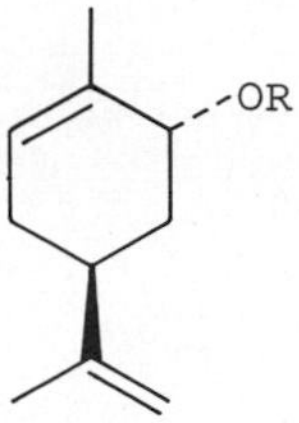

ROCH₂

287

R = H, (+)- in EtOH
R = Ac, (±)- in Ac₂O

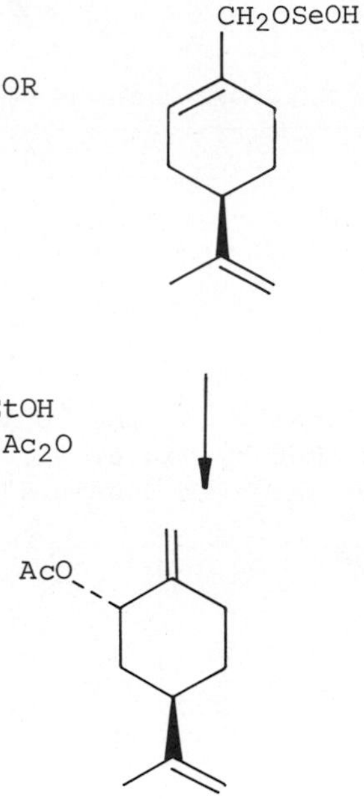

CH₂OSeOH

AcO

289

discovered by Naves and Grampoloff in gingergrass oil
(*Cymbopogon densiflorus*) and to which they erroniously attributed the *cis*-structure.[353] The correct structure of this
alcohol (289) was given by Schroeter, who also synthesized it
as one of the products resulting from the sensitized photochemical oxidation of (+)-limonene.[297][298]
    An example of a (+)-menth-1-ene (290) oxidation during
which chirality is lost is the tristriphenylphosphinechloro-
rhodium catalyzed oxidation that leads to racemic carvotanacetone (291) and piperitone (250), presumably via the symmetrical intermediate (292).[354] Similar intermediates have
been proposed by Schenck et al. to account for racemic products in the autoxidation of limonene (as opposed to the sensitized photoxidation).[355]

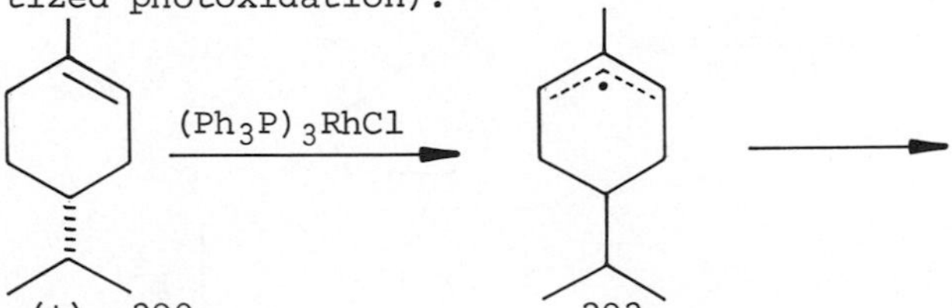

(Ph₃P)₃RhCl

(+) 290   292

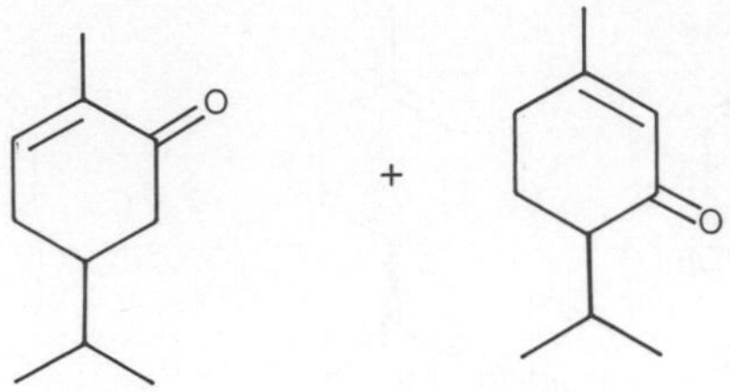

291   250

Finally, of the "building block" type syntheses of carvone, that by Vig et al.[356] is probably the most interesting,
and is shown in Scheme 43.

Scheme 43

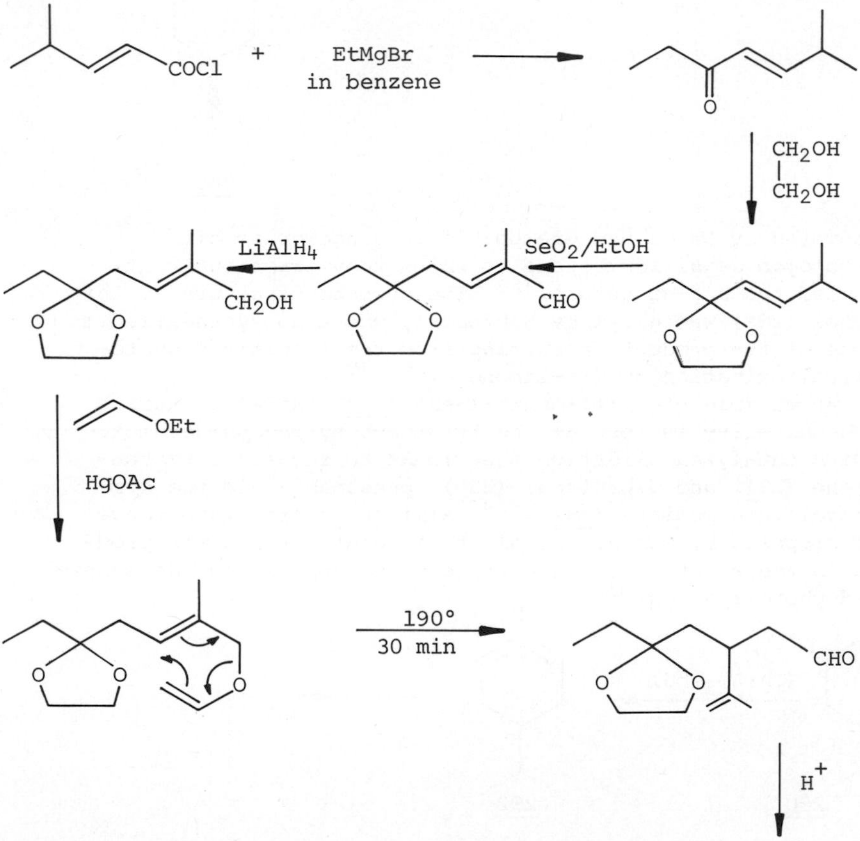

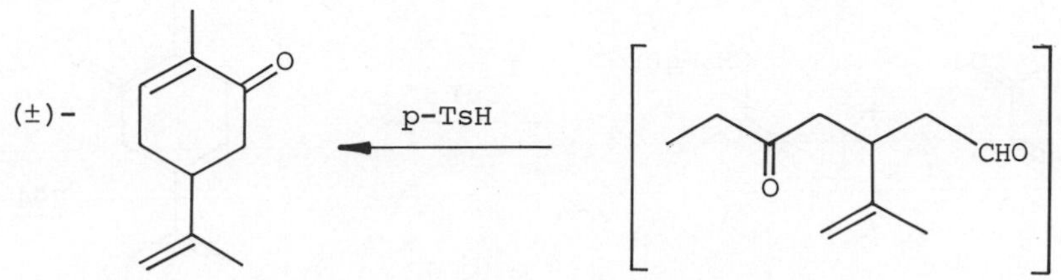

Dihydrocarveols are generally made by reduction of carvone (281); sodium in ethanol reduction gives all four of them, with the all equatorial one in largest amount[357] (the names under the formulas are taken by analogy with the carvomenthol names-- see also, Ref. 341).

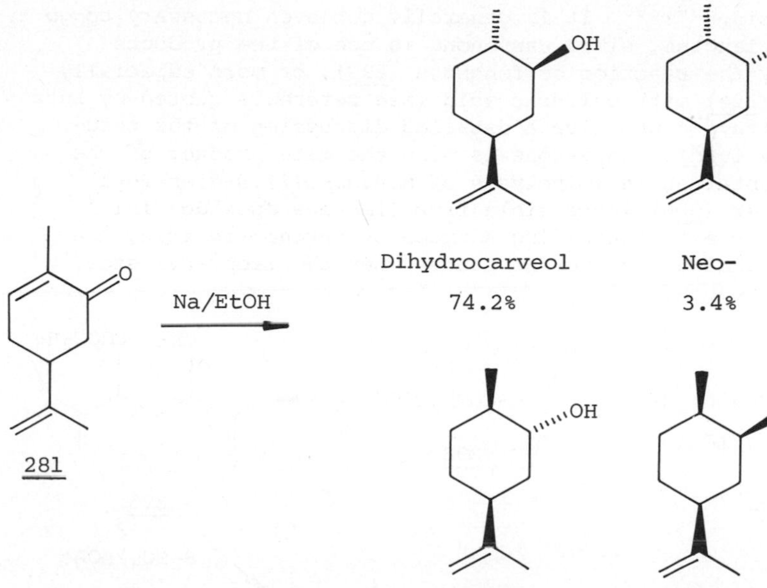

Dihydrocarveol   Neo-
74.2%   3.4%

Iso-   Neoiso-
9.3%   8.7%

Dihydrocarvone can be made from these alcohols, of course, but has also been synthesized in several ways, the most straightforward being again by Vig et al.,[358] who treated 2-methylcyclohex-5-enone (293) with isopropenyl magnesium bromide in the presence of cuprous iodide to obtain a 60% yield of the desired product (294).

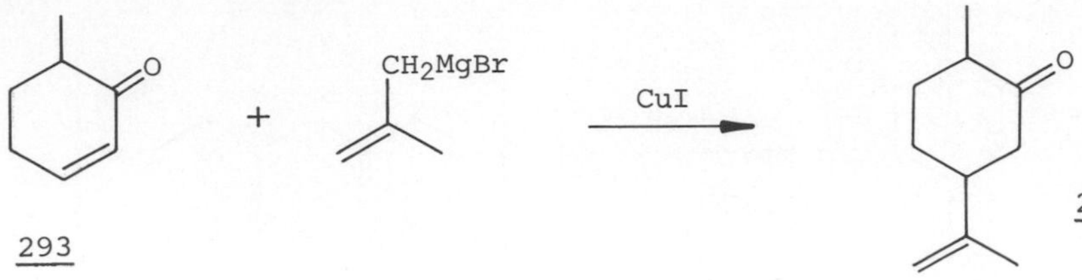

293   294

Dihydrocarvone

Carvenone (295) is easily synthesized from the compound obtained by reaction of the Mannich base methiodide (297) derived from butanone, and the β-ketoester (298). This compound (296) can be cyclized to carvenone with sulfuric acid in acetic acid.[359,360] It is generally not even necessary to go to these lengths, since carvenone is one of the products formed by the reaction of fenchone (299), or more especially camphor (224) with sulfuric acid (see reference quoted by Lutz and Roberts,[361] who give a detailed discussion of the reactions involved). Carvenone is also the main product of the mixture obtained on solvolysis of mentha-1(7),8-dien-2-yl ethyl ether (300) [accessible from limonene epoxide (191)] with acetic acid containing a trace of perchloric acid, the reaction taking a different course when the isoprenyl side chain is reduced.[362]

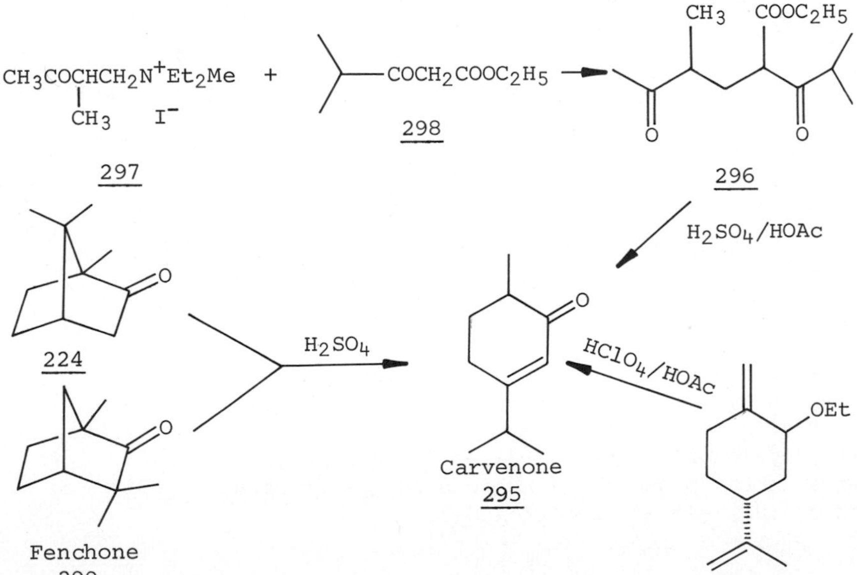

$CH_3COCHCH_2N^+Et_2Me$ +
|
CH_3 I^-

297

$COCH_2COOC_2H_5$

298

CH_3 $COOC_2H_5$

296

$H_2SO_4/HOAc$

224

H_2SO_4

Fenchone
299

Carvenone
295

$HClO_4/HOAc$

OEt

300

The 3-Oxygenated Menthanes

In dealing with thymol (301) and its reduced natural products
(listed in Scheme 44) it is convenient to include the derivatives of 3,9-dioxygenated menthanes, since these are usually

Scheme 44

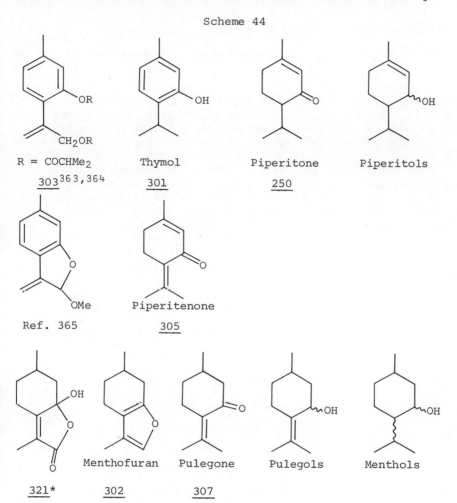

R = COCHMe₂

303³⁶³,³⁶⁴   Thymol   Piperitone   Piperitols

                    301   250

Ref. 365   Piperitenone

                    305

321*   Menthofuran   Pulegone   Pulegols   Menthols

            302   307

*Two isomers (at least) of this substance have been isolated
from natural products (see Refs. 366 and 367).

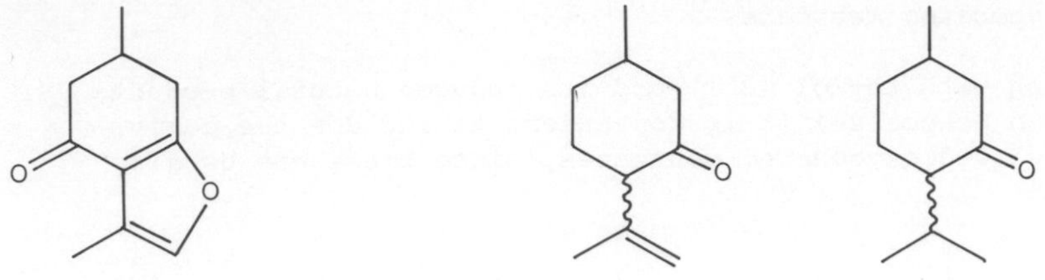

Evodone[365]

332

Isopulegone

Menthones

found in cyclized form, menthofuran (302) being the best known member of the group. Some of these substances, however, have been found by Bohlmann et al. in various *Helenium* species and in *Doronicum austriacum* as uncyclized esters (e.g., 303).[363,364]

Many of these 3-oxygenated menthane derivatives have been synthesized by cyclization procedures from open chain acids. One of the earliest of these to be examined in detail was the cyclization of geranic acid (304) to piperitenone* (305)[369,370] but similar cyclizations of citronellic acid (306) to pulegone (307),[371,372] lavandulic acid (308) to piperitenone (305),[372] and dihydrolavandulic acid (309) to piperitone (250)[372] have also been described. These cyclizations occur generally with a small amount of sulfuric acid in acetic anhydride, but certain cyclizations of open chain aldehydes occur simply on heating, and the relationship between *trans*-citral, isopiperitenol, and *cis*-citral, for example, have been discussed by Ohloff[99] (see also in the section on the open chain aldehydes, above).

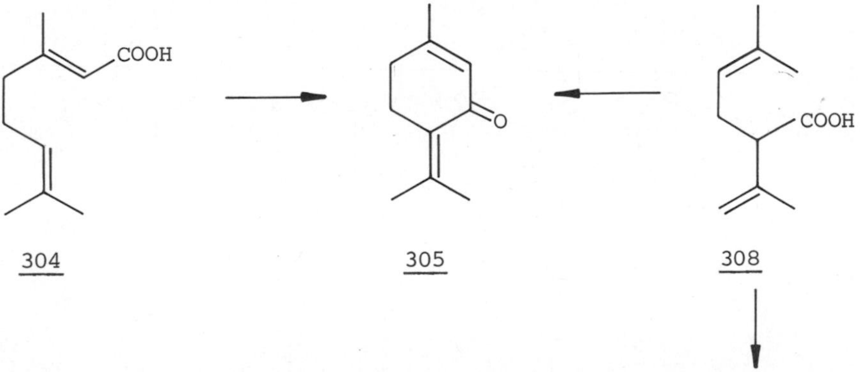

304   305   308

*This cyclization gives other products of considerable interest, see Beereboom[368] and the section on filifolone (below).

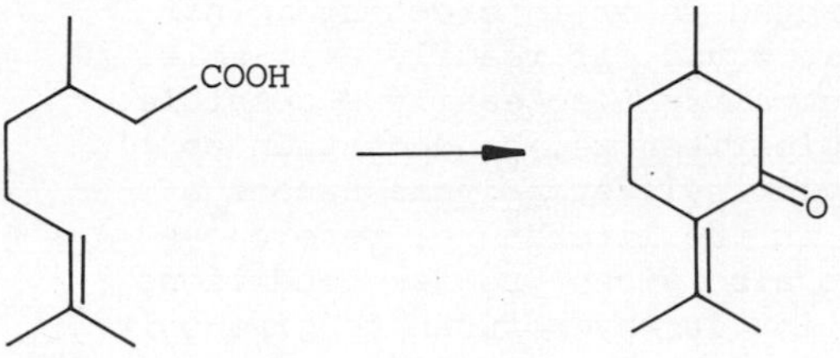

306   307 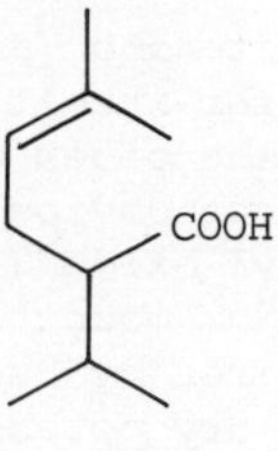   309

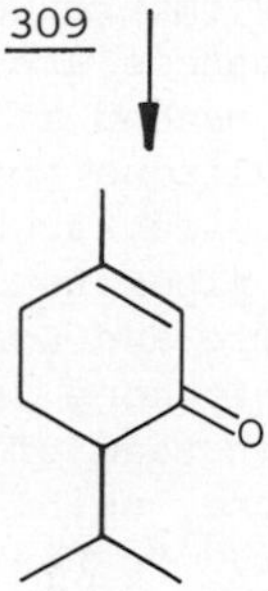

250

An interesting cyclization of citronellol is that described by Van Bruggen; when treated with a t-butyl peroxide,
the alcohol forms citronellyl radicals that cyclize to a mixture of menthols.[373]

Thymol (301). Thymol and its methyl ether both occur naturally; the phenol is very widely distributed, and is of some
commercial importance. It is usually synthesized by reaction
of *m*-cresol (310) with propylene in a Friedel-Crafts-type
reaction, and the principal difficulty is in separating the
thymol from the different isomers that are formed in the reaction.[337] Thymol is a well-known example of a cryptophenol
(one that is not soluble in aqueous alkali, nor methylated
with dimethyl sulfate in base, see Ref. 374), and this fact
can sometimes be used in its purification.

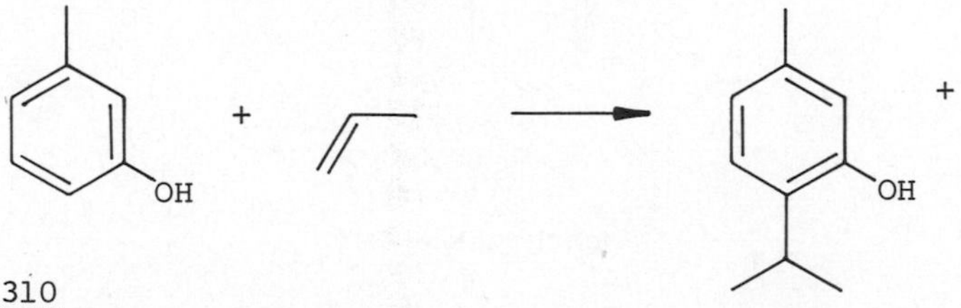

310   301

+ Other
  isomers

Piperitenone (305). This compound is of interest again not only for itself, but because it would, if readily available, make the other more reduced compounds also easily accessible. The most obvious way to make piperitenone, condensation of mesityl oxide (311) with methyl vinyl ketone, has caused a certain amount of controversy in the literature; references are given by Beereboom,[368] who also gives precise conditions using dry potassium hydroxide in tetrahydrofuran to give only 22% formation of the interfering isoxylitone (312). The two substances are difficult to separate by distillation, and the best method is that of Naves and Papazian, using the fact that isoxylitone forms a water soluble bisulfite compound.[375] This synthesis, as well as the preparation of menthone and menthol from piperitenone has also been more recently studied by Ohshiro and Doi.[376] Another method that raises the amount of piperitenone in the mixture to 50% used mesityl oxide (311), and instead of methyl vinyl ketone, the Mannich base (313) of acetone, using Triton B as the condensing agent in boiling ether.[377] When precautions are not taken in this base-catalyzed condensation, some idea of the complex isoxylitone-like products that are formed can be gained from the work of Furth and Wiemann.[378]

311   or $Et_2NCH_2CH_2COCH_3$

305

312

Pd/C

"Lippione"   Menthones

314

Piperitenone 1,2-epoxide (<u>314</u>, "lippione") is also a constituent of various *Mentha* species, and has been synthesized from pulegone (<u>307</u>).[379]   (See also the section on dioxygenated *p*-menthanes for compounds related to <u>314</u>.)

Piperitone (250), cis- *and* trans-*Piperitol (315a* and *315b)*. Syntheses of piperitone (<u>250</u>) are generally variations of the original one of Walker,[380] that consisted in allowing 4-chlorobutan-2-one (<u>316</u>, a methyl vinyl ketone equivalent) to react with isopropyl ethyl acetoacetate (<u>317</u>, R = Et) when the piperitone system (<u>318</u>) is formed directly, the free ketone being obtained after hydrolysis in base.   Lawesson et al.[381] used the t-butyl acetoacetate (<u>317</u> R = t-Bu), adding methyl vinyl ketone by base catalysis, and then obtaining the piperitone by pyrolysis.   Finally, Stepanov and Myrcina used a somewhat similar method to Walker's, but starting with the appropriate malonic ester.[382]

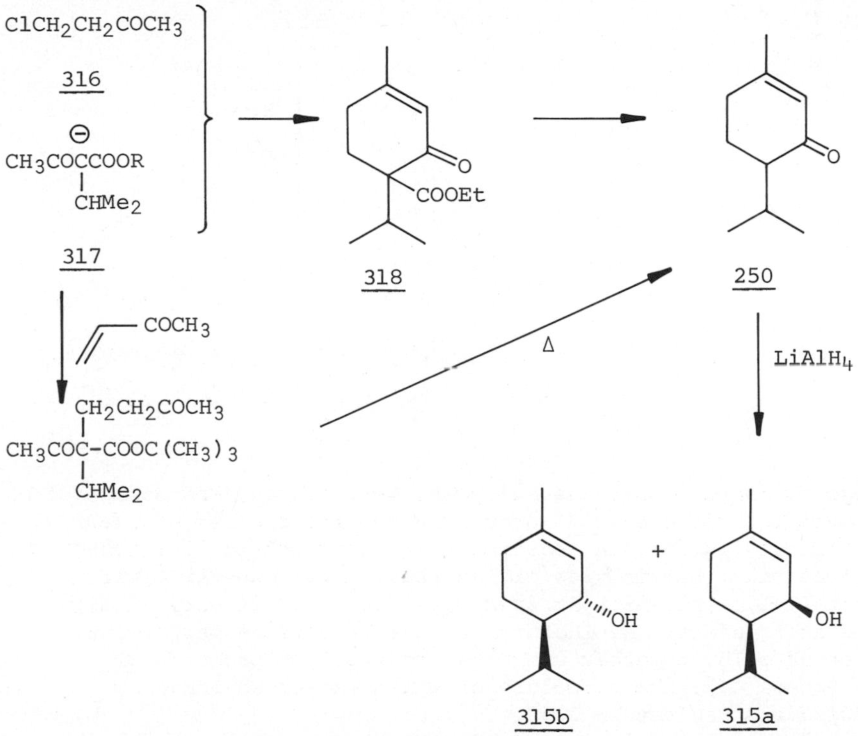

The corresponding alcohols are obtained by metal hydride reduction of piperitone, the *cis*- and *trans*-isomers are separ-

able by distillation,[383] but care must be taken, and the
author has occasionally encountered seemingly inexplicable dehydrations (probably catalyzed by traces of acid) occurring
on storage, particularly of *trans*-piperitol. (In this connection, see the discussion in Ref. 384 concerning the preparation and decomposition of the piperitols.) A different preparation of piperitone utilized Birch reduction of thymol
methyl ether (319), followed by dilute oxalic acid cleavage
of the enol ether formed (320).[384]

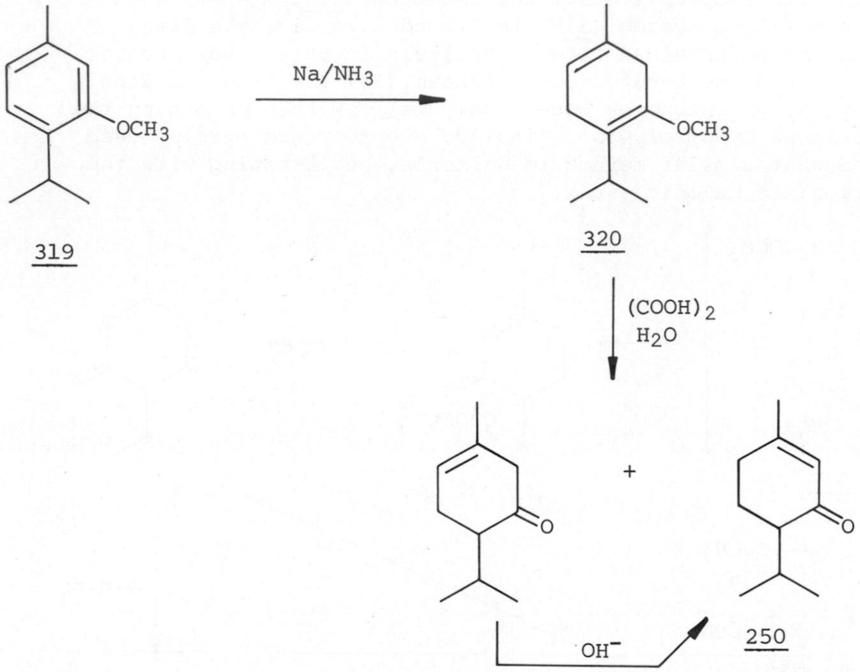

*Pulegone, Isopulegone, the Pulegols, Menthofuran, and Related
Compounds.* Pulegone (307) occurs in several species of plant,
particularly in certain *Mentha* spp.; isopulegol is also found
in plant oils, but in view of its ready interconversion with
citronellal, with which it is always found, it is very likely
to be an artefact. Menthofuran occurs in various peppermint
and other oils, together with its autoxidation product, 321
(see Scheme 44), the structure of which was established by
Woodward and Eastman in 1950.[385]

    In view of the large supplies of natural pulegone, there
has not been extensive effort expended on syntheses; it can
be made from 3-methylcyclohexanone (322) by preparing the

β-ketoester (323) (with diethyl oxalate), protecting the ketone group as the ethylene ketal (324), then adding the methyl
groups with excess methyl magnesium iodide and hydrolysis with
concentrated hydrochloric acid. The mixture of pulegone (307)
and isopulegone (325) thus obtained can be converted into
practically pure pulegone with sodium ethoxide in ethanol.[386]
Ohloff et al. have shown that pyrolysis causes the opposite
rearrangement to take place, isoisopulegone (325a) being formed
in 38% yield from pulegone.[387]

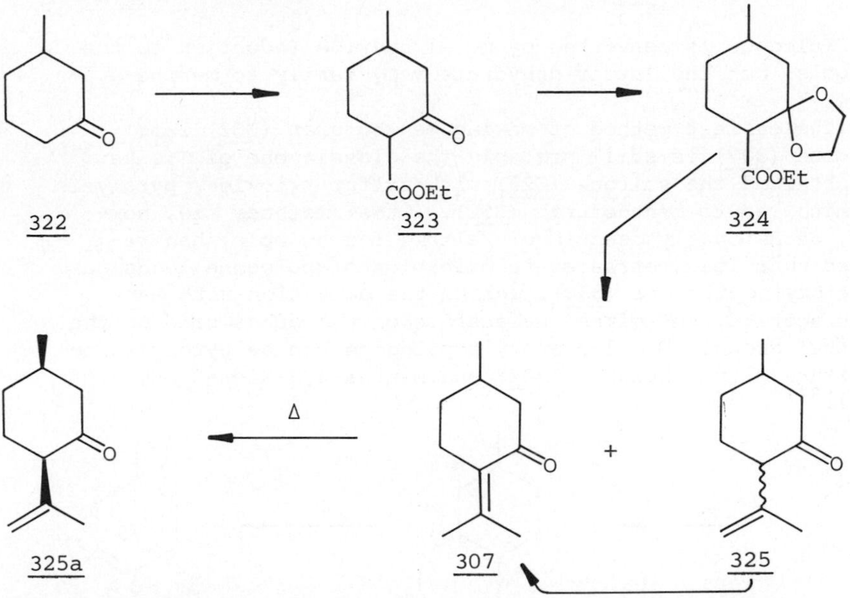

322   323   324

325a   307   +   325

Pulegone can also be prepared from aromatic sources.
Wolinsky has shown how reduction of the cresolic acid methyl
ester (326) leads to a mixture of cyclohexanol esters (327)
that are converted to a mixture of glycols (328) with methyl
lithium. Oxidation of the secondary alcohol (Jones reagent)
and distillation of the product with a trace of iodine yields
pulegone.[388]

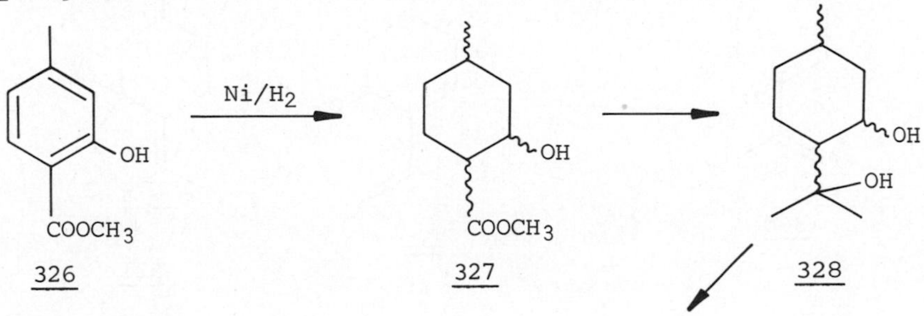

326   327   328

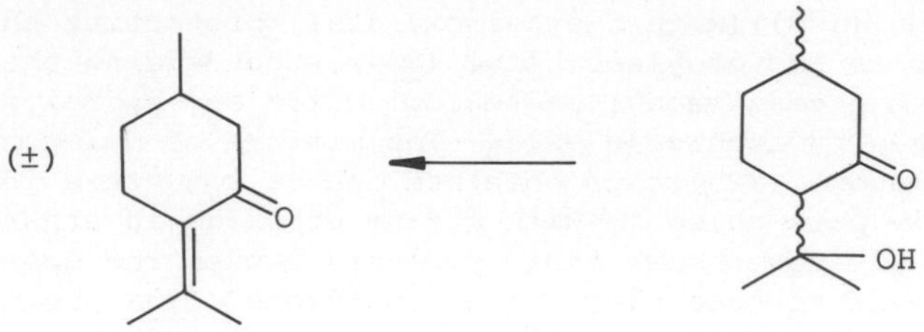

307

Pulegone is converted by metal hydride reduction to the pulegols, but the latter dehydrate very easily to mentha-3,8-diene.[389]

The easiest method of making menthofuran (302) from pulegone (307) is still probably the classic one of Treibs,[390] who obtained the sultone (329) with sulfur trioxide, pyrolysis of which led to menthofuran (302). Other methods are, however, occasionally described; Zalkow, for example, has reported that lead tetraacetate oxidation of pulegone leads to 4-acetoxyisopulegone (330), unlike the oxidation with mercuric acetate that gives the acetate on the other side of the carbonyl group. The 5-acetoxyisopulegone can be pyrolyzed to a mixture of menthofuran (302) and mentha-4,8-dien-3-one (331).[391]

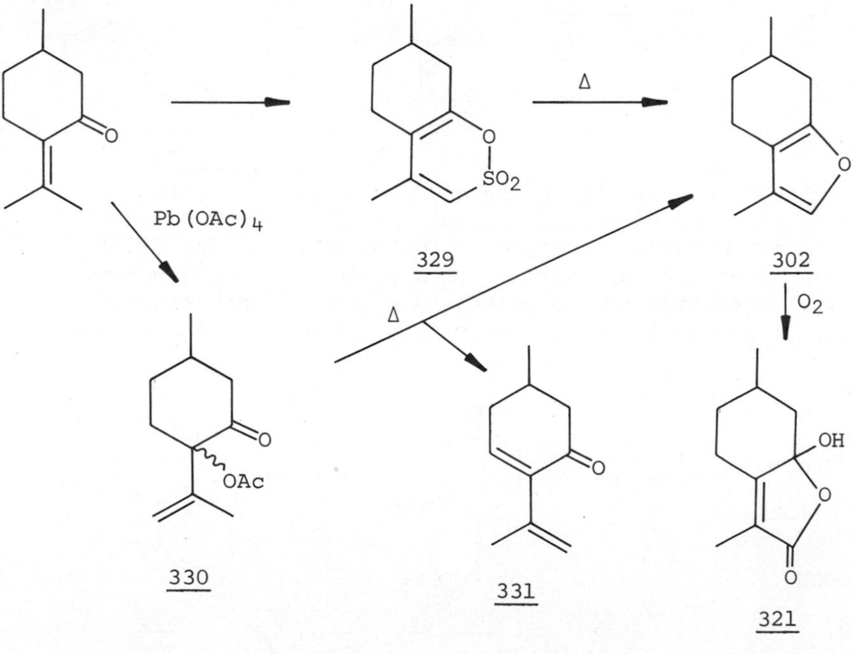

329

302

330

331

321

The light-induced oxidation of menthofuran to the hydroxy-butenolide (321) has been worked out in detail by Schulte-Elte[392] (see, also, Ref. 393).

Evodone (332), related to menthofuran, has been isolated from the leaves of *Evodia hortensis*,[366,394] and there is a report of an abortive synthesis, that started from the Birch reduction product (334) of 3,5-dimethoxytoluene (333). Con-densation of (334) with β-bromopropionaldehyde acetal using potassium amide in liquid ammonia was described, but such bromoacetals are well known to be extremely reluctant to react in this type of condensation.[395] At any rate the treatment of the supposed product (316) with hydrochloric acid did not lead to anything that could be correlated with evodone.[396]

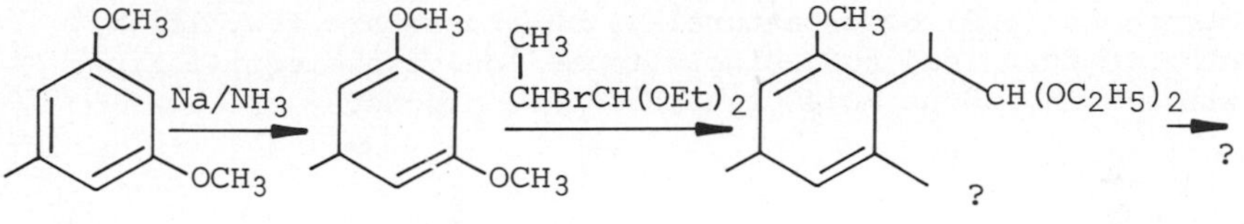

333   334   335

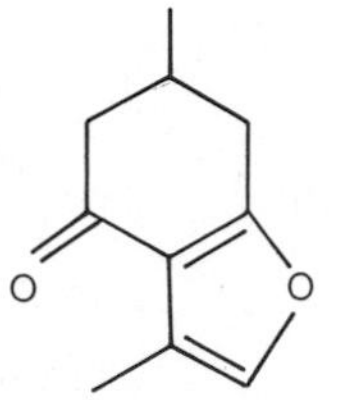

Evodone

332

2-Hydroxy-3-methylene-6-methylbenzofuran. Bohlmann has found this benzofuran in *Helenium* species, and has synthesized its methyl ether (336) from the 6-methylcoumaranone (337) by the route shown in Scheme 45.[365]

Scheme 45

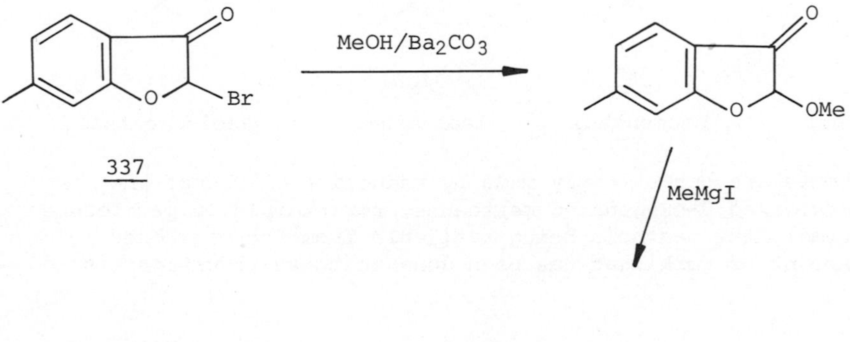

337

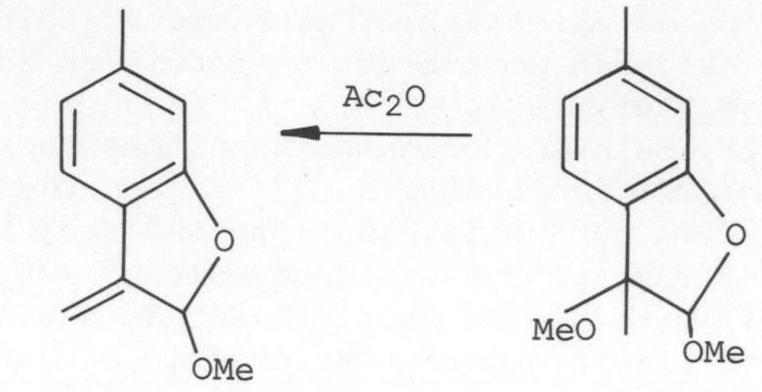

336

Menthones and Menthols. These compounds are very important
constituents of various mint oils. Both menthone (338) and
isomenthone (339) occur naturally, but there are few, if any,
syntheses that lead to a single isomer only, the equilibrium
mixture being about 60:40 menthone:isomenthone.

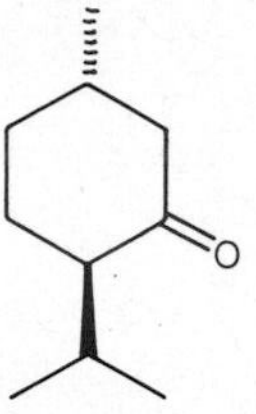

338	339
Menthone	Isomenthone

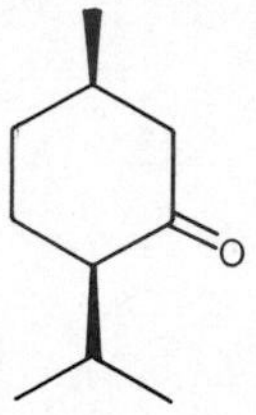

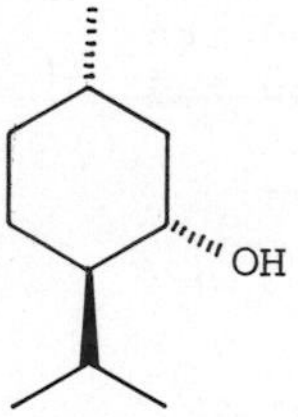

342	343	340	341
Menthol	Neomenthol	Isomenthol	Neoisomenthol

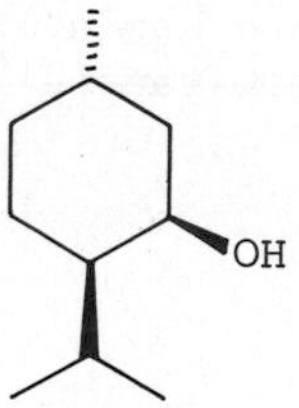

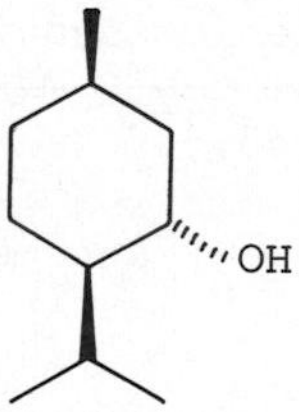

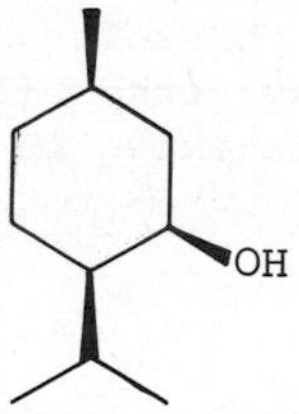

Menthones are conveniently made by reduction of one of the
more oxidized 3-oxygenated menthanes, for example, piperitenone
or thymol, the menthols being available from the menthones.
The amount of work that has been done on these reductions is

enormous; some idea of it is obtained by consulting Ref. 397. It is sufficient to say that catalytic reduction of thymol completely to the alcohols generally results in a high yield of isomenthol (340); typical figures are those obtained with Raney cobalt:[398] 84.7% isomenthol (340), 9.8% neoisomenthol (341), and 2.7% each of menthol (342) and neomenthol (343). Reductions of menthone that do not cause extensive enolization, for example, metal hydride reductions in ether, etc., lead to mixtures of menthol (342) and neomenthol (343), while isomenthone (339) gives iso- (340) and neoisomenthols (341). Menthone and isomenthone are readily separated by gas chromatography, but the separation of the four menthols is more difficult, the best column is probably Hyprose.[399] Piperitol is at the same oxidation stage as menthone, and Bhati has shown how distillation of piperitol with potassium hydroxide, a reaction based on a discovery of Ohloff,[40] leads to (±) menthone.[400]

The 7-Oxygenated Menthanes

The naturally occurring 7-oxygenated menthanes are shown in Scheme 46, and the oxidation-reduction relationships between

Scheme 46

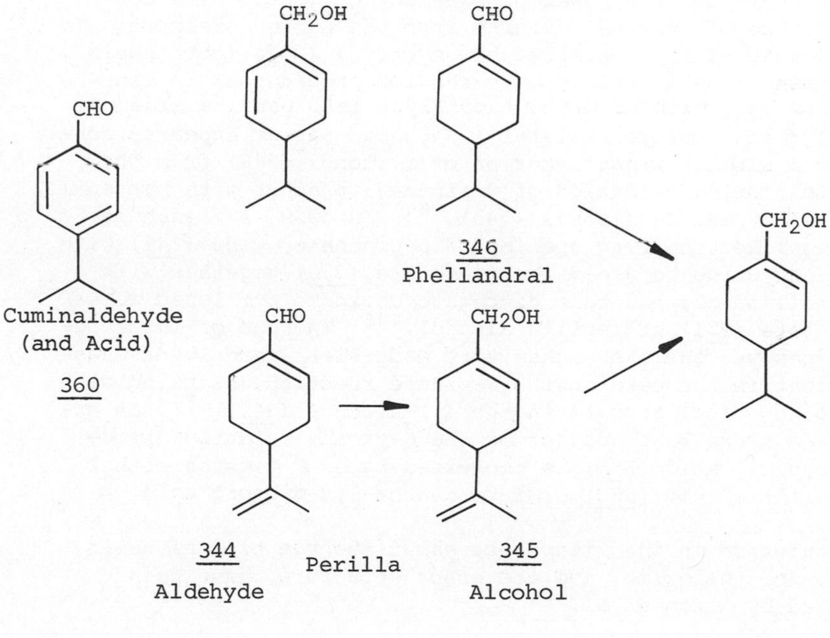

Cuminaldehyde
(and Acid)

360

346
Phellandral

344
Perilla
Aldehyde

345
Alcohol

CHO   CH₂OH

356   358

CHO   CHO

357   359

them can be deduced from the foregoing sections. The aromatic
compounds have been known since the last century, and little
new concerning their total synthesis has appeared in the past
20 years.

Perilla aldehyde (344) and alcohol (345), and the reduced
phellandral (346) are, however, of considerable interest, not
only for their own syntheses, but also because the α-oxime* of
perilla aldehyde is a powerful sweetening agent.[402] A convenient type of synthesis starts from the pinene skeleton. In
1947, Toshio et al. described how myrtenal (347) (obtainable
from α-pinene by a variety of oxidation procedures) is converted by heat with or without catalyst into perilla aldehyde
(344),[403] and four years later a Japanese patent appeared concerning a similar transformation of myrtenol (348) from the
selenium dioxide oxidation of α-pinene, together with myrtenal
(347), into perilla alcohol (345).[404] In 1959, a French patent described the ring opening of β-pinene epoxide (349) with
acetic anhydride to yield the diacetate (350) together with
some phellandral, and this diacetate could be pyrolyzed to
the acetate (351) of perilla alcohol.[405] In none of these de-
scriptions was the stereochemistry made clear, and Keenan has
shown that in the case of the β-pinene reaction, starting with
(-)-β-pinene (14) results in the (-)-alcohol (345).[406] It has
also been shown that another of the β-pinene oxidation products, nopinic acid (352) is converted by acid fission either
into perillic acid (353) or into mentha-1,3-dienoic acid

*The confusion in the literature about the *syn* or *anti* nomen-
clature for the oxime, and its exact structure, have been
clarified by Acton et al.[401]

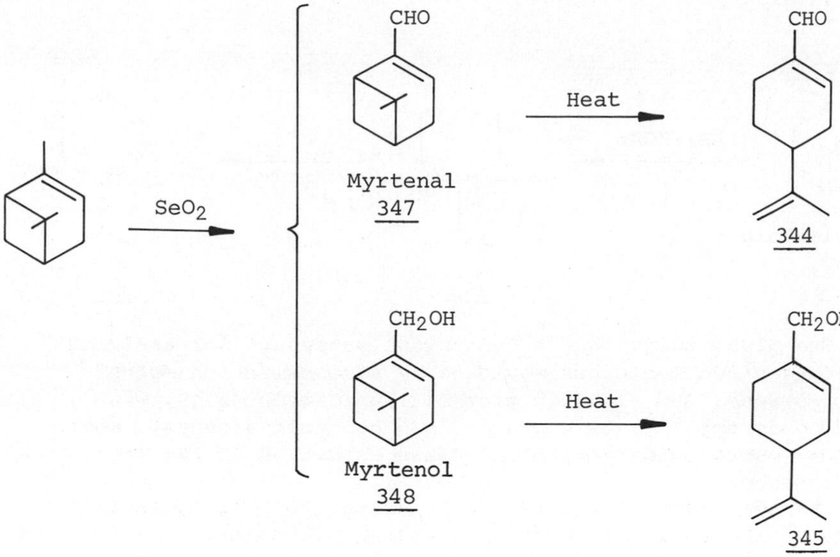

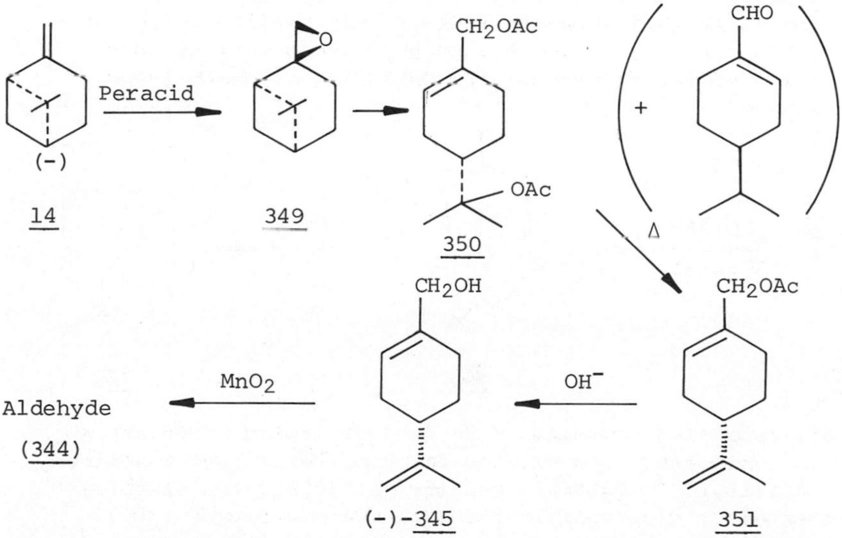

(354) according to the conditions.[407,408] The moving of the C-8 double bond into the ring is common to all doubly un-saturated monoterpenes having an isopropenyl group (as we have mentioned above in the case of limonene), and perilla com-pounds are not exception, as the varying products of this

nopinic acid reaction show, and as has been shown to occur with
perilla aldehyde.[409]

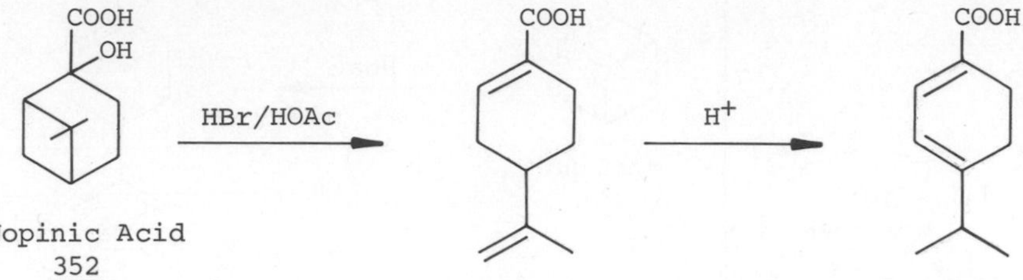

Nopinic Acid
352

353   354

One might think that a convenient procedure for arriving
at 7-oxygenated menthanes would be by some oxidation method
from p-cymene, and although microbial oxidation (e.g., with
Pseudomonas spp.[410]) is known to lead to cumin aldehyde, most
methods (photo-oxidation, etc.) cause oxidation in the isopropyl group.

The oxidation of one of the products (355) of menth-1-
ene (247) photo-oxidation[298] with sodium bichromate in sulfuric acid leads to phellandral (346),[411] but this reaction
might not be so easily accomplished in the perilla series in
view of the lability of the isopropenyl group to acid (see,
also, Ref. 345 for differences in oxidation of menth-1-ene
and limonene).

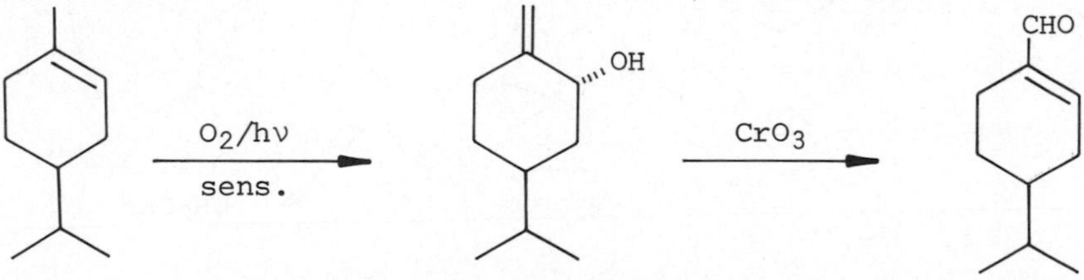

247   355   346

The remaining compounds (356 to 359) listed above are
found in cumin seeds and oil and Varo and Heinz have described
their lability.[412] Mentha-1,4-dien-7-al (356), for example,
is converted by disproportionation to cumin aldehyde (360)
and menth-3-en-7-al (359) merely by injection into a gas
chromatograph. Mentha-1,3-dien-7-al (357) has already been
mentioned as arising from perilla aldehyde with acid.[409]

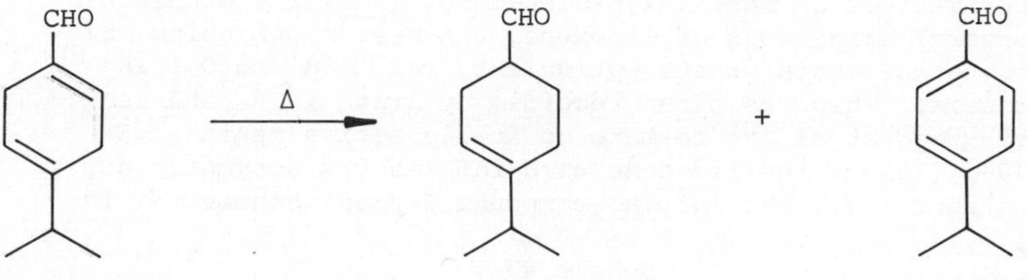

356   359   +   360

The 9-Oxygenated Menthanes

It is a curious fact that most of the compounds of this group
have only relatively recently been discovered in natural
materials; indeed, Gildermeister-Hoffmann lists none at all.
Hunter and Moshonas discovered two of the alcohols [(361) and
(362)] in cold pressed Valencia orange oil,[413] the diol uroterpinol (363) was found[414,415] and identified[416] as the
glycoside (364) in human urine, probably from dietary limonene
metabolism,* and the two possible diastereomers of meth-1-en-9-
al (365) have been found in rose oil.[417]   In addition to the
synthesis of these natural products that is mentioned below,
the series of papers by Camps and Pascual[418-419a] should be mentioned; they describe the synthesis of a number of saturated
and unsaturated menthane 9- (and some 7-) alcohols, not yet
reported in nature together with their stereochemistry.

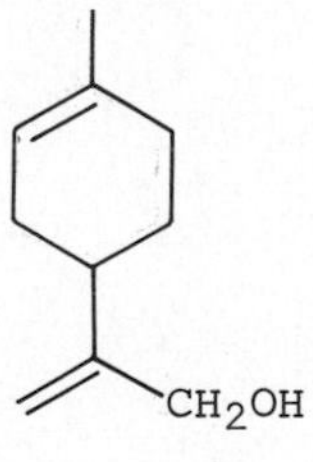

361 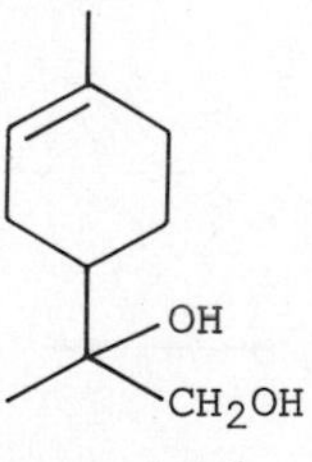   Uroterpinol   364
                     363

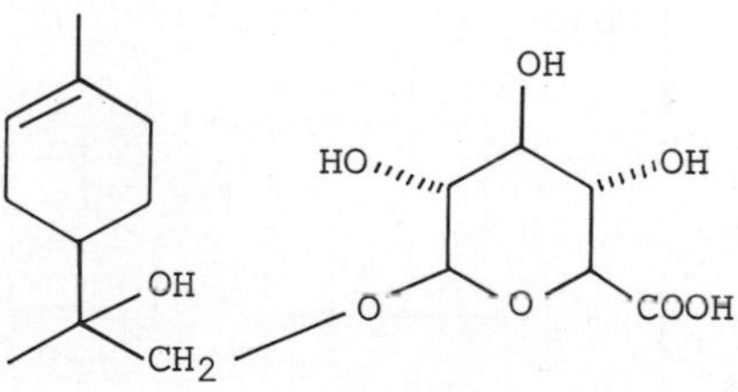

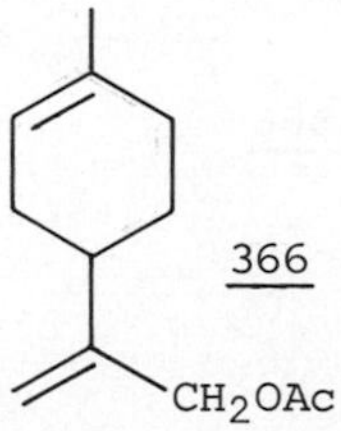

366 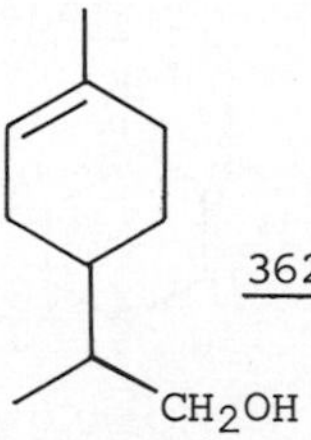   362 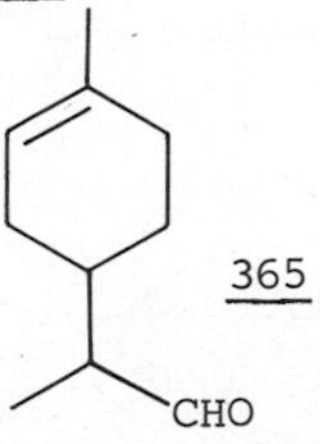   365

*The natural compound is not optically pure, probably because
dietary limonene is not optically pure.[416]

Derivatives of mentha-1,8-dien-10-ol (<u>361</u>) are accessible
from several oxidations of limonene, the best known being that
of lead tetraacetate or its equivalent, red lead (Pb_3O_4) in
acetic acid. This was first recorded by Aratani,[420] and was
used by Ruegg et al.[421] to make optically active mentha-1,8-
dien-10-ol, since the limonene used retains its assymetry during the reaction. The initial compound formed (Scheme 47) is

Scheme 47

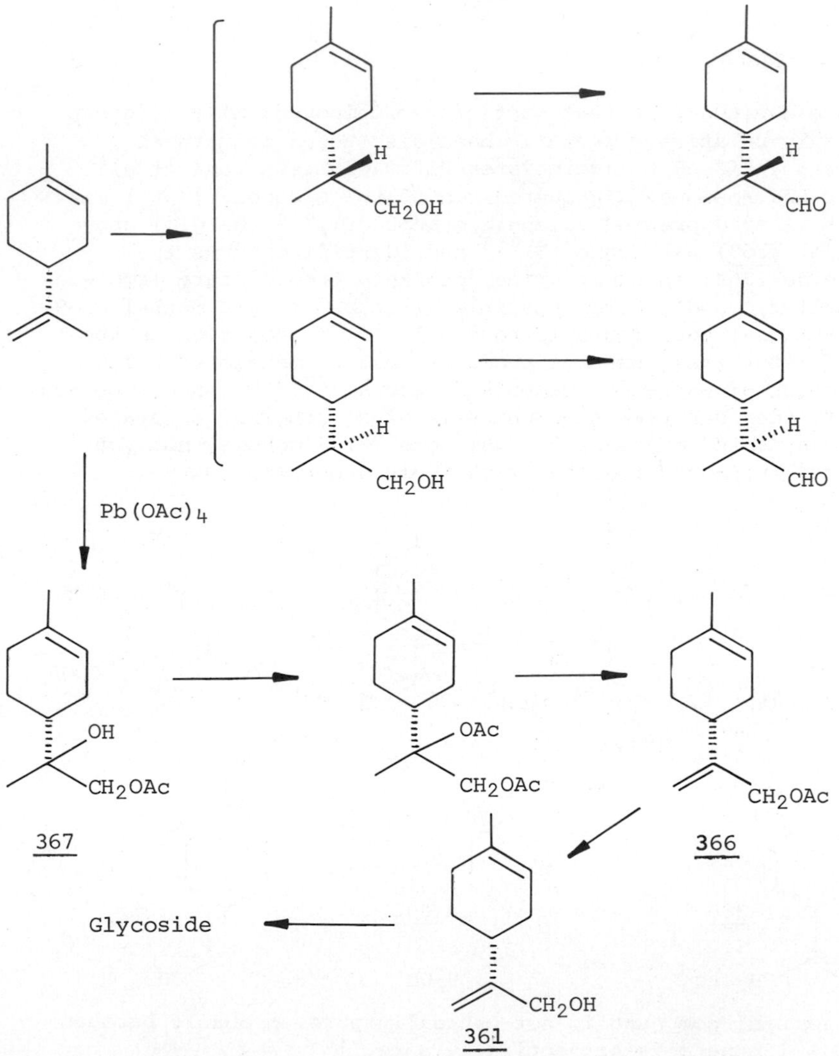

the 8,9-diol in the form of its monoacetate (367) and it was
this diol that Dean et al. used for making the glycoside,
364.[416]   Acetylation of the other hydroxyl group (tertiary)
and pyrolysis leads to the dienyl acetate (366), from which
the free alcohol is obtained in the usual way.   The mono-
acetate of the dienol is also one of the main compounds obtained by selenium dioxide oxidation of limonene in acetic
anhydride,[288] and is itself a natural product, occurring in
the oil of *Citrus natsudaidi*,[422] *C. unishu*,[423] and Valencia
orange oil.[424]

A detailed discussion of the stereochemistry of the 9-
oxygenated menth-1-enes has been given by Ohloff et al.,[417]
and this is very important, since early publications on this
topic were obscure.   Albaigès et al.[418] pointed out that there
were two diastereomers of menth-1-en-9-ol (362), and a mixture
of the two can be obtained by hydroboration[425-427] or by addition of aluminum alkyls[428] to limonene, during which the
chirality center at C-4 is, of course, maintained.   The two
isomers can be separated by fractional crystallization of the
3,5-dinitrobenzoates, and from the pure alcohols it is possible
to make the two isomers of the aldehyde.[417]

Di- and Polyoxygenated Menthanes, Cineols, Pinol, etc.

Some of the dioxygenated menthanes have already been described,
notably the 3,9-diols and their dehydration products (menthofuran derivatives), and the 8,9-diols.   Some of the remaining
natural products of this type are given in Scheme 48.   Loss of

Scheme 48

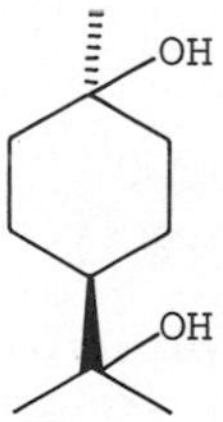

373

trans-Terpin

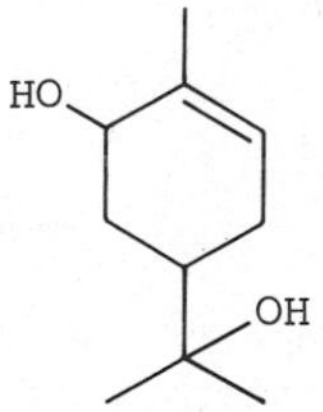

370

Sobrerol

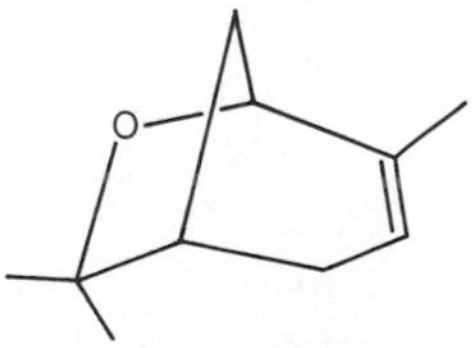

371

Pinol

(not naturally occurring)

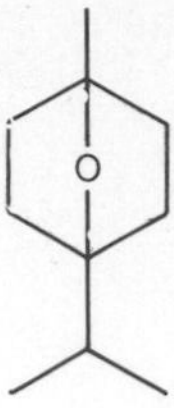

369   368   376   374

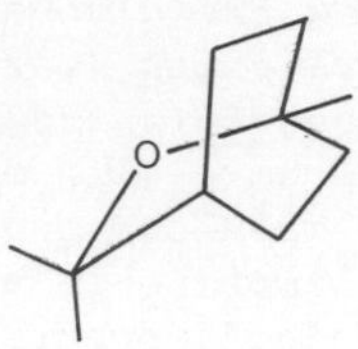

1,4-Cineol   1,8-Cineol   Mullilam Diol   Ascaridole

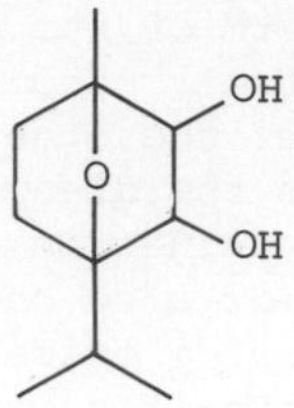

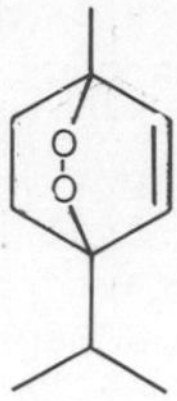

380   381

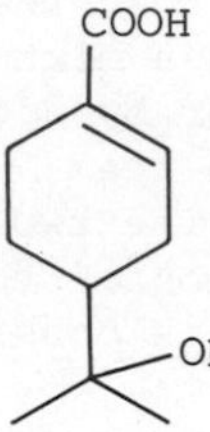

377

Oleuropic Acid

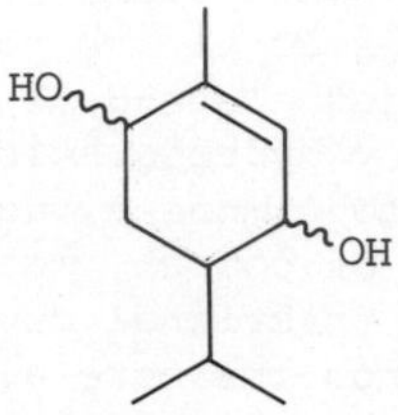

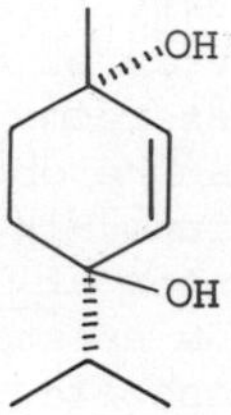

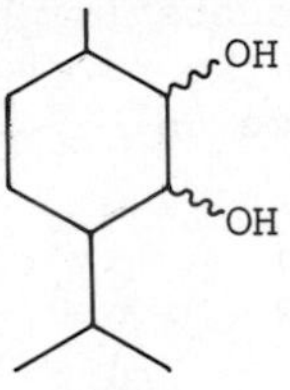

384 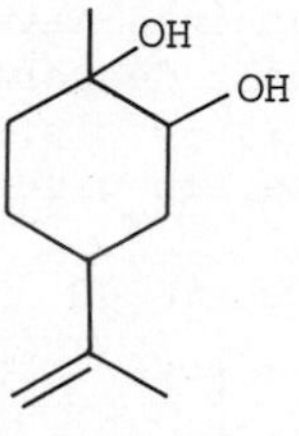   378 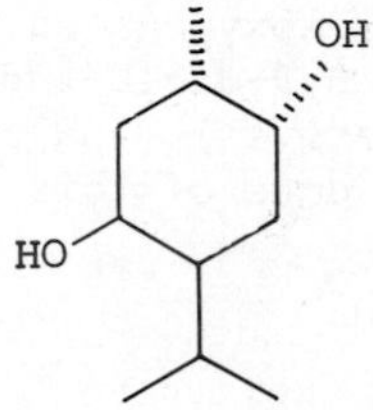   379

                                  Menthane-2,5-diol

water between the hydroxyl groups of the well-known hydration
products of pinene, terpin hydrate, or of the hydroxyl groups
in menthane-1,4-diol gives the cineols, 1,8-(368) and 1,4-
(369), respectively. In the same way, sobrerol (370), known
since the eighteenth century, loses water to give pinol (371).*
This list is by no means exhaustive, and there is some question as to how "authentic" some natural products of this
nature really are. For example, it is known that terpinolene
(51) is autoxidized to menth-3-ene-1,2-trans-8-triol (372),[429]

*Pinol itself is not known as a natural product.

and this may indeed crystallize in bottles of terpinolene (in
which it is not very soluble). Thus, if it were to be found
in a natural product containing at the same time terpinolene,

51   372

the question would always arise concerning the way the natural
product had been stored. A further difficulty is the thermo-
lability of certain compounds. For many years the authenticity
of *trans*-terpin (373) was in doubt, but it was recently identified as certainly as is possible at present in *Illicum verum*
and in fennel oil (*Foeniculum vulgare* var. dulce M),[430] although it loses water at 70°-80°--to give, of course, a variety
of the well-known terpineols we have already discussed.
Peyron et al., who made this observation, also refer to the
difficulty of extraction on account of the water solubility of
trans-terpin; 1 g dissolves in 32 g of water at 100°.[430] The
cineols are fairly commonly encountered in natural products,
for example 1,4-cineol (369) is a vital component for the
flavor of lime juice (*Citrus medica* L., *var. acida*)[431] and
1,8-cineol is one of the main oxygen-containing terpenes of
various *Melaleucia* species (tea trees).[432] Finding ethers and
unsaturated alcohols rather than diols makes one wonder not
only how authentic known polyoxygenated compounds are, but how
many have been missed by loss of water or water solubility
during workup. In this section, the discussion will be limited
to those compounds mentioned in Scheme 48.

The compounds of the first line of the list (371 to 373)
do not merit a long discussion here, since their synthesis has
been known for a long time (see, e.g., Ref. 433). 1,4-Cineol
(369) has generally been synthesized in the past by reduction
of the natural product, ascaridole* (374) [obtainable by
photo-oxidation of α-terpinene (226)] to *cis*-1,4-terpin (375),

*An interesting new source of singlet oxygen is made from
phosphite esters and ozone. Its structure is reported to be
$(RO)_3P<^O_O>O$, and it gives ascaridole in 60% yield from α-
terpinene.[434]

and treatment of the latter with oxalic acid,[435] but this reaction is by no means as clean as Wallach evidently believed, and careful purification of the cineol, obtained in only moderate yield, is necessary.[436] A mixture of both cineols (1,4- and 1,8-) is obtained directly from isoprene when this is treated at 30° in a nitrogen atmosphere with 50% sulfuric acid. A 23% yield of distillable product is obtained, consisting of 28% 1,4-cineol (369) and 25% of 1,8-cineol (368).[437] In addition to the classical methods of making 1,8-cineol, the method of hydroboration in the presence of mercuric acetate[438] when applied to α-terpineol (39) yields 90% of 1,8-cineol (368).[439]

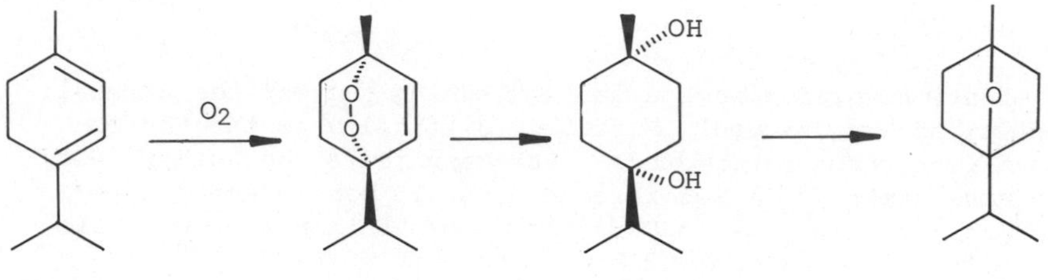

226   374   375   369

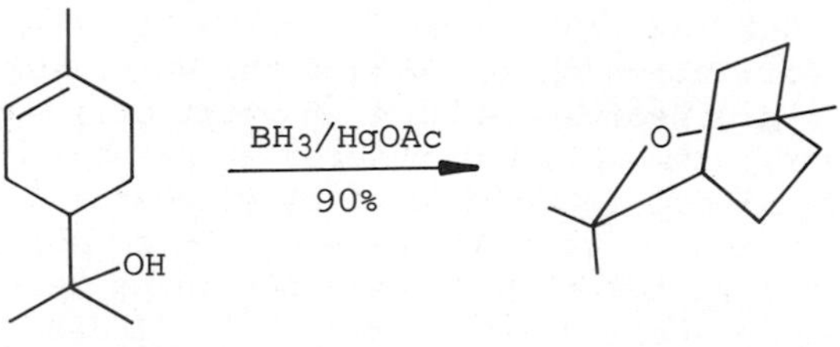

39   368

"Mullilam diol" is the name that Mathur et al. gave to the 1,4-cineol-2,3-diol (376) that they isolated from *Zanthoxylium rhetsa*.[440] The complete stereochemistry is not certain, but the diol arrangement is *trans*. The compound was obtained many years previously as a product from the peracid oxidation of sabinene (233).[441] An isomer with a *cis*-diol arrangement has been made using the reduction of ascaridole with ferrous sulfate.[442]

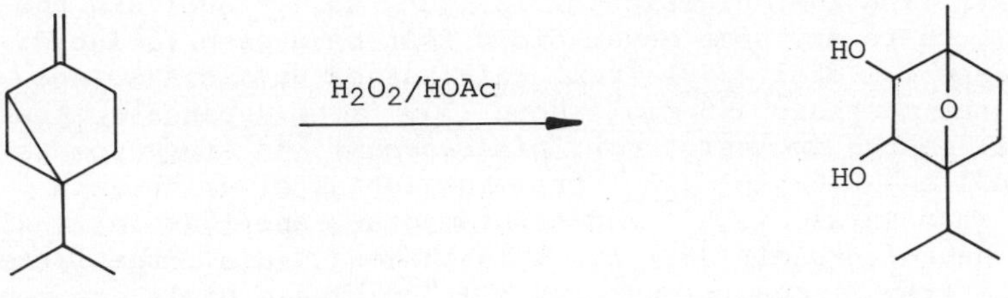

233   →   H₂O₂/HOAc   →   376

Oleuropic acid (377) was isolated from the hydrolyzate of a bitter principle from various parts of the olive tree (*Olea europaea*), and after initial assignment of an incorrect structure,[443] the correct formula was determined by Mechoulam et al., and the compound synthesized from cyclohexanone-4-carboxylic acid by the route shown in Scheme 49,[444] and by Herz and Wahlborg from (-)-β-pinene (14),[408] the latter leading to optically active oleuropic acid.

Scheme 49

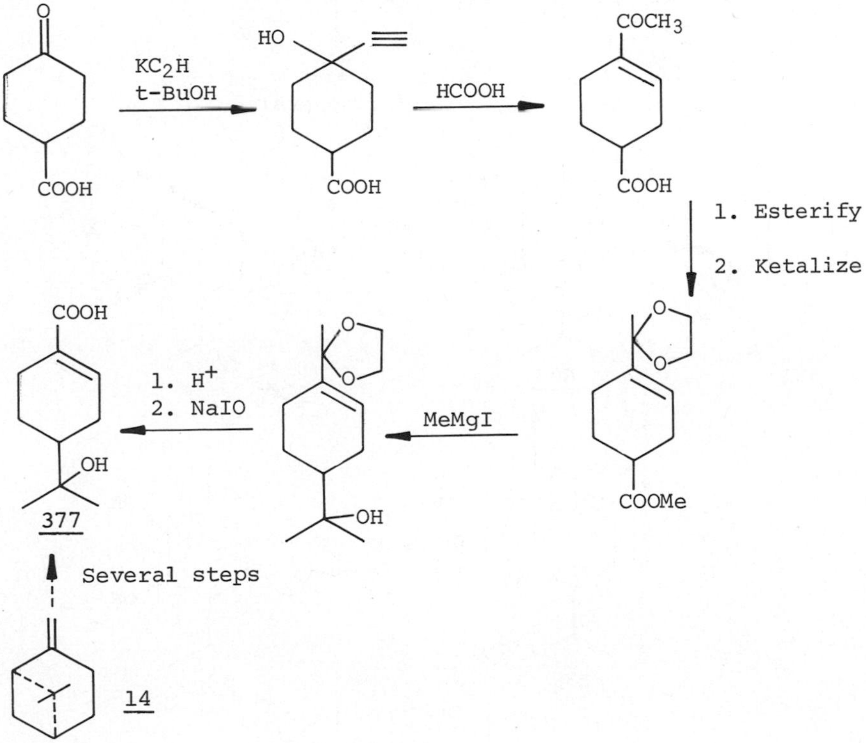

Apart from the "classic" diols (such as sobrerol and the terpins), there are some newer diols that have been isolated: menth-8-ene-1,2-diol (378) from cold-pressed Valencia orange oil,[413] the menthane 2,5-diol shown (379) from Japanese pepper-mint oil (*Mentha arvensis, var. piperasceus*),[445] some isomers of menth-1-ene-3,6-diol (380) from *Eucalyptus dives*[446] and Chinese star anise oil,[447] and *trans*-menth-2-ene-1,4-diol (381) (-)-menthane-1,3-diol (382) and a menthane-2,3-diol, the latter three all from Mitcham peppermint oil.[448]   These diols are gen-erally synthesized by some means of oxidation of suitable hydrocarbons; for example, the peracid oxidation of α-phel-landrene (228) gives a mixture of three possible isomers of the menth-1-ene-3,6-diols (all of which are present in *Euca-lyptus dives*, and the structures of which have been estab-lished.[446]   Piperitone epoxide (+)-(383), on the other hand,

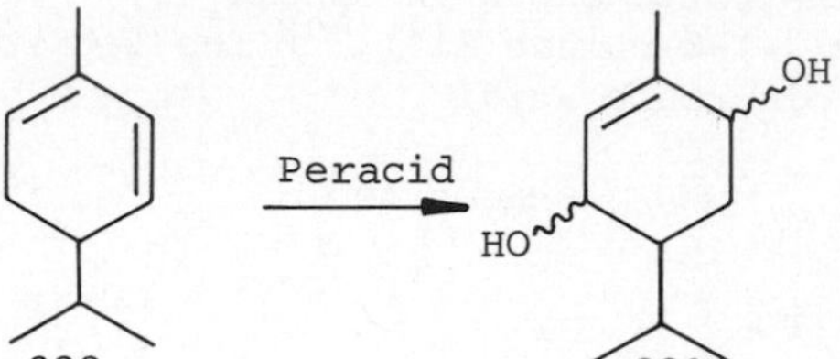

228   →   Peracid   →   380
(3 isomers)

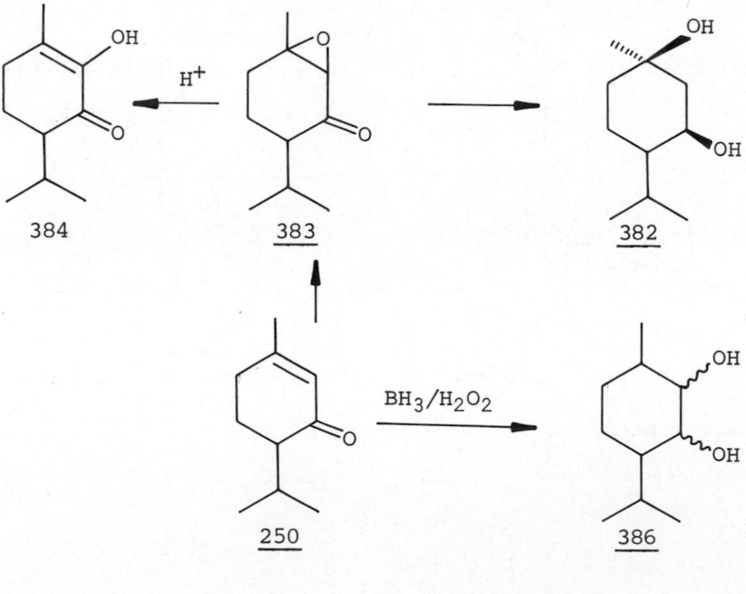

384   ←   H+   383   →   382

250   BH₃/H₂O₂   →   386

can be converted by lithium aluminum hydride reduction followed by catalytic reduction into the naturally occurring men-
thane-1,3-diol (382).[448] Piperitone epoxide (383) is a natural product, too, and is readily converted by acid catalysis
into one of the more "classic" dioxygenated menthanes, diosphenol (384).[449] The more unsaturated diosphenolene (385) is
similarly available from piperitenone oxide ("lippione," 314,
see above),[450] or by aluminum chloride catalyzed ring opening
of verbenone epoxide (386), that leads first to isopiperitenone epoxide (387) and then with acid, to diosphenolene
(385).[451]

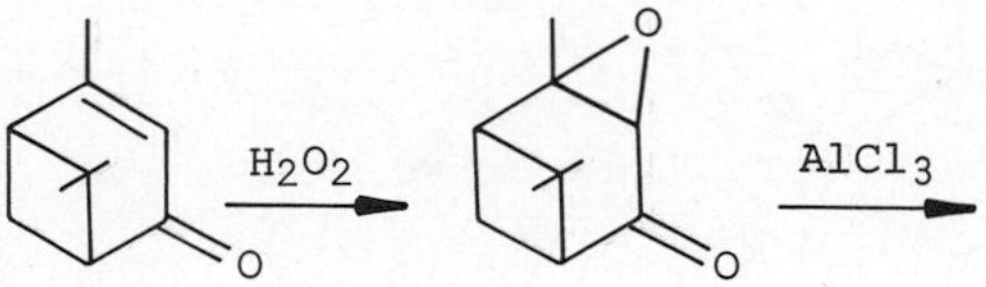

386

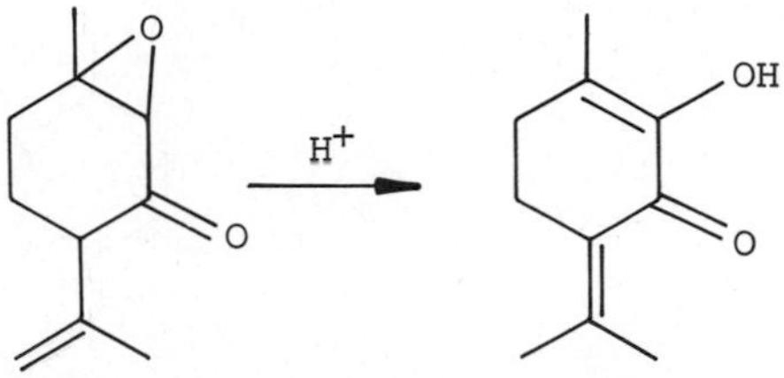

387   385

    The saturated menthane-1,2-diols are obtained by hydra-
tion of limonene epoxides (see the comprehensive article of
Royls and Leffingwell[301]), but the paper on the isolation[413]
does not make it clear precisely which isomers were found.
Finally, Klein and Dunkelblum have synthesized two menthane
trans-2,3-diols (386) from the hydroboration of piperitone
(250),[452] but they do not appear to be the same as the men-
thane-2,3-diol in Mitcham peppermint oil,[448] since the melting points reported are completely different.

9.   THE *m*-MENTHANES

It was considered for a long time that the only *m*-menthane

derivatives encountered in nature were artifacts arising from
the acid-catalyzed ring opening of Δ^3-carene (164) via sylvestrene dihydrochloride (388). Nevertheless, Bardyshev et al.
have isolated m-menthenol (389) from the high boiling fraction of Russian turpentine[453]--admittedly one that contains
Δ^3-carene. However this may be, the synthesis of this type
of compound is generally from sylvestrene. For example, the
dihydrochloride (cis + trans, 388), on treatment with soap,
water, and steam, yields 30.7% of hydrocarbons, and 22.5% of
m-menth-1-en-8-ol (390) 36% of which is "natural" (+)-m-menth-
5-en-8-ol (389), formerly known as "sylveterpineol").[454]

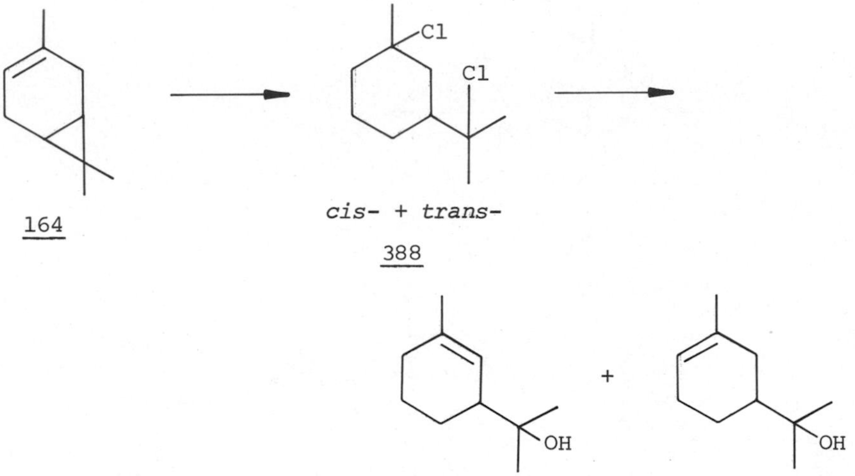

164

cis- + trans-

388

(-)-390   (+)-389

10. 1,1,2,3-TETRAMETHYLCYCLOHEXANES

A. Safranal (391)

1-Formyl-2,6,6-trimethylcyclohexa-1,3-diene (safranal, 391) is
a breakdown product of the bitter principle of safran, and can
be considered as being related either to the ionones, where
the side chain has been lost, or to the monoterpenes as a dehydrogenated cyclocitral. Indeed, its synthesis is generally
from β-cyclocitral (392) by oxidation with selenium dioxide[455]
or N-bromosuccinimide.[456]

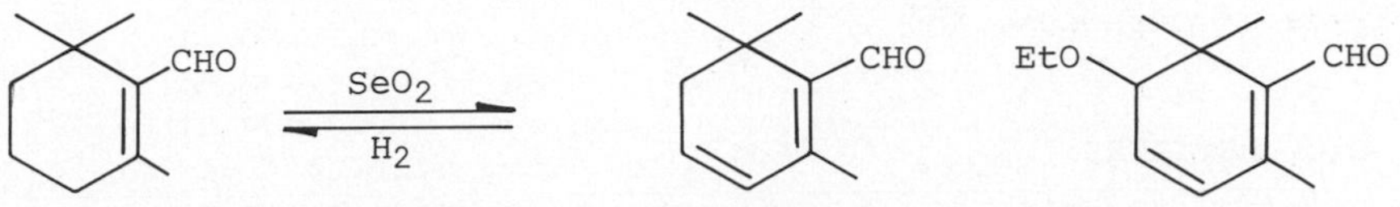

392   SeO₂ / H₂ →   Safranal   391   EtO...   393

An ethoxysafranal (393) has been isolated from a species of Mexican compositae, *Piqueria trinervia*, Cav.,[457] but no synthesis is reported.

B. Karahana Ether (394)

This substance (394), isolated from Japanese hops ("Shinshu-wase"),[458] can be considered as a 2,2,6-trimethylbenzyl derivative, although it is at the same time a tetrahydrofuran. It has been synthesized by Coates and Melvin from geraniol, the acetate of which was first cyclized with benzoyl peroxide, cupric benzoate, and cupric chloride in acetonitrile following the technique of Breslow et al.[459] The resulting mixture was hydrolyzed to the corresponding diols, from which the desired *cis*-diol (395) was separated by careful chromatography. Reaction of this diol with one equivalent of *p*-toluenesulfonyl chloride in pyridine at room temperature yielded racemic karahana ether. Coates and Melvin suggest that this may also represent the biogenetic pathway for the formation of natural karahana ether.[460]

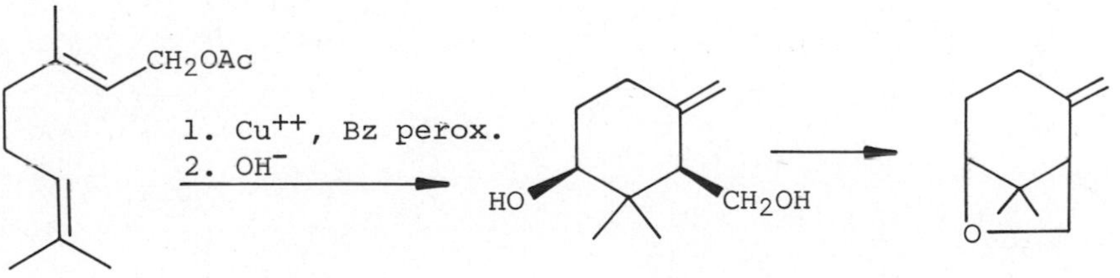

1. Cu⁺⁺, Bz perox.
2. OH⁻

395   Karahana Ether 394

11. THE *o*-MENTHANES

Although a number of *o*-menthane derivatives have been isolated as natural products,[457,461,462] none of these has been synthesized. Entry to certain *o*-menthatrienes is possible by pyrolysis of verbenene.[463]

12.   CYCLOHEPTANES

A.   Thujic Acid (396), Shonanic Acid (397), Eucarvone (398), and Karahanaenone (399)

These compounds are derivatives of 1,1,4-trimethylhepta-2,4,6-triene, and until relatively recently, there was a lingering doubt as to their structure, largely owing to certain of their reactions that could be associated with derivatives of bicyclo[4.1.0]heptanes (carenes). The doubts were finally resolved in the case of thujic acid (396) by measurement of the NMR spectrum[464] and the crystal structure of the p-bromophenacyl ester.[465]   Thujic acid occurs in the wood oils of various Thuja and other species, including Libocedrus formosana in which it is accompanied by a reduced form[466] (called shonanic acid when it was first isolated,[467] the cycloheptadiene structure of which was attributed first by Erdtman.[468] Although there appears to be no total synthesis of thujic acid, its relation to shonanic acid has been discussed by Pasto, who showed that the reduction of thujic acid gives initially a dihydrocompound (400), that is rearranged by base to a mixture of three further acids (401a, 401b, and 401c), none of which is identical to shonanic acid.

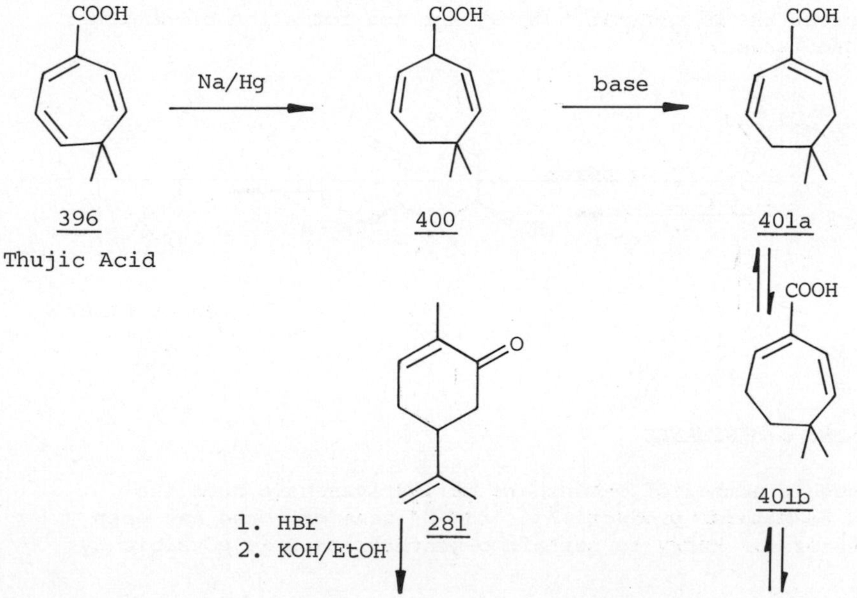

396
Thujic Acid

400

401a

401b

281

1. HBr
2. KOH/EtOH

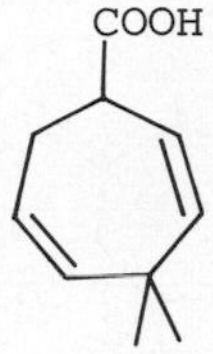

397

Shonanic Acid

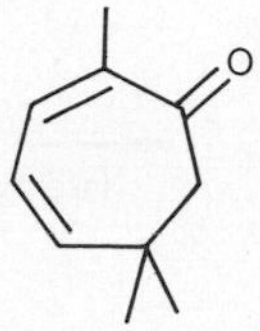

398

Eucarvone

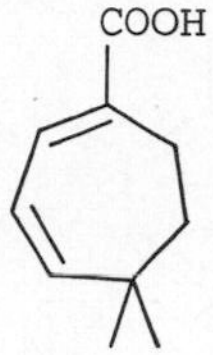

401c

Eucarvone (398) was reported in the older literature to occur in *Asurum Sieboldii,* var. *seoulensis,*[469] but this is the only report of its occurrence as a natural product. It is nevertheless readily available by the synthesis of Bayer from carvone (281),[470] and a study of its various reactions, including a consideration of its relationship with the carane derivatives, has been given by Corey and Burke.[471] The total synthesis of eucarvone has been carried out by Barnes and Houlihan (Scheme 50), who obtained an appreciable amount of carvacrol in the last dehydrobromination stage.[472] Eucarvone is also obtained in low yield by dehydration of 2-hydroxyisopinocamphone (402).[473]

Scheme 50

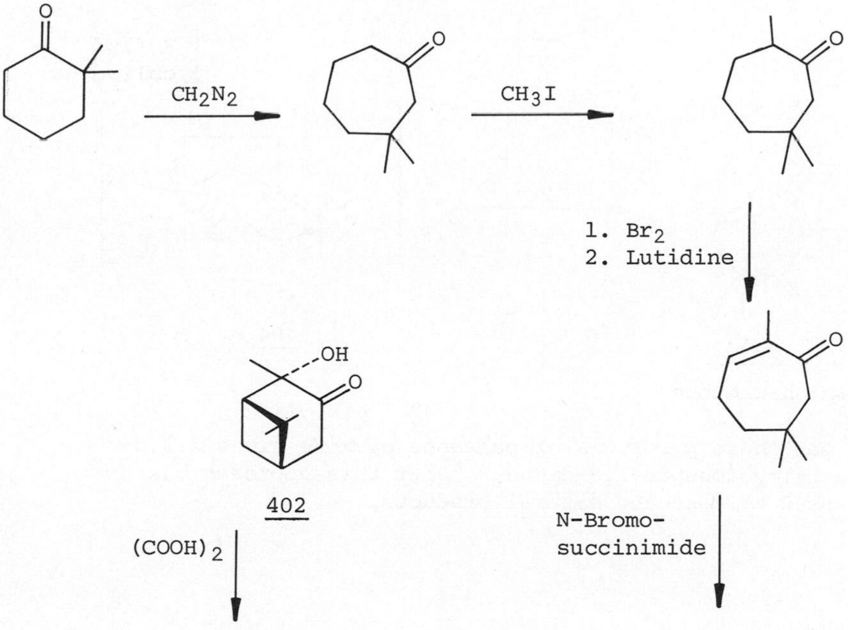

CH$_2$N$_2$

CH$_3$I

1. Br$_2$
2. Lutidine

402

(COOH)$_2$

N-Bromo-succinimide

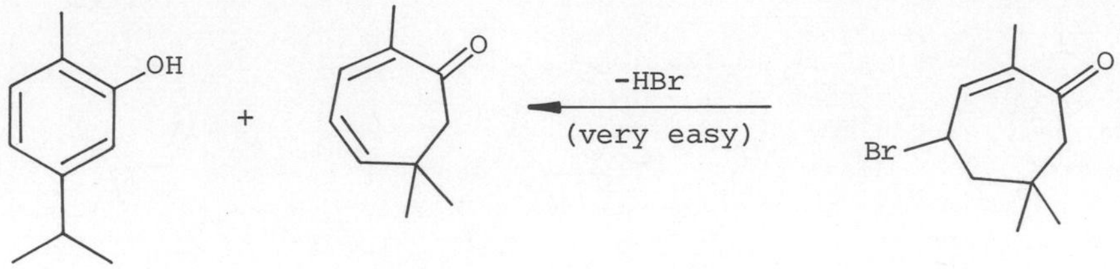

278   398

Karahanaenone (2,2,5-trimethylcyclohept-4-enone, 399)
was isolated from hop oil in 1968,[458] and its synthesis followed shortly afterwards by Demole and Enggist, who treated
linalool (6) with N-bromosuccinimide, when they obtained the
furan (403) that on refluxing in collidine lost hydrogen
bromide. The intermediate (404) was not isolated, but converted directly by heat to karahanaenone (399).[474] A full
description of this synthesis has also appeared.[475]

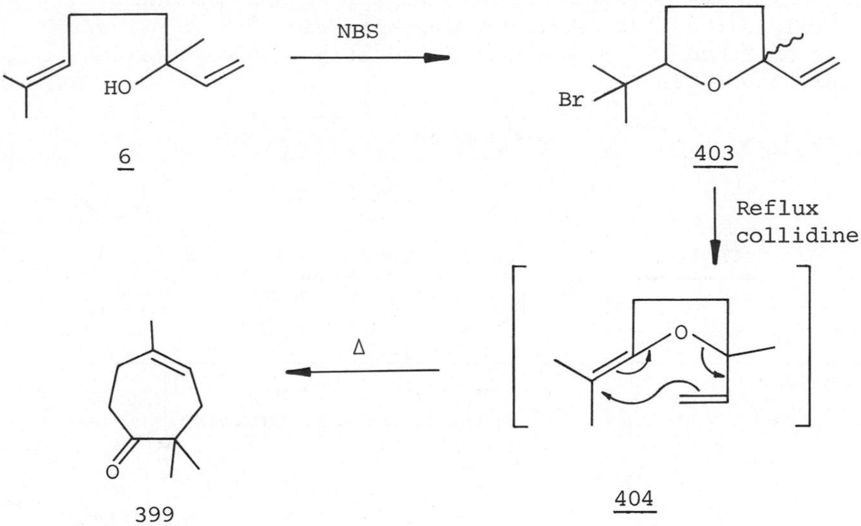

6   403

399
Karahanaenone

404

Gas phase pyrolysis of pulegone epoxide gives 2,2,5-
trimethylcyclohepta-1,3-dione,[476] but this approach has not
been used to make any natural products.

B.   Nezukone and The Thujaplicins

These compounds are not isoprenoid, but can clearly be derived
from the isopropylbicyclo[3.1.0]hexane system of the thujanes
or the bicyclo[4.1.0]heptane system of the caranes (see below).
Nezukone (405) is, in fact, found in the Japanese nezuko tree
(*Thuja standishii*) together with α- and β-thujaplicin (406a
and 406b).[477]
     Syntheses of these molecules are not numerous, but Birch
has made nezukone from the sodium in ammonia reduction (407)
product of *p*-isopropylanisole, using addition of dichlorocarbene (to 408) then treatment with silver borofluoride to
give the product (405).[478]   The tropolones are not conveniently treated in a chapter on monoterpene synthesis, but it
might be mentioned that α-thujaplicin (406a) has been synthesized by ter Borg et al. from 2-chlorotropone (409) by reaction of isopropylmagnesiumbromide, followed by heat.[479]

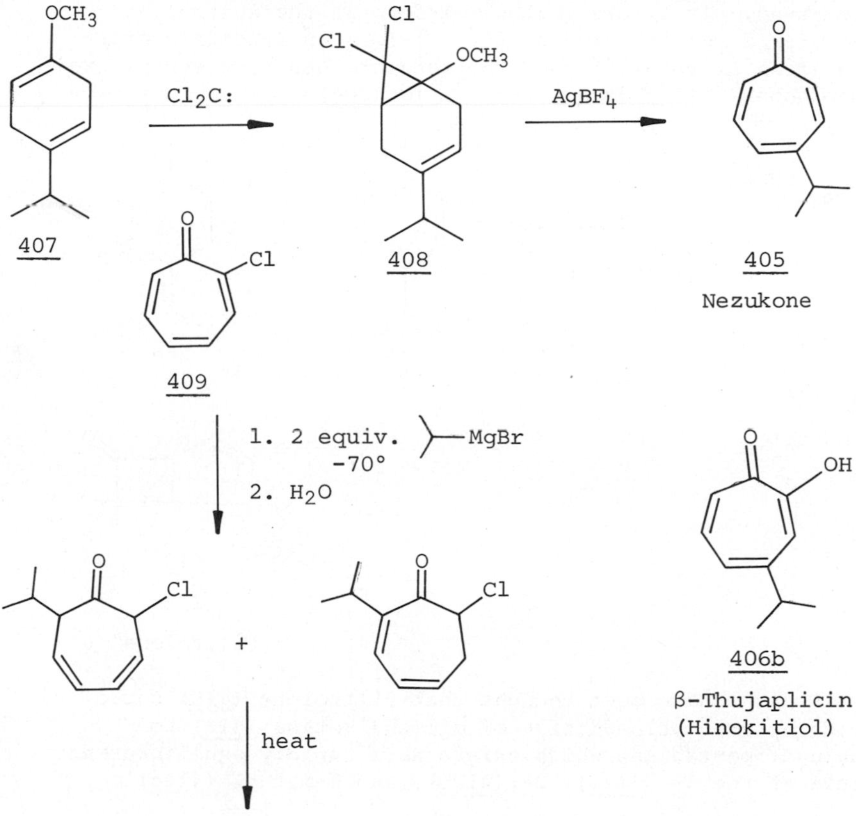

407

409

405

Nezukone

406b

β-Thujaplicin
(Hinokitiol)

1. 2 equiv. —MgBr
   −70°
2. H₂O

heat

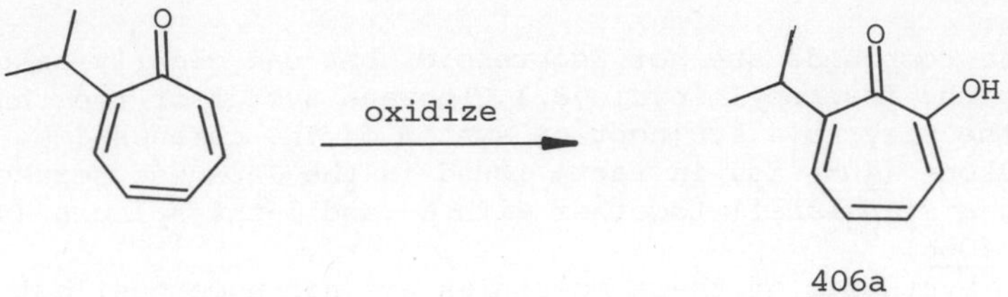

oxidize

406a

α-Thujaplicin

13. BICYCLO[3.2.0]HEPTANES

A. Filifolone

2,6,6-Trimethylbicyclo[3.2.0]hept-2-en-7-one (filifolone, 410)
is the only naturally occurring representative of this class
of compound. It occurs in the (+)-form in the Australian
plant *Zieria smithii*, and as the (-)-form in *Artemisia filifolia* (from Arizona),[480] and the racemate has been synthesized
by Beereboom[481,482] from geranic acid (304).

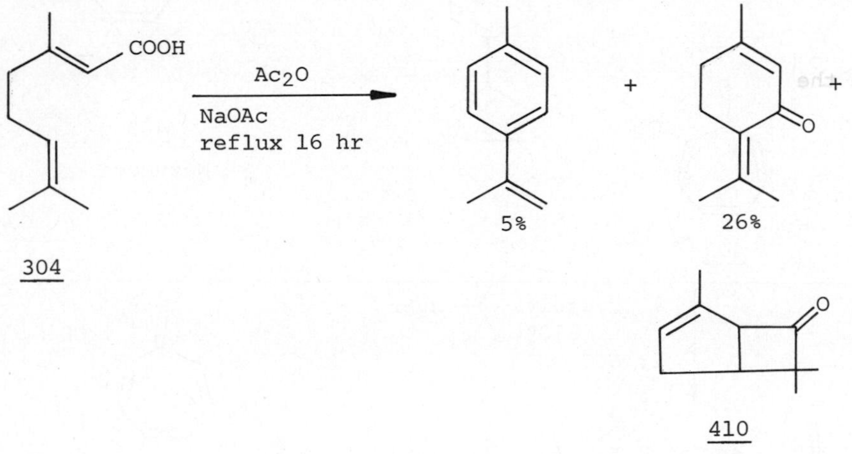

Ac_2O

NaOAc
reflux 16 hr

304

5%

+

26%

+

410

28%

Filifolone

It might have been thought that filifolone (410) could
be made by the cycloaddition of dimethylketene (411) to
methylcyclopentadiene which exists as a rapidly equilibrating
mixture of the 1- (412a), 2- (412b), and 5-methyl (412c)

isomers.[483,484] Huber and Dreiding have shown, however, that
this reaction leads mainly to the isomeric bicyclo[3.2.0]-
heptenone (413), together with a second isomer (414) and small
amounts of unidentified byproducts.[485] Filifolone is obtained
from chrysanthenone (see bicyclo[3.1.1]heptanes below) by the
action of acetic acid.[485a]

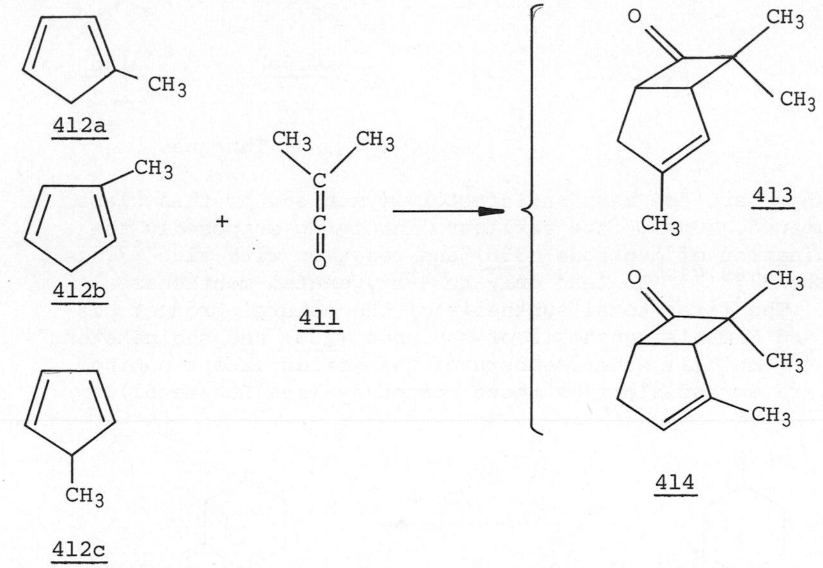

14. BICYCLO[3.1.0]HEXANES

This class of compounds includes sabinene, 2-thujene, umbellulone, and related substances. The parent hydrocarbon, 1-
isopropyl-4-methylbicyclo-[3.1.0]hexane (415) is called thujane, both stereoisomers of which can be made by carbene addition to pulegene (416)[486] accessible from pulegone (307) via
a Favorskii rearrangement.[253,487]

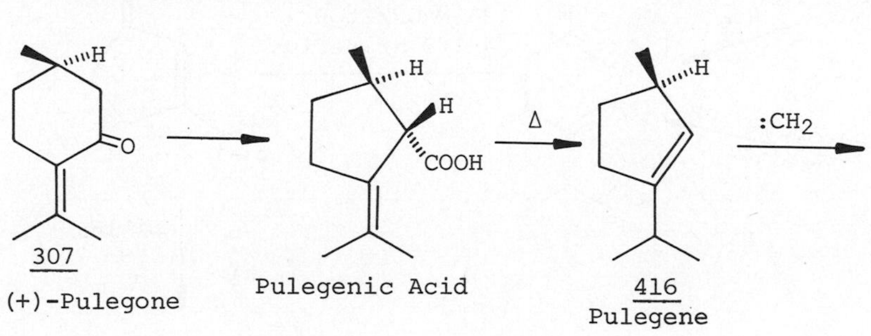

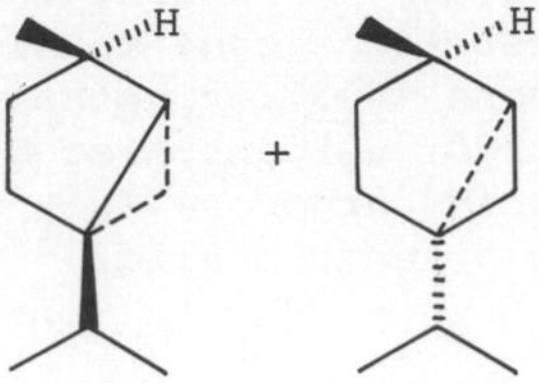

415a   415b

cis   *trans*

Thujanes

    There are not many satisfactory syntheses of this class
of compound. One of the earliest, involving supposedly the
dibromination of menthone (338) and reaction with zinc[488] has
been shown[489,490] to lead only to 4-oxygenated menthones
(417). The first total synthesis of the natural products is
Fanta and Erman's synthesis of sabinene (233) and sabinaketone
(418),[491] but all other members of the series except sabinol
(419) are accessible from these compounds (see Scheme 51).

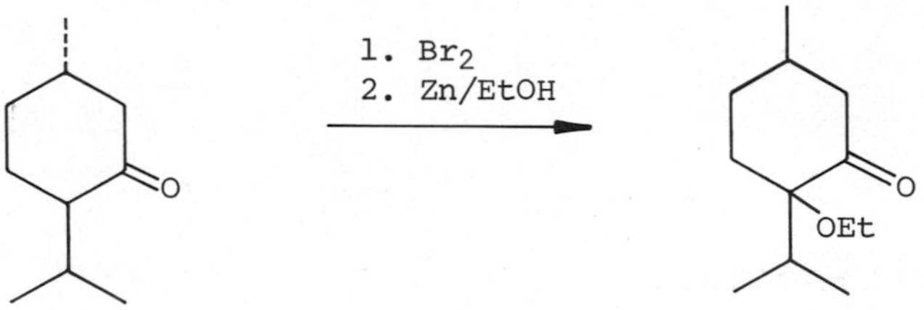

1. Br$_2$
2. Zn/EtOH

338   417

Scheme 51

O$_3$, reduce

Hydroboration
Oxidize

2% NaOH/EtOH
4 1/2 hr reflux

LiAlH$_4$

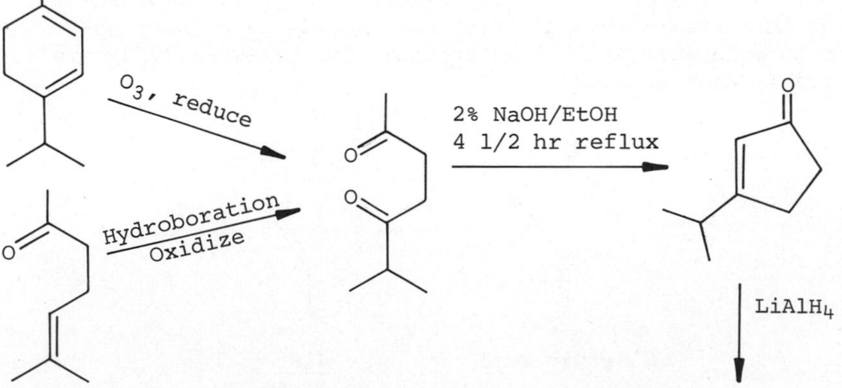

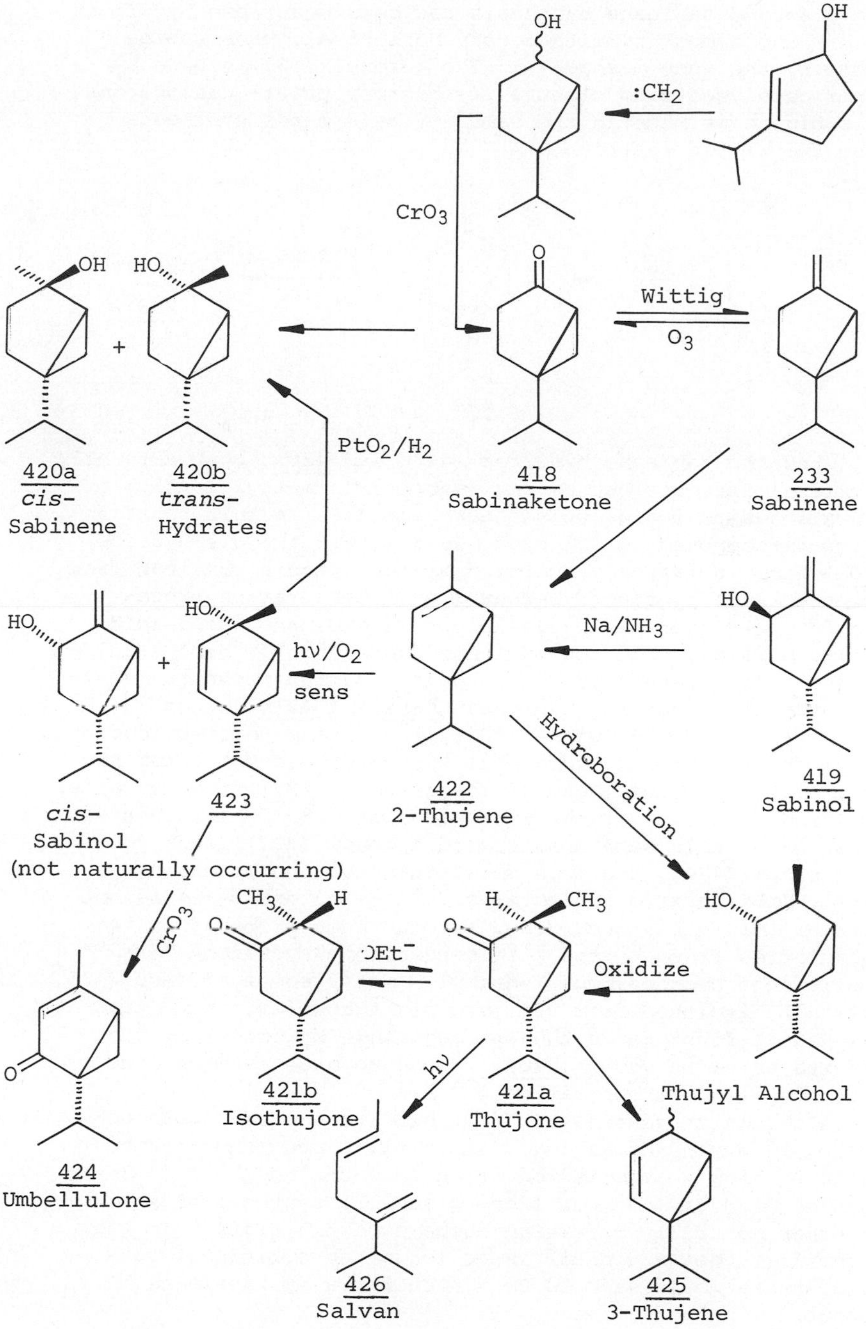

$\underline{420a}$
cis-
Sabinene

$\underline{420b}$
trans-
Hydrates

$\underline{418}$
Sabinaketone

$\underline{233}$
Sabinene

CrO$_3$

:CH$_2$

Wittig
O$_3$

PtO$_2$/H$_2$

cis-
Sabinol
(not naturally occurring)

$\underline{423}$

hν/O$_2$
sens

Na/NH$_3$

$\underline{422}$
2-Thujene

$\underline{419}$
Sabinol

Hydroboration

CrO$_3$

$\underline{424}$
Umbellulone

CH$_3$ H

$\underline{421b}$
Isothujone

OEt$^-$

H   CH$_3$

$\underline{421a}$
Thujone

hν

Oxidize

HO

Thujyl Alcohol

$\underline{426}$
Salvan

$\underline{425}$
3-Thujene

A second sabinene synthesis has been described by Vig et
al.,[492] and almost simultaneously Mori et al. made sabina
ketone by the same method.[493] The ketone (418) is made by
copper catalyzed ring closure of the appropriate diazoketone,
and sabinene is made in the same way as Erman's synthesis
using the Wittig reaction:

418

Thus the sabinene hydrates (420a and 420b, both naturally
occurring) are obtained by the reaction of methyl lithium or
methylmagnesium iodide on sabinaketone.[494] In this reaction,
the *trans*-compound (420b) predominates, but the *cis*-hydrate
(420a) forms a larger proportion of the mixture obtained from
sabinene (233) by the oxymercuration-demercuration procedure.[495] 2-Thujene is accessible from sabinene using acid
catalysis (i.e., with ion exchange resins[496]) or base catalysis (potassium t-butoxide[497]). Hydroboration techniques make
the naturally occurring thujones (421a and 421b) accessible[486],[497] from 2-thujene (422), from which photo-oxidation
yields *trans*-sabinene hydrate as the major product after reduction of the double bond of the alcohol (423) produced after
bisulfite reduction of the hydroperoxide.[486],[498] Chromic acid
oxidation of this same unsaturated alcohol (423) leads to
umbellulone (424), the main constituent of the mountain
laurel, *Umbellularia californica*.[498] 3-Thujene (425) is not
certainly naturally occurring, but has been prepared in two
laboratories from thujone p-toluenesulfonylhydrazone.[499],[500]
Finally, the hydrocarbon, 2-methyl-3-methylenehept-5-ene (426,
"salvan"), isolated from various *Salvia* species,[501],[502] and
believed at first to be thujane but shown by Brieskorn and
Dalferth to be as shown (426),[503] can be made from thujone by
photochemical decarbonylation.[504]
   Although *trans*-sabinol (419) and its acetate (both occurring in *Juniperus sabina*, "savin oil") are converted by
sodium in liquid ammonia reduction to 2-thujene,[486],[505] there
is no method yet known of linking sabinol synthetically with
the other naturally occurring members of the group. In view
of the fact that savin oil is no longer an article of com-
merce, there would seem to be a need for a synthesis of this
alcohol.

15. BICYCLO[2.2.1]HEPTANES

Among its naturally occurring members, this group of compounds includes norbornanes with the following methyl substitutions: 2,2,3-(camphene, 427), 1,7,7-(borneols, camphor, 224, etc.), 1,3,3-, 2,5,5-, and 2,7,7- (fenchane derivatives), but syntheses of recent date are too rare to warrant subdivision of the section. This was not always the case, however, and in the days when the structure of camphor was still of vital importance (one might recall that Bredt's rule dates from this time), it was hardly possible to open a volume of Liebig's Annalen without finding some attempt at some partial structure of the series. Today, the emphasis has passed to mechanistic studies, although even here the unique rearrangements associated with the bicyclo[2.2.1]heptane structure are historically firmly anchored to the fascinating and frequently inexplicable facts that the chemists of the late nineteenth century were discovering. An additional spur to work in this group of compounds was always the unique properties of camphor, a compound known from the dawn of civilization by the odor it imparted to woods and oils that contained it, and later because of the abnormally high cryoscopic constant responsible for the exceptional properties observed when physical measurements became fashionable as new instrumentation was developed.*

The first total synthesis of the group was the celebrated one of camphor by Komppa[507,508] and, apart from historical interest, there was a small point that remained unresolved until recently, namely, the exact structure of the product that Komppa claimed to have obtained from compound 428, and which he wrote as 429. This aspect was recently examined by Agharomurthy and Lewis,[509] who found that Komppa's supposition of a C-methylated rather than an O-methylated compound was correct (although both 428 and 429 are, in fact, entirely in the enol form--a fact that Komppa certainly suspected), and therefore that the camphor synthesis, which depends on precisely this point, is quite correct.

*One might still make a case out for trying camphor in any new physical technique--the use of solvent dependency of NMR chemical shifts was first demonstrated with camphor,[506] because it is a very readily available rigid molecule having sterically well-defined substituents in fixed positions from carbonyl group.

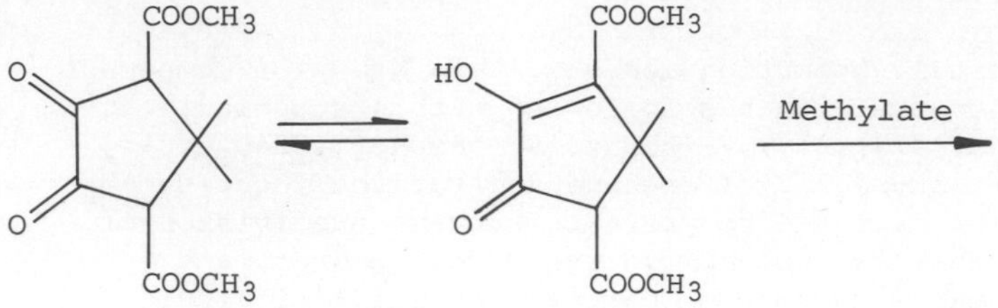

428

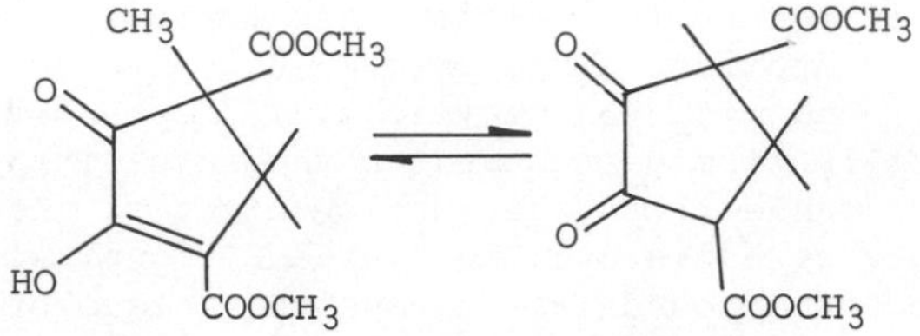

429

The more recent syntheses of bicyclo[2.2.1]heptanes use the Diels-Alder reaction most frequently, and have as their ancestor, so far as natural products are concerned, the synthesis of bornyl acetate (430) and camphor (224) from 1,1,2-trimethylcyclopentadiene (431) and vinyl acetate followed by reduction. Epicamphor (432) is also obtained from this reaction.[510]

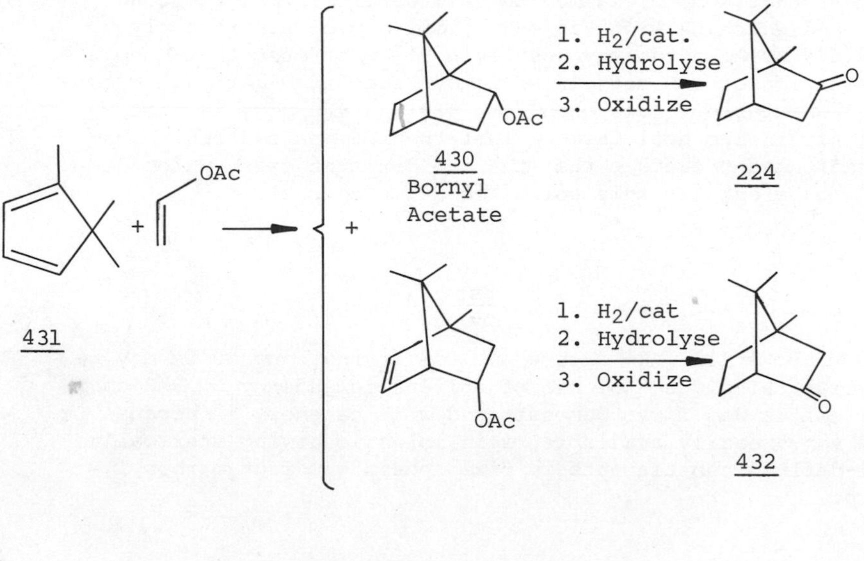

Similar techniques (Scheme 52) were used by Vaughan and

Scheme 52

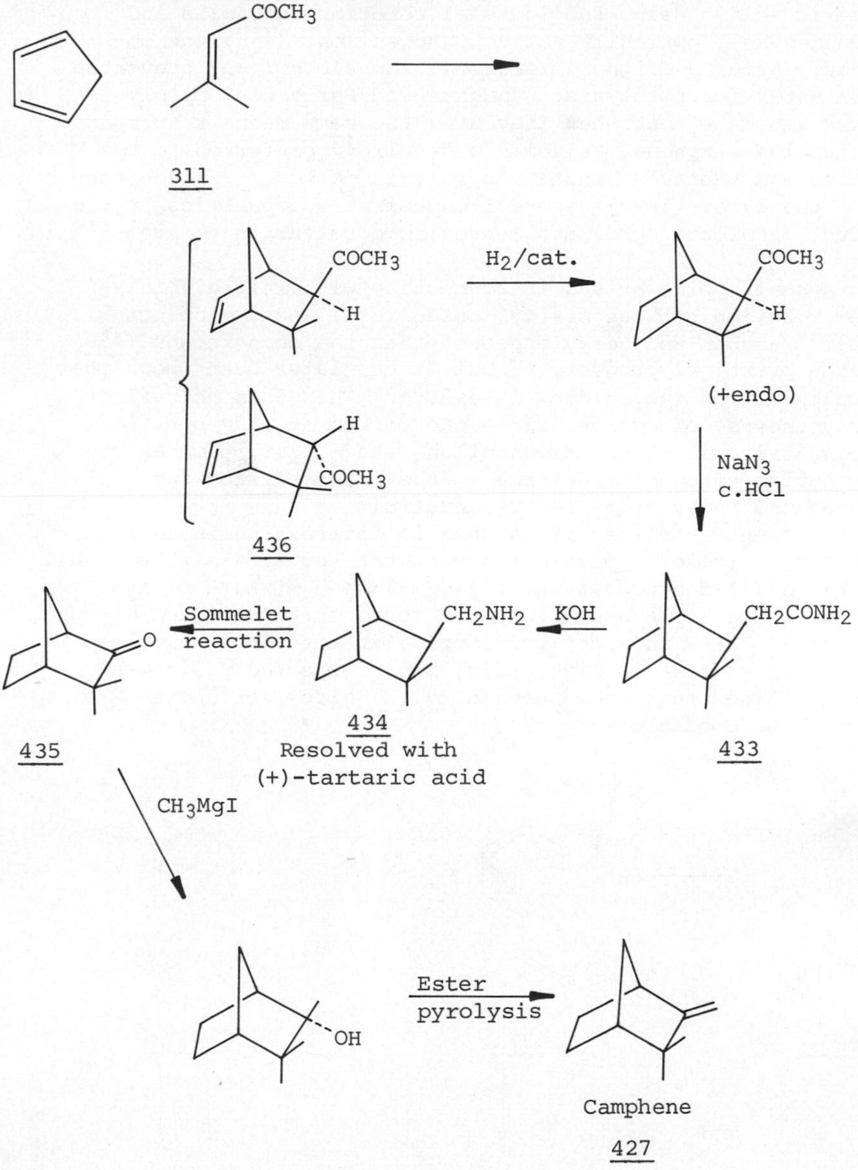
COCH3
311
COCH3
H
H
COCH3
436
H2/cat.
COCH3
H
(+endo)
NaN3
c.HCl
CH2CONH2
433
KOH
CH2NH2
434
Resolved with
(+)-tartaric acid
Sommelet
reaction
O
435
CH3MgI
OH
Ester
pyrolysis
Camphene
427

Perry in order to make optically active camphene.[511] The
acetamide (433), obtained by the route shown (Scheme 52) was
resistant to acid hydrolysis, but potassium hydroxide converted it to camphenylamine (434) that was resolved with (+)-
tartaric acid. Using the Sommelet reaction (hexamine and
formaldehyde), optically active camphenilone (435) was obtained. After a Grignard reaction, the alcohol was converted
to an ester for pyrolysis. Vaughan and Perry used pyrolysis
of the xanthate, but when they used the same means to prepare
[13]C-labeled camphene, Friedman and Wolf[512] preferred to use
Roberts and Yancey's benzoate pyrolysis method,[513] since they
found there were less traces of undesirable impurities in the
product (xanthate pyrolysis leaves some sulfur in the camphene).

A small point arises in connection with initial Diels-
Alder reaction between mesityl oxide (311) and cyclopentadiene. Vaughan and Perry supposed that the endo-ketone (436)
was the principal product,[514] but it has later been shown that
actually 60% of the product is exo.[515] This does not affect
the synthesis, of course, since the carbon at this position
becomes trigonal at the camphenilone stage, but it makes the
attribution of endo-structures to the intermediates (as
Vaughan and Perry did[511]) very doubtful.

A recent synthesis of camphor is interesting in that it
purports to follow a possible biogenetic route. Fairlie et al.
converted (+)-dihydrocarvone [(+)-294] to a mixture of its
enol acetates (437a and 437b), and found that treatment of one
of them (437a) with boron trifluoride in methylene chloride
gave camphor (224) in high yield, but unexpectedly, it was
racemic. The other enol acetate (437b) gives carvenone (295)
under these conditions.[516]

(+)-294   437a   437b

                           BF$_3$ BF$_3$

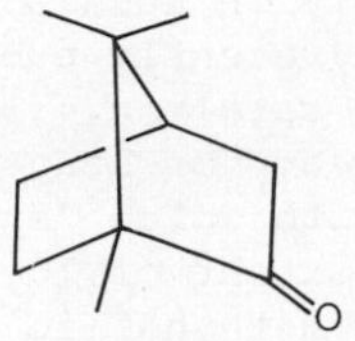

(±)-224

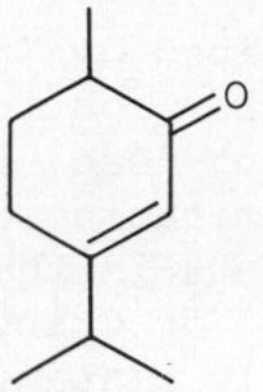

295

Camphor is produced on a very large scale commercially, but using pinene as starting material, that is, it is once again only a partial synthesis. The route followed[517] (Scheme 53) is generally using the Wagner-Meerwein rearrangement,

Scheme 53

Possible Industrial Routes to Camphor

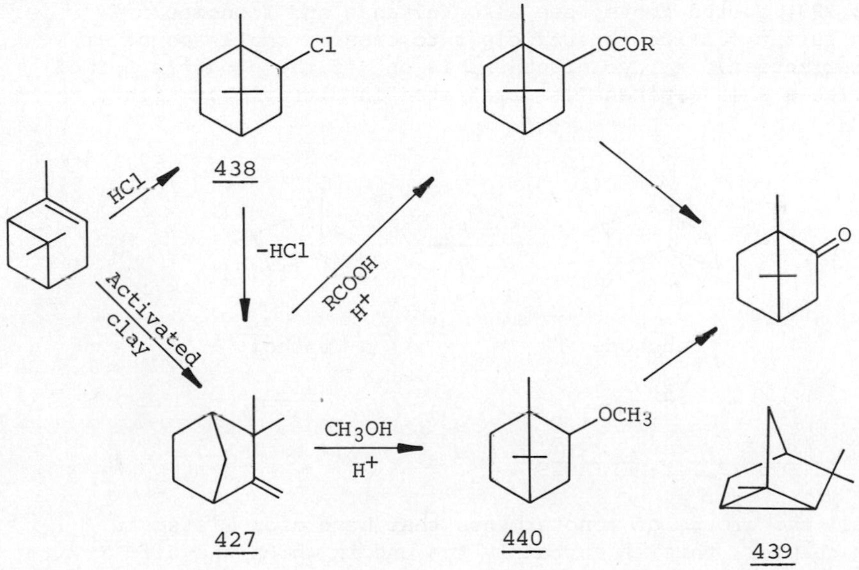
Cl
OCOR
HCl
438
-HCl
RCOOH
H+
Activated clay
O
CH3OH
H+
OCH3
427
440
439
Tricyclene

originally to bornyl chloride (438), which loses hydrogen chloride to give camphene (427). The latter can be converted via an isobornyl ester to camphor. There are many modifications of this route; one of the oldest is perhaps the direct catalytic conversion of pinene to camphene. This was first

described by Tishchenko and Rudakov in 1933,[518] but similar
work has continued, especially in Russia, right up to the
present (see, for example, a recently published patent that
uses titanium dioxide as the catalyst[519]). There are also
variants in the camphene to camphor stage, for example, camphene (or tricyclene, 439, with which "synthetic" camphene is
usually contaminated to the extent of 10-20%) can be converted
to 2-methoxybornane (440) by methanol in the presence of a
methanol-wetted strong cation exchange resin in the H-form at
60°, and 2-methoxybornane can be oxidized to camphor by a
mixture of nitrogen dioxide and oxygen at 10°-15°.[520]

The first total synthesis of fenchone (299) was that of
Ruzicka,[521] but since 1935, when Komppa and Klami published
an improved method,[522] there has been no further synthetic
work carried out. α-Fenchol (441), the main product from the
metal hydride[523] and metal-ammonia[524] reductions of fenchol*
is, together with its acetate, also a widely distributed natural product, and is usually obtained in small amounts during
manipulations of α-pinene that involve acids (in addition to
Ref. 289, quoted above, see also Valkanas and Iconomou[525]);
even sulfur dioxide is sufficient to cause a small amount of
isomerization.[526] Cyclofenchene is one of the products formed
in traces when α-pinene is irradiated in ultraviolet light.[527]

Fenchone   α-Fenchol

299   441

16.   BICYCLO[3.1.1]HEPTANES

Of all the groups of monoterpenes that have been listed in
this chapter, this is certainly the one in which the difference between availability as natural products and as synthetic products is the greatest. α-Pinene (17) is probably
the most common of the monoterpene hydrocarbons, yet it has
never been synthesized directly. β-Pinene (14) was synthesized only relatively recently by irradiation of myrcene (12),

*But not aluminum isopropylate reduction which leads to the
other isomer.[524]

which Crowley found to product mainly the cyclobutene (442),
β-pinene being formed in only 10% yield[528,529] (the sensitized
irradiation of myrcene follows a different path[530]). This
nevertheless represents a total synthesis of the whole group
of compounds, since they are all available from β-pinene or
α-pinene, which can be made from β-pinene by lithium in ethylenediamine;[531] or by the action of iron pentacarbonyl (the
latter gives a 97% optical yield of (-)-α-pinene, 17a, from
(-)-β-pinene, 14a).[532]

Actually, most monoterpene synthesis starting from the
pinenes* require the slightly less common β-pinene, and this
is also obtainable from α-pinene by hydroboration and pyrolysis
of the organoborane.[534]

A brief summary of reactions leading from the pinenes to
other members of the pinane group that are naturally occurring
is given below in Scheme 54. The lead tetraacetate reaction
from α-pinene (17) passes through the stage of the 2-acetate
(443), leading finally to *trans*-verbenyl acetate (444).[534a]
From the latter compound, verbenone (445) and *cis*-verbenol
(446) are obtainable. The stereochemistry of these compounds
is fairly evident from the reactions involved, and has been
taken as a matter of course in most laboratories concerned
with monoterpenes, but this has not prevented several detailed
discussions of the stereochemistry of *cis*- and *trans*-
verbenol derivatives from appearing in recent years (see, e.g.,
Ref. 93, 535-537).

*For a review of the various other monoterpenes available from
the pinenes, see Ref. 533.

Scheme 54

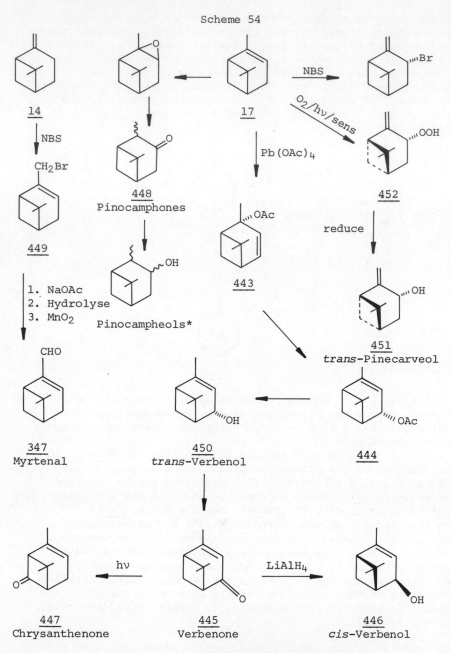

14

NBS

CH$_2$Br

449

1. NaOAc
2. Hydrolyse
3. MnO$_2$

448
Pinocamphones

OH

Pinocampheols*

347
Myrtenal

CHO

17

Pb(OAc)$_4$

OAc

443

450
trans-Verbenol

NBS

O$_2$/hν/sens

Br

OOH

452

reduce

OH

451
trans-Pinecarveol

OAc

444

447
Chrysanthenone

hν

445
Verbenone

LiAlH$_4$

446
cis-Verbenol

OH

*For a full description of the structures and spectra of the
different pinocampheol isomers, see Teisseire.[537]

Hurst and Whitham have shown how verbenone is converted into cyrhsanthenone (447) by ultraviolet irradiation,[538] the yield of which can reach 67%.[539]   The naturally occurring pinocamphones (448) can be prepared from α-pinene epoxide by conventional ring opening reactions of epoxides.[540,542]   The reaction of β-pinene with N-bromosuccinimide yields the 7-brominated α-pinene derivative (449) that can be converted into myrtenal (347).[543]   The radical oxidation of α-pinene with t-butyl perbenzoate in the presence of cuprous bromide yields 30% of alcohols, 30% of this mixture being trans-verbenol (450) and 19% trans-pinocarveol (451);[544] but one of the most effective methods for making trans-pinocarveol (451, that occurs naturally in the essential oil of camomile[545]) is by sensitized photoxygenation of α-pinene,[546] which leads to 95% trans-pinocarveyl hydroperoxide (452) and 0.56 cis-pin-3-ene 2-hydroperoxide.[547]

17. BICYCLO[4.1.0]HEPTANES

The most important natural product of this group is Δ^3-carene (164), the main constituent of some terpentines, notably from Eastern Europe and India (Pinus longifolia, from Indian Chir trees). Since this is readily accessible, most work in this group has been that of conversion of Δ^3-carene to some other more useful compound, but there are a few syntheses of the skeleton that will be briefly mentioned here.

Δ^2-Carene (453) also occurs naturally, and the two car-enes are interconvertible with ethylenediaminolithium;* Ohloff et al. found that the acid catalyzed isomerizations reported earlier were erronious.[548] Most of the syntheses of the structure use an old reaction of Kishner, who found that pulegone (307) can be converted to the imidazoline (454) with hydrazine, then heat in the presence of base, and copper sulfate converts this to trans-carane (455).[550 551] The same reaction carried out on piperitenone (305) results in the synthesis of (±)-Δ^2-carene (453).[552] Although not a natural product, caran-2-one is readily synthesized from dihydrocarvone hydrobromide with alcoholic potassium hydroxide,[553] and has also been made more recently by total synthesis from 4-methylpent-3-enyl bromide (456) following Scheme 55.[554]

*This isomerization also occurs under hydrogenation conditions.[549]

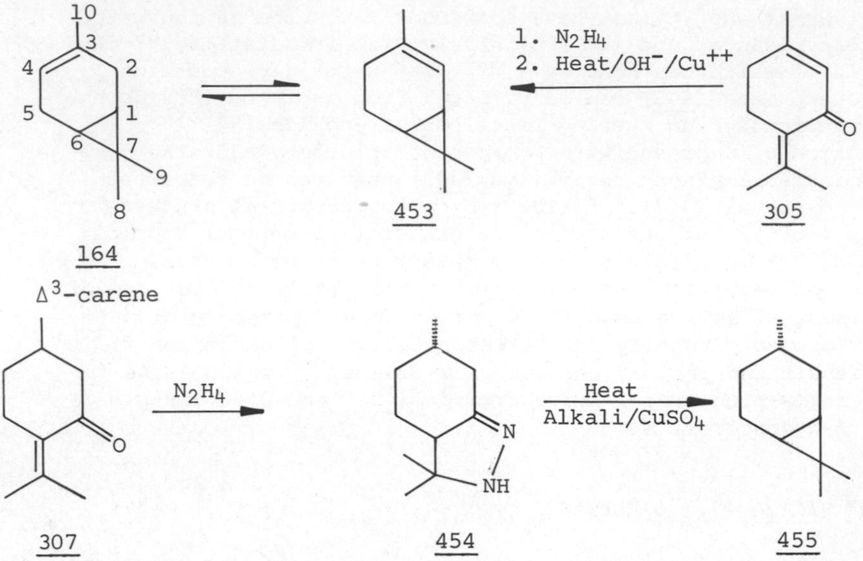
10
3
4
2
5
1
6
7
9
8
164
Δ3-carene
1. N2H4
2. Heat/OH−/Cu++
453
305
307
N2H4
N
NH
454
Heat
Alkali/CuSO4
455

Scheme 55

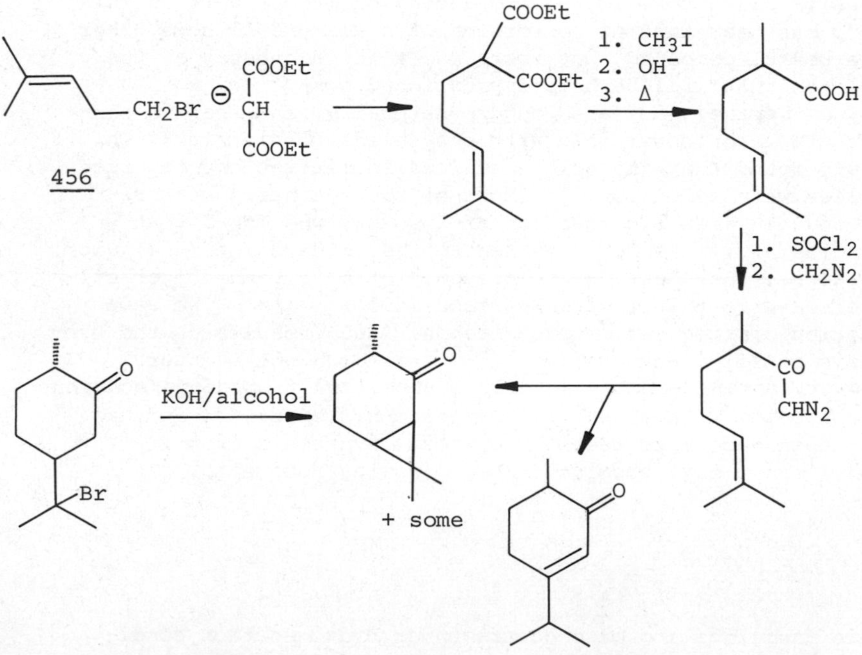
CH2Br
456
COOEt
CH
COOEt
COOEt
COOEt
1. CH3I
2. OH−
3. Δ
COOH
1. SOCl2
2. CH2N2
CO
CHN2
O
KOH/alcohol
O
Br
+ some
O

18.   FURAN MONOTERPENES

A.   3-Methyl-2-substituted and 3-Substituted Furans

Apart from the menthofuran derivatives (see above), the following furan monoterpenes with a 3-substituent occur in nature
(the reference given under each formula refers to the correct
structural identification of the compound):

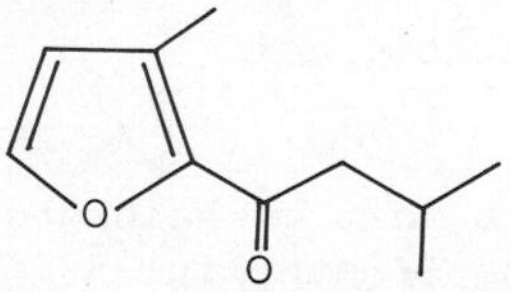

457

Elsholtzione[555]
(*Elsholtzia cristata*)
and others

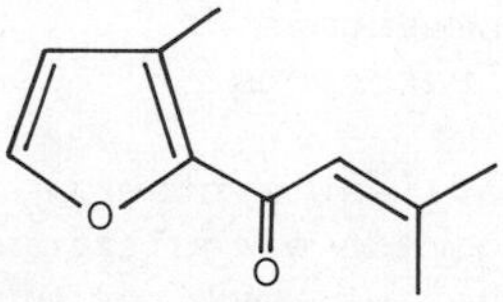

458

Naginata Ketone[556,557]
Perilla frutescens,
& *Elsholtzia cristata*

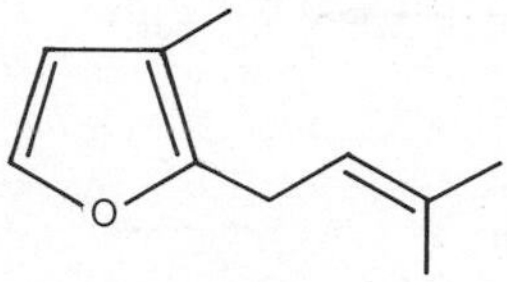

459

Rose Furan[558]
Rose oil

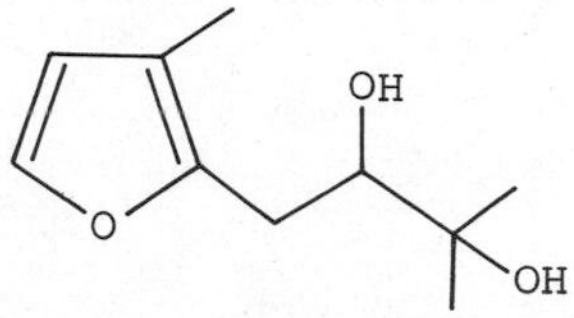

Elsholtzidiol[559]
Elsholzia densa
(synthesis not yet reported)

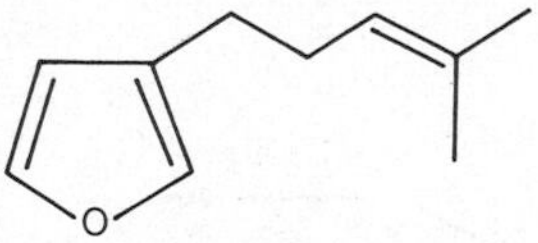

460

Perillene[560]
Perilla frutescens
and elsewhere

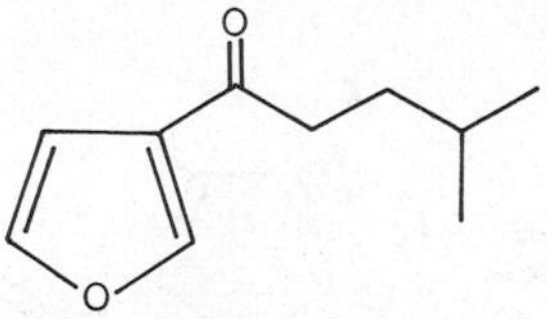

461

Perilla Ketone[561]
P. frutescens

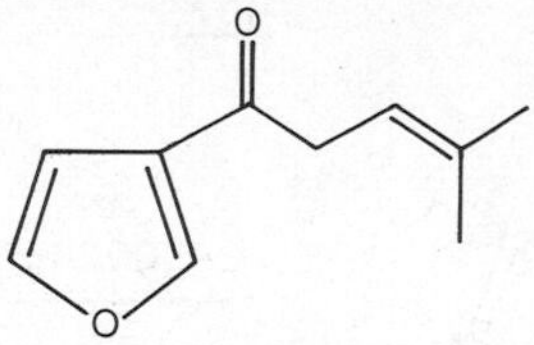

462

Egomaketone[562]
P. frutescens
(synthesis not yet
reported)

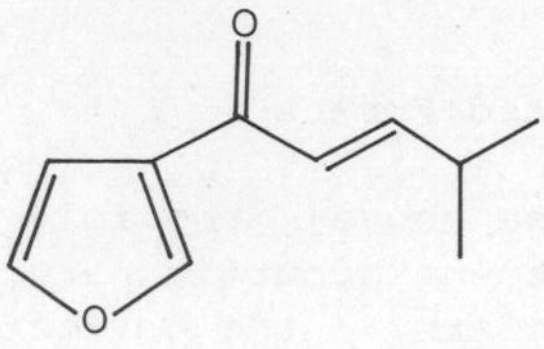

463

Isoegomaketone[563]
P. frutescens

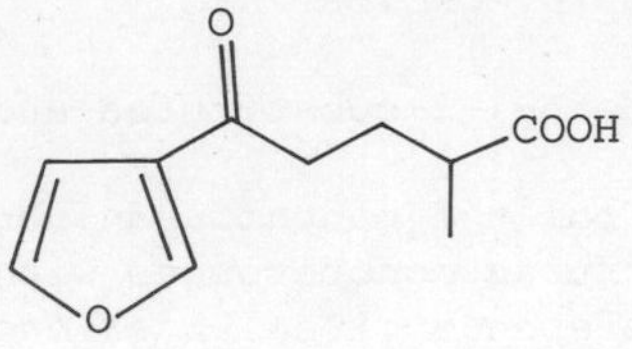

464

Batatic Acid[564]

The earliest synthesis in the group is that of Reichstein et al., who made elsholtzione by converting 3-methylfuran (465) to the 2-formyl compound (466) with hydrogen cyanide, then dehydrating the oxime of this with acetic anhydride and carrying out a Grignard reaction on the nitrile (467), thereby obtaining the saturated ketone (457).[565] Naginata ketone (458)

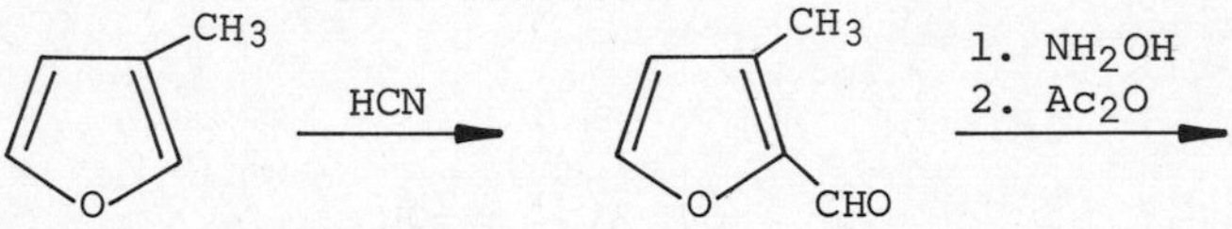
CH3
HCN
CH3
CHO
1. NH2OH
2. Ac2O

465 466

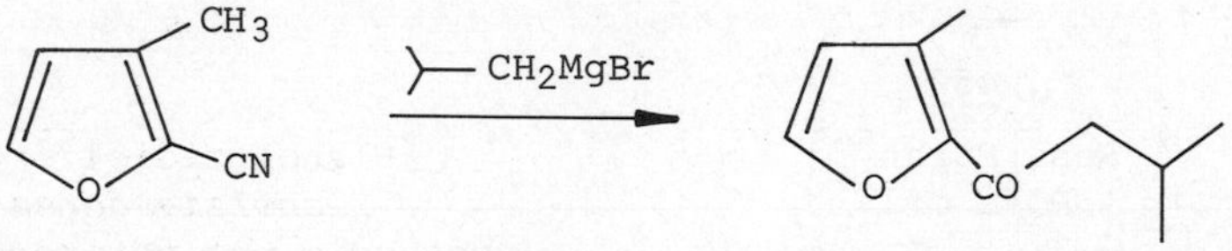
CH3
CN
CH2MgBr
CO

467 457

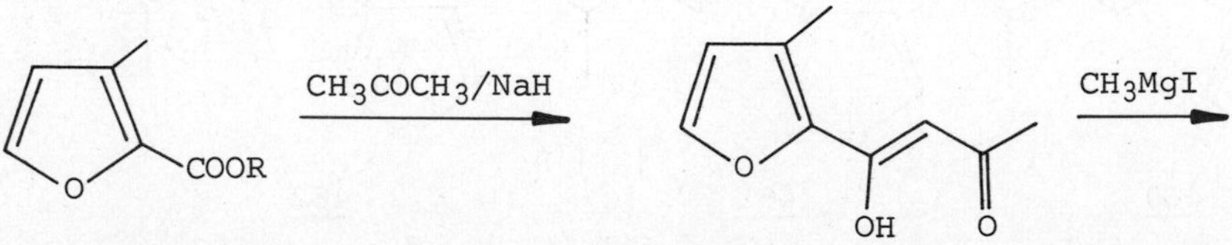
COOR
CH3COCH3/NaH
OH
O
CH3MgI

468

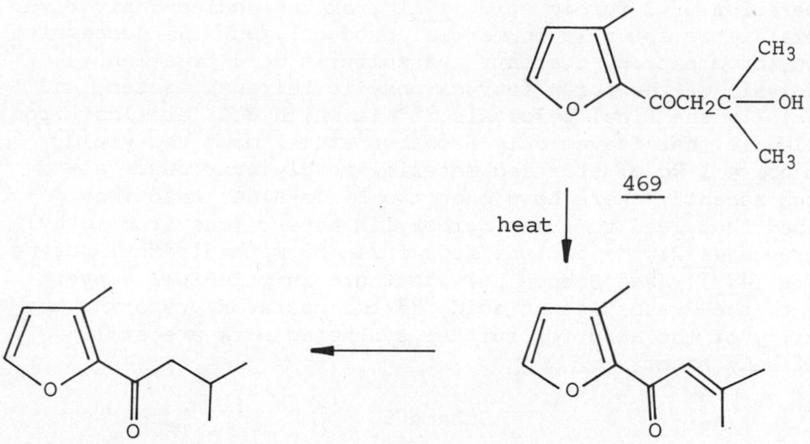

469

heat

457   458

that had already been reduced to elsholtzione (457) catalytically,[556,557] was synthesized very simply by Büchi et al., using
the reaction of 3-methyl-2-furoic ester (468) with acetone in
the presence of sodium hydride, followed by reaction with
methyl magnesium iodide and dehydration by heating the tertiary alcohol resulting (469).[558] Büchi's aim, however, was
to synthesize rose furan (459), and reduction of the carbonyl
group of naginata ketone proved to be an insuperable obstacle,
so for rose furan another synthesis was carried out.   Mer-
curation of 3-methylfuran and replacement of the mercur-
ichloro group by lithium led to 2-lithiated 3-methylfuran
(470), that could be reacted with prenyl bromide to give rose
furan (459).[558]

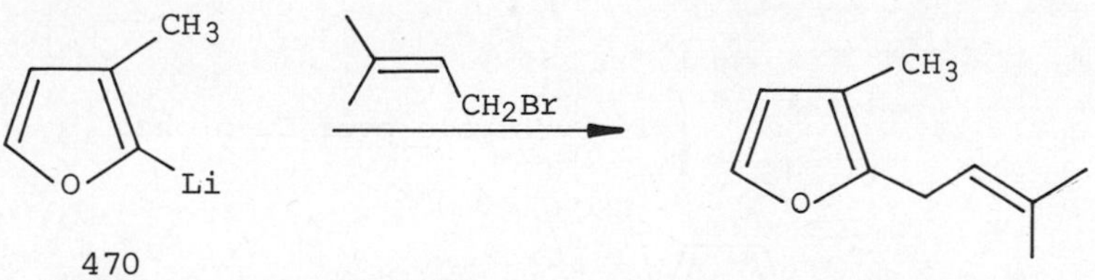

470

459

The starting material for the remaining furans listed
above, that are substitued only in the 3-position, is generally a 3-furoic acid derivative, and therein lies one of the
chief difficulties of synthesis on any scale above that of the

research laboratory. The syntheses usually start with the
preparation of 3-furoic acid (471) from the sodium derivative
of oxalacetic ester (a commercial product), but the successive
steps (bromination, concentrated sulfuric acid treatment,
hydrolysis of the furan tetracarboxylic tetraethylester, and
especially the final pyrolysis,[566] in which $3CO_2$ are lost from
a molecule that leaves only 5 carbon atoms) make the yield
from above 1 kg of starting material rarely over 50g.   Although recently there have been two Diels-Alder reactions described that lead to furan carboxylic esters [one from methyl
2-furoate (472),[567] and one from furan[568] with dimethyl butyne
dioate (473)] (see Scheme 56), that are in principal convertible to the 3-substituted acid,[569] the costs of preparing any
quantity of the acid for further synthetic work are still
relatively high.

Scheme 56

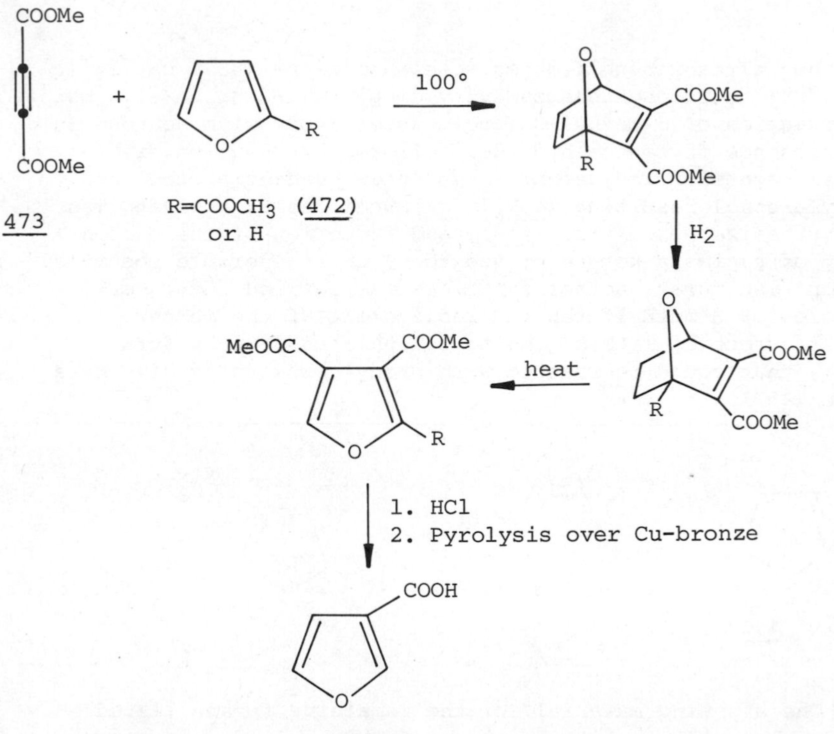

Once the acid has been obtained, the routes to the various 3-substituted furans are relatively straightforward. For example, the synthesis of perillene (460),[570] involves heating 3-furylmethanol (474) with 1-ethoxy-2-methylbutadiene in the presence of mercuric acetate to form the diene ether; the

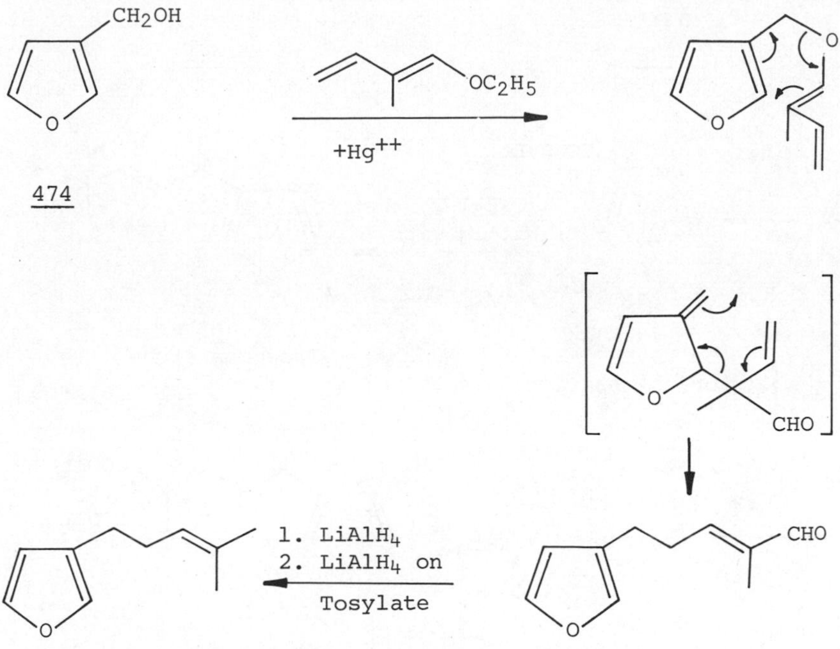

latter then rearranges in a double Claisen-Cope reaction to yield an aldehyde (475) that can be converted, via lithium aluminum hydride reduction of the tosylate of the corresponding alcohol, to perillene (460). Perilla ketone (461) was synthesized by Matsuura[571] as follows: 3-furoyl chloride (476) was allowed to react with the organocadmium compound obtained from isopentylmagnesium halide (Scheme 57):

Scheme 57

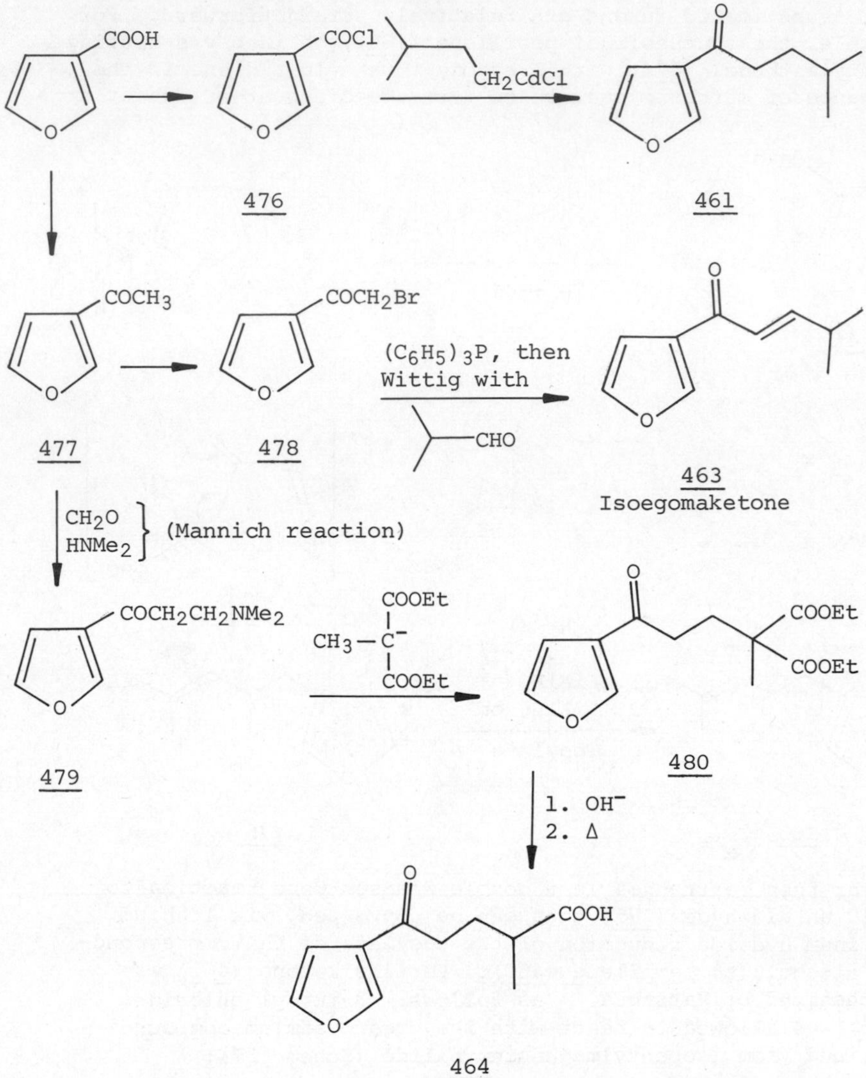

476

461

477   478

(C₆H₅)₃P, then
Wittig with

463
Isoegomaketone

479

CH₂O
HNMe₂ } (Mannich reaction)

480

1. OH⁻
2. Δ

464

The corresponding dehydro compound, isoegomaketone (463) was also made from 3-furoic acid. The latter was converted to 3-acetylfuran (477), which was then brominated (to 478). The phosphorane, made from the bromide (478) and triphenylphosphine, was converted by the Wittig reaction to isoegomaketone (463).[572] The isomer, egomaketone (462), does not appear to have been synthesized.

The synthesis of batatic acid (464) also starts from 3-acetylfuran, which is converted into its Mannich base (479) with formaldehyde and dimethylamine, and this base is then condensed with the sodium derivative of diethyl methylmalonate, after which the usual hydrolysis and pyrolysis of the malonic ester (480) leads to batatic acid (464).[568,573]

B. 2,5,5-Substituted Tetrahydrofurans

"Linalool oxides" have been known for many years to occur in certain essential oils, but the definite structure was established only in 1963 by Felix et al.[574] Similar results were published by Klein et al.[575] The former paper in particular gives a full discussion of the oxidation products of linalool obtained by the reaction with peracids, and suggests that the name linalool oxide be reserved for the *cis-* and *trans-*isomers of formula 481. This peracid oxidation leads first to a diastereomeric pair of unstable 6,7-epoxides (482) that are converted by heat or acid to the linalool oxides (481a and 481b). There is a small amount of the corresponding tetrahydropyran derivatives (483) formed at the same time. Linalool oxide can also be made from citral diepoxide (484)[576] using the Wharton reaction (hydrazine on an epoxyketone[577]), or from geranyl acetate epoxide (485) with acid in an inert solvent.[578] Whether the native product in the plant is the initially formed epoxide (482) or the tetrahydrofuran (481) is not certain. It has been maintained that the compound is not formed from the oxidation of linalool by air,[579] but practically all samples of linalool that the author has examined contain traces of linalool oxide (cf. Ref. 580), and Ohloff has indicated[581] that whereas rose oxide occurs in the plant as a single isomer, linalool oxide is always present as a mixture of *cis-* and *trans-* forms.

2-(But-2-en-2-yl)-5,5-dimethyltetrahydrofuran (486) was isolated by Strickler and Kováts from lime oil, and is formed by treatment of linalool (6) with dilute sulfuric acid, together with the corresponding tetrahydropyran (487).[582]

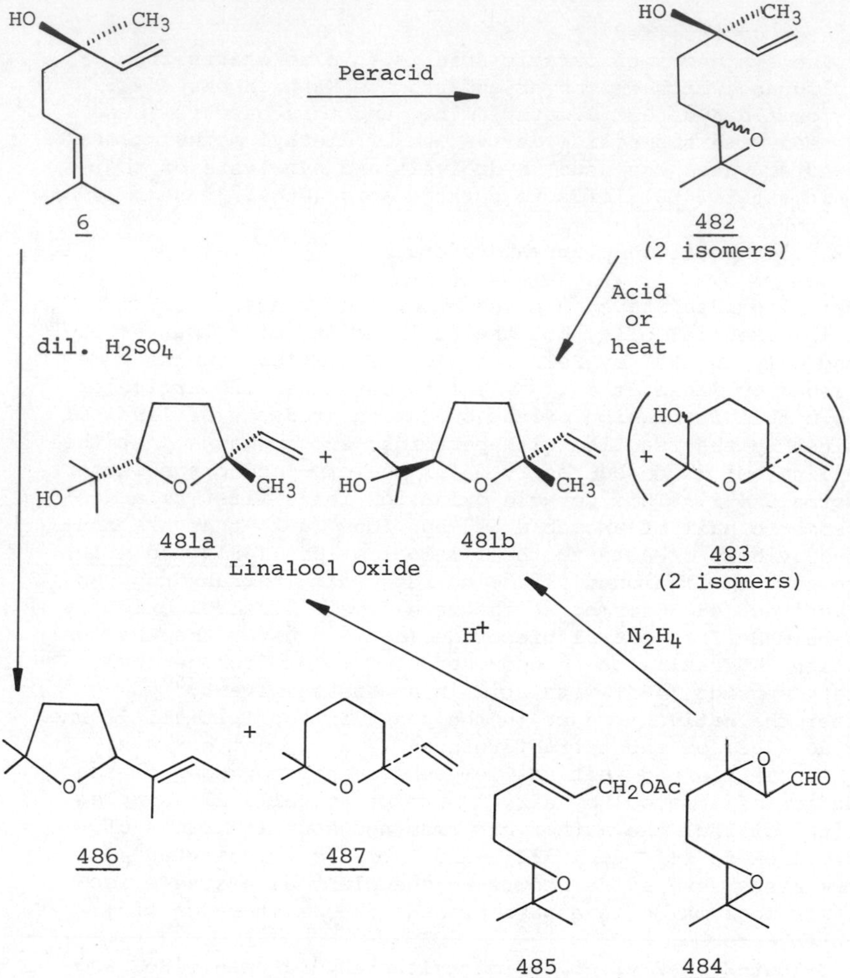

$\underline{6}$ Peracid   → $\underline{482}$
(2 isomers)

Acid
or
heat

dil. H_2SO_4

$\underline{481a}$ + $\underline{481b}$ $\left(+ \quad \underline{483} \right)$
Linalool Oxide   (2 isomers)

H^+ N_2H_4

$\underline{486}$ + $\underline{487}$ $\underline{485}$ $\underline{484}$

19. OXETONES

$\underline{488}$ $\underline{489}$

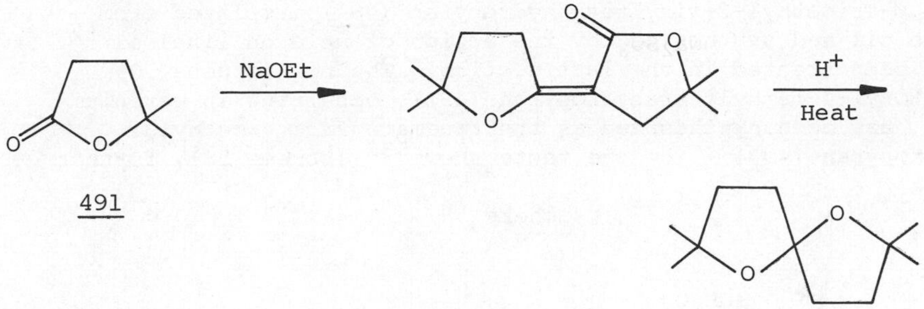

491

490

Although not strictly monoterpenes, we might include here the two spirodihydrofuran derivatives (488 and 489) that have been isolated from hop oil by Naya and Kotaka. The fully hydrogenated derivative (490) was synthesized by reaction of sodium ethylate on 4-methyl-4-hydroxypentenoic acid lactone (491) followed by heating with acid.[583]

20. TETRAHYDROPYRANS

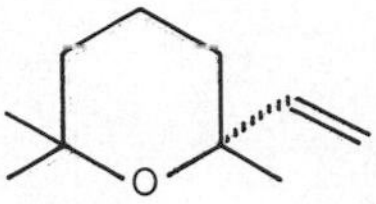

487

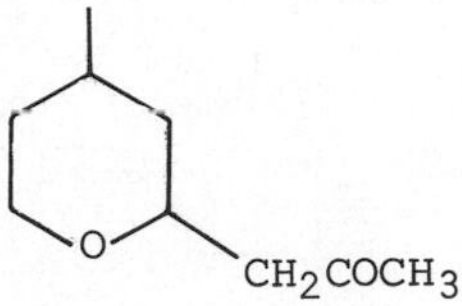

492
2-Acetonyl-4-Methyltetra-
hydropyran

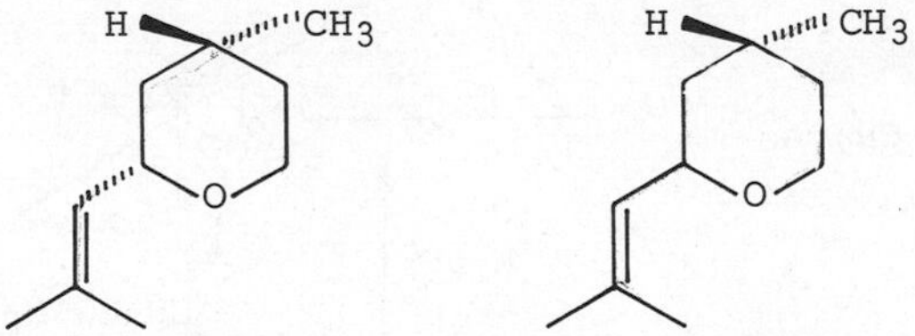

cis-
496a

trans-
496b

Rose Oxide

2,6,6-Trimethyl-2-vinyltetrahydropyran (<u>487</u>), isolated from
lime oil and synthesized by the action of acid on linalool[582]
has been treated in the last section. The nor-terpene, 2-
acetonyl-4-methyltetrahydropyran (<u>492</u>), occurring in geranium
oil, has been synthesized as the racemate from 4-methyldi-
hydropyran (<u>493</u>)[584] by the route shown[585] (Scheme 58), further

Scheme 58

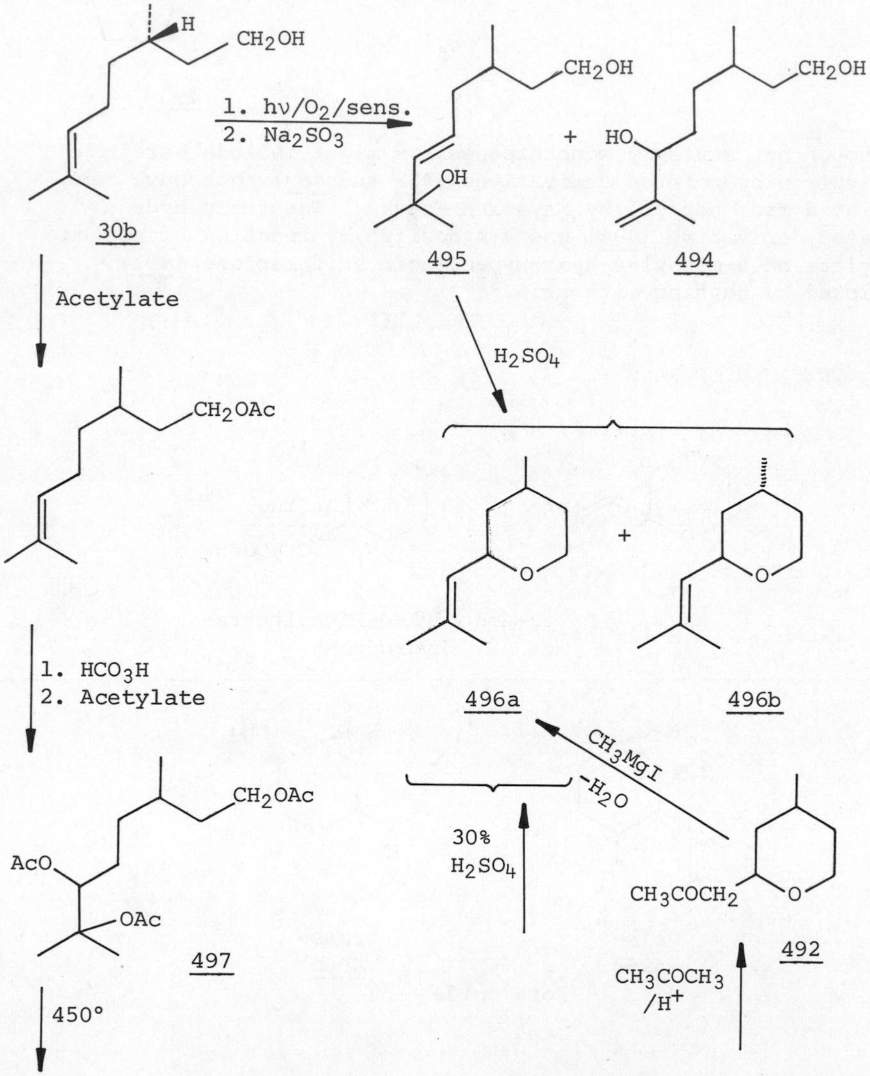

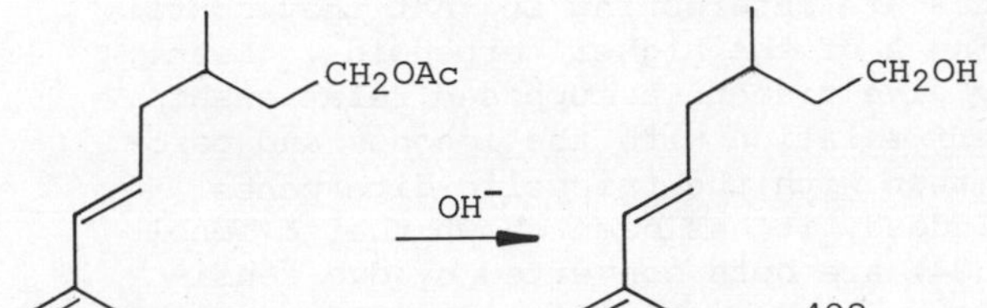

498 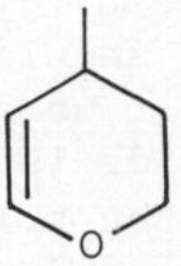   493

reaction of methylmagnesium iodide and dehydration leading to rose oxide.[586] The latter substance, occurring in both rose[587] and geranium[588] oils, is, however, much more conveniently made by the sensitized photooxidation of (-)-citronellol (30b).[589, 590] After reduction of the intermediate hydroperoxides, the mixture is treated with sulfuric acid at room temperature, under which conditions the secondary alcohol (494, 40% of the mixture) is unchanged, while the tertiary alcohol (495, 60% of the mixture) is converted to a 1:1 mixture of natural cis-(496a) and trans- (496b) rose oxides. Other oxidative techniques have been carried out on citronellol; for instance, acetylating the performic acid oxidation product of citronellyl acetate yields a triacetate (497) that can be converted by pyrolysis and hydrolysis to the diene alcohol (498). Treatment of the latter with 30% sulfuric acid yields a 9:1 mixture of cis- and trans-rose oxides.[591]

21.   HEXAHYDROBENZOFURAN-2-ONES

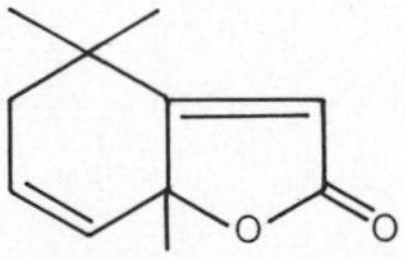

| 499 | 500 | 501 |
| Actinidiolide | Dihydroactinidiolide | Loliolide |

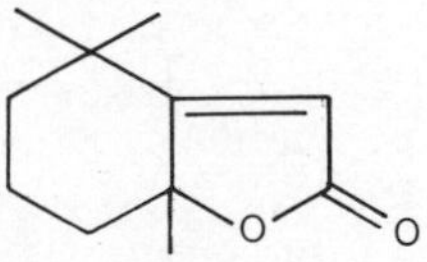

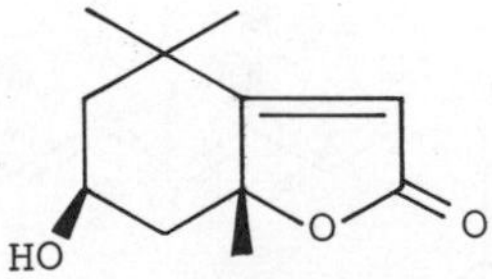

502
Actinidol

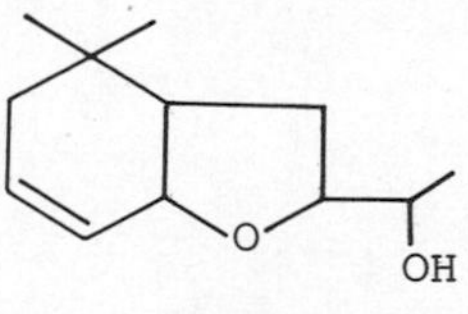

(for position of disubstituted double bond, see Ref. 592).

These terpenoid compounds are interesting in that they contain
trimethylcyclohexane ring A of the higher terpenoids. Lest
this should immediately give rise to a supposed relationship,
let it be said that their relation with the ionones and caro-
tenoids is much closer than with the tricyclic diterpenes
and the triterpenes. Indeed, it has been shown that β-ionol
(503) and β-carotene (504) are both converted by dye sensitized photo-oxidation into dihydroactinidiolide (500) and the
allene (505).[593]

Actinidiolide (499) and actinidol (502) have been found
in *Actinidia polygama*,[594] the dihydrocompound (500) in several
plants (a list of which is given by Demole et al.[595]) and
loliolide has also been isolated from several sources (cf.
Ref. 596). The first three of the compounds have been synthesized as racemates from homosafranic acid (506) by Demole
and Enggist,[592] by the routes shown in Scheme 59. Sakan had
already used β-cyclohomogeranic acid (507) to synthesize (±)-
dihydroactinidiolide.[594]

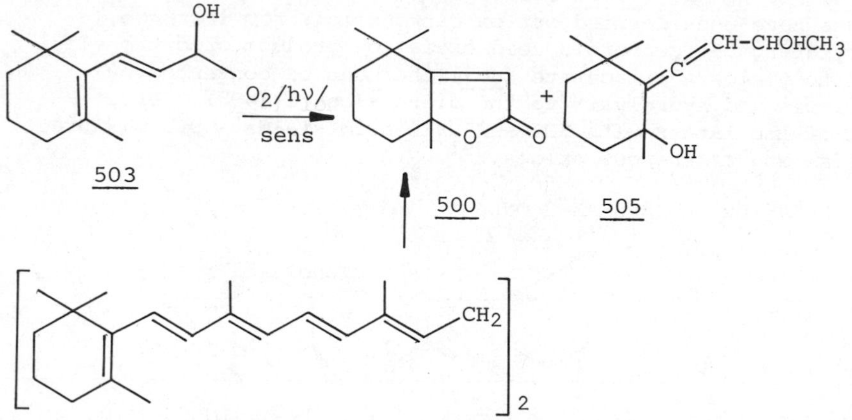

Scheme 59

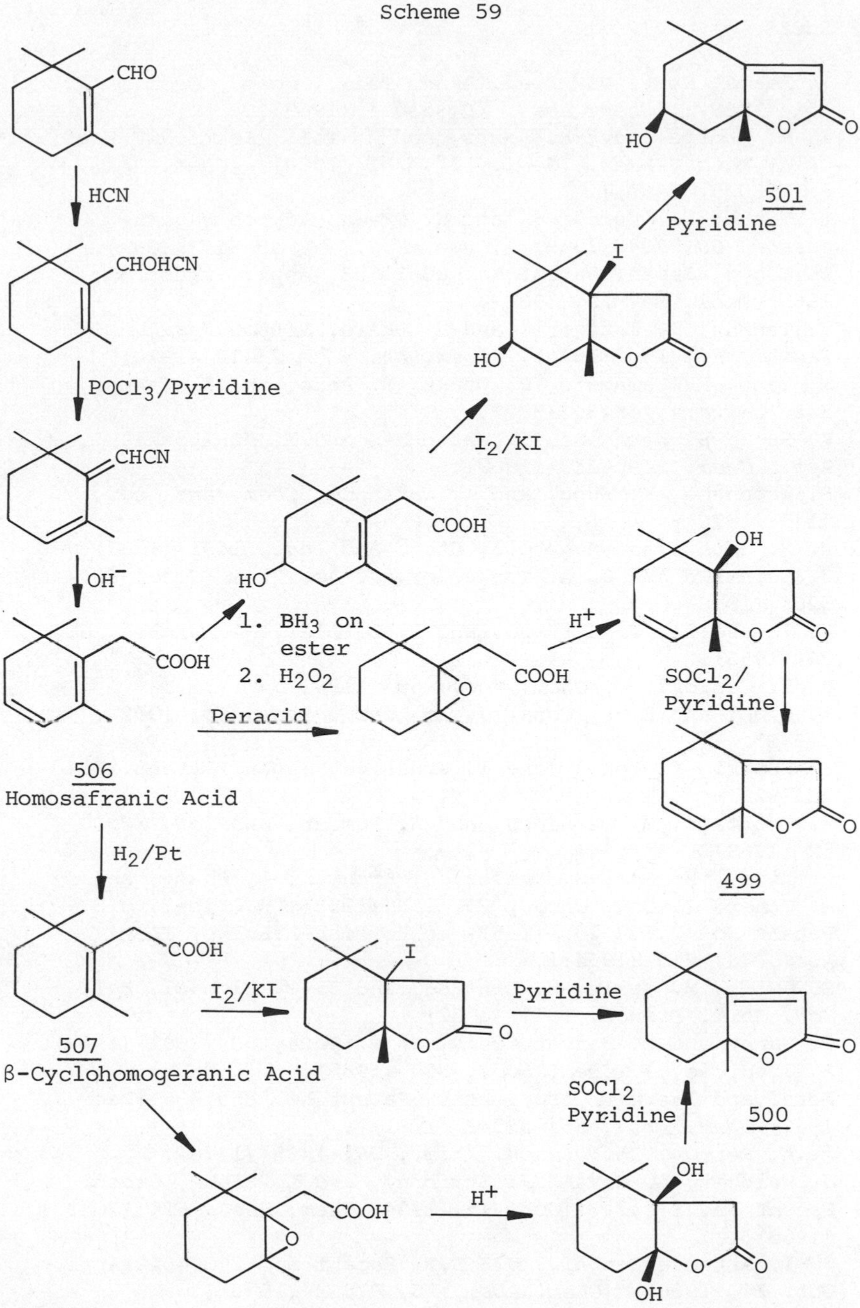

CHO
HCN
CHOHCN
POCl3/Pyridine
CHCN
OH⁻
COOH
506
Homosafranic Acid
1. BH3 on ester
2. H2O2
HO
COOH
I2/KI
I
HO
O
O
Pyridine
HO
O
O
501
Peracid
O
COOH
H+
OH
O
O
SOCl2/Pyridine
O
O
499
H2/Pt
COOH
507
β-Cyclohomogeranic Acid
I2/KI
I
O
O
Pyridine
O
O
500
SOCl2 Pyridine
O
COOH
H+
OH
O
O
OH

172   The Synthesis of Monoterpenes

REFERENCES

1. K. Laats, *Eesti NSV Tead. Akad. Toim., Keem., Geol., 17,*
   355 (1968); [*Chem. Abs., 70,* 53034 (1969)].
2. I. B. Kudryavtsev, K. Laats, and M. Tali, *Eesti NSV Tead.
   Akad. Toim., Keem., Geol., 17,* 361 (1968); [*Chem. Abs.,
   70,* 53035 (1969)].
3. J. Tanaka, T. Katagiri, and H. Okawa, *Nippon Kagaku
   Zasshi, 90,* 204 (1969); [*Chem. Abs.,70,* 58034 (1969)].
4. G. Kogyo, Japan. Patent No. 68 11893 (Appl. August 30,
   1965, Publ. May 20, 1968).
5. J. Tanaka, T. Katagiri, and H. Okawa, *Nippon Kagaku
   Zasshi, 91,* 156 (1970); [*Chem. Abs.,73,* 25672 (1970)].
6. M. Yang, K. Yamamoto, N. Otake, M. Ando, and K. Takase,
   Tet. Letters, 3843 (1970).
7. K. Suga, S. Watanabe, T. Watanabe, and M. Kuniyoshi, *J.
   Appl. Chem., 19,* 318 (1969).
8. S. Watanabe, K. Suga, and T. Watanabe, *Chem. and Ind.,*
   1145 (1970).
9. A. F. Thomas and M. Stoll, *Chem. and Ind.,* 1491 (1963).
10. J. Meinwald and J. A. Yankeelov, *J. Am. Chem. Soc., 80,*
    5266 (1958).
11. P. Teisseire, P. Bernard, and B. Corbier, *Recherches, 6,*
    30 (1956).
12. M. F. Carroll, *J. Chem. Soc.,* 507 (1941).
13. W. Kimel and A. C. Cope, *J. Am. Chem. Soc., 65,* 1992
    (1943).
14. S. Futaki, Y. Yonea, and T. Kunshige, Japan. Patent. No.
    23,782.
15. W. Hoffmann, H. Pasedach, and H. Pommer, *Annalen, 729,*
    52 (1969).
16. W. Kimel, N. Sax, S. Kaiser, G. Eichman, G. Chase, and
    A. Ofner, *J. Org. Chem., 23,* 153 (1958); W. Kimel, U.S.
    Patent No. 2,628,250 [1953, to Hoffman-LaRoche, *Chem.
    Abs., 48,* 710 (1954)].
17. S. Julia, M. Julia, H. Linarès, and J.-C. Blondel, *Bull.
    Soc. Chim. France,* 1947 (1962).
18. G. Saucy and R. Marbet, Helv. *Chim. Acta, 50,* 2091 (1967).
19. A. J. Ultée, *J. Chem. Soc.,* 530 (1948).
20. Metal and Thermit Corp., Brit. Patent No. 855,696 [Dec. 7,
    1960, *Chem. Abs., 55,* 23342 (1961)].
21. A. A. Petrov, *Zh. Obshch. Khim., 28,* 1435 (1958).
22. J. Weichet, L. Novak, J. Stribrny, and L. Blaha, Czech.
    Patent No. 112,243 [Oct. 15, 1964, *Chem. Abs., 62,* 13049
    (1965)].
23. V. I. Artem'ev et al., U.S.S.R. Patent No. 268,404 (Appl.
    Oct. 24, 1966); [*Chem. Abs., 73,* 87432 (1970)].

24. L. A. Goldblatt and S. Palkin, *J. Am. Chem. Soc.*, *63*, 3517 (1941).
25. E. L. Patton, *Amer. Perfumer*, *56*, 118 (1950).
26. B. M. Mitzner, E. T. Theimer, L. Steinbach and J. Wolt, *J. Org. Chem.*, *30*, 646 (1965).
27. K. J. Crowley, *J. Org. Chem.*, *33*, 3679 (1968).
28. Y.-R. Naves, *Helv. Chim. Acta*, *28*, 1220, 1231 (1945).
29. R. L. Blackmore, U.S. Patent No. 3,231,485 (Oct. 25, 1962).
30. W. F. Erman, *J. Am. Chem. Soc.*, *89*, 3828 (1967).
31. G. Frank, *J. Chem. Soc. (B)*, 130 (1968).
32. P. J. Kropp, *J. Am. Chem. Soc.*, *91*, 5783 (1969).
33. R. Mayer, K. Bochow, and W. Zieger, *Z. Chem.*, *4*, 348 (1964).
34. D. V. Banthorpe and D. Whittaker, *Quart. Rev. (London)*, *20*, 373 (1966).
35. L. Ruzicka and V. Fornasir, *Helv. Chim. Acta*, *2*, 182 (1919).
36. B. A. Arbusow and W. S. Abramow, *Ber.*, *67*, 1942 (1934).
37. Y.-R. Naves and F. Bondavelli, *Helv. Chim. Acta*, *48*, 563 (1965).
38. G. Ohloff, J. Seibl, and E. sz. Kováts, *Annalen*, *675*, 83 (1964).
39. B. M. Mitzner, S. Lemberg, and E. T. Theimer, *Can. J. Chem.*, *44*, 1090 (1966).
40. G. Ohloff, *Chem. Ber.*, *90*, 1554 (1957).
41. A. Bhati, *Perf. and Ess. Oil Rec.*, *54*, 376 (1963).
42. O. P. Vig, B. Vig, R. K. Khetarpal, and R. C. Anand, *Ind. J. Chem.*, *7*, 450 (1969).
43. U. G. Nayak, Sukh Dev, and P. C. Guha, *J. Ind. Chem. Soc.*, *29*, 23 (1952).
44. Sukh Dev (personal communication).
45. A. D. Dembitskii, R. A. Yurina, L. A. Ignatova, and M. I. Goryaev, *Khim. Prir. Soedin*, *4*, 251 (1968); [*Chem. Abs.*, *71*, 3489 (1969)].
46. A. D. Dembitskii, R. A. Yurina, L. A. Ignatova, and M. I. Goryaev, *Izv. Akad. Nauk Kaz. S.S.S.R.*, *Ser. Khim.*, *19*, 49 (1969).
47. E. Klein and W. Rojahn, *Chem. Ber.*, *97*, 2700 (1964).
48. Y. Ogata, *J. Chem. Soc. Japan*, *63*, 417, 419 (1942).
49. K. H. Schulte-Elte and M. Gadola, *Helv. Chim. Acta*, *54*, 1095 (1971).
50. H. Pines, N. E. Hoffman, and V. I. Ipatieff, *J. Am. Chem. Soc.*, *76*, 4412 (1954).
51. R. Rienäcker and G. Ohloff, *Angew. Chem.*, *73*, 240 (1961).
52. A. D. Dembitskii, R. A. Yurina, and M. I. Goryaev, *Khim. Prir. Soedin.*, *5*, 443 (1969); [*Chem. Abs.*, *72*, 82892

(1970)].

53.   J. S. Sörensen and N. A. Sörensen, *Acta Chem. Scand.*, *8*, 284 (1954).

54.   P. Nayler and M. C. Whiting, *J. Chem. Soc.*, 4006 (1954).

55.   W. Treibs and D. Merkel, in *Die Ätherischen Öle* (Gildermeister-Hoffmann), Vol. IIIa (Akademie-Verlag, Berlin, 1960), p. 500.

56.   P. de Mayo, in *The Chemistry of Natural Products* (edited by K. W. Bentley) (Interscience, New York, 1959), Vol. II, p. 40.

57.   H. E. Eschinazi, *J. Org. Chem.*, *26*, 3072 (1961).

58.   Y.-R. Naves and C. Frei, *Helv. Chim. Acta*, *46*, 2551 (1963).

59.   *Chem. Abs.*, Subject Index, 1967.

60.   W. E. Wright, J. C. Benstead, and J. D. Shimmin, U.S. Patent No. 3,324,160 (April 12, 1963, to Shell Oil Co.).

61.   R. Heilmann and R. Glenat, *Bull. Soc. Chim. France*, 1586 (1955).

62.   B. N. Joshi, R. Seshadri, K. K. Chakravarti, and S. C. Bhattacharyya, *Tetrahedron*, *20*, 2911 (1964).

63.   P. Barbier, *Compt. Rend. Acad. Sci.*, *114*, 674 (1892). Linalool is referred to as "licaréol" in this early paper.

64.   H. Normant, *Ind. Parfumérie*, *11*, 172 (1956).

65.   E. Tomikashi, *Koryo*, *45*, 12 (1957).

66.   A. Boake Roberts and Co., Ltd., Brit. Patent No. 896,262 (Sept. 15, 1968).

67.   G. Ohloff and E. Klein, *Tetrahedron*, *18*, 37 (1962).

68.   P. S. Wharton and D. H. Bohlen, *J. Org. Chem.*, *26*, 3615 (1961).

69.   G. V. Nair and G. D. Pandit, *Tet. Letters*, 5097 (1966).

70.   G. V. Nair and G. D. Pandit, Brit. Patent No. 1,082,364 (March 18, 1964, to Unilever, Ltd.).

71.   S. K. Pradhan and V. M. Girijavollabhan, *Tet. Letters*, 3103 (1968).

72.   K. Suga, S. Watanabe, and I. Okoshi, *Bull. Chem. Soc. Japan*, *39*, 1335 (1966).

73.   E. Jirát and F. Vonášek, Czech. Patent No. 95,472 [June 15, 1960: *Chem. Abs.*, *55*, 6378 (1961)].

74.   K. Kogami, J. Kumanotani, and T. Kuwata, *Perf. and Ess. Oil Rec.*, *58*, 872 (1967).

75.   P. Nayler, Brit. Patent No. 878,680 [May 7, 1959, to Distillers Co., Ltd., *Chem. Abs.*, *56*, 8862 (1962)].

76.   T. Kuwata, K. Kumano, and K. Kunio, Jap. Patent No. '65 8334 [July 22, 1961, to T. Hawegawa Co., Ltd., *Chem. Abs.*, *63*, 5689 (1965)].

77.   W. Treibs and D. Merkel, in *Die Ätherischen Öle*

(Gildemeister-Hoffmann) (Akademie-Verlag, Berlin, 1960), Vol. IIIa, p. 572.

78. G. W. Pigulevski and G. B. Trojan, *Izv. Akad. Nauk S.S.S.R.*, 401 (1950).

79. L. Ruzicka and H. Schinz, *Helv. Chim. Acta*, *23*, 959 (1940).

80. A. F. Thomas, B. Willhalm, and R. Müller, *Org. Mass Spec.*, *2*, 223 (1969).

81. J. W. K. Burrell, R. F. Garwood, L. M. Jackman, E. Oskay, and B. C. L. Weedon, *J. Chem. Soc.*, *(C)*, 2144 (1966).

81a. Y. Yukawa, T. Hanafusa, and K. Fujita, *Bull. Chem. Soc. Japan*, *37*, 158 (1964).

82. J. Tanaka, T. Katagiri, and T. Takeshita, *Nippon Kagaku Zasshi*, *89*, 65 (1968).

83. A. R. Battersby, R. T. Brown, J. A. Knight, J. A. Martin, and A. O. Plunkett, *Chem. Comm.*, 346 (1966).

84. P. Loew, H. Goeggel, and D. Arigoni, *Chem. Comm.*, 347 (1966).

85. E. S. Hall, F. McCapra, T. Money, K. Fukumoto, J. R. Hanson, B. S. Mootoo, G. T. Phillips, and A. I. Scott, *Chem. Comm.*, 348 (1966).

86. R. C. Haley, J. A. Miller, and H. C. S. Wood, *J. Chem. Soc. (C)*, 264 (1969).

87. V. Prelog and E. Watanabe, *Annalen*, *603*, 1 (1957).

88. T. Yoshida, H. Kawamura, and A. Komatsu, *Agr. Biol. Chem.*, *33*, 343 (1969).

89. Y. Nakatani, S. Sato, and T. Yamanishi, *Agr. Biol. Chem.*, *33*, 967 (1969).

90. T. Yamanishi, M. Nose, and Y. Nakatani, *Agr. Biol. Chem.*, *34*, 599 (1970).

91. T. Matsuura and Y. Butsugan, *Nippon Kagaku Zasshi*, *89*, 513 (1968).

92. R. M. Silverstein and J. O. Rodin, *Science*, *154*, 509 (1966).

93. C. A. Reece, J. O. Rodin, R. G. Brownlee, W. G. Duncan, and R. M. Silverstein, *Tetrahedron*, *24*, 4249 (1968).

94. E. J. Corey and D. Seebach, *Angew. Chem. Int. Ed.*, *4*, 1075 (1965).

95. D. Seebach, *Synthesis*, *1*, 17 (1969).

96. R. C. Krug and T. F. Yen, *J. Org. Chem.*, *21*, 1082 (1956).

97. A. Verlay, *Bull. Soc. Chim. France*, 849 (1928).

98. V. G. Cherkaev, A. A. Bag, and S. A. Prepelkina, *Trudy Vsesoyuz Nauch Inst. Sintheticheskikh i Natural. Dushistykh Veshchesv.* 35 (1954); [*Chem. Abs.*, *53*, 18082 (1959)].

99. G. Ohloff, *Tet. Letters*, 10 (1960).

100. V. Grignard and J. Dœuvre, *Compt. Rend. Acad. Sci.*, *190*, 1164 (1930).

101. P. Joseph-Nathan and A. Manjarrez, *Rev. Soc. Quím. Mex*, *11*, 116 (1967).

102. M. Ohtsuru, M. Teraoka, K. Tori, and K. Takeda, *J. Chem. Soc. (B)*, 1033 (1967).

103. W. Treibs and D. Merkel, in *Die Ätherischen Öle* (Gildermeister-Hoffmann) (Akademie-Verlag, Berlin, 1963), Vol. IIIc, p. 89.

104. I. Majewska, *Tluszcze i Srodki Piorace*, *8*, 296 (1964); [*Chem. Abs.*, *63*, 12695 (1965)].

105. T. Hashimoto and T. Shibutani, Japan. Patent No. '65 15,077 (Dec. 12, 1961, to Toyotama Perfumery Co.); [*Chem. Abs.*, *63*, 16124 (1965)].

106. K. Kogami and J. Kumanotani, *Bull. Chem. Soc. Japan*, *41*, 2508 (1968).

107. K. Nakagawa, R. Konaka, and T. Nakata, *J. Org. Chem.*, *27*, 1597 (1962).

108. I. N. Nazarov, S. M. Makin, V. B. Mochalin, and D. V. Nazarova, *Zhur. Obshch. Khim.*, *29*, 3965 (1959).

109. M. Montavon, H. Lindlar, R. Marbet, R. Ruegg, G. Ryser, G. Saucy, P. Zeller, and O. Isler, *Helv. Chim. Acta*, *40*, 1250 (1957).

110. L. S. Povarov, *Russ. Chem. Rev.*, 639 (1965).

111. G. I. Samokhvalov, L. A. Vakulova, T. V. Men, L. T. Zhikhareva, V. I. Koltunova, and N. A. Preobrazhenskii, *Zh. Obshch. Khim.*, *29*, 2575 (1959); U.S.S.R. Patent No. 118,496 (1959).

112. A. F. Thomas, *J. Am. Chem. Soc.*, *91*, 3281 (1969).

113. G. Saucy, R. Marbet, H. Lindlar, and O. Isler, *Helv. Chim. Acta*, *42*, 1945 (1959).

114. I. Sausa, L. Gazo, and J. Morvic, Czech. Patent No. 126,763 (Nov. 29, 1965); [*Chem. Abs.*, *70*, 37941 (1969)].

115. J. Redel and P. Raymond, *Compt. Rend. Acad. Sci.*, *255*, 1127 (1962).

116. K. Suga and S. Watanabe, Japan. Patent No. '64 3011 (Nov. 30, 1961); [*Chem. Abs.*, *60*, 15921 (1964)].

117. T. G. H. Jones and F. B. Smith, *J. Chem. Soc.*, 2530 (1925); 2767 (1926).

118. E. E. Boehm, V. Thaller, and M. C. Whiting, *J. Chem. Soc.*, 2535 (1963).

119. P. Teisseire and B. Corbier, *Recherches*, *17*, 5 (1969).

120. O. P. Vig, K. L. Matta, and I. Raj, *J. Ind. Chem. Soc.*, *41*, 752 (1964).

121. O. P. Vig, K. L. Matta, M. S. Bhatia, and R. Anand, *Ind. J. Chem.*, *8*, 107 (1970).

122. H. Lindlar, *Helv. Chim. Acta*, *35*, 446 (1952).

123. R. B. Bates and S. K. Paknikar, *Tet. Letters,* 1453 (1965).

124. L. Crombie, R. P. Houghton, and D. K. Woods, *Tet. Letters,* 4553 (1967).

125. A. F. Thomas and B. Willhalm, *Tet. Letters,* 3775 (1964).

126. O. N. Devgan, M. M. Bokadia, A. K. Bose, M. S. Tibbets, G. K. Trivedi, and K. K. Chakravarti, *Tet. Letters,* 5337 (1967); *Tetrahedron, 25,* 3217 (1969).

127. Y. Chrétien-Bessière, L. Peyron, L. Bénezet, and J. Garnero, *Bull. Soc. Chim. France,* 2018 (1968).

128. Y. Asahina and S. Takagi, *J. Pharm. Soc. Japan, 464,* 837 (1920).

129. F. Bohlmann and W. Thefeld, *Chem. Ber., 102,* 1698 (1969).

130. B. Willhalm and A. F. Thomas, *Chem. Comm.,* 1380 (1969).

131. K. Yano, S. Hayashi, T. Matsuura, and A. W. Burgstahler, *Experientia, 26,* 8 (1970).

132. H. Schinz and C. F. Seidel, *Helv. Chim. Acta, 25,* 1572 (1942).

133. A. E. Wick, D. Felix, K. Steen, and A. Eschenmoser, *Helv. Chim. Acta, 47,* 2425 (1964).

134. W. Sucrow, *Angew. Chem. Int. Ed., 7,* 629 (1968).

135. B. M. Trost and R. LaRochelle, *Tet. Letters,* 3327 (1968).

136. A. F. Thomas, *Chem. Comm.,* 1054 (1970).

137. J. N. Hines, M. J. Peagram, G. H. Whitham, and M. Wright, *Chem. Comm.,* 1593 (1968).

138. A. F. Thomas, *Chimia, 24,* 452 (1970); and unpublished work.

139. W. Sucrow and W. Richter, *Tet. Letters,* 3675 (1970).

140. W. Sucrow, *Tet. Letters,* 4725 (1970).

141. J. Colonge and P. Dumont, *Bull. Soc. Chim. France, 11,* 125 (1944).

142. L. Ruzicka and T. Reichstein, *Helv. Chim. Acta, 19,* 646 (1936).

143. T. Takemoto and T. Nakajima, *Yakugaku Zasshi, 77,* 1344 (1957); [*Chem. Abs., 52,* 4479 (1958)].

144. A. R. Battersby and D. A. Rowlands (personal communication).

145. A. Eschenmoser, H. Schinz, R. Fischer, and J. Colonge, *Helv. Chim. Acta, 34,* 2329 (1951).

146. J. E. Baldwin, R. E. Hackler, and D. P. Kelly, *Chem. Comm.,* 537 (1968).

147. G. M. Blackburn, W. D. Ollis, J. D. Plackett, S. Smith, and I. O. Sutherland, *Chem. Comm.,* 186 (1968).

148. V. Rautenstrauch, *Chem. Comm.,* 4 (1970).

149. J. E. Baldwin, J. DeBernardis, and J. E. Patrick, *Tet.*

Letters, 353 (1970).

150. H. Kwart and R. K. Miller, *J. Am. Chem. Soc., 76,* 5403 (1954).

151. W. Giersch and K. H. Schulte-Elte (personal communication).

152. M. Julia and M. Baillarge, *Bull. Soc. Chim. France,* 734 (1966).

153. W. Sucrow, *Tet. Letters,* 1431 (1970).

154. W. Sucrow, *Chem. Ber., 103,* 3771 (1970).

155. H. Schinz and G. Shäppi, *Helv. Chim. Acta, 30,* 1483 (1947).

156. H. Grütter and H. Schinz, *Helv. Chim. Acta, 35,* 1656 (1952).

157. A. Pfau and P. L. Plattner, *Helv. Chim. Acta, 15,* 1250 (1932).

158. K. Brack and H. Schinz, *Helv. Chim. Acta, 34,* 2009 (1951).

159. M. Matsui and B. Stalla-Bourdillon, *Agr. Biol. Chem. (Tokyo), 32,* 1246 (1968).

160. K. C. Brannock, H. S. Pridgen, and B. Thompson, *J. Org. Chem., 25,* 1815 (1960).

161. P. Teisseire and M. Rinaldi, *Recherches, 13,* 4 (1963).

162. K. Takabe, T. Katagiri, and J. Tanaka, *Nippon Kagaku Zasshi, 90,* 943 (1969); [*Chem. Abs., 72,* 42628 (1970)]; see also R. Maurin and M. Bertrand, *Bull. Soc. Chim. France,* 2356 (1972).

163. A. F. Thomas (unpublished work).

164. S. M. Baba, H. H. Mathur, and S. C. Bhattacharyya, *Tetrahedron, 22,* 903 (1966).

165. T. Sakai, S. Eguchi, and M. Ohno, *J. Org. Chem., 35,* 790 (1970).

166. M. K. Logani, I. P. Varshney, R. C. Pandey, and Sukh Dev, *Tet. Letters,* 2645 (1967).

167. S. M. Dixit, A. S. Rao, and S. K. Paknikar, *Chem. and Ind.,* 1256 (1967).

168. U. Steiner and H. Schinz, *Helv. Chim. Acta, 34,* 1508 (1951).

169. C. Ferrero and H. Schinz, *Helv. Chim. Acta, 39,* 2109 (1956).

170. L. Crombie and M. Elliott, in *Progress in the Chemistry of Natural Products,* edited by L. Zechmeister (Springer-Verlag, Vienna, 1961), Vol. 19, p. 120.

171. H. Staudinger and L. Ruzicka, *Helv. Chim. Acta, 7,* 177 (1924).

172. R. Yamamoto, *J. Chem. Soc. Japan, 44,* 311, 1070 (1923).

173. H. Staudinger, O. Muntwyler, L. Ruzicka, and S. Seibt, *Helv. Chim. Acta, 7,* 390 (1924).

174. I. G. M. Campbell and S. H. Harper, *J. Chem. Soc.*, 283 (1945).

175. T. Matsumoto, A. Nagai, and Y. Takahashi, *Bull. Chem. Soc. Japan, 36,* 481 (1963).

176. L. Crombie, S. H. Harper, and K. C. Sleep, *J. Chem. Soc.*, 2743 (1954).

177. S. Julia, M. Julia, and G. Linstrumelle, *Bull. Soc. Chim. France,* 3499 (1966).

178. T. Hanafusa, M. Ohnishi, M. Mishima, and Y. Yukawa, *Chem. & Ind.*, 1050 (1970).

179. M. Matsui and H. Yoshioka, Japan. Patent No. 65 6,457 (April 23, 1963, to Sumimoto Chemical Co., Ltd.); [*Chem. Abs., 63,* 1822 (1965)].

180. M. Matsui and M. Uchiyama, *Agr. Biol. Chem., 26,* 532 (1962).

181. S. Julia, M. Julia, and G. Linstrumelle, *Bull. Soc. Chim. France,* 2693 (1964).

182. S. Julia and G. Linstrumelle, *Bull. Soc. Chim. France,* 3490 (1966).

183. G. Widmark, *Ark. Kem., 11,* 195 (1957).

184. M. Delépine, *Bull. Soc. Chim. France,* 1369 (1936).

185. M. Julia, S. Julia, and B. Cochet, *Bull. Soc. Chim. France,* 1476 (1964).

186. M. Julia, S. Julia, and B. Cochet, *Bull. Soc. Chim. France,* 1487 (1964).

187. J. Martel and C. Huynh, *Bull. Soc. Chim. France,* 985 (1967).

188. Yu. P. Volkov and L. N. Khachatur'yan, *Zh. Obshch. Khim., 37,* 2358 (1967).

189. Yu. P. Volkov and L. N. Khachatur'yan, *Zh. Org. Khim., 4,* 1398 (1968).

190. H. Yoshioka, M. Matsui, Y. Yamada, and H. Sakimoto, *Ind. Chim. Belge, 32* (special no.), Compt. Rend. 36ème Cong. Int. de Chimie Industrielle, III, 890 (1967).

191. F. W. Semmler and H. Schiller, *Ber., 60,* 1951 (1927).

192. K. Ueda and M. Matsui, *Agr. Biol. Chem., 34,* 119 (1970).

193. L. Crombie, C. F. Doherty, and G. Pattenden, *J. Chem. Soc. (C),* 1076 (1970).

194. L. Velluz, J. Martel, and G. Nominé, *Compt. Rend. Acad. Sci. (C), 268,* 2199 (1969).

195. J. Martel and J. Buenida, I.U.P.A.C. Meeting, Riga, U.S.S.R., 1970, Abstracts E 112, p. 572.

196. K. Ueda and M. Matsui, *Agr. Biol. Chem., 34,* 1115 (1970).

197. J. H. Tumlinson, D. D. Hardee, R. C. Gueldner, A. C. Thompson, P. A. Hedin, and P. Minyard, *Science, 166,* 1010 (1969).

198. R. Zurflüh, L. L. Dunham, V. L. Spain, and J. B. Siddall, *J. Am. Chem. Soc., 92,* 425 (1970).
199. K. Kafuku, T. Nozoe, and C. Hata, *Bull. Chem. Soc. Japan, 6,* 40, 111 (1931).
200. T. Ishiguro, N. Koga, and K. Nara, *J. Pharm. Soc. Japan,* 77, 566 (1957).
201. A. F. Thomas, *Perf. and Ess. Oil Rec., 56,* 301 (1965).
202. K. Sisido, S. Kurozumi, K. Utimoto, and T. Isida, *J. Org. Chem., 31,* 2795 (1966).
203. S. A. Achmad and G. W. K. Cavill, *Austral. J. Chem., 16,* 858 (1963).
204. H. Strickler, G. Ohloff, and E. sz. Kováts, *Tet. Letters,* 649 (1964).
205. R. Thomas, *Tet. Letters,* 544 (1961).
206. E. Wenkert, *J. Am. Chem. Soc., 84,* 98 (1962).
207. E. Leete and J. N. Wemple, *J. Am. Chem. Soc., 88,* 4743 (1966).
208. H. Strickler, G. Ohloff, and E. sz. Kováts, *Helv. Chim. Acta, 50,* 759 (1967).
209. R. C. Cookson, J. Hudec, S. A. Knight, and B. Whitear, *Tet. Letters,* 79 (1962).
210. O. Wallach, *Annalen, 323,* 333 (1902).
211. S. Forsén and T. Norin, *Tet. Letters,* 4183 (1966).
212. M. P. Hartshorn, D. N. Kirk, and A. F. A. Wallis, *J. Chem. Soc.,* 5494 (1964).
213. T. Ikeda and K. Wakatsuki, *J. Chem. Soc. Japan, 57,* 425 (1936).
214. Y. Sebe and T. Naito, *J. Taiwan Pharm. Assoc., 2,* 23 (1950); [*Chem. Abs., 45,* 6163 (1951)].
215. T. Ikeda and S. Takada, *J. Chem. Soc., Japan, 57,* 71 (1937).
216. E. sz. Kováts and H. Strickler, *J. Gas Chrom., 3,* 244 (1965).
217. L. M. Roth and T. Eisner, *Ann. Rev. Entomol., 7,* 107 (1952).
218. M. Pavan, *Ricerce Sci., 20,* 1853 (1950).
219. A. F. Thomas, in *Specialist Report on Terpenes,* edited by K. H. Overton (Chemical Society, London, 1971).
220. S. M. McElvain, R. D. Bright, and P. R. Johnson, *J. Am. Chem. Soc., 63,* 1558 (1941).
221. J. Meinwald, *J. Am. Chem. Soc., 76,* 4571 (1954).
222. R. B. Bates and C. W. Sigel, *Experientia, 19,* 564 (1963).
223. T. Sakan, A. Fujino, F. Murai, A. Suzuki, and Y. Butsugan, *Bull. Chem. Soc. Japan, 33,* 1737 (1960).
224. T. Sakan, S. Isoe, S. B. Hyeon, R. Katsumura, T. Maeda, J. Wolinsky, D. Dickerson, M. R. Slabaugh, and D. Nelson,

Tet. Letters, 4097 (1965).

225. M. Pavan, *Ricerce Sci., 19,* 1011 (1949).

226. R. Fusco, R. Trave, and A. Vercellone, *Chim. e Ind. (Milano), 37,* 251 (1955); *37,* 958 (1955).

227. F. Korte, J. Falbe, and A. Zschocke, *Tetrahedron, 6,* 201 (1959).

228. K. J. Clark, G. I. Fray, R. H. Jaeger, and R. Robinson, *Tetrahedron, 6,* 217 (1959).

229. G. W. K. Cavill, D. L. Ford, and H. D. Locksley, *Austral. J. Chem., 9,* 288 (1956).

230. T. Sakan, F. Murai, Y. Hayashi, Y. Honda, T. Shono, M. Nakajima, and M. Kato, *Tetrahedron, 23,* 4635 (1967).

231. S. B. Hyeon, S. Isoe, and T. Sakan, *Tet. Letters,* 5325 (1968).

232. D. J. McGurk, J. Frost, G. R. Waller, E. J. Eisenbraun, K. Vick, W. A. Drew, and J. Young, *J. Insect Physiol., 14,* 841 (1968); [*Chem. Abs., 69,* 41988 (1968)].

233. G. W. K. Cavill and H. Hinterberger, *Austral. J. Chem., 13,* 514 (1960).

234. J. Meinwald, M. S. Chadha, J. J. Hurst, and T. Eisner, *Tet. Letters,* 29 (1962).

235. G. W. K. Cavill and F. B. Whitfield, *Austral. J. Chem., 17,* 1260 (1964).

236. W. R. Dunstan and F. W. Short, *Pharm. J. and Trans., 14,* 3 (1883).

237. A. R. Battersby, R. S. Kapil, and R. Southgate, *Chem. Comm.,* 131 (1968).

238. G. Büchi, J. A. Carlson, J. E. Powell, Jr., and L.-F. Tietze, *J. Am. Chem. Soc., 92,* 2165 (1970).

239. C. Djerassi, J. D. Gray, and F. Kincl, *J. Org. Chem., 25,* 2174 (1960).

240. C. Djerassi, T. Nakano, A. N. James, L. H. Zalkow, E. J. Eisenbraun, and J. N. Shoolery, *J. Org. Chem., 26,* 1192 (1961).

241. G. Büchi, B. Gubler, R. S. Schneider, and J. Wild, *J. Am. Chem. Soc., 89,* 2776 (1967).

242. W. H. Tallent, *Tetrahedron, 20,* 1781 (1964).

243. L. H. Briggs, B. F. Cain, P. W. LeQuesne, and J. N. Shoolery, *J. Chem. Soc.,* 2595 (1965).

244. G. Büchi and R. E. Manning, *Tet. Letters,* 5 (1960).

245. T. Sakan and K. Abe, *Tet. Letters,* 2471 (1968).

246. P. W. Thies, *Tetrahedron, 24,* 313 (1968).

247. M. Bridel, *Compt. Rend. Acad. Sci., 176,* 1742 (1923).

248. H. Inouye, T. Arai, Y. Miyoshi, and Y. Yaoi, *Tet. Letters,* 1031 (1963).

249. J. Wolinsky and D. Nelson, *Tetrahedron, 25,* 3767 (1969).

250. S. Isoe, T. Ono, S. B. Hyeon, and T. Sakan, *Tet. Letters,*

5319 (1968).

251. R. Trave, A. Marchesini, and L. Garanti, *Gazz. Chim. Ital.*, *98*, 1132 (1968).

252. M. Romaňuk, V. Herout, F. Šorm, Y.-R. Naves, P. Tullen, R. B. Bates, and C. W. Sigel, *Coll. Czech. Chem. Comm.*, *29*, 1048 (1964).

253. S. A. Achmad and G. W. K. Cavill, *Proc. Chem. Soc.*, 166 (1963).

254. S. A. Achmad and G. W. K. Cavill, *Austral. J. Chem.*, *18*, 1989 (1965).

255. G. W. K. Cavill and C. D. Hall, *Tetrahedron*, *23*, 1119 (1967), and references quoted therein.

256. W. Reusch and P. Mattison, *Tetrahedron*, *24*, 4933 (1968).

257. J. Wolinsky, M. R. Slabaugh, and T. Gibson, *J. Org. Chem.*, *29*, 3740 (1964).

258. J. Wolinsky, T. Gibson, D. Chan, and H. Wolf, *Tetrahedron*, *21*, 1247 (1965).

259. R. Robinson, R. H. Jaeger, and K. J. Clark, U.S. Patent No. 3,010,997 (June 23, 1958, to Shell Chemicals).

260. G. W. K. Cavill and F. B. Whitfield, *Austral. J. Chem.*, *17*, 1245 (1964).

261. F. Korte, K. H. Büchel, and A. Zschocke, *Chem. Ber.*, *94*, 1952 (1961).

262. K. Sisido, K. Utimoto, and T. Isida, *J. Org. Chem.*, *29*, 3361 (1964).

263. K. Sisido, K. Inomata, T. Kageyama, and K. Utimoto, *J. Org. Chem.*, *33*, 3149 (1968).

264. A. R. Battersby, R. T. Brown, R. S. Kapil, J. A. Martin, and A. O. Plunkett, *Chem. Comm.*, 890 (1966).

265. P. Loew and D. Arigoni, *Chem. Comm.*, 137 (1968).

266. A. R. Battersby, R. S. Kapil, J. A. Martin, and L. Mo, *Chem. Comm.*, 133 (1968).

267. A. R. Battersby and B. Gregory, *Chem. Comm.*, 134 (1968).

268. P. L. Lentz, Jr., and M. G. Rossmann, *Chem. Comm.*, 1269 (1969).

269. B. D. Challand, H. Hikino, G. Kornis, G. Lange, and P. de Mayo, *J. Org. Chem.*, *34*, 794 (1969).

270. R. U. Lemieux and C. Brice, *Can. J. Chem.*, *33*, 1701 (1967).

271. D. H. R. Barton, H. P. Faro, E. P. Serebryakov, and N. F. Woolsey, *J. Chem. Soc.*, 2438 (1965).

272. H. Schmidt, *Chem. Ber.*, *80*, 528, 533 (1947).

273. J. Wolinsky and W. Barker, *J. Org. Chem.*, *82*, 636 (1960).

274. A. F. Thomas, *Helv. Chim. Acta*, *55*, 815 (1972).

275. G. Ciamician and P. Silber, *Ber.*, *43*, 1341 (1910).

276. M. P. Hartshorn, D. N. Kirk, and A. F. A. Wallis, *J. Chem. Soc.*, 5494 (1964).

277. J. B. Lewis and G. W. Hedrick, *J. Org. Chem.*, *30*, 4271 (1965).

278. O. P. Vig, K. L. Matta, A. Lal, and I. Raj, *J. Ind. Chem. Soc.*, *41*, 142 (1964).

279. R. B. Bates, E. S. Caldwell, and H. P. Klein, *J. Org. Chem.*, *34*, 2615 (1969).

280. V. I. Shabalina, A. D. Dembitskii, and M. I. Goryaev, *Tr. Inst. Khim. Nauk, Akad. Nauk Kaz. S.S.S.R.*, *19*, 17 (1967).

281. L. S. Ivanova, A. G. Borokovskaya, and G. A. Rudakov, *Zh. Org. Khim.*, *3*, 2162 (1967).

282. B. Mitzner and E. T. Theimer, *J. Org. Chem.*, *27*, 3359 (1962).

283. I. I. Bardyshev, V. M. Ya. Shashkina, and V. I. Kulikov, *Zhur. Org. Khim.*, *2*, 1039 (1966).

284. H. Kuczyński and A. Zabża, *Rocz. Chem.*, *40*, 643 (1966).

285. D. S. Deorha and S. P. Sareen, *Rec. Trav. Chim.*, *84*, 137 (1965).

286. J. Garnero, L. Benezet, L. Peyron and Y. Chrétien-Bessière, *Bull. Soc. Chim.*, 4679 (1967).

287. A. J. Birch and G. Subba Rao, *Austral. J. Chem.*, *22*, 2037 (1969).

288. A. F. Thomas and W. Bucher, *Helv. Chim. Acta*, *53*, 771 (1970).

289. J. J. Loori and A. R. Cover, *J. Food Sci.*, *29*, 576 (1964).

290. O. Zeitschel and H. Schmidt, *Ber.*, *60*, 1327 (1927).

291. I. I. Bardyshev and R. I. Livshits, *Zhur. Prikl. Khim.*, *25*, 1289 (1952).

292. G. H. Keats, *J. Chem. Soc.*, 2003 (1937).

293. H. van Bekkum, D. Medema, P. E. Verkade, and B. M. Wepster, *Rec. Trav. Chim.*, *81*, 269 (1962).

294. M. Winter and P. Enggist (personal communication).

295. K. Stephan and J. Helle, *Ber.*, *35*, 2147 (1902).

296. H. B. Henbest and R. S. McElhinney, *J. Chem. Soc.*, 1834 (1959).

297. G. Schroeter, dissertation (Göttingen, 1962).

298. G. O. Schenck, O.-A. Neumüller, G. Ohloff, and S. Schroeter, *Annalen*, *687*, 26 (1965).

299. P. J. Kropp, *J. Org. Chem.*, *35*, 2435 (1970).

300. E. Klein and G. Ohloff, *Tetrahedron*, *19*, 1091 (1963).

301. E. E. Royals and J. C. Leffingwell, *J. Org. Chem.*, *31*, 1937 (1966).

302. J. C. Leffingwell and R. E. Shackelford, *Tet. Letters*, 2003 (1970).

303. C. H. Brieskorn and H. Brunner, *Planta Med. Suppl.*,

96 (1967).

304. O. Wallach and R. Heyer, *Annalen*, *362*, 280 (1908).

305. B. M. Mitzner, E. T. Theimer, and S. Lemberg, *Can. J. Chem.*, *41*, 2097 (1963).

306. R. L. Frank and J. B. McPherson, Jr., *J. Am. Chem. Soc.*, *71*, 1387 (1949).

307. A. Bayer, *Ber.*, *27*, 443 (1894).

308. R. M. Bowman, A. Chambers, and W. R. Jackson, *J. Chem. Soc. (C)*, 1296 (1966).

309. O. Wallach, *Annalen*, *239*, 12 (1887).

310. A. J. Birch, *J. Chem. Soc.*, 593 (1946).

311. M. D. Soffer and M. A. Jevnik, *J. Am. Chem. Soc.*, *77*, 1003 (1955).

312. G. S. K. Rao and Sukh Dev, *J. Ind. Chem. Soc.*, *33*, 539 (1956).

313. M. D. Soffer and A. C. Williston, *J. Org. Chem.*, *22*, 1254 (1957).

314. G. Mukherji, B. K. Ganguly, R. C. Banerjee, D. Mukherji, and J. C. Bardhan, *J. Chem. Soc.*, 2407 (1963).

315. G. Stork, A. Brizzolara, J. Szmuszkovicz, and R. Terrell, *J. Am. Chem. Soc.*, *85*, 207 (1963).

316. K. G. Lewis and G. J. Williams, *Tet. Letters*, 4573 (1965).

317. A. F. Thomas (unpublished work).

318. M. D. Soffer and G. E. Günay, *Tet. Letters*, 1355 (1965).

319. O. Wallach, *Annalen*, *356*, 217 (1907).

320. T. Sakai, K. Yoshihara, and Y. Hirose, *Bull. Chem. Soc. Japan*, *41*, 3348 (1968).

321. N. Shinoda, M. Shiya, and K. Nishimura, *Agr. Biol. Chem.*, *34*, 234 (1970).

322. E. Klein and W. Rojahn, *Tetrahedron*, *21*, 2173 (1965).

323. J. Leffingwell, France Patent Appl. No. 2,003,498 (March 8, 1968, to Reynolds Tobacco Co.); [*Chem. Abs.*, *72*, 100934 (1970)].

324. E. Klein, H. Farnow, and W. Rojahn, *Dragoco Rep.*, *12*, 3 (1965).

325. E. Klein and W. Rojahn, *Dragoco Rep.*, *14*, 231 (1967).

326. Y. Sakuda, *Bull. Chem. Soc. Japan*, *42*, 3348 (1969).

327. E. N. Trachtenberg and J. R. Carver, *J. Org. Chem.*, *35*, 1646 (1970).

328. B. M. Mitzner, S. Lemberg, V. Mancini, and P. Barth, *J. Org. Chem.*, *31*, 2419 (1966).

329. B. M. Mitzner and S. Lemberg, *J. Org. Chem.*, *31*, 2022 (1966).

330. G. Ohloff, *Arch. Pharm.*, *287*, 258 (1954).

331. D. Merkel, in *Die Ätherischen Ölen* (Gildermeister-Hoffmann) (Akademie-Verlag, Berlin, 1962), Vol. IIIb,

p. 70.

332. M. Bukala, B. Burczyk, S. Kucharski, and J. Rulinska, *Chem. Stosow. Ser. A*, *12*, 371 (1968); [*Chem. Abs.*, *70*, 68536 (1969)].

333. Z. Chabudiński and J. Kudik, *Roczniki Chem.*, *39*, 1833 (1965).

334. H. Boardman, *J. Am. Chem. Soc.*, *84*, 1376 (1962).

335. A. F. Thomas, *Helv. Chim. Acta*, *48*, 1057 (1965).

336. W. O. Kermack, *J. Chem. Soc.*, 2285 (1924).

337. D. Merkel, in *Die Ätherischen Ölen* (Gildermeister-Hoffmann) (Akademie-Verlag, Berlin, 1966), Vol. IIId, p. 404.

338. E. C. Taylor, H. W. Atland, R. H. Danforth, C. McGillivray, and A. McKillop, *J. Am. Chem. Soc.*, *92*, 3520 (1970).

339. E. E. Royals and S. E. Horne, Jr., *J. Am. Chem. Soc.*, *73*, 5856 (1951).

340. B. S. Tyagi, B. B. Ghatge, and S. C. Bhattacharyya, *J. Org. Chem.*, *27*, 1430 (1962).

341. S. H. Schroeter and E. L. Eliel, *J. Org. Chem.*, *30*, 1 (1965).

342. A. Manjarrez and V. Mendoza, *Perf. and Ess. Oil Rec.*, *58*, 23 (1967).

343. V. K. Honwad, E. Siscovic, and A. S. Rao, *Ind. J. Chem.*, *5*, 234 (1967).

344. W. G. Dauben, M. Lorber, and D. S. Fullerton, *J. Org. Chem.*, *34*, 3587 (1969).

345. W. Treibs and H. Bast, *Annalen*, *561*, 165 (1949).

346. E. E. Royals and L. L. Harrell, Jr., *J. Am. Chem. Soc.*, *77*, 3405 (1955).

347. F. Humbert and G. Guth, *Bull. Soc. Chim. France*, 2867 (1966).

348. I. C. Nigam and L. Levi, *Can. J. Chem.*, *46*, 1944 (1968).

349. N. P. Damodaran and Sukh Dev, *Tet. Letters*, 1941 (1963).

350. S. M. Linder and F. P. Greenspan, *J. Org. Chem.*, *22*, 949 (1957).

351. K. Wiberg and S. D. Nielsen, *J. Org. Chem.*, *29*, 3353 (1964).

352. J. P. Schaefer, B. Horvath, and H. P. Klein, *J. Org. Chem.*, *33*, 2647 (1968).

353. Y.-R. Naves and V. Grampoloff, *Bull. Soc. Chim. France*, 37 (1960).

354. J. E. Baldwin and J. C. Swallow, *Angew. Chem. Int. Ed. Engl.*, *8*, 601 (1969).

355. G. O. Schenck, O.-A. Neumüller, G. Ohloff, and S. Schroeter, *Annalen*, *687*, 26 (1965).

356. O. P. Vig, S. D. Sharma, S. Chander, and I. Raj, *Ind. J.*

Chem., *4*, 275 (1966).

357. B. M. Lawrence and J. W. Hogg, *Perf. and Ess. Oil Rec.*, *59*, 515 (1968).

358. O. P. Vig, O. P. Chugh, and K. L. Matta, *J. Ind. Chem. Soc.*, *45*, 748 (1968).

359. H. Henecka, *Chem. Ber.*, *82*, 112 (1949).

360. J. K. Roy, *Sci. and Culture (India)*, *19*, 156 (1953).

361. R. P. Lutz and J. D. Roberts, *J. Am. Chem. Soc.*, *84*, 3715 (1962).

362. A. Kergomard, J. C. Tardivat, H. Tantou, and J. P. Vuillerme, *Tetrahedron*, *26*, 2883 (1970).

363. F. Bohlmann, U. Niedballa, and J. Schulz, *Chem. Ber.*, *102*, 864 (1969).

364. F. Bohlmann and C. Zdero, *Tet. Letters*, 3375 (1970).

365. F. Bohlmann, J. Schulz, and U. Bühmann, *Tet. Letters*, 4703 (1969).

366. A. J. Birch and R. W. Richards, *Austral. J. Chem.*, *9*, 241 (1956).

367. K. J. Crowley, *J. Chem. Soc.*, 4254 (1964).

368. J. J. Beereboom, *J. Org. Chem.*, *31*, 2026 (1966).

369. C.-A. Vodoz, thesis (E.T.H., Zürich, 1949).

370. C. Balant, C.-A. Vodoz, H. Kappeler, and H. Schinz, *Helv. Chim. Acta*, *34*, 722 (1951).

371. J. C. Bardhan and S. C. Bhattacharyya, *Chem. and Ind.*, 800 (1951).

372. W. Kuhn and H. Schinz, *Helv. Chim. Acta*, *36*, 161 (1953).

373. E. Van Bruggen, *Rec. Trav. Chim.*, *87*, 1134 (1968).

374. R. B. Woodward and T. Singh, *J. Am. Chem. Soc.*, *72*, 494 (1950).

375. Y.-R. Naves and G. Papazian, *Helv. Chim. Acta*, *25*, 1023 (1942).

376. S. Ohshiro and K. Doi, *Yakugaku Zasshi*, *88*, 417 (1968).

377. H. Ueda, K. Takeo, P. Tsai, and C. Tatsumi, *Agric. Biol. Chem.*, *29*, 374 (1965).

378. B. Furth and J. Wiemann, *Bull. Soc. Chim. France*, 1819 (1965).

379. R. H. Reitsema, *J. Am. Chem. Soc.*, *78*, 5022 (1956).

380. J. Walker, *J. Chem. Soc.*, 1585 (1935).

381. S.-O. Lawesson, E. H. Larsen, and H. J. Jakobsen, *Rec. Trav. Chim.*, *83*, 464 (1964).

382. F. N. Stepanov and R. A. Myrcina, *Zh. Obshch. Khim.*, *34*, 3092 (1964).

383. A. K. Macbeth, B. Milligan, and J. S. Shannon, *J. Chem. Soc.*, 901 (1953).

384. A. F. Thomas, B. Willhalm, and J. H. Bowie, *J. Chem. Soc. (B)*, 392 (1967).

385. R. B. Woodward and R. H. Eastman, *J. Am. Chem. Soc.*, *72*,

399 (1950).

386. C. Black, G. L. Buchanan, and A. W. Jarvie, *J. Chem. Soc.*, 2971 (1956).

387. G. Ohloff, J. Osiecki, and C. Djerassi, *Chem. Ber.*, *95*, 1400 (1962).

388. J. Wolinsky, M. Senyek, and S. Cohen, *J. Org. Chem.*, *30*, 3207 (1965).

389. A. K. Macbeth and J. S. Shannon, *J. Chem. Soc.*, 4748 (1952).

390. W. Treibs, *Ber.*, *70*, 85 (1937).

391. L. H. Zalkow and J. W. Ellis, *J. Org. Chem.*, *29*, 2626 (1964).

392. K. H. Schulte-Elte, thesis (Göttingen, 1961).

393. G. O. Schenck, *Strahlentherapie*, *115*, 518 (1961).

394. H. Stetter, and R. Lauterbach, *Chem. Ber.*, *93*, 603 (1960).

395. J. Hesse, *Ber.*, *30*, 1438 (1897).

396. B. I. Nurunnabi, *Pakistan J. Sci. Ind. Res.*, *4*, 37 (1961).

397. D. Merkel, in *Die Ätherischen Ölen* (Gildemeister-Hoffmann) (Akademie-Verlag, Berlin, 1962), Vol. IIIb, p. 25-33. It should be noted that the formulas of (+)-menthone and (+)-isomenthone are interchanged in this reference.

398. W. J. Houlihan and D. R. Moore, U.S. Patent No. 3,405,185 (Oct. 4, 1961, to Universal Oil Products); [*Chem. Abs.*, *70*, 87994 (1969)].

399. D. R. Moore and A. D. Kossoy, *Analyt. Chem.*, *33*, 1437 (1961).

400. A. Bhati, *Perf. and Ess. Oil Rec.*, *54*, 448 (1963).

401. E. M. Acton, H. Stone, M. A. Leaffer, and S. M. Oliver, *Experientia*, *26*, 473 (1970).

402. S. Furukawa and Z. Tomizawa, *J. Chem. Ind. (Tokyo)*, *23*, 342 (1920).

403. A. Toshio, T. Takaaki, I. Kazutomo, and M. San'ichirô, *Science (Japan)*, *17*, 241 (1947); [*Chem. Abs.*, *45*, 1976d (1951)].

404. G. Taga and T. Naki, Japan. Patent No. 53 2979 (June 26, to Arakawa Forest Products Chem. Co.); [*Chem. Abs.*, *49*, 4023b (1955)].

405. A. Kergomard, S. Philibert-Bigou, and M. T. Geneix, French Patent No. 1,183,849 (July 15, 1959); [*Chem. Abs.*, *55*, 27404 (1961)].

406. T. R. Keenan, B.S. thesis (Massachussetts Inst. of Technology, 1966) (unpublished); quoted in G. Büchi, W. Hofheinz, and J. V. Paukstelis, *J. Am. Chem. Soc.*,

91, 6473 (1969).

407. A. Kergomard, *Bull. Soc. Chim.,* 1161 (1957).

408. W. Herz and H. J. Wahlborg, *J. Org. Chem., 27,* 1032 (1962).

409. H. Kayahara, H. Ueda, I. Ichimoto, and C. Tatsumi, *J. Org. Chem., 33,* 4536 (1968).

410. R. I. Leavitt, *J. Gen. Microbiol., 49,* 411 (1967).

411. F. Porsch, *Dragoco Report,* 59 (1964); [*Chem. Abs., 40,* 16009 (1964).

412. P. T. Varo and D. E. Heinz, *J. Agr. Food Chem., 18,* 239 (1970).

413. G. L. K. Hunter and M. G. Moshonas, *Analyt. Chem., 37,* 378 (1965).

414. A. P. Wade, G. S. Wilkinson, F. M. Dean, and A. W. Price, *Biochem. J., 101,* 727 (1966).

415. R. Tschesche, I. Duphorn, and K. Gelissen, *Z. Physiol. Chem., 345,* 100 (1966).

416. F. M. Dean, A. W. Price, A. P. Wade, and G. S. Wilkinson, *J. Chem. Soc. (C),* 1893 (1967).

417. G. Ohloff, W. Giersch, K. H. Schulte-Elte, and E. sz. Kováts, *Helv. Chim. Acta, 52,* 1531 (1969).

418. J. Albaigès, J. Castells, and J. Pasual, *J. Org. Chem., 31,* 3507 (1966).

418a. F. Camps, J. Castells and J. Pascual, *J. Org. Chem., 31,* 3510 (1966).

419. F. Camps and J. Pascual, *An. Real Soc. Espan. Fis. y Quím., 64,* 167 (1968).

419a. F. Camps, *J. Org. Chem., 33,* 2466 (1968).

420. T. Aratani, *J. Chem. Soc. Japan, 80,* 528 (1959).

421. R. Ruegg, A. Pfiffner, and M. Montavon, *Recherches,* 3 (1966).

422. Y. Ohta and Y. Hirose, *Agr. Biol. Chem., 30,* 1196 (1966).

423. Y. Kita, Y. Nakatani, A. Kobayashi, and T. Yamanishi, *Agric. Biol. Chem., 33,* 1559 (1969).

424. M. G. Moshonas and E. D. Lund, *J. Food Sci., 34,* 502 (1969).

425. H. C. Brown and G. Zweifel, *J. Am. Chem. Soc., 83,* 1241 (1961).

426. R. Dulou and Y. Chrétien-Bessière, *Bull. Soc. Chim. France,* 1362 (1959).

427. K. H. Schulte-Elte and G. Ohloff, *Helv. Chim. Acta, 50,* 153 (1967).

428. K. Ziegler, F. Krupp, and K. Zosel, *Annalen, 629,* 241 (1960).

429. E. J. Lorand and J. E. Reese, *J. Am. Chem. Soc., 72,* 4596 (1950).

430. L. Peyron, L. Benezet, D. de Dortan, J. Garnero, *Bull.*

Soc. Chim. France, 339 (1969).

431. E. sz. Kováts, *Helv. Chim. Acta*, *46*, 2705 (1963).

432. A. R. Penfold, F. R. Morrison, and H. G. McKern, *Perf. and Ess. Oil Rec.*, *40*, 149 (1949).

433. D. Merkel, in *Die Ätherischen Ölen* (Gildermeister-Hoffmann) (Akademie-Verlag, Berlin, 1962), Vol. IIIb, p. 120-128.

434. R. W. Murray and M. L. Kaplan, *J. Am. Chem. Soc.*, *91*, 5358 (1969).

435. O. Wallach, *Annalen*, *392*, 59 (1912).

436. A. F. Thomas and B. Willhalm, *Helv. Chim. Acta*, *47*, 475 (1964), and unpublished work.

437. T. Aikawa, Y. Shiihara, H. Sano, and H. Izumitani, Japan. Patent No. 68 24,186 (Sept. 11, 1965); [*Chem. Abs.*, *70*, 58058 (1969)].

438. H. C. Brown and P. Geohegan, *J. Am. Chem. Soc.*, *89*, 1522 (1967).

439. J. M. Coxon, M. P. Hartshorn, J. W. Mitchell, and K. E. Richards, *Chem. and Ind.*, 652 (1968).

440. R. K. Mathur, S. K. Ramaswamy, A. S. Rao, and S. C. Bhattacharyya, *Tetrahedron*, *23*, 2495 (1967).

441. G. G. Henderson and A. Robertson, *J. Chem. Soc.*, 1849 (1923).

442. D. Brown, B. T. Davis, T. G. Halsall, and A. R. Hands, *J. Chem. Soc.*, 4492 (1962).

443. B. Shasha and Y. Leibowitz, *Nature*, *184*, 2019 (1959).

444. R. Mechoulam, N. Danieli, and Y. Mazur, *Tet. Letters*, 709 (1962).

445. T. Hashizume and I. Sakata, *Tet. Letters*, 3355 (1967).

446. A. Blumann, E. W. Della, C. A. Henrick, J. Hodgkin, and P. R. Jefferies, *Austral. J. Chem.*, *15*, 290 (1962).

447. K. Ishibashi, J. Katsuhara, K. Hashimoto, and M. Kobayashi, *Kogo Kagaku Zasshi*, *70*, 1195 (1967); [*Chem. Abs.*, *67*, 111321 (1967).

448. J. Katsuhara, K. Ishibashi, K. Hashimoto, and M. Kobayashi, *Kogyo Kagaku Zasshi*, *69*, 1170 (1966); [*Chem. Abs.*, *65*, 17006 (1966)].

449. R. H. Reitsema and V. J. Varnis, *J. Am. Chem. Soc.*, *78*, 3792 (1956).

450. Y.-R. Naves [*Helv. Chim. Acta*, *49*, 2012 (1966)] gives relevant literature.

451. J. A. Retamar, V. R. Medel, O. A. Arpesella, A. Orlando, and D. A. De Iglesias, *Arch. Bioquim. Quím. Farm.*, *14*, 139 (1968).

452. J. Klein and E. Dunkelblum, *Tetrahedron*, *24*, 5701 (1968). The diol structures are written identically in this paper; they should clearly have epimeric methyl groups

at C-1.

453.   I. I. Bardyshev, I. V. Gorbacheva, and A. L. Pertsovskii,
       Vestsi Akad. Navuk Belarus. S.S.R., Ser. Khim. Navuk,
       102 (1969); [*Chem. Abs., 72,* 3575 (1970).

454.   I. I. Bardyshev, I. V. Gorbacheva, and A. L. Pertsovskii,
       Vestsi Akad. Navuk Belarus. S.S.R., Ser. Khim. Navuk,
       47 (1969); [*Chem. Abs., 71,* 3487 (1969)].

455.   R. Kuhn and G. Wendt, *Chem. Ber., 69,* 1549 (1936).

456.   M. Mousseron-Canet, J.-C. Mani, and J.-L. Olivé, *Compt.
       Rend. Acad. Sci., 262,* 1725 (1966).

457.   J. Romo, A. Romo de Vivar, L. Quijano, T. Rios, and E.
       Diaz, *Revista Latinoamericana de Quím., 1,* 73 (1970).

458.   Y. Naya and M. Kotake, *Tet. Letters,* 1645 (1968).

459.   R. Breslow, J. T. Groves, and S. S. Olin, *Tet. Letters,*
       4717 (1966).

460.   R. M. Coates and L. S. Melvin, Jr., *J. Org. Chem., 35,*
       865 (1970).

461.   Y.-R. Naves, *Bull. Soc. Chim. France,* 1871 (1959).

462.   F. Bohlmann and C. Zdero, *Tet. Letters,* 2419 (1969).

463.   A. F. Thomas, B. Willhalm, and G. Ohloff, *Helv. Chim.
       Acta, 52,* 1249 (1969).

464.   J. Gripenberg, *Acta Chem. Scand., 10,* 487 (1956).

465.   R. E. Davis and A. Tulinsky, *Tet. Letters,* 839 (1962).

466.   T.-B. Lo and Y.-T. Lin, *J. Chinese Chem. Soc. (Formosa),
       Ser. II, 3,* 30 (1956).

467.   N. Ichikawa, *Bull. Soc. Chem. Japan, 12,* 267 (1937).

468.   H. Erdtman, in *Progress in Organic Chemistry,* edited by
       J. W. Cook (Academic, New York, 1952), Vol. 1, p. 51.

469.   T. Kaku and T. Kondo, *J. Pharm. Soc. Japan, 51,* 3, 112
       (1931).

470.   A. von Bayer, *Ber., 31,* 2067 (1898).

471.   E. J. Corey and H. J. Burke, *J. Am. Chem. Soc., 78,* 174
       (1956).

472.   R. A. Barnes and W. J. Houlihan, *J. Org. Chem., 26,* 1609
       (1961).

473.   T. Suga, K. Mori, and T. Matusuura, *J. Org. Chem., 30,*
       669 (1965).

474.   E. Demole and P. Enggist, *Chem. Comm.,* 264 (1969).

475.   E. Demole and P. Enggist, *Helv. Chim. Acta, 54,* 456
       (1971).

476.   W. Reusch, D. F. Anderson, and C. K. Johnson, *J. Am.
       Chem. Soc., 90,* 4988 (1968).

477.   Y. Hirose, B. Tomita, and N. Nakatsuka, *Tet. Letters,*
       5875 (1966).

478.   A. J. Birch and R. Keeton, *J. Chem. Soc. (C),* 109 (1968).

479.   A. P. ter Borg, R. van Helden, and A. F. Bickel, *Rec.
       Trav. Chim., 81,* 591 (1962).

480. R. B. Bates, M. J. Onore, S. K. Paknikar, C. Steelink, and E. P. Blanchard, *Chem. Comm.*, 1037 (1967).

481. J. J. Beereboom, *J. Am. Chem. Soc.*, *85*, 3525 (1963).

482. J. J. Beereboom, *J. Org. Chem.*, *30*, 4230 (1965).

483. S. McLean and P. Haynes, *Tetrahedron*, *21*, 2313 (1965).

484. V. A. Mironov, E. V. Sobolev, and A. N. Elizarova, *Tetrahedron*, *19*, 1939 (1963).

485. U. A. Huber and A. S. Dreiding, *Helv. Chim. Acta, 53*, 495 (1970).

485a. W. F. Erman, *J. Am. Chem. Soc.*, *91*, 779 (1969).

486. G. Ohloff, G. Uhde, A. F. Thomas, and E. sz. Kováts, *Tetrahedron*, *22*, 309 (1966).

487. J. Wolinsky, H. Wolf, and T. Gibson, *J. Org. Chem.*, *28*, 274 (1963).

488. P. C. Guha and B. Nath, *Ber.*, *70*, 931 (1937).

489. W. G. Dauben, A. C. Albrecht, E. Hoerger, and H. Takimoto, *J. Org. Chem.*, *23*, 457 (1958).

490. A. K. Bose and M. S. Tibbetts, *Tetrahedron*, *23*, 457 (1958).

491. W. I. Fanta and W. F. Erman, *J. Org. Chem.*, *33*, 1656 (1968).

492. O. P. Vig, M. S. Bhatia, K. C. Gupta, and K. L. Matta, *J. Ind. Chem. Soc.*, *46*, 991 (1969).

493. K. Mori, M. Ohki, and M. Matsui, *Tetrahedron*, *26*, 2821 (1970).

494. J. W. Daly, F. C. Green, and E. H. Eastman, *J. Am. Chem. Soc.*, *80*, 6330 (1958).

495. G. F. Russell and W. G. Jennings, *J. Agr. Food Chem.*, *18*, 733 (1970).

496. V. I. Shabalina, A. D. Dembitskii, and M. I. Goryaev, *Zhur. Obshch. Khim.*, *34*, 3855 (1964).

497. S. P. Acharya, H. C. Brown, A. Suzuki, S. Nozawa, and M. Itoh, *J. Org. Chem.*, *34*, 3855 (1969).

498. E. Klein and W. Rojahn, *Chem. Ber.*, *98*, 3045 (1965).

499. J. W. Wheeler, R. H. Chung, Y. N. Vaishov, and C. C. Shroff, *J. Org. Chem.*, *34*, 545 (1969).

500. J. E. Baldwin and H. C. Karuss, Jr., *J. Org. Chem.*, *35*, 2426 (1970).

501. H. Seyler, *Ber.*, *35*, 550 (1902).

502. R. Salgues, *Compt. Rend. Acad. Sci.*, *243*, 177 (1956).

503. C. H. Brieskorn and S. Dalferth, *Annalen*, *676*, 171 (1964).

504. R. H. Eastman, J. E. Starr, R. S. Martin, and M. K. Sakata, *J. Org. Chem.*, *28*, 2162 (1963).

505. A. J. Birch, *J. Chem. Soc.*, 811 (1945).

506. N. S. Bhacca and D. H. Williams, *Tet. Letters*, 3127 (1964).

507. G. Komppa, *Ber.*, *36*, 4332 (1903).
508. G. Komppa, *Annalen*, *370*, 209 (1909).
509. K. Aghoramurthy and P. M. Lewis, *Tet. Letters*, 1415 (1968).
510. K. Alder and E. Windmuth, *Annalen*, *543*, 41 (1939).
511. W. R. Vaughan and R. Perry, Jr., *J. Am. Chem. Soc.*, *75*, 3168 (1953).
512. L. Friedman and A. P. Wolf, *J. Am. Chem. Soc.*, *80*, 2424 (1958).
513. J. D. Roberts and J. A. Yancey, *J. Am. Chem. Soc.*, *75*, 3165 (1953).
514. W. R. Vaughan and R. Perry, Jr., *J. Am. Chem. Soc.*, *74*, 5355 (1952).
515. A. F. Thomas and B. Willhalm, *Helv. Chim. Acta*, *50*, 826 (1967).
516. J. C. Fairlie, G. L. Hodgson, and T. Money, *Chem. Comm.*, 1196 (1969).
517. D. Merkel, in *Die Ätherische Öle* (Gildemeister-Hoffmann) (Akademie-Verlag, Berlin, 1963), Vol. IIIc, p. 310.
518. V. E. Tishchenko and G. A. Rudakov, *Zhur. Prikl. Khim.*, *6*, 691 (1933).
519. S. Ya. Korotov, V. A. Vyrodov, E. A. Afanas'eva, A. I. Kolesov, Z. L. Maslakova, T. D. Oblivantseva, P. K. Chirkov, P. I. Zhuravlev, and O. I. Minaeva, U.S.S.R. Patent No. 238,541 (Appl. Oct. 20, 1962); [*Chem. Abs.*, *71*, 39216 (1969)].
520. B. J. Kane and R. M. Albert, U.S. Patent No. 3,383,422 (Appl. Jan. 27, 1965); [*Chem. Abs.*, *69*, 44063 (1968)].
521. L. Ruzicka, *Ber.*, *50*, 1362 (1917).
522. G. Komppa and A. Klami, *Ber.*, *68*, 2001 (1935).
523. P. Hirsjärvi, *Ann. Acad. Sci. Fennicae, Ser. A II*, 81 (1957).
524. A. Coulombeau and A. Rassat, *Bull. Soc. Chim. France*, 3338 (1965).
525. G. Valkanas and N. Iconomou, *Helv. Chim. Acta*, *46*, 1089 (1963).
526. B. Burczyk and M. Bukala, *Chem. Stosowana*, *7*, 245 (1963).
527. R. Mayer, K. Bochow, and W. Zieger, *Z. Chem.*, *9*, 348 (1964).
528. K. J. Crowley, *Proc. Chem. Soc.*, 334 (1962).
529. K. J. Crowley, *Proc. Chem. Soc.*, 245 (1962).
530. R. H. S. Liu and G. S. Hammond, *J. Am. Chem. Soc.*, *89*, 4936 (1967).
531. B. N. Joshi, R. Seshadri, K. K. Chakravarti, and S. C. Bhattacharyya, *Tetrahedron*, *20*, 2911 (1964).
532. P. A. Spanninger and J. L. von Rosenburg, *J. Org. Chem.*, *34*, 3658 (1969).

533. Y.-R. Naves, *Russ. Chem. Rev.*, *37*, 779 (1968).
534. H. C. Brown and M. V. Bhatt, *J. Am. Chem. Soc.*, *82*, 2074 (1960).
534a. G. H. Whitham, *J. Chem. Soc.*, 2232 (1961).
535. S. S. Poddubnaya, V. G. Cherkaev, and T. A. Rudol'fi, *Akad. Nauk. Belorussk. S.S.R. Tsentr. Nauchn.-Tekhn. Soveshch Gorki*, 209 (1963); [*Chem. Abs.*, *62*, 11855 (1965)].
536. M. A. Cooper, J. R. Salmon, D. Whittaker, and U. Scheidegger, *J. Chem. Soc. (B)*, 1259 (1967).
537. P. Teisseire, *Recherches*, *17*, 37 (1969).
538. J. J. Hurst and G. H. Whitham, *J. Am. Chem. Soc.*, *82*, 2864 (1960).
539. W. F. Erman, *J. Am. Chem. Soc.*, *89*, 3828 (1967).
540. I. C. Nigam and L. Levi, *Can. J. Chem.*, *46*, 1944 (1968).
541. B. A. Arbuzov, Z. G. Isaeva, and I. S. Andreeva, *Izvest. Akad. Nauk S.S.S.R.*, *Ser. Khim.*, 838 (1965).
542. F. J. Chloupek and G. Zweifel, *J. Am. Chem. Soc.*, *29*, 2092 (1964).
543. G. Zweifel and C. C. Whitney, *J. Org. Chem.*, *31*, 4178 (1966).
544. H. Heikman, P. Baeckström, and K. Torssell, *Acta Chem. Scand.*, *22*, 2034 (1968).
545. Y. Chrétien-Bessière, L. Peyron, L. Benezet, and J. Garnero, *Bull. Soc. Chim. France*, 381 (1970).
546. G. O. Schenck, H. Eggert, and W. Denk, *Annalen*, *584*, 177 (1953).
547. G. Helms, thesis (Göttingen, 1961) (unpublished).
548. G. Ohloff, K. H. Schulte-Elte, and W. Giersch, *Helv. Chim. Acta*, *48*, 1665 (1965).
549. W. Cocker, P. V. R. Shannon, and P. A. Staniland, *J. Chem. Soc. (C)*, 41 (1966).
550. N. Kishner and A. Zavadovski, *J. Russ. Phys. Chem. Soc.*, *43*, 1132 (1911); *43*, 1554 (1911).
551. B. Ramamoorthy and G. S. K. Rao, *Tet. Letters*, 5147 (1967).
552. Y.-R. Naves, *Helv. Chim. Acta*, *25*, 732 (1942).
553. A. Bayer, *Ber.*, *27*, 1919 (1894).
554. F. Medina and A. Manjarrez, *Tetrahedron*, *20*, 1807 (1964).
555. Y. Asahina and Y. Murayama, *Arch. Pharm.*, *252*, 435 (1914).
556. Y. Fujita and T. Ueda, *Chem. and Ind.*, 236 (1960).
557. Y.-R. Naves and P. Ochsner, *Helv. Chim. Acta*, *50*, 406 (1960).
558. G. Büchi, E. sz. Kováts, P. Enggist, and G. Uhde, *J. Org. Chem.*, *33*, 1227 (1968).

559. V. N. Vashist and C. K. Atal, *Experientia, 26,* 817 (1970).

560. H. Kondo and H. Suzuki, *Ber., 69,* 2459 (1936).

561. R. Goto, *J. Pharm. Soc. Japan, 57,* 17 (1937).

562. T. Ueda and Y. Fujita, *Chem. and Ind.,* 1618 (1962).

563. H. Ito, *J. Pharm. Soc. Japan, 84,* 1123 (1964).

564. T. Kubota and K. Naya, *Chem. and Ind.,* 1618 (1962).

565. T. Reichstein, H. Zschokke, and A. Goerg, *Helv. Chim. Acta, 14,* 1277 (1931).

566. E. Sherman and E. D. Amstutz, *J. Am. Chem. Soc., 72,* 2195 (1950).

567. L. Mavoungou-Gomès, *Bull. Soc. Chim. France,* 1764 (1967).

568. M. J. Cook and E. J. Forbes, *Tetrahedron 24,* 4501 (1968).

569. T. Reichstein, A. Grüssner, K. Schindler, and E. Hardmeier, *Helv. Chim. Acta, 16,* 276 (1933).

570. A. F. Thomas and M. Ozainne, *J. Chem. Soc. (C),* 220 (1970).

571. T. Matsuura, *Bull. Chem. Soc. Japan, 30,* 430 (1957).

572. R. A. Massy-Westropp and G. D. Reynolds, *Austral. J. Chem., 19,* 891 (1966).

573. T. Kubota, *Tetrahedron, 4,* 68 (1958).

574. D. Felix, A. Melera, J. Seibl, and E. sz. Kováts, *Helv. Chim. Acta, 46,* 1513 (1963).

575. E. Klein, H. Farnow, and W. Rojahn, *Tet. Letters,* 1109 (1963).

576. G. V. Nair and G. D. Pandit, Brit. Patent No. 1,108,208.

577. P. S. Wharton and D. H. Bohlen, *J. Org. Chem., 26,* 3615 (1961).

578. G. V. Nair and G. D. Pandit, Brit. Patent No. 1,122,593 (Appl. May 24, 1965).

579. Y.-R. Naves, *Helv. Chim. Acta, 28,* 1231 (1945).

580. B. Willhalm, A. F. Thomas, and M. Stoll, *Acta Chem. Scand., 18,* 1573 (1964).

581. G. Ohloff (personal communication).

582. H. Strickler and E. sz. Kováts, *Helv. Chim. Acta, 49,* 2055 (1966).

583. Y. Naya and M. Kotaka, *Tet. Letters,* 1715 (1967).

584. W. E. Parham and H. E. Holmquist, *J. Am. Chem. Soc., 73,* 913 (1951).

585. Y.-R. Naves, P. Ochsner, A. F. Thomas, and D. Lamparsky, *Bull. Soc. Chim. France,* 1608 (1963).

586. Y.-R. Naves and P. Ochsner, *Helv. Chim. Acta, 45,* 397 (1962).

587. C. F. Seidel, D. Felix, A. Eschenmoser, K. Biemann, E. Palluy, and M. Stoll, *Helv. Chim. Acta, 44,* 598 (1961).

588. Y.-R. Naves, D. Lamparsky, and P. Ochsner, *Bull. Soc. Chim. France,* 645 (1961).

589. G. Ohloff, E. Klein, and G. O. Schenck, *Angew. Chem.*, *73*, 578 (1961).

590. G. Ohloff, in *Fortschritte der chemischen Forschung* (Springer-Verlag, Berlin, 1969), Vol. 12, p. 185, lists a number of further references relevant to this reaction.

591. E. H. Eschinazi, *J. Org. Chem.*, *35*, 1097 (1970).

592. E. Demole and P. Enggist, *Helv. Chim. Acta*, *51*, 481 (1968).

593. S. Isoe, S. B. Hyeon, and T. Sakan, *Tet. Letters*, 279 (1969).

594. T. Sakan, S. Isoe, and S. B. Hyeon, *Tet. Letters*, 1623 (1967).

595. E. Demole, P. Enggist, and M. Stoll, *Helv. Chim. Acta*, *52*, 24 (1969).

596. R. Hodges and A. L. Porte, *Tetrahedron*, *20*, 1463 (1964).

TOTAL SYNTHESIS OF SESQUITERPENES

Clayton H. Heathcock

Department of Chemistry
University of California
Berkeley, California

1.	Introduction	199
2.	Acyclic Sesquiterpenes	200
	A. Farnesol, Nerolidol	200
	B. Juvenile Hormone	207
	C. Sinensals	222
	D. Torreyal, Neotorreyol, Dendrolasin, Ipomeamarone (Ngaione)	227
3.	Monocarbocyclic Sesquiterpenes	233
	A. Sesquiterpenes Related to Bisabolene	233
	B. β-Bisabolene, γ-Bisabolene, Lanceol, Isobisabolene, Bisabolol	234
	C. Curcumenes, Zingiberene	241
	D. ar-Turmerone	247
	E. Nuciferal	250
	F. Cryptomerion	251
	G. Todomatuic acid, Juvabione	253
	H. Perezone	262
	I. Bilobanone	263
	J. Furoventalene	264
	K. Elemanes	265
	L. Elemane (Tetrahydroelemene)	266
	M. Tetrahydrosaussaurea Lactone, Saussaurea Lactone	267
	N. Tetrahydroelemol, β-Elemene, Elemol	269
	O. Furopelargone-A, and Furopelargone-B	274
	P. Nootkatin, Procerin	276
	Q. Germacrane, Dihydrocostunolide	277
	R. Humulene	280

4. Bicarbocyclic Sesquiterpenes, Hydronaphthalenes   282
   A. Eudesmanes   282
   B. α-Cyperone, β-Cyperone   285
   C. α-, β-, and γ-Eudesmol   289
   D. α- and β-Selinene   296
   E. Costol, Costal, Costic Acid   299
   F. α- and β-Agarofuran, Nor-ketoagarofuran   300
   G. Epi-γ-Selinene   304
   H. Chamaecynone, 4α-Hydroxyisochamaecynone   305
   I. Occidol, Occidentalol   308
   J. Atractylon and Lindesterene   310
   K. Santonin   315
   L. Artemisin   324
   M. Alantolactone, Isoalantolactone, Telekin   326
   N. Cadinanes; Calamenene, ε-Cadinene,
      Vetacadinol, Veticadinene   330
   O. Drimanes; Drimenol (Bicyclofarnesol),
      Drimenin, Farnesiferol A (Biogenetic Routes)   338
   P. Drimanes; Drimenin, Isodrimenin,
      Conterifolin, Drimenol, Isoiresin, Winterin   345
   Q. Valeranone   353
   R. Eremophilanes; Isonootkatone (α-Vetivone),
      Nootkatone, Valerianol, Valencene, Eremoligenol,
      Eremophilene, Fukinone   361
   S. Tricyclic Sesquiterpenes Having a Decalin
      Nucleus with an Additional Cyclopropane Ring   380
5. Other Bicarbocyclic Sesquiterpenes   395
   A. Guaiazulenes; Bulnesol, α-Bulnesene, and
      Kessane   395
   B. Guaianolides; Arborescin, Geigerin,
      Desacetoxymatricarin, and Achillin   412
   C. Tricyclic Hydroazulenes Containing
      a Cyclopropane Ring; Aromadendrene,
      Cyclocolorenone   417
   D. Cyperolone   422
   E. β-Himachalene, Widdrol   424
   F. Sesquicarene, Sirenin   428
   G. α- and β-cis-Bergamotene   446
   H. Chamigrene   449
   I. Cuparene, β-Cuparenone, Aplysin,
      Debromoaplysin   453
   J. Carabrone   459
   K. Helminthosporal   463
   L. β-Vetivone, Hinesol, Epihinesol
      (Agarospirol)   466
   M. Caryophyllene, Isocaryophyllene   474
   N. α-Santalol, β-Santalol, α-Santalene, β-Santalene   481

6. Tricarbocyclic Sesquiterpenes   492
    A. Patchouli Alcohol, α-Patchoulene,
        β-Patchoulene   492
    B. Seychellene   499
    C. Cedrol, Cedrene   504
    D. Epizizanoic Acid   514
    E. Longifolene   517
    F. Copaene, Ylangene   520
    G. Sativene, Cyclosativene   525
    H. Culmorin   528
    I. α- and β-Bourbonene   530
    J. Illudin M   534
    K. Tricyclic Rearrangement Products   540

1. INTRODUCTION

The group of naturally occurring substances containing 15 carbon atoms and derivable biogenetically from mevalonic acid via
farnesyl pyrophosphate[1] (sesquiterpenes), offers a truly remarkable variety of structural goals to the synthetic chemist.
Within this rather restricted class of organic compounds, one
finds substances with structures ranging from the commonplace
to the exotic.[2] Acyclic, monocyclic, bicyclic, tricyclic, and
even tetracyclic compounds are represented. Various sesquiterpenes are known which contain three-, four-, five-, six-,
seven-, nine-, ten-, and eleven-membered carbon cycles. A
high degree of stereochemical subtlety is encountered, as members of the class are known which possess as many as eight
asymmetric carbon atoms. Because of this wide diversity of
structural types, both skeletal and stereochemical, the sesquiterpene field is an excellent arena for the testing and re-
fining of new synthetic methods and concepts.
    In this chapter, we document the activity in this area
through the middle of 1970. We have attempted complete coverage of all sesquiterpene total syntheses. Formal total syntheses are included when the precursor natural material had
been previously prepared by total synthesis (in optically
active form, if relevant) and when fairly extensive structure
modification is involved. Thus, the mere interrelation of
sesquiterpenes by minor chemical changes will not be covered,
even if the starting material happens to have been prepared
previously by total synthesis. Likewise, the total synthesis
of a given sesquiterpene does not warrant the additional inclusion of all prior chemical transformations of that substance which lead to other sesquiterpenes.
    We have not attempted to be exhaustive in our coverage of

peripheral studies, such as the synthesis of model compounds, hydrogenation products, and degradation products. These topics have been introduced mainly when they offer special insight into the problems encountered with specific structural types, or when complete solutions to various synthetic problems have not yet been achieved. We have included several total syntheses of materials which are not actually natural products, mainly tricyclic materials resulting from acid-catalyzed rearrangements of sesquiterpenes (α-caryophyllene alcohol, isolongifolene, clovene). This decision is less arbitrary than it appears, since there is growing realization that many compounds long regarded as of natural origin are in fact artifacts of the isolation process (*inter alia*, elemol, elemene).[3] In any event, the main thrust of the chapter justifies the inclusion of such topics, since they typify the approaches which have classically been taken toward sesquiterpene synthesis.

2.   ACYCLIC SESQUITERPENES

A.   Farnesol, Nerolidol

The sesquiterpene alcohols farnesol and nerolidol occur in nature mainly as the *trans,trans* and *trans*-isomerides (1 and 2, respectively), although *cis,trans*-farnesol (3) is reported to occur naturally.[4] Ruzicka's synthesis of farnesol and nerolidol reported in 1923,[5] was the first sesquiterpene total

| 1 | 2 | 3 |

synthesis. Scheme 1 outlines Ruzicka's synthesis. The problem

Scheme 1. Ruzicka's Synthesis of Farnesol and Nerolidol

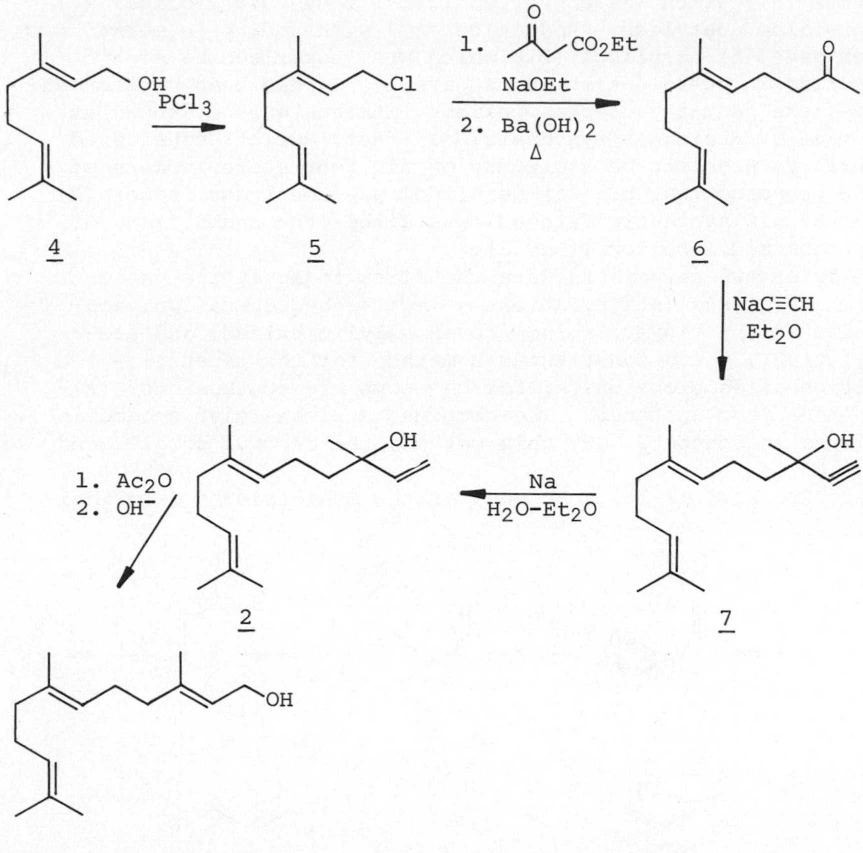

of establishing a uniform stereochemistry at the 6,7-double bond, which was then unknown, was circumvented by starting with natural geraniol 4, which had been prepared by total synthesis via linalool (8)[6] or citral (9),[7] although probably

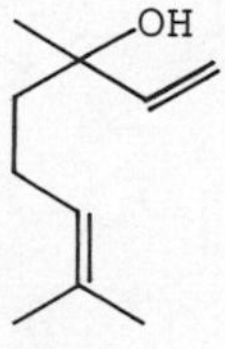

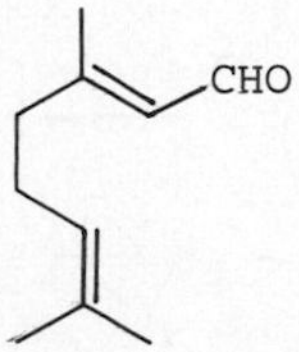

not in a stereochemically homogeneous state. Standard conversion of geraniol (4) via geranyl chloride (5) afforded geranyl
acetone (6), which was converted into 1,2-dehydronerolidol (7)
with sodium acetylide. Reduction of 7 with sodium in moist
ether gave (±)-nerolidol (2), which was isomerized by acetic
anhydride in petroleum ether to give 1. As has been mentioned
above, the geometry of the olefinic linkages was not known at
the time of Ruzicka's synthesis. In fact, Ruzicka considered
natural farnesol to be a mixture of all four stereoisomers of
1 and proposed that his (±)-nerolidol was the *trans* isomer (2)
and that his synthetic farnesol was either the *trans,trans* or
cis,trans modification (1 or 3).

Isler and co-workers were the first to apply the basic
Ruzicka sequence (ethynylation, selective hydrogenation, conversion of the vinyl carbinol to an allylic halide, and acetonylation), which constitutes a method for the repetitive
addition of isoprene units, for the complete synthesis of (±)-
nerolidol from acetone.[8] The combined Ruzicka-Isler scheme is
outlined in Scheme 2. By this method, the central double bond

Scheme 2. Ruzicka-Isler Scheme for the Synthesis of Nerolidol

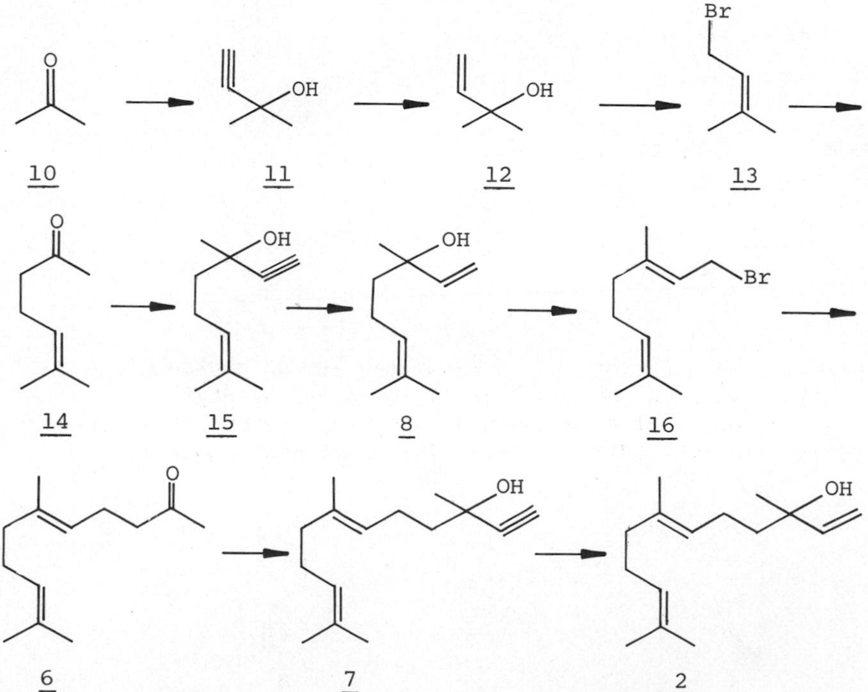

was clearly not introduced stereospecifically, and dienone 6
was separated from its isomer via its semicarbazone derivative.
It was later reported that this method of synthesis gives a
mixture of *trans* and *cis* nerolidols containing 65% of the *trans*
isomer.[9]

Nazarov, Gussev and Gunar reported a thorough study of
reaction conditions in the Ruzicka-Isler scheme and introduced
several modifications to facilitate the large-scale application of the sequence.[10]   The Nazarov synthesis of (±)-nerolidol
and farnesol is given in Scheme 3.   The major modification

Scheme 3.   Nazarov Modification of the Ruzicka-Isler Synthesis

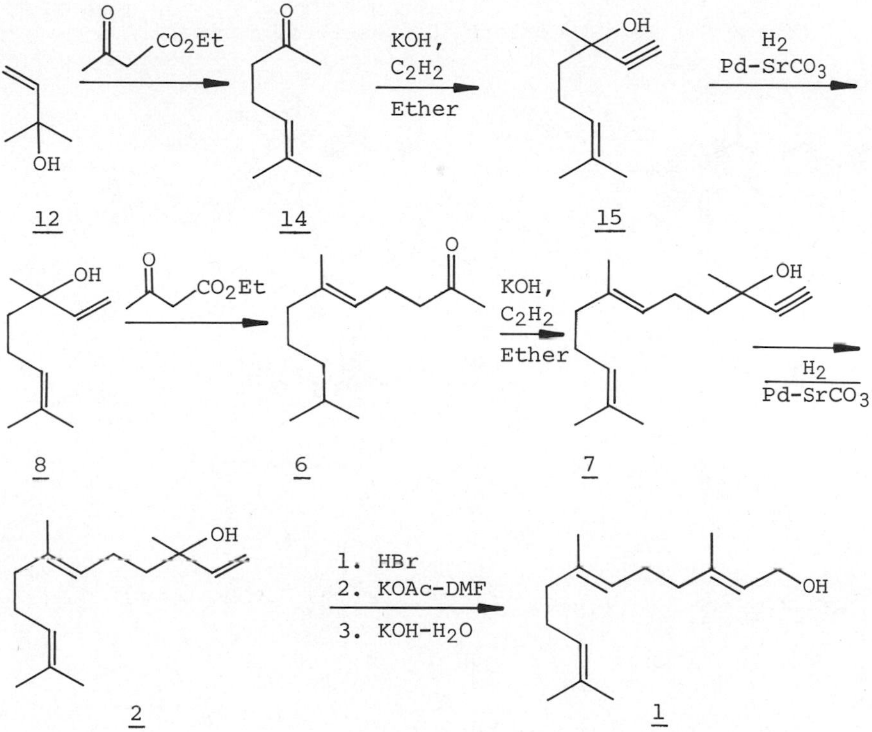

introduced was the direct acetonylation of allylic alcohols
12 and 8, which was accomplished by heating a mixture of the
appropriate alcohol and acetoacetic ester at approximately
200° for several hours.   Alternatively, the transformation of
12 to 14 or 8 to 6 was carried out by reaction of the allylic
alcohol with gaseous hydrogen bromide, followed by condensation of the resulting halide with sodioacetoacetic ester and

saponification, or by reaction of the allylic alcohol with diketene,[11] followed by pyrolysis of the resulting aceto-acetic ester at 170-200°.[10] The authors report complete stereochemical homogeneity in their synthetic products, re-gardless of the method of synthesis, but no definitive evidence is presented to reinforce this assertion.

The basic Ruzicka method of converting geranyl chloride into (±)-nerolidol and farnesol has been reported subsequently by other workers,[12,13] although no further details on stereo-specificity were given.

Radio-labeled farnesol has been synthesized from *trans*-geranyl acetone (6) by Popjak, Cornforth, and co-workers, by the method given in Scheme 4.[14] The *cis,trans* and *trans,trans*

Scheme 4. Popjak-Conforth Synthesis of Farnesol

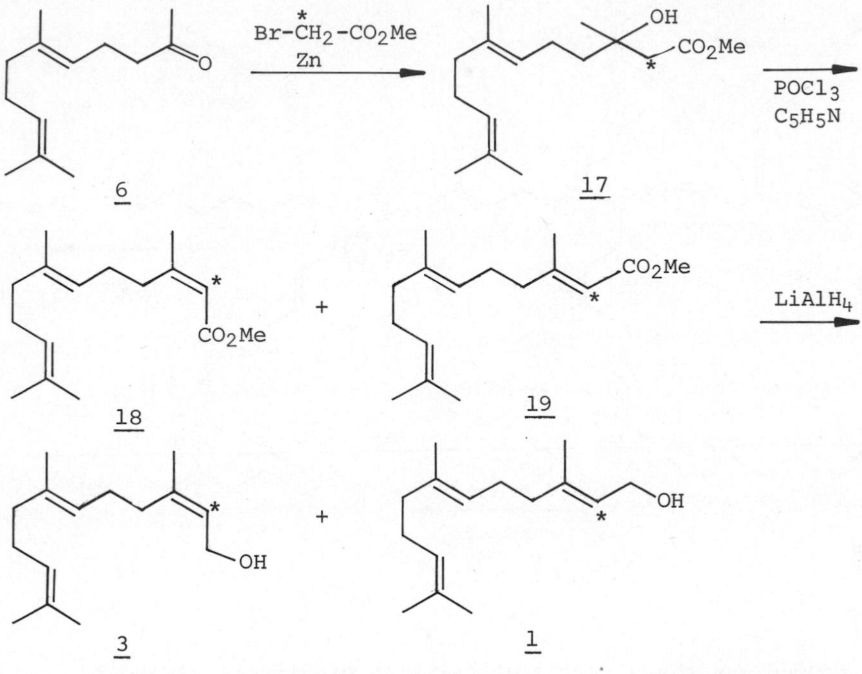

methyl farnesates (18 and 19) were obtained in a ratio of 1:1 and were separated, prior to their reduction, by preparative vapor phase chromatography (vpc).

Julia, Julia, and Guegan introduced an elegant method for the construction of polyisoprenoid chains which utilizes methylcyclopropyl ketone (20) as the basic building unit.[15]

The application of the Julia scheme for the synthesis of (±)-nerolidol is outlined in Scheme 5. Cyclopropyldimethylcarbinol

Scheme 5.   Julia Synthesis of Nerolidol

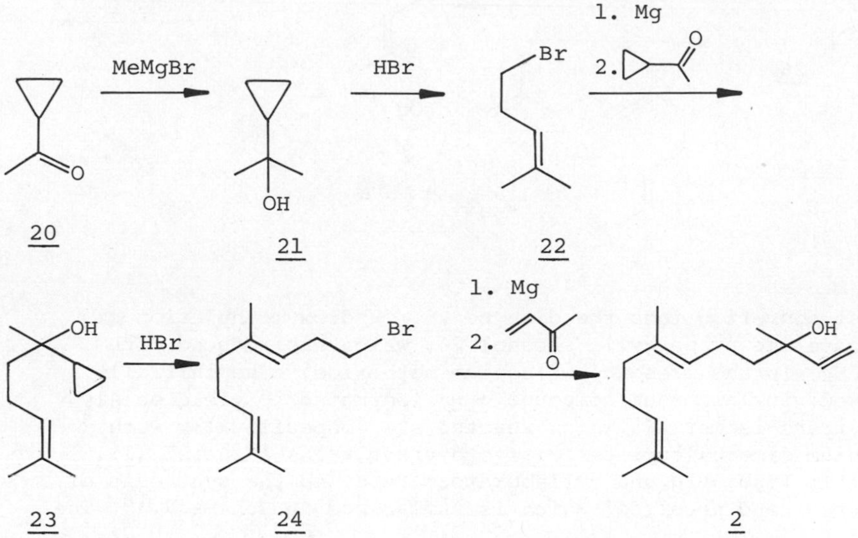

(21) reacts with 48% HBr to yield the homoallylic bromide 22, the Grignard reagent of which reacts with methyl cyclopropyl ketone to yield alcohol 23. Opening of 23 in the same manner gives the *trans* olefin 24, along with approximately 25% of its *cis* isomer.[16]

A clearly stereospecific synthesis of farnesol was provided in 1967 by Corey and co-workers.[17] The Corey synthesis (Scheme 6) begins with *trans*-geranyl acetone (6), which was

Scheme 6.   Corey Synthesis of *trans,trans*-Farnesol

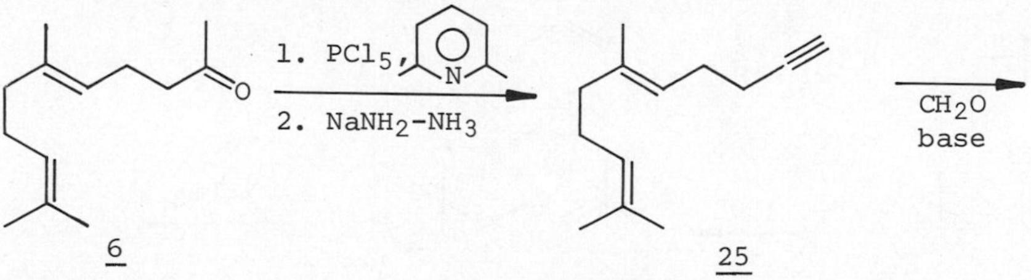

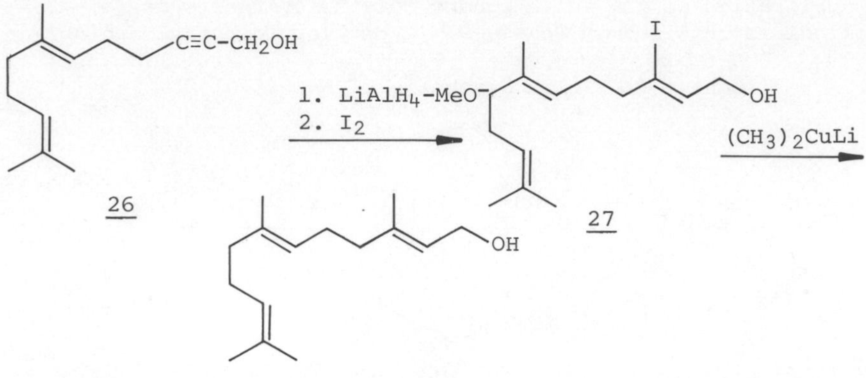

first converted into the dienyne 25. Hydroxymethylation of
25 gave the propargylic alcohol 26, which was reduced with
LiAlH$_4$, in the presence of sodium methoxide. The initially
formed vinylaluminum compound was iodinated to yield solely
the *trans*-isomer 27, which reacted stereospecifically with
lithium dimethylcuprate to afford *trans,trans*-farnesol (1).

In 1969, Vig and collaborators reported the synthesis of
farnesol and nerolidol which is summarized in Scheme 7.[18] The

Scheme 7. Vig Synthesis of Farnesol

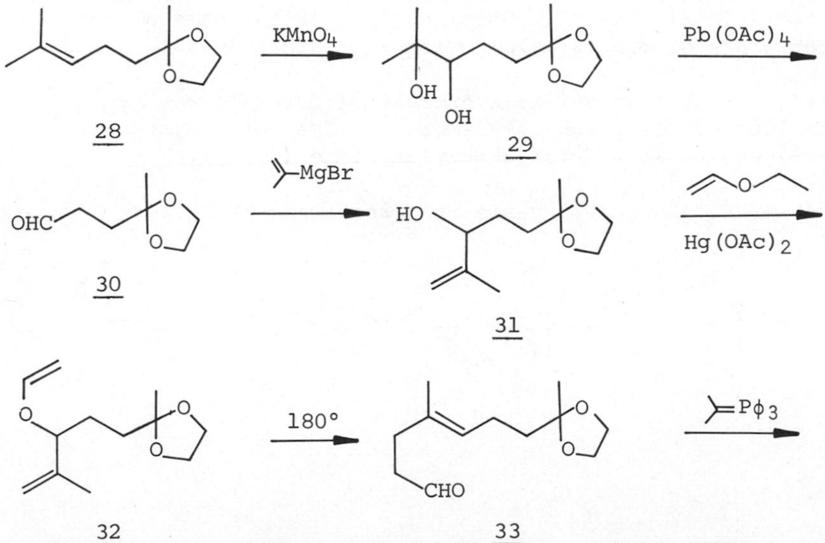

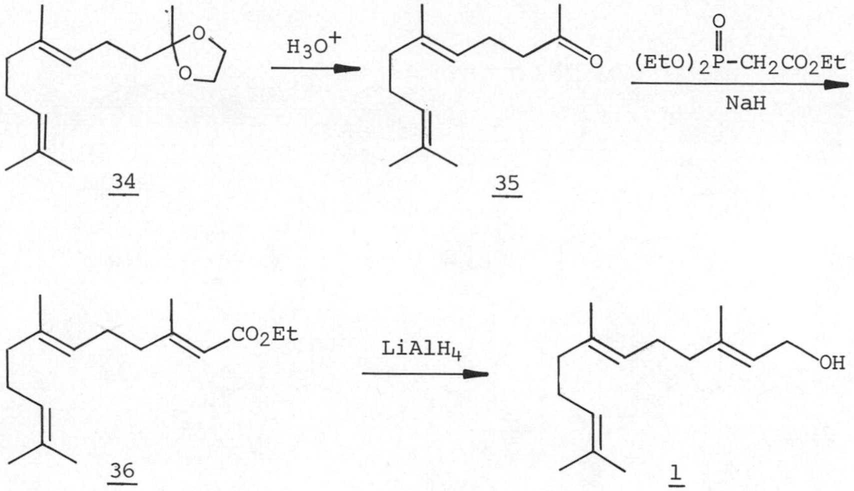

key step is utilization of the Claisen rearrangement to establish geometry at the central double bond. It is claimed that
the reaction leads *exclusively* to the *trans* isomer (33).
Stereospecificity was not obtained in the modified Wittig reaction, although ethyl *trans*-farnesate (36) predominates.

B. Juvenile Hormone

Juvenile hormone, the hormonal substance responsible for ar-
resting development at the pupal stage in the Cecropia moth,
is not actually a sesquiterpene. However, its close structural
resemblance to the acyclic sesquiterpenes warrants its inclusion in the present discussion. The hormone was identified as
methyl *trans,trans,cis*-10-epoxy-7-ethyl-3,11-dimethyl-2,6-
tridecadienoate (37) by Röller, Dahm, Sweeley, and Trost in
1967.[19] Since that time, no less than ten syntheses of the
substance have appeared.[20-29]

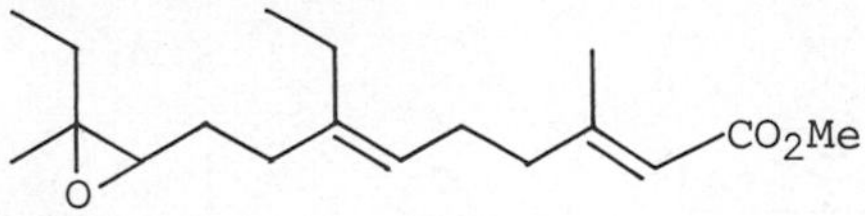

The first synthesis, reported in 1967 by the Wisconsin
group of Dahm and Röller (zoology) and Trost (chemistry),[20]
was nonstereoselective, affording mixtures of isomers at each
stage where new stereochemistry is introduced (Scheme 8).

Scheme 8. Juvenile Hormone--The Wisconsin Synthesis

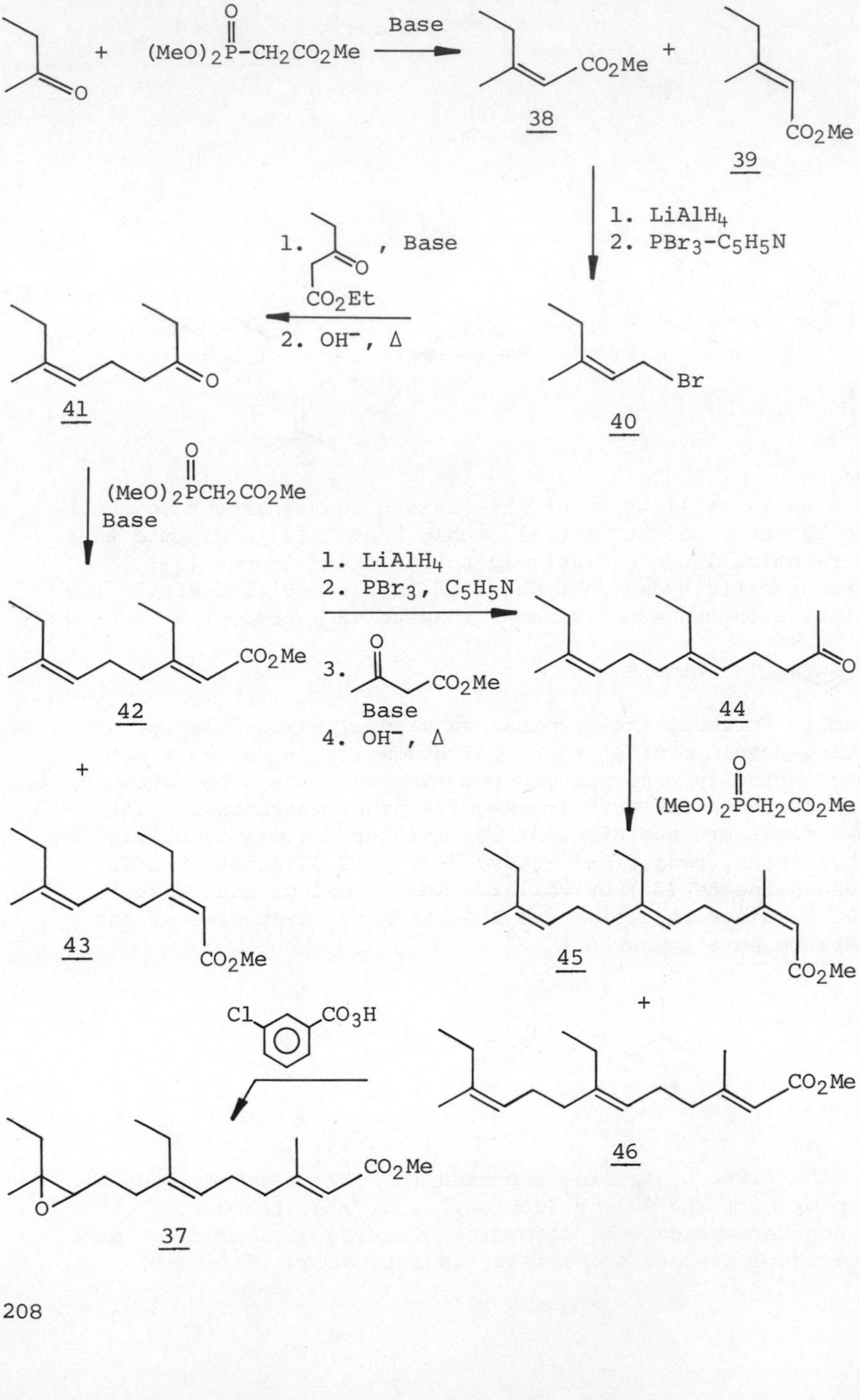

208

Although synthetically "inelegant," such an approach offers
the important advantage of making unnatural isomers available
for physiological screening. Methyl *cis*-3-methyl-2-pentenoate
(38) was prepared in 11% yield by treating 2-butanone with the
sodium salt of trimethylphosphonoacetate.   The corresponding
trans isomer (39) was also produced, in 6% yield. After fractionation of the mixture of esters, the pure *cis* ester was
reduced to the allylic alcohol, which was converted into *cis*
bromide 40 by PBr$_3$ in pyridine. Alkylation of ethyl 3-oxo-
pentanoate with 40, followed by alkaline hydrolysis and decarboxylation afforded enone 41. Application of the Emmons
reaction to 41 gave the *trans,cis*- and *cis,cis*-esters 42 and
43 in yields of 27% and 12%, respectively. Repetition of the
above sequence (LiAlH$_4$, PBr$_3$, alkylation of acetoacetic ester)
on the *trans,cis*-isomer 42 gave dienone 44 in 23% yield.   When
44 was submitted to the Emmons reaction, the *cis,trans,cis*-
and *trans,trans,cis*-C$_{17}$-methyl esters 45 and 46 were produced
in yields of 5% and 25%, respectively. Epoxidation of 46 with m-
chloroperbenzoic acid gave racemic juvenile hormone (37) in 40%
yield, along with 10% of the 6,7-epoxy compound and 10% of the 6,7,-
10,11-bisepoxide.
   Three ingenious stereoselective syntheses of juvenile
hormone quickly issued from Harvard (Corey), Stanford (Johnson), and Syntex (Siddall and Edwards). Corey's synthesis,
outlined in Scheme 9,[21] handled the stereochemistry of the

Scheme 9.   Juvenile Hormone--Harvard Synthesis

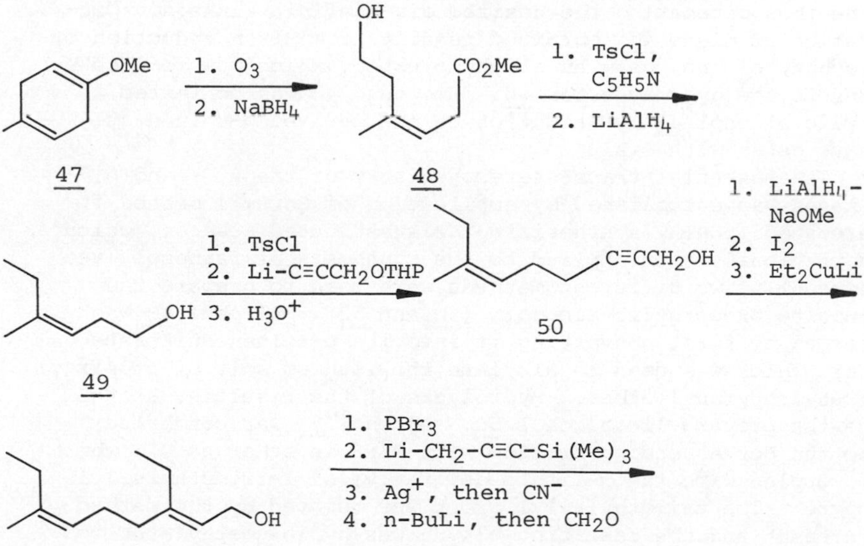

1. LiAlH$_4$-
   NaOMe

2. I$_2$

3. (Me)$_2$CuLi

C≡CCH$_2$OH

52

1. MnO$_2$-
   hexane

2. MnO$_2$-NaCN
   MeOH

CH$_2$OH

53

1. HOBr

2. i-PrO$^-$

CO$_2$Me

46

CO$_2$Me

37

potential epoxide ring by selective scission of a cyclohexa-
diene thus generating the desired *cis*-olefinic linkage.   Ozonization of diene 47, obtained readily from Birch reduction of
p-methoxytoluene, gave an aldehydo ester, which was reduced by
NaBH$_4$ to the hydroxy ester 48. Compound 48 was converted to
allylic alcohol 49 by reduction of the derived p-toluenesulfonate ester with LiAlH$_4$.

     The essential *trans*-stereochemistry of the 2,3- and 6,7-
linkages was established by application of Corey's method for
stereospecifically synthesizing trisubstituted alkenes, which
had previously been applied to the synthesis of farnesol (see
Scheme 6).   Two different methods were used to prepare the
requisite propargylic alcohols (50 and 52).   Alcohol 49 was
extended by first converting it into its p-toluenesulfonate
ester, which was used to alkylate the lithium salt of propargyl
tetrahydropyranyl ether.   Hydrolysis of the resulting acetal
gave the propargylic alcohol 50.   Alcohol 51 was converted
into the corresponding bromide with PBr$_3$ in ether at 0°, which
was coupled with the 3-lithio derivative of 1-trimethylsilyl-
propyne.   The trimethylsilyl group was removed by the method
of Arens[30] and the resulting alkyne was hydroxymethylated to

give 52. No comment was made as to how much stereochemical homogeneity was lost in the PBr₃ and coupling steps, the implication being that the process is stereospecific.

Allylic alcohol 53, resulting from a second application of the Corey sequence, was oxidized directly to methyl ester 46 by an unusual method developed for oxidation of allylic alcohols to α,β-unsaturated esters without *cis,trans* isomerization. The terminal olefinic linkage was selectively epoxidized by van Tamelen's method,[31] affording racemic juvenile hormone (52% yield).

The highly imaginative Syntex synthesis (Scheme 10) led

Scheme 10.   Juvenile Hormone--Syntex Synthesis

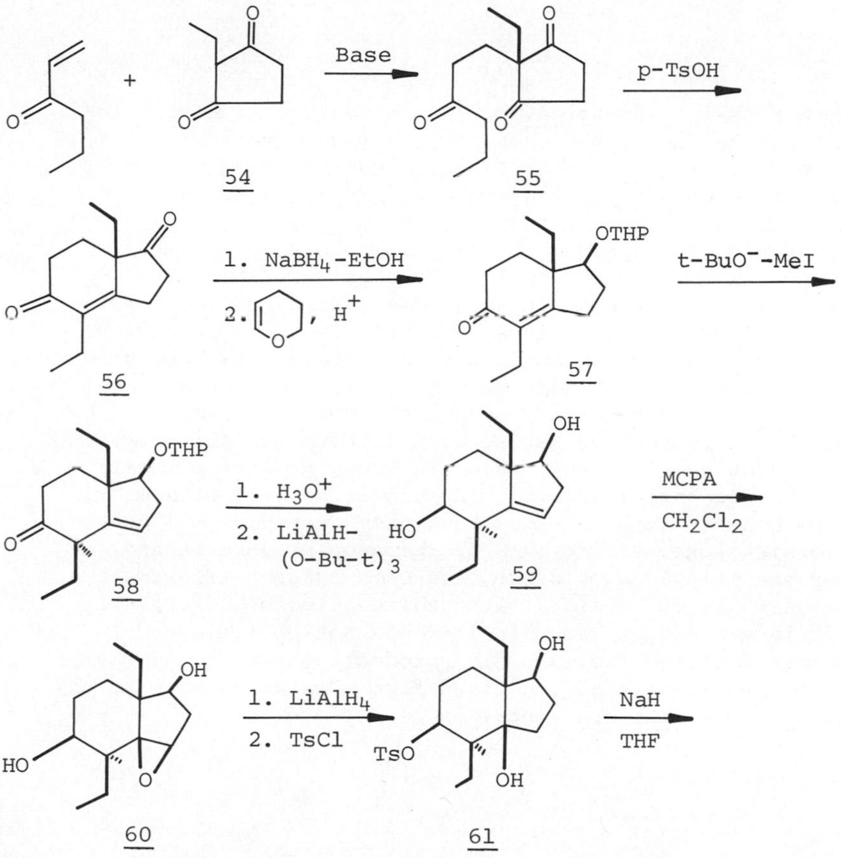

to the Wisconsin dienone 44.[22] The geometry of the two olefinic linkages was established by Grob fragmentation[32] of 1,3-
diol monotosylates 61 and 63. Consideration of the known
steric course of similar fragmentations[33] required that the
ethyl group be *cis* to the tosylate group in both 61 and 63.
The synthesis of compound 61 is a masterpiece of asymmetric
induction. The five centers of asymmetry were introduced
sequentially, as outlined in the scheme, following sound principles of kinetic control in each case. In two steps (56 →
57 and 57 → 58), stereospecificity was obtained by taking advantage of the steric hindrance exerted by the angular ethyl
group. The second secondary hydroxyl group was formed by reducing the corresponding ketone with lithium tri-*tert*-butoxy-
aluminum hydride, a reagent which is known to give predominately the equatorial alcohol. Epoxidation of 59 with m-
chloroperbenzoic acid in ether gave only α-epoxide 64, due to
the aforementioned steric bulk of the angular substituent.
However, in methylene chloride, the same oxidant afforded the
β-epoxide 60 in 50% yield. Although the stereochemistry of
the tertiary hydroxyl group in 61 would not be reflected in
the geometry of the fragmentation product, it was found that
the epimeric alcohol (65), derived from epoxide 64 gave mainly
oxetane 66 under the fragmentation conditions.

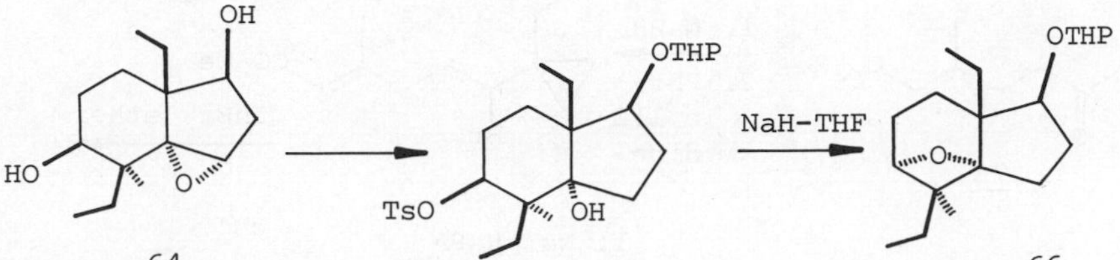

64   65   66

The stanford group, headed by Johnson, established the
stereochemistry at the epoxide ring and the two double bonds
by three different methods of stereoselective synthesis.[23]
The starting point was 1-ethyl-1-acetylcyclopropane (70).
This ketone was carbonated and the resulting keto ester was
alkylated with methyl trans-4-bromo-3-methylbutanoate (71).
This allylic halide, which represents the eventual 2,3-double
bond in juvenile hormone, was prepared by the nonspecific
bromination of β,β-dimethylacrylic acid (67). A mixture of
cis and trans isomers (68 and 69) is produced, but on workup,
the cis bromo acid spontaneously lactonizes, leaving the desired trans isomer as the only acidic material.

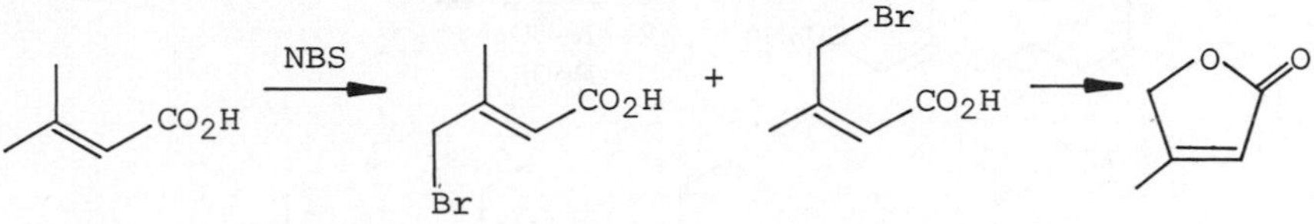

67   68   69

After hydrolysis, decarboxylation and re-esterification,
the unsaturated cyclopropyl ketone 73 was obtained. This substance is the key to the Stanford synthesis. Johnson had previously developed a highly stereoselective modification[16] of
the Julia olefin synthesis.[15] The method involves the

Scheme 11. Juvenile Hormone--Stanford Synthesis

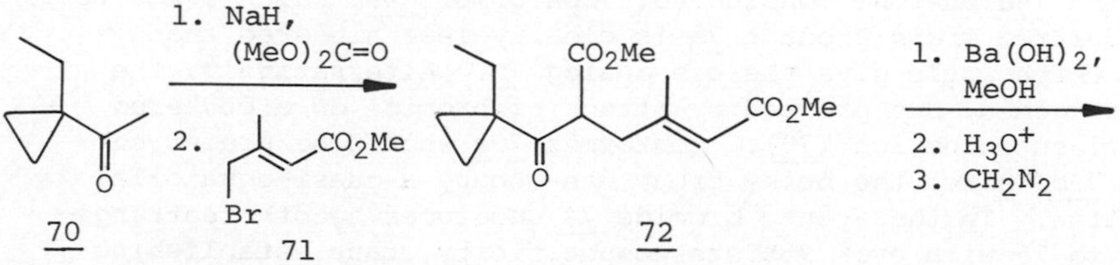

70   71   72

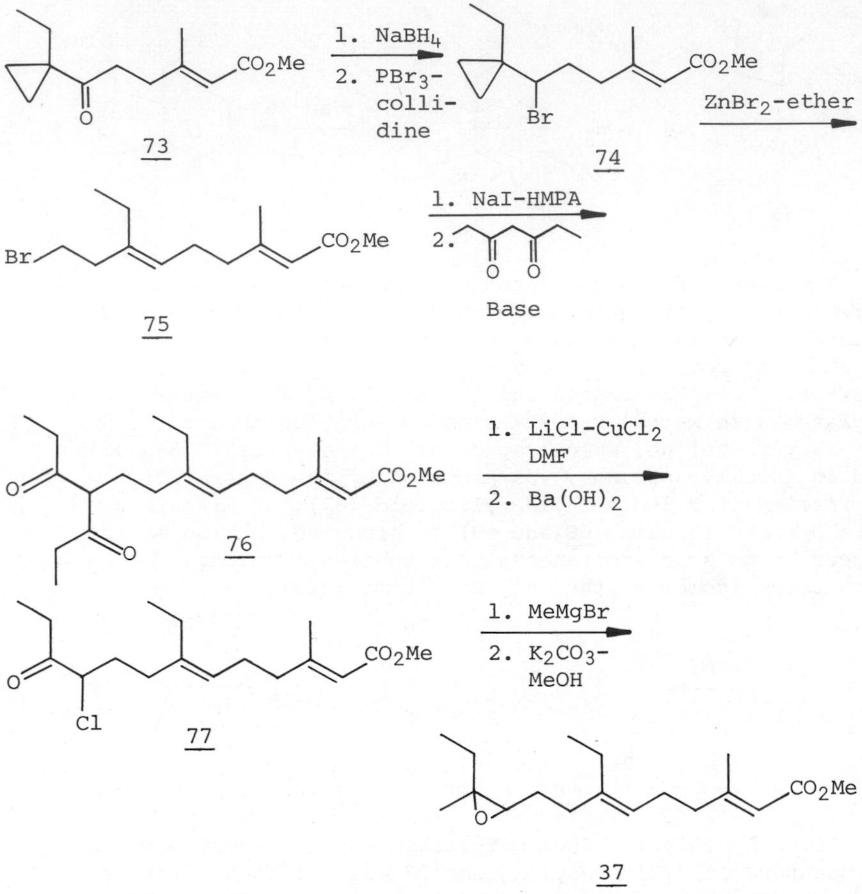

73 → 1. $NaBH_4$ 2. PBr_3-collidine → **74** → $ZnBr_2$-ether →

75 → 1. NaI–HMPA 2. (diketone) Base →

76 → 1. $LiCl$–$CuCl_2$ DMF 2. $Ba(OH)_2$ →

77 → 1. $MeMgBr$ 2. K_2CO_3–MeOH →

37

rearrangement of a cyclopropylcarbinyl bromide (74) to a homo-allyl bromide (75) under the influence of zinc bromide in ether. If one assumes that ring opening is concomitant with ionization, and that an antiperiplanar arrangement of the C–Br and breaking cyclopropyl C–C bond is necessary, then conformers 74a and 74b must be considered. Conformer 74a, which leads to the desired *trans* product 75 is clearly less hindered than 74b, which would give the *cis* analog 78. Alternatively, the rearrangement may proceed by attack of bromide on a puckered bicyclobutonium ion (79). Conformer 79a should be preferred over 79b, since the bulky group can occupy a quasi-equatorial position. In the event, bromide 74 undergoes smooth rearrangement to 75 with over 95% stereospecificity, thus establishing the desired stereochemistry at the eventual 6,7 double bond.

74a 75

74b 78

74

79a → 75 79b → 78

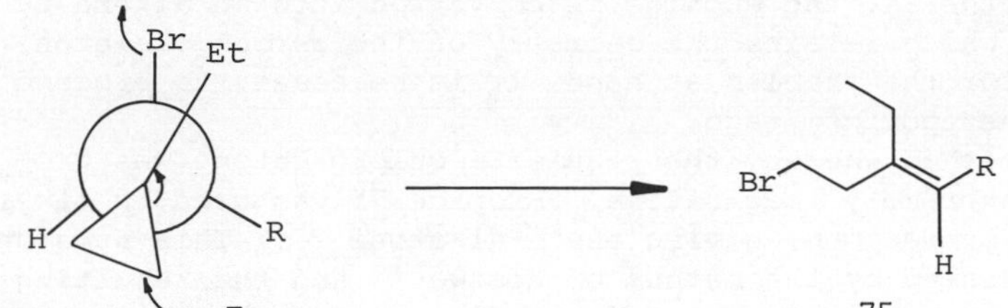

The stage was now set for the introduction of the epoxide moiety of juvenile hormone. In this respect, the Stanford synthesis differs from the earlier syntheses, which relied on selective epoxidation in a pentultimate stage to achieve this functionality. The plan was to utilize Cornforth's method for the stereoselective synthesis of trisubstituted olefins.[34] The Cornforth method involves the formation of a chlorohydrin by addition of a Grignard reagent to an α-chloro ketone. One of the two possible diastereomers is usually produced in predominance, in a predictable manner.[35] Therefore, the epoxide which results from base catalyzed cyclization is produced stereoselectively, by asymmetric induction in the first step

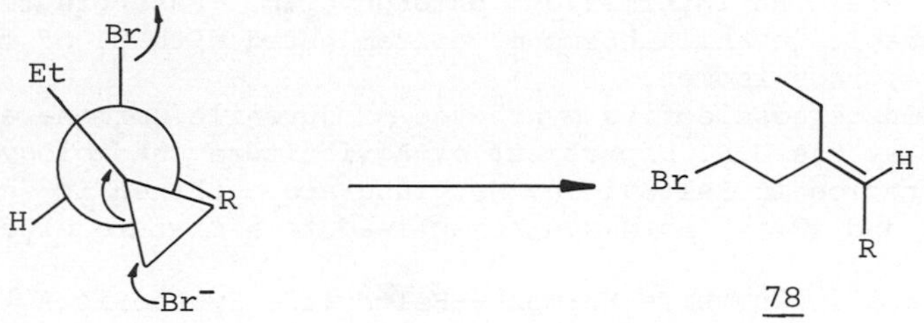

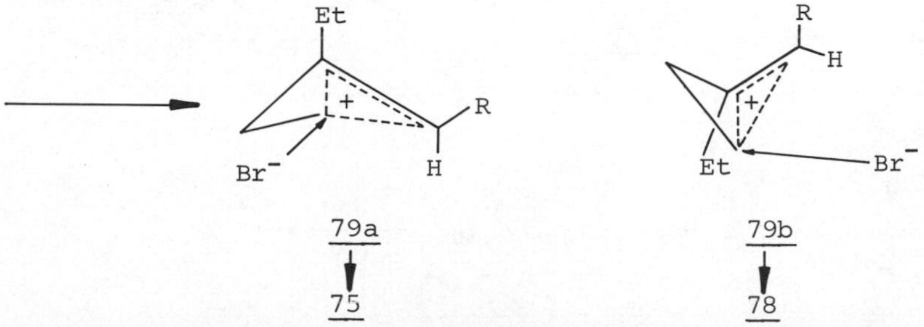

of the sequence. In applications of the Cornforth scheme to
olefin synthesis, the epoxide is converted into an alkene by
a process which retains the geometry of the carbon skeleton.
However, for the problem at hand, it is necessary to proceed
only to the epoxide stage.

The method whereby the requisite chloro ketone was produced is extremely imaginative. Bromide 75 was used to alkylate 3,5-heptanedione, giving the β-diketone 76. This substance
was chlorinated by the method of Kosower[36] and the resulting
chloro diketone was deacylated with barium hydroxide in ethanol. Chloro ketone 77 was treated with methylmagnesium bromide at -75° (it having previously been found that low temperatures greatly enhance the stereoselectivity of the process)[16] to yield an intermediate chlorohydrin. Base treatment
yielded racemic juvenile hormone, contaminated with 8% of the
trans,trans,trans isomer.

Two nonstereoselective syntheses of juvenile hormone ac-
complished at the U.S. Department of Agriculture Entomology
Research Station in Beltsville, Maryland are outlined in
Schemes 12 and 13.[24] Both syntheses lead to a mixture of all

Scheme 12.   Juvenile Hormone--Beltsville Synthesis A

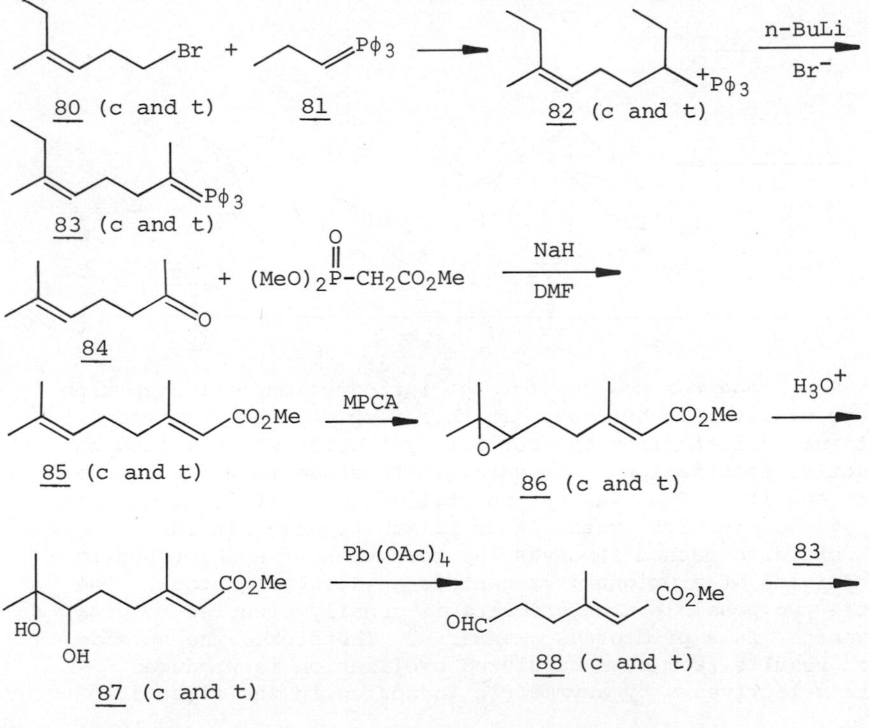

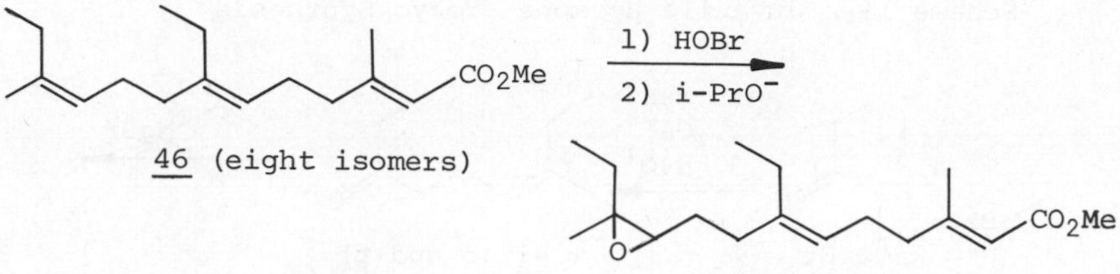

46 (eight isomers)

1) HOBr
2) i-PrO⁻

37 (eight isomers)

Scheme 13.   Juvenile Hormone--Beltsville Synthesis B

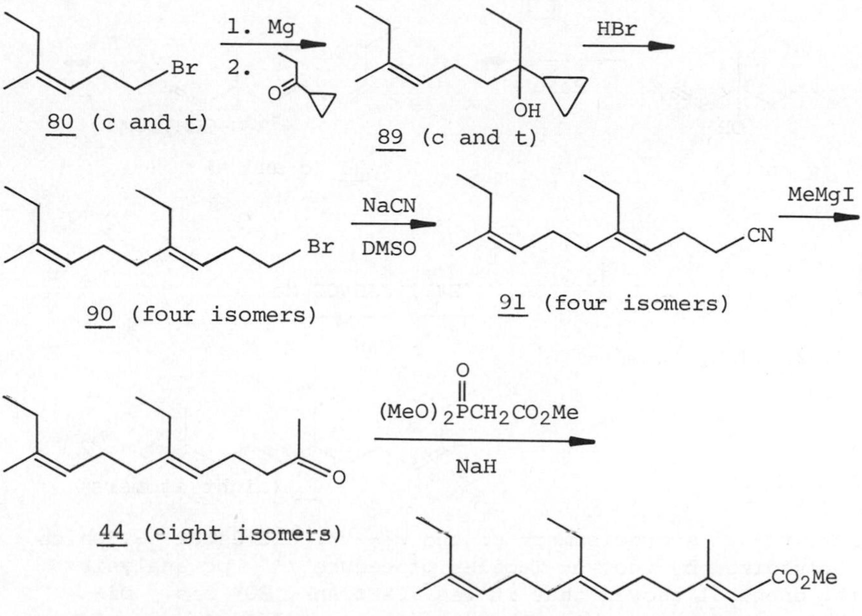

80 (c and t)

1. Mg
2.

89 (c and t)

HBr

90 (four isomers)

NaCN
DMSO

91 (four isomers)

MeMgI

44 (cight isomers)

$$(MeO)_2 \overset{\overset{\displaystyle O}{\|}}{P} CH_2 CO_2 Me$$

NaH

46 (eight isomers)

eight stereoisomers of the natural hormone. Although this
method is clearly unsuitable for preparing a single isomeride
in good yield, it does make active material fairly readily
available for entomological research.

Two further nonstereoselective syntheses of the hormone,
originating in the Department of Agricultural Chemistry at the
University of Tokyo[25] and the Department of Chemistry at the
University of New Brunswick[26] are summarized in Schemes 14 and
15. The Tokyo synthesis is patterned after the classical
Ruzicka-Isler farnesol synthesis (see Scheme 2), as modified
by Kimel.[11,37] The synthesis leads, is a straightforward

Scheme 14.   Juvenile Hormone--Tokyo Synthesis

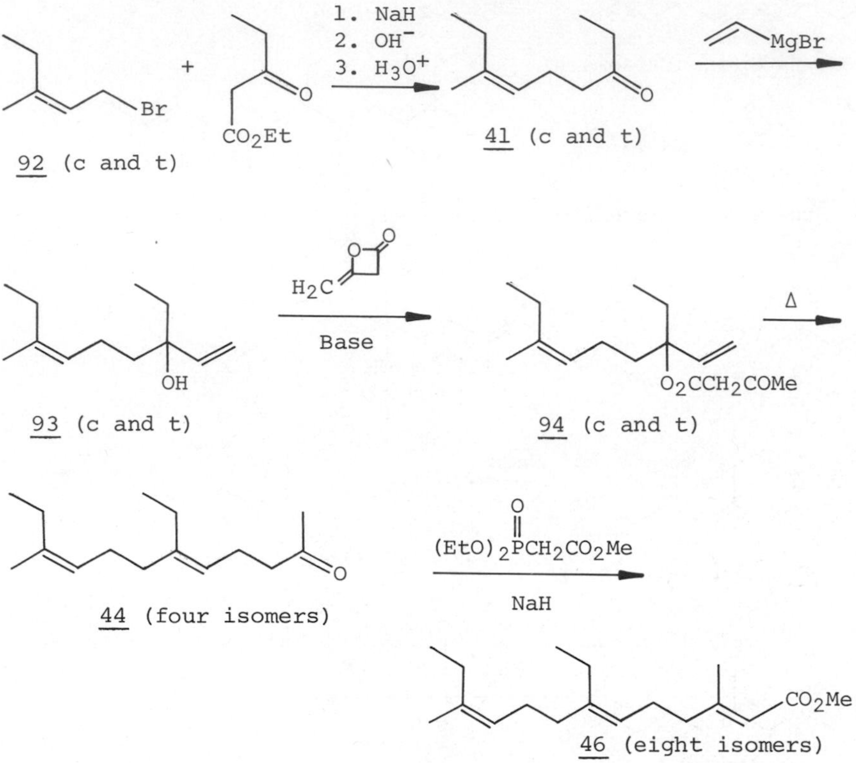

1. NaH
2. OH⁻
3. H₃O⁺

92 (c and t)   + CO_2Et →   41 (c and t)   →   MgBr

93 (c and t)   OH   →   Base   →   94 (c and t) O_2CCH_2COMe →   Δ

44 (four isomers)

$$(EtO)_2 \overset{O}{\overset{\|}{P}} CH_2CO_2Me$$

NaH

CO_2Me

46 (eight isomers)

manner, to all stereoisomers of the C_{17}-trienic ester 46, which
was epoxidized by the van Tamelen procedure.[31] Vpc analysis
of the enone 41 showed that it was 70% *trans*, 30% *cis*. Dienone 44 was found to have the composition: 16% *cis,cis*; 35%
cis,trans; 17% *trans,cis*; 32% *trans,trans*.

The New Brunswick synthesis (Scheme 15) utilizes the
Claisen rearrangement to prepare unsaturated aldehyde 97. This
method had been introduced earlier by Saucy and Marbet for
terpene synthesis.[38] The Claisen method gives a mixture of
stereoisomeric γ,δ-unsaturated aldehydes containing 60%
and 40% *cis* isomers. Vpc analysis of dienone 44 revealed it
to be a mixture of 15% *cis,cis*; 23% *trans,cis*; 25% *cis,trans*;
and 37% *trans,trans* isomers (see above). By preparative vpc
at various stages, all four isomers of dienone 44 were obtained in a pure state. The *trans,cis* isomer was converted
into racemic juvenile hormone by the method of Dahm, Trost,
and Röller (see Scheme 8).

Scheme 15.   Juvenile Hormone--New Brunswick Synthesis

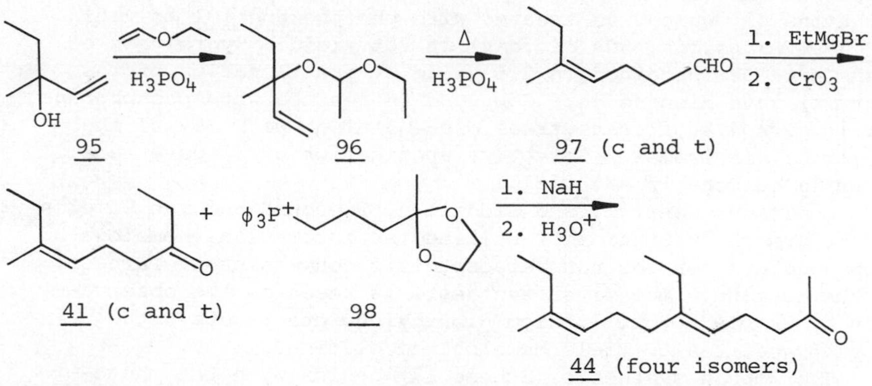

41 (c and t)   + ϕ_3P^+ 98

1. NaH
2. H_3O^+

44 (four isomers)

Schultz and Sprung, at Schering AG in Berlin, have reported an economical synthesis of the hormone, which gives a mixture of all isomers (Scheme 16).[27]   Unsaturated ketal 100

Scheme 16.   Juvenile Hormone--Schering Synthesis

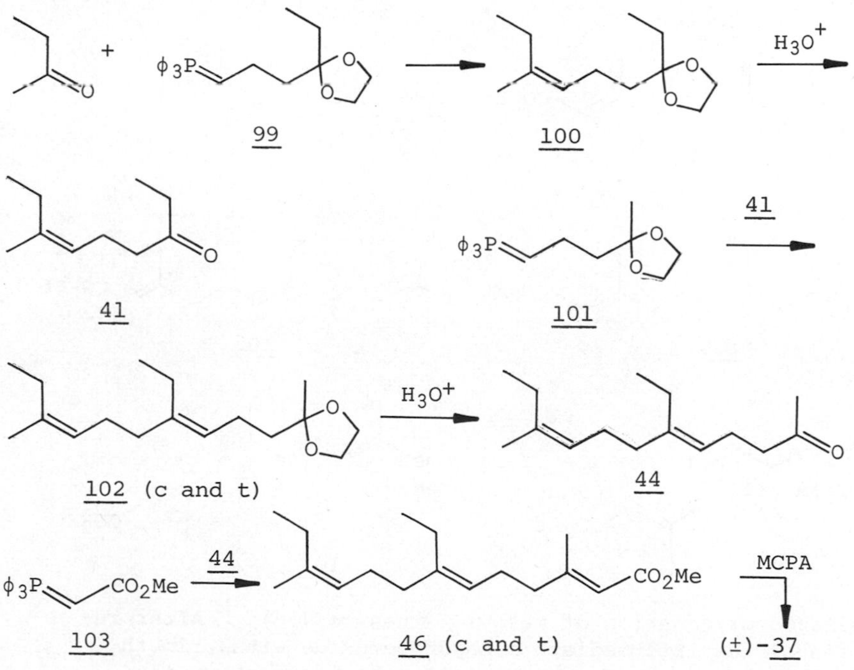

99   100 H_3O^+

41   ϕ_3P 101   41

102 (c and t) H_3O^+ 44

ϕ_3P CO_2Me 44   46 (c and t) CO_2Me MCPA

103   46 (c and t)   (±)-37

was obtained in 69% yield by coupling phosphorane 99 with 2-butanone. The *cis* isomer (63% of the mixture) was converted to ketone 41, which was treated with phosphorane 101 to yield 102 (50% *trans,cis*; 50% *cis,cis*) in 70% yield. Hydrolysis of *trans,cis*-102 (obtained in 29% yield by fractionation of the mixture) gave dienone 44. Compound 44 reacted with phosphorane 103 to give 41% of *trans,trans,cis*-46, along with 35% of the *cis,trans,cis* isomer. Selective epoxidation of 46 gave racemic juvenile hormone in 43% yield.

Anderson, Henrick, and Siddall at Zoecon[28] and van Tamelen and McCormick at Stanford[29] utilized an interesting homologation sequence for the nonstereospecific conversion of farnesol to the hormone. The first synthesis is based on the observation by Crabbé[39] that lithium dimethylcuprate reacts with allylic acetates to yield homologated olefins.

The Zoecon synthesis (Scheme 17) begins with the photo-

Scheme 17.   Juvenile Hormone--Zoecon Synthesis

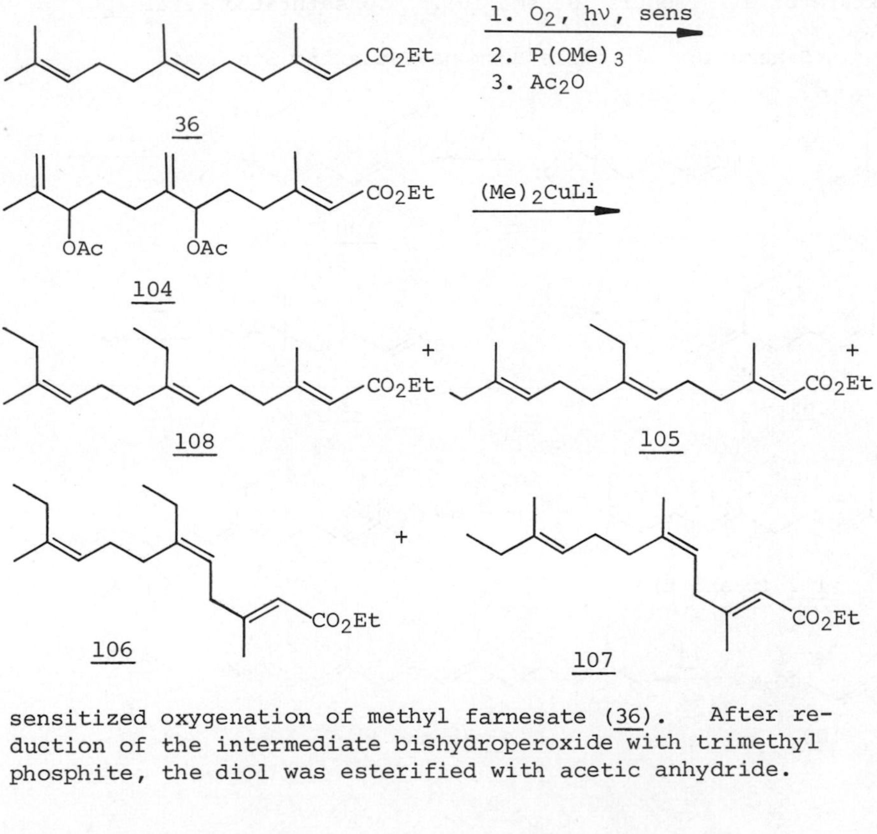

sensitized oxygenation of methyl farnesate (36). After reduction of the intermediate bishydroperoxide with trimethyl phosphite, the diol was esterified with acetic anhydride.

Compound 104 reacted with lithium dimethylcuprate in ether to give trienes 107, 105, and 106 in yields of 76, 14, and 8% respectively. When the reaction was carried out in tetrahydrofuran, the mixture of triene esters contained "appreciable amounts" of isomer 108. The corresponding methyl ester (46) had previously been converted to juvenile hormone.

Van Tamelen and McCormick (Scheme 18) converted farnesyl

Scheme 18. Juvenile Hormone--van Tamelen Synthesis

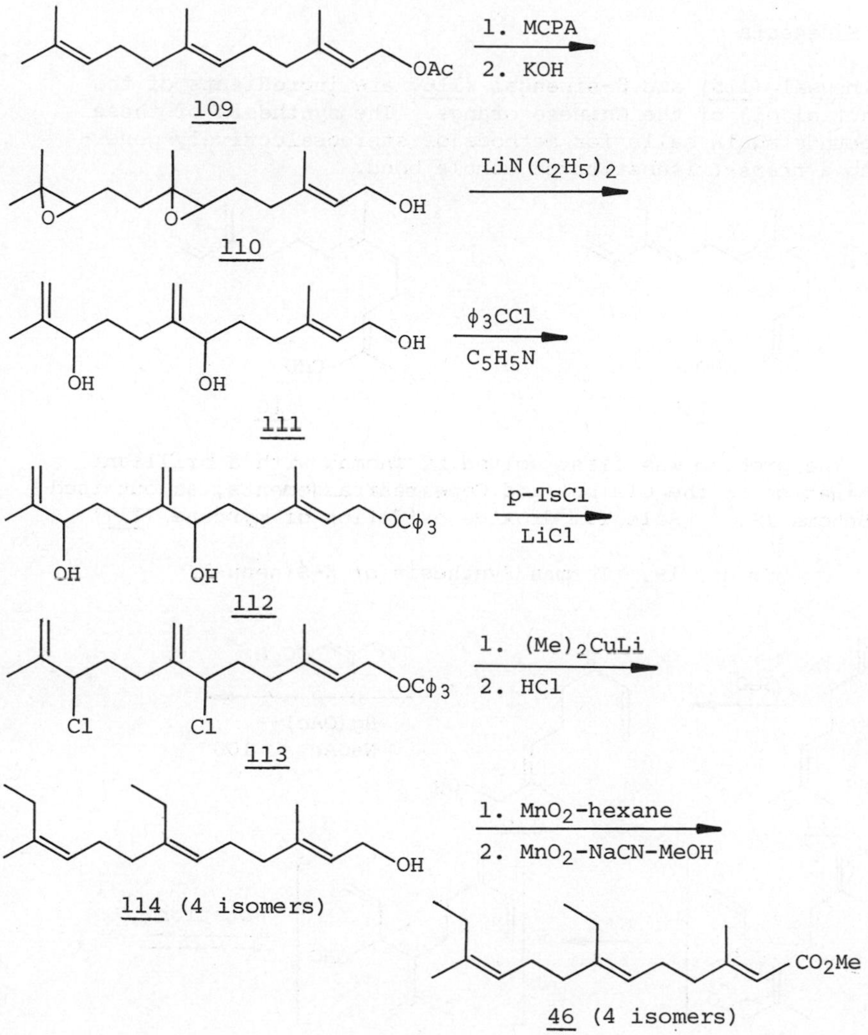
OAc
1. MCPA
2. KOH
109
O
O
OH
$LiN(C_2H_5)_2$
110
OH
OH
OH
ϕ_3CCl
C_5H_5N
111
$OC\phi_3$
OH
OH
p-TsCl
LiCl
112
Cl
Cl
$OC\phi_3$
1. $(Me)_2CuLi$
2. HCl
113
OH
1. MnO_2-hexane
2. MnO_2-NaCN-MeOH
114 (4 isomers)
CO_2Me
46 (4 isomers)

acetate (<u>109</u>), via bis-epoxide <u>110</u> into the tris-allylic triol <u>111</u>. The primary hydroxyl was protected by the trityl group, and the two secondary alcohols were replaced by chlorine. Alkylation of bis-chloride <u>113</u> with lithiumdimethylcuprate gave, after deprotection of the primary hydroxyl, a mixture of geometric isomers of trienol <u>114</u>. This was oxidized by Corey's method[40] to give a mixture of isomers of <u>46</u>. The isomer distribution was determined by glpc to be 40% *trans,cis,trans*; 20% *trans,trans,trans*; 20% *trans,cis,cis*; and 20% *trans,trans, cis* (desired isomer).

C. Sinensals

α-Sinensal (<u>115</u>) and β-sinensal (<u>116</u>) are ingredients of the essential oil of the Chinese orange. The synthesis of these compounds again calls for methods of stereoselectively generating a *trans*-trisubstituted double bond.

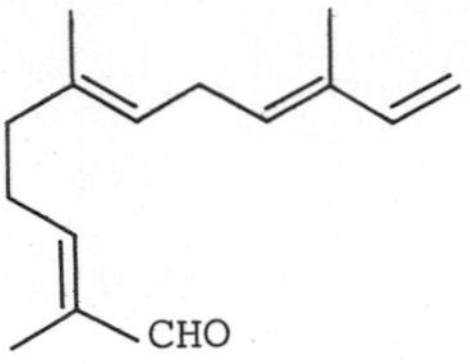

<u>115</u> 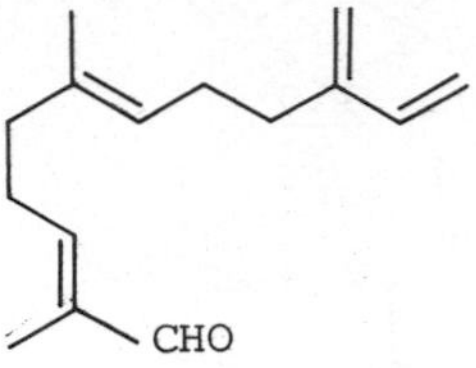   <u>116</u>

The problem was first solved by Thomas with a brilliant combination of the Claisen and Cope rearrangements, as outlined in Scheme 19.[41] Selenium dioxide oxidation of myrcene (<u>117</u>)

Scheme 19.   Thomas Synthesis of β-Sinensal

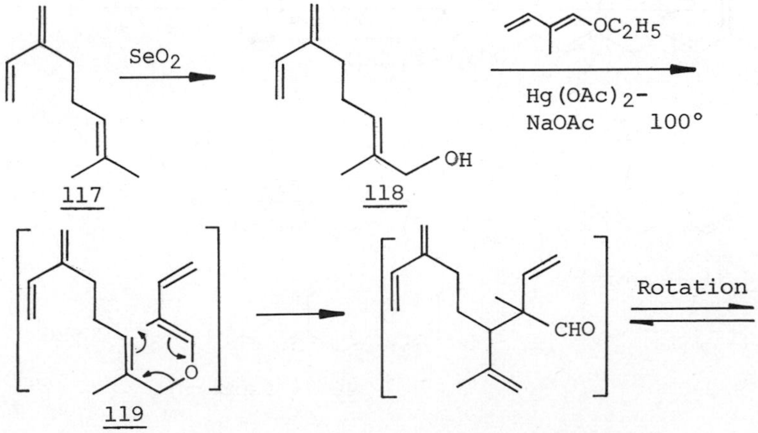

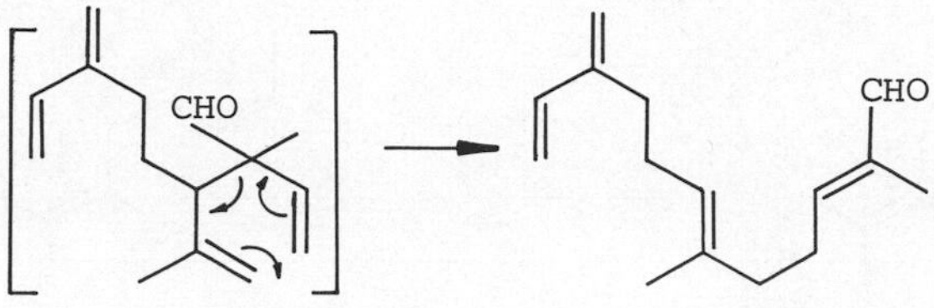

gave allylic alcohol 118, along with the corresponding *trans*-aldehyde, which could be reduced to 118. When a mixture of 118 and 1-ethoxy-2-methyl-1,3-butadiene was heated with mercuric acetate and sodium acetate at 98° for 50 hr, β-sinensal was produced in 43% yield. The transformation presumably occurs in the following manner: mercuric ion catalyzed trans-etherification, yielding 119, which undergoes a Claisen rearrangement to yield 120, which undergoes a second thermal reorganization to form 116.

Stereochemically, the Thomas synthesis seems to produce only *trans*,*trans*-β-sinensal, the natural isomer. This result can be rationalized in terms of the known stereochemistry of the Cope rearrangement.[42] Using a chairlike, four-center transition state, the Claisen rearrangement (119 → 120) can be written as

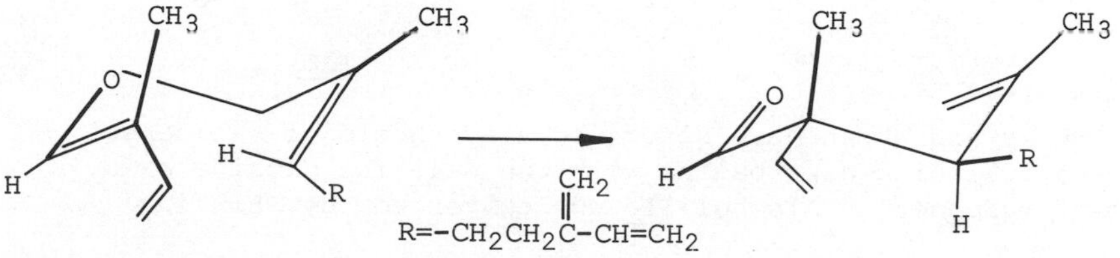

Diastereomer 120 can rotate into a suitable conformation for subsequent Cope rearrangement in two ways:

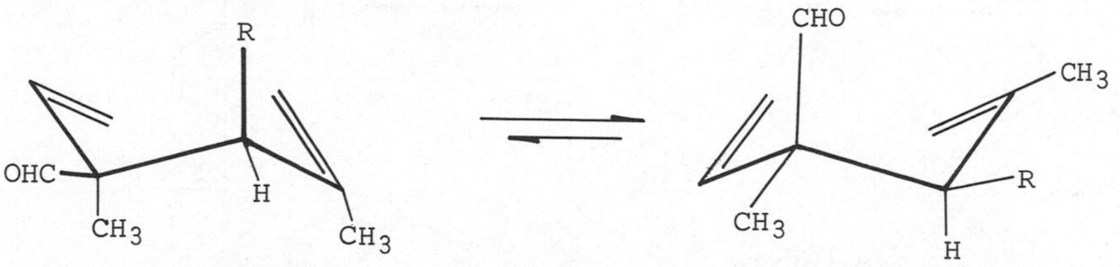

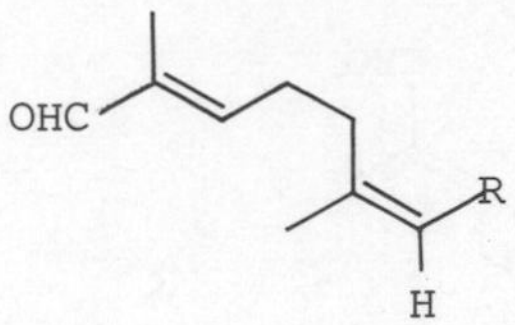

121

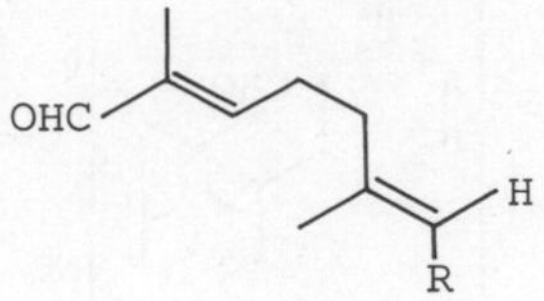

122

Conformer 120a should give *trans,cis*-isomer (121), while conformer 120b should give a *cis,trans*-isomer (122). Furthermore,
conformer 120b should be preferred, on the grounds that the
bulky substituents (CH$_3$ and R) occupy pseudo-equatorial positions. Thus, the 7,8 double bond should be *trans*. However,
it is quite likely that the 2,3 double bond would isomerize to
the more stable *trans* arrangement during the reaction.

Alternatively, diastereomer 123 might be produced in the
Claisen rearrangement. The preferred conformer of this isomer would lead directly to *trans,trans*-β-sinensal.

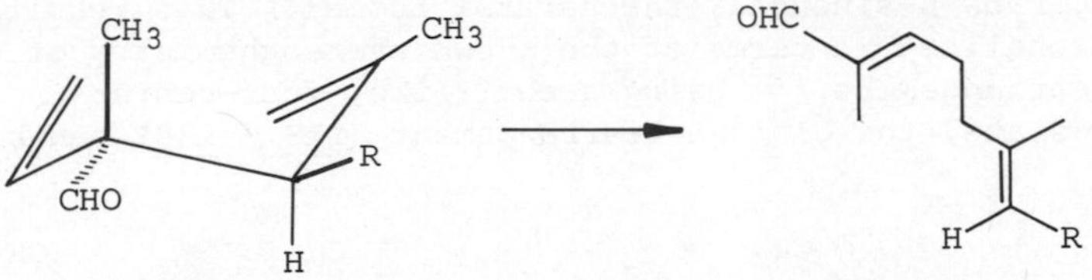

123   116

Büchi and Wüest have described a synthesis of β-sinensal
(Scheme 20) which also begins with the selenium dioxide oxidation of myrcene.[43] Alcohol 118 was converted, by PBr$_3$ in

Scheme 20. Büchi-Wüest Synthesis of β-Sinensal

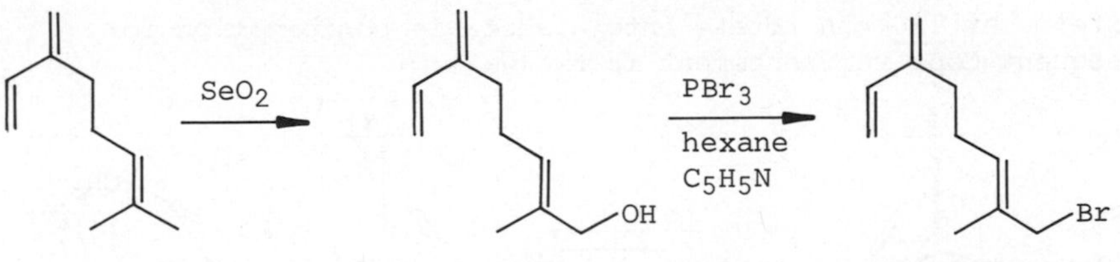

117   118   124

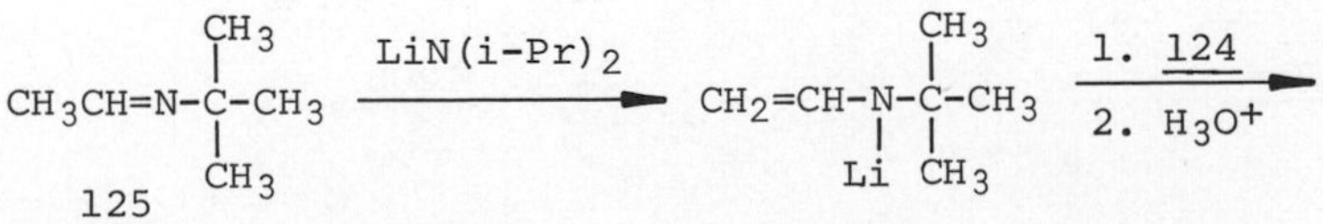

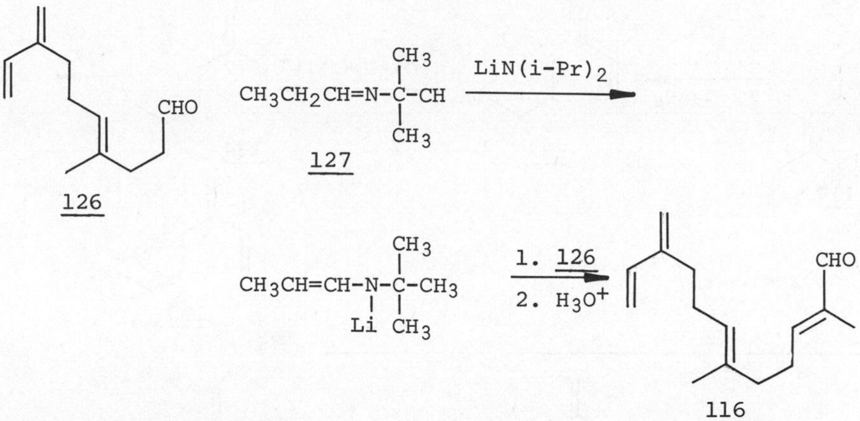

126 127 116

hexane, to allylic bromide 124. This halide was used to alkylate the lithium salt of Schiff's base 125 (Stork's method),[44] yielding the C-12 aldehyde 126 after hydrolysis. Compound 126 was treated with the lithium salt of Schiff's base 127, in an application of Wittig's "directed aldol condensation"[45] to yield β-sinensal.

Bertele and Schudel have recorded a synthesis of β-sinensal which is summarized in Scheme 21.[46] Ozonolysis of

Scheme 21. Bertele-Schudel Synthesis of β-Sinensal

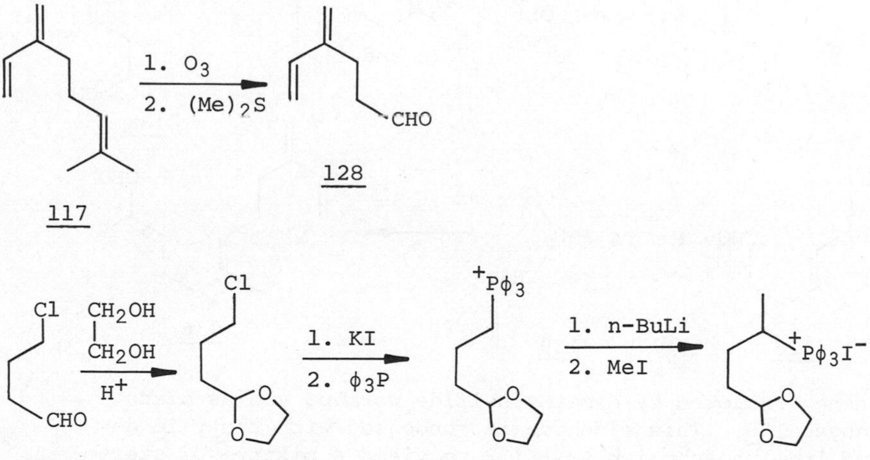

117 128 129 130 131 132

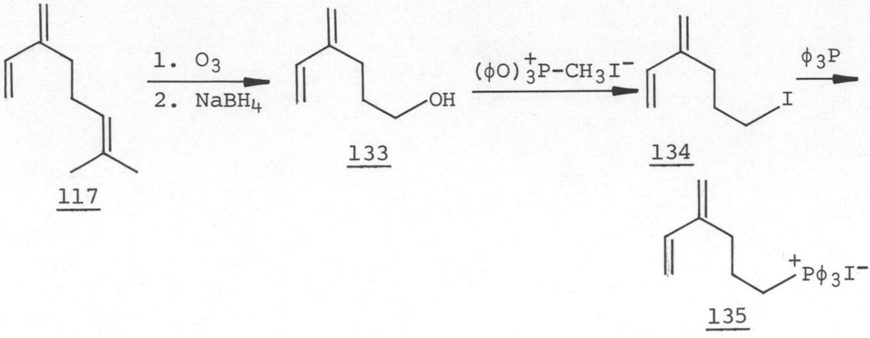

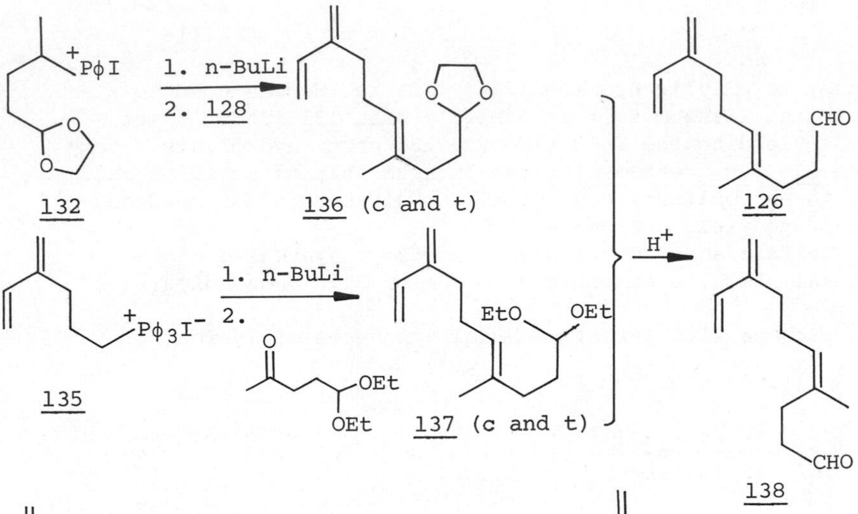

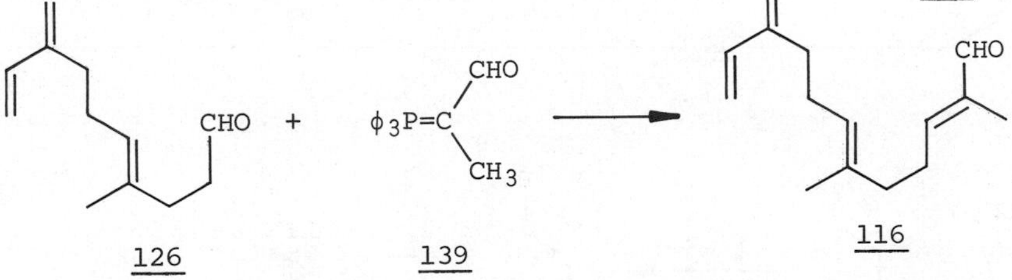

myrcene, followed by dimethylsulfide workup, yields diene aldehyde **128**. This aldehyde is condensed with the ylid derived from phosphonium salt **132** to yield a mixture of stereoisomeric triene acetals **136** containing 40% *cis-* and 60% *trans-*isomers. Alternatively, the myrcene ozonide may be reduced with sodium borohydride to afford alcohol **133**. Subsequent conversion of this alcohol into phosphonium salt 135, followed by condensation of the derived ylid with 5,5-diethoxy-2-pentanone yields a mixture of isomeric triene acetals 137 in a ratio of 30% *trans* and 70% *cis*. The acetal mixtures 136 and 137 were separated by preparative vpc, the major isomer being readily obtained in each case. Hydrolysis of purified *trans*-136 and *cis*-137 afforded 126 and 138, respectively. The stereochemistry was established for the two isomers by an examination of the NMR spectra. Condensation of 126 with ylid 139 yielded *trans,trans*-β-sinensal, identical with the natural material.

The same authors synthesized α-sinensal from intermediate 126 by the route shown in Scheme 22. Condensation of 126 with

Scheme 22.   Bertele-Schudel Synthesis of α-Sinensal

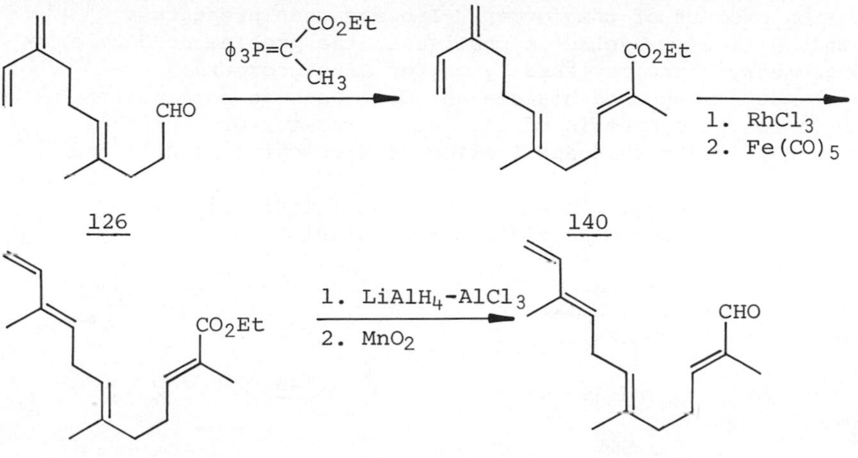

1-carbethoxyethyldinetriphenylphosphorane gave the tetraene ester 140. Treatment of this ester with rhodium chloride in ethanol[47] rearranged the 1,3-diene system to a mixture of 9,10 *cis,trans* isomers (70% *trans*). When the mixture was heated with iron pentacarbonyl at 180°, the proportion of *trans*-isomer increased to 95%.[48] Reduction of the ester, followed by oxidation of the resulting allylic alcohol gave α-sinensal.

D.   Torreyal, Neotorreyol, Dendrolasin, Ipomeamarone (Ngaione)

Torreyal (141), neotorreyol (142), dendrolasin (143), and ipomeamarone (ngaione, 144) are sesquiterpene substances containing a β-alkylated furan ring. The double bonds in 141–143 are *trans*, and the synthetic problem is therefore similar

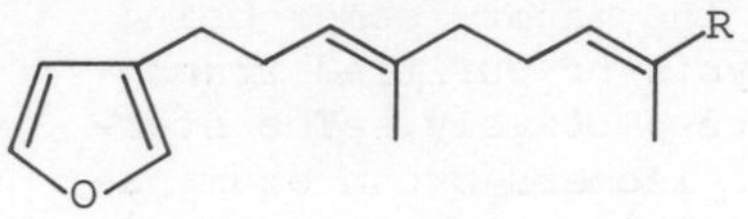

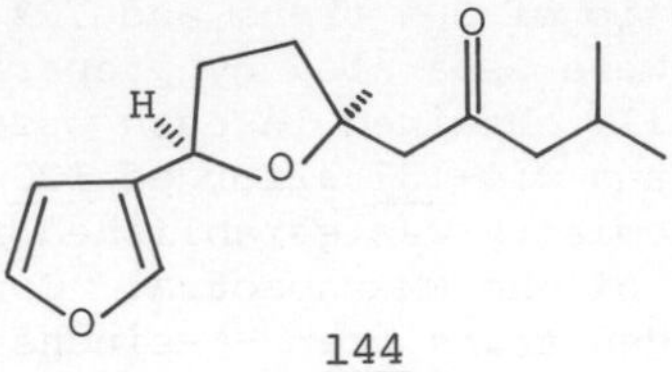

141   R=CHO
142   R=CH$_2$OH
143   R=CH$_3$

144

to that faced in the synthesis of farnesol, juvenile hormone or sinensal. In fact, the structural similarity between β-sinensal and torreyal is striking, the latter being simply an oxidation product of the former. Ipomeamarone presents a somewhat different synthetic challenge, the problem of double bond geometry being replaced by one of ring geometry.

Thomas has applied his elegant Claisen-Cope combination sequence to the synthesis of 141–143 from 3-furfuryl alcohol (Scheme 23).[49] In this application, the reaction is utilized

Scheme 23.   Thomas Synthesis of Torreyal,
Neotorreyol, and Dendrolasin

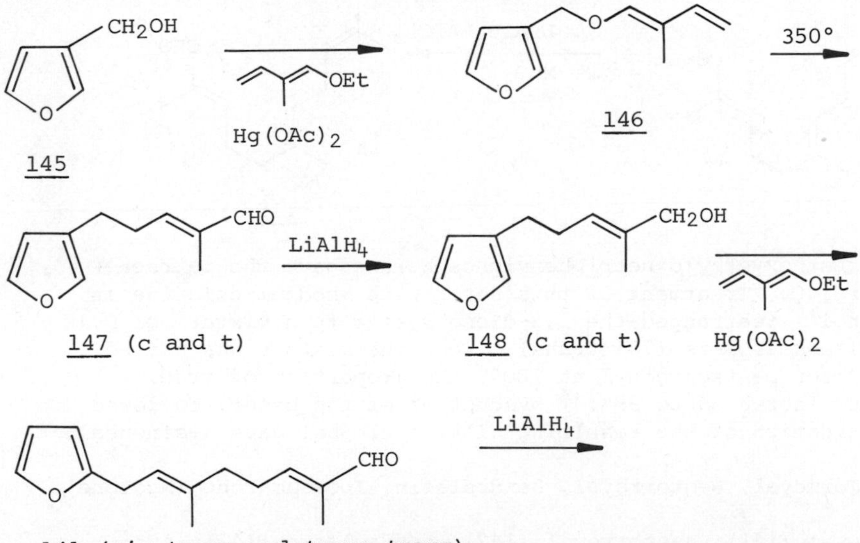

145

146

147 (c and t)

148 (c and t)   Hg(OAc)$_2$

141 (*cis,trans* and *trans,trans*)

as a method for the repetitive addition of two isoprene units, after reduction of the initially formed aldehyde (147). The

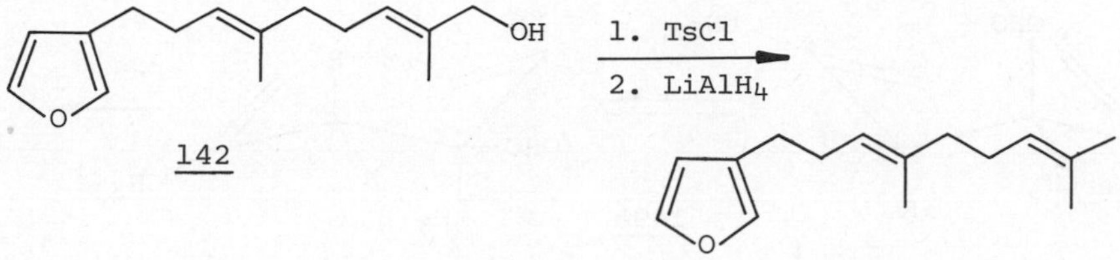

142

1. TsCl

2. LiAlH₄

143

synthetic torreyal was converted, via neotorreyol, into dendrolasin by the indicated sequence. Stereochemically, the
Thomas synthesis gives aldehydes _147_ in a *trans:cis* ratio of
2:1. However, application of the sequence to the derived alcohol mixture _148_ gave only *trans,trans-* and *cis,trans-* torreyal in a 4:1 ratio, the former being identical with the
natural substance.

The conversion of both *cis-* and *trans-148_ to a compound
(_141_) containing a *trans-*6,7-double bond deserves comment.
The stereospecific conversion of *trans-148_ to 6,7-*trans-141_ is
understandable in terms of the conformational arguments presented previously (p. 223). The 6,7-*cis* isomer of _148_ can
react with the ethoxydiene to give two ethers (_148a_ and _148a_),
which should give diastereomers _149a_ and _149b_, respectively.
The preferred "chairlike" conformations of both diastereomers

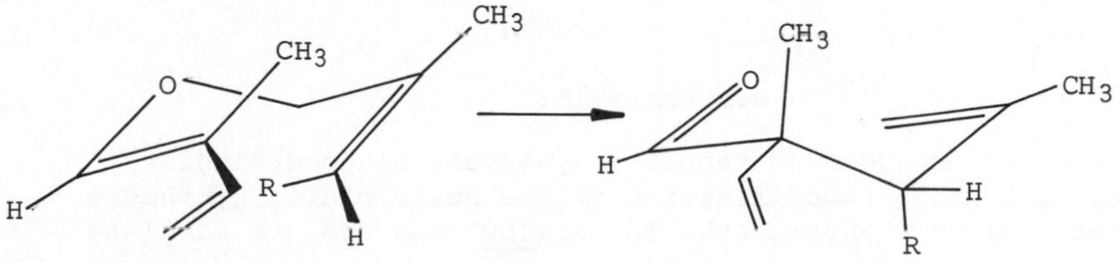

148a

149a

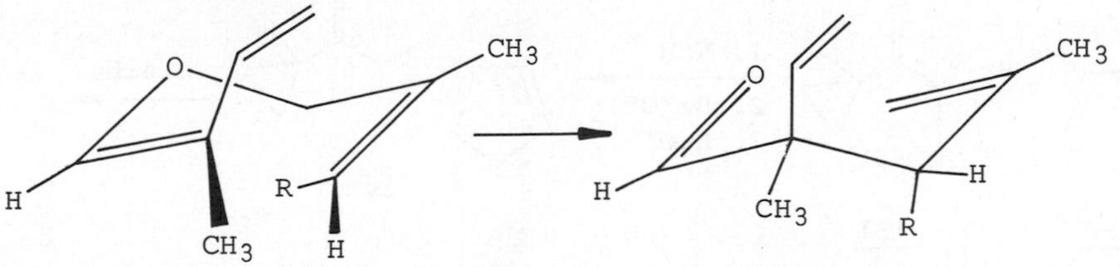

148b

149b

_149a_ and _149b_ (e.g., the conformations having two of the groups
CH₃, vinyl, and R in pseudo-equatorial positions) lead to isomers of _141_ containing a *trans-*6,7-double bond.

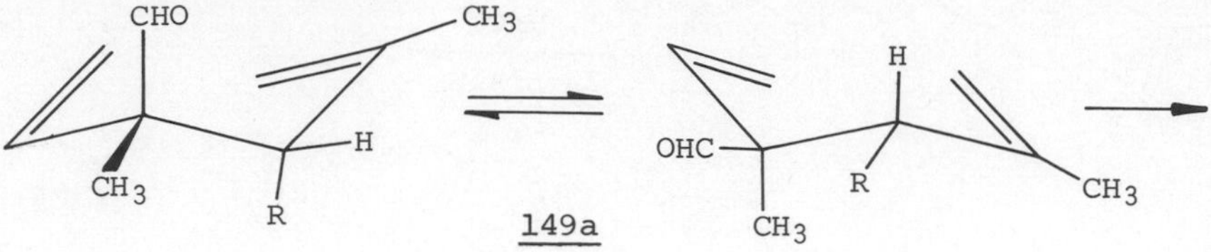

149a

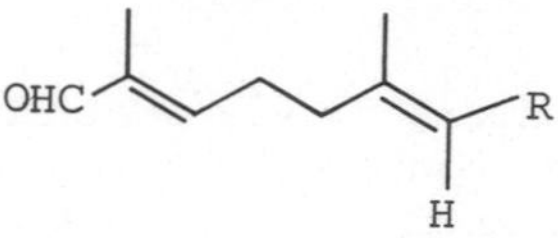

trans,trans-141

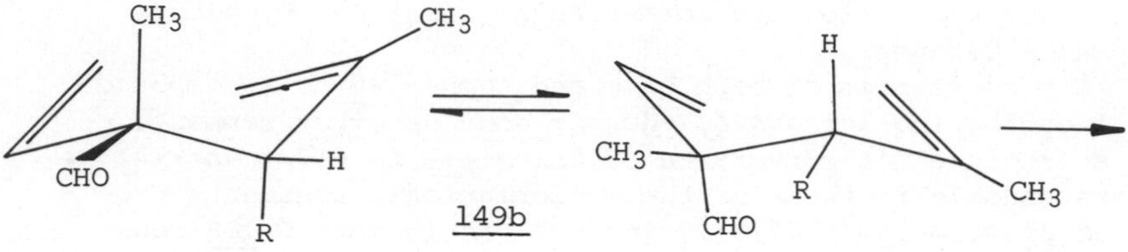

149b

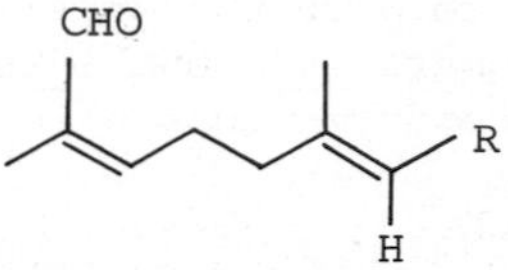

cis,trans-141

Parker and Johnson report a synthesis of dendrolasin based on Johnson's modification of the Julia olefin synthesis (Scheme 24).[50] 3-Bromomethylfuran (150) was used to alkylate

Scheme 24.   Parker-Johnson Synthesis of Dendrolasin

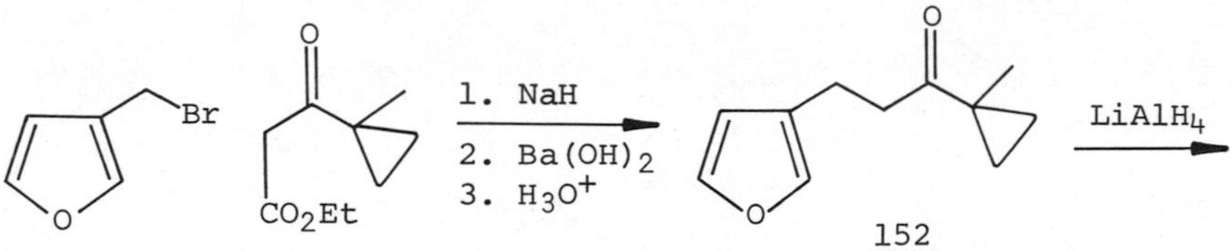

1. NaH
2. Ba(OH)$_2$
3. H$_3$O$^+$

152

LiAlH$_4$

150   151

153

1. PBr$_3$
2. ZnBr$_2$

154

1. NaCN-DMSO
2. NaOH

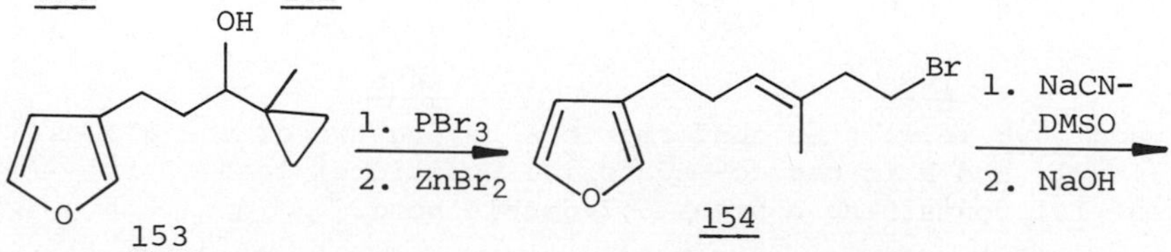

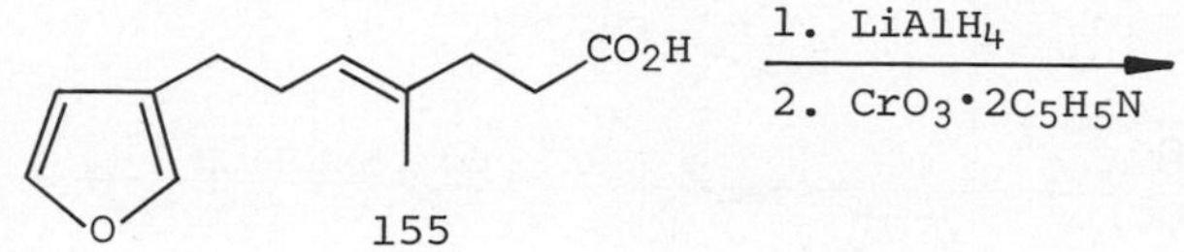

155

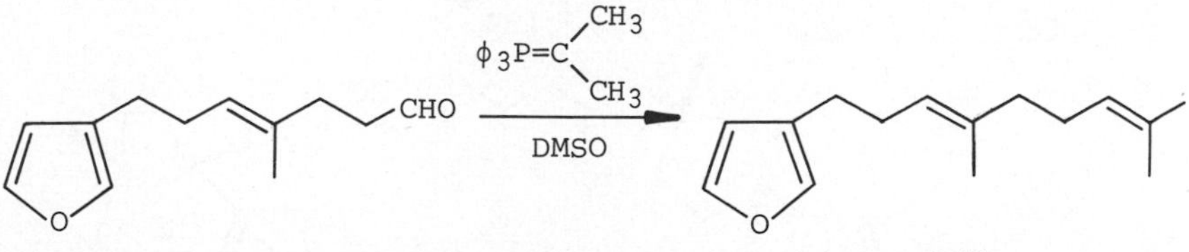

156   143

β-keto ester 151. After hydrolysis and decarboxylation, cyclopropyl ketone 152 was reduced to carbinol 153. This material
was submitted to the two-stage PBr₃-ZnBr₂ treatment to yield
homoallylic bromide 154 (p. 214). After nitrile chain extension, the remaining three carbons were added by a Wittig reaction of isopropylidinetriphenylphosphorane on aldehyde 156.
As in Johnson's juvenile hormone synthesis (see Scheme 11),
the central double bond was greater than 95% *trans*.

In 1956, Kubota and Matsuura communicated a synthesis of
(±)-ipomeamarone (144) which is summarized in Scheme 25.[51]

Scheme 25.   Kubota-Matsuura Synthesis of Ipomeamarone

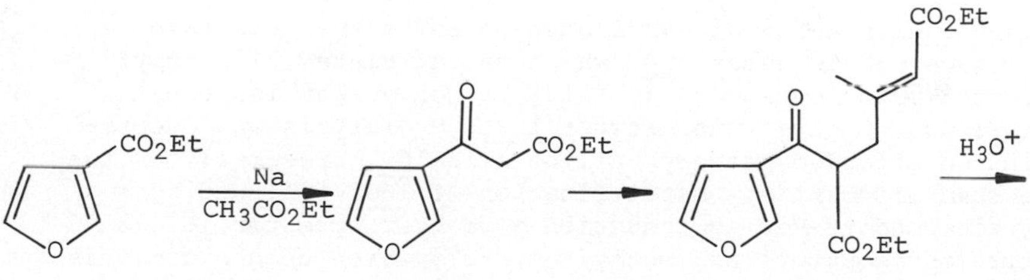

157   158   159   160   161

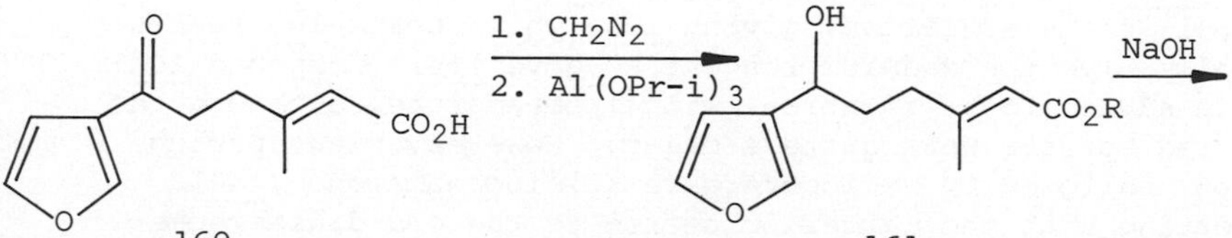

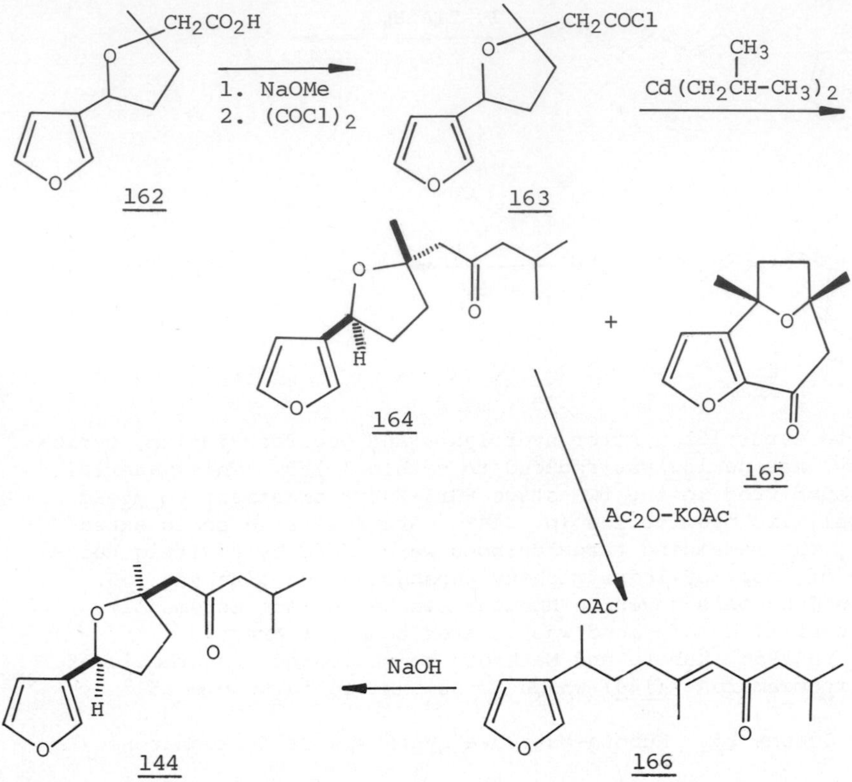

Claisen condensation of ethyl acetate and ethyl 3-furoate
(157) gave β-keto ester 158, which was alkylated with ethyl
4-bromo-3-methylbutanoate (p. 213), a convenient isoprene
equivalent, to give keto diester 159. Hydrolysis and decarboxylation afforded the crystalline acid 160, presumably as the
pure *trans*-isomeride. Esterification of 160, followed by
Meerwein-Pondorf-Verley reduction gave hydroxy ester 161 (a
mixture of isopropyl and methyl esters), which upon hydrolysis
afforded a mixture of diastereomeric acids 162. The derived
acid chloride mixture 163 reacted with isobutylcadmium to
yield 164 and 165. Presumably *cis*-163 undergoes intramolecular
Friedel-Crafts acylation, giving 165, while *trans*-163 reacts
normally with the cadmium reagent to give 164. Compound 164
was found not to be identical with ipomeamarone, although the
infrared spectra were quite similar. However, ring opening
to 166, followed by reclosure gave (±)-ipomeamarone (144),
suggesting that the latter substance is the *cis*-diastereomer.
Full details of this synthesis have not yet appeared, so that

the stereoselectivity in the ring opening-reclosure sequence
cannot be assessed.

3. MONOCARBOCYCLIC SESQUITERPENES

A. Sesquiterpenes Related to Bisabolene

A large number of sesquiterpenes are related to the bisabolenes
(2 and 3) which probably arise in vivo by the cyclization of
cis-Δ^2-farnesyl pyrophosphate (1).[1] The members of this class

$$\underline{1} \qquad\qquad \underline{2} \qquad\qquad \underline{3}$$

which have been synthesized are α-curcumene (4), ar-tumerone
(5), nuciferal (6), β-curcumene (7), γ-curcumene (8), α-
zingiberene (9), β-bisabolene (2), γ-bisabolene (3), isobisabolene (10), lanceol (11), α-bisabolol (12), cryptomerion (13),
todomatuic acid (14), juvabione (15), dehydrojuvabione (16),
and bilobanane (17).

Most of the syntheses in this area begin with a preformed
six-membered ring and center about various approaches to the
addition of the C-8 side chain. Stereochemistry is not a
typical problem in this series; only three of the compounds
synthesized (3, 6, 11) are capable of *cis-trans* isomerism.
Diastereomers are possible with α-zingiberene (9), todomatuic
acid (14), juvabione (15), and dehydrojuvabione (16). However,
the syntheses to date have not reckoned with this difficulty.
The recorded progress in the area consists mainly of various
approaches *to* the carbon skeleton and methods for controlling
the oxidation level of various positions on the basic skeleton.

$$\underline{4} \qquad\qquad \underline{5} \qquad\qquad \underline{6} \qquad\qquad \underline{7}$$

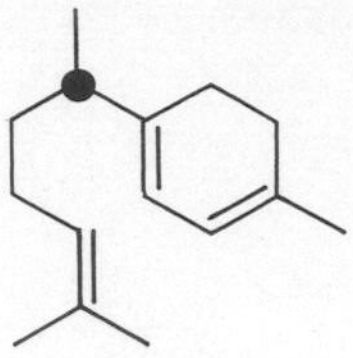

8 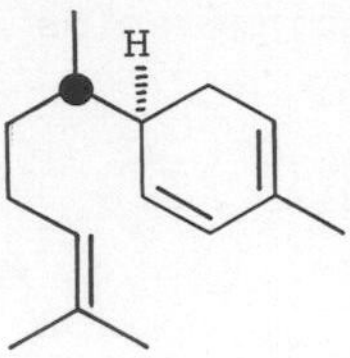   9 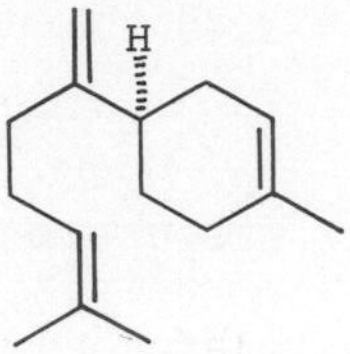   2 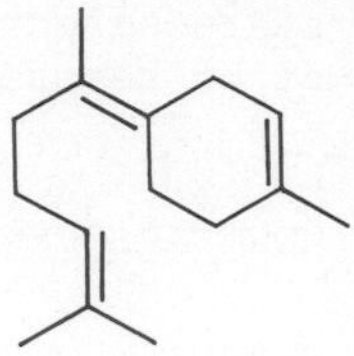   3

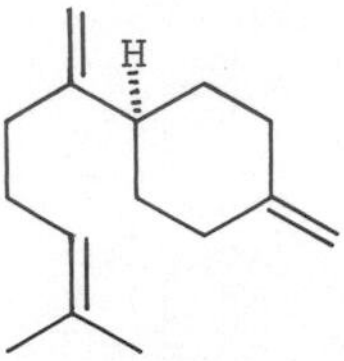

10 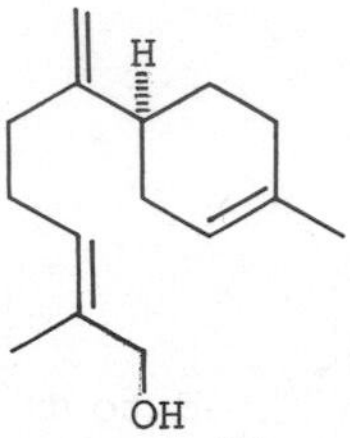   11 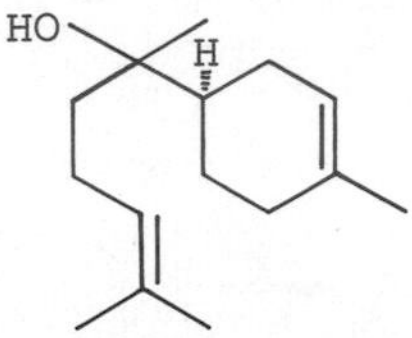   12 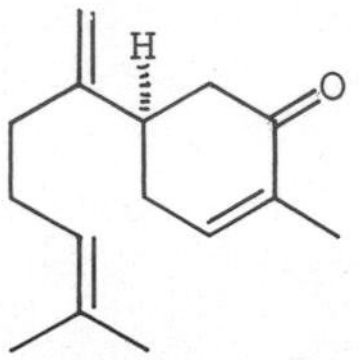   13

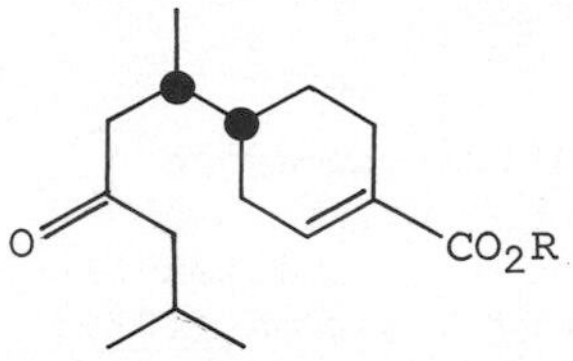

14   R=H   16   17
15   R=CH$_3$

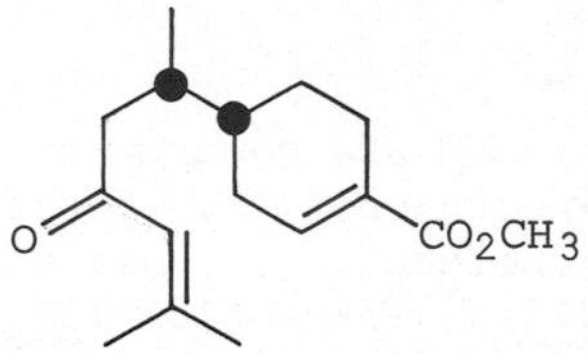

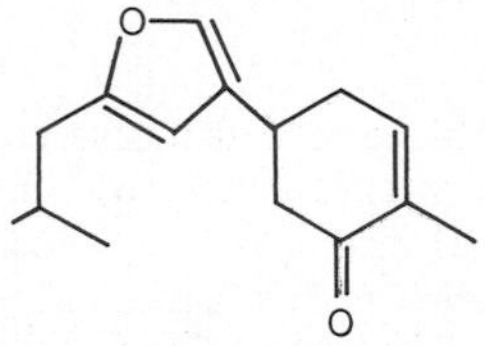

B. β-Bisabolene, γ-Bisabolene, Lanceol, Isobisabolene, Bisabolol

In a pioneering paper in the area of sesquiterpene synthesis, Ruzicka reported in 1925 that (±)-nerolidol (II-2, p. 200) is dehydrated by acetic anhydride to yield a hydrocarbon mixture, from which he succeeded in preparing a crystalline trishydrochloride, m.p. 79°-80°.[52] This substance was identical with the trishydrochloride obtained from natural bisabolene. Treatment of the synthetic trishydrochloride with sodium acetate in acetic acid gave synthetic bisabolene, "...wie natürlick zu erwarten war, mit dem analog aus Naturprodukten isolierten Kohlenwasserstoff identisch."[52]

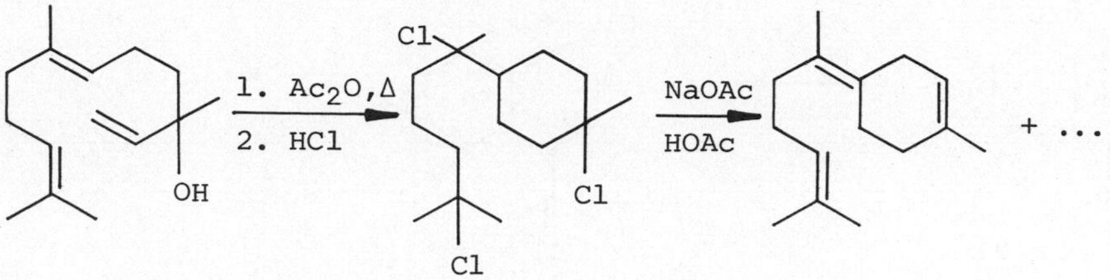

$\underline{2}$ $\underline{18}$ $\underline{3}$

Seven years later, Ruzicka reported a second synthesis of γ-bisabolene, which is summarized in Scheme 1.[53]   1-methyl-

Scheme 1.   Ruzicka's Second Bisabolene Synthesis

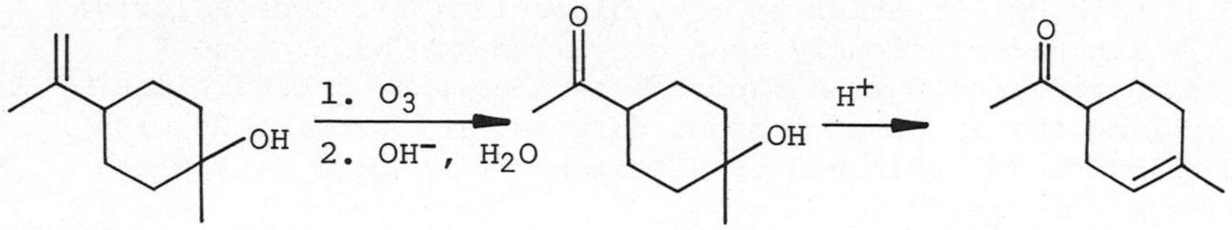

$\underline{19}$ $\underline{20}$ $\underline{21}$

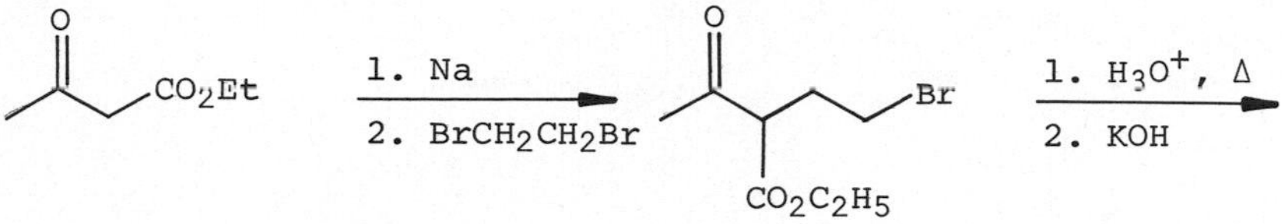

$\underline{22}$

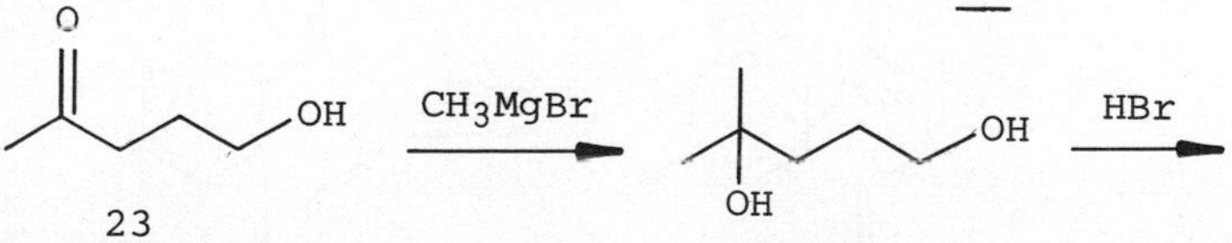

$\underline{23}$ $\underline{24}$

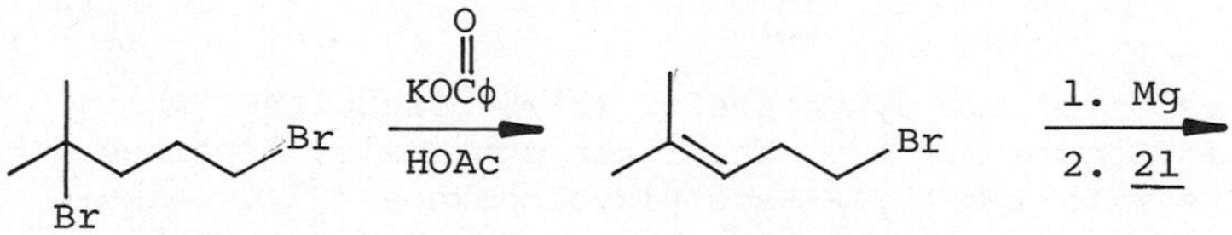

$\underline{25}$ $\underline{26}$

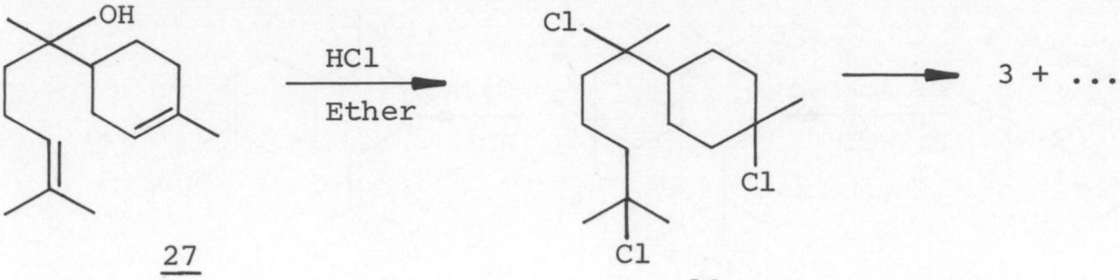

4-acetylcyclohexene (21), was synthesized by ozonolysis-dehydration of β-terpineol (19). To this ketone was added the
Grignard reagent derived from 2-methyl-5-bromo-2-pentene (26),
which was prepared in a straightforward manner beginning with
acetoacetic ester. The resulting alcohol, "bisabolol" (27),
reacted with·HCl in ether to give bisabolene trishydrochloride
(18), which had previously been converted to "bisabolene."

· Manjarrez and Guzman report a synthesis of β-bisabolene
by coupling the Grignard reagent derived from bromide 26 with
acid chloride 28, followed by Wittig methylenation of the resulting ketone 29.[54]

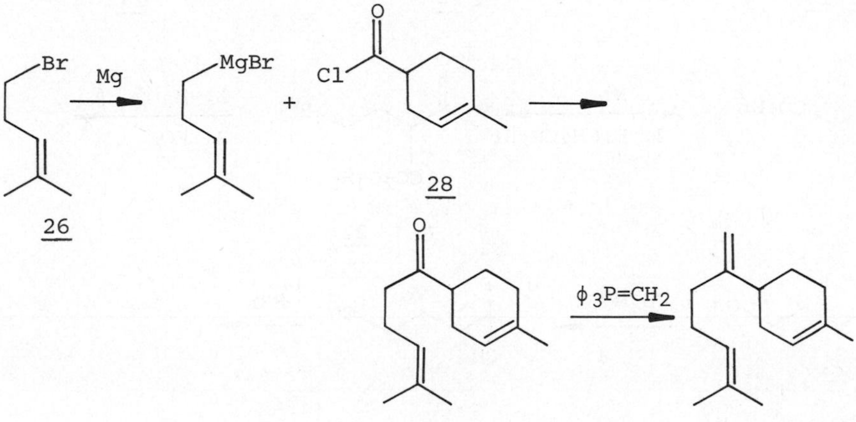

Vig has reported two syntheses of (±)-β-bisabolene, which
are outlined in Scheme 2.[55,56] The first synthesis, reported
in 1966, begins with 1-methyl-4-acetylcyclohexene (21). Alkylation of the derived β-keto ester 30 with 4-bromo-2-methyl-
2-butene gave 31, which was converted into dienone 32 by hydrolysis and decarboxylation. Wittig methylenation afforded
β-bisabolene. The second Vig synthesis is longer and was actually accomplished only as an adjunct to a synthesis of

Scheme 2. Vig's β-Bisabolene Syntheses

First Synthesis

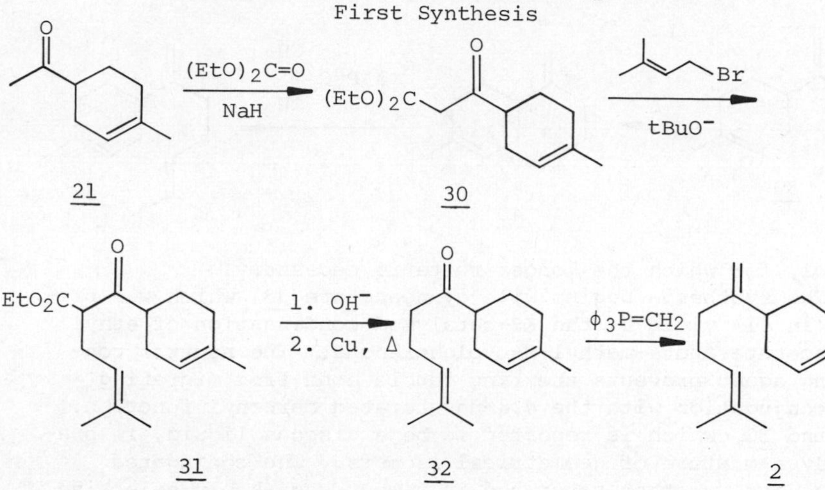

$\underline{21}$ → (EtO)$_2$C=O / NaH → $\underline{30}$ → with Br-containing reagent / tBuO$^-$

$\underline{31}$ → 1. OH$^-$ 2. Cu, Δ → $\underline{32}$ → ϕ_3P=CH$_2$ → $\underline{2}$

Second Synthesis

$\underline{33}$ (CO$_2$Et, NC) → H$_2$-Pd/C → $\underline{34}$ (CO$_2$Et, NC) → NaBH$_4$ →

$\underline{35}$ (CN, OH) → NaOH / Δ → $\underline{36}$ (HO$_2$C) → EtOH / H$^+$ →

$\underline{37}$ (EtO$_2$C) → LiAlH$_4$ → $\underline{38}$ (OH) → vinyl ethyl ether / Hg(OAc)$_2$ →

237

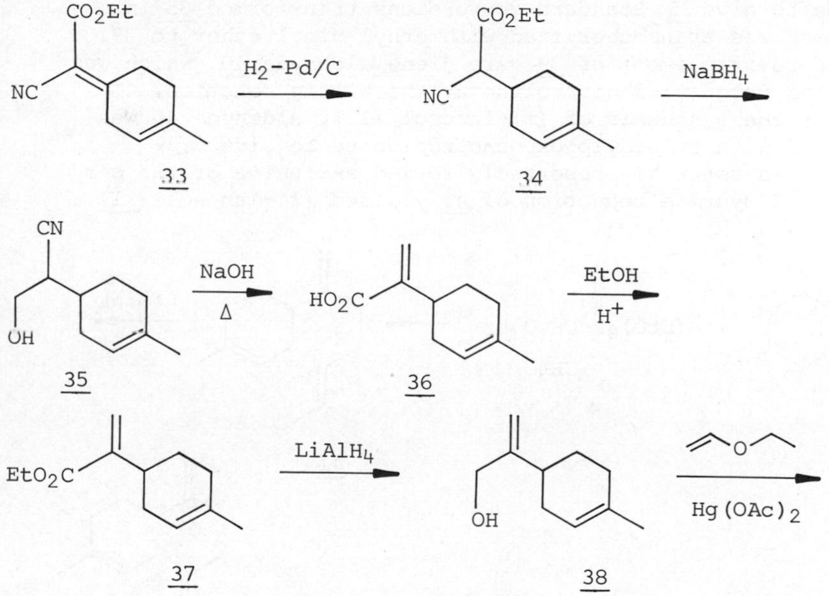

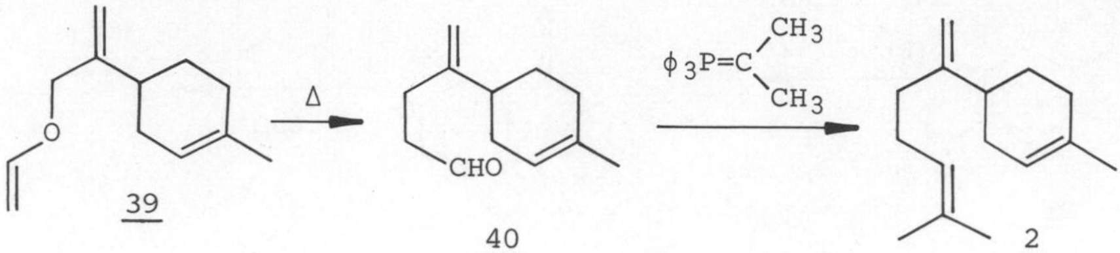

lanceol, for which the longer route is necessary.

The synthesis begins with cyanoacetate 33, which was prepared in 41% yield by the KF-catalyzed condensation of ethyl
cyanoacetate and 4-methyl-3-cyclohexenone. The neutral condensing agent prevents the ring double bond from migrating
into conjugation with the α,β-unsaturated carbonyl function.
Compound 33, which is reported to be a viscous liquid, is presumably a mixture of geometrical isomers. The conjugated
double bond was then saturated in a most remarkably selective
hydrogenation to give 34, which was reduced with sodium borohydride to give 35. Standard conversions transformed 35 into
38, which was transetherified with ethyl vinyl ether to 39.
Claisen rearrangement of 39 gave diene aldehyde 40, which was
converted into (±)-β-bisabolene by the Wittig reaction.

For the synthesis of (±)-lanceol (11), aldehyde 40 was
condensed with triethylphosphonopropionate to give α,β-
unsaturated ester 41, reportedly formed exclusive of the cis
isomer.[56] Hydride reduction of 41 yielded (±)-lanceol (11).

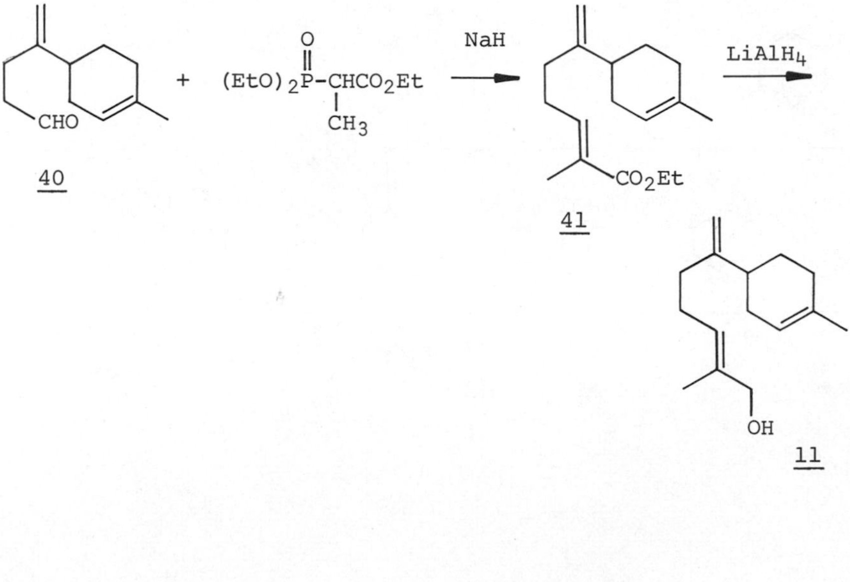

(±)-Lanceol had previously been synthesized by Manjarrez, Rios, and Guzman[57] by a route which unambiguously determined the geometry (then unknown) of the olefinic linkage. The synthesis (Scheme 3) starts from enone 21 (prepared by Diels-Alder

Scheme 3.   Manjarrez-Rios-Guzman Synthesis of (±)-Lanceol

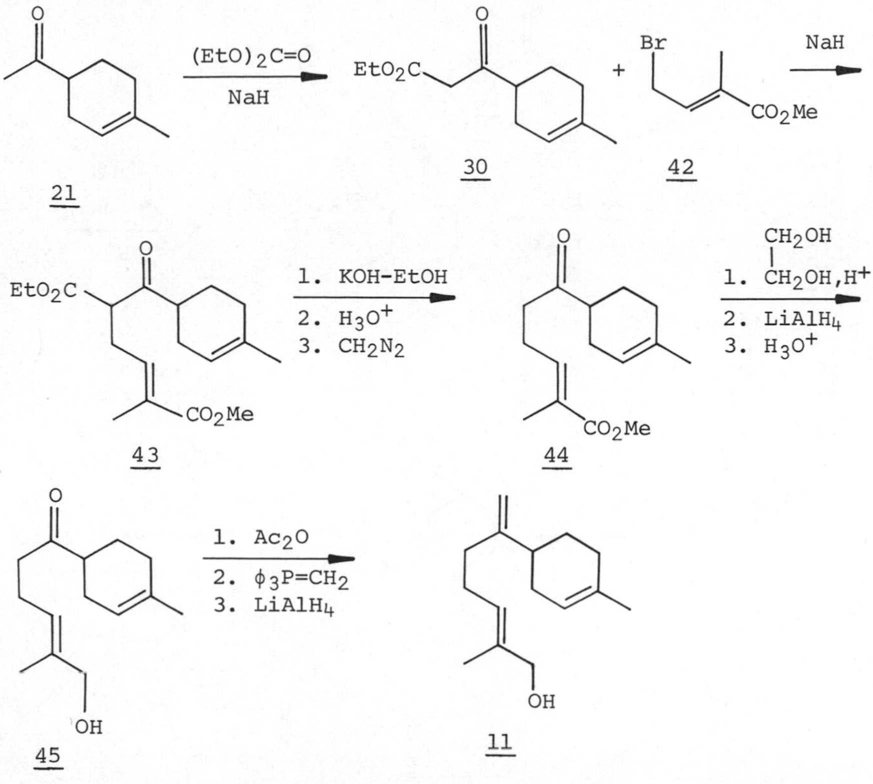

condensation of isoprene and methyl vinyl ketone). Carbethoxylation gave 30, which was alkylated with methyl γ-bromotiglate, of known *trans*-geometry, to yield keto diester 43. Hydrolysis, decarboxylation and re-esterification then afforded 44. After protecting the ketonic carbonyl as its dioxolane derivative, the ester grouping was reduced, leading to keto alcohol 45. Wittig methylenation of the derived acetate gave, after removal of the acetyl protecting group (±)-lanceol (11).

Vig's group has reported a synthesis of isobisabolene (10) which is somewhat different from other work in the class, since the six-membered ring is not the starting point (Scheme 4).[58]   Methyl heptenone (II-14) was carbethoxylated to yield

Scheme 4. Vig's Isobisabolene Synthesis

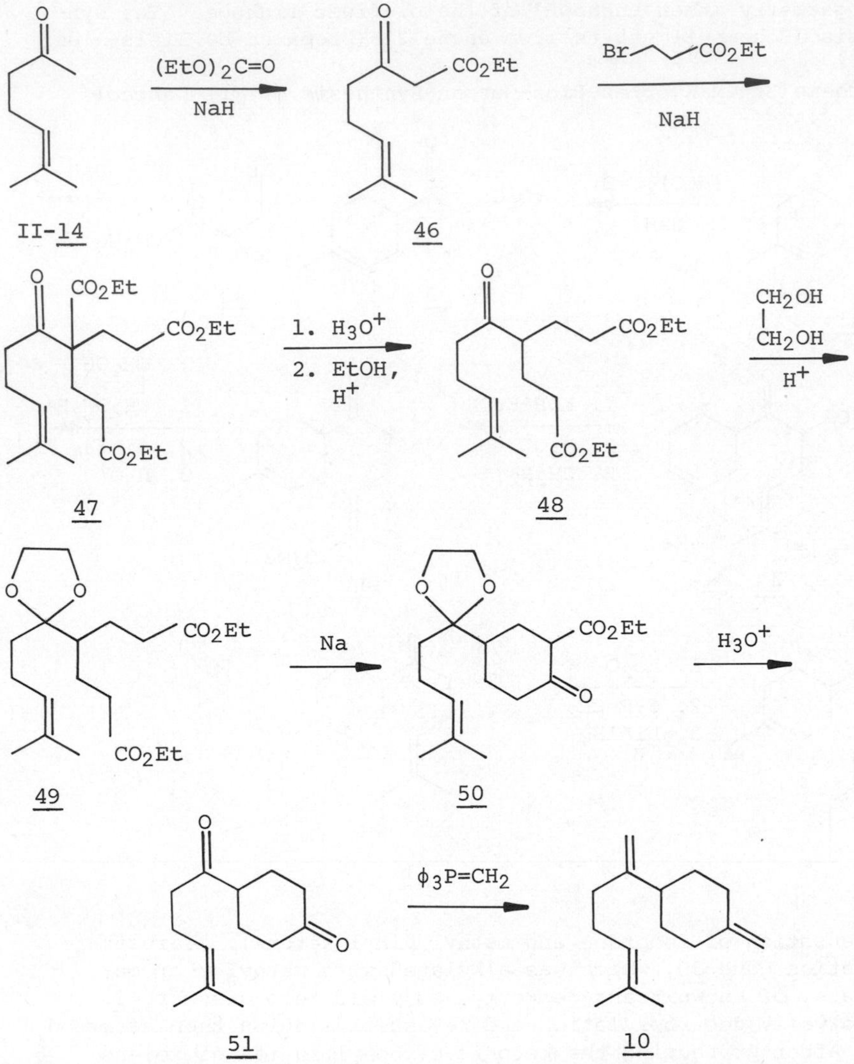

β-keto ester <u>46</u>. No comment is made regarding the apparent
clean selectivity for the methyl group. However, all of the
intermediates in the synthetic sequence are liquids, and no
crystalline derivatives are reported. It is therefore likely
that some isomeric materials may be present as side products.
Keto ester <u>46</u> was alkylated with two equivalents of ethyl

β-bromopropionate, giving keto triester $\underline{47}$, which was hydrolyzed, decarboxylated, and re-esterified to obtain keto diester $\underline{48}$.   After ketalization, a Dieckman cyclization yielded
$\underline{50}$, which was hydrolyzed to dione $\underline{51}$.   The synthesis of (±)-
isobisabolene ($\underline{10}$) was completed by a double Wittig reaction.

Kuznetsov and Myrsina report that (±)-bisabolol ($\underline{12}$) may
be prepared from tetrahydro-*p*-toluonitrile ($\underline{52}$) by treating it
with 4-methyl-3-pentenylmagnesium bromide.   The resulting dienone ($\underline{32}$) was treated with methylmagnesium bromide to give
(±)-bisabolol ($\underline{12}$), presumably a mixture of diastereomers.[59]

C.   Curcumenes, Zingiberene

Perhaps the simplest sesquiterpene related to the bisabolenes,
from a synthetic standpoint, is α-curcumene ($\underline{4}$).   The substance occurs in nature along with its double bond isomer ($\underline{53}$),
and the isolated mixture of $\underline{4}$ and $\underline{53}$ has been referred to as
(-)-α-curcumene.   To avoid confusion we shall refer to $\underline{4}$ as
α-curcumene and $\underline{53}$ as *iso*-α-curcumene.   Most synthetic work
in the area has been directed toward the synthesis of a mix-
ture of $\underline{4}$ and $\underline{53}$ which approximates the composition of the
naturally derived mixture.

In 1940, Simonsen, Carter, and Williams synthesized α-
curcumene and *iso*-α-curcumene by the route shown in Scheme 5.[60]
Friedel-Crafts acylation of toluene by glutaric anhydride
gives keto acid $\underline{54}$ in low yield.   This substance was esterified and treated with methylmagnesium iodide to give crystal
line acid $\underline{55}$, probably the *trans* isomer.   Reduction of $\underline{55}$

Scheme 5.   Simonsen Synthesis of α-Curcumene

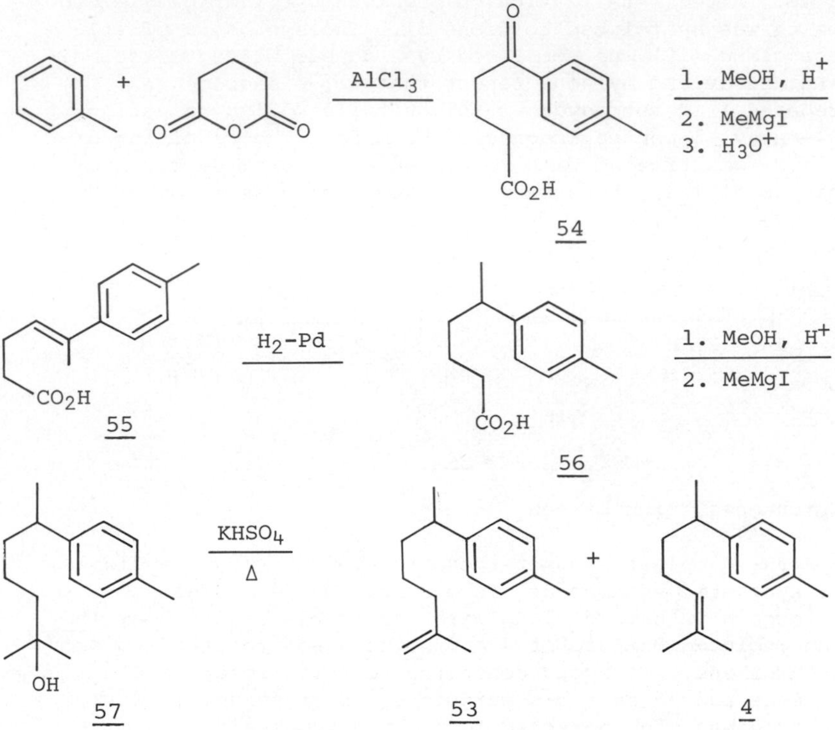

yielded 56. Due to the low yield in the first stage, this acid was also prepared by nitrile chain extension of ethyl γ-p-tolyl-n-valerate. Addition of Grignard reagent to the corresponding methyl ester gave an alcohol (57), which was dehydrated to a mixture of 4 and 53. On the basis of ozonization studies, it was concluded that the latter isomer predominated, while the naturally isolated mixture is mostly the isopropylidine isomer (4).

Birch and Mukherji, in a 1949 paper pointing out the utility of dissolving metal reductions in synthesis, record syntheses of α-, β-, and γ-curcumene (Scheme 6).[61] Condensation of p-tolylmagnesium bromide with methylheptenone (II-14) gave alcohol 58. Stepwise hydrogenolysis and reduction of the aromatic ring led to (±)-α-curcumene (4) and (±)-β-curcumene (7). For the synthesis of γ-curcumene (8), methylheptenone was added to the Grignard reagent derived from p-bromoanisole. The resulting alcohol was readily dehydrated upon workup with

Scheme 6.   Birch Synthesis of the Curcumenes

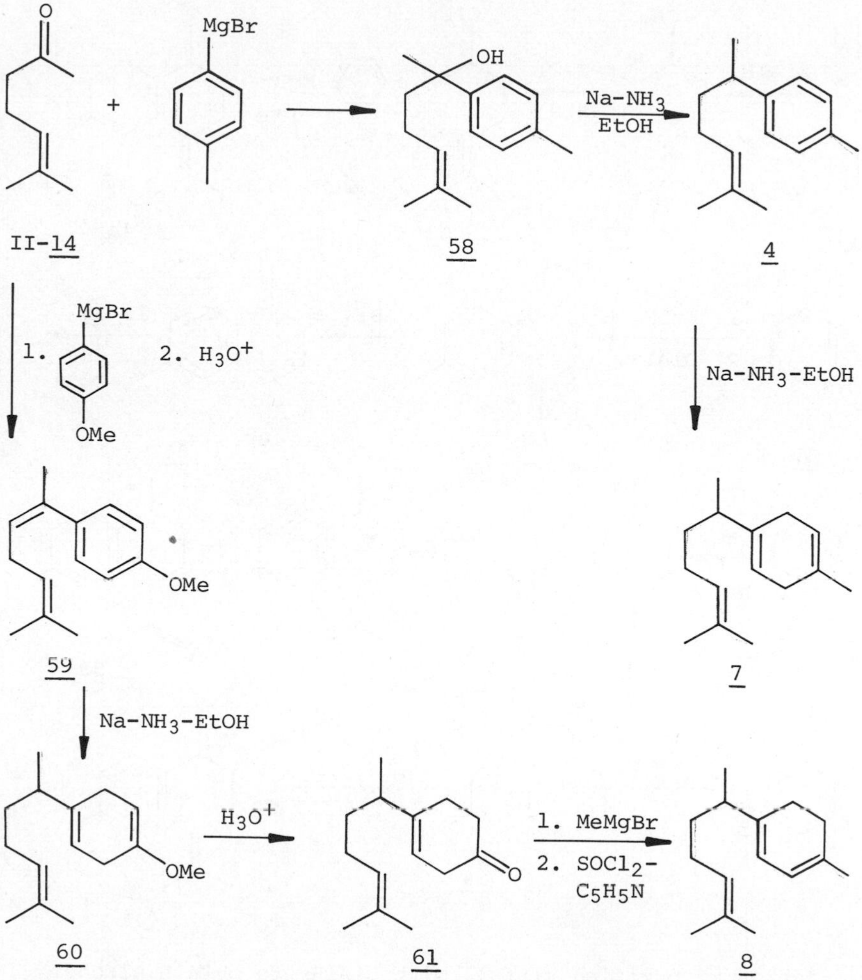

II-14   58   4

aqueous ammonium chloride.   Diene 59 was reduced to the tetrahydro stage, and the resulting enol ether 60 was hydrolyzed to
give dienone 61. Compound 61 was formulated as the β,γ-
isomer because of its lack of absorption in the 220-280 mμ
region. Addition of methylmagnesium bromide to 61 and dehydration of the resulting alcohol gave (±)-γ-curcumene (8).

In three 1965 papers, Rao and Honwad reported the syntheses of α-curcumene outlined in Scheme 7.[62,63] In the
simplest,[62] Rao reduced methylheptenone (obtained by reverse

Scheme 7.   Rao's α-Curcumene Syntheses

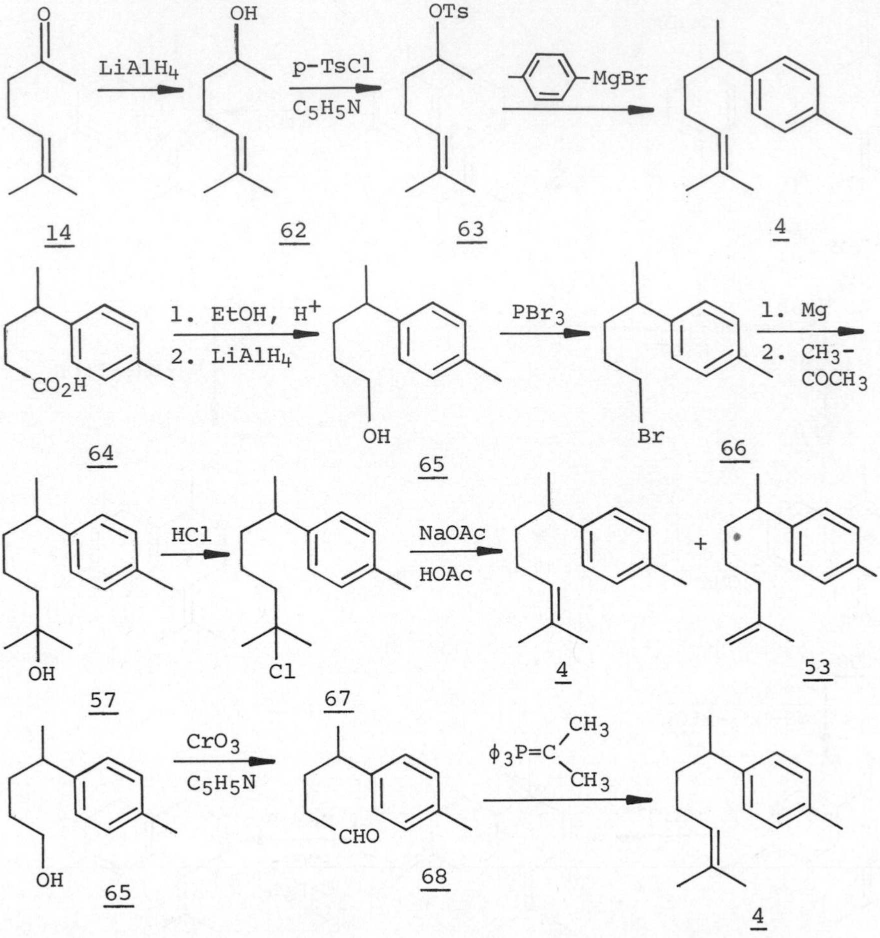

aldolization of purified citral) with LiAlH₄ to obtain alcohol
62. The derived p-toluenesulfonate ester was treated with p-
tolylmagnesium bromide to give (±)-α-curcumene (**4**) in approximately 10% yield.

In conjunction with a study which established the absolute configuration of (-)-α-curcumene,[63] Honwad and Rao synthesized a mixture of (±)-α-curcumene and (±)-*iso*-α-curcumene
(**53**) from Rupe's γ-p-tolyl-n-valerate (**64**).[64] After esterification and hydride reduction, alcohol **65** was obtained. The
Grignard reagent derived from the corresponding bromide was
added to acetone to yield alcohol **57**. This alcohol was

converted into chloride 67, which was dehydrochlorinated with
sodium acetate in acetic acid to yield 4 and 53.

In a separate communication,[63] it was reported that alcohol 65 may be oxidized by Sarett's reagent to yield aldehyde
68. Treatment of 68 with isopropylidinetriphenylphosphorane
gives (±)-α-curcumene (4).

Although Birch[61] and Rao[62,63] had prepared pure (±)-α-
curcumene, uncontaminated with (±)-iso-α-curcumene, Vig was
the first to synthesize each isomer in a pure state (Scheme
8).[65] Rupe's acid (64)[64] was converted, by hydride reduction

Scheme 8. Vig's Synthesis of (±)-α-Curcumene and
(±)-iso-α-Curcumene

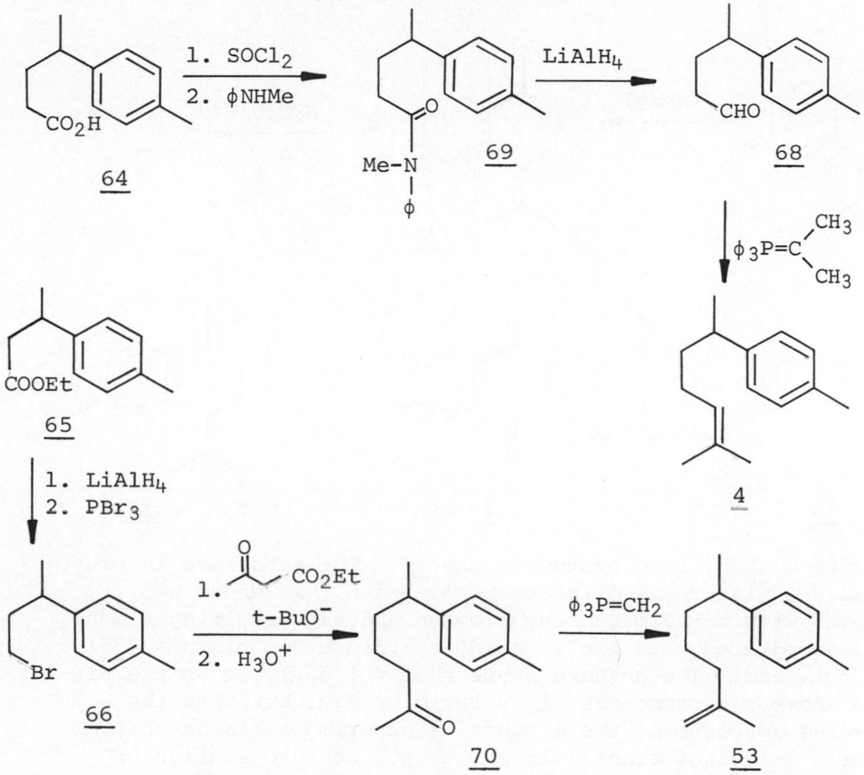

of its N-methylanilide (69) into aldehyde 68. Alternatively,
reduction of the ethyl ester 65 gave an alcohol, which was
converted into bromide 66. Acetoacetic ester synthesis on 66
gave methyl ketone 70. Appropriate Wittig reactions on 68 and
70 gave the isomeric α-curcumenes.

In 1968, Joshi and Kulkarni reported a synthesis of (-)-zingiberene (9) and (-)-α-curcumene (4) from (+)-citronellal (Scheme 9).[66] Citronellal (71) was converted, by Stork's

Scheme 9.   Joshi-Kulkarni Synthesis of α-Curcumene

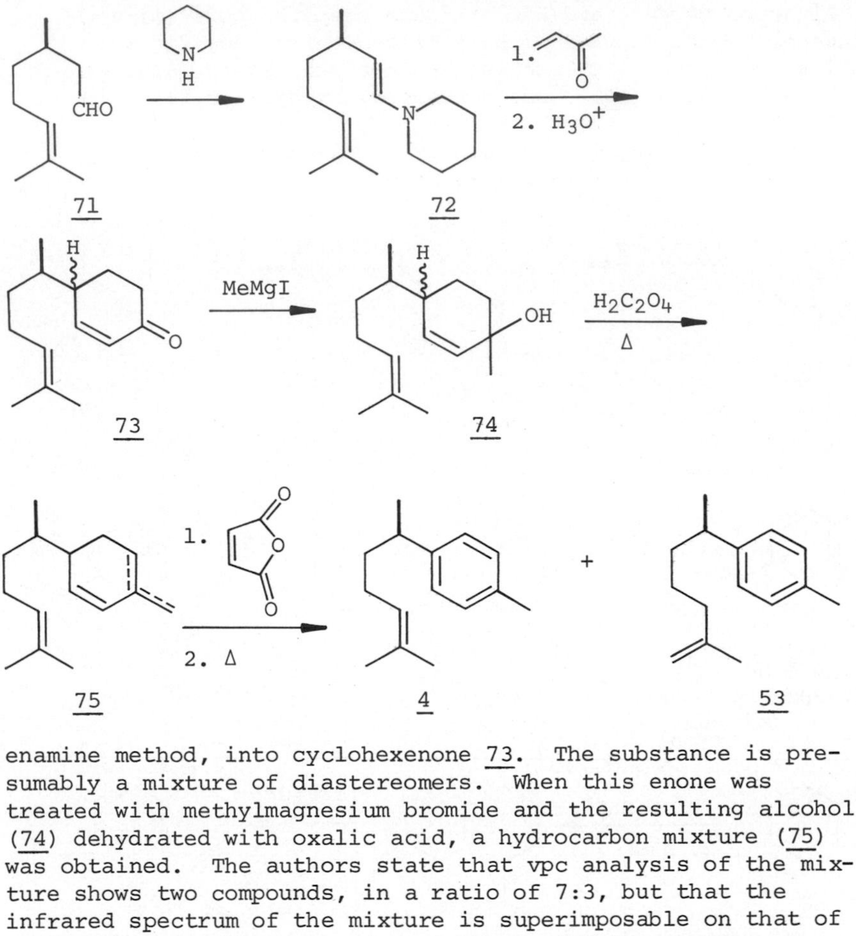

enamine method, into cyclohexenone 73. The substance is presumably a mixture of diastereomers. When this enone was treated with methylmagnesium bromide and the resulting alcohol (74) dehydrated with oxalic acid, a hydrocarbon mixture (75) was obtained. The authors state that vpc analysis of the mixture shows two compounds, in a ratio of 7:3, but that the infrared spectrum of the mixture is superimposable on that of natural (-)-zingiberene. In any event, dehydrogenation of mixture 75, by pyrolysis of its maleic anhydride adduct over a free flame, gave a mixture of 4 and 53. Although this synthesis of the curcumene isomers is probably valid, the uncertainty surrounding the relative stereochemistry of the two asymmetric centers in the Joshi-Kulkarni "zingiberene" remains. In an probability, the actual synthetic material was a mixture of all possible isomers.

Other nonstereospecific syntheses of zingiberene had previously been reported by Mukherji and Bhattacharyya[67] and Banerjee.[68] For the Mukherji-Bhattacharyya synthesis (Scheme 10), the starting point was the enone 61 utilized by Mukherji

Scheme 10. Mukherji-Bhattacharyya Synthesis of Zingiberene

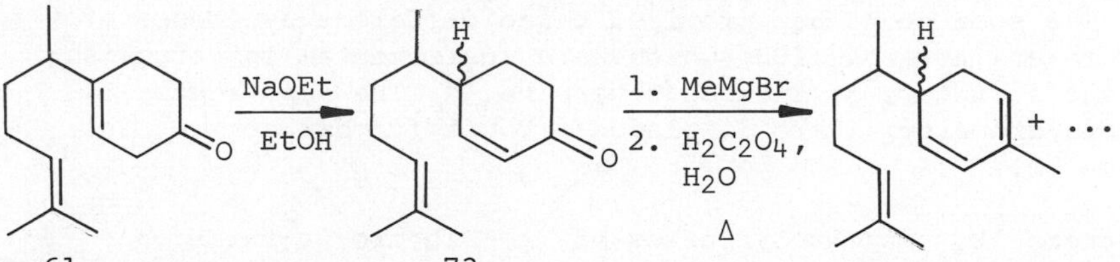

61   73   9

and Birch in the synthesis of γ-curcumene (Scheme 6). From this point, the synthesis is identical to that of Joshi and Kulkarni, which it predates by 15 years. The synthetic mixture gave an ultraviolet spectrum quite similar to that of isolated zingiberene. Banerjee's 1962 synthesis is also identical to the earlier Mukherji-Bhattacharyya synthesis.

D. ar-Turmerone

Rupe has synthesized (±)-ar-turmerone (5) as outlined in Scheme 11.[69,70] Reformatsky reaction of ethyl bromoacetate

Scheme 11. Rupe's Synthesis of (±)-ar-turmerone

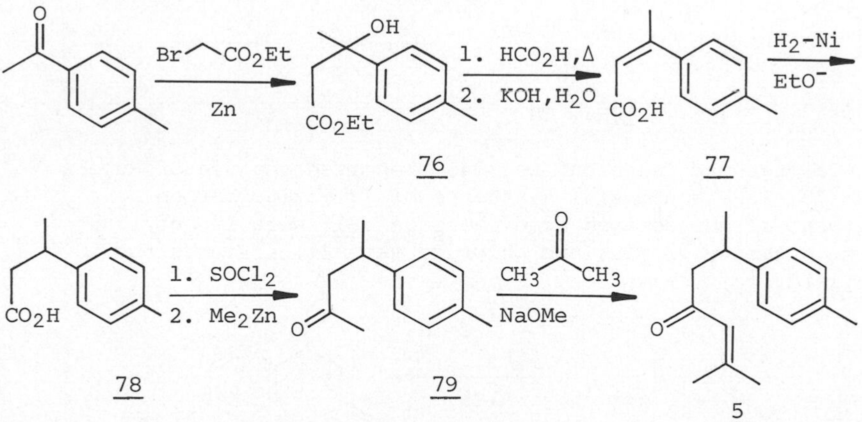

76   77

78   79   5

and p-methylacetophenone gave crystalline alcohol 76, which
was dehydrated by formic acid to the dimethyl cinnamic acid
77. Hydrogenation over nickel gave acid 78. Treatment of the
derived acid chloride with dimethylzinc gave, in quantitative
yield, (±)-curcumone (79), itself isolable from natural
sources. When the synthetic curcumone was condensed with acetone, in the presence of sodium methoxide, a very low yield of
(±)-ar-turmerone (5) was obtained.

The same group has provided three different syntheses of
dihydro-ar-turmerone (80) which were instrumental in establishing the structure of the sesquiterpene.[70] The Rupe routes to
the dihydromaterial are displayed without further comment in
Scheme 12.

Scheme 12.   Rupe's Syntheses of (±)-Dihydro-ar-turmerone

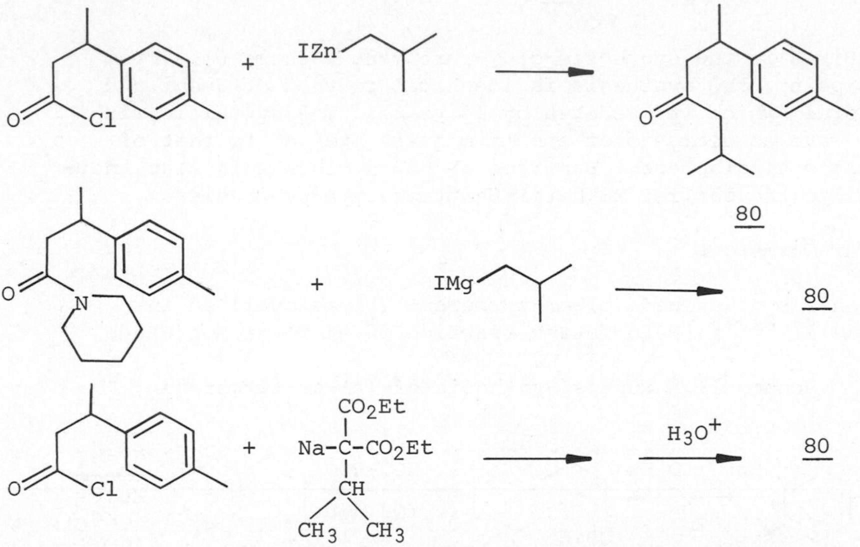

Colonge and Chambion, in 1946, reported the use of Rupe's
acid (78) in a successful synthesis of (±)-ar-turmerone.
Treatment of the derived acid chloride (81) with isobutylene
in the presence of aluminum chloride gave (±)-ar-turmerone in
50% yield.[71]

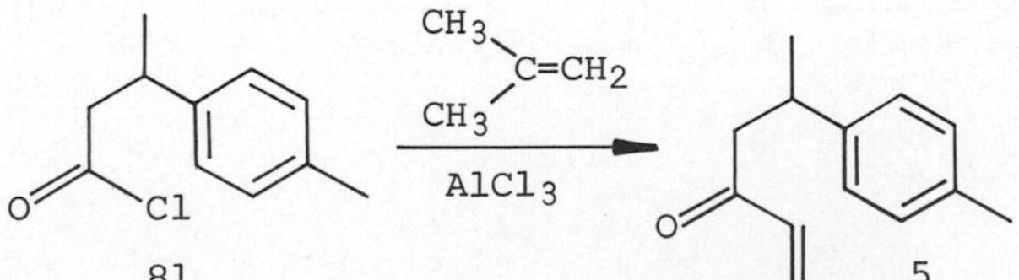

In 1947, Mukherji provided an "unambiguous" synthesis,[72] starting with aldehyde 82 (Scheme 13), which is prepared from p-methyl-acetophenone. Compound 82 was reduced with zinc in acetic acid to alcohol 83. The Grignard reagent derived from the corresponding bromide (84) was added to the piperidide of 3-methyl-2-butenoic acid to obtain (±)-ar-turmerone.

Scheme 13. Mukherji's Synthesis of (±)-ar-turmerone

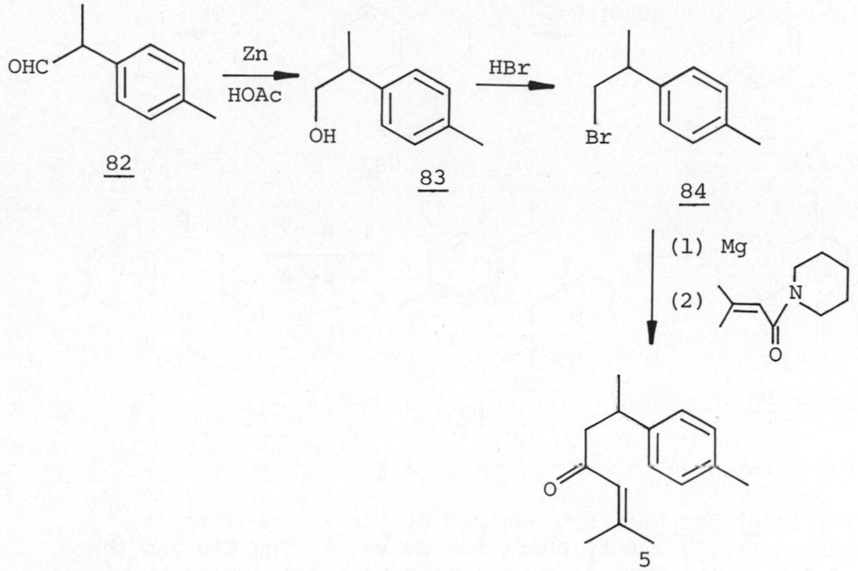

Twelve years later, Gandhi, Vig, and Mukherji provided another "unambiguous" synthesis of (±)-ar-turmerone, outlined in Scheme 14.[73]

Scheme 14. Gandhi-Vig-Mukherji Synthesis of (±)-ar-Tumerone

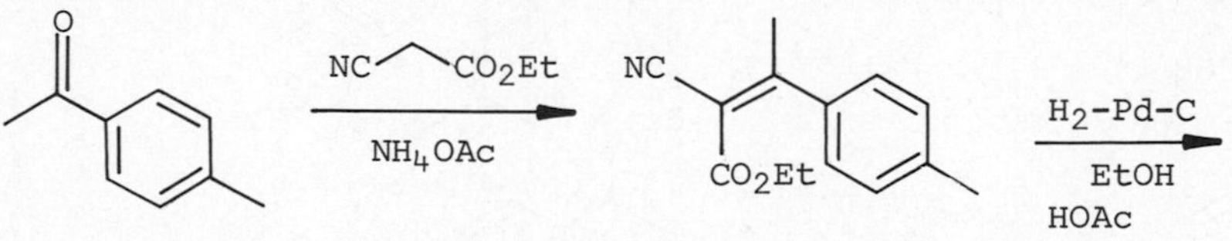

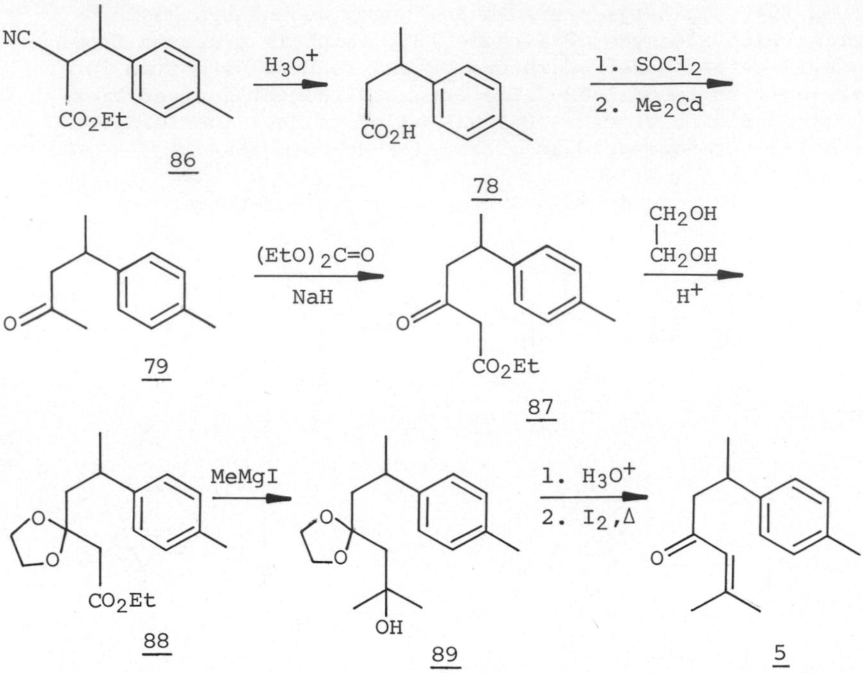

E. Nuciferal

Nuciferal (6) has been synthesized by Vig and co-workers,[74] Honwad and Rao,[63] and by Büchi and Wüest.[75] The Vig and the Honwad-Rao syntheses are identical (Scheme 15) and begin with

Scheme 15. Vig-Honwad-Rao Synthesis of Nuciferal

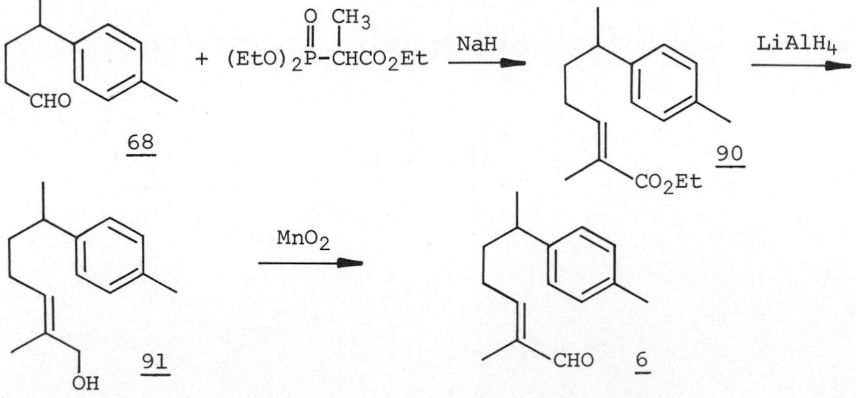

aldehyde 68, which was the key intermediate in the synthesis
of α-curcumene (Schemes 7 and 8). Condensation of this aldehyde with triethylphosphonopropionate gave exclusively the
trans ester 90, which was reduced by LiAlH₄ to allylic alcohol
91. Oxidation of 91 with manganese dioxide gave (±)-nuciferal
    The Büchi-Wüest synthesis (Scheme 16) also proceeds via

Scheme 16. Büchi-Wüest Nuciferal Synthesis

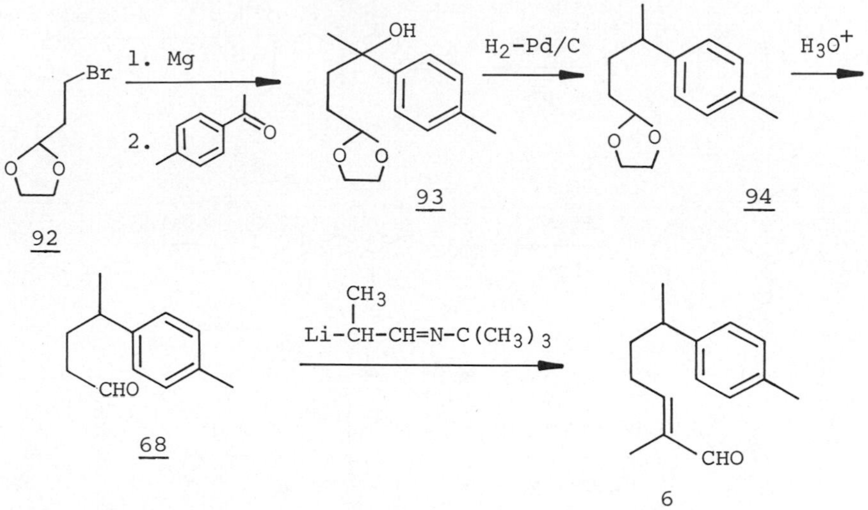

aldehyde 68, which was prepared by adding the Grignard reagent from bromide 92 to p-methylacetophenone, hydrogenolysis
of the resulting benzylic alcohol (93) and hydrolysis of the
acetal grouping. The remaining three carbons were added
stereospecifically by a Wittig directed aldol condensation (cf.
Büchi's synthesis of sinensal, Scheme 20), yielding racemic
nuciferal.

F. Cryptomerion

Vig and c-workers have reported the synthesis of (±)-crypto-
merion (13) which is outlined in Scheme 17.[76] The theme of
this synthesis, utilization of the Claisen rearrangement to
elaborate a five-carbon aldehydic side-chain which is converted
into the requisite isopropylidine moiety by a Wittig reaction,
is similar to the approach used in the synthesis of β-bisabolene (Scheme 2). 6-Methyl-2-cyclohexenone (96), obtained by
Birch reduction of o-cresol methyl ether (95) was converted
into keto diester 97 by Michael addition of diethyl malonate.

Scheme 17. Vig's Synthesis of (±)-Cryptomerion

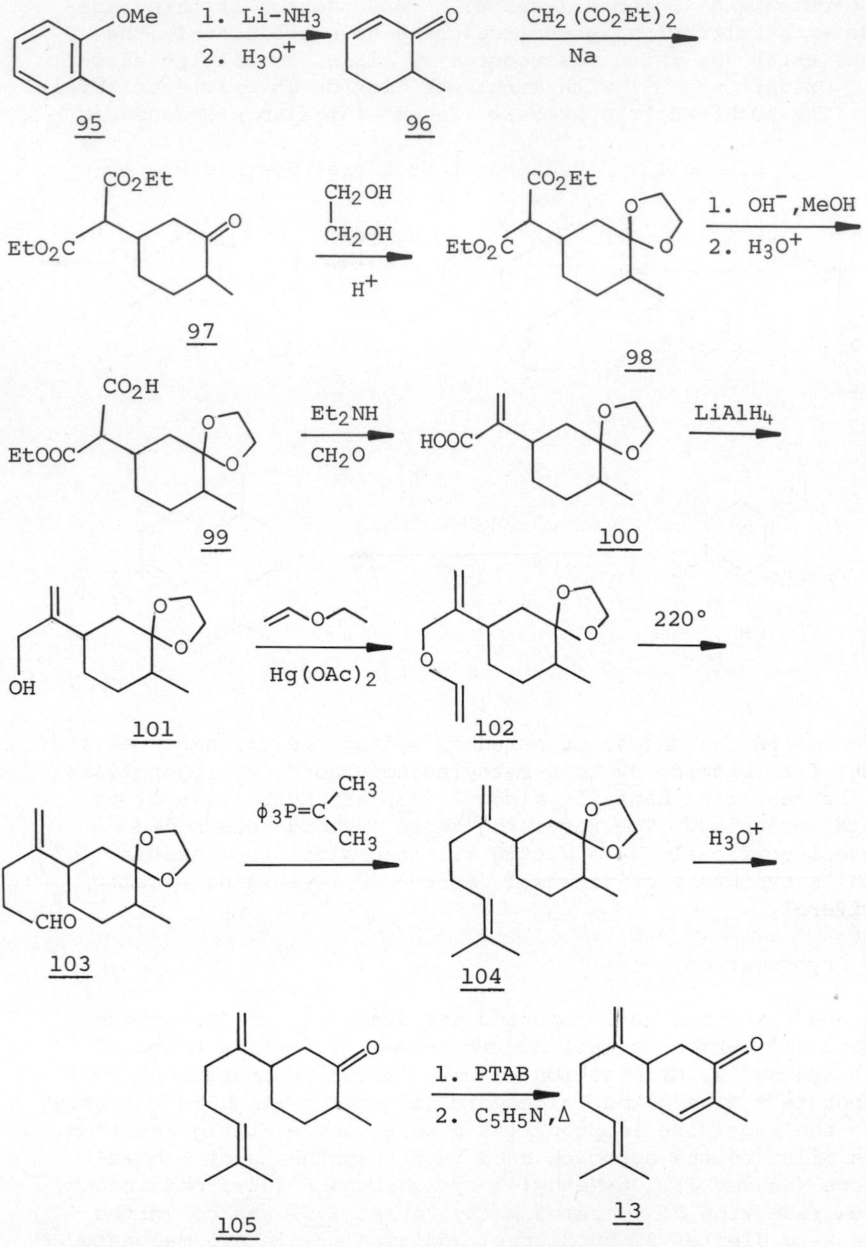

After ketalization, the diester was selectively hydrolyzed to give acid 99. This material was subjected to a Mannich reaction to give unsaturated ester 100, which was reduced with LiAlH₄ to give allylic alcohol 101. After chain extension via vinyl ether 102, the resulting aldehyde (103) was treated with isopropylidinetriphenylphosphorane to give diene 104. The protecting group was removed by hydrolysis and the resulting ketone brominated with phenyltrimethylammonium perbromide (PTAB). Dehydrobromination afforded (±)-cryptomerion (13). The selectivity in the last two stages of the synthesis is remarkable. No isomeric products are reported to be formed in either the bromination or dehydrobromination steps.

G.   Todomatuic Acid and Juvabione

Todomatuic acid was first isolated in 1940 from a byproduct of the sulfite pulp industry in Japan.[77] The correct structure (14) was assigned in 1941 by Momose.[78] In more recent years, the methyl ester of todomatuic acid ("juvabione," 15) and the related dehydrojuvabione (16) have been isolated from balsam fir and shown to have high juvenile hormone activity.[79,80]

    An early synthesis of (±)-desoxotodomatuic acid (106)[81] has been followed by four syntheses of juvabione (15)[82-85]

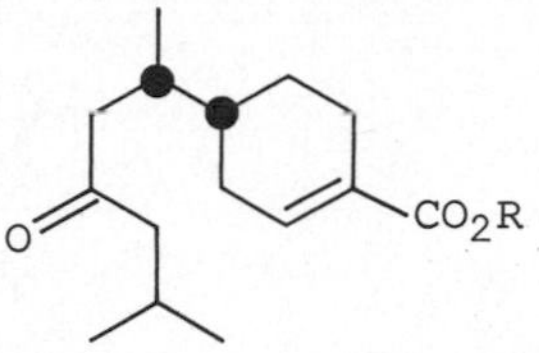

14: R=H
15: R=CH₃   106 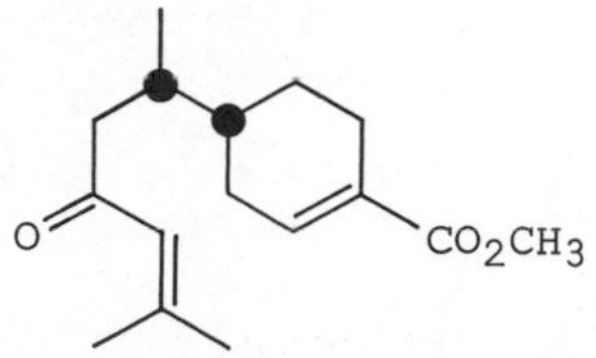   16

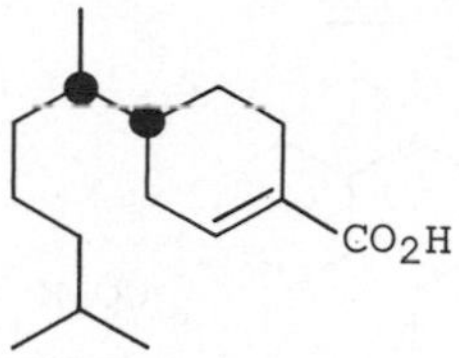

and a nonstereoselective synthesis of dehydrojuvabione (16).[86]

    The Nakazaki-Isoe synthesis of (±)-desoxotodomatuic acid (106) carried out in order to confirm Momose's structure assignment, is outlined in Scheme 18.[81] Anisole was acylated

Scheme 18. Nakazaki-Isoe Synthesis of (±)-Desoxytodomatuic Acid

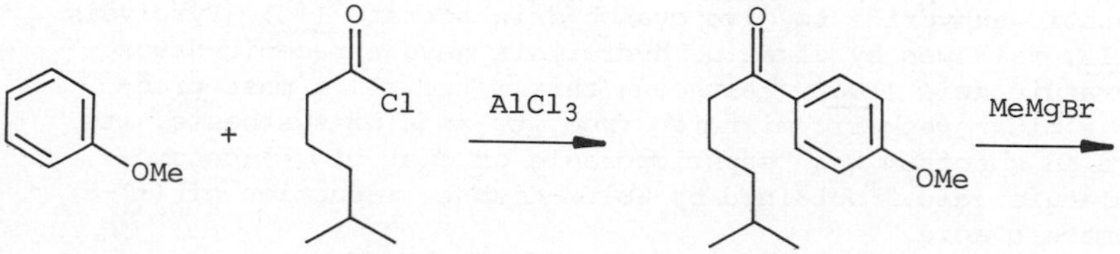

107

108 109

110 111

112 113

106

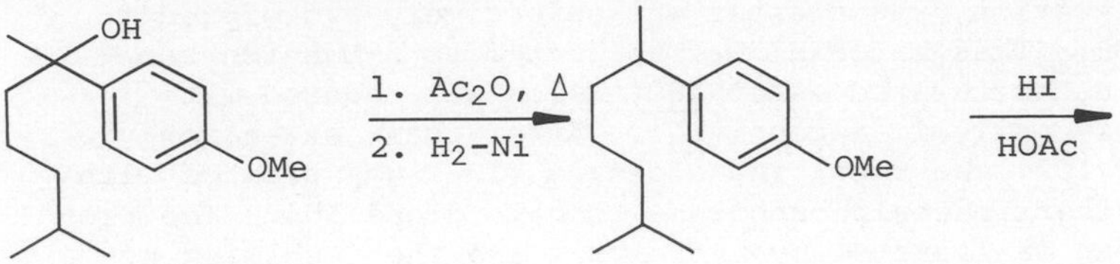

with 5-methylhexanoyl chloride to yield ketone 107, which was treated with methylmagnesium bromide to afford alcohol 108. This alcohol was dehydrated and the resulting alkene hydrogenated to give phenyl ether 109. Demethylation gave phenol 110, which was hydrogenated to give alcohol 111 (diastereomeric mixture?). Oxidation of 111 gave 112, which reacted with KCN in acetic anhydride to give cyanohydrin acetate 113. Pyrolysis of 113, followed by alkaline hydrolysis gave a racemic desoxotodomatuic acid (106). Although this material is most probably a diastereomeric mixture, from its mode of synthesis, its infrared spectrum was "superimposable on that of (+)-desoxotodomatuic acid," obtained by Wolff-Kishner reduction of (+)-todomatuic acid.

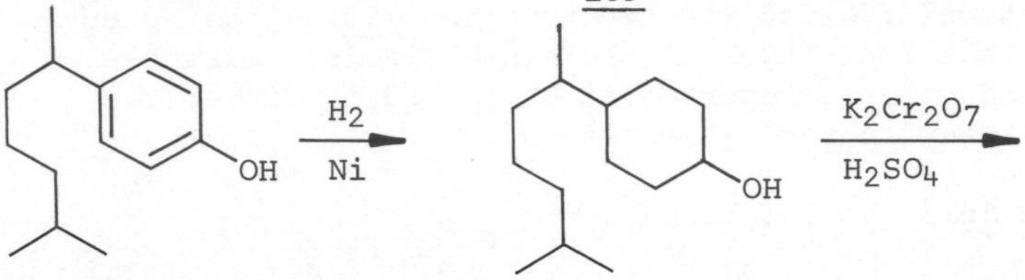

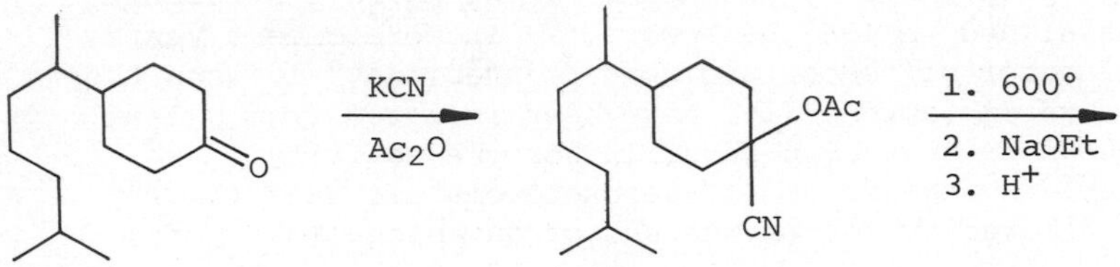

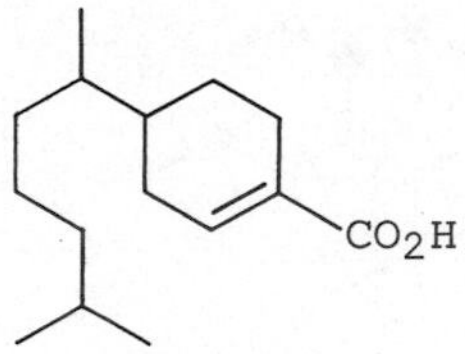

The Mori-Matsui synthesis of todomatuic acid and juvabione (Scheme 19) follows an outline similar to most work on

Scheme 19.   Mori-Matsui Synthesis of (±)-Todomatuic Acid
and (±)-Juvabione

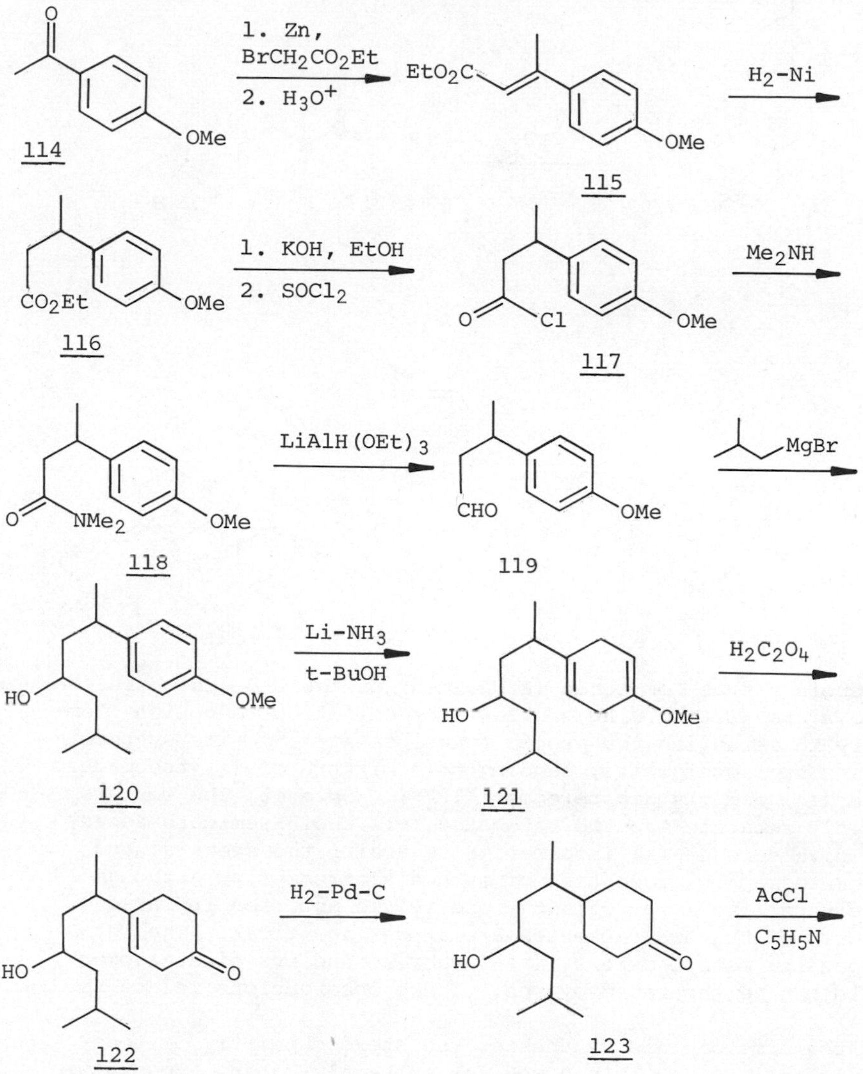

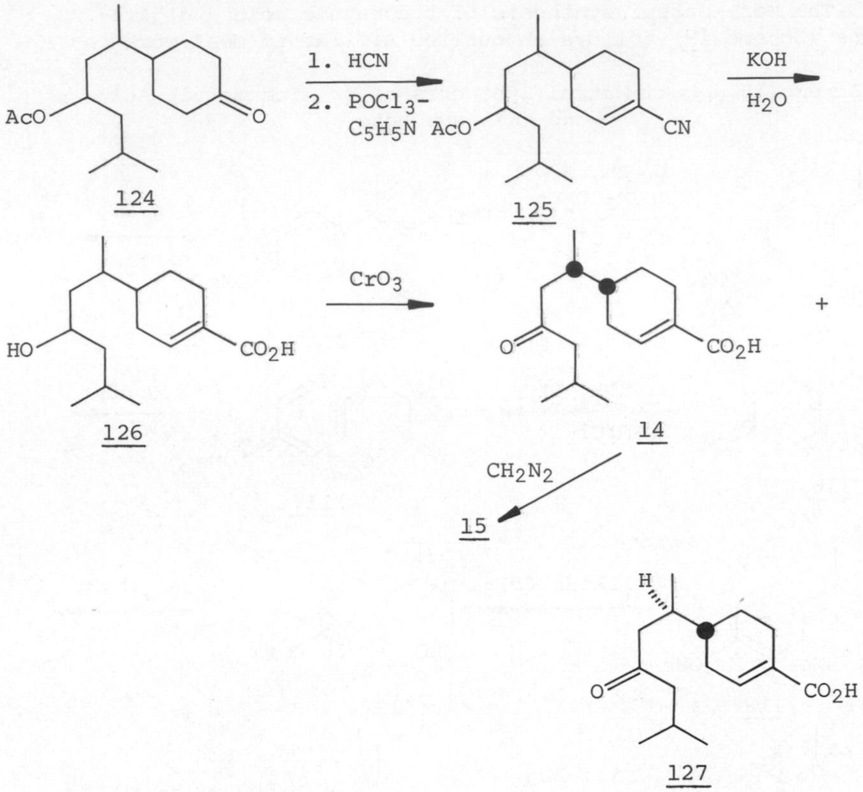

bisabolane-type compounds (attachment of the C-8 sidechain onto a preexisting ring, followed by oxidation-reduction chemistry to establish the proper functionality).[82] The synthesis is not stereoselective, leading to a mixture of (±)-todomatuic acid (14) and its diastereomer (127*). However, the workers cleanly separate the two racemates (via their semicarbazones) and give some useful information regarding the spectral and chromatographic properties of such diastereomers. Although 14 is crystalline (m.p. 65°-66°) and 127 is not, the infrared spectra of the two racemates are almost identical. The corresponding methyl esters, (±)-juvabione and its diastereomer, could not be separated by vpc. These observations relate to

*At the time of this synthesis, the stereochemistry of todomatuic acid and juvabione was not correctly known. Todomatuic acid had been incorrectly assigned structure 127 on the basis of molecular rotation.[81] The correct stereochemistry was later deduced by the Hoffman-LaRoche group.[84b]

the stereochemical homogeneity of the synthetic zingiberine of Joshi and Kulkarni (Scheme 9), Mukherji and Bhattacharyya (Scheme 10) and the synthetic desoxo-todomatuic acid of Naka-zaki and Isoe (Scheme 18). In these various cases, liquid products were obtained which were assayed by infrared spectra and vpc. If the observations regarding the diastereomeric todomatuic acids and juvabiones are extended to these similar systems, then one expects little difference in the spectral and chromatographic properties of such diastereomers. It is therefore highly likely that the above-mentioned syntheses are nonstereoselective.

Shortly after the Mori-Matsui synthesis was announced, Ayyar and Rao reported an essentially identical synthesis (Scheme 20).[86] Ayyar and Rao obtained the racemate melting at

Scheme 20.   Ayyar-Rao Synthesis of (±)-Todomatuic Acid
and (±)-Juvabione

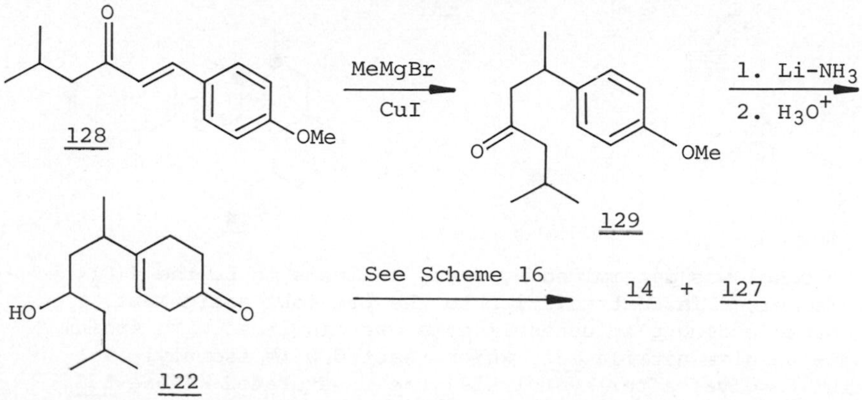

65°-66°, corresponding to the relative stereochemistry in (+)-todomatuic acid, by regeneration of the acid from its purified S-benzylthiouronium salt. They did not isolate the other pure racemate, although it was undoubtedly produced by their route.

In 1968, a group at Hoffmann-LaRoche, headed by Beverly A. Pawson, reported an elegant transformation of R-(+)-limonene (130) into (+)-todomatuic acid and (+)-juvabione, which estab-lished the absolute configuration in the two sesquiterpenoids as 4(R),8(S).[84] The synthesis, outlined in Scheme 21, begins with the selective hydroboration of R-(+)-limonene (130) with disiamylborane. The resulting mixture of 4(R),8(R)- and 4(R),8(S)-p-menth-1-3-en-9-ols was purified by recrystalliza-tion of the corresponding 3,5-dinitrobenzoates. The less solu-ble ester was hydrolyzed to give the unsaturated alcohol 131.

Scheme 21.   Hoffman-LaRoche Synthesis of (±)-Todomatuic
Acid and (±)-Juvabione

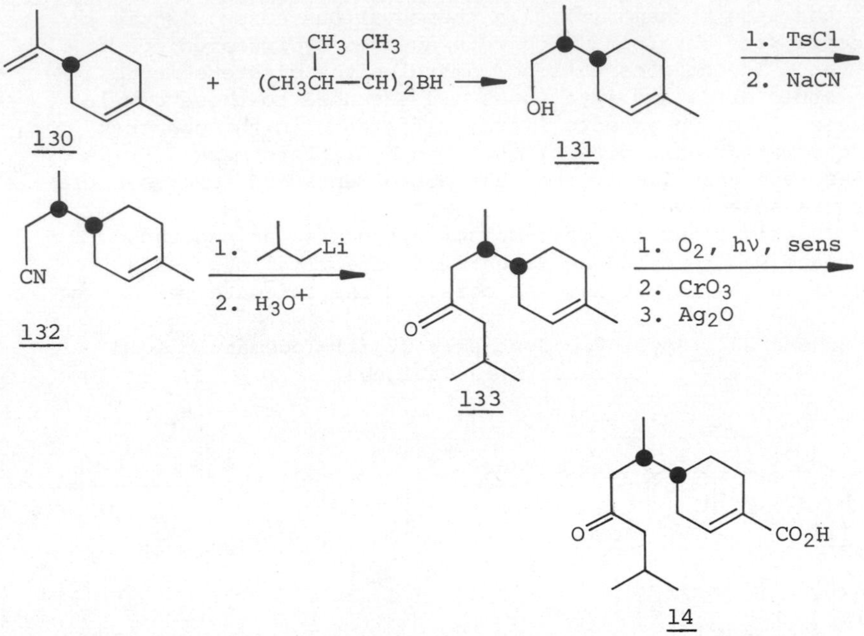

This alcohol was determined by x-ray analysis to be the 4(R),
8(R) isomer,[87] in contradiction to the previous assignment.[88]
The corresponding p-toluenesulfonate was displaced with sodium
cyanide to give nitrile 132, which reacted with isobutyl-
lithium to give, after hydrolysis, the unsaturated ketone 133.
This substance was oxidized successively with singlet oxygen,
chromic acid, and silver oxide to yield (+)-todomatuic acid
(m.p. 64.0°-65.5°), which yielded (+)-juvabione on esterification.   The exact mechanism of the oxidation sequence is obscure, and probably involves the following steps:

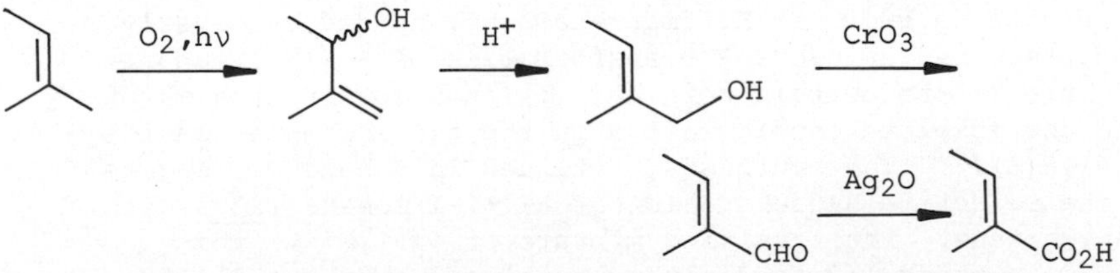

The juvabione synthesis reported by Birch and co-workers

is outlined in Scheme 22.[85]   The relative stereochemistry at

Scheme 22.   Birch's Juvabione Synthesis

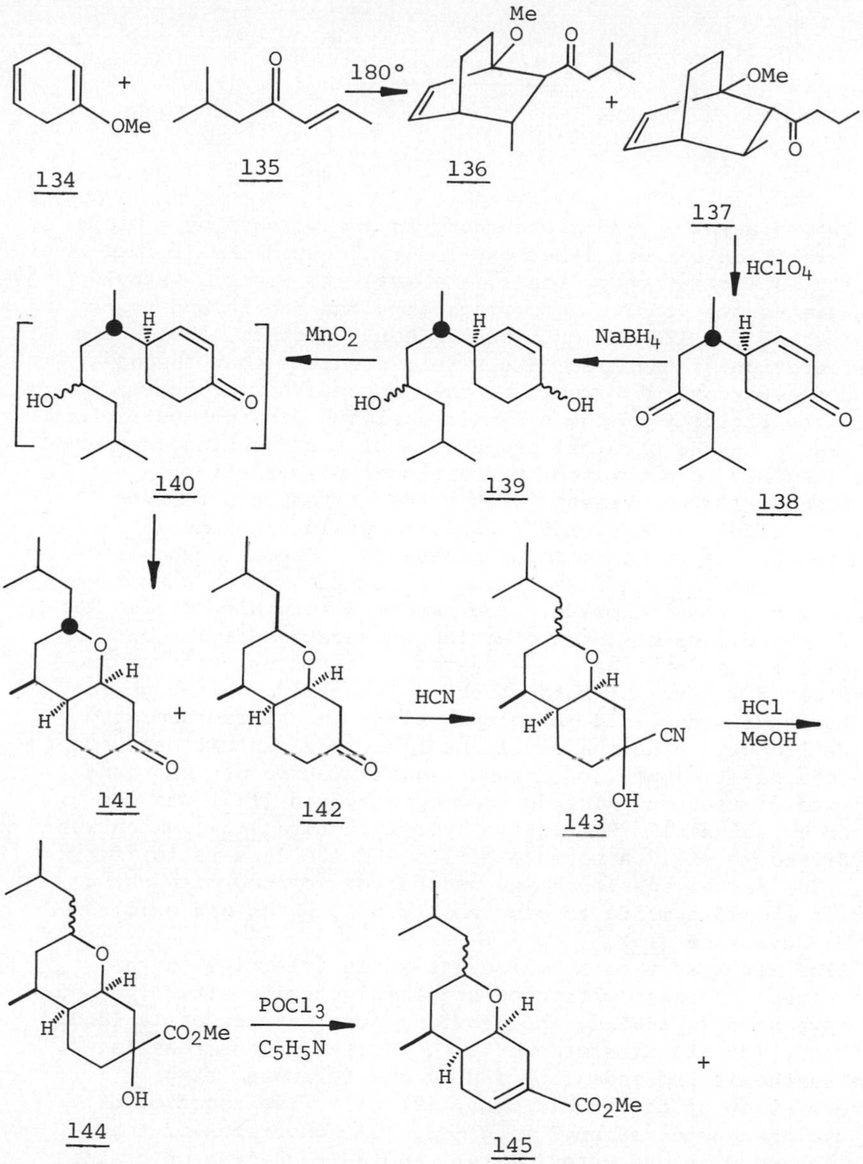

the two adjacent asymmetric carbons is established by a Diels-Alder reaction between 1-methoxy-1,3-cyclohexadiene (formed by in situ isomerization of the 1,4-isomer) and *trans*-6-methyl-hept-2-en-4-one (135). Unformtunately, the "*anti*" and "*syn*" adducts 136 and 137 are obtained in equal amounts, although a consideration of the Alder-Stein rule predicts that the desired diastereomer 137 would be the preponderant product. However, the rigidity of the bicyclic skeleton provides sufficient difference in the physical properties of the two diastereomers that they may be separated by fractional distillation.

Isomer 137 is treated with acid, whereupon a type of "reverse aldol condensation" occurs to yield diketone 138, as a stereochemically homogeneous substance. Sodium borohydride reduction gave 139, a diastereomeric mixture, which was oxidized by manganese dioxide. The expected keto alcohol 140 was not obtained, due to intramolecular conjugate addition of the hydroxyl group onto the enone system. The actual product isolated was a mixture of keto ethers 141 and 142. Although both diastereomers can yield juvabione, since the stereochemistry of the isobutyl side-chain will be lost in a pentultimate step, only the major isomer (of unknown configuration at that center) was carried on. The derived cyanohydrin (143) was converted by methanolic HCl into a hydroxy ester (144), which was dehydrated to yield a mixture of 145 and 146 in a ratio of 1:2. The latter substance was reductively opened with calcium in liquid ammonia to give an alcohol, which was oxidized to (±)-juvabione (15).

The Birch synthesis, while not stereoselective in the first step, proceeds with good stereoselectivity after 136 and 137 have been separated, and provides pure samples of 14 (and, in theory, of its diastereomer). A shorter, nonstereoselective synthesis proceeds from 138 by the following steps: (1) hydrogenation of the double bond, (2) selective reaction of the cyclohexanone carbonyl with HCN, (3) conversion of the nitrile group into a methyl ester, and (4) dehydration of the α-hydroxy ester with POCl$_3$.

It is interesting that, although the synthesis provides authentic juvabione in a stereorational manner, the authors did not recognize that the accepted stereochemistry of the hormone was in error. The oversight apparently came about through an error in formulating the ring opening reaction of intermediate 137. In their communication, Birch and co-workers formulate the reaction as follows:

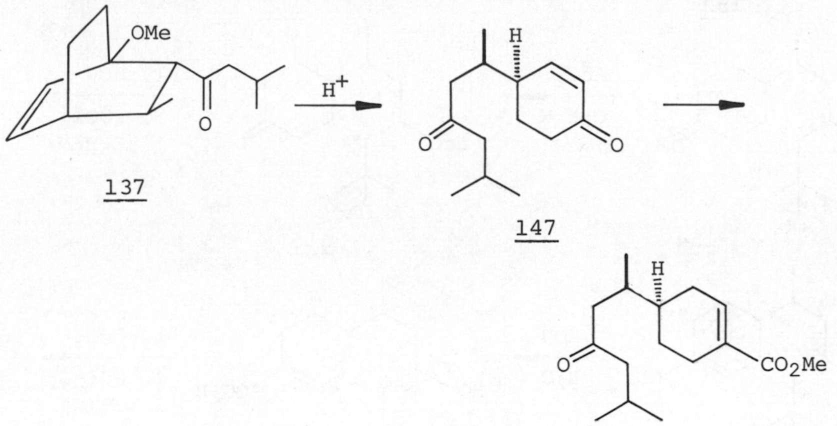

137

147

148

Intermediate 147 would indeed have given a juvabione of structure 148.

The Mori-Matsui synthesis of (±)-dehydrojuvabione (16) is outlined in Scheme 23.[86] The synthesis proceeds in a straight-

Scheme 23.   Mori-Matsui Synthesis of (±)-Dehydrojuvabione

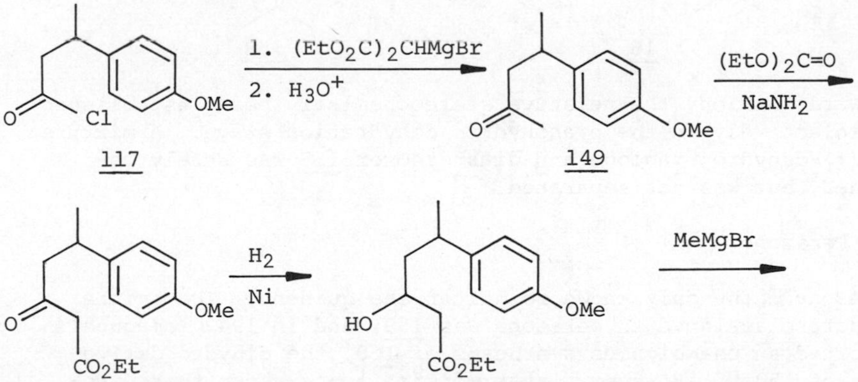

117

149

150

151

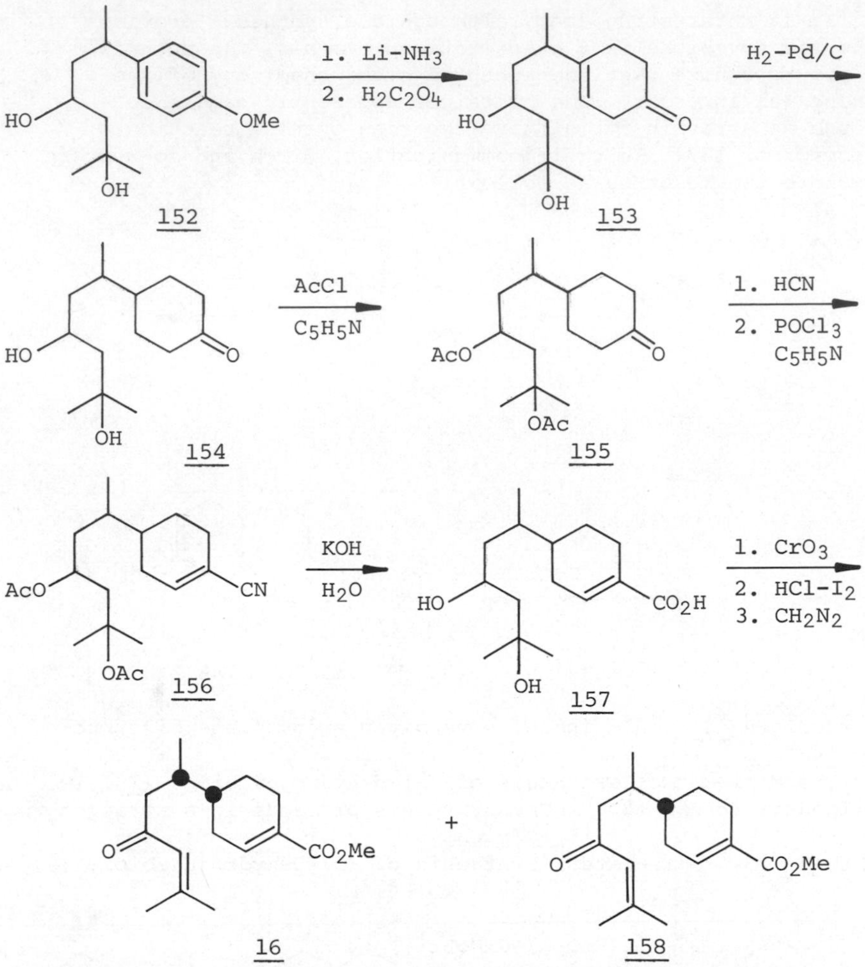

forward fashion, the relative stereochemistry being established
nonselectively in the cyanohydrin dehydration stage. A mixture
of (±)-dehydrojuvabione and diastereomer 158 was surely obtained, but was not separated.

H.   Perezone

Perezone is the only known sesquiterpene quinone. The initial
structure assigned to perezone was 159, and in 1942 Yamaguchi
reported an unambiguous synthesis of 160, the dihydro derivative of 159.[89] However, later work[90,91] has shown that perezone is correctly represented by structure 161.

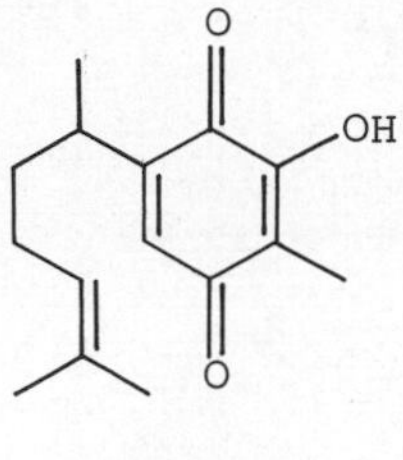

159 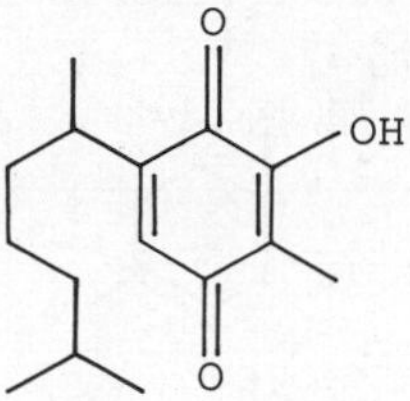   160 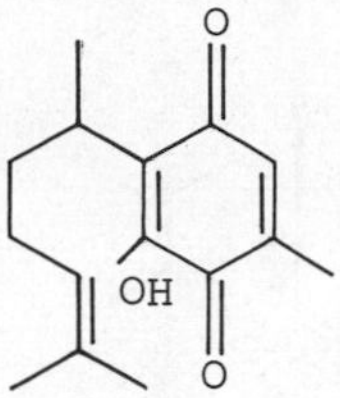   161

In 1965, Cortes, Salmon, and Walls reported an inefficient synthesis of (±)-perezone, which is outlined in Scheme
24.[92] The lithio derivative of 3,5-dimethoxytoluene (162) was

Scheme 24.   Walls-Salmon-Cortes Synthesis of (±)-Perezone

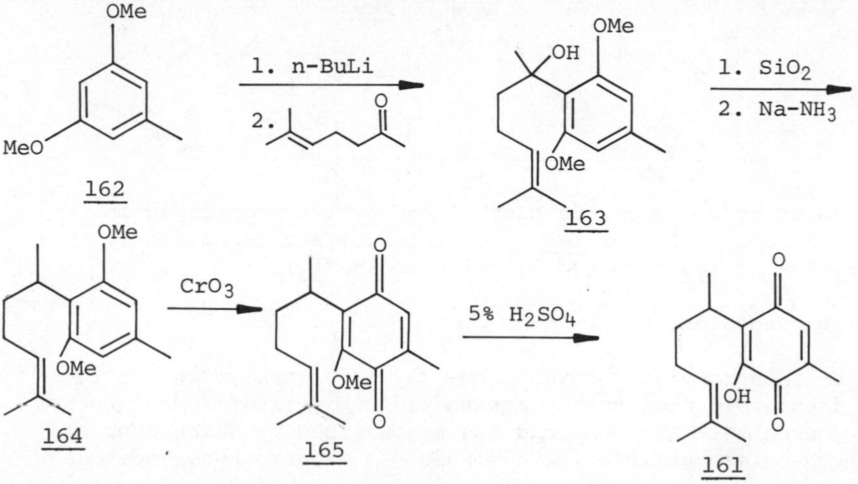

162   163

164   165

161

treated with 6-methylhept-5-en-2-one to give alcohol 163 in
18% yield. This benzylic alcohol was dehydrated (silica gel)
and the resulting alkene reduced to yield the diether
164. This substance was oxidized, in 7% yield to the methyl
ether of perezone (165), which was demethylated in 4% yield by
stirring with dilute sulfuric acid.

I. Bilobanone

Bilobanone (17) has been synthesized in optically active form
by Büchi and Wüest (Scheme 25).[93] (+)-Carvone (166) was oxidized by selenium dioxide in ethanol to keto aldehyde 167 in
60% yield. Compound 167 reacted with 3-methyl-1-butynyl-

Scheme 25. Büchi-Wüest Bilobanone Synthesis

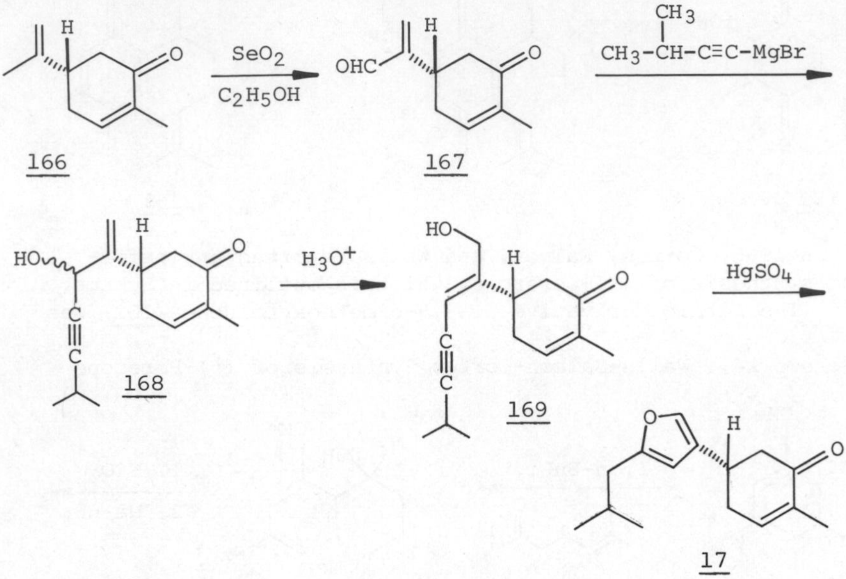

magnesium bromide to give alcohol 168, which rearranged in
mineral acid to isomer 169. Mercuric sulfate catalyzed cyclization of 169 afforded (+)-bilobanone (17).

J. Furoventalene

The C-15 benzofuran furoventalene (177) is interesting in that
its isoprenoid skeleton is apparently not derived from farnesyl
pyrophosphate. The compound was synthesized by Weinheimer and
Washecheck as outlined in Scheme 26.[94]   m-bromophenoxyacetone

Scheme 26. Weinheimer-Washecheck Synthesis of Furoventalene

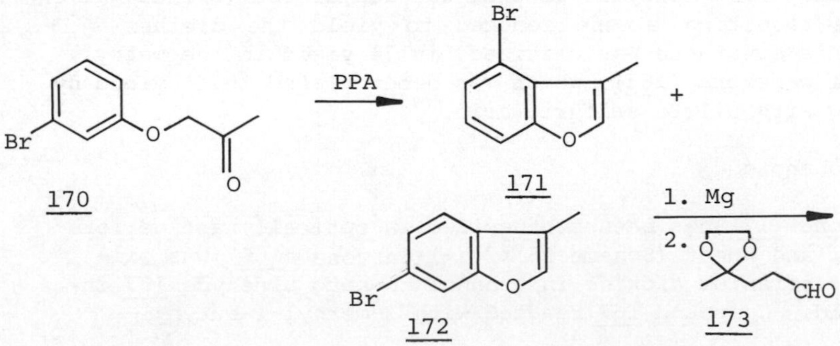

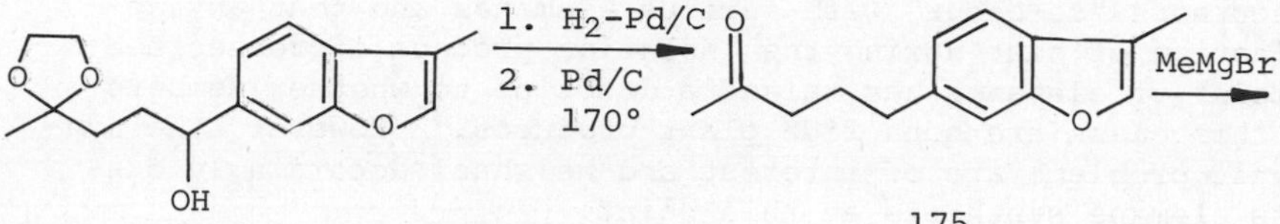

(170) was cyclized with polyphosphoric acid to give a mixture
of bromobenzofurans 171 and 172. The minor isomer, 172, was
converted into the corresponding Grignard reagent, which was
added to 4,4-ethylenedioxypentanal (173) to give alcohol 174.
Compound 174 was hydrogenolyzed, with concommitant reduction
of the furan ring. After dehydrogenation, ketone 175 was obtained (the ketal having been hydrolyzed during the hydrogenolysis step). Compound 175 was condensed with methylmagnesium bromide and the resulting alcohol dehydrated. The
product of the dehydration was furoventalene (177), its terminal double bond isomer and a tricyclic alkylation product.

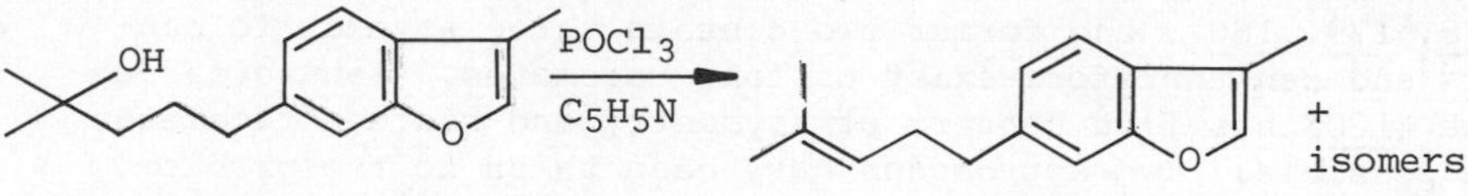

K. Elemanes

Sesquiterpenes of the elemane class [i.e., β-elemene (178),
elemol (179), saussurea lactone (180)] were originally thought
to arise by in vivo Cope rearrangement of cyclofaresyl cation.[1]
However, the observation that the actual cyclodecadiene

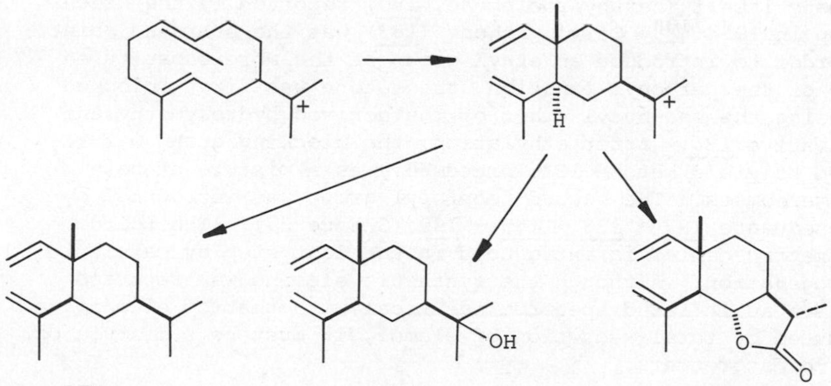

precursors "co-occur" with various elemanes and that any application of heat during the isolation process increases the "yield" of elemanes has raised a doubt as to whether members of this class are bona fide plant products.[3]   However, the synthetic problems are of interest and we shall accordingly discuss elemane synthesis at this point.

Of the compounds in the group which have been synthesized (178, 179, 180), the former two contain three asymmetric centers and can therefore exist as four racemates. Saussurea lactone (180) has five centers of asymmetry and sixteen racemates are possible. Two approaches have been taken to the problem of elaborating the gross skeleton. Tetrahydrosaussurea lactone (181),[95] saussurea lactone,[96] β-elemene,[97] and elemol[97] have all been prepared by relay synthesis from santonin (182), in which most of the relative stereochemistry has been established. Both β-elemene[98] and elemol[99] have been synthesized

181

182

in a nonstereosepcific coupling of 1,10-dibromo-2,8-decadienes. Direct stereorational routes to members of the group have not yet been found.

L. Elemane (Tetrahydroelemene)

The first synthetic work on the elemanes was the synthesis of elemane itself (hexahydroelemene, 190) reported by the Czech group in 1954.[100] Carvomenthone (183) was the starting point. In order to introduce an ethyl group at the more substituted side of the carbonyl function, the ketone was first "blocked" by using the *sec*-butyl ether of the derived hydroxymethylene derivative 184. After ethylation, the blocking group was removed to yield ketone 187, presumably as a mixture of both diastereomers. The second isopropyl group was introduced by the sequence 187 → 188 → 189 → 190 (Scheme 27). The third asymmetric center is introduced in the last step by catalytic hydrogenation. Although the synthetic elemane was reported to have an infrared spectrum identical with that of elemane produced by total reduction of elemol, it must be a mixture of four diastereomers.

Scheme 27. Synthesis of Elemane, Sykora-Cerny-Herout-Sorm

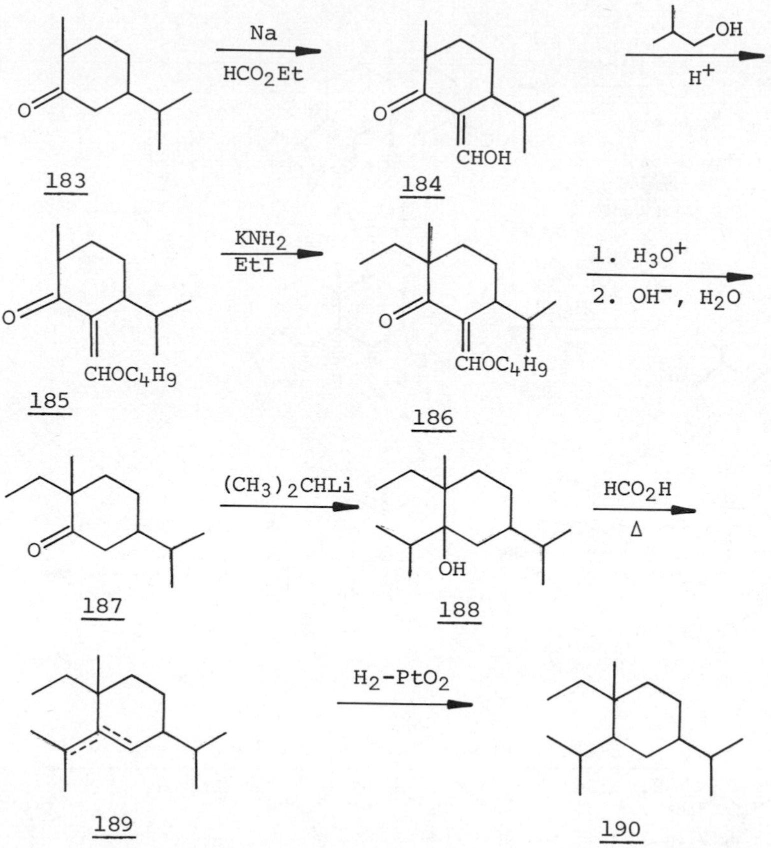

M.   Tetrahydrosaussaurea Lactone, Saussaurea Lactone

The next success in the elemane area came in 1963, when Simonovic, Rao, and Bhattacharyya reported the synthesis of tetrahydrosaussaurea lactone (181) from santonin (Scheme 28).[95]
This is a classic example of a relay synthesis, in light of the
fact that santonin had previously been synthesized in optically
active form (see p. 322). Hydrogenation of santonin (182)
over palladized carbon gives tetrahydro isomers 191 and 192 in
22% and 56% yields, respectively. Isomer 192, "α-tetrahydro-
santonin," had previously been converted into keto-oxide 195
by the indicated route.[101] The A ring was cleaved by an adaptation of Johnson's method (ozonolysis of the derived benzyl-
idine or furfurilidine derivative).[102] The resulting diacid

Scheme 28.   Synthesis of Tetrahydrosaussurea Lactone--
               Simonovic-Rao-Bhattacharyya

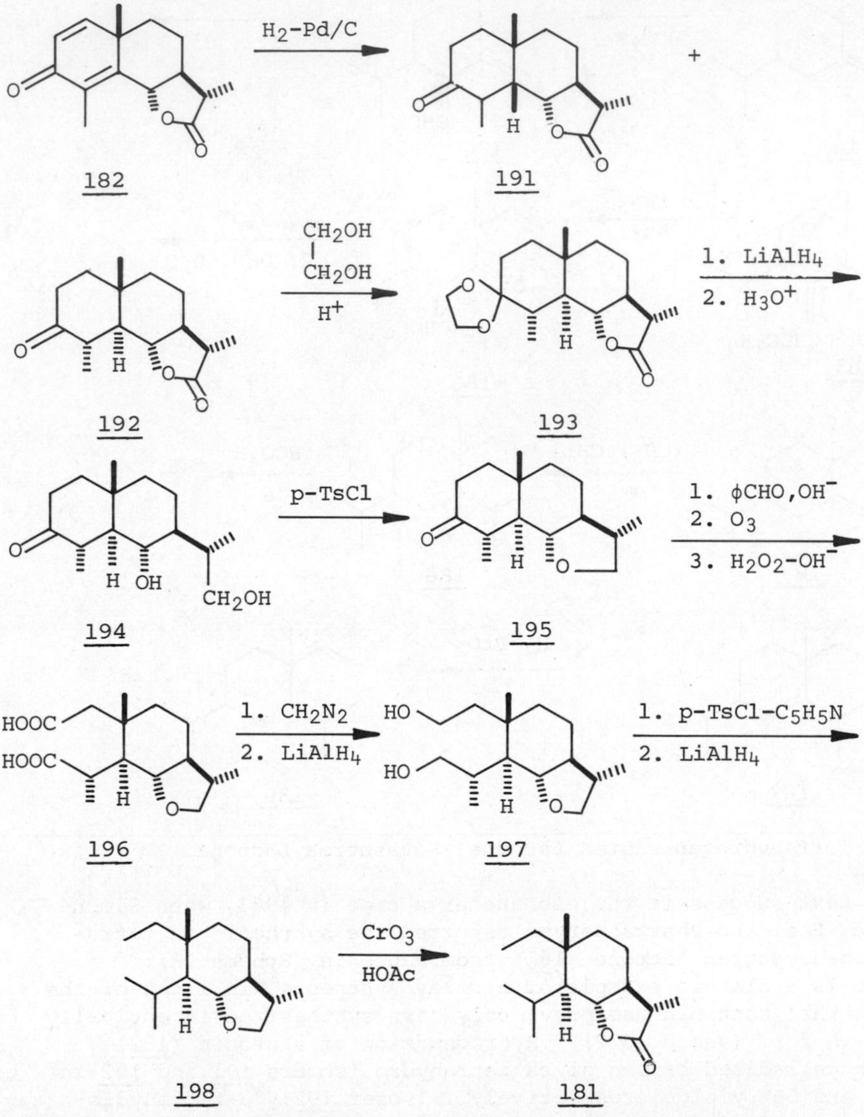

was converted to its diester and reduced with hydride to give
diol 197. This was converted into a ditosylate, which was reduced by hydride to oxide 198. Oxidation of 198 gave tetrahydrosaussurea lactone in high yield.

In a subsequent paper, Honwad, Siscovic, and Rao reported the conversion of intermediate 197 into saussurea lactone itself (Scheme 29).[96] Diol 197 was converted via its ditosylate,

Scheme 29. Honwad-Siscovic-Rao Synthesis of Saussurea Lactone

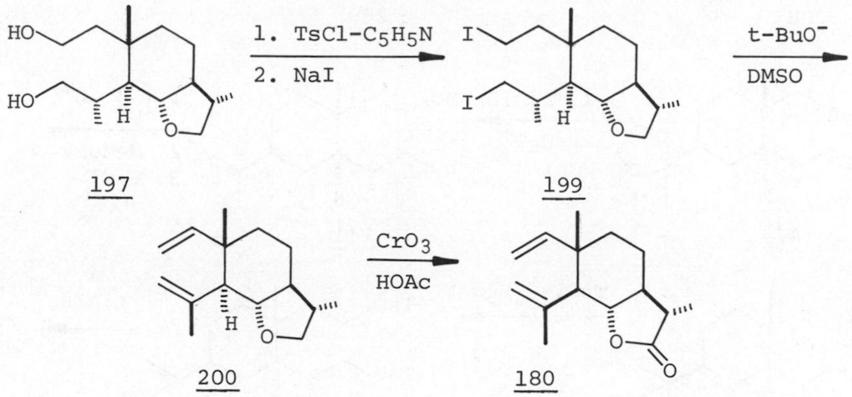

197

199

200

180

into iodide 199. This was dehydrohalogenated by potassium t-butoxide in DMSO to yield oxide 200. Oxidation of 200 with chromium trioxide in acetic acid gave saussurea lactone in 15% yield.

N. Tetrahydroelemol, β-Elemene, Elemol

A formal total synthesis of tetrahydroelemol (214), is available through the combined efforts of Halsall, Theobald, and Walshaw in England and Rao's group in India (Scheme 30). The

Scheme 30. Synthesis of Tetrahydroelemol
(Halsall-Theobald-Rao)

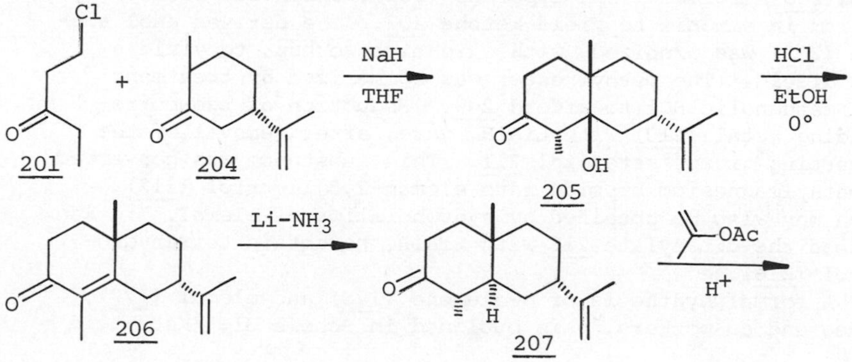

201

204

205

206

207

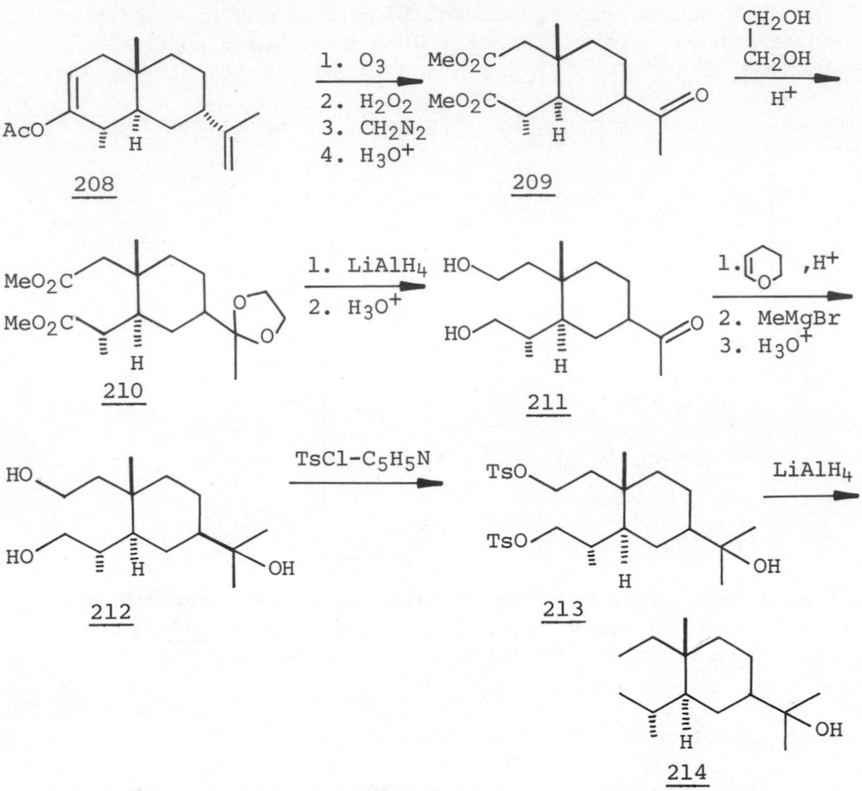

British group prepared eleman-2,3,11-triol (212) starting from
(-)-dihydrocarvone (204).[103] Alkylation of this substance
with 1-chloro-3-pentanone gave ketol 205. The relative stereochemistry of the angular methyl group and the isopropenyl
group is established in the alkylation stage. Dehydration of
ketol 205 afforded 7-epi-cyperone (206), which was reduced by
lithium in ammonia to yield ketone 207. The derived enol acetate (208) was ozonized, with oxidative workup, to yield a
keto diacid. The methyl ester was epimerized by treatment
with methanolic HCl to afford 209. Reduction of the corresponding ketal (210) with LiAlH₄ gave, after removal of the
protecting group, keto diol 211. This substance was converted
by methylmagnesium bromide into eleman-2,3,11-triol (212),
which may also be obtained by hydroboration of elemol.[103] Rao
reduced the ditosylate 213 with LiAlH₄ to obtain tetrahydroelemol (214).[104]

A formal synthesis of β-elemene (178) and elemol (179),
by Rao and co-workers,[97] is outlined in Scheme 31. Keto-acid

Scheme 31.   Rao's Synthesis of β-Elemene and Elemol

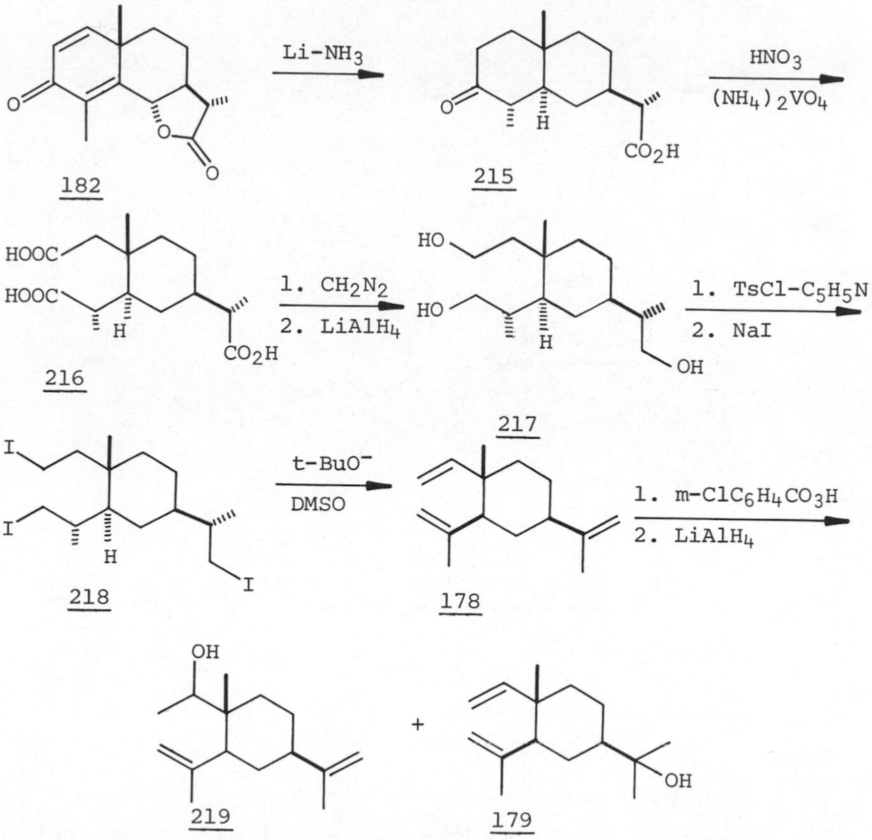

215, prepared from α-santonin (182) by the route shown,[105] was oxidized by nitric acid and ammonium vanadate to yield triacid 216. The corresponding triester was reduced to eleman-2,3,12-triol (217). The primary triiodide 218 was obtained by sodium iodide displacement on the tritosylate. Base catalyzed dehydroiodination of 218 gave β-elemene (178). Selective epoxidation, followed by hydride reduction, gave elemol (179) and secondary alcohol 219.

Vig has reported a synthesis of β-elemene,[98] based on Corey's observation that 1,10-dibromo-2,8-decadienes cyclize under the influence of nickel carbonyl.[106] The synthesis (Scheme 32) begins with unsaturated ester 220. The requisite ten-carbon chain is built up in a clever fasion, utilizing a Claisen rearrangement and two modified Wittig reactions to add six carbons. The first Wittig reaction (223 → 224) is stereo-

Scheme 32. Vig's Synthesis of β-Elemene

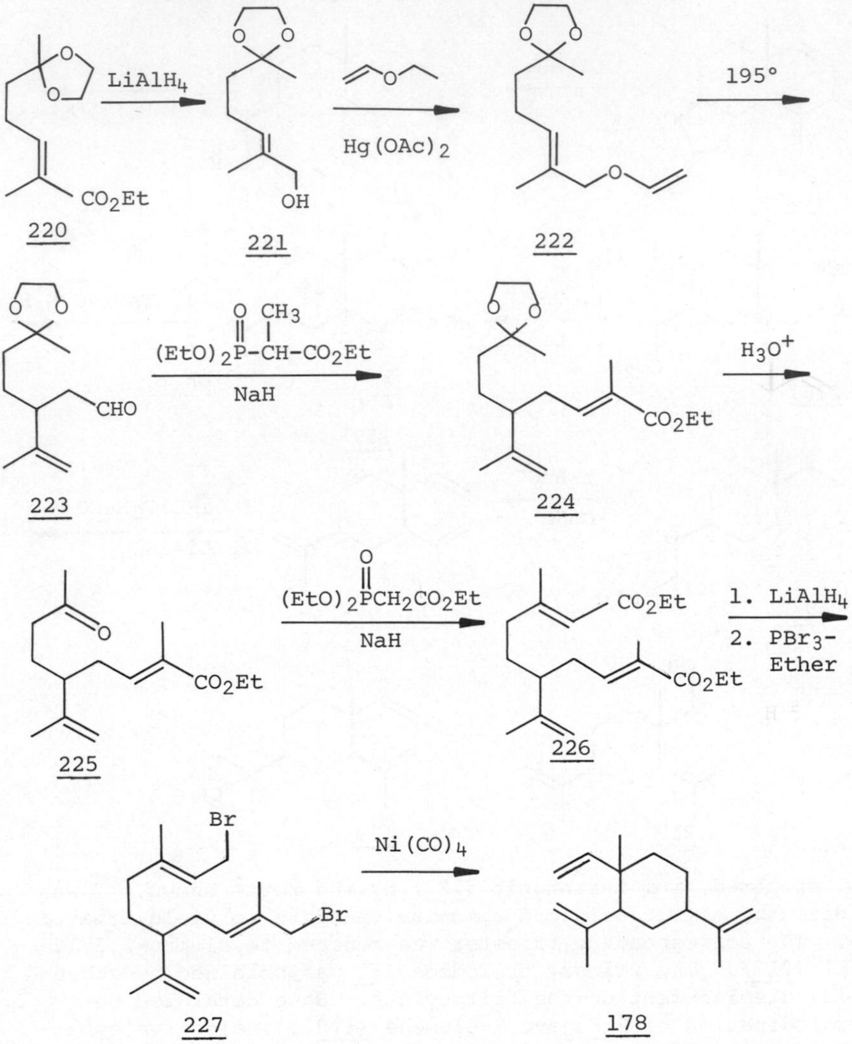

specific, yielding only the *trans* isomer 224. The second
Wittig, using triethylphosphonoacetate (225 → 226) yields
trans-trans isomer 226, along with some *trans-cis* isomer. The
mixture of isomers was used to complete the synthesis. Vig
reports that his synthetic β-elemene is only one of the four
possible racemates, as judged by the fact that the synthetic
hydrocarbon has IR and NMR spectra which are identical with
those of natural β-elemene. In this respect, the work differs

from the observations of Corey and Broger (below).

Corey and Broger prepared (±)-elemol by a route which also involved nickel carbonyl catalyzed coupling of a bis-allylic 1,10-dibromide.[99] The requisite dibromide (235) was prepared as outlined in Scheme 33. The interesting isoprene

Scheme 33. Corey's Synthesis of (±)-Elemol

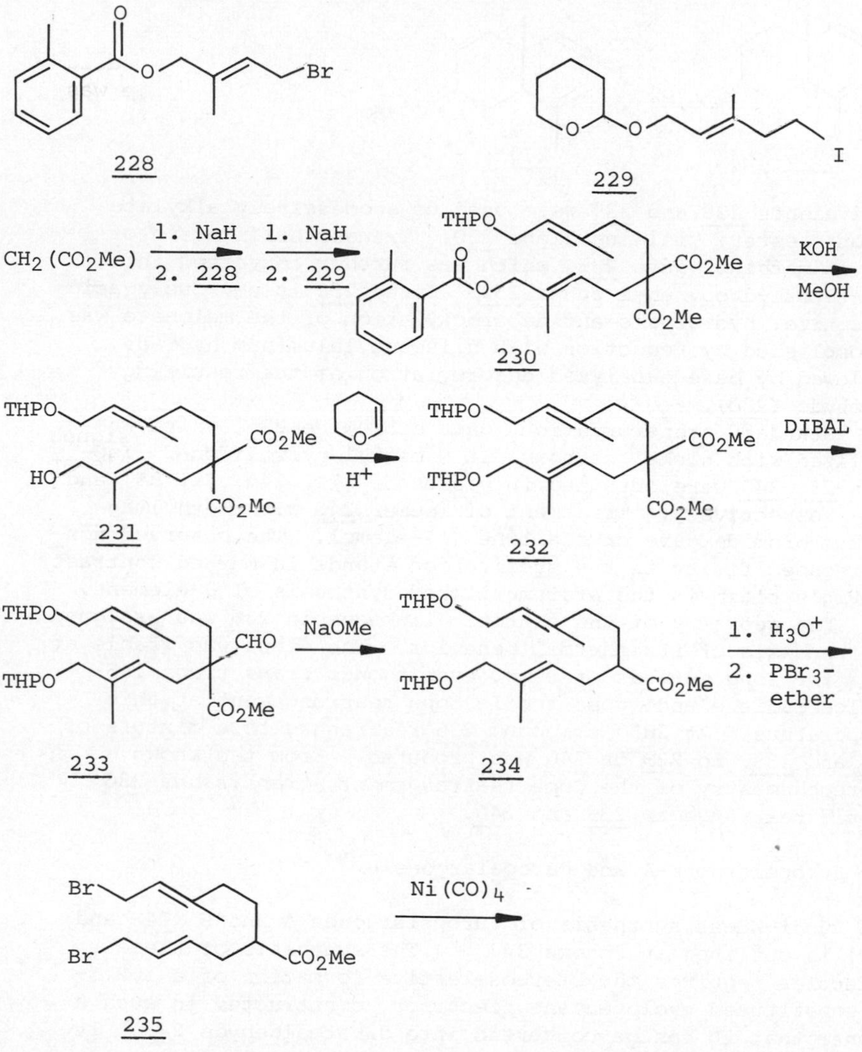

equivalents 228 and 229 were used to successively alkylate
malonic ester, yielding diene 230. Transesterification of
230 in methanol gave 231, which was further converted into
bis-tetrahydropyranyl ether 232. Since 232 is obviously acid
sensitive, hydrolysis and decarboxylation of the malonate was
accomplished by reduction with diisobutylaluminum hydride,
followed by base catalyzed deformylation of the resulting
aldehyde (233).

Standard transformations gave dibromide 235, which was
cyclized with nickel carbonyl in N-methyl-pyrrolidone. Isomers 236-240 were obtained in yields of 11%, 24%, 3%, 3%, and
24%, respectively. Treatment of isomer 240 with methylmagnesium bromide gave crystalline (±)-elemol. The observed non-
stereospecificity in the cyclization stands in marked contrast
to Vig's claim in the aforementioned synthesis of β-elemene.

The geometry of the olefinic linkages in 236 was assigned
on the basis of its thermal behavior. The diene was stable at
100° for long periods of time, even though *trans,trans*-1,5-
cyclodecadiene undergoes facile Cope rearrangement at this
temperature. At 285° compound 236 rearranged to a mixture of
237 and 238; no 239 or 240 was produced. From the known
stereochemistry of the Cope rearrangement, *trans,trans*-236
should rearrange to 239 and 240.

O. Furopelargone-A and Furopelargone-B

The Büchi-Wüest synthesis of furopelargones A and B (248 and
249) is outlined in Scheme 34.[107] The synthesis of these
molecules requires the stereoselective formation of a 1,2,3-
trisubstituted cyclopentane precursor, constructed in such a
manner that it can be converted into diketoaldehyde 247. It
was anticipated that 247 would undergo dehydration to yield
furopelargone-A (248), rather than the more strained system

Scheme 34.   Büchi-Wüest Synthesis of the Furopelargones

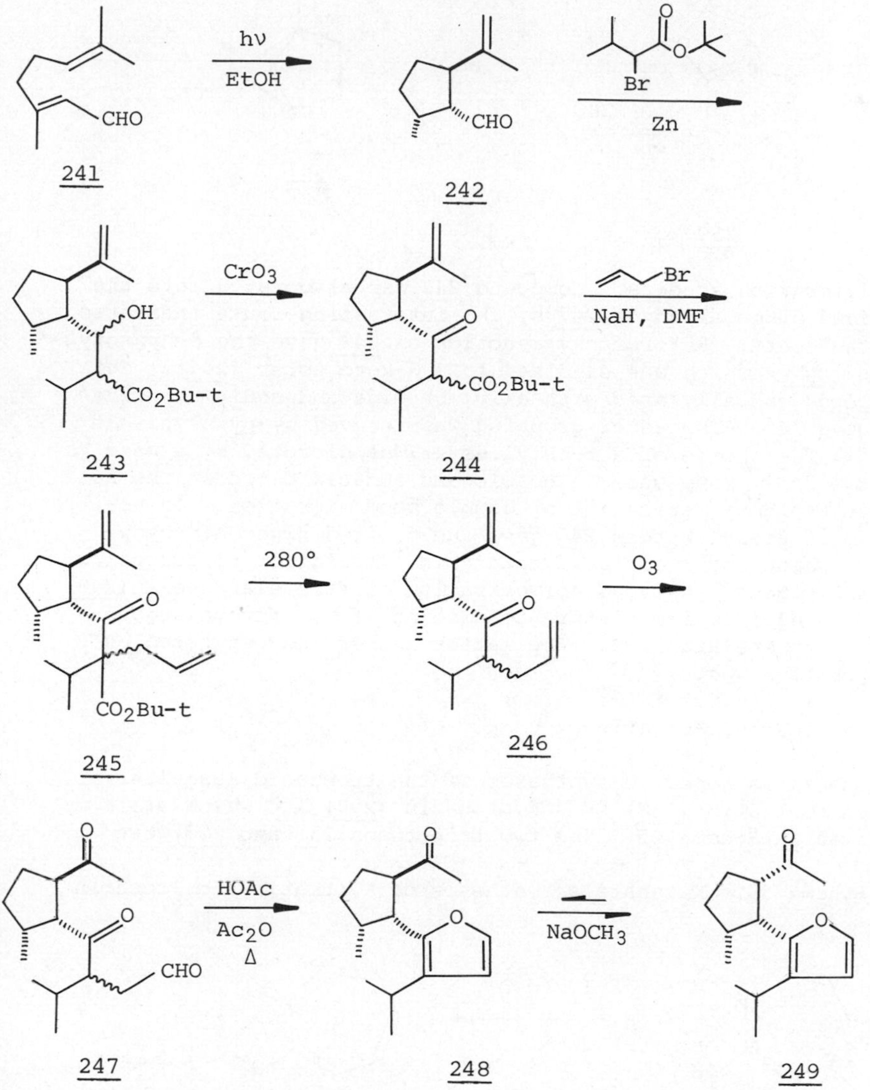

250. The initial objective was achieved by taking advantage
of the well known photocyclization of citral (241). Photo-
citral-A (242) is produced in 20% yield, along with photo-
citral-B (251), when citral is irradiated in alcoholic solution. The stereochemistry of 242 was assigned on the basis of

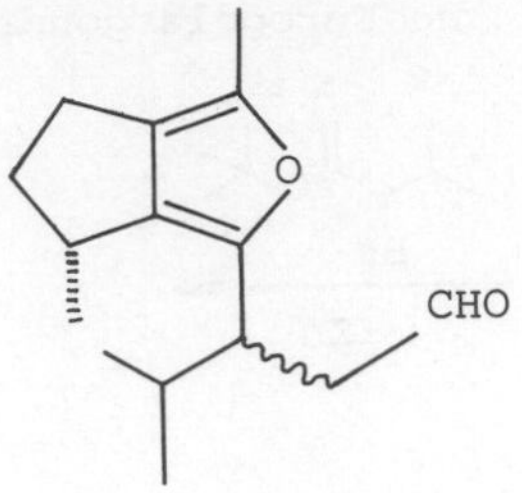

250

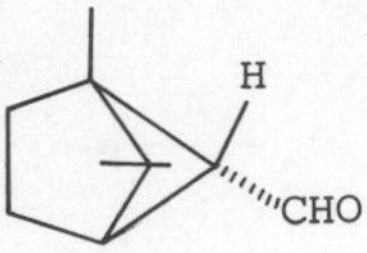

251

equilibration studies. Compound 242 was elaborated into the
desired diketoaldehyde 247 by the interesting route indicated
in the chart. Reformatsky reaction on 242 gave the β-hydroxy
ester 243, which was oxidized to a β-keto ester (244). This
compound was allylated with allyl bromide and sodium hydride
to give 245. The ester grouping was removed by pyrolysis at
280°. The choice of a t-butyl ester was dictated by a need to
remove this group under nonbasic and nonacidic conditions so
as to avoid epimerization or double bond migration. Ozonolysis of dienyl ketone 246 gave the desired diketoaldehyde
(247) which underwent acid catalyzed cyclization to furopelargone-A. Base catalyzed epimerization of furopelargone-A (248)
gave an equilibrium mixture consisting of 94% furopelargone-A
and 6% furopelargone-B. The latter isomer was separated by
preparative vpc.

P. Nootkatin, Procerin

Kitahara has reported syntheses of the troponoid sesquiterpenes nootkatin (258)[108] and procerin (259),[109] which are summarized in Scheme 35. The two tropotones 252 and 253 were

Scheme 35. Kitahara's Synthesis of Nootkatin and Procerin

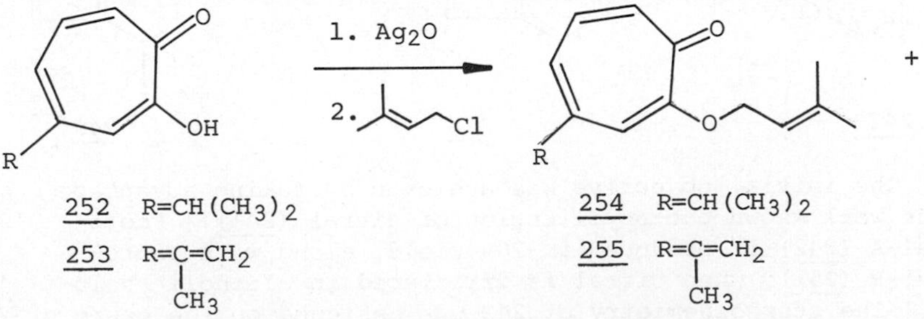

252	R=CH(CH₃)₂
253	R=C=CH₂
	CH₃

254	R=CH(CH₃)₂
255	R=C=CH₂
	CH₃

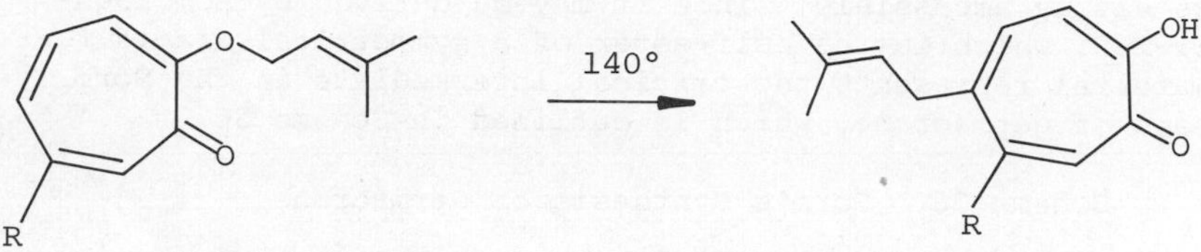

140°

256   R=CH(CH₃)₂

257   R=C=CH₂
           |
          CH₃

258   R=-CH(CH₃)₂

259   R=-C=CH₂
            |
           CH₃

alkylated with β,β-dimethylallyl chloride to give mixtures of
tropotone ethers. Pyrolysis of the ether mixture gave nootkatin (258) or procerin (259) in 5-7% yield.

Q.   Germacrane, Dihydrocostunolide

The first synthetic success in the germacrane area was Sorm's
synthesis of germacrane itself (260), which was reported in
1958.[110] Although not itself a natural product, the substance
may be considered the parent hydrocarbon of the germacrane
family. The only other synthesis in this area is Corey's
relay synthesis of dihydrocostunolide (261) from santonin.[111]

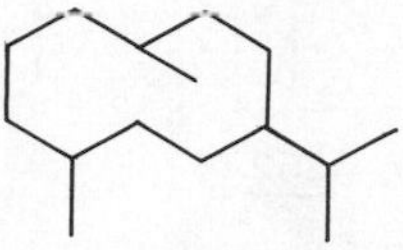

260

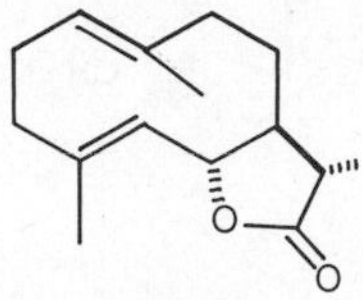

261

     The primary problem which must be faced in the synthesis
of germacranes is elaboration of the cyclodecane ring. This
task, for which there are few reliable methods available, is
amplified by the substitution pattern in more complicated members of the class.
     One of the few reliable methods for preparing medium
rings is the acyloin condensation. An analysis of germacrane
indicates that it may be obtained by reducing the acyloin derived from diesters 262 and 263. Intermediate 262 appears to

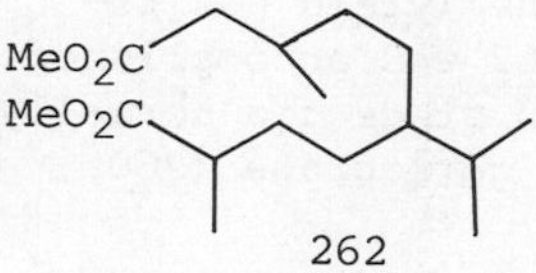

262

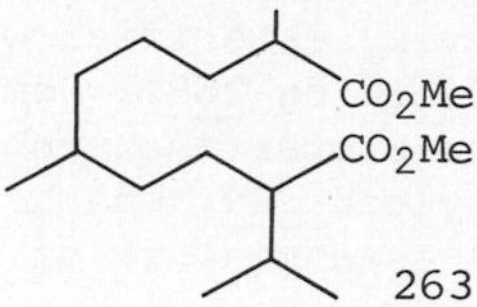

263

be more easily accessible, since it may be derived by homologation of 268, which is the half-ester of a symmetrical diacid. This material represents the critical intermediate in the Sorm synthesis of germacrane, which is outlined in Scheme 36.

Scheme 36. Sorm's Synthesis of Germacrane

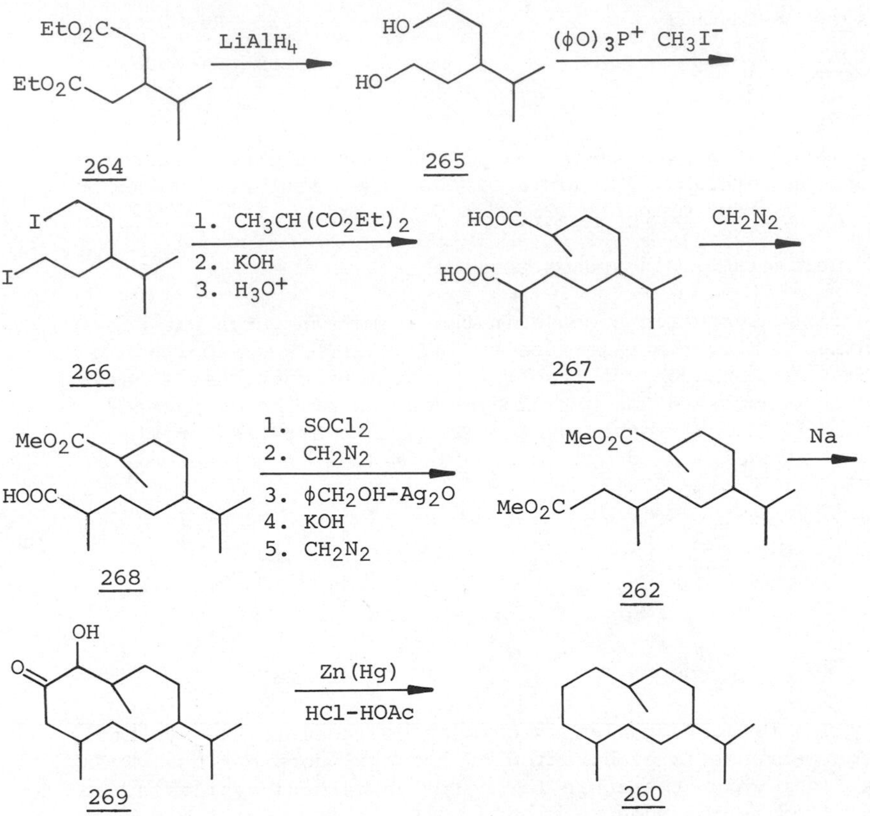

Diethyl β-isopropylglutarate (264) was reduced to diol 265, which was converted into a diiodide (266) with triphenoxyphosphonium methiodide. Compound 266 was used to alkylated two moles of diethyl methylmalonate, yielding a tetraester which was hydrolyzed and decarboxylated. The resulting diacid (267) was selectively esterified with diazomethane (yield not reported), affording 268. Homologation to 262 was accomplished by the Arndt-Eistert sequence. Cyclization proceeded normally, yielding acyloin 269, which was reduced to germacrane (260), undoubtedly as a mixture of diastereomers.

Corey's synthesis of dihydrocostunolide (261) is outlined in Scheme 37. The overall concept embodies a scheme for

Scheme 37. Corey's Synthesis of Dihydrocostunolide

182   270

192   271

272   273

274   275

261

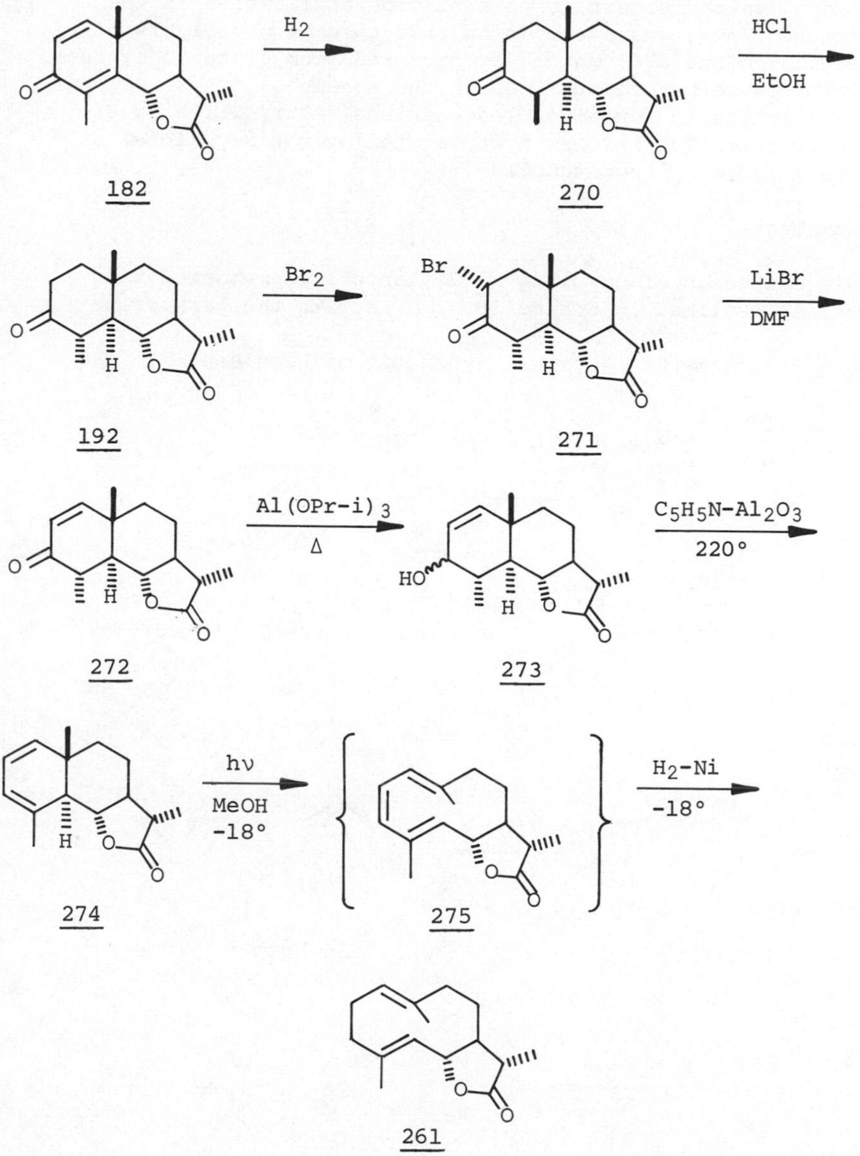

elaboration of a medium ring which has great promise, but has not yet been used extensively in natural product synthesis. The basic idea involves the scission of the central bond in a 1,2-fused bicyclic system, leading to a new ring corresponding in size to the periphery of the old bicyclic compound. The technique, which is certain to find wide application in the germacrane class, was executed in this case by photolysis of the crucial diene 274, which was prepared from santonin by the straightforward method outlined in the scheme.

The initially formed cyclodecatriene (275), which is exceedingly thermolabile, was hydrogenated over Raney nickel at -18° to yield dihydrocostunolide.

R. Humulene

Corey's synthesis of humulene (292), an eleven-membered ring triene, is outlined in Scheme 38.[112] As with the germacrane

Scheme 38. Corey's Synthesis of Humulene

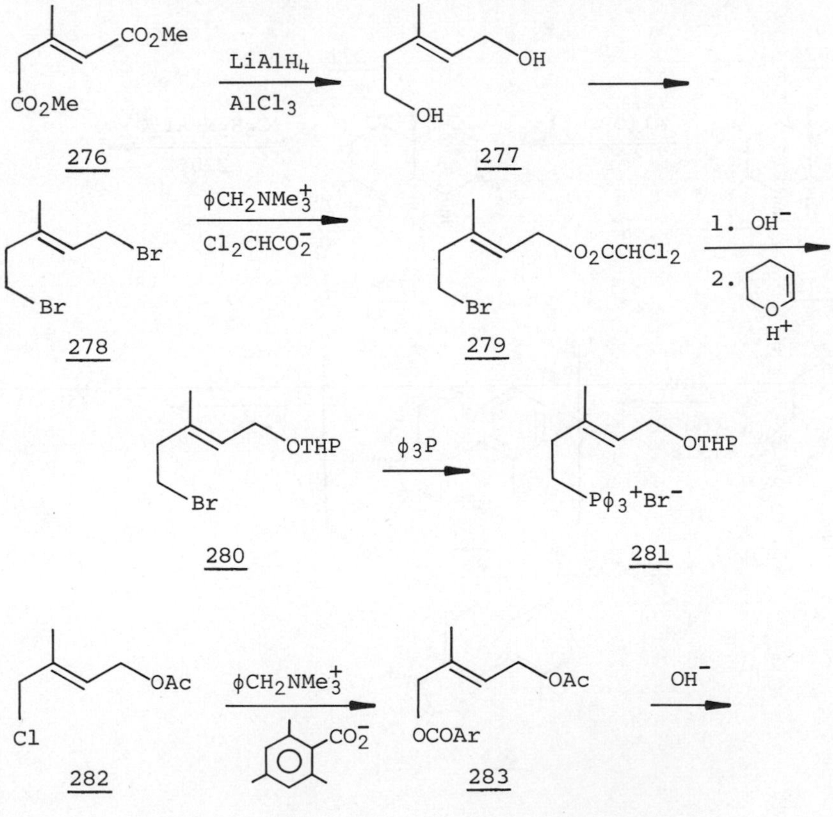

284 285 286

287

281

288

289 290

291 292

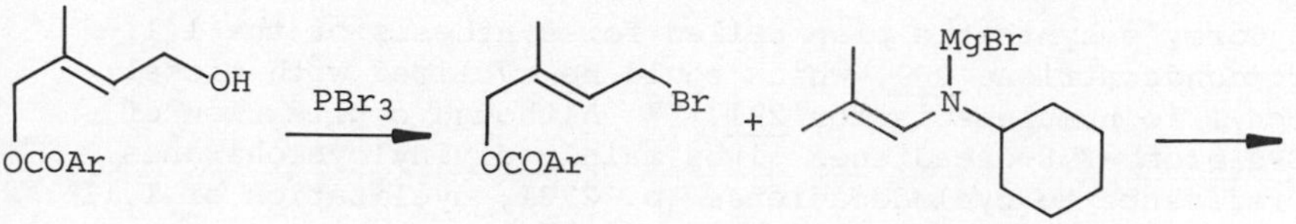

family, the chief synthetic problem here is construction of the eleven-membered ring. Double bond geometry, which appears to be an added complication, is actually no problem, since humulene is known to be the most stable of the various isomers. For this reason, one may design a synthesis leading to any of the isomers and then equilibrate in the final step.

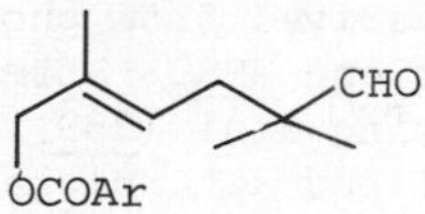

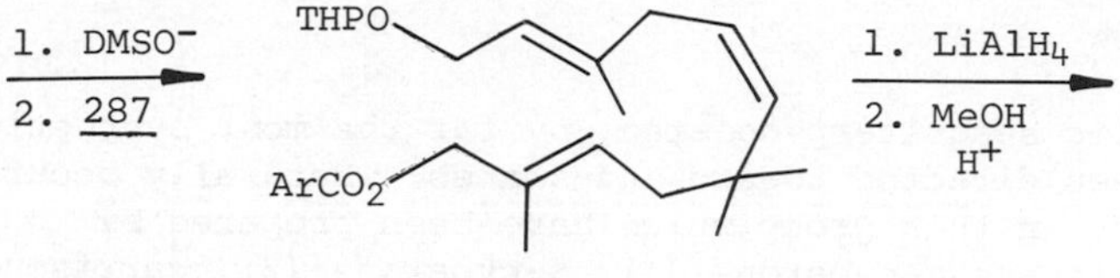

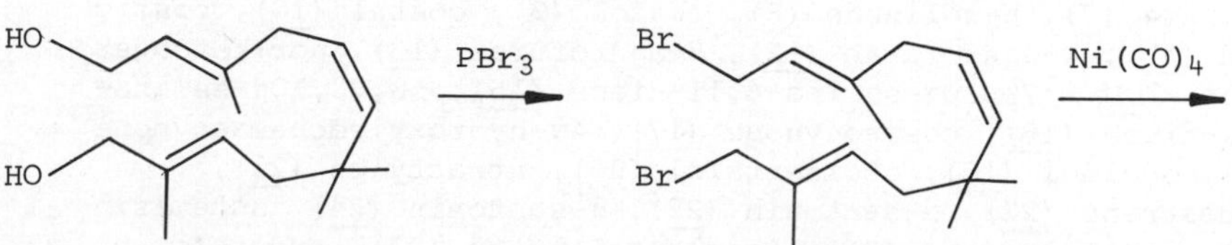

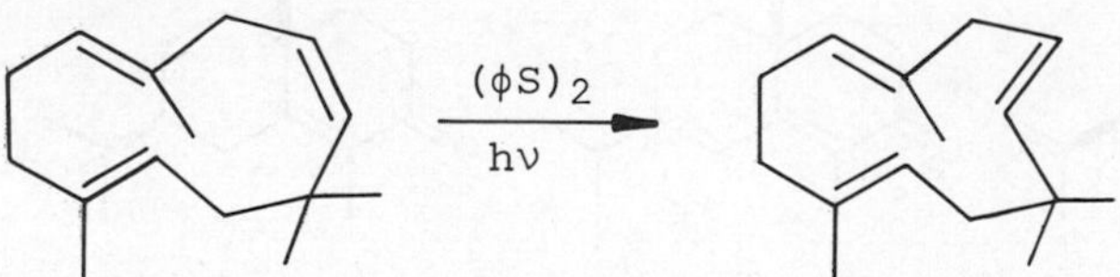

Corey's synthetic plan called for synthesis of the 1,11-dibromoundecatriene 289, which could be cyclized with nickel carbonyl to humulene isomer 291.[106] Although cyclization of 1,10-dibromo-2,8-decadienes gives mainly divinylcyclohexanes in preference to cyclodecadienes (p. 273), cyclization of 1,11-dibromo-2,9-undecadienes yields predominately the cycloundeca-diene. The requisite dibromoundecatriene was prepared by a Wittig reaction of the ylid derived from phosphonium salt 281 and aldehyde 287, yielding triene 288. After removal of the protecting groups, the resulting diol (289) was converted to dibromide 290, which was cyclized with nickel carbonyl in N-methylpyrrolidone. The initial product, 4,5-cis-humulene (291), was isomerized to humulene by irradiation with diphenyl-disulfide in cyclohexane.

4. BICARBOCYCLIC SESQUITERPENES, HYDRONAPHTHALENES

A. Eudesmanes

In the bicyclic sesquiterpene area, by far the most synthetic effort has been directed toward eudesmanes. Naturally occur-ring compounds in this group which have been prepared by total synthesis are α-cyperone (1), β-cyperone (2), carrisone (3), α-eudesmol (4), β-eudesmol (5), γ-eudesmol (6), α-selinene (7), β-selinene (8), costol (9), costal (10), costic acid (11), α-agarofuran (12), β-agarofuran (13), nor-ketoagaro-furan (14), 7β,10α-selina-4,11-diene (15), 5β,7β,10α-selina-3,11-diene (16), chamaecynone (17), 4α-hydroxyisochamaecynone (18), occidol (19), occidentalol (20), atractylon (21), lindestrene (22), α-santonin (23), β-santonin (24), astemisin (25), alantolactone (26), isoalantolactone (27), and telekin (28).

1   2   3

4   5   6   7

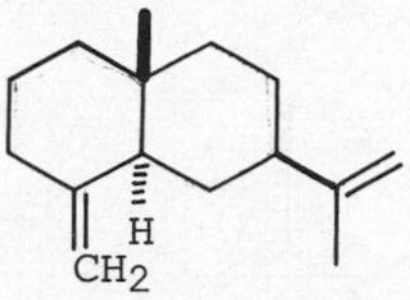

8

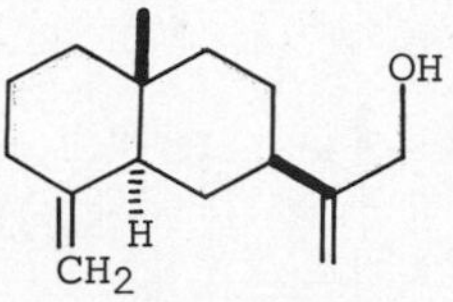

9

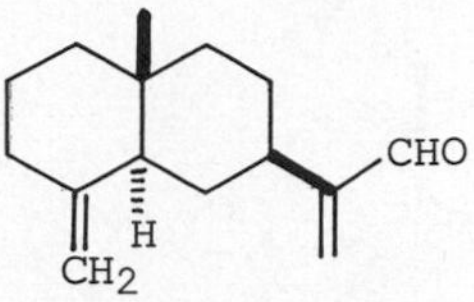

10

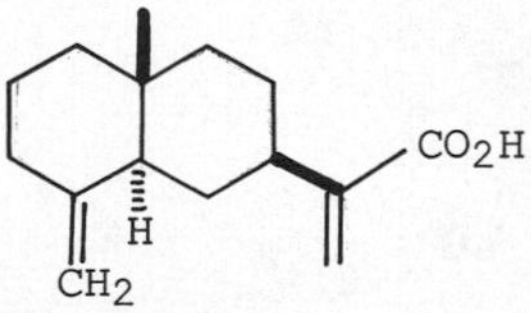

11

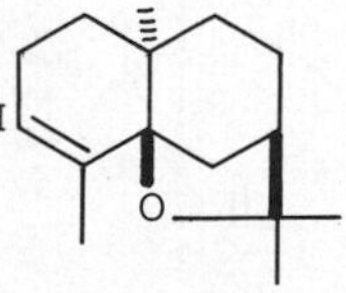

12

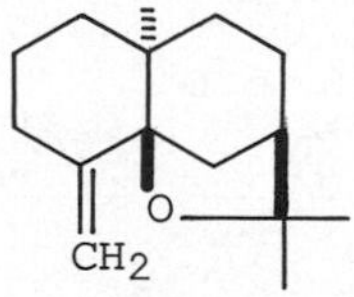

13

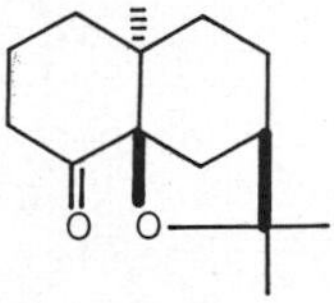

14

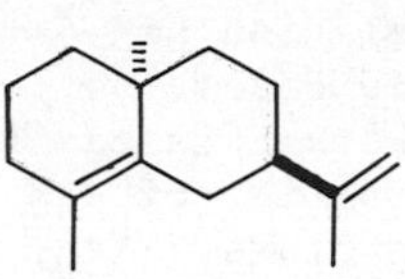

15

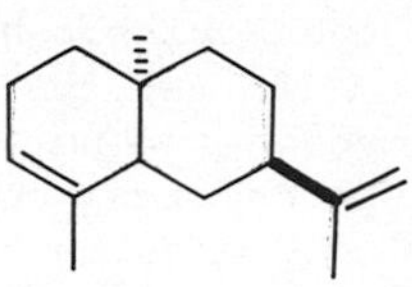

16

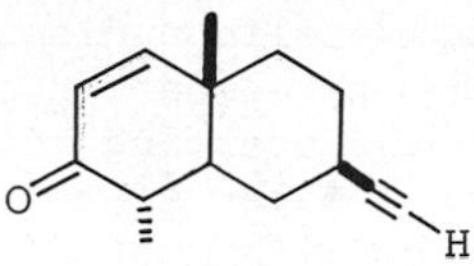

17

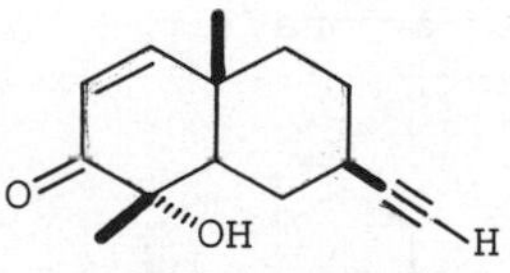

18

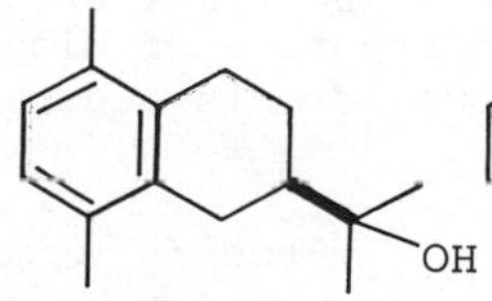

19

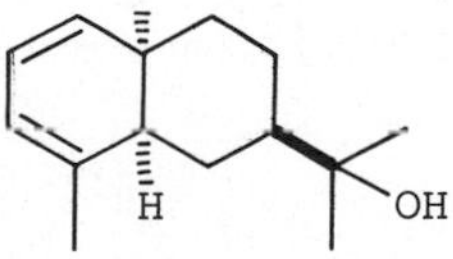

20

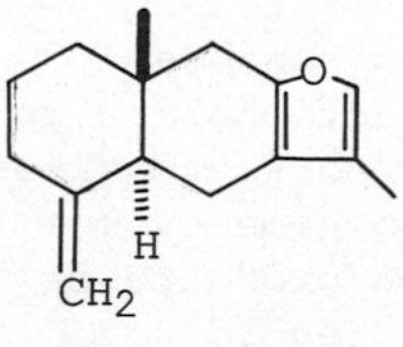

21

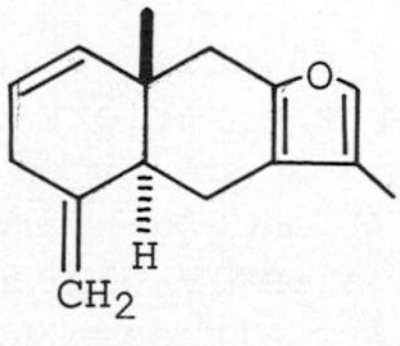

22

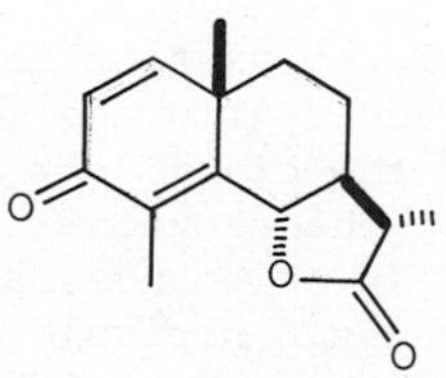

23

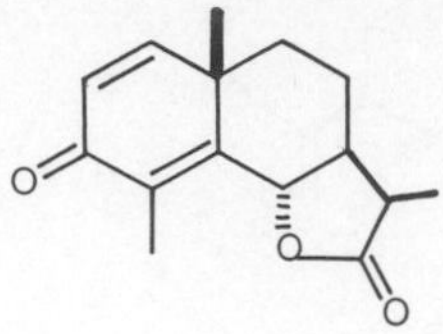

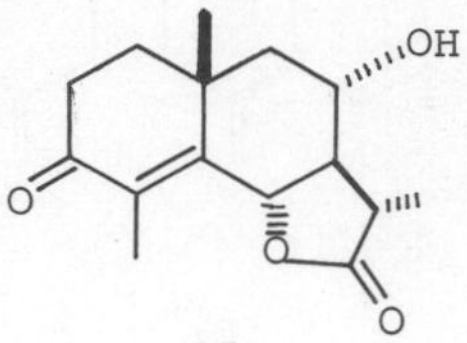

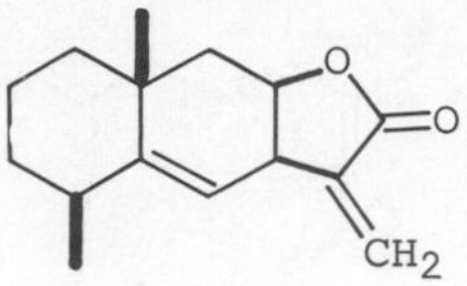

24   25   26

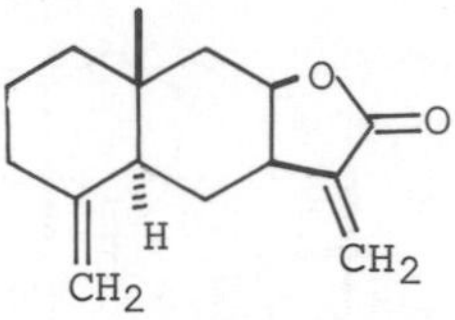

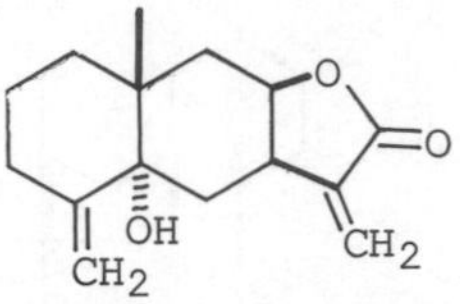

27   28

Almost invariably, workers in the field have constructed
the 9-methyldecalin nucleus by the Robinson annelation sequence. The three carbon side chain may, in some cases, be
present in the starting methylcyclohexanone; in other cases it
is added at a later stage. Stereochemical control is not a
major problem with most of the compounds in this class. Many
of the eudesmanes have a *trans* decalin nucleus, with the three
carbon function in an equatorial position, so that thermodynamic control may be exerted. Those members which have the
angular methyl group and the three carbon group *trans* have
been synthesized by way of 7-epi-cyperone (31), the kinetically formed isomer in condensation of dihydrocarvone (29)
with ethyl vinyl ketone (30), or its equivalent.

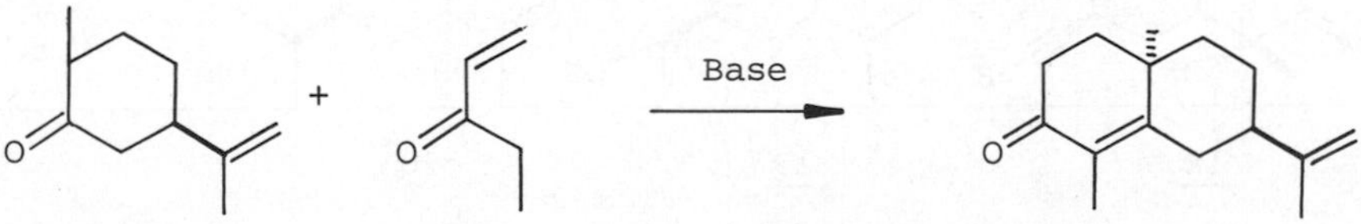

29   30   Base   31

The wide variation in synthetic approaches has been
mainly aimed at allowing for the variety of functionality in
members of the class. It should be noted that nor-ketoagarofuran (14), chamaecynone (17) and 4α-hydroxyisochamaecynone
(18) are C-14 compounds, apparently degradation products of
sesquiterpenes. Occidol (19), although nonisoprenoid, is
probably a rearrangement product of occidentalol.

B. α-Cyperone, β-Cyperone

In the earliest paper on a bicyclic sesquiterpene synthesis, Adamson, McQuillin, Robinson, and Simonsen reported the synthesis of "substances structurally identical with α- and β-cyperones."[113] The synthesis involved condensation of (-)-dihydrocarvone (32) with the methiodide of 1-diethylamino-pentan-3-one (33). The resulting product, ketol 34* was dehydrated with sodium ethoxide in benzene to give "α-cyperone." Later work has shown that this product is actually a mixture of compounds 35 and 36 with the former predominating.

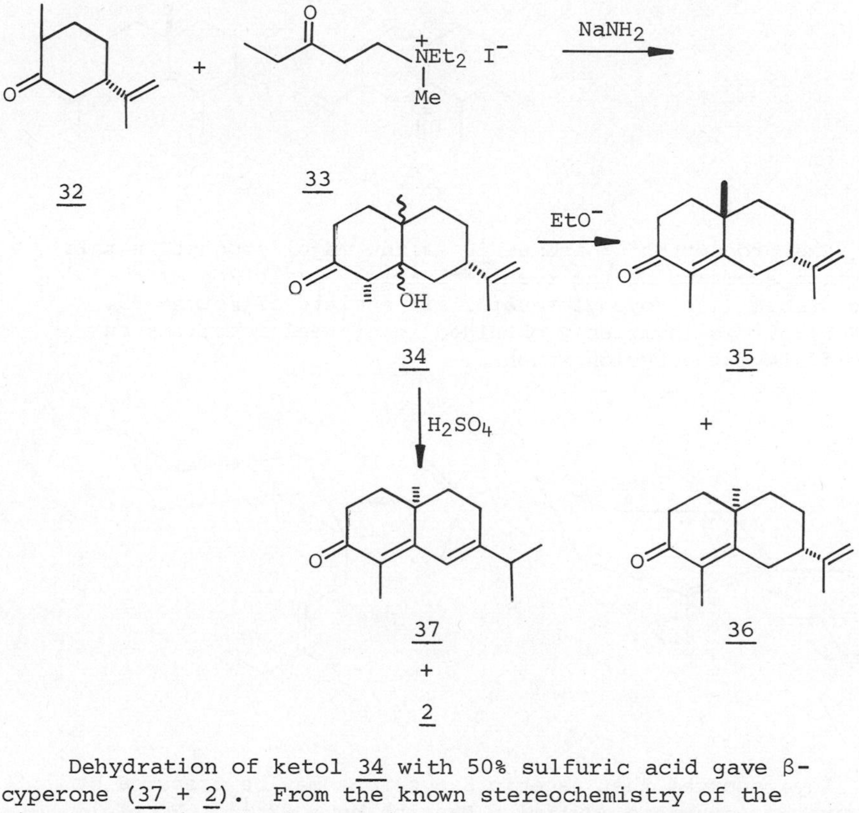

Dehydration of ketol 34 with 50% sulfuric acid gave β-cyperone (37 + 2). From the known stereochemistry of the Michael reaction of this series, it may be concluded that this sample was a mixture of approximately 80% 37 and 20% 2
Howe and McQuillin later reinvestigated this work.[114]

*The authors erroneously considered this substance to be the uncyclized diketone.

(+)-Dihydrocarvone (29) was condensed with compound 33 to give a mixture consisting mainly of ketol 38 and α-cyperone (1). After purification via its oxime, α-cyperone was obtained in 3% yield.

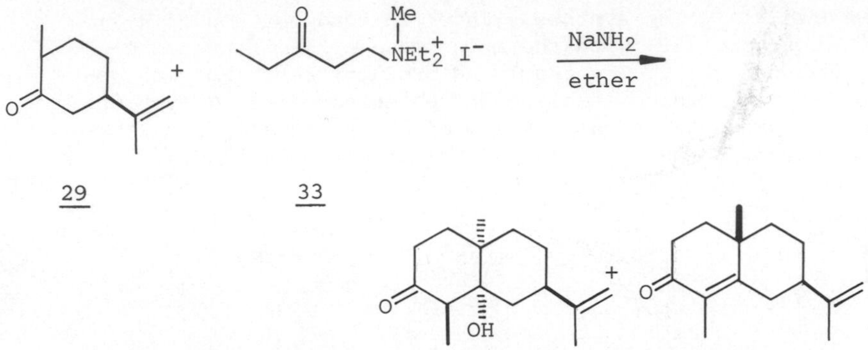

29   33

38   1

The production of isomer 38 as the major product in this reaction seems to be the result of axial alkylation of the more stable (isopropenyl equatorial) enolate of ketone 29. This result is invariably obtained in Michael reactions on such dialkylated cyclohexanones.

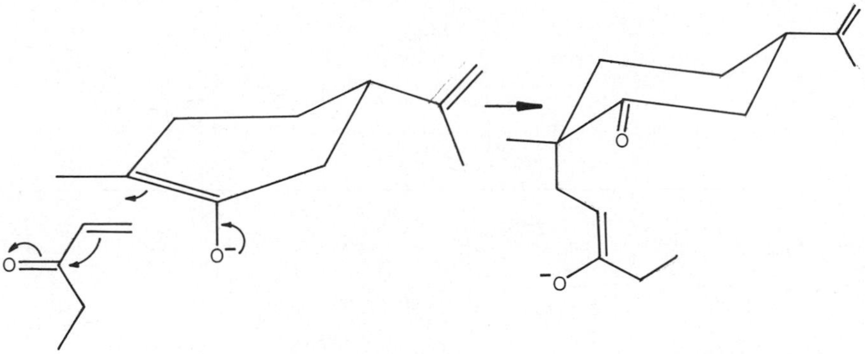

Roy reported that racemic β-cyperone may be prepared by condensing carvenone (39) with Mannich base 33.[115] No yield is given.

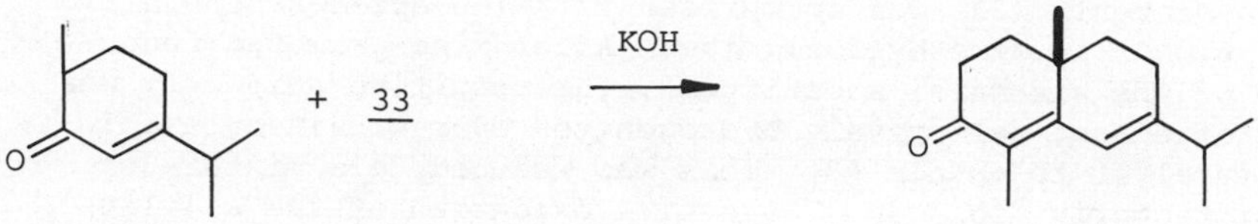

Since the yield of α-cyperone which may be obtained by such direct annelations is low, Piers developed a synthesis of the terpene from α-santonin (3-23) which is readily available commercially.[116]   The Piers synthesis is outlined in Scheme 1.

Scheme 1.   Piers' Synthesis of (−)-Cyperone (Relay)

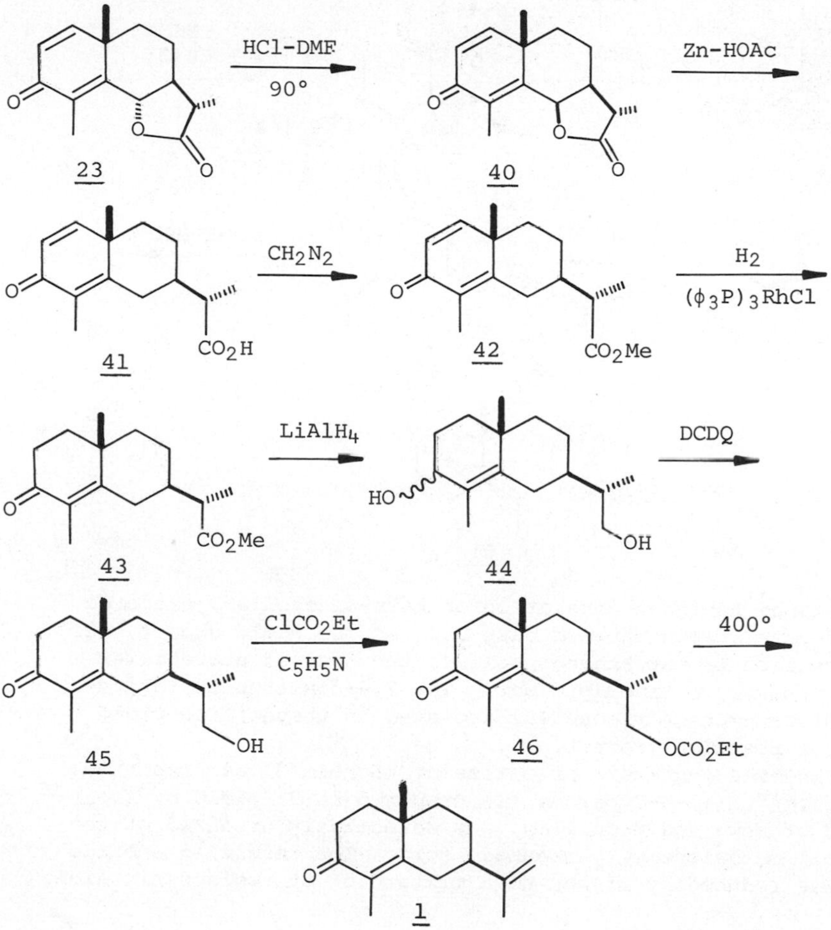

(-)-α-Santonin (23) was epimerized to (-)-6-epi-α-santonin (40) by HCl in dimethylformamide. After zinc promoted reduction of the six-axial substituent, the resulting acid (41) was esterified and selectively hydrogenated over Wilkinson's soluble catalyst to obtain 43. This was reduced to a mixture of diastereomeric diols 44. Selective oxidation of the allylic hydroxyl, with dichlorodicyanoquinone, gave enone-alcohol 45. The resulting carbonate ester (46) was pyrolyzed at 400° to effect elimination. α-Cyperone (1) was obtained in 61% yield, accompanied by 32% of enone-alcohol 45.

The first synthesis of carissone (3) was reported by Mukherji, Singh, and Vig in 1960 (Scheme 2).[117] The synthesis

Scheme 2.   Mukherji-Singh-Vig Synthesis of Carissone

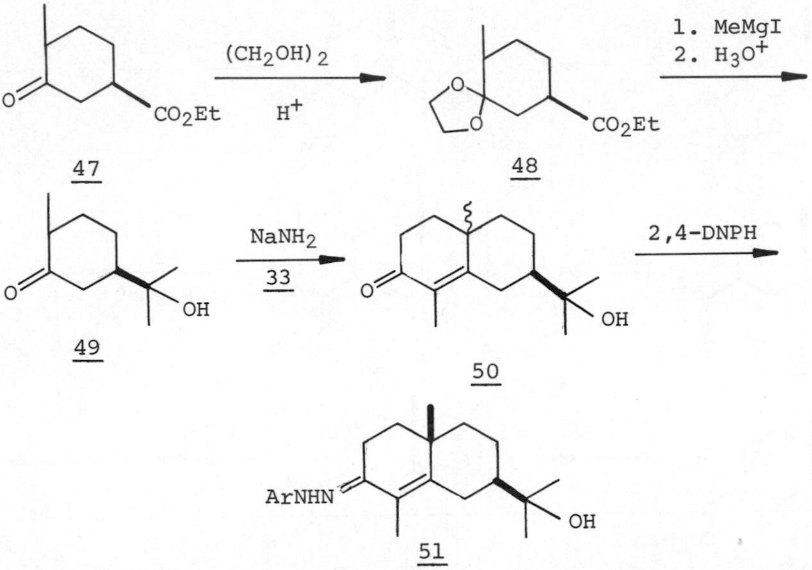

proceeds by Robinson annelation on keto-alcohol 49, prepared as indicated, with Mannich base 33. As one would expect, the condensation is not stereospecific, the desired stereosomer being probably a minor product. The 2,4-dinitrophenylhydrazone of racemic carrisone was isolated in unspecified yield from the reaction product.

Pinder's synthesis of carissone (Scheme 3) was reported in 1961.[118] (+)-α-Cyperone (1) prepared in 3% yield by the method of Howe and McQuillin, was selectively oxidized at the terminal methylene by perbenzoic acid. The resulting epoxide (52) was reduced by LiAlH₄ to a mixture of stereoisomeric diols

Scheme 3.   Pinder's Synthesis of (+)-Carissone

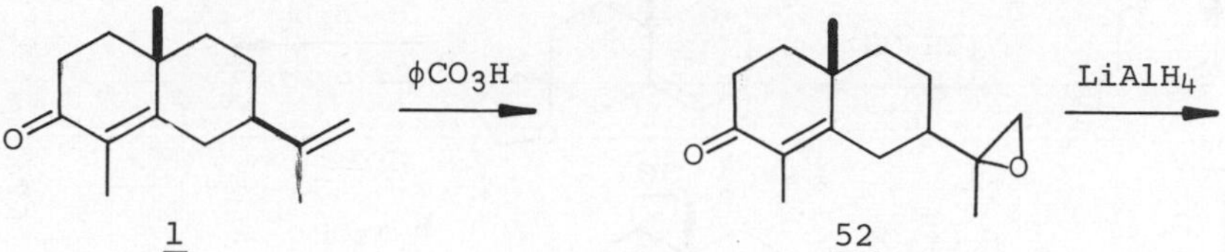

ϕCO_3H

$\underline{1}$

$\underline{52}$

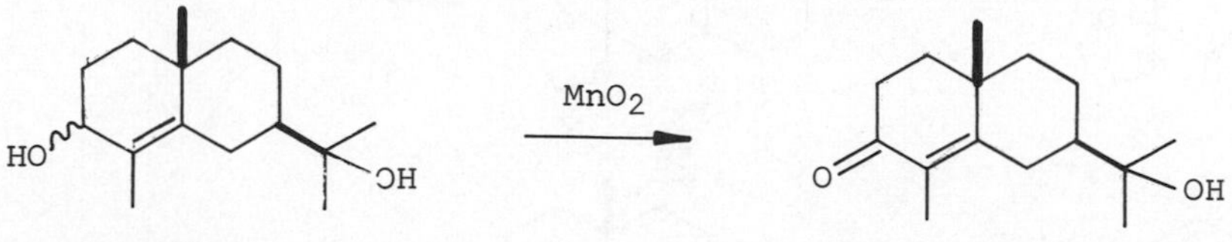

LiAlH$_4$

$\underline{53}$

MnO$_2$

$\underline{3}$

(53) which was oxidized with manganese dioxide to afford (+)-carissone (3).

C.   α-, β-, and γ-Eudesmol

Pinder and Williams reported the first synthesis of a eudesmol in 1963.[118b] Their synthetic (+)-carissone was converted into its dithioketal (54) which was desulfurized with Raney nickel to yield (+)-γ-eudesmol (6).

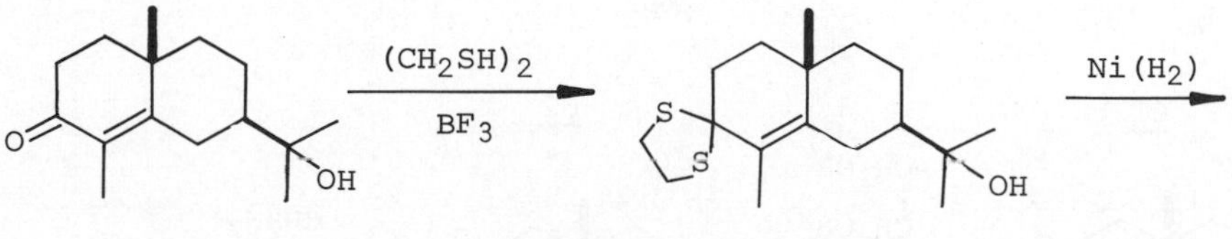

$(CH_2SH)_2$

BF$_3$

$\underline{3}$

$\underline{54}$

Ni(H$_2$)

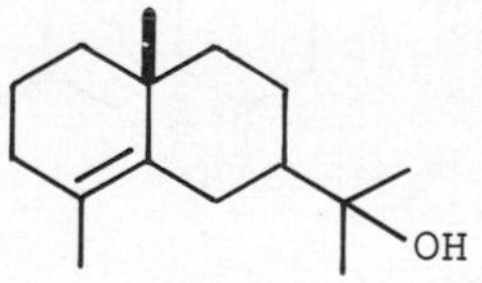

$\underline{6}$

Marshall's synthesis of β-eudesmol is outlined in Scheme 4.[119] Octalone 55 was ketalized, with concomitant double bond migration, to afford unsaturated ketal 56. Although it was not possible to drive this reaction to completion, crystalline ketal 56 could be obtained in 40% yield. Hydroboration of 56, by Brown's method,[120] gave a mixture containing alcohol

Scheme 4. β-Eudesmol--Marshall's Synthesis

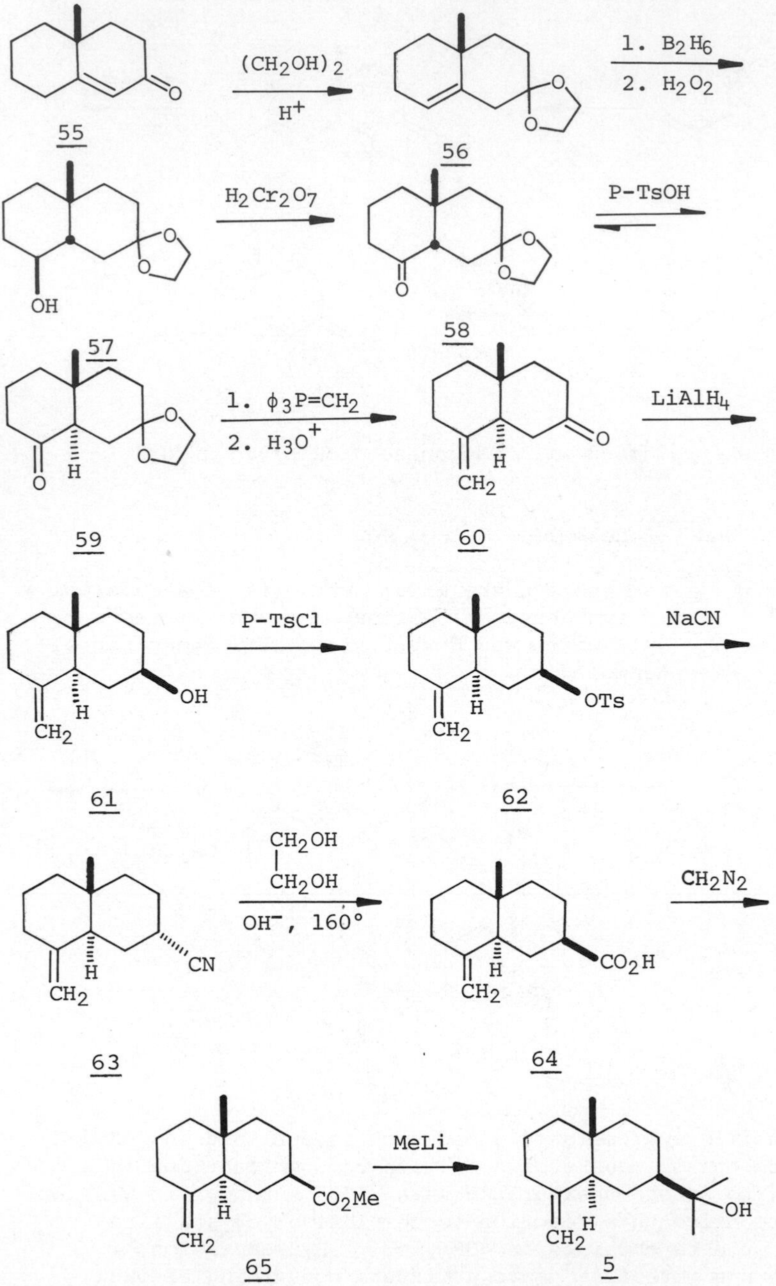

57.* Oxidation of this mixture with Jones reagent, followed by direct crystallization, allowed the production of crystalline keto-ketal 58 in 30% yield. When treated with p-toluenesulfonic acid in refluxing toluene, equilibrium was reached between *cis* ketone 58 and its *trans* counterpart 59 with the latter predominating (*trans/cis* = 3/1). Direct crystallization gave 59 in 65% yield. Both 58 and 59 reacted with methylenetriphenylphosphorane in DMSO to give predominately the *trans* fused product. After hydrolysis, the crystalline unsaturated ketone 60 was obtained (47% from 58 44% from 59). Lithium aluminum hydride reduction gave predominately the equatorial alcohol 61 (isolated in 83% yield), which was converted into its p-toluenesulfonate ester (62). When 62 was treated with sodium cyanide in N-methylpyrrolidone, nitrile 63 was produced in 64% yield. This was hydrolyzed, with concomitant epimerization, to equatorial acid 64 (65%). The corresponding ester (65) was treated with methyllithium to obtain (±)-β-eudesmol (5).

Pinder's synthesis of β-eudesmol (Scheme 5)[122] begins with a Robinson annelation of (-)-dihydrocarvone (32) and 1-diethylaminobutan-3-one methiodide. As discussed earlier, the major product in such reactions is that produced by axial alkylation on the more stable conformer of the corresponding enolate. Therefore, the resulting ketol (67), isolated in 25% yield, has the undesired *trans* disposition of the angular methyl and isopropenyl groups. In order to render C-7 epimerizable, ketol 67 was ozonized to diketone 68. This material was

*Careful analysis of this reaction indicated the following products:

Ring-opening of ethylene ketals by diborane had been observed previously.[121] (However, see p. 405.)

Scheme 5. β-Eudesmol--Pinder's Synthesis

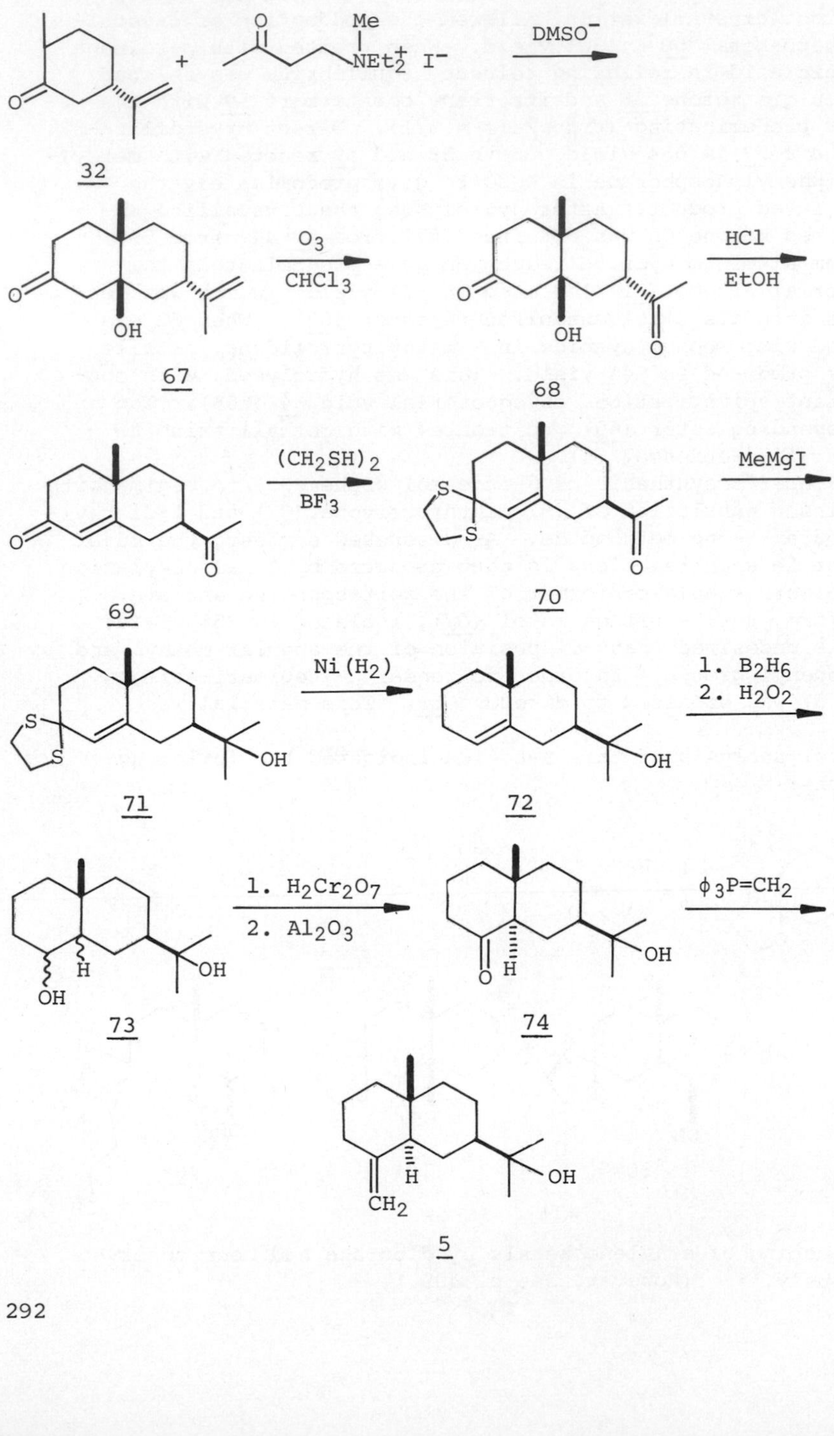

dehydrated, with simultaneous epimerization, to obtain endione
69. The unsaturated carbonyl group of 69 reacted preferentially with ethanedithiol, affording crystalline 70 in 30% yield.
Raney nickel desulfurization of the derived dimethyl carbinol
(71) gave nor-γ-eudesmol (72). This was converted into (+)-
β-eudesmol in the straightforward manner shown.

An efficient stereoselective synthesis of β-eudesmol,
suitable for preparing the material in quantity, was reported
by Heathcock and Kelly (Scheme 6) in 1968.[123] The synthesis

Scheme 6. β-Eudesmol--Heathcock-Kelly Synthesis

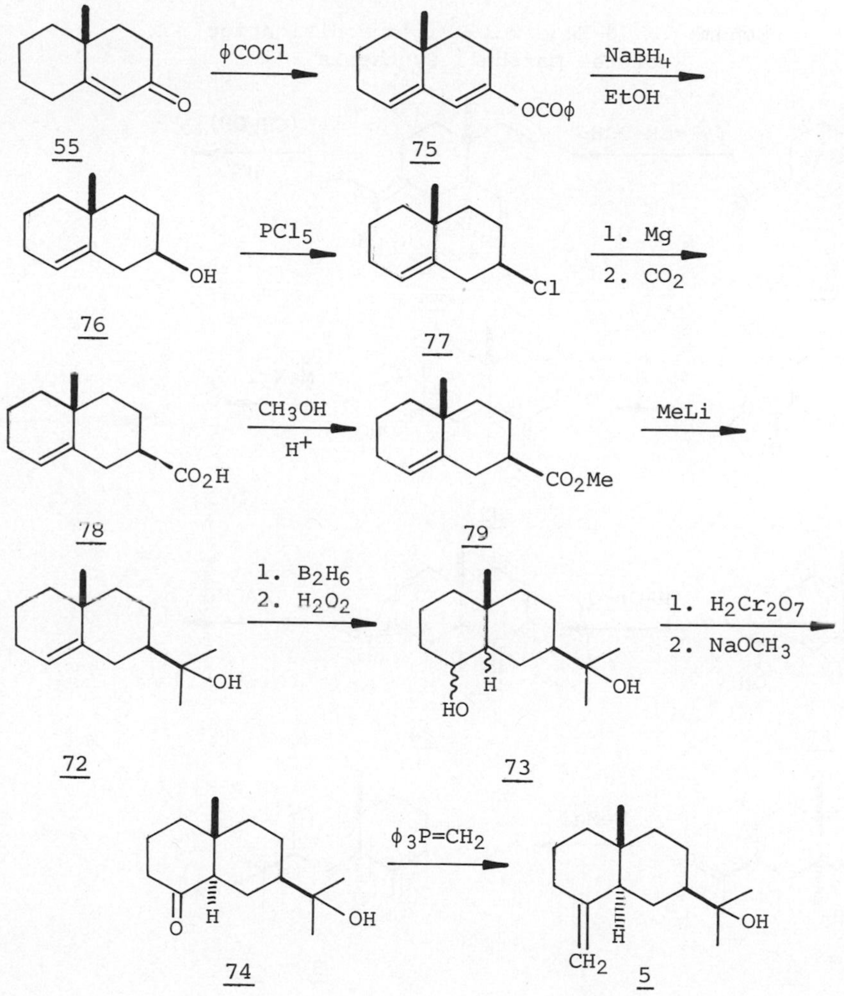

begins with methyl octalone 55, which was deconjugated by treating the derived enol benzoate (75) with sodium borohydride in ethanol. The resulting unsaturated alcohol reacted with PCl_5 to give chloride 77. Carbonation of the corresponding Grignard reagent gave exclusively the equatorial acid 78, which was converted, via methyl ester 79, into (±)-nor-γ-eudesmol (72). The remainder of the synthesis is identical to that of Pinder in the optically active series. The overall yield for the ten steps was 10%.

Vig and co-workers have proposed a modification of Marshall's synthesis (Scheme 7) in which the isopropanol side

Scheme 7.   β-Eudesmol--Vig's Modification of the Marshall Synthesis

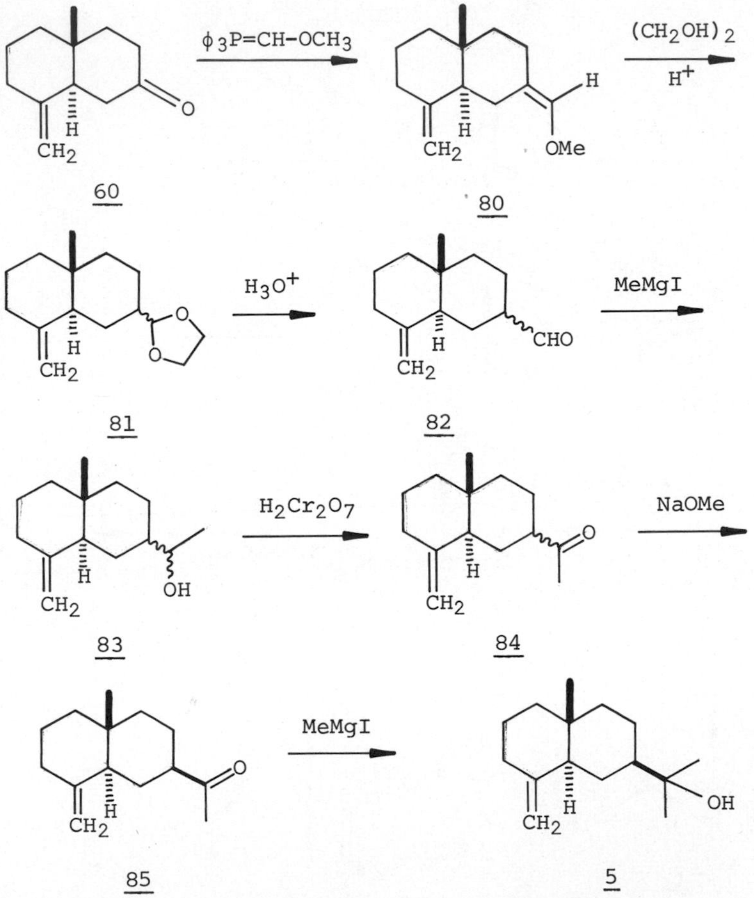

chain is introduced by a Wittig reaction.[124]   The route appears
to offer little advantage over the original Marshall method.
        Marshall has reported a synthesis of γ-eudesmol,[125] which
begins with a key intermediate in his alantolactone synthesis
(Scheme 8).   Unsaturated ketone 86 was carbonated to give

Scheme 8.   Marshall's Synthesis of γ-Eudesmol

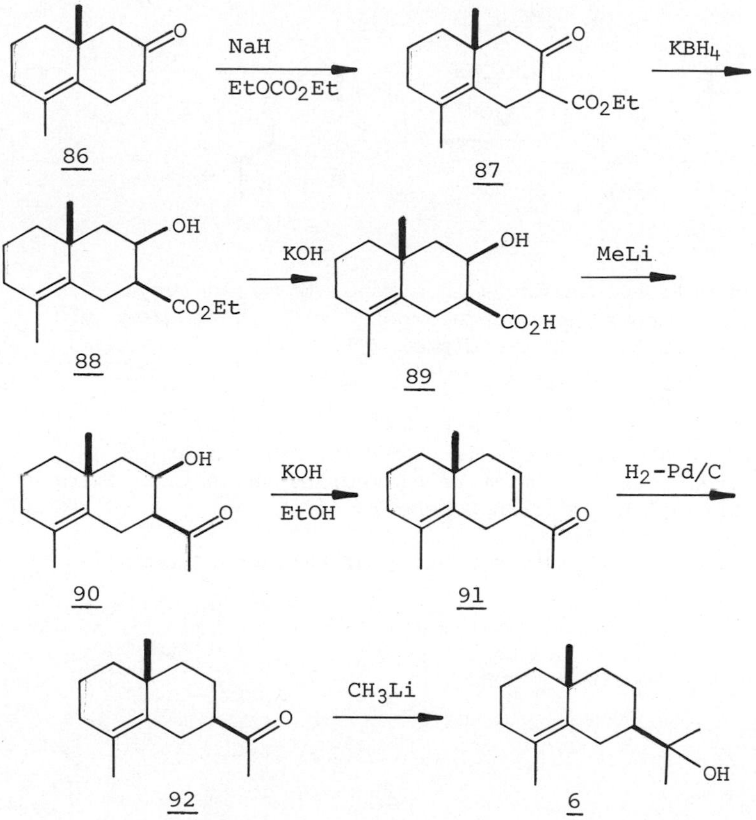

β-keto ester 87, which was selectively reduced to β-hydroxy
ester 88.   Treatment of the corresponding hydroxy acid with
methyllithium gave hydroxy ketone 90, which was smoothly dehydrated with alcoholic base to enone 91.   This was hydrogenated to 92, which reacted with methyllithium to afford
(±)-γ-eudesmol.
        Pinder reports a synthesis of α-eudesmol (4) from caris-
sone (3) which is outlined in Scheme 9.[122b]   Birch reduction
of carissone gave trans-dihydrocarissone (93), which reacted

Scheme 9. Pinder's Synthesis of α-Eudesmol

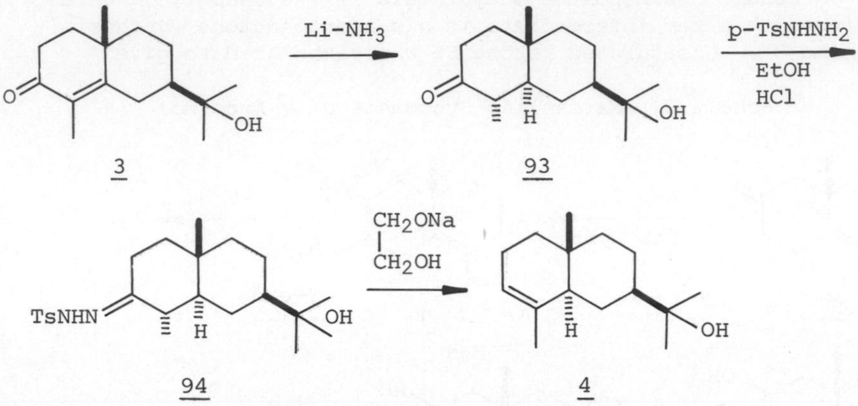

$$ \underline{3} \quad \xrightarrow{\text{Li-NH}_3} \quad \underline{93} \quad \xrightarrow[\substack{\text{EtOH} \\ \text{HCl}}]{\text{p-TsNHNH}_2} $$

$$ \underline{94} \quad \xrightarrow[\text{CH}_2\text{OH}]{\text{CH}_2\text{ONa}} \quad \underline{4} $$

with p-toluenesulfonylhydrazine in ethanol to yield tosylhydrazone 94. Compound 94 was decomposed with the sodium salt
of ethylene glycol to give α-eudesmol (4).

D.   α- and β-Selinene

An early attempt at the synthesis of β-selinene (Banerjee,
1960), which resulted in a mixture of various selinene isomers
is outlined without comment in Scheme 10.[126]

Scheme 10. Banerjee's Synthesis of Selinene Isomers

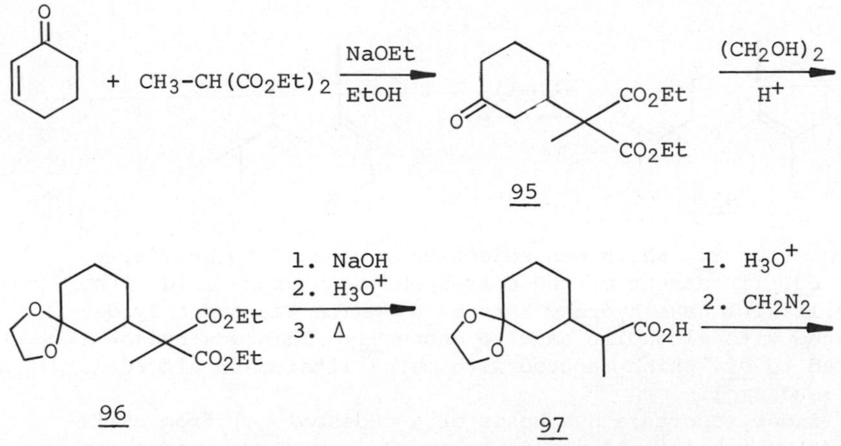

$$ + \; \text{CH}_3\text{-CH(CO}_2\text{Et)}_2 \quad \xrightarrow[\text{EtOH}]{\text{NaOEt}} \quad \underline{95} \quad \xrightarrow[\text{H}^+]{\text{(CH}_2\text{OH)}_2} $$

$$ \underline{96} \quad \xrightarrow[\substack{\text{2. H}_3\text{O}^+ \\ \text{3. }\Delta}]{\text{1. NaOH}} \quad \underline{97} \quad \xrightarrow[\text{2. CH}_2\text{N}_2]{\text{1. H}_3\text{O}^+} $$

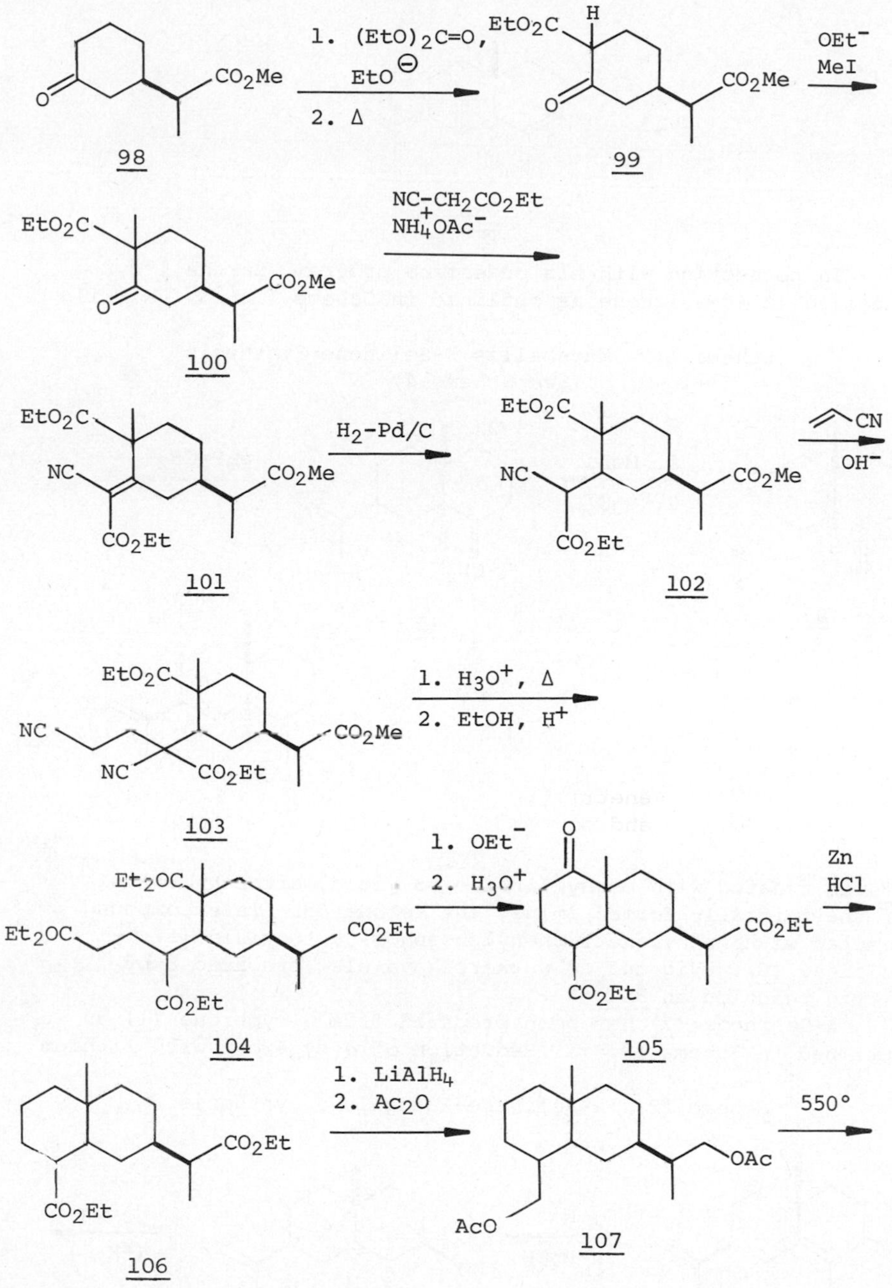

$$\underset{\underline{98}}{}\quad\xrightarrow[\text{2. }\Delta]{\begin{array}{c}\text{1. (EtO)}_2\text{C=O,}\\ \text{EtO}^{\ominus}\end{array}}\quad\underset{\underline{99}}{}\quad\xrightarrow[\text{MeI}]{\text{OEt}^-}$$

$$\underset{\underline{100}}{}\quad\xrightarrow{\begin{array}{c}\text{NC-CH}_2\text{CO}_2\text{Et}\\ +\\ \text{NH}_4\text{OAc}^-\end{array}}$$

$$\underset{\underline{101}}{}\quad\xrightarrow{\text{H}_2\text{-Pd/C}}\quad\underset{\underline{102}}{}\quad\xrightarrow[\text{OH}^-]{\diagup\!\!\diagdown\text{CN}}$$

$$\underset{\underline{103}}{}\quad\xrightarrow[\text{2. EtOH, H}^+]{\text{1. H}_3\text{O}^+,\ \Delta}$$

$$\underset{\underline{104}}{}\quad\xrightarrow[\text{2. H}_3\text{O}^+]{\text{1. OEt}^-}\quad\underset{\underline{105}}{}\quad\xrightarrow[\text{HCl}]{\text{Zn}}$$

$$\underset{\underline{106}}{}\quad\xrightarrow[\text{2. Ac}_2\text{O}]{\text{1. LiAlH}_4}\quad\underset{\underline{107}}{}\quad\xrightarrow{550°}$$

297

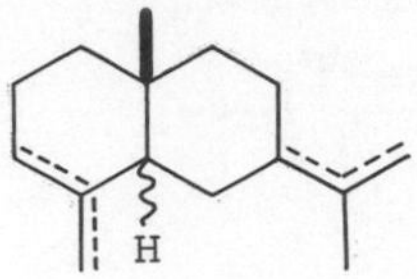

108

In connection with his eudesmane program, Marshall synthesized (±)-β-selinene as outlined in Scheme 11.[119b] Nitrile

Scheme 11.   Marshall's β-Selinene Synthesis
(See Scheme 4)

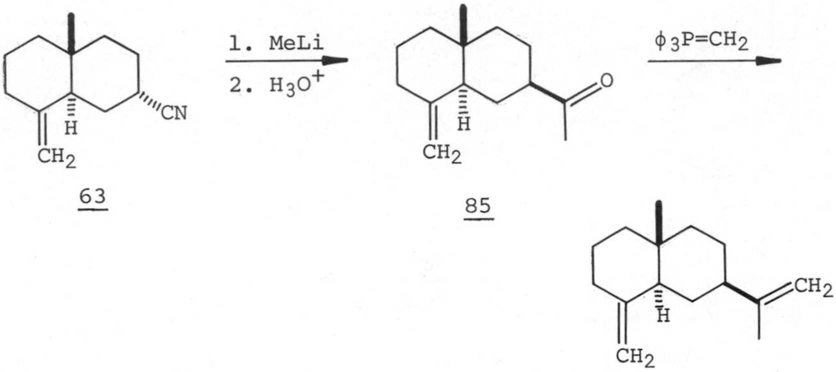

63 was treated with methyllithium to yield, after hydrolysis
of the initially formed imine, the ketone 85. This compound
reacted with methylenetriphenylphosphorane to yield (±)-β-
selinene (8). Vig and co-workers have also prepared 8 by
Wittig reaction on 85.[124]

α-Selinene (7) has been prepared from α-cyperone (1) as
outlined in Scheme 12.[127] Reduction of α-cyperone with lithium

Scheme 12.   α-Selinene--Bangalore Synthesis

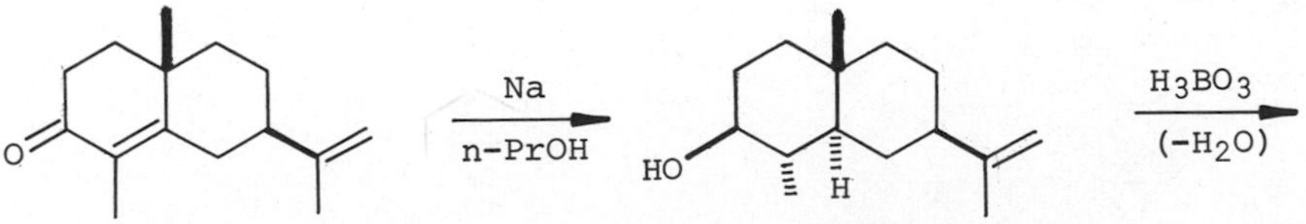

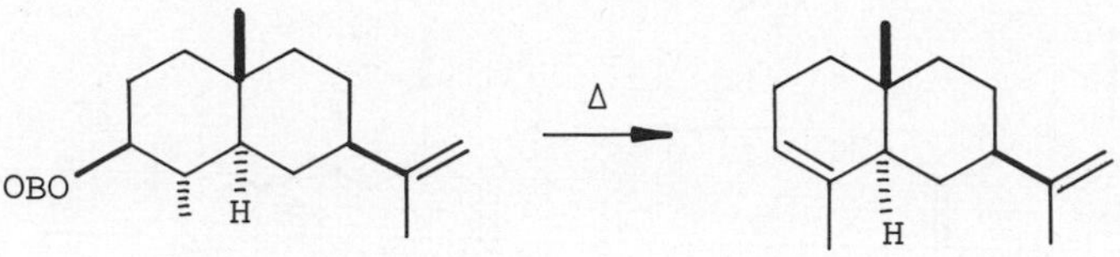

110 7

in ammonia containing ethanol or by sodium and n-propanol gave dihydrocyperol (109). In order to effect *cis* elimination of the equatorial hydroxyl and its neighboring tertiary hydrogen, the corresponding metaborate ester (110) was pyrolized by heating over a free flame. The resulting hydrocarbon mixture contained predominately α-selinene.

E. Costol, Costal, Costic Acid

These related compounds were prepared by Marshall as outlined in Scheme 13.[119b,128] The equatorial tosylate 62 was converted

Scheme 13. Marshall's Synthesis of (±)-Costol (9), (±)-Costol (10), and (±)-Costic Acid (11)

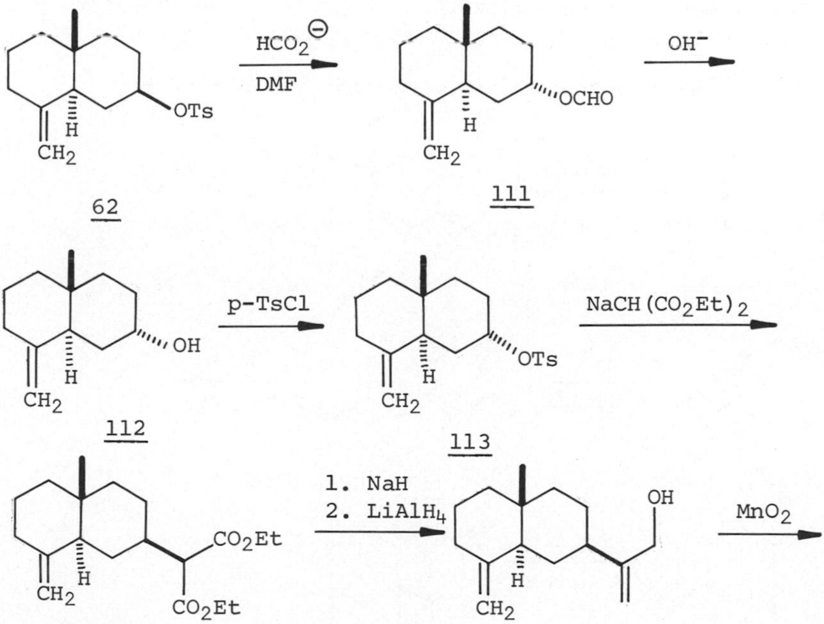

62 111 112 113 114 9

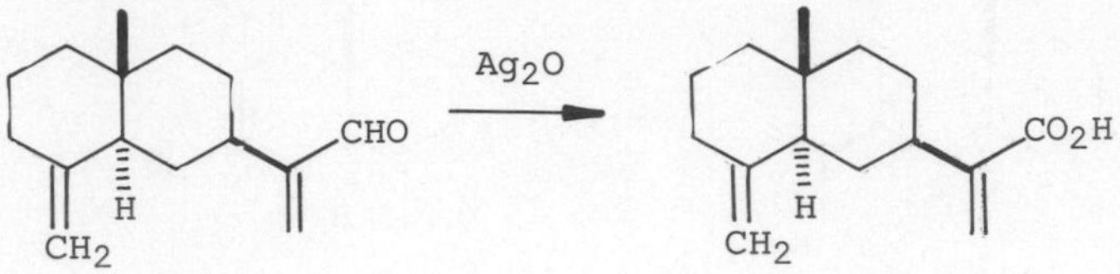

$$10 \xrightarrow{Ag_2O} 11$$

into axial tosylate 113 by the three-step sequence shown. Alkylation of malonic ester with tosylate 113 gave diester 114. Reduction of the corresponding sodium salt with lithium aluminum hydride gave a 3:1 mixture of (±)-costol (9) and dihydrocostol. Oxidation of the allylic hydroxyl with activated manganese dioxide gave (±)-costal (10) and further oxidation of this substance with silver oxide gave (±)-costic acid (11).

F. α- and β-Agarofuran, Nor-ketoagarofuran

In the agarofurans (12-14), the angular methyl group, and the three-carbon side chain are *trans*. Therefore, the readily available 7-epi-α-cyperone (31) may be used as a starting point.

In a synthesis designed to ascertain the relative stereochemistry of the molecule, Barrett and Büchi prepared α-agarofuran from compound 31 as shown in Scheme 14.[129] Reduction of

Scheme 14. Barrett-Büchi Synthesis of α-Agarofuran

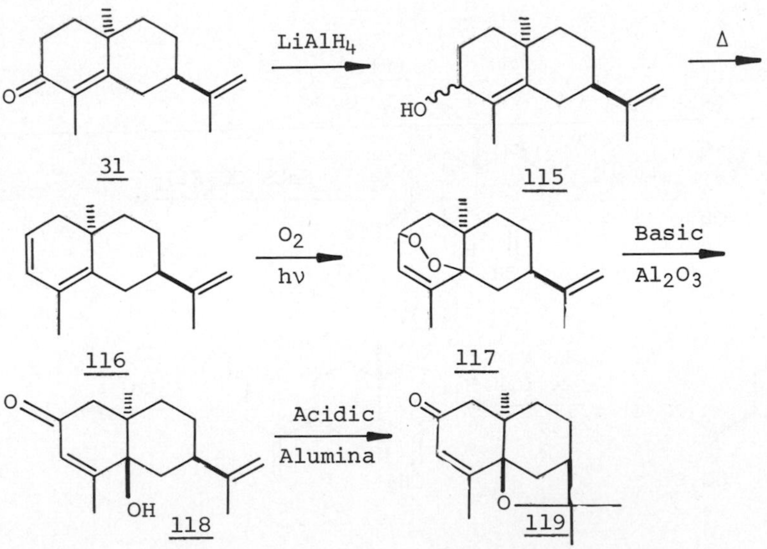

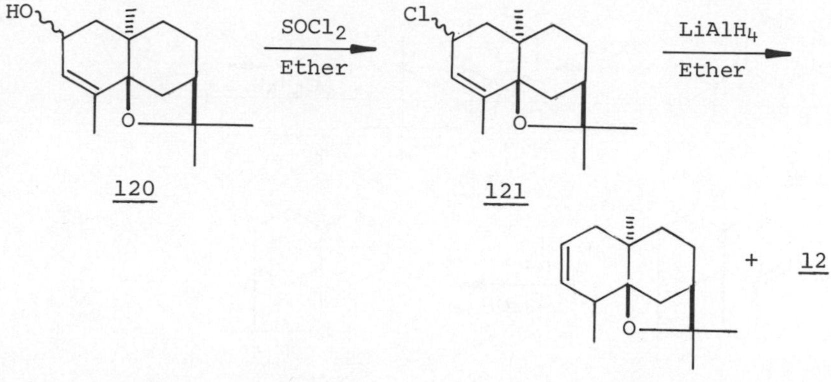

120   121

122   + 12

7-epi-α-cyperone (31) with LiAlH₄ gave a crystalline alcohol (stereochemistry undetermined) in 80% yield. Pyrolytic dehydration of this alcohol (115) gave a mixture of diene 116 and its heteroannular isomer in a ratio of 3:1. Sensitized photooxygenation of 116 gave peroxide 117, the singlet oxygen apparently adding predominately from the less hindered side of the molecule. When peroxide 117 was treated with basic alumina, it was transformed into the crystalline keto-alcohol (118), which readily cyclized to keto-ether 119 upon exposure to acid-washed alumina. Compound 119 was reduced by LiAlH₄ to a mixture of allylic alcohols 120. This mixture was deoxygenated by reducing the derived chlorides 121 with LiAlH₄ α-agarofuran (12) and its double bond isomer (122) were produced in equal amounts.

Marshall and Pike synthesized α- and β-agarofuran as shown in Scheme 15.[130] 7-epi-α-cyperone was epoxidized to 123

Scheme 15. Marshall's Synthesis of the Agarofurans

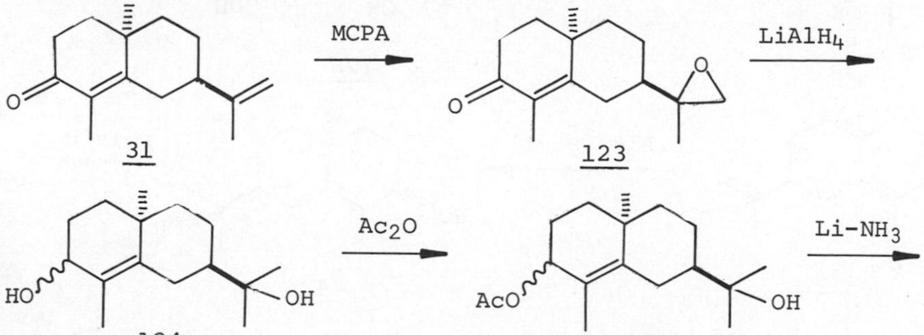

31   123

124   125

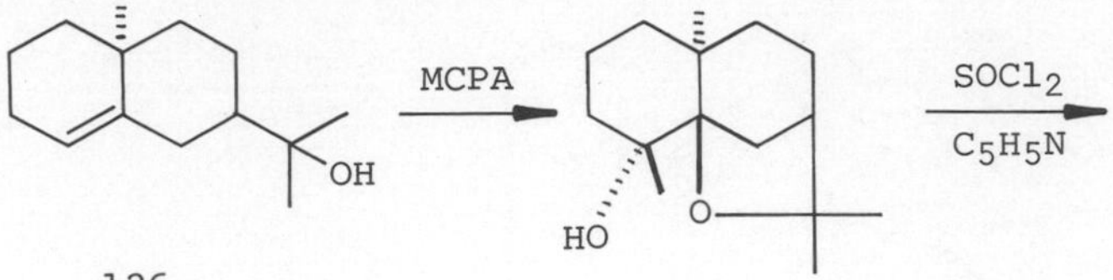

126 127

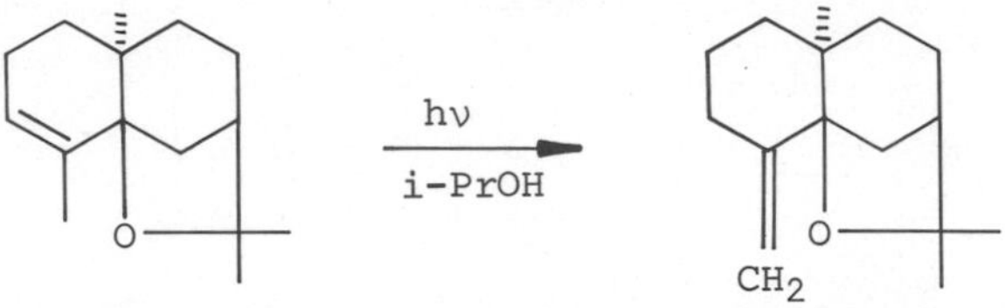

12 13

(probably a mixture of diastereomers) which was reduced to a mixture of isomeric diols (124). The mixture was not separated but was directly acetylated and the mixture of acetates (125) reductively cleaved with lithium in ammonia to 7-epi-γ-eudesmol (126). Treatment of 126 with m-chloroperbenzoic acid gave directly the naturally occurring 4-hydroxydihydroagarofuran (127), which was dehydrated by thionyl chloride in pyridine to α-agarofuran (12). When 12 was irradiated in isopropyl alcohol with xylene as sensitizer, β-agarofuran (13) was produced.

The most efficient synthesis of α-agarofuran yet reported is that of Deslongchamps and co-workers (Scheme 16).[131] Ketol

Scheme 16. Deslongchamp's Synthesis of α-Agarofuran

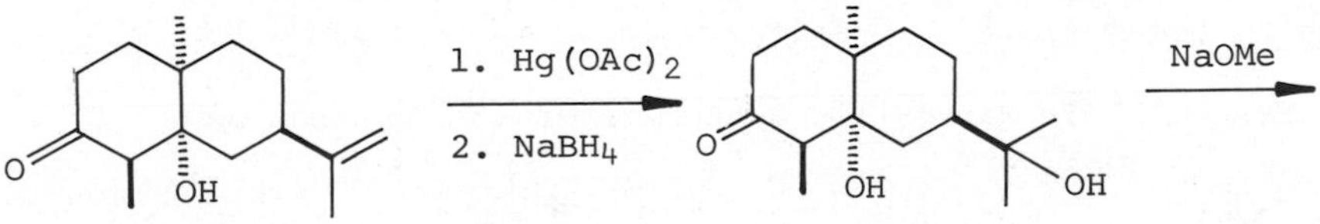

38 128

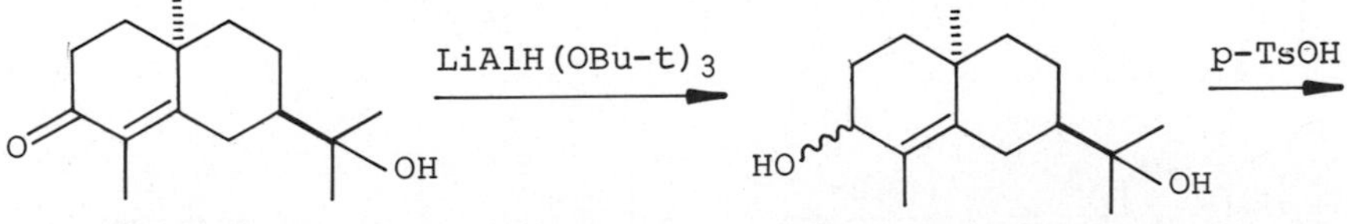

129 124

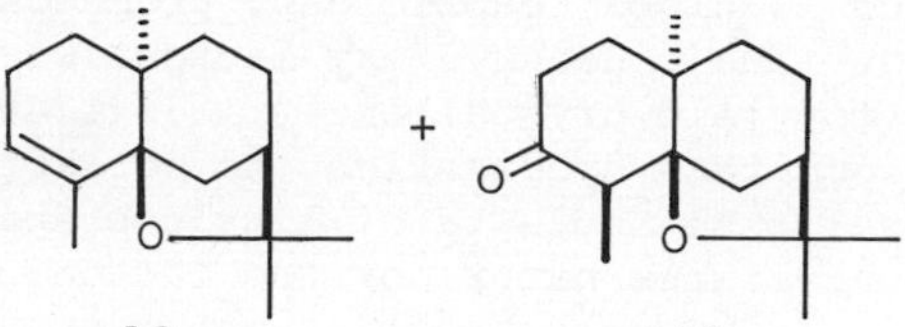

12   **130**

38 (p. 286) was converted into keto diol **128** by oxymercuration-demercuration. Base catalyzed dehydration yielded 7-
epi-carrisone (**129**), which was reduced to a mixture of unsaturated diols (**124**). Treatment of **124** with p-toluenesulfonic acid in benzene gave α-agarofuran in 80% yield, along
with 20% of keto-ether **130**.

     Kelly and Heathcock reported the synthesis of nor-keto-
agarofuran (**14**) which is outlined in Scheme 17.[132] Unsaturated

Scheme 17. Heathcock-Kelly Synthesis of Nor-Ketoagarofuran

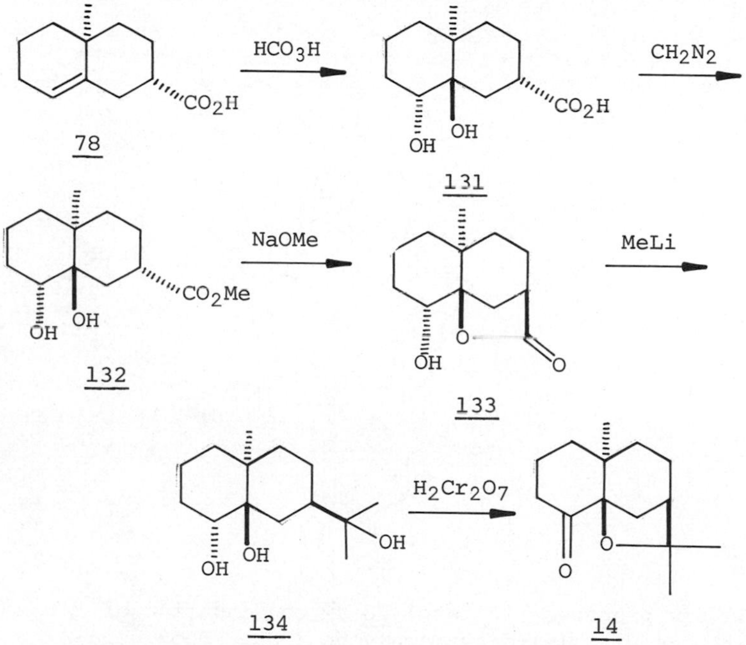

acid **78**, used in the synthesis of β-eudesmol (Scheme 6), was
oxidized by performic acid to dihydroxy acid **131**. The corresponding methyl ester **132** was epimerized and lactonized under
carefully controlled conditions to obtain hydroxy lactone **133**.

This reaction, which is highly capricious, proceeds in yields
ranging from 10-70%. The crude hydroxy lactone was treated
with methyllithium to yield crystalline triol (134), which was
oxidized, with concommitant cyclization, to nor-ketoagarofuran
(14). Although novel, the route is of little preparative
value, due to the unreliable nature of the lactonization step.

G. Epi-γ-Selinene

Like the agarofurans and occidentalol, epi-γ-selinene (15) and
epi-α-selinene (16) have the angular methyl and the three-
carbon side chain *trans*. As mentioned earlier (p. 286), this
stereochemistry may be established kinetically by Robinson
annelation of ethyl vinyl ketone (or its equivalent) and dihydrocarvone. As part of the structure proof, Klein and Rojahn
synthesized isomer 15 as outlined in Scheme 18.[133] Epi-α-

Scheme 18. Klein-Rojahn Synthesis of Epi-γ-Selinene

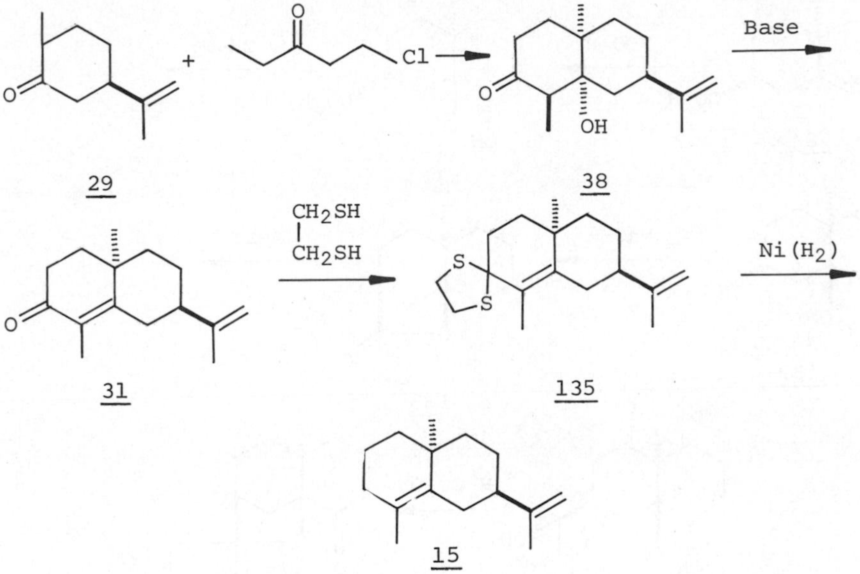

cyperone (31) was prepared via ketol 38 by condensation of 1-
chloro-3-pentanone with (+)-dihydrocarvone (29). Deoxygenation was accomplished by Raney nickel desulfurization of the
corresponding dithioketal (135).

H.   Chamaecynone and 4α-Hydroxyisochamaecynone

The chamaecynones are a group of bicyclic nor-sesquiterpenes which have been isolated from the essential oil of the Benihi tree.   Chamaecynone (17) was synthesized in 1967 by the Sendai group of Nozoe, Asao, Ando, and Takase.[134]   The starting point was α-santonin (23) which was converted via its C-6 epimer into keto-acid 41.   Compound 41 was hydrogenated, in the presence of 1% NaOH, to give saturated ketone 136.   The configuration at C-4, which is epimerizable, is the more stable one. Compound 136 was subjected to chlorodecarboxylation (Kochi's modification of the Hunsdiecker reaction) to obtain chloride 137, undoubtedly a mixture of diastereomers.   Since the next step planned involved treatment with strong base, and aldol condensations must be avoided, the carbonyl group was temporarily reduced, affording a mixture of diastereomeric chloro-alcohols 138.   Base catalyzed dehydrochlorination gave unsaturated alcohol mixture 139, which was brominated and dehydrobrominated to a mixture of acetylenic alcohols 141. Compound 141 was oxidized to ketone 142.   Compound 142 was brominated and the resultant mixture of bromoketones 143 was dehydrobrominated to give chamaecynone (17) in 38% yield. Epimerization at C-4 apparently occurs during the bromination stage.

Scheme 19.   Chamaecynone--Sendai Synthesis

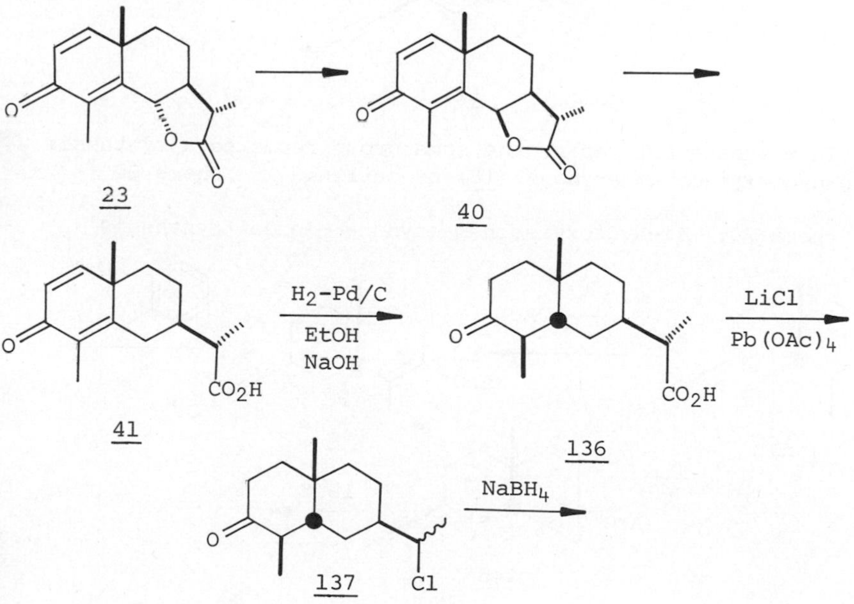

23

40

H$_2$-Pd/C

EtOH
NaOH

41

136

LiCl

Pb(OAc)$_4$

137   Cl

NaBH$_4$

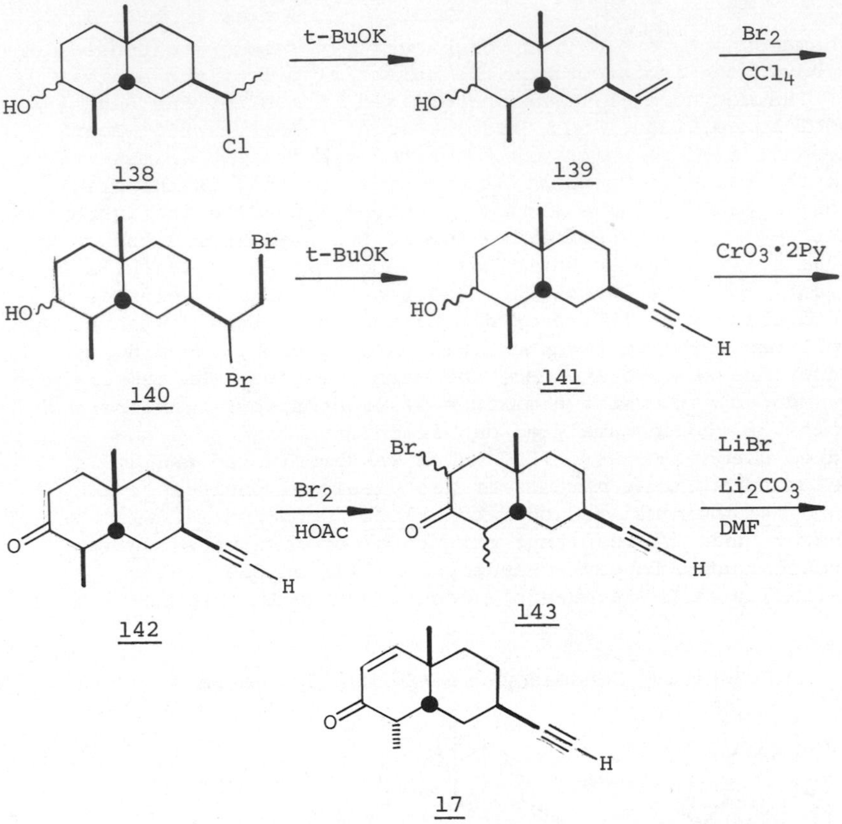

In a subsequent paper, the same group reported a synthesis of 4α-hydroxyisochamaecynone (18) as outlined in Scheme 20.[135]

Scheme 20.   4α-Hydroxyisochamaecynone--Sendai Synthesis

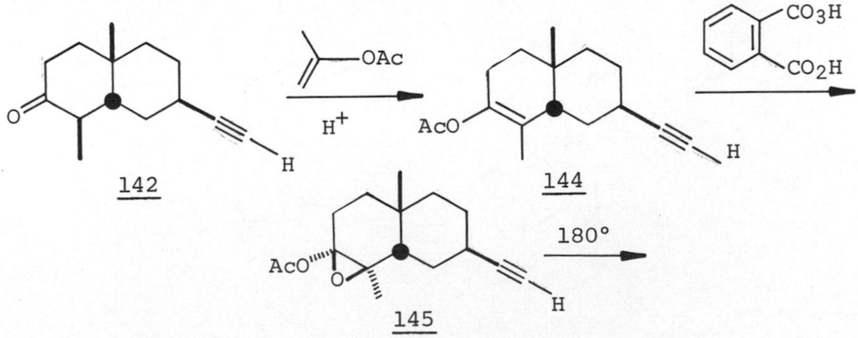

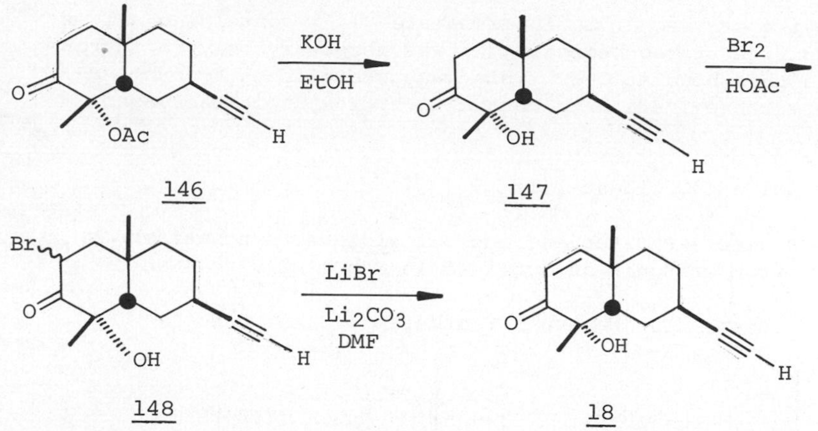

146   147

148   18

The synthesis, which was carried out in order to determine the stereochemistry at C-4, began with acetylenic ketone 142. This material was converted into enol acetate 144, which was epoxidized to oxide 145. Because of the folded nature of the *cis*-octalin 144, the oxygen is delivered stereospecifically from the convex surface of the molecule. Pyrolytic rearrangement of 145 gave keto acetate 146, which was hydrolyzed to hydroxy ketone 147. The isomeric hydroxy ketone 149 was prepared by direct basic hydrolysis of 145. The formation of isomers 147 and 149 may be depicted mechanistically as follows:

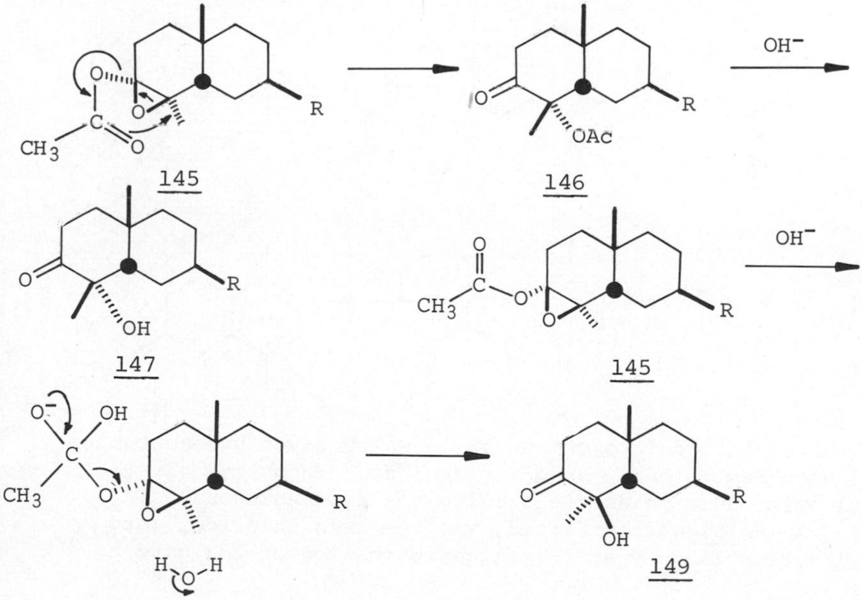

145   146   147   145   149

Having arrived at an intermediate (147) containing all of the requisite stereochemistry, it was necessary only to introduce a double bond at C-1. This was accomplished by bromination-dehydrobromination to obtain 4α-hydroxyisochamaecynone (18).

I. Occidol and Occidentalol

The rearranged sesquiterpene occidol (19) was synthesized by Nakazaki from santonin as outlined in Scheme 21.[136] Santonin

Scheme 21. Nakazaki Synthesis of (+)-Occidol

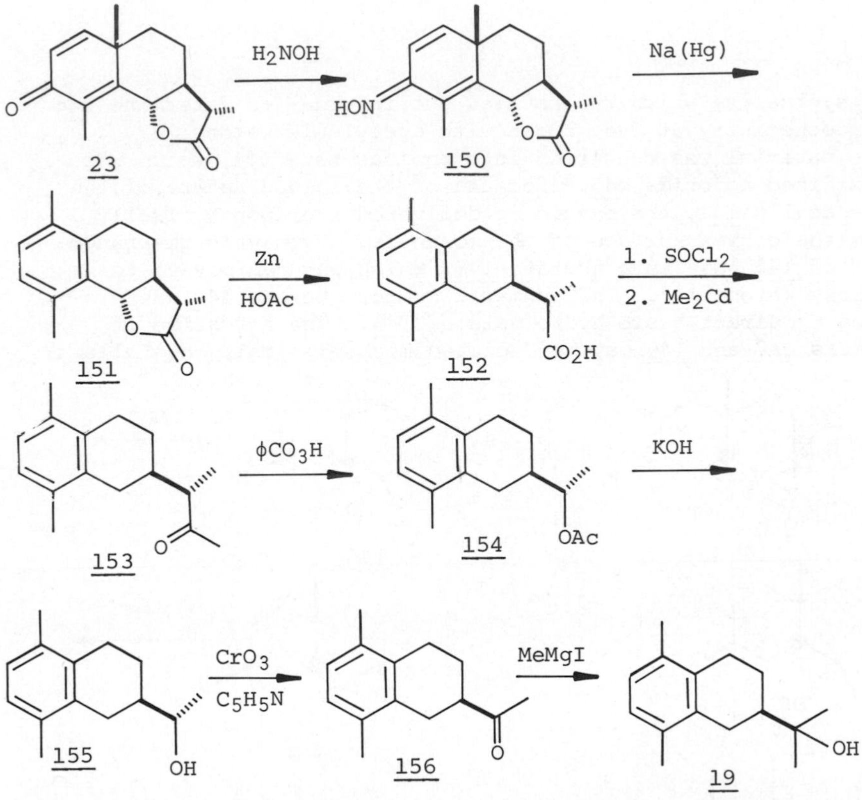

(23) was converted into oxime (150), which gives hyposantonin (151) upon reduction with sodium amalgam. Reductive cleavage of 151 (zinc dust in acetic acid) gives hyposantonous acid (152). Acid 152 was converted, via the acid chloride, into methyl ketone 153. Baeyer-Villiger oxidation of 153 gave an

acetate (154), which was hydrolyzed to an alcohol (155). Compound 155 was oxidized to ketone 156, which gave (+)-occidol when treated with methylmagnesium iodide.

Occidentalol is an interesting eudesmane in that the decalone system is *cis*-fused and the angular methyl and three-carbon groups are *trans*. The stereochemistry must be introduced solely by kinetic methods. Hortmann has stated in a footnote[137] that occidentalol (20) can be obtained, along with isomer 158, via photolysis of "*trans*-occidentalol (157)." However, the synthesis of 157 has not yet been reported.

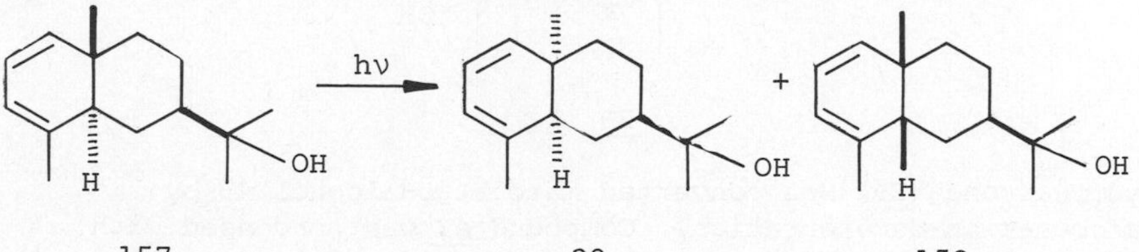

157   20   158

This reaction, which must proceed by closure of an intermediate cyclodecatriene, is similar to the key step in the Corey-Hortmann synthesis of dihydrocustunolide (p. 279).

Heathcock and Amano have recently reported a synthesis of (+)-occidentalol which is summarized in Scheme 22.[138]   (+)-

Scheme 22.   Heathcock-Amano Synthesis of Occidentalol

29   49

128   129

159   160

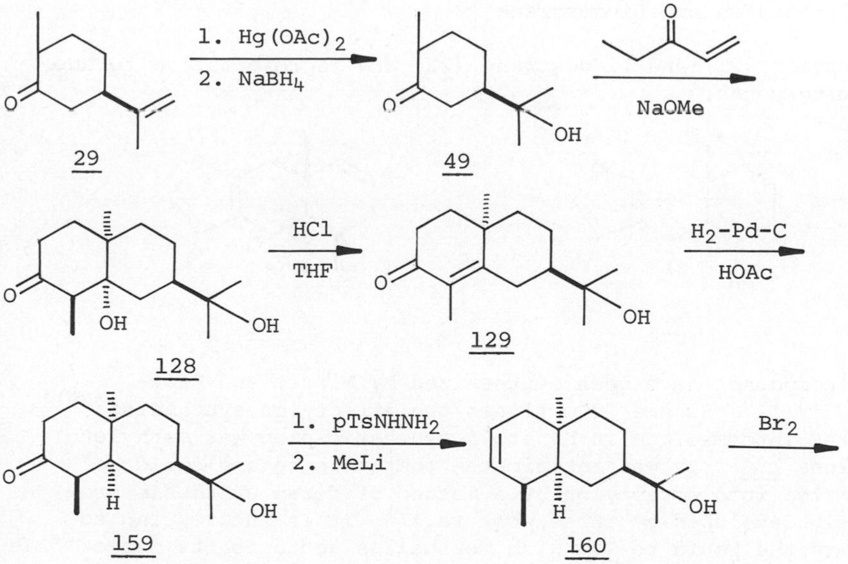

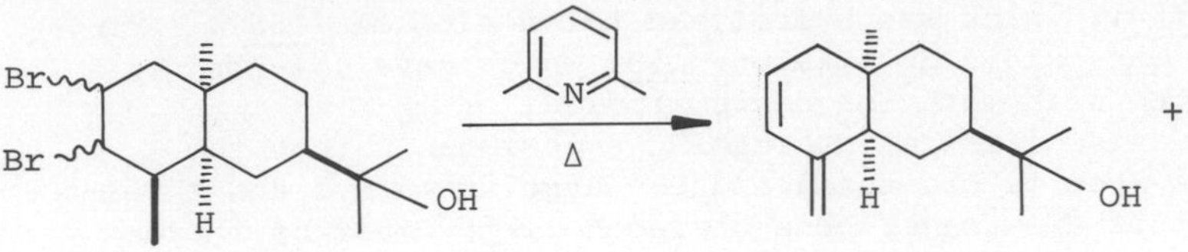

161

162

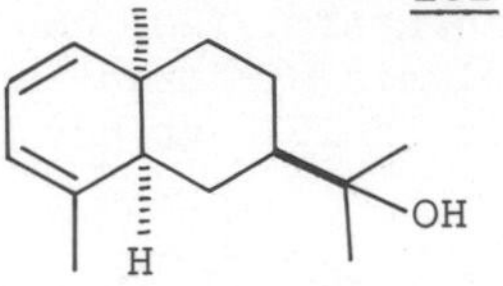

20

Dihydrocarvone (29) was converted into keto-alcohol 49 by oxymercuration-demercuration. Compound 49 was condensed with ethyl vinyl ketone (sodium methoxide in ether) to obtain keto diol 128, in which the angular methyl and isopropanol groups are *trans*. Dehydration of 128 gave epi-carissone (129). Compound 129 was hydrogenated over palladized carbon in acetic acid to give a 4:1 mixture of *cis*-dihydrocarissone (159) and its *trans* isomer. Ketone 159 was subjected to Shapiro's modification of the Bamford-Stevens reaction to yield olefin 160. Bromination-dehydrobromination gave a mixture of occidentalol (20) and its isomer 162.

J. Atractylon and Lindestrene

Atractylon (21) and lindestrene (22) are representative furano-sesquiterpenes.

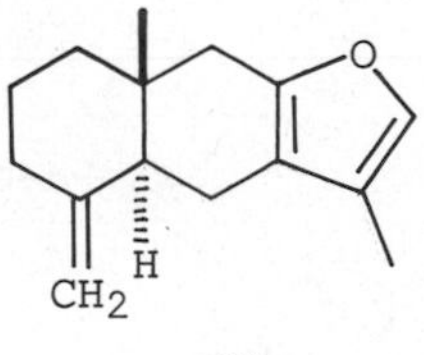

21

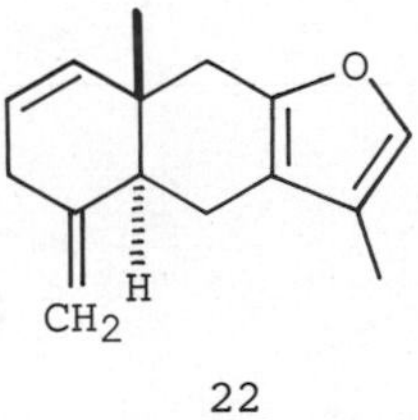

22

Both compounds have been synthesized by Minato and Naga-saki.[139,140] Scheme 23 outlines the atractylon synthesis.[139] The key intermediate in Minato's synthetic plan was methylene decalone 176. It was anticipated that this ketone could be converted into atractylon by a method of furan synthesis previously developed by these workers.[141] It is interesting to compare the route to 176 with Marshall's route to its isomer

Scheme 23. Minato-Nagasaki Synthesis of Atractylon

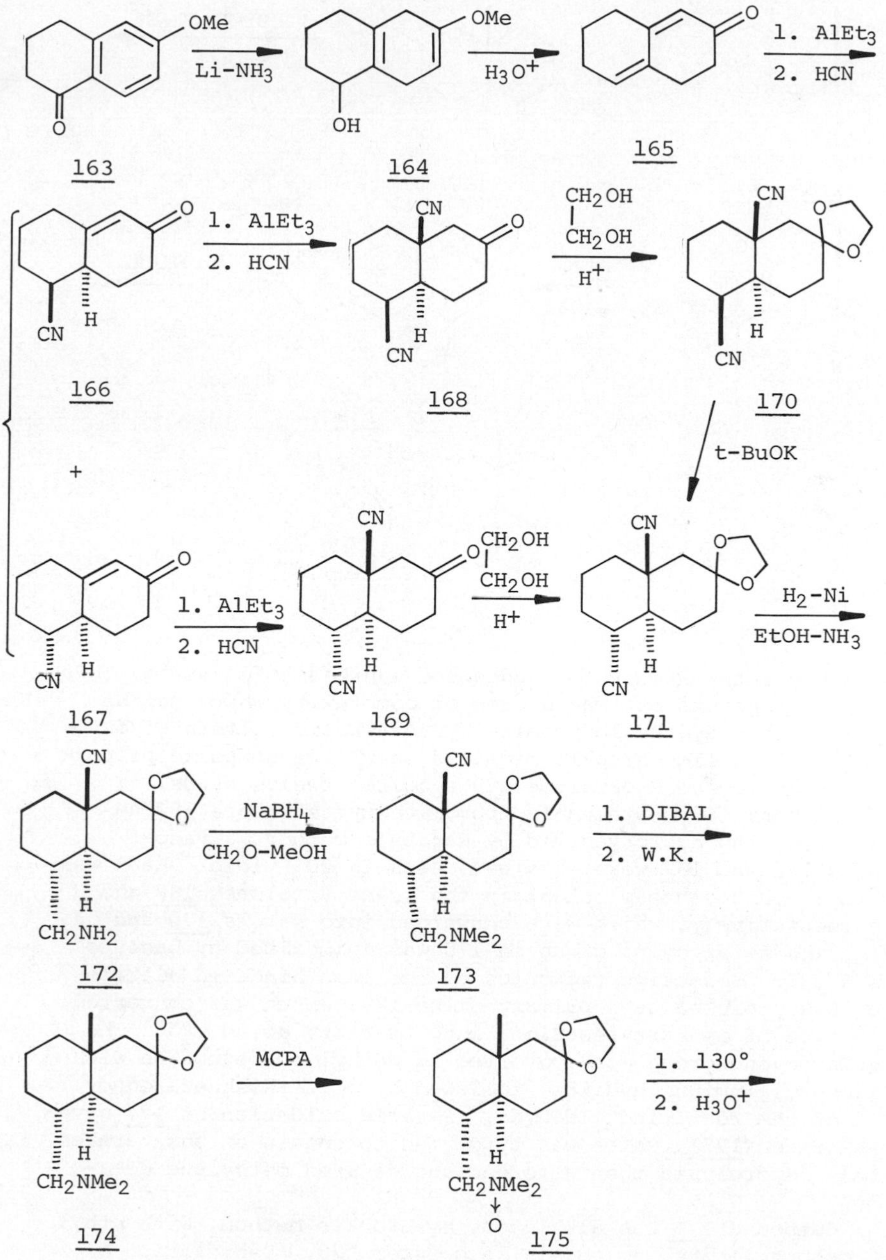

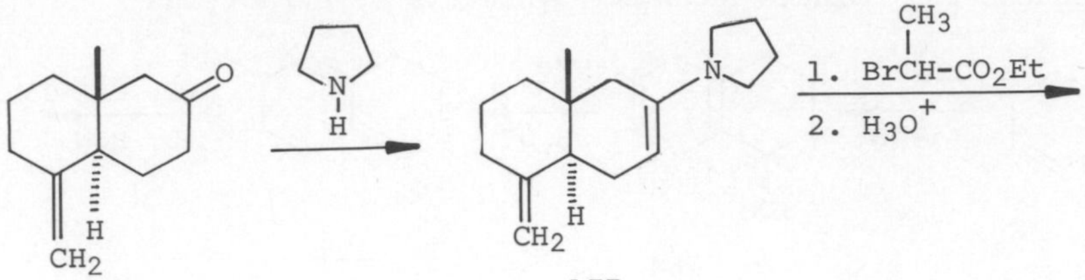

176

177

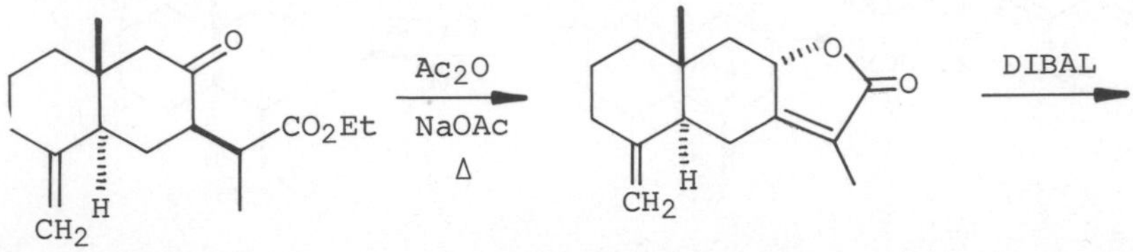

178

179

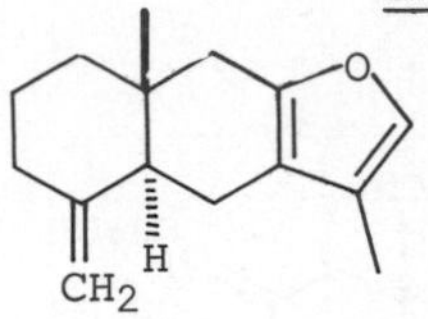

21

60, a key intermediate for eudesmane synthesis (p. 290). This comparison points out the degree of complexity which can be imparted to a synthetic problem by an awkward pattern of functionality. While Marshall required only five steps to prepare 60, the analogous 2-decalone 179 requires twelve steps.

Dienone 165, obtainable from methoxy tetralone 163 as indicated, was hydrocyanated by Nagata's method. Cyano-enones 166 and 167 were obtained, each in 20% yield. Each was further hydrocyanated to obtain the trans decalones 168 and 169 respectively, which were converted into ketols 170 and 171. The axial cyano group in 170 was epimerized by base to yield 171. Selective reduction of the less hindered nitrile over Raney nickel gave primary amine 172, which was converted, by a type of Leukart reaction, into tertiary amine 173. The angular cyano group was converted to methyl by reduction with diisobutylaluminum hydride, followed by Wolff-Kishner reduction of the resulting aldehyde. Peracid oxidation of 174 gave an N-oxide (175), which was pyrolyzed to obtain an unsaturated ketal. Hydrolysis then afforded the desired methylene decalone.

Compound 176 was alkylated, by Stork's method, with ethyl α-bromopropionate, to obtain keto ester 178. Dehydration of this substance afforded butenolide <u>179</u>, which was reduced with diisobutylaluminum hydride to atractylon (<u>21</u>).

The Minato-Nagasaki synthesis of lindestrene is outlined in Scheme 24.[140] The starting point was the well-known

Scheme 24. Minato-Nagasaki Synthesis of (±)-Lindestrene

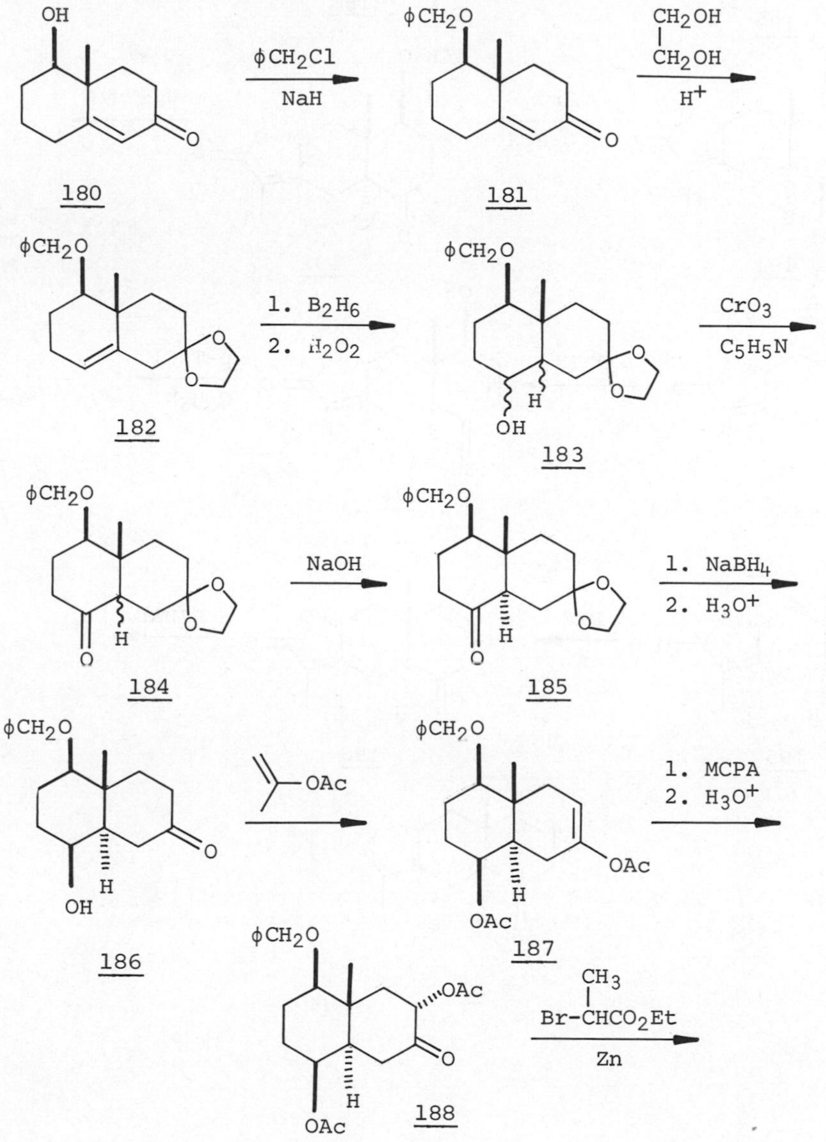

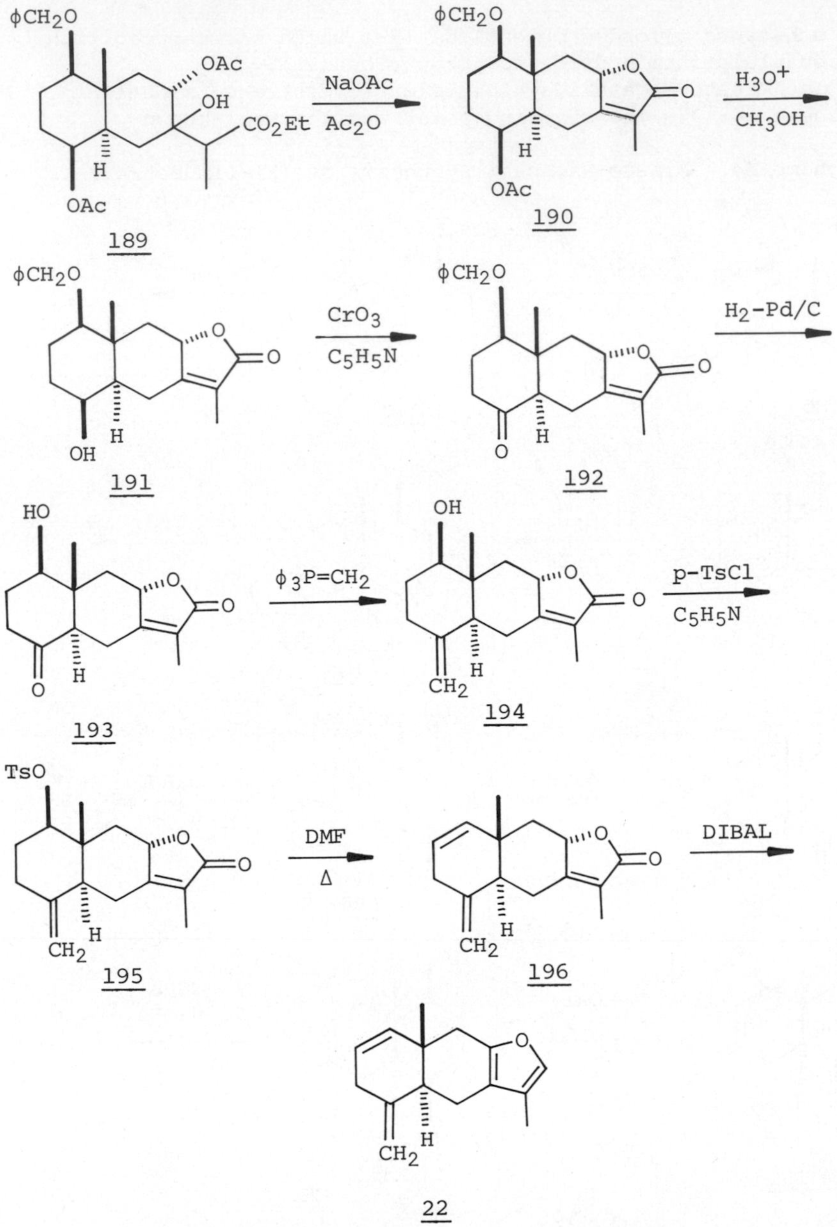

φCH₂O → structures with reagents

ϕCH_2O

NaOAc, Ac₂O

ϕCH_2O

H_3O^+ / CH_3OH

189

OAc, OH, CO₂Et, OAc, H

190

OAc, H

ϕCH_2O

CrO₃ / C₅H₅N

ϕCH_2O

H_2-Pd/C

191

OH, H

192

O, H

HO

$\phi_3P=CH_2$

OH

p-TsCl / C₅H₅N

193

O, H

194

CH₂, H

TsO

DMF Δ

DIBAL

195

CH₂, H

196

CH₂, H

22

CH₂, H

314

unsaturated hydroxy-ketone 180.[142] The secondary alcohol was
protected by base catalyzed benzylation.* Ketalization of 181
gave 182, which was converted into a mixture of diastereomeric
alcohols by hydroboration. Oxidation with Sarrett's reagent
gave a mixture of decalones, which was epimerized by alcoholic
base. The resulting trans-decalone 185 was reduced and the
dioxolane grouping hydrolyzed to obtain hydroxy ketone 186.

When 186 was treated with isopropenyl acetate, the diacetate 187 was produced. Peracid oxidation, followed by acidic
hydrolysis gave 188, which reacted in the Reformatsky reaction
to yield 189. Sodium acetate catalyzed dehydration gave
butenolide 190, which was hydrolyzed to alcohol 191. Sarrett
oxidation of the secondary hydroxyl gave 192, which was hydrogenolyzed to 193. The methylene group was introduced by a
Wittig reaction, yielding 194. The second double bond was
formed by solvolysis of tosylate 195 in dimethylformamide.†
The resulting triene (196) was reduced by diisobutyluminum
hydride to (+)-lindestrene (22).

K. Santonin

α-santonin (23) and β-santonin (24) are two of the most well-
known and earliest studied sesquiterpenes. Many years of
structural investigation were culminated in 1930 when Clemo,

23   24

Haworth, and Walton proposed the correct structures for the
santonin isomers.[145] In an outstanding fraud, an Indian group
claimed in 1943 that they had effected the total synthesis of

---

*Recent attempts to repeat this reaction have been unsuccess-
ful.[143] This is not surprising, due to the well known pro-
pensity of this type of compound to undergo vinylogous retro-
aldol condensations.[144] However, compound 182, prepared by
another route[143] was found to have the same physical properties as those reported by Minato and Nagasaki.
†This elimination is most remarkable, in light of the finding
that trans-decalyl tosylates of this type undergo predominate
skeletal rearrangement under these conditions.[143]

optically active santonin without the assistance of any di-
symmetric reagents.[146] This claim was vigorously disputed by
various workers.[147]

Total synthesis of the santonin isomers was pursued by a
number of groups on account of the powerful anthelmintic activity possessed by the compounds. Success was eventually
realized by a group headed by Abe at the Takeda Pharmaceutical
Laboratories in Osaka. In addition to synthesizing the natural
santonin isomers 23 and 24, the Abe group prepared the unnatural isomers (±)-santonin A (197), (±)-santonin B (198),
(±)-santonin C (199), and (±)-santonin D (200).[148]*

197   198

199   200

Because of the interest in santonin chemistry, and because natural (−)-α-santonin itself has been used as the starting point in many formal total syntheses of other sesquiterpenes (see Schemes 28, 29, 31, and 37 in section 3; Schemes 1,
19, 20, and 21 in section 4; and Schemes 6, 7, 9, and 11 in
section 5), we shall give a full account of the synthetic
studies carried out by the Takeda group.

Scheme 25 outlines the synthesis of (±)-santonin A

*For the sake of convenience, the racemic santonin isomers are
written in the enantiomeric form with the angular methyl group
β. At the time of the Takeda synthesis, no definitive information on the stereochemistry at C-11 (see formula 197) was
available, and for some time this configuration was in dis-
pute.[149] The configuration was defined independently in 1962
by three groups.[150]

Scheme 25. Takeda Synthesis of (±)-Santonin A, (±)-Santonin B, and (±)-Santonin D

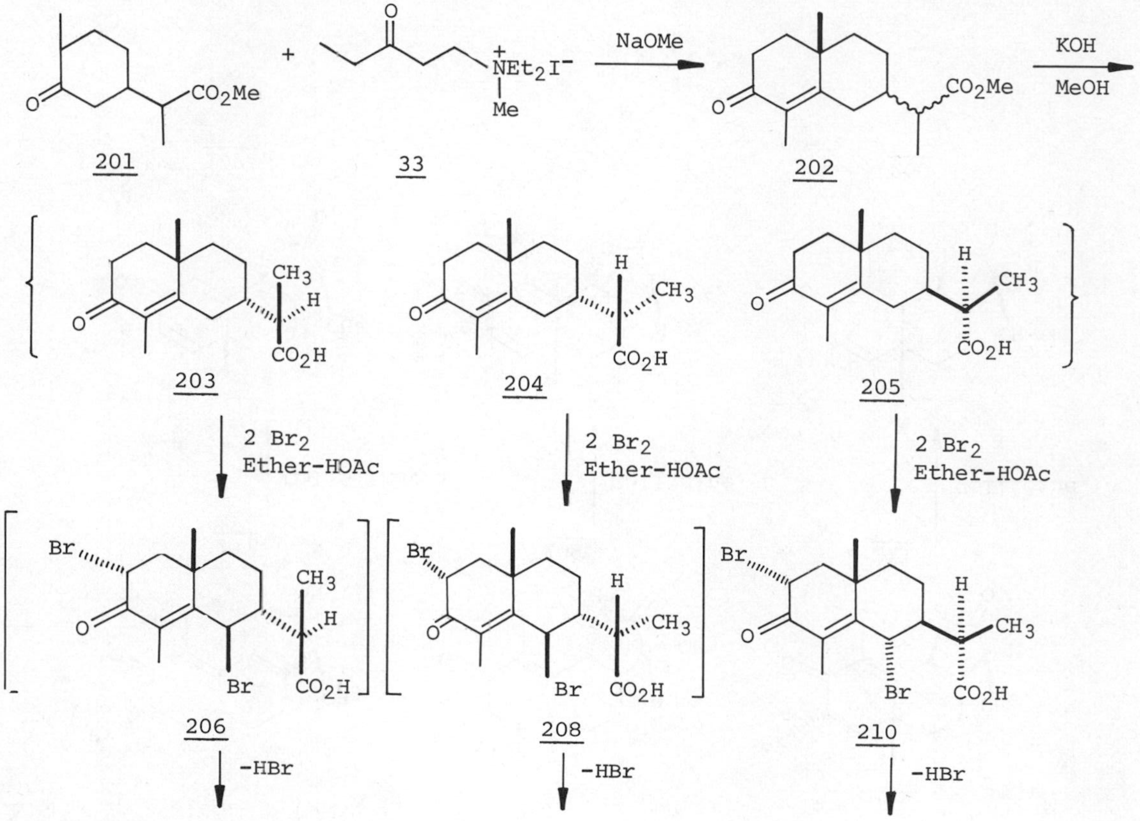

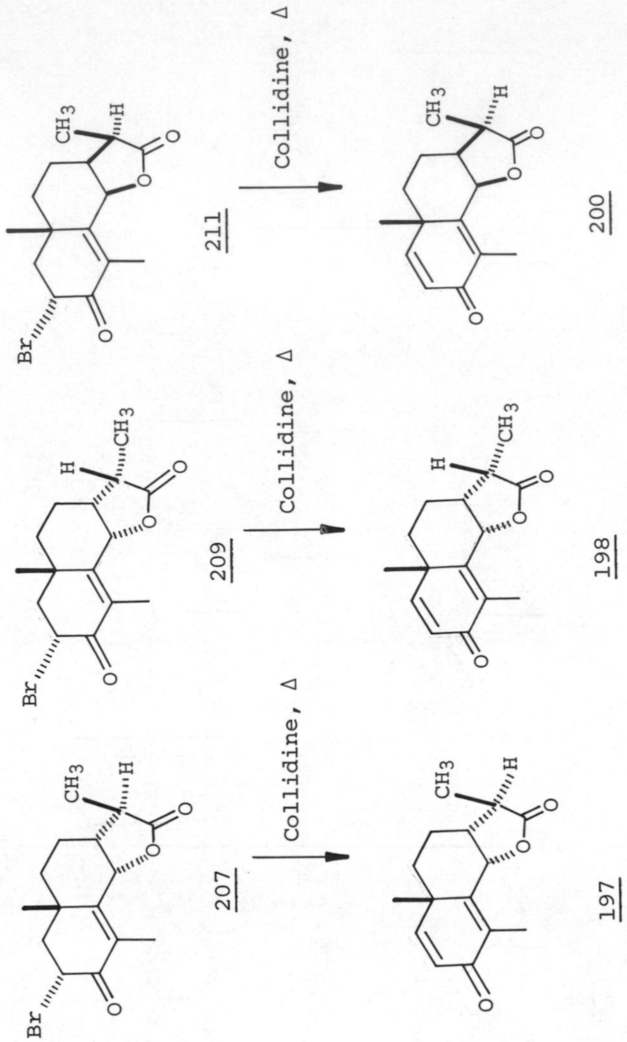

Br
CH3
H
207
Collidine, Δ
197
Br
H
CH3
209
Collidine, Δ
198
Br
CH3
H
211
Collidine, Δ
200

(197),[148a] (±)-santonin B (198),[148a] and (±)-santonin D
(200).[148b] Keto-ester 201 was condensed with 1-diethylamino-
pentan-3-one methiodide (33) in a Robinson annelation to yield
a mixture of diastereomeric octalones 202 in 51% yield. Alkaline hydrolysis of the mixture gave a mixture of acids.
Fractional crystalline yielded (±)-A-acid (203) in 21% yield
and (±)-B-acid (204) in 16% yield. Bromination of A-acid
(203) with two mole-equivalents of bromine in ether-acetic
acid gave a bromo-lactone 207, probably via the initially
formed dibromo-acid 206. Dehydrobromination of 207 yielded
(±)-santonin A (197). Analogous treatment of B-acid (204)
yielded (±)-santonin B (198), via intermediates 208 and 209.

     After removal of the crystalline A- and B-acids, the re-
maining mother liquors were brominated in a similar manner.
Careful examination of reaction mixture afforded the dibromo-
acid 210 and the bromo-lactone 211 in low yield. Both 210 and
211, upon refluxing in collidine, yielded (±)-santonin D.
From these results, it may be inferred that D-acid (205) is
produced, along with isomers 203 and 204, as a minor product
of the Robinson annelation.

     The fourth unnatural santonin isomer containing a *cis*-
fused lactone ring, (±)-santonin C (199), was obtained, along
with (±)-santonin D, by the route shown in Scheme 26. Hexalone

Scheme 26.   Takeda Synthesis of (±)-Santonin C
             and (±)-Santonin D

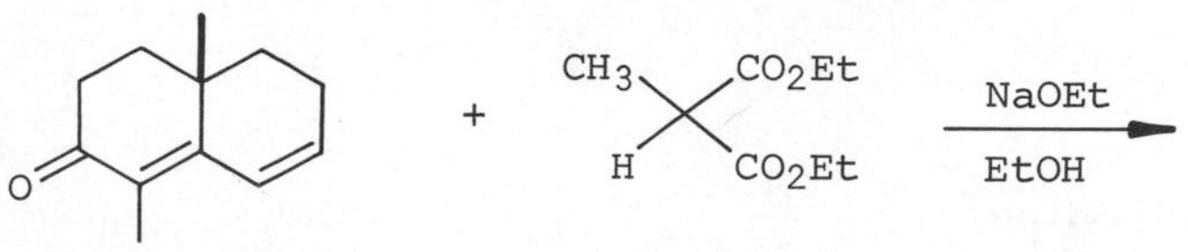

212

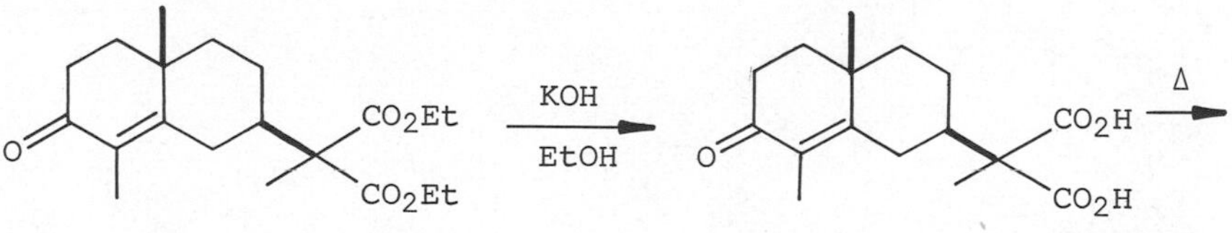

213   214

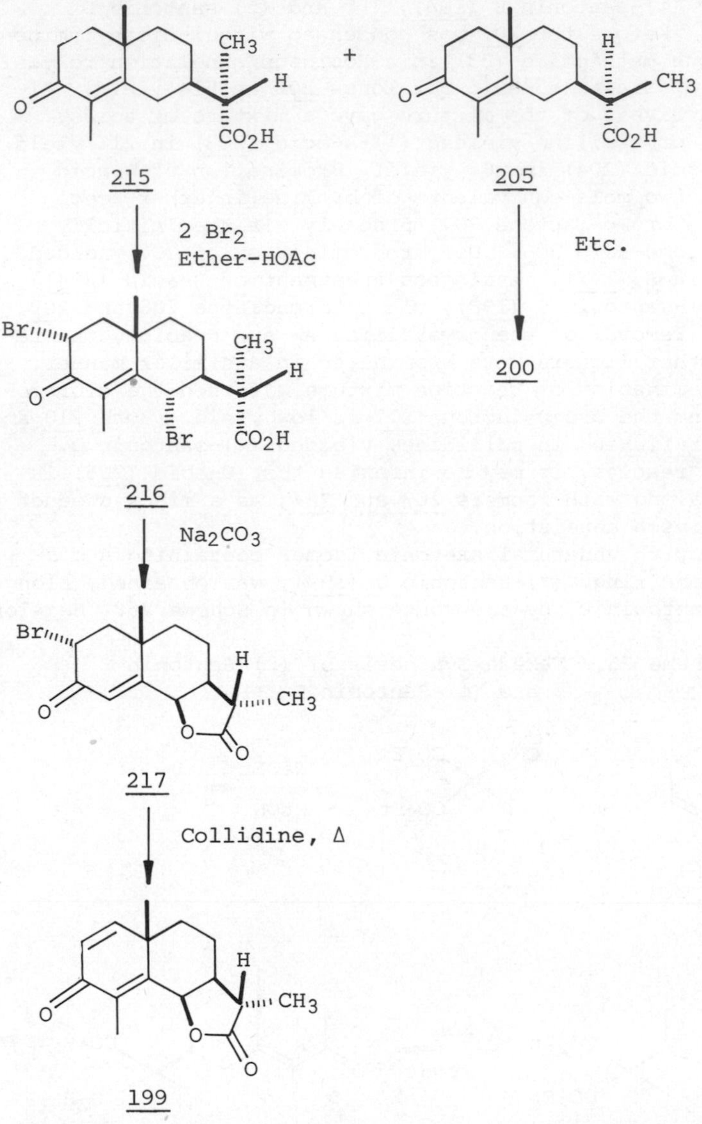

215

205

2 Br$_2$
Ether-HOAc

Etc.

216

200

Na$_2$CO$_3$

217

Collidine, Δ

199

212 underwent 1,6-Michael addition with the sodium salt of
diethyl methylmalonate to give *exclusively* the adduct 213, in
which the incoming group had occupied the equatorial position
(69% yield of crystalline product). Alkaline hydrolysis gave
214, which was decarboxylated to a mixture of keto-acids.
From this mixture, C-acid (215) and D-acid (205) were isolated

(in low yield) by fractional crystallization.

C-acid (215) underwent bromination to give dibromo-acid 216. This substance was lactonized to bromo-lactone 217 only upon treatment with sodium carbonate. Further dehydrobromination afforded (+)-santonin C (199). D-acid (205) had already been converted into (+)-santonin D (200).

Several noteworthy points emerge from a study of these syntheses. Firstly, we see again the tendency for Robinson annelation on cyclohexanones of the type 201 to give predominately octalones in which the angular methyl group and the three-carbon side chain are *trans* (cf. p. 286). Thus, acids 203 and 204, both having the three-carbon group in the unnatural axial configuration are the predominant products. As would be expected, essentially no stereospecificity is seen at the side-chain position (C-11), with 203 and 204 being produced in roughly equal amounts.

The stereochemistry of the lactone ring is, in each case, established by lactonization of a γ-bromo acid. It is likely that the allylic bromination occurs to give an axial bromide initially (see 206 and 208). In the case of 206 and 208, backside displacement by the carboxyl group can occur immediately, before the bromine is epimerized to the more stable equatorial configuration, giving lactones 207 and 209, respectively. In the case of C-acid (215) or D-acid (205), backside displacement on an axial bromide must proceed through a strained transition state. In these cases, epimerization of the initial bromide occurs, and dibromo-acids 216 and 210 may even be isolated. Lactonization is now more difficult, as the carboxyl group must approach from an axial direction.

The stereospecificity of the Michael reaction (212 → 213) is remarkable. Hexalone 212 reacts with diethyl malonate to give adducts 218 and 219 in variable amounts, depending upon the reaction conditions. With potassium t-butoxide in t-butanol at 25° for 10 days, crystalline 218 was obtained in

51% yield. None of isomer 219 was isolated under these conditions. When the reaction was carried out at reflux for 10 hr,
crystalline isomer 219 was isolated in 39% yield. Examination
of the mother liquors in this case revealed that isomer 218
was a minor product.[151]   These data suggest that compound 218
is a kinetic product (axial alkylation) and 219 is a thermodynamic product (equatorial alkylation). However, the conversion of 212 to 213 was carried out under mild conditions
(sodium ethoxide in ethanol at room temperature) which should
favor the kinetic product.

Scheme 27 outlines the Takeda synthesis of α-santonin and

Scheme 27.   Takeda Synthesis of α-Santonin and β-Santonin

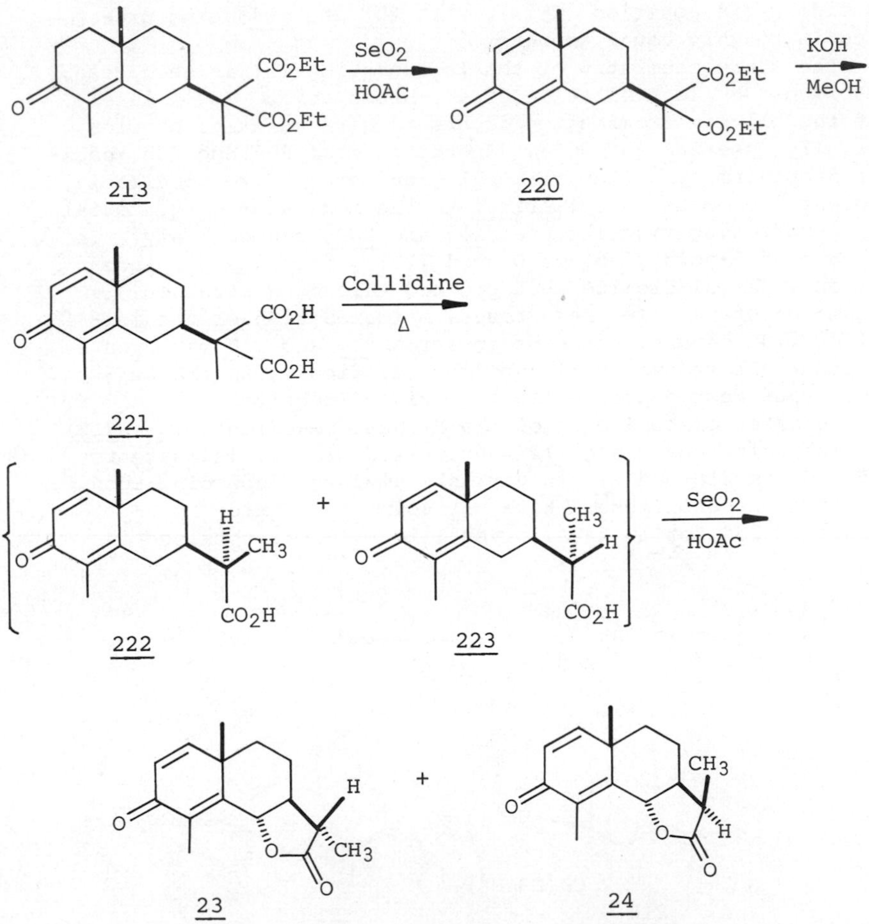

213

220

221

222

223

23

24

β-santonin.[152] Enone 213 was oxidized with selenium dioxide
in acetic acid to a dienone 220, which was hydrolyzed to a diacid 221. Decarboxylation of diacid 221 gave a mixture of
mono-acids 222 and 223. Hydroxylation of this mixture, again
with selenium dioxide in acetic acid, gave a mixture of (±)-
α-santonin (23) and (±)-β-santonin (24). The intermediate diacid 221 was resolved as its brucine of quinine salt. The
levorotatory antipode of 221 was carried through the same sequence to afford (-)-α-santonin and (-)-β-santonin.
    Scheme 28 outlines an alternative synthesis by Abe's

Scheme 28. Takeda Synthesis of α-Santonin

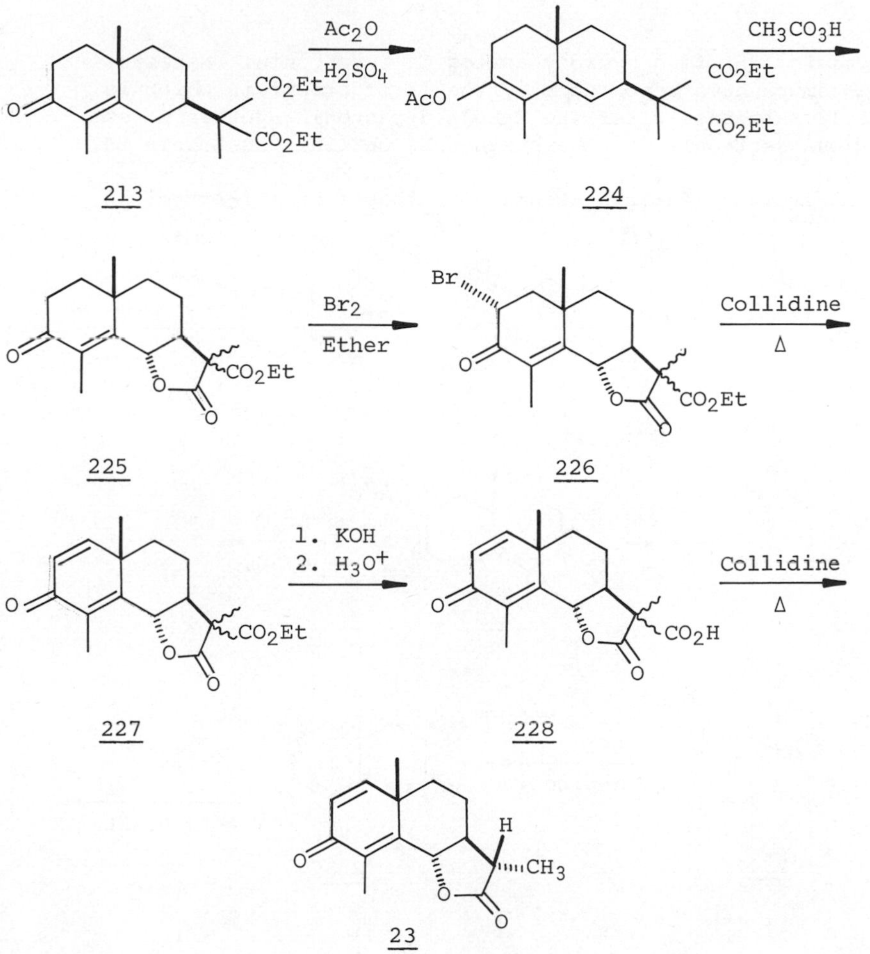

group which led only to α-santonin.[152] Keto-diester 213 was
converted into the dienyl acetate 224 by acetic anhydride.
Peracid oxidation occurred from the face of the molecule opposite to the methyl group, yielding (presumably) an alcohol,
which lactonized. The isolated product was the *trans* fused
lactone 225. Bromination of 225 gave a bromo-ketone 226,
which was dehydrobrominated with collidine to afford 11-
carbethoxy santonin (227). Alkaline hydrolysis, followed by
acidification, gave acid 228, which decarboxylated to give
only (±)-α-santonin (23). Compound 228 was resolved as its
brucine salt and the levorotatory antipode decarboxylated to
yield (-)-α-santonin.

L.   Artemisin

Artemisin (25) is a hydroxy analog of α-santonin. Nakazaki
and Naemura have reported a synthesis of artemisin which is
closely patterned after the Takeda α-santonin synthesis (see
previous section).[153] The synthesis, outlined in Scheme 29,

Scheme 29. Nakazaki-Naemura Synthesis of (±)-Artemisin

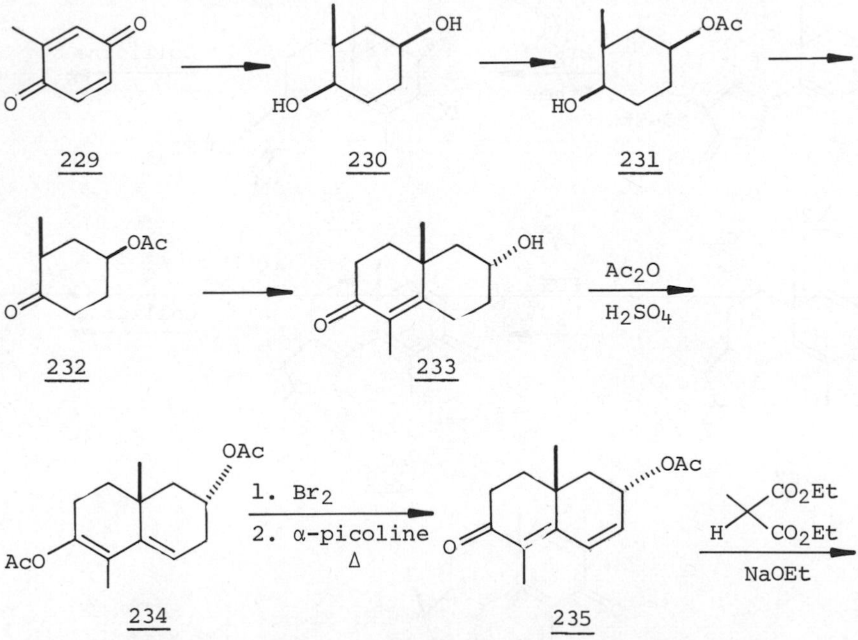

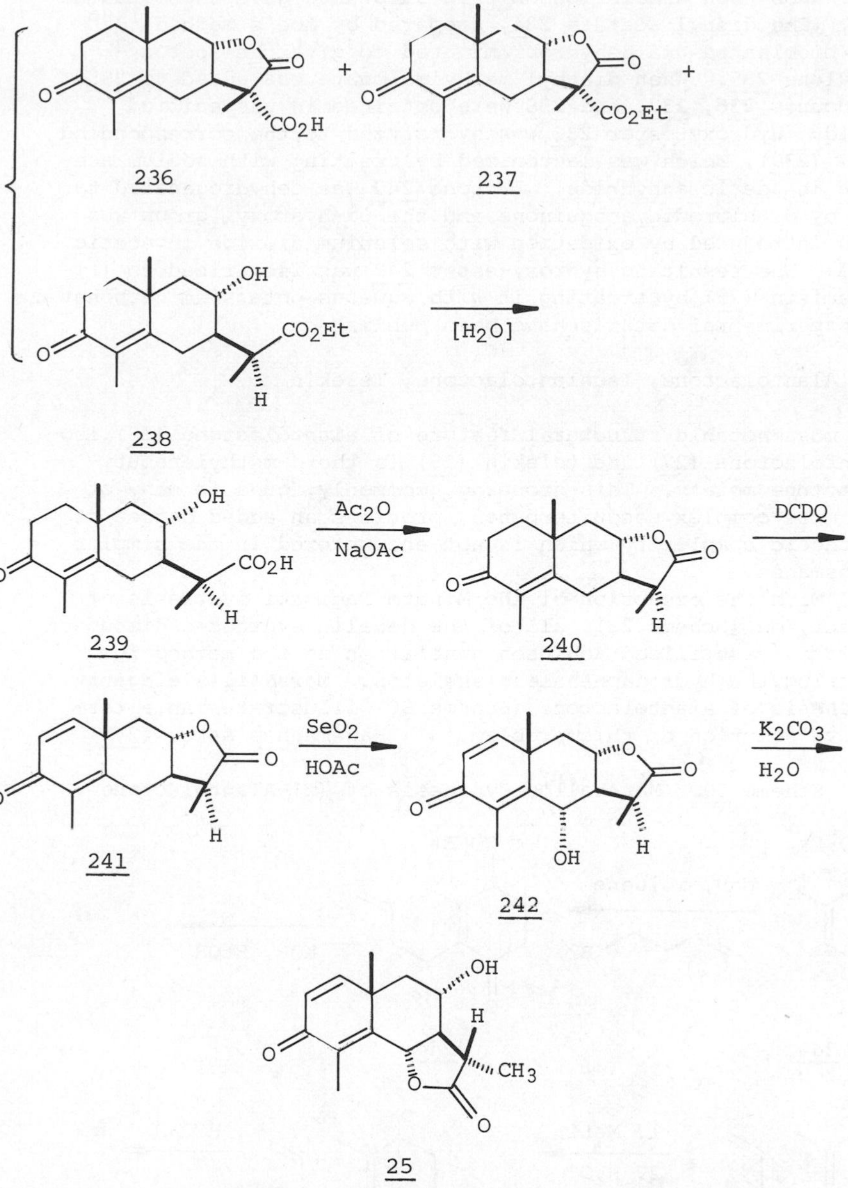

begins with p-toluoquinone (<u>229</u>), which was hydrogenated to give the *cis* glycol <u>230</u>. Selective acetylation of <u>230</u>, afforded monoacetate <u>231</u>, which was oxidized to keto-acetate

232. Robinson annelation on this substance gave the octalone
233. The dienyl acetate 234, prepared by Abe's method,[148a]
was brominated and dehydrobrominated to give the acetoxy-
hexalone 235. When diethyl methylmalonate was added to 235,
compounds 236, 237, and 238 were obtained in unspecified
yield. Hydroxy-ester 238 was hydrolyzed to the corresponding
acid (239), which was lactonized by treating with sodium acetate in acetic anhydride. Lactone 240 was dehydrogenated to
241 by dichlorodicyanoquinone and the 6α-hydroxyl group was
then introduced by oxidation with selenium dioxide in acetic
acid. The resulting hydroxy-ester 242 was isomerized to (±)-
artemisin (25) by treating it with aqueous potassium carbonate.
No experimental details have been published.

M. Alantolactone, Isoalantolactone, Telekin

The most notable structural feature of alantolactone (26) isoalantolactone (27) and telekin (28) is the α-methylenebutyrolactone moiety. This grouping, commonly found in many of
the more complex sesquiterpenes, provides an added degree of
synthetic complexity which is not encountered in the simpler
eudesmanes.

   With the exception of the Minato-Nagasaki synthesis of
atractylon (Scheme 23), all of the decalin syntheses discussed
so far have utilized Robinson annelation as the method for
securing the hydronaphthalene skeleton. Marshall's elegant
synthesis of alantolactone (Scheme 30) illustrates an alternative solution of this problem.[154] Hagemann's ester (243)

Scheme 30. Marshall's Synthesis of (±)-Alantolactone

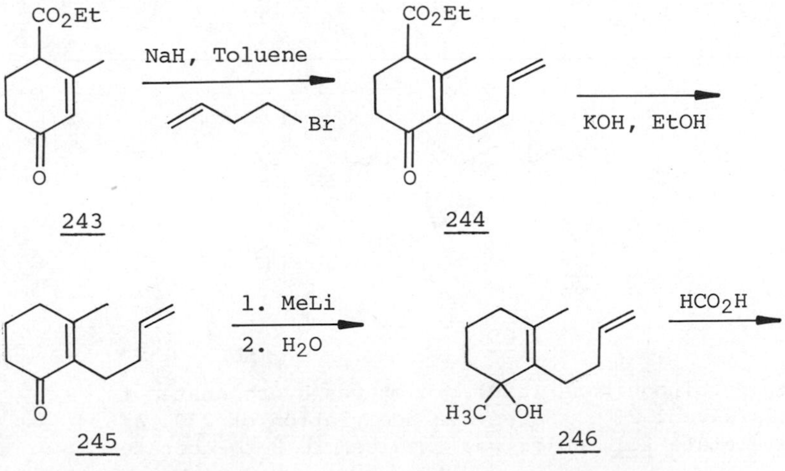

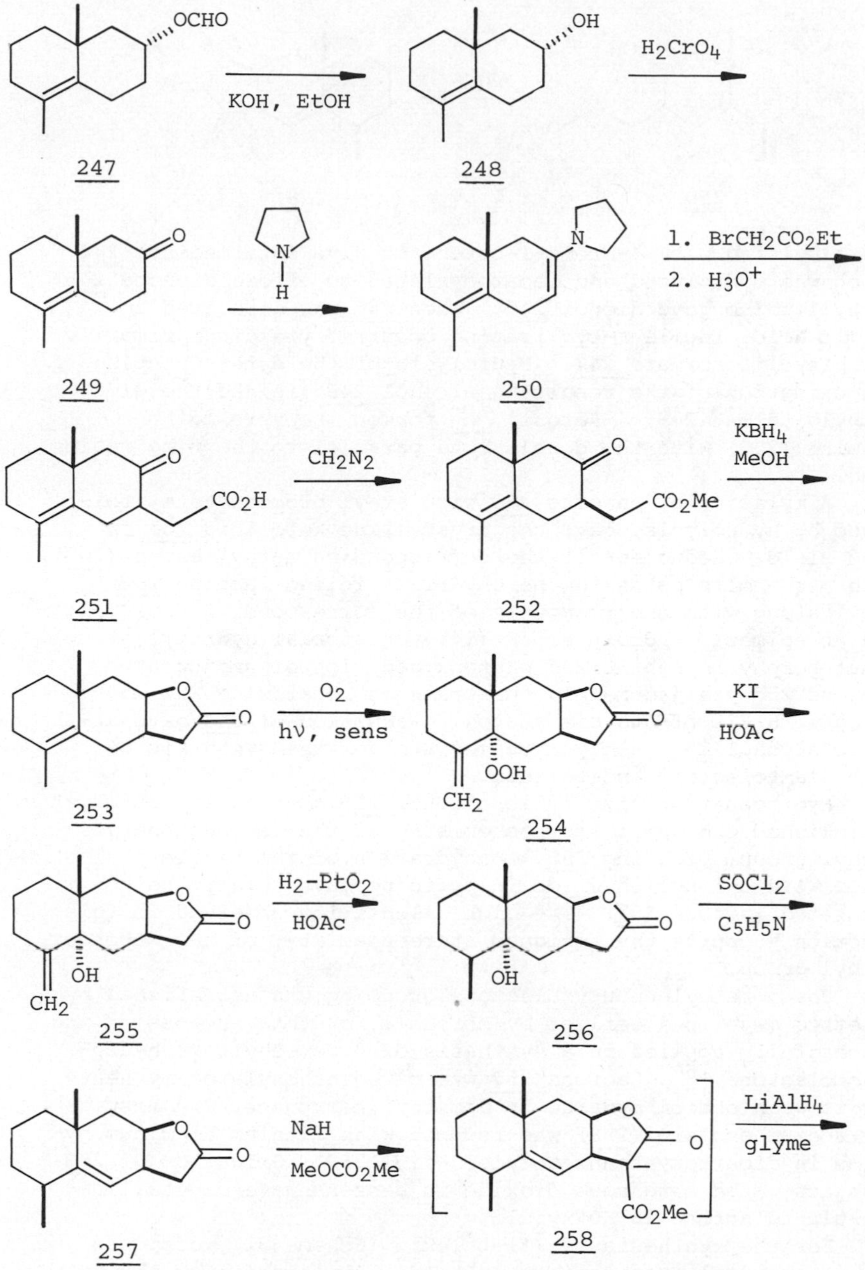
OCHO
KOH, EtOH
OH
H_2CrO_4
247
248
O
N
H
N
1. $BrCH_2CO_2Et$
2. H_3O^+
249
250
O
CO_2H
CH_2N_2
O
CO_2Me
KBH_4
MeOH
251
252
O
O
O_2
hν, sens
O
O
OOH
CH_2
KI
HOAc
253
254
O
O
OH
CH_2
H_2-PtO_2
HOAc
O
O
OH
$SOCl_2$
C_5H_5N
255
256
O
O
NaH
$MeOCO_2Me$
O
O
CO_2Me
$LiAlH_4$
glyme
257
258

259   26

was alkylated with 4-bromo-1-butene to give intermediate 244,
which was hydrolyzed and decarboyxlated to afford dienone 245.
Methyllithium gave dienol 246. When 246 was dissolved in
formic acid, facile π-cyclization occurred yielding primarily
the bicyclic formate 247. Hydrolysis of the ester grouping
and oxidation of the resulting alcohol 248 yielded the di-
methyloctalone 249. Ketone 249 formed a pyrrolidine
enamine (250) with the double bond parallel to the ring fusion
bond.

     Alkylation of enamine 250 with ethyl bromoacetate, fol-
lowed by hydrolysis, gave the crystalline keto acid 251 in
good yield. Reduction of the corresponding methyl ester (252)
with methanolic potassium borohydride yielded lactone 253
(74%) along with minor amounts of the corresponding diol (11%)
and an epimeric hydroxy ester (14%, equatorial hydroxyl group).
Hematoporphyrin sensitized photooxygenation of unsaturated
lactone 253 yielded mainly the *trans* hydroperoxide 254 (53%)
along with 22% of its *cis* analog. Reduction of 254 gave allylic alcohol 255, a key intermediate for the synthesis of
both alantolactone and telekin.

     Hydrogenation of allylic alcohol 255 over Adam's catalyst
established the axial stereochemistry of the new secondary
methyl group, yielding 256. Dehydration of the tertiary alcohol with thionyl chloride in pyridine gave solely the unsaturated lactone 257. The high selectivity observed in this
reaction supports the assigned stereochemistry of the secondary
methyl group.

     The α-methylenebutyrolactone grouping was established by
a method developed earlier by Marshall for this purpose[155] and
successfully applied in a synthesis of 4-demethyltetrahydro-
alantolactone.[156] Lactone 257 was carbomethoxylated by heating it with sodium hydride in dimethyl carbonate. The resulting sodium enolate (258) was reduced with lithium aluminum hydride in dimethoxyethane to yield diol 259. Oxidation of 259
with activated manganese dioxide in benzene gave crystalline
(±)-alantolactone in 80% yield.

     For the synthesis of (±)-telekin (Scheme 31), compound
255 is an ideal intermediate.[154b,157] Application of the α-
methylenebutyrolactone synthesis to 255 (via 260 and 261) gave

Scheme 31.   Marshall's Synthesis of (±)-Telekin

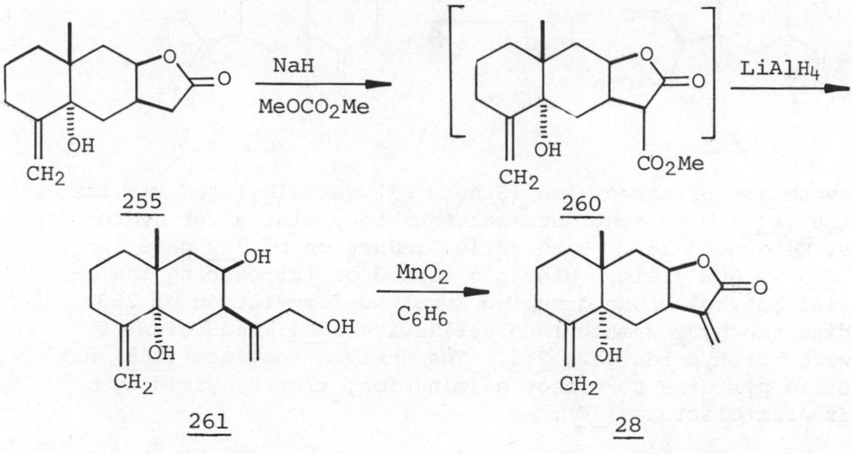

crystalline (±)-telekin (28).

Minato's synthesis of (±)-isoalantolactone (27) is outlined in Scheme 32.[158]   Methylene decalone, an intermediate in

Scheme 32.   Minato's Synthesis of (±)-Isoalantolactone

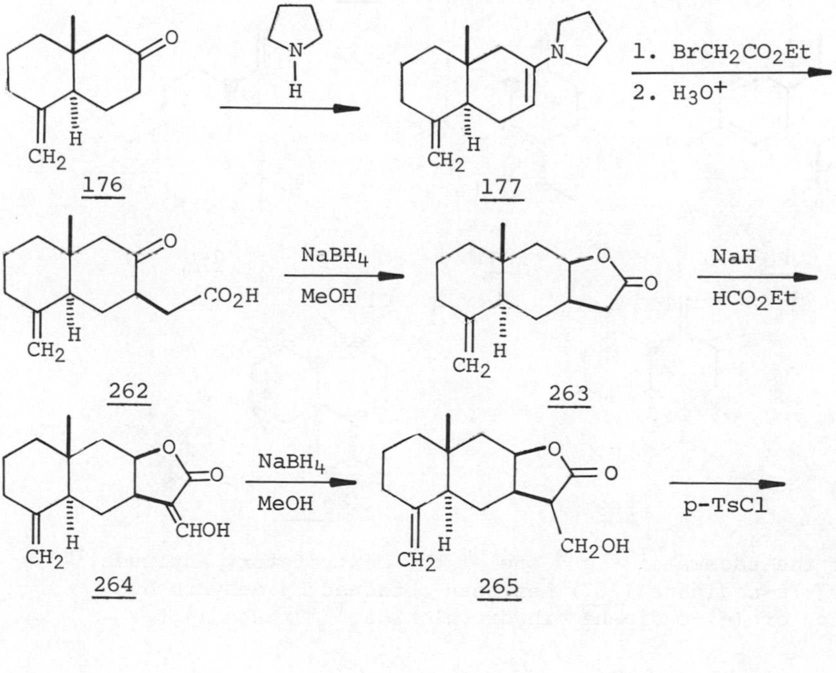

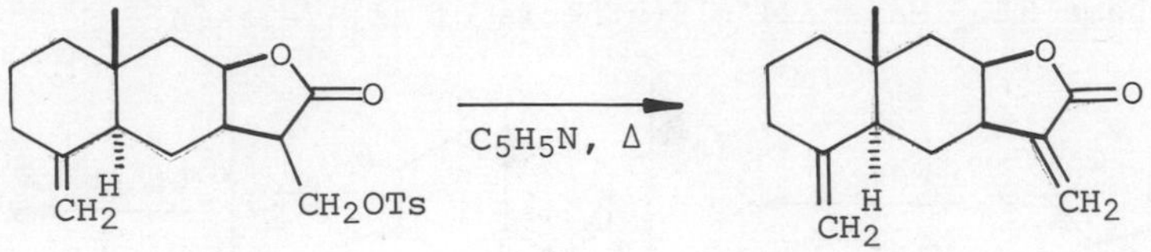

266   27

the synthesis of atractylon (Scheme 23) was alkylated via its
enamine (177) with ethyl bromoacetate to yield, after hydrolysis, keto acid 262.   Borohydride reduction of 262 gave lactone 263 in 66% yield.   Minato's method of introducing the α-
methylenebutyrolactone grouping involved formylation of 263,
yielding the hydroxymethylene derivative 264.   Reduction of
264 with borohydride gave 265.   The derived tosylate (266) was
heated in pyridine to effect elimination, thereby yielding
(±)-isoalantolactone (27).

N.   Cadinanes; Calamenene, ε-Cadinene, Veticadinol, Veticadinene

The only naturally occurring cadinanes which have been synthesized are calamenene (266),[159] ε-cadinene (267),[160] veticadinol (268),[161] and veticadinene (269).[162]   In addition two
syntheses of cadinene dihydrochloride (270) have appeared,

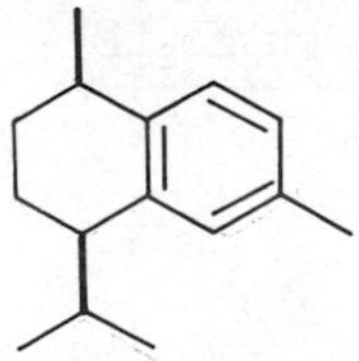

266 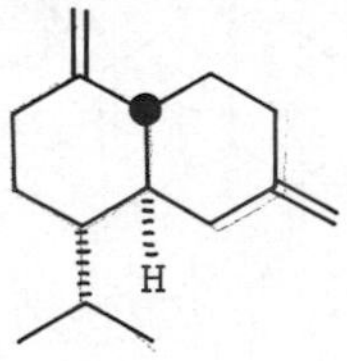   267 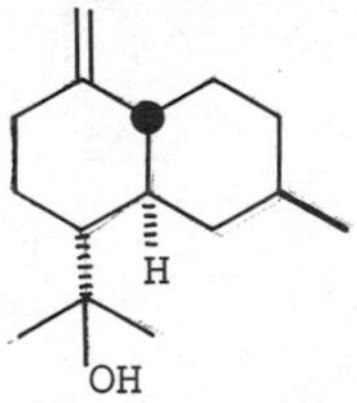   268

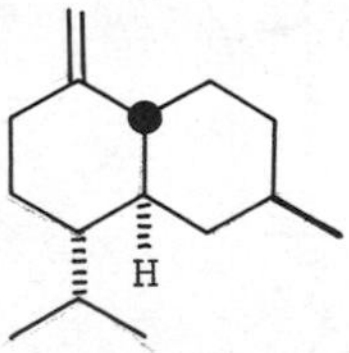

269 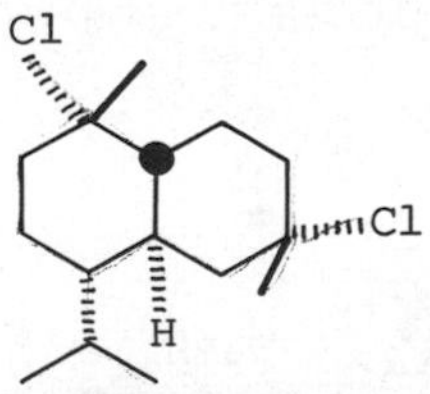   270

one of the racemate[163] and one of the dextrotatory antipode.[164]
Since (-)ε-cadinene (267) has been obtained by dehydrochlorination of (-)-cadinene dihydrochloride,[165] these latter

syntheses may be considered formal total syntheses of 267.

The Ladwa-Joshi-Kulkarni synthesis of (+)-calamenene, the optical antipode of the more commonly occurring terpene, is outlined in Scheme 33.[159] (-)-Menthone (271) was converted

Scheme 33. Ladwa-Joshi-Kulkarni Synthesis of Calamenene

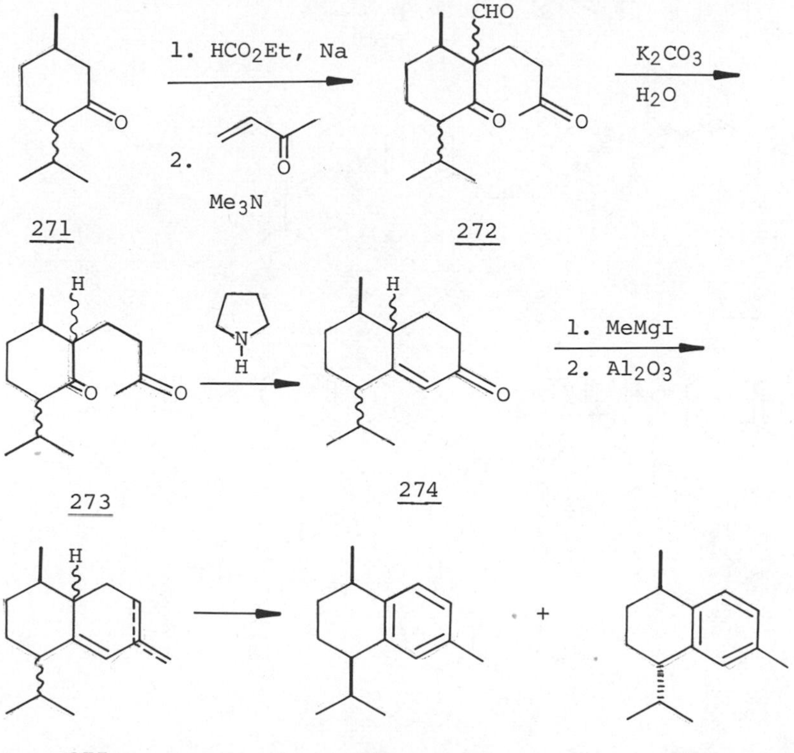

271 272 273 274 275 266 276

into diketoaldehyde 272 by the method of Corey and Nozoe.[166] Deformylation with aqueous potassium carbonate gave dione 273, which was cyclized with pyrrolidine to enone 274. Compound 274 was methylated and dehydrated, affording a mixture of dienes (275, *exo*:*endo* ratio = 1:1). Selenium dioxide dehydrogenation of this mixture gave a mixture of (+)-calamenene (266, 60%) and its diastereomer (276, 40%).

Rao, Rao, and Dev at the Indian Institute of Science in Bangalore, reported a synthesis of (±)-cadinene dihydrochloride in 1960.[163] The starting material, 4-isopropyl-6-methoxy-1-tetralone (281) had been previously prepared by

Bardhan and Mukherji (Scheme 34)[167] and by Rao and Dev (Scheme 35).[168] Although the latter synthesis is more lengthy, it is

Scheme 34. Bardhan-Mukherji Synthesis of Tetralone 281

<u>277</u>   <u>278</u>

<u>279</u>
<u>280</u>

<u>281</u>

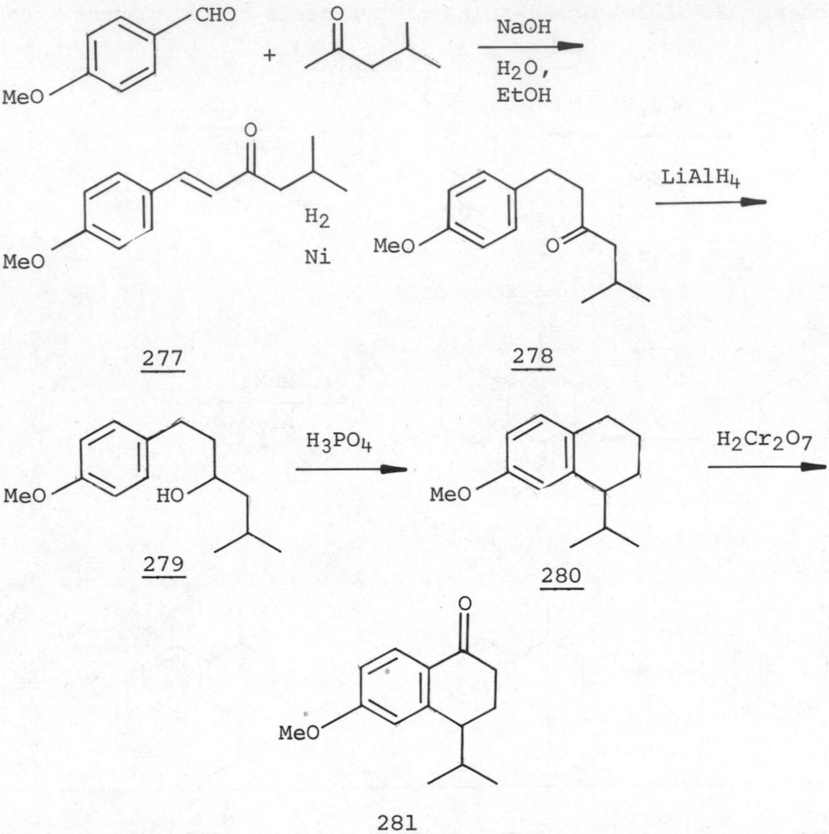

Scheme 35. Rao-Dev Synthesis of Tetralone 281

<u>282</u>   <u>283</u>

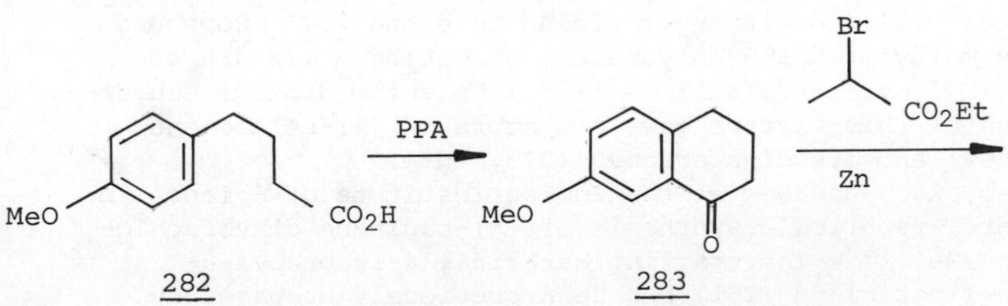

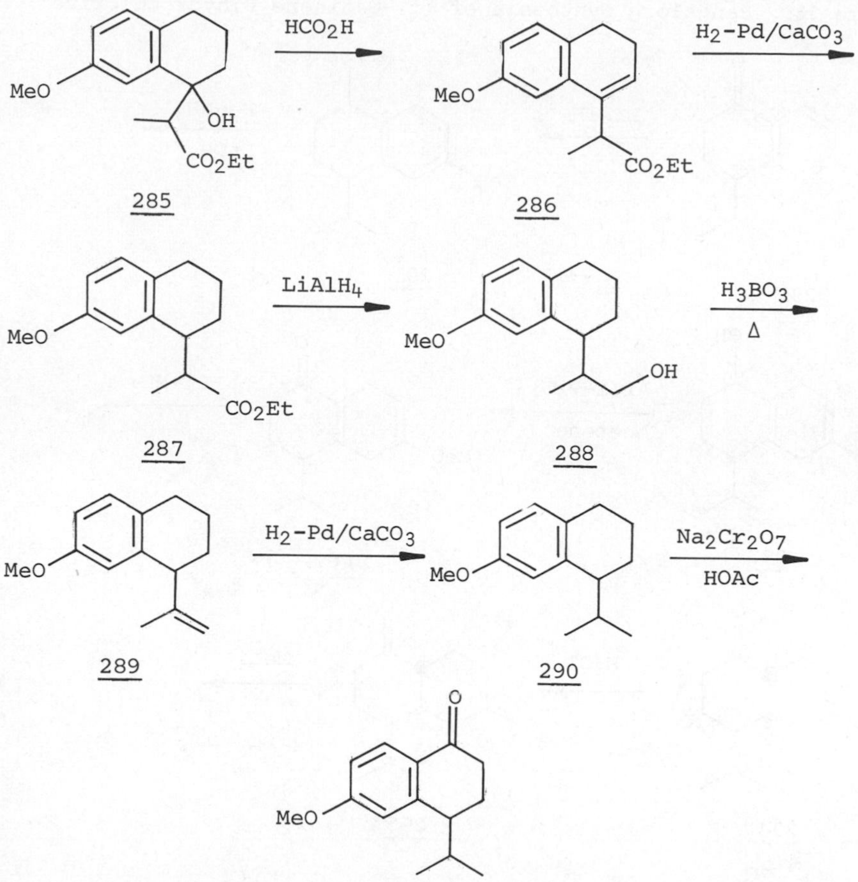

reported that tetralone 281 may be obtained in 45% yield based on acid 282.

For the synthesis of (±)-cadinene dihydrochloride, (Scheme 36),[163] the Bangalore group first reduced tetralone 281 to the secondary alcohol 291. Birch reduction of 291 afforded 292, which was submitted to Oppenauer oxidation to obtain enone 293. Lithium-ammonia reduction of 293 gave compound 294, which was hydrolyzed to diketone 295. Compound 295 was treated with excess methyllithium to obtain a gummy diol, which reacted with HCl in ether to give crystalline (±)-cadinene di-dihydrochloride (270), identical with that obtained from (±)-δ-cadinene.[169]

The stereochemistry of the three centers in crucial

Scheme 36.   Bangalore Synthesis of (±)-Cadinene Dihydrochloride

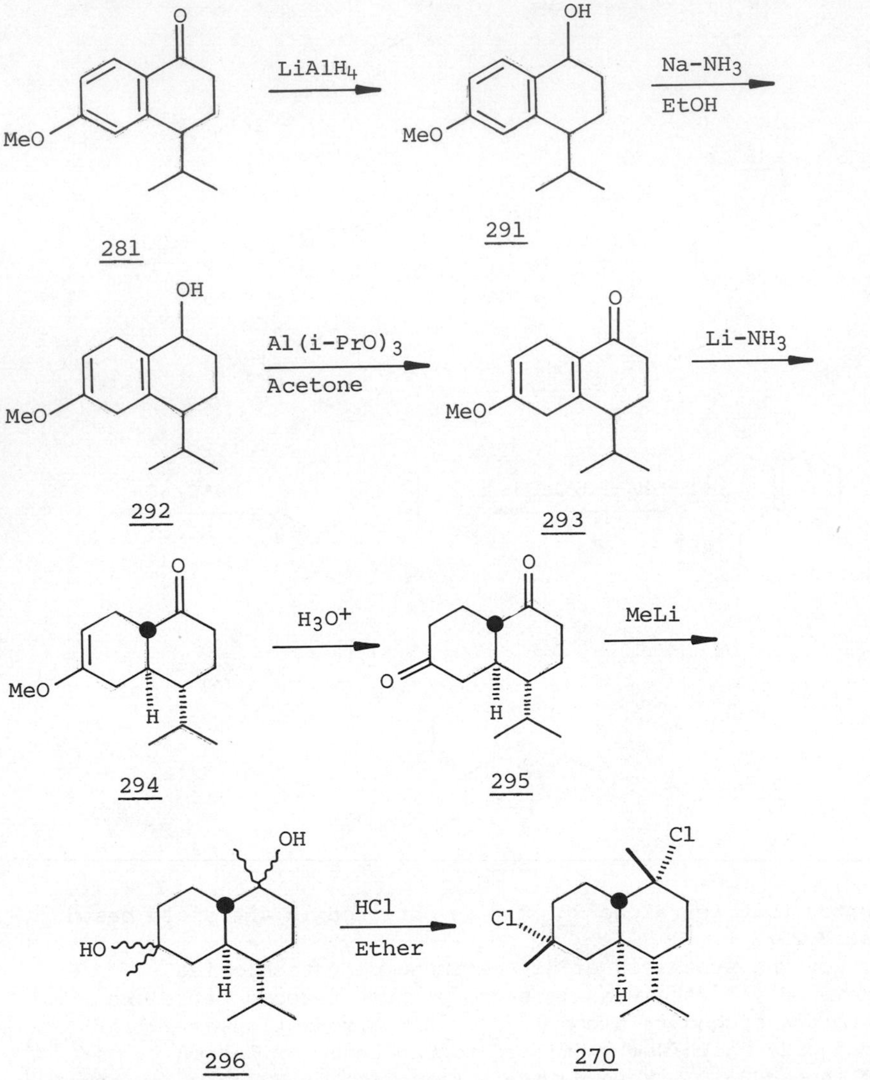

intermediate 295 is established in the metal-ammonia reduction of 293. Mechanistic considerations[170] predict that the more stable product (294) will be formed predominately. In any event, diketone 295 was identical by infrared spectroscopy with a sample of the levorotatory diketone, obtained by degradation of ε-muurolene (297)* by Herout and Santary.[171]

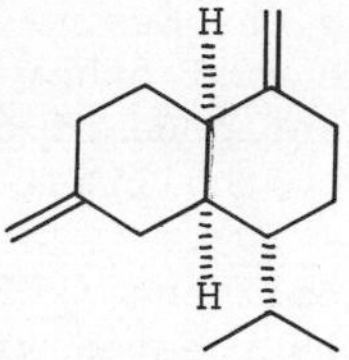

297

In 1965, Soffer reported a more direct synthesis of (+)-cadinene dihycrochloride, which is outlined in Scheme 37.[164]

Scheme 37. Soffer's Synthesis of (+)-Cadinene Dihydrochloride

298   299   300

295   270

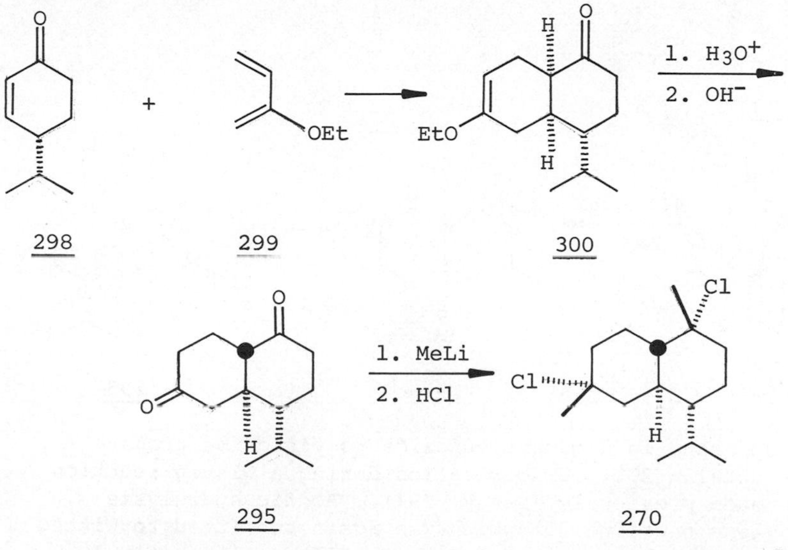

(-)-Cryptone (298), obtained from synthetic (±)-cryptone[172] by resolution of its p-carboxyphenylhydrazone (with quinine), reacted with 2-ethoxy-1,3-butadiene (299) to give predominately

*Before 1964, the term "ε-cadinene" was erroneously applied to structure 297. Westfelt showed that compound 297 in fact possesses a cis-ring fusion and found that under traditional hydrochlorination conditions compound 297 yields (-)-cadinene dihydrochloride.[165] However, since the ring juncture point adjacent to the carbonyl group in 295 is epimerizable, ozonolytic degradation of 297 can still yield 295

the octalone 300 (unspecified yield). It is expected that the
diene would attack mainly from the side of 298 *trans* to the
isopropyl group, thus assuring the desired stereochemistry in
adduct 300. Hydrolysis of the enol ether and basic epimerization gave the dextrorotatory antipode of dione 295. This
dione was converted, first by methyllithium, then by HCl, into
(+)-cadinene dihydrochloride (270).

   Soffer's synthesis of ε-cadinene (267) is outlined in
Scheme 38.[160] Octalone 300 was treated with methylenetri-

Scheme 38.   Soffer's Synthesis of (+)-ε-Cadinene

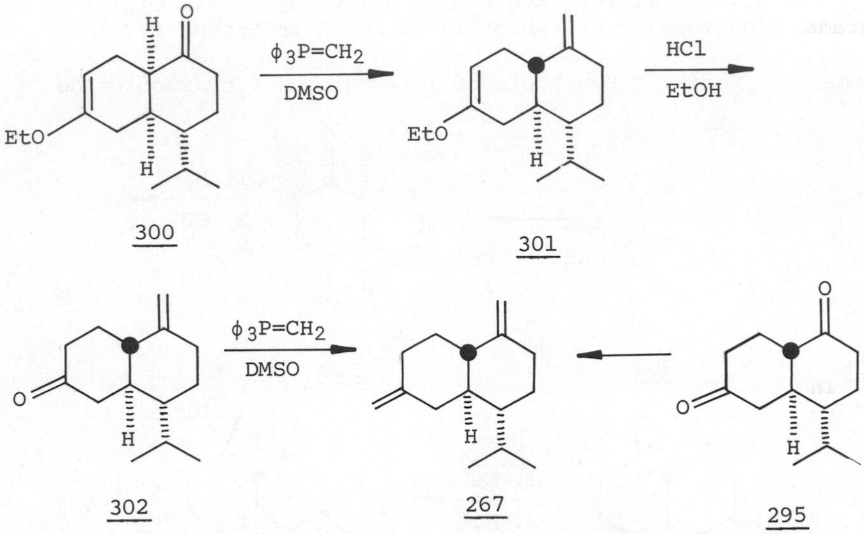

phenylphosphorane in dimethyl sufoxide to yield the *trans*-
methylene octalin 301. Epimerization during a Wittig reaction
had been noted previously (see p. 291). Acidic hydrolysis
gave methylene decalone 301 which was again submitted to Wittig
methylenation to obtain (+)-ε-cadinene (267). Alternatively,
267 could be obtained by the direct *bis*-methylenation of dione
295.

   Vig and co-workers reported a synthesis of "veticadinol"
(268), outlined in Scheme 39.[161] Relative stereochemistry
would be established in the hydrogenation of compound 304 and
in the acid catalyzed esterification leading to 308. Since
none of the intermediates in Scheme 39 were crystalline, and
since no comparison was made with authentic veticadinol, there
is considerable doubt about the stereochemical homogeneity of
the final product. It is probably a mixture of 268 and isomers

Scheme 39. Vig's Synthesis of "(±)-Veticadinol"

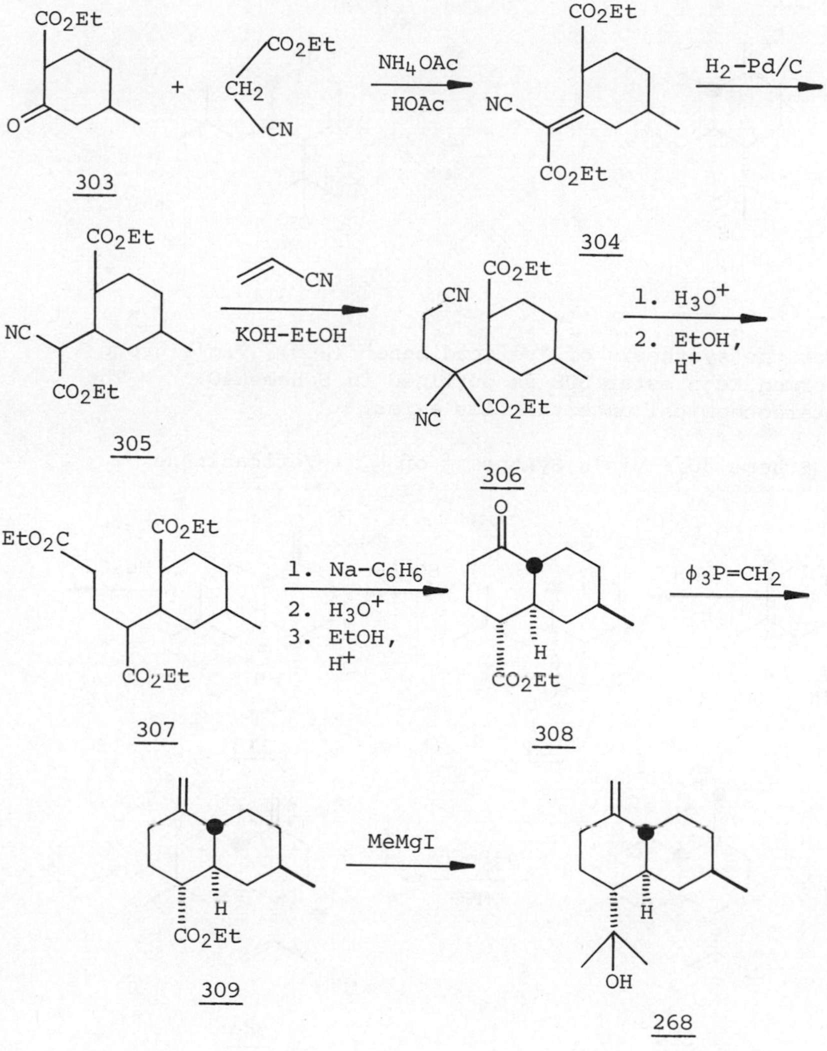

303 + CH$_2$ (CO$_2$Et, CN) $\xrightarrow[\text{HOAc}]{\text{NH}_4\text{OAc}}$ 304 $\xrightarrow{\text{H}_2-\text{Pd/C}}$

305 $\xrightarrow[\text{KOH-EtOH}]{\text{CN}}$ 306 $\xrightarrow[\substack{\text{2. EtOH,} \\ \text{H}^+}]{\text{1. H}_3\text{O}^+}$

307 $\xrightarrow[\substack{\text{2. H}_3\text{O}^+ \\ \text{3. EtOH,} \\ \text{H}^+}]{\text{1. Na-C}_6\text{H}_6}$ 308 $\xrightarrow{\phi_3\text{P}=\text{CH}_2}$

309 $\xrightarrow{\text{MeMgI}}$ 268

337

310 and 311.

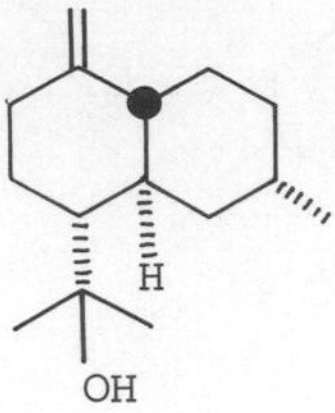

310

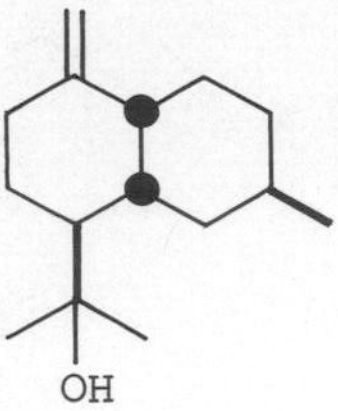

311

For the synthesis of "veticadinene" (269), Vig's group transformed keto ester 308 as outlined in Scheme 40.[162] The same stereochemical uncertainties exist.

Scheme 40. Vig's Synthesis of "(±)-Veticadinene"

308

312

313

314

269

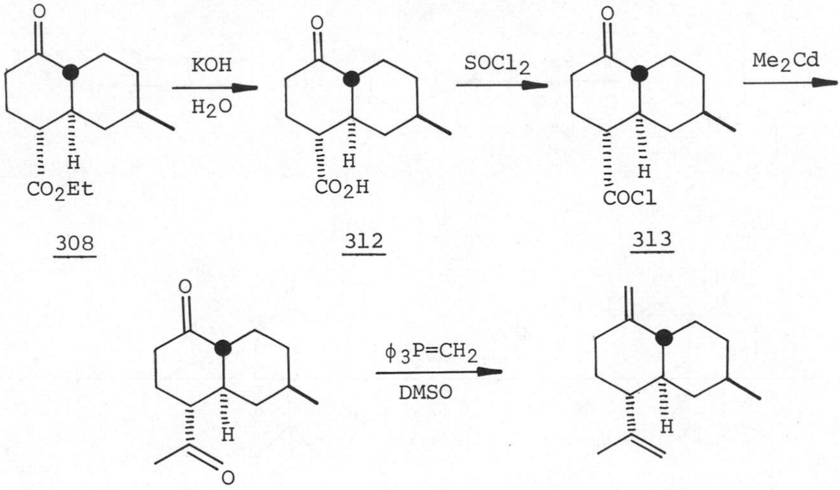

O. Drimanes; Drimenol (Bicyclofarnesol), Drimenin, Farnesi-ferol A (Biogenetic Routes)

The drimane group of sesquiterpenes possesses a bicyclofarnesol skeleton with substitution similar to that typically found in the di- and triterpene families. The probable biogenesis of drimenol (315) from farnesol has stimulated a good deal of study of this and related cyclizations in vitro.

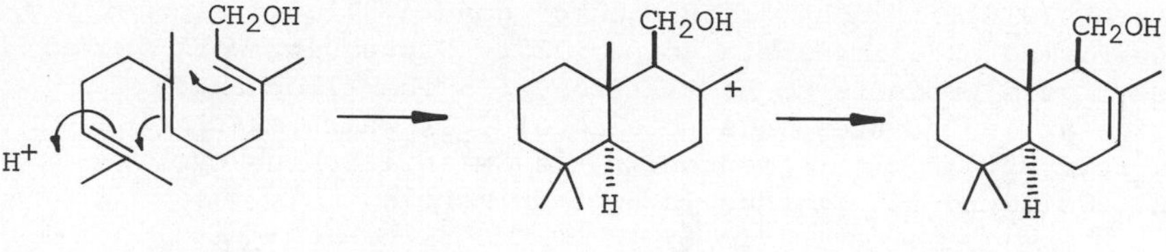

315

Drimanic sesquiterpenes which have actually been synthe-sized by this type of approach include drimenol (315),[173,174] the corresponding acid, "bicyclofarnesic acid" (316, not natu-rally occurring),[173,175] drimenin (317),[176] and farnesiferol A (318).[177] In this section we shall discuss these "biogeneti-cally styled" drimane syntheses. In the next, we shall dis-cuss other syntheses of these and related drimanes.

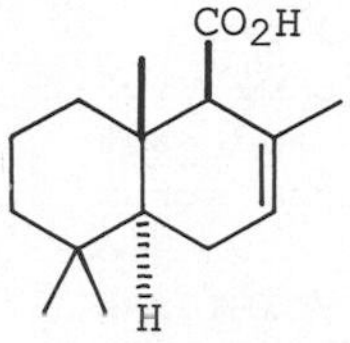

316 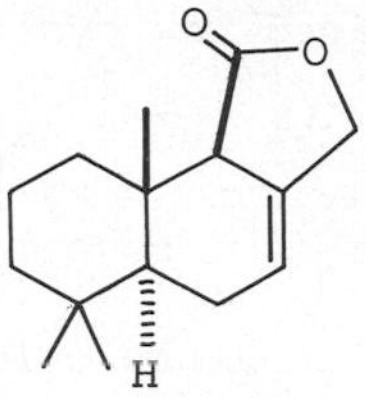   317 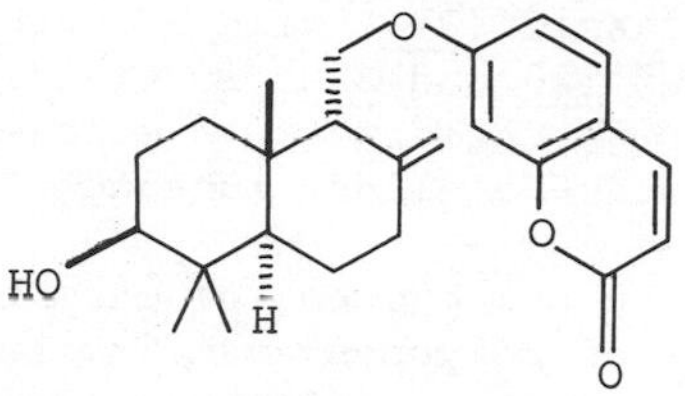   318

The first work on the cyclization of farnesic acid (319) was reported by Caliezi and Schinz in 1949.[173a] From the reaction of acid 319 in a formic acid-sulfuric acid mixture, they iso-lated a crystalline acid (subsequently shown to be bicyclo-farnesic acid, 316)[175b] and a "liquid acid" (subsequently shown to be a mixture).[175] Compound 316 was obtained in 10%

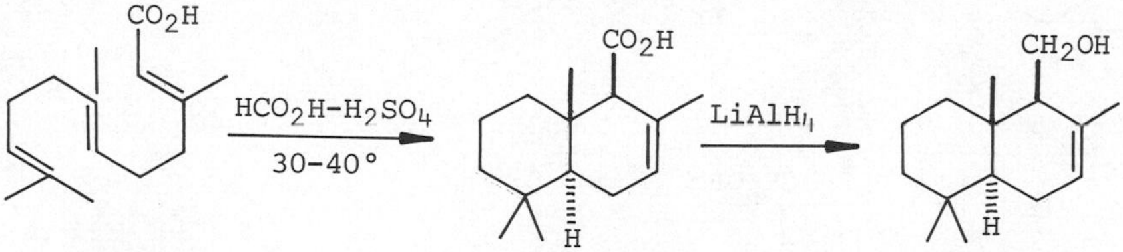

319   316   315

yield.   Reduction of 316 with lithium aluminum hydride gave a racemic alcohol, m.p. 64-65°, subsequently shown to be (±)-drimenol (315).[175] The "liquid acid" gave a "liquid alcohol," which formed an allophanate, m.p. 193°. Subsequent work showed these liquid products to be mixtures.[174] The allophanate melting at 193° corresponds to alcohol 321, which must arise from acid 320, a minor product in the Caliezi-Schinz cyclization. Compound 321 is thus (±)-epi-drimenol.

320   321

Stork and Burgastahler examined the cyclization of far-nesic acid (319) with BF_3 etherate.[175a] They isolated two acids, 316 and 320, in yields of 35% and 2-5%, respectively. Although these workers erroneously concluded that 316 and 320 were *cis*-decalins, subsequent work[175b] revealed the true struc-trues.

In the process of examining in vitro oxidative cycliza-tions of polyenes, van Tamelen and co-workers carried out the sequence of reactions outlined in Scheme 41, which led from

Scheme 41.   Van Tamelen Synthesis of (±)-Drimenol
and (±)-Epi-Drimenol

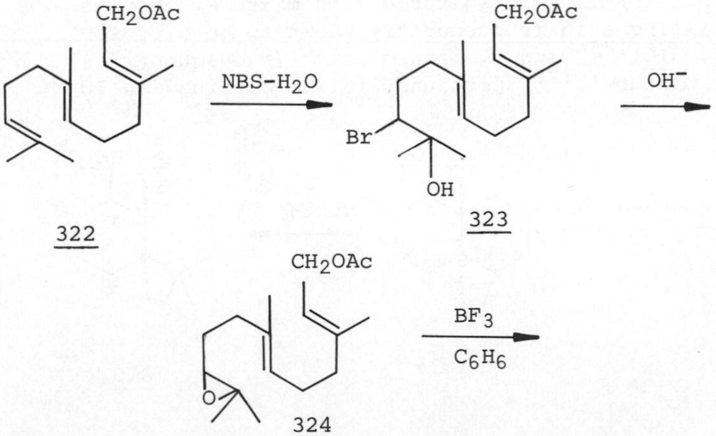

322   323

324

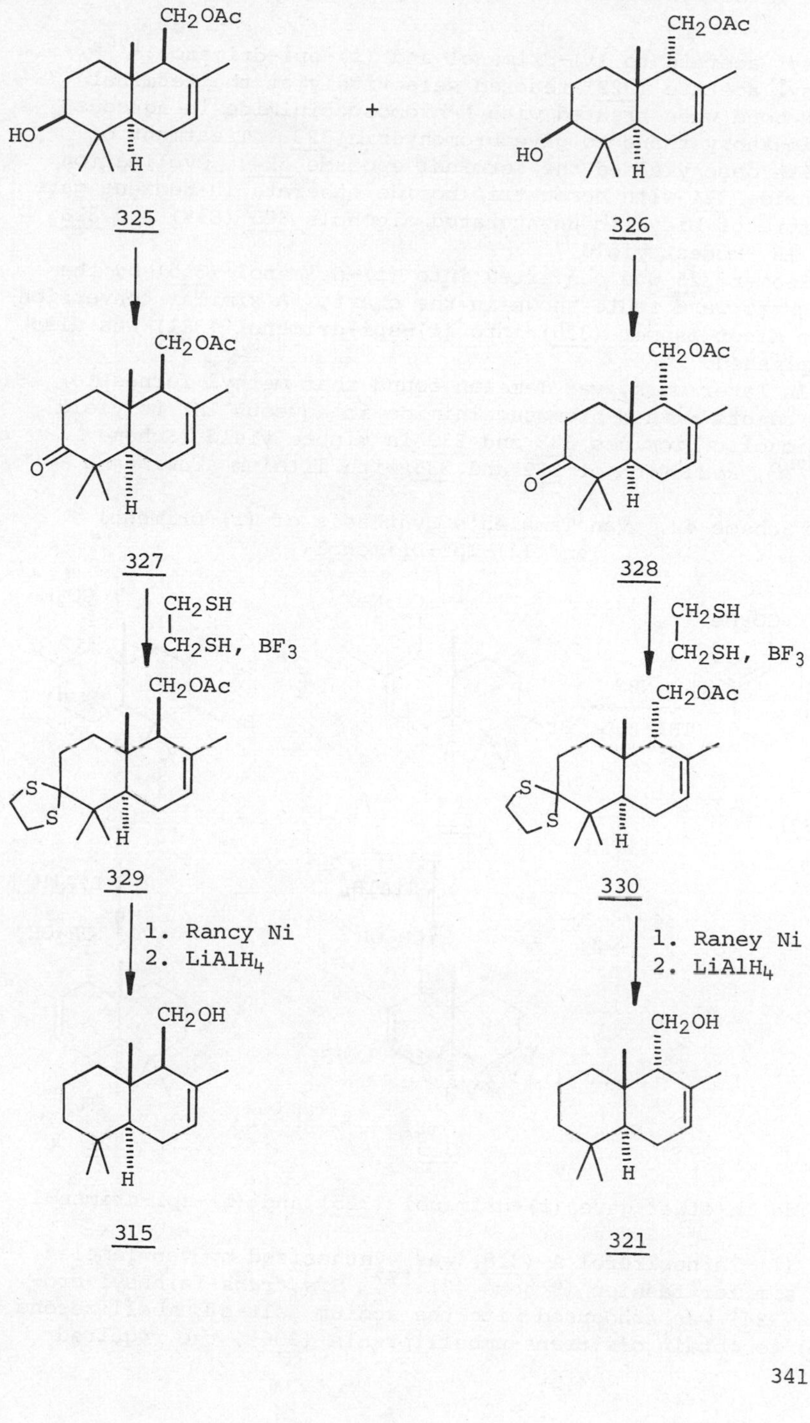
CH2OAc
HO
H
325
+
CH2OAc
HO
H
326
CH2OAc
O
H
327
CH2OAc
O
H
328
CH2SH
CH2SH, BF3
CH2OAc
S
S
H
329
CH2SH
CH2SH, BF3
CH2OAc
S
S
H
330
1. Rancy Ni
2. LiAlH4
CH2OH
H
315
1. Raney Ni
2. LiAlH4
CH2OH
H
321

farnesyl acetate to (±)-drimenol and (±)-epi-drimenol.[174a]
Farnesyl acetate (322) reacted selectively at the terminal
double bond when treated with N-bromosuccinimide in aqueous
1,2-dimethoxyethane to give bromohydrin 323. Treatment of
323 with base yielded the terminal epoxide 324. Cyclization
of epoxide 324 with boron trifluoride etherate in benzene gave
a mixture of bicyclic unsaturated alcohols 325 (85%) and 326
(15%) in "modest yield."

Isomer 325 was converted into (±)-drimenol (315) by the
straightforward route shown in the chart. A similar conversion
of the minor isomer (326) into (±)-epi-drimenol (321) was also
accomplished.

In later work, van Tamelen found that methyl farnesate
(331) reacts with N-bromosuccinimide in aqueous THF to yield
the bicyclic bromides 332 and 333 in minute yield (Scheme
42).[174b] Reduction of 332 and 333 with lithium aluminum

Scheme 42. Van Tamelen's Synthesis of (±)-Drimenol
and (±)-Epi-Drimenol

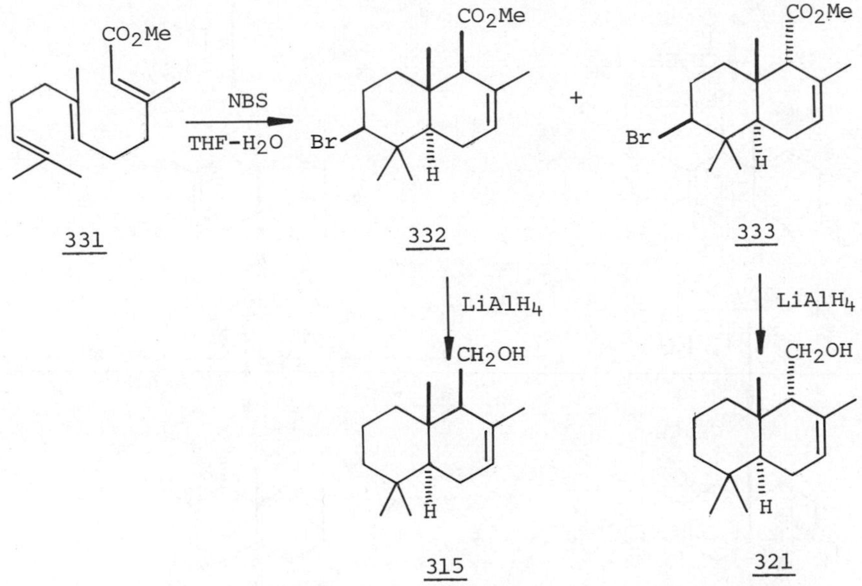

hydride in ether gave (±)-drimenol (315) and (±)-epi-drimenol
(321).

(±)-Farnesiferol A (318) was synthesized by van Tamelen
in a similar fashion (Scheme 43).[177] Cis,trans-farnesyl bromide (334) was condensed with the sodium salt of umbelliferone
(335) to obtain cis,trans-umbelliprenin (336), the required

Scheme 43. Van Tamelen Synthesis of (±)-Farnesiferol A

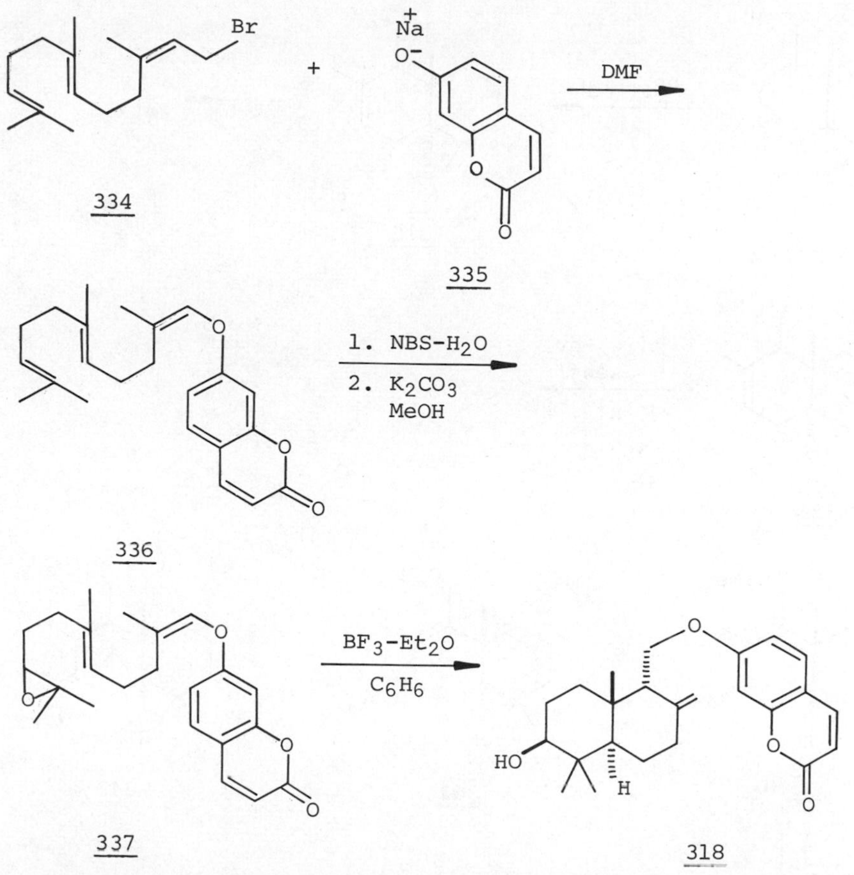
Br
Na
O
+
DMF
O
O
334
335
O
1. NBS-H2O
2. K2CO3
MeOH
O
O
336
O
BF3-Et2O
C6H6
O
O
O
O
HO
H
O
O
337
318

starting material. Terminal epoxidation yielded the oxide 337, which was cyclized by boron fluoride etherate in benzene to a mixture containing approximately 2% of (±)-farnesiferol A (318).

The last biogenetically styled synthesis to be discussed in this section is Kitahara's synthesis of (±)-drimenin (317), outlined in Scheme 44.[176] Monocyclofarnesic acid (338) had previously been cyclized to drimanic acid (316) by Stork and Burgstahler.[175a] Kitahara and co-workers increased the yield of this cyclization to 55% by carrying out the reaction with boron fluoride in a mixture of ether and benzene at 30°. The methyl ester (339) derived from drimanic acid was photooxygenated in the presence of hematoporphyrin or Rose Bengal.

Scheme 44.   Kitahara's Synthesis of (±)-Drimenin

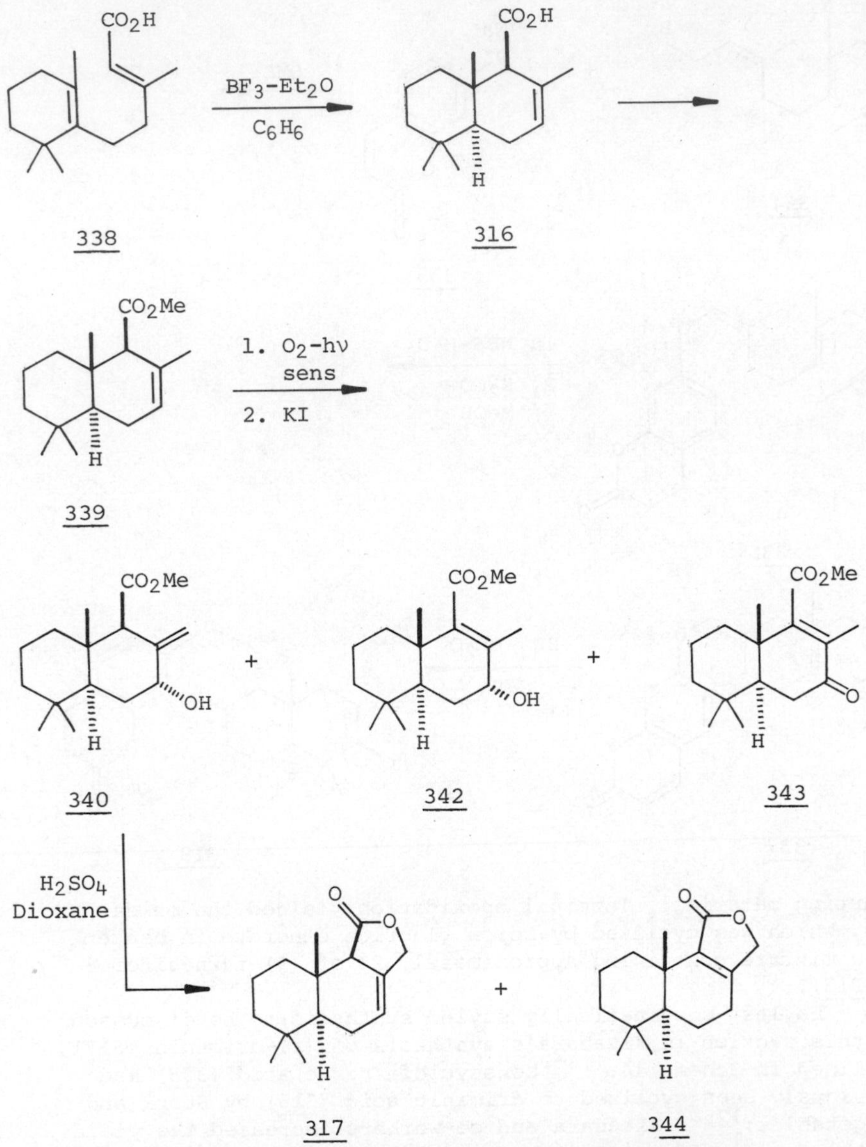

After decomposition of the peroxides with potassium iodide, compounds 340, 341, and 342 were obtained in yields of 33, 23, and 33%, respectively. Hydrolysis and concommittant lactonization occurred when hydroxy ester 340 was treated with sulfuric acid in dioxane, yielding (±)-drimenin (317, 69%) and (±)-isodrimenin (344, 11%).

P. Drimanes; Drimenin, Isodrimenin, Conterifolin, Drimenol, Isoiresin, Winterin

The syntheses of drimenin (317), isodrimenin (344), conterifolin (345), and drimenol (315) by Wenkert are interesting, in that the starting material, drimic acid (346) is actually a degradation product of various diterpenes.[178] The Wenkert synthesis of (±)-drimenin is outlined in Scheme 45.

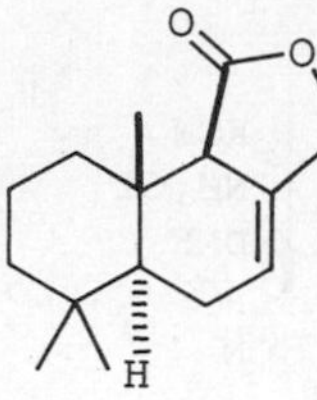

317 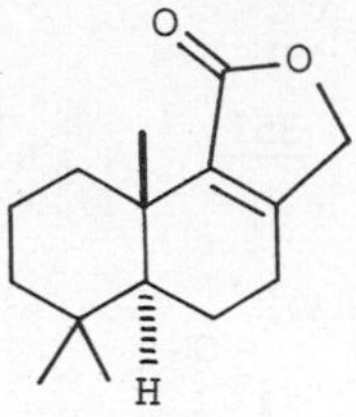   344 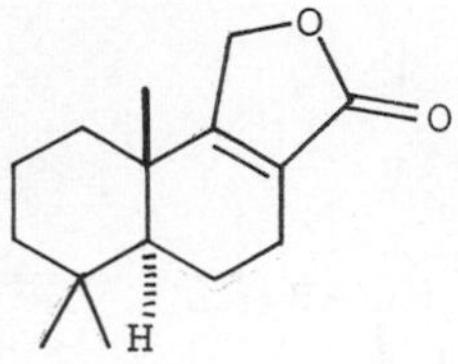   345

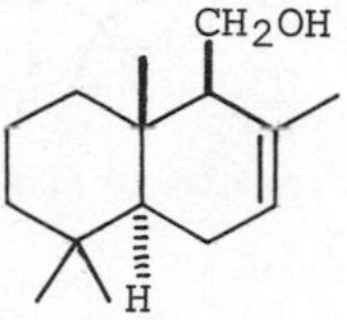

315 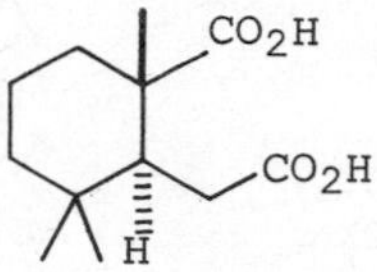   346

Scheme 45. Wenkert's Synthesis of (±)-Drimenin

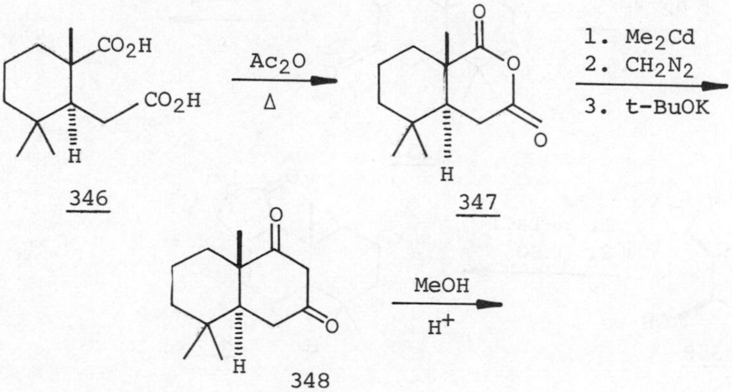

346   347

348

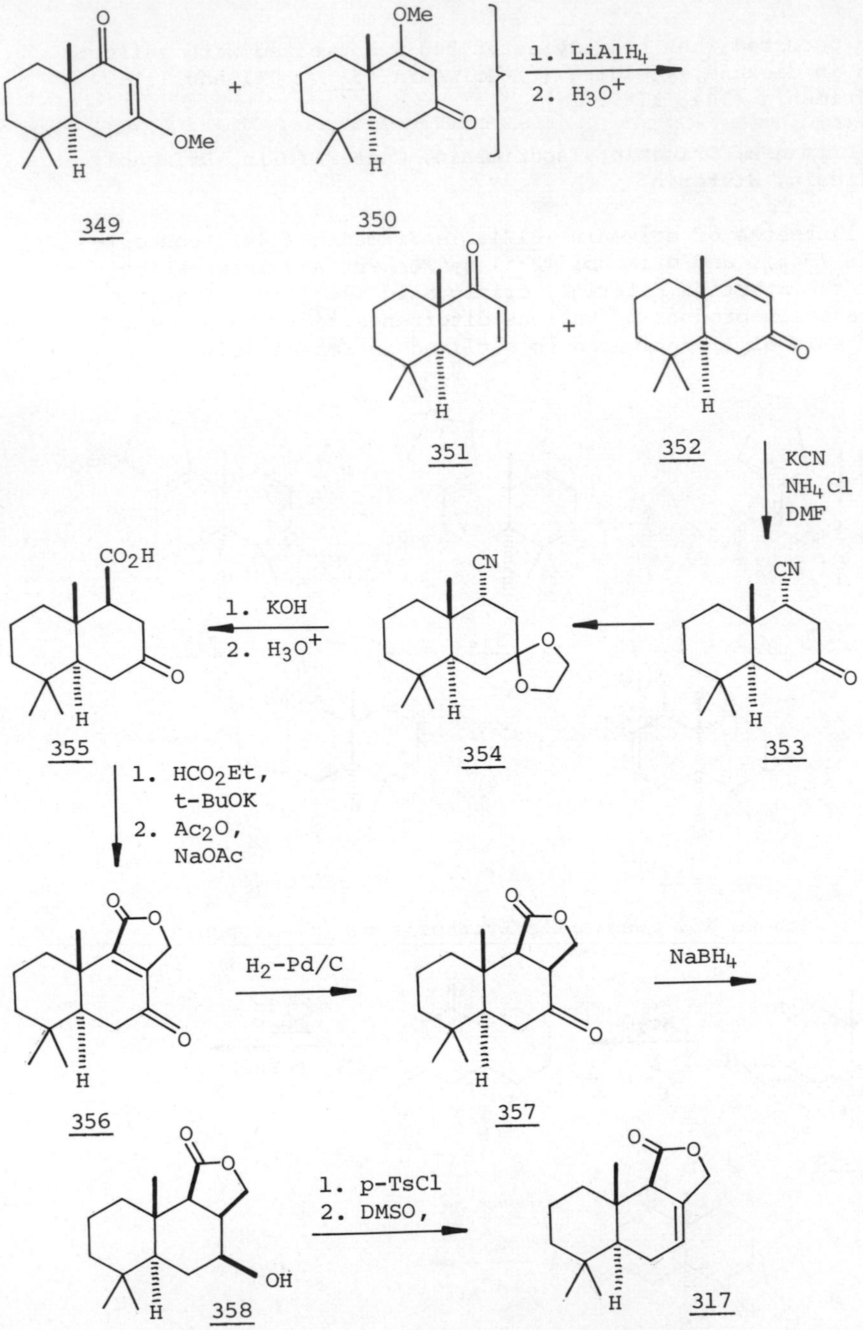

349
350
1. $LiAlH_4$
2. H_3O^+
351
352
KCN
NH_4Cl
DMF
353
354
1. KOH
2. H_3O^+
355
1. HCO_2Et, t-BuOK
2. Ac_2O, NaOAc
356
H_2-Pd/C
357
$NaBH_4$
358
1. p-TsCl
2. DMSO,
317

Drimic anhydride 347, obtained by heating drimic acid
with acetic anhydride, was treated with dimethylcadmium. The
resulting keto acid was methylated and the crude product cyclized with base to yield the 1,3-dione 348. When 348 was
treated with acidic methanol, the enol ethers 349 and 350 were
obtained in 84 and 13% yields, respectively. Lithium aluminum
hydride reduction, followed by acidic hydrolysis, converted
349 and 350 into enones 351 and 352. The major product 352
was hydrocyanated by Nagata's method. The resulting cyano
ketone (353), shown to have an axial cyano group, was converted
into ketal 354.

Vigorous hydrolysis of the nitrile occurred with prior
epimerization to the equatorial position. After deketalization, keto acid 355 was produced in good yield. Formylation
of keto acid 355 gave a hydroxymethylene derivative, which was
lactonized by treatment with sodium acetate in acetic anhydride. The resulting product, 7-ketoisodrimenin (356), was
reduced to the dihydro analog 357, which was further reduced
with sodium borohydride to alcohol 358. When the derived p-
toluenesulfonate ester was heated in dimethyl sulfoxide, (±)-
drimenin (317) was obtained.

Wenkert's further conversions of (±)-drimenin into (±)-
isodrimenin, (±)-conterifolin, and (±)-drimenol are outlined
in Scheme 46. (±)-Isodrimenin, obtained by isomerization of
(±)-drimenin, was reduced by lithium aluminum hydride to diol
359. Manganese dioxide oxidation of 359 yielded (±)-conteri-
folin (345), along with some (±)-isodrimenin. Diol 360, obtained by hydride reduction of (±)-drimenin, was acetylated
and the resulting diacetate 361 was reduced by lithium in

Scheme 46. Wenkert's Conversions of (±)-drimenin into
    (±)-Isodrimenin, (±)-Conterifolin, and (±)-Drimenol

317   344   359

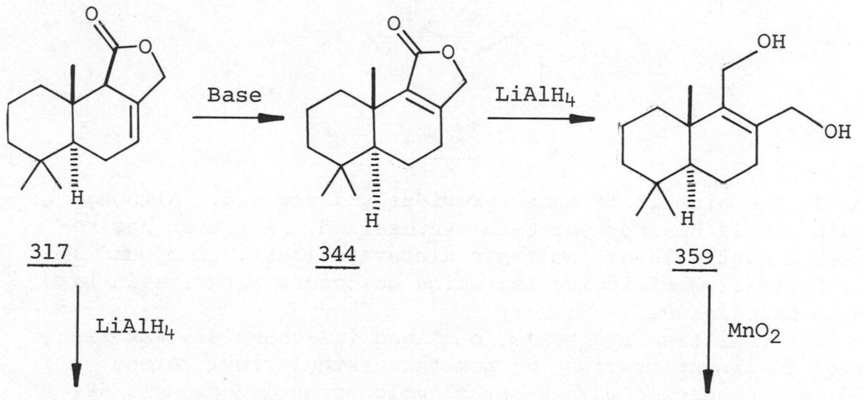

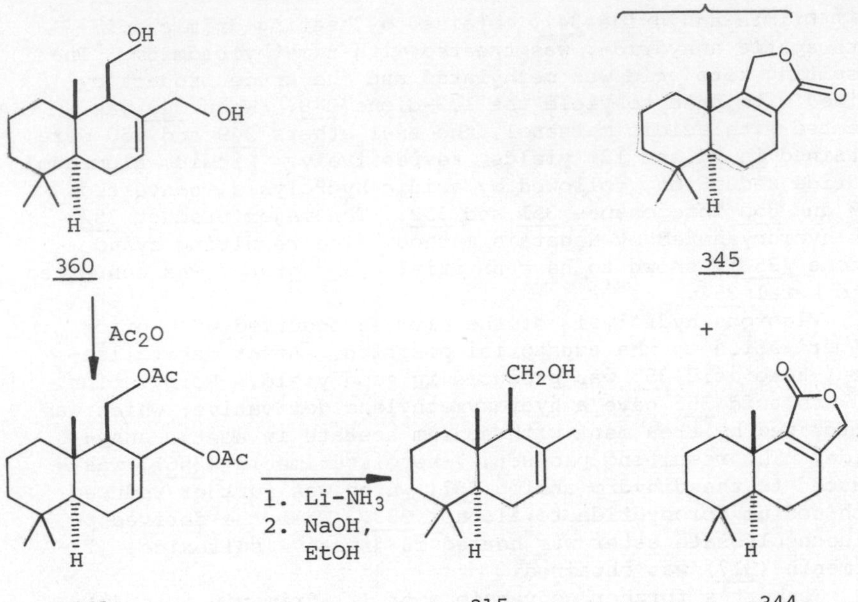

360

Ac₂O

361

1. Li-NH₃
2. NaOH, EtOH

315

344

345

+

ammonia.   Subsequent alkaline hydrolysis yielded (±)-drimenol (315).

The drimanic sesquiterpene iresin (362) was the first sesquiterpene in which the bicyclofarnesol skeleton was demonstrated.   Synthetically, it presents a much greater challenge

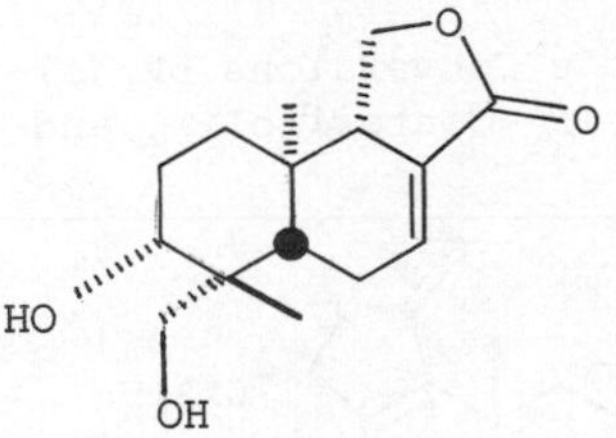

362

than do the simpler drimanes previously discussed.   Although iresin itself has not yet been synthesized, Pelletier has reported a synthesis of isoiresin diacetate (386).   Compound 386 is a derivative of isoiresin, which co-occurs with iresin in *Iresin celosioides*.

The Pelletier synthesis, outlined in Scheme 47, was reported in two stages.[179]   Carbomethoxymethyl vinyl ketone (363) was condensed with 2-methylcyclohexane-1,3-dione (364)

Scheme 47. Pelletier's Synthesis of (±)-Isoiresin Diacetate

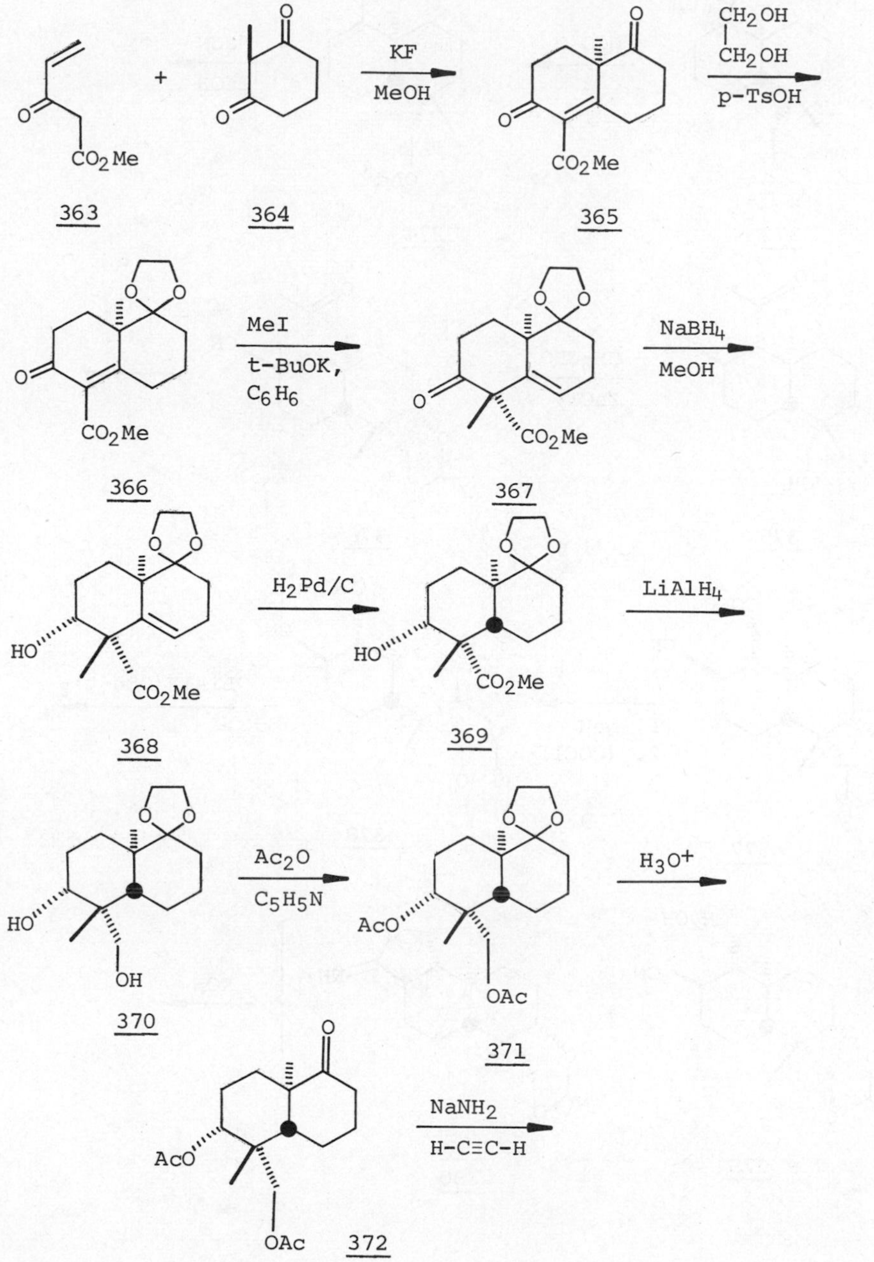

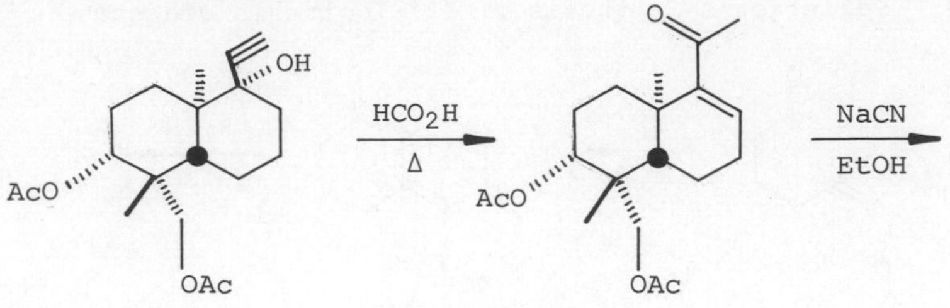

373 374

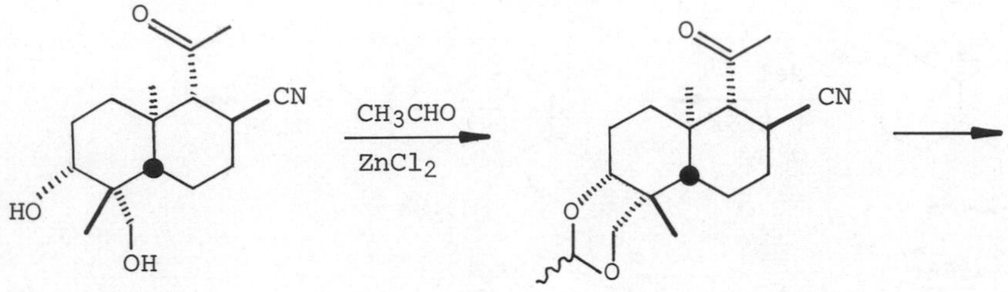

375 376

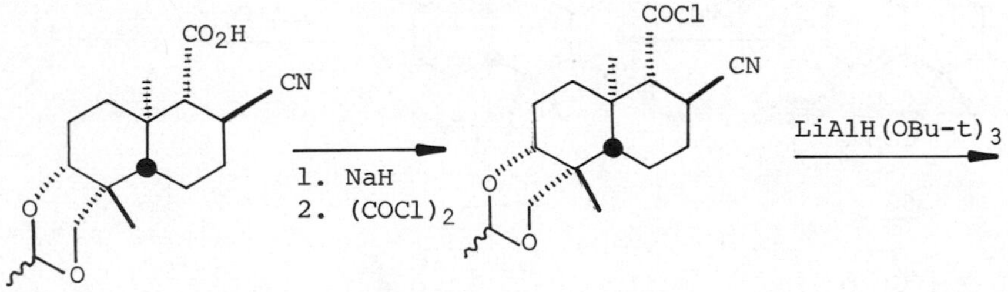

377 378

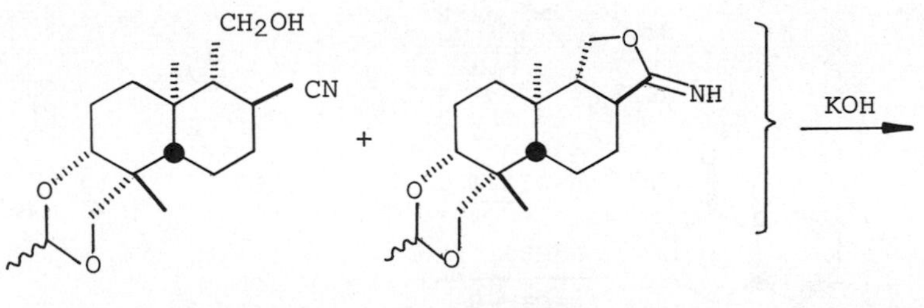

379 380

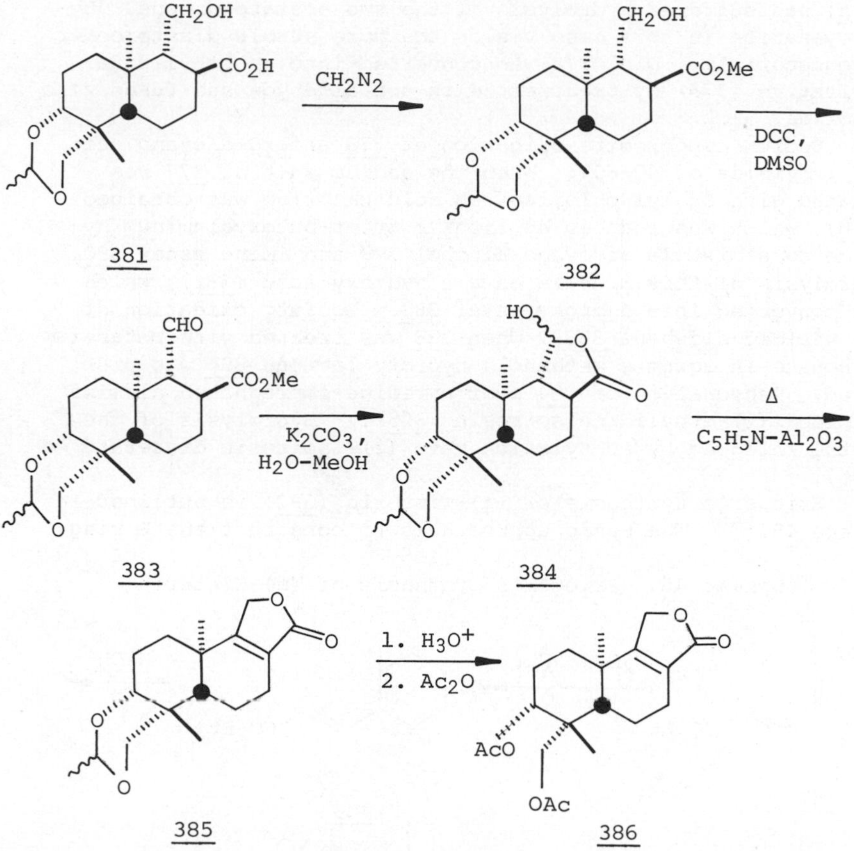

under the influence of anhydrous potassium fluoride in methanol, yielding the diketo ester 365 in 30-50% yield. Selective ketalization of 365 afforded ketal 366, which was methylated stereospecifically to obtain 367. After reducing the ketone with borohydride, the resulting hydroxy ester 368 was catalytically reduced to yield mainly the *trans*-fused bicycle 369. Lithium aluminum hydride reduction of hydroxy ester 369 yielded diol 370, which was converted into diacetate 371. Hydrolysis of the ketal grouping gave keto diacetate 372.[179a]

When keto diester 372 was condensed with the sodium salt of acetylene, ethynyl carbinol 373 was the main product (81%). Rupe rearrangement of 373 gave enone 374 in 48% yield (along with 12% of the corresponding α,β-unsaturated aldehyde). Hydrocyanation of 374 occurred when it was heated with a large excess of sodium cyanide in ethanol. The resulting product

(375) had suffered hydrolysis of the two acetate groups. Hydrocyanation in this case yields the more stable diastereomer (diequatorial). Diol 375 was converted into an ethylidine derivative (376) by treatment with acetaldehyde and fused zinc chloride.

Sodium hypobromite oxidation of 376 afforded cyano acid 377 in yields of 40-60%. When the sodium salt of 377 was treated with oxalyl chloride, an acid chloride was obtained (378), which was reduced by lithium tri-t-butoxyaluminum hydride to a mixture of cyano alcohol 379 and imino ester 380. Hydrolysis of this mixture gave a hydroxy acid (381), which was converted into hydroxy ester 382. Moffatt oxidation of 382 yielded aldehyde 383. When 383 was treated with potassium carbonate in aqueous methanol, hydroxy lactone 384 was produced. Dehydration of 384 over pyridine-impregnated alumina afforded (±)-ethylidineisoiresin (385). Hydrolysis of the acetal followed by acetylation gave (±)-isoiresin diacetate (386).

Brieger's synthesis of (±)-winterin (392) is outlined in Scheme 48.[180] The basic approach is to construct the B ring

Scheme 48. Brieger's Synthesis of (±)-Winterin

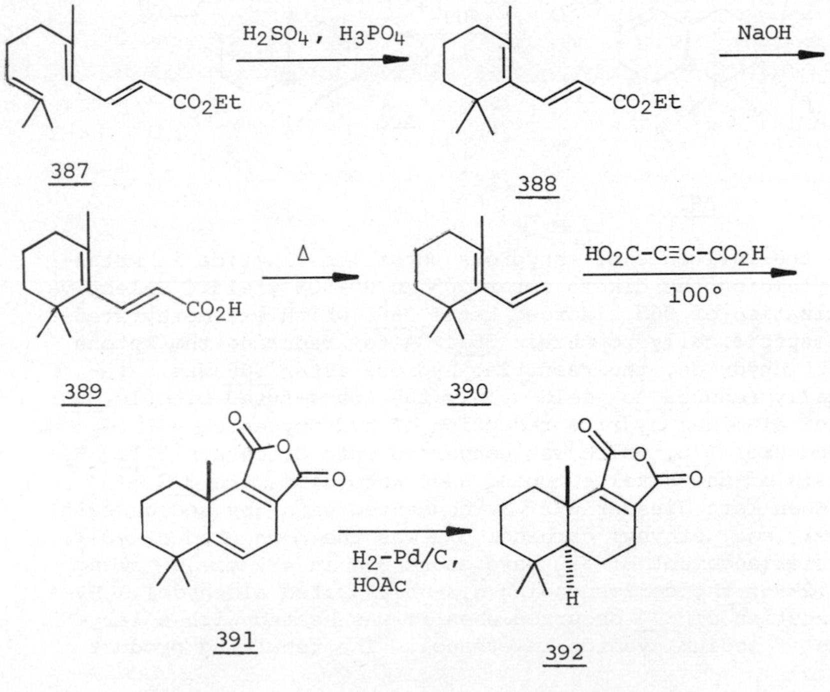

by a Diels-Alder reaction. The requisite diene, 1-vinyl-
2,6,6-trimethylcyclohexene (390), was prepared by thermal decarboxylation of β-cyclocitrylidineacetic acid (389). Royals
had previously shown that this acid may be produced from ethyl
citrylidineacetate (387) by π-cyclization, followed by alkaline
hydrolysis.[181] When diene 390 was condensed with acetylene-
dicarboxylic acid, adduct 391 was obtained in 4% yield. Catalytic hydrogenation gave (±)-winterin (392), contaminated with
its dihydro derivative.

Q. Valeranone

Valeranone (393) is a member of an interesting class of sesquiterpenes. While isoprenoid, the carbon skeleton cannot be

393

derived from farnesol. Since the C-10 angular methyl group
and the three carbon side chain at C-7 are *trans*, the most
promising synthetic approach to the skeleton is via 7-epi-
cyperone (31) or an analog thereof (see p. 284). The most
interesting synthetic challenge is the second angular methyl

31

group at C-9, which must be introduced stereospecifically *cis*
to the C-10 methyl.
     Several interesting solutions to this problem have been
explored. In Schemes 49 and 50 are outlined Marshall's initial synthesis of the antipode of natural valeranone. Robinson annelation of methyl vinyl ketone and (-)-dihydrocarvone
(32) gave ketol 67, which was hydrogenated to ketol 394.
Acid-catalyzed dehydration afforded enone 395, which possesses
the proper relative stereochemistry at the two asymmetric
centers.
     Initial attempts to introduce the second angular methyl

Scheme 49. Marshall's Valeranone Synthesis—First Variation

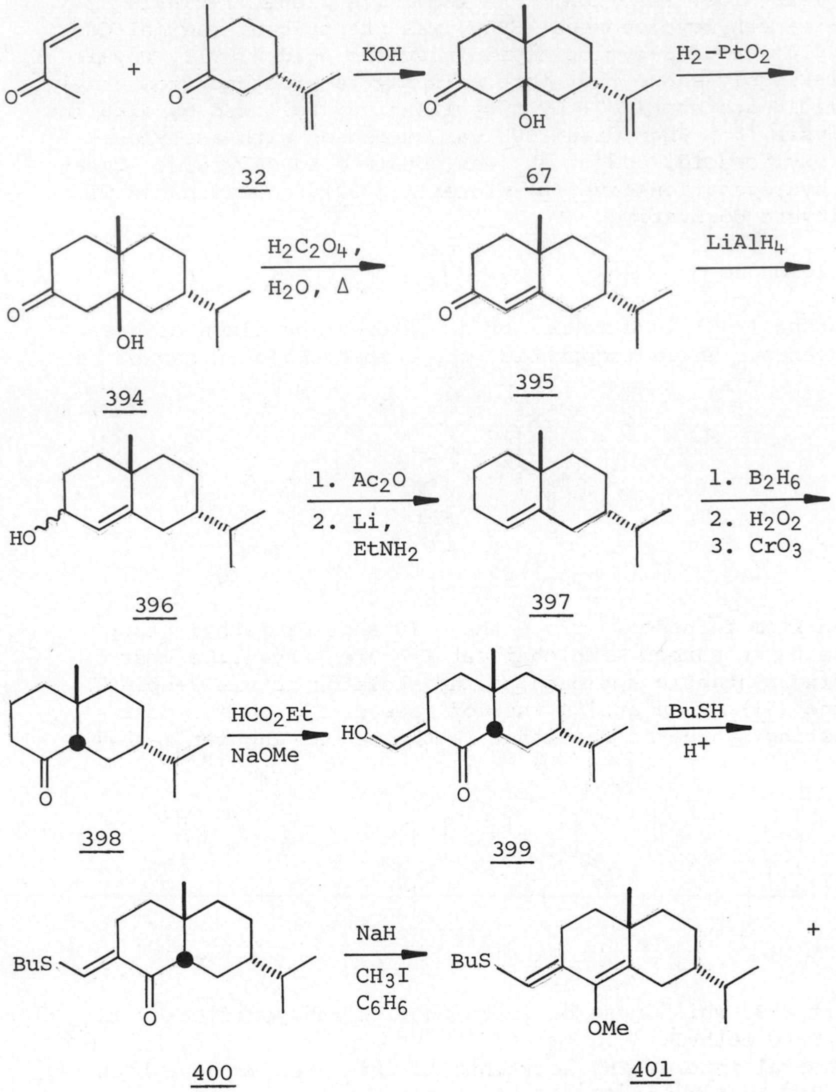

32   67

394   395

396   397

398   399

400   401

354

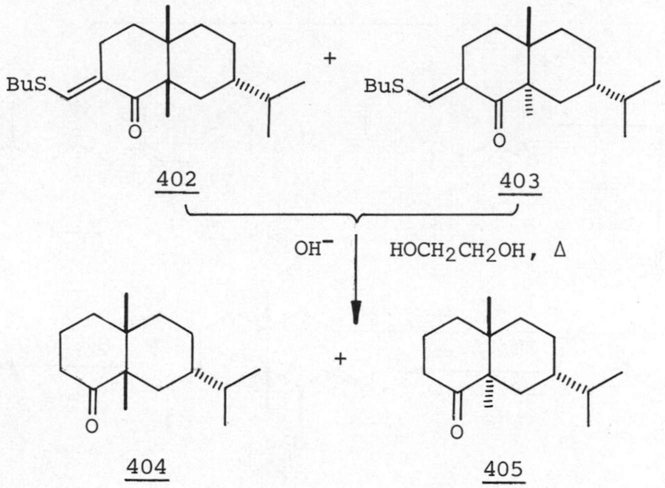

Scheme 50.   Marshall's Valeranone Synthesis--Second Variation

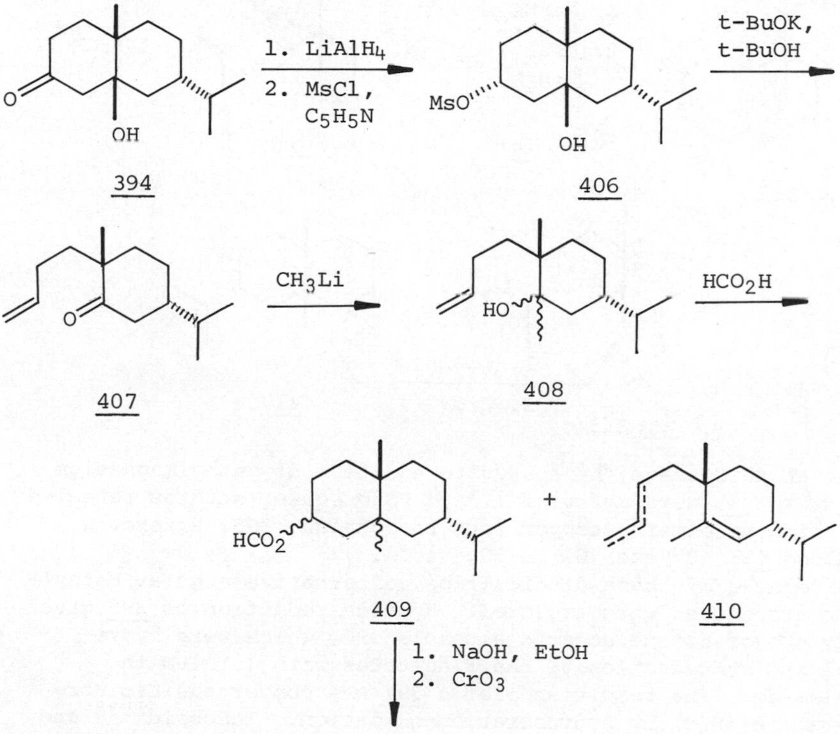

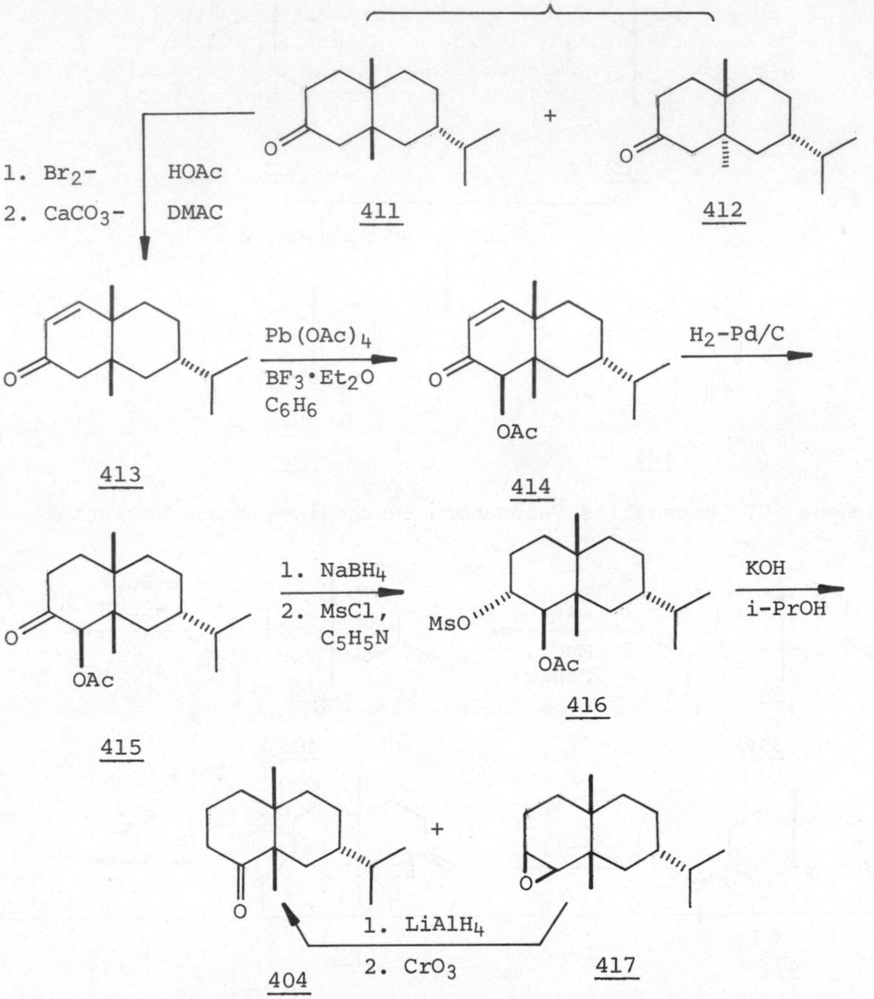

1. Br$_2$- | HOAc
2. CaCO$_3$- | DMAC

411   +   412

413

Pb(OAc)$_4$
BF$_3$·Et$_2$O
C$_6$H$_6$

OAc

414

H$_2$-Pd/C

OAc

415

1. NaBH$_4$
2. MsCl, C$_5$H$_5$N

MsO
OAc

416

KOH
i-PrOH

+

404

1. LiAlH$_4$
2. CrO$_3$

417

group at this stage, by conjugate addition of methylmagnesium iodide to 395 were unfruitful.[182b] Subsequent studies revealed that lithium dimethylcopper adds to octalone 395, affording decalone 411 (Scheme 50) in 10% yield.

Because of these difficulties, alternative angular methylation procedures were explored. Hydride reduction of 395 gave a mixture of diastereomeric alcohols 396, which were hydrogenolyzed by reduction of their acetates with lithium in ethylamine. The resulting olefin 397 was converted into norvaleranone (398) by hydroboration-oxidation. Theobald[183a] and

and an Indian group[183b] have also reported the synthesis of
ketone 398, in essentially the same manner.

The problem now becomes one of stereospecific angular
methylation. To whatever extent angular alkylation occurs,
the stereochemistry is predictable. *Trans* alkylation can only
occur through one of the highly strained transition states
398a or 398b. In one case (398a), the incoming methyl group

398a   398b

398c

encounters severe hindrance due to the axial isopropyl group.
If ring B is adjusted into a twist conformation to avoid this
interaction, a price in transition state energy must be paid.
The transition state leading to *cis* alkylation (398c) is not
similarly encumbered.

However, it had been known for some time that 1-decalone
undergoes predominant alkylation not in the angle, but at
C-2.[184] It thus became necessary to temporarily block C-2, in
order to force alkylation at the angle. This was readily accomplished by treating the derived hydroxymethylene derivative
399 with *n*-butyl mercaptan. The resulting *n*-butylthiomethylene
derivative 400 was alkylated with sodium hydride and methyl
iodide in benzene. Under these conditions, enol ether 401 and
C-alkylated diastereomers 402 and 403 were obtained in yields
of 56, 18, and 2%, respectively. Vigorous base catalyzed hydrolysis of the mixture of 402 and 403 gave (+)-valeranone
(404) and its stereoisomer 405 in a ratio of 92:8.

An alternative solution to the angular methyl problem was

explored and is summarized in Scheme 50. Ketol 394 was reduced and the resulting diol converted into mono-methane-
sulfonate 406. Fragmentation of 406 yielded the butenyl
cyclohexanone 407. Treatment of ketone 407 with methyllithium
yielded an alcohol (408), which was treated with formic acid
at room temperature. There was obtained a mixture of bicyclic
formates (409) in 40-50% yield, along with some monocyclic
diene (410). Again, the stereochemistry of the cyclization
process is predictable, albeit only by analogy. Since cyclization of diones such as 418 is known to yield solely *cis*
decalones (e.g., 67), it is reasonable that cation 408a would
cyclize in a similar steric sense.

418   67

408a   409

    In the event, when the diastereomeric mixture 409 was
hydrolyzed and oxidized, *cis* and *trans* ketones 411 and 412
were obtained in a ratio of 87:13. In order to convert 411
into (+)-valeranone, ketone transposition must be accomplished.
It is a measure of the difficulty of this process that Marshall and co-workers were forced to use nine separate synthetic stages to accomplish it.
    In order to block undesired reactions at C-3 (compound
411 forms a 1:1 mixture of isomeric enol acetates), bromination-dehydrobromination was first carried out. The resulting
enone (413) was acetoxylated with lead tetraacetate and boron
fluoride in benzene, to afford acetoxy enone 414 in 21% yield.

After catalytic hydrogenation, the acetoxy ketone was reduced
and the new hydroxyl esterified with methanesulfonyl chloride.
Basic hydrolysis gave a mixture of (+)-valeranone (404) and
oxide 417 in a ratio of 16:84. Oxide 417 was reduced and the
resulting secondary alcohol oxidized to (+)-valeranone.
    Wenkert has reported a highly imaginative solution to the
angular methyl problem posed by valeranone.[185] The Wenkert
synthesis, outlined in Scheme 51, employs 1,4-dimethoxy-2-

Scheme 51. Wenkert's Synthesis of (-)-Valeranone

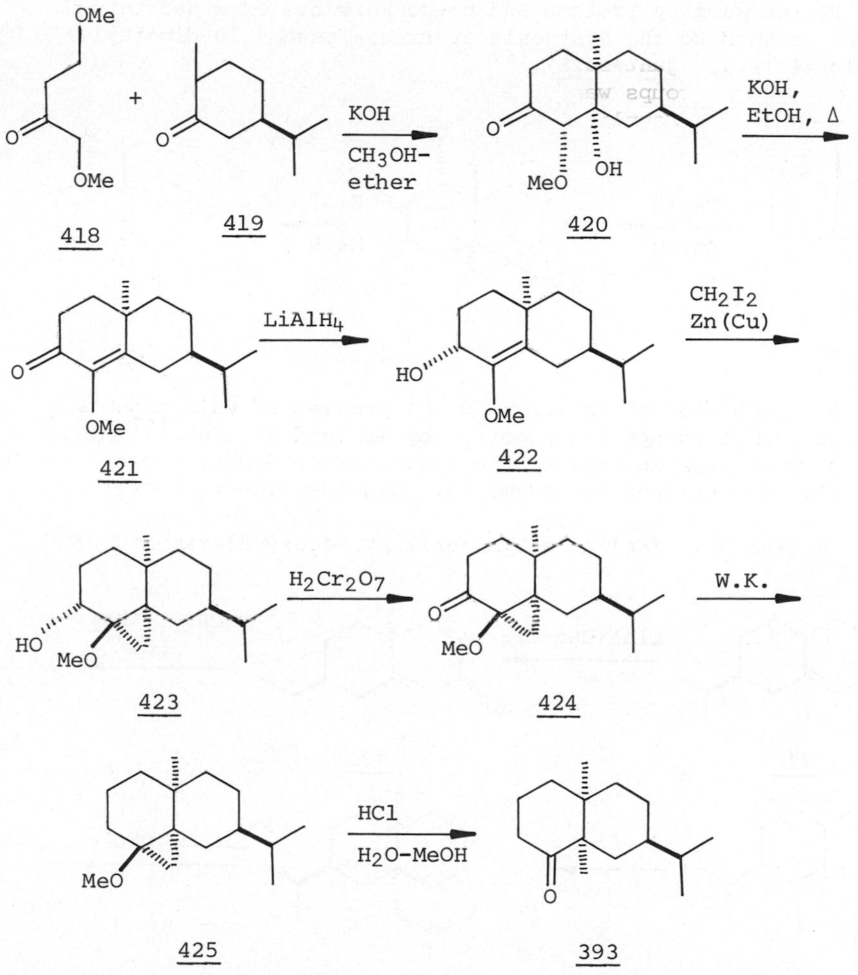

butanone (418) as an in situ source of methoxymethyl vinyl
ketone in a Robinson annelation with (+)-carvomenthone (419).
The resulting ketol (420) was dehydrated with ethanolic potassium hydroxide to obtain methoxy enone 421. Hydride reduction
of 421 yielded predominantly equatorial alcohol 422.

Simmons-Smith methylenation of 422 proceeded stereospecifically, with the resulting methylene bridge being *cis* to
the secondary hydroxyl group.[186] Jones oxidation gave ketone
424, which was reduced by the Wolff-Kishner method to ether
425. Compound 425 underwent specific electrophilic cleavage
as expected, yielding (-)-valeranone.

Recent work by Ireland and co-workers has extended this
elegant method to the synthesis of *trans*-fused 9,10-dimethyl-
decalones (e.g., 426 → 428).[187]

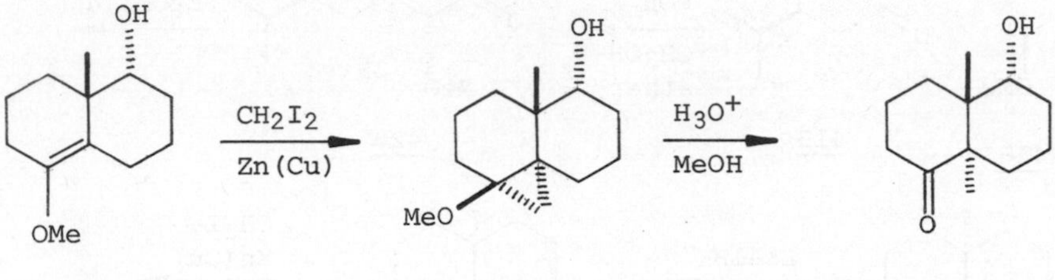

426   427   428

A second elegant solution to the problem of valeranone's
angular methyl groups was reported by Ireland in 1968.[188] The
method, which was applied to the synthesis of desoxy derivative 434, is outlined in Scheme 52. Hydride reduction of

Scheme 52. Ireland's Synthesis of Desoxyvaleranone

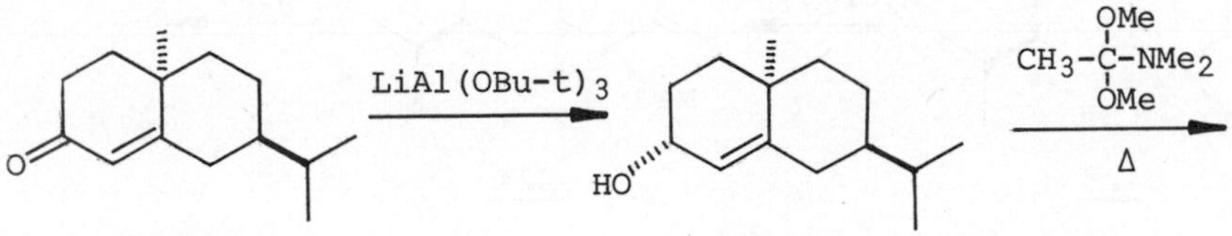

429   430

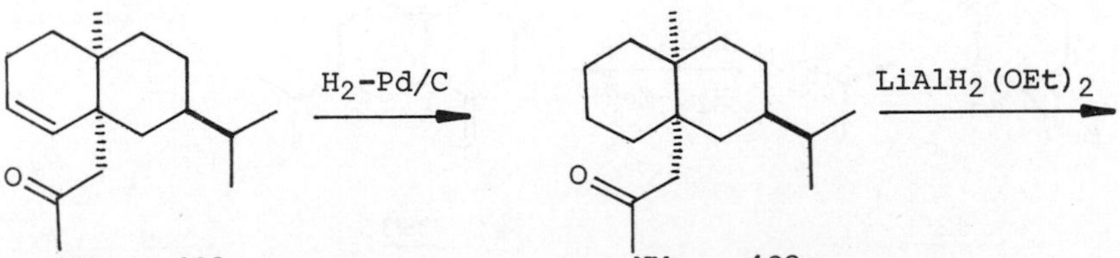

431   432

Rh(Pϕ_3)$_3$Cl

C$_6$H$_6$

433   434

enone 429 (Scheme 49) afforded the equatorial allylic alcohol
430 specifically (86% yield). Eschenmoser's variant of the
Claisen vinyl ether rearrangement[189] allowed the conversion of
430 into unsaturated amide 431. The stereochemistry of the
new angular substituent is decided by the geometry of the
transition state; it must be attached to the same face of the
molecule as the departing oxygen.

Lithium diethoxyaluminohydride reduction of amide 432
gave aldehyde 433, which was decarbonylated by treatment with
tristriphenylphosphinechlororhodium(I) chloride in benzene.
The resulting product (434) is the desoxy analog of (+)-
valeranone.

An interesting photochemical rearrangement discovered by
Marshall offers a further solution to the foregoing synthetic
problem, although it has not yet been applied to the synthesis
of a natural product. Upon irradiation in a mixture of t-
butyl alcohol and acetic or formic acid, olefin 435 is converted into isomer 436 in yields of 40-50%.[190]

hν

t-BuOH-HCO$_2$H

435   436

R. Eremophilanes; Isonootkatone (α-Vetivone), Nootkatone,
Valerianol, Valencene, Eremoligenol, Eremophilene, Fukinone

The eremophilane family represents an interesting nonisoprenoid
group of compounds believed to arise by cationic rearrangement
of a eudesmanoid precursor. Members of the group are all

characterized by *cis*-vicinal methyl groups at carbons 1 and 9, although the three-carbon side chain at C-7 may be oriented either *cis* or *trans* to the angular methyl.

Successful syntheses have been recorded for isonootkatone (α-vetivone, 437),[191] nootkatone (438),[192,193] valencene (439),[194] valerianol (440),[194] eremophilene (441),[194] eremoligenol (442),[194] and fukinone (443).[195] In addition, syntheses of tetrahydroeremophilone (444),[196] eremophil-3,11-diene (445),[197] and dehydrofukinone (446)[198] will be discussed in this section. Tricyclic eremophilanes will be taken up in Sec. 4-S.

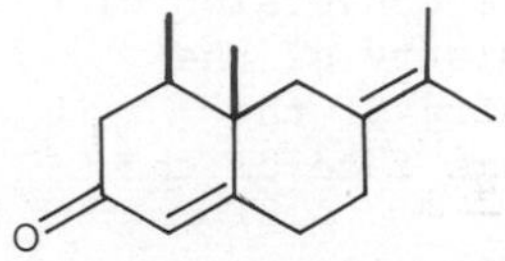

437

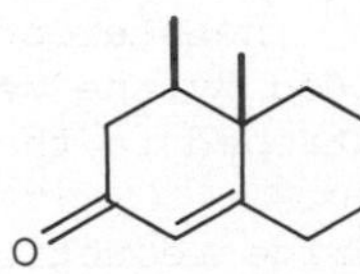

438

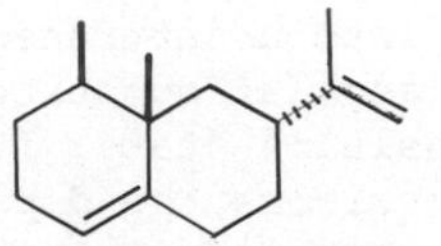

439

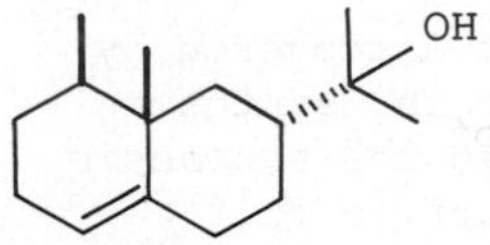

440

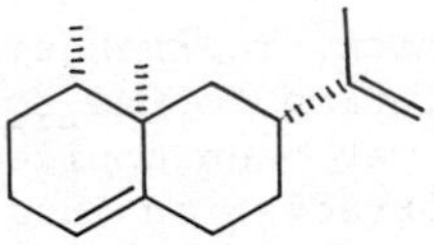

441

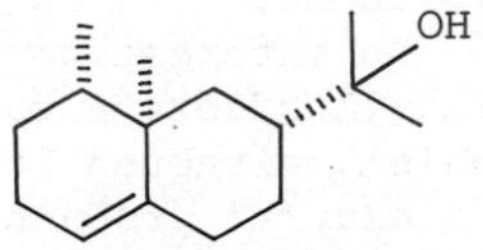

442

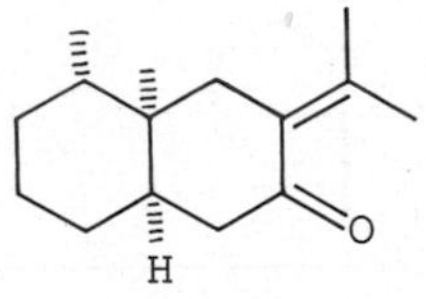

443

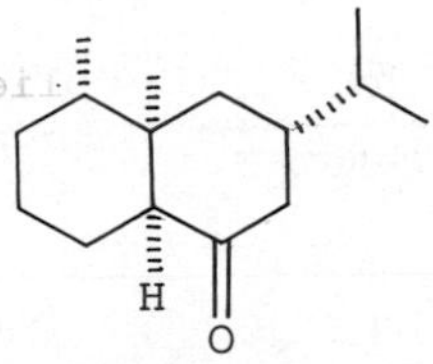

444

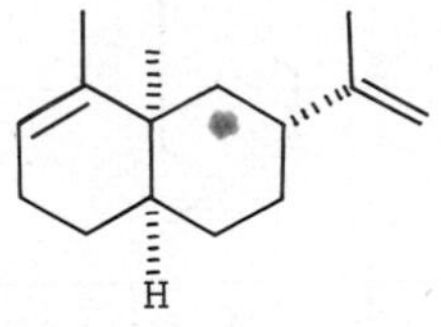

445

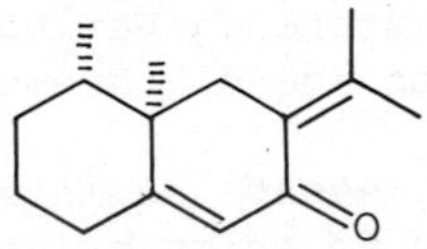

446

The synthetic problems associated with this group are several. The problem of the *cis*-vicinal methyl groups has been approached in a variety of ways, and still may not be

considered to have been adequately solved. With the "normal"
eremophilanes (C-9 methyl and C-7 isopropyl *cis*), the stereochemistry at C-7 is the less stable; either the three-carbon
chain must be axial or ring B must adopt a boat conformation.
The main synthetic obstacle is a rather subtle one. The arrangement of side-chain groups, combined with functional pat-
terns, is not easily accommodated with the standard repertoire
of synthetic methods.

The synthetically simplest target, and the first member
to yield to synthesis, is isonootkatone (437). Here the only
synthetic problems are the carbon skeleton itself and the *cis*
disposition of the two methyl groups. Marshall's synthesis is
summarized in Scheme 53. Diethyl isopropylidene-malonate
(447) was reduced to diol 448, which reacted with phosphorous
tribromide in ether-hexane-pyridine to give dibromide 449.
Compound 449 was used to alkylate two moles of diethyl malonate
and after hydrolysis and decarboxylation, diacid 450 was obtained. Dieckmann cyclization of ester 451 afforded keto
ester 452.

Compound 452 was used in a Robinson annelation with *trans*-
3-penten-2-one to obtain octalone 453. Although full experimental details have not yet been published, octalone 453 is
apparently the major isomer produced in this reaction. A
related reaction (Scheme 54) used in the synthesis of nootkatone gave a *cis,trans* ratio of 3:1. This result is interesting, in light of other work with 2-methylcyclohexan-1,3-dione
(364), where the *cis* isomer 453 is the minor product when
annelation is accomplished with KOH-pyrrolidine[199] or on the
pyrrolidine enamine of 364 in benzene.[200] At best, isomers
452 and 453 may be produced in equal amounts, when the pyrrolidine enamine of 364 is used in dimethylformamide.[200]*

364

452

453

*As Coates has pointed out, the stereochemistry in Marshall's
annelation is determined in the *Michael* reaction of 452 with
pentenone. In the annelations with cyclic 1,3-diketones such
as 364, the stereochemistry is probably set in the *aldol* condensation of the intermediate triketone.[214b]

With the production of 453, the problem now reduces to one of redox chemistry, reduction of the angular carbomethoxyl group to methyl. This was accomplished in straightforward fashion, after protection of the ketonic carbonyl, as outlined in Scheme 53.

Scheme 53. Marshall's Synthesis of (±)-Isonootkatone

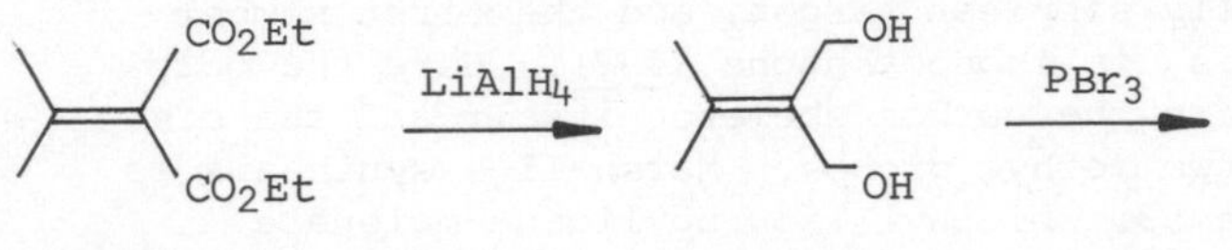

447 448

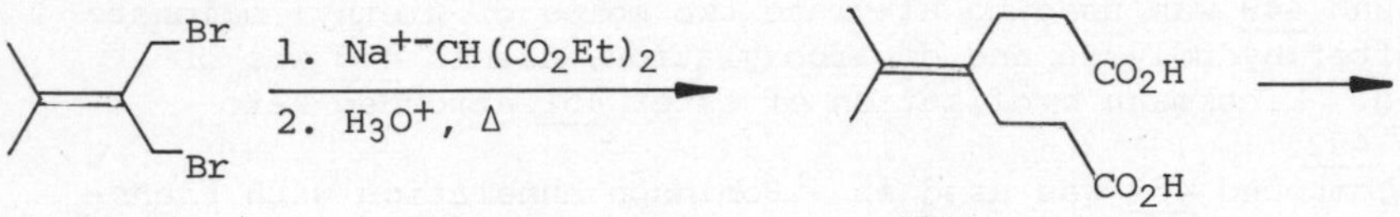

449 450

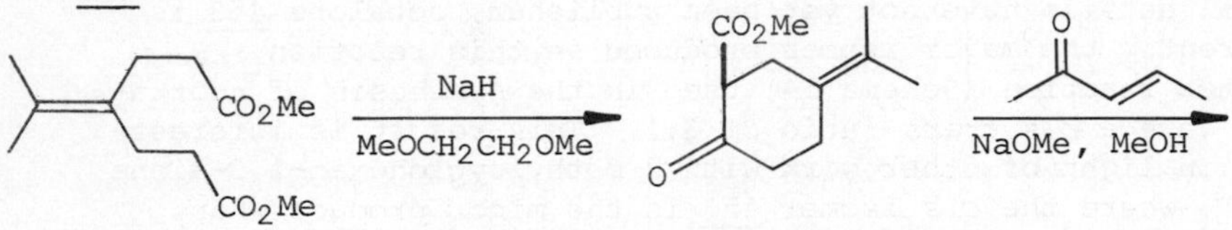

451 452

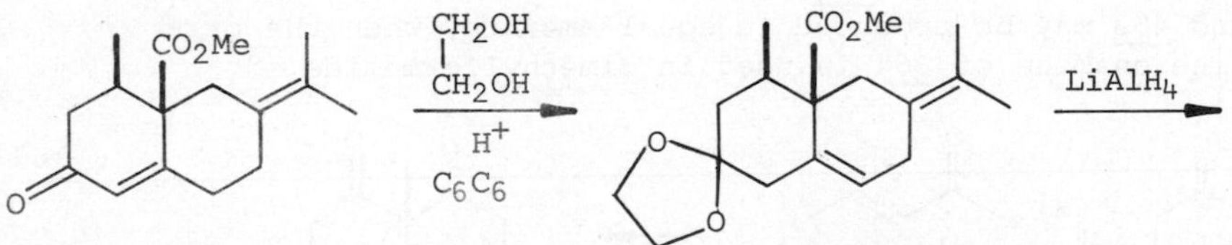

453 454

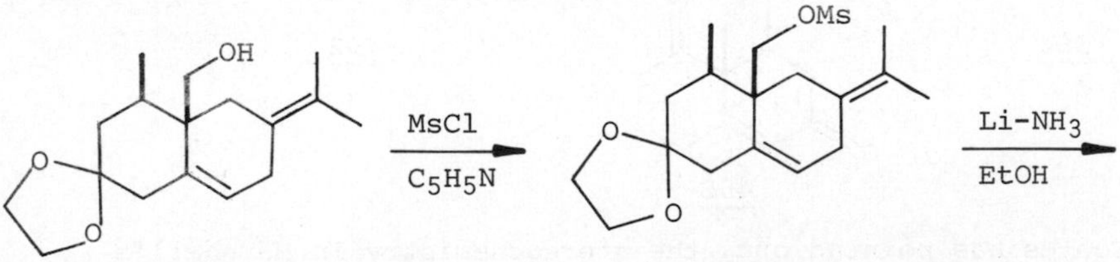

455 456

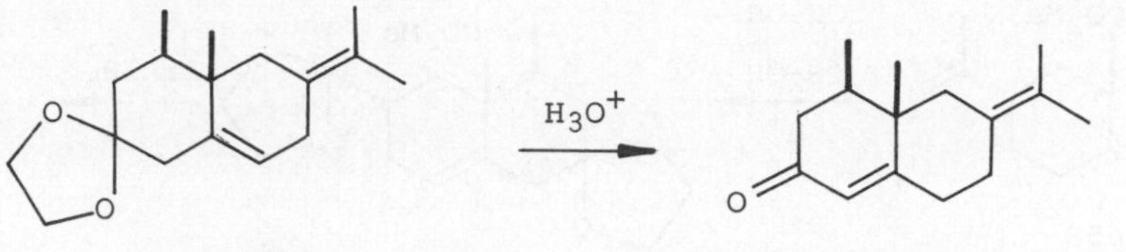

457   437

Nootkatone (438) has been synthesized by Schudel[192] and by Marshall.[193] A reported synthesis by Pinder[201] has been retracted.[202] Because of its similarity to the preceding synthesis, Marshall's synthesis will be discussed first (Scheme 54). Dimethyl γ-ketopimelate (454), when treated with ethylidenetriphenylphosphorane in dimethyl sulfoxide, undergoes Wittig reaction and Dieckmann cyclization, affording β-keto ester 455 as a 1:1 mixture of geometric isomers. Robinson annelation with trans-3-penten-2-one (potassium t-amyloxide in t-amyl alcohol) gave octalone 456 and its trans isomer (3:1 ratio) in 75% yield.

Epoxidation of 456 gave an oxide (457), which was rearranged to methyl ketone 458 when treated with BF₃ etherate. After protection of the two ketonic carbonyl groups, the carbomethoxy group was reduced to methyl by hydride reduction, Moffatt oxidation of the resulting primary alcohol (460) to an aldehyde (461), and Wolff-Kishner reduction of the aldehyde to methyl (462). Acid catalyzed deprotection yielded a dione (463) from which (±)-nootkatone was produced by selective Wittig methylation.

The Givaudan synthesis (Schudel),[192] summarized in Scheme 55, begins with 4-acetyl-1-ethoxycyclohexene (464). Wittig

Scheme 54. Marshall's (±)-Nootkatone Synthesis

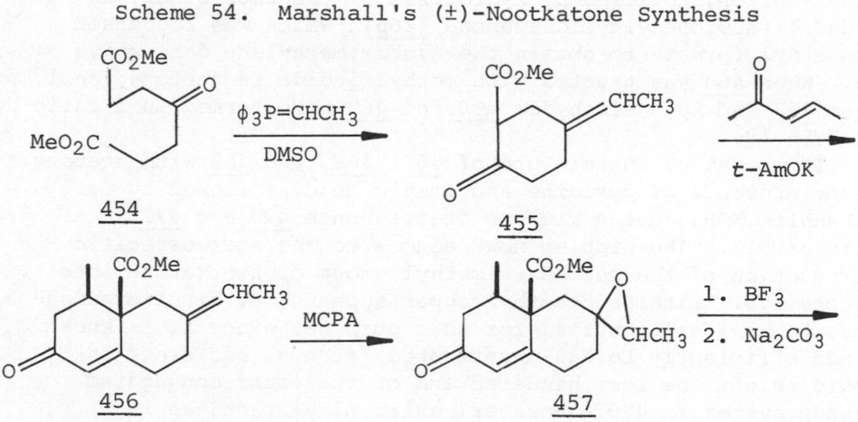

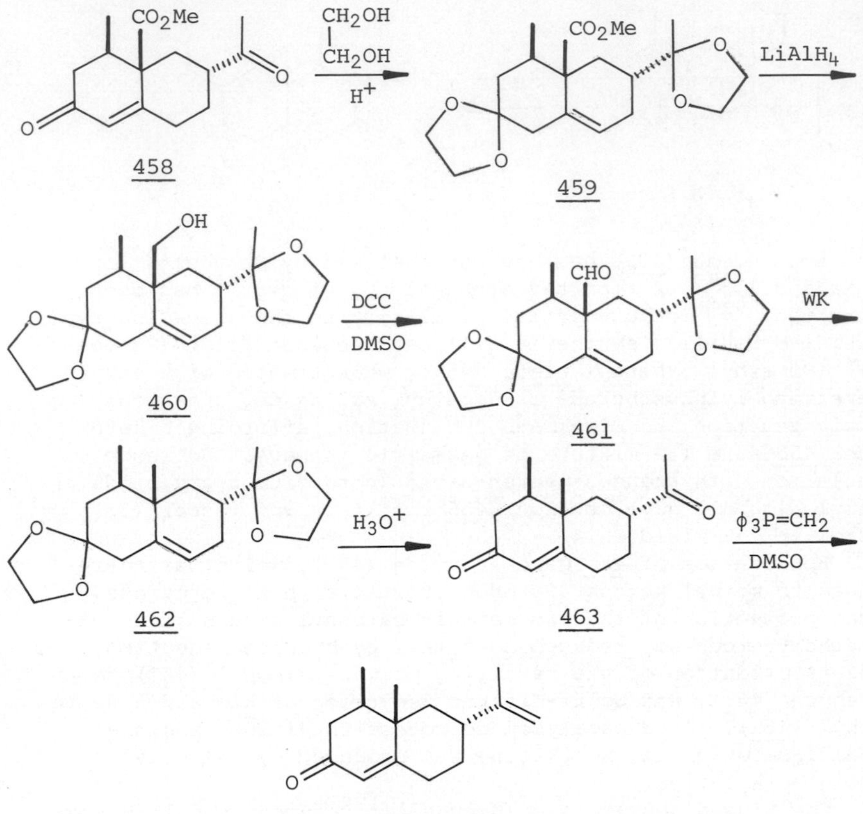

methylenation, followed by hydrolysis of the enol ether, afforded 4-isopropenylcyclohexanone (465), which was condensed
with ethyl formate to obtain the hydroxymethylene derivative
466. When 466 was treated with methyl iodide in acetone, enol
ether 467 and keto aldehydes 468 and 469 were formed in a ratio
of 27:55:18.

Treatment of the mixture of 467, 468, and 469 with acetone
in the presence of pyridine and acetic acid, followed by
methanolic KOH, gave a mixture of trienones 470 and 471 in a
ratio of 5:1. The problem now reduces to the stereospecific
introduction of the secondary methyl group of nootkatone into
dienone 470. Lithium dimethylcopper appears, at first examination, to be ideally suited for this purpose, since it is known
to add efficiently to α,β-unsaturated ketones, and since it
should select the less hindered arm of the cross-conjugated
dienone system in 470. However, axial alkylation, as is

usually encountered with such reactions, would lead to a *trans* disposition of the two methyl groups. In the event, dienone 470 did react smoothly with lithium dimethylcopper, affording 4-epinootkatone (472) in 85% yield. Compound 472 was dehydrogenated with dichlorodicyanoquinone (DCDQ) to dehydronootkatone (473), but this compound could not be selectively reduced at the 3,4-double bond.

In order to facilitate reduction of this linkage, keto ester 474 was prepared by condensing the mixture 467-469 with methyl acetoacetate (see Scheme 55). Compound 474 added lithium dimethylcopper cleanly, affording keto ester 475, which was again dehydrogenated with DCDQ. The 3,4-double bond was reduced smoothly with sodium borohydride in pyridine (Michael

Scheme 55.   Givaudan Synthesis of (±)-Nootkatone

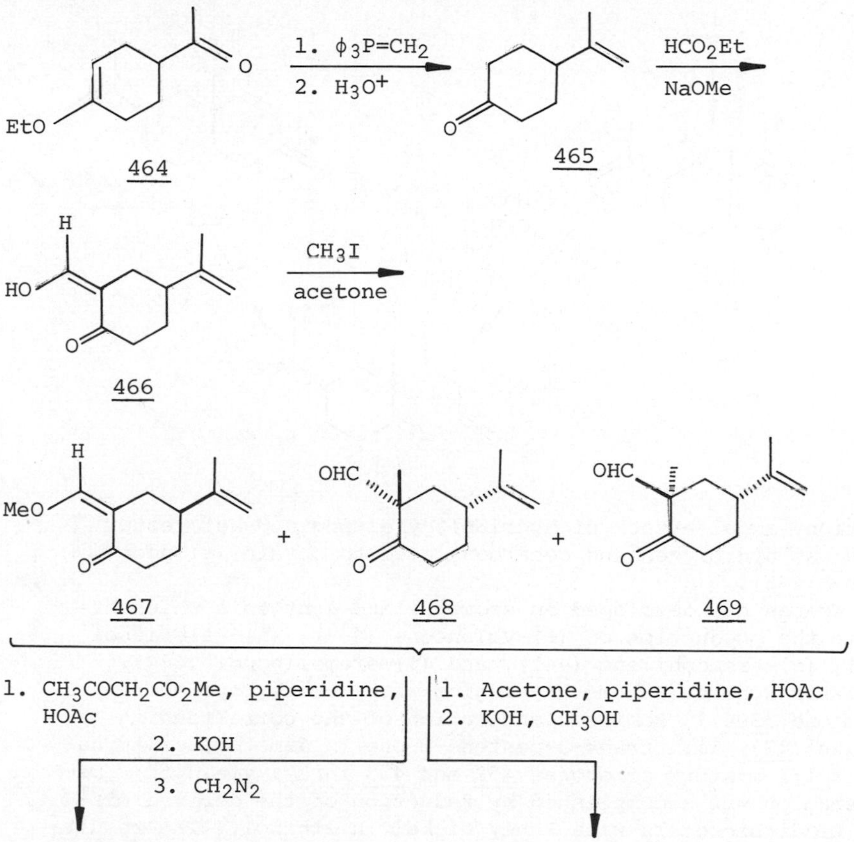

1. $\phi_3P=CH_2$
2. H_3O^+

HCO_2Et

NaOMe

464

465

CH_3I

acetone

466

MeO

467   +   468   +   469

1. $CH_3COCH_2CO_2Me$, piperidine, HOAc

2. KOH

3. CH_2N_2

1. Acetone, piperidine, HOAc
2. KOH, CH_3OH

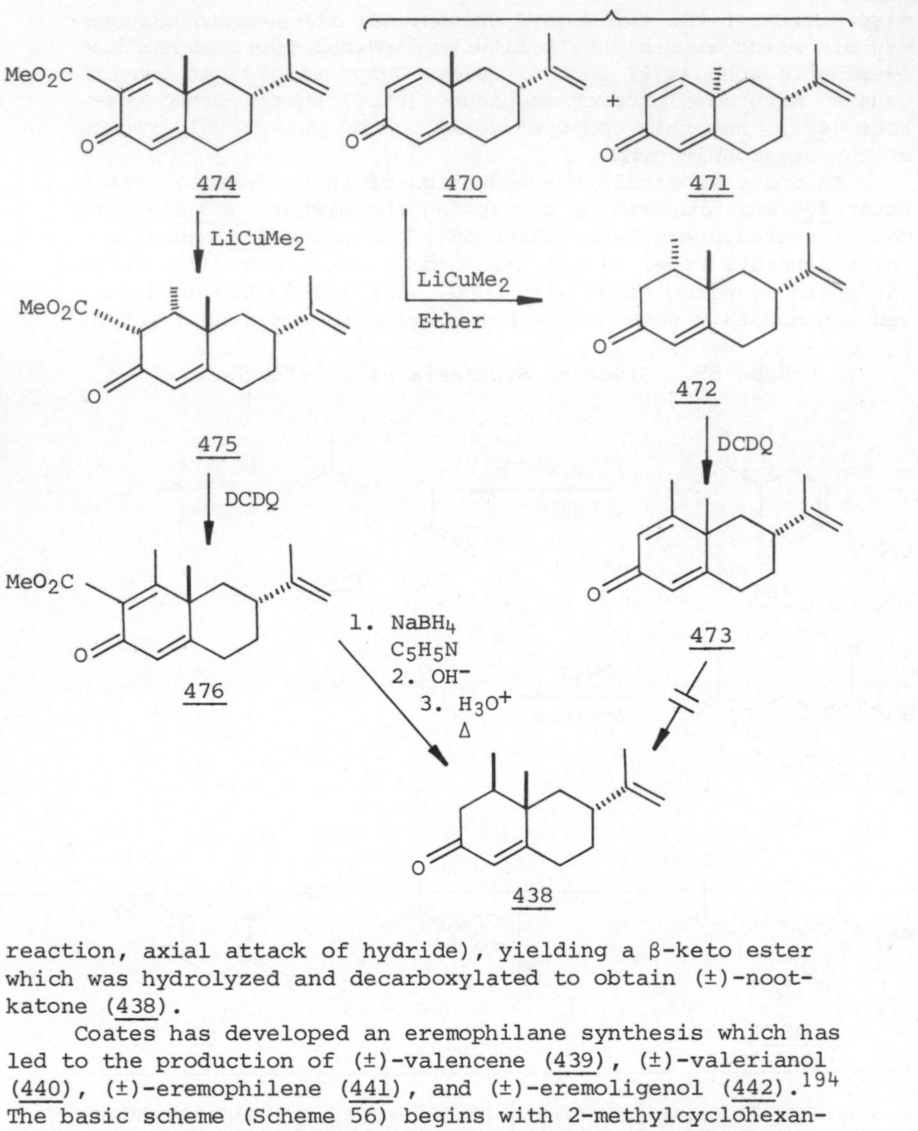

MeO₂C — 474

470 + 471

LiCuMe₂

MeO₂C — 475

472

LiCuMe₂ / Ether

476

473

438

1. NaBH₄ C₅H₅N
2. OH⁻
3. H₃O⁺ Δ

reaction, axial attack of hydride), yielding a β-keto ester which was hydrolyzed and decarboxylated to obtain (±)-nootkatone (438).

Coates has developed an eremophilane synthesis which has led to the production of (±)-valencene (439), (±)-valerianol (440), (±)-eremophilene (441), and (±)-eremoligenol (442).[194] The basic scheme (Scheme 56) begins with 2-methylcyclohexan-1,3-dione (364). Robinson annelation of the corresponding enamine (477) with *trans*-3-pentene-2-one in dimethylformamide gave a 1:1 mixture of enones 452 and 453 in 27% yield.[200] Deoxygenation was accomplished by reduction of the derived mixture of dithioketals with Raney nickel in ethanol. The resulting mixture of octalones 478 and 479 was separated by fractional distillation.[203]

Compound 479, which has the desired *cis* relationship between the two methyl groups, was carbethoxylated and the resulting β-keto ester was alkylated with chloromethyl methyl
ether in hexamethylphosphoramide. Under these conditions, O-
alkylation occurred to the complete exclusion of C-alkylation.
Reduction of enol ether 481 with lithium in ammonia afforded
the unsaturated ester 482 as the sole product in 61% yield.
That the less stable (axial) isomer was produced was shown by
sodium ethoxide-catalyzed equilibration of 482 to a more stable
isomer (483).

The reduction of 481 to 482 is an interesting and poten-
tially useful process which has been extensively studied by
Coates.[204] The probable course of events is outlined below:

481

482

The stereochemistry is determined in protonation of the final
enolate. Evidently, kinetic control occurs, with the proton
approaching from the less hindered face of the molecule.

Standard methods were then used to convert 482 into (±)-
eremoligenol (442) and (±)-eremophilene (441). Similarly, the
more stable ester (483) was converted into (±)-valerianol
(440) and (±)-valencene (439).

Piers' synthesis of fukinone (443) is outlined in Scheme
57.[195] The initial phase of the synthesis, synthesis of
octalone 489 was reported earlier.[205] The approach taken to
the *cis*-vicinal methyl substituents here is interesting and
certainly different from those previously discussed. Previous
work, both by Piers and by Ourisson,[206] had shown that 2,3-
dimethylcyclohexanone (498) may be used in a Robinson annelation

Scheme 56. Coates' Synthesis of (±)-Valencene, (±)-Valerianol,
            (±)-Eremophilene, and (±)-Eremoligenol

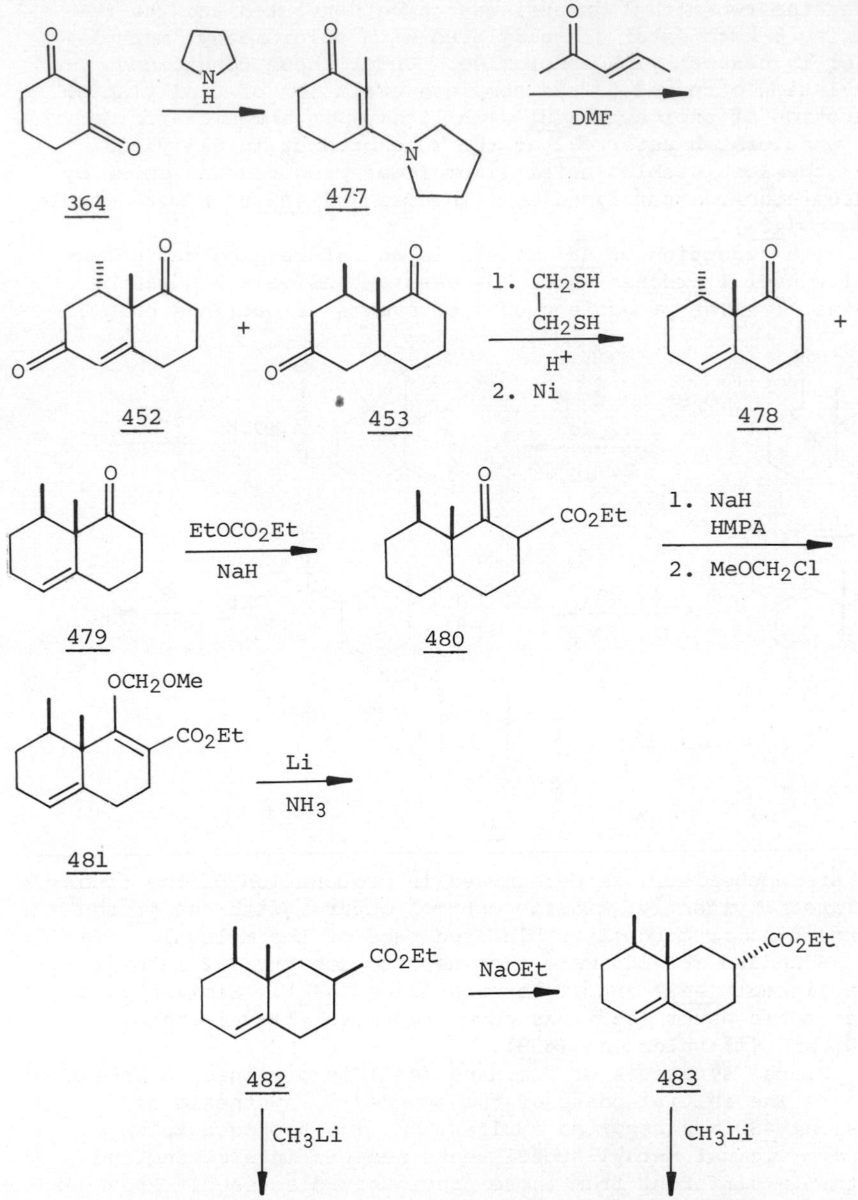

364

47·7

452   +   453

1. CH₂SH
   |
   CH₂SH
   ──────
   H⁺
2. Ni

478   +

479

EtOCO₂Et
────────
  NaH

480

1. NaH
   HMPA
─────────
2. MeOCH₂Cl

481

Li
────
NH₃

482   NaOEt   483

CH₃Li   CH₃Li

370

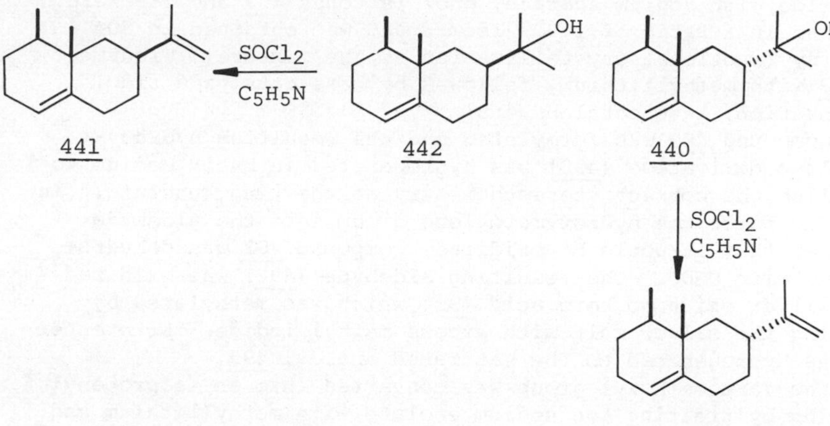

441   442   440

439

with methyl vinyl ketone. However, Piers found that the yield
of adduct is low (15%) and that a 3:2 ratio of stereoisomers
499 and 500 is obtained.

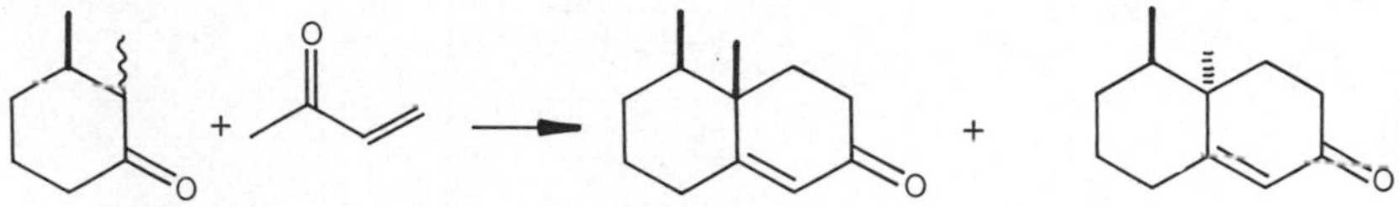

498   499   500

However Piers had further noted that alkylation of 2,3-
dimethyl-6-n-butylthiomethylenecyclohexanone (484) with
methallyl chloride produced a mixture of ketones 501 and 502
in high yield in a ratio of 4:1. This finding suggested an
obvious method for circumventing the current problem.

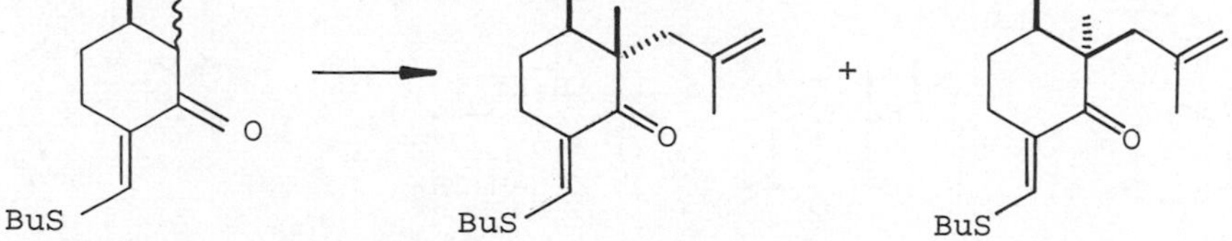

484   501   502

Ketone 484 was alkylated with ethyl 3-bromopropionate to
obtain a mixture of keto esters (485) which was hydrolyzed to a
keto acid mixture (486). When 486 was refluxed in acetic

anhydride with sodium acetate, enol lactones 487 and 488 were
obtained in a ratio of 1:9. Isomer 488 was obtained in 80%
yield by fractional crystallization of the mixture. Treatment
of 488 with methyllithium, followed by base-catalyzed aldol
condensation, gave octalone 489.

Compound 489 was formylated and the resulting hydroxymethylene derivative (490) was hydrogenated in basic medium to
establish the correct stereochemistry at the ring juncture. In
order to force the hydroxymethylene group into the aldehyde
form, so that it could be oxidized, compound 491 was dehydrogenated with DCDQ. The resulting aldehyde (492) was oxidized
with silver oxide to keto acid 493, which was methylated by
treating the silver salt with excess methyl iodide. Keto ester
494 was hydrogenated to the saturated analog, 495.

The carbomethoxyl group was converted into an isopropenyl
grouping by treating the sodium enolate with methyllithium and
dehydrating the resulting keto alcohol with thionyl chloride

Scheme 57. Piers' Synthesis of (±)-Fukinone

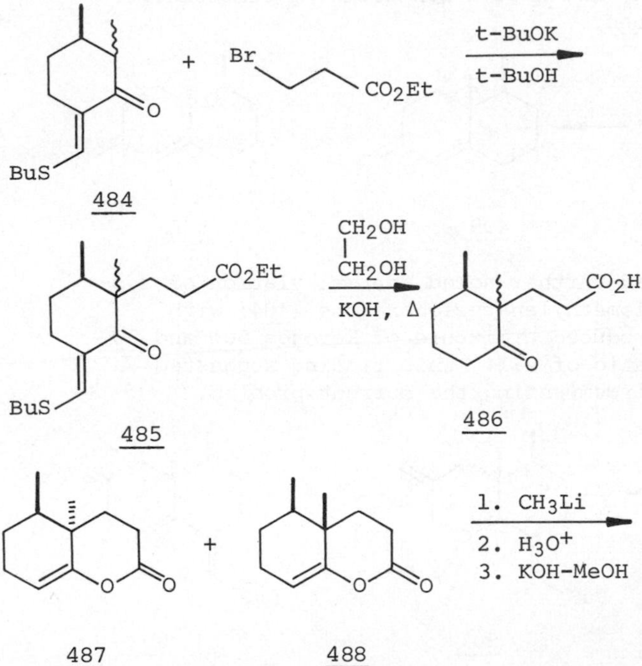

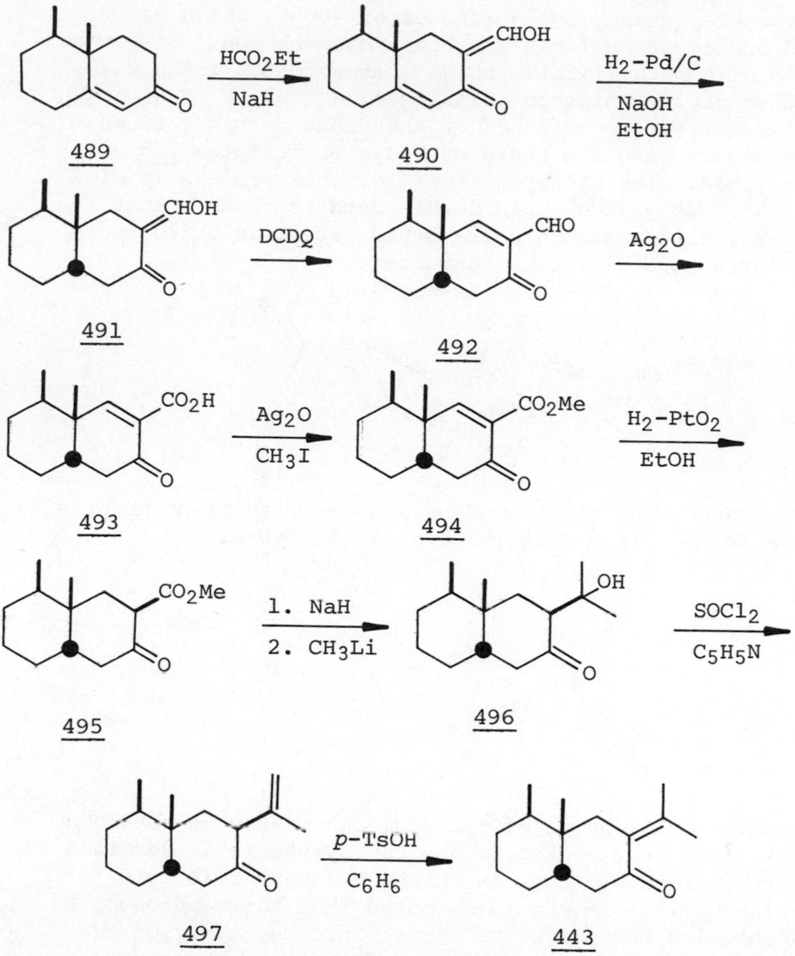

in pyridine. The product, isofukinone (497), was isomerized to (±)-fukinone (443) by p-toluenesulfonic acid in benzene.

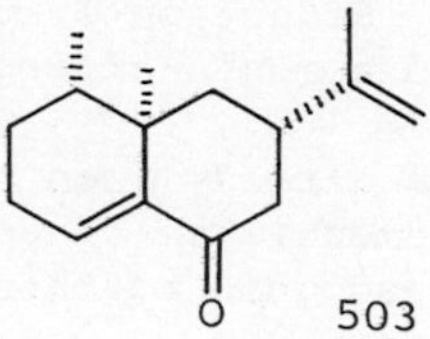

Although eremophilone (503) has not been successfully synthesized, Brown has reported a synthesis of its tetrahydro derivative (Scheme 58).[196] Alkylation of keto ester 504 with *trans*-5-bromo-2-pentene gave a keto ester (505), which was hydrolyzed and decarboyxlated to obtain dienone 506. Compound 506 reacted with methyllithium to give an alcohol, which was dehydrated to yield a mixture of trienes.

π-Cyclization, accomplished by anhydrous formic acid at room temperature, gave a mixture of bicyclic formates 508 and 509 in 67% yield. The stereochemistry of this process is of interest. Since the side-chain double bond is *trans*, either 508 or 509 may arise through a transition state in which the incipient A-ring has chairlike character:

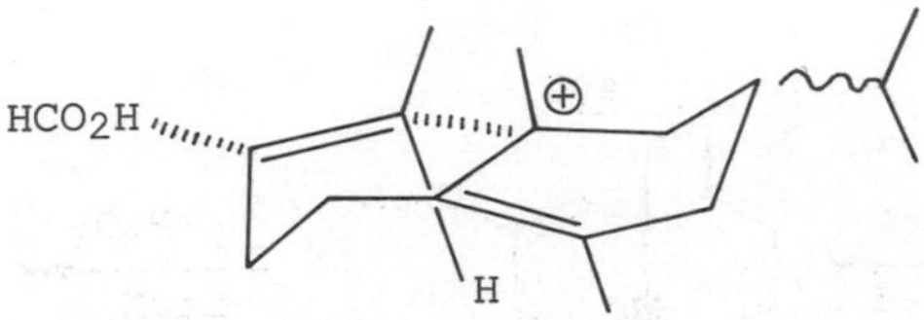

On the other hand, the corresponding isomers with *trans* methyls can only be formed via a boatlike transition state:

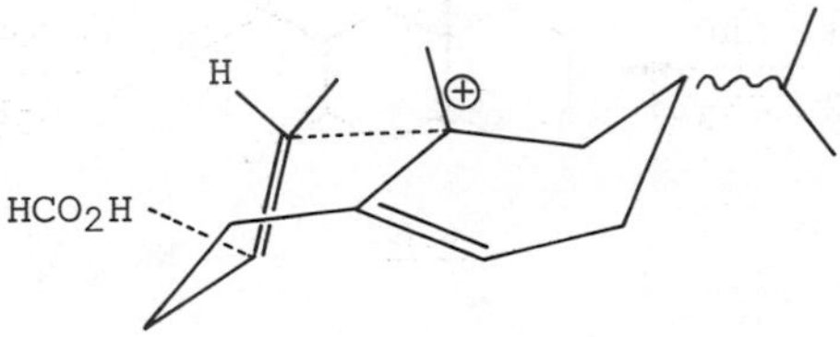

In any event, should the undesired stereochemistry be obtained, the situation may be remedied, since the synthesis is destined to pass through intermediates in which this center is epimerizable (see 512). It should be noted that this approach, while guaranteeing the proper relative stereochemistry of the two methyl groups, offers no control over the three-carbon side chain.

The mixture of 508 and 509 was reductively cleaved to a mixture of alcohols 510 and 511, which were isolated in a ratio of 3:2. The minor isomer was oxidized and the resulting ketone 512 reduced to octalin 513. Photo-oxygenation gave a mixture of diastereomeric allylic alcohols 514, which was carefully ozonized to obtain ketols 515. Reduction of the derived acetate 516 with calcium in liquid ammonia afforded (±)-tetrahydroeremophilone (444).

Eremophilene (441), as has already been noted, is one of the simplest members of the eremophilane class. Originally, structure 445 was incorrectly assigned to this substance.[207]

Scheme 58. Brown's Synthesis of (±)-Tetrahydroeremophilone

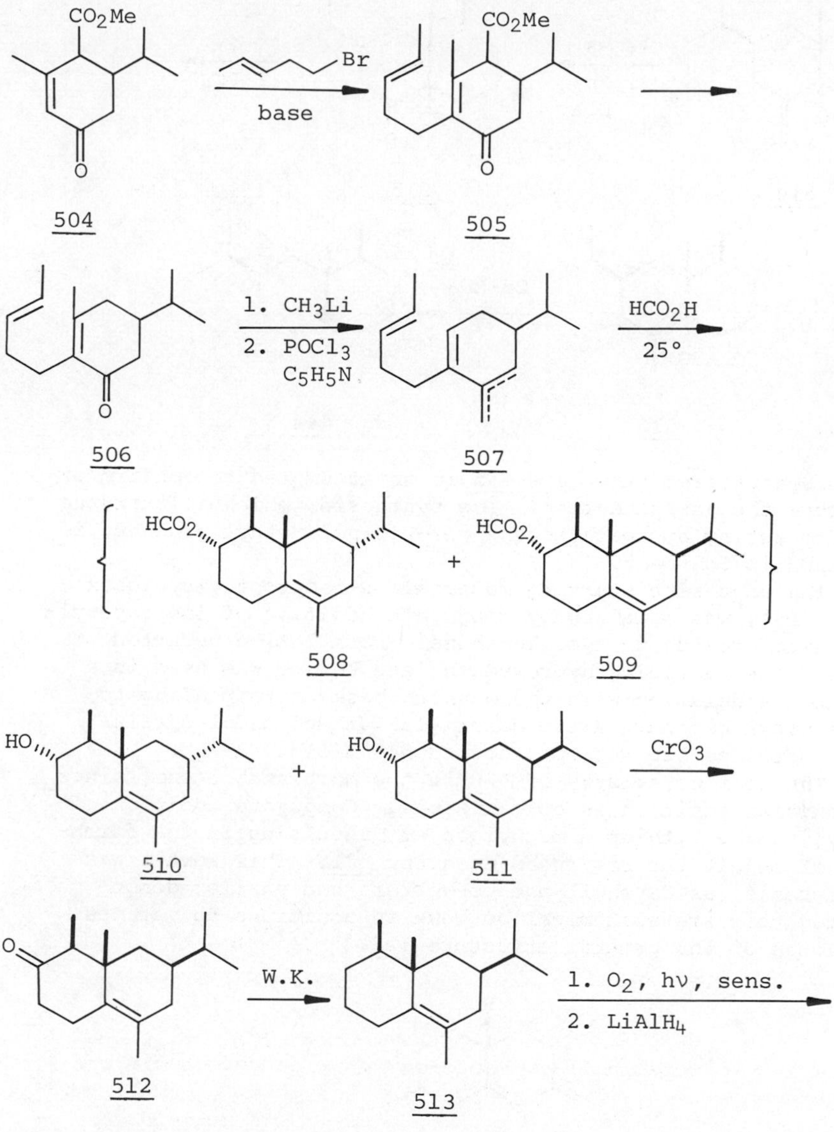

504

505

506

1. CH₃Li
2. POCl₃
   C₅H₅N

507

HCO₂H
25°

508 + 509

510 + 511

CrO₃

512

W.K.

513

1. O₂, hν, sens.
2. LiAlH₄

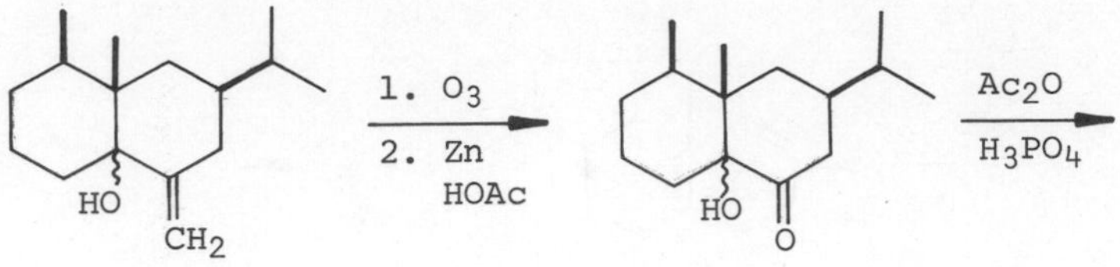

HO
CH₂

514

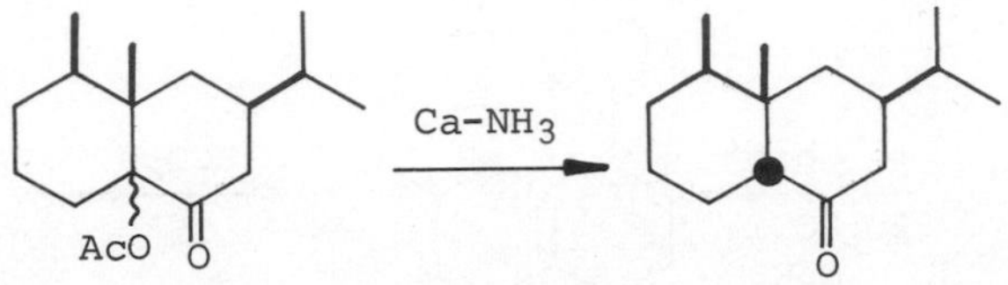

1. O₃
2. Zn
   HOAc

HO
O

515

Ac₂O
H₃PO₄

AcO O

516

Ca-NH₃

O

444

Piers synthesized structure 445 in an attempted to confirm or disprove the assignment.[197] The synthesis, which illustrates an interesting approach to the cis-vicinal methyl problem, is outlined in Scheme 59.

The requisite starting material, 3-isopropenylcyclohexanone (518), was obtained by conjugate addition of isopropenylmagnesium bromide to cyclohexenone. Formylation occurred at C-6 and the resulting hydroxymethylene ketone was used in a Robinson annelation with the Mannich base corresponding to ethyl vinyl ketone. After deformylation and aldol cyclization, octalone 520 was obtained in 64% yield.

The stereochemistry of 520 is the more stable one, since the angular position is epimerizable. Conjugate addition of methyl, using lithium dimethylcopper, resulting in the formation of solely the cis-fused decalone 521. This result was predictable, as Marshall and co-workers had earlier demonstrated this stereochemical outcome in additions to various octalones of the general structure 522.[182b,208]

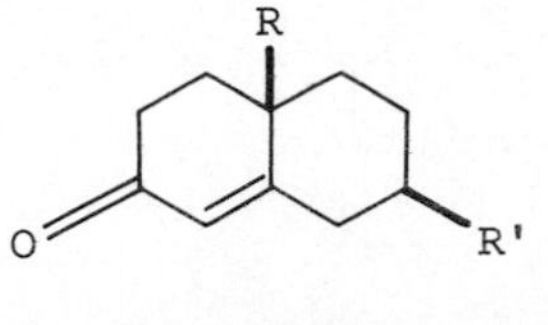

R

O   R'

522

Finally, Bamford-Stevens reaction on decalone 521 yielded the desired compound, (±)-eremophil-3,11-diene (445), which was found to be different from the natural substance.

Scheme 59. Piers' Synthesis of (±)-Eremophil-3,11-diene

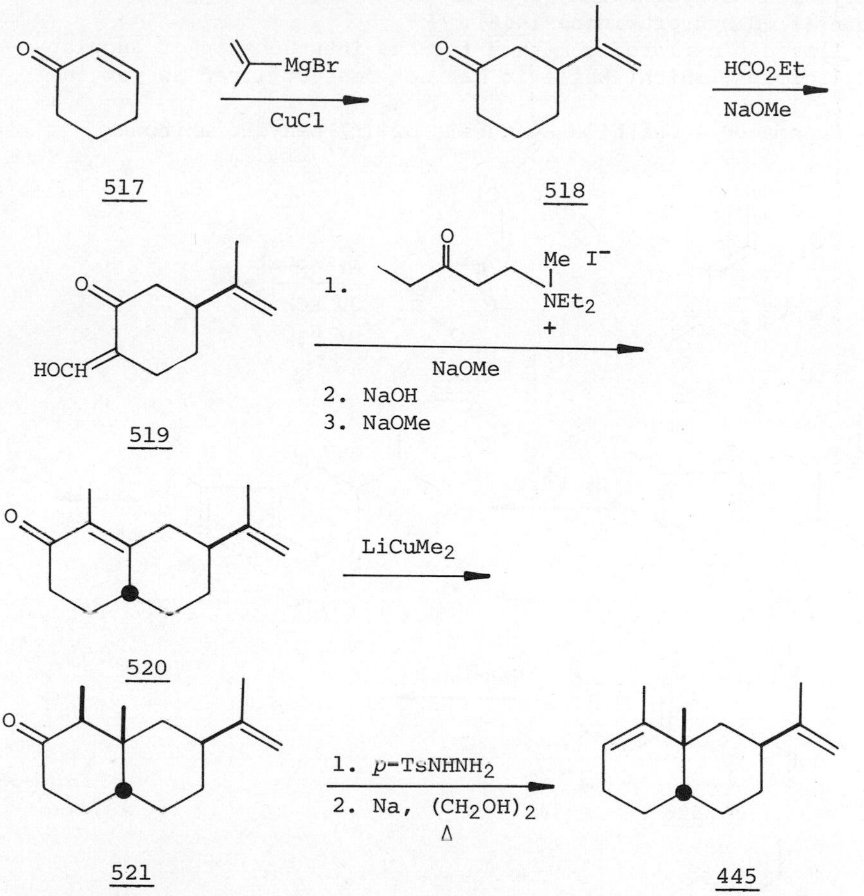

Dehydrofukinone (446) was prepared by Okashi (Scheme 60)[198] by a route involving Stork's isoxazole annelation method.[209] Hagemann's ester (243) was alkylated with 4-chloromethyl-3,5-dimethylisoxazole (523). After hydrolysis and decarboxylation, enone 524 was obtained. Hydrogenation yielded saturated ketone 525, which was protected at C-6 by the isopropoxymethylene group. Alkylation with methyl iodide gave 527, which was deprotected by acid hydrolysis. The cis isomer 528 was isolated in 40% yield after chromatography. Alkylation of the isoxazole nitrogen, followed by ring cleavage, hydrolysis and aldol cyclization, gave acetyl octalone 529, which exists in the enolic form indicated. This substance was

converted into an enol ether (530), which was treated successively with methyllithium, aqueous acid, and POCl$_3$-pyridine to
obtain (±)-dehydrofukinone (446).

Sims has explored a method for the introduction of angular
methyl groups, which, while it has not been utilized as yet in

Scheme 60.   Ohashi's Synthesis of (±)-Dehydrofukinone

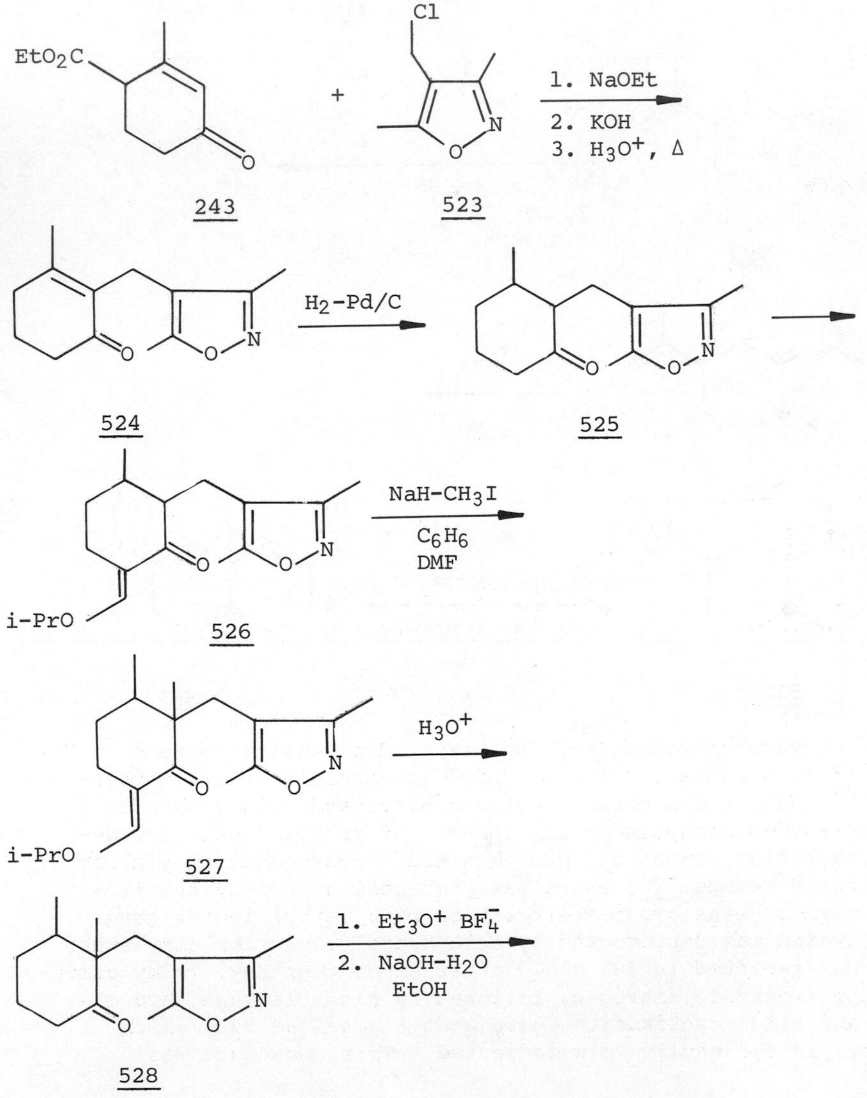

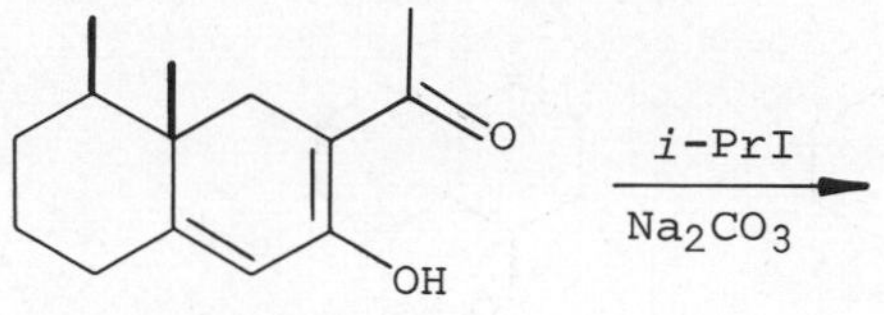
O
OH
i-PrI
Na2CO3

529

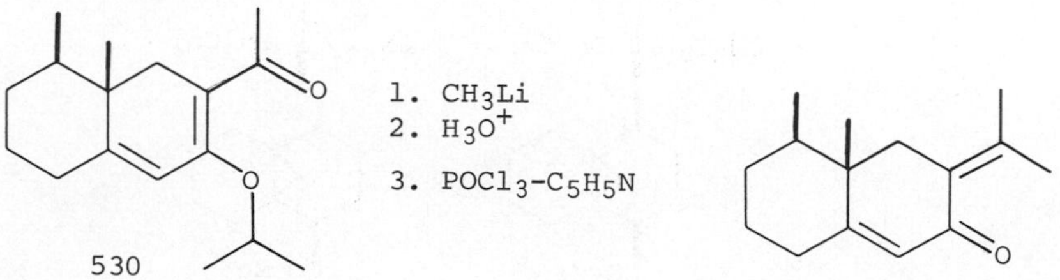
O
O
1. CH3Li
2. H3O+
3. POCl3-C5H5N
O

530

446

a natural product synthesis, offers promise as an alternative solution to the *cis*-vicinal methyl group problem.[210] Scheme 61 outlines the synthesis of octalins 540 and 541 by the Sims method. 4-methyl-1-naphthoic acid (531) was converted into the isomeric tetrahydro acids 532 and 533 (532:533 = 2:1) by Birch reduction. The isomeric acids were separated by fractional crystallization. Hydrogenation of the corresponding methyl esters 534 and 535 gave octalins 536 and 537, respectively. Simmons-Smith methylenation gave, in each case, a

Scheme 61. Sims' Method of Angular Methyl Indroduction

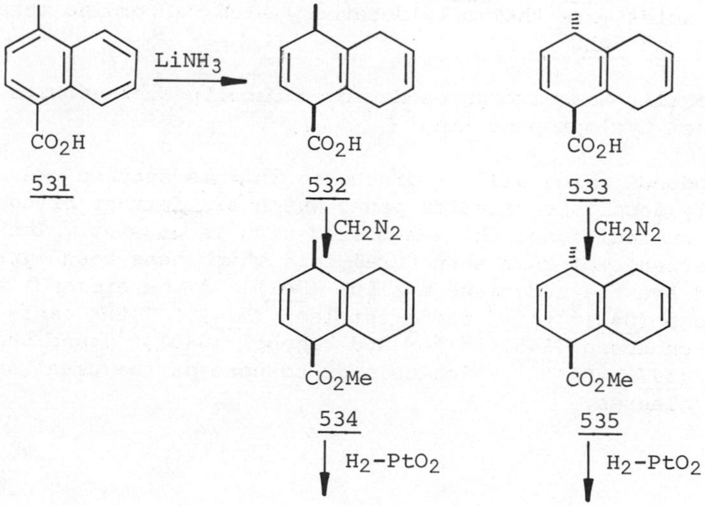
LiNH3
CO2H
531
CO2H
532
CO2H
533
CH2N2
CH2N2
CO2Me
534
CO2Me
535
H2-PtO2
H2-PtO2

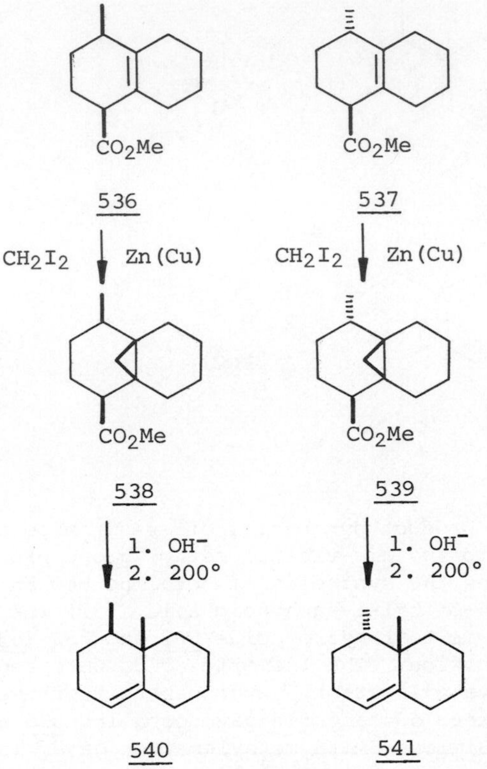
CO2Me
CO2Me
536
537
CH2I2
Zn(Cu)
CH2I2
Zn(Cu)
CO2Me
CO2Me
538
539
1. OH⁻
2. 200°
1. OH⁻
2. 200°
540
541

single isomer (538 and 539). After saponification, the resulting acids were thermally decarboxylated, affording octalins 540 and 541.

S. Tricyclic Sesquiterpenes Having a Decalin Nucleus with an Additional Cyclopropane Ring

The compounds which will be discussed in this section are basically decalinic sesquiterpenes which are further bridged so that an additional three-membered ring is present. The sesquiterpenes of this structural type which have been synthesized are the eudesmane maaliol (542),[211] the eremophilanes aristolone (543)[212,213] and aristolene (544),[214] the cadinanes β-cubebene (545)[215,216] and cubebol (546),[216] and thujopsene (547),[217,218] which belongs to none of the usual decalinic classes.

542   543   544

545   546   547

The synthetic problems associated with these compounds closely parallel the problems encountered in the respective bicyclic series, complicated by the necessity of introducing the cyclopropane ring.

Büchi's synthesis of maaliol (542) is outlined in Schemes 62 and 63.[218] The starting point was 7-epi-cyperone (31, see p. 284), which smoothly added HBr in acetic acid to yield the bromide 548. Base catalyzed dehydrobromination then afforded the tricyclic enone 549 in 62% overall yield. With the basic carbon skeleton in hand, the problem now becomes one of redox chemistry. The carbonyl group must be removed, and, in a formal sense, water must be added to the double bond in one specific orientation and *cis* to the cyclopropane ring. As will be seen in the sequel, these problems were not adequately solved, although two different routes leading from enone 549 to maaliol in poor yield were developed.

Wolff-Kishner reduction of enone 549 led to a mixture of olefins 550 and 551 (550:551 = 15:85) in 64% yield. Selenium dioxide oxidation of the mixture gave unsaturated aldehyde 552 in 6% yield. A second Wolff-Kishner reduction on 552 gave a mixture of olefins 550, 551, and 553. Osmium tetroxide oxidation of this mixture gave a mixture of diols from which β-maalidiol (554) was isolated in 20% yield. Reduction of the derived mono-p-toluenesulfonate (555) with lithium aluminum hydride gave, probably via an epoxide intermediate, maaliol (542).

Scheme 62. Büchi's Synthesis of Maaliol--First Variation

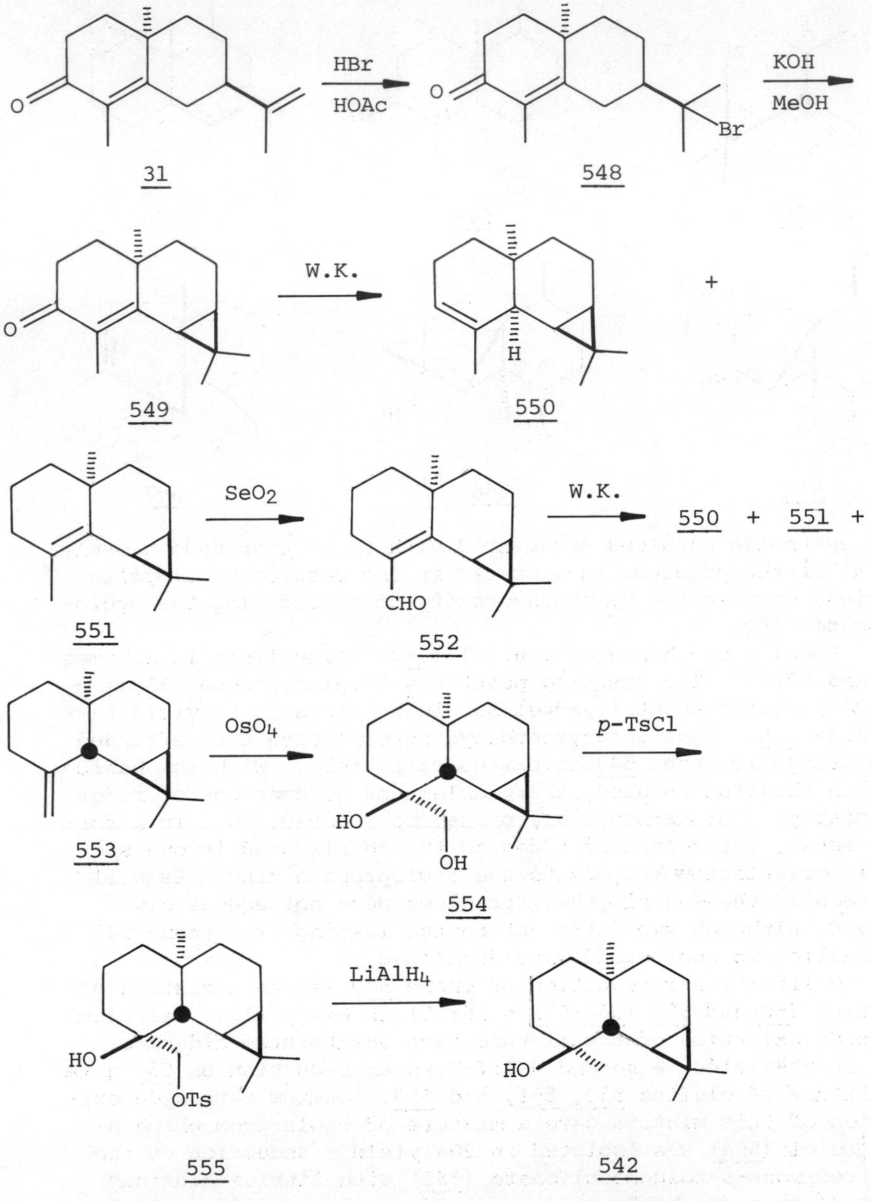

An alternative route (Scheme 63) begins with lithium-ethanol-ammonia reduction of the tricyclic enone 549. From a complex mixture of products, alcohol 556 was isolated in 28% yield. The derived acetate 557 was pyrolyzed, whereupon a 1:1 mixture of olefins 558 and 559 was obtained in 74% yield. Osmium tetroxide oxidation gave iso-α-maalidiol (560) in 5% yield and this diol was oxidized by Sarrett's procedure to the

Scheme 63.   Büchi's Synthesis of Maaliol--Second Variation

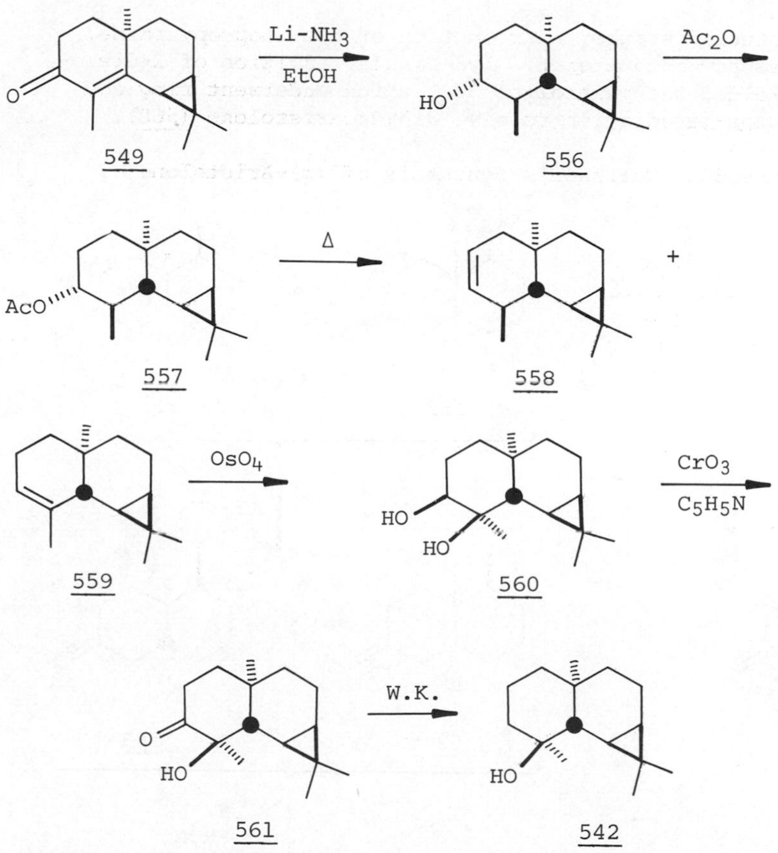

acyloin 561 in 25% yield. Wolff-Kishner reduction of 561 gave maaliol in 18% yield.

The first reported aristolone (543) synthesis was that of Ourisson, which is outlined in Scheme 64.[212] The plan was to introduce the isopropylidine bridge by addition of a carbene or carbenoid species to an octalone such as 566.

2,3-Dimethylcyclohexanone reacted with methyl vinyl ketone
to give isomers 499 and 500 in a ratio of 3:2 (see Piers, p.
371 for yield). Lithium-ammonia reduction of the mixture gave
a mixture of decalones 562 and 563. Although this double bond
is present in the eventual product, it was temporarily saturated, apparently to avoid the complication to selective reaction in a dienone such as 569.* After bromination of the mixture, the crystalline bromo ketone 564 was separated from the
undesired isomer 565. Dehydrobromination then afforded octalone 566.

The crucial stages, introduction of the isopropylidine
group, were now encountered. 1,3-Dipolar addition of 2-diazo-
propane yielded the pyrazoline 567, which underwent ring contraction upon irradiation to give dihydroaristolone (568).

Scheme 64. Ourisson's Synthesis of (±)-Aristolone

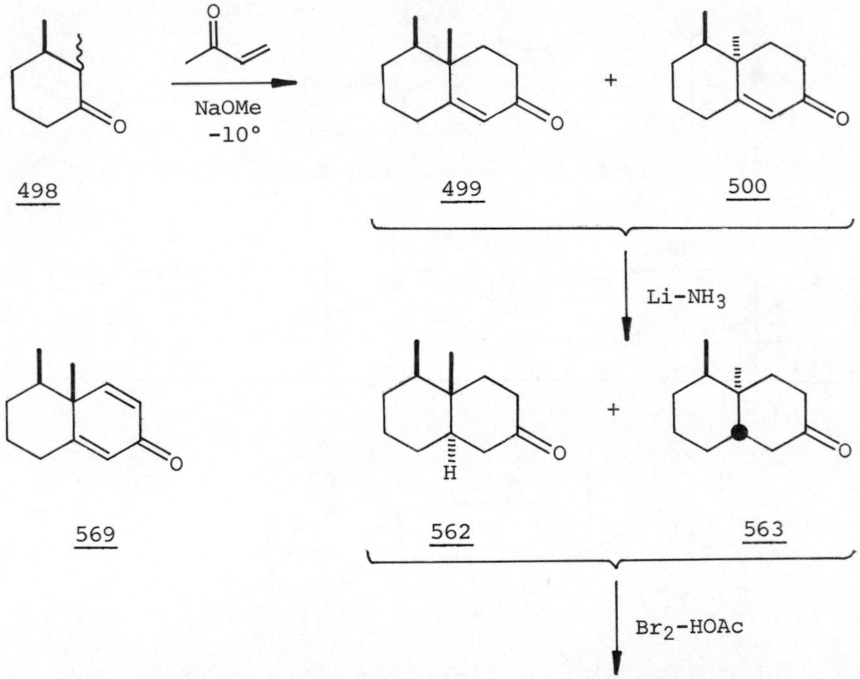

498

NaOMe
-10°

499   +   500

Li-NH₃

569   562   H   +   563

Br₂-HOAc

*However, such selective reaction would probably succeed; see
Scheme 55.

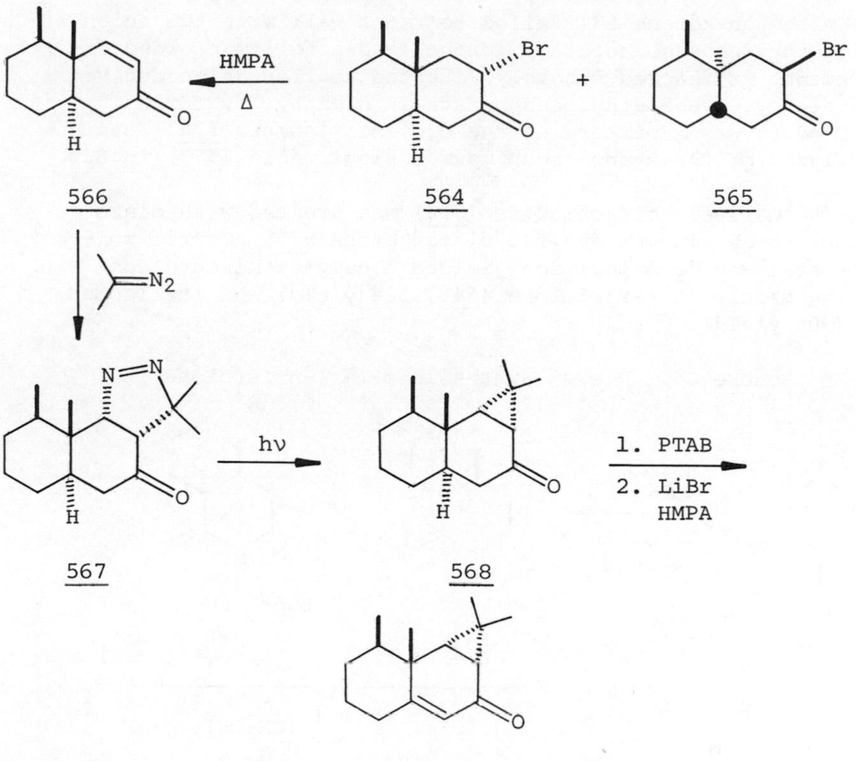

566   564   565

567   568

543

Bromination and dehydrobromination of this substance led to (±)-aristolone (543).

Piers' synthesis of aristolone (Scheme 65) incorporates several novel features. The cis-vicinal methyl groups were established by alkylating the 6-butylthiomethylene derivative of 2,3-dimethylcyclohexanone (484) with methylallyl chloride (see p. 371). After deprotection of the resulting diastereomeric mixture, isomers 569 and 570 were obtained in a ratio of 4:1. Equilibration of 569 with p-toluenesulfonic acid in benzene yielded a complex mixture from which isomer 571 was isolated by fractional distillation in 22% yield.*

The synthetic plan called for the introduction of a three-carbon diazo ketone chain at the carbonyl position. Insertion of the derived α-keto carbene into the methylpropenyl double bond would then yield the aristolone skeleton, in a sequence reminiscent of Büchi's thujopsene synthesis (Scheme 70).

*See Ref. 213, footnote 2 for an improvement in this process.

Although ketone 571 failed to react well with the anion
of triethyl phosphonoacetate, probably due to its crowded environment, it reacted smoothly with the smaller anion derived
from diethyl cyanomethylphosphonate in dimethyl sulfoxide.
The product was a mixture of the olefinic isomers 572. Basic
hydrolysis of the cyano group gave a single acid (573) in 82%
yield.

The derived acid chloride (574) was treated with diazomethane to obtain the desired diazo ketone 575. Cupric sulfate-catalyzed decomposition yielded a complex mixture containing mainly (±)-aristolone (543, 30% yield) and its isomer
576 (20% yield).

Scheme 65. Piers' Synthesis of (±)-Aristolone

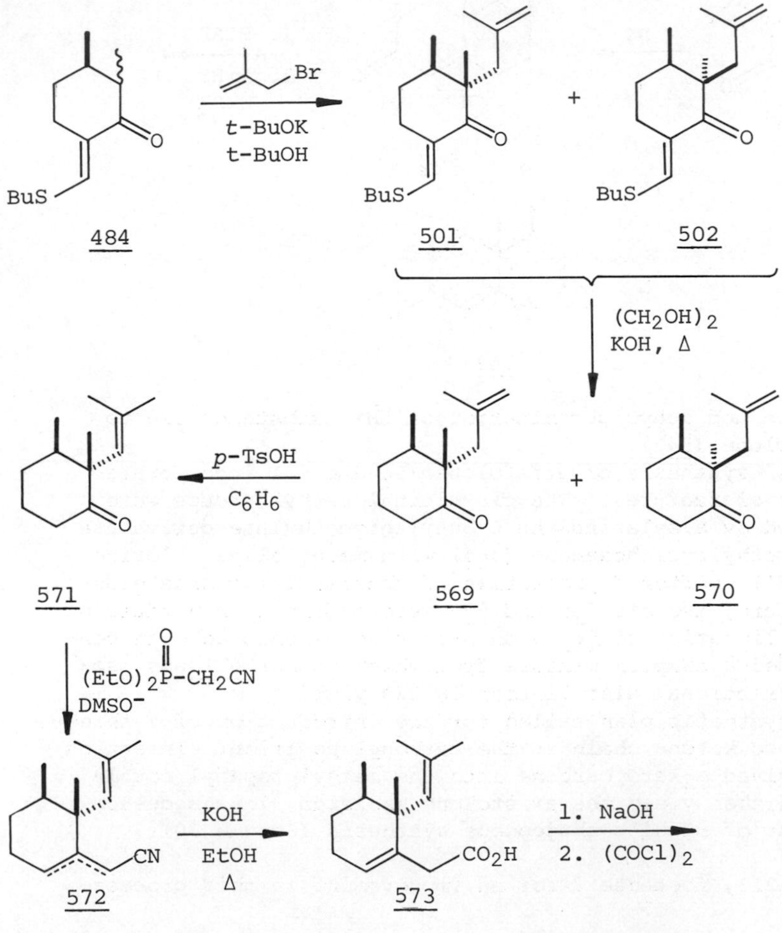

574   575

543   +   576

For the synthesis of (±)-aristolene (calarene), 544, Coates made use of unsaturated keto ester 480, a crucial intermediate in the synthesis of several bicyclic eremophilanes (see Scheme 66).[214] The synthetic plan called for introduction of the isopropylidine group by pyrolysis of a pyrazoline ring, first employed by Kishner and Zavodorsky in the synthesis of carane, and later used by Ourisson in the synthesis of (±)-aristolone (Scheme 64).

Keto ester 480 reacted with methyllithium to give a hydroxy ketone (577), which was dehydrated by methanolic HCl to obtain dienone 578. This substance reacted with hydrazine in ethanol to yield the desired pyrazoline 579. Decomposition was effected by heating with powdered potassium hydroxide at 250°. The product was nearly pure (±)-aristolene (544). A similar decomplisition of pyrazoline 580 afforded two stereo-isomers, 581 and 582 in a ratio of 3:1. The stereochemistry

580   581   +   582

of the cyclopropane ring probably reflects the thermodynamic stability of the pyrazoline ring. In both cases, the major decomposition product has stereochemistry corresponding to that of the more stable pyrazoline isomer. However, the production of substantial amounts of 582 in the decomposition of 580 indicates that, in some unknown manner, the stereochemistry of the

Scheme 66.   Coates' Synthesis of (±)-Aristolene (Calarene)

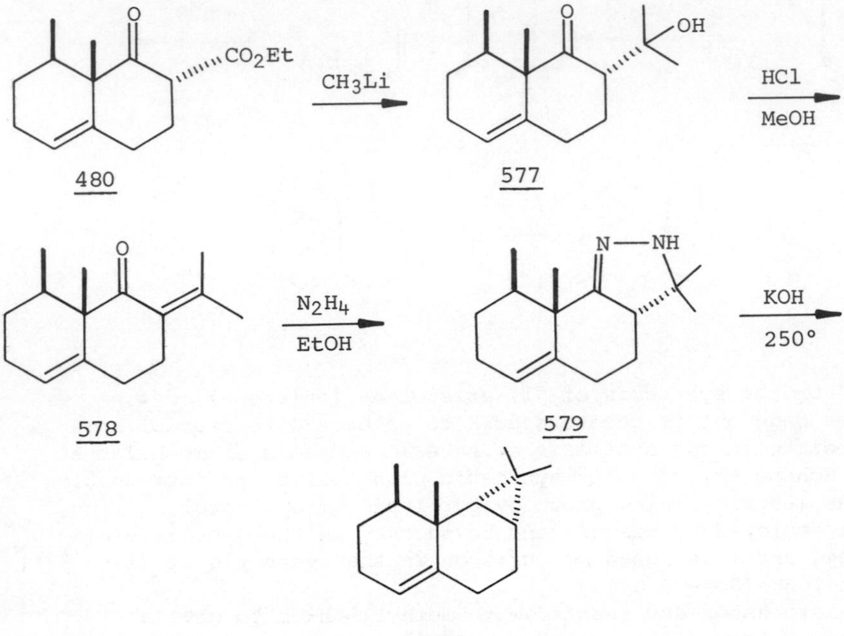

480   577

578   579

544

secondary methyl group influences the stereochemical outcome
of the reaction.

The structures of β-cubebene and cubebol (545 and 546) are
interesting in that the cyclopropane ring is found *within* the
periphery of the basic cadinane skeleton. The bicyclo[3.1.0]-
hexane arrangement found in the B and C rings immediately suggests a synthetic route. Stork had previously explored the
intramolecular insertion of α-keto carbenes into olefinic
linkages as a means of generating such structures.[219] The
crucial intermediate is thus unsaturated diazoketone 583.

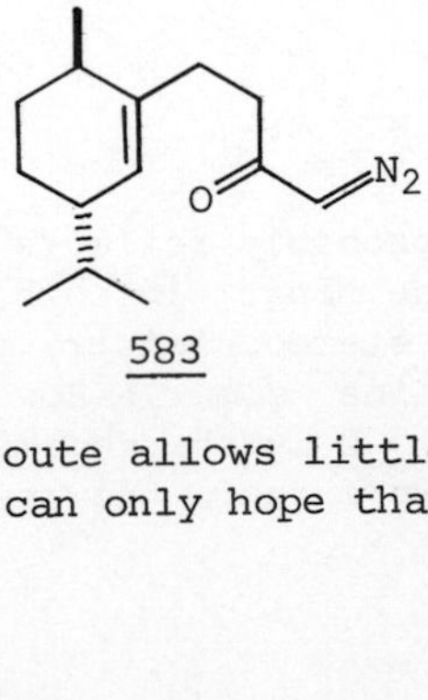

583

Unfortunately, such a route allows little chance for stereochemical control. One can only hope that the carbene will add

predominately *trans* to the larger isopropyl group, in which
case the major cyclopropyl ketone would have the proper stereochemistry.

Piers' synthesis of (±)-β-cubebene is outlined in Scheme
67.[215] A mixture of (±)-menthone and (±)-isomenthone (584)
was formylated and the resulting hydroxymethylene derivative
was converted into a mixture of *n*-butylthiomethylene compounds
585. Sodium borohydride reduction of the ketone, followed by
acidic hydrolysis, gave a mixture of aldehydes 586. Reduction
with borohydride yielded allylic alcohols 587. While the isomer ratio at this point is not stated, one expects a predominance of the isomer having the two alkyl groups *trans*.

Alcohol 587 was converted into bromide 588 and this bromide was used to alkylate carbethoxymethylenetriphenylphos-
phorane. The resulting phosphonium salt was subjected to
vigorous hydrolysis to obtain a mixture of unsaturated acids
589. The desired diazoketone, as a mixture of *cis* and *trans*
isomers (590), was prepared from 589 in the normal manner.
Cupric sulfate-catalyzed decomposition gave three isomeric
cyclopropyl ketones 591, 592, and 593, in a ratio of 2:3:5.
The minor isomer was converted by a Wittig reaction into (±)-
β-cubebene (545). Although the stereochemistry of isomers
592 and 593 remains obscure, the major isomer should be

Scheme 67. Piers' Synthesis of (±)-β-Cubebene

584

585

586

587

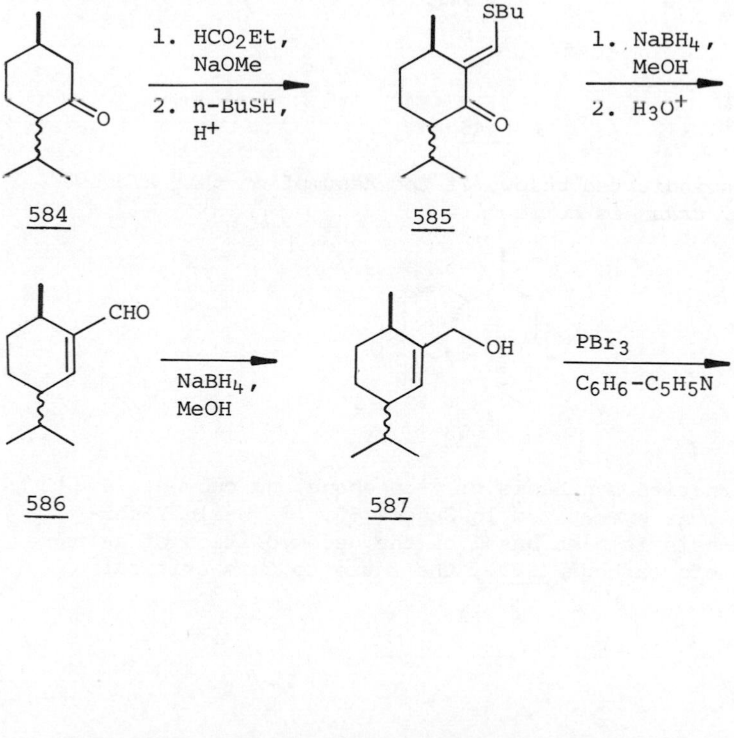

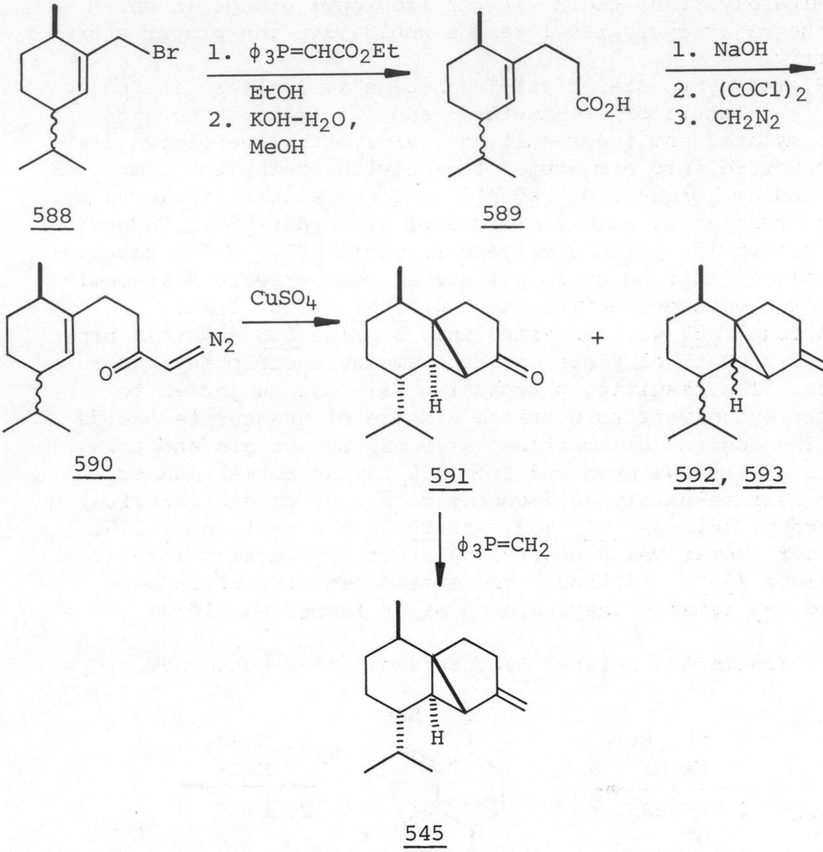

588   589

590   591   +   592, 593

$\phi_3 P=CH_2$

545

formulated as indicated below, if the assumption that alcohol 587 is mostly *trans* is correct.

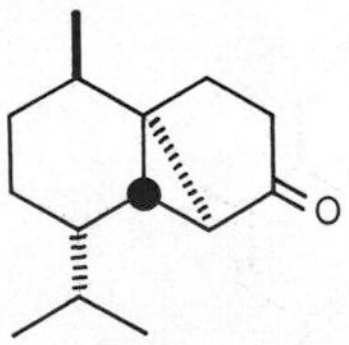

593 ?

An alternative synthesis of β-cubebene and cubebol, due to Yoshikoshi, is summarized in Scheme 68.[216] While Yoshikoshi's synthesis is also based on the decomposition of an unsaturated α-keto carbene (599), the route to this critical

intermediate is quite different. The starting point was (-)-
trans-caran-2-one (594), which reacted with allylmagnesium
bromide to give alcohol 595 (stereochemistry unspecified).
Hydroboration-oxidation gave a diol (596), which was further
oxidized to the spiro-lactone 597. Pyrolysis of this substance, in the presence of pyridine, yielded the dienic acid
598 in which the methyl and isopropenyl groups are presumably
trans, as in the cubebenes.

    After preparation of the diazoketone 599, copper catalyzed
decomposition gave tricyclic ketones 600, 601, and 602 in 11,
13, and 1% overall yields, based on spiro-lactone 597. Isomer 600 was hydrogenated to the nor-ketone 591, which gave β-
cubebene (545) upon treatment with methylenetriphenylphosphorane. Addition of methylmagnesium iodide to 591 gave cubebol

Scheme 68. Yoshikoshi's Synthesis of β-Cubebene and Cubebol

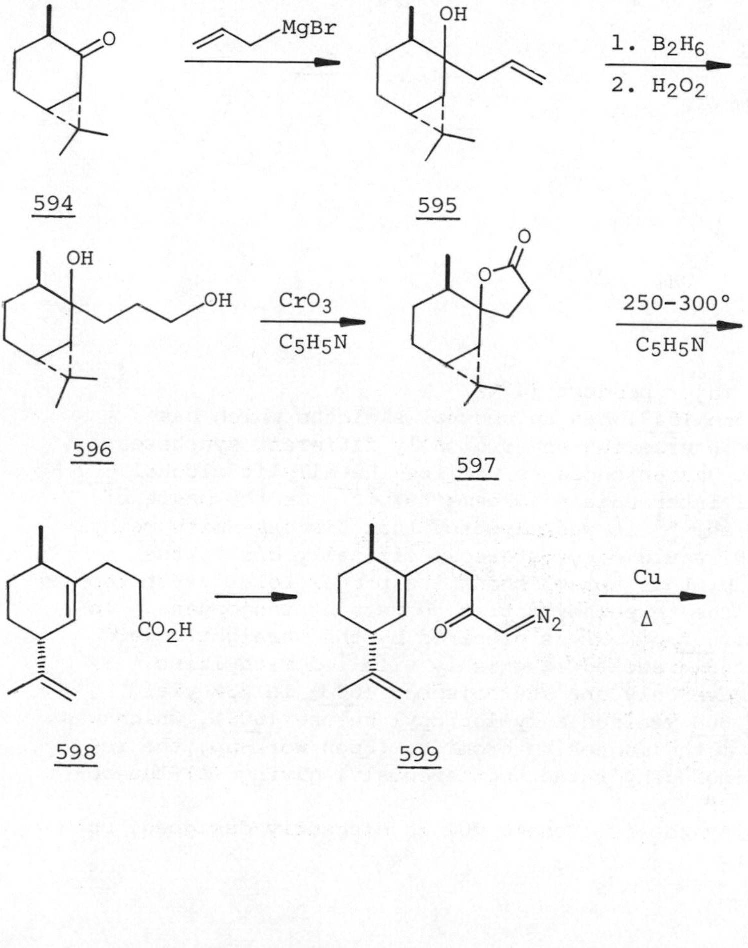

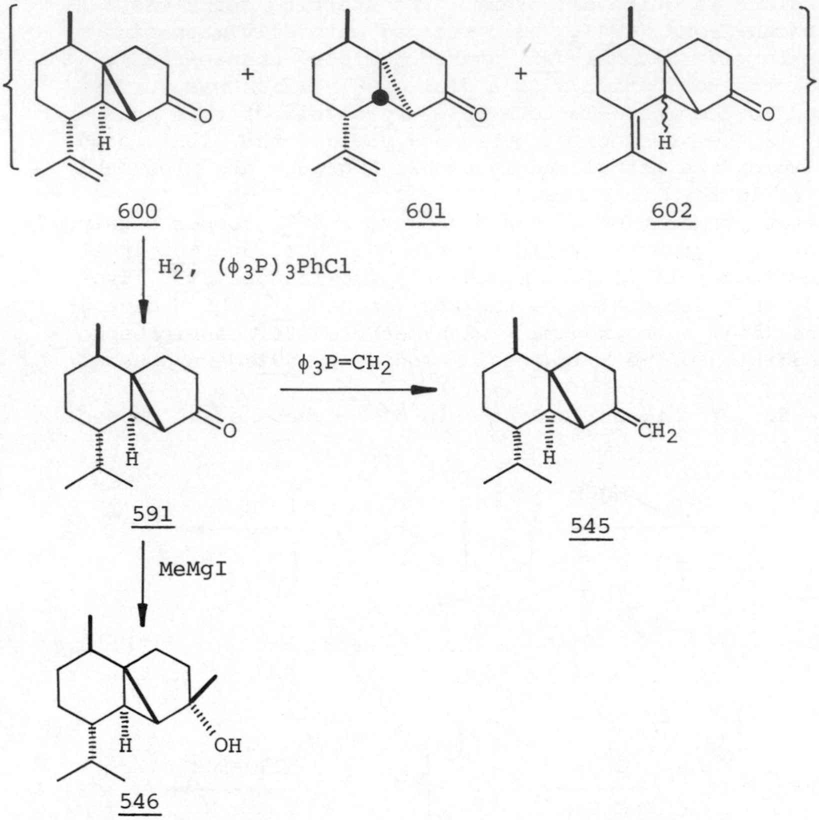

H
O
+
O
+
H
O
600
601
602
H₂, (ϕ₃P)₃PhCl
ϕ₃P=CH₂
H
O
H
CH₂
591
545
MeMgI
H
OH
546

(546) as the major product (47%).

Thujopsene (547) has an unusual skeleton which has prompted two interesting and radically different syntheses. In the first, Dauben chose to utilize the allylic alcohol 607 as a critical intermediate (Scheme 69).[217] On the basis of earlier studies,[186] it was expected that Simmons-Smith methylenation of 607 would occur stereospecifically *cis* to the hydroxyl group, thus establishing the proper relative stereochemistry in the three asymmetric centers of thujopsene. In fact, alcohol 607, which was prepared by the straightforward route indicated, reacted sluggishly with iodomethylzinc iodide, but gave only one stereoisomer (608) in 23% yield. Oxidation of 608 yielded a cyclopropyl ketone (609), which was treated with methylmagnesium bromide. Upon work-up, the resulting carbinol dehydrated spontaneously, giving (±)-thujopsene (547).

Büchi's synthesis (Scheme 70) is elegantly designed, but

Scheme 69.   Dauben's Synthesis of (±)-Thujopsene

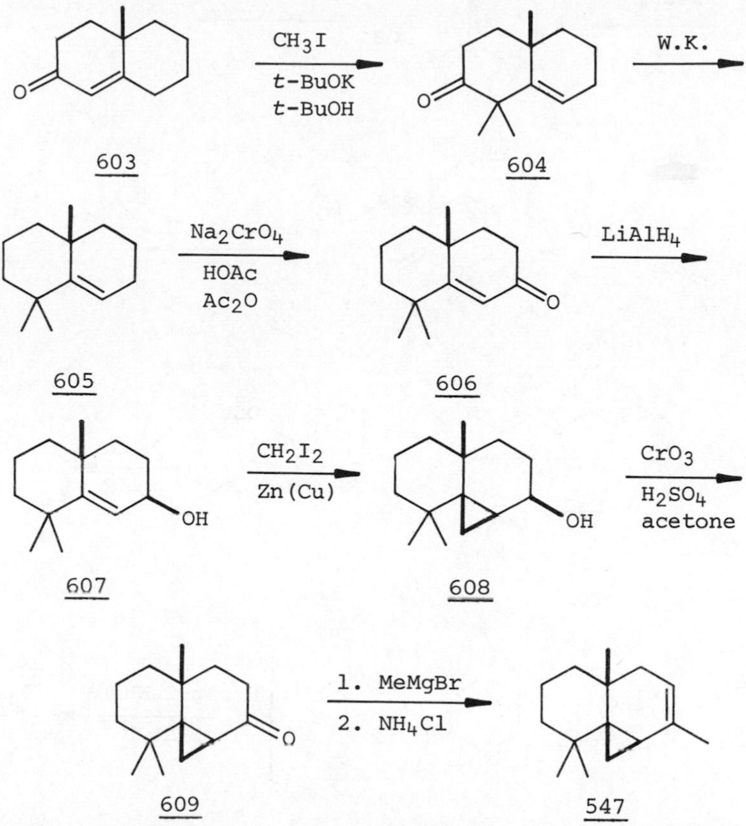

603   604

605   606

607   608

609   547

suffers from an extremely low yield in the final step.   β-
Cyclocitral (610) was reduced to β-cyclogeraniol (611) which
was converted into vinyl ether 612.   Pyrolytic rearrangement
of this material yielded aldehyde 613.   This substance, as its
ethoxy acetal (614), was allowed to react with ethyl propenyl
ether to obtain the mixed acetal 615.   When 615 was heated
with sodium acetate in acetic acid, a mixture of unsaturated
aldehydes 616 and 617 was obtained in 76% yield.   Although one
isomer greatly predominated (12:1 ratio), they were not separated and their stereochemistry was not defined.   The corresponding p-toluenesulfonylhydrazones (618 and 619) were irradiated, as their sodium salts, to give a complex mixture of
products.   The hydrocarbon fraction resulting from this reaction (18%)

Scheme 70. Büchi's Synthesis of (±)-Thujopsene

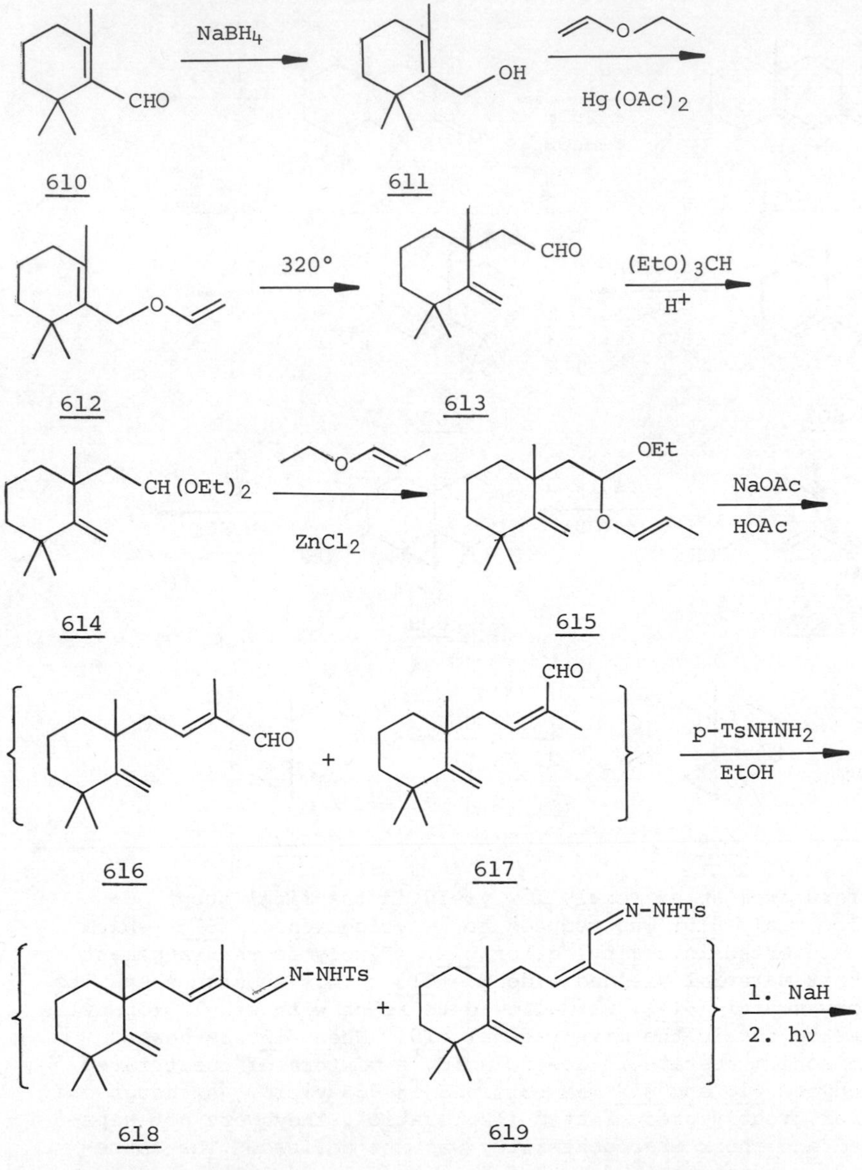

610

611

612

613

614

615

616   +   617

618   +   619

620   547

consisted primarily of cyclopropene 620 (10%) and (±)-thujop-
sene (547, 4%). Five unidentified hydrocarbons were also observed as minor products.

5.   OTHER BICARBOCYCLIC SESQUITERPENES

A.   Guaiazulenes; Bulnesol, α-Bulnesene, and Kessane

In contrast to the synthetic success which has been realized
in the decalinic area (see Sec. 4), only a few hydroazulenic
sesquiterpenes have yielded to total synthesis. There are two
reasons for this disparity. While there are many reliable
methods for the formation of the hydronaphthalene nucleus
(Robinson annelation, reduction of naphthalenes, Diels-Alder
cycloaddition, π-cyclization), analogous methods for the generation of hydroazulenes are rarer. Complicating this dearth of
methods for elaboration of the basic skeleton is the poorly
developed state of conformational analysis in cyclopentanes
and cyclohexanes. Thus, one may confidently design stereoselective syntheses in the decalin area, in which relative
stereochemistry is established by either kinetic or thermodynamic methods. Stereoselective design in the hydroazulene
area is much more difficult.

As a consequence of these factors, syntheses in the
hydroazulene area have uniformly been accomplished by the rearrangement of a bicyclic precursor possessing a skeleton
which is more amenable to stereoselective synthesis. In this
section we shall discuss the synthesis of bulnesol (1),[220,222]
α-bulnesene (2),[221,223] and kessane (3).[224] In the following
section, we will take up the syntheses of the quaianolides

1   2   3

arborescin (4), geigerin (5), desacetoxymatricarin (6), and achillin (7), all of which have been prepared from santonin derivatives.

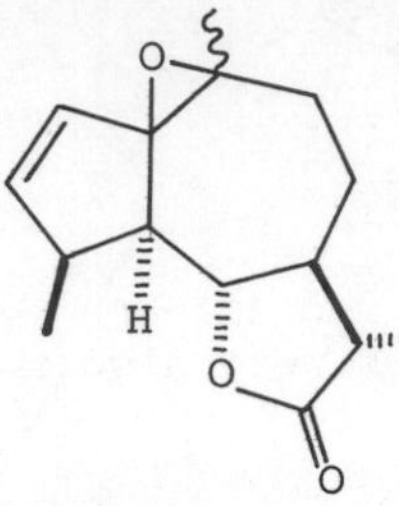

4

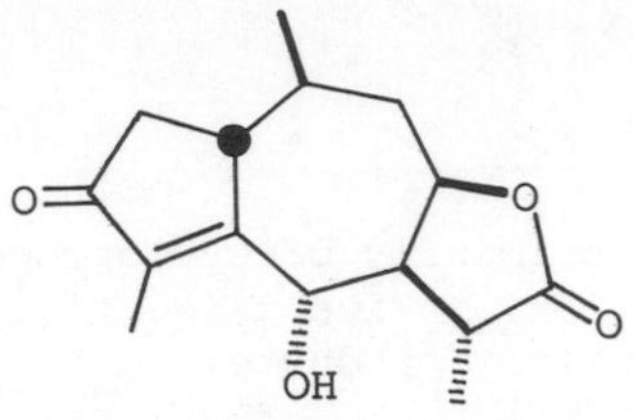

5

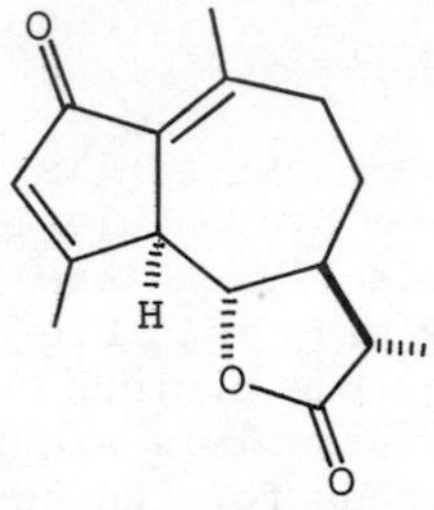

6

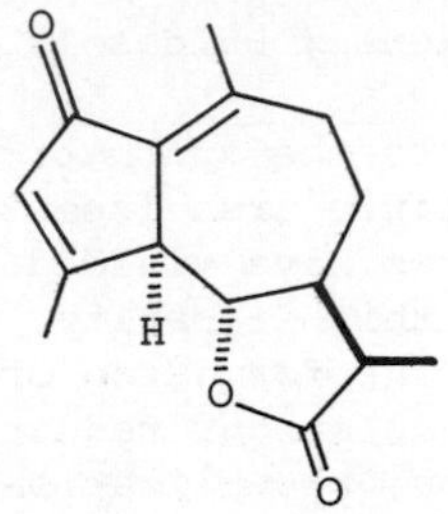

7

Marshall's synthesis of (±)-bulnesol, which proceeds via bicyclo[4.3.1]decane intermediates, is outlined in Scheme 1.[220] The synthesis, which is beautifully designed, was based upon the expectation that a bicyclo[4.3.1]decane derivative such as 8 would solvolyze with rearrangement to the hydroazulene 9. Preliminary studies demonstrated the feasibility of the

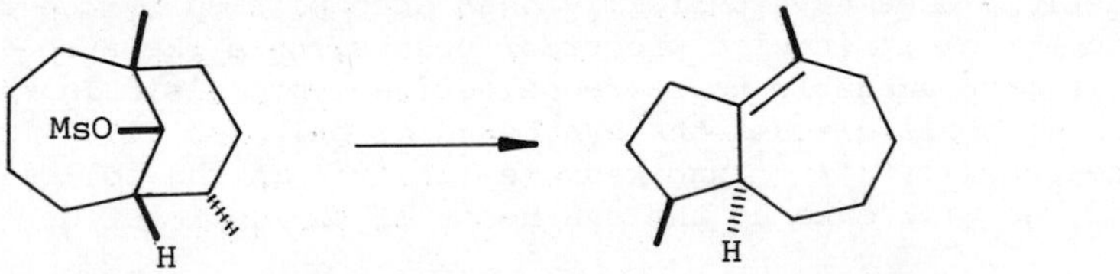

8

9

approach.[225] With this knowledge, the problem becomes one of stereospecifically synthesizing an analog of compound 8 which contains the necessary functionality for incorporation of the isopropylol group. The rearrangement product would then

possess the correct stereochemistry (11). Less desirably, a
substrate with the stereochemistry indicated in 12 might be

10   11

employed. The initial product (13) might be converted by epimerization to 11, which could then be converted into bulnesol.
A consideration of models suggested that the equilibrium
13 ⇌ 11 (X = carbomethoxy) should lie strongly to the right.

12   13

The starting point for Marshall's synthesis was 4-carbethoxycyclo-
hexanone (14) Scheme 1. The derived ethylene ketal was reduced by
lithium aluminum hydride to primary alcohol 15. Treatment of
the corresponding methanesulfonate with sodium p-chlorophen-
oxide gave ether 16a, which was hydrolyzed to keto ether 17a.
    Compound 17a reacted smoothly with ethyl diazoacetate in
the presence of boron trifluoride to give the ring-expanded
β-keto ester 18a in 79% yield. Compound 18a reacted with
methyl vinyl ketone to yield exclusively isomer 19a. Acid
catalyzed cyclization, followed by ester hydrolysis, afforded
the crystalline keto acid 20a in 65% yield.
    At this point, comment on the choice of the grouping
CH$_2$OAr as the potential isopropylol group is in order. Preliminary experiments on the related cycloheptanone 33 showed
that intramolecular cyclization competed with the desired
Michael alkylation. The only product isolated in attempted
alkylation of 33 was the keto ester 35, which presumably arose
by fragmentation of diketone 34. In order to avoid this complication, it was decided to replace the carbethoxy group with
a protected methylol group. Other considerations prompted the

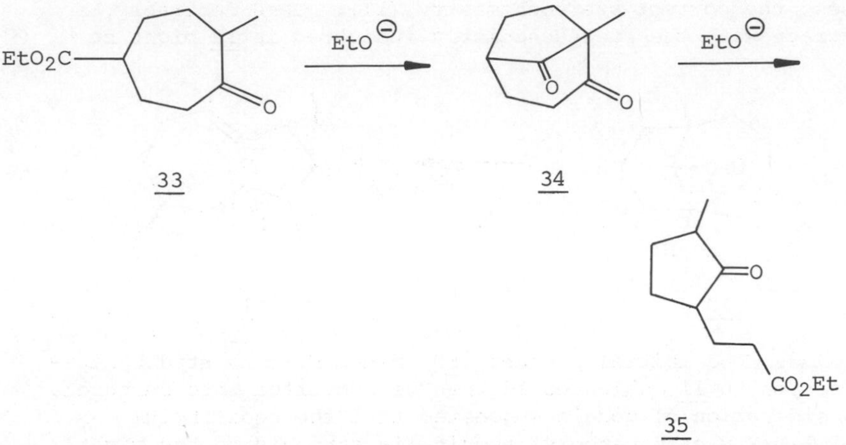

33   34

35

replacement of the α-methyl group by an ester grouping, which
could later be reduced to the desired methyl substituent. The
first protecting group explored was phenoxy. However, it was
found that this group was incompatible with the conditions required for cyclization of the derived diketo ester. The problem seemed to lie in sulfonation of the phenoxy ring. It was
therefore reasoned that the incorporation of a deactivating
group in this ring would stabilize it with respect to this unwanted side reaction. As was found in the sequel, this pre-
diction was admirably borne out.

The remarkable stereospecificity observed in the allkyla-
tion of β-keto ester 18a is interesting. If one assumes a
"reactant-like" transition state, the stereochemistry is ex-
plicable in terms of steric hindrance to approach of the reagent (below). Unfortunately, the resulting product (19a) has

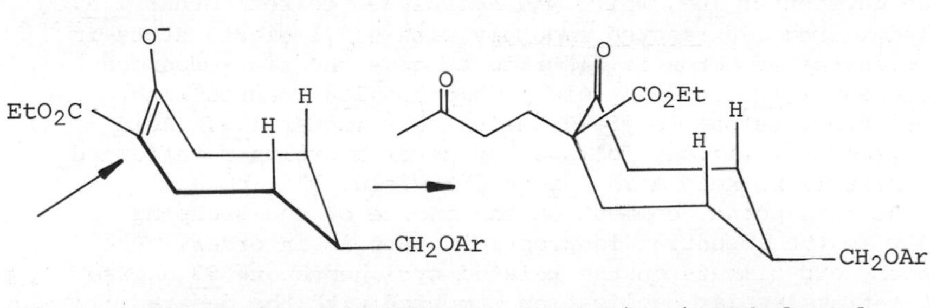

the incorrect geometry for direct conversion into bulnesol.
However, as noted previously, this situation may be remedied
at a later stage.

With the desired bicyclo[4.3.0]decane skeleton in hand,
Marshall and Partridge turned to the problem of further

elaboration of compound 20a into a suitable precursor for the synthesis of bulnesol. The first task, reduction of the bridgehead carboxyl to methyl, was accomplished in a relatively straightforward manner. Hydride reduction of keto acid 20a was accomplished by partial removal of the aromatic chlorine. The resulting mixture of diols 21 was converted into the corresponding p-toluenesulfonates 22, which were reduced with lithium aluminum hydride to hydroxy ethers 23. Birch reduction of 23, followed by hydrolysis of the resulting enol ethers, gave the single crystalline diol 24. The stereochemistry of compound 24 was confirmed by oxidation to keto acid 36, which was converted into lactone 37. Reduction of 37, for which the

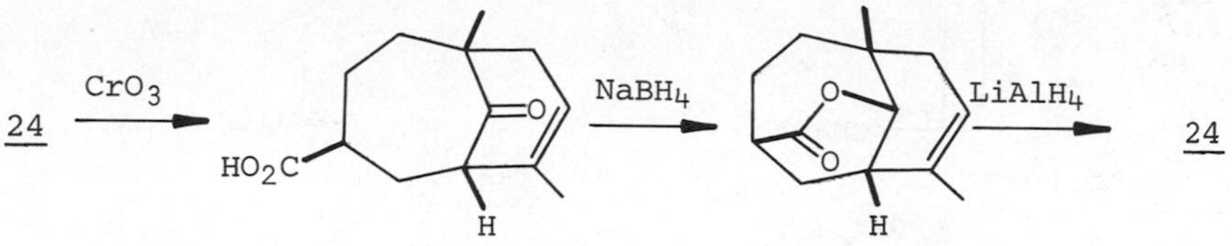

stereochemistry is fixed, led to the same diol (24).*

*The stereochemistry of the secondary alcohol (*syn* to the four carbon bridge) is, of course, critical. In order for concerted rearrangement to occur in the desired sense, the leaving group must be *anti*-coplanar with the migrating bond:

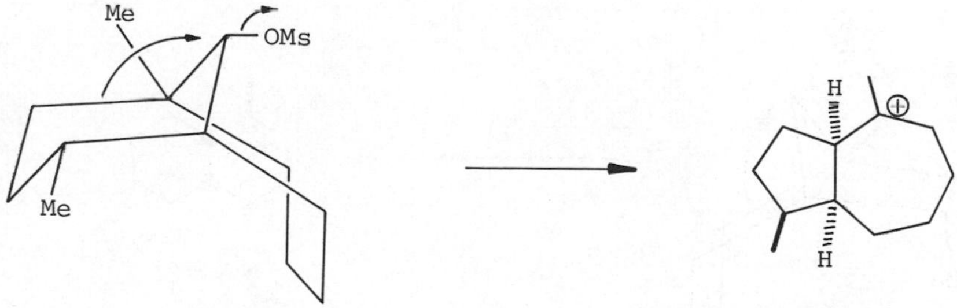

The alternative isomer would solvolyze to a decalinic system.

After selective acetylation of the primary hydroxyl, the double bond was reduced catalytically to afford compound 26. At this stage, the relative stereochemistry at two of bulnesol's asymmetric centers is fixed. Model studies had shown that hydrogen is delivered predominately from the *exo* face of the bicyclic system, leading to an equatorial methyl group, as in 26.

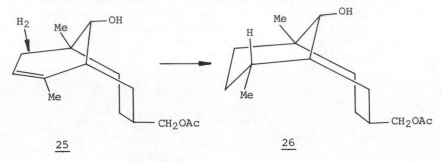

25   26

Solvolysis of methanesulfonate 27 occurred with rearrangement as planned, affording the octahydroazulene 28 in 92% yield. After removal of the acetyl group, the methylol group was oxidized to carboxyl. The resulting acid (30) gave a methyl ester (31), which was equilibrated with methanolic sodium methoxide. Isomer 32 predominated at equilibrium, as predicted (32:31 = 7:3). The major isomer was separated by

(footnote continued)

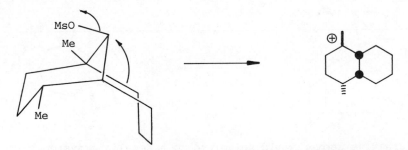

The stereochemistry at this center is established in the hydride reduction of keto acid 20a and may be rationalized on steric grounds. Preliminary studies had shown that simpler ketones analogous to 20a are reduced either by hydride or sodium in alcohol to a single alcohol. Apparently, the four carbon bridge effectively blocks approach to that side of the carbonyl group.

preparative glpc and treated with methyllithium to obtain (±)-bunesol (1).

Scheme 1. Marshall's Synthesis of (±)-Bulnesol

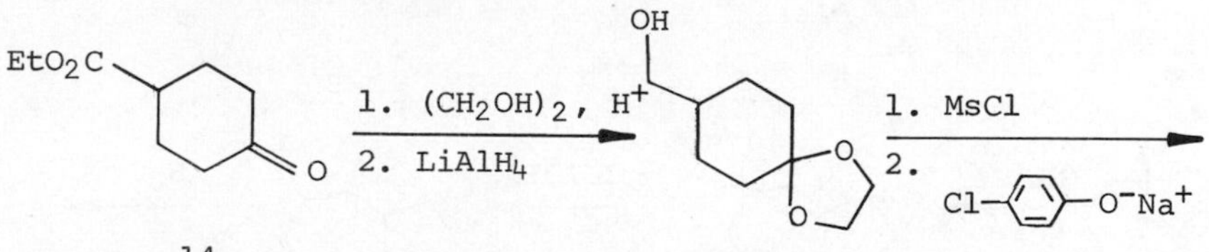

14 15

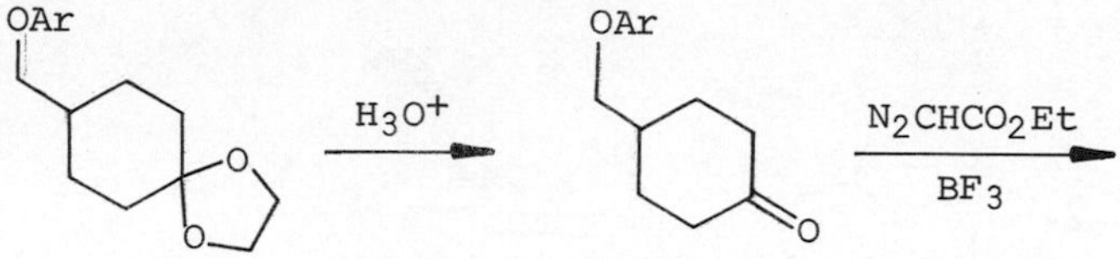

16 17a

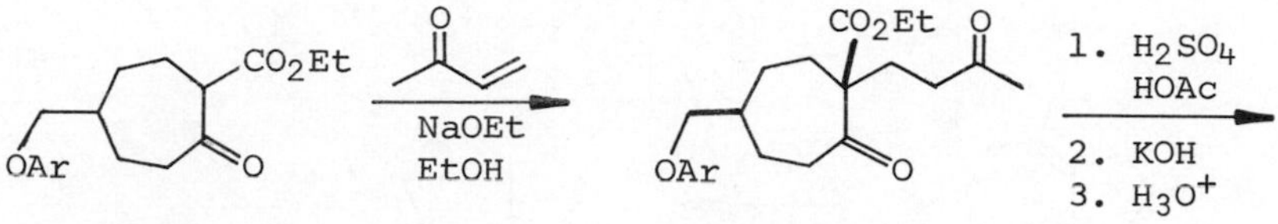

18a 19a

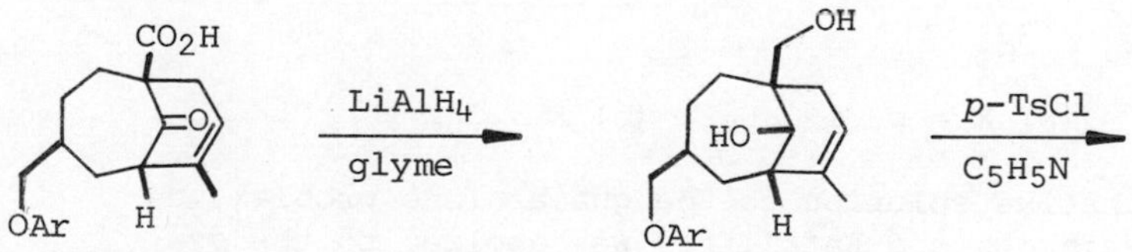

20a 21a, 21b

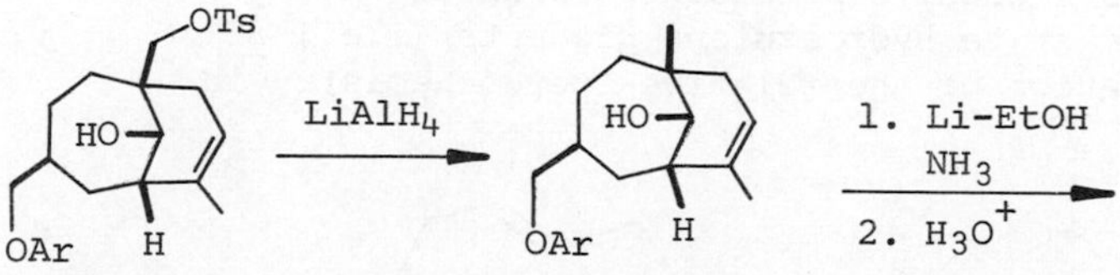

22a, 22b 23a, 23b

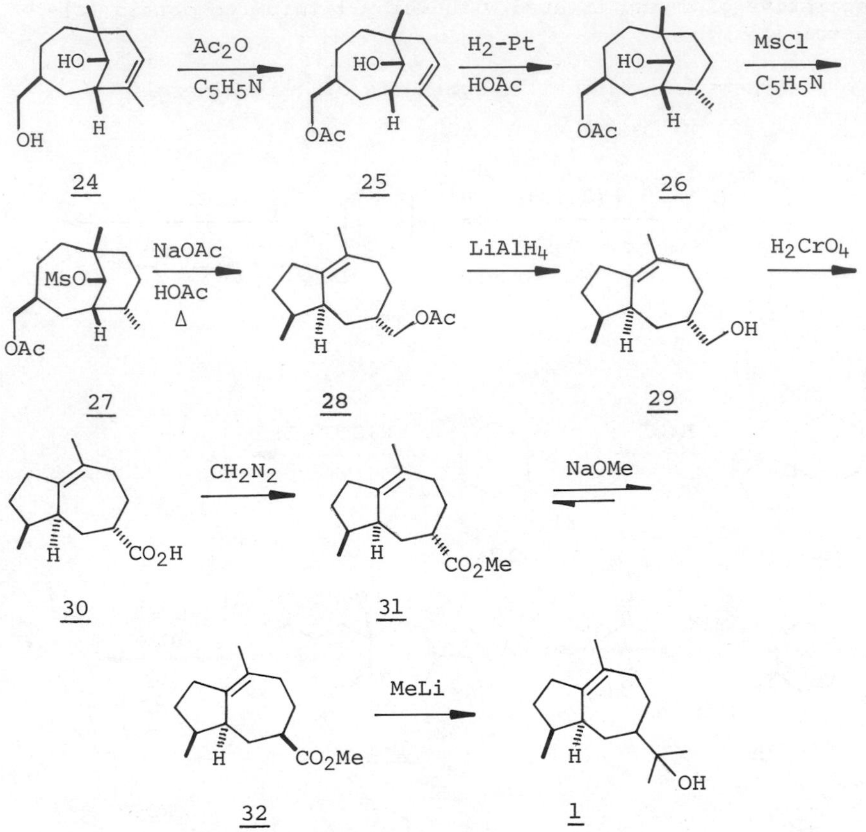

(a) X = p-Cl-C$_6$H$_4$-; (b) X = C$_6$H$_5$-

An alternative solution to the guaiazulene problem, developed by Heathcock and Ratcliffe, was applied to the synthesis of bulnesol (1) and α-bulnesene (2).[221] The Heathcock-Ratcliffe approach was based on the discovery that both *cis*- and *trans*-9-methyl-1-decalyl *p*-toluenesulfonates undergo solvolytic rearrangement to hydroazulene products (i.e., 38 → 39).[226] Consideration of the relative stereochemistry in

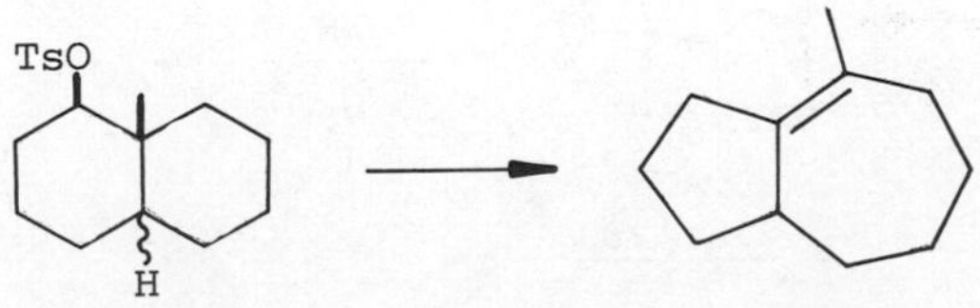

bulnesol reveals three decalyl tosylates which might yield
this type of product upon solvolytic rearrangement (40-42).
The known stereochemistry of the Wagner-Meerwein rearrangement places an added synthetic constraint on the problem. In
cis-decalin 40, the conformer which has the tosylate group and
the migrating bond *anti* and coplanar (conformer 40a) is hope-

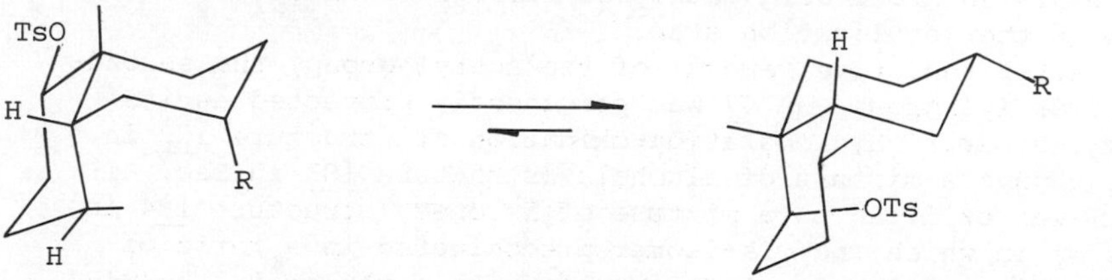

lessly crowded. The alternative chair-chair conformer (40b)
would predominate heavily. Thus, this material would probably *not* yield a hydroazulene on solvolysis. The *cis*-fused
decalyl tosylate 41a, however, is expected to exist predominately as conformer 41a, in which the requisite *anti*-coplanar
relationship is present. Finally, the *trans*-fused decalyl
tosylate 42 is restricted to one low-energy conformation, in
which the tosylate group and migrating bond are *anti*-coplanar.
The choice of synthetic target for this approach is thus
limited to tosylates 41 and 42.

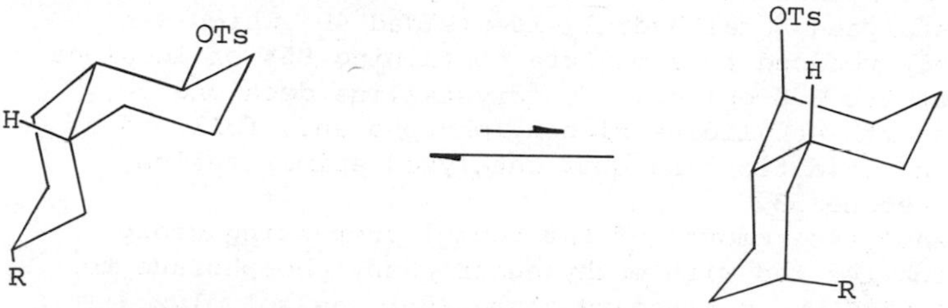

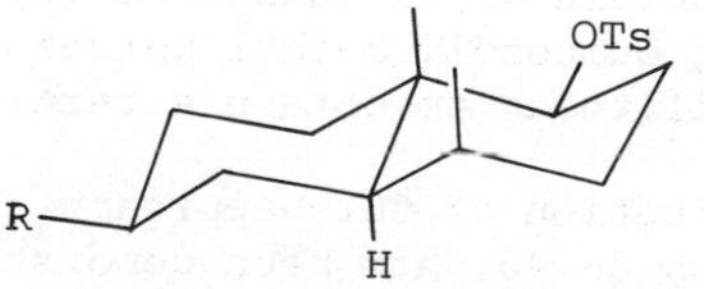

The Heathcock-Ratcliffe synthesis of bulnesol follows
the *trans*-decalin route (42). An alternative application of
this approach, by Yoshikoshi,[222] adopts a route proceeding

through *cis*-decalin intermediates (41).*

Schemes 2 and 3 outline the Heathcock-Ratcliffe synthesis of (±)-α-bulnesene (2) and (±)-bulnesol (1). Acetylation of unsaturated keto alcohol (structure 180 in Sec. 4) gave acetate 45, which was ketalized, with concommittant double bond migration, to give 46. The acetyl protecting group was necessary to avoid acid catalyzed retrograde aldol condensation in the ketalization step.

After reductive removal of the acetyl group, the secondary hydroxyl group in 47 was permanently protected as its benzyl ether.† Hydroboration-oxidation of structure 182 in Sec. 4 gave a mixture of alcohols (structure 183 in Sec. 4) which was oxidized to a mixture of ketones (structure 184 in Sec. 4) in which the *cis*-isomer predominated in a ratio of 2:1. This mixture reacted directly with methylenetriphenylphosphorane in dimethylsulfoxide to give *only* the crystalline *trans*-fused methylene decalin 48 in 87% yield.

Acid-catalyzed ketal hydrolysis yielded 49, which was quantitatively reduced to a mixture containing 85% of decalone 50 and 15% of its C-8 epimer. The crystalline decalone 50, when treated with ethylidinetriphenylphosphorane, followed by hydroboration, oxidation and acid catalyzed epimerization, gave methyl ketone 53.

Hydrogenolytic removal of the benzyl protecting group gave 54, which reacted with methylenetriphenylphosphorane to give alcohol 55. The derived tosylate (56) was solvolyzed in buffered acetic acid to obtain (±)-α-bulnesene (2).

For the synthesis of bulnesol, methyl ketone (53) was first treated with methyllithium to obtain a tertiary alcohol

---

*The third possible application of this approach, proceeding through intermediates such as 40, has been demonstrated by Hendrickson in the rearrangement of compound 43, leading to compounds in the pseudoquaianolide series.[227]

†Although Minato and Nagasaki report that keto alcohol (structure 180 in Sec. 4) can be benzylated directly,[140] Heathcock and Ratcliffe were unable to repeat the reaction (see p. 313).

Scheme 2. Heathcock-Ratcliffe Synthesis of (±)-α-Bulnesene

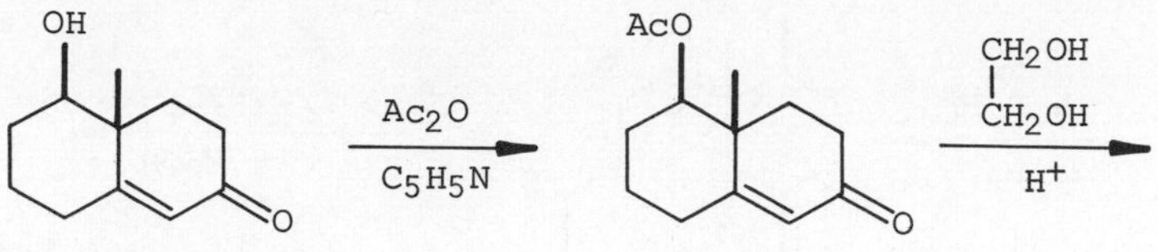

180 (sec. 4)

45

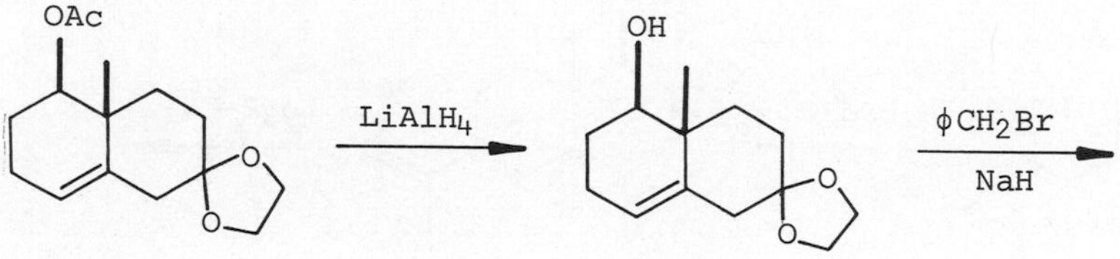

46

47

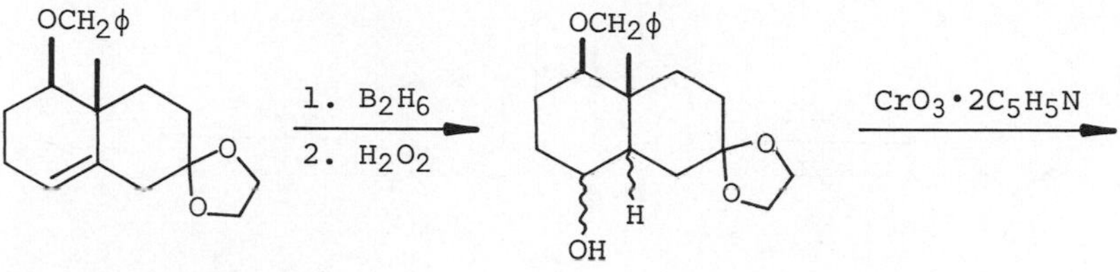

182 (sec. 4)

183 (sec. 4)

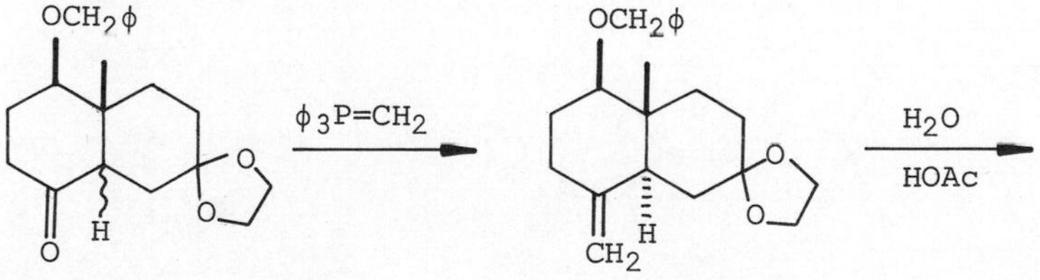

184 (sec. 4)

48

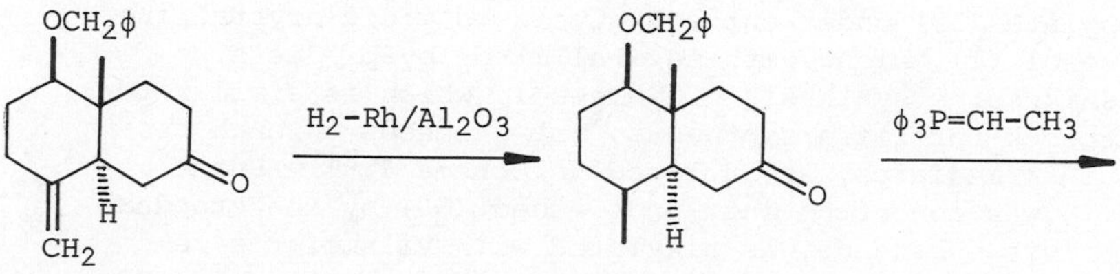

49

50

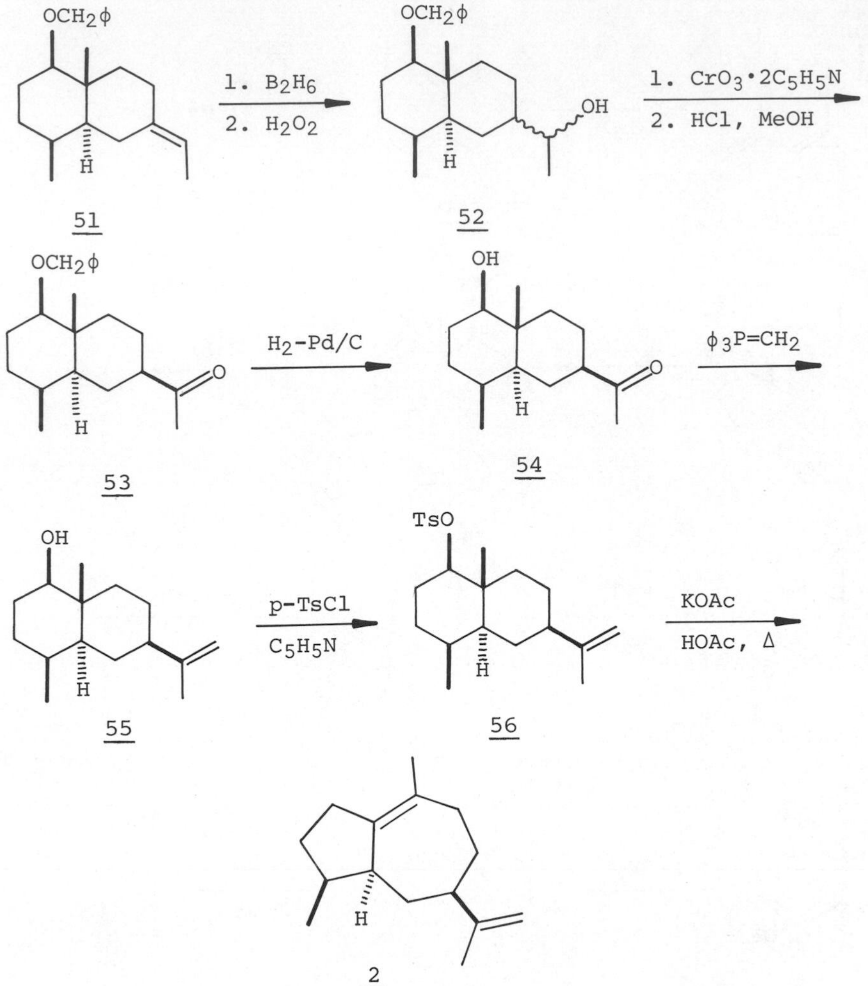

$$51 \xrightarrow[\text{2. } H_2O_2]{\text{1. } B_2H_6} 52 \xrightarrow[\text{2. HCl, MeOH}]{\text{1. } CrO_3 \cdot 2C_5H_5N}$$

$$53 \xrightarrow{H_2-Pd/C} 54 \xrightarrow{\phi_3P=CH_2}$$

$$55 \xrightarrow[\text{C}_5H_5N]{p-TsCl} 56 \xrightarrow[\text{HOAc, } \Delta]{KOAc}$$

2

(57), which was hydrogenolyzed to diol 58. The corresponding mono-tosylate (59) underwent solvolysis to yield crystalline (±)-bulnesol (1), along with several minor by-products.

Yoshikoshi's synthesis of bulnesol, which is similar to the Heathcock-Ratcliffe synthesis, but proceeds through *cis*-decalin intermediates, is outlined in Scheme 4.[222] Enone acetate 45 was converted into enol ether (60) by the standard method. This substance was alkylated with Vilsmeier's reagent to give immonium salt 61 which was reduced to amine 62. The corresponding methiodide (63) was hydrogenolyzed by refluxing it in ethanol with freshly prepared Raney nickel.

Scheme 3. Heathcock-Ratcliffe Synthesis of (±)-Bulnesol

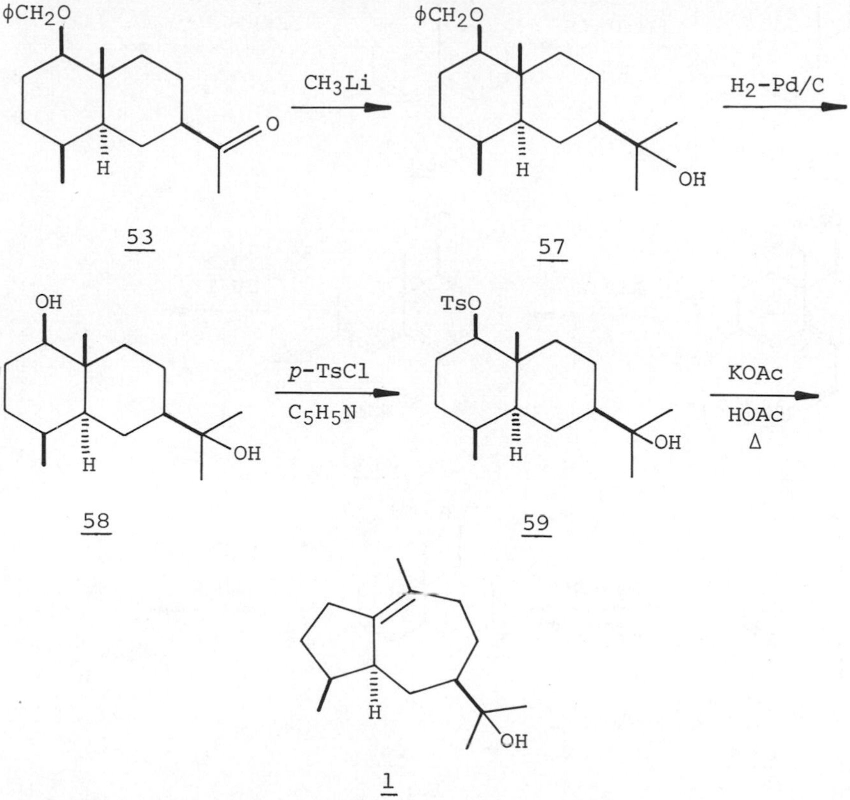

53

57

58

59

1

Hydrolysis of the resulting enol ether (64) gave compound 65, in which the stereochemistry at C-8 is the more stable one (equatorial methyl). Although hydroxy enone 65 gave a mixture of *cis* and *trans* decalones on hydrogenation, the corresponding acetate (66) reacted with hydrogen in the presence of palladium on strontium carbonate to give almost solely the *cis* decalone 67.

The derived tetrahydropyranyl ether 69 was reduced with lithium aluminum hydride to give equatorial alcohol 70, whose tosylate (71) reacted with cyanide, affording nitrile 72. Basic hydrolysis, hydrolysis of the protecting group, and esterification then yielded hydroxy ester 73. The corresponding tosylate (74) solvolyzed with rearrangement to give ester 32, which had previously been converted by Marshall into (±)-bulnesol (1).

Scheme 4. Yoshikoshi's Synthesis of (±)-Bulnesol

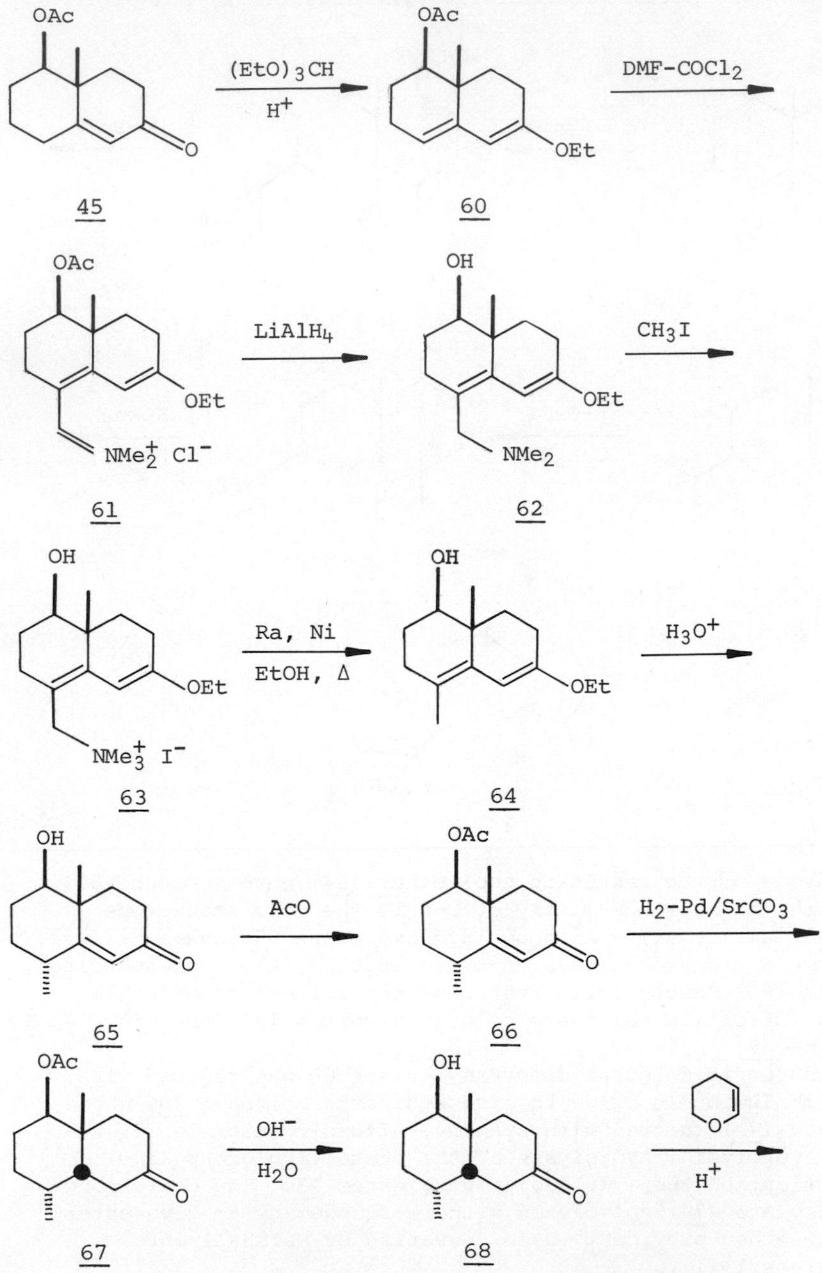

408

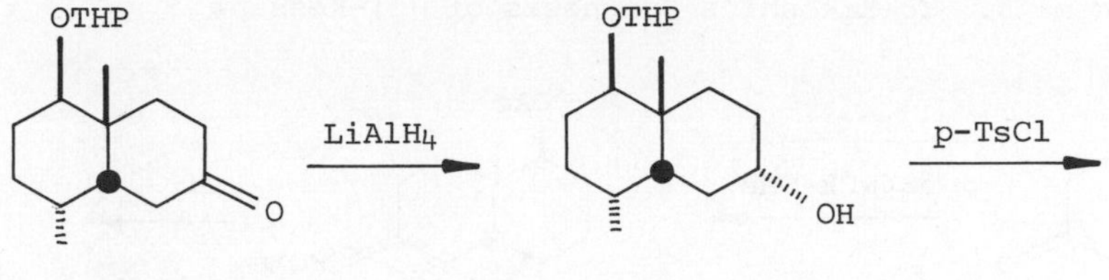

A modification of this basic scheme has recently allowed the synthesis of the bicyclic hydroazulenic ether kessane (**3**). This synthesis, also from Yoshikoshi's laboratory, is outlined in Scheme 5.[227] The *cis*-fused decalone **67** was treated with methoxymethylenetriphenylphosphorane to give enol ether **75**. Acid catalyzed hydrolysis gave a mixture of aldehydes (**76**), which was oxidized by Jones reagent to a 2:3 mixture of epimeric acids **77** and **78**. The major isomer (**78**) was esterified and the resulting ester treated with methyllithium to obtain diol **79**. The corresponding monomesylate (**80**) was

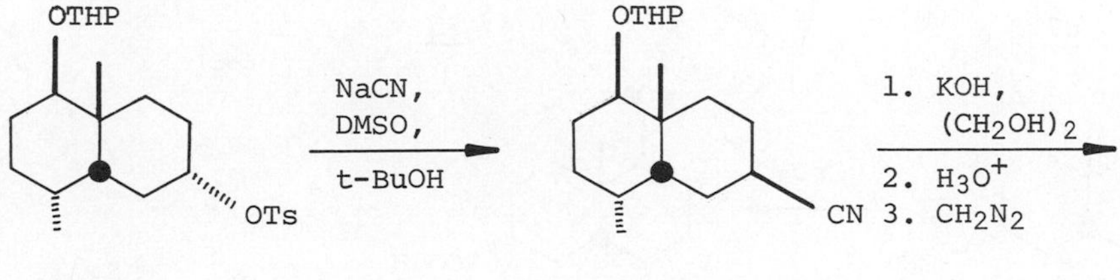

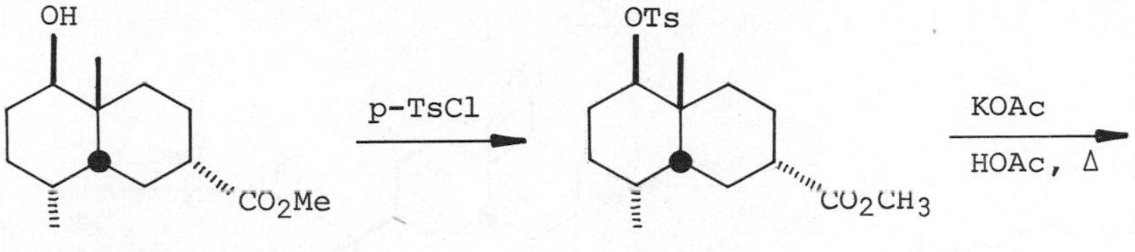

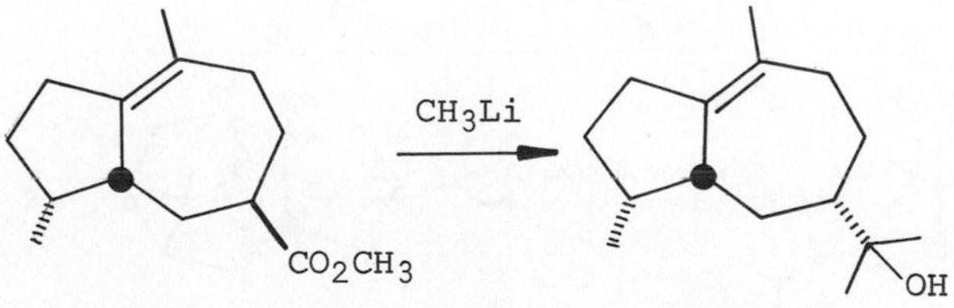

Scheme 5. Yoshikoshi's Synthesis of (±)-Kessane

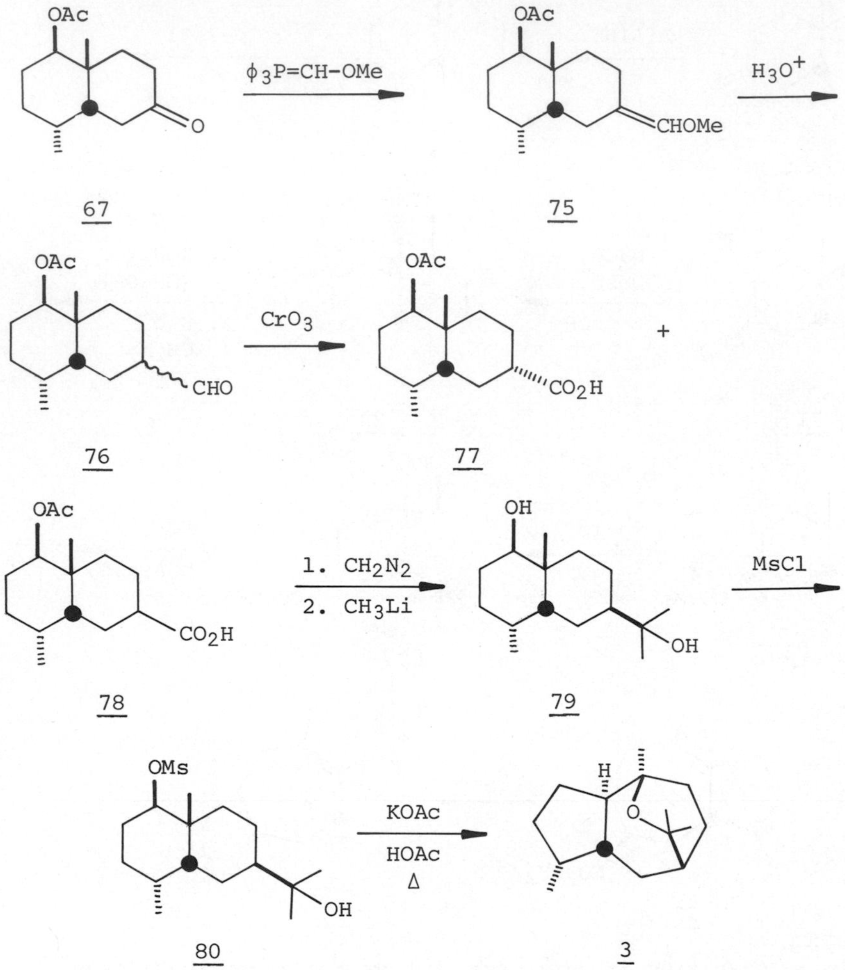

solvolyzed in buffered acetic acid, affording a ternary mixture from which (±)-kessane (3) was isolated in 30% yield.

Piers has reported a relay synthesis of α-bulnesene, beginning with the santonin derivative 42 (Sec. 4) (see p.
287).[223] Irradiation of dienone 42 (Sec. 4) in aqueous acetic
acid yielded the photo-product 81, which was acetylated to 82.
Catalytic hydrogenation of 82 gave saturated ketone 83. After
reduction of the ketonic carbonyl with sodium borohydride, the
resulting mixture of epimeric alcohols (84) was treated with

p-toluenesulfonyl chloride in pyridine to effect dehydration.
Compound 85 was obtained in 55% yield. Although 85 gave a
mixture of isomers when hydrogenated over platinum oxide, the
analogous reduction catalyzed by the homogeneous catalyst
tris(triphenylphosphine)rhodium(I) chloride was completely
selective, yielding isomer 86.

Hydride reduction of 86 gave a diol (87) which was
treated with methyl chlorocarbonate to form the carbonate
ester at the less-hindered primary position. Dehydration of
the tertiary alcohol gave 88. Pyrolysis of 88 at 400° gave a

Scheme 6. Piers' Synthesis of α-Bulnesene (Relay)

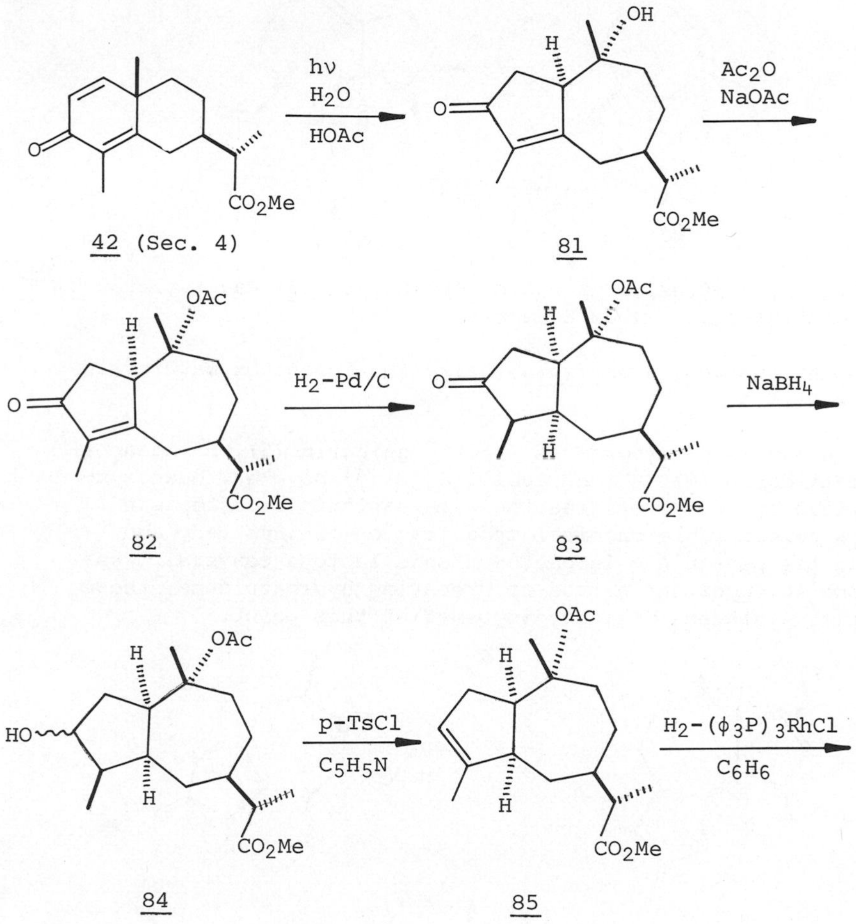

42 (Sec. 4)

81

82

83

84

85

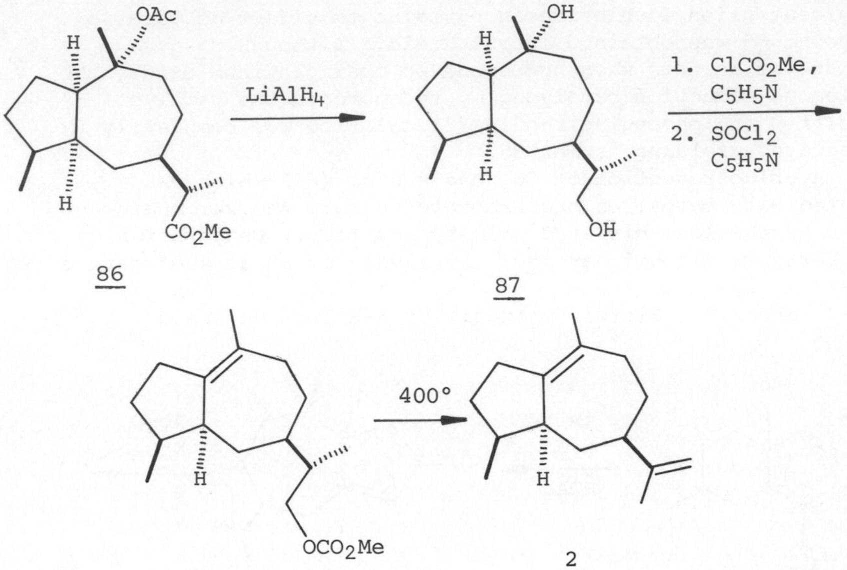

mixture of olefins, from which α-bulnesene (2) was isolated by preparative glpc.   (See Scheme 6.)

B.   Guaianolides; Arborescin, Geigerin, Desacetoxymatricarin, and Achillin

The quaianolides arborescin (4),[228] geigerin (5),[229] desacetoxymatricarin (6),[230] and achillin (7)[231] have all been synthesized by routes originating with santonin or artemisin. Since considerable chemical modification is involved, and since the santonin → isophotosantonic lactone conversion has become an important method of preparing hydroazulenes, these partial syntheses will be discussed at this point.

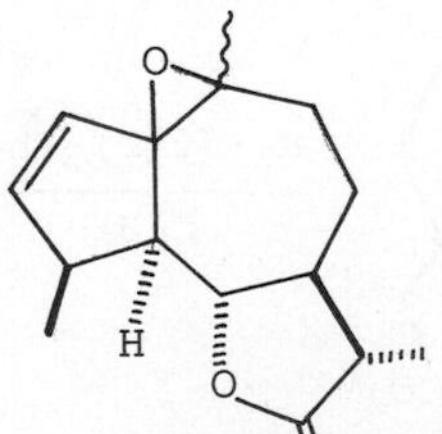

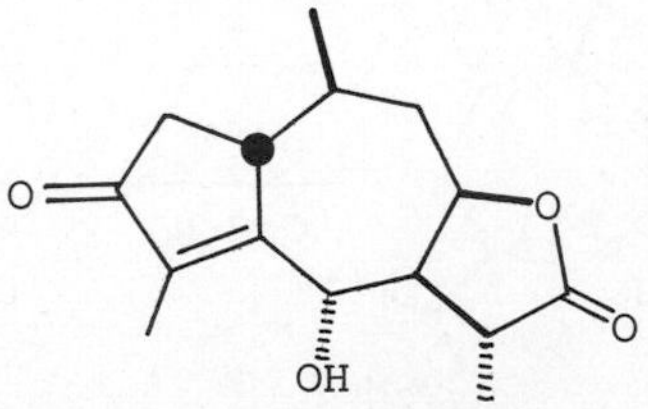

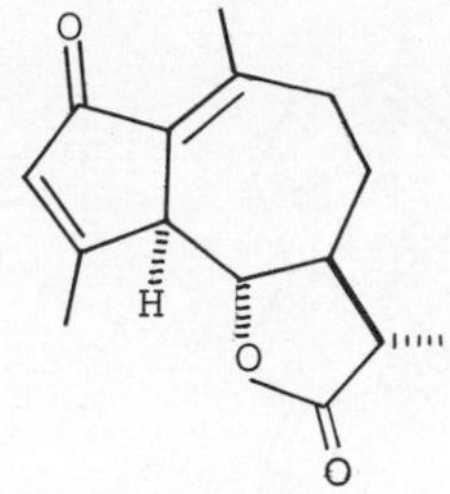

6

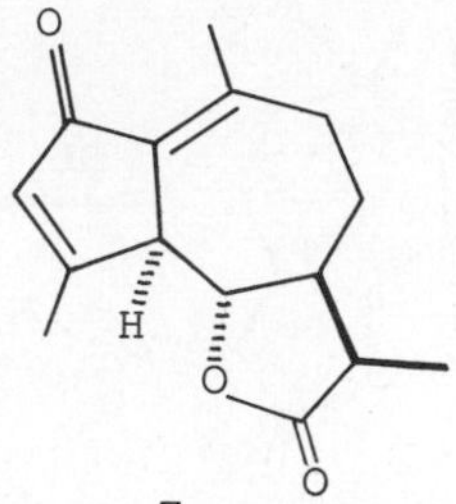

7

The synthesis of arborescin (4) by Suchy, Herout, and Sorm[228] is outlined in Scheme 7. α-Santonin (structure 23, Sec. 4) was converted into O-acetyldihydroisophotosantonic lactone (90) by Barton's procedure.[232] Borohydride reduction of 90 yielded a mixture of alcohols (91), shown later by White and co-workers to be a 1:2:4:5 mixture, epimeric at the new hydroxyl and the secondary methyl group.[230] The derived benzoate mixture (92) was treated with BF₃ to eliminate acetic acid. The resulting olefin (93) was epoxidized to yield an ether

Scheme 7. Arborescin--Czechoslovak Synthesis (Relay)

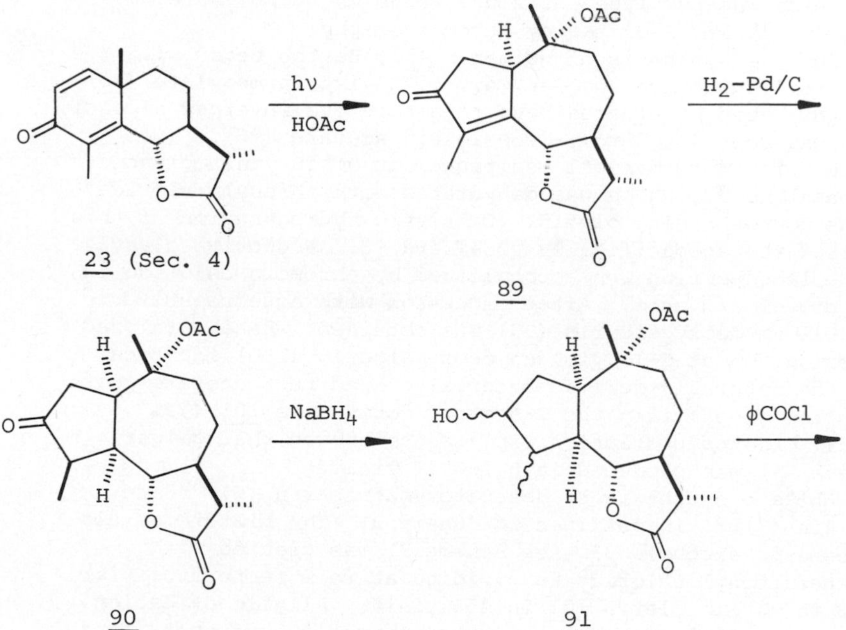

23 (Sec. 4)

89

90

91

92 → BF₃ → 93 → φCO₃H →

94 → 210° → 4

(94) which was pyrolyzed at 210°. From the pyrolysate, arborescin (4) was isolated by chromatography.

For the synthesis of geigerin (5), Barton began with
artemisin (structure 25, Sec. 4).   Geigerin mesylate (95)
was hydrolyzed by aqueous base to obtain the inverted alcohol,
which was acetylated to γ-isogeigerin acetate (96). (See
Scheme 8.) Photochemical rearrangement of 96 yielded the
hydroazulene 97, which was dehydrated with thionyl chloride
to the corresponding olefin. Catalytic hydrogenation of this
occurred stereospecifically to afford 98. Reductive cleavage
of the lactone ring was accomplished by chromous chloride in
1N hydrochloric acid. After digestion with aqueous sulfuric
acid, 11-epideoxygeigerin (99) was obtained. Base catalyzed
epimerization at C-11 yielded deoxygeigerin (100) identical
with the naturally derived material. Lead tetraacetate acet-
oxylation gave mainly the 2-acetoxy derivative 101 (72%
yield). Isotopic dilution experiments showed that geigerin
acetate (5) was produced in 0.1-0.3% yield.

White's synthesis of desacetoxymatricarin (6)[230] and
achillin (7)[231] is outlined in Scheme 9. The mixture of diastereomeric alcohols 91 (see Scheme 7) was treated with
methanesulfonyl chloride in pyridine at room temperature for
24 hr to obtain olefin 102 in 45% yield. Allylic oxidation
of this material (t-butyl chromate in acetic acid containing

Scheme 8. Barton's Synthesis of Geigerin (Relay)

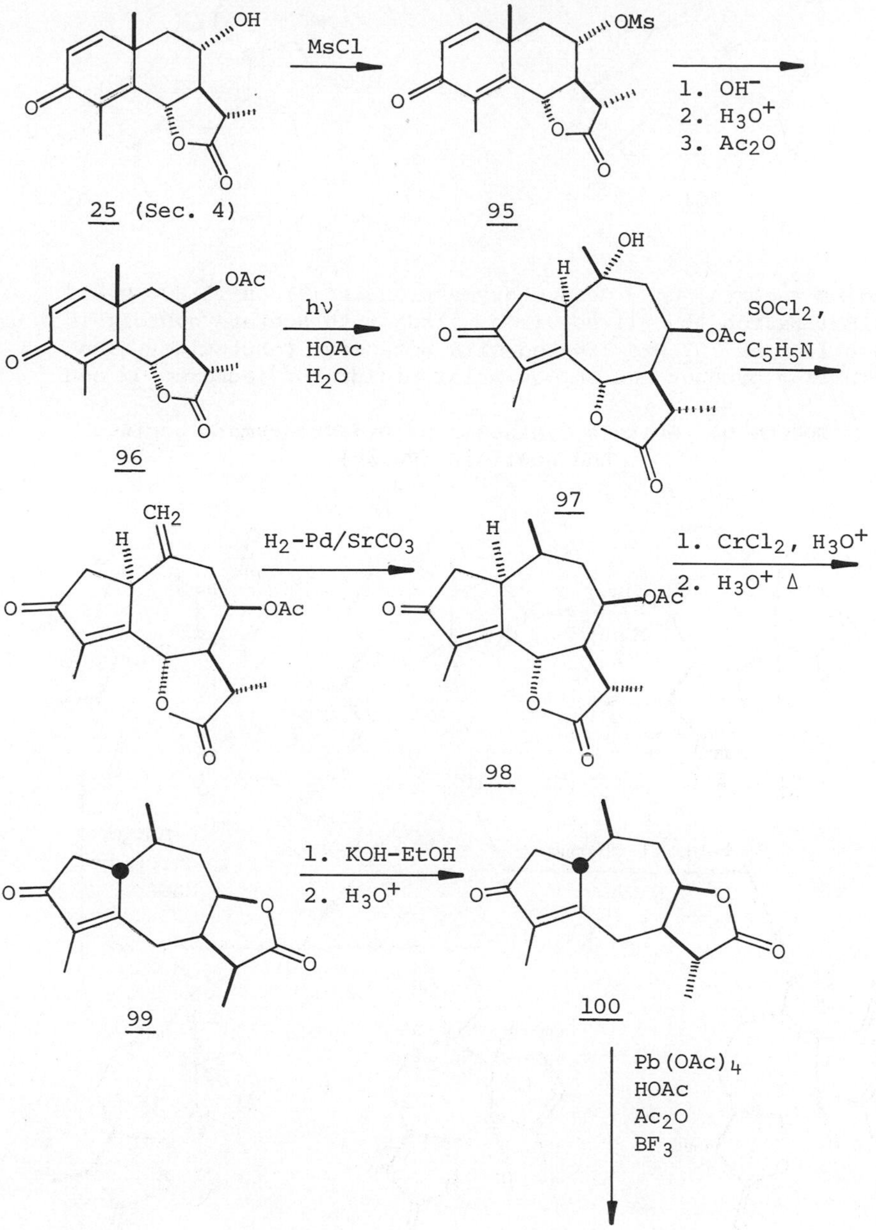

25 (Sec. 4)   95

96   97

98

99   100

415

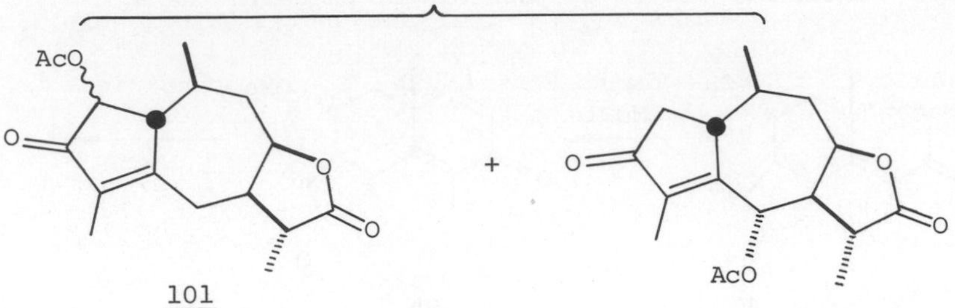

101 5

sodium acetate) gave desacetoxymatricarin (5) in 15-20% yield. Epimerization at C-11 occurred, along with acetate hydrolysis when lactone 102 was treated with potassium *t*-butoxide. The resulting product was an equimolar mixture of isomers 103 and

Scheme 9. White's Synthesis of Desacetoxymatricarin and Achillin (Relay)

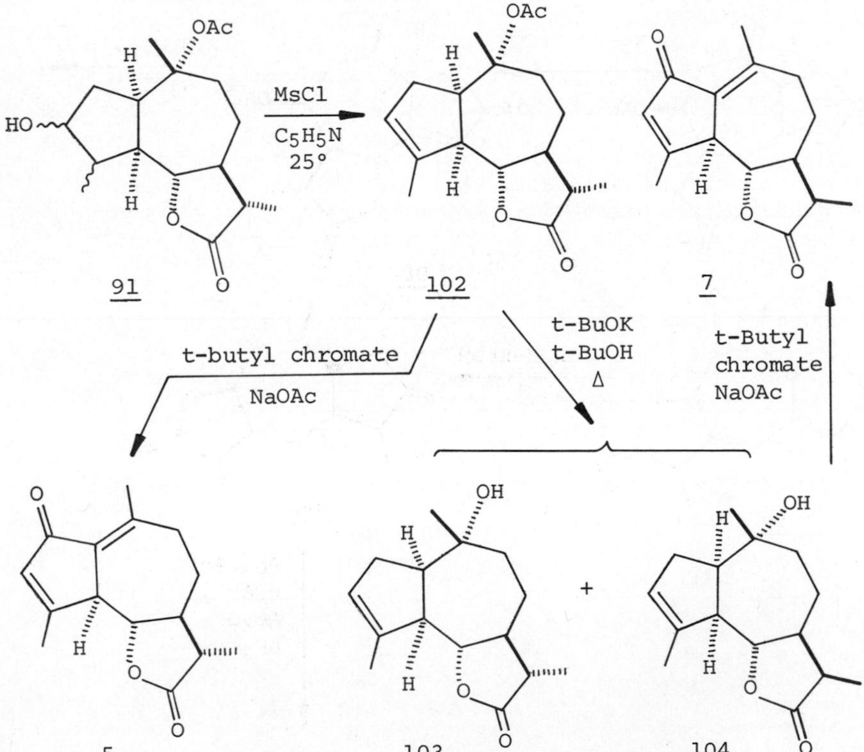

91 102 7 5 103 104

<u>104</u>. Allylic oxidation of <u>104</u> furnished achillin in 5% yield.

C. Tricyclic Hydroazulenes Containing a Cyclopropane Ring;
Aromadendrene, Cyclocolorenone

The tricyclic hydroazulenes aromadendrene (<u>105</u>) and cyclocolorenone (<u>106</u>) present synthetic challenges which are quite
similar to those encountered in the simpler hydroazulenes.
The major synthetic task here is the stereospecific elaboration

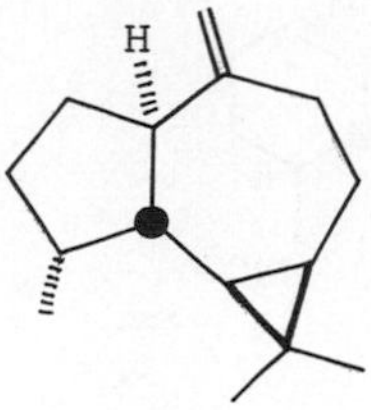

<u>105</u> 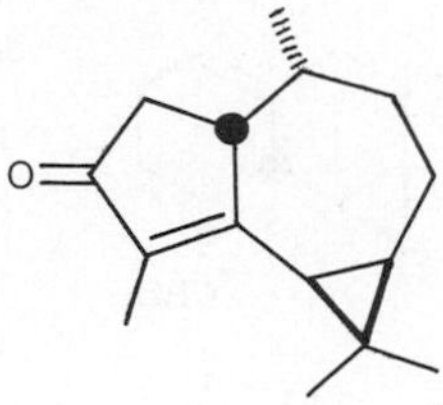   <u>106</u>

of the hydroazulene nucleus. In this section, we discuss
Büchi's elegant synthesis of (-)-aromadendrene, the enantiomer
of <u>105</u>, as well as the synthesis of an epimer of compound <u>106</u>,
by Büchi and Narang and Dutta.

Büchi's aromadendrene synthesis, outlined in Scheme 10,
begins with (-)-perillaldehyde (<u>107</u>).[233] The cyclopropane
ring was formed by addition of HBr, followed by base catalyzed dehydrobromination. The product, unsaturated aldehyde
<u>109</u>, was condensed with methylenetriphenylphosphorane to ob-
tain diene <u>110</u>. Compound <u>110</u> reacted with acrolein to give a
mixture of aldehydes <u>111</u> and <u>112</u> in a ratio of 85:15. The
more abundant isomer <u>111</u> was converted into hydrocarbon <u>113</u>
by standard methods.

Osmium tetroxide hydroxylation of <u>113</u> gave a single diol
(<u>114</u>), which formed a mono-p-toluenesulfonate ester (<u>115</u>).
Compound <u>115</u> readily underwent pinacol rearrangement when

Scheme 10. Büchi's Synthesis of (-)-Aromadendrene

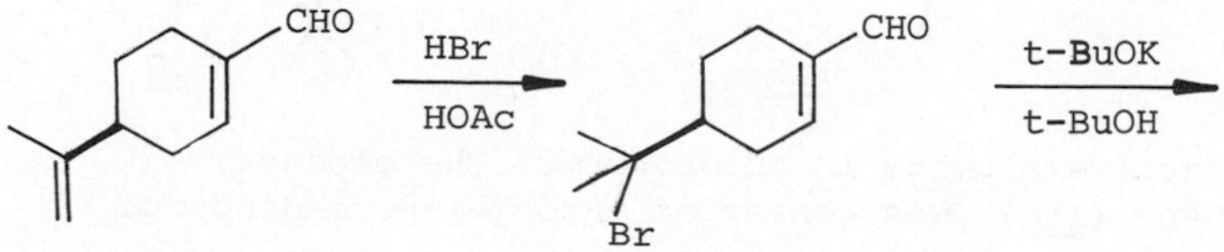

<u>107</u>   <u>108</u>

109 110

111 112

113 114

115 116 117

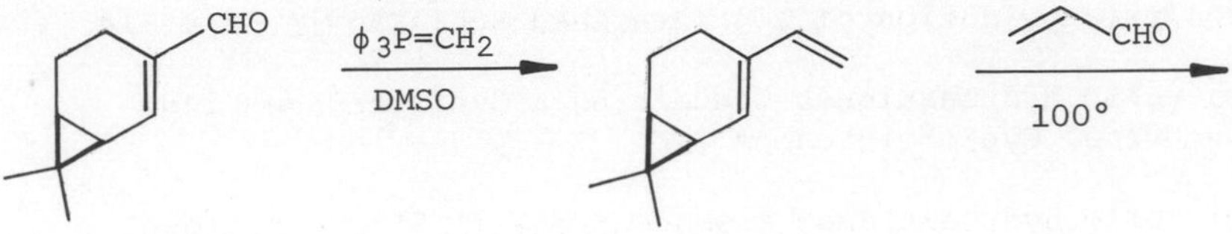

treated with active alumina in chloroform. The product, apoaromadendrene (116), was converted into (-)-aromadendrene (117) by a Wittig reaction.

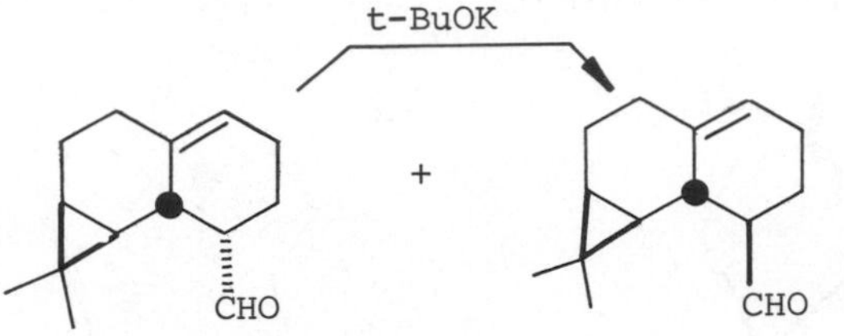

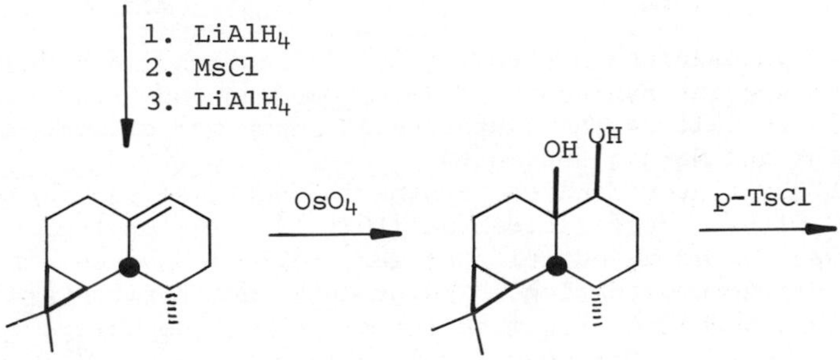

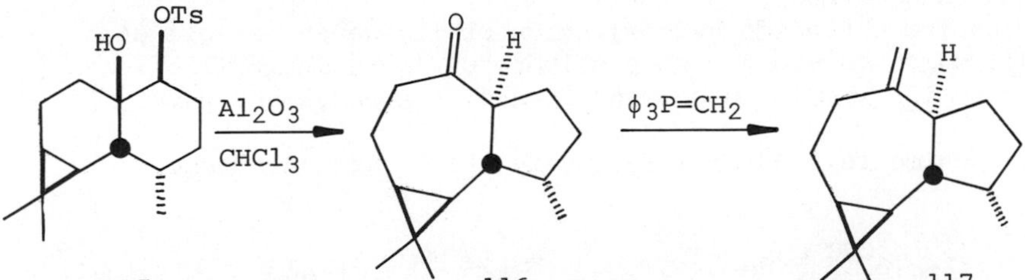

Büchi's synthesis of epicyclocolorenone (124) is outlined

in Scheme 11.[234]   O-acetyl photosantonic lactone (89) was reduced with chromous chloride in acetic acid to dienone 118.
Partial hydrogenation gave mainly a dihydro isomer 119 whose
methyl ester was reduced with lithium aluminum hydride to a
diol.   Reoxidation with dichlorodicyanobenzoquinone gave hydroxy enone 121.   The derived p-bromobenzenesulfonate was
treated with dimethylamine to obtain an amino ketone (122).
Cope elimination of the corresponding N-oxide gave dienone

Scheme 11.   Büchi's Synthesis of Epicyclocolorenone (Relay)

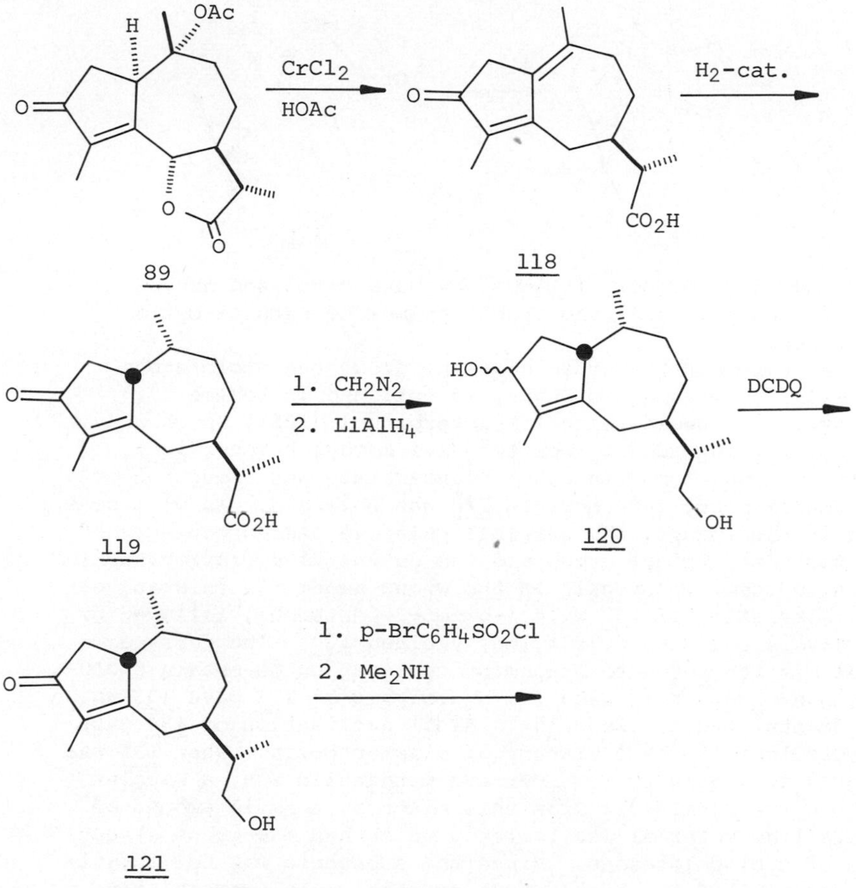

123. Addition of HBr, followed by base catalyzed dehydrobromination gave 124, the stable epimer of natural cyclocolorenone.

A sterorandom synthesis of the gross cyclocolorenone
skeleton, by Narang and Dutta, is outlined in Scheme 12.[235]
The synthesis begins with (±)-α-terpineol (125), which was
oxidatively degraded to homoterpenyl methyl ketone (126).
After condensation with ethyl cyanoacetate and reduction of
the double bond, intermediate 128 was obtained. As will be
seen in the sequel, the critical relative stereochemistry of
the secondary methyl group and the potential cyclopropane ring
is introduced, apparently in the wrong sense, in this stage.

Alkylation of 128 with 1-bromo-3-pentanone, followed by
hydrolysis and decarobyxlation, yielded 130. The derived
ketal was subjected to Dieckmann cyclization to obtain cycloheptanone 132. Mild alkaline hydrolysis of 132 gave 133 which
was deketalized to dione 134. Aldol cyclization of 134 gave
hydroazulene 135 as a mixture of diastereomers. When 135 was
treated successively with HBr and methanolic KOH, a complex
mixture was obtained. From this mixture, a small amount of
crystalline material was isolated which had the gross structure of cyclocolorenone. Since the substance was not identical with either cyclocolorenone or epicyclocolorenone, the
cyclopropane ring and the secondary methyl group must be cis.
The configuration at the angle is unspecified.

Scheme 12. Narang-Dutta Synthesis of Cyclocolorenone Isomer

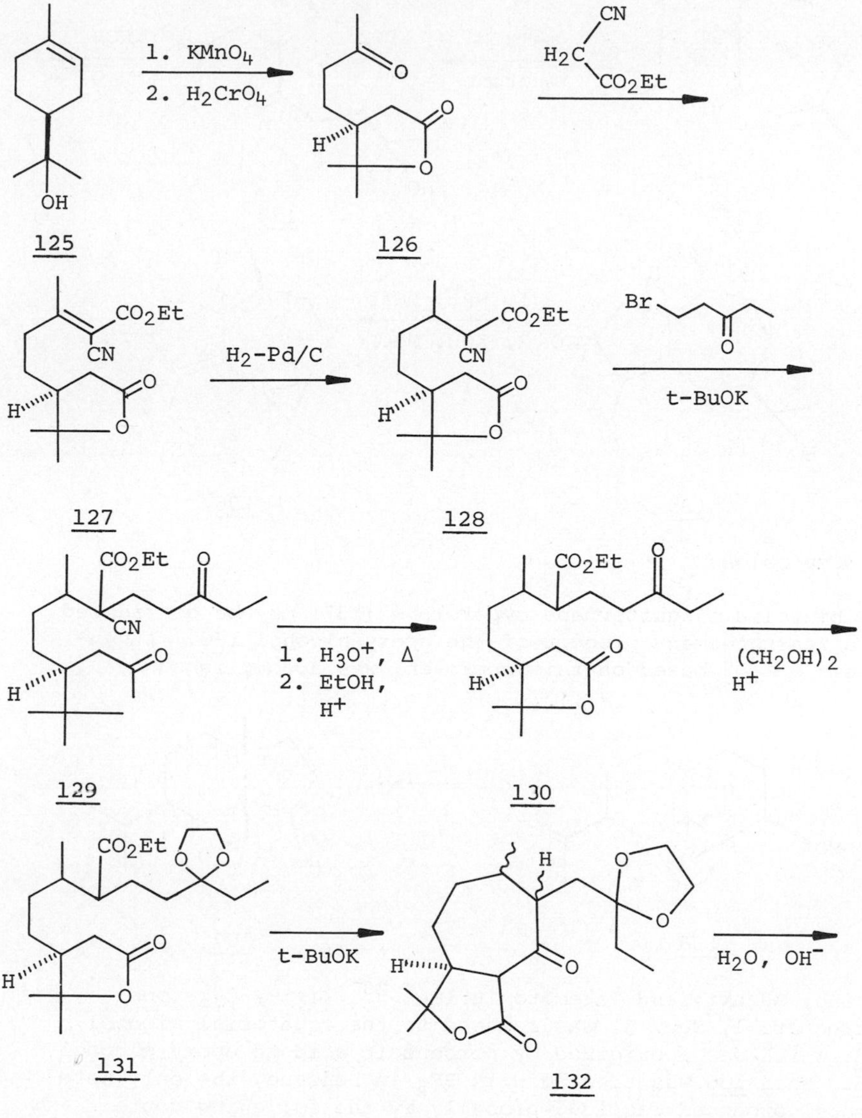

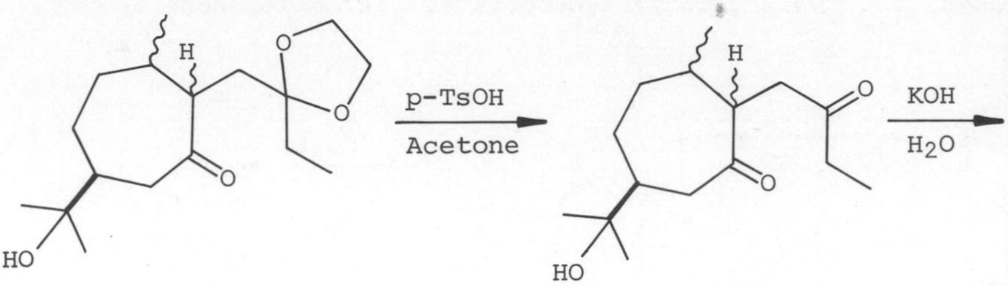

133   134

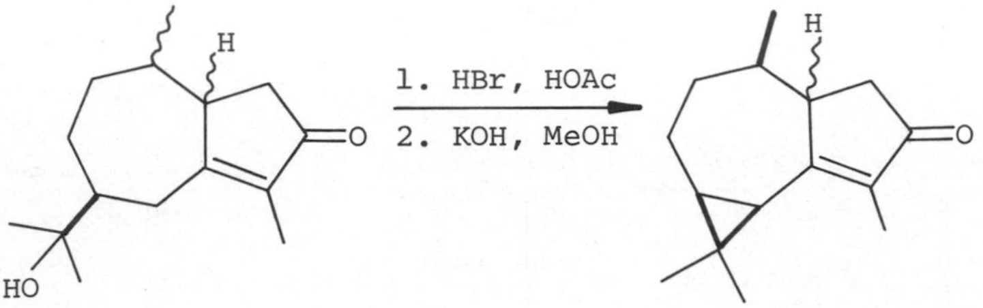

135   136

D. Cyperolone

The bicyclic sesquiterpene cyperolone (137) may be considered as a rearrangement product of the epoxy alcohol 136. A synthesis of 135 based on this approach, was accomplished by

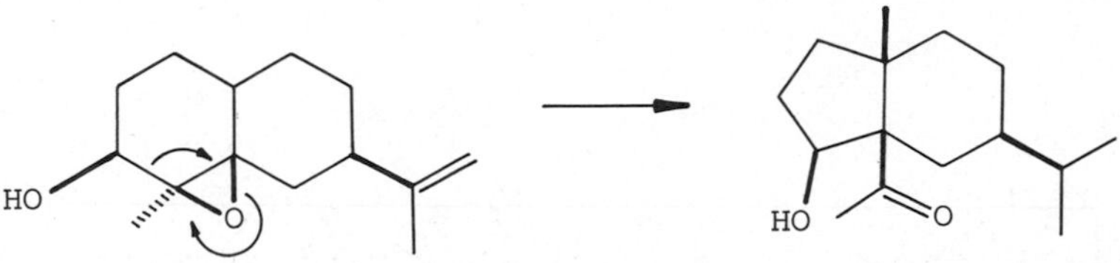

136   137

Hikino, Suzuki, and Takemoto in 1966.[236]   (+)-α-Cyperone (structure 1, Sec. 4) was reduced to the equatorial alcohol 138, which was epoxidized by perbenzoic acid to epoxy alcohol 136. When 136 was treated with BF$_3$ in benzene, the only keto alcohol produced was 139, probably by the following route:

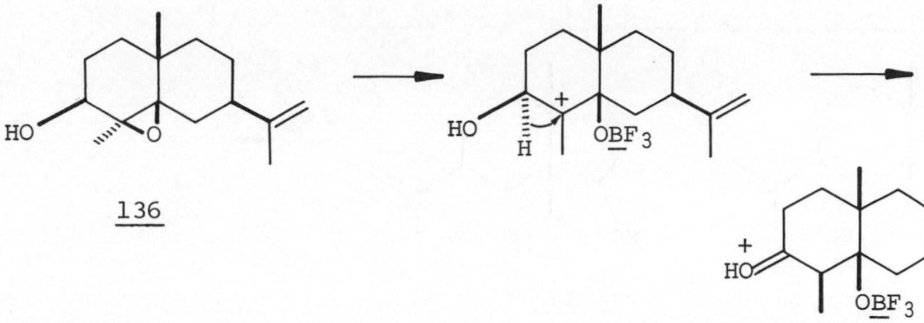

136

Oxidation of 136 gave epoxy ketone 140, which was rearranged with BF₃ in benzene. Diketone 141 was produced in
25% yield. Hydride reduction gave a mixture of diols 142 and
143 (34%). Selective acetylation of 143 gave 144 in 49% yield,

Scheme 13. Hikino-Suzuki-Takemoto Synthesis of Cyperolone

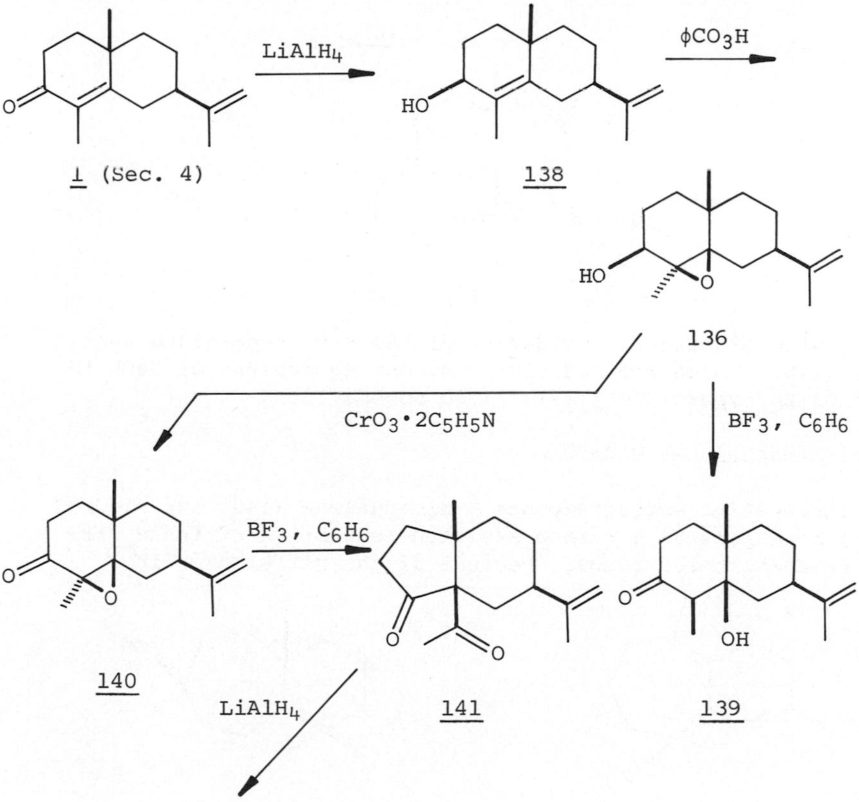

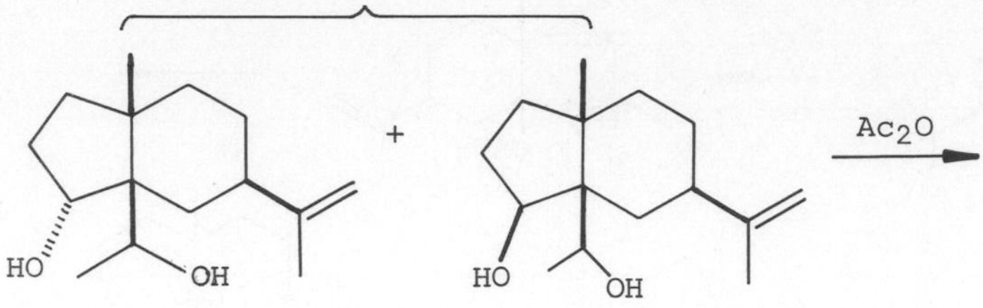

142   143

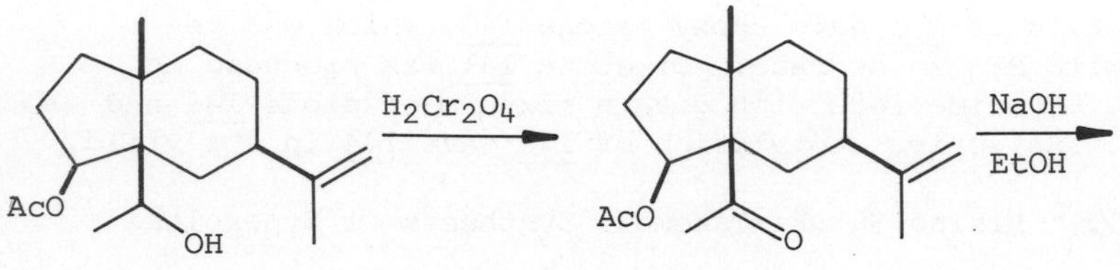

144   145

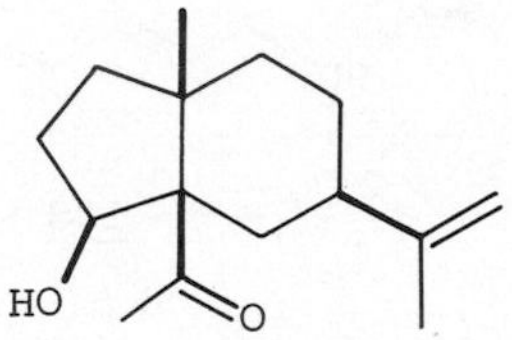

137

along with diacetate. Oxidation of 144 gave cyperolone acetate (145), which was hydrolyzed by one equivalent of NaOH in
ethanol to cyperolone (137). (See Scheme 13.)

E. β-Himachalene, Widdrol

The interesting sesquiterpenes β-himachalene (146) and widdrol
(147) both possess a carbon skeleton consisting of fused six-
and seven-membered rings. Because of the differences in

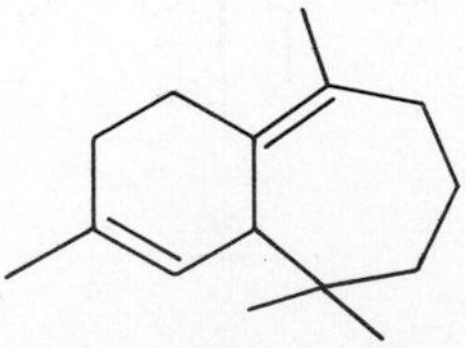

146 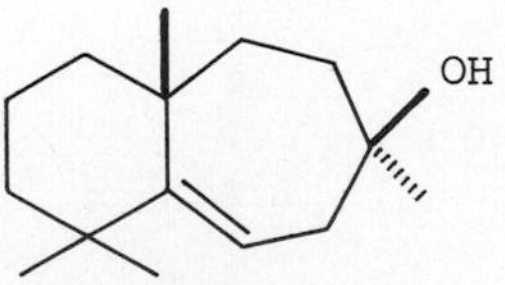   147

substitution, the two compounds present quite different synthetic challenges. De Mayo's approach to the β-himachalene skeleton was based on his discovery that acetylacetone undergoes photochemical cycloaddition with cyclohexene, yielding an intermediate cyclobutanol which can fragment to a δ-diketone:[237]

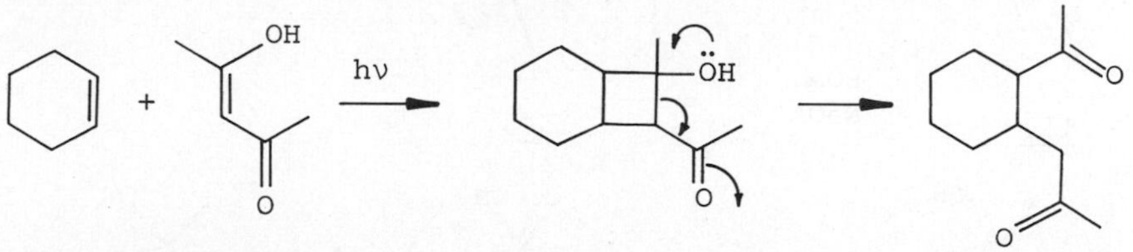

The logical extension, use of a cyclic β-diketone, led to the formation of bicyclic substances.[238]

De Mayo's application of this approach to the synthesis of β-himachalene is outlined in Scheme 14.[238] The ketal of

Scheme 14. De Mayo's Synthesis of (±)-β-Himachalene

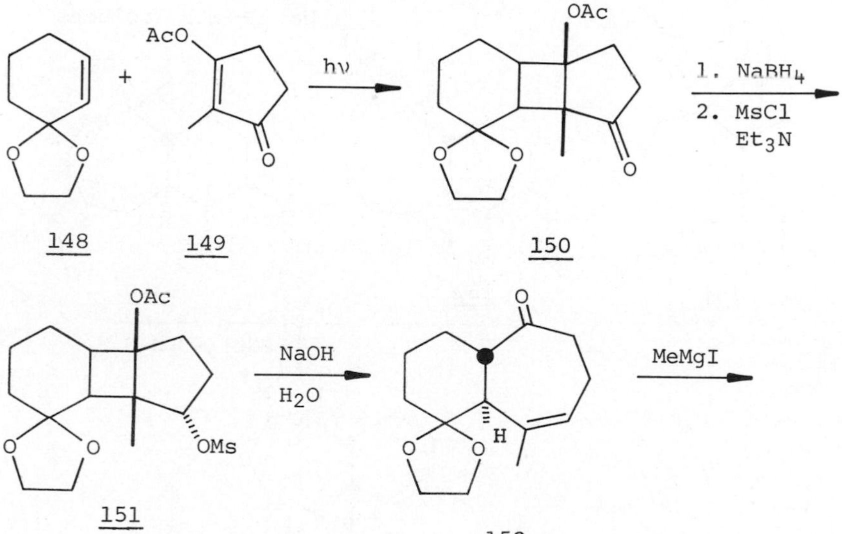

148 149 150

151

152

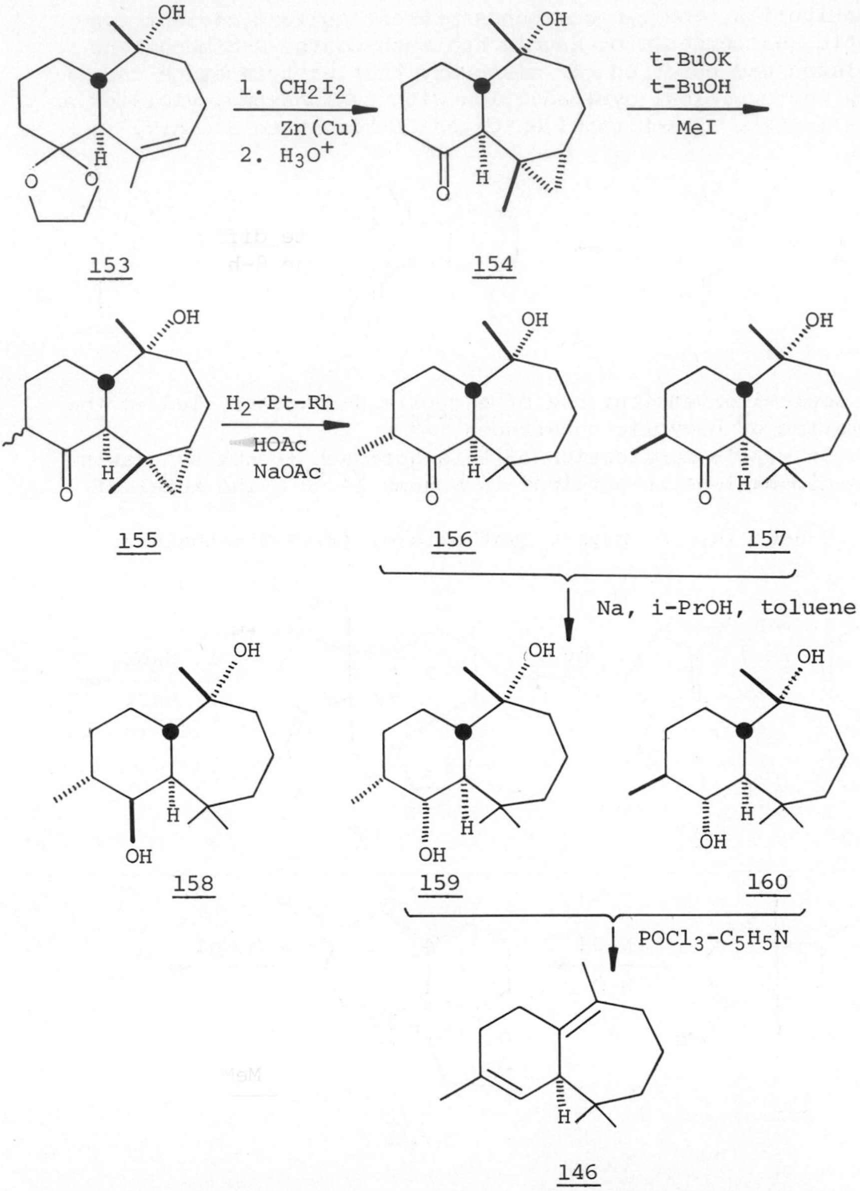

cyclohexenone (148) was photolyzed in the presence of the enol
acetate of 2-methyl-1,3-cyclopentanedione (149). The photo-
adduct 150 was obtained in 35% yield. Although the full
stereochemistry of this adduct is not known, the five-four

ring fusion is probably *cis*. Models indicated that reducing
agents should attack the carbonyl group from the side *cis* to
the angular methyl group. This relative stereochemistry, although lost in the ultimate product, is critical, since only
such an arrangement can lead to concerted fragmentation. The
alternative disposition (methyl and leaving group *cis*) would,
in theory, yield a *trans*-cycloheptene. The high activation
energy associated with such a process would probably preclude
fragmentation.

In the event, borohydride reduction of 150 gave a single
alcohol, which reacted with methanesulfonyl chloride in triethylamine to give mesylate 151. When treated with dilute
alkali, 151 underwent hydrolysis and fragmentation to give
152. Under these conditions, the more stable ring fusion is
expected to result. Subsequent results revealed that the expected *trans* fusion was present. Upon treatment of 152 with
methylmagnesium iodide, an alcohol (153) was produced which
reacted in a Simmons-Smith reaction to yield the tricyclic substance 154, after deketalization.

Since β-himachalene has only one asymmetric center, the
stereochemistry at the five centers in 154 is important only
insofar as they influence the course of further transformations. Since the synthetic plan called for introduction of
the tetrasubstituted double bond of β-himachalene by some type
of dehydration of the tertiary alcohol, a *trans* arrangement of
the hydroxyl group and the adjacent angular hydrogen seemed
ideal. The stereochemistry shown for alcohol 153 was assigned
on the basis of molecular models, assuming attack by Grignard
reagent to the less hindered face of the carbonyl group in
152. In light of the dehydration results (vide infra), this
assignment might be reconsidered.

Methylation of 154 gave 155 as a mixture of diastereomers.
Hydrogenolysis over a platinum-rhodium catalyst afforded a
mixture of keto alcohols 156 and 157. When this mixture was
reduced by sodium and isopropyl alcohol in toluene, diols
158-160 were obtained in yields of 40, 18, and 26%, respectively. Upon dehydration with phosphorous oxychloride in
pyridine, isomers 159 and 160 gave a complex mixture of hydrocarbons, containing 8% of (±)-β-himachalene (146). The major
products in each case were isomers 161 and 162, *trans*-α-
himachalene and *trans*-γ-himachalene.

161   162

Widdrol (147) was synthesized by Enzell as outlined in Scheme 15.[239] Trimethyloctalone structure 605, sec. 4 (see Scheme 69, Sec. 4) was ring expanded with diazomethane and

Scheme 15. Enzell's Synthesis of (±)-Widdrol

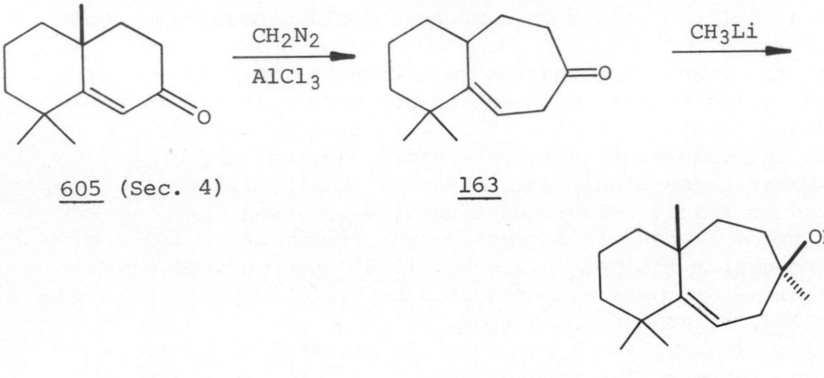

605 (Sec. 4)   163

147

aluminum chloride to give a complex mixture of ketones, from which the desired product 163 was isolated in 15% yield. This material reacted with methyllithium to give (±)-widdrol in 25% yield.

F. Sesquicarene, Sirenin

Sesquicarene (164), the sesquiterpene analog of Δ^2-carene and sirenin (165), an interesting plant hormone, have elicited wide-spread attention. Four syntheses of the former and five syntheses of the latter will be discussed in this section.

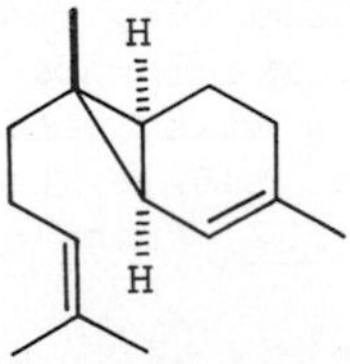

164 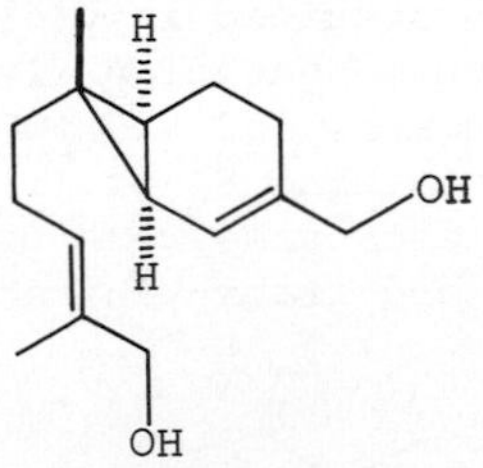   165

The bicyclo[4.1.0]heptane skeleton of sesquicarene and sirenin strongly invites a synthesis based on intramolecular carbene insertion (see Sec. 4-S, cubebene, thujopsene, aristolone):[219]

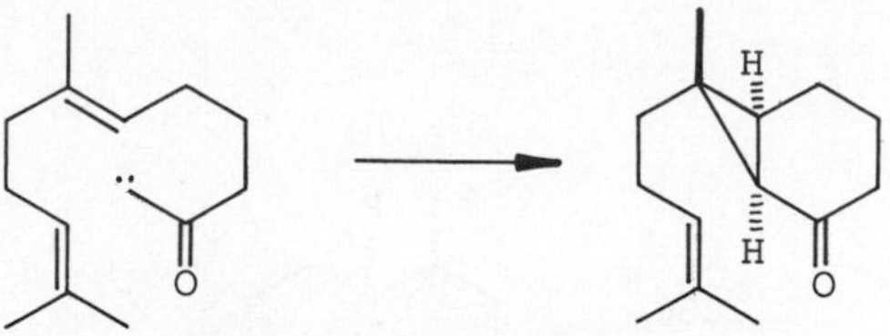

All nine syntheses thus far recorded take this basic approach. Because of the known stereospecificity of this process,[240] the geometry of the trisubstituted double bond will be mirrored in the three asymmetric centers of the cyclization product.   A *trans* disposition will afford "natural" stereochemistry, while a *cis* arrangement would lead to isosesquicarene or isosirenin.

Corey's sixteen-stage route to sesquicarene from geranyl bromide (structure 16, Sec. 2) is outlined in Scheme 16.[241] Alkylation of the lithium salt of propargyl tetrahydropyranyl

Scheme 16.   Corey's Synthesis of (±)-Sesquicarene

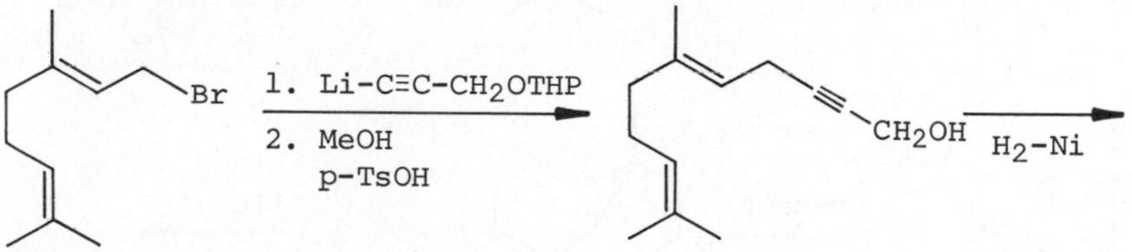

1. Li—C≡C—CH$_2$OTHP

2. MeOH
   p—TsOH

16 (Sec. 2)

166

H$_2$—Ni

CrO$_3$·2C$_5$H$_5$N

CH$_2$Cl$_2$

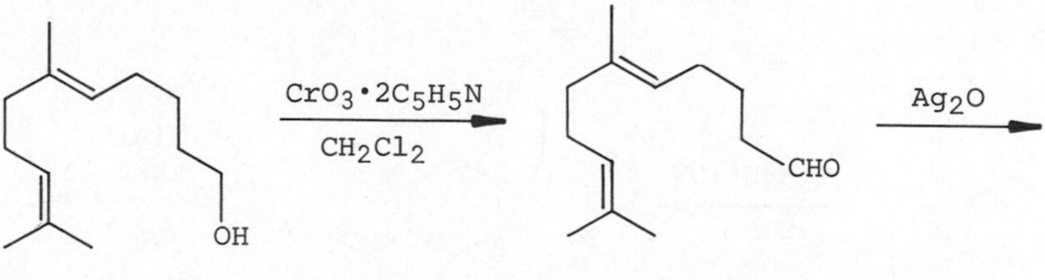

167

168

Ag$_2$O

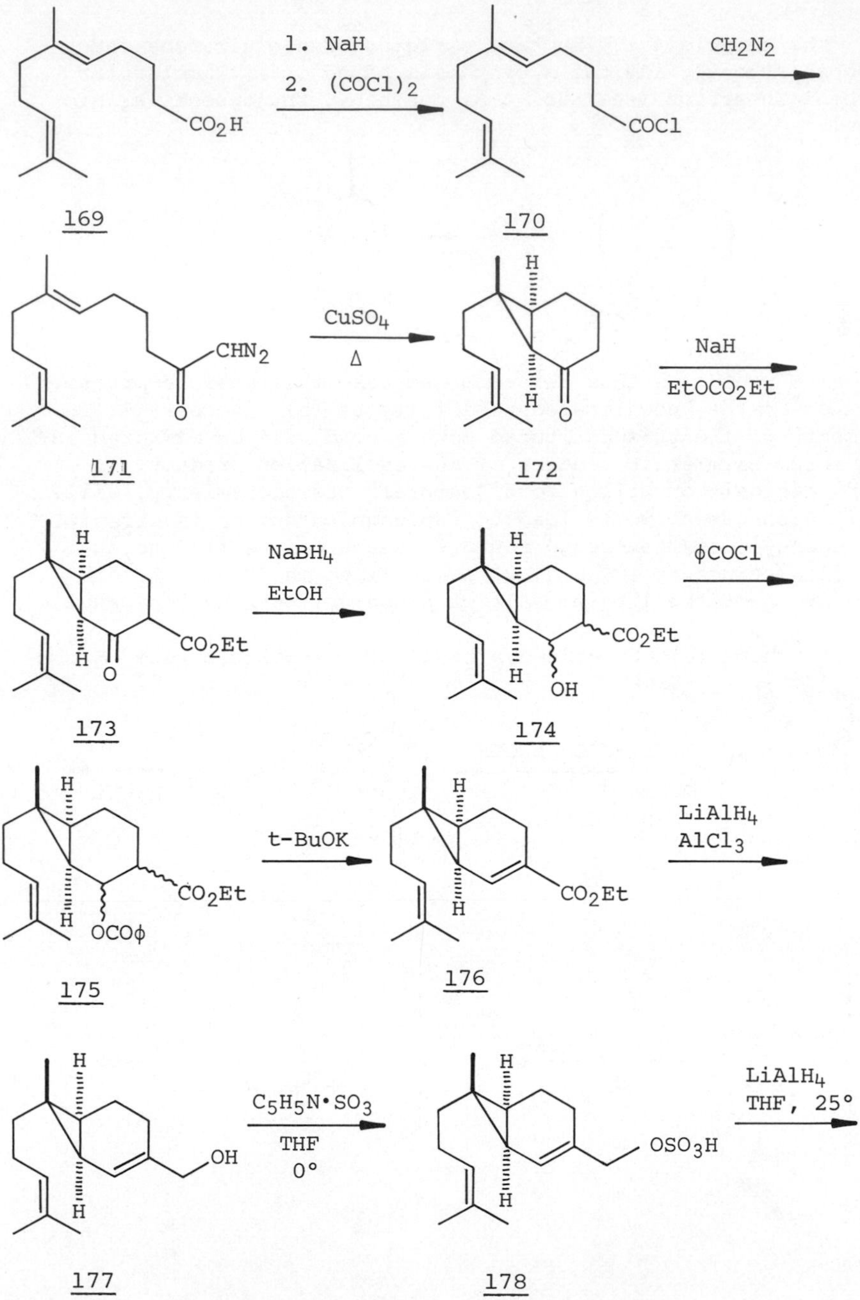

1. NaH
2. $(COCl)_2$

CO_2H

$COCl$

169

170

CH_2N_2

CHN_2

O

171

$CuSO_4$
Δ

H

H
O

172

NaH
$EtOCO_2Et$

H

H
O
CO_2Et

173

$NaBH_4$
EtOH

H

H
CO_2Et
OH

174

$\phi COCl$

H

H
CO_2Et
$OCO\phi$

175

t-BuOK

H

H
CO_2Et

176

$LiAlH_4$
$AlCl_3$

H

H
OH

177

$C_5H_5N\cdot SO_3$
THF
0°

H

H
OSO_3H

178

$LiAlH_4$
THF, 25°

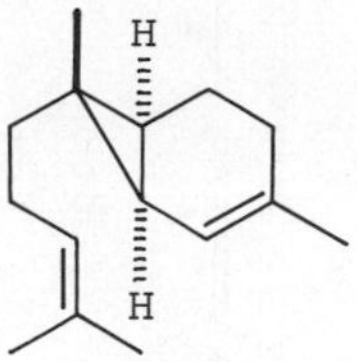

164

ether with structure 16, Sec. 2, gave, after hydrolysis, the propargylic alcohol 166. Reduction of the triple bond, followed by a two-stage oxidation, gave acid 168. The corresponding diazoketone, 171, prepared in the standard manner, was decomposed by heating in cyclohexane with cupric sulfate. Bicyclic ketone 172 was obtained, as the only isomer, in 60% yield from acid 169.

From this point, one must introduce the potential vinyl methyl group adjacent to the carbonyl group and change the carbonyl group into a double bond. The route chosen by Corey, although lengthy, was designed so as to be applicable in a synthesis of sirenin (vide infra). Carboxylation occurred when 172 was treated with sodium hydride in diethyl carbonate. The resulting β-keto ester (173) was reduced to a mixture of β-hydroxy esters (174), which gave the corresponding benzoates (175). Base catalyzed elimination afforded unsaturated ester 176, which was reduced to allylic alcohol 177. Because of the sensitivity of 177, its hydrogenolysis was accomplished by an interesting new method. Successive treatment with sulfur trioxide-pyridine complex, followed by lithium aluminum hydride, gave (±)-sesquicarene (164).

The Mori-Matsui (Scheme 17)[242] and Coates (Scheme 18)[242] syntheses of sesquicarene are similar in their general outline.

Scheme 17. Mori-Matsui Synthesis of (±)-Sesquicarene

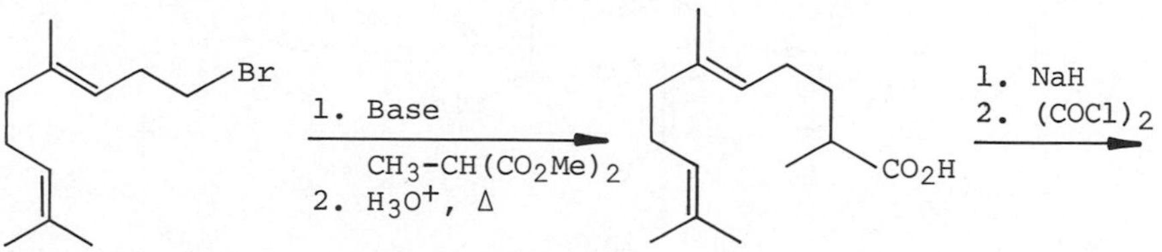

24 (Sec. 2)   180

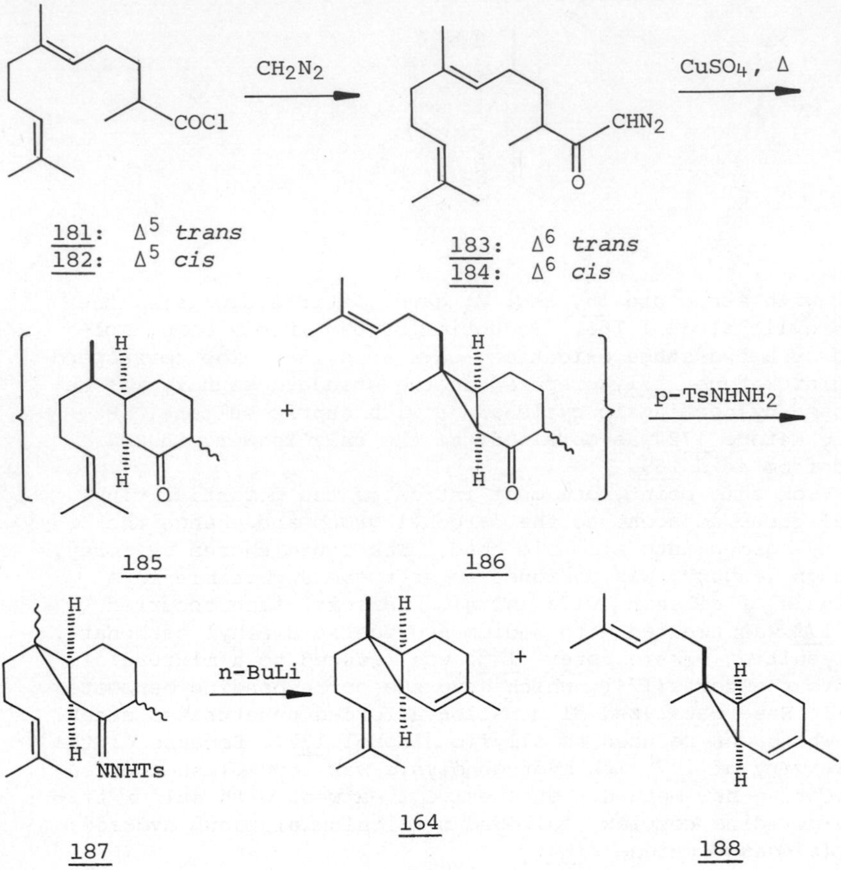
COCl
CH_2N_2
CHN_2
O
$CuSO_4$, Δ
181: Δ^5 *trans*
182: Δ^5 *cis*
183: Δ^6 *trans*
184: Δ^6 *cis*
H
H
O
+
H
H
O
p-$TsNHNH_2$
185
186
H
H
NNHTs
n-BuLi
H
H
+
H
H
187
164
188

Scheme 18. Coates' Synthesis of (±)-Sesquicarene

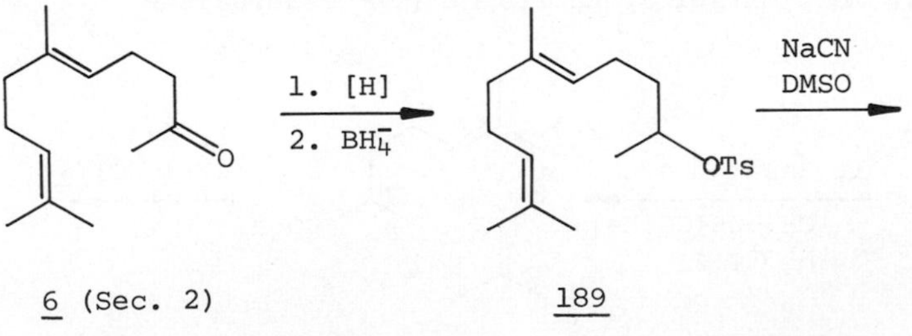
O
1. [H]
2. BH_4^-
OTs
NaCN
DMSO

6 (Sec. 2)

189

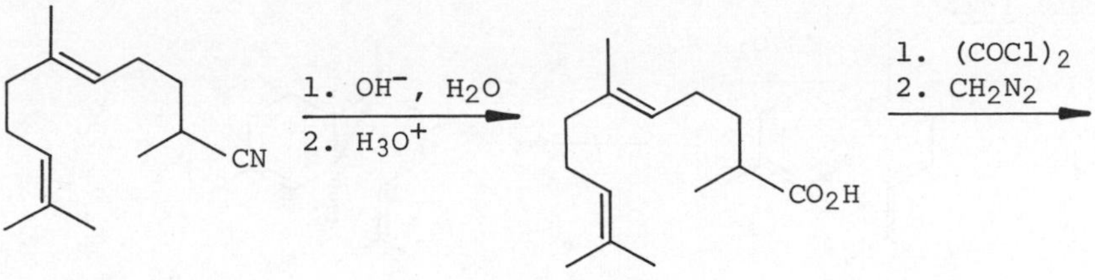

190

180

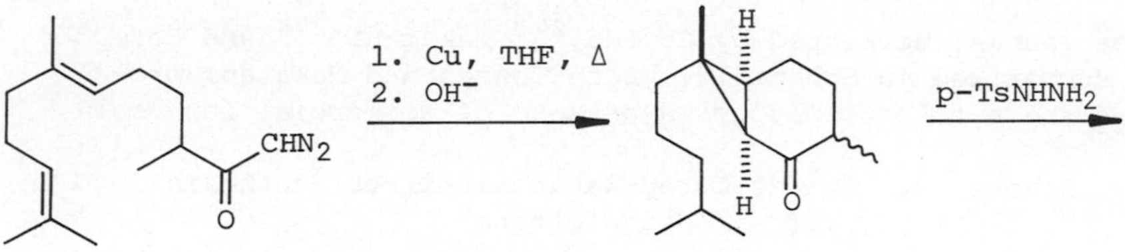

183

184

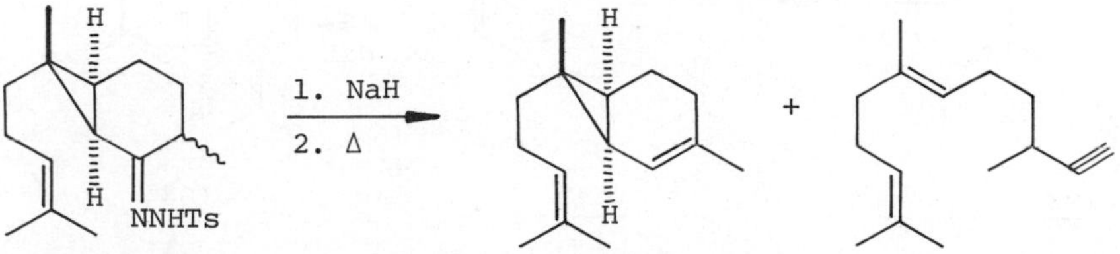

186

164

191

Both proceed through the diazo ketone 183, which is arrived at in different routes. The Mori-Matsui route is nonselective; *trans*:*cis* ratios of 2.5:1 were observed at the trisubstituted double bond. Because of this mixture, Mori and Matsui obtained not only sesquicarone (185), but its C-7 isomer 186 in a ratio of 2.4:1. Coates, working with a pure isomer of diazo ketone 183, obtained only 184 (as a mixture of methyl epimers). In both syntheses, the double bond was introduced by Bamford-Stevens reaction on *p*-toluenesulfonyl hydrozone 186. Mori and Matsui decomposed 186 with *n*-butyllithium and obtained 164 and 188 in approximately 20% yield. Coates pyrolyzed the sodium salt of 186 and obtained (±)-sesquicarene (164) and the fragmentation product 191 in 15% and 53% yields, respectively.

Several groups have explored direct routes to (±)-sesquicarene which are based on intramolecular insertion of carbene 192.

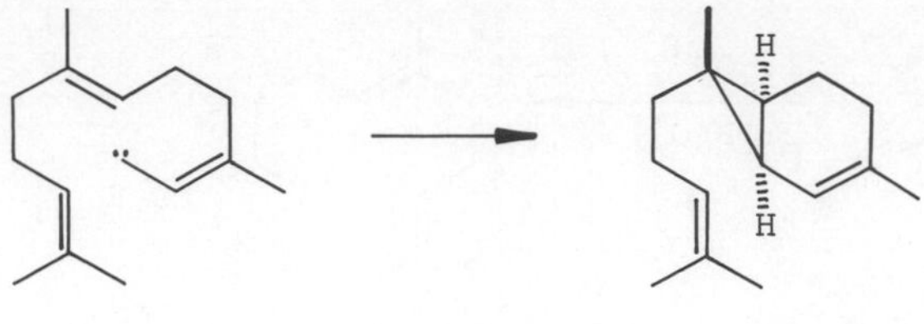

192   164

These routes, developed by Coates,[243] Nakatani,[244] and Corey[245] are summarized in Scheme 19. Both Coates and Nakatani prepared the *p*-toluenesulfonyl hydrazone of commercial farnesal

Scheme 19.   Coates-Corey-Nakatani Direct Synthesis
             of (±)-Sesquicarene

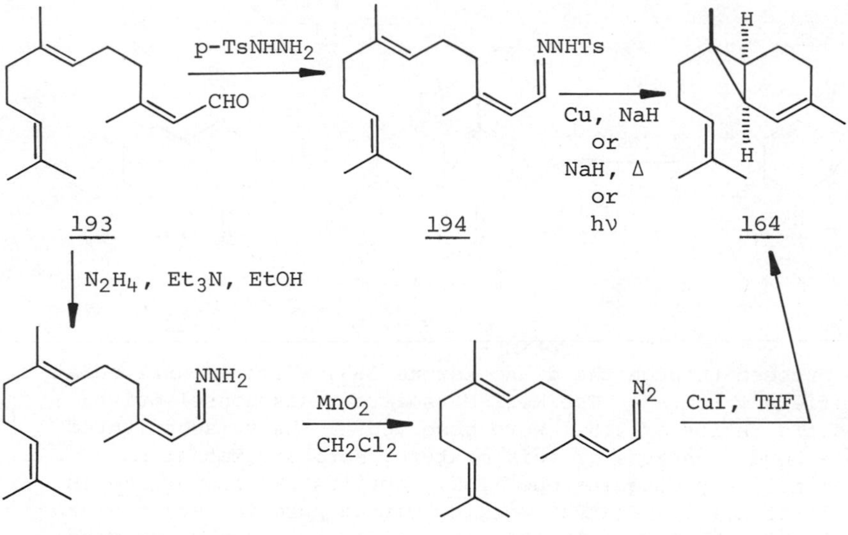

(mixture of *cis,trans* and *trans,trans*). Coates decomposed the tosylhydrazone with copper in the presence of sodium hydride. (±)-Sesquicarene was obtained in 5% yield. Nakatani carried out the cyclization by photolysis or pyrolysis of the sodium salt of 194. (±)-Sesquicarene and its C-7 epimer were obtained in 5-15% yield.

Corey fractionated commercial farnesol to obtain only the *cis,trans* isomer. Oxidation gave *cis,trans* farnesal (193), which formed the hydrazone (195) with hydrazine and triethylamine in ethanol. Oxidation of 195 with excess activated manganese dioxide gave the diazo compound 196. Cuprous iodide-catalyzed decomposition of 196 gave (±)-sesquicarene in 25% yield, based on *cis,trans*-farnesol. In a later study, Corey reported that mercuric iodide catalyzes the conversion of both 196 and its *trans,trans* isomer to sesquicarene.[250] Using this modification, the commercially available mixture of *trans,trans*- and *cis,trans*-farnesol may be used. The yield reported for cyclization of the mixture of diazoketones to (±)-sesquicarene is 60%.[250]

As with sesquicarene, all of the sirenin syntheses are based on intramolecular carbene insertion. Rapoport's synthesis, which was designed to yield both (±)-sirenin and (±)-isosirenin for biological testing, is outlined in Scheme 20.[246] Methyl heptenone (structure 14, Sec. 2) was condensed with the ylid derived from phosphonium salt 197 to yield a mixture of acids 198 and 169 in a ratio of 3:2. The minor isomer, shown to possess a *trans* double bond, was converted by the normal method (see Scheme 16) into bicyclic ketone 172. The major acid (198) with a *cis* double bond, yielded bicyclic ketone 200. Rapoport's method for changing the carbonyl group into the required vinylic hydroxymethyl function, was identical to that used by Corey for sesquicarene (Scheme 16), except that pivaloate, rather than benzoate, was eliminated.

Selenium dioxide oxidation of 204 gave a mixture of aldehyde and allylic alcohol. Oxidation of the entire product with MnO_2 yielded only aldehydo ester 205. When compound 205 was reduced with $LiAlH_4$-$AlCl_3$ at 0°, (±)-sirenin was obtained. A similar sequence of reactions converted bicyclic ketone 200 into (±)-isosirenin. (See Scheme 20.)

Scheme 20. (±)-Sirenin--Rapoport's Synthesis

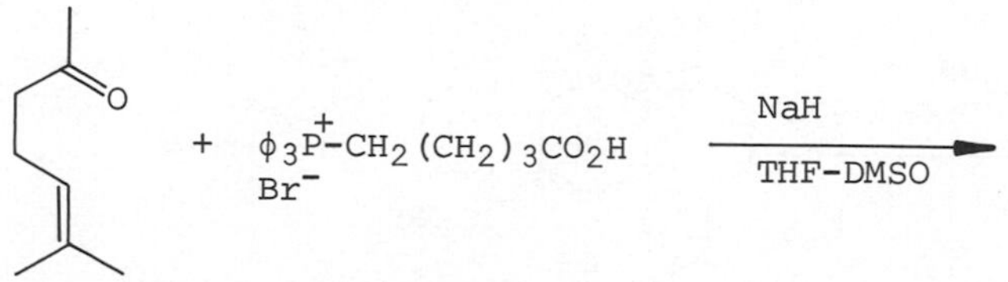

14 (Sec. 2)   197

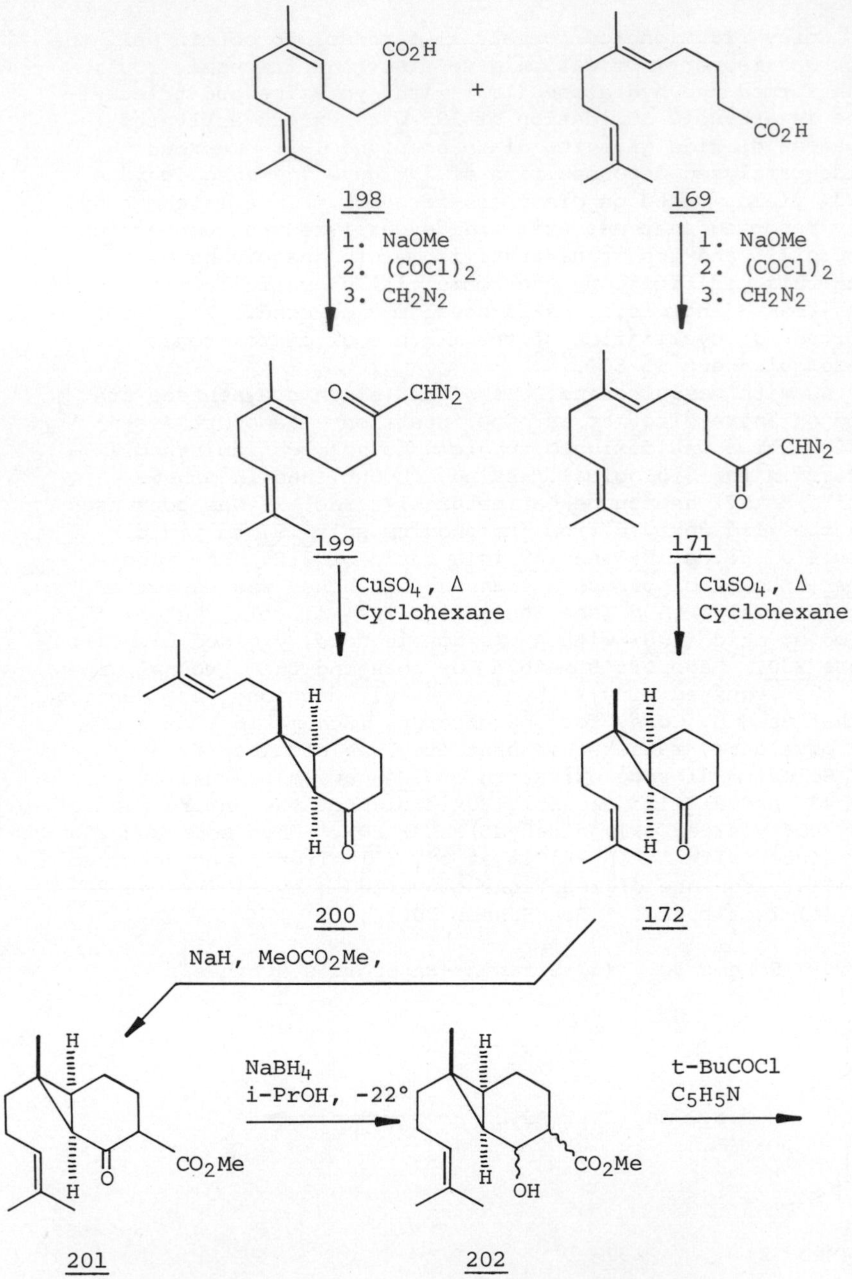
CO2H
+
CO2H
198
169
1. NaOMe
2. (COCl)2
3. CH2N2
1. NaOMe
2. (COCl)2
3. CH2N2
CHN2
O
CHN2
O
199
171
CuSO4, Δ
Cyclohexane
CuSO4, Δ
Cyclohexane
H
H
O
H
H
O
200
172
NaH, MeOCO2Me,
H
H
O
CO2Me
201
NaBH4
i-PrOH, -22°
H
H
OH
CO2Me
202
t-BuCOCl
C5H5N

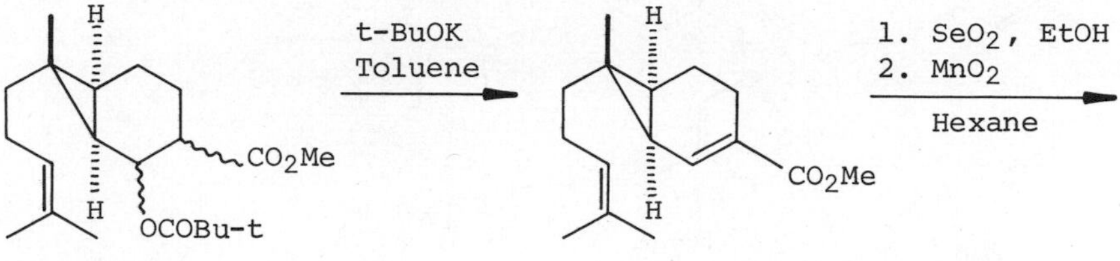
H
H
CO2Me
OCOBu-t
t-BuOK
Toluene
H
H
CO2Me
1. SeO2, EtOH
2. MnO2
Hexane

203 204

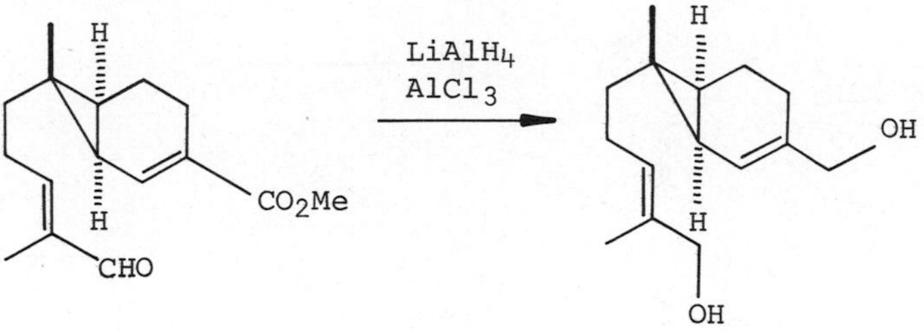
H
H
CO2Me
CHO
LiAlH4
AlCl3
H
H
OH
OH

205 165

Corey's first synthesis of (±)-sirenin (Scheme 21)[247] differs from Rapoport's in that the side chain is functionalized *before* the cyclization step. A high price is paid for this novelty, which is apparently unnecessary. Corey's synthesis, while stereoselective (Scheme 21) or stereospecific (Scheme 22), requires either 23 or 25 steps, with one isomer separation required in either case. Allylic oxidation of dienic ester 176 (as in Rapoport's synthesis, Scheme 20; 204 → 205), synthesized stereospecifically by Corey for the synthesis of sesquicarene (Scheme 16), would yield (±)-sirenin in 15 steps.

In order to establish the desired *trans*,*trans* geometry in diazo ketone 217, Corey explored two routes. The more efficient route, although not stereospecific, is outlined in

Scheme 21. (±)-Sirenin--Corey's First Synthesis

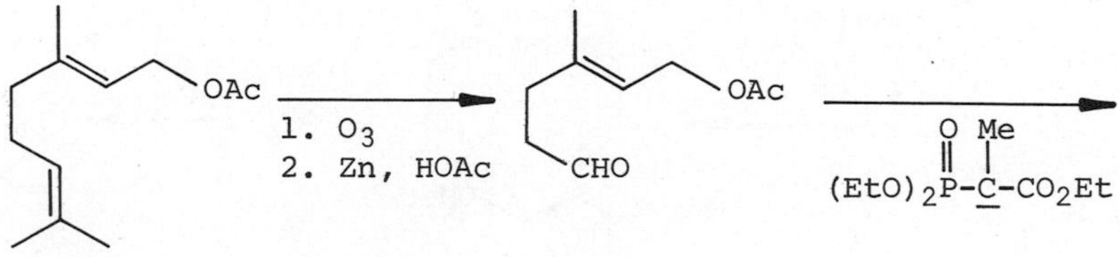
OAc
1. O3
2. Zn, HOAc
OAc
CHO
(EtO)2P-C-CO2Et
O Me

206 207

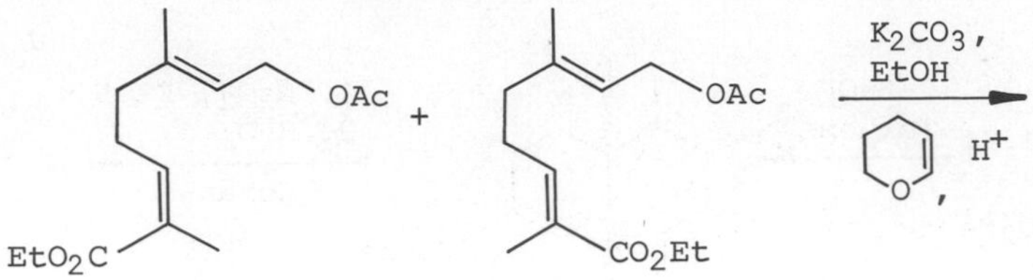

208 209

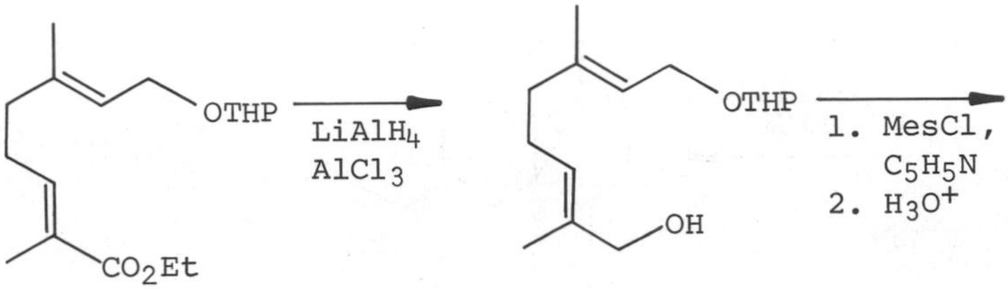

210 211

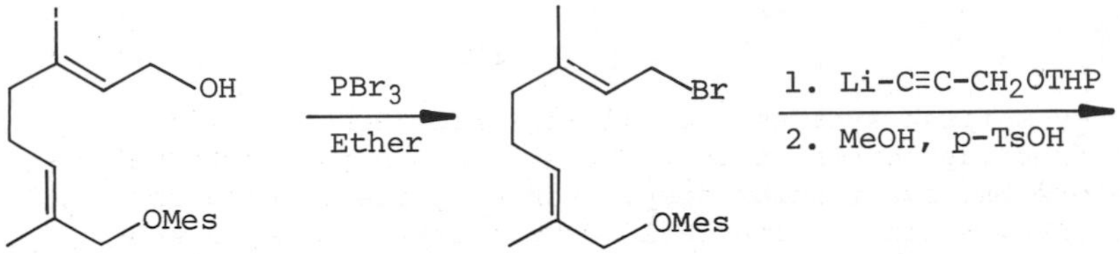

212 213

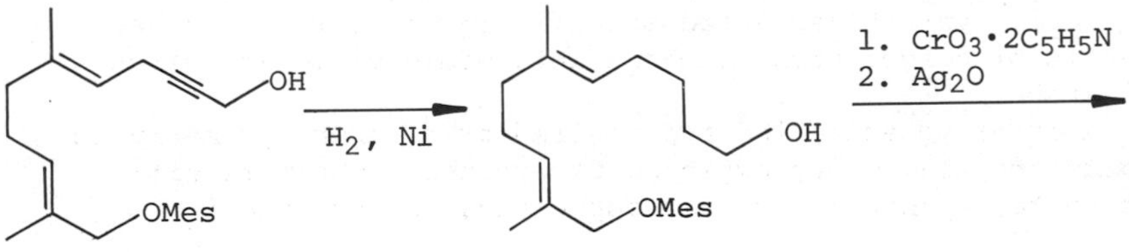

214 215

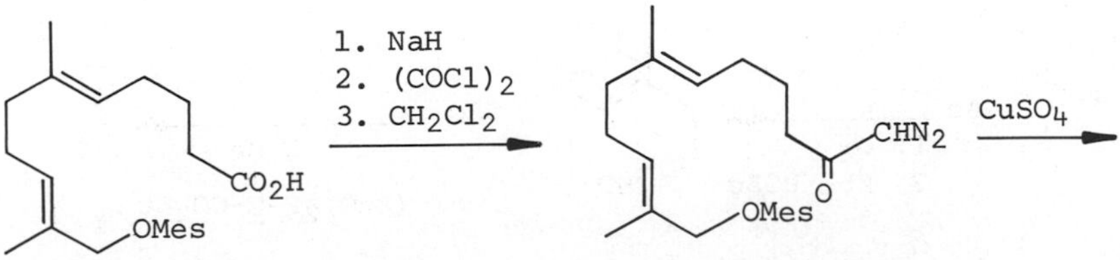

216 217

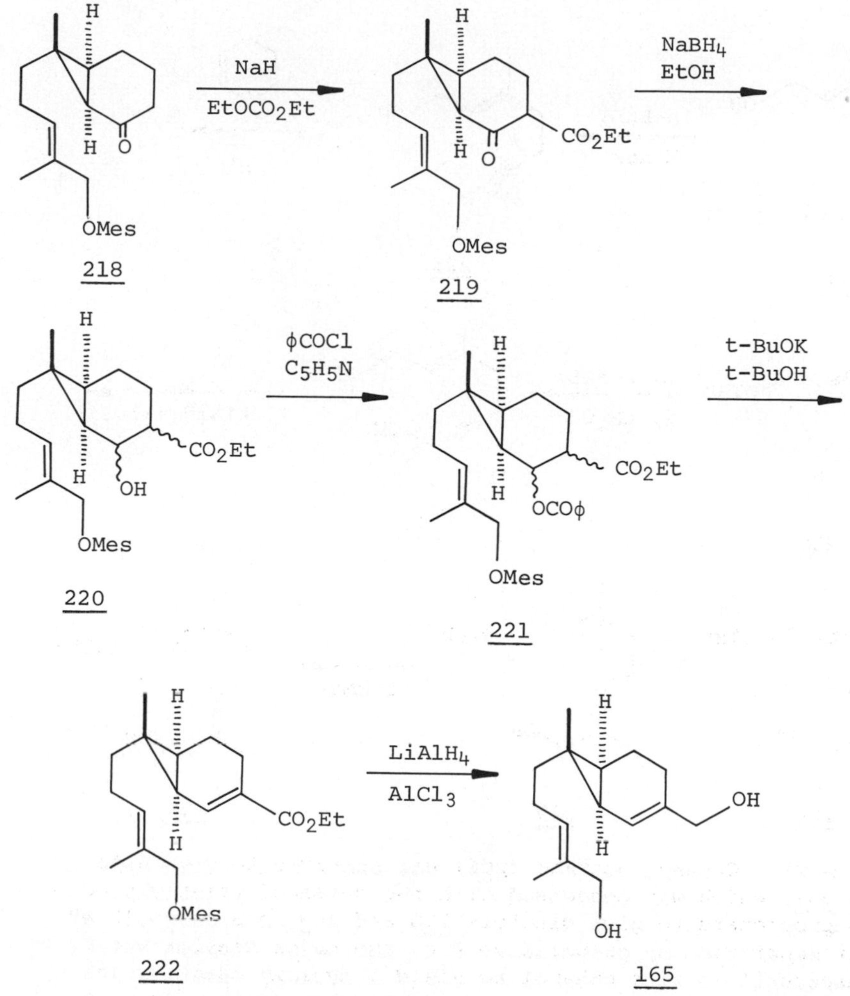

Scheme 22. Corey's Alternative Synthesis of Hydroxy Ether 211

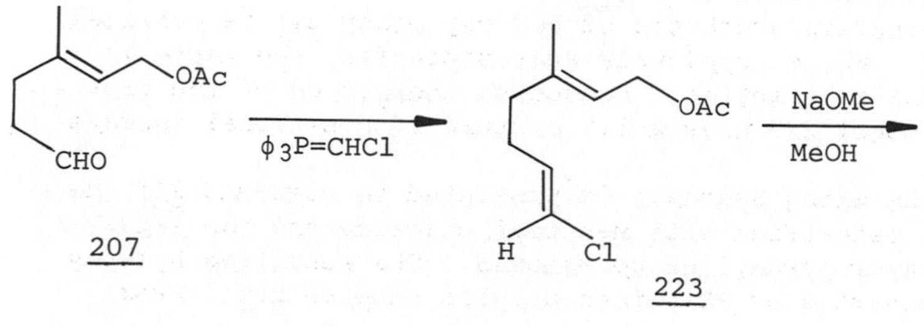

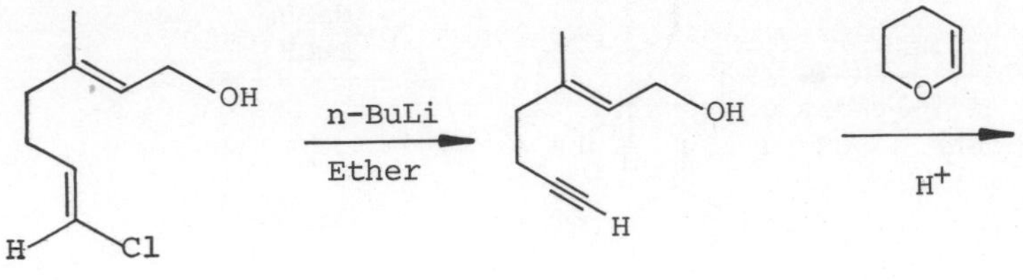

224 225

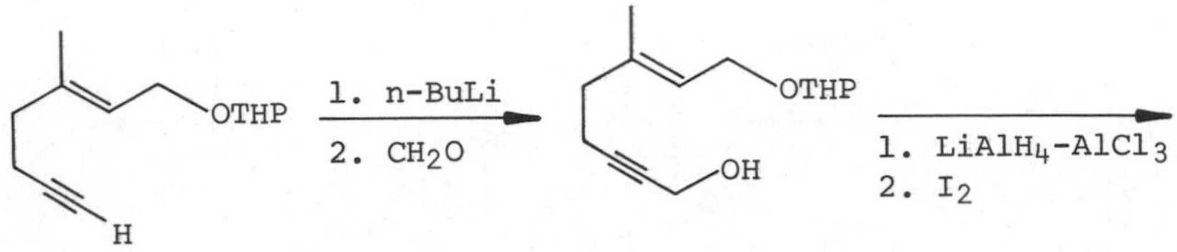

226 227

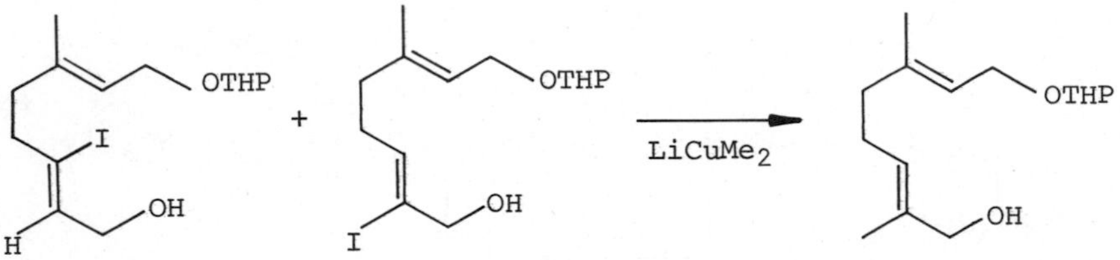

228 229 211

Scheme 21. Geranyl acetate (206) was ozonized to give aldehyde 207, which was condensed with the anion of triethylphosphonopropionate to give diesters 208 and 209 in a ratio 12:88. After separation by preparative tlc, the major diester was transesterified with ethanol to yield a hydroxy ester, which was protected as its tetrahydropyranyl ether (210). The α,β-unsaturated ester function was then reduced by $LiAlH_4$-$AlCl_3$ to give the differentiated diol 211.

An alternative synthesis of hydroxy ether 211 is outlined in Scheme 22. While completely stereospecific, the route is less efficient in practice. Reductive iodination of the propargyllic alcohol 227 gave a 1:2 mixture of positional isomers 228 and 229.

Since the wrong hydroxyl is protected in compound 211 the free end was esterified with mesitoyl chloride and the acid-labile tetrahydropyranyl group removed. The resulting hydroxy ester was converted by PBr_3 into allylic bromide 213. From this point, the synthesis closely parallels Corey's sesqui-carene synthesis (Scheme 16).

Grieco, at Columbia University, synthesized sirenin by the method shown in Scheme 23.[248] As in Rapoport's synthesis,

Scheme 23.   (±)-Sirenin--Grieco's Synthesis

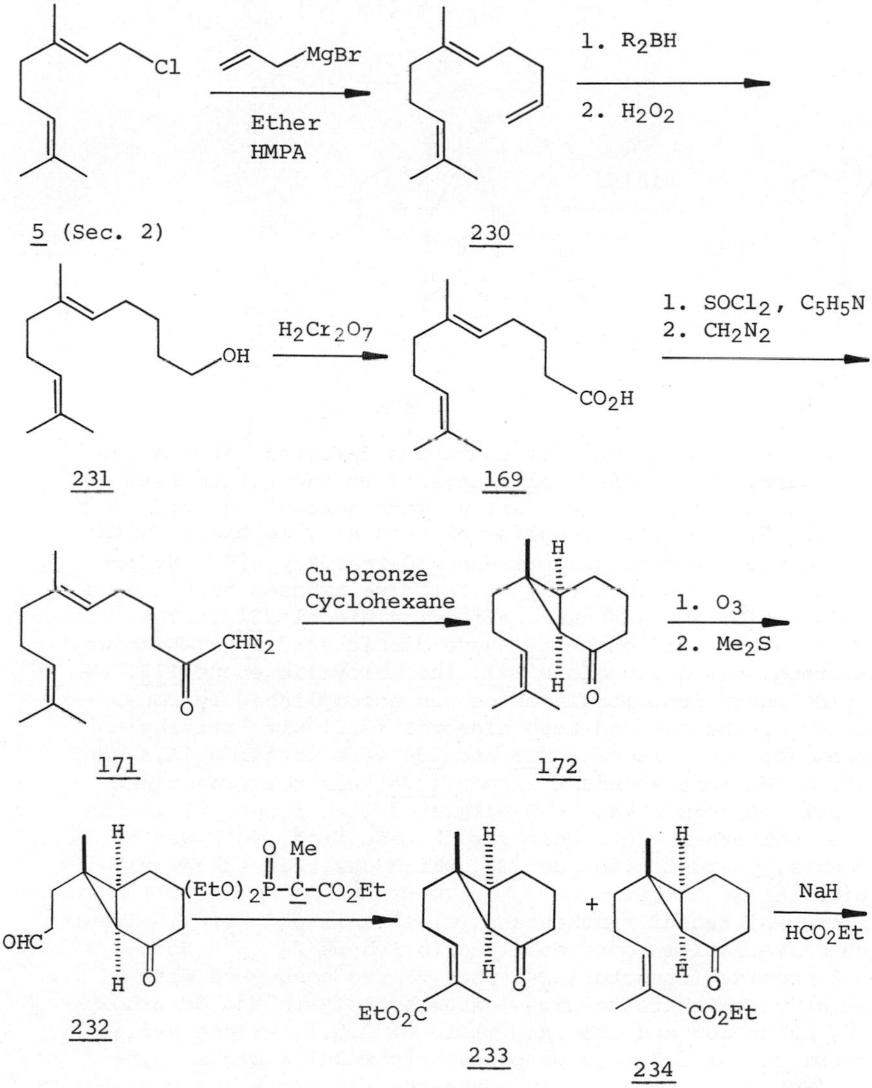

5 (Sec. 2)   230

231   169

171   172

232   233   234

235

236

237

165

functionalization of the side chain was deferred until after ring closure. Stereochemical integrity at the trisubstituted double bond was assured by starting with geranyl chloride (structure 5, Sec. 2). Coupling of this allylic halide with allylmagnesium bromide gave triene 230 in 90% yield. Hydroboration occurred selectively at the more exposed double bond when disiamylborane was used, affording dienol 231 in 70% yield. Jones oxidation of 231 gave dienic acid 169, which was transformed, via diazoketone 171, into bicyclic enone 172.

Side chain functionalization was accomplished by condesnation of the derived keto aldehyde (232) with triethylphosphonoacetate. Isomers 233 and 234 were obtained in a ratio of 8:92. The more abundant isomer (234) was formylated and the resulting formyl ketone O-alkylated with isopropyl iodide to give enol ether 236. Borohydride reduction, followed by hydrolysis, yielded aldehyde 237, which was reduced to (±)-sirenin.

Mori and Matsui synthesized (±)-sirenin and (±)-isosirenin by the nonselective route outlined in Scheme 24.[249] Homogeranyl bromide (structure 24, Sec. 2) was condensed with sodiodiethyl malonate to give, after hydrolysis and decarboxylation, acids 169 and 198 in a ratio of 2.5:1. Since halide, structure 24, Sec. 2, was prepared by the Julia olefin synthesis (Scheme 5, Sec. 2), it contained the *trans* and *cis* isomers in approximately this ratio. From this acid mixture, the synthesis follows a line similar in most respects to the

Scheme 24.   (±)-Sirenin--Mori-Matsui Synthesis

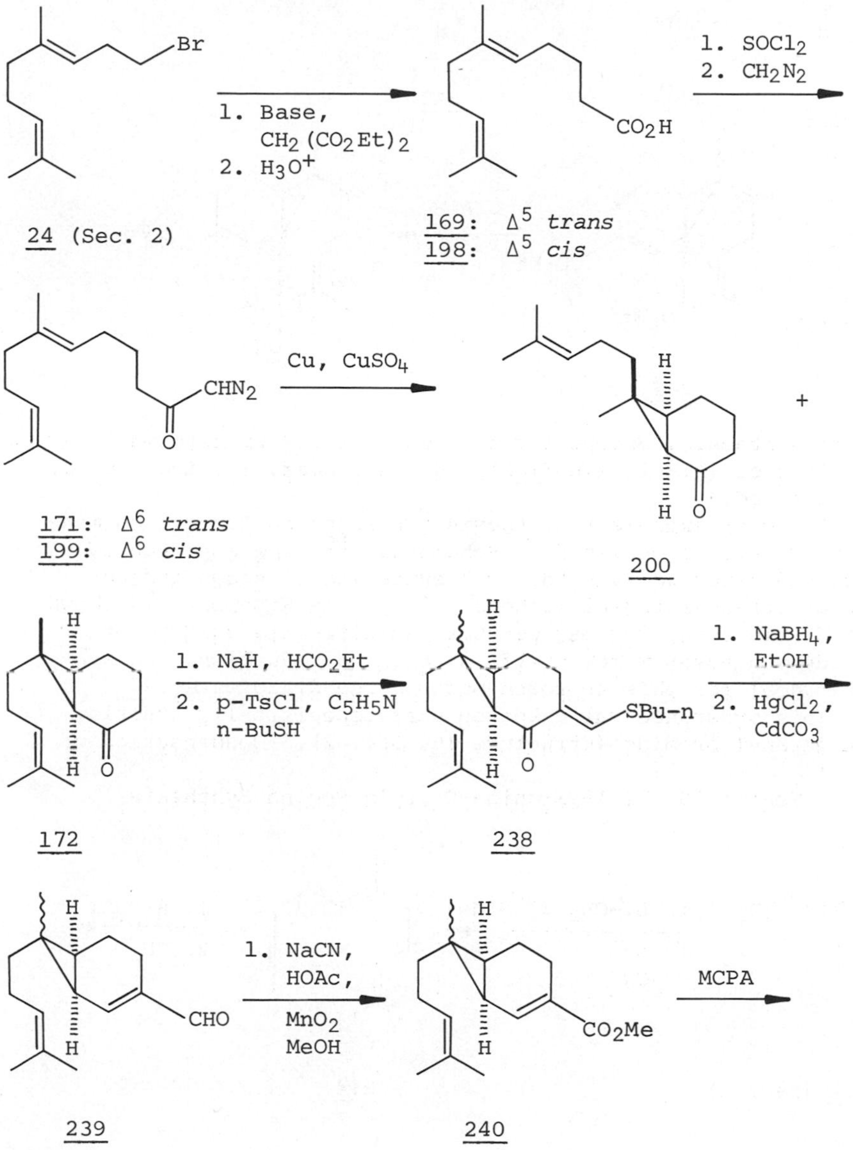

24 (Sec. 2)

1. Base,
   $CH_2(CO_2Et)_2$
2. H_3O^+

$\underline{169}$: Δ^5 *trans*
$\underline{198}$: Δ^5 *cis*

1. $SOCl_2$
2. CH_2N_2

$\underline{171}$: Δ^6 *trans*
$\underline{199}$: Δ^6 *cis*

Cu, $CuSO_4$

$\underline{200}$

+

172

1. NaH, HCO_2Et
2. p-TsCl, C_5H_5N
   n-BuSH

238

1. $NaBH_4$,
   EtOH
2. $HgCl_2$,
   $CdCO_3$

239

1. NaCN,
   HOAc,
   MnO_2
   MeOH

240

MCPA

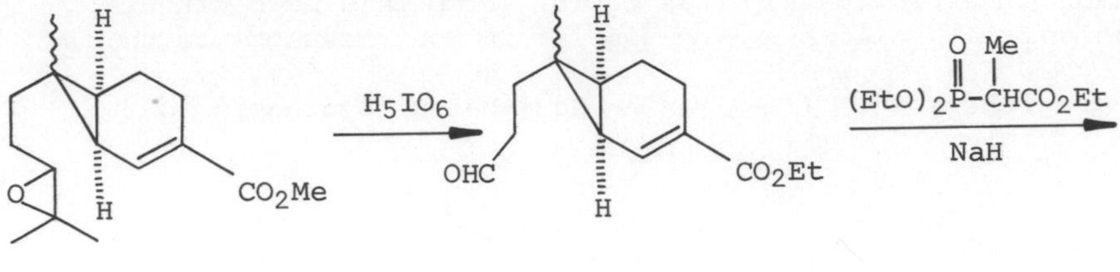

241   242

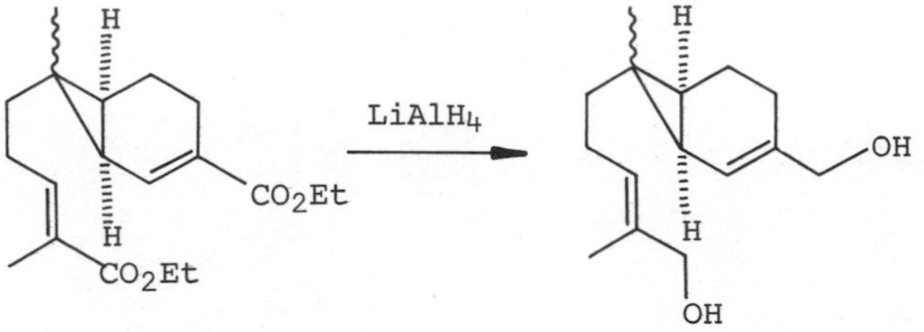

243   165

Grieco synthesis, except for the order of the transformations.
A mixture of (±)-sirenin (165) and its isomer, (±)-isosirenin,
was obtained.

The most expedient synthesis yet reported for the hormone,
due to Corey, is outlined in Scheme 25.[250] The approach is
patterned after Corey's "direct" synthesis of sesquicarene
from *cis,trans*-farnesal (Scheme 19). Since Rapoport had shown
that dienic ester 204 may be stereospecifically oxidized to
an aldehydo ester which yields sirenin on reduction, the synthetic goal for this approach becomes the diazo ester 250.

This substance was prepared stereospecifically starting
with geranyl bromide (structure 16, Sec. 2). Condensation of

Scheme 25. (±)-Sirenin--Corey's Second Synthesis

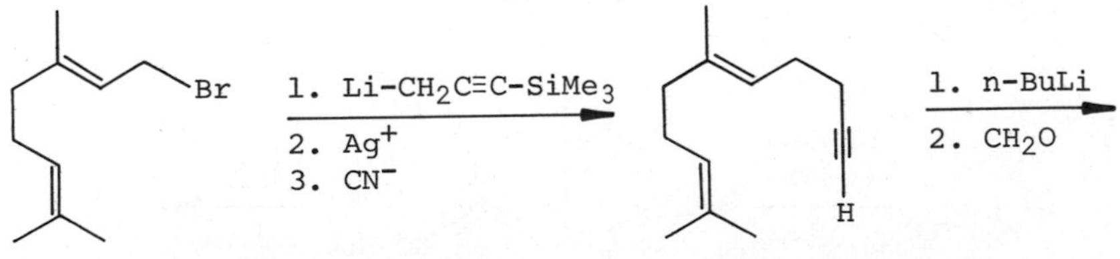

16 (Sec. 2)   214

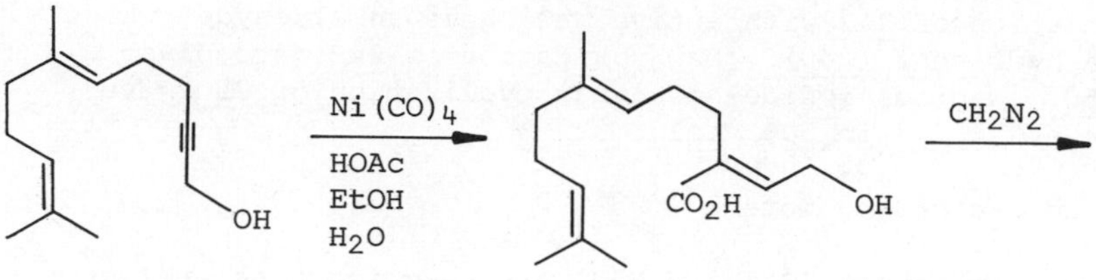

245

246

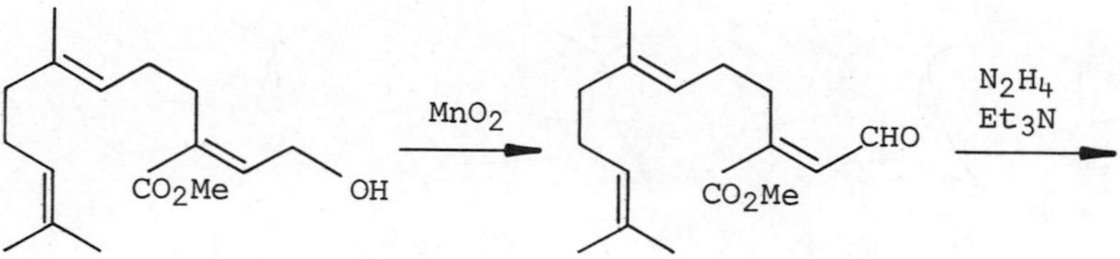

247

248

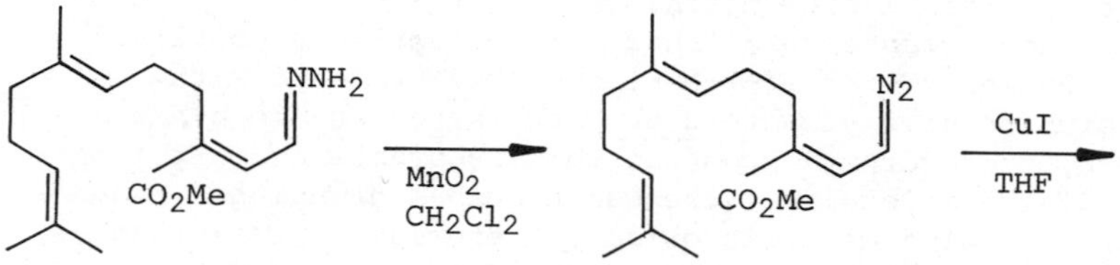

249

250

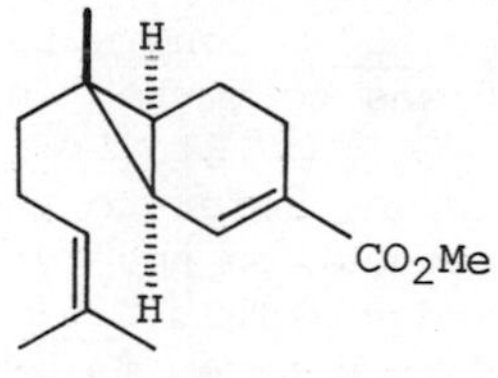

204

structure 16, Sec. 2, with lithio-1-trimethylsilylpropyne gave,
after treatment with Ag$^+$ and CN$^-$, dienyne 244.[251]   The derived
lithium salt was treated with paraformaldehyde to give propargyl alcohol 245, which was carboxylated in a most interesting reaction with nickel carbonyl in HOAc-EtOH-H$_2$O to yield
hydroxy acid 246.   Methyl ester 247 was isolated in 30% yield
based on alcohol 245.

Oxidation of 247 with active MnO$_2$ gave an aldehyde, which formed a hydrazone (249). MnO$_2$ oxidation of 249 gave diazo ester 250. Cuprous iodide-catalyzed cyclization of 250 gave bicyclic ester 204 in 50% yield.

G.   α- and β-cis-Bergamotene

The cis-bergamotenes* (251 and 252) are sesquiterpene analogs of the monoterpenes α- and β-pinene (253 and 254). This analogy suggests a method of synthesis, if one of the methyl groups

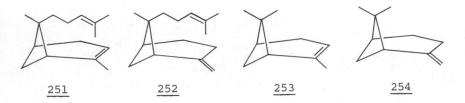

| 251 | 252 | 253 | 254 |

in pinene can be functionalized, and elaborated into the required methylpentenyl side chain.

Gibson and Erman accomplished this objective as outlined in Schemes 26 and 27.[255] (+)-Nopinone (257) reacted with methyllithium to give alcohol 258. Because of steric hindrance to approach, the hydroxyl group in this substance is on the same side of the molecule as the gem-dimethyl grouping. When 258 was treated with mercuric oxide and bromide in refluxing pentane, tricyclic ether 259 was produced in 81% yield.

Oxidation of ditertiary ether 259 afforded lactone 260 in 72% yield. Hydride reduction, followed by selective acetylation, gave hydroxy ester 261. Dehydration of this material yielded olefins 262 and 263, along with ether 259, in a ratio of 15:4:1. Isomer 262 was converted into tosylate 264 and

*The prefixes cis and trans refer to the disposition of the more complicated substituent on the cyclobutane ring, relative to the three-carbon bridge. Thus, 251 is α-cis-bergamotene[252] and 252, not yet found in nature, is β-cis-bergamotene. The corresponding trans isomers, 255[253] and 256,[254] are both natural products.

| 255 | 256 |

Scheme 26. Gibson-Erman Synthesis of (+)-α-*cis*-Bergamotene

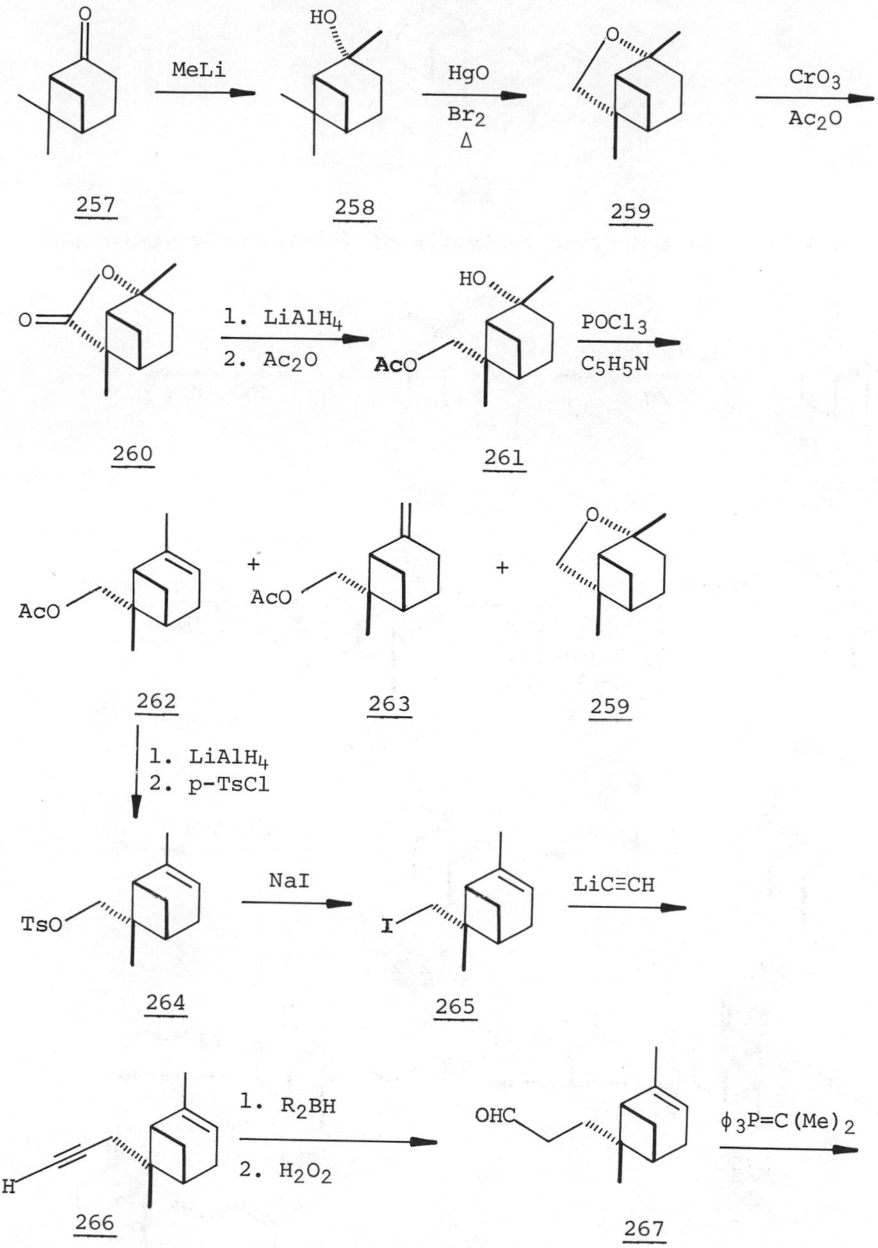

447

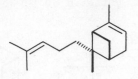

<u>251</u>

Scheme 27.   Gibson-Erman Synthesis of (+)-*cis*-β-Bergamotene

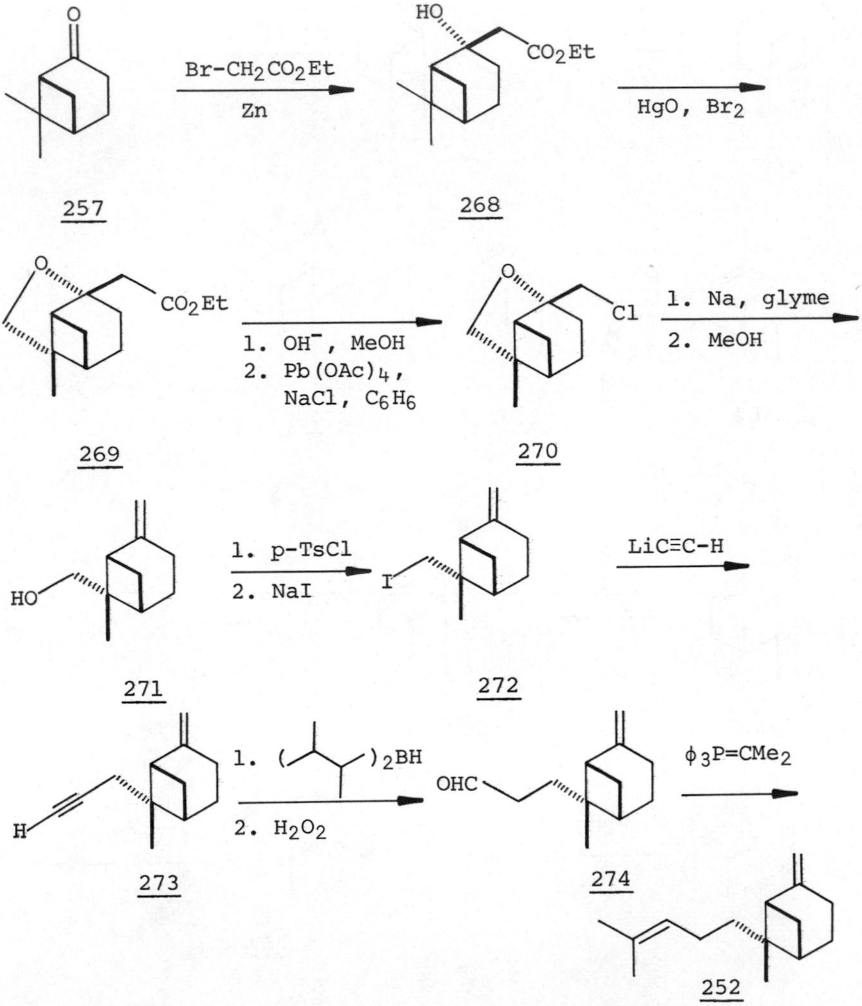

<u>257</u>   Br-CH₂CO₂Et / Zn →   <u>268</u>   HgO, Br₂ →

<u>269</u>   1. OH⁻, MeOH 2. Pb(OAc)₄, NaCl, C₆H₆ →   <u>270</u>   1. Na, glyme 2. MeOH →

<u>271</u>   1. p-TsCl 2. NaI →   <u>272</u>   LiC≡C-H →

<u>273</u>   1. (   )₂BH 2. H₂O₂ →   <u>274</u>   φ₃P=CMe₂ →   <u>252</u>

448

thence into iodide 265. When iodide 265 was coupled with the
ethylenediamine complex of lithium acetylide, alkyne 266 was
obtained in 50% yield. Hydroboration, with disiamylborane,
gave aldehyde 267. This unstable aldehyde was condensed directly with isopropylidenetriphenylphosphorane to obtain (+)-
α-cis-bergamotene (251) in 5% yield, based on acetylene 266.

For production of the β isomer 252, a more efficient
method for introduction of the exocyclic double bond was developed. Reformatsky reaction of 257 gave hydroxy ester 268
in 57% yield. Cyclization of this material, again with HgO
and Br$_2$, gave ether 269. The corresponding acid was degraded
by a Kochi-Hunsdiecker reaction to chloro ether 270. When
this material was treated with sodium, fragmentation occurred
to yield unsaturated alcohol 271 in 48% yield. The side chain
was elaborated, as before, to give (+)-cis-β-bergamotene (252).

H. Chamigrene

Chamigrene (275) and α-chamigrene (276) are interesting spiro-
sesquiterpenes which possibly represent a link (as ion 277) in
the biogenesis of thujopsene (structure 547, Sec. 4) and
cuparene (278)[256] Yoshikoshi's synthesis of (±)-chamigrene[251]
and (±)-α-chamigrene[258] is outlined in Scheme 28.

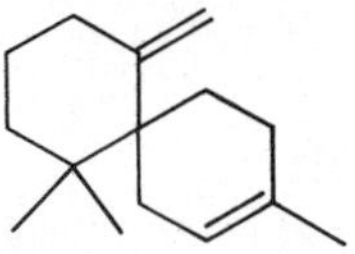

275 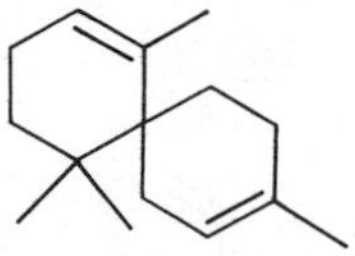   276 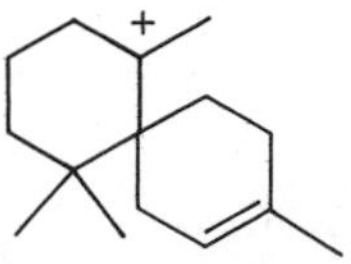   277

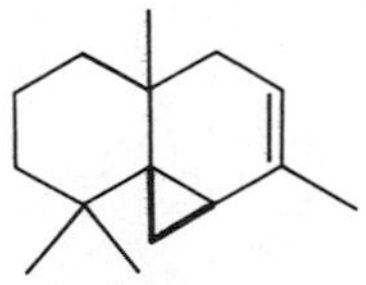

547 (Sec. 4) 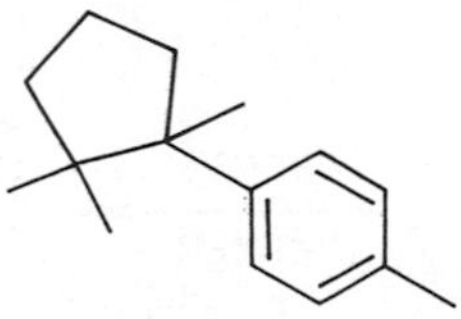   278

Scheme 28. Yoshikoshi's Synthesis of (±)-Chamigrene and (±)-α-Chamigrene

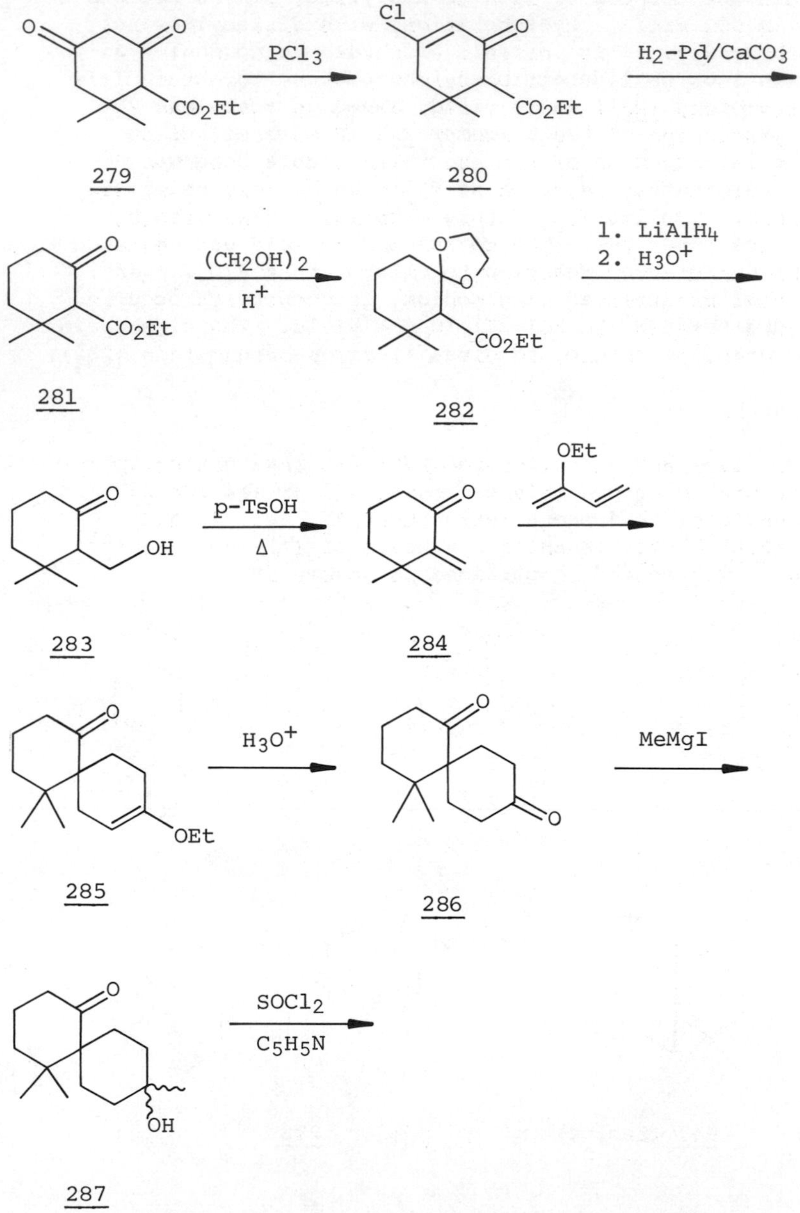

279 PCl$_3$ 280 H$_2$–Pd/CaCO$_3$

281 $\dfrac{\text{(CH}_2\text{OH)}_2}{\text{H}^+}$ 282 1. LiAlH$_4$ 2. H$_3$O$^+$

283 $\dfrac{\text{p-TsOH}}{\Delta}$ 284 OEt

285 H$_3$O$^+$ 286 MeMgI

287 $\dfrac{\text{SOCl}_2}{\text{C}_5\text{H}_5\text{N}}$

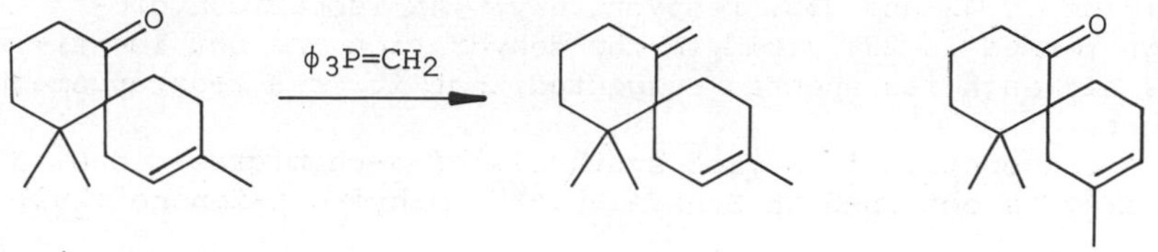

Diketo ester 279, obtained in 94% yield from mesityl oxide and malonic ester, was converted by known procedures[259] into enone 284. Diels-Alder reaction of 284 with 2-ethoxy-1,3-butadiene gave a single keto ether (285) in 20% yield. After hydrolysis, diketone 286 was treated with excess methylmagnesium iodide to give a mixture of epimeric keto alcohols (287), which was dehydrated to unsaturated ketone 288. Enone 284 reacted with isoprene to give 288 and its isomer 289 in 20% and 12% yields, respectively. The crystalline isomer 288 could be obtained by seeding the mixture of isomers with pure 288.

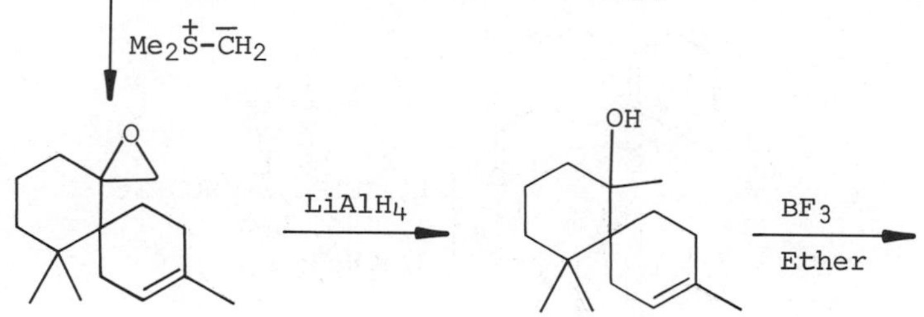

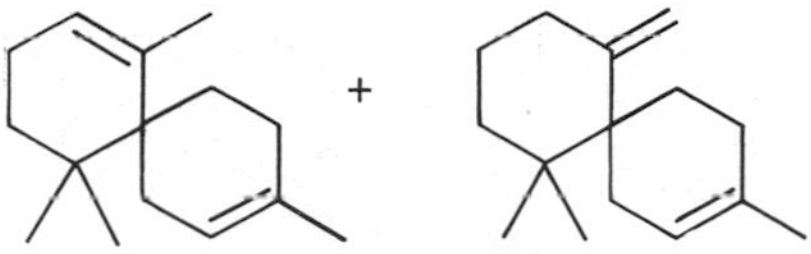

Although the hindered carbonyl group in 288 did not react with methyllithium or methylmagnesium iodide, it reacted with methylenetriphenylphosphorane in dimethyl sulfoxide at 60° for 48 hr to give chamigrene (275) in 70% yield. The epoxide 290, formed from 288 and excess dimethylsulfonium methylide, was reduced by LiAlH₄ to a single unsaturated alcohol (291) (stereochemistry undetermined). Dehydration (BF₃ in ether) of 291 gave (±)-chamigrene (275) and (±)-α-chamigrene (276)

in yields of 42 and 35%, respectively. An isomeric hydrocarbon formed in 23% yield in the dehydration was not identified, although its spectra suggested that it is a rearrangement product.

A biogenetically styled synthesis of α-chamigrene, due to Kitahara, is outlined in Scheme 29.[260] Dihydro-β-ionone (292)

Scheme 29. Kitahara's Synthesis of (±)-α-Chamigrene

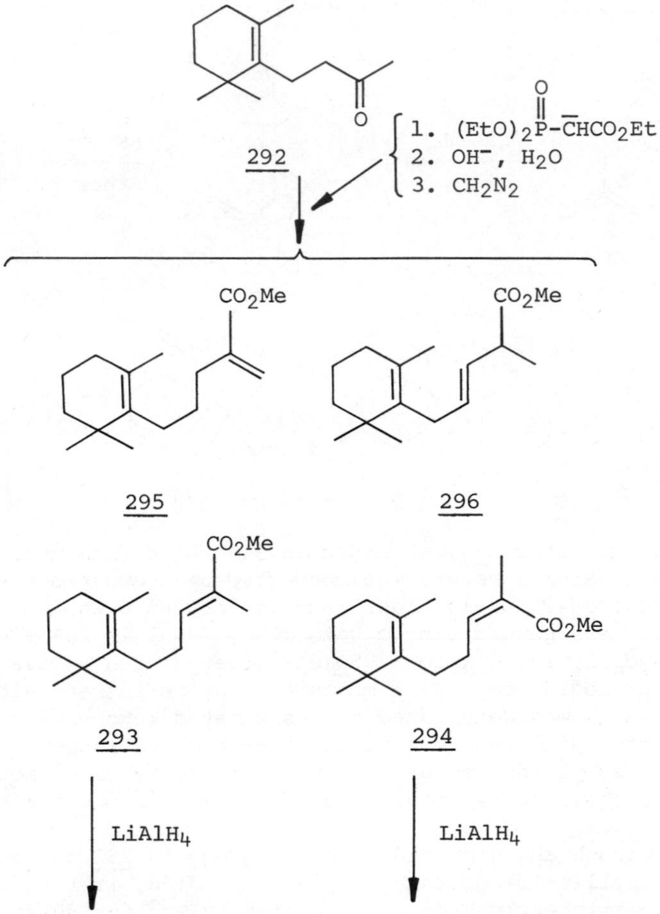
292
1. $(EtO)_2P(O)\text{-}\bar{C}HCO_2Et$
2. OH^-, H_2O
3. CH_2N_2
CO_2Me
CO_2Me
295
296
CO_2Me
CO_2Me
293
294
$LiAlH_4$
$LiAlH_4$

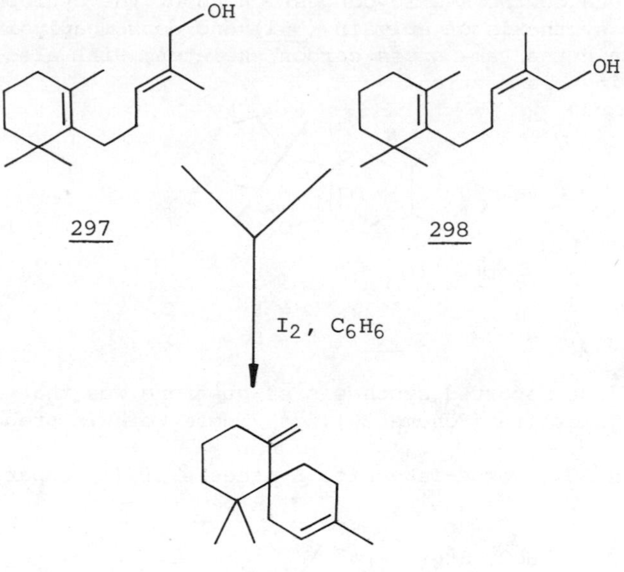

297

298

I_2, C_6H_6

275

was condensed with triethylphosphonoacetate to give a mixture of esters, which was hydrolyzed. After crystallization of *trans*-monocyclofarnesic acid (acid corresponding to 293) in 30% yield, the remaining acids were esterified and separated by chromatography. Esters 293-296 were obtained in relative yields of 25, 22, 33, and 20%, respectively. Esters 293 and 294 were reduced to *trans*- and *cis*-monocyclofarnesol (297 and 298). Treatment of either isomer with iodine in benzene solution gave a mixture of hydrocarbons containing 25% (±)-α-chamigrene (275).

I. Cuparene, β-Cuparenone, Aplysin, Debromoaplysin

Cuparene (299) and β-cuparenone (300) are typical members of a class of bicyclic sesquiterpenes produced by the order *Cupressales*. The compounds present an interesting synthetic

299

300

challenge due to the steric congestion about the cyclopentane
ring. The synthesis of aplysin (301) and debromoaplysin (302),
which contain the same gross carbon skeleton, will also be
discussed in this section.

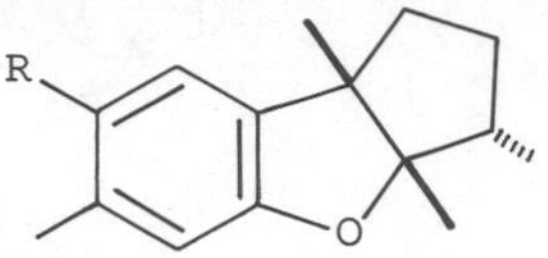

301: X = Br
302: X = H

The first reported synthesis of cuparene was that of
Nozoe and Takeshita (Scheme 30).[261] These workers prepared

Scheme 30. Nozoe-Takeshita Synthesis of (±)-Cuparene

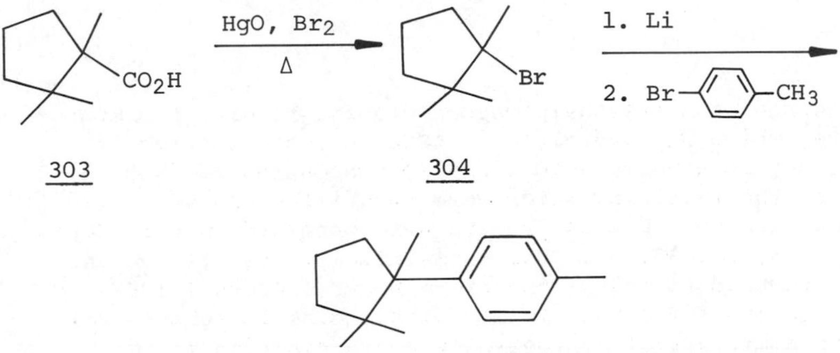

303   304

299

1-bromo-1,2,2-trimethylcyclopentane (304) by Hunsdiecker
degradation of camphonanic acid (303). They reported that the
corresponding lithio derivative coupled with p-bromotoluene to
give (±)-cuparene (299). Since the yield is not specified,
and since the experimental details of this synthesis have not
been forthcoming after ten years, it is probably not an especially good preparation of the material.

A more plausible synthesis of 299 is outlined in Scheme
31.[262] 3-Methylcyclohex-2-enone and toluene were condensed
in a Friedel-Crafts reaction to give the bicyclic ketone 305.
The corresponding furfurylidine derivative (306) was methylated to give 307, which was ozonized to the trimethyl-p-
tolyladipic acid 308. Dieckmann cyclization of the

Scheme 31. Raphael's Synthesis of (±)-Cuparenone
and (±)-Cuparene

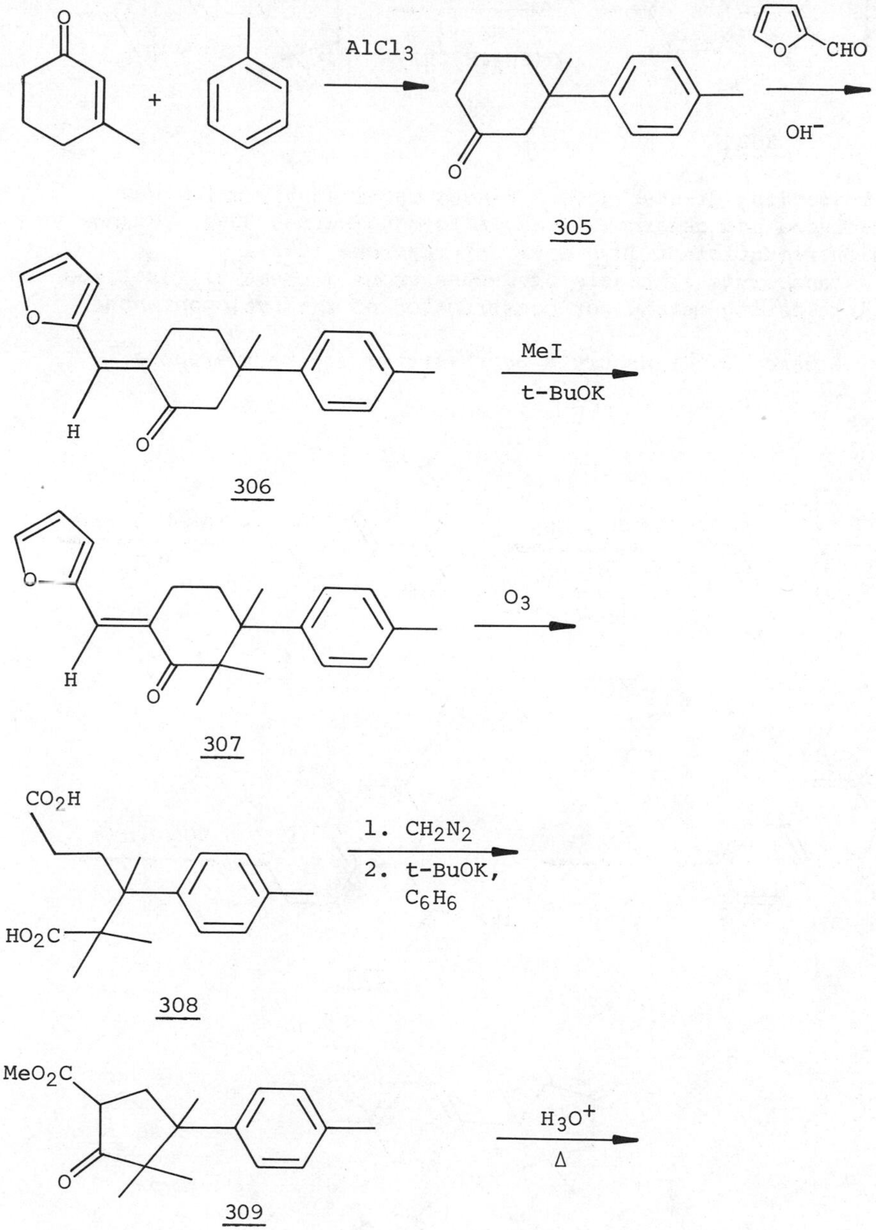

305

306

307

308

309

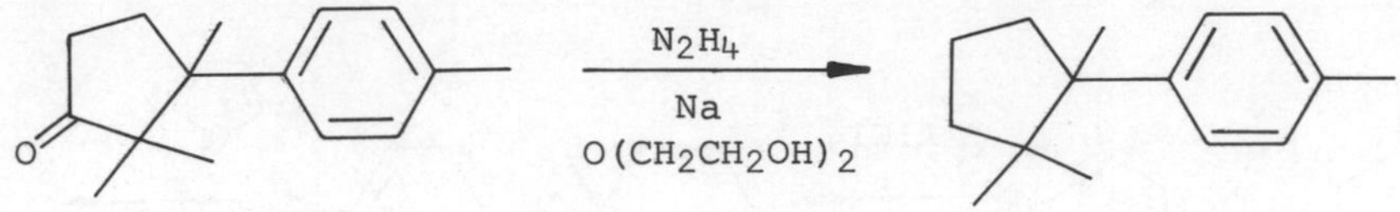

309a   299

corresponding diester gave a β-keto ester (309), which was
hydrolyzed and decarboxylated, affording ketone 309a. Huang-
Minlon reduction of 309a gave (±)-cuparene (299).
     Lansbury's synthesis of β-cuparenone (Scheme 32) utilizes
an interesting method for construction of the cyclopentanone

Scheme 32. Lansbury's Synthesis of (±)-β-Cuparenone

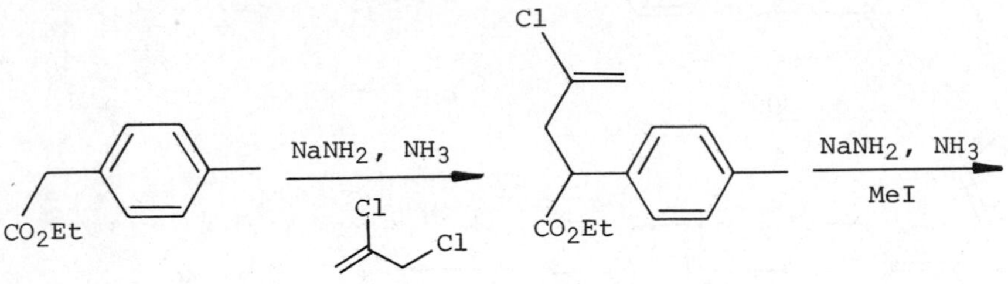

310   311

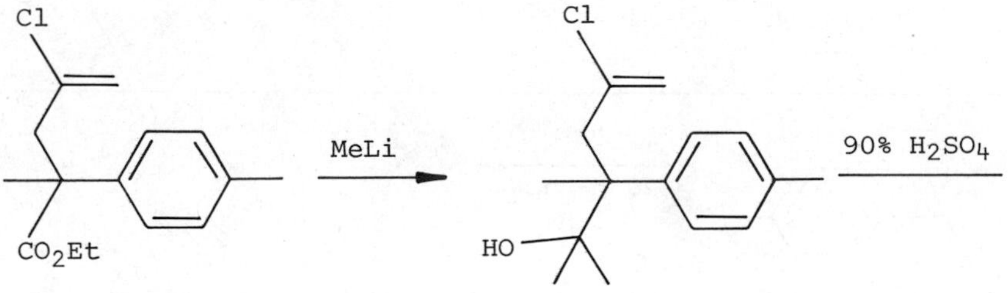

312   313

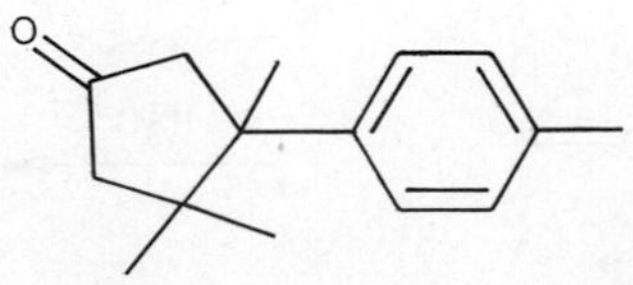

300

ring.[263] Ethyl p-tolylacetate (310) was alkylated with 2,3-dichloropropene and then with methyl iodide to obtain ester 312 (54% overall yield). Compound 312 reacted with methyllithium to give an alcohol (313), which was treated with 90% sulfuric acid. (±)-β-cuparenone (300) was produced in 34% yield. The cyclization may be depicted as follows:

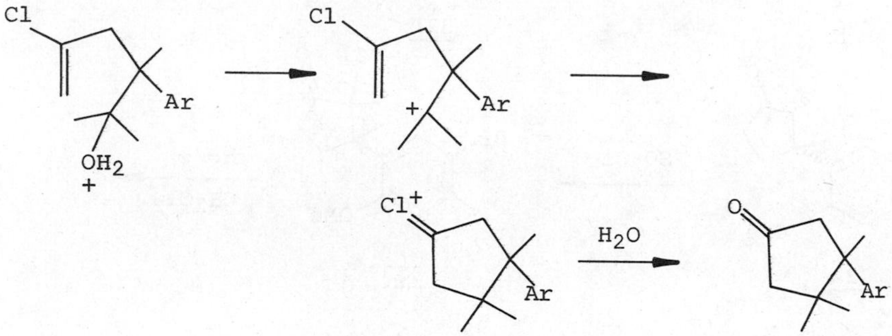

Hirata's synthesis of (±)-aplysin (301) and (±)-debromoaplysin (302) is outlined in Scheme 33.[264] 4-Bromo-3-methyl-anisole (314) was converted into the 6-lithio derivative by

Scheme 33.   (±)-Aplysin--Hirata's Synthesis

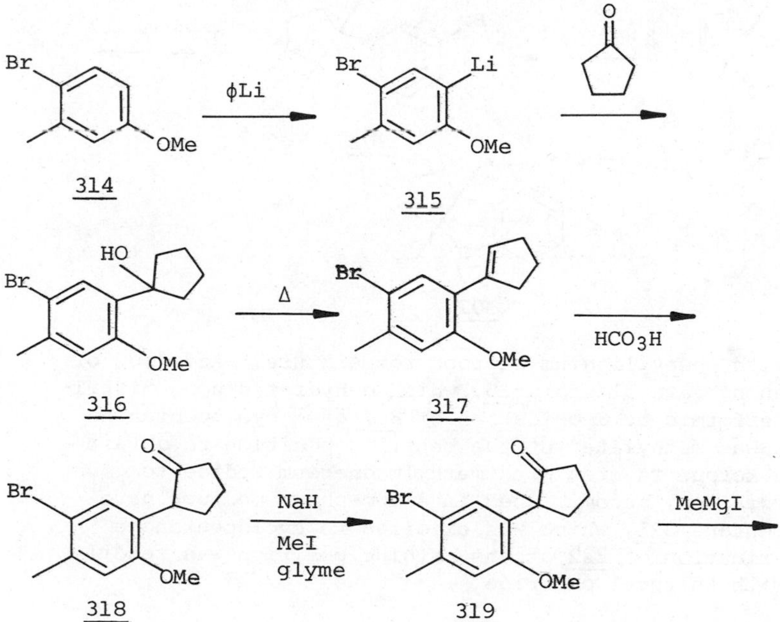

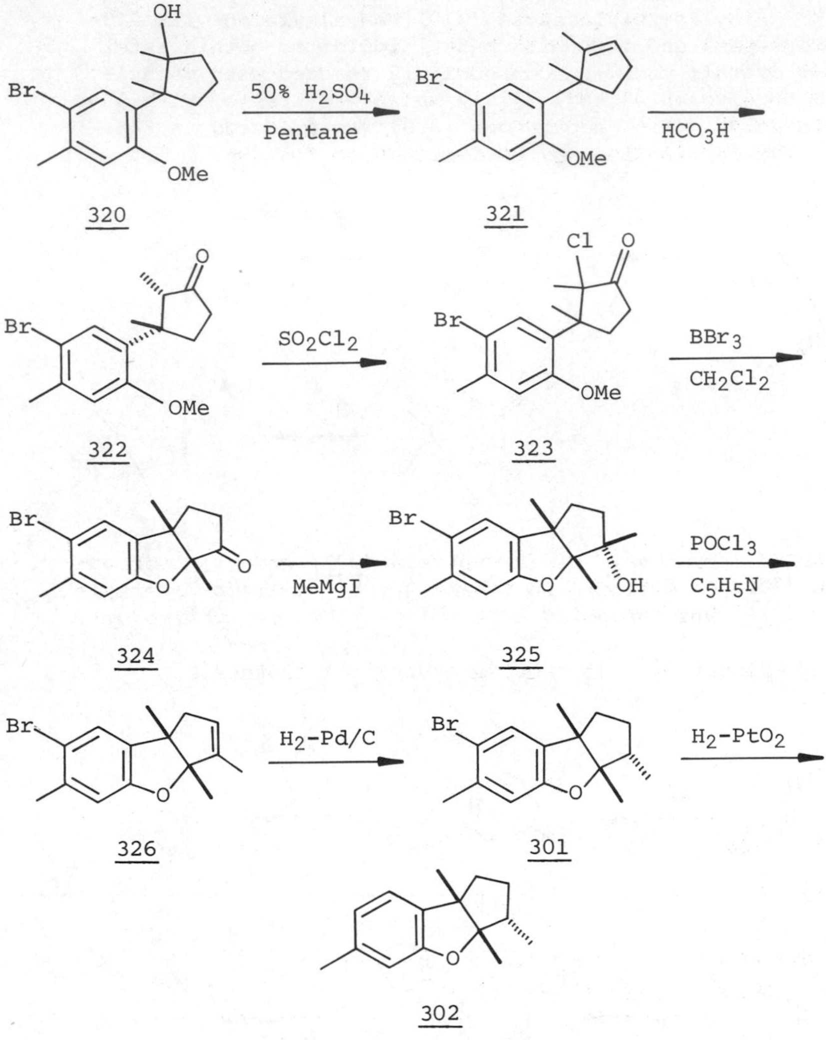

reaction with phenyllithium at room temperature. Addition of
cyclopentanone gave alcohol 316, which dehydrated upon distillation. Performic acid oxidation of 317 gave cyclopentanone
318, which was methylated at the benzylic position to obtain
319. This ketone reacted with methylmagnesium iodide to give
320. Dehydration, accomplished in a two-phase medium, gave
the cyclopentene 321, which was oxidized to cyclopentanone
322. Chlorination of 322 at the methine position was readily
achieved with sulfuryl chloride.

The aryl methyl ether was cleaved, with concomitant cyclization, affording the tricyclic keto ether 324 in 20% yield.
This ketone reacted with methylmagnesium iodide to give an
alcohol (probably 325) which was dehydrated to 326. Stepwise
hydrogenation gave (±)-aplysin (301) and (±)-debromoaplysin
(302).

J. Carabrone

The sesquiterpene lactone carabrone (327) may be viewed as a
guaianolide which has suffered scission of the five-membered
ring. With five asymmetric centers on the simple carbocyclic

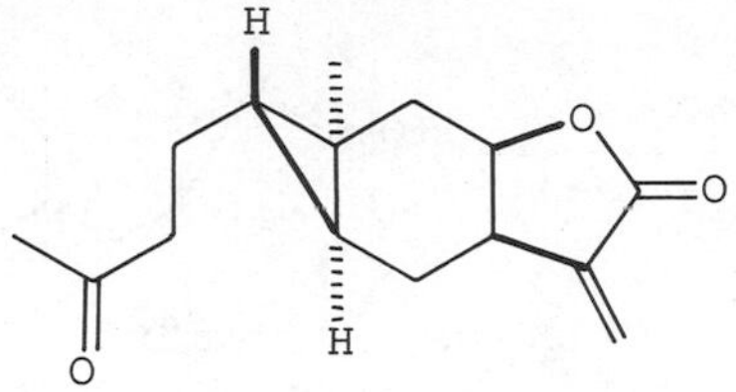

327

nucleus, carabrone presents an interesting problem in stereorational synthesis. Although this goal remains to be achieved,
Minato has reported the lengthy synthesis outlined in Scheme
34.[265]

Scheme 34. Minato's Synthesis of (±)-Carabrone

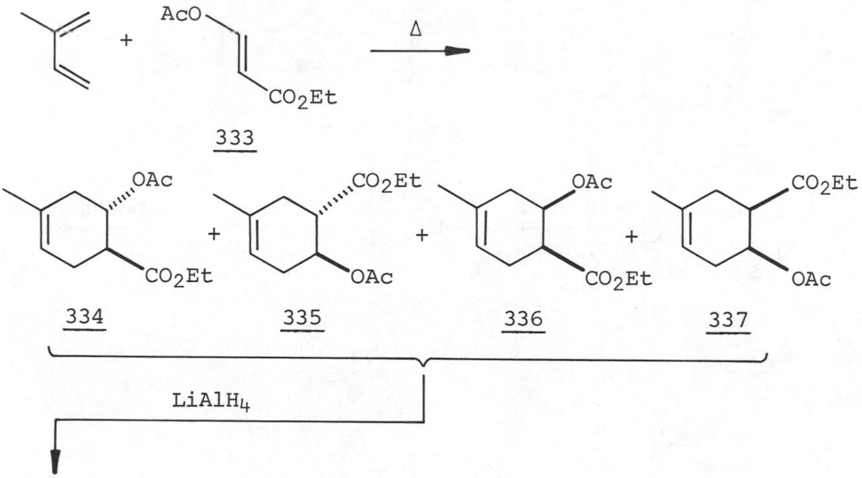

333

334   335   336   337

LiAlH$_4$

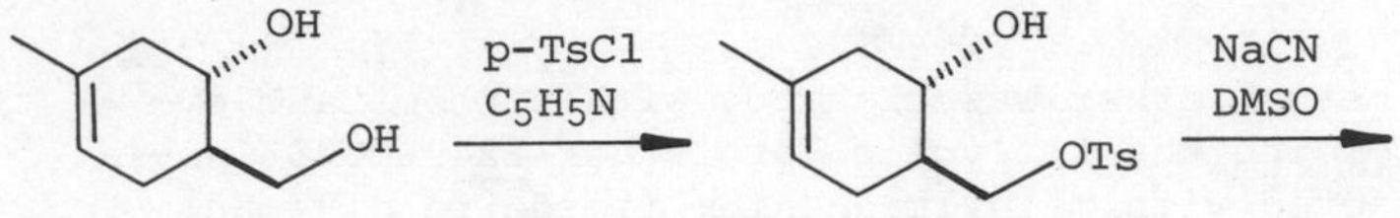

$$\underline{338} \xrightarrow[\text{C}_5\text{H}_5\text{N}]{\text{p-TsCl}} \underline{339} \xrightarrow[\text{DMSO}]{\text{NaCN}}$$

$$\underline{340} \xrightarrow[\text{EtOH}]{\text{NaOH}} \underline{341} \xrightarrow{200°}$$

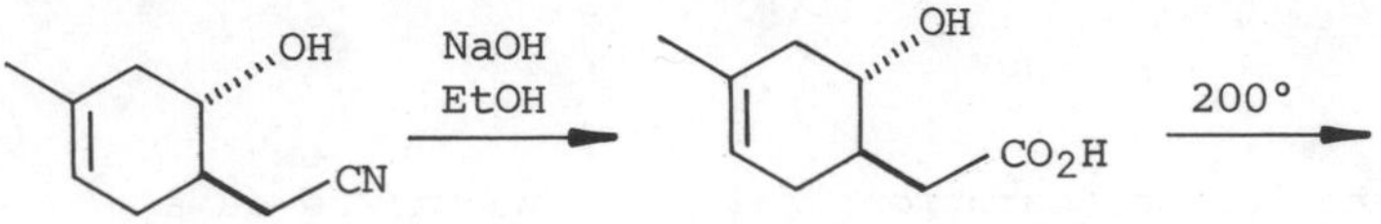

$$\underline{332} \xrightarrow[\text{2. K}_2\text{CO}_3\text{, MeOH}]{\text{1. N}_2\text{CHCO}_2\text{Et,Cu}} \underline{342} \xrightarrow{195°}$$

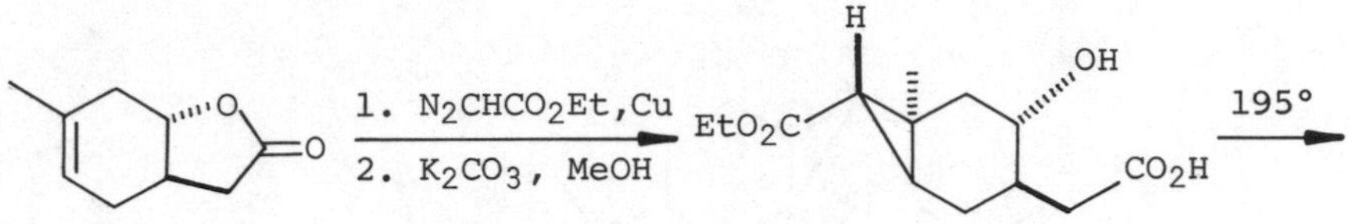

$$\underline{343} \xrightarrow[\text{2. H}_3\text{O}^+]{\text{1. NaOH}} \underline{344} \begin{array}{l} \text{1. NaHCO}_3 \\ \text{2. (COCl)}_2 \\ \text{3. n-BuSH} \\ \text{4. Ni} \end{array}$$

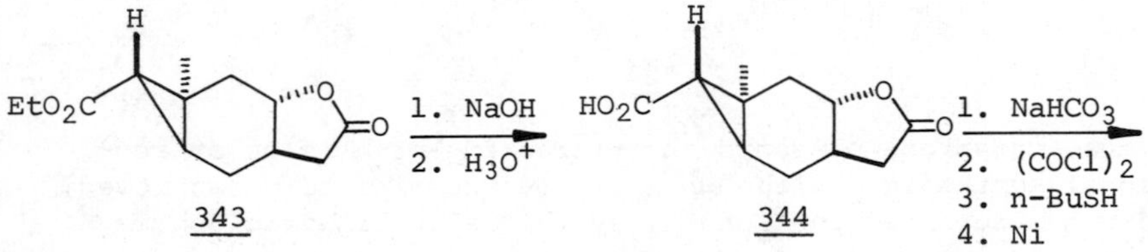

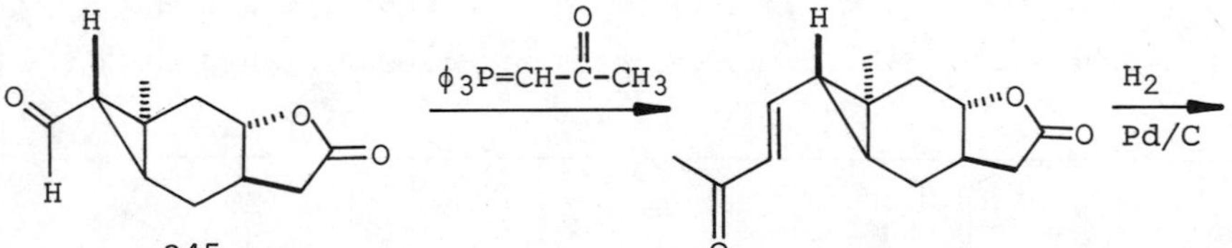

$$\underline{345} \xrightarrow{\phi_3\text{P}=\text{CH-C-CH}_3} \underline{346} \xrightarrow[\text{Pd/C}]{\text{H}_2}$$

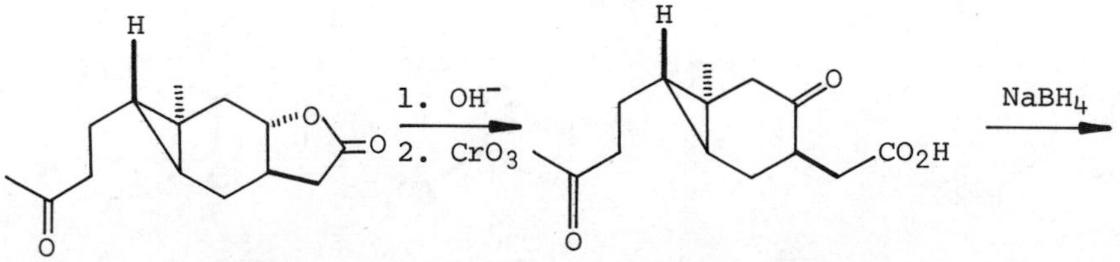

$$\underline{347} \xrightarrow[\text{2. CrO}_3]{\text{1. OH}^-} \underline{348} \xrightarrow{\text{NaBH}_4}$$

460

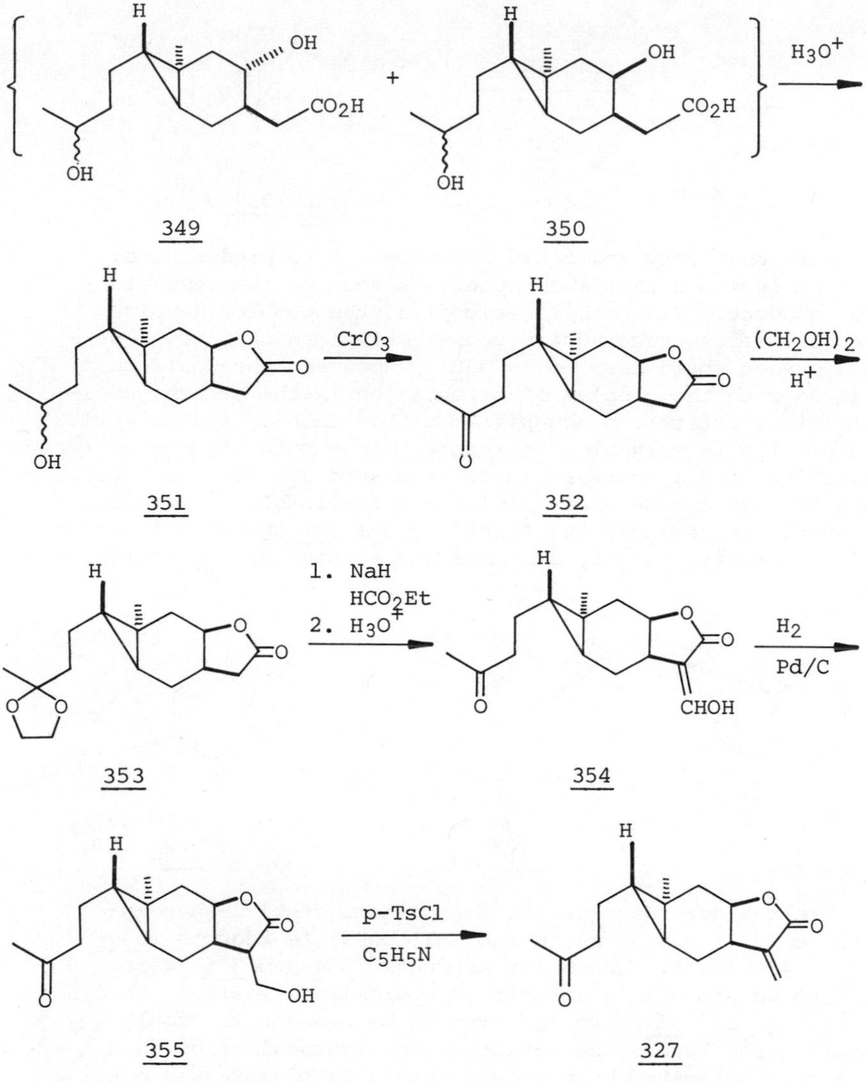

349   350

351   352

353   354

355   327

Minato's synthetic plan called for construction of the bicyclo[4.1.0]heptane skeleton by adding carbethoxycarbene to a methylcyclohexene (i.e., 328 → 329).

328

329

The stereochemistry indicated was expected to predominate,
since it is known that such reactions lead to the less hindered product. Given this basic plan, the problem devolves
into one of constructing the proper methylcyclohexene. The
ideal target would seem to be 330.* However, one would then
be faced with the problem of orientation in the carbene addition, since attack can occur from either face of the molecule.
This problem is solvable, in theory, since *both* lactone centers
can be inverted by conversion to keto-acid 331 (at which point
the acetic acid side-chain would be epimerizable). However,
for synthetic reasons, the *trans*-lactone 332 appeared to be an
easier synthetic target, and this was adopted as the initial
goal.

330

331

332

Diels-Alder reaction of isoprene and ethyl β-acetoxyacrylate (333) gave a mixture of all possible adducts (334-
337) in 48% yield. The major products (334 and 335) were
found to be present in a ratio of approximately 3:1. The *cis*
isomers 336 and 337 were inferred to be present in "small
amounts" from later observations. The synthesis from this
point is complicated by the fact that this mixture was not
separated, but was carried through as such. In order to avoid
confusion, we shall illustrate the synthesis with the major
isomer only (334). Hydride reduction of the mixture afforded
diol 338, which was converted, via mono-tosylate 339 into
hydroxy nitrile 340. Hydrolysis of 340 gave a hydroxy acid

*The methylene group was to be introduced at a later stage by
Minato's method, see sec. 4-M.

(341), which lactonized on heating at 200°, affording the desired *trans*-lactone 332.

When this lactone was treated with ethyl diazoacetate and copper powder in diglyme at 170°, and the resulting crude product hydrolyzed with methanolic potassium hydroxide, a mixture of monocarboxylic acids was obtained. At this point, the crude product was treated with HCl in ether to give a lactone fraction and an acid fraction in a ratio of 1:9. The lactone fraction contained *cis* lactones derived from the minor Diels-Alder adducts 336 and 337. When the acid fraction was heated at 195°, a mixture of lactones was obtained in 53% yield. Chromatographic analysis indicated that two compounds were present in a ratio of 85:15. Structure 343 was assigned to the major isomer on the basis of spectral arguments.

Alkaline hydrolysis of 343, followed by acid catalyzed lactonization gave the crystalline acid lactone 344 in 41% yield. From this point the synthesis proceeds with a single stereoisomer. Acid 344 was converted, via its acid chloride, into a butyl thioester, which was desulfurized by Raney nickel to aldehyde 345. Wittig condensation of 345 with acetyl-methylenetriphenylphosphorane gave enone 346, which was reduced to the saturated keto lactone 347.

At this point it was necessary to invert the lactone C-O bond in order to establish the proper stereochemistry at this center. This was accomplished by alkaline hydrolysis of 347, followed by oxidation with Jones reagent. The oily keto acid 348 was reduced by sodium borohydride to a mixture of diol acids (349 and 350). Upon treatment with sulfuric acid at room temperature, one of these isomers (350) lactonized, giving *cis*-lactone 351 (27% yield) and unreacted *trans*-hydroxy acid 349.

Oxidation of 351 gave keto lactone 352, which now has all of the proper stereochemistry for conversion into carabrone. After protection of the side-chain carbonyl, the methylene group was introduced by Minato's method (see sec. 4-M), yielding finally (±)-carabrone (327).

K. Helminthosporal

Corey's synthesis of the interesting fungal toxin helmintho-sporal (356) is summarized in Scheme 35.[266] The compound has four asymmetric centers, of which two (C-1 and C-5) are fixed

356

relative to one another on geometrical grounds. The stereochemistry at C-8 is no problem. Since helminthosporal is
known to resist acid catalyzed epimerization, the formyl group
must occupy the most stable configuration. Thus, the only
stereochemical feature which must be reckoned with in a synthetic plan is the disposition of the isopropyl group on the
cyclohexane ring relative to the two-carbon bridge.

The nature of the two-carbon bridge in 356 suggests a
method for its formation. If one considers 359 as an aldolization product of keto aldehyde 358 then the requisite intermediate might be available through oxidative cleavage of the
bicyclo[3.3.1]nonene 357.

357   358

359

The synthetic plan was reduced to practice as shown in
the scheme. (-)-Carvomenthone (structure 419, Sec. 4) was
activated at C-6 by conversion into its α-hydroxymethylene
derivative (360). Condensation of 360 with methyl vinyl ketone, followed by deformylation, yielded dione 361. Treatment
of 361 with BF₃ in methylene chloride gave bicyclic ketones
362 and 363 in yields of 8 and 32%, respectively. Although
bicyclo[3.3.0]nonanes such as 362 and 363 are normally observed
only as minor side-products in the base-catalyzed Robinson
annelation sequence,[267] acid catalyzed hydrolysis-cyclization
of compound 368 had been previously reported to yield mainly
the bridged product 369.[268]

368   369

The relative stereochemistry of the isopropyl group and the potential two-carbon bridge is established, stereoselectively, in this step. Although both substituents α to the carbonyl in 361 are epimerizable, cyclization of 361a, with an axial isopropyl group in the transition state, is less favorable than cyclization of 361b.

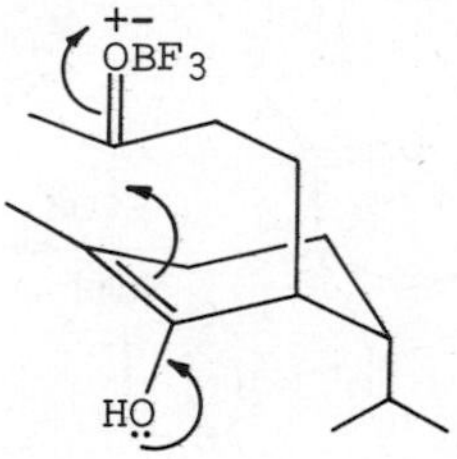

361a   361b

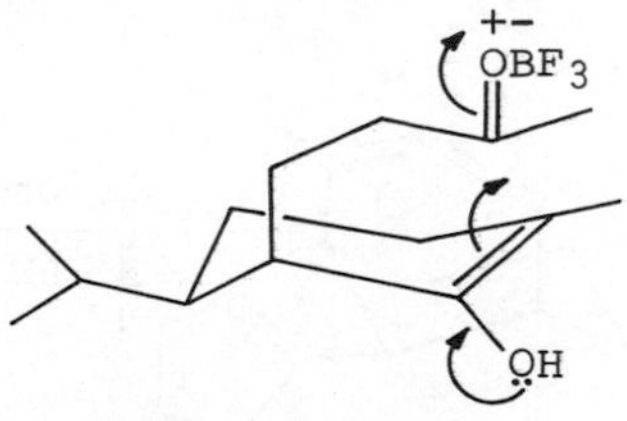

362   363

After separation of the two isomers (362 and 363) via their semicarbazone derivatives, compound 363 was condensed with methoxymethylenetriphenylphosphorane to give an enol ether (364), which was transformed into ethylene acetal 365. Cleavage of the double bond yielded keto aldehyde 366, which

Scheme 35. Corey's Synthesis of (−)-Helminthosporal

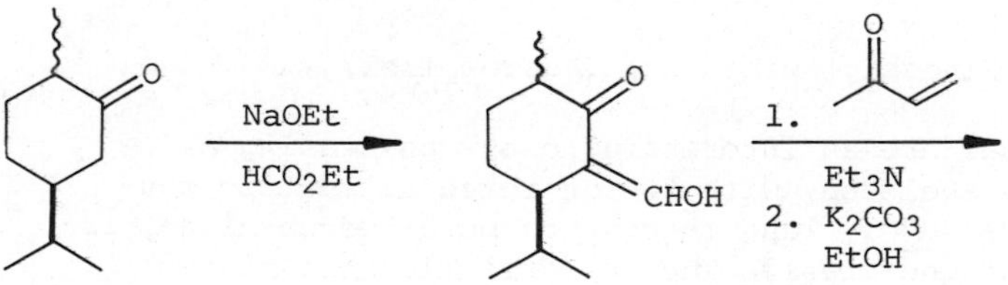

419 (Sec. 4)　　　　　　　360

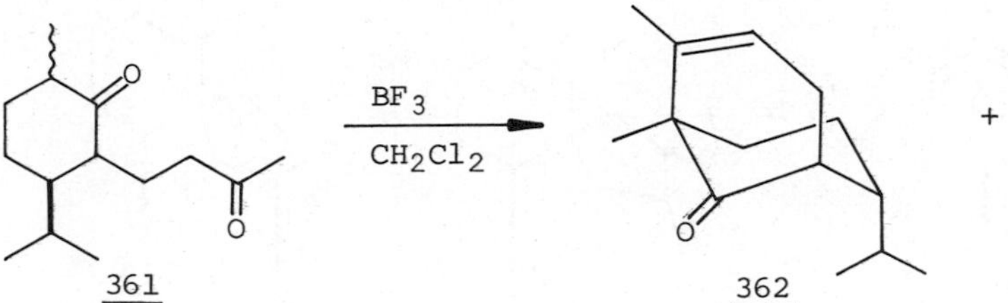

361　　　　　　　　362

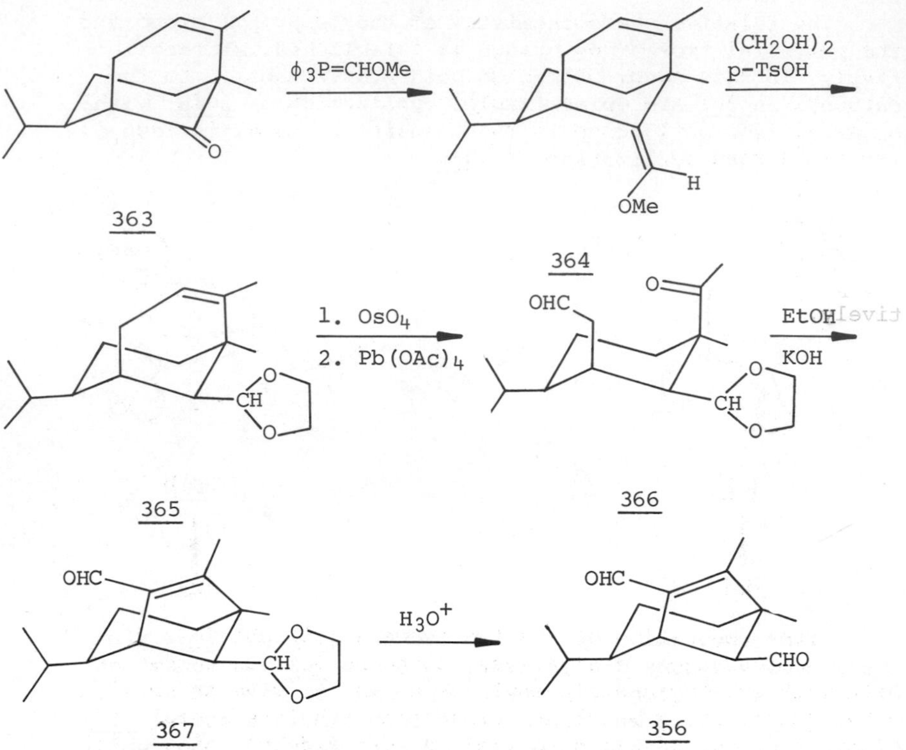

363

364

365

366

367

356

was aldolized with ethanolic base. Hydrolysis of the resulting acetal (367) yielded (-)-helminthosporal (356). (See
Scheme 35.)

L.   β-Vetivone, Hinesol, Epihinesol (Agarospirol)

The spirovetivanes are an interesting class possessing a
spiro[4.5]decane skeleton, with two or three asymmetric centers. β-Vetivone (370), long thought to be a hydroazulene, is
the key member of the class. The epimeric alcohols hinesol
(371) and agarospirol (372) contain an additional chiral center, and thus present more challenging synthetic tasks.

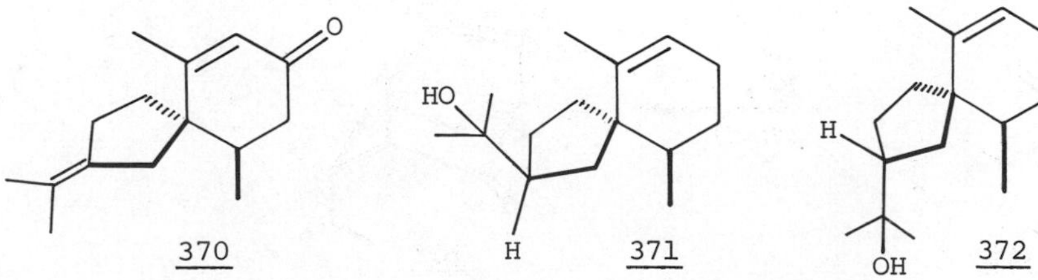

370

371

372

One obvious approach to compounds of this class is through the spiro[4.5]decane 375, obtained by Kropp as a solvolysis product of the tricyclic enone 374.[269] Enone 374 is a photo-chemical rearrangement product of the dienone 373. From the stereochemical relationships which had been worked out for the related santonin-lumisantonin rearrangement,[270] compound 375 is known to have the stereochemistry indicated in Scheme 36.

Scheme 36.   Marshall's Synthesis of (±)-β-Vetivone

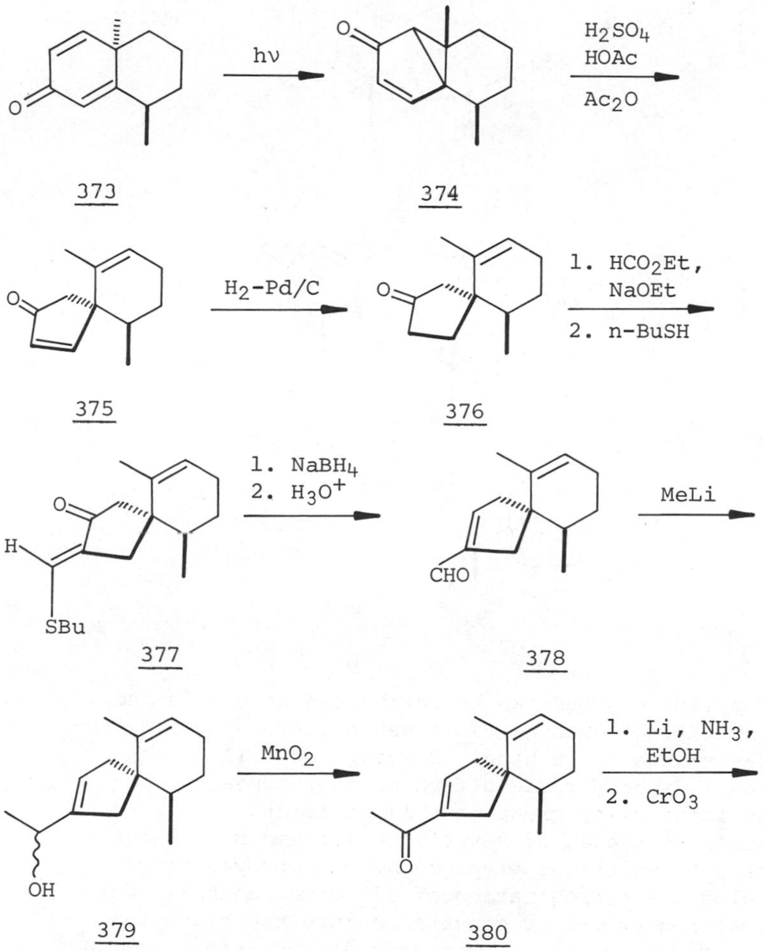

373   374

375   376

377   378

379   380

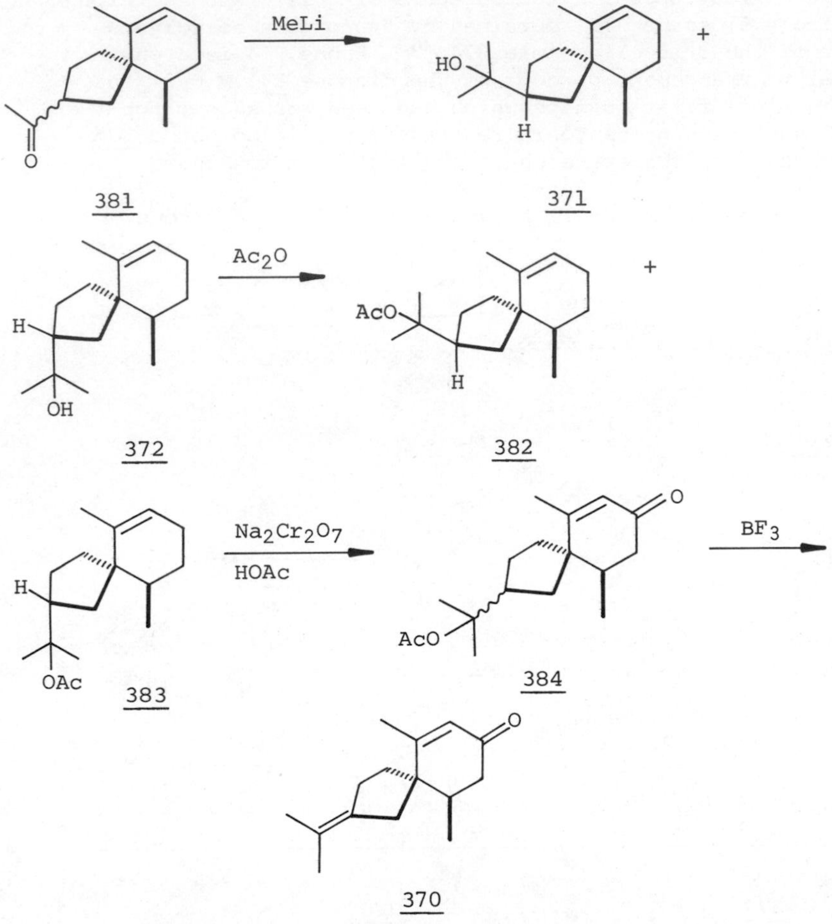

381

371

372

382

383

384

370

If an isopropylidine group can be introduced at C-3 in the dihydro derivative of dienone 375, then a stereospecific synthesis of β-vetivone is in hand. However, as will be seen from the sequel, such a route offers no stereochemical control over an isopropylol group at this position.

Marshall's synthesis of β-vetivone (Scheme 36) begins with dienone 375, which was prepared by an improved procedure.[271] Selective hydrogenation of 375 gave enone 376 which was functionalized at C-3 by conversion into the n-butyl-thiomethylene derivative 377. Borohydride reduction, followed by acid hydrolysis, yielded aldehyde 378. This substance was treated with methyllithium to obtain 379, a diastereomeric mixture, which was oxidized to dienone 380.

Dissolving metal reduction of 380 gave, after re-oxidation, the enone 381 a 3:2 mixture of epimers. This mixture reacted with methyllithium to give hinesol (371) and epihinesol (372)* in a ratio of 5:3. Although 371 and 372 are not separable, the corresponding acetates (382 and 383) could be separated by preparative glpc. Allylic oxidation of the mixture gave keto ester mixture 384, which was dehydrated by a known method to (±)-β-vetivone (370).

Although this route produces synthetic β-vetivone in a stereorational manner, it provides no synthetic proof of the stereochemistry of hinesol. In order to provide this evidence, Marshall carried out the synthesis in Scheme 37.[272] The

Scheme 37. Marshall's Synthesis of (±)-Hinesol

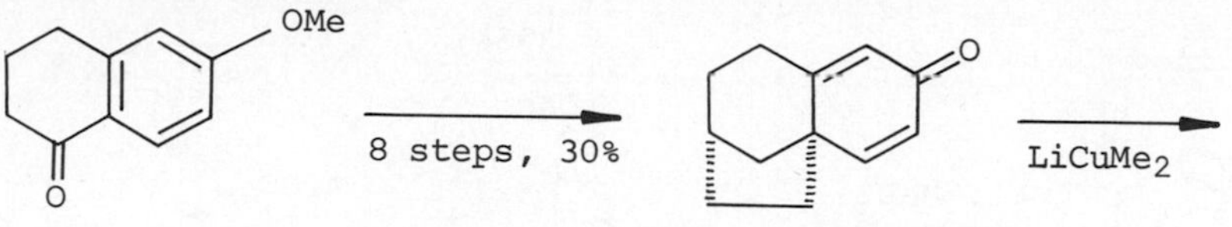

385

$\xrightarrow{\text{8 steps, 30%}}$

386

$\xrightarrow{\text{LiCuMe}_2}$

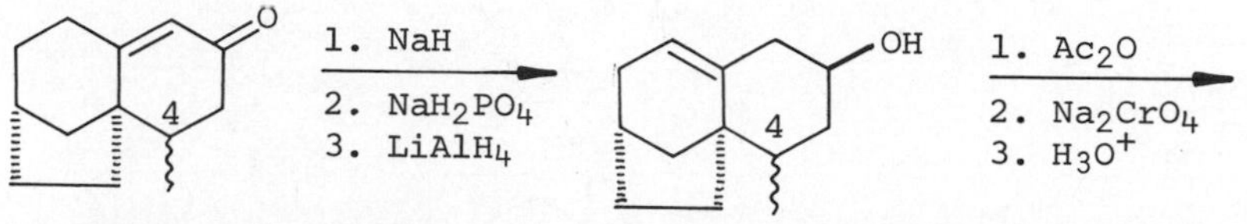

387: C-4 Me β (20%)
388: C-4 Me α (80%)

1. NaH
2. NaH$_2$PO$_4$
3. LiAlH$_4$

389: C-4 Me β (20%)
390: C-4 Me α (80%)

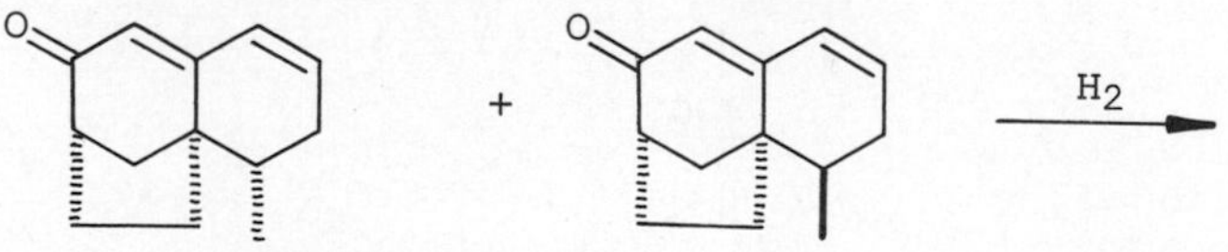

1. Ac$_2$O
2. Na$_2$CrO$_4$
3. H$_3$O$^+$

391 (80%)

+

392 (20%)

$\xrightarrow{\text{H}_2}$

*Compound 372 (epihinesol) is very probably agarospirol. Although samples of agarospirol are no longer available for comparison, Deslongchamps reports that his epihinesol (Scheme 38) which is identical to Marshall's, gave an epoxide which is spectrally identical to that derived from natural agarospirol.

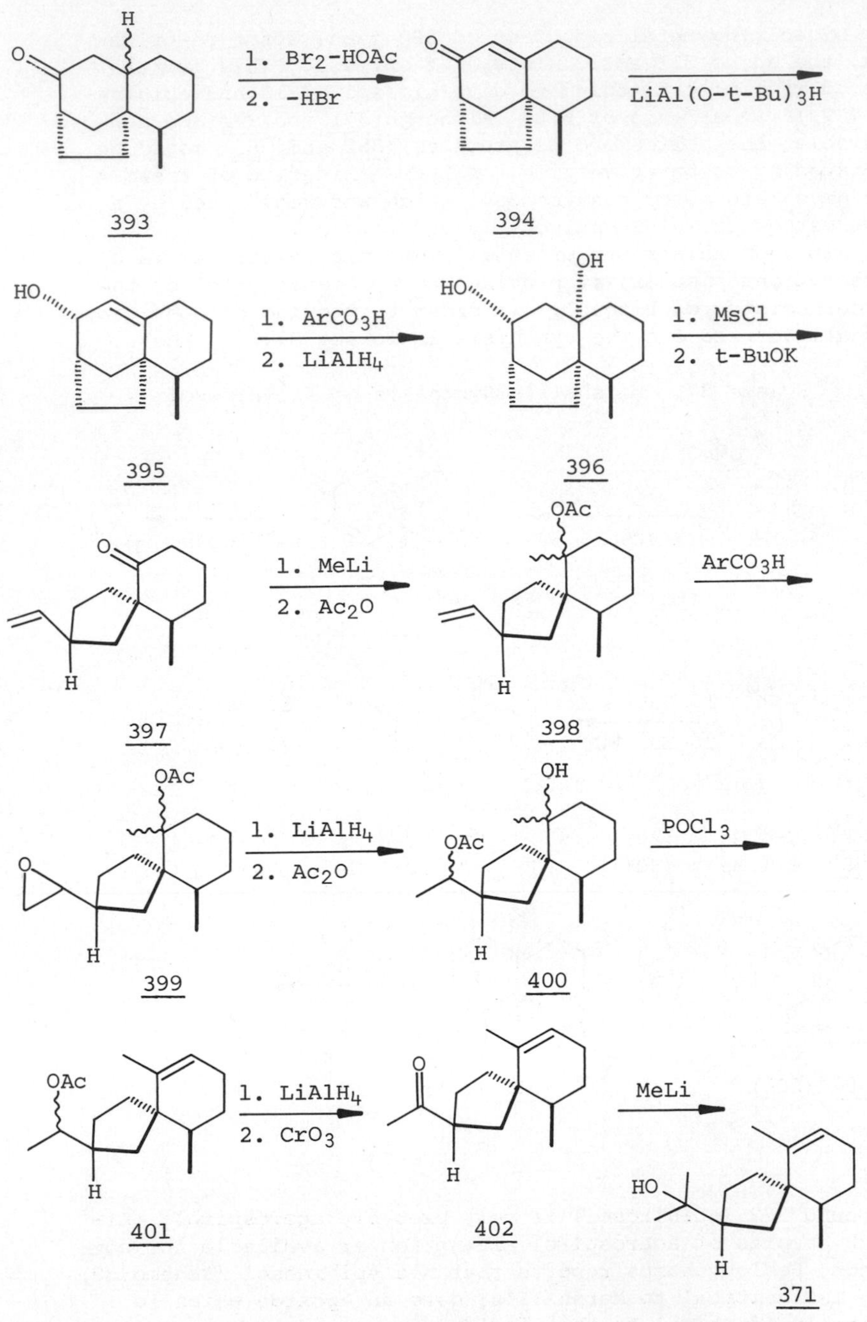

1. Br2-HOAc
2. -HBr
LiAl(O-t-Bu)3H
393
394
1. ArCO3H
2. LiAlH4
1. MsCl
2. t-BuOK
395
396
1. MeLi
2. Ac2O
ArCO3H
397
398
1. LiAlH4
2. Ac2O
POCl3
399
400
1. LiAlH4
2. CrO3
MeLi
401
402
371

synthesis, while providing hinesol in a stereochemically unambiguous manner, it is not stereoselective. Tricyclic ketone
386 was prepared in eight steps from 6-methoxy-1-tetralone
(385) in a modification of the route originally employed by
Masamune for the production of this substance from 6-methoxy-
2-tetralone.[273]

Addition of lithium dimethylcopper to the more accessible
double bond of dienone 386 yielded a mixture of enones 387 and
388 in a ratio of 1:4. The stereochemistry of the adducts was
established firmly by x-ray analysis of a derivative of the
minor (desired) isomer. The mixture of enones was converted
into allylic alcohols 389 and 390 by reduction of the corresponding β,γ-unsaturated isomers. Allylic oxidation of the
corresponding homoallylic acetates, followed by elimination of
acetic acid, yielded a major and a minor dienone (391 and 392),
which were separated by preparative glpc. The minor isomer
was used in the remaining seventeen steps needed for the production of synthetic hinesol.

Catalytic hydrogenation gave a mixture of isomeric ketones (393), which was converted into enone 394 by bromination-dehydrobromination. Hydride reduction of 394 yielded
allylic alcohol 395. Reduction of the corresponding epoxide
gave diol 396. Grob fragmentation of the derived monomethane-
sulfonate afforded the spiro[4.5]decane derivative 397.

Compound 397 reacted with methyllithium to give a mixture
of alcohols, which was acetylated to obtain unsaturated acetates 398. The corresponding epoxide (399) was reduced with
LiAlH₄ and the product acetylated to obtain hydroxy ester 400
(a mixture of four diastereomers). Dehydration of this substance yielded unsaturated ester 401. The alcohol resulting
from LiAlH₄ reduction of 401 was oxidized to ketone 402, of
unambiguous stereochemistry. Treatment of 402 with methyllithium gave (±)-hinesol (371).

Deslongchamps has reported the elegant synthesis of
agarospirol outlined in Scheme 38.[274] The readily available

Scheme 38. Deslongchamps' Synthesis of Agarospirol

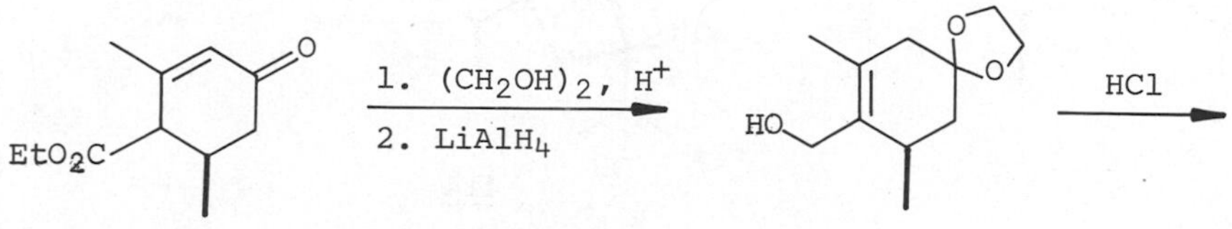

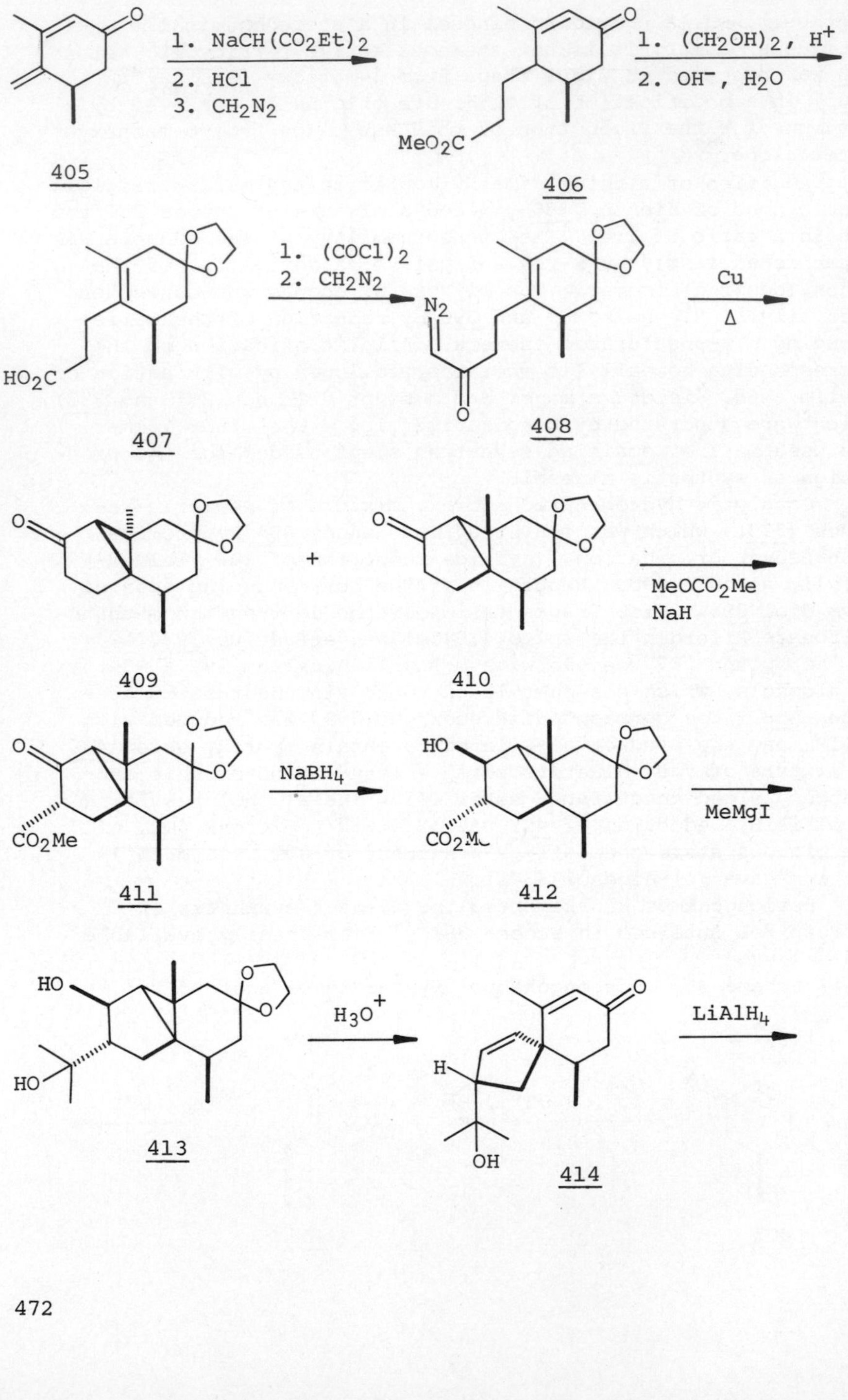
1. NaCH(CO2Et)2
2. HCl
3. CH2N2
405
MeO2C
406
1. (CH2OH)2, H+
2. OH−, H2O
HO2C
407
1. (COCl)2
2. CH2N2
N2
408
Cu
Δ
409
+
410
MeOCO2Me
NaH
CO2Me
411
NaBH4
HO
412
MeMgI
HO
413
H3O+
H
OH
414
LiAlH4

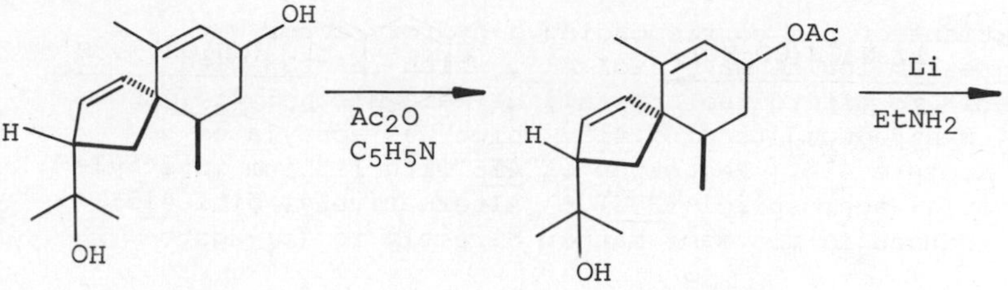

415 → (Ac$_2$O, C$_5$H$_5$N) → 416 → (Li, EtNH$_2$) → 372

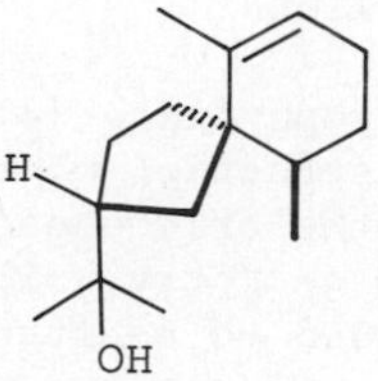

enone ester 403 when ketalized and then reduced, afforded
allylic alcohol 404. Hydrolysis of the ketal grouping occurred
with concomittant dehydration of yield dienone 405. This
material underwent 1,6 addition of sodio diethyl malonate to
yield, after hydrolysis, decarboxylation and esterification,
the enone ester 406. Compound 406 was obtained from enone
ester 403 in 47% overall yield. Ketalization again occurred
with the normal double bond shift to yield a ketal ester,
which was saponified to obtain ketal acid 407.

The diazo ketone 408, obtained in the usual manner, was
decomposed in the presence of copper powder to yield a mixture
of tricyclic ketones 409 and 410 in a ratio of 1:9. The mixture of ketones 409 and 410 was converted into β-keto ester
411. (For the sake of convenience, we illustrate in the
scheme with the major isomer only. Compounds 411-413 were
each contaminated by 10% of an isomer derived from the minor
tricyclic ketone 409.) The relative stereochemistry of agarospirol is established at this stage. Nonbonded interactions
cause the carbomethoxyl group to be *anti* in the bicyclo[3.1.0]-
hexane system.

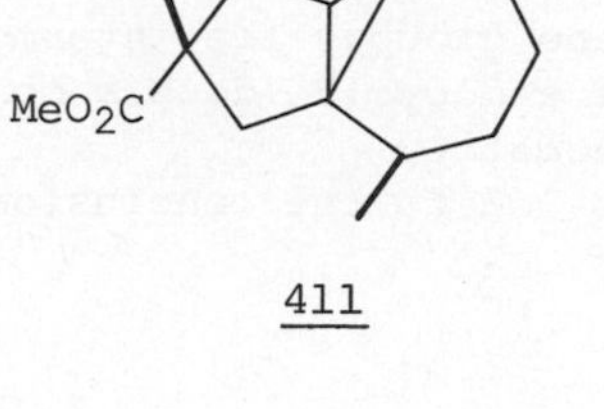

411

Treatment of the corresponding β-hydroxy ester with methylmagnesium iodide gave diol 413, which reacted with aqueous acid to afford the crystalline keto alcohol 414. Compound 414 was reduced to diol 415, which was acetylated to the mono acetate 416. Reduction of 416 with lithium in ethylamine gave (±)-agarospirol (372).* Alternatively, diol 415 could be reduced in the same manner directly to (±)-agarospirol.

M. Caryophyllene, Isocaryophyllene

Caryophyllene (417) and isocaryophyllene (418) pose three major synthetic problems: construction of the bicyclo[7.2.0]-undecane skeleton, control of the bridgehead stereochemistry, and control of the double bond geometry. In Corey's approach to the problem,[275] it was decided to construct a tricyclic

417   418

precurser and generate the nine-membered ring by scission of the internal bond of a hydrindane (i.e., 419 → 420).

419   420

Given this basic approach, the problem reduces to the synthesis of an intermediate with the gross skeleton of 419 functionalized in such a way as to allow the fragmentation and leave functionality adjacent to the cyclobutane ring for introduction of the methylene group. The intermediate should also allow formation of the endocyclic double bond, and should provide control over its geometry.

Various considerations led to the conclusion that the

*See footnote, p. 469.

tricyclic diol monotosylates 421 and 423 are ideal intermediates for the purpose. Base catalyzed Grob fragmentation would lead to the bicyclic ketones 422 and 424, respectively.

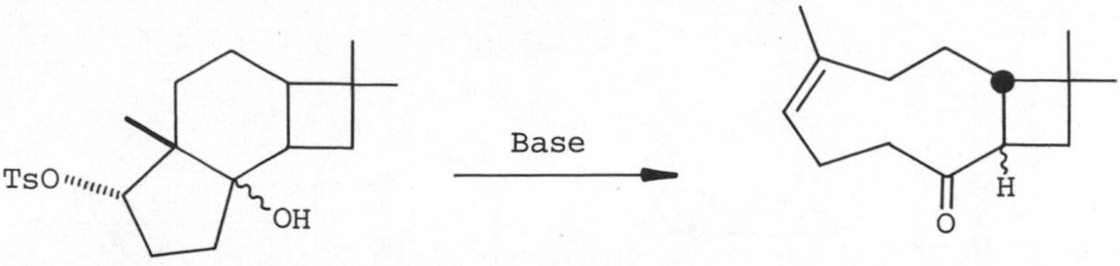
TsO
OH
Base
H
O

421 422

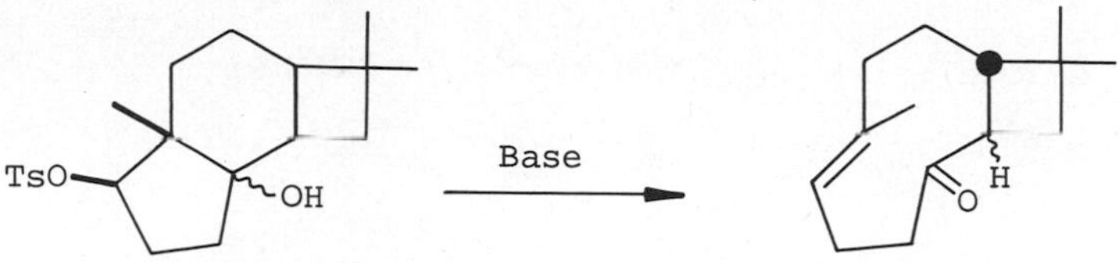
TsO
OH
Base
H
O

423 424

Assuming a concerted fragmentation, the geometry of the new double bond is fixed by the relative stereochemistry of the angular methyl group and the leaving group. The products 422 and 424 possess functionality for introduction of the characteristic methylene group and the carbonyl group should allow equilibration of the ketones to the more stable *trans* ring juncture.

Scheme 39 illustrates the application of this approach to the synthesis of (±)-isocaryophyllene. Photoaddition of isobutylene and cyclohexenone gave, after homogenization of the

Scheme 39. Corey's Synthesis of (±)-Isocaryophyllene

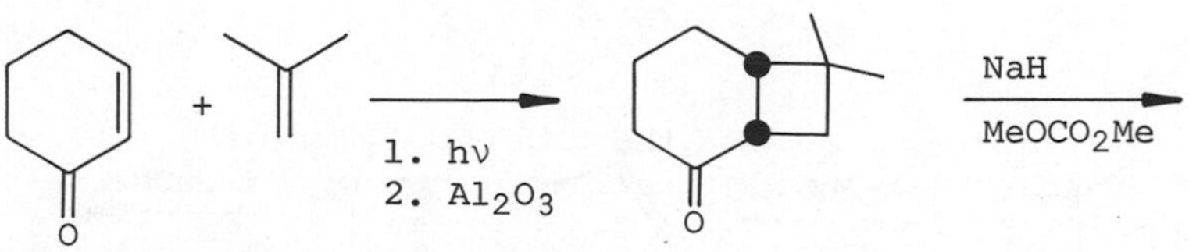
O
+
1. hν
2. Al2O3
O
NaH
MeOCO2Me

425

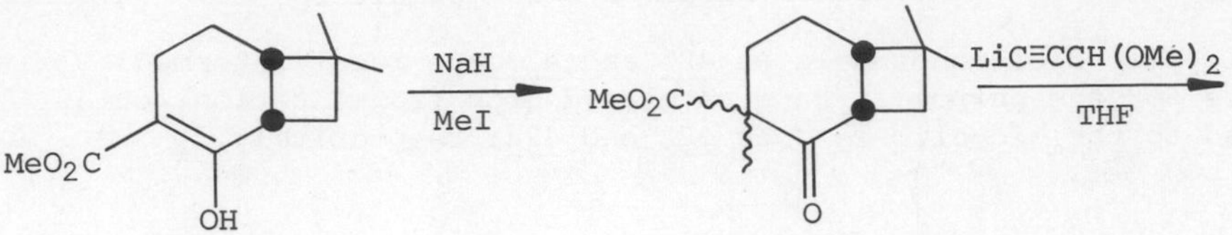

426 427

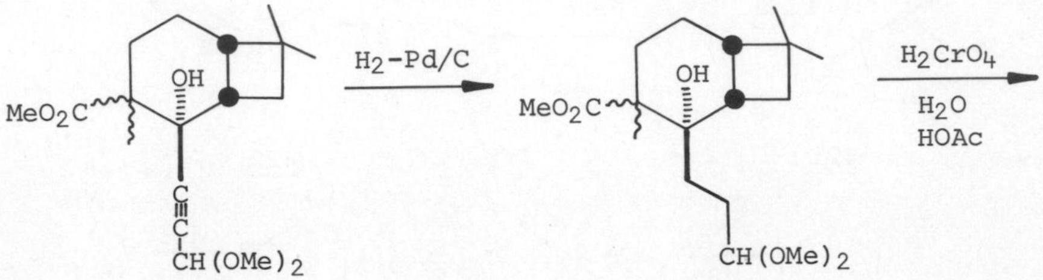

428 429

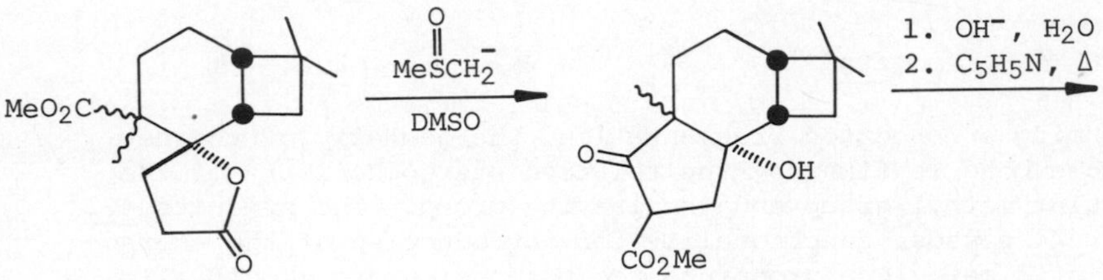

430 431

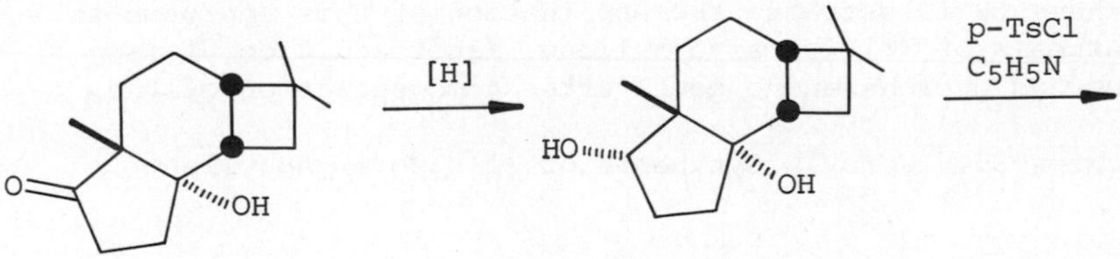

432 433

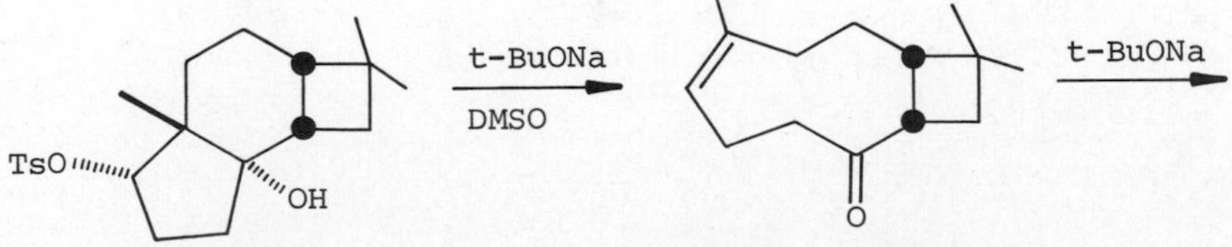

434 435

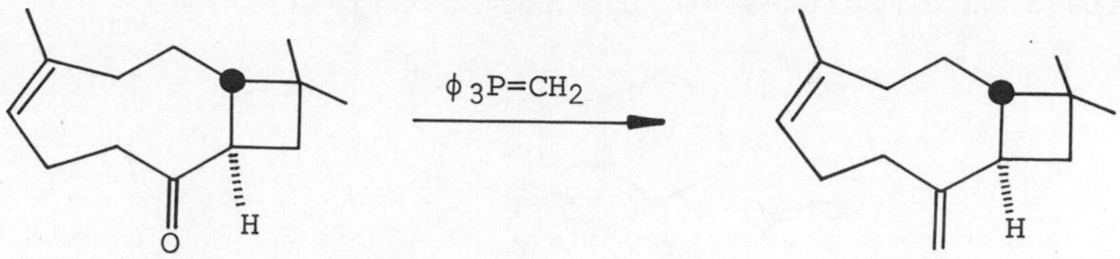

436   418

ring juncture, the bicyclo[4.2.0]octanone 425 in 35-45% yield.
This material was converted into the enolic β-keto ester 426
in the normal manner (86%). Methylation of 426 gave a mixture
of diastereomers 427 in a ratio of 3:1 (94% yield). Although
the isomers were not separated, and their structures not rig-
orously proven, it seems likely that the major isomer has the
stereochemistry depicted in 427a. Structure 427a would result
from methylation from the less hindered convex face of the bicyclic system.

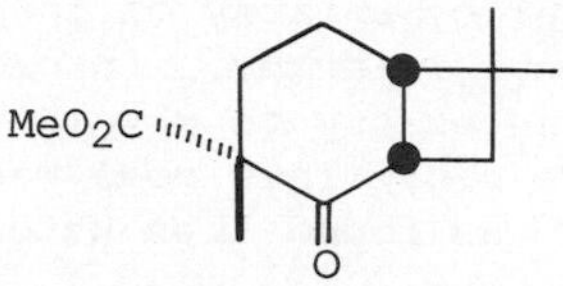

427a

Addition of the lithium salt of propargaldehyde dimethyl
acetal to the mixture 427 gave a mixture of ethynyl carbinols
428, which was reduced to 429. Oxidation of 429 with chromic
acid in aqueous acetic acid gave the spiro-lactone 430, still
a mixture of diastereomers, in 86% yield. The stereochemistry
of the new center is based on the assumption that the lithium
acetylide adds from the less hindered face of 427.

Dieckmann cyclization of 430 gave a β-keto ester (431),
which was saponified and decarboxylated to give a major (m.p.
126-127°) and a minor (m.p. 140-141.5°) tricyclic keto alcohol. Although no information pertaining to relative stereochemistry of the angular methyl group was adduced by Corey,
for the reason given above, the structure shown (432) seems
most reasonable to this writer. Fortunately, this stereochemical point is not critical, since the geometry of the
double bond to be formed by fragmentation depends *only* on the
relative disposition of this methyl group and the vicinal
leaving group (see above). For purposes of clarity, we shall
illustrate in the sequel with the methyl group β oriented.
The alternative structure, 437, could lead, through the

appropriate intermediates, to the same ultimate product.

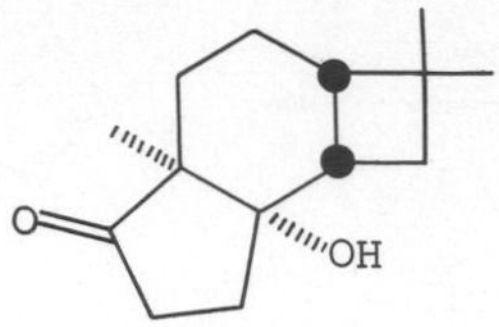

437

Reduction of the carbonyl group in 432 with various
agents [NaBH$_4$, LiAlH$_4$, LiAl(O-Bu-t)$_3$H, Na-H$_2$O/ether] led to a
single diol (433). This compound formed a mono-tosylate
(434), which reacted with sodium t-butoxide in DMSO to yield,
initially, the *cis*-fused unsaturated ketone, 435. Equilibration of this substance yielded the *trans* isomer 436 which reacted with methylenetriphenylphosphorane to yield (±)-iso-
caryophyllene.

The obtention of isocaryophyllene from diol 434 demon-
strates that the relative stereochemistry of the secondary
hydroxyl and the angular methyl is *trans*. In order to secure
caryophyllene itself, it is necessary to obtain the isomer of
434 in which these groups are *cis*. This goal was finally
achieved (Scheme 40) when it was found that ketol 432 is

Scheme 40.   Corey's Synthesis of (±)-Caryophyllene

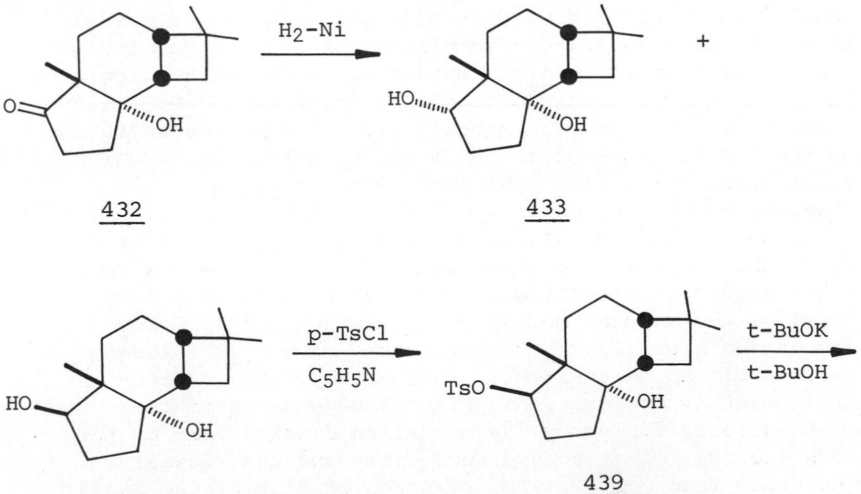

H_2-Ni

432   433

+

p-TsCl
C$_5$H$_5$N

438

439

t-BuOK
t-BuOH

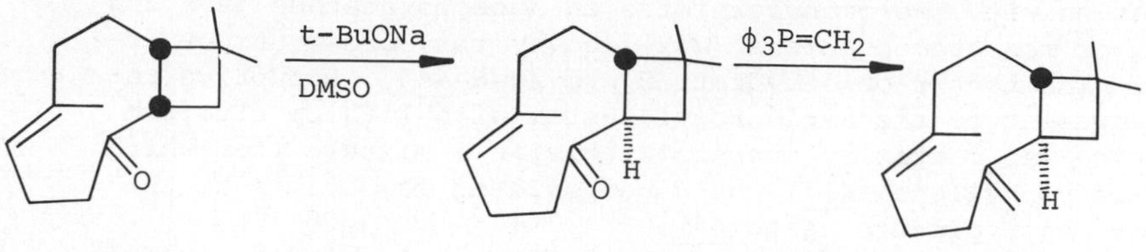

440   441   417

reduced by hydrogen in the presence of Raney nickel to a 1:1
mixture of diol 433 and its isomer 438. Compound 438 gave a
mono-tosylate (439), which on reaction with potassium t-butoxide in *t*-butanol initially gave the *cis*-fused ketone, 440
which on equilibration as above gave the *trans* isomer 441.
Wittig methylenation provided (±)-caryophyllene.

An approach starting from the macrocycle humulene (structure 279, Sec. 3) is described by Sutherland[276] (Scheme 41).

Scheme 41. Sutherland's Synthesis of (±)-Caryophyllene
from Humulene

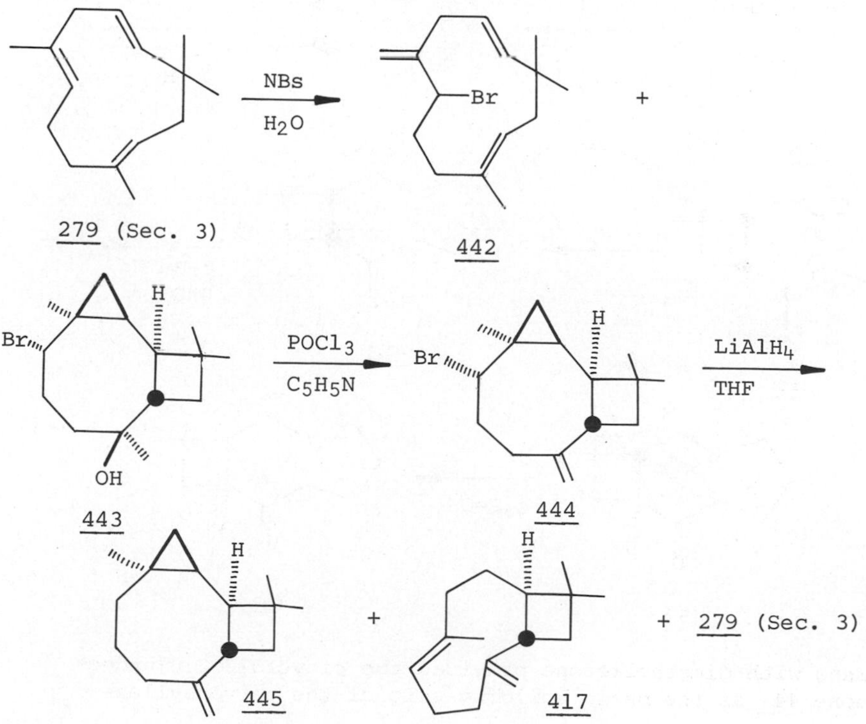

279 (Sec. 3)   442

443   444

445   417   + 279 (Sec. 3)

Reaction with N-bromosuccinimide in aqueous acetone gave 20%
each of monobromohumulene (442) and hydroxy-bromo compound
443. The latter on dehydration provided 444. Reductive rearrangement of the cyclopropyl system with lithium aluminum
hydride was partially successful giving a mixture from which
(±)-caryophyllene (417) could be isolated in 30% yield, the
other major product being 445.

A novel approach starting from 1,2-cyclononadiene (446)
has been reported[277] (Scheme 42). Reaction of the cyclic

Scheme 42. Gras-Maurin-Bertrand Synthesis of
5,6-Dihydro-Nor-Caryophyllene

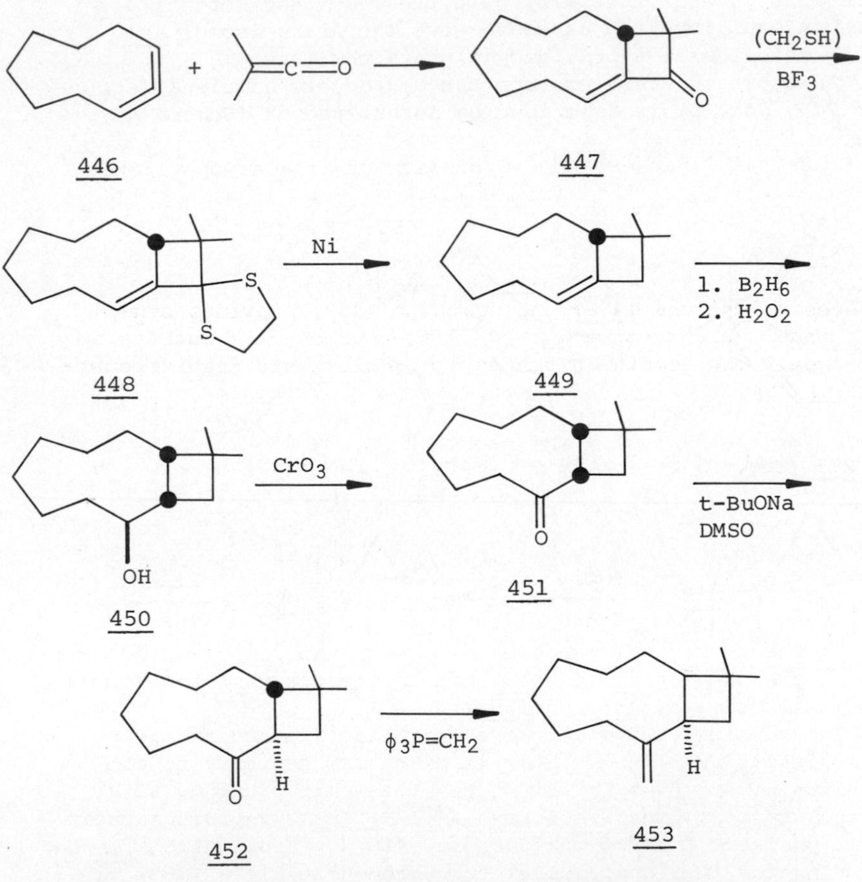

446   447

448   449

450   451

452   453

allene with dimethylketone provided the bicyclo[7.2.0]undec-
eneone 447 as the basic building unit of the caryophyllene

skeleton. Thioketal formation to <u>448</u> followed by reduction
gave the olefin <u>449</u>. Hydroboration to <u>450</u>, followed by oxidation gave the unstable *cis* ketone <u>451</u> which on equilibration
yielded the *trans* ketone <u>452</u>. A wittig reaction provided 5,6-
dihydronorcaryophyllene <u>453</u>.

N. α-Santalol, β-Santalol, α-Santalene, β-Santalene

The characteristic odor of East Indian Sandalwood oil, highly
prized in perfumery, is due mainly to the companion alcohols
α- and β-santalol (<u>454</u> and <u>455</u>), which occur in the oil along
with their hydrocarbon analogs α- and β-santalene (<u>456</u> and
<u>457</u>). The relationship of these sesquiterpenes to the mono-

<u>454</u>: X = OH   <u>455</u>: X = OH
<u>456</u>: X = H   <u>457</u>: X = H

terpenes tricyclene (<u>458</u>) and camphene (<u>459</u>) provides synthetic analogy for construction of the carbocyclic nucleus in
each case. The remaining synthetic problems are the introduc-

<u>458</u>   <u>459</u>

tion of the isopentenyl substituent, with the proper orientation in the case of <u>455</u> and <u>457</u> and the geometry of the double
bond in the case of <u>454</u> and <u>455</u>.
    The simplest target, from a synthetic standpoint, is
α-santalene (<u>456</u>), since there is no stereochemistry to con-
tend with. Corey was the first to solve this problem, with
the synthesis outlined in Scheme 43.[278] *Trans*-π-bromocamphor
(<u>462</u>), prepared from α-bromocamphor (<u>460</u>) by bromination-
debromination, was converted into π-bromotricyclene (<u>463</u>) via
the diazo compound. The corresponding Grignard reagent reacted with γ,γ-dimethylallyl mesitoate to give (+)-α-santalene

Scheme 43.   Corey's Synthesis of (±)-α-Santalene

(456) in 18% yield. The remaining material was mostly the
crystalline bi-π-tricyclyl (465). The isopentenyl side-chain

could not be efficiently introduced by treatment of 464 with
γ,γ-dimethylallyl bromide, since the resulting α-santalene was
accompanied by up to 50% of the isomer 466.

In 1962, Corey reported an elegant, stereospecific synthesis of (±)-β-santalene (457), and its epimer, (±)-epi-β-
santalene (472), which he showed to be a companion of β-santalene in sandalwood oil (Scheme 44).[279] Norbornanone (467) was

Scheme 44. Corey's Synthesis of (±)-β-Santalene
and (±)-epi-β-Santalene

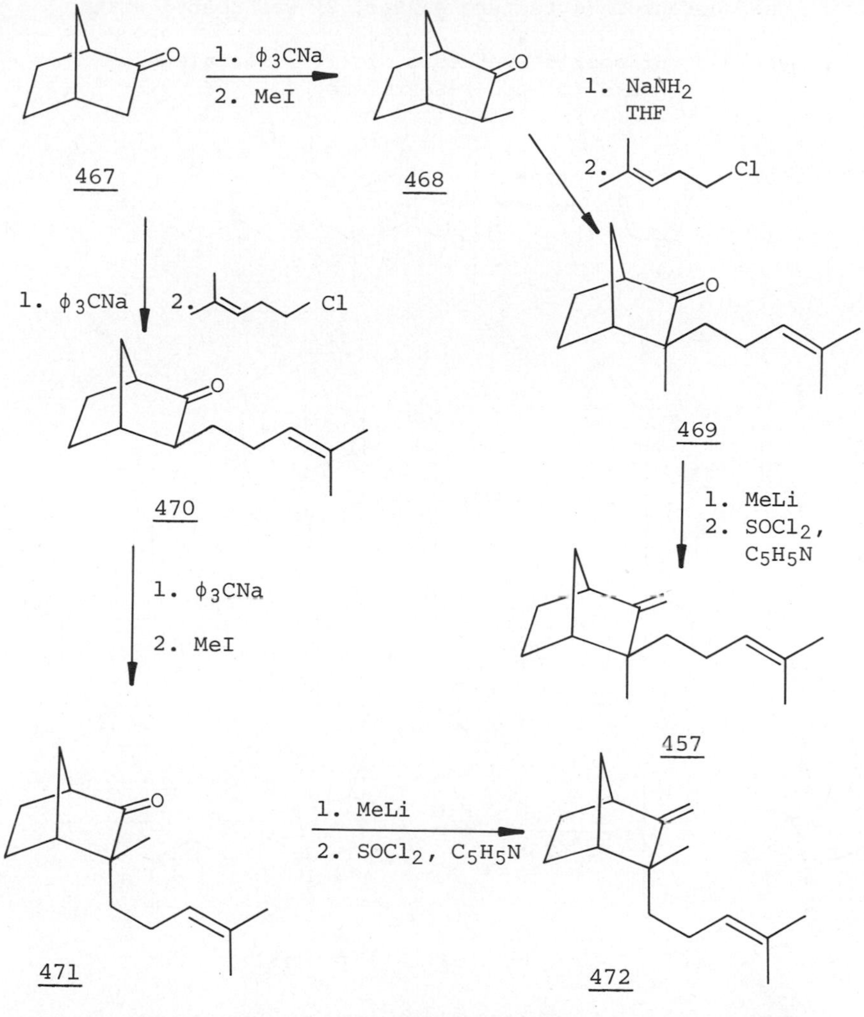

methylated to give the *exo* derivative 468. Careful analysis
showed that the *exo-endo* alkylation ratio was greater than
30:1. Alkylation of 468 with 2-methyl-5-chloro-2-pentene
yielded solely ketone 469. The tertiary alcohol formed on
reaction of 469 with methyllithium was dehydrated to obtain
(±)-β-santalene (457). By reversing the order of introduction
of the two side-chains, the isomer 472 was prepared, also
stereospecifically.

Brieger has reported an interesting, although inefficient, synthesis of a mixture of the two β-santalene isomers (Scheme 45).[280] When geraniol (structure 4, Sec. 2) was heated with

Scheme 45.   Brieger's Synthesis of (±)-β-Santalene
and (±)-epi-β-Santalene

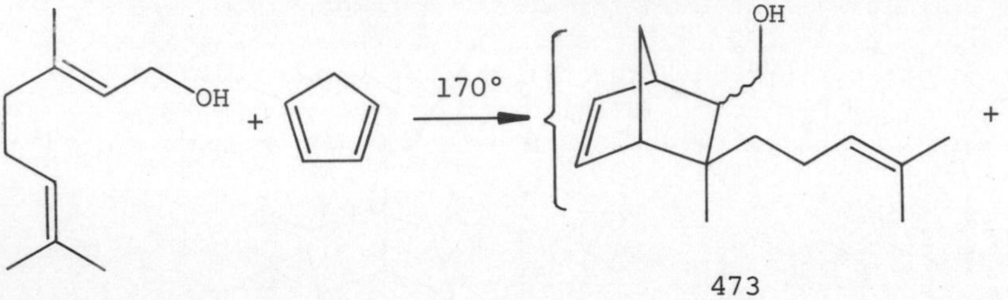

4 (Sec. 2)

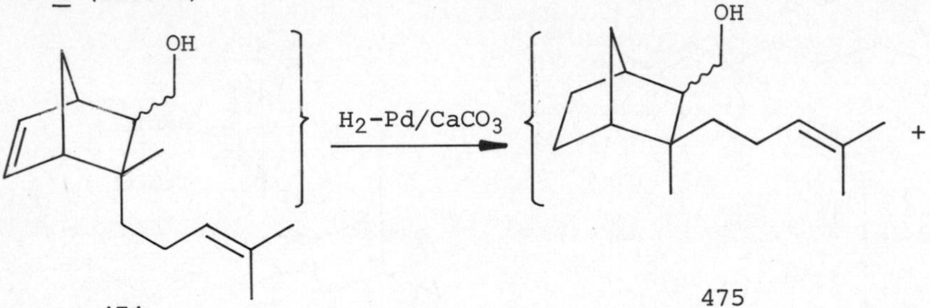

474

475

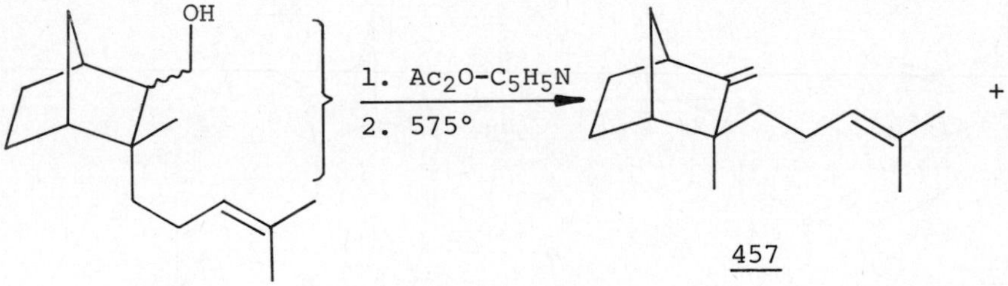

476

457

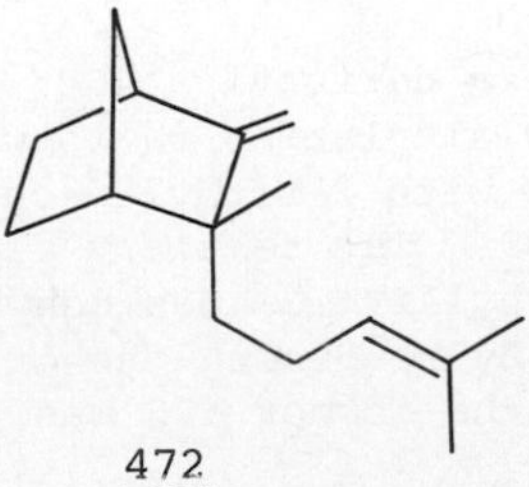

472

cyclopentadiene at 170° for two days, a complex mixture of
adducts and polymers was produced. Fractional distillation of
the mixture yielded a mixture of 473 and 474 in 4% yield.
After saturation of the more accessible double bond, the corresponding acetates were pyrolyzed to yield a 3:2 mixture of
(±)-β-santalene (457) and (±)-epi-β-santalene (472).

In a synthesis which was first announced in 1947, and
finally revealed in 1967, Bhattacharyya prepared α-santalene
from π-bromotricyclene (462) as outlined in Scheme 46.[281]   The

Scheme 46.   Bhattacharyya's Synthesis of (±)-α-Santalene

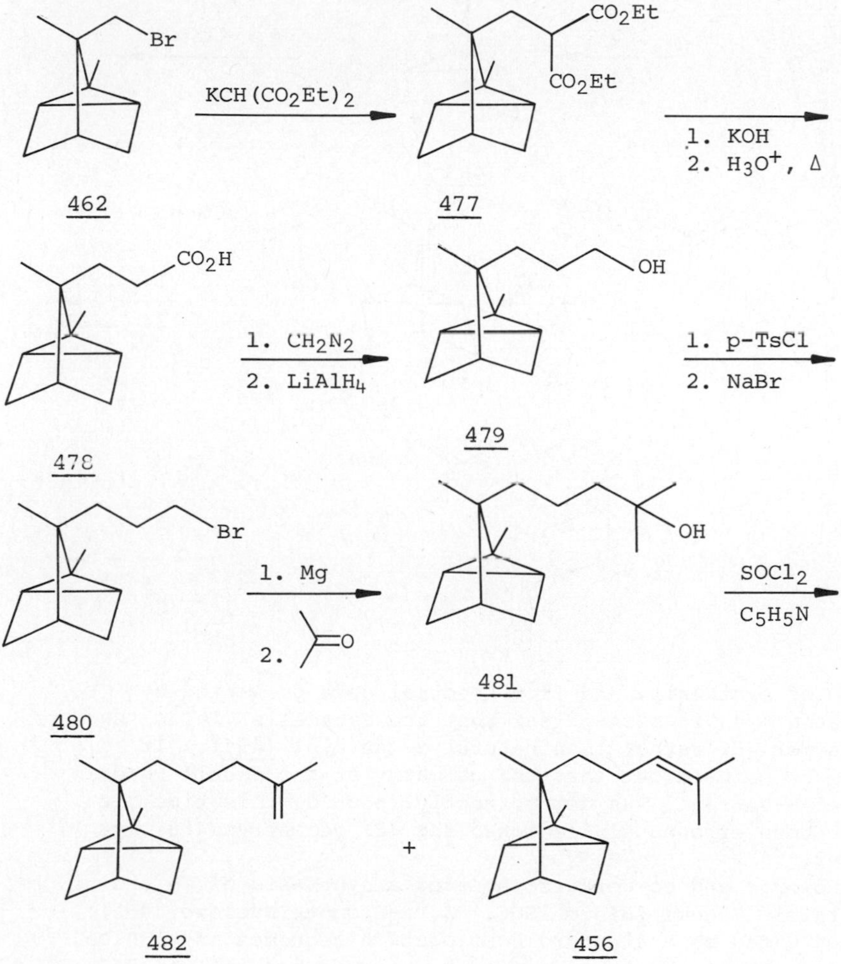

ten-stage conversion, which affords a mixture of (+)-α-santa-
lene (456) and its isomer 482 offers little advantage over
Corey's one-step conversion, even though the latter proceeds
in only 18% yield.

In the same 1967 paper, Bhattacharyya reported a synthesis
of "(+)-α-santalol."[281b] The synthesis, which proceeds through
tricycloekasantalal (483), is outlined in Scheme 47. From the

Scheme 47. Bhattacharyya's Synthesis of (+)-(E)-α-Santalol

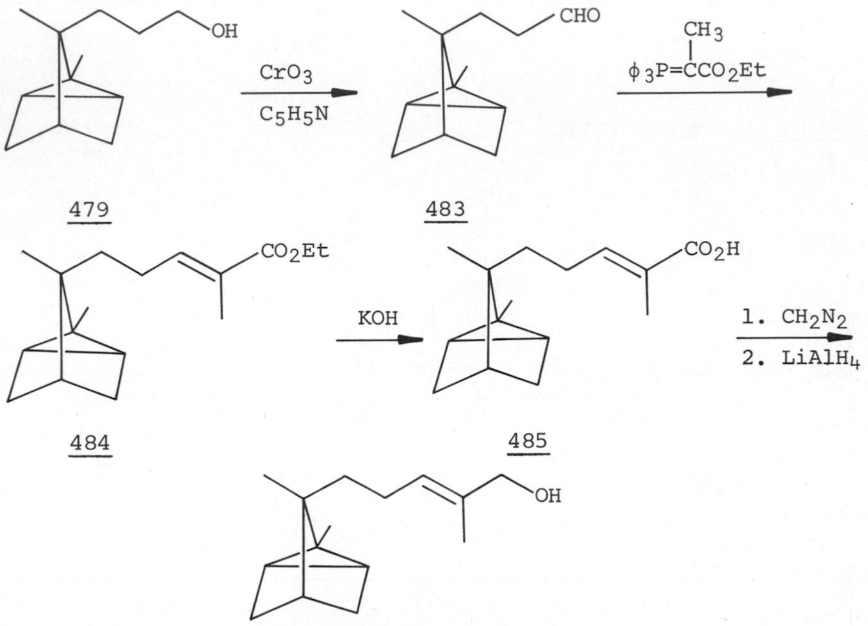

method of synthesis, and from spectral data presented by
Bhattacharyya, it seems clear that the synthesis yields the
(E)-isomer 486 rather than natural α-santalol (454). It
should be pointed out that the geometry of the double bond in
natural α-santalol was not correctly known at this time, it
having been erroneously assigned the (E) geometry (486) by
Brieger.[282]

Colonge and co-workers reported a synthesis of α- and
β-santalol (Scheme 48) in 1966.[283] π-Bromotricyclene (462)
was converted by a standard homologation sequence into chloride 490. This substance formed a Grignard reagent which

Scheme 48.   Colonge's Synthesis of α- and β-Santalol (?)

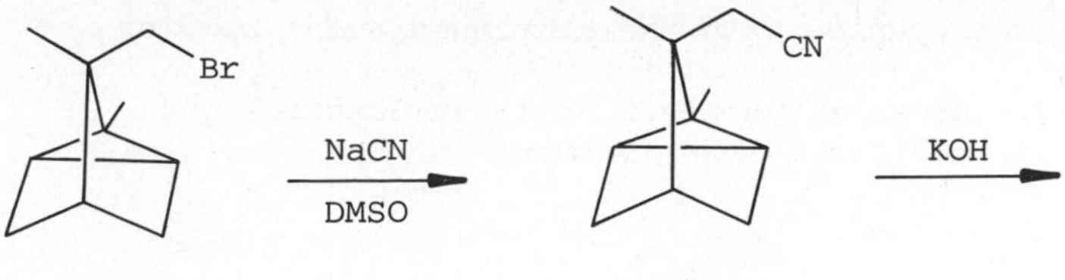

462   487

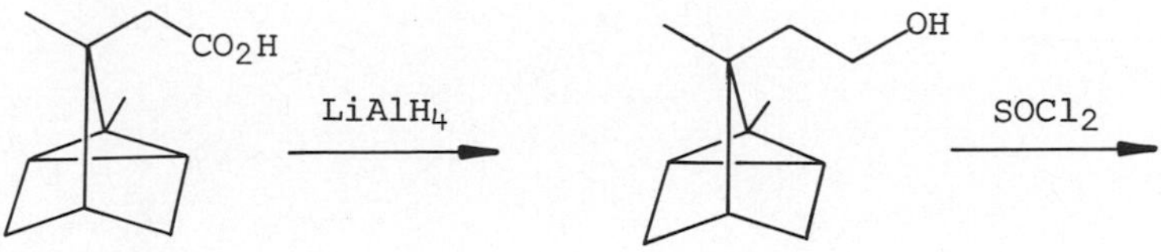

488   489

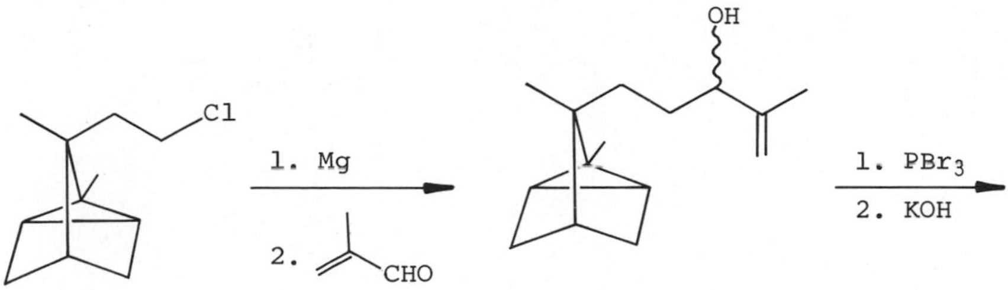

490   491

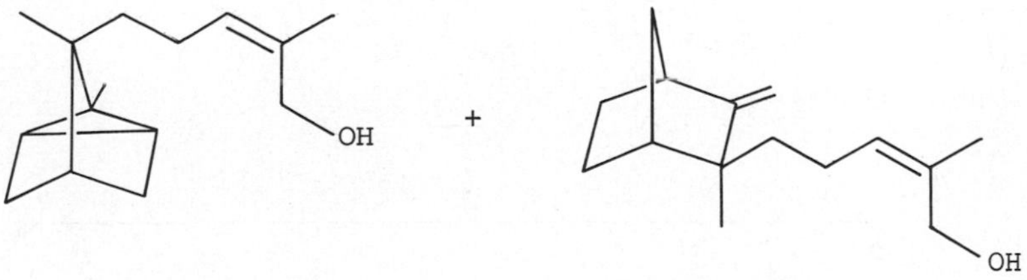

454 (+ isomer?)   455 (+ isomer?)

reacted with methacrolein to yield alcohol 491. Upon treatment of 491 with PBr₃, followed by KOH, a mixture of α-santalol
(454) and β-santalol (455) was isolated. The stereochemical
integrity of the products has been questioned.[284]

In 1967, Erman and co-workers, at the Procter and Gamble
Laboratories in Cincinnati, Ohio, provided conclusive evidence
that natural α-santalol indeed has the (Z) geometry indicated

in 454. The Erman synthesis, which provides both the (E) (486) and (Z) (454) isomers, is outlined in Scheme 49.[285] π-Bromo-tricyclene (462) was coupled with the ethylenediamine complex

Scheme 49.   Erman's Synthesis of (+)-α-Santalol and Its (+)-(E) Isomer

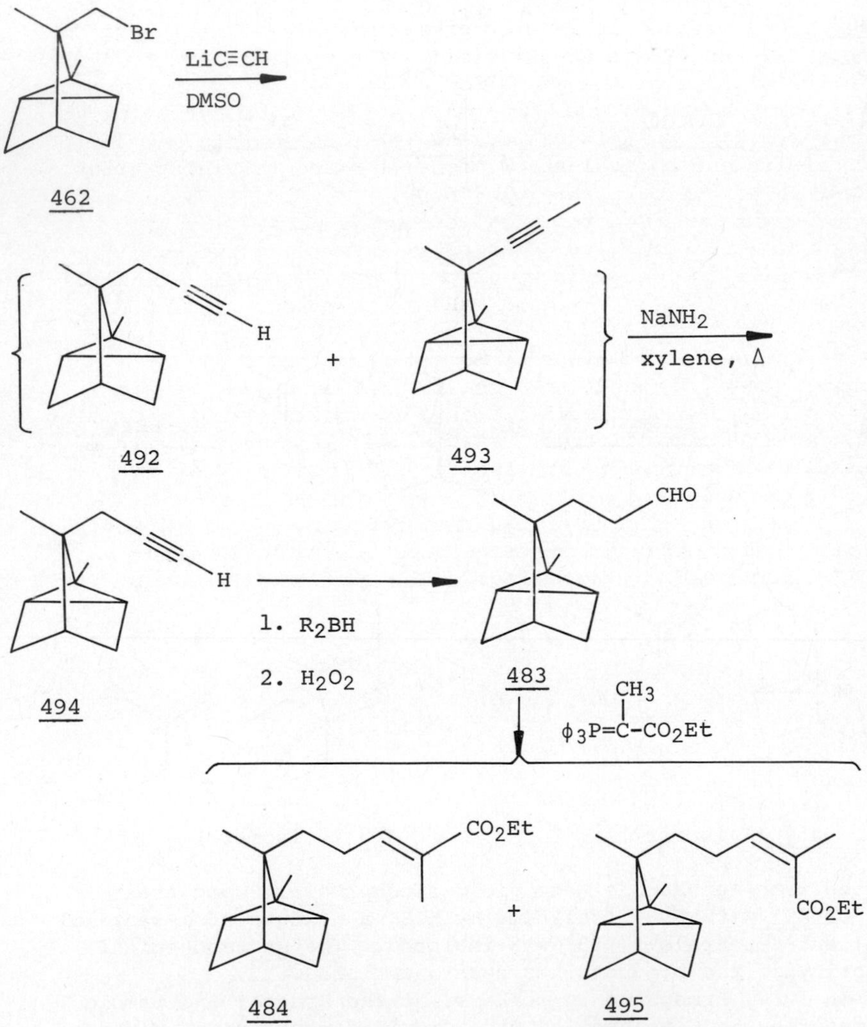

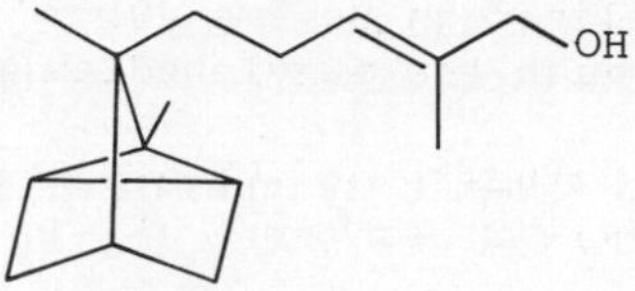

486 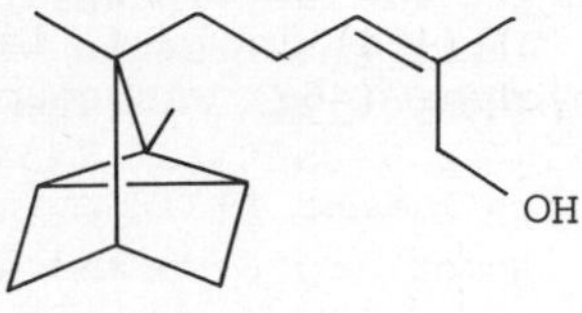   454

of lithium acetylide in HMPA to give a mixture of acetylenic
isomers 492 and 493 in 80-90% yield. The ratio of 492 to 493
was highly variable, the amount of 493 increasing with reaction
time. Homogenization was achieved, in 63% yield, by refluxing
the mixture with sodamide in xylene. Terminal acetylene 494
reacted with one equivalent of disiamylborane to yield, after
oxidation, tricycloekasantalal (483).

This aldehyde reacted with (carbethoxyethylidine)tri-
phenylphosphorane to give unsaturated esters 484 and 495 in
a ratio of 5:1. The geometry of the double bond was, in each
case, assigned on the basis of NMR spectroscopy. Reduction of
the major ester (484) gave (+)-(E)-2-santalol (486). Analogous reduction of the minor isomer (495) afforded (+)-α-
santalol (454) identical with the natural material. Analysis
of eight East Indian sandalwood oils and one Australian sandalwood oil, all from different sources, revealed that isomer 454
is a constituent of all, and that the (E) isomer (486) is a
constituent of none.

By a simple (though elegant) modification of the Wittig
reaction, Corey achieved a stereospecific synthesis of α-
santalol (Scheme 50).[286] Tricycloekasantalal (483) was treated

Scheme 50. Corey's Synthesis of (+)-α-Santalol

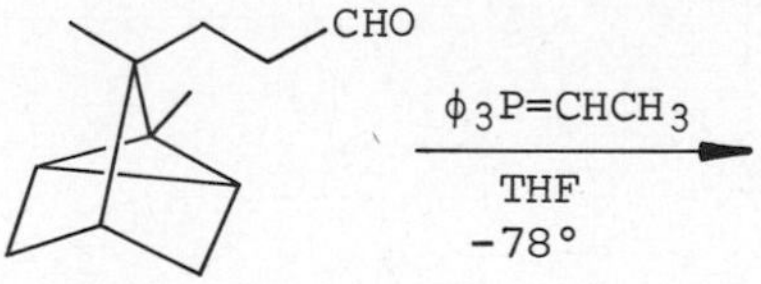

CHO

φ₃P=CHCH₃
———————————→
THF
-78°

483

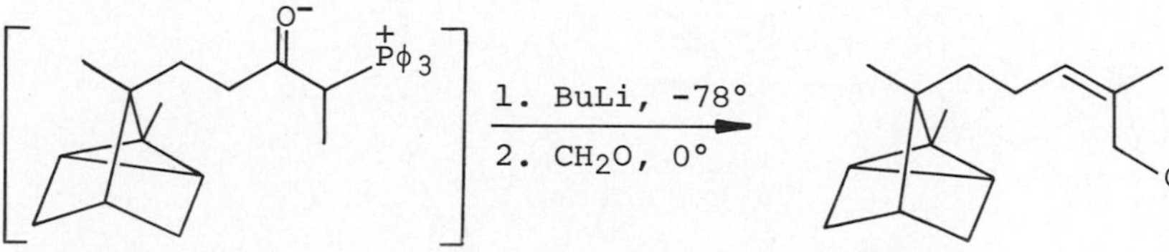

O⁻   +
     Pφ₃

1. BuLi, -78°
———————————→
2. CH₂O, 0°

OH

496   454

with ethylidinetriphenylphosphorane at -78° to give the betaine
496. This was treated with *n*-butyllithium at -78°, then paraformaldehyde at 0°. The product obtained on work-up consisted
solely of α-santalol (454).

In Scheme 51 is outlined Erman's synthesis of the (E) and
(Z) isomers of β-santalol (505 and 455), the latter of which

Scheme 51. Erman's Synthesis of (±)-β-Santalol
and Its (±)-(E) Isomer

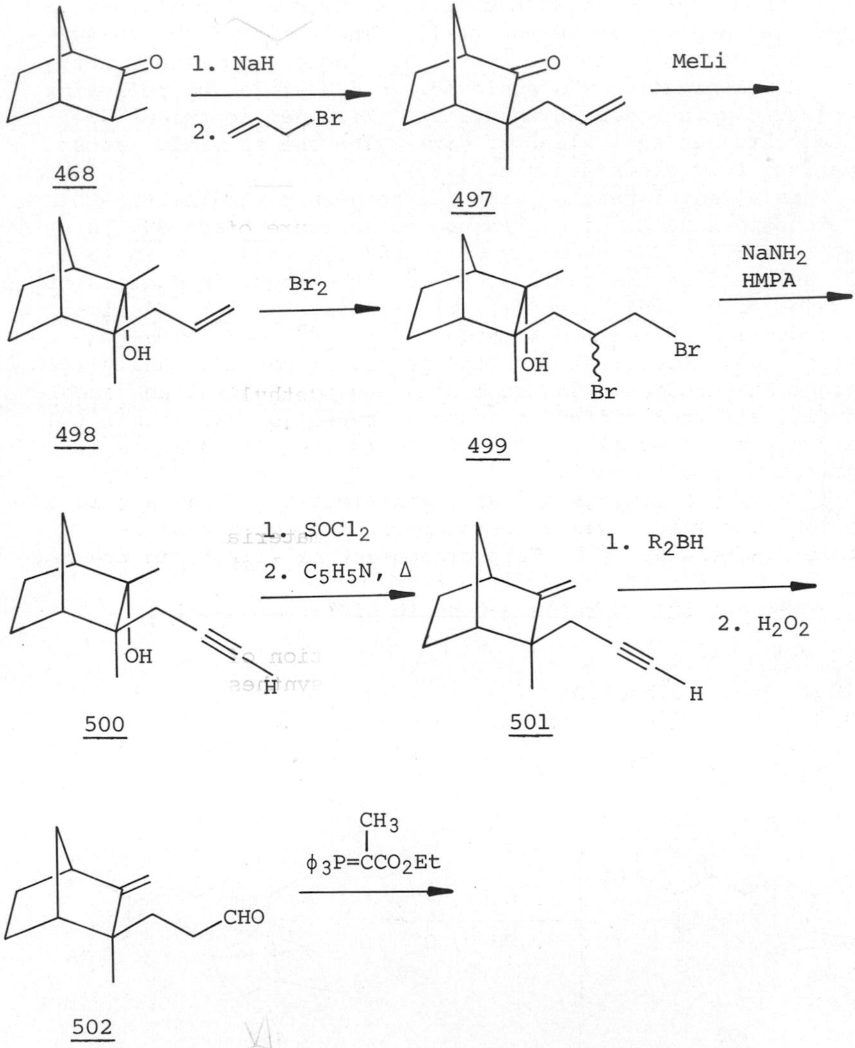

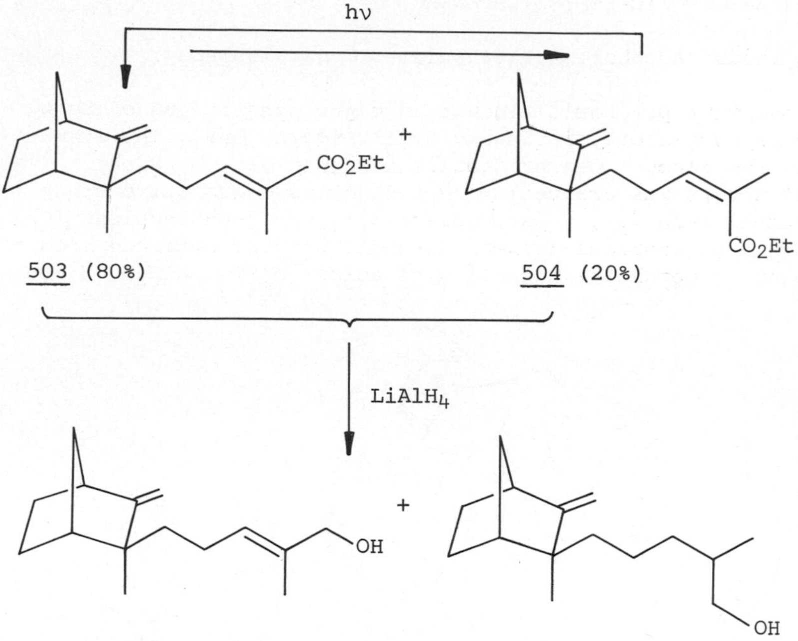

503 (80%)   504 (20%)

505   455

is identical with the natural terpene.[287] The synthesis
closely parallels Erman's synthesis of the α-santalol isomers
(Scheme 49). Exo-3-methylnorbornanone (468) was alkylated with
allyl bromide to afford, stereospecifically, the ketone 497.
This reacted with methyllithium to give alcohol 498, which was
brominated and then dehydrobrominated to give the hydroxy
acetylene 500. Dehydration yielded 501, which was converted
into unsaturated aldehyde 502 by Erman's method (see Schemes
26, 27, and 49). Aldehyde 502 reacted with (carbethoxyethylidine)triphenylphosphorane to give unsaturated esters 503 and
504 (18% yield) in a ratio of 5:1. Ultraviolet irradiation
changed the 503:504 ratio to 2:1. When the mixture was reduced, a mixture of (±)-(E)-β-santalol (505 major product) and
(±)-β-santalol (455, minor product) was produced. Corey's
modification (Scheme 50) could probably be applied to aldehyde
502, providing a stereospecific synthesis of 455.

6.   TRICARBOCYCLIC SESQUITERPENES

A.   Patchouli Alcohol, α-Patchoulene, β-Patchoulene

The history of patchouli alcohol, a major constituent of East
Indian patchouli oil, is indeed an intriguing one. The structure of the alcohol was originally assigned as 1 by Büchi.[288]
This structure was arrived at by a rigorous degradative study
of α-patchoulene (2), γ-patchoulene (3), and β-patchoulene (4).
Although β-patchoulene (4) was the sole product obtained from
treatment of patchouli alcohol with acids (H_2SO_4, HCO_2H, H_3BO_3,

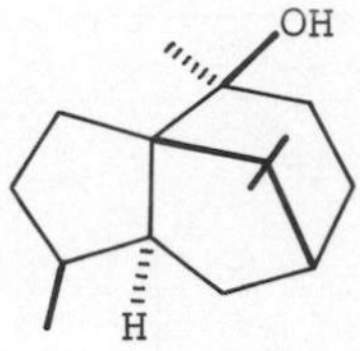

1

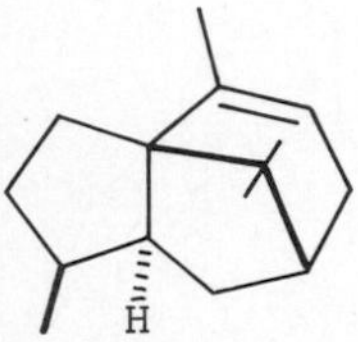

2

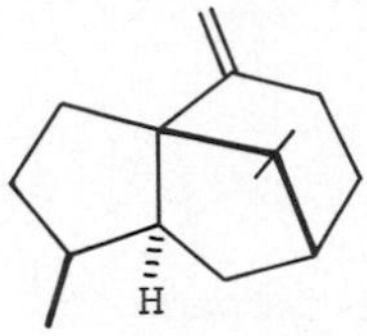

3

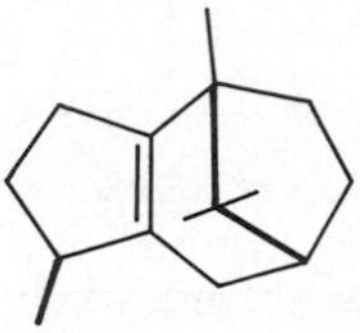

4

I_2), a mixture of 2, 3, and 4 in a ratio of 52:46:4 was obtained on pyrolysis of patchouli alcohol acetate. When
patchouli alcohol was dehydrated with $POCl_3$-pyridine, the
ratio obtained was 2:3:4 = 78:7:7. Furthermore, both 2 and 3
were shown to give 4, on acid treatment.

Assuming that 4 must arise by a Wagner-Meerwein rearrangement, and that the $POCl_3$-pyridine and, particularly, the acetate pyrolysis dehydrations proceed with no rearrangement,
patchouli alcohol was thought to be 1. The assumption was
strengthened when α-patchoulene (2) was converted into patchouli alcohol by a sequence interpreted by Büchi as:[288]

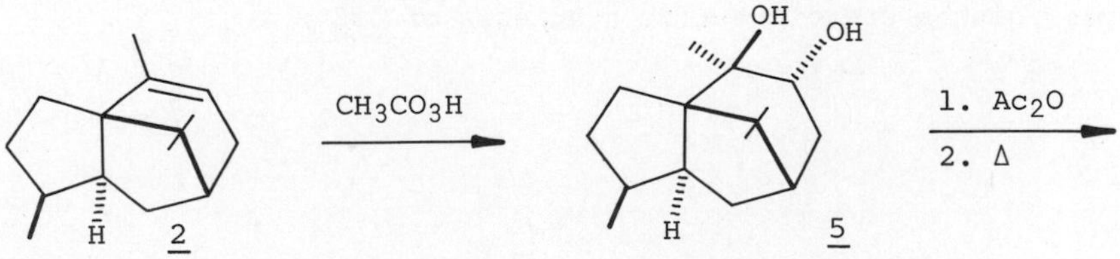

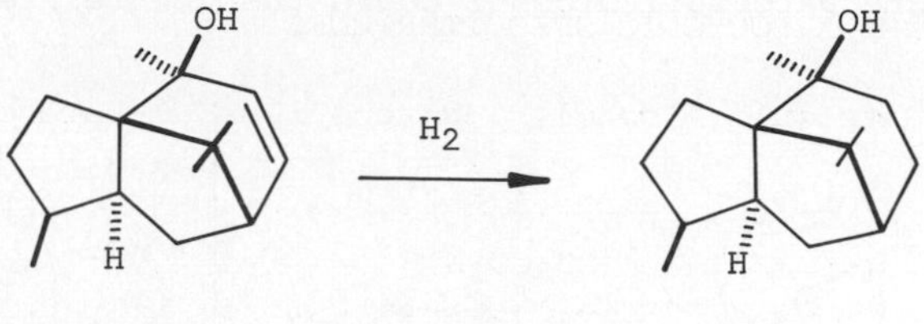

6   1

With this structure in mind, Büchi set out to synthesize
patchouli alcohol via α-patchoulene (2), which he had previously converted into patchouli alcohol by the steps outlined
above. The Büchi synthesis, which is outlined in Scheme 1,
will be discussed shortly.[289] It was only *after* the synthesis
had been completed (presumably confirming structure 1 for
patchouli alcohol), that Dunitz and co-workers completed an
x-ray investigation of the patchouli alcohol diester of
chromic acid, with the intent of determining the Cr-O-Cr bond
angle.[290] Surprisingly, it was found that patchouli alcohol
has structure 7, rather than 1.

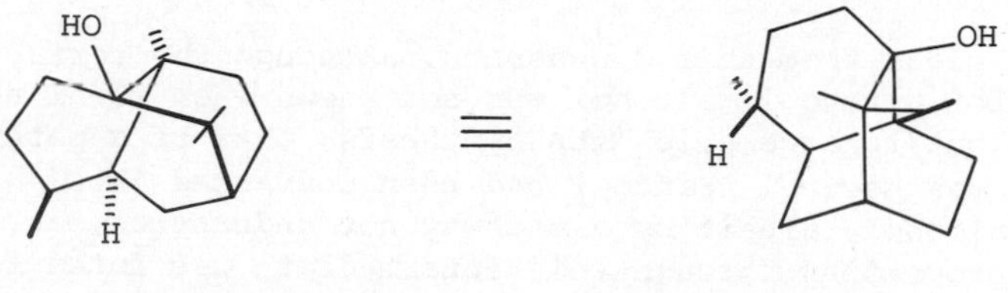

7   7

It then became clear that rearrangement had occurred in
pyrolysis of patchouli alcohol acetate (8) to α-patchoulene:

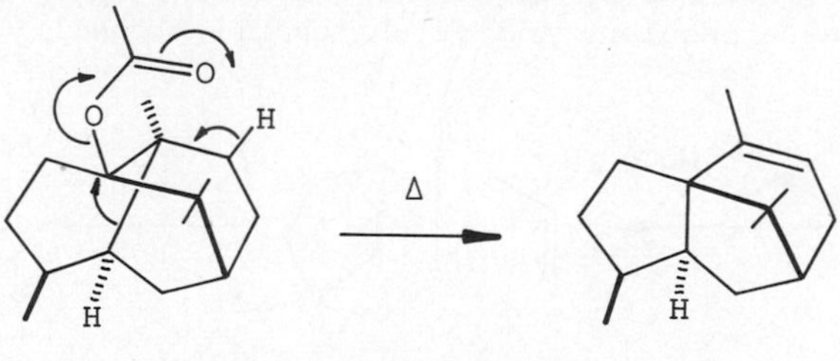

8   2

Amusingly, the exact reverse rearrangement had occurred in the
peracetic acid oxidation of α-patchoulene. The conversion of

2 into patchouli alcohol could then be reinterpreted correctly as:

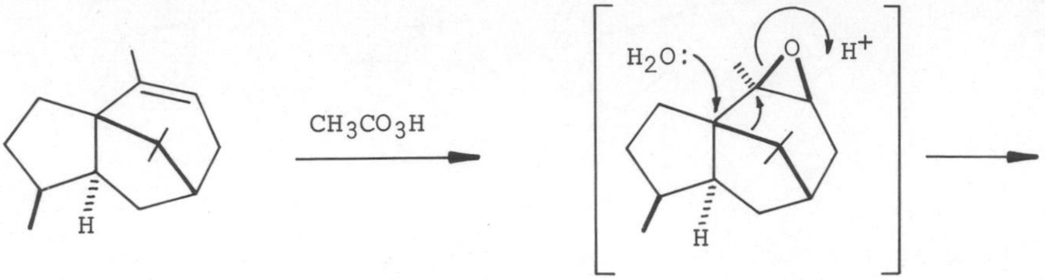

2

9

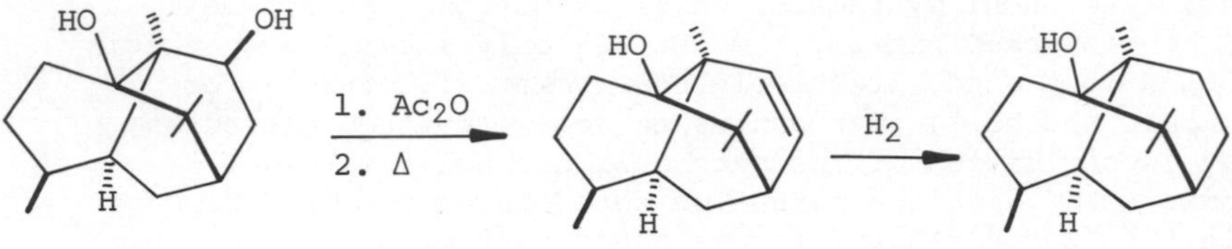

10
("5")

1. Ac$_2$O
2. Δ

11
("6")

H$_2$

7
("1")

As is clear from this discussion, although the correct structure for patchouli alcohol was not known when Büchi actually set out to accomplish its synthesis, that of α-patchoulene (2) was secure. Since 2 had been converted into patchouli alcohol, albeit by a pathway not understood, any synthesis proceeding through this intermediate was fated for success, even though the structure of the final product would remain obscure. Fortunately, Büchi elected such a synthetic plan, which is outlined in Scheme 1.[289]

Scheme 1.   Büchi's Synthesis of (-)-β-Patchoulene,
(+)-α-Patchoulene and (-)-Patchouli Alcohol

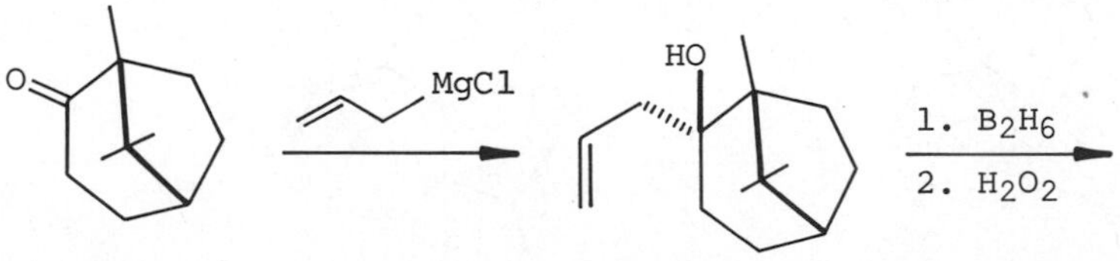

12

13

1. B$_2$H$_6$
2. H$_2$O$_2$

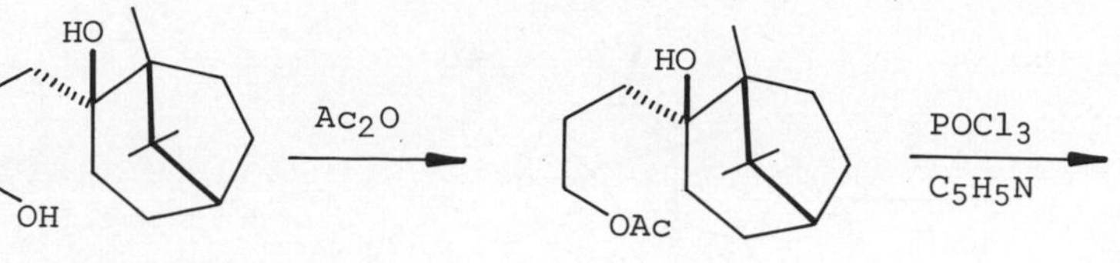

14 15

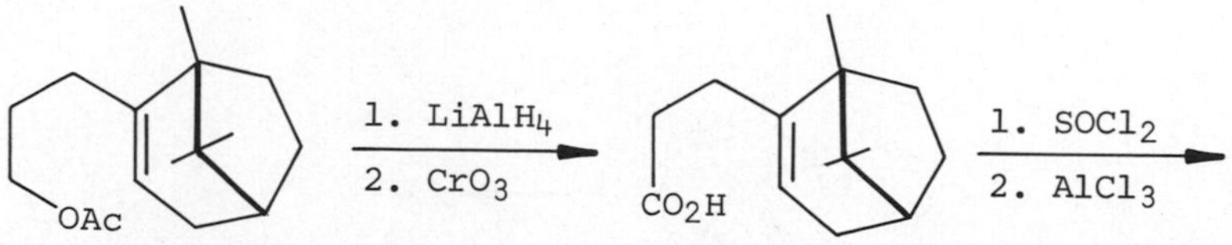

16 17

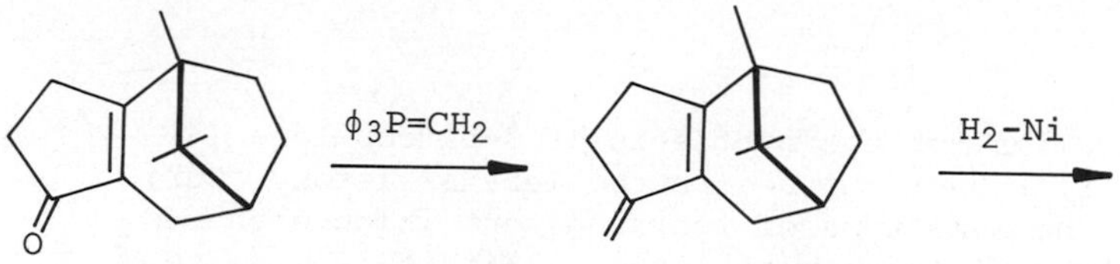

18 19

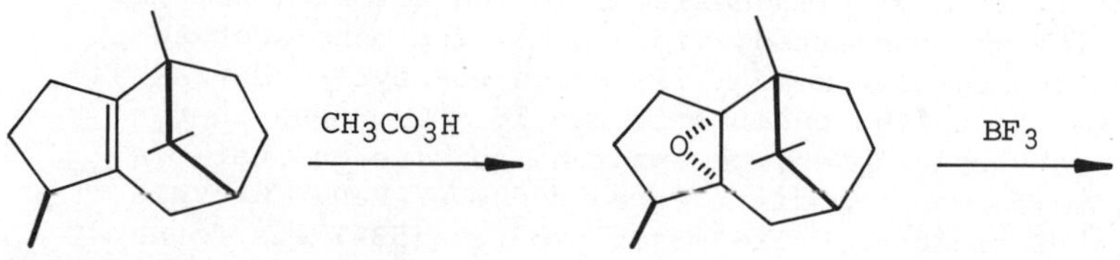

4 20

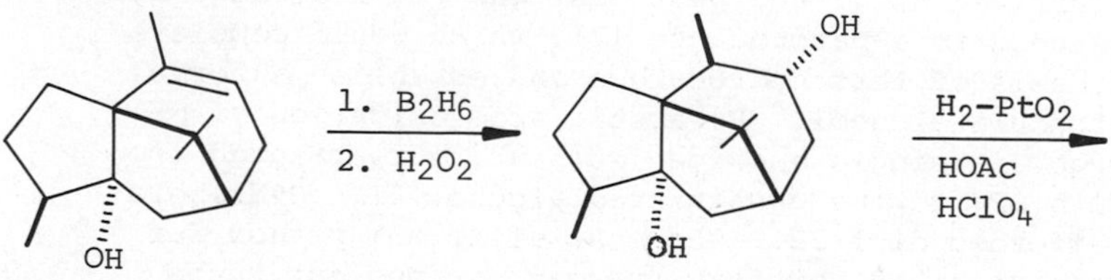

21 22

OH

OAc

Ac$_2$O

350°

H

H

23

24

7

H

2

The plan called for the synthesis of β-patchoulene (4), which could be rearranged to a α-patchoulene skeleton. Curiously, this is the same type of rearrangement which obscured the true structure of patchouli alcohol. With this goal in mind, homo-camphor (12) becomes an ideal starting point. Treatment of 12 with allylmagnesium chloride afforded the alcohol 13, which was converted, via diol 14 and mono-acetate 15, into the unsaturated acid (17), which was cyclized via its acid chloride to the cyclopentenone 18. Compound 18 reacted with methylenetriphenylphosphorane to give an unstable diene (19), which was immediately reduced over Raney nickel to a mixture of products. The major product (53%) was found to be β-patchoulene (4), thus completing the total synthesis of this natural product.

Having arrived at 4, the next goal was the rearrangement of this substance to α-patchoulene (2), which would complete a total synthesis of that hydrocarbon and establish a formal route to patchouli alcohol. Peracetic acid oxidation of β-patchoulene gave a single epoxide (20), which rearranged upon treatment with BF$_3$ to the unsaturated alcohol 21. Hydroboration of 21 afforded diol 22. The most efficient method for removal of the unwanted tertiary hydroxyl turned out to be catalytic hydrogenation in the presence of perchloric acid. Compound 23 was obtained in 31% yield. The reaction probably proceeds by prior dehydration to 25, which is reduced to 23. Pyrolysis of acetate 24, obtained by acetylation of alcohol 23, afforded a mixture of olefins consisting mainly (80%) of α-patchoulene (2). In light of the prior conversion of 2 to

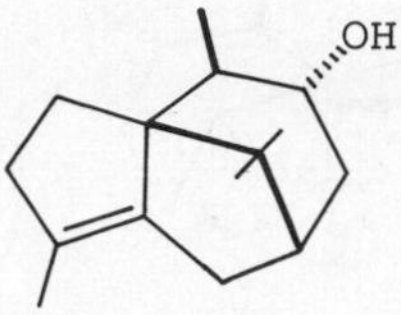

25

patchouli alcohol, a formal total synthesis of the alcohol was
established.

After the structure of patchouli alcohol became known,[290]
rational syntheses not proceeding through the patchoulenes
could be planned. Such a route, due to Danishevsky, is outlined in Scheme 2.[291] Danishevsky's synthetic plan called for

Scheme 2. Danishevsky's Synthesis of (±)-Patchouli Alcohol

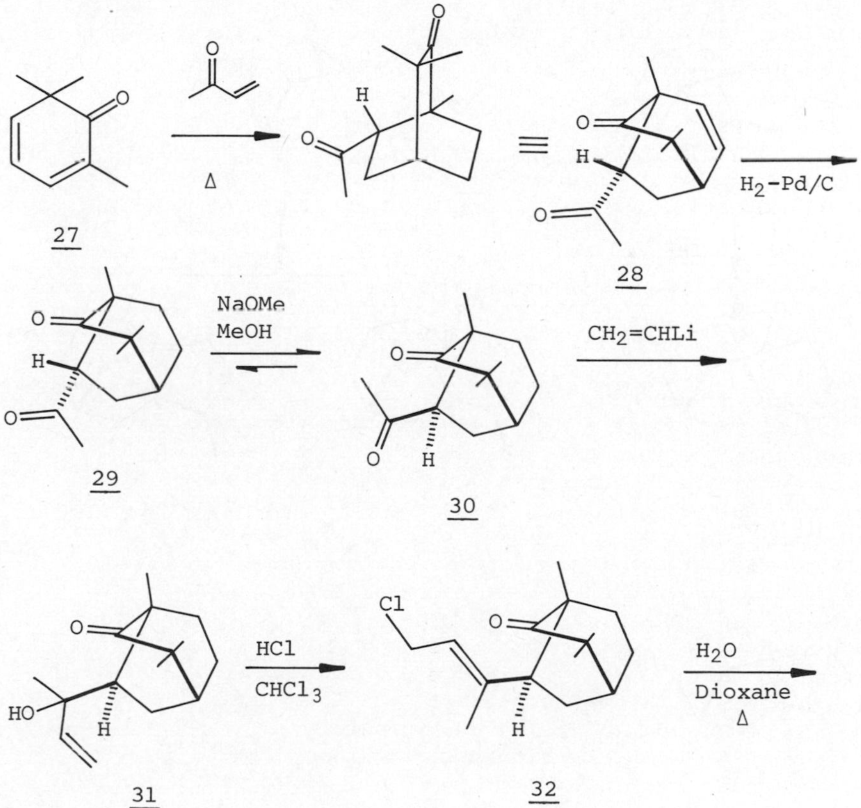

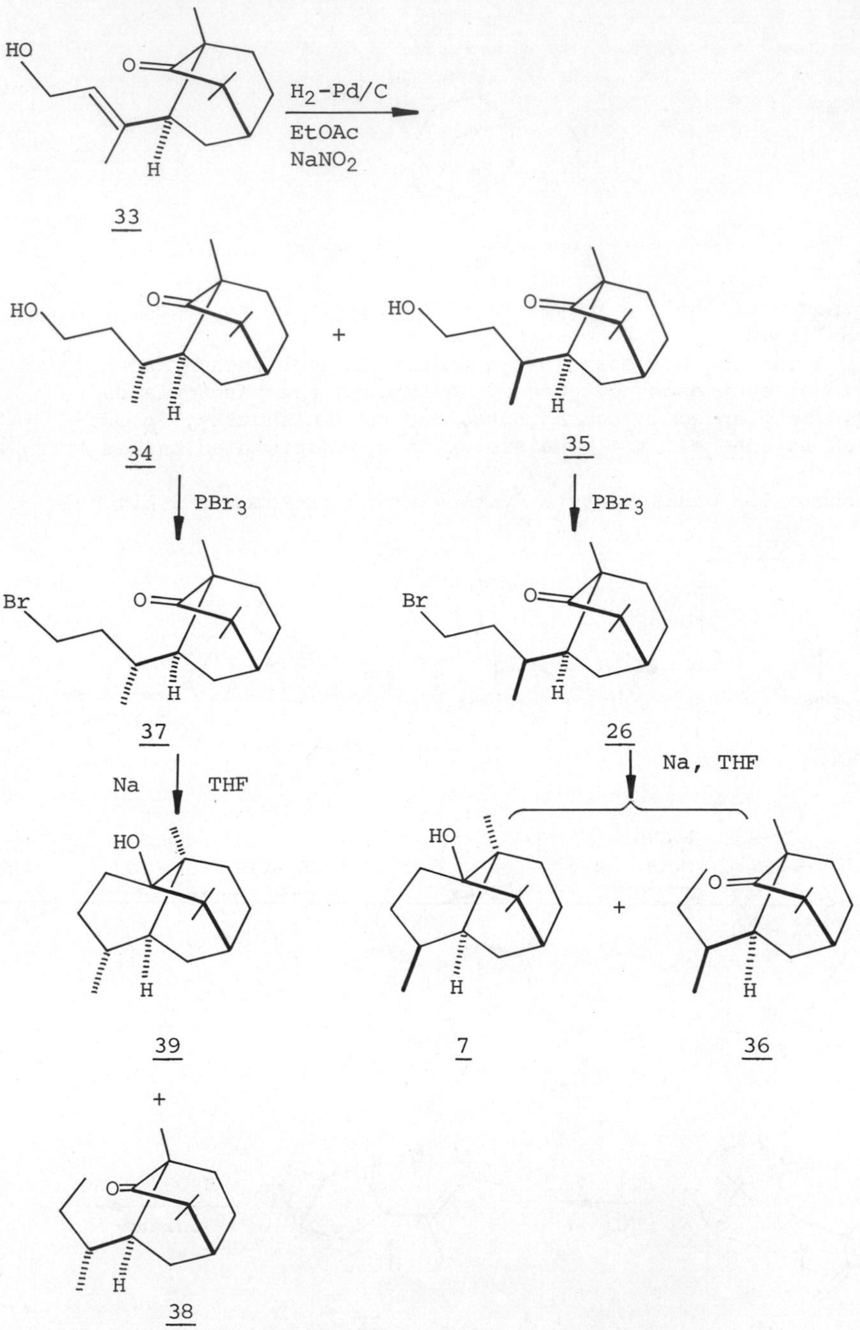
HO
O
H
H2-Pd/C
EtOAc
NaNO2
33
HO
O
H
+
HO
O
H
34
35
PBr3
PBr3
Br
O
H
Br
O
H
37
26
Na
THF
Na, THF
HO
H
HO
H
+
O
H
39
7
36
+
O
H
38

the prior construction of the bicyclo[2.2.2]octane moiety (rings B and C), followed by closure of ring A by reductive alkylation of an ε-bromo ketone (i.e., 26 → 7).

The requisite bicyclo[2.2.2]octanone was constructed starting with the trimethylcyclohexadienone 27, which underwent Diels-Alder addition with methyl vinyl ketone to afford dione 28 in 93% yield. After saturation of the double bond, the acetyl group was epimerized with base. The equilibrium ratio of 29:30 was 3:7. By crystallization of the major isomer (30) and equilibration of the mother liquors, isomer 30 was obtained in 65% yield. Dione 30 reacted with vinyllithium at the more accessible carbonyl to yield alcohol 31, which yielded allylic chloride 32 upon treatment with HCl in chloroform.

Since catalytic reduction of the double bond in 32 was accompanied by considerable hydrogenolysis, the halide was solvolyzed to obtain alcohol 33. Hydrogenation of this material gave two epimers, 34 and 35, which were separated by chromatography. Isomer 35 gave a bromide (26), which was treated with sodium in THF. The product, obtained in 55% yield, was an equimolar mixture of saturated ketone 36 and (±)-patchouli alcohol (7). Analogous treatment of the isomeric bromide 37 gave almost exclusively saturated ketone 38, and only a trace of (±)-epi-patchouli alcohol (39).

B. Seychellene

The tricyclic hydrocarbon seychellene (40), isolated from patchouli oil, may arise biogenetically by rearrangement of patchouli alcohol. The hydrocarbon has recently been synthesized, in a perfectly stereorational route, by Piers.[292]

Analysis of the tricyclic skeleton reveals various ways in which it may be constructed by one-ring closure in a bicyclic **precursor**. One method, proceeding through a bicyclo-[2.2.2]octane was illustrated by Danishevsky in his patchouli alcohol synthesis. Such a route suffers since it offers no control over the stereochemistry of the secondary methyl group, this center being in a side chain prior to closure of the final ring. A method of synthesis proceeding through intermediates which contain this center in a ring, preferably a six-membered ring, would be more desirable, since control over its stereochemistry could then be exerted.

Four routes to <u>40</u> are possible in which the immediate bicyclic **precursor** is a decalin, with the secondary methyl group in one ring. Of these potential decalinic routes, c is preferable, since one of the termini of the missing bond is *adjacent* to the methylene group.

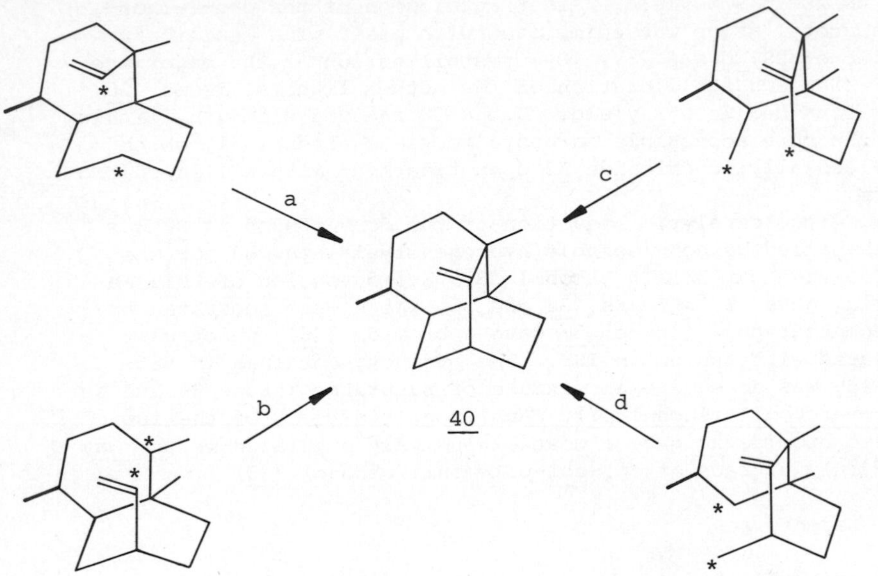

If one recognizes that a methylene group is synthetically equivalent to a carbonyl group, then a logical synthetic target is the keto tosylate <u>41</u>, which could undergo intramolecular alkylation to the tricyclic ketone <u>42</u>.

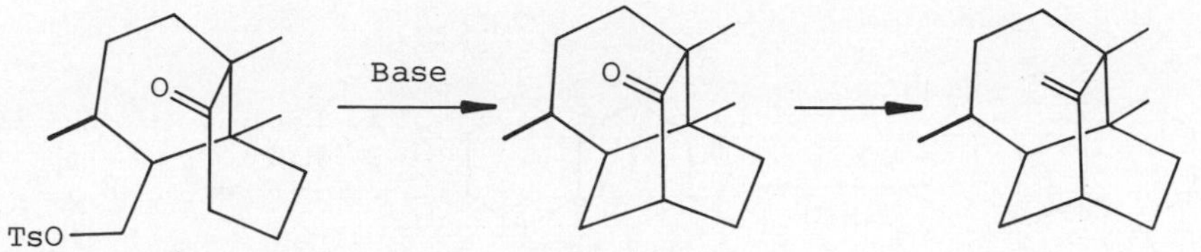

41 42 40

The construction of keto tosylate 41 and its elaboration into (±)-seychellene is outlined in Scheme 3.[292] The readily

Scheme 3. Piers' Synthesis of (±)-Seychellene

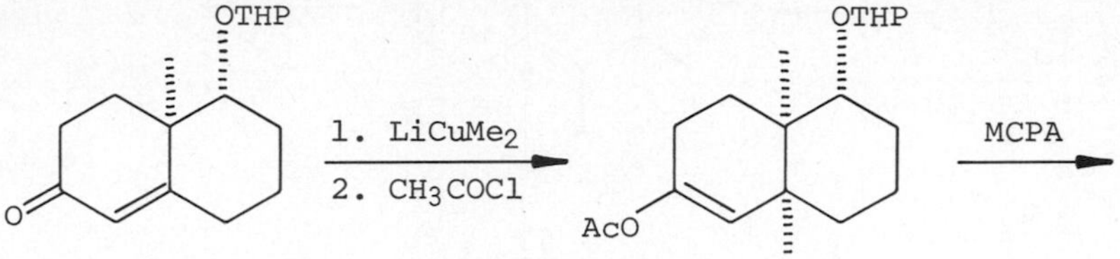

43 44

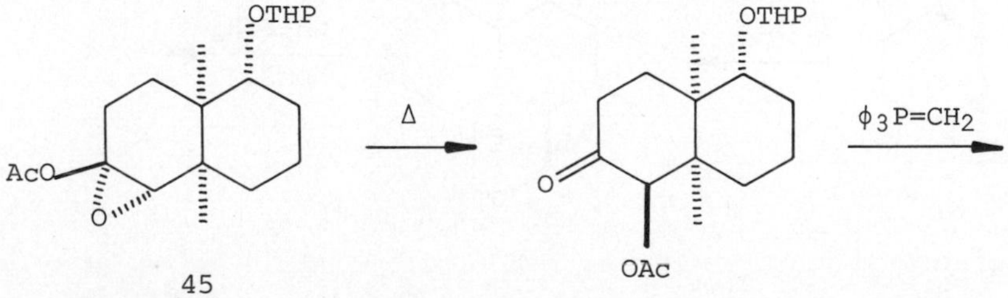

45 46

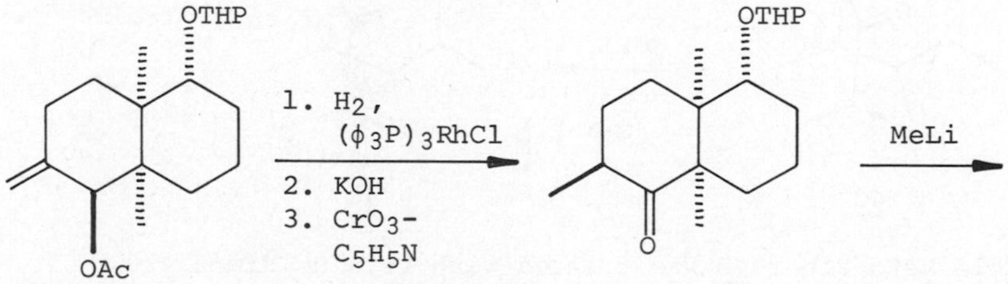

47 48

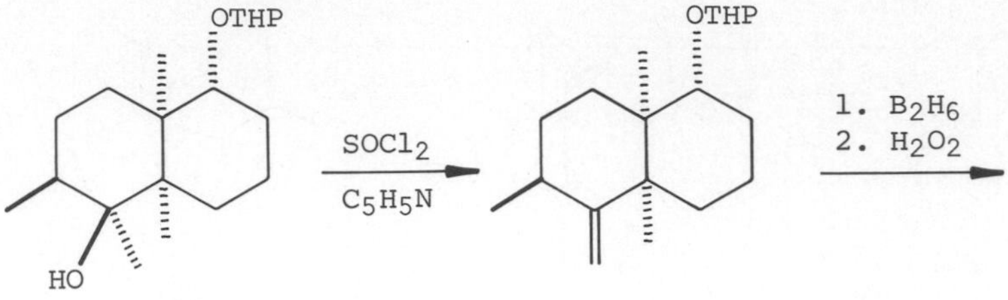

available keto ether 43 was treated with lithium dimethylcopper. The resulting enolate was quenched with acetyl chloride to obtain the enol acetate 44. The process is stereospecific, leading to compound 44 in 78% yield. Epoxidation,
followed by thermal rearrangement of the acetoxy epoxide,

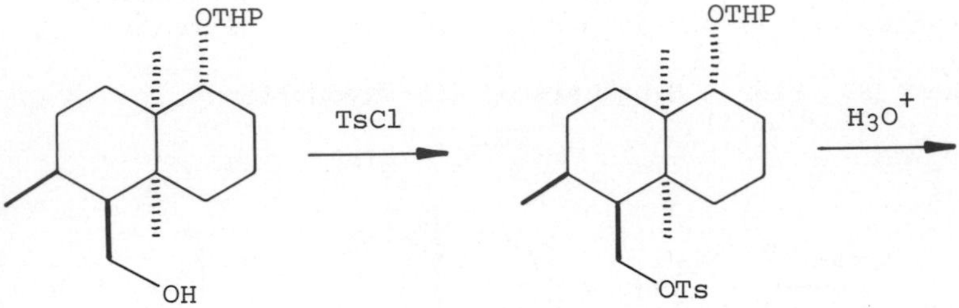

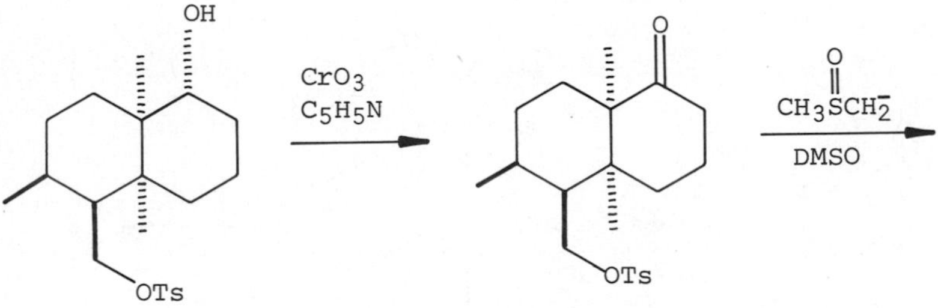

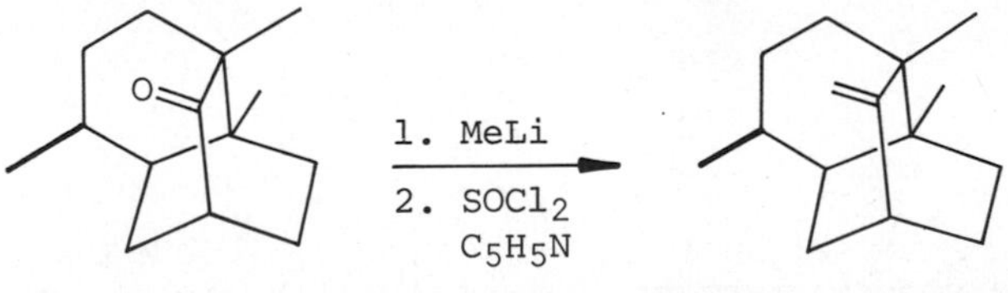

yielded keto acetate 46. This material was treated with methylenetriphenylphosphorane to give an allylic acetate (47), which was subjected successively to hydrogenation, saponification, and oxidation to obtain ketone 48.

It is in the hydrogenation step that the relative stereochemistry at the secondary methyl group is probably established. Apparently, compound 47 exists in the conformation 47a, with hydrogen being delivered exclusively from the less hindered face. The point is of little practical importance, however, since this center is epimerizable in ketone 48. Analysis reveals that isomer 48 should be more stable than its

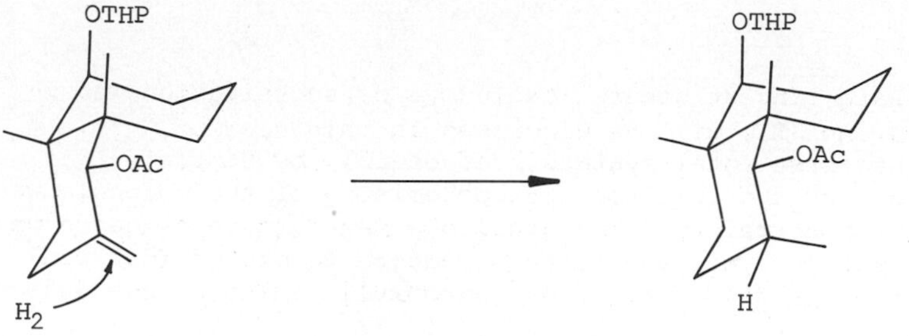

47a

epimer 49. In fact, ketone 48 was unchanged upon treatment with base.

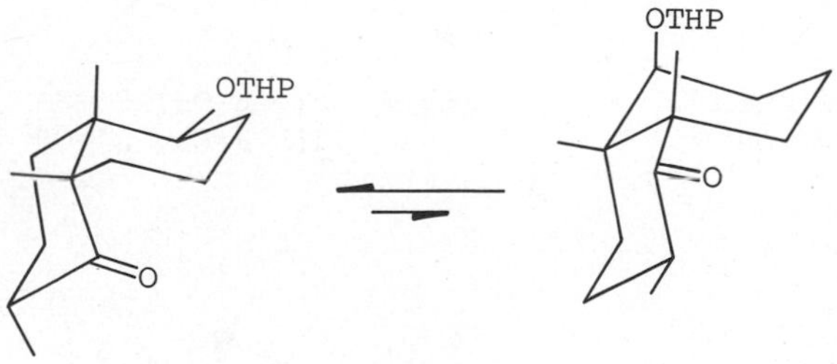

48   49

Ketone 48 reacted with methyllithium to give an alcohol (50), which was dehydrated to obtain the exocyclic olefin 51. Hydroboration occurred from the more accessible convex face, leading to primary alcohol 52 which was transformed into tosylate 53. After removal of the protecting group and oxidation, the desired keto tosylate 41 was in hand. Base catalyzed cyclization indeed yielded tricyclic ketone 42, which was transformed into (±)-seychellene (40).

C. Cedrol, Cedrene

The tricyclic alcohol cedrol (55) and its companion hydrocarbon α-cedrene (56) are constituents of cedar wood oil. The synthesis of these tricyclic sesquiterpenes, which possess the

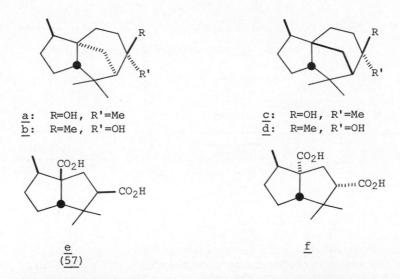

55   56

same basic ring structure (exclusive of substitution) as α-patchoulene (2), will be discussed in this section.
The first total synthesis of cedrol, by Stork,[293] was begun before the complete stereochemistry of the molecule was known with certainty. The available degradative evidence was sufficient to limit the stereostructure to one of four possibilities (a-d). Norcedrenedicarboxylic acid, a degradation

a: R=OH, R'=Me   c: R=OH, R'=Me
b: R=Me, R'=OH   d: R=Me, R'=OH

e
(57)   f

product of the natural product, thus becomes e or f. In order to clarify this stereochemical ambiguity by synthesis, it is necessary to design a route which will lead to structures a-d in an unambiguous manner.
At the outset, Stork decided to focus his attention on

norcedrenedicarboxylic acid for two reasons. Firstly, there are only two, rather than four, possible structures for this substance. On biogenetic grounds, one of these seemed more likely than the other (e, 57). Second, it seemed likely that this diacid would represent a suitable intermediate for subsequent elaboration into cedrol itself. Naturally derived norcedrenedicarboxylic acid could then be used as a relay for completing the synthesis of cedrol.

The reasoning which led to the synthetic attack on diacid 57 was an intuitive one. It was reasoned that "...(structure e) appeared a priori rather more likely since the trans fusion of the A/B system in (f), while not impossible, implies a degree of strain which would be surprising on the basis of biogenetic speculations."[293b] As will be seen, this intuition turned out to be correct.

As a starting point for the synthesis, Stork chose the known cyclopentanone diester 58, which was alkylated with benzyl α-bromopropionate to give the triester 59) (Scheme 4).

Scheme 4. Stork's Synthesis of Cedrol

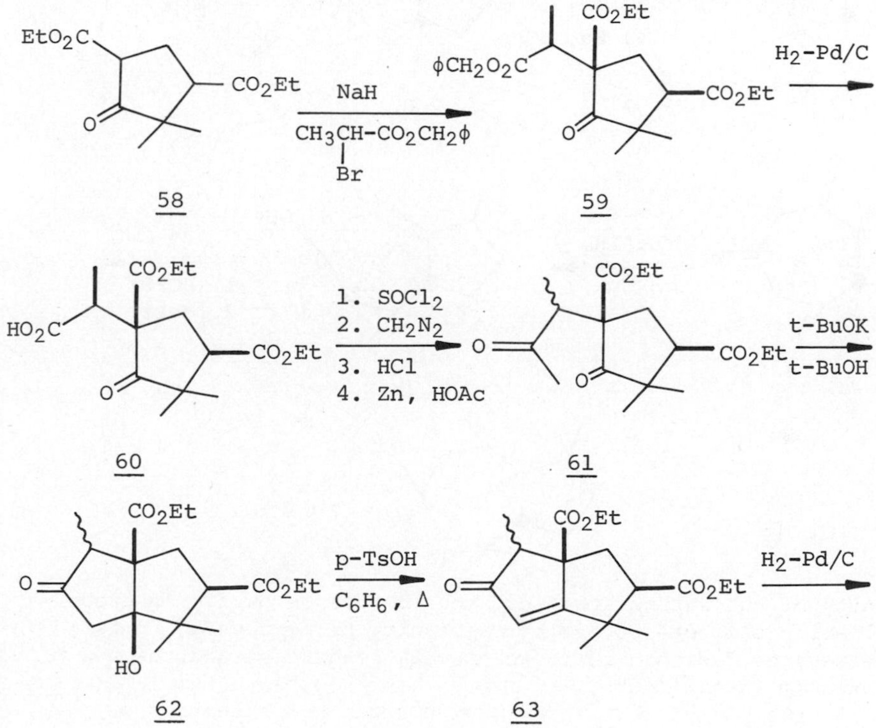

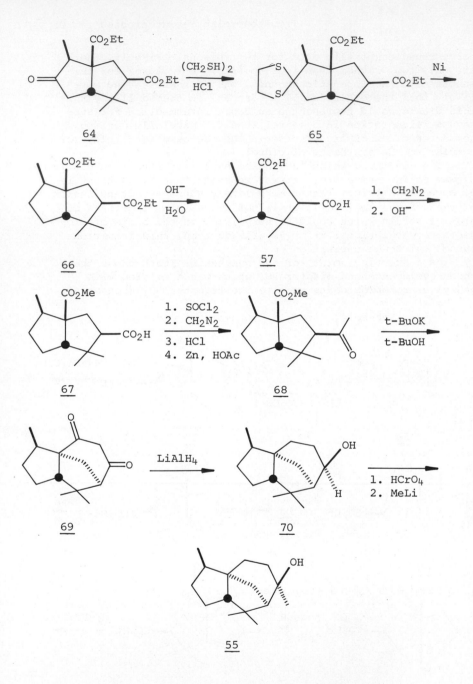

64

65

66

57

67

68

69

70

55

Catalytic hydrogenolysis gave the crystalline keto acid 60 in 40% overall yield. At this point, we shall comment on the stereochemistry of keto acid 60. Although three asymmetric centers are present the only critical feature is the *cis* disposition of the two carbethoxy groups. The secondary methyl group, which is adjacent to a carbonyl function is epimerizable. The stereoselectivity of the alkylation step (58 → 59) is difficult to assess. Although crude keto triester 59 was obtained in 90% yield (after distillation), no evidence was obtained relative to its stereochemical homogeneity. Upon hydrogenolysis, acid 60 (presumably the major isomer) was obtained as the only crystalline product in 45% yield. While alkylation probably proceeds mainly *trans* to the C-4 carbethoxy group, other isomers of 59 may be formed in the reaction.

Keto acid 60 was transformed into diketone 61 by an adaptation of the Arndt-Eistert method (via the acid chloride, diazoketone and chloro ketone) in 79% overall yield. The product obtained (61) was a mixture of epimers at the secondary methyl group. Closure of ring-A was accomplished by brief exposure of 61 to *t*-BuOK in *t*-BuOH. The resulting aldol (62) was dehydrated by acid to yield the cyclopentenone 63 in 80% overall yield.

Catalytic hydrogenation of 63 yielded the crystalline keto diester 64. The same product was obtained by lithium-ammonia reduction of 63, a process expected to yield the more stable *cis* ring fusion. The carbonyl group was removed via its dithio ketal 65, yielding diester 66. Saponification of 66 yielded (±)-norcedrenedicarboxylic acid (57), thus confirming the stereochemistry assigned to the secondary methyl group and establishing the complete stereochemistry of 55. After resolution of 57 via its quinine salt, the remaining synthesis was accomplished using naturally derived 57.

Compound 57 was converted, by partial saponification of the dimethyl ester, into the half-ester 67. The secondary carboxyl of 67 was transformed into acetyl by the diazoketone route. Claisen cyclization of 68 gave β-diketone 69, which possesses the essential skeleton of cedrol. Lithium aluminum hydride reduction of 69 gave (remarkably) the secondary alcohol 70 in good yield. Oxidation of 70 gave a ketone which reacted with methyllithium to afford cedrol (55).

The probable biogenesis of cedrol and α-cedrene (below) prompted Corey[294] and Lawton[295] to design syntheses styled on such a cationic cyclization.

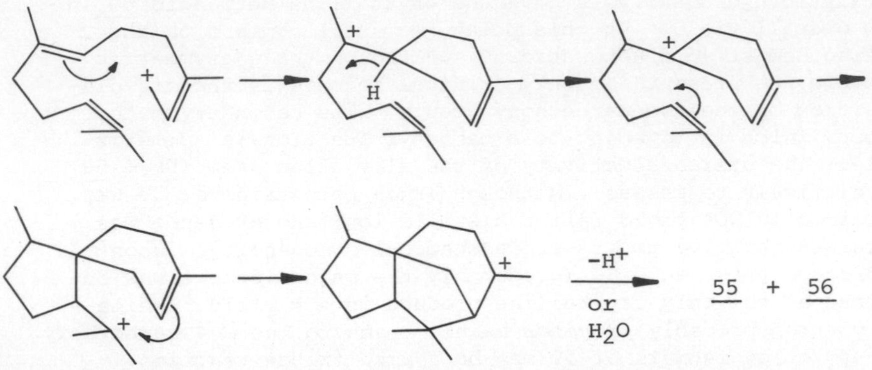

Both groups focus their attention on cyclization of spiro[4.5]-decanes (last step of proposed biosynthesis).

Corey synthesized the crucial spirodienone 80 as outlined in Scheme 5.[294] *p*-Methoxyacetophenone (71) was converted, by

Scheme 5. Synthesis of Spirodienone 80 (Corey)

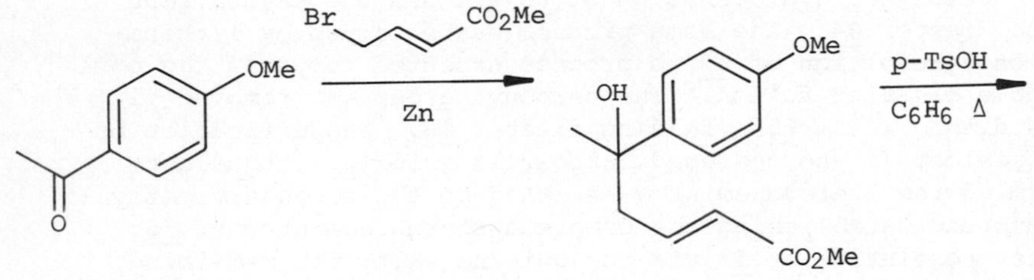

71 72

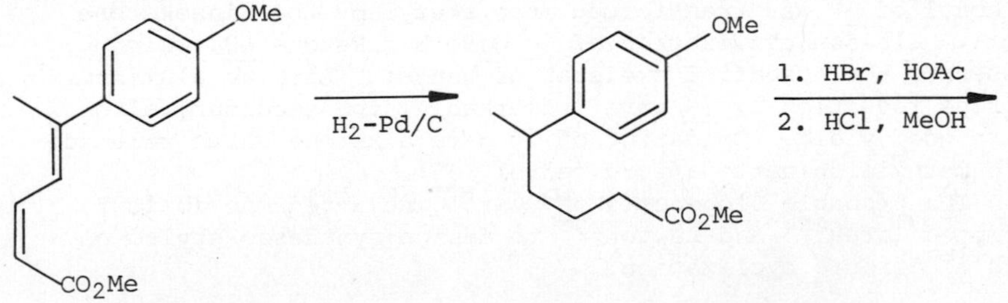

73 74

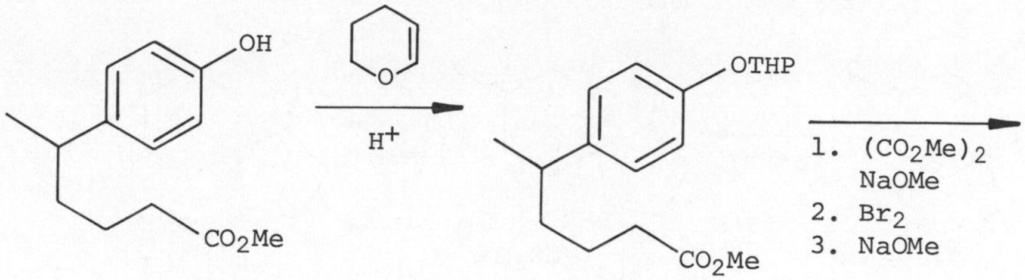

Reformatsky reaction with ethyl γ-bromocrotonate, into hydroxy ester $\underline{72}$.   Dehydration, followed by hydrogenation gave ester $\underline{74}$.   After demethylation and re-esterification, the phenolic hydroxyl was protected as its tetrahydropyranyl ether.   Bromination α to the ester group in $\underline{76}$ was achieved by oxalylation, bromination of the resulting sodio derivative and deoxalylation.   The resulting α-bromo ester ($\underline{77}$) was deprotected to yield the phenolic bromo ester $\underline{78}$.   Base catalyzed cyclization gave an equimolar mixture of the *cis*- and *trans*-isomers $\underline{79}$ and $\underline{80}$, which was homogenized upon further treatment with base to obtain isomer $\underline{80}$.

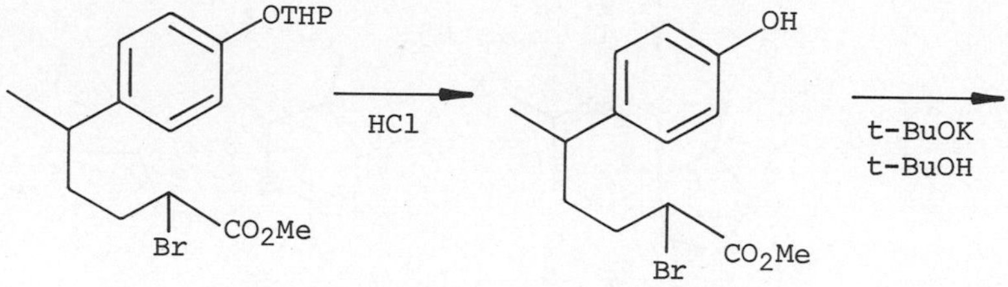

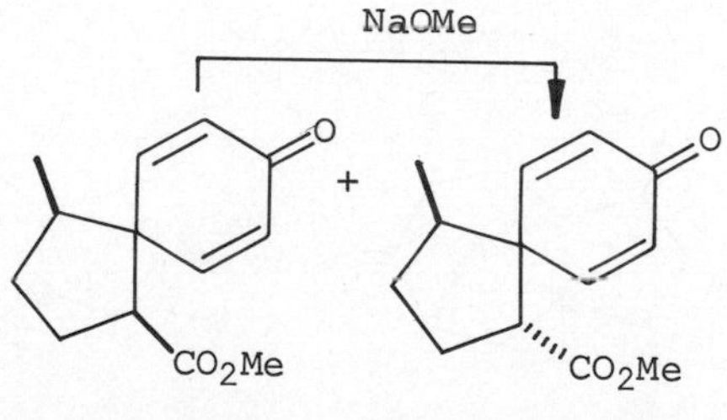

Several methods which were used to elaborate intermediate $\underline{80}$ are summarized in Scheme 6.   Catalytic hydrogenation yielded

Scheme 6. Conversion of Spirodienone $\underline{80}$ into (±)-Cedrol and (±)-α-Cedrene (Corey)

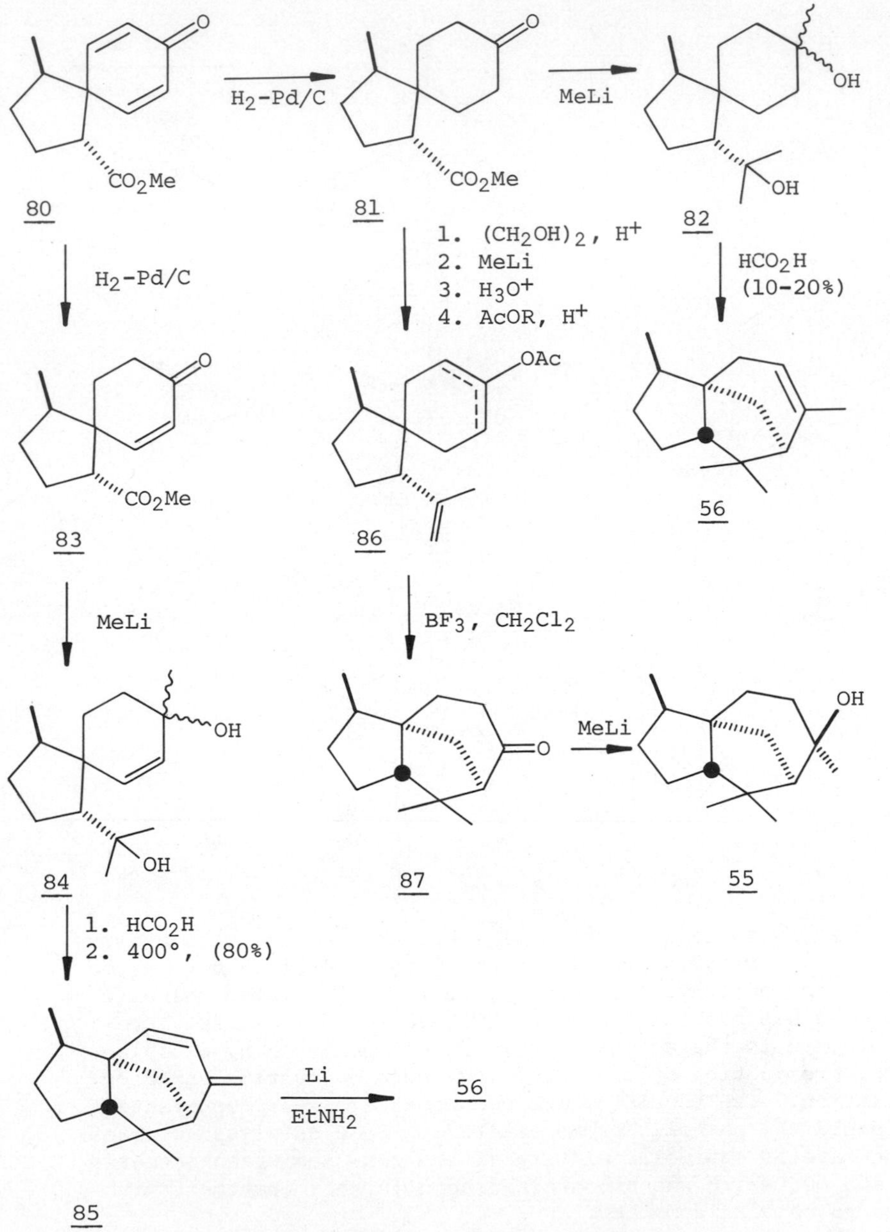

$\underline{80}$

H$_2$-Pd/C

$\underline{81}$

MeLi

$\underline{82}$

1. (CH$_2$OH)$_2$, H$^+$
2. MeLi
3. H$_3$O$^+$
4. AcOR, H$^+$

HCO$_2$H
(10-20%)

H$_2$-Pd/C

$\underline{83}$

$\underline{86}$

$\underline{56}$

MeLi

BF$_3$, CH$_2$Cl$_2$

$\underline{84}$

$\underline{87}$

MeLi

$\underline{55}$

1. HCO$_2$H
2. 400°, (80%)

$\underline{85}$

Li
EtNH$_2$

$\underline{56}$

the tetrahydro keto ester 81, which reacted with methyllithium
to afford a mixture of diastereomeric diols (82). When this
mixture was exposed to formic acid at room temperature, (±)-α-
cedrene (56) was produced in 10-20% yield.

Partial hydrogenation of 80 was selective, yielding a
single dihydro derivative (83). Enone 83 reacted with methyllithium to give a mixture of diols (84), which was treated
with formic acid. The product, a mixture of tricyclic formates, was pyrolyzed briefly at 400° to obtain diene 85 in
80% yield. Reduction of 85 with lithium in ethylamine afforded
exclusively (±)-α-cedrene (56).

A third route to the cedrene skeleton proceeded through
the enol acetate 86, prepared in the manner indicated. Compound 86 underwent cyclization to tricyclic ketone 87 when
treated with BF₃. Treatment of 87 with methyllithium yielded
(±)-cedrol (55).

The stereochemistry of these cyclization reactions is
interesting. The isomeric enol acetates 86a and 86b can, in
theory, lead to isomeric ketones 87 and 88. As seen, ketone
87 possesses the stereochemistry of cedrol.

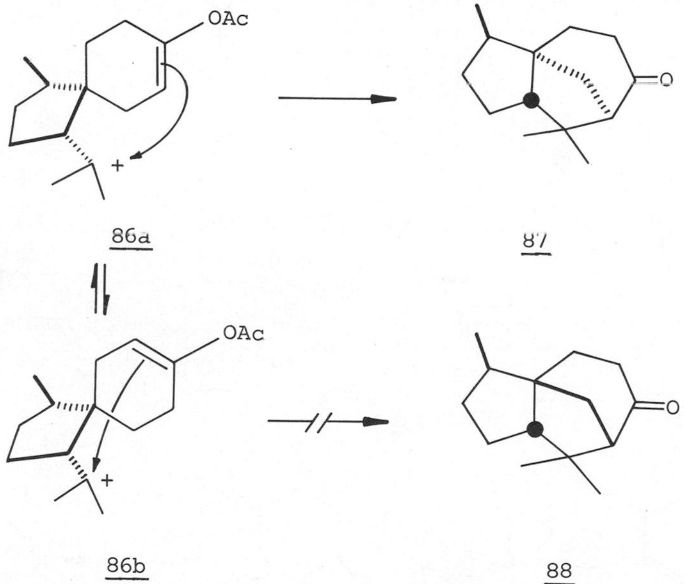

Isomer 88 would lead to an isomer of natural cedrol. However,
the strain inherent in compound 88 probably prevents its formation, even though ion 86b is doubtless present. Since ions
86a and 86b are in equilibrium, and only the former cyclizes,

the process is quite efficient. Similar arguments can be ad-
vanced for the cyclization of diols 82 and 84, although the
disparity in yields is inexplicable.

    Lawton's synthesis, which was similarly patterned, is
outlined in Scheme 7.[295] Diethyl p-benzyloxyphenylmalonate

Scheme 7. Lawton's Synthesis of (±)-α-Cedrene

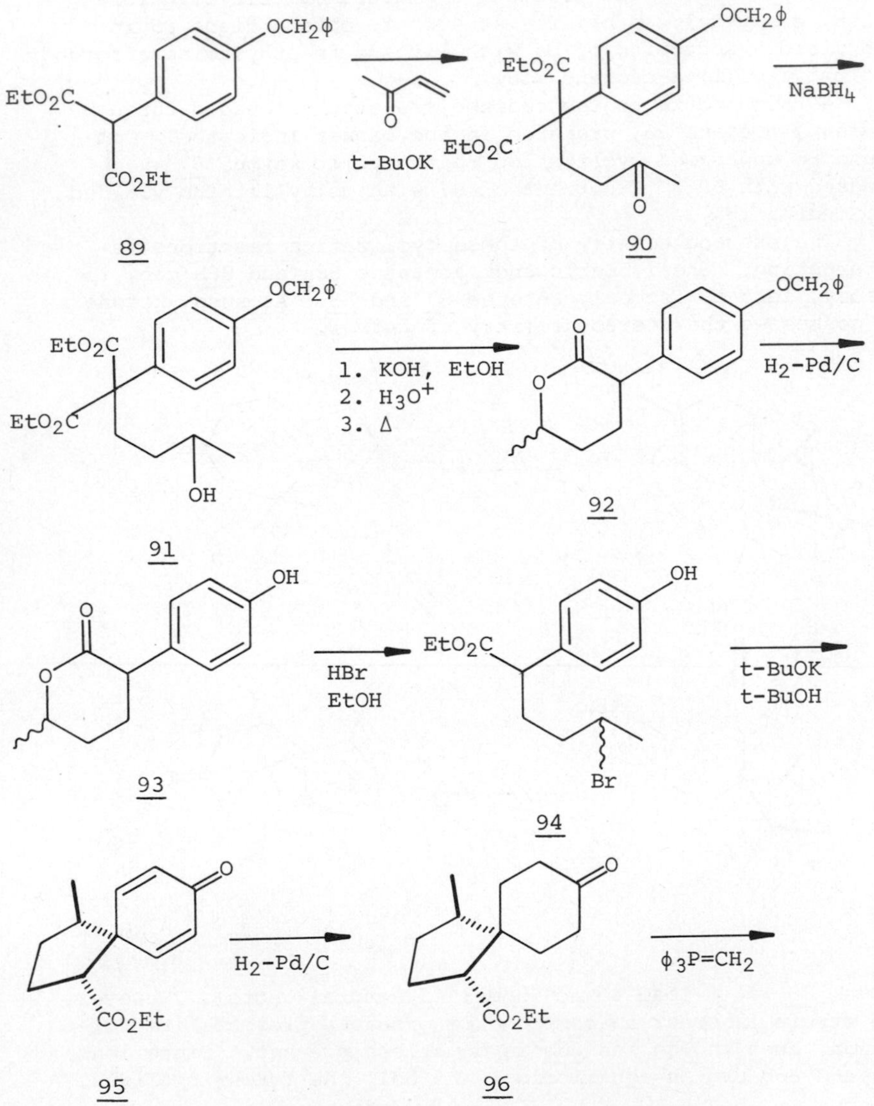

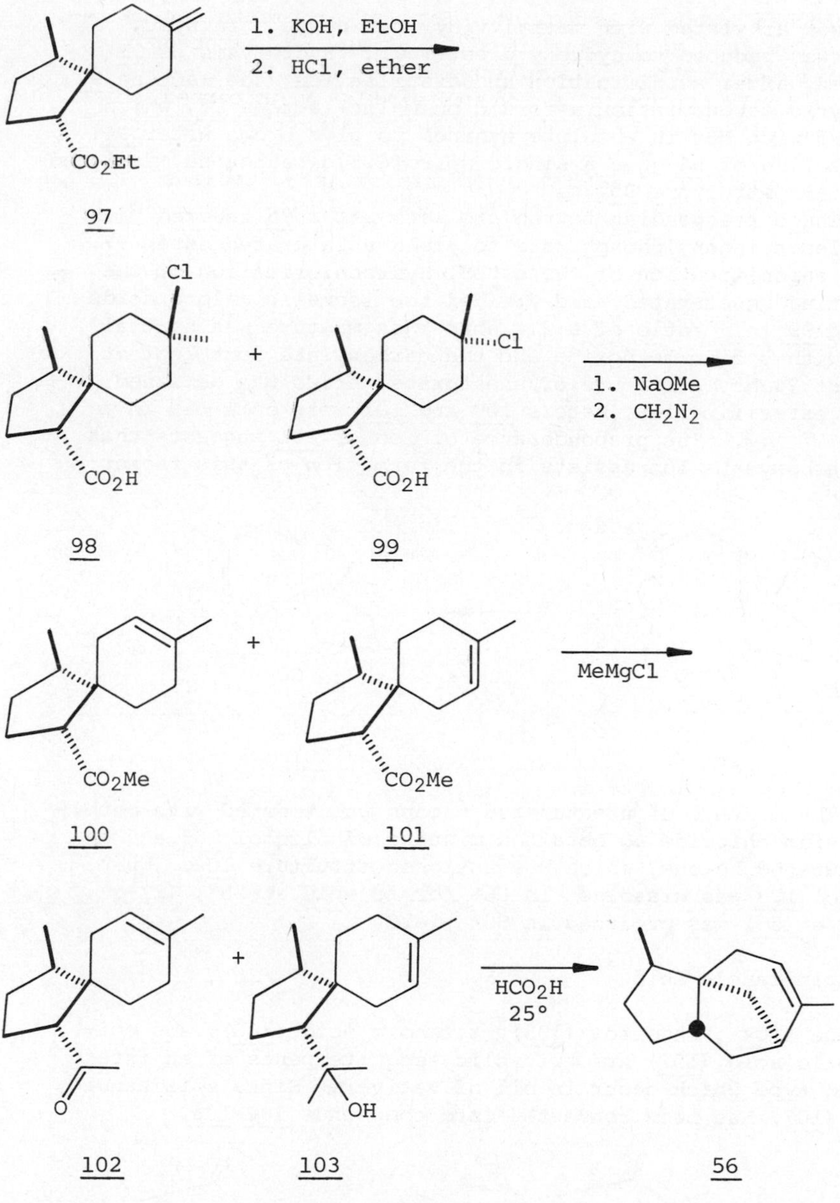

97

1. KOH, EtOH
2. HCl, ether

98 + 99

1. NaOMe
2. CH$_2$N$_2$

100 + 101

MeMgCl

102 + 103

HCO$_2$H
25°

56

(89) was alkylated with methyl vinyl ketone to obtain 90, which was reduced to hydroxy diester 91. Hydrolysis of 91 yielded, after acidification and distillation, the lactone 92. Catalytic debenzylation afforded phenolic lactone 93, which reacted with HBr in absolute ethanol to give bromo ester 94. Cyclization of 94 gave a single spiro[4.5]decadienone, assigned the *trans* structure 95.

The corresponding tetrahydro keto ester 96 reacted with methylenetriphenylphosphorane to yield unsaturated ester 97. After saponification of the ester, hydrochlorination of the resulting unsaturated acid yielded the isomeric chloro acids 98 and 99 in a ratio of 3:2. When this mixture was neutralized with sodium methoxide and the carboxylate ions kept at 35° for 24 hr, a mixture of unsaturated acids was obtained. After esterification, esters 100 and 101 were obtained in a ratio of 1:3. The preponderance of isomer 101 suggests that the carboxylate ion assists in the formation of this isomer:

101

The mixture of unsaturated esters was treated with methyl-magnesium chloride to obtain a mixture of alcohol 103 and an unsaturated ketone, which was assigned structure 102. When alcohol 103 was dissolved in 88% formic acid at 25°, (±)-α-cedrene (56) was produced in 80% yield.

D. Epizizanoic Acid

Zizaene (104), khusimol (105), zizanoic acid (106), and epi-zizanoic acid (107) are tricyclic sesquiterpenes of an interesting type which occur in oil of vetiver. Since epizizanoic acid (107) has been converted into compounds 104-106,

104   HO   105

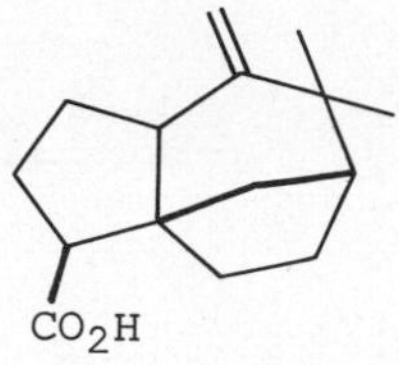

106

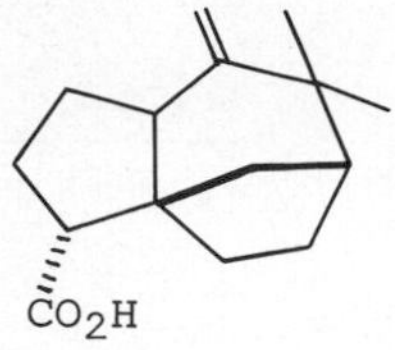

107

Yoshikoshi's synthesis of 107 (Scheme 8) constitutes a formal synthesis of the others.[296]

Scheme 8.   Yoshikoshi's Synthesis of Epizizanoic Acid

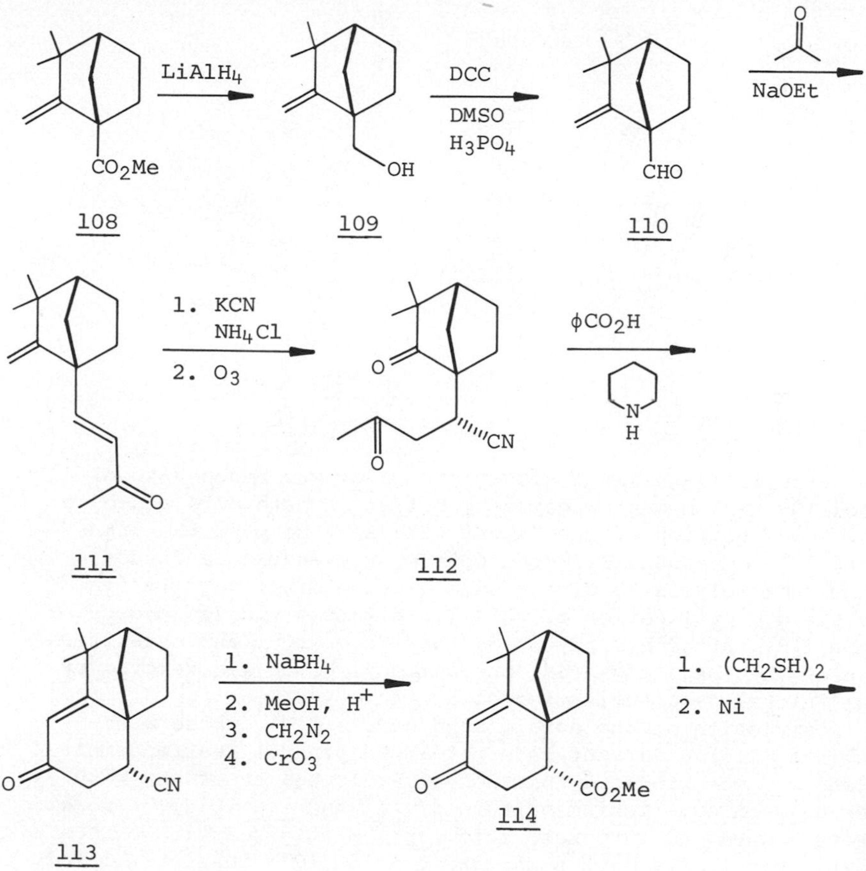

108

109

110

111

112

113

114

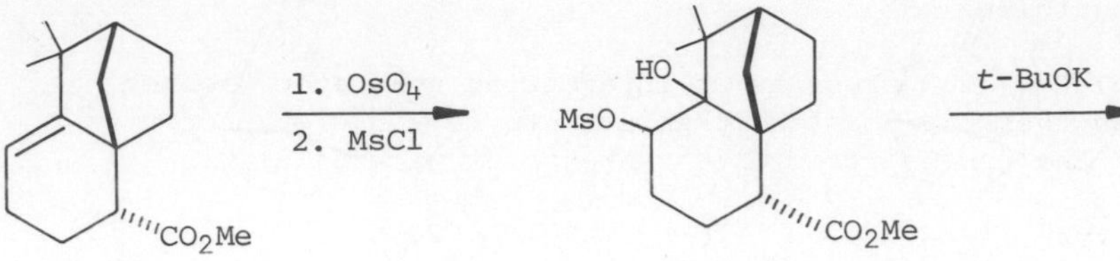

115   **116**

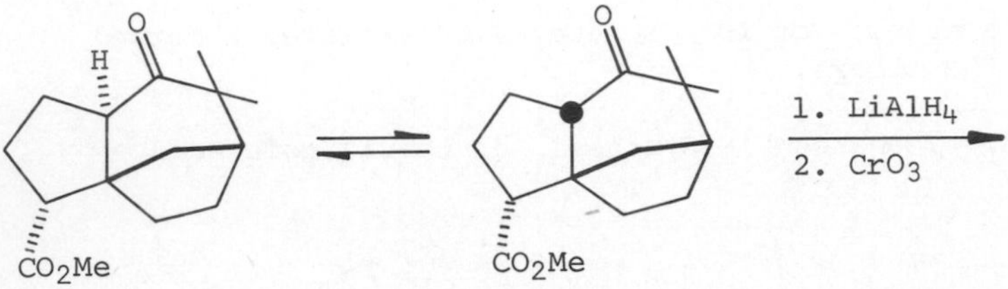

117   **118**

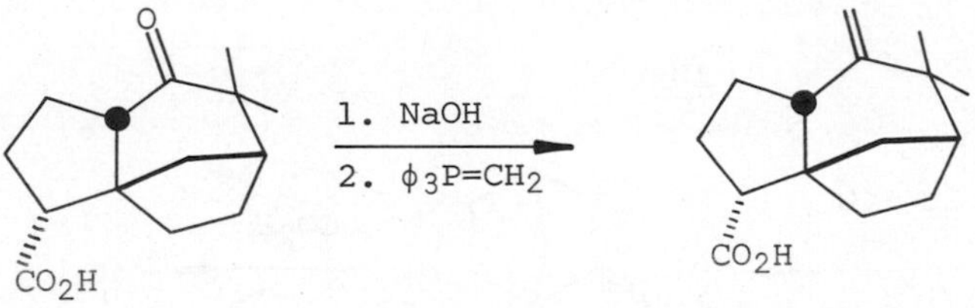

119   **107**

Methyl (+)-camphenecarboxylate (108) was reduced to alcohol 109, which was oxidized by Moffatt's method to aldehyde
110. Condensation of the latter with acetone gave the *trans*
enone 111. Compound 111 underwent hydrocyanation and subsequent ozonolysis to give a single cyano diketone (112) in
45% yield. Cyclization of this 1,5-diketone yielded the cyano
enone 113. After hydrolysis of the cyano group and esterification of the resulting acid, the carbonyl group was removed by
Raney nickel desulfurization of a dithio ketal.

Osmylation of the double bond gave a diol, whose monomesylate (116) underwent base catalyzed pinacol rearrangement.
Ketone 117 was produced initially. Prolonged treatment with
base gave an equilibrium mixture of 117 and 118 (2:3). Isomer
118 was converted into keto acid 119. Wittig reaction on the
sodium salt of 119 gave epizizanoic acid (107) in approximately
10% yield.

E.   Longifolene

Longifolene (120) presents an interesting synthetic problem.
Although there is no stereochemistry to reckon with, construction of the unusual tricyclic skeleton is no trivial task.

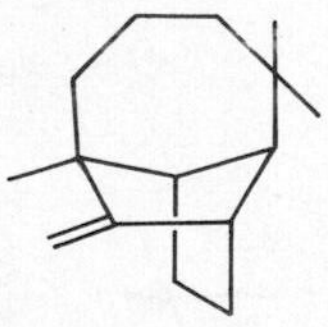

120

After consideration of various routes to the skeleton of
120,[297] Corey adopted an approach based on internal Michael
alkylation of the bicyclic enedione 121.

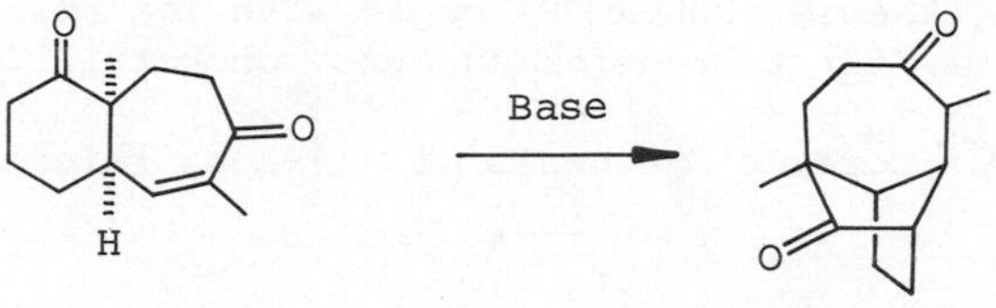

After construction of the basic network, it would then be
necessary to introduce the third methyl group, remove the
superfluous carbonyl group and transform the remaining carbonyl into a methylene group.   The indicated intramolecular
Michael reaction had ample precedent in the known conversion
of santonin (structure 23 in Sec. 4) into santonic acid (123),
which presumably occurs by a similar mechanism:

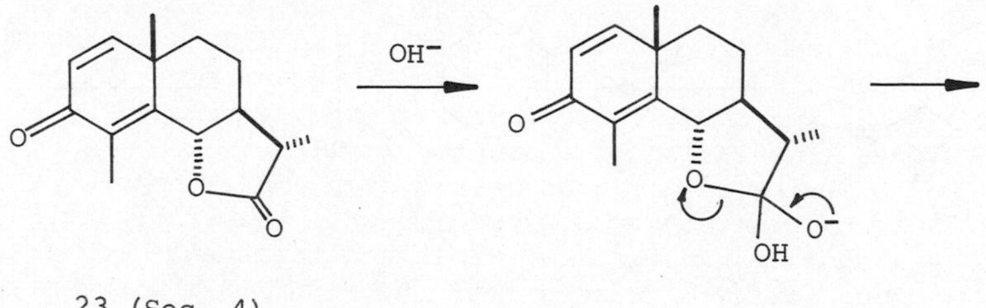

23 (Sec. 4)

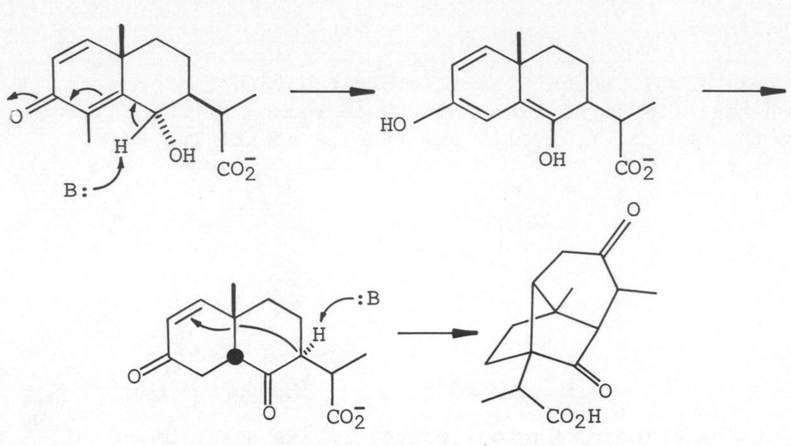

<u>123</u>

The Corey synthesis (Scheme 9) began with the unsaturated dione <u>124</u> which was first transformed into monoketal <u>125</u>.

Scheme 9. Corey's Synthesis of (±)-Longifolene

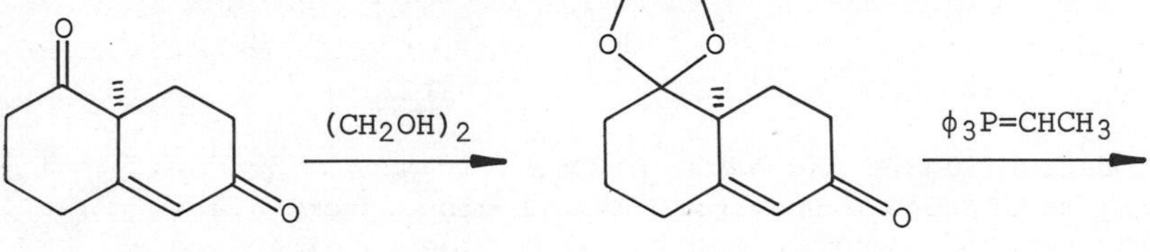

<u>124</u> <u>125</u>

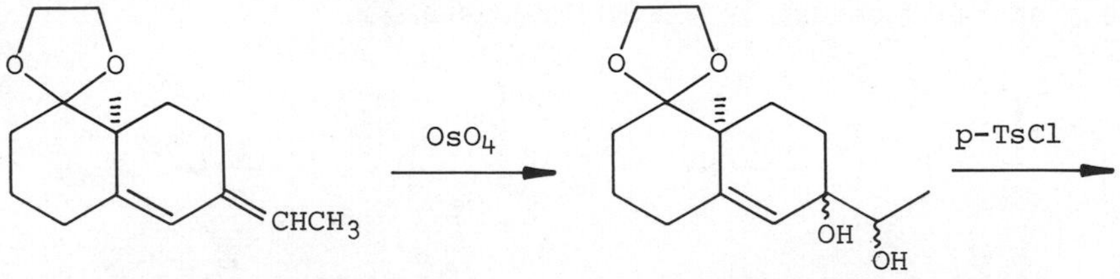

<u>126</u> <u>127</u>

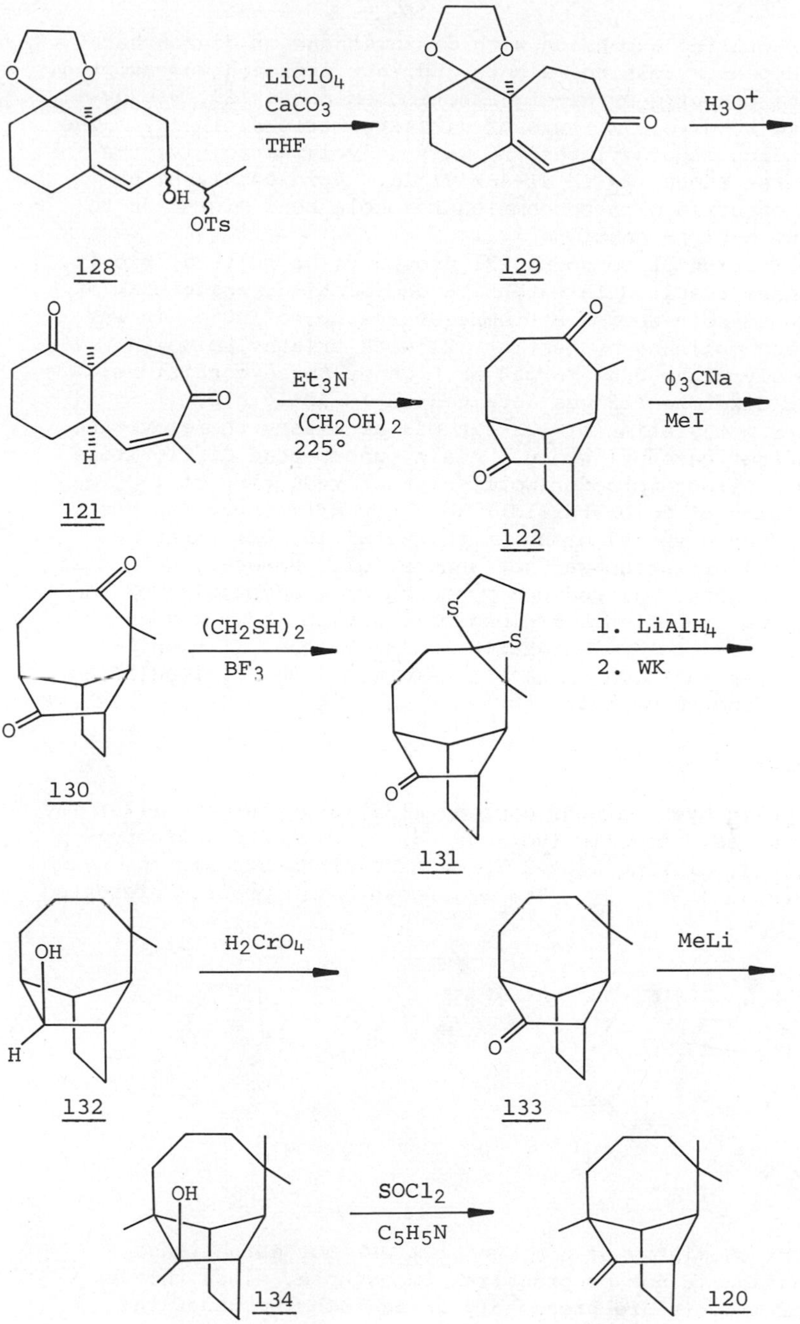
LiClO$_4$
CaCO$_3$
THF
128
129
H$_3$O$^+$
121
Et$_3$N
(CH$_2$OH)$_2$
225°
122
ϕ$_3$CNa
MeI
130
(CH$_2$SH)$_2$
BF$_3$
131
1. LiAlH$_4$
2. WK
132
H$_2$CrO$_4$
133
MeLi
134
SOCl$_2$
C$_5$H$_5$N
120

Since direct ring expansion with diazomethane or diazoethane
did not appear promising, a more indirect approach was adopted.
Diene 126, prepared by Wittig reaction on enone 125, was hydroxylated to give a mixture of diastereomeric diols 127. The
corresponding monotosylates 128 were solvolyzed to give the
unconjugated enone 129 in 41-48% yield. Acid catalyzed hydrolysis occurred with concomitant double bond migration to
afford the desired enedione 121.*

     Cyclization of compound 121 proved to be quite difficult.
Strong bases resulted in extensive degradation, while weak
bases caused no reaction at temperatures up to 100°. It was
finally accomplished by heating 121 with triethylamine in
ethylene glycol at 225° for 24 hr. Under these conditions,
tricyclic diketone 122 was obtained in 10-20% yield.

     After methylation of 122, it was necessary to remove the
less hindered carbonyl group. Again, unexpected difficulties
intruded. Direct methods (Wolff-Kishner reduction of 130, desulfurization of thioketal 131) were not effective. Moreover,
even when the carbonyl group in thioketal 131 was first reduced, desulfurization was not successful. However, when this
hydroxy thioketal was reduced by Georgian's adaptation of the
Wolff-Kishner reaction, (±)-longicamphenylol (132) was obtained in good yield. Oxidation of 132 yielded (±)-longi-
camphenylone (133), which was transformed into (±)-longifolene
(120) by standard methods.

F.   Copaene, Ylangene

The tricyclic hydrocarbons copaene (135) and ylangene (136) may
be regarded as tricyclic isoprenologs of α-pinene (structure
253, Sec. 5), or ring-closed forms of α-trans-bergamotene
(structure 255, Sec. 5). The major synthetic problem presented

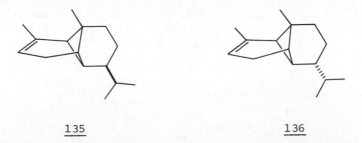

135   136

*The stereochemistry of compound 121 was not established.
This question is not of practical importance, since the cis
and trans isomers are presumably in equilibrium under the
conditions of the Michael reaction.

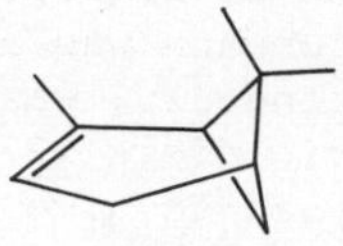

253 (Sec. 5) 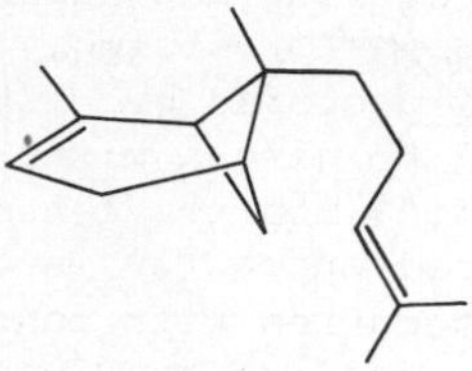   255 (Sec. 5)

by these sesquiterpenes is construction of the interesting tricyclic nucleus. Of subsidiary importance is the stereochemistry of the isopropyl group, relative to the functionality in
the other six-membered ring. An ideal candidate for further
elaboration into 135 and 136 is the tricyclic intermediate
137, which possesses the basic nucleus and has differentiated
oxygen functions. For conversion of 137 into copaene or

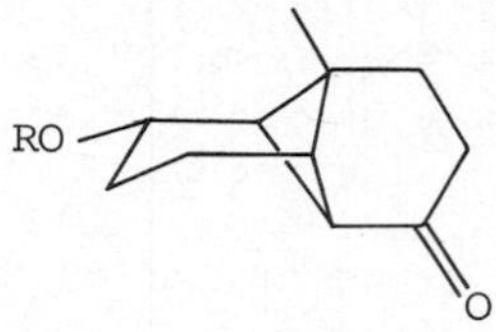

137

ylangene, two further changes would be required. The carbonyl
group must be converted into an isopropyl group, and the masked
hydroxyl group must be changed into vinyl methyl. Both operations should be routine.

    Analysis of the carbon network suggests that it may be
produced by one-bond closure of a cis-decalin precursor. Since
there are four bonds common to all three rings, there are four
possible precursors (a-d). There are various reasons for
choosing an approach based on a type a intermediate. First,
one of the new-bond termini is adjacent to a carbonyl group.
Thus, intramolecular alkylation might be used for ring closure.
Second, methyl decalones of this general type are readily
available by Robinson annelation.

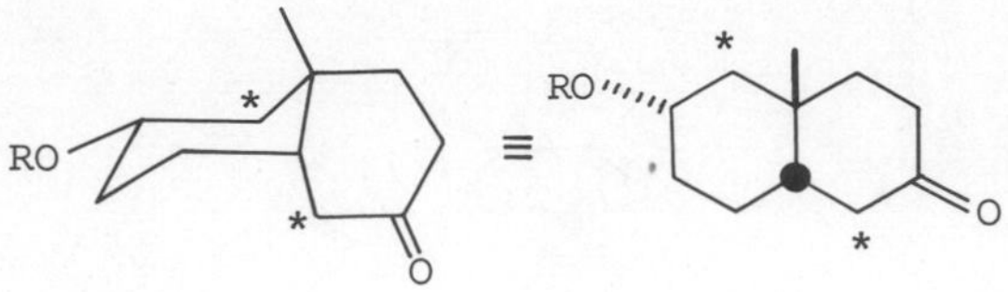

a

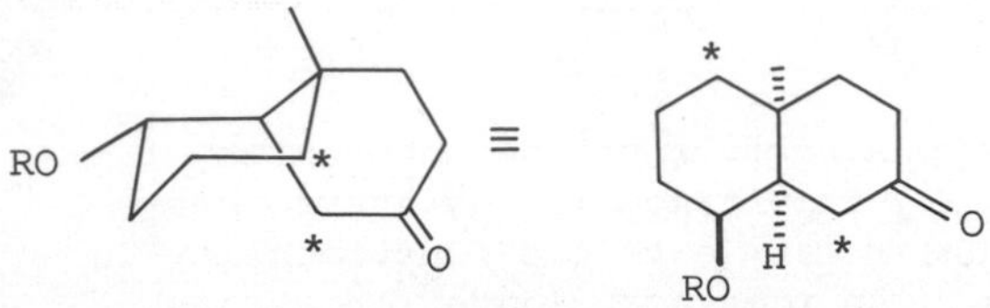

b

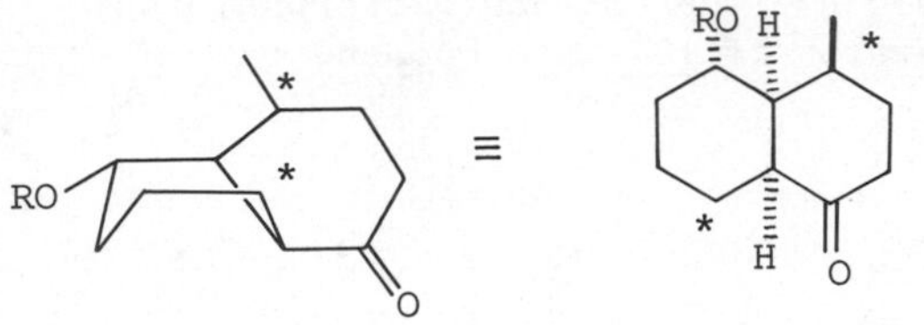

c

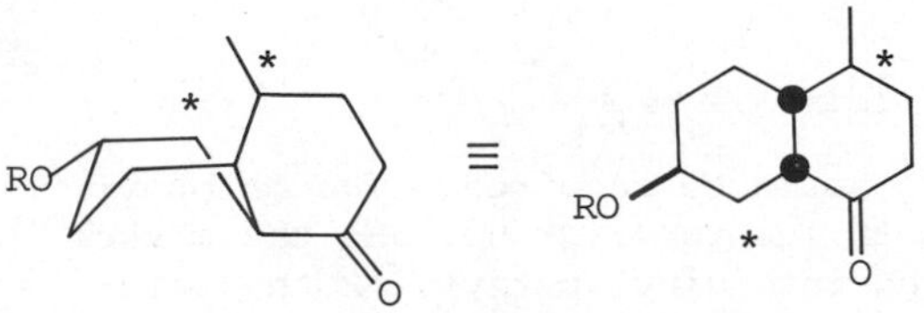

d

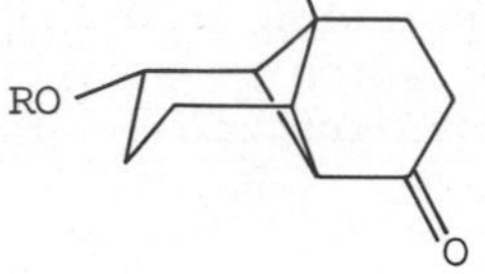

137

The Heathcock synthesis, based on this general synthetic plan, is summarized in Scheme 10.[298] Unsaturated keto alcohol structure 180, Sec. 4, was converted into tosylate 138, which was hydrogenated to yield the *cis*-fused keto tosylate 139. Pyridine-catalyzed dehydrotosylation gave the octalone 140. After protection of the carbonyl group, the double bond was oxidized by peracid to obtain oxide 142. The oxidation is stereospecific, attack occurring only from the more exposed convex side of the unsaturated ketal 141.

The epoxide ring was opened by reaction with sodium benzylate in benzyl alcohol at 200°. The resulting alcohol 143 was changed to tosylate and the ketal grouping hydrolyzed to yield intermediate 145. Keto tosylate 145 reacted with methylsulfinyl carbanion in dimethylsulfoxide, affording the

Scheme 10. Heathcock's Synthesis of
(±)-Copaene and (±)-Ylangene

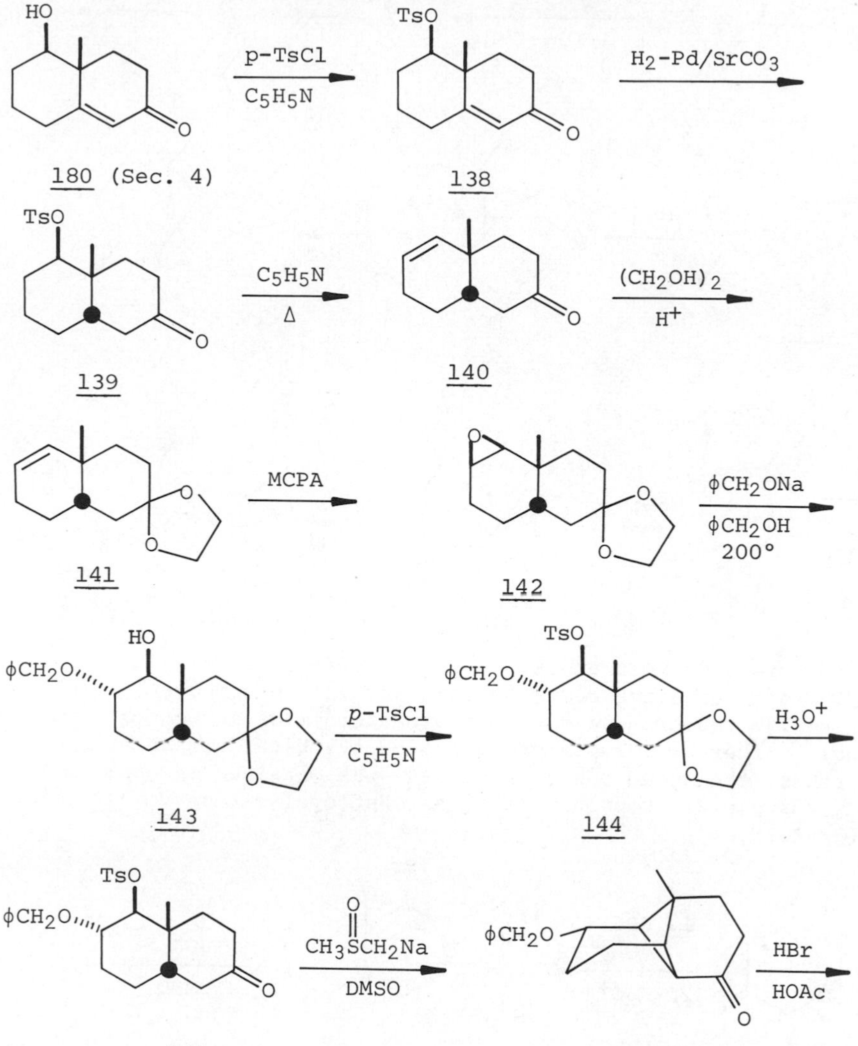

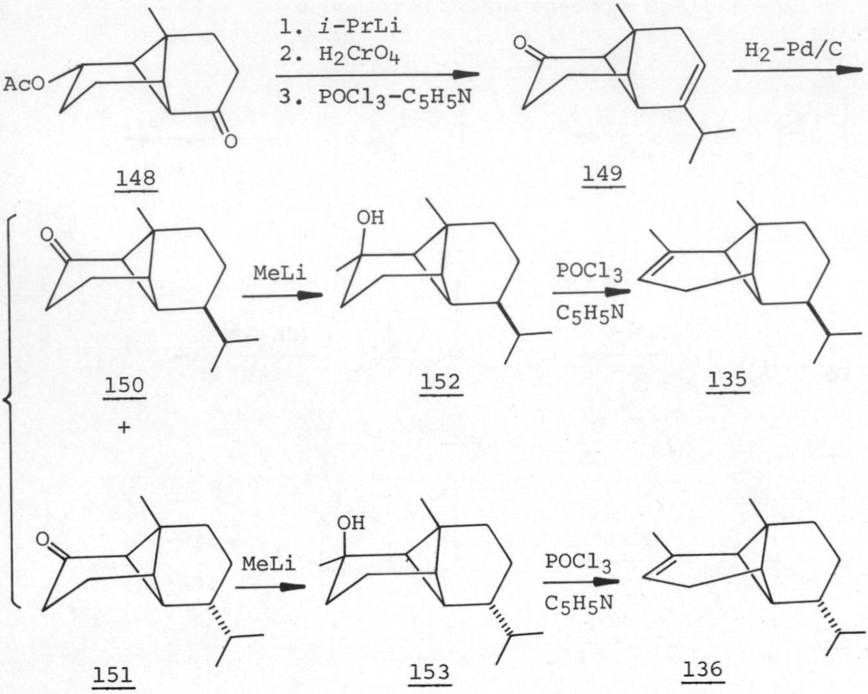

desired tricyclic keto ether 146 in 96% yield.

In an alternative route to this crucial intermediate, octalone 140 was treated with N-bromosuccinimide in benzyl alcohol. After hydrolysis of some dibenzyl ketal and benzyl enol ether, the bromo benzyl ether 147 was obtained in approximately 40% yield. Compound 147 underwent cyclization to 146 in 50% yield.

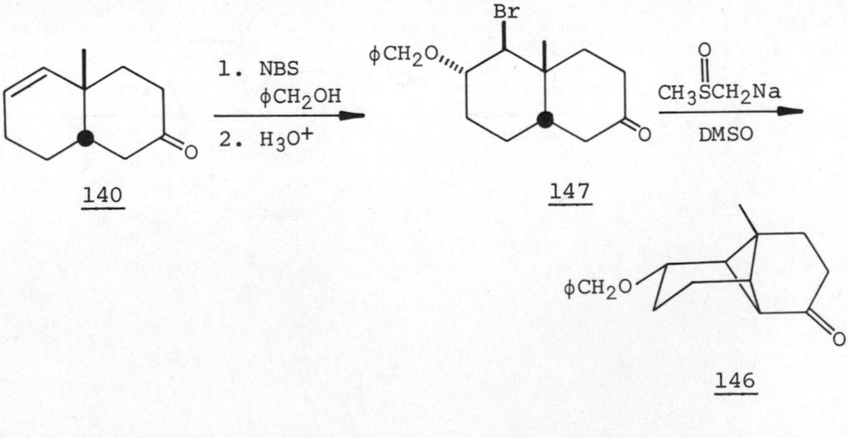

Keto ether 146 was debenzylated by treatment with HBr in glacial acetic acid. Keto acetate 148 was obtained in 55% yield. Treatment of 148 with isopropyllithium gave a mixture of diastereomeric secondary-tertiary diols, which was oxidized to a mixture of keto alcohols. Dehydration of the latter mixture gave unsaturated ketone 149 in 17% yield for the three steps.

At this point, the relative stereochemistry was established. Catalytic hydrogenation of 149 gave a mixture of saturated ketones 150 and 151, which were separated by preparative glpc. Isomer 150 was converted, via alcohol 152, into (±)-copaene (135). Isomer 151 was transformed in a similar manner into (±)-ylangene (136). The proportion of 150 and 151 formed in the hydrogenation reaction was found to be condition dependent. Thus, when 149 was reduced in ethyl acetate or methanol, isomers 150 and 151 were produced in a ratio of 3:7. When the reaction was carried out in hexane, the 150:151 ratio was 6:4. When enone 149 was first reduced to alcohol 154 and the latter hydrogenated, a mixture of isomeric secondary alcohols was obtained. Jones oxidation of this mixture

149   154

gave ketones 150 and 151 in a ratio of 9:1. Thus, by choosing the proper hydrogenation conditions, the synthesis is stereoselective for either (±)-copaene or (±)-ylangene.

G. Sativene, Cyclosativene

Sativene and cyclosativene have been shown to possess structures 155 and 156, respectively. For the synthesis of 155

155

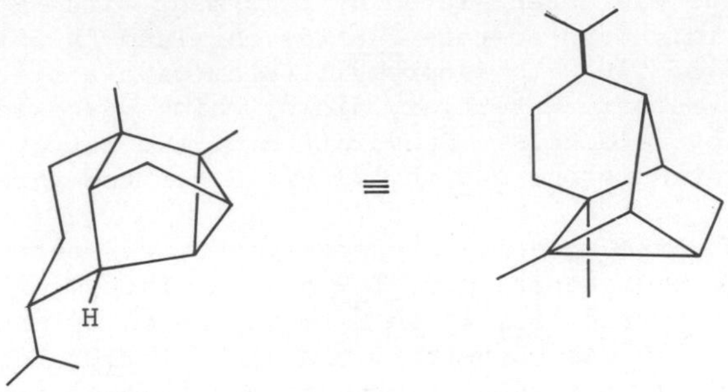

156

McMurry chose a route based on intramolecular alkylation of an enolate ion. The ideal candidate for such a cyclization is keto tosylate 157. The McMurry synthesis is outlined in Scheme 11.[299a]

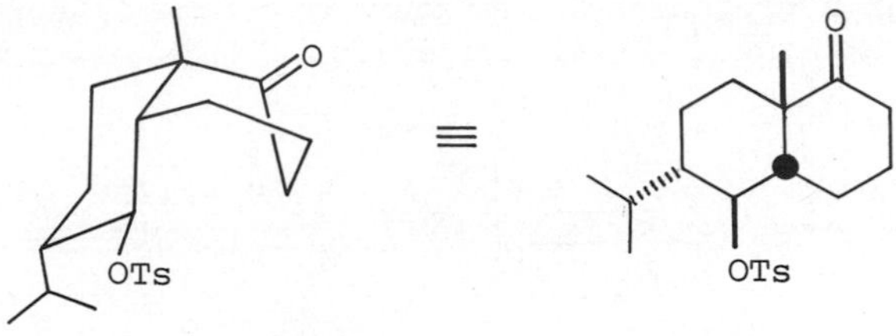

157

Scheme 11.   McMurry's Synthesis of (±)-Sativene and (±)-Cyclosativene

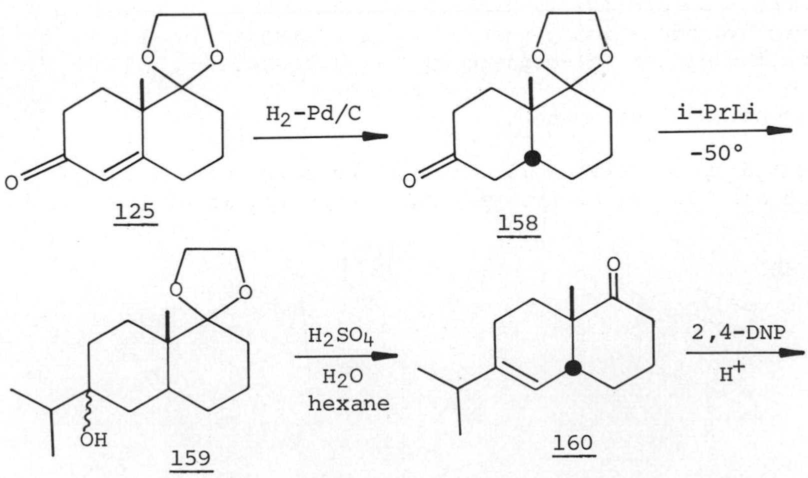

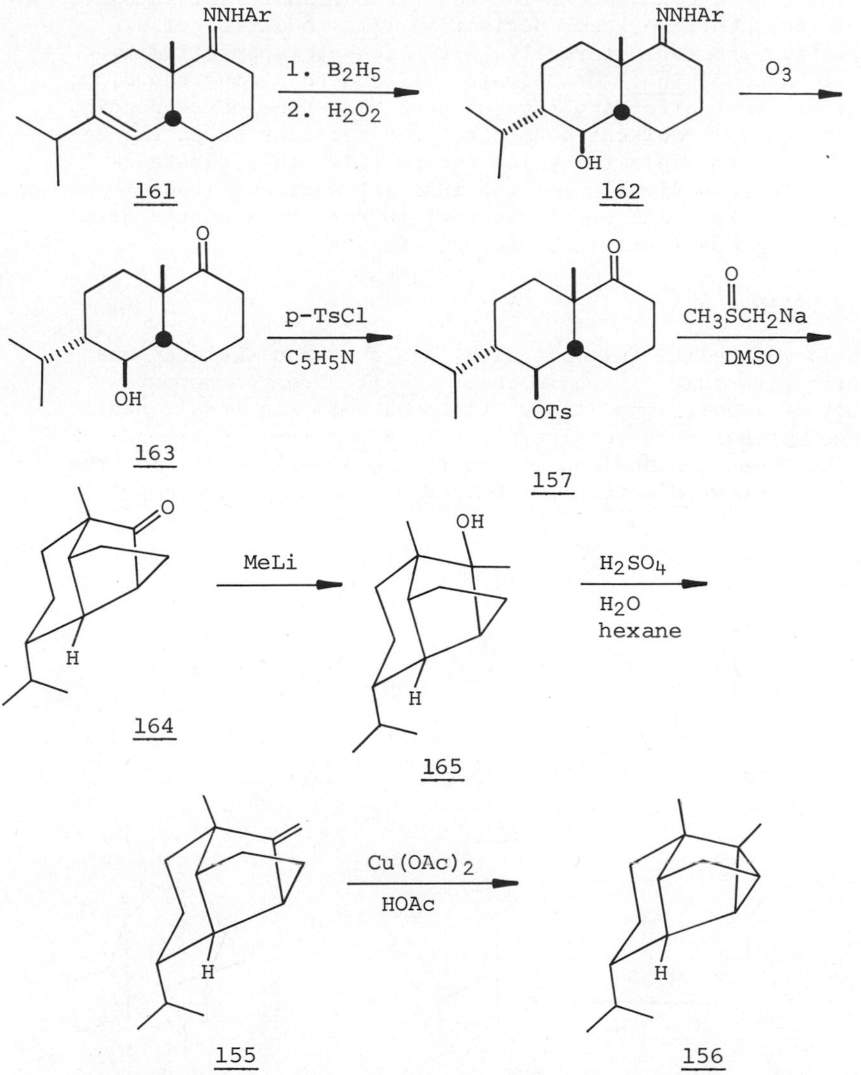

Unsaturated keto ketal 125 was hydrogenated to give the cis-fused decalone 158, along with approximately 5% of its trans isomer. Compound 158 reacted with isopropyllithium at -50° to give a mixture of alcohols (159), which was dehydrated by treatment with aqueous sulfuric acid. Unsaturated ketone 160 was obtained in 80% yield, based on ketone 158.

Since diborane reacted with the carbonyl group faster

than the double bond, enone 160 was first transformed into its
2,4-dinitrophenylhydrazone derivative 161. Addition of diborane then proceeded normally, with high stereospecificity,
to yield alcohol 162. Ozonolysis sufficed to remove the blocking group, thus affording keto alcohol 163. The corresponding
tosylate (157) underwent base catalyzed cyclization in the desired manner to yield tricyclic ketone 164. This substance
was transformed, via alcohol 165 into (±)-sativene (155).
Treatment of 155 with cupric acetate in refluxing acetic acid
gave (±)-cyclosativene (156) in 32% yield.[299b]

H. Culmorin

The mold metabolite culmorin (166) has a carbon skeleton enantiomeric with that of longiborneol (167), a rearrangement
product of longifolene (120). Although culmorin itself has
not been prepared by total synthesis, the related diketone
(168) has been synthesized as the racemate by Roberts.[300] The
optically active diketone is reduced by sodium in isopropyl

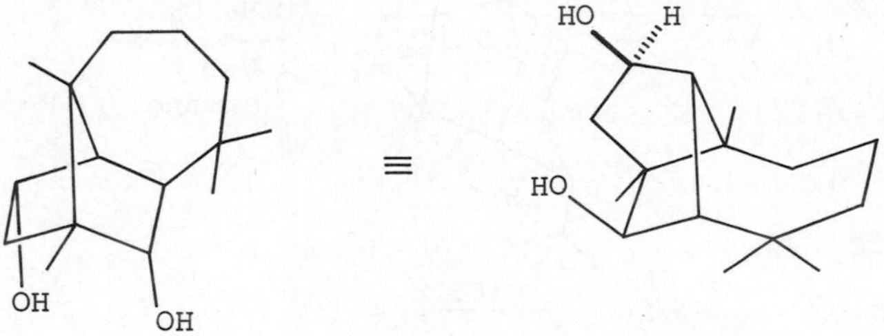

166

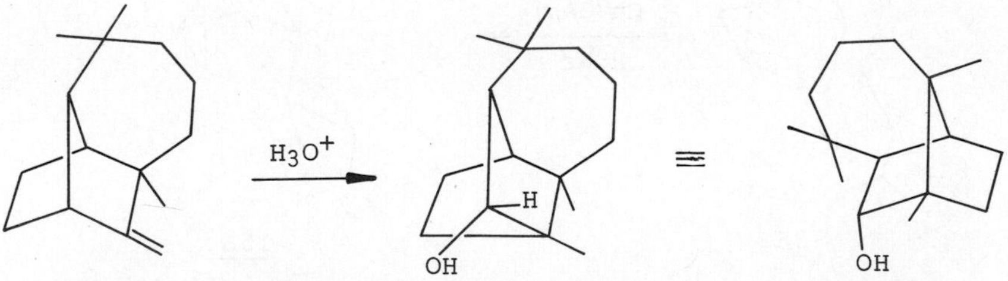

120   167

alcohol to a mixture of culmorin (166) and the isomeric diol
169 in a ratio of 3:2.[301]

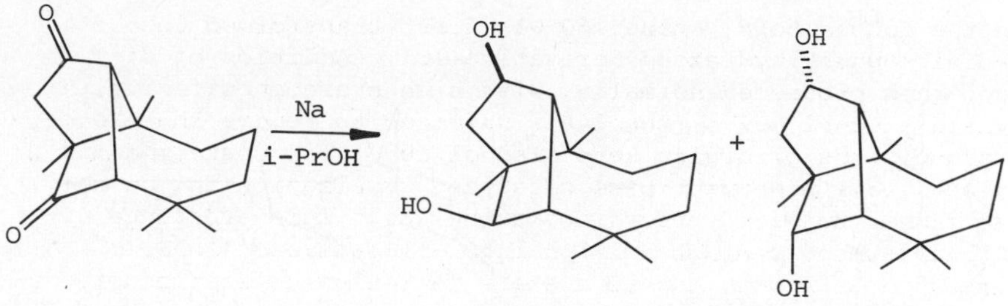

168   166   169

Roberts' synthesis of (±)-culmorin diketone is summarized
in Scheme 12.   Tetrahydroeucarvone (<u>170</u>) was alkylated with

Scheme 12.   Roberts' Synthesis of (±)-Culmorin Diketone

170   171

172   173

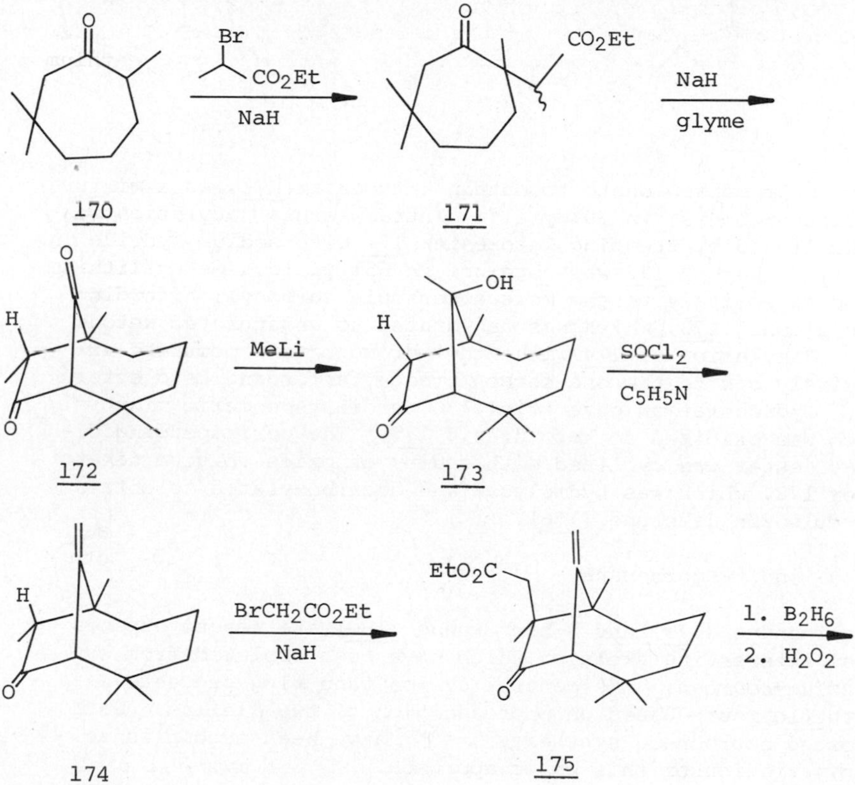

174   175

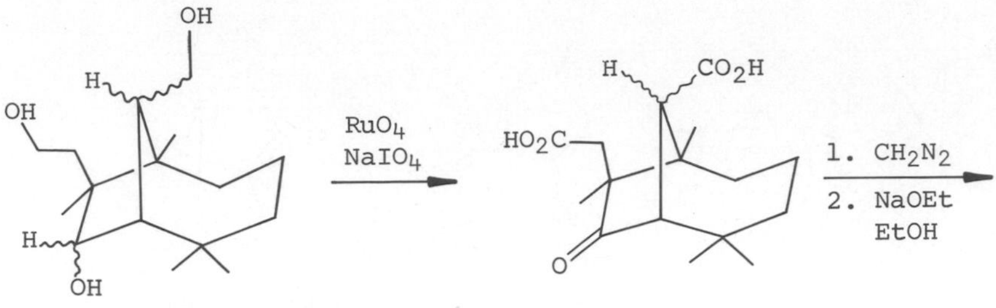

176   177

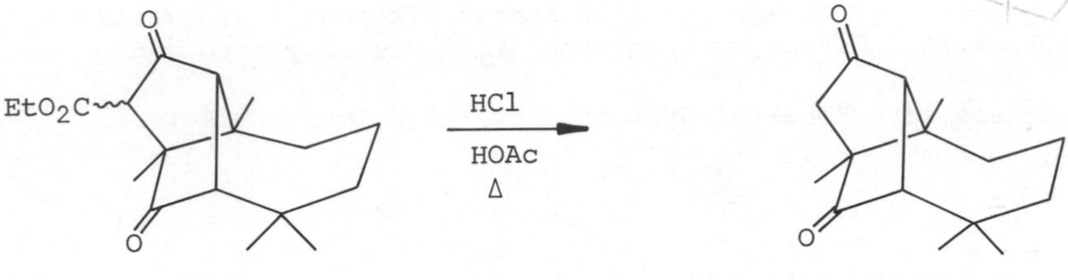

178   166

ethyl α-bromopropionate to obtain keto ester 171, as a mixture
of diastereomers, in 30% yield. Intramolecular acylation was
accomplished by treating keto ester 171 with sodium hydride in
glyme. Diketone 172 was obtained in 65% yield. Methyllithium
added selectively to the more accessible carbonyl, affording
keto alcohol 173, which was dehydrated to unsaturated ketone
174. Alkylation of 174 with ethyl bromoacetate occurred exclusively cis to the one-carbon bridge, affording keto ester
175. Hydroboration gave triol 176 (a diastereomeric mixture),
which was oxidized to keto diacid 177. The corresponding dimethyl ester was cyclized with sodium ethoxide to give diketo
ester 178, which was hydrolyzed and decarboxylated to obtain
(±)-culmorin diketone (166).

I.   α- and β-Bourbonene

α-Bourbonene (179) and β-boubornene (180) are sesquiterpenes
of an interesting skeleton which have been isolated from
Geranium bourbon. The central cyclobutane ring suggests a
synthetic route based on photoaddition of two olefins. Both
recorded bourbonene syntheses[302,303] have been accomplished
by a variation of this basic approach.

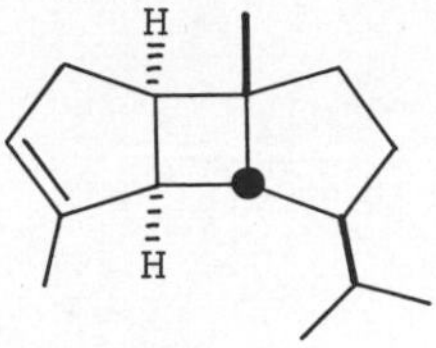

179 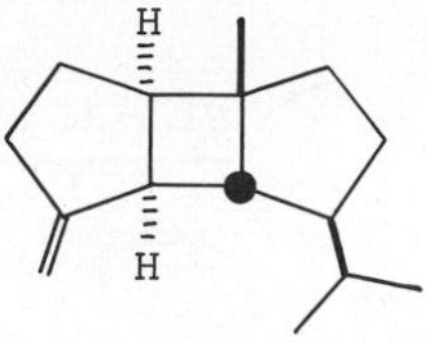   180

White's synthetic plan (Scheme 13) called for the synthesis of 1-methyl-3-isopropylcyclopentene (181). Photoaddition of this olefin and cyclopentenone in a well-documented
reaction,[304] should give tricyclic ketones 182 and/or 183.
Although the *cis-anti-cis* relative stereochemistry about the

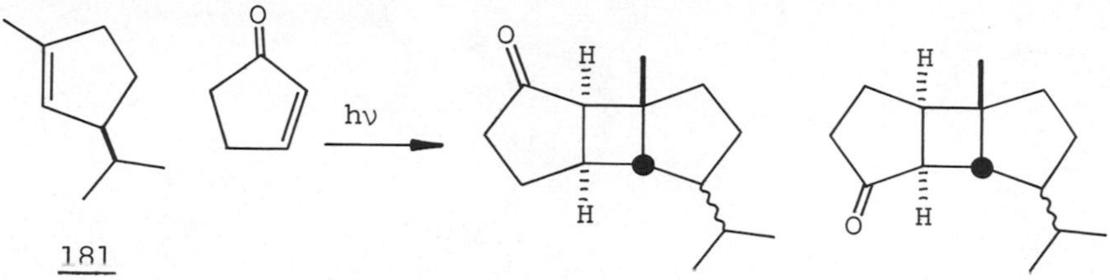

181

182   183

cyclobutane ring could be confidently predicted on the basis
of analogy,[304] the orientation (head-to-head, 182, or head-to-
tail, 183) could not. The relative stereochemistry of the
isopropyl group is also difficult to predict in such a path,
although one would probably expect addition to occur predominately *trans* to this group.

Scheme 13.   White's Synthesis of (±)-α-Bourbonene
and (±)-Bourbonene

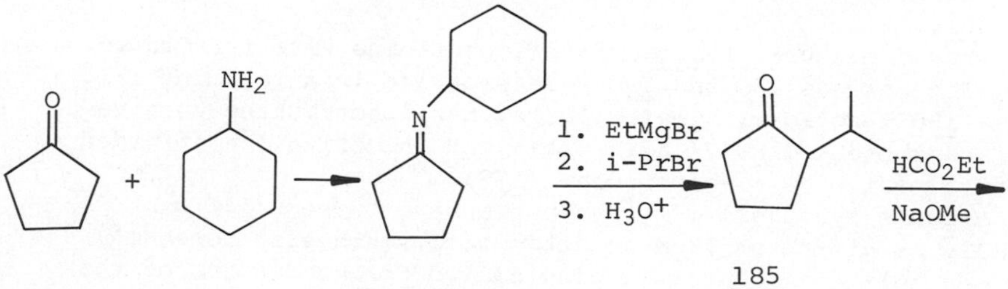

184   185

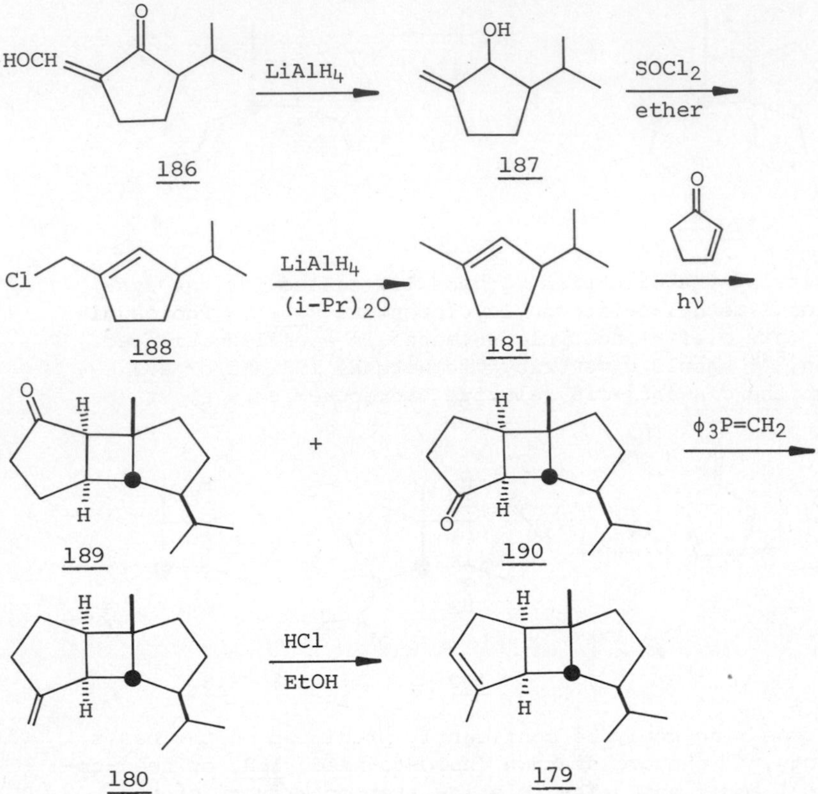

The cyclohexylimine of cyclopentanone (184) was alkylated with isopropyl bromide to give, after hydrolysis, 2-isopropyl-cyclopentanone (185). The corresponding α-hydroxymethylene derivative (186) was reduced by LiAlH₄ to allylic alcohol 187 (70%). Alcohol 187 reacted with thionyl chloride in ether to give rearranged chloride 188, which was reduced to cyclopentene 181.

When a mixture of 181 and cyclopentenone were irradiated, tricyclic ketones 189 and 190 were produced in a ratio of 1:1. Isomer 190 reacted with methylenetriphenylphosphorane to give (±)-β-bourbonene (180). Acid catalyzed isomerization afforded the more stable (±)-α-bourbonene (179).

Brown's synthesis of (±)-α-bourbonene (Scheme 14) was actually on off-shoot from an attempt to synthesize copaene (see sec. 6). The synthetic plan called for production of the diester 191. Photocyclization of 191 could lead to a copaene precurser (192) or a bourbonene precurser (193). As will be seen, diester 191 affords exclusively 193.

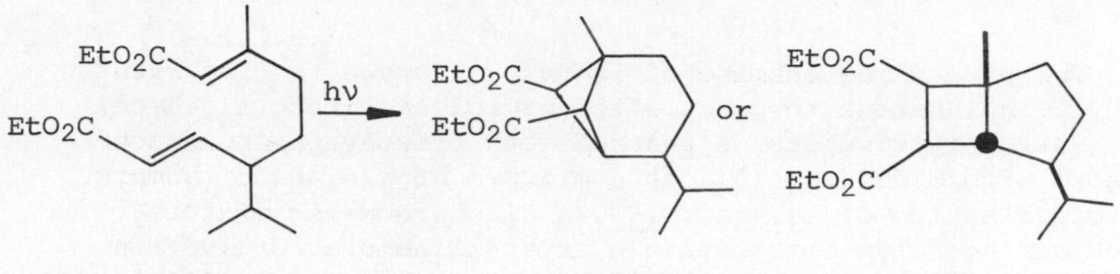

Scheme 14. Brown's Synthesis of (±)-α-Bourbonene

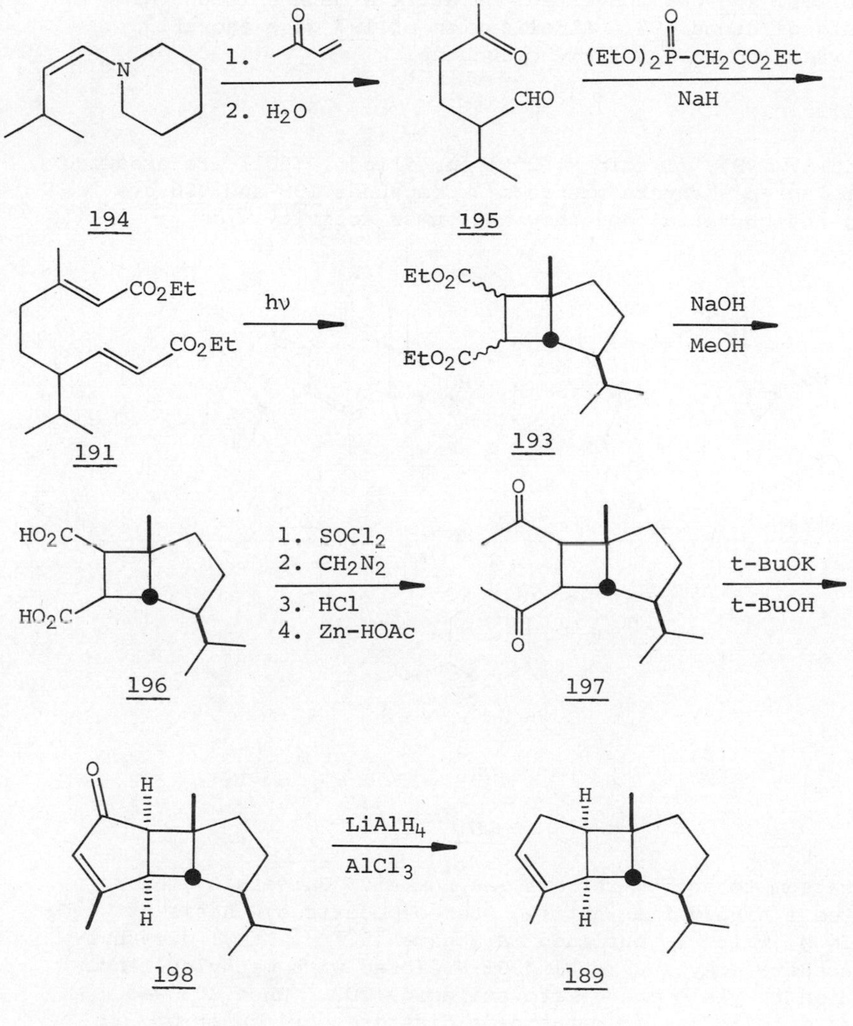

The piperidine enamine of isovaleraldehyde reacted with methyl vinyl ketone to give, after hydrolysis, keto aldehyde 195 (79%). This substance reacted with triethylphosphonoacetate to afford diester 191, as a mixture of cis,trans isomers. Photocyclization of 191 gave 193, a diastereomeric mixture, which was hydrolyzed to obtain a crystalline diacid (196) in 53% yield. Although the full stereochemistry of 196 is not known, the ring fusion must be cis and the isopropyl group must be cis to the angular methyl group.

Diacid 196 was converted, by Stork's method (see Scheme 4), into diketone 197. Aldolization of 197 gave enone 198, which was reduced to (±)-α-bourbonene.

J.   Illudin M

Illudin S (199), illudin M (200) and illudol (201) are produced by the Jack-o'-lantern mushroom. Compounds 199 and 200 are toxic, antibacterial and show antitumor activity.

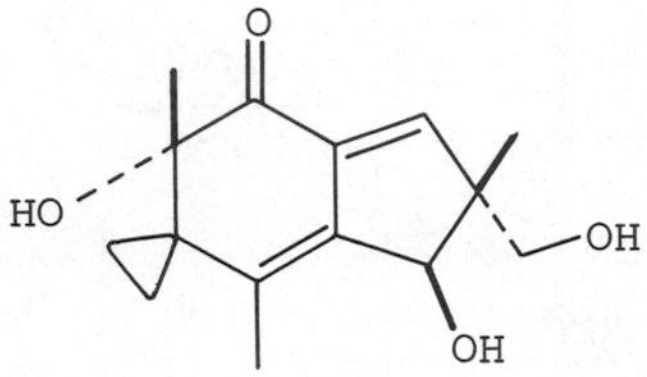

199 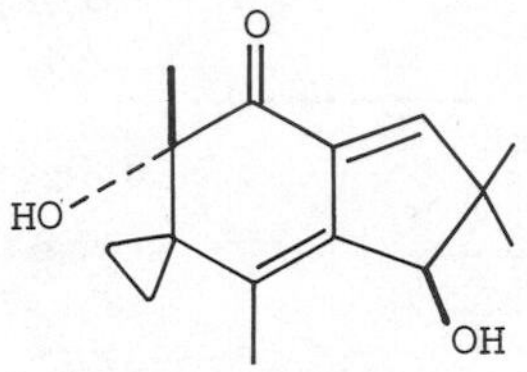   200

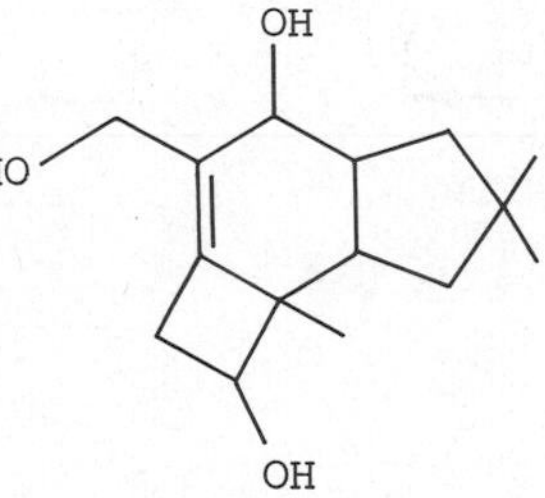

201

Matsumoto and co-workers, at Hokkaido University, have reported a highly imaginative, stereospecific synthesis of illudin M, which is outlined in Scheme 15.[305] Ethyl dimethylacetoacetate ethylene ketal (202) reacted with methylsulfinyl carbanion to yield the β-keto sulfoxide 203. When 203 was treated with iodine in methanol a diastereomeric pair of

Scheme 15. Illudin M--First Hokkaido Synthesis

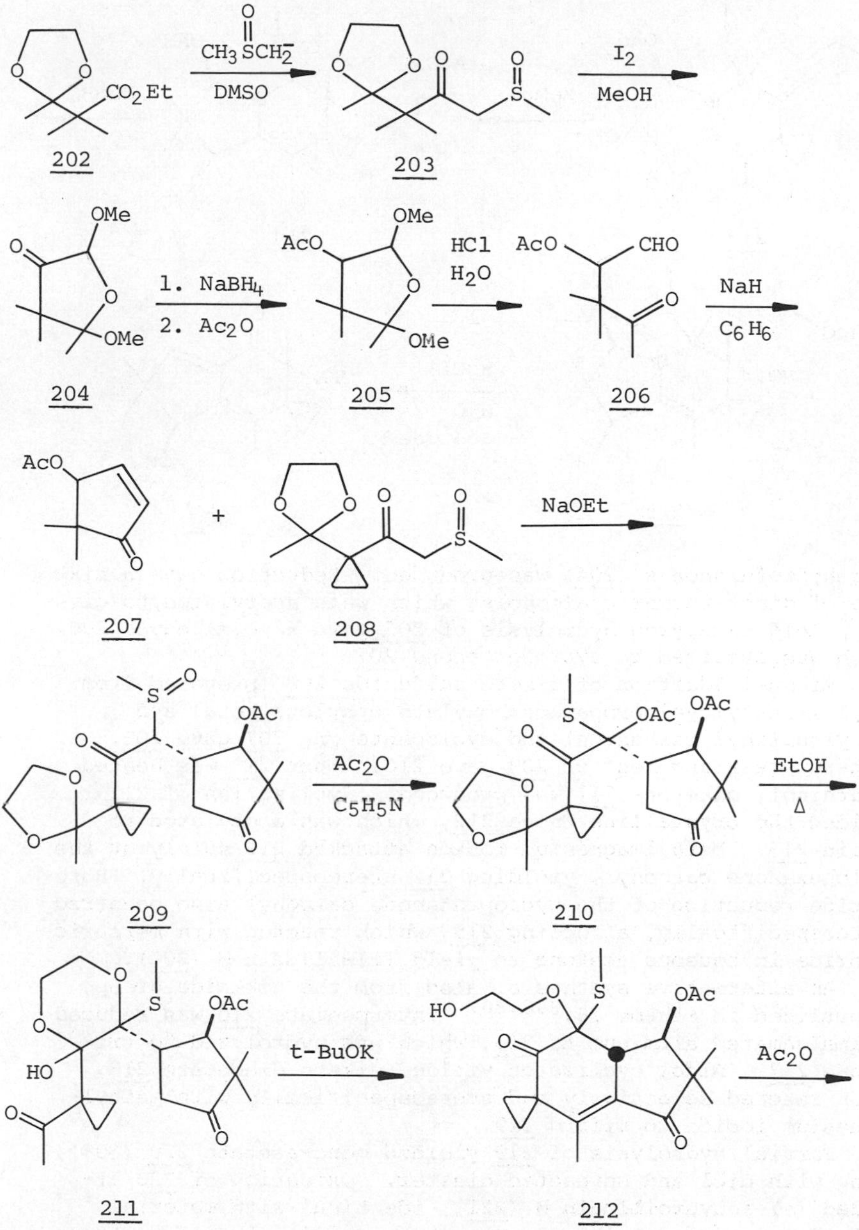

213 → (MeMgI) → 214 → (NaBH₄)

215 → (HgCl₂, H₂O, Acetone) → 200

tetrahydrofuranones (204) was produced. Reduction gave a mixture of diastereomeric alcohols, which were acetylated to give
205. Acid catalyzed hydrolysis of 205 gave keto aldehyde 206,
which was cyclized to cyclopentenone 207.

Michael addition of β-keto sulfoxide 208 (prepared from
ethyl 1-acetylcyclopropanecarboxylate ethylene ketal and
methylsulfinyl carbanion) and cyclopentenone 207 gave 209.
Pummerer rearrangement of 209 gave 210. When 210 was heated
in ethanol, diketone 211 was produced. Cyclization of 211
yielded the crystalline enone 212, which was acetylated to
obtain 213. Methylmagnesium iodide attacked 213 solely at the
cyclohexanone carbonyl, yielding 214 stereospecifically. Borohydride reduction of the cyclopentanone carbonyl also occurred
stereospecifically, affording 215, which reacted with mercuric
chloride in aqueous acetone to yield (±)-illudin M (200).

An alternative synthesis, also from the Hokkaido group,
is outlined in Scheme 16.[306,307] Intermediate 210 was reduced
by amalgamated aluminum to 216, which was hydrolyzed to triketone 217. Aldol cyclizaton yielded diketo diacetate 218,
which reacted selectively and stereospecifically with methylmagnesium iodide to afford 219.

Partial hydrolysis of 219 yielded mono-acetate 220 (30%),
along with diol and unreacted diester. Oxidation of 220 afforded (±)-dehydroilludin M (221), identical with material
derived from natural illudin M. The remaining transformations
were carried out on optically active 221. Hydride reduction
gave a triol (222), which formed a diacetate (223). Selective

Scheme 16.   Illudin M--Second Hokkaido Synthesis

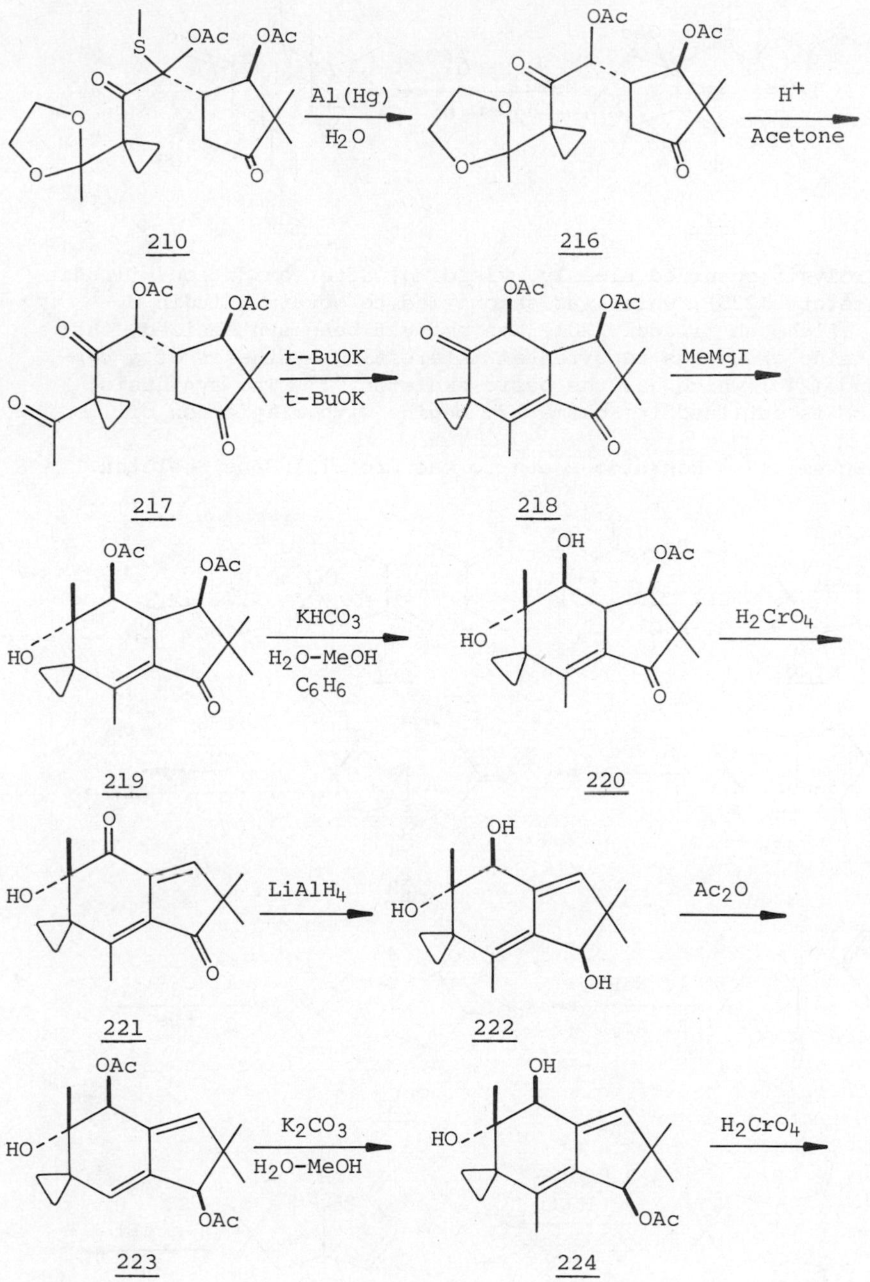

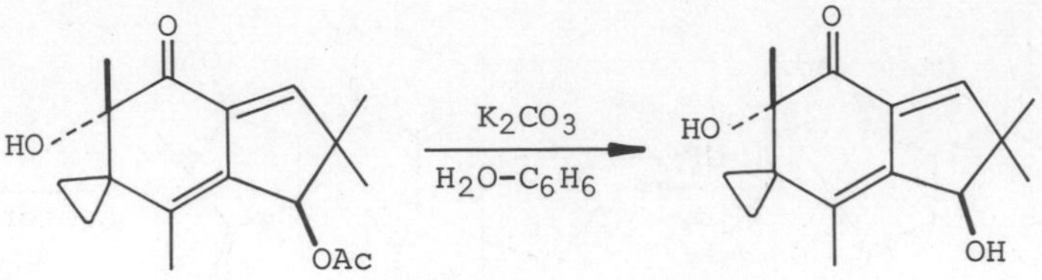

225   200

hydrolysis occurred cleanly, yielding, after oxidation, illudin
M acetate (225), which was saponified to obtain illudin M.
      Although illudol (201) has not yet been synthesized, the
Hokkaido group has reported an interesting synthesis of a compound (240) which has the basic skeleton.[308] The synthesis,
which is outlined in Scheme 17, begins with alkylation of

Scheme 17. Hokkaido Route to the Protoilludane Skeleton

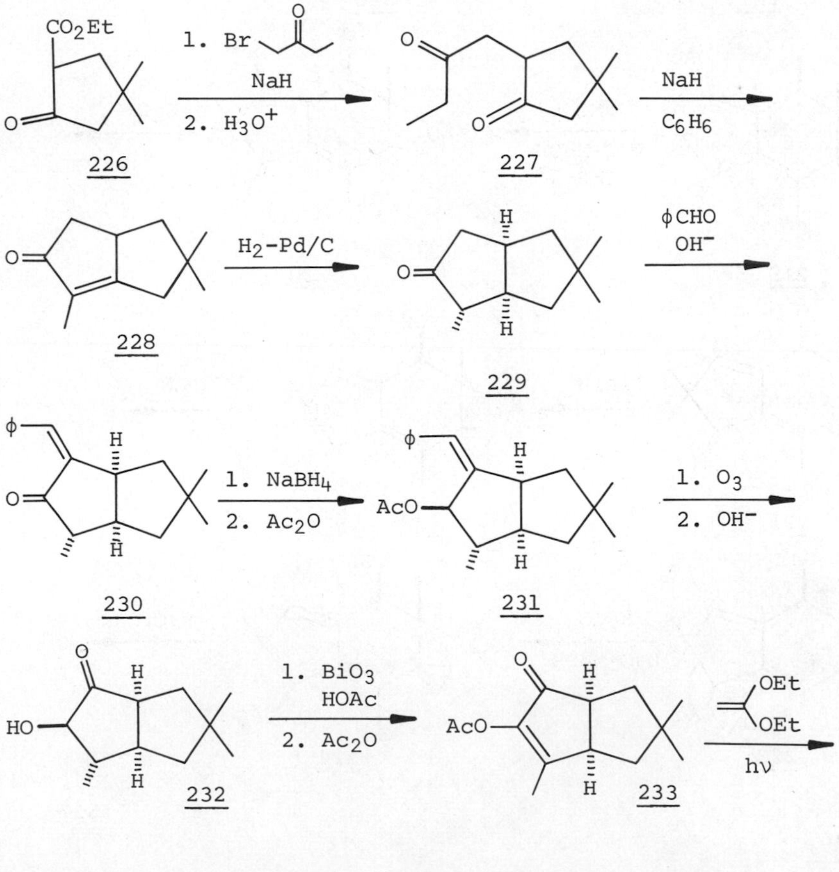

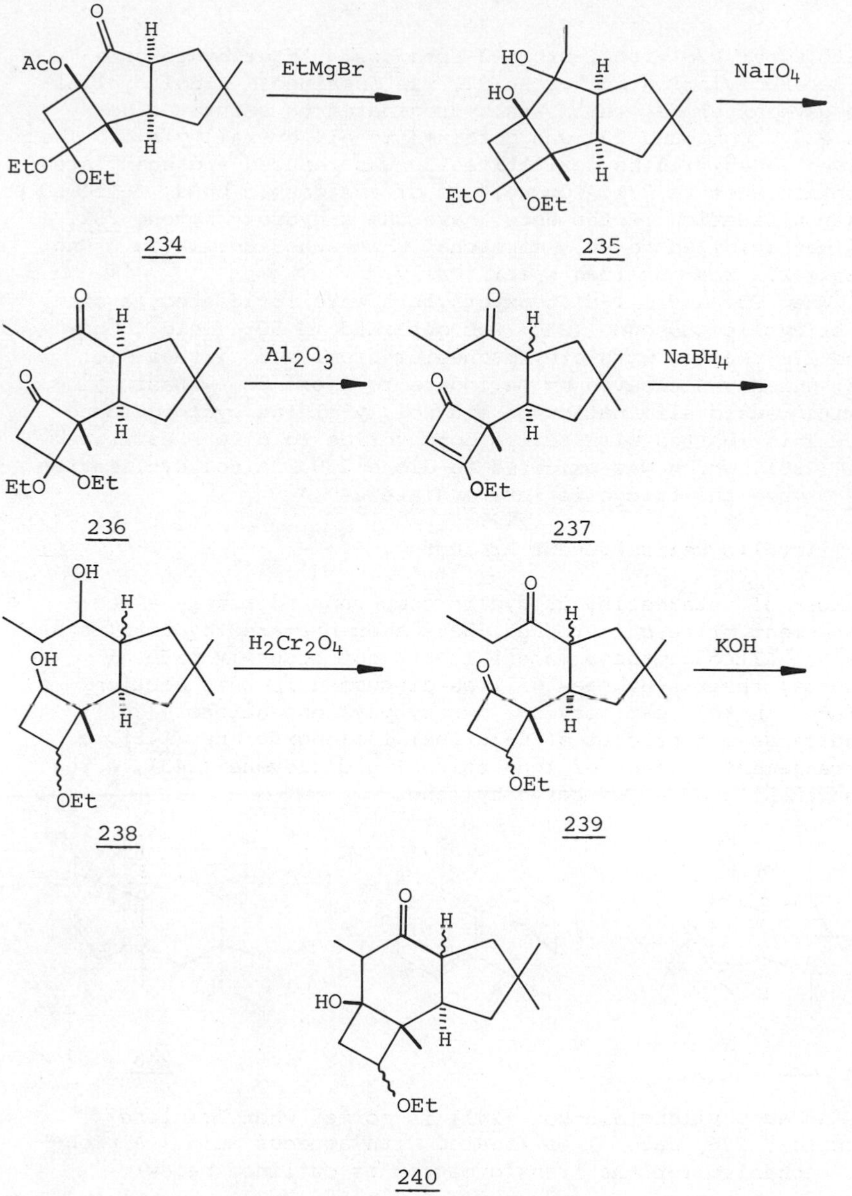
EtMgBr
NaIO4
Al2O3
NaBH4
H2Cr2O4
KOH
AcO
EtO
OEt
HO
OH
234
235
236
237
238
239
240

β-keto ester 226 with 1-bromo-2-butanone. After hydrolysis
and decarboxylation, diketone 227 was obtained. Aldol cyclization gave an enone, which was hydrogenated to saturated ketone 229. Compound 229 was obtained in 41% overall yield. The
derived α-benzylidine derivative 230 was reduced and acetylated
to obtain acetate 231. Ozonolysis of the double bond, followed
by saponification of the ester gave the α-hydroxy ketone 232.
This was oxidized to an α-diketone, which was acetylated. Enol
acetate 233 was obtained specifically.

When 233 and 1,1-diethoxyethylene were irradiated, a single tricyclic compound (234) was obtained in 50% yield. Compound 234 reacted with ethylmagnesium bromide to give a diol
(235) which was cleaved by periodate to dione 236. Basic
alumina caused elimination of ethanol, yielding cyclobutenone
237. This reacted with sodium borohydride to give a saturated
diol (238), which was oxidized to dione 239. Aldol cyclization
of 239 gave the tricyclic intermediate 240.

K. Tricyclic Rearrangement Products

A number of interesting tricyclic compounds, derived by rearrangement of sesquiterpenes, have been prepared by total synthesis. Although these materials are not actually natural
products, their syntheses will be discussed in this section.
Included in this category are α-caryophyllene alcohol (241),
a rearrangement product of humulene; isolongifolene (242), a
rearrangement product of longifolene; and clovene (243), a rearrangement product of caryophyllene.

241   242   243

α-Carophyllene alcohol (241) is formed when humulene
(structure 276, Sec. 3) is treated with aqueous acid. A probable mechanism for the transformation is outlined below:

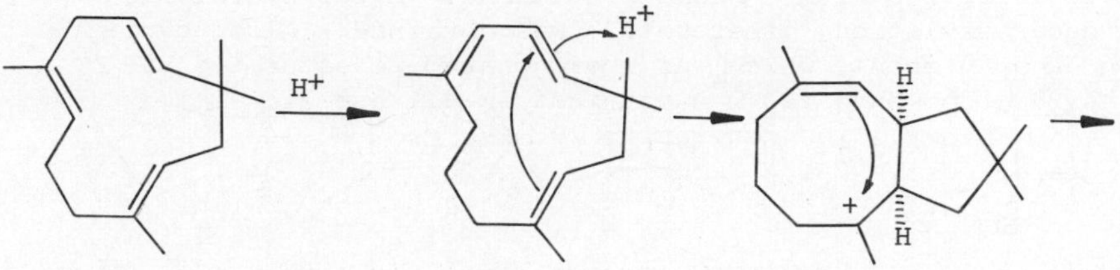

276 (Sec. 3)

244

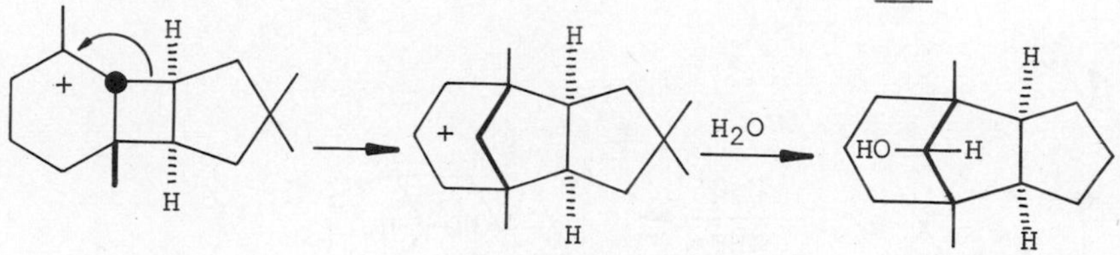

245

241

Although compound 241 itself is rather complicated from a synthetic standpoint, Corey reasoned that tertiary alcohols corresponding to ions 244 and 245 might be more easily derivable. In particular, alcohol 247, which might derive from tricyclic ketone 246 represents an attractive synthetic intermediate. Corey's synthesis of α-caryophyllene alcohol, based on this approach, is outlined in Scheme 18.[309] Photoaddition

Scheme 18. Corey's Synthesis of α-Caryophyllene Alcohol

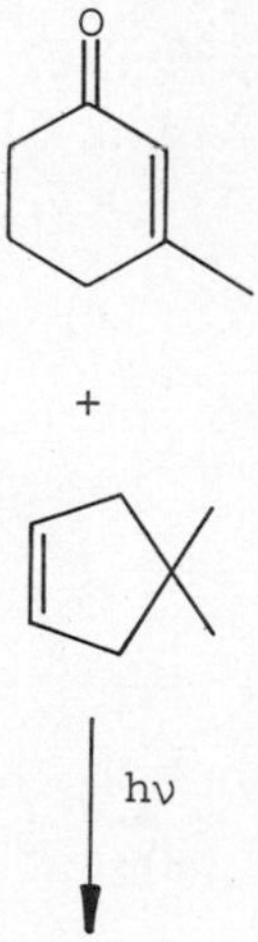

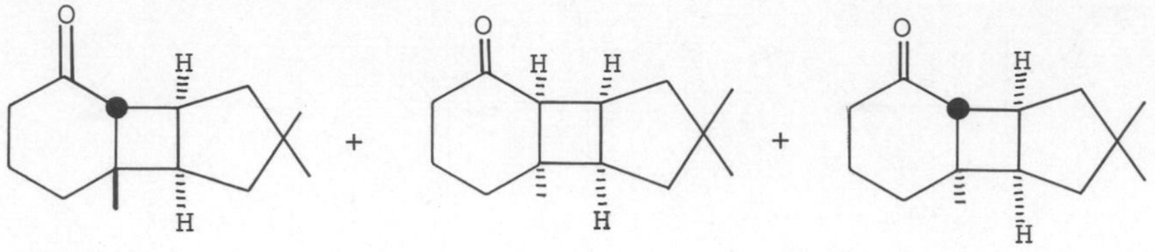

246   248   249

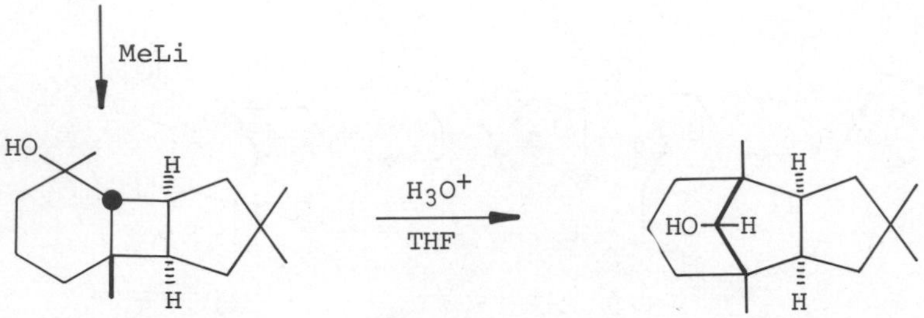

MeLi

247   241

of 3,3-dimethylcyclopentene and 3-methylcyclohexenone gave
three isomeric tricyclic adducts (246, 248, and 249) in a ratio
of 72:12:14. The major isomer 246 was purified by preparative
glpc and treated with methyllithium to yield the crystalline
tertiary alcohol 247. When alcohol 247 was treated with sulfuric acid in aqueous THF, α-caryophyllene alcohol was obtained
in 50% yield.

Isolongifolene (242), a rearrangement product of longifolene, has been synthesized by Dev by the route outlined in
Scheme 19.[310] The synthesis consists of fusing the C-ring

Scheme 19. Dev's Synthesis of (±)-Isolongifolene

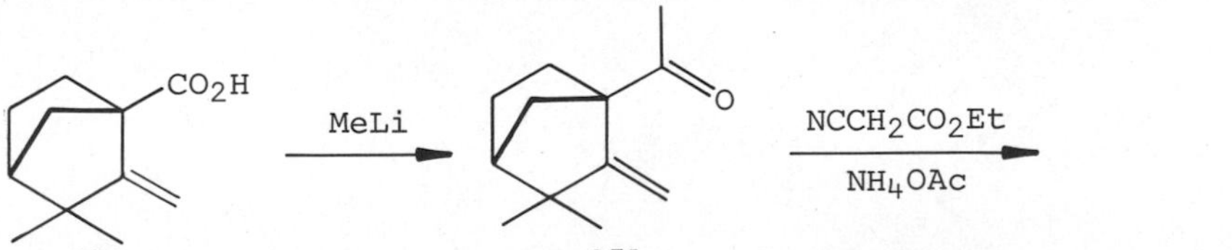

250   251

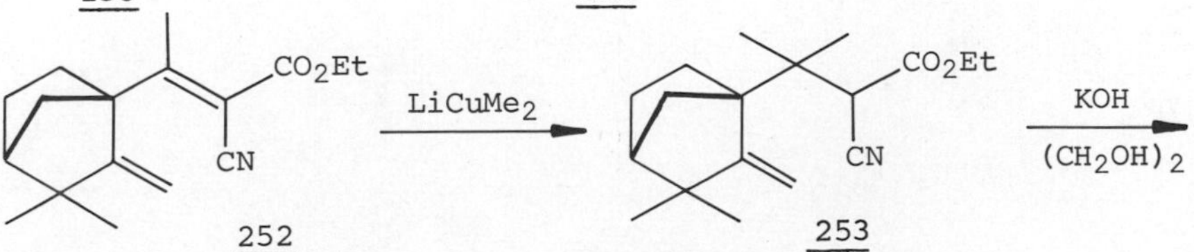

252   253

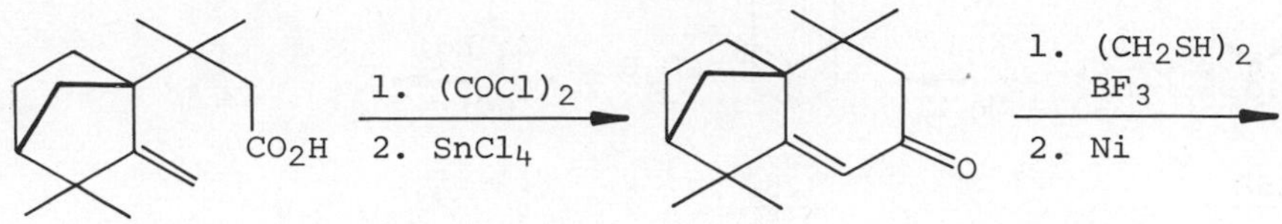

254   255

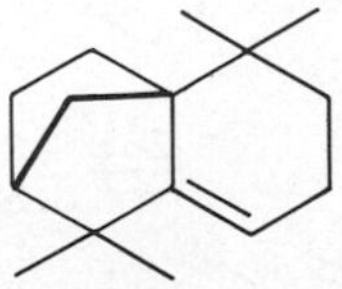

242

onto camphene. Dev's starting material, (±)-camphene-1-carb-
oxylic acid, obviously possesses the necessary functionality
for this purpose. For a synthesis of epizizanoic acid based
on a similar approach, see Scheme 8.

Unsaturated acid 250 was converted by methyllithium to
enone 251, which was condensed with ethyl cyanoacetate to obtain intermediate 252. Conjugate addition of lithium dimethylcopper yielded 253, which was hydrolyzed to unsaturated acid
254. Ring closure was effected by treating the derived acid
chloride with stannic chloride. Removal of the carbonyl group
then afforded (±)-isolongifolene (242). The eight-step synthesis was accomplished in 17% overall yield.

Clovene (243) is one of the products obtained when caryophyllene is treated with acid. It has an interesting carbon
skeleton, with a cyclopentene ring fused onto a bicyclo[3.3.1]-
nonane system. Raphael has disclosed a synthesis of clovene.[311] The Raphael synthesis is outlined in Scheme 20.
Since complete details are still lacking, the synthesis is
simply presented in the chart without further comment.

Scheme 20. Raphael's Synthesis of (±)-Clovene

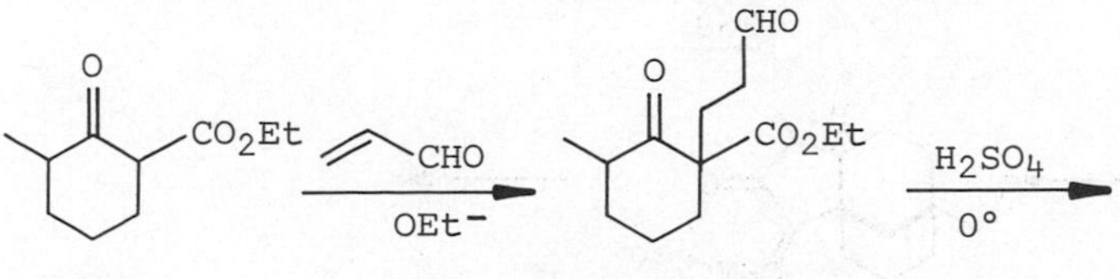

256   257

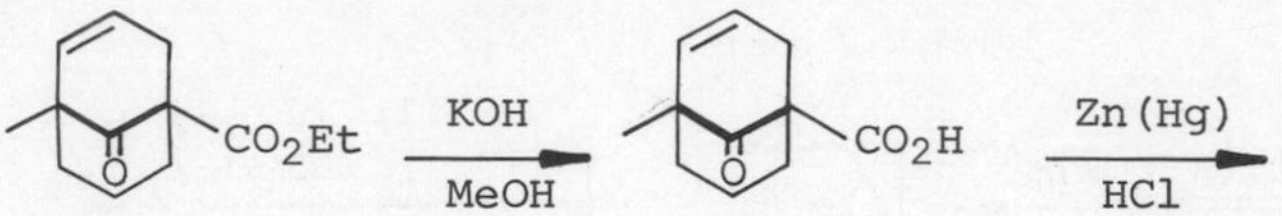

258

259

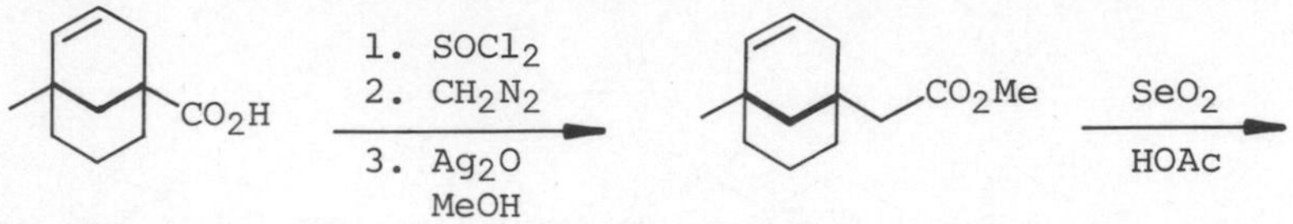

260

261

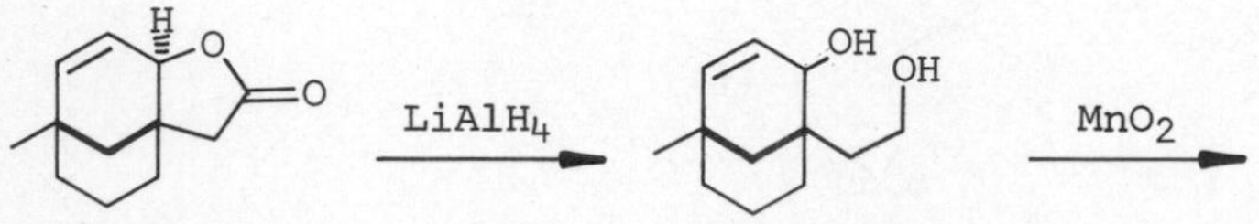

1. SOCl₂
2. CH₂N₂
3. Ag₂O
   MeOH

262

263

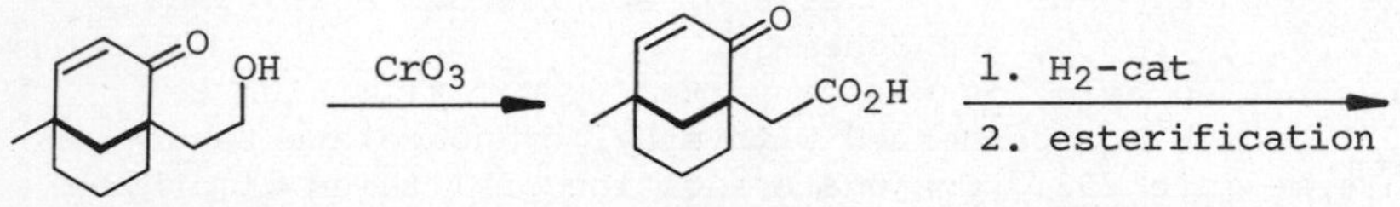

264

265

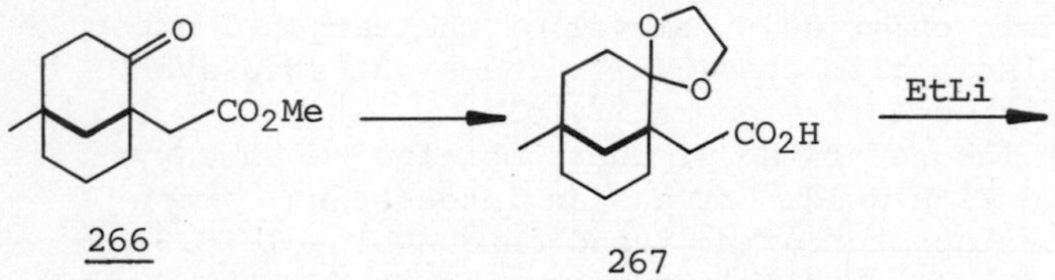

266

267

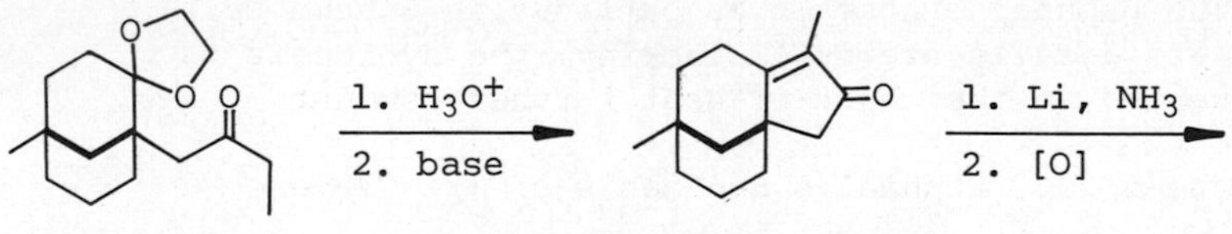

268

268

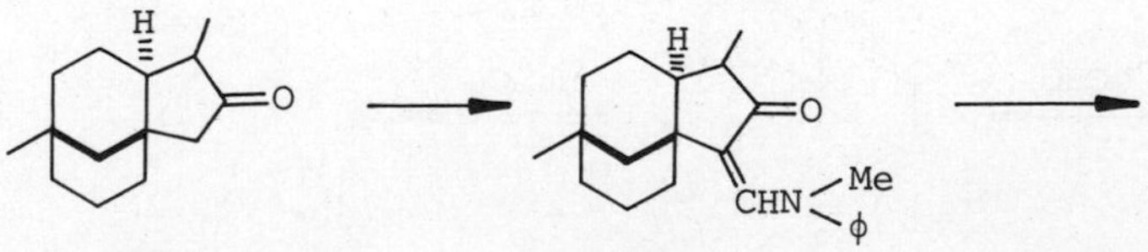

269

270

544

271   LiAlH₄   272

OCO₂Me   Δ

273   243

REFERENCES

1. (a) W. Parker, J. S. Roberts and R. Ramage, *Quart. Rev.*, *21*, 331 (1967); (b) R. B. Clayton, *ibid.*, *19*, 168 (1965); (c) J. H. Richards and J. B. Hendrickson, *Biosynthesis of Steroids, Terpenes and Acetogins* (Benjamin, New York, 1964).

2. (a) G. Ourisson, S. Munavalli, and C. Ehret, *Data Relative to Sesquiterpenoids* (Pergamon Press, Oxford, 1966); (b) A. R. Pinder, *The Chemistry of the Terpenes* (Wiley and Sons, Inc., New York, 1970); (c) J. L. Simonsen and D. H. R. Barton, *The Terpenes*, Vol. III, 2nd ed. (Cambridge University Press, Cambridge, England, 1960); (d) J. L. Simonsen and W. C. J. Ross, *The Terpenes*, Vol. V, addenda to Vol. III (with P. DeMayo) (Cambridge University Press, Cambridge, England, 1957); (e) P. DeMayo, *Mono- and Sesquiterpenoids* (Interscience, New York, 1959); (f) F. Sorm and L. Dolejs, *Guaianolides and Germacranolides* (Holden-Day, Inc., San Francisco, 1966).

3. R. V. Jones and M. D. Sutherland, *Chem. Commun.*, 1229 (1968); (b) K. Morikawa and Y. Hirose, *Tet. Letters*, 1799 (1969); (c) A. S. Rao, A. P. Sadgopal and S. G. Bhattacharyya, *Tetrahedron*, *13*, 319 (1961); (d) K. Takeda, H. Minato and M. Ishikawa, *J. Chem. Soc.*, 4578 (1964).

4. R. B. Bates, D. M. Gale and B. J. Gruner, *J. Org. Chem.*, *28*, 1086 (1963).

5. L. Ruzicka, *Helv. Chim. Acta.*, *6*, 492 (1923).

6. G. Bouchardat, *Compt. Rend.*, *116*, 1253 (1893); (b) F. Tiemann and F. W. Semmler, *Ber.*, *26*, 2708 (1893); (c) K. Stephan, *J. Prakt. Chem.*, *58*, 109 (1898); (d) L. Ruzicka and V. Fornasir, *Helv. Chim. Acta.*, *2*, 182 (1919).

7. (a) F. Tiemann, *Ber.*, *31*, 808 (1898); (b) P. Barbier and
    L. Bouveault, *Compt. Rend.*, *122*, 1422 (1896).

8. O. Isler, R. Ruegg, L. Chopard-dit-Jean, H. Wagner, and
    K. Bernhard, *Helv. Chim. Acta*, *39*, 897 (1956).

9. A. Ofner, W. Kimel, A. Holmgren, and F. Forrester, *ibid.*,
    42, 2577 (1959).

10. I. N. Nazarov, B. P. Gusev, and V. I. Gunar, *Zhur. Obshch.
    Khim. S.S.S.R.*, *28*, 1444 (1958).

11. P. Chaleyer, *Perfumery. Essent. Oil Rec.*, *49*, 17 (1958).

12. N. I. Skvortsova, V. Y. Tokareva, and V. N. Belov, *Zhur.
    Obshch. Khim. S.S.S.R.*, *29*, 3113 (1959).

13. E. Yu. Shvarts and A. A. Petrov, *ibid.*, *30*, 3598 (1960).

14. G. Popjak, J. W. Cornforth, R. H. Cornforth, R. Ryhage,
    and D. S. Goodman, *J. Biol. Chem.*, *237*, 56 (1962).

15. M. Julia, S. Julia, and R. Guégan, *Bull. Soc. Chim.
    France*, 1072 (1960).

16. S. F. Brady, M. A. Ilton, and W. S. Johnson, *J. Amer.
    Chem. Soc.*, *90*, 2882 (1968).

17. E. J. Corey, J. A. Katzenellenbogen, and G. H. Posner,
    ibid., *89*, 4245 (1967).

18. O. P. Vig, J. C. Kapur, C. K. Khurana, and B. Vig, *J.
    Indian Chem. Soc.*, *46*, 505 (1969).

19. H. Roller, K. H. Dahm, C. C. Sweeley, and B. M. Trost,
    Angew. Chem., *79*, 190 (1967).

20. K. H. Dahm, B. M. Trost, and H. Roller, *J. Amer. Chem.
    Soc.*, *89*, 5292 (1967).

21. E. J. Corey, J. A. Katzenellenbogen, N. W. Gilman, S. A.
    Roman, and B. E. Erickson, *ibid.*, *90*, 5618 (1968).

22. R. Zurfluh, E. N. Wall, J. B. Siddall, and J. A. Edwards,
    ibid., *90*, 6224 (1968).

23. W. S. Johnson, T. Li, D. J. Faulkner, and S. F. Campbell,
    ibid., *90*, 6225 (1968).

24. B. H. Braun, M. Jacobson, M. Schwarz, P. E. Sonnet, N.
    Wakabayashi, and R. M. Waters, *J. Econom. Entomol.*, *61*,
    866 (1968).

25. K. Mori, B. Stalla-Bourdillon, M. Ohki, M. Matsui, and
    W. S. Bowers, *Tetrahedron*, *25*, 1667 (1969).

26. J. A. Findlay and W. D. Mackay, *Chem. Commun.*, 733 (1969).

27. H. Schulz and I. Sprung, *Angew. Chem. Internal. Edn.
    Engl.*, *8*, 271 (1969).

28. R. J. Anderson, C. A. Henrick, and J. B. Siddall, *J. Amer.
    Chem. Soc.*, *92*, 735 (1970).

29. E. E. van Tamelen and J. P. McCormick, *ibid.*, *92*, 737
    (1970).

30. H. M. Schmidt and J. F. Arens, *Rec. Trav. Chim.*, *86*, 1138
    (1967).

31. E. E. van Tamelen, M. A. Schwartz, E. J. Hessler, and A.
    Stone, *Chem. Commun.*, 409 (1966).

32. C. A. Grob, H. R. Kiefer, H. Lutz, and H. Wilkins, *Tet. Letters*, 2901 (1964).

33. P. S. Wharton, *J. Org. Chem.*, *26*, 4781 (1961).

34. J. W. Cornforth, R. H. Cornforth, and K. K. Mathew, *J. Chem. Soc.*, 112, 2539 (1959).

35. D. J. Cram and F. A. Abd Elhafez, *J. Amer. Chem. Soc.*, *74*, 5828 (1952).

36. E. M. Kosower, W. J. Cole, G. S. Wu, D. E. Cardy, and G. Meisters, *J. Org. Chem.*, *28*, 630 (1963).

37. (a) W. Kimel, J. D. Surmatis, J. Weber, G. O. Chase, N. W. Sax, and A. Ofner, *ibid.*, *22*, 1611 (1957); (b) W. Kimel, N. W. Sax, S. Kaiser, G. G. Eichmann, G. O. Chase, and A. Ofner, *ibid.*, *23*, 153 (1958).

38. G. Saucy and R. Marbet, *Helv. Chim. Acta*, *50*, 2099 (1967).

39. P. Rona, L. Tökes, J. Tremble, and P. Crabbé, *Chem. Commun.*, 43 (1969).

40. E. J. Corey, N. W. Gilman, and B. E. Ganem, *J. Amer. Chem. Soc.*, *90*, 5616 (1968).

41. (a) A. F. Thomas, *Chem. Commun.*, 947 (1967); (b) *J. Amer. Chem. Soc.*, *91*, 3281 (1969).

42. W. von E. Doering and W. R. Roth, *Tetrahedron*, *18*, 67 (1962).

43. G. Büchi and H. Wüest, *Helv. Chim. Acta*, *50*, 2440 (1967).

44. G. Stork and S. R. Dowd, *J. Amer. Chem. Soc.*, *85*, 2178 (1963).

45. G. Wittig and H. D. Frommeld, *Chem. Ber.*, *97*, 3548 (1964).

46. E. Bertele and P. Schudel, *Helv. Chim. Acta*, *50*, 2445 (1967).

47. M. S. Lemberg, French Patent 1456900 (September 19, 1966).

48. G. F. Emerson, J. E. Mahler, R. Kochhar, and R. Petit, *J. Org. Chem.*, *29*, 3620 (1964).

49. A. F. Thomas, *Chem. Commun.*, 1657 (1968).

50. K. A. Parker and W. S. Johnson, *Tet. Letters*, 1329 (1969).

51. T. Kubota and T. Matsuura, *Chem. & Ind.*, 521 (1956).

52. L. Ruzicka and E. Capato, *Helv. Chim. Acta*, *8*, 259 (1925).

53. L. Ruzicka and M. Liguori, *ibid.*, *15*, 3 (1932).

54. A. Manjarrez and A. Guzman, *J. Org. Chem.*, *31*, 348 (1966).

55. O. P. Vig, K. L. Matta, G. Singh, and I. Raj, *J. Indian Chem. Soc.*, *43*, 27 (1966).

56. O. P. Vig, J. P. Salota, B. Vig, and B. Ram, *Indian J. Chem.*, *5*, 475 (1967).

57. A. Manjarrez, T. Rios, and A. Guzman, *Tetrahedron*, *20*, 333 (1964).

58. O. P. Vig, I. Raj, J. P. Salota, and K. L. Matta, *J. Indian Chem. Soc.*, *46*, 205 (1969).

59. K. V. Kuznetsov and R. A. Myrsina, *Dopov. Akad. Nauk. Ukr. RSR, Ser. B.*, *21*, 810 (1969) [*Chem. Abs.*, *72*, 21789 (1970)].

60. F. D. Carter, J. L. Simonsen, and H. O. Williams, *J. Chem. Soc.*, 451 (1940).

61. A. J. Birch and S. M. Mukherji, *ibid.*, 2531 (1949).

62. A. S. Rao, *Indian J. Chem.*, *3*, 419 (1965).

63. (a) V. Honwad and A. S. Rao, *Tetrahedron*, *21*, 2593 (1965); (b) *Current Sci. (India)*, *34*, 534 (1965).

64. H. Rupe and A. Steinbach, *Ber. Dtsch. Chem. Ges.*, *44*, 584 (1911).

65. O. P. Vig, J. P. Salota, and B. Vig, *Indian J. Chem.*, *4*, 323 (1966).

66. (a) G. D. Joshi and S. N. Kulkarni, *Indian J. Chem.*, *3*, 91 (1965); (b) G. D. Joshi and S. N. Kulkarni, *Indian J. Chem.*, *6*, 127 (1968).

67. N. K. Bhattacharyya and S. M. Mukherji, *Science and Culture*, *16*, 269 (1950); (b) S. M. Mukherji and N. K. Bhattacharyya, *J. Amer. Chem. Soc.*, *75*, 4698 (1953).

68. R. C. Banerjee, *J. Sci. Ind. Res. (India)*, *21B*, 285 (1962).

69. H. Rupe and F. Wiederkehr, *Helv. Chim. Acta.*, *7*, 654 (1924).

70. H. Rupe and A. Gassmann, *ibid.*, *19*, 569 (1936).

71. J. Colonge and J. Chambion, *Compt. Rend.*, *222*, 557 (1946).

72. S. M. Mukherji, *J. Indian Chem. Soc.*, *24*, 341 (1947).

73. R. P. Gandhi, O. P. Vig, and S. M. Mukherji, *Tetrahedron*, *7*, 736 (1959).

74. O. P. Vig, B. Vig, and I. Raj, *J. Indian Chem. Soc.*, *42*, 673 (1965).

75. G. Büchi and H. Wüest, *J. Org. Chem.*, *34*, 1122 (1969).

76. O. P. Vig, J. C. Kapoor, J. Puri, and S. D. Sharma, *Indian J. Chem.*, *6*, 60 (1968).

77. R. Tuchihashi and T. Hanzawa, *J. Chem. Soc. Japan*, *61*, 1041 (1940).

78. T. Momose, *J. Pharm. Soc. Japan*, *61*, 288 (1941).

79. W. S. Bowers, H. M. Fales, M. J. Thompson, and E. C. Uebel, *Science*, *154*, 1020 (1966).

80. V. Cerny, L. Doljes, L. Labler, F. Sorm, and K. Slama, *Tet. Letters*, 1053 (1967).

81. M. Makazaki and S. Isoe, *Bull. Chem. Soc. Japan*, *34*, 741 (1961); *36*, 1198 (1963).

82. K. Mori and M. Matsui, *Tet. Letters*, 2515 (1967); (b) *Tetrahedron*, *24*, 3127 (1968).

83. K. S. Ayyar and G. S. K. Rao, *Tet. Letters*, 4677 (1967); (b) *Can. J. Chem.*, *46*, 1467 (1968).

84. B. A. Pawson, H. C. Cheung, S. Gurbaxani, and G. Saucy, *Chem. Commun.*, 1057 (1968); (b) *J. Amer. Chem. Soc.*, *92*, 366 (1970).

85. A. J. Birch, P. L. Macdonald, and V. H. Powell, *Tet. Letters*, 351 (1969).

86. K. Mori and M. Matsui, *ibid.*, 4853 (1967).

87. J. F. Blount, B. A. Pawson, and G. Saucy, *Chem. Commun.*, 715, 1016 (1969).

88. K. H. Schulte-Elte and G. Ohloff, *Helv. Chim. Acta.*, *49*, 2150 (1966).

89. K. Yamaguchi, *J. Pharm. Soc. Japan, 62,* 491 (1942).

90. D. A. Archer and R. H. Thompson, *Chem. Commun.*, 354 (1965).

91. E. R. Wagner, R. D. Moss, R. M. Brooker, J. P. Heeschen, W. J. Potts, and M. L. Dilling, *Tet. Letters,* 4233 (1965).

92. E. Cortes, M. Salmon, and F. Walls, *Bol. Inst. Quim. Univ. Nacl. Autom. Mex., 17,* 19 (1965).

93. G. Büchi and H. Wüest, *J. Org. Chem., 34,* 857 (1969).

94. A. J. Weinheimer and P. H. Washecheck, *Tet. Letters,* 3315 (1969).

95. D. M. Simonovic, A. S. Rao, and S. C. Bhattacharyya, *Tetrahedron, 19,* 1061 (1963).

96. V. K. Honwad, E. Siscovic, and A. S. Rao, *ibid., 23,* 1273 (1967).

97. L. J. Patel and A. S. Rao, *Tet. Letters,* 2273 (1967).

98. O. P. Vig, K. L. Matta, J. C. Kapur, and B. Vig, *J. Indian Chem. Soc., 45,* 973 (1968).

99. E. J. Corey and E. A. Broger, *Tet. Letters,* 1779 (1969).

100. V. Sykora, J. Cerny, V. Herout, and F. Sorm, *Coll. Czech. Chem. Commun., 19,* 566 (1954).

101. H. Matsumura, I. Iwai, and E. Ohki, *J. Pharm. Soc. Japan, 75,* 687 (1955); (b) H. Ogura, *J. Org. Chem., 25,* 679 (1960).

102. W. S. Johnson, B. Bannister, R. Pappo, and J. E. Pike, *J. Amer. Chem. Soc., 78,* 6354 (1956).

103. T. G. Halsall, D. W. Theobald, and K. B. Walshaw, *J. Chem. Soc.,* 1029 (1964).

104. L. J. Patil, K. S. Kulkarni, and A. S. Rao, *Indian J. Chem., 4,* 400 (1966).

105. K. S. Kulkarni and A. S. Rao, *Tetrahedron, 21,* 1167 (1965).

106. E. J. Corey and E. K. W. Wat, *J. Amer. Chem. Soc., 89,* 2757 (1967).

107. G. Büchi and H. Wüest, *J. Amer. Chem. Soc., 87,* 1589 (1965).

108. Y. Kitahara and M. Funamizu, *Bull. Chim. Soc. Japan, 31,* 782 (1958).

109. Y. Kitahara and T. Kato, *ibid., 37,* 895 (1964).

110. M. Suchy and F. Sorm, *Coll. Czech. Chem. Commun., 23,* 2175 (1958).

111. (a) E. J. Corey and A. G. Hortmann, *J. Amer. Chem. Soc., 85,* 4033 (1963); (b) *87,* 5736 (1965).

112. E. J. Corey and E. Hamanaka, *ibid.*, *89*, 2758 (1967).

113. P. S. Adamson, F. J. McQuillin, R. Robinson, and J. L. Simonsen, *J. Chem. Soc.*, 1576 (1937).

114. R. Howe and F. J. McQuillin, *ibid.*, 2423 (1955).

115. J. K. Roy, *Chem. and Ind.*, 1393 (1954).

116. E. Piers and K. F. Cheng, *Can. J. Chem.*, *46*, 377 (1968).

117. S. M. Mukherji, S. Singh, and O. P. Vig, *Sci. and Culture (Calcutta)*, *25*, 533 (1960).

118. (a) A. R. Pinder and R. A. Williams, *Chem. and Ind.*, 1714 (1961); (b) *J. Chem. Soc.*, 2773 (1963).

119. (a) J. A. Marshall and M. T. Pike, *Tet. Letters*, 3107 (1965); (b) J. A. Marshall, M. T. Pike, and R. D. Carroll, *J. Org. Chem.*, *31*, 2933 (1966).

120. H. C. Brown, K. H. Murray, L. J. Murray, J. A. Snover, and G. Zweifel, *J. Amer. Chem. Soc.*, *82*, 4233 (1960).

121. M. Nussim, T. Mazur, and F. Sondheimer, *J. Org. Chem.*, *29*, 1120 (1964).

122. (a) R. P. Houghton, D. C. Humber, and A. R. Pinder, *Tetrahedron Letters*, 353 (1966); (b) D. C. Humber, A. R. Pinder, and R. A. Williams, *J. Org. Chem.*, *32*, 2335 (1967).

123. C. H. Heathcock and T. R. Kelly, *Tetrahedron*, *24*, 1801 (1968).

124. O. P. Vig, R. C. Anand, B. Kumar, and S. D. Sharma, *J. Indian Chem. Soc.*, *45*, 1033 (1968).

125. J. A. Marshall and M. T. Pike, *Tet. Letters*, 4989 (1966).

126. D. K. Banerjee and P. S. Halwe, *J. Indian Chem. Soc.*, *37*, 669 (1960).

127. G. L. Chetty, G. S. K. Rao, S. Dev, and D. K. Banerjee, *Tetrahedron*, *22*, 2311 (1966).

128. J. A. Marshall and R. D. Carroll, *Tet. Letters*, 4223 (1965).

129. H. C. Barrett and G. Büchi, *J. Amer. Chem. Soc.*, *89*, 5665 (1967).

130. J. A. Marshall and M. T. Pike, *J. Org. Chem.*, *33*, 435 (1968).

131. A. Asselin, M. Mongrain, and P. Deslongchamps, *Can. J. Chem.*, *46*, 2817 (1968).

132. C. H. Heathcock and T. R. Kelly, *Chem. Commun.*, 267 (1968).

133. E. Klein and W. Rojahn, *Tet. Letters*, 279 (1970).

134. T. Nozoe, T. Asao, M. Ando, and K. Takase, *ibid.*, 2821 (1967).

135. M. Ando, T. Asao, and K. Takase, *ibid.*, 4689 (1969).

136. M. Nakazaki, *Chem. and Ind.*, 413 (1962).

137. A. G. Hortmann, *Tet. Letters*, 5785 (1968).

138. C. H. Heathcock and Y. Amano, *Can J. Chem.*, *50*, 340 (1972).

139. (a) H. Minato and T. Nagasaki, *Chem. Commun.*, 377 (1965);

(b) *J. Chem. Soc. (C)*, 1866 (1966).

140. (a) H. Minato and T. Nagasaki, *Chem. Commun.*, 347 (1966);
(b) *J. Chem. Soc. (C)*, 621 (1968).

141. H. Minato and T. Nagasaki, *ibid.*, 377 (1966).

142. C. B. C. Boyce and J. S. Whitehurst, *J. Chem. Soc.*, 2680 (1960).

143. C. H. Heathcock, R. Ratcliffe, and J. Van, *J. Org. Chem.*, *37*, 1796 (1972).

144. S. Swaiminathan, J. P. John, and J. Ramachandron, *Tet. Letters*, 729 (1962).

145. G. R. Clemo, R. D. Haworth, and E. Walton, *J. Chem. Soc.*, 1110 (1930).

146. K. Paranjape, N. L. Phalnikar, B. V. Bhide, and K. S. Nargund, *Rasayanam, 1*, 233 (1943) [*Chem. Abs.*, *38*, 4266 (1944)]; *Nature, 153*, 141 (1944).

147. (a) J. W. Cornforth, R. H. Cornforth, and M. J. S. Dewar, *Nature, 153*, 317 (1944); (b) J. M. O'Gorman, *J. Amer. Chem. Soc.*, *66*, 1041 (1944); (c) G. R. Clemo, W. Cocker and S. Hornsby, *J. Chem. Soc.*, 616 (1946); (d) A. L. Wilds and C. Djerassi, *J. Amer. Chem. Soc.*, *68*, 1715 (1946); (e) R. B. Woodward and T. Singh, *ibid.*, *72*, 494 (1950).

148. Y. Abe, T. Harukawa, H. Ishikawa, T. Miki, M. Sumi and T. Toga, *ibid.*, *75*, 2567 (1953); (b) *78*, 1416 (1956).

149. R. B. Woodward and P. Yates, *Chem. and Ind.*, 1391 (1954); (b) W. Cocker and T. B. H. McMurry, *Tetrahedron, 8*, 181 (1960); (c) Y. Abe, T. Miki, M. Sumi, and T. Toga, *Chem. and Ind.*, 953 (1956).

150. (a) J. D. M. Asher and G. A. Jim, *Proc. Chem. Soc.*, 111 (1962); (b) D. H. R. Barton, T. Miki, J. T. Pinkey, and R. J. Wells, *ibid.*, 112 (1962); (c) M. Nakazaki and H. Arakawa, *ibid.*, 151 (1962).

151. M. Yanagita, S. Inayama, M. Hirakura, and F. Seki, *J. Org. Chem.*, *23*, 690 (1958).

152. Y. Abe, T. Harukawa, H. Ishikawa, T. Miki, M. Sumi, and T. Toga, *J. Amer. Chem. Soc.*, *78*, 1422 (1956).

153. M. Nakazaki and K. Naemura, *Tet. Letters*, 2615 (1966).

154. (a) J. A. Marshall and N. Cohen, *J. Amer. Chem. Soc.*, *87*, 2773 (1965); (b) J. A. Marshall, N. Cohen, and A. R. Hochstetler, *ibid.*, *88*, 3408 (1966).

155. J. A. Marshall and N. Cohen, *J. Org. Chem.*, *30*, 3475 (1965).

156. J. A. Marshall, N. Cohen, and K. R. Arenson, *ibid.*, *30*, 762 (1965).

157. J. A. Marshall and A. R. Hochstetler, *Tet. Letters*, 55 (1966).

158. (a) H. Minato and I. Horibe, *Chem. Commun.*, 531 (1965);
(b) *J. Chem. Soc. (C)*, 1575 (1967).

159. P. D. Ladwa, G. D. Joshi, and S. N. Kulkarni, *Chem. and Ind.*, 1601 (1968).
160. M. D. Soffer and L. A. Burk, *Tet. Letters*, 211 (1970).
161. O. P. Vig, K. L. Matta, G. Singh, and I. Raj, *J. Indian Chem. Soc.*, *43*, 605 (1966).
162. O. P. Vig, A. Lal, T. R. Malhotra, and K. L. Matta, *ibid.*, *45*, 48 (1968).
163. (a) M. V. R. K. Rao, G. S. K. Rao, and S. Dev, *Tet. Letters*, No. 27, 27 (1960); (b) *Tetrahedron*, *22*, 1977 (1966).
164. M. D. Soffer, G. E. Günay, O. Korman, and M. B. Adams, *Tet. Letters*, 389 (1963); (b) M. D. Soffer and G. E. Günay, *ibid.*, 1355 (1965).
165. L. Westfelt, *Acta. Chem. Scand.*, *18*, 572 (1964); *20*, 2852 (1966).
166. E. J. Corey and S. Nozoe, *J. Amer. Chem. Soc.*, *87*, 5728 (1965).
167. J. C. Bardhan and D. N. Mukherji, *J. Chem. Soc.*, 4629 (1956).
168. G. S. K. Rao and S. Dev, *J. Indian Chem. Soc.*, *34*, 255 (1957).
169. (a) R. O. Hellyer and H. H. G. McKern, *Austral. J. Chem.*, *9*, 547 (1956); (b) R. P. Hilderbrand and M. D. Sutherland, *ibid.*, *12*, 678 (1959).
170. G. Stork and S. D. Darling, *J. Amer. Chem. Soc.*, *86*, 1761 (1964).
171. V. Herout and F. Santavy, *Coll. Czech. Chem. Commun.*, *19*, 118 (1954).
172. M. D. Soffer and M. Jevnik, *J. Amer. Chem. Soc.*, *77*, 1003 (1955).
173. (a) A. Caliezi and H. Schinz, *Helv. Chim. Acta*, *32*, 2556 (1949); (b) *33*, 1129 (1950); (c) *35*, 1637 (1952).
174. (a) E. E. van Tamelen, A. Storni, E. J. Hessler, and M. Schwartz, *J. Amer. Chem. Soc.*, *85*, 3295 (1963); (b) E. E. van Tamelen and E. J. Hessler, *Chem. Commun.*, 411 (1966).
175. (a) G. Stork and A. W. Burgstahler, *J. Amer. Chem. Soc.*, *77*, 5068 (1955); (b) P. A. Stadler, A. Eschenmoser, H. Schinz, and G. Stork, *Helv. Chim. Acta*, *40*, 2191 (1957).
176. Y. Kitahara, T. Kato, T. Suzuki, S. Kanno, and M. Tanemura, *Chem. Commun.*, 342 (1969).
177. E. E. van Tamelen and R. M. Coates, *ibid.*, 413 (1966).
178. E. Wenkert and D. P. Strike, *J. Amer. Chem. Soc.*, *86*, 2044 (1964).
179. (a) S. W. Pelletier, R. W. Chappell, and S. Prabhakar, *Tet. Letters*, 3489 (1966); *J. Amer. Chem. Soc.*, *90*, 2889 (1968); (b) S. W. Pelletier and S. Prabhakar, *ibid.*, *90*, 5318 (1968).
180. G. Brieger, *Tet. Letters*, 4429 (1965).

181. E. E. Royals, *J. Amer. Chem. Soc.*, *69*, 841 (1947).

182. (a) J. A. Marshall, W. I. Fanta, and G. L. Bundy, *Tet. Letters*, 4807 (1965); (b) J. A. Marshall, W. I. Fanta, and H. Roebke, *J. Org. Chem.*, *31*, 1016 (1966); (c) J. A. Marshall and G. L. Bundy, *Tet. Letters*, 3359 (1966); (d) J. A. Marshall, G. L. Bundy, and W. I. Fanta, *J. Org. Chem.*, *33*, 3913 (1968).

183. (a) D. W. Theobald, *Tet. Letters*, 969 (1966); (b) T. R. Govindachari, N. Viswanathan, B. R. Pai, P. S. Santhanam, and M. Srinivasin, *Indian J. Chem.*, *6*, 475 (1968).

184. J. W. Cook and C. A. Lawerence, *J. Chem. Soc.*, 817 (1937).

185. E. Wenkert and D. A. Berges, *J. Amer. Chem. Soc.*, *89*, 2507 (1967).

186. W. G. Dauben and G. H. Berezin, *ibid.*, *85*, 468 (1963).

187. R. E. Ireland, D. R. Marshall, and J. W. Tilley, *ibid.*, *92*, 4754 (1970).

188. D. J. Dawson and R. E. Ireland, *Tet. Letters*, 1899 (1968).

189. A. E. Wick, D. Felix, K. Steen, and A. Eschenmoser, *Helv. Chim. Acta*, *47*, 2425 (1964).

190. J. A. Marshall and A. R. Hochstetler, *Chem. Commun.*, 732 (1967).

191. J. A. Marshall, H. Faubl, and T. M. Warne, Jr., *ibid.*, 753 (1967).

192. M. Pesaro, G. Bozzato, and P. Schudel, *ibid.*, 1152 (1968).

193. J. A. Marshall and R. A. Ruden, *Tet. Letters*, 1239 (1970).

194. (a) R. M. Coates and J. M. Shaw, *J. Org. Chem.*, *35*, 2597 (1970); (b) *Tet. Letters*, 5405 (1968).

195. E. Piers and D. R. Smillie, *J. Org. Chem.*, *35*, 3997 (1970).

196. S. Murayama, D. Chan and M. Brown, *Tet. Letters*, 3715 (1968).

197. (a) E. Piers and R. J. Keziere, *ibid.*, 583 (1968); (b) *Can. J. Chem.*, *47*, 137 (1969).

198. M. Ohashi, *Chem. Commun.*, 893 (1969).

199. R. L. Hale and L. H. Zalkow, *ibid.*, 1249 (1968).

200. R. M. Coates and J. M. Shaw, *ibid.*, 47 (1968).

201. H. C. Odom and A. R. Pinder, *ibid.*, 26 (1969).

202. H. C. Odom, A. K. Torrence, and A. R. Pinder, "Synthetic Studies in the Eremophilane Sesquiterpene Group," presented at the Symposium on Synthetics and Substitutes for the Food Industry, American Chemical Society, Division of Agricultural and Food Chemistry, 158th National A.C.S. Meeting, September 8-12, 1969, New York, Abstract 48 (unpublished).

203. R. M. Coates and J. E. Shaw, *Chem. Commun.*, 515 (1968).
204. R. M. Coates and J. E. Shaw, *J. Org. Chem.*, *35*, 2601 (1970).
205. E. Piers, R. W. Britton, and W. de Waal, *Can. J. Chem.*, *47*, 4307 (1969).
206. G. Berger, M. Franck-Neumann, and G. Ourisson, *Tet. Letters*, 3451 (1968).
207. J. Hockmannova and V. Herout, *Coll. Czech. Chem. Commun.*, *29*, 2369 (1964).
208. J. A. Marshall and H. Roebke, *J. Org. Chem.*, *33*, 840 (1968).
209. (a) G. Stork, S. Danishevsky, and M. Ohashi, *J. Amer. Chem. Soc.*, *89*, 5459 (1967); (b) G. Stork and J. E. McMurry, *ibid.*, *89*, 5463 (1967).
210. J. J. Sims and L. H. Selman, *Tet. Letters*, 561 (1969).
211. R. B. Bates, G. Büchi, T. Matsuura, and R. R. Shaffer, *J. Amer. Chem. Soc.*, *82*, 2327 (1960).
212. C. Berger, M. Franck-Neumann, and G. Ourisson, *Tet. Letters*, 3451 (1968).
213. E. Piers, R. W. Birtton, and W. de Waal, *Can. J. Chem.*, *47*, 831 (1969).
214. (a) R. M. Coates and J. E. Shaw, *Chem. Commun.*, 515 (1968); (b) *J. Amer. Chem. Soc.*, *92*, 5657 (1970).
215. E. Piers, R. W. Britton, and W. de Waal, *Tet. Letters*, 1251 (1969).
216. A. Tanaka, H. Uda, and A. Yoshikoshi, *Chem. Commun.*, 308 (1969).
217. W. G. Dauben and A. C. Ashcraft, *J. Amer. Chem. Soc.*, *85*, 3673 (1963).
218. G. Büchi and J. D. White, *ibid.*, *86*, 2884 (1964).
219. G. Stork and J. Ficini, *ibid.*, *83*, 4678 (1961).
220. (a) J. A. Marshall and J. J. Partridge, *ibid.*, *90*, 1090 (1968); (b) *Tetrahedron*, *25*, 2159 (1969).
221. C. H. Heathcock and R. Ratcliffe, *J. Amer. Chem. Soc.*, *93*, 1746 (1971).
222. M. Kato, H. Kosugi, and A. Yoshikoshi, *Chem. Commun.*, 185 (1970).
223. E. Piers and K. F. Cheng, *ibid.*, 562 (1969).
224. M. Kato, H. Kosugi, and A. Yoshikoshi, *ibid.*, 934 (1970).
225. J. A. Marshall and J. J. Partridge, *Tet. Letters*, 2545 (1966).
226. C. H. Heathcock and R. Ratcliffe, *Chem. Commun.*, 994 (1968).
227. J. B. Hendrickson, C. Ganter, D. Dorman, and H. Link, *Tet. Letters*, 2235 (1968).
228. M. Suchy, Y. Herout, and F. Sorm, *Coll. Czech. Chem. Commun.*, *29*, 1829 (1964).

229. D. H. R. Barton, J. T. Pinhey, and R. J. Wells, *J. Chem. Soc.*, 2518 (1964).

230. E. H. White, S. Eguchi, and J. N. Marx, *Tetrahedron, 25,* 2099 (1969).

231. J. N. Marx and E. H. White, *ibid., 25,* 2117 (1969).

232. D. H. R. Barton, *Helv. Chim. Acta, 42,* 2604 (1959).

233. G. Büchi, W. Hofheinz, and J. V. Paukstelis, *J. Amer. Chem. Soc., 88,* 4113 (1966); *91,* 6473 (1969).

234. G. Büchi and H. J. E. Loewenthal, *Proc. Chem. Soc.,* 280 (1962).

235. S. A. Narang and P. C. Dutta, *J. Chem. Soc.,* 1119 (1964).

236. H. Hikino, N. Suzuki, and T. Takemoto, *Chem. Pharm. Bull., 14,* 1441 (1966).

237. P. de Mayo and H. Takeshita, *Can. J. Chem., 41,* 440 (1963).

238. (a) B. D. Challand, G. Kornis, G. L. Lange, and P. de Mayo, *Chem. Commun.,* 704 (1967); (b) B. D. Challand, H. Hikino, G. Kornis, G. Lange, and P. de Mayo, *J. Org. Chem., 34,* 794 (1969).

239. C. Enzell, *Tet. Letters,* 185 (1962).

240. G. Stork and M. Gregson, *J. Amer. Chem. Soc., 91,* 2373 (1969).

241. E. J. Corey and K. Achiwa, *Tet. Letters,* 1837 (1969).

242. K. Mori and M. Matsui, *ibid.,* 2729 (1969).

243. (a) R. M. Coates and R. M. Freidinger, *Chem. Commun.,* 871 (1969); (b) *Tetrahedron, 26,* 3487 (1970).

244. Y. Nakatani and T. Yamanishi, *Agr. Biol. Chem., 33,* 1805 (1969).

245. E. J. Corey and K. Achiwa, *Tet. Letters,* 3257 (1969).

246. (a) J. J. Plattner, U. T. Bhalerao, and H. Rapoport, *J. Amer. Chem. Soc., 91,* 4933 (1969); (b) U. T. Bhalerao, J. J. Plattner, and H. Rapoport, *ibid., 92,* 3429 (1970).

247. E. J. Corey, K. Achiwa, and J. A. Katzenellenbogen, *ibid., 91,* 4318 (1969).

248. P. A. Grieco, *ibid., 91,* 5660 (1969).

249. (a) K. Mori and M. Matsui, *Tet. Letters,* 4435 (1969); (b) *Tetrahedron, 26,* 2801 (1970).

250. E. J. Corey and K. Achiwa, *ibid.,* 2245 (1970).

251. E. J. Corey and H. A. Kirst, *ibid.,* 5041 (1968).

252. C. J. Muller and W. G. Jennings, *J. Agr. Food Chem., 15,* 762 (1967).

253. V. Herout, V. Ruzicka, M. Vrany, and F. Sorm, *Coll. Czech. Chem. Commun., 15,* 373 (1950).

254. K. S. Kulkarni, S. K. Paknikar, and S. C. Bhattacharyya, *Tetrahedron, 22,* 1917 (1966).

255. (a) T. W. Gibson and W. F. Erman, *Tet. Letters,* 905 (1967); (b) *J. Amer. Chem. Soc., 91,* 4771 (1969).

256. S. Ito, K. Endo, T. Yoshida, M. Yatagai, and M. Kodama,

Chem. Commun., 186 (1967).

257. A. Tanaka, H. Uda, and A. Yoshikoshi, *ibid.*, 188 (1967).

258. A. Tanaka, H. Uda, and A. Yoshikoshi, *ibid.*, 56 (1968).

259. (a) D. Stauffacher and H. Schinz, *Helv. Chim. Acta, 37,* 1227 (1954); (b) U. Steiner and B. Willhalm, *ibid., 35,* 1752 (1952).

260. S. Kanno, T. Kato, and Y. Kitahara, *Chem. Commun.*, 1257 (1967).

261. T. Nozoe and H. Takeshita, *Tet. Letters,* No. 23, 14 (1960).

262. (a) W. Parker, R. Ramage, and R. A. Raphael, *Proc. Chem. Soc.,* 74 (1961); (b) *J. Chem. Soc.,* 1558 (1962).

263. P. T. Lansbury and F. R. Hilfiker, *Chem. Commun.,* 619 (1969).

264. K. Yamada, H. Yazawa, D. Uemura, M. Toda, and Y. Hirata, *Tetrahedron, 25,* 3509 (1969).

265. (a) H. Minato and I. Horibe, *Chem. Commun.,* 358 (1967); (b) *J. Chem. Soc. (C),* 2131 (1968).

266. (a) E. J. Corey and S. Nozoe, *J. Amer. Chem. Soc., 85,* 3527 (1963); (b) *87,* 5733 (1965).

267. W. S. Johnson, J. J. Korst, R. A. Clement, and J. Dutta, *ibid., 82,* 614 (1960).

268. S. Julia, *Bull. Soc. Chim. France, 21,* 780 (1954).

269. P. J. Kropp, *J. Amer. Chem. Soc., 87,* 3914 (1965).

270. D. H. R. Barton, P. de Mayo, and M. Shafig, *J. Chem. Soc.,* 140 (1958).

271. (a) J. A. Marshall and P. C. Johnson, *Chem. Commun.,* 391 (1968); (b) *J. Org. Chem., 35,* 192 (1970).

272. J. A. Marshall and S. F. Brady, *Tet. Letters,* 1387 (1969).

273. S. Masamune, *J. Amer. Chem. Soc., 83,* 1009 (1961).

274. M. Mongrain, J. Lafontaine, A. Belanger, and P. Deslong-champs, *Can. J. Chem., 48,* 3273 (1970).

275. (a) E. J. Corey, R. B. Mitra, and H. Uda, *J. Amer. Chem. Soc., 85,* 362 (1963); (b) *86,* 485 (1964).

276. J. M. Greenwood, J. K. Sutherland, and A. Torre, *Chem. Commun.,* 410 (1965).

277. J. L. Gras, R. Maurin, and M. Bertrand, *Tet. Letters,* 3533 (1969).

278. E. J. Corey, S. W. Chow, and R. A. Scherrer, *J. Amer. Chem. Soc., 79,* 5773 (1957).

279. E. J. Corey, R. Hartmann, and P. A. Vatakencherry, *ibid., 84,* 2611 (1962).

280. G. Brieger, *Tet. Letters,* 1949 (1963).

281. (a) S. C. Bhattacharyya, *Sci. and Cult., 13,* 208 (1947); (b) S. Y. Kamat, K. K. Chakravart, and S. C. Bhattacharyya, *Tetrahedron, 23,* 4487 (1967).

282. G. Brieger, *Tet. Letters,* 2123 (1963).

283. J. Colonge, G. Descotes, Y. Bahurel, and A. Memet, *Bull. Soc. Chim. France,* 374 (1966).

284. See Ref. 285, footnote 2.

285. R. G. Lewis, D. H. Gustafson, and W. F. Erman, *Tet. Letters,* 401 (1967).

286. E. J. Corey and H. Yamamoto, *J. Amer. Chem. Soc., 92,* 226, 3523 (1970).

287. H. C. Kretschmer and W. F. Erman, *Tet. Letters,* 41 (1970).

288. G. Büchi, R. E. Erickson, and N. Wakabayashi, *J. Amer. Chem. Soc., 83,* 927 (1961).

289. (a) G. Büchi and W. D. MacLeod, Jr., *ibid., 84,* 3205 (1962); (b) G. Büchi, W. D. MacLeod, Jr., and J. Padilla O., *ibid., 86,* 4438 (1964).

290. M. Dobler, J. D. Dunitz, B. Gubler, H. P. Weber, G. Büchi, and J. Padilla O., *Proc. Chem. Soc.,* 383 (1963).

291. S. Danishevsky and D. Dumas, *Chem. Commun.,* 1287 (1968).

292. E. Piers, R. W. Britton and W. de Waal, *ibid.,* 1069 (1969).

293. (a) G. Stork and F. H. Clarke, Jr., *J. Amer. Chem. Soc., 77,* 1072 (1955); (b) *83,* 3114 (1961).

294. E. J. Corey, N. N. Girotra, and C. T. Mathew, *ibid., 91,* 1557 (1969).

295. T. G. Crandall and R. G. Lawton, *ibid., 91,* 2127 (1969).

296. F. Kido, H. Uda, and A. Yoshikoshi, *Chem. Commun.,* 1335 (1969).

297. (a) E. J. Corey, M. Ohno, P. A. Vatakencherry, and R. B. Mitra, *J. Amer. Chem. Soc., 83,* 1251 (1961); (b) E. J. Corey, M. Ohno, R. B. Mitra, and P. A. Vatakencherry, *ibid., 86,* 478 (1964).

298. (a) C. H. Heathcock, *ibid., 88,* 4110 (1966); (b) C. H. Heathcock, R. A. Badger, and J. W. Patterson, Jr., *ibid., 89,* 4133 (1967).

299. (a) J. E. McMurry, *ibid., 90,* 6821 (1968); (b) *Tet. Letters,* 55 (1969).

300. B. W. Roberts, M. S. Poonian, and S. C. Welch, *ibid., 91,* 3400 (1969).

301. D. H. R. Barton and N. H. Werstiuk, *J. Chem. Soc., C,* 148 (1968).

302. (a) J. D. White and D. N. Gupta, *J. Amer. Chem. Soc., 88,* 5364 (1966); (b) *90,* 6171 (1968).

303. M. Brown, *J. Org. Chem., 33,* 162 (1968).

304. P. E. Eaton, *Accs. Chem. Res., 1,* 50 (1968).

305. T. Matsumoto, H. Shirahama, A. Ichihara, H. Shin, S. Kagawa, F. Sakan, S. Matsumoto, and S. Nishida, *J. Amer. Chem. Soc., 90,* 3280 (1968).

306. T. Matsumoto, H. Shirahama, A. Ichihara, H. Shin, S. Kagawa, S. Nishida, *Tet. Letters,* 1925 (1968).

307.   T. Matsumoto, H. Shirihama, A. Ichihara, H. Shin, S.
       Kagawa, and F. Sakan, *ibid.*, 1171 (1970).
308.   S. Kagawa, S. Matsumoto, S. Nishida, S. Yu, J. Morita,
       A. Ichihara, H. Shirahama, and T. Matsumoto, *ibid.*, 3913
       (1969).
309.   (a) E. J. Corey and S. Nozoe, *J. Amer. Chem. Soc.*, *86*,
       1652 (1964); (b) *87*, 5733 (1965).
310.   R. R. Sobti and S. Dev, *Tet. Letters*, 2893 (1967); (b)
       Tetrahedron, *26*, 649 (1970).
311.   (a) R. D. H. Murry, W. Parker, R. A. Raphael, and D. B.
       Jhaveri, *Tetrahedron*, *18*, 55 (1962); (b) P. Doyle, I. R.
       MacLean, W. Parker, and R. A. Raphael, *Proc. Chem. Soc.*,
       239 (1963).

THE SYNTHESIS OF TRITERPENES

J. W. ApSimon and J. W. Hooper*

*Department of Chemistry, Carleton University
Ottawa, Ontario, Canada*

1. Introduction 559
2. Squalene and Related Compounds 560
   A. Squalenoid Compounds by the Coupling of Two
      Identical Units 560
   B. Squalenoid Compounds by the Coupling of Two
      Different Units 562
   C. Stepwise Syntheses of Squalene 562
   D. Malabaricanediol 45 566
   E. Ambrein 56 569
3. Triterpenes with Steroidal Ring Systems 569
   A. Lanosterol and Congeners (Scheme 2) 569
   B. Cycloartenol 60 572
4. Nonsteroidal Polycyclic Triterpenes 574
   A. The Onocerins and Related Compounds 574
   B. The Amyrin Group 594
   C. Stereorational Syntheses 610
5. Presqualene Alcohol 633
Acknowledgments 635

1. INTRODUCTION

The triterpenes constitute a large terpenoid class[1] in which
relatively little synthetic work had been reported until recently. This is undoubtedly a reflection of their complexity
and the stereochemical problems inherent in the elaboration of

*Present address: Bristol Laboratories of Canada, Candiac,
P.Q., Canada

such molecules. This chapter is divided into three sections:
squalene 1 and related compounds; triterpenes with steroidal
ring systems (e.g., lanosterol 2); and nonsteroidal polycyclic
triterpenes (e.g., α-onocerin 3, β-amyrin 4, hydroxyhopanone
5, and alnusenone 6).

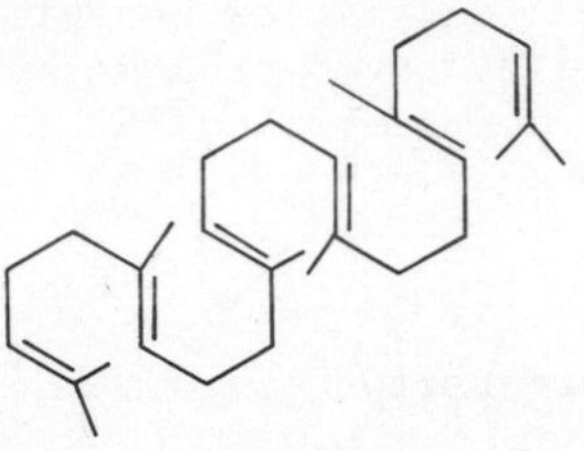

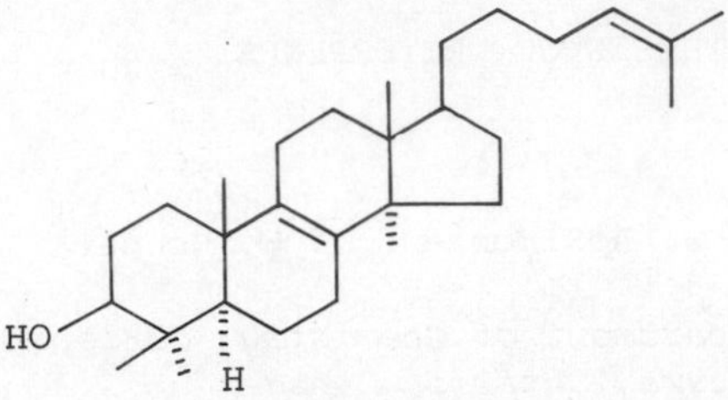

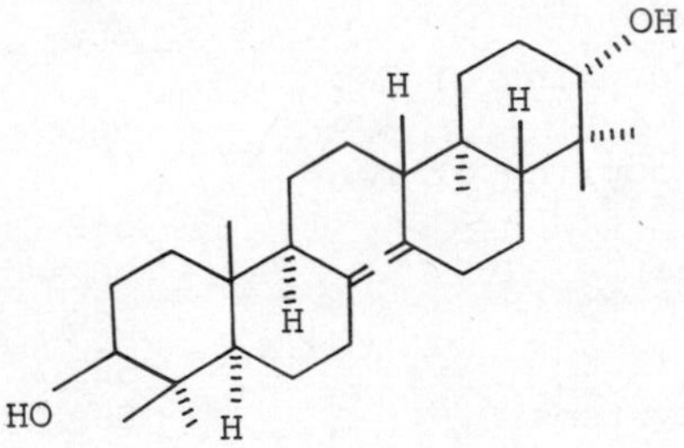

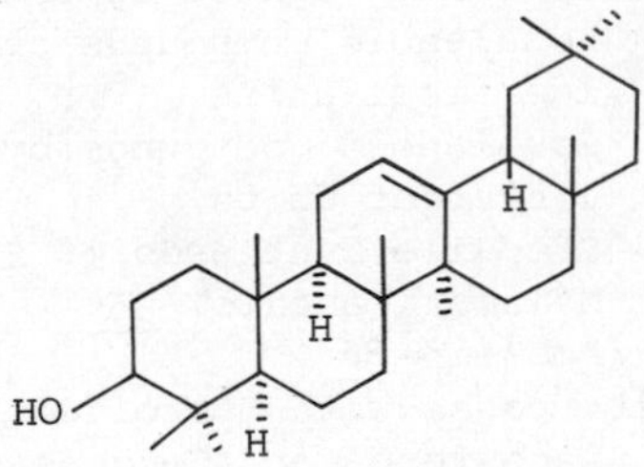

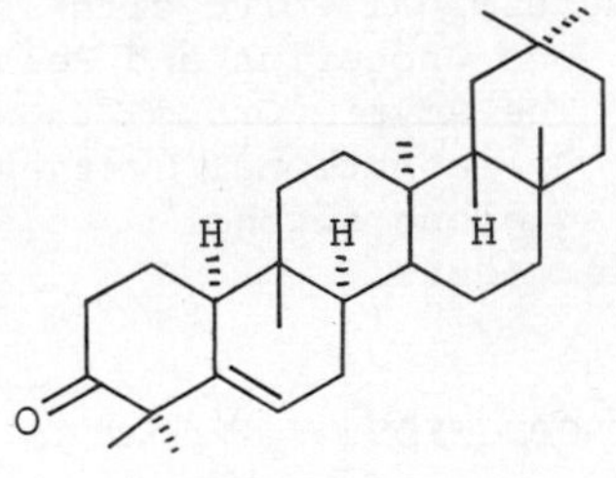

2.   SQUALENE AND RELATED COMPOUNDS

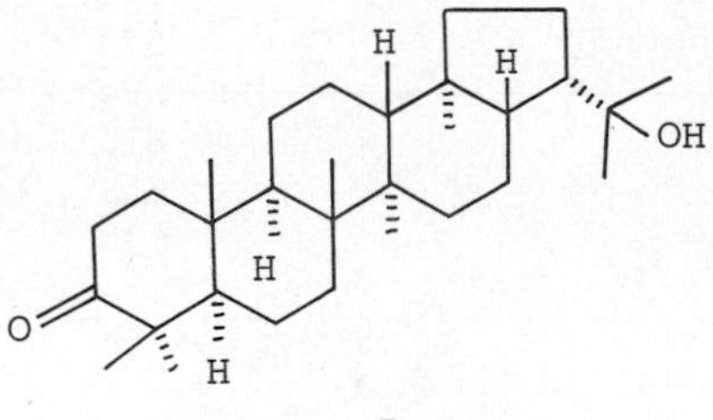

A.   Squalenoid Compounds by the Coupling of Two Identical Units

Squalene 1 occupies a key position in the biosynthetic pathway
leading to triterpenoids and steroids.[2] Early synthetic work

centered on the coupling of two molecules of farnesyl bromide
$7^{3,4}$ with lithium to give a low yield of all *trans*-squalene 1
as the thiourea adduct.   Recently,[5] a more efficient method
has been developed and the analog 8 was coupled using nickel
tetracarbonyl to yield 10,15-bis-norsqualene in high yield and
with 90% stereoselectivity.   Another generally applicable
coupling reaction involving the reaction of allylic or benzylic
alcohols (e.g., 9 to give 10) with a reagent formed from methyl
lithium and titanium trichloride has also been reported.[6]
This could presumably be applied to squalene synthesis.   No

7 R=CH$_3$
8 R=H   9   10

comment has appeared as to the mechanism but the reaction can
be used for "cross coupling" so that alcohols 11 and 12
yielded ketal 13, which was subsequently transformed into 15-
norsqualene.   10-norsqualene and 10'-norsqualene were also
prepared from similar precursors.   Various other modified
squalenes, including 20,21-dehydrosqualene[7] and 1-norsqualene
(*cis* and *trans*)[8] have been obtained by degradation of squalene
and resynthesis.   The various synthetic modified squalenes
were used in highly significant triterpene and steroid biosynthesis investigations.[9]

11   12   13

B. Squalenoid Compounds by the Coupling of Two Different Units

The use of the Wittig reagent 14 has been demonstrated by four groups[10] who condensed it with geranyl acetone 15 and obtained a mixture of isomers from which squalene was isolated in low

$Ph_3P=CHCH_2CH_2CH=PPh_3$

14   15

yield. Recent refinements in the stereochemical control of the Wittig reaction[11] would seem to make this route worthy of reexamination.

The coupling of two different C-15 units has been relatively unexplored until recently when Biellman[12] reported a method based on the alkylation of carbanions of the type 16 stabilized by a double bond and an adjacent sulphur atom. These α-carbanions appear to suffer very little isomerization of the double bond(s) during reaction. Thus the carbanion 17, derived from farnesyl phenyl thioether, was reacted with farnesyl bromide to yield the coupled product 18 subsequently desulphurized by lithium in ethylamine to squalene in 60% overall yield. The method is general for a wide variety of substrates and would appear to be the most useful to date.

Another route[13] utilized the sigmatropic rearrangement of allylic sulphonoum ylids, as in 19 → 20, a method suggested as a possible biosynthetic route to squalene. Thus, farnesyl nerolidyl sulphide was reacted with benzyne to give the ylid 21 which rearranged immediately to 12-phenylthiosqualene 18. No yield was reported for this sequence. The coupling of phosphorous ylids and allylic halides and related reactions[14] are important in the synthesis of 1,5-dienes. Thus, the ylid obtained from the tributylphosphonium bromide 22 was reacted with farnesyl bromide 7 to yield the phosphonium salt 23 readily converted to squalene in 65% overall yield.[14]

C. Stepwise Syntheses of Squalene

The stepwise synthesis of squalene and similar acyclic terpenoids requires a highly stereoselective pathway to *trans*-trisubstituted double bonds, an area which has recently enjoyed a resurgence of attention with the discovery of terpenoid insect hormones.[15]

The first highly stereoselective route to squalene was

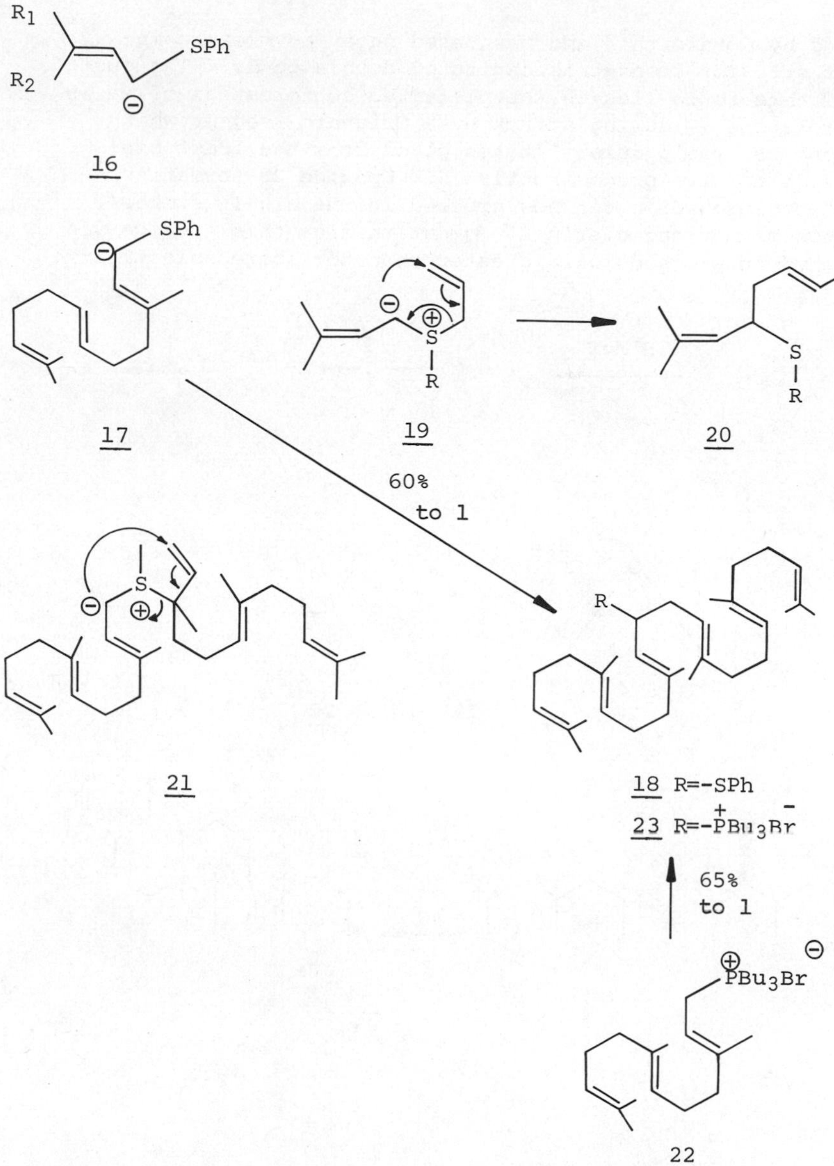

16

17

19 → 20

60%
to 1

21

18 R=-SPh
23 R=-PBu₃Br

65%
to 1

22

reported by Cornforth[16] and was based on an asymmetric induction step for *trans*-trisubstituted double bonds. The success of this route lies in the preferred conformation of an
α-chloroketone 24 during attack by a Grignard reagent which,
according to Cram's rule,[17] takes place from the least hindered side to give predominantly 25. Epoxide 26 formation
from the chlorohydrin can now proceed in one highly stereoselective manner and olefin 27 formation from this epoxide via
an iodohydrin proceeds with greater than 80% stereoselectivity.

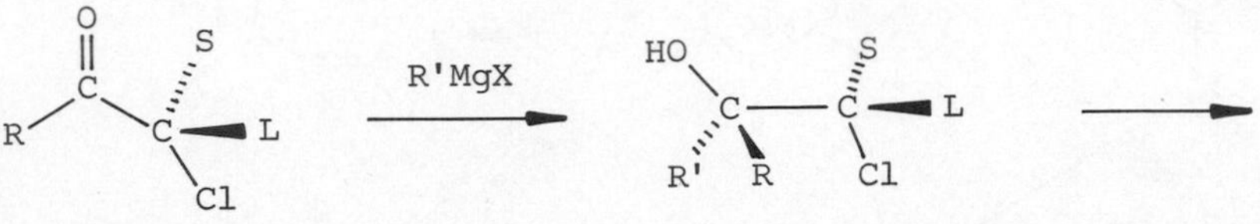

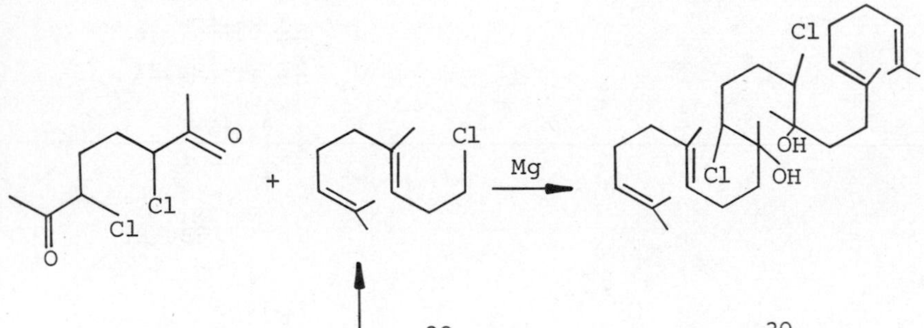

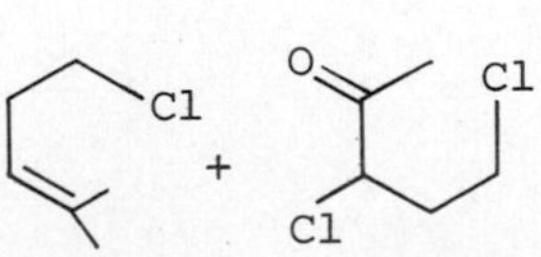

Using this principle the chloroketone 28 was reacted with the Grignard reagent from 29 to yield chlorohydrin 30, of correct geometry for final conversion into squalene. Chloride 29 was prepared in a similar way from 31 and 32. The overall yield of squalene was 18-20% and a stereoselectivity of >70% per double bond was achieved. More recently, Brady, Ilton, and Johnson[18] reported even greater stereoselectivity by working at lower temperatures.

An elegant stepwise total synthesis of squalene has been accomplished[19] with about 98% stereoselectivity for each double bond formed. The basis of this method is a version of the Claisen rearrangement in which a vinyl ether of the type 33 is formed in situ and thermally rearranged to 34, thus providing a *trans*-trisubstituted double bond with pendant functionality available for further modification. Similar highly stereoselective Claisen type processes have recently been reported[20] and the success of these methods probably depends on the nonbonded interactions developing in a chairlike six-membered transition state which favors 35 → 36, leading to a *trans*-trisubstituted double bond, over 37 → 38 which would lead to a *cis*-double bond.

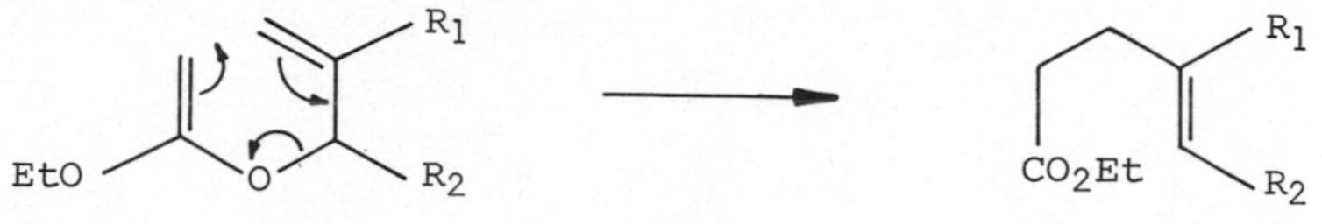

33   34

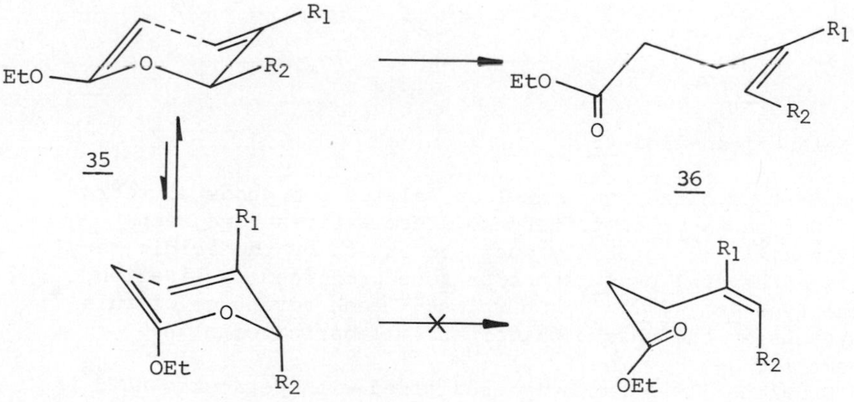

35

36

37   38

The synthesis of squalene itself[19] started with succinaldehyde 39 which, on reaction with 2-propenyllithium, gave
the diene diol 40 as a mixture of *dl* and *meso* forms.   On
treatment with ethyl orthoformate and propionic acid followed
by pyrolytic rearrangement, the diene ester 41 was formed with
a purity of 97%.   Transformation into the corresponding dialdehyde and repetition of the above sequence of reactions
provided the tetraene diol 42 and, in turn, the tetraene diester 43 (R=CO$_2$Et) containing about 9% of isomers other than
all *trans*.   Crystallization of the corresponding alcohol 43
(R=CH$_2$OH) allowed removal of unwanted isomers.   Reaction of
43 (R=CHO) with isopropylidenetriphenylphosphorane afforded
squalene 1 with 96% purity with respect to double bond isomers.   The use of 3-methoxyisoprene 44[19,20a] in a similar
Claisen rearrangement is apparently equally efficient.

39   40   41

42   43   44

D.   Malabaricanediol 45

Attempts to cyclize squalene 1 or related compounds to tetracyclic or pentacyclic triterpenoid derivatives have been
fruitless,[9,21,22] probably due[23] to attack by the acidic reagents at several centers and/or to a preferred cyclization
of the type 46 → 47.[9,24]   Sharpless[23] took advantage of this
to synthesize the unique triterpene malabaricanediol 45
(Scheme 1).
     Squalene 1 was randomly epoxidized with peracetic acid in
dichloromethane to give a mixture which, on controlled hydrolysis, afforded a mixture of 1, squalene-2,3-diol 48, and
the isomeric epoxides 49 and 50.   The internal epoxides 49

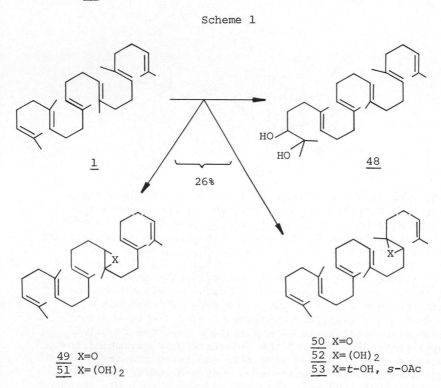

46   47

Scheme 1

1

48

26%

49 X=O
51 X=(OH)$_2$

50 X=O
52 X=(OH)$_2$
53 X=t-OH, s-OAc

41% from 50

55

54

45

and 50 were separated from diol 48 by formation of their
thiourea clathrates, and from squalene by chromatography.
Hydrolysis of the mixture of 49 and 50 gave the corresponding
erythro-diols 51 and 52 which were separated by chromatography
on silver nitrate impregnated silica gel. The mono-acetate
53 of diol 52 reacted with NBS in t-butanol to give (mainly)
the cyclic bromoether 54, thus protecting the oxygen functions
in this region. Terminal epoxide formation via the bromohydrin (NBS/H_2O and then K_2CO_3) followed by treatment with
zinc dust and a catalytic quantity of acetic acid and subsequent basic hydrolysis of the acetate gave the required
epoxy-erythro-diol 55. If the reductive elimination of the
bromoether were performed with zinc and large amounts of
acetic acid then the epoxide group was reduced to the olefin
(cis and trans), and this reaction was shown to apply to other
aliphatic epoxides.

Lewis acid catalyzed cyclization of epoxydiol 55 did not
afford detectable amounts of malabaricanediol 45, which was
unstable under these conditions. It was reasoned that the use
of an acid strong enough to protonate epoxides but not olefins
was needed. Picric acid was used since it is fairly acidic
and also hindered, and therefore less likely than other pro-
tonic acids to act as a nucleophile. This gave a 16% yield of
a mixture of four isomeric diols. The corresponding trimethyl-
silyl ethers were separated by preparative g.l.c. and dl-
malabaricanediol 45 was obtained after hydrolysis.

E.   Ambrein 56

Ambrein is obtained from ambergris and has not as yet been synthesized, although its dehydration product ambertriene 57 has been obtained[25] from the degradation product ambreinolide 58 and dihydro-γ-ionone 59, both of which have been independently synthesized.[26]

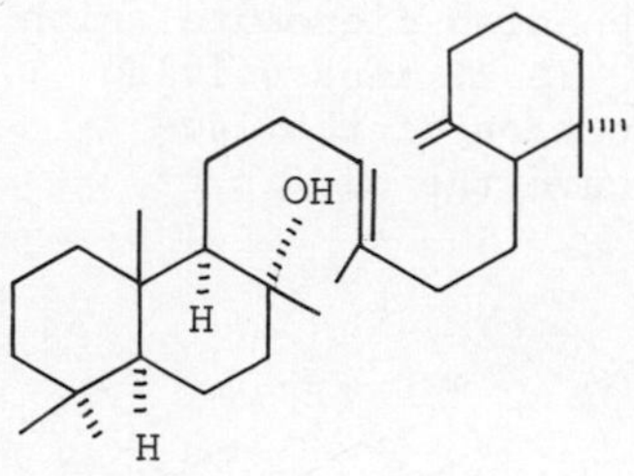

56 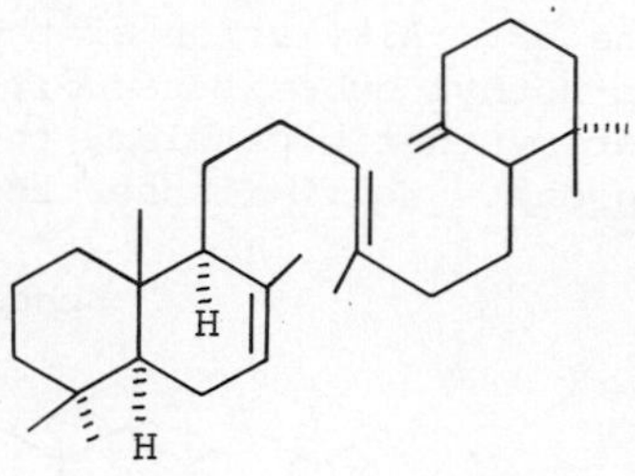   57

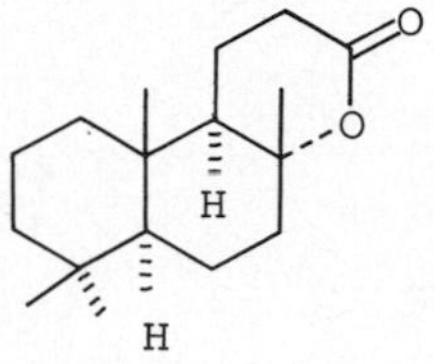

58 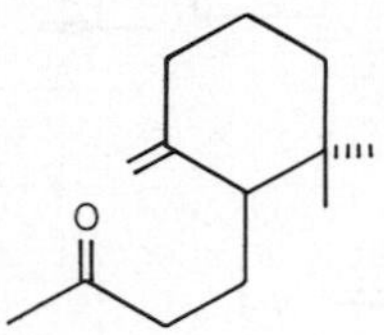   59

3.   TRITERPENES WITH STEROIDAL RING SYSTEMS

Compounds belonging to this group include lanosterol 2 and the variation on its methyl substitution pattern, cycloartenol 60.

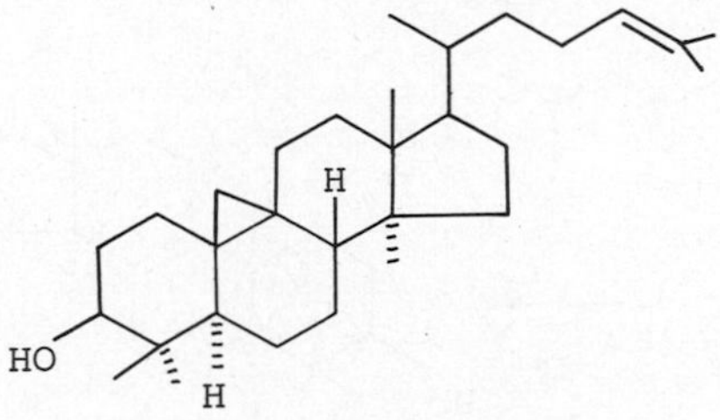

60

Most of the syntheses depend on the conversion of cholesterol 61 (already totally synthesized[27]) to lanosterol 2, which has been used for the (formal total) synthesis of several compounds.

A.   Lanosterol and Congeners (Scheme 2)

Cholesterol 61 was transformed by unexceptional methods[28] to the enone 62. Alkylation of the corresponding dienolate anion gave the methyl substituted β,γ-enone 63 (63%) thus solving the major synthetic problem, the introduction of the 14α-methyl group. Wolff-Kishner reduction gave the olefin 64

Scheme 2

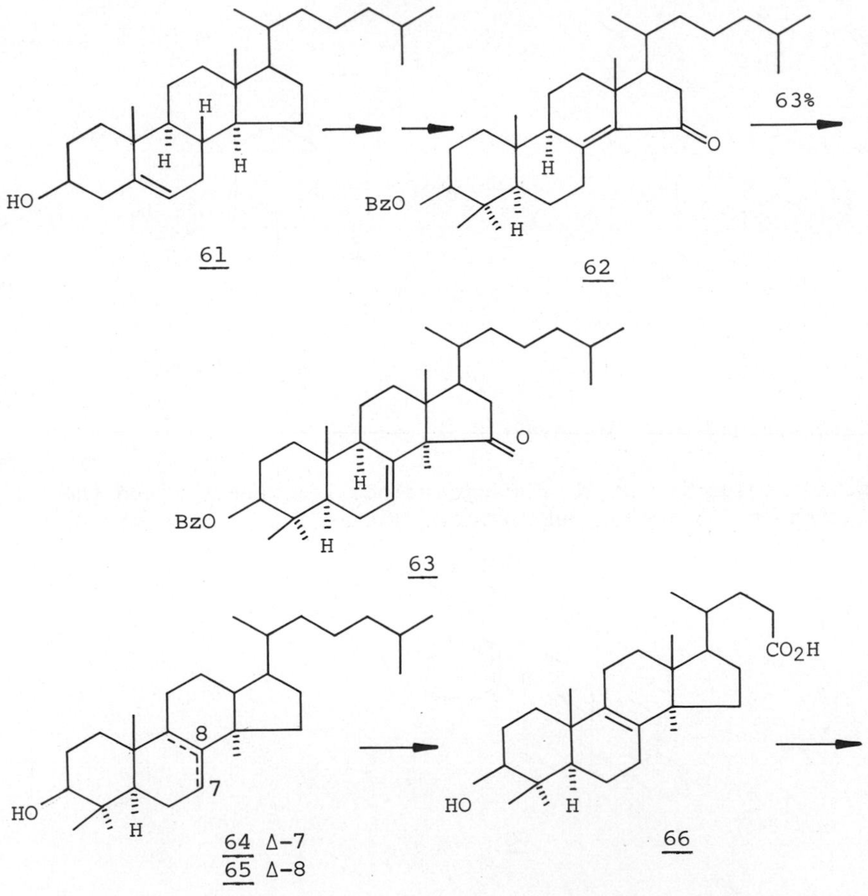

61

62

63%

63

64 Δ-7
65 Δ-8

66

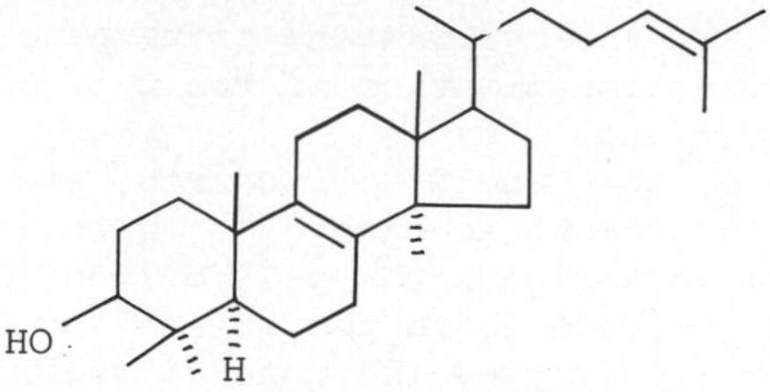

2

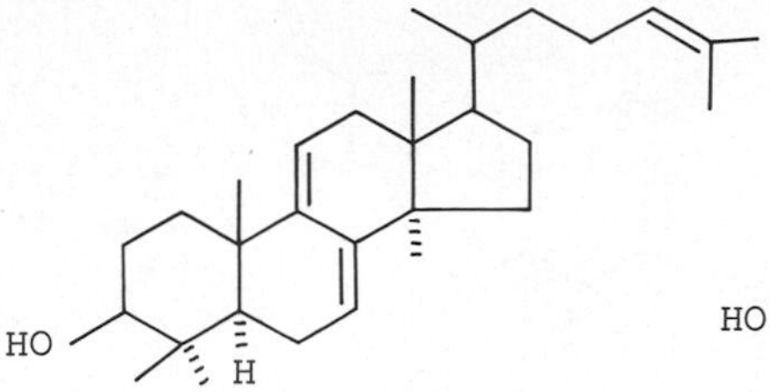

67
68 Side Chain Saturated

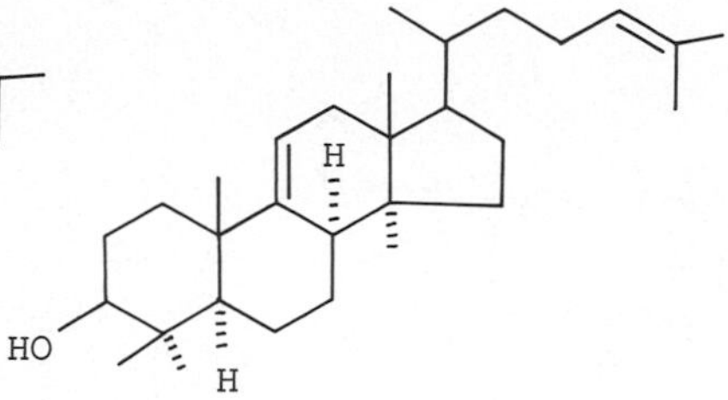

69
70 Side Chain Saturated

which is in acid catalyzed equilibrium with lanostenol 65. There remained the problem of introducing the double bond into the side chain and this was achieved by degradation to 66 and resynthesis to give lanosterol 2. Simple transformations of lanosterol have led to total syntheses of agnosterol 67,[29] dihydroagnosterol 68,[30] parkeol 69,[31] and dihydroparkeol 70.[32]

Van Tamelen has recently announced[87] total syntheses of 24,25-dihydrolanosterol, 24,25-dihydro-$\Delta^{13(17)}$-protosterol, isoeuphenol, (-)-isotirucallol and parkeol via the intermediacy of the nonenzymic cyclization of the epoxides 70a or their C-3 epimers.

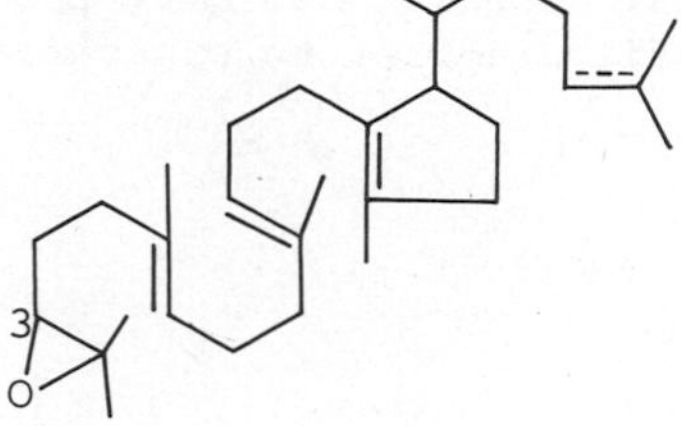

70a

B.   Cycloartenol <u>60</u>

The published approaches to the cycloartane triterpenes all
depend upon photolytic functionalization of the C-19 methyl
group of lanostane derivatives.

Thus, cycloartane <u>71</u>, the parent hydrocarbon, was obtained from 11β-hydroxylanostanyl acetate <u>72</u>. Photolysis of
the nitrite ester of the latter gave the C-19 oximino compound
<u>73</u> which was dehydrated and reduced to the amine <u>74</u>. Deamina-
tion of this gave 3β,11α-dihydroxycycloartane <u>75</u> readily converted to <u>71</u>.[33]

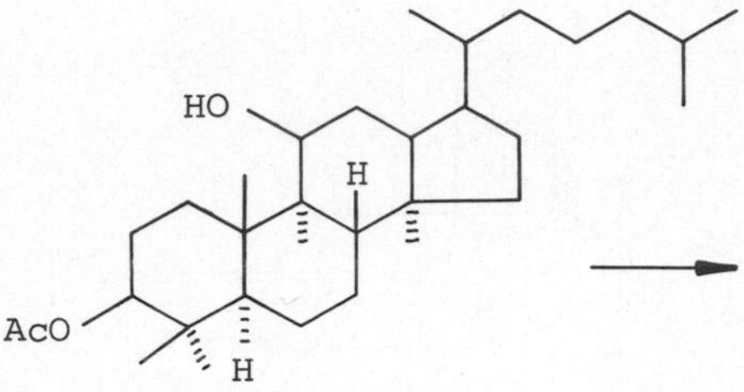

<u>72</u> 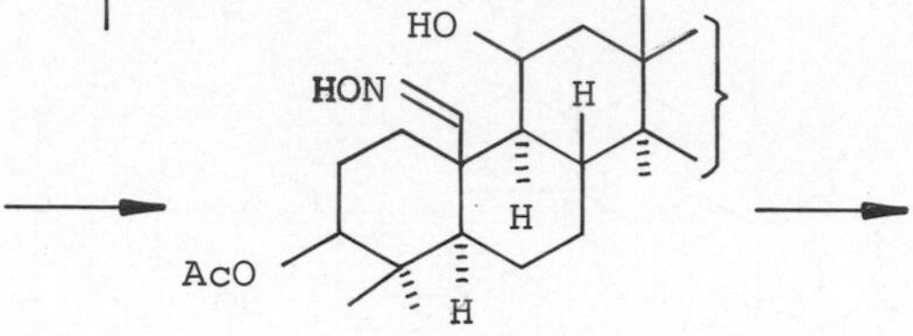   <u>73</u>

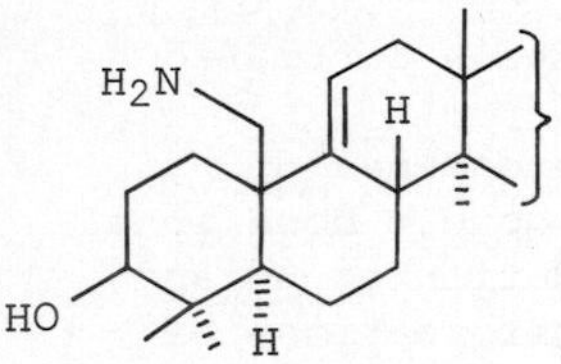

<u>74</u>

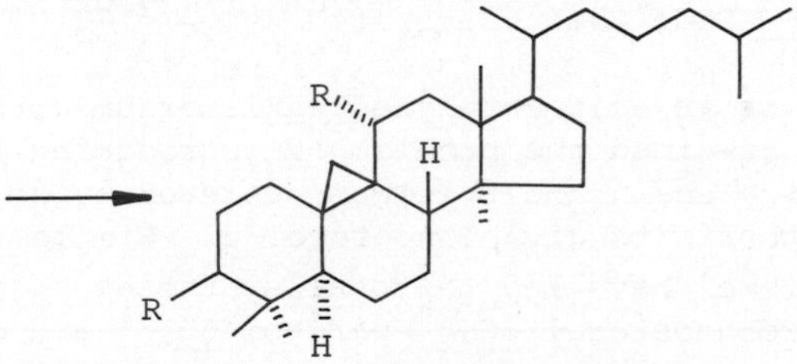

<u>71</u> R=H
<u>75</u> R=OH

Shaffner et al.[34] photolyzed 11-oxolanostenol <u>76</u> and
obtained the diol <u>77</u> which was oxidized with lead tetraacetate to the hemiacetal <u>78</u>. This, on treatment with methanesulphonyl chloride under basic conditions, gave the cyclopro-
panoketone <u>79</u> which was readily reduced to cycloartanol <u>80</u>.

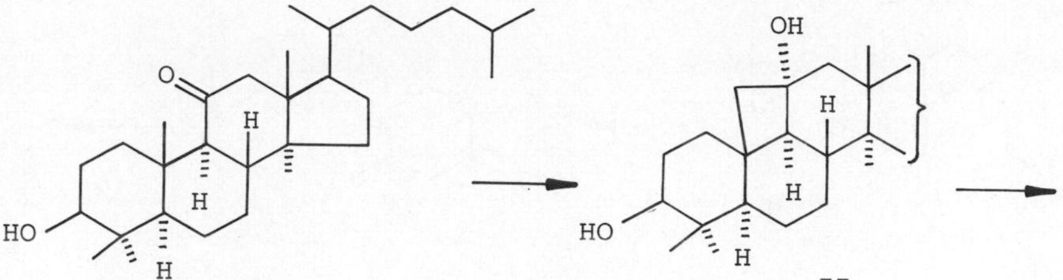

76   77

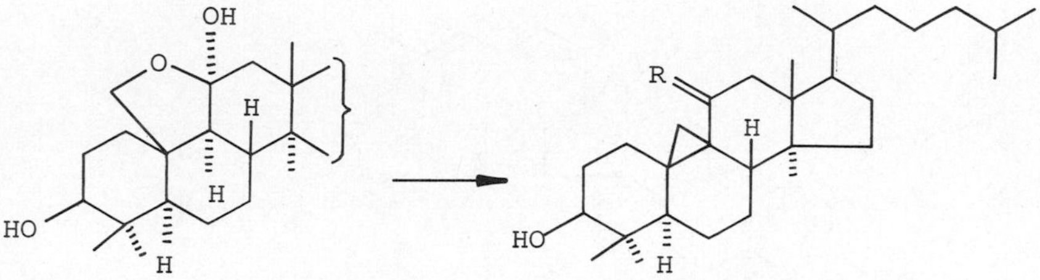

78

79 R=O
80 R=H$_2$

Barton et al.[35] have synthesized cycloartenol 60 itself
(Scheme 3). Thus, reduction of the 11-keto compound 81, available from lanosterol 2 by a simple series of reactions, gave
the 11β-hydroxy compound 82 Photolysis of the corresponding
nitrite ester 83 in the presence of iodine gave the C-19 iodo
compound 84 and thence the iodoketone 85. Treatment of this
with base then gave 11-oxocycloartenyl benzoate 86 which suffered loss of the C-11 oxygen function on reduction with
lithium aluminum hydride to give cycloartenol 60, isolated as
its acetate 87.

Scheme 3

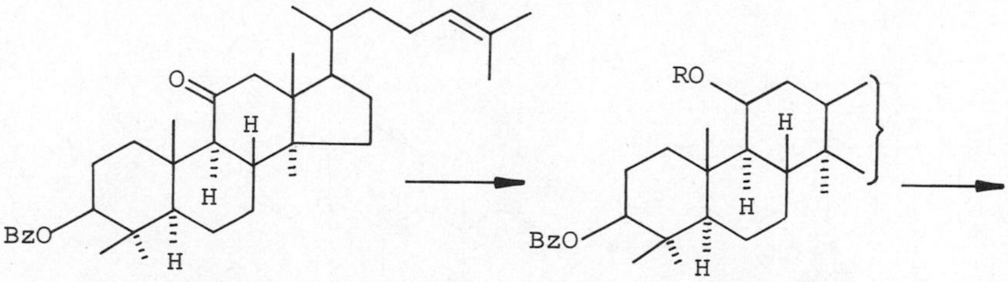

81

82 R=H
83 R=ONO

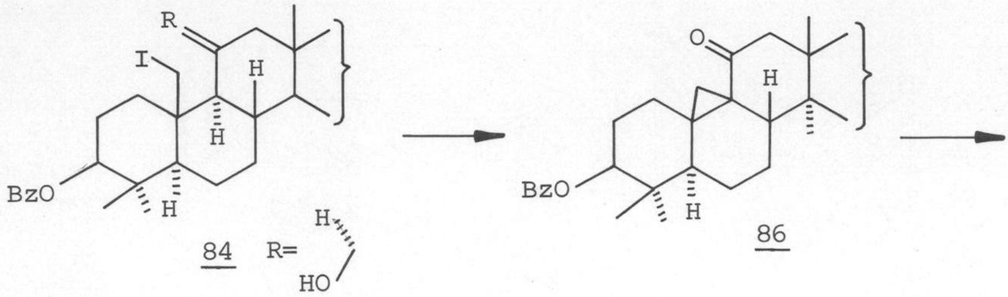

84 R=

85 R=O

86

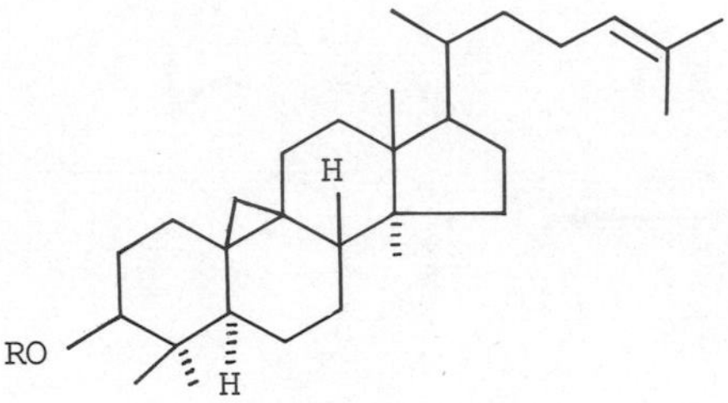

60 R=H
87 R=Ac

4. NONSTEROIDAL POLYCYCLIC TRITERPENES

A. The Onocerins and Related Compounds

α-Onocerin was shown by Barton and Overton[36] to have the symmetrical structure 3 and to undergo acid catalyzed isomerization to β-onocerin 88 and subsequent cyclization to γ-onocerin
89. These reactions form the basis for several total syntheses in which α-onocerin, whose total synthesis is described
below, was used as a natural relay.

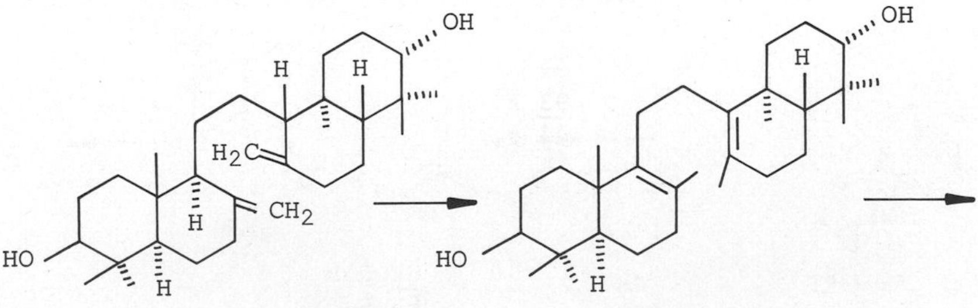

3

88

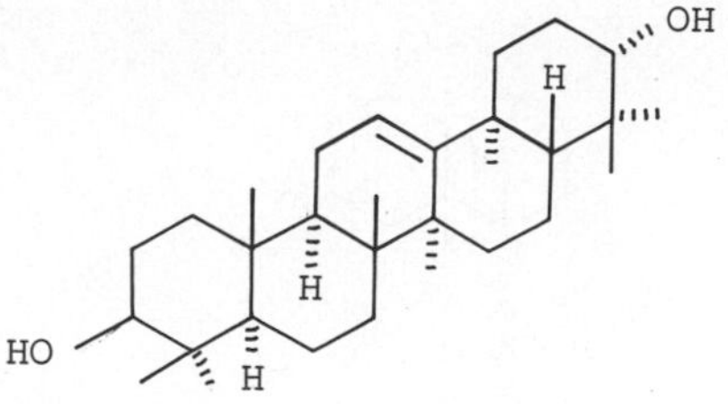

<u>89</u>

Dinoronocerane <u>90</u> *(Scheme 4)*

Sondheimer and Elad[37] have synthesized this degradation product of α-onocerin <u>3</u>. Reduction of the enedione <u>91</u> gave a mixture of diols <u>92</u> which, on oxidation with manganese dioxide,
afforded the hydroxyenone <u>93</u>. Alkylation of the corresponding
benzoate <u>94</u> gave the β,γ-unsaturated ketone <u>95</u> which was successively reduced by the Wolff-Kishner method and hydrogenated
to give the *trans*-decalol <u>96</u>. The corresponding ketone <u>97</u> was
reacted with sodium acetylide and the acetylenic alcohol <u>98</u> so
formed was converted to its disodium salt and reacted with a
further equivalent of ketone <u>97</u> to give a mixture of *meso* (m.p.
210°) <u>99</u> and *dl* (m.p. 190°) <u>100</u> forms of the diols. These
could not be produced directly by reaction of two equivalents
of ketone <u>97</u> with acetylene dimagnesium bromide. Attack on
ketone <u>97</u> was stereospecific in both reactions, proceeding by
axial addition to the less hindered α side. The assignment of

Scheme 4

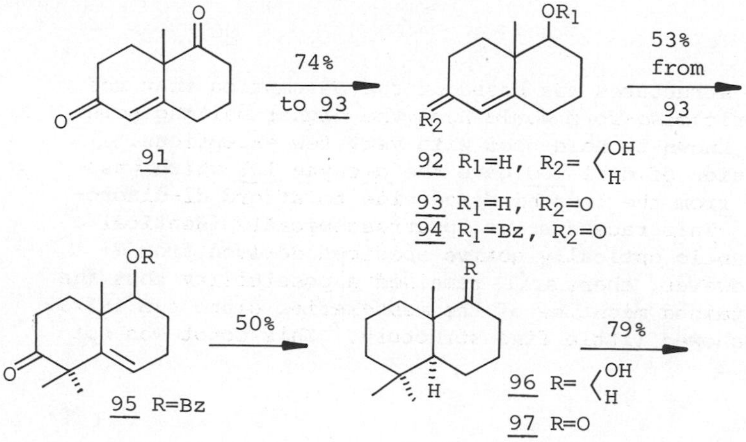

<u>91</u>   74%
to <u>93</u>

<u>92</u> R₁=H, R₂= OH/H
<u>93</u> R₁=H, R₂=O
<u>94</u> R₁=Bz, R₂=O

53%
from
<u>93</u>

<u>95</u> R=Bz   50%

<u>96</u> R= OH/H
<u>97</u> R=O

79%

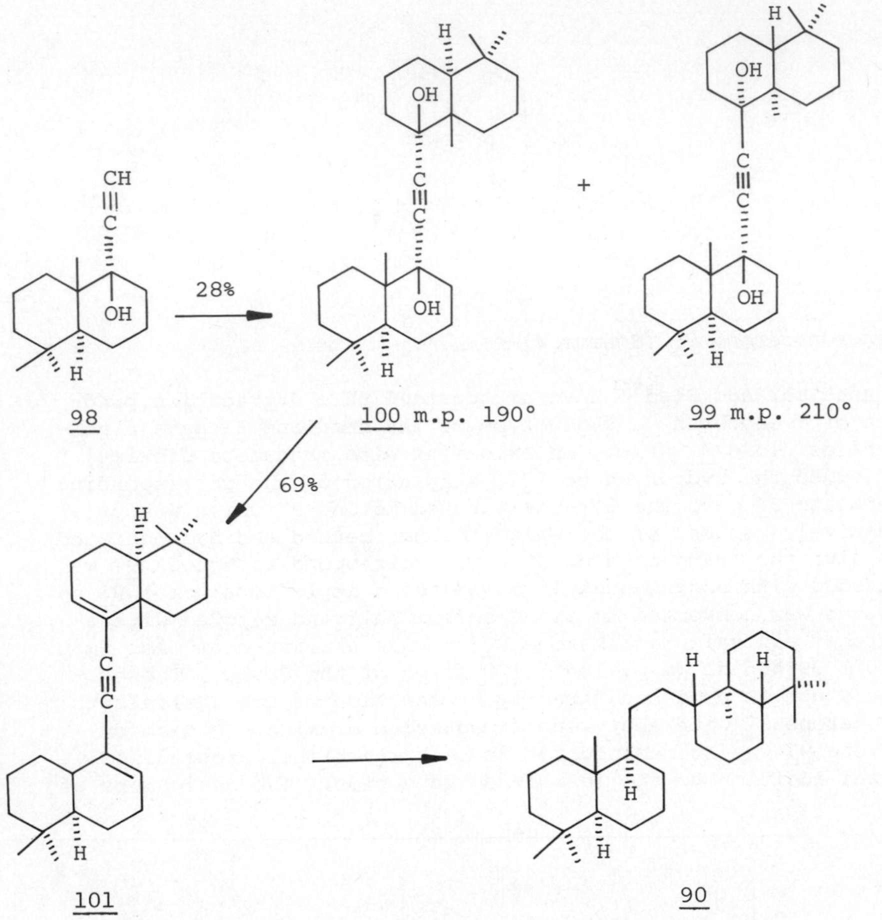

98   28% →   100 m.p. 190°   +   99 m.p. 210°

69% →

101   →   90

dl- and meso-structures was based on the assumption that the centrosymmetric meso-form would have the higher melting point, a phenomenom known to hold good with very few exceptions.

Dehydration of diol 100 gave the dienyne 101 which was hydrogenated from the less hindered side to afford dl-dinoronocerane 90. This racemate was spectroscopically identical with an authentic optically active specimen derived from α-onocerin. However, there still remained a possibility that the substance obtained might be of the meso-series since the infra-red spectra showed little fine structure. This doubt was not resolved.

The Onoceradienes 102, 103 and 104 (Schemes 5 and 6)

Two different approaches[38,39] to the onoceradienes (i.e., onocerins lacking the C-3 and C-3' hydroxyl groups) have led to the production of the same key intermediate 105.

Eschenmoser and co-workers[38] (Scheme 5) used keto-ester 106, obtained from the triene-acid 107 by acid catalyzed cyclization, to prepare the enone 108, two different routes being employed. Reduction of 106 with lithium aluminum hydride and sodium hydride gave the allylic alcohol 109 which afforded 108 on oxidation. Alternatively, reduction of 106 with lithium aluminum hydride alone gave a diol, the monomesylate of which 110 was oxidized to the ketomesylate which afforded the enone 108 on treatment with base.

Scheme 5

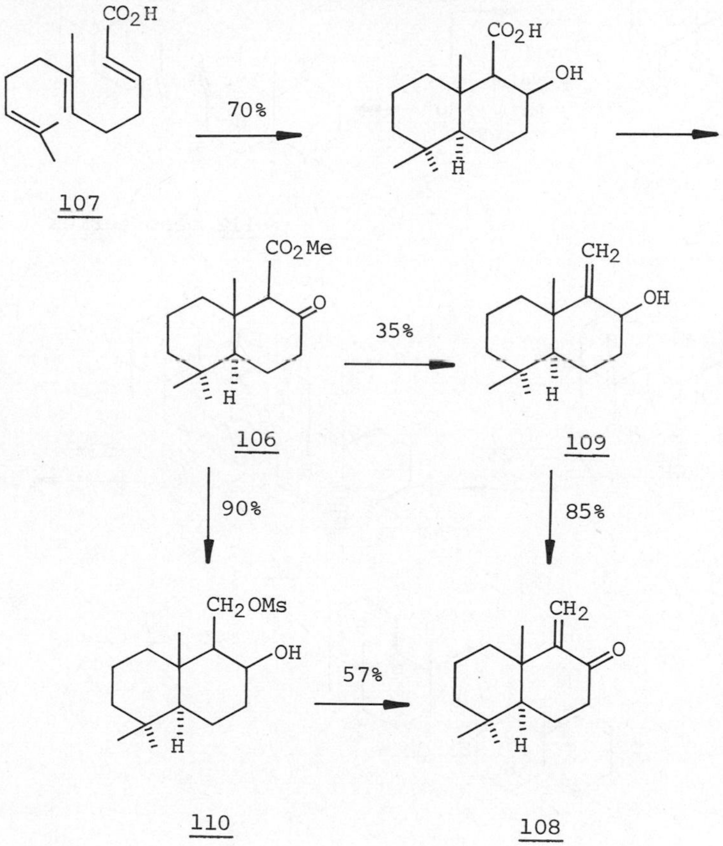

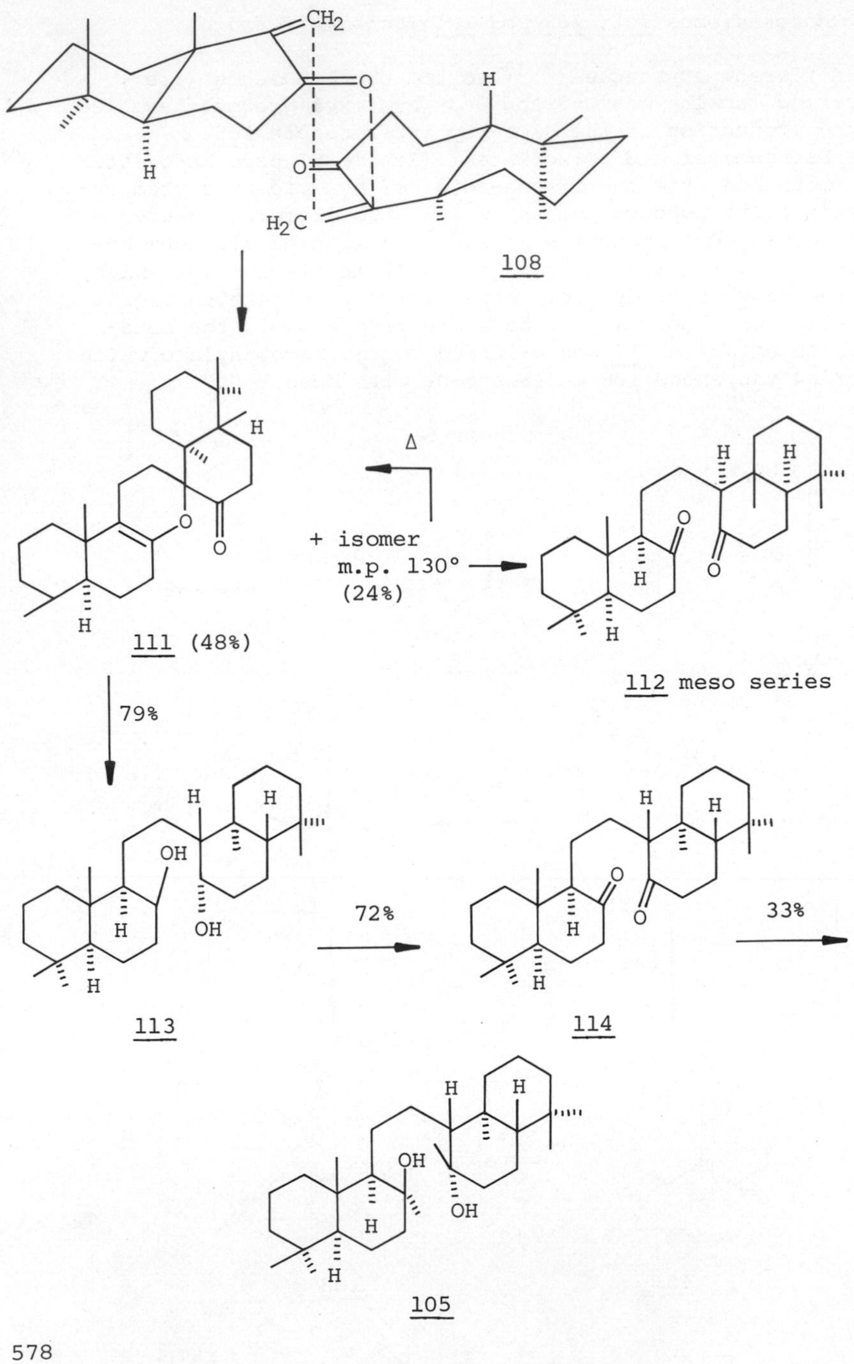

108

111 (48%)

Δ

+ isomer
m.p. 130°
(24%)

112 meso series

79%

113

72%

114

33%

105

When 108 was heated in xylene it underwent a Diels-Alder reaction to give two dimeric products m.p. 172° and m.p. 130° of which the former led to a racemic onocerin derivative and therefore had structure 111 while the other belonged to the *meso*-series and led to diketone 112. Thus the dimers were the products of endo addition. Reduction of the dimer m.p. 172° with sodium in propanol gave the diol 113 which was resolved via its (-)-menthyloxyacetate. Oxidation of the (+)-diol 113 gave bisnoronoceradione 114 identical with the product obtained from an ozonolysis of α-onoceradiene 102. Finally, reaction of diketone 114 with methylmagnesium iodide gave the diol 105 by attack from the less hindered side, and this was identical with a sample prepared by Corey[39] and described below.

Corey and Sauers (Scheme 6)[39] also utilized a dimerization, the Kolbé electrolysis of carboxylic acids, in their approach. Sclareol 115, which has been totally synthesized,[40] was oxidized to the *trans*-lactone 116 which could be isomerized by acid to the more stable *cis*-lactone 117. Both lactones have the same absolute configuration as the onocerins. Hydrolysis of the lactones gave the corresponding acids 118 and 119. The acetate 120 of acid 118 was also prepared from sclareol via the acetoxy aldehyde 121. Electrolytic oxidative decarboxylation (Kolbé coupling) of the ammonium salts of the hydroxy acids 118 and 119 gave the dimeric diols 122 (17%) and 123 (12%), respectively, each accompanied by the ketone 124 (ca. 35%). The latter was probably formed by fission of the C-8 to C-9 bond in the intermediate 125 although an ionic mechanism cannot be excluded. The diol 123 is identical with that produced by Eschenmoser and co-workers[38] described above.

Electrolysis of the acetate 120 gave the dimeric diol 122 in improved yield (34%) and none of the ketone 124 was formed.

Dehydration of diol 122 (hydroxyl groups equatorial) with phosphorous oxychloride and pyridine gave α-onoceradiene 102 while similar treatment of diol 123 (hydroxyl groups axial and *trans*-coplanar with the C-9 protons) gave β-onoceradiene 103. Treatment of either of the tetracyclic diols 122, 123 with perchloric acid in benzene gave γ-onocerene (pentacyclo-squalene) 104, identical with a sample derived from natural α-onocerin.

The Onocerins

The syntheses described so far relate to substances *derived* from natural products. Stork et al.[41] achieved the first synthesis of a naturally occurring nonsteroidal polycyclic triterpene (α-onocerin 3). The Stork et al. approach utilized the Kolbé electrolysis method used earlier by Corey[39] to good

Scheme 6

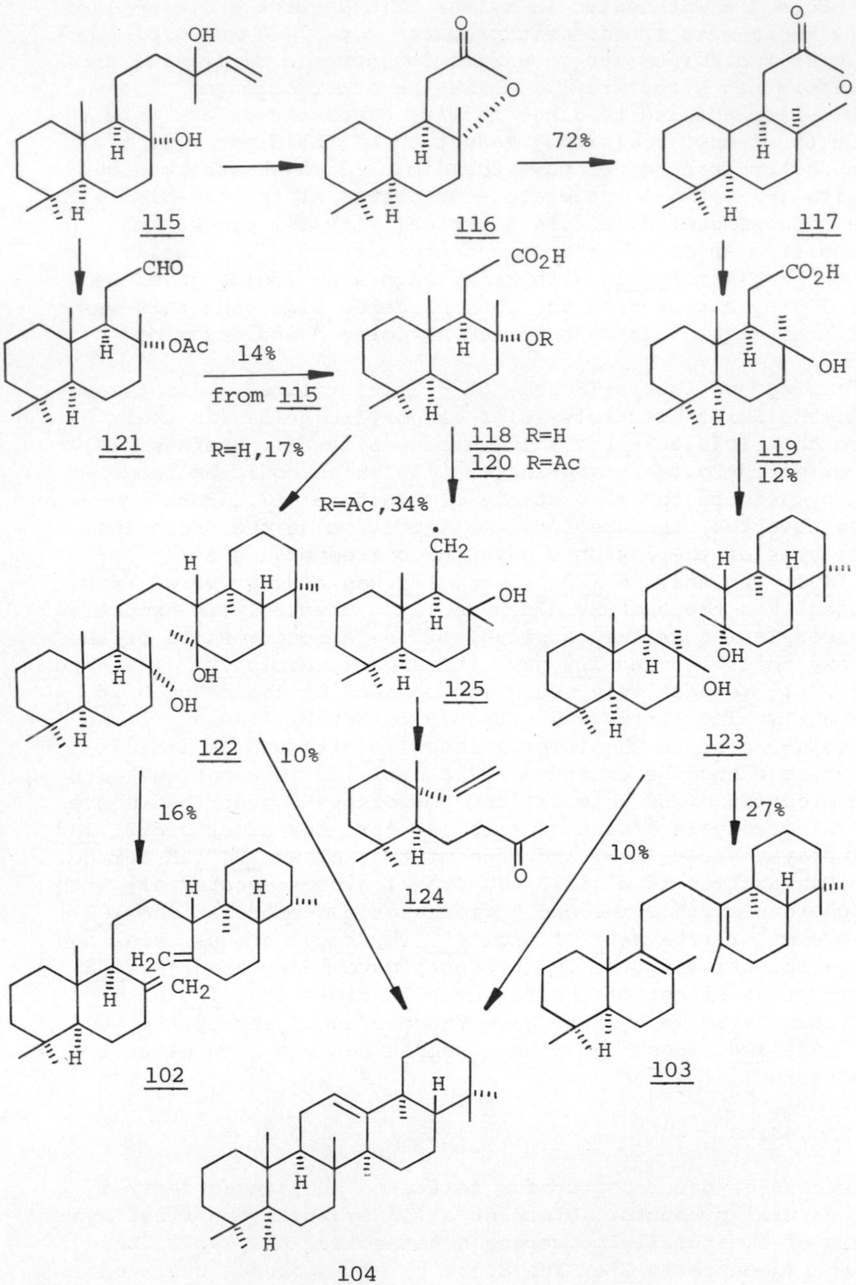

OH
OH
H
H
115
O
O
H
H
116
72%
O
O
H
H
117
CHO
OAc
H
H
121
14%
from 115
CO2H
OR
H
H
118 R=H
120 R=Ac
CO2H
H
OH
H
119
12%
R=H,17%
R=Ac,34%
H
H
OH
H
OH
H
122
CH2
OH
H
H
125
H
H
OH
H
OH
H
123
10%
16%
H
O
124
10%
27%
H
H
H2C
CH2
H
H
102
H
H
103
H
H
H
104

effect.   A simple model compound 126 for the ultimate substrate (keto-acid 127) was coupled to diketone 128.   Keto-acid
127 was first synthesized by Stork and later produced by other
workers[42,43] using different methods.

Stork's Approach to Acid 127 (Scheme 7).[41]   Reaction of 6-
methoxy-1-tetralone 129 with methyl magnesium iodide gave
olefin 130 by spontaneous dehydration of the tertiary benzylic
alcohol.   Epoxidation of this olefin and acid catalyzed

Scheme 7

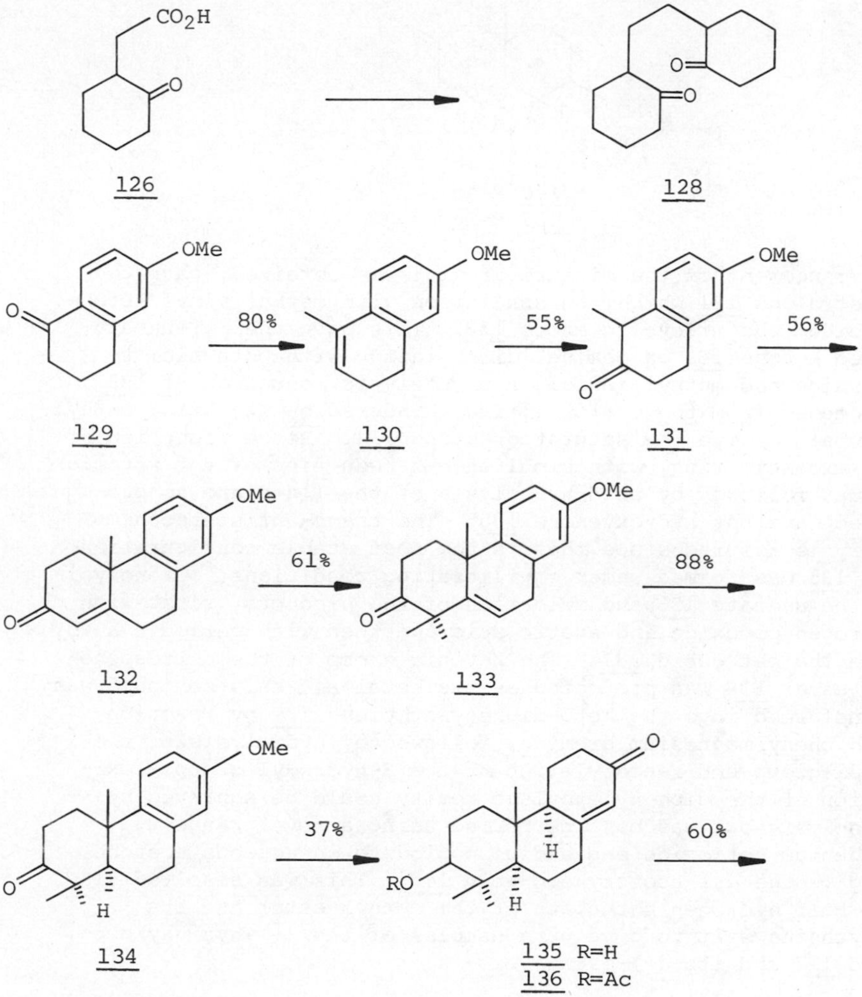

137 R=H
138 R=Me

139

140 R=Ac
127 R=H

rearrangement of the mixture of diols so obtained, gave the β-tetralone 131 which, on annelation with methyl vinyl ketone, afforded the tricyclic enone 132. This gave the β,γ-unsaturated ketone 133 on geminal dialkylation with potassium t-butoxide and methyl iodide, and catalytic reduction of 133 proceeded from the α side (β face hindered by two axial methyl groups) to give the saturated ketone 134. Birch reduction of the aromatic ring, with simultaneous reduction of the ketonic group, followed by acid hydrolysis of the 1,4-diene so produced gave the hydroxyenone 135. The *trans-anti*stereochemistry is assured since this is the most stable configuration and 135 was formed under equilibrating conditions. Ozonolysis of the acetate 136 and oxidation of the products, first with hydrogen peroxide and acetic acid and then with periodic acid, gave the keto-acid 137. The ketonic group of the corresponding ester 138 was protected as the ketal and this compound was transformed into the keto-diphenylethylene 139 by reaction with phenylmagnesium bromide, followed by hydrolysis of the ketal group and reacetylation of the 3-hydroxyl group. Oxidation of the diphenylethylene moiety could be achieved by ozonolysis but was best performed using a novel reagent, ruthenium tetroxide and sodium periodate in aqueous acetone, to give the *d,l*-acetoxyketo-acid 140. This was resolved via the half hydrogen phthalate of the methyl ester and its strychnine salt to give pure samples of the (-)-hydroxyketo-acid 127 and its (+)-isomer.

Sondheimer's Approach to Acid 127 (Scheme 8).[42]   The starting
material, the hydroxyenone 95 (R=H) (page 575) was hydrogenated over platinum to give the diol 141 with a *trans*-fused ring
system.   Chromium trioxide oxidation of diol 141 could be controlled to give the desired ketol 142 together with unchanged
141, diketone 143, and ketol 144.   The unwanted products of
oxidation could be reduced back to diol 128 and thus recycled.

Scheme 8

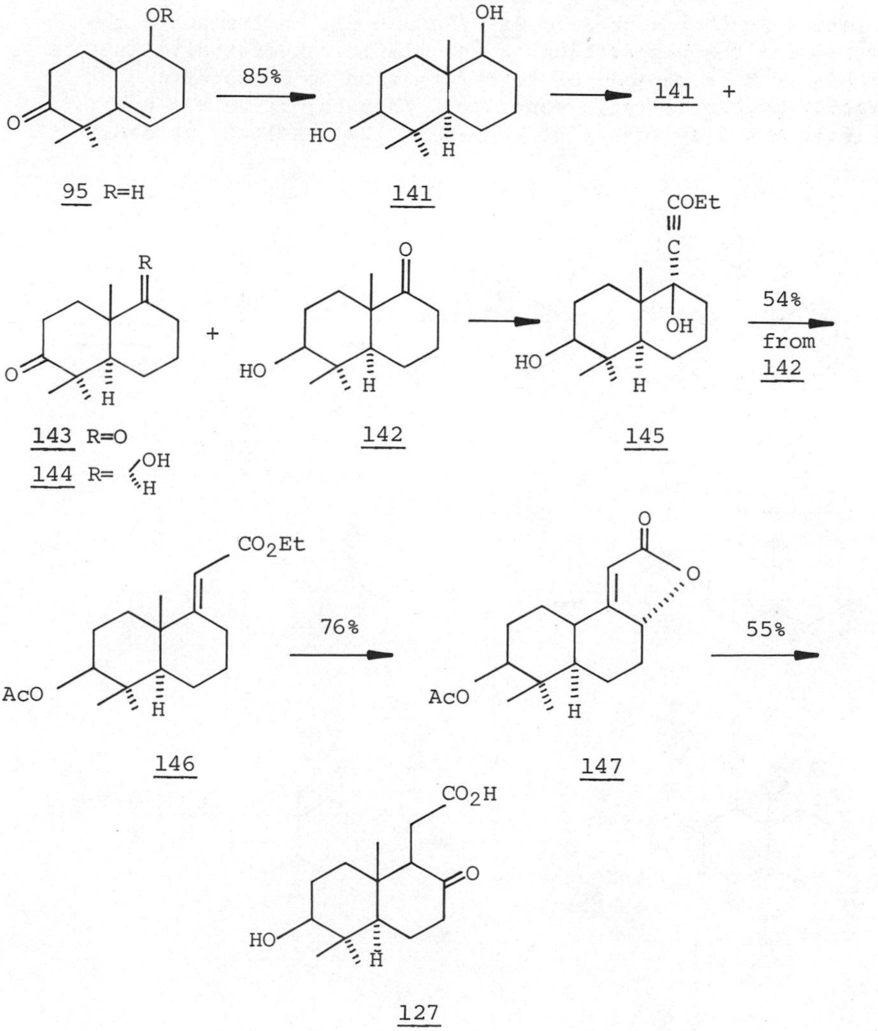

Reaction of ketol 142 with lithium ethoxyacetylide gave the acetylenic diol 145 which afforded the α,β-unsaturated ester 146 on acid treatment followed by acetylation. Allylic oxidation of this ester with selenium dioxide (assumed to occur from the less hindered side) served to introduce the needed oxygen function at C-8. The unsaturated lactone 147 so formed was treated with strong base when double bond migration, hydrolysis of both ester groups and ketonization of the enol formed gave the required acid 127, identical with that prepared by Stork.

Ireland's Synthesis of Acid 127 (Scheme 9).[43] Ireland's objective was the preparation of the olefinic hydroxyacid 148, which should be capable of transformation to α-onocerin 3 directly by electrolytic coupling. This objective was not achieved and a synthesis of keto-acid 127 resulted instead.

Scheme 9

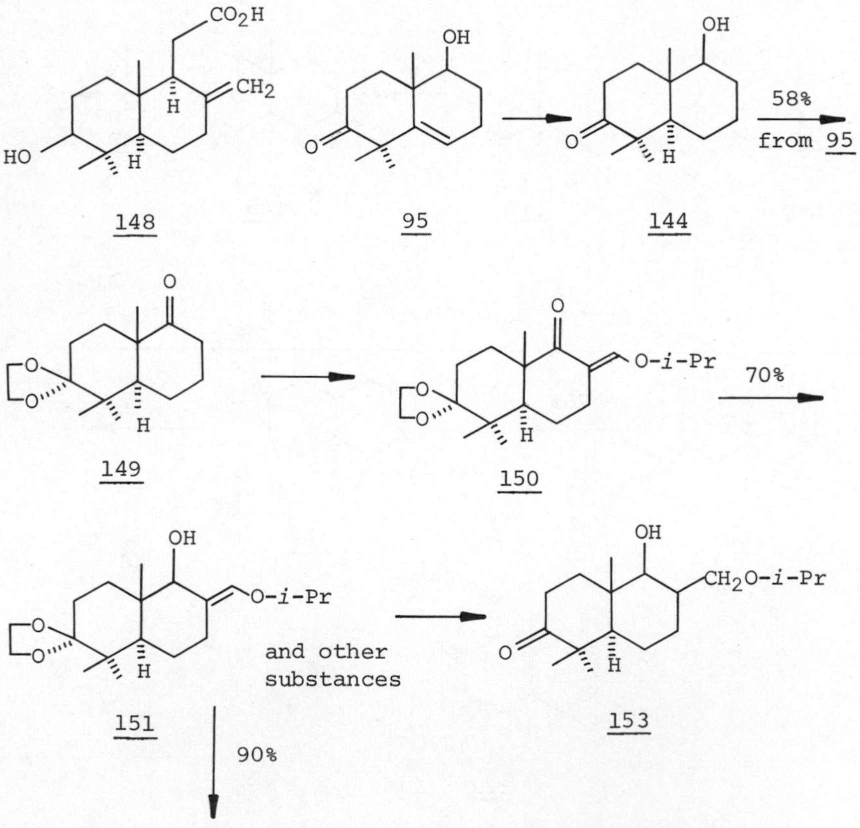

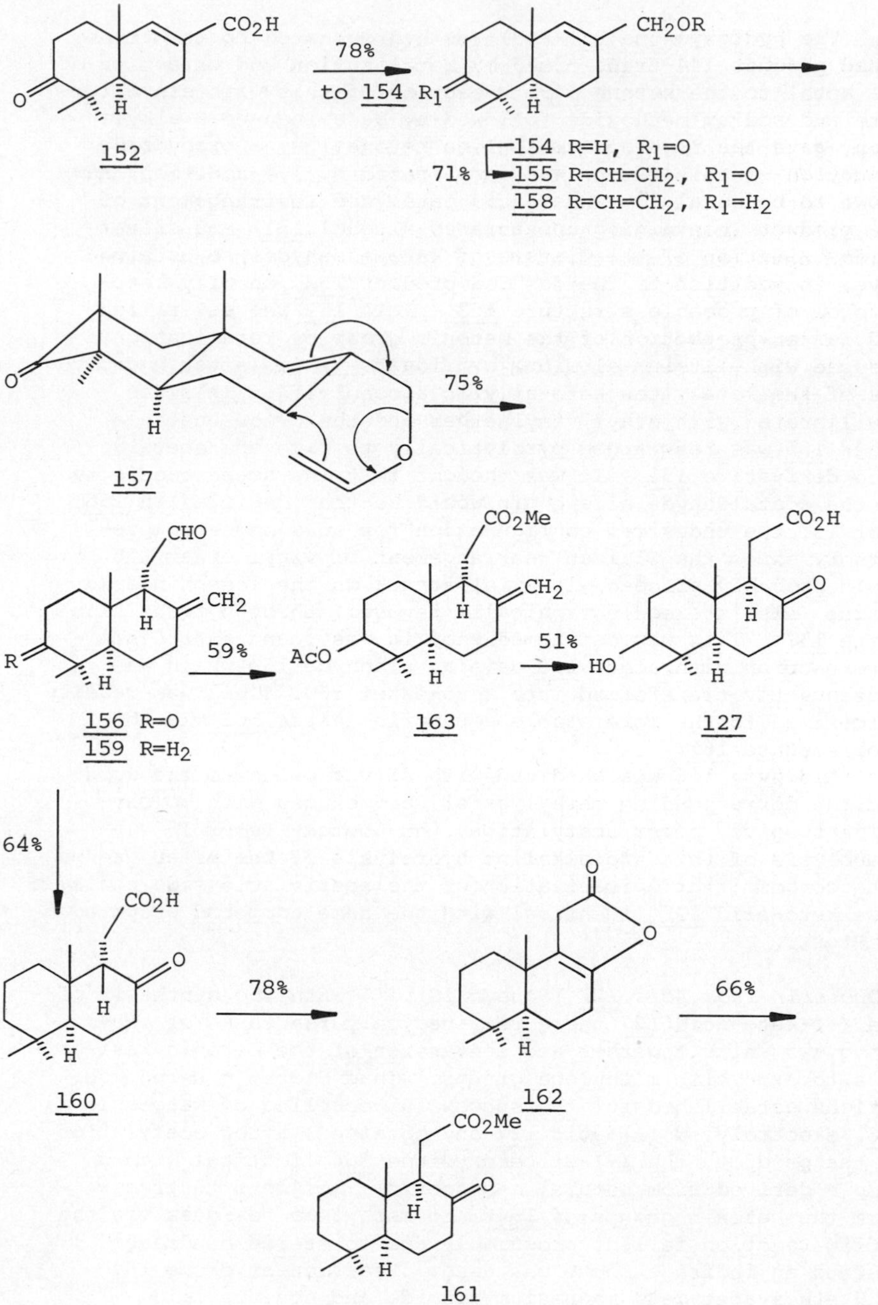

152

78%
to 154   R₁=

154 R=H, R₁=O
155 R=CH=CH₂, R₁=O
158 R=CH=CH₂, R₁=H₂

71%

157

75%

156 R=O
159 R=H₂

163

127

59%

51%

64%

160

78%

162

66%

161

The hydroxyenone 95 (R=H) was hydrogenated to the *trans*-fused product 144 transformed by ketalization and oxidation of the ketal to the ketone 149. Reaction of this with ethyl formate and sodium methoxide followed by base induced 0-alkylation, gave the *iso*propoxymethylene ketone 150. Borohydride reduction of this ketone was complicated by 1,4-addition, now known to be usual.[44] Thus, acid catalyzed rearrangement of the product (containing unsaturated alcohol 151) and silver oxide oxidation of the mixture of keto-aldehydes so obtained gave, in addition to the desired product 152, an oily keto-alcohol of probable structure 153. Acid 152 was esterified and, after protection of the ketonic group by ketalization, reduced with lithium aluminum hydride to yield, after hydrolysis of the ketal, the keto-allylic alcohol 154. This was equilibrated with ethyl vinyl ether and the crude unstable ether 155 was rearranged pyrolytically to give the acetaldehyde derivative 156. It was thought that the stereochemistry at the acetaldehyde side chain would be that depicted in 156, that is, the undesired configuration for an α-onocerin precursor, since the Claisen rearrangement of vinyl ether 156 would probably proceed via axial attack on the less hindered α side (also stereoelectronically favored) through transition state 157. This was confirmed when it was found that a similar reaction with ether 158 gave an aldehyde 159 which was subsequently transformed into keto-ester 160. This was readily epimerized to the more stable equatorial ester 161 via the enol acetate 162.

Aldehyde 156 was oxidized with silver oxide to the acid and the corresponding methyl ester was reduced with borohydride to give, after acetylation, the acetoxy ester 163. Ozonolysis of this and alkaline hydrolysis of the ester groups, with concomitant epimerization of the acetic acid side chain, gave keto-acid 127, identical with the same compound prepared by Stork.

α-Onocerin from Acid 127 (Scheme 10).[41] With the synthesis of the (-)-keto-acid 127 there remained only the tasks of coupling two units together and conversion of the ketonic residues to exocyclic methylene groups. Thus Stork,[41] using conditions established for the successful coupling of keto-acid 126, electrolyzed (-)-acid 127 and obtained, after acetylation of the product, the (-)-diacetoxydione 164 identical with a sample derived from natural α-onocerin. Attempts to transform the ketonic groups of 164 into methylene residues via the Wittig reaction failed, presumably due to steric hindrance. Instead an indirect route was used. Reaction of dione 164 with ethoxyacetylene magnesium bromide and aqueous acid

Scheme 10

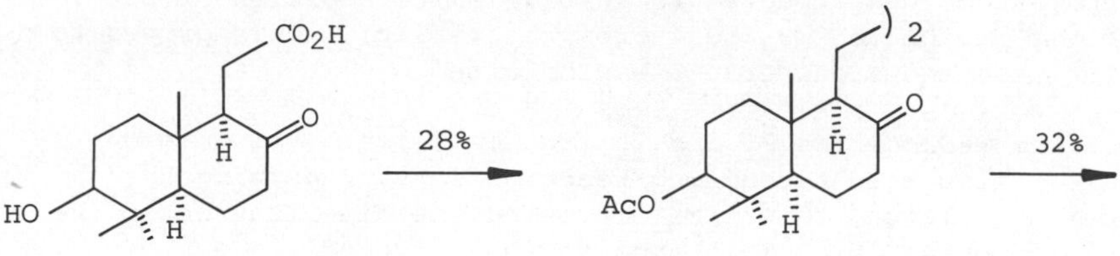

(−)-acid
127

164

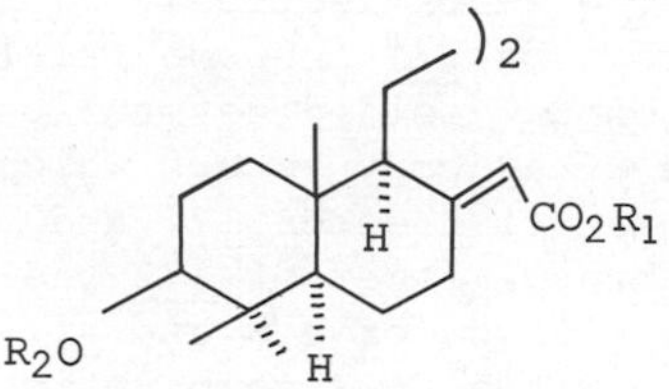

165 R$_1$=Et, R$_2$=Ac
166 R$_1$=R$_2$=H

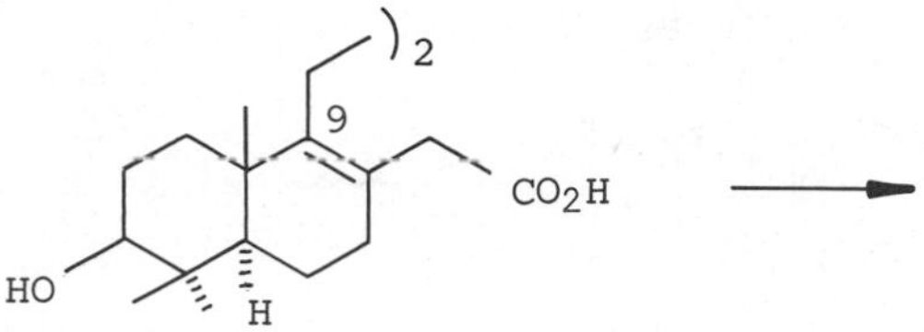

167

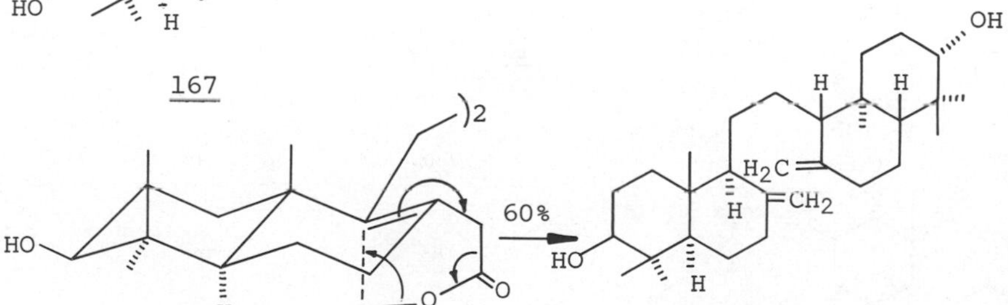

168

3

catalyzed rearrangement of the diacetylenic diol so obtained gave the diacetoxy diester 165. The corresponding dihydroxy diacid 166 was then thermally decarboxylated. The mechanism of this reaction requires the intermediacy of the β,γ-unsaturated isomer and it is known that in the onocerin system the isomer shown 167 is favored. Thus, the asymmetric center at C-9 is destroyed during this reaction. However, it was anticipated that α-onocerin 3 would nevertheless be produced (rather

than an isomer) since the transition state 168 for the decarboxylation was expected to involve axial addition of the
carboxyl proton to C-9 (best orbital overlap). This proved to
be the case and α-onocerin 3 was obtained.

The β and γ-Onocerins 75 and 76 (Scheme 11).[45] Van Tamelen
et al.[45] also used a coupling reaction in synthesizing the
onocerin skeleton, this time employing the dimerization of the
allylic bromide 169. This bromide was produced via a biogenetic type cyclization of an epoxydiene. Thus, methyl
trans-trans-farnesate 170 was selectively epoxidized at the
terminal double bond to give 171. Boron trifluoride catalyzed
cyclization of this epoxide yielded several partially cyclized
products and a mixture of hydroxy esters which was shown to
contain >90% of the equatorial ester 172 and ca. 2% of the
axial ester 173. The benzyl ether of 172 was isomerized with
base to an equilibrium mixture of α,β and β,γ-unsaturated
esters. These were separated by a combination of chromatography

Scheme 11

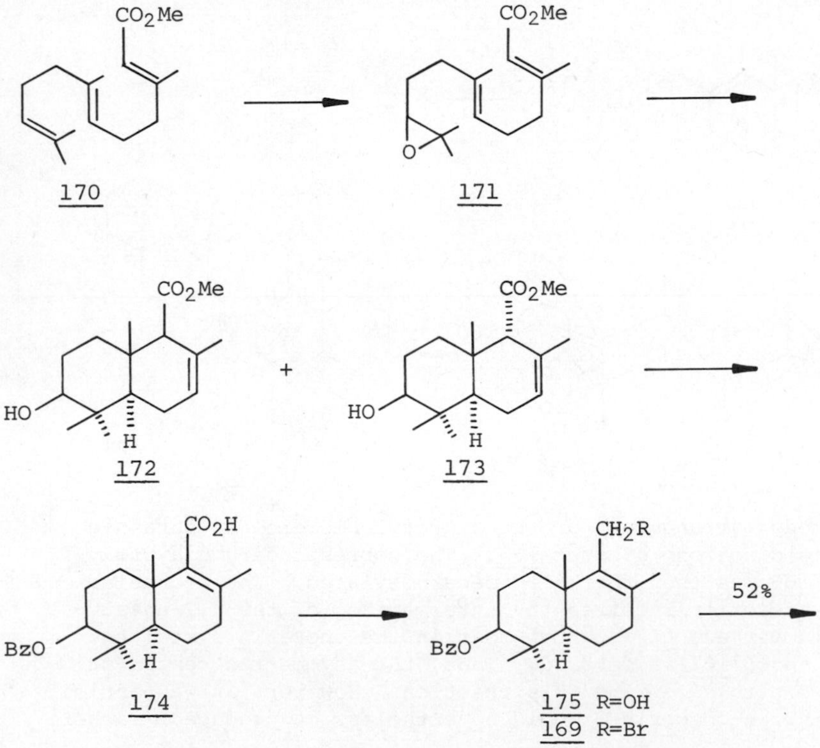

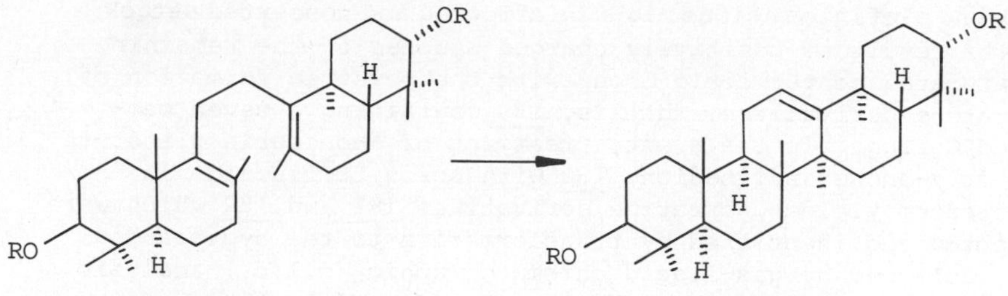

176 R=Ac
88 R=H

177 R=Ac
89 R=H

and selective hydrolysis (formic acid/sulphuric acid) of the
α,β-unsaturated ester, to give the acid 174. This was resolved through its brucine salt and the (+)-isomer was reduced to the allylic alcohol 175 which was readily transformed
into the corresponding bromide 169 using hydrobromic acid.
Coupling of the bromide with magnesium in ether and cleavage
of the benzyloxy groups using sodium in liquid ammonia gave,
after acetylation, (+)-β-onocerin diacetate 176. Acid treatment of this afforded γ-onocerin diacetate 177. The conversions of the diacetates 176 and 177 to β-onocerin 88 and γ-
onocerin 89 are well known.

Serratenediol 178

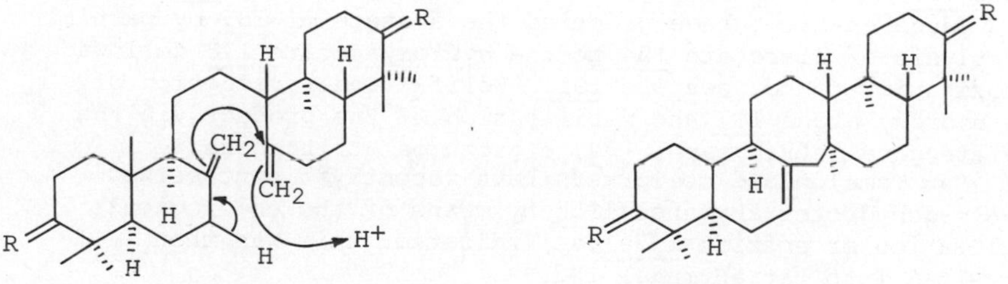

179 R=H, OAc
180 R=O

178 R=H, OH
181 R=H, OAc
182 R=O
183 R=H$_2$

As previously mentioned, derivatives of α-onocerin rearrange
when treated with acid catalysts, first to β-onocerin types
and subsequently to the pentacyclic γ-onocerin series. The
α-onocerin → β-onocerin transformation obviously involves protonation of both of the exocyclic methylene groups. Tsuda et
al.[46] reasoned that if a bulky acid catalyst were used then

only one olefinic residue might be affected and concerted attack
of the resultant positively charged species on the remaining
unsaturated center could occur with the possible formation of
serratene derivatives, that is, 178 containing a seven mem-
bered C ring. In the event, treatment of α-onocerin diacetate
179 or α-onocerindienedione 180 with boron trifluoride in
chloroform yielded serratene derivatives 181 and 182 which were
detected and identified by transformation to the hydrocarbon
183, followed by gas-liquid chromatographic (g.l.c.) analysis.
β and γ-onocerin derivatives were also formed in these reactions. Serratenedione 182 was isolated by fractional crystallization and shown to be identical with a sample derived from
the natural product, serratenediol 178, to which it could be
converted by reduction thus completing the synthesis.

Tetrahymanol and the Hopane Group (Schemes 12 and 13)

Swiss workers[47] were the first to take advantage of the transformation of α-onocerin to pentacyclic γ-onocerin for the synthesis of triterpenoid derivatives of the hopane series. Previously, Barton and Overton[36] had shown that γ-onocerin diacetate 177 gave the keto-diacetate 184 (Scheme 12) on treatment with peracid. Jëger et al.[47] reduced this ketone (Wolff-
Kishner) to the diacetate 185 and then partially oxidized the
corresponding diol 186 to the hydroxyketone 187. Treatment of
this with Fullers earth in boiling xylene gave hop-17(21)-
enone 188, a degradation product of hydroxyhopanone 189.
     Tsuda et al.[48] have prepared the acetate of 187 by partial
hydrolysis of diacetate 184 to the hydroxyacetate 176 followed
by oxidation to ketoacetate 191. Wolff-Kishner reduction of
the hydroxyketone 187 and purification of the product via the
acetate gave tetrahymanol 192, a protozoal triterpene.
     Van Tamelen and co-workers have recently[88] synthesized
dl-Δ[12]-dehydrotetrahymanol 195b by means of the abiological
cyclization of epoxide 195a as indicated; this was then
converted into tetrahymanol 192.

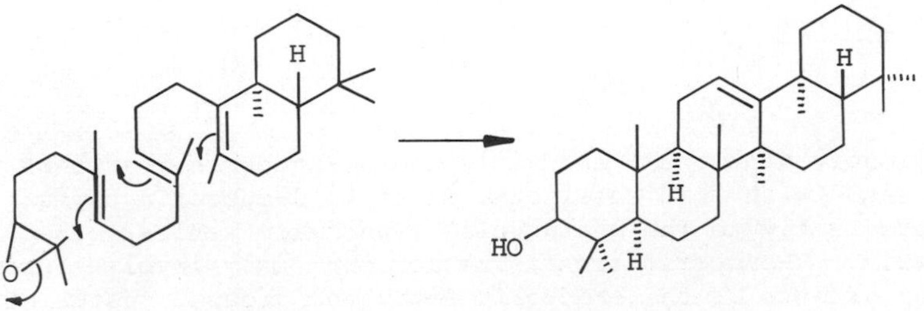

195a   195b

Scheme 12

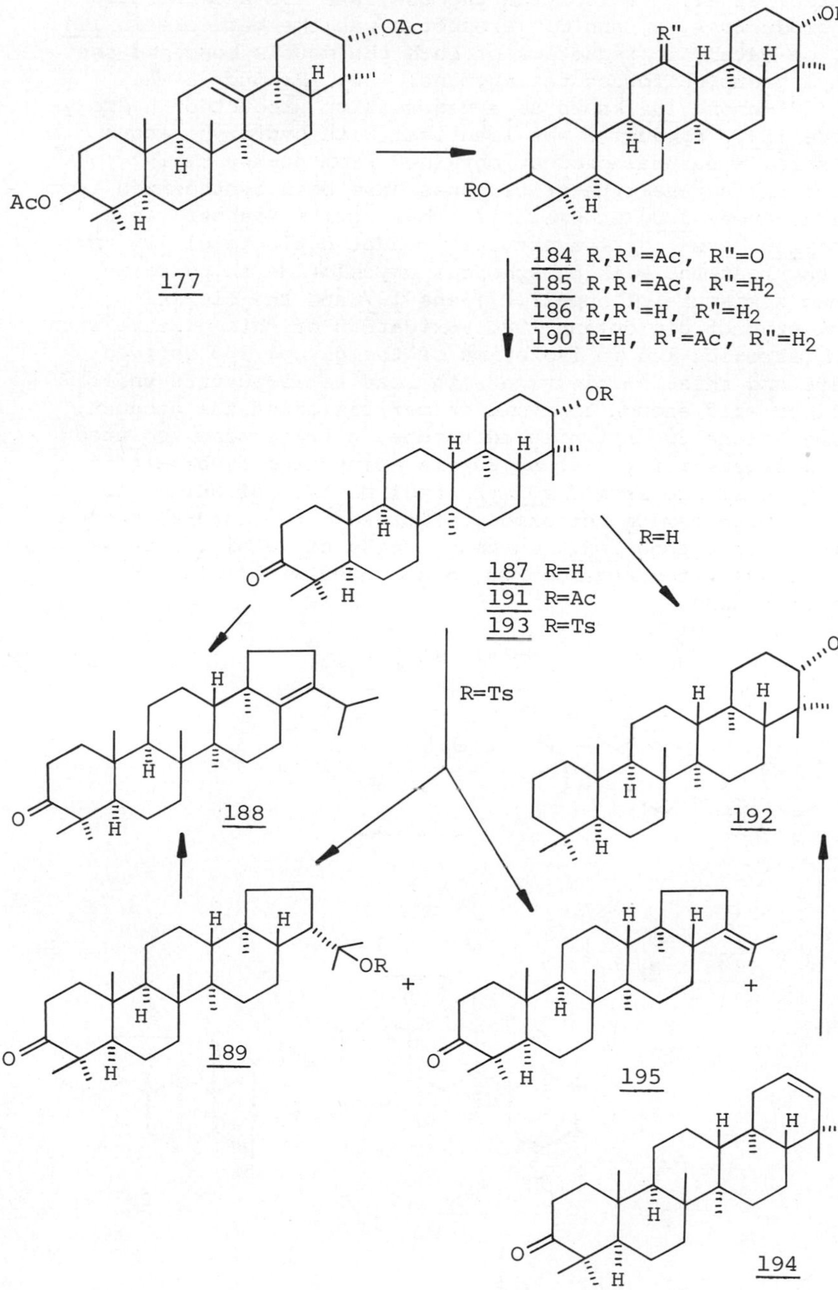

OAc
AcO
177
OR'
R"
RO
184 R,R'=Ac, R"=O
185 R,R'=Ac, R"=H2
186 R,R'=H, R"=H2
190 R=H, R'=Ac, R"=H2
OR
187 R=H
191 R=Ac
193 R=Ts
R=H
R=Ts
188
OH
192
OR
189
195
194

Kishi et al.[49] solvolyzed the tosylate 193 and isolated three products A, B, and C. Product A was the keto-olefin 194 which, on catalytic reduction of both the double bond and the carbonyl group, afforded tetrahymanol 192. Product B was hop-21(22)-enone 195 known as a degradation product of hydroxy-hopanone 189. Product C was identical with hydroxyhopanone 189 itself, a natural product obtained from dammar resin.

Several hopane type triterpenes have been synthesized from hydroxyhopanone 189 (Scheme 13). Thus, Wolff-Kishner reduction of 189 gave[50,51] the natural product diplopterol 196 and this, on treatment with phosphorous oxychloride in pyridine, afforded a mixture of hop-21(22)-ene 197 and the olefinic natural product diploptene 198. Oxidation of this mixture with osmium tetroxide led to isolation of the glycol 199 derived from 198 and this, on cleavage with lead tetra-acetate under conditions mild enough to avoid epimerization of the product, gave the ketone 200 known as adiantone, a triterpene isolated from a Japanese fern. Ketone 200 is epimerized by base[51,52] to the more stable isomer 201 the enol acetate of which, on oxidation with osmium tetroxide, afforded[52] the natural product hydroxyadiantone 202. Chromatography of ketol 202 on alumina gives ketohakonanol 203, a compound with which it co-occurs in nature.

Scheme 13

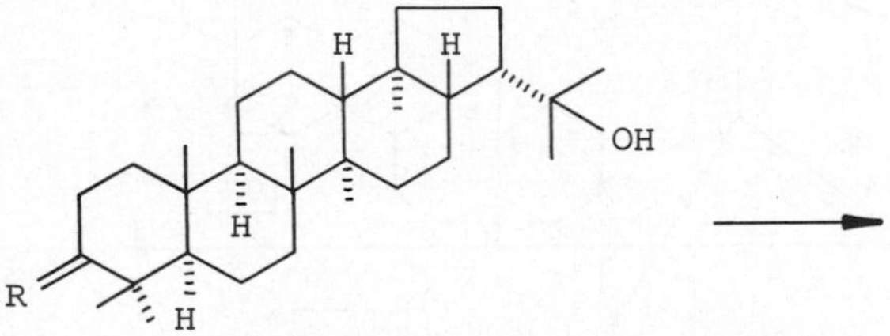

189 R=O
196 R=H$_2$

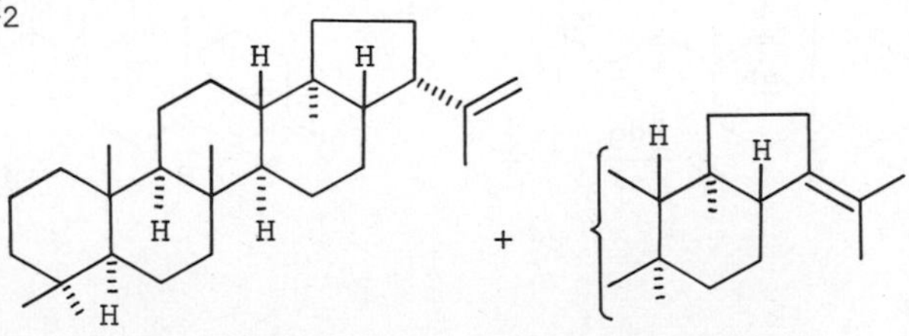

196   +   197

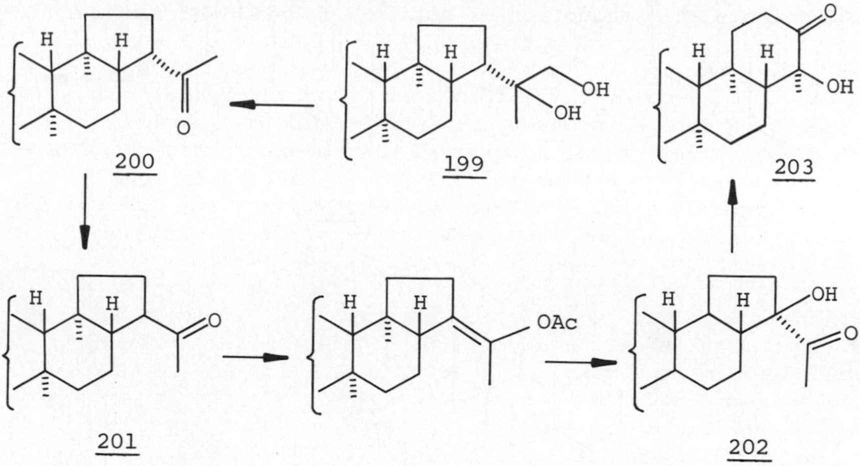

Recently, Iguchi and Kakisawa[53] achieved a synthesis of
the natural product fern-8-ene 204 (Scheme 14). γ-Onocerin
diacetate 177 was partially deacetylated with methylmagnesium
iodide to give a mixture of hydroxyacetates 205 and 206. The
mixture of isomers was not separated until the penultimate
stage but since both series underwent identical reactions until
then, only the structures leading to fern-8-ene are described
and depicted. Thus, oxidation of 205 to the ketone followed
by Wolff-Kishner reduction gave the hydroxyolefin 207 which,
on treatment with phosphorous pentachloride, afforded the rearranged diene 208. Further rearrangement using Fullers earth
in refluxing xylene gave a mixture from which the diene 209
was isolated as a pure compound. This gave fern-8-ene on
Birch reduction.

Scheme 14

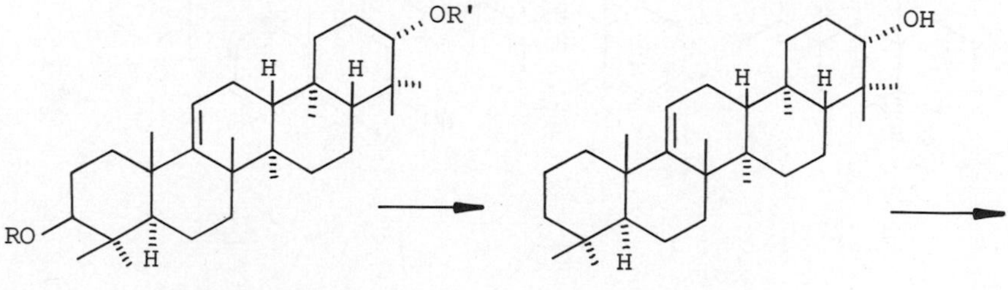

177 R=R'=Ac
205 R=H, R'=Ac
206 R=Ac, R'=H

207

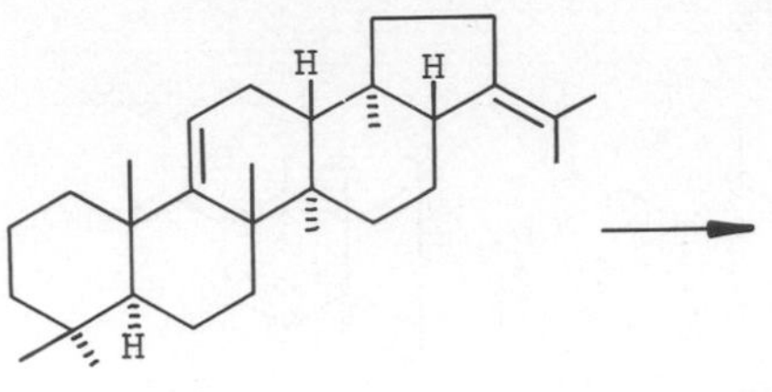

208 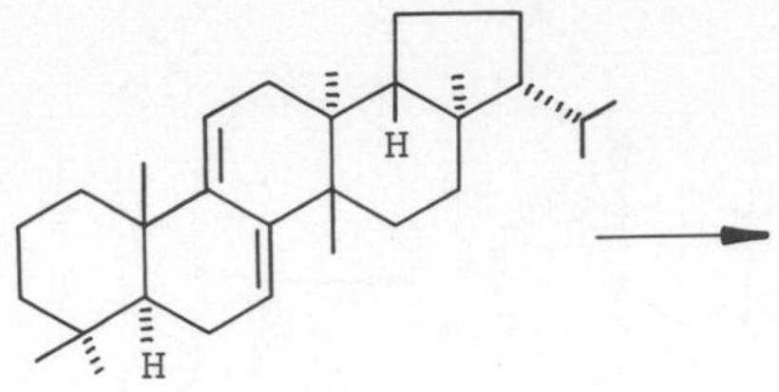   209

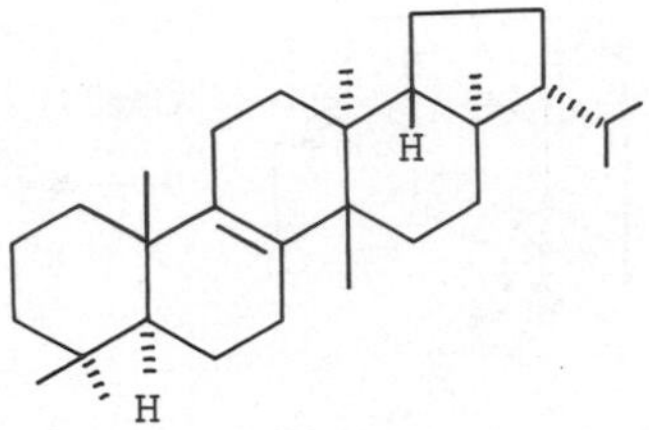

204

B.   The Amyrin Group

Total synthesis of triterpenes of the amyrin group, including
those of natural products which can be derived from β-amyrin 4
by 1,2-methyl and/or hydrogen shifts (e.g., taraxerol 210 and
multifluorenol 211), are based on three syntheses of olean-
13(18)-ene 212, and a later partial synthesis of β-amyrin itself from olefin 212 by Barton.[54]

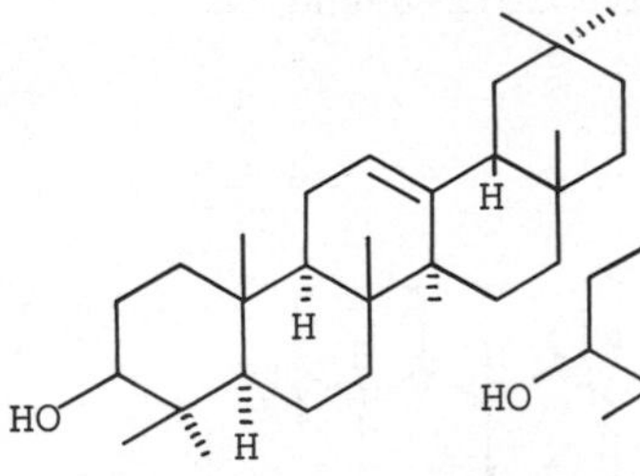

4 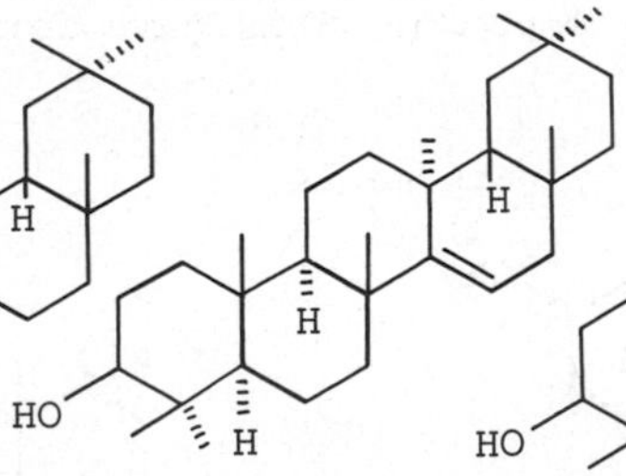   210 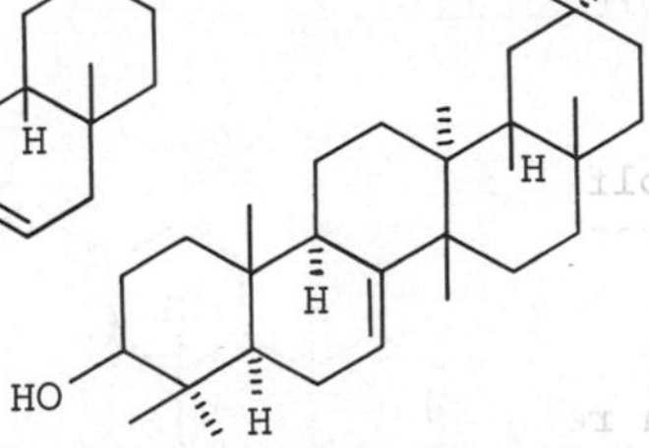   211

Olean-13(18)-ene 212 and Related Compounds

All three syntheses of this type have involved a preformed AB/DE tetracyclic ring system which was then cyclized by acid catalysis. Corey et al.[55] and Barltrop et al.[56] have produced substances formulated as the same tetracyclic intermediate 226 by different routes, and each group transformed this compound into an olean-13(18)-ene derivative, also by different methods. Sondheimer and Ghera[57] used procedures reminiscent of those already described in their syntheses of onocerin derivatives.

Olean-11,12,13,18-diene 214 (Scheme 15). Corey[55] synthesized this derivative, which had previously been transformed into olean-13(18)-ene, using a rings ABE → rings ABDE → amyrin approach. The key intermediates, the AB rings precursor 215 and the E ring precursor 216, were produced as follows.

The starting material (+)-ambreinolide 58, a degradation product of ambrein 56 (p. 569), has been totally synthesized in the racemic form (26a). Saponification of lactone 58 and subsequent methylation and dehydration of the tertiary alcohol gave the olefinic ester 217 which, on reduction with lithium aluminum hydride, afforded the alcohol 218 and then the tosylate and the (+)-bromide 215. The acid 219 was also prepared from the aldehyde 220 obtained from sclareol. Thus, aldehyde 220 was converted to the nitrile 221 via the alcohol and brosylate, the acetoxy group was extruded by refluxing in quinoline, and hydrolysis of the unsaturated nitrile gave acid 219.

The Michael adduct 222 from diketone 223 and methyl acrylate was saponified and cyclodehydration of the acid gave enol-lactone 224. Hydrogenation of the olefinic bond was accompanied by hydrogenolysis of the lactonic oxygen to give a keto-acid which afforded the racemic enol lactone 216 on recyclization.

Reaction of the Grignard reagent derived from (+)-bromide 215 with the racemic enol-lactone 216 gave the crude diketones 225 as an oily mixture of stereoisomers. Intramolecular aldol condensation was achieved by treatment with potassium *t*-butoxide and the mixture of conjugated ketones 226 was readily separated from substantial amounts of by-products, although separation of the two isomers could not be achieved. Treatment of ketones 226 with methyl lithium gave a mixture of tertiary alcohols 227 which was used directly for the acid catalyzed cyclization. This critical step was expected to proceed via initial dehydration to give, for example, the triene 228 which could then protonate to form various cations in equilibrium, including the species 229. A driving force

Scheme 15

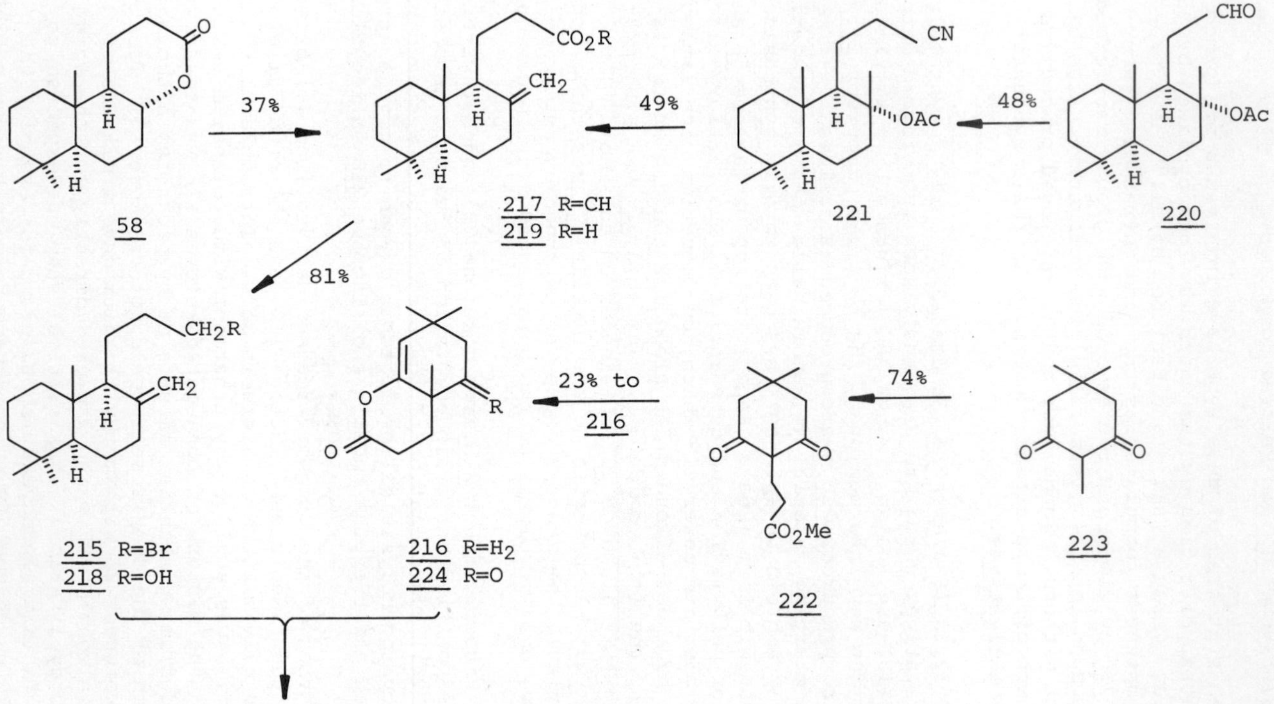
CHO
OAc
CN
OAc
CO_2R
CH_2
37%
49%
48%
58
217 R=CH
219 R=H
221
220
81%
CH_2R
CH_2
R
23% to
216
74%
CO_2Me
215 R=Br
218 R=OH
216 R=H_2
224 R=O
222
223

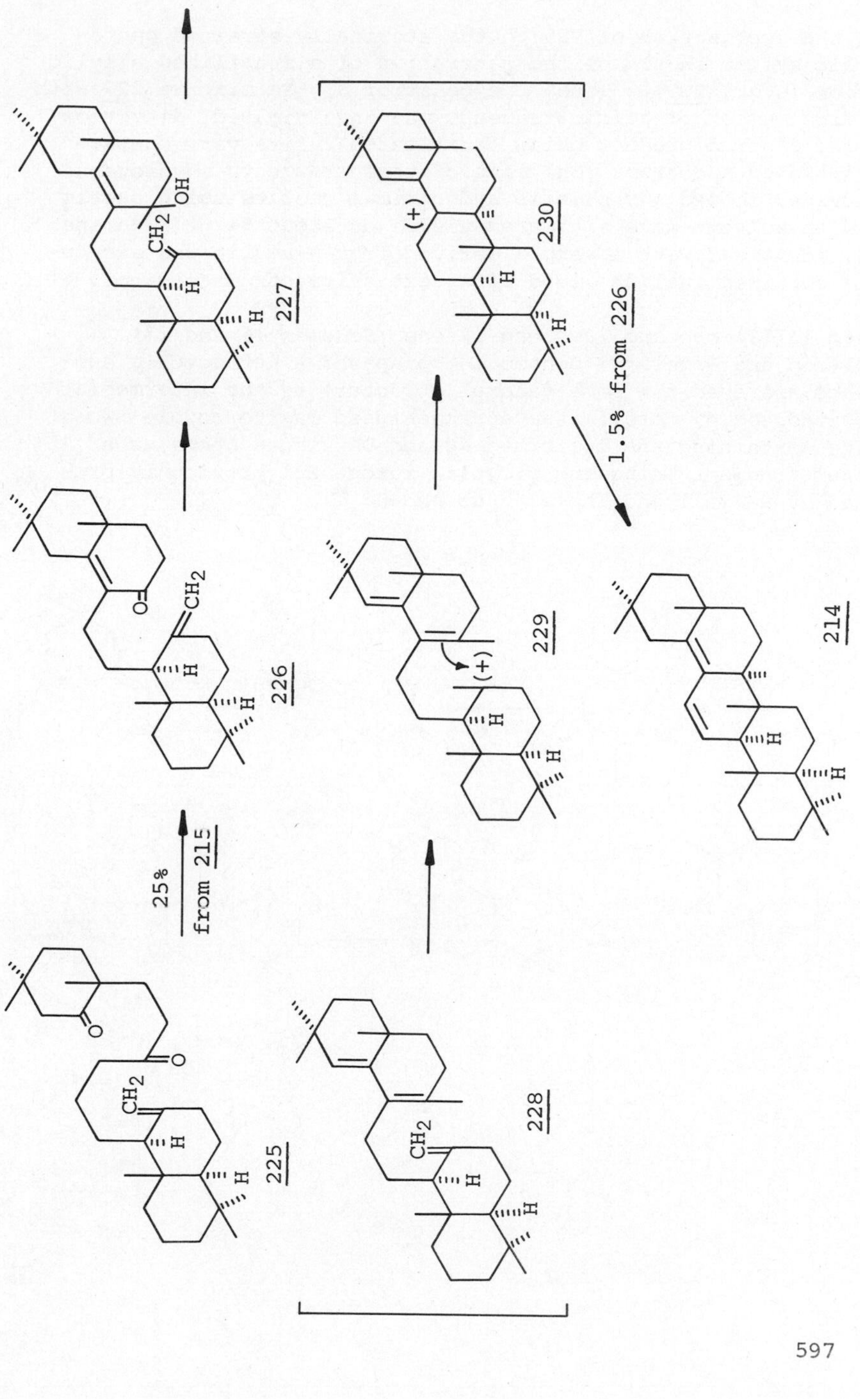

225
25% from 215
226
227
228
229
230
1.5% from 226
214
CH2
OH
(+)

for the cyclization of 229 to the sterically strained pentacyclic system should be the generation of a stabilized allylic cation (e.g., 230). Thus, the behavior of the mixture 227 with a large number of acidic reagents was investigated, detection of the desired product being facilitated by its very characteristic uv spectrum. Only one of these reagents, hydrogen chloride, showed any promise and optimum results using acetic acid as solvent gave estimated yields of about 5%. (-)-diene 214, identical with a sample derived from β-amyrin was eventually obtained in 1.5% yield after extensive chromatography.

Olean-13(18)-ene and 18-Olean-12-ene (Schemes 16 and 17). Barltrop and Rogers[56] (Scheme 16) prepared a tetracyclic substance assigned the same general structure as the intermediate 226 produced by Corey. The approach used was to couple two units containing the preformed AB and DE ring systems, the latter fragment being the bicyclic ketone 231 previously produced by Halsall and Thomas[58] as shown.

Scheme 16

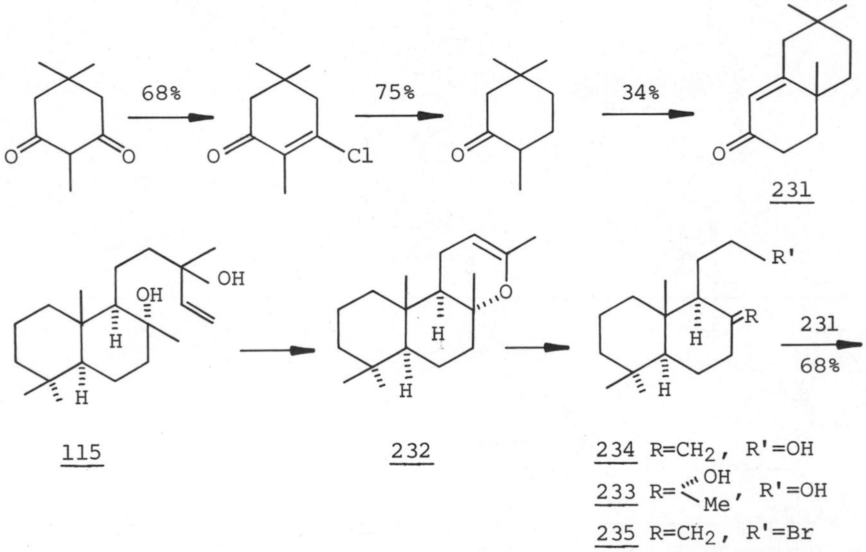
68%
75%
34%
O
O
O
Cl
O
O
231
OH
OH
H
H
O
H
H
R'
R
H
H
231
68%
115
232
234 R=CH2, R'=OH
233 R=OH, Me, R'=OH
235 R=CH2, R'=Br

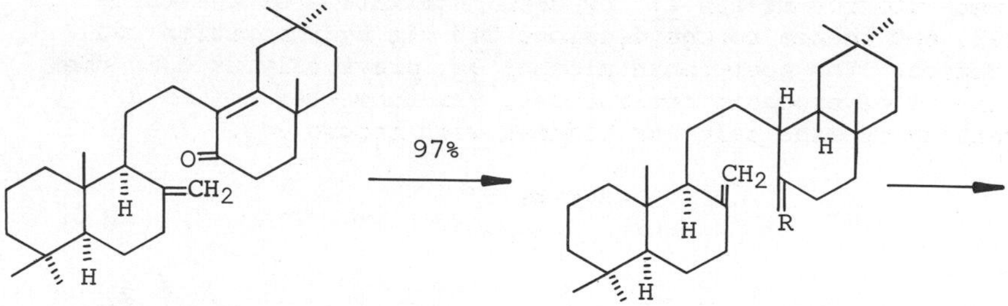

236   97% →   237 R=O
                        238 R= $\langle$ OH / Me

239
241 17α-Me

+

212
240 17α-Me

The diterpene sclareol 115, which has been totally synthesized (26c), was oxidized under carefully controlled conditions to give the enol ether 232 which, on ozonolysis and
hydride reduction of the mixture of acids and aldehydes produced, afforded the diol 233 and the olefinic alcohol 234. On
treatment with phosphorous oxybromide both of these gave the
(+)-olefinic bromide 235, the precursor of rings A and B.

Alkylation of the anion of racemic enone 231 with (+)-
bromide 235 gave a mixture of epimeric dienones 236 which on
Birch reduction afforded the mixture of olefinic ketones 237.
These were converted by methyl lithium to the mixture of alcohols 238 which was cyclodehydrated with aluminum chloride to
give a product containing about 35% of pentacyclic hydrocarbons. Preparative g.l.c. of this material afforded a sample
of mixed crystals of (-)-olean-13(18)-ene 212 and (-)-18α-
olean-12-ene 239, identical with the same substance obtained
from β-amyrin. The other pentacyclic substances were assigned
the 17α-structures 240 and 241.

Sondheimer and Ghera's[57] synthesis (Scheme 16) of the
tital compounds also started from enone 231 which was trans-

formed into the olefin 242 by desulphurization of the thioketal, and thence to the decalone 243 via hydroboration and
oxidation.   The acetylenic alcohol 98, previously used in the
synthesis of onocerin derivatives, was converted to its dimagnesium bromide salt and reacted with ketone 243.   The diol

Scheme 17

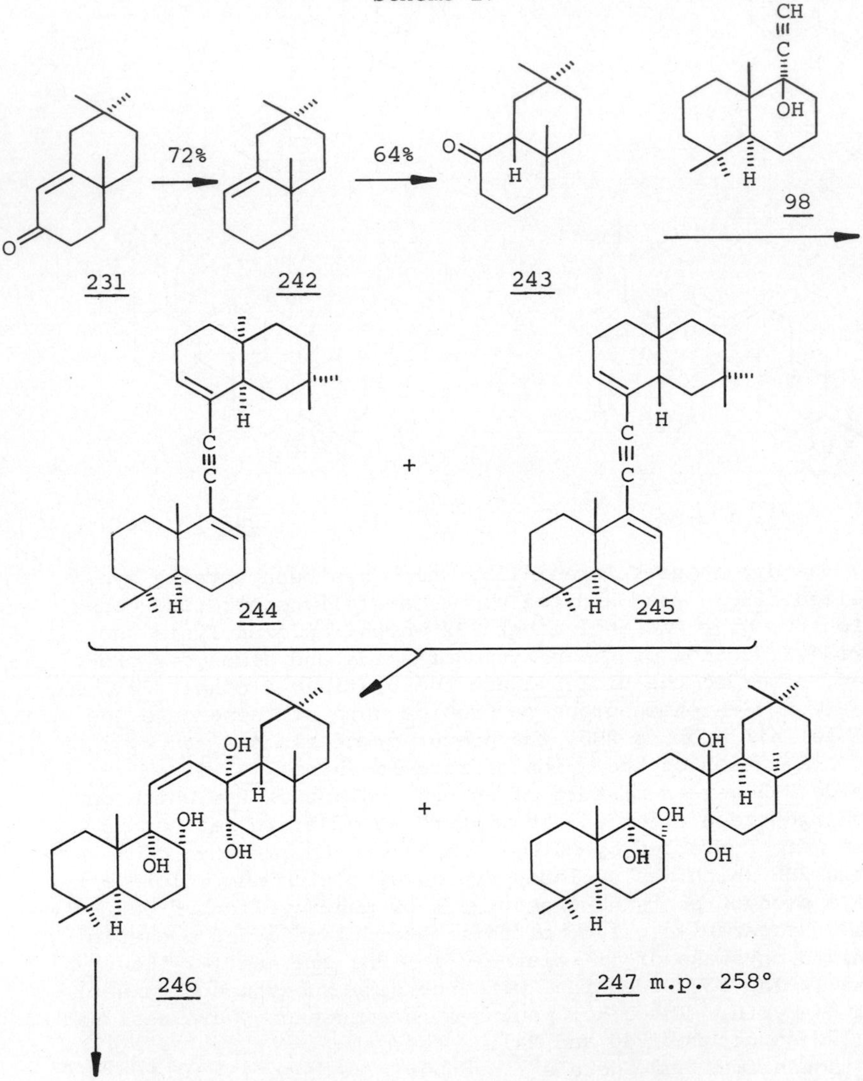

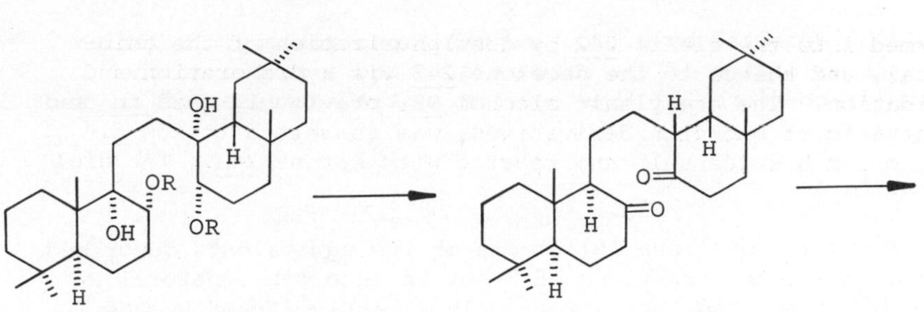

248 R=H m.p. 245°
249 R=Ac

250

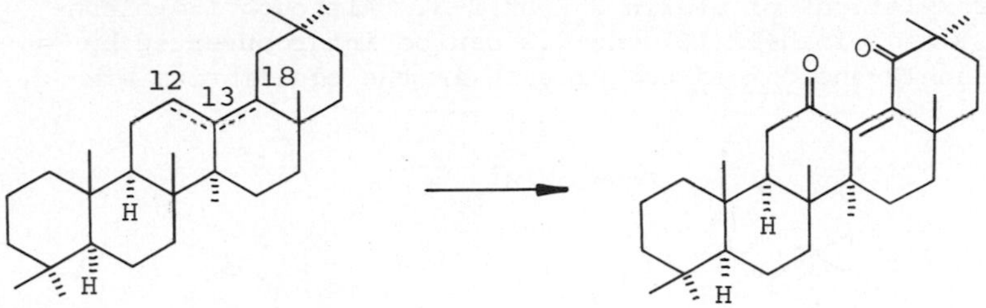

239 Δ-12,18α-H
212 Δ-13(18)

251

produced was dehydrated to give a mixture of the dienynes 244
and 245. Formation of the required trisubstituted double bond
isomers (shown) was predicted for the *cis*-fused decalone on
the basis of conformational analysis. It was thought that the
trans-fused isomer of 243 would have yielded compounds containing tetrasubstituted double bonds.

Osmium tetroxide catalyzed oxidation of the double bonds
was followed by hydrogenation over platinum. The latter reaction proceeded differently for the two isomers 244 and 245
giving compounds later identified as olefinic tetra-ol 246 and
saturated tetra-ol 247 m.p. 258°. These were readily separated and the olefinic compound 246 was further hydrogenated
to the isomer of 247, tetra-ol 248 m.p. 245°. Both saturated
tetra-ols gave diacetates confirming the presence of trisubstituted double bonds in the dienynes 244 and 245. Zinc catalyzed Serini reaction of the diacetate 249 gave, after base
catalyzed equilibration, a mixture presumed to be mainly the
isomer 250. Treatment of this with methylmagnesium bromide
and cyclodehydration of the ditertiary alcohol obtained, using
perchloric acid and acetic anhydride in benzene, gave a complex mixture in which the presence of oleanenes 212 and 239

could be detected by g.l.c. and mass spectroscopy. Selenium
dioxide oxidation of the mixture afforded a sample of the
known diene-dione 251 thus confirming the syntheses of 212 and
239.

β-*Amyrin* 4

The syntheses of olean-13(18)-ene or its equivalents described
above, and transformations of β-amyrin into other natural
products described later, were only recently linked by the
elegant series of reactions achieved by Barton et al.[54] involving, typically, a photolytic process as the key step.

Barton's synthesis (Scheme 18) was in two phases: the
formation of olean-12-ene 252 from 18α-olean-12-ene 239 and
the "hydroxylation" of olefin 252 at C-3. Although 18α-olean-
12-ene 239 and olean-13(18)-ene 212 can be interconverted by
acid, olean-12-ene 252 is not present in the equilibrium mixture.

Scheme 18

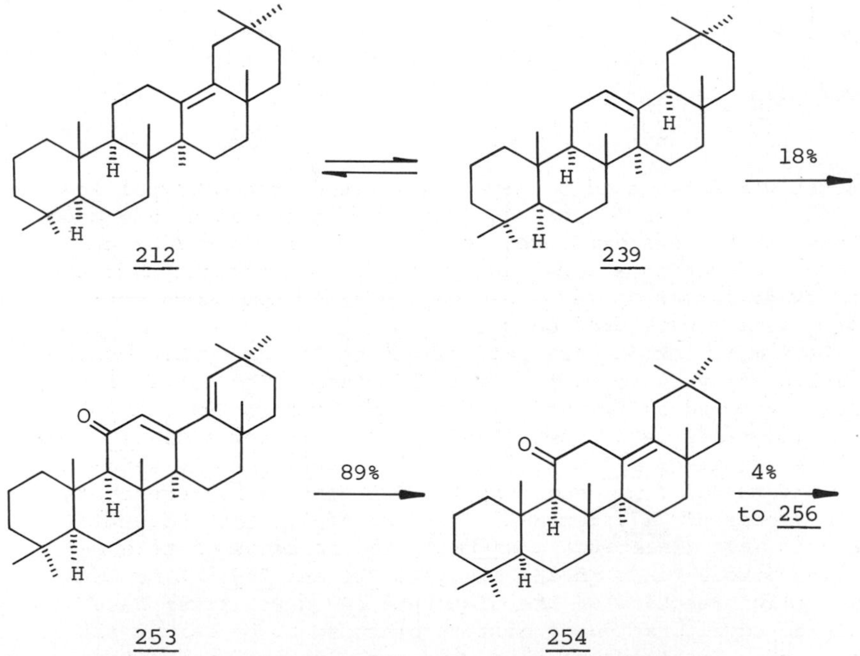

212   239   18%

253   89%   254   4%
                                       to 256

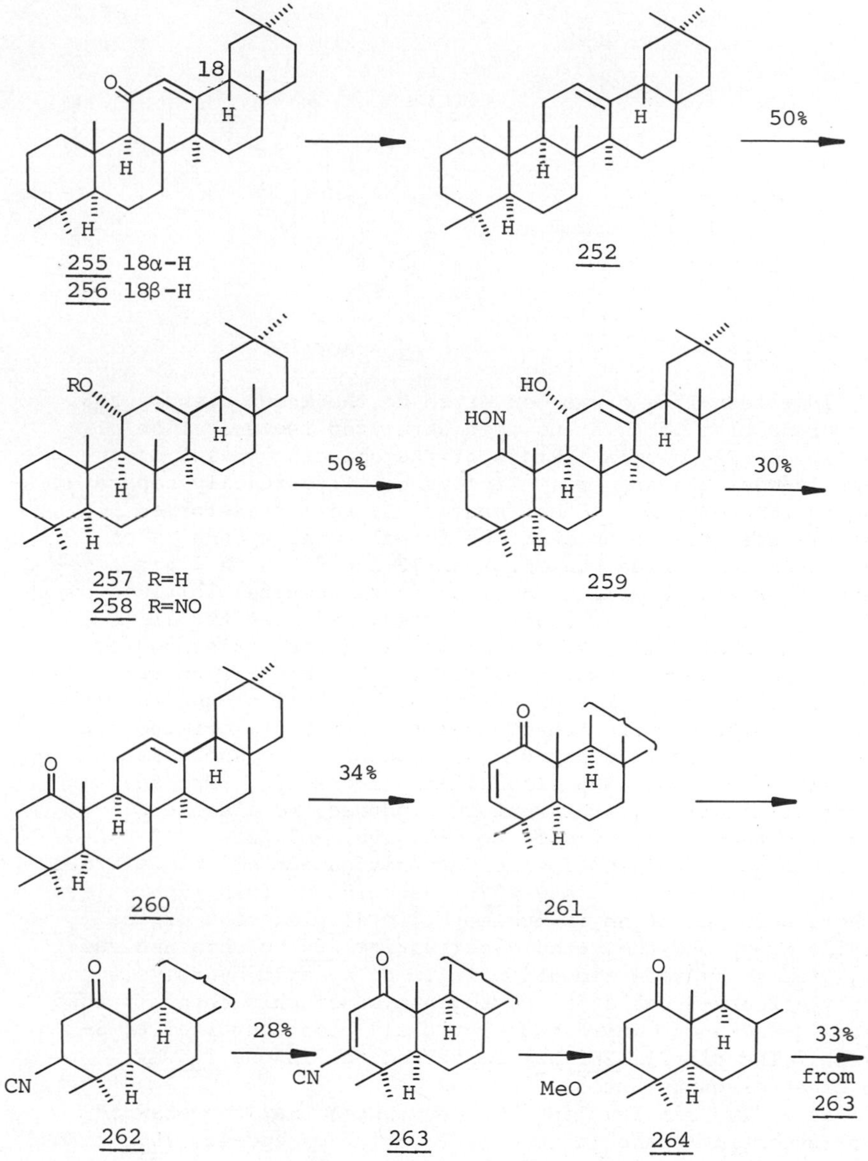

255 18α-H
256 18β-H

252

50%

RO

257 R=H
258 R=NO

50%

HON
HO

259

30%

260

34%

261

262

CN

28%

263

CN

264

MeO

33%
from
263

265   4 β-Amyrin

   18α-olean-12-ene was converted to the known dienone 253.
Birch reduction of this and acid catalyzed isomerization of
the product 254 gave a mixture of the epimeric (C-18) olean-
12-en-11-ones 255 and 256. Since these were readily separated
and the 18β compound 256 had previously been transformed into
olean-12-ene, this completed the formal total synthesis of
that compound. Oxidation of olean-12-ene 252 with a novel reagent, N-bromosuccinimide *and* lead tetra-acetate, followed by
basic hydrolysis of the acetate formed, afforded the 11α-
allylic alcohol 257 (50%). Photolysis of the corresponding
nitrite 258 gave the oximino-alcohol 259, this Barton reaction
being facilitated by the proximity of the C-11 oxygen function
and C-1. Nitrous acid treatment transformed the oximino function into a carbonyl group and subsequent hydrogenolysis of
the allylic alcohol gave olean-12-ene-1-one 260. Bromination/
dehydrobromination of this ketone produced the dienone 261
which smoothly added cyanide ion to give a mixture of 3-cyano-
1-ketones 262. Bromination/dehydrobromination of this mixture
then gave the β-cyano-enone 263. Reaction of this with excess
methoxide ion resulted in nucleophilic displacement of the
nitrile group and the methoxy derivative 264 so obtained was
subjected to hydride reduction followed by acid hydrolysis to
give the 2-ene-3-one 265. Hydrogenation of this then afforded
olean-12-ene-3-one which had previously been converted to β-
amyrin. The nitrile 263 was converted to dienone 265 by
two other elegant routes.
   Recently, van Tamelen[89] has announced that the stannic
chloride-nitromethane induced cyclization of epoxide 265a leads
to δ-amyrin 212 (3β-OH) in 8% yield. In view of existing conversions of the natural compound, this experiment also leads
to formal total syntheses of β-amyrin 4 and germanicol 265a.

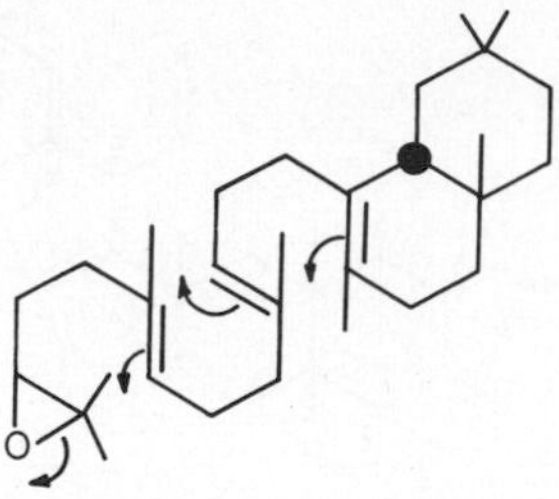

265a

Taraxerol 210 (Scheme 19)

Part of the structural elucidation of taraxerol 210 by Spring
and co-workers[59] involved the partial synthesis of this triterpene from β-amyrin. Thus, the acetate 266 was oxidized to
the ketone 267, using hydrogen peroxide/acetic acid, and this
on bromination/dehydrobromination gave the unsaturated ketone
268. A further bromination/dehydrobromination proceeded with
methyl migration, presumably via a C-13 carbonium ion, to
yield the dienone 269 with a taraxerol skeleton. This compound has been prepared by other methods from β-amyrin.[60]
Birch reduction of the conjugated double bond followed by a
Wolff-Kishner reduction under forcing conditions gave taraxerol 210.

Scheme 19

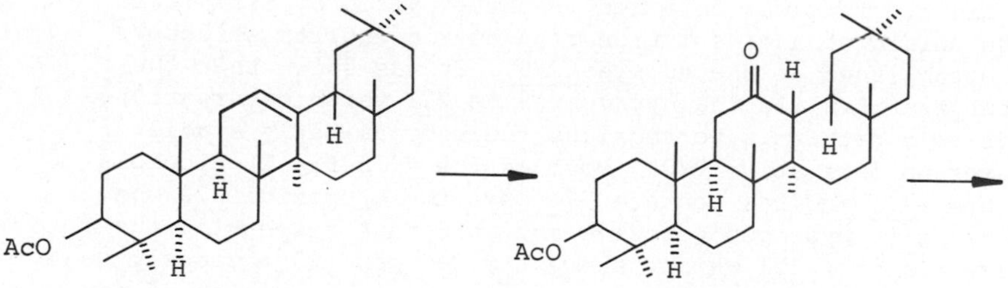

266

267

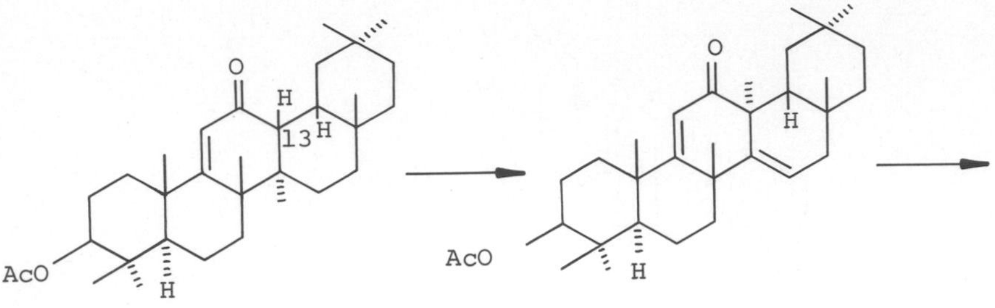

268

269

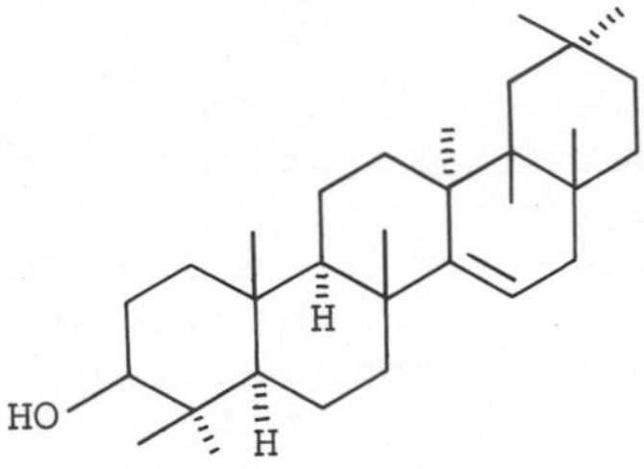

210

Multiflorenol 211 (Scheme 20)

Corey et al.[61] noted that ethanol solutions of β-amyrin deposit a crystalline oxidation product on prolonged exposure to air and light. This was obtained in better yield by irradiation of an acidic ethanol solution of β-amyrin and its structure was established as the epoxide 270. It was shown that this probably arises via the hydroperoxide 271 which can rearrange under acid catalysis to the less thermodynamically stable taraxerane derivative 270. *Both* of the epimeric 11-alcohols 272 and 273, and the acetate 274 gave the α-epoxide 270 when treated with hydrogen peroxide and acid thus supporting the intermediacy of 271 which, since the C-O bond is obviously cleaved during the reaction, probably arises by attack of peroxide on the allylic carbonium ion 275. It was suggested that these reactions may parallel the postulated natural processes whereby the less stable triterpene skeletal types are derived from the amyrins, or their ions, by an oxidative process. The oxidation is said to "power" the rearrangement(s) to less stable structures. The taraxerol synthesis described already, involving a bromination with methyl migration, falls in the

Scheme 20

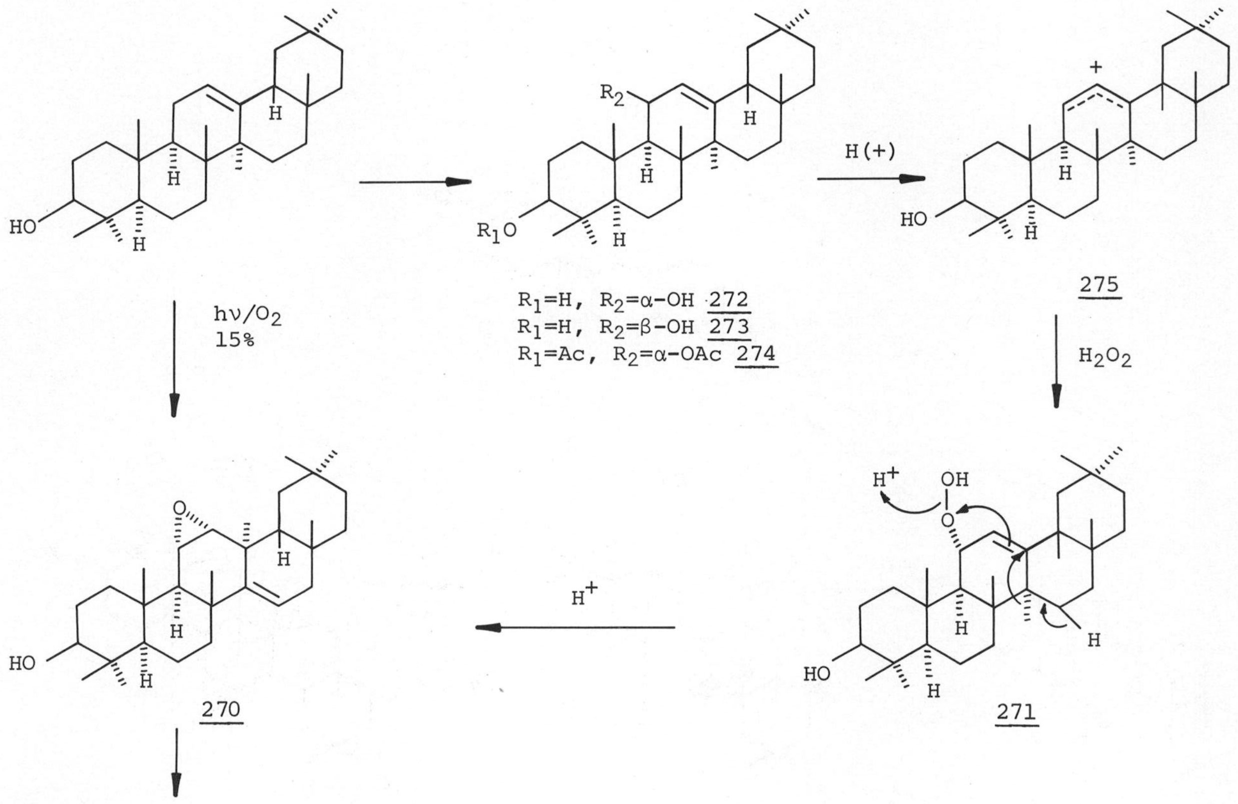

HO
H
hν/O_2
15%
R_2
R_1O
R_1=H, R_2=α-OH 272
R_1=H, R_2=β-OH 273
R_1=Ac, R_2=α-OAc 274
H(+)
+
275
H_2O_2
O
270
H^+
OH
271

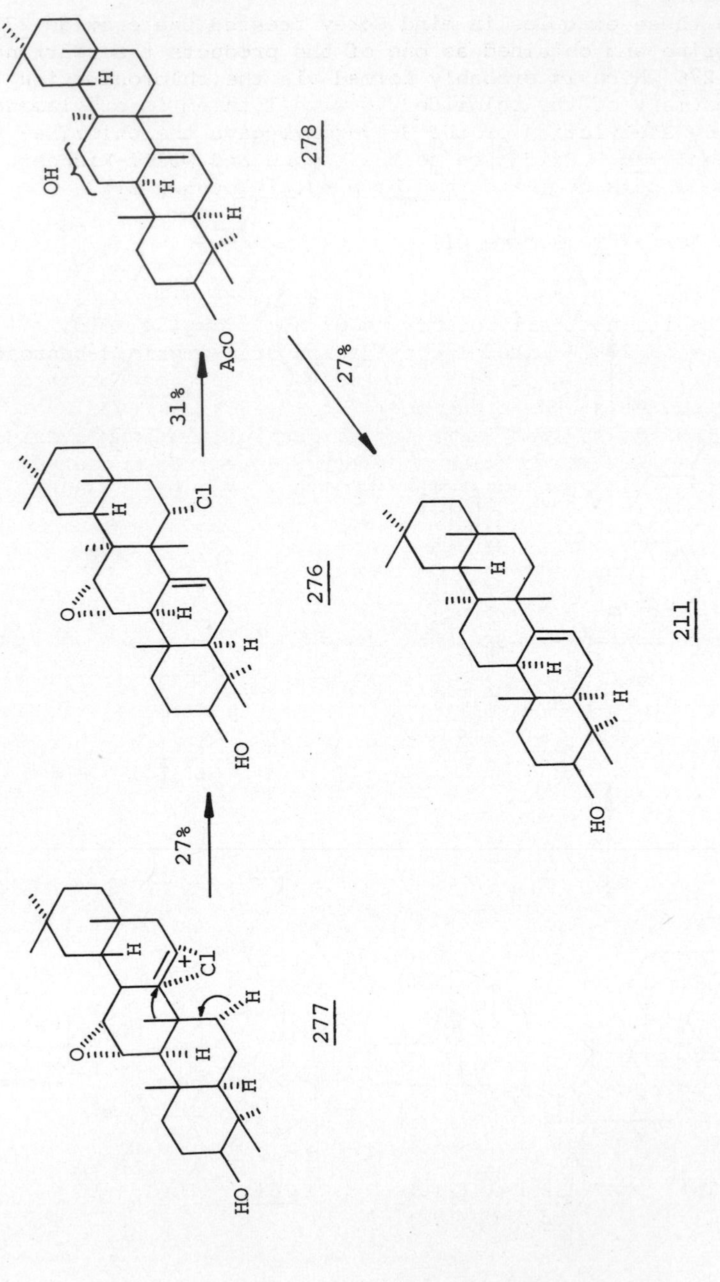
277
27%
276
31%
278
27%
211

same category.

With these examples in mind Corey treated the epoxide 270
with chlorine and obtained as one of the products the rearranged
chloride 276 which is probably formed via the chloronium ion
277. Treatment of the chloride 276 with lithium in ethylamine
followed by acetylation of the 3-hydroxyl gave the chlorine-
free alcohol 278. Oxidation to the ketone and Wolff-Kishner
reduction of this compound then gave multiflorenol 211.

Oleanolic Acid 279 (Scheme 21)

Barton[62] again utilized a photolytic step in the conversion of
β-amyrin to its naturally occurring C-28 carboxylic acid,
oleanolic acid 279. Careful epoxidation of β-amyrin-3-benzoate
gave the 12,13-β-epoxide 280. Previous workers had obtained
only the product of rearrangement of this epoxide, the 12-
ketone. Epoxide 280 was inert toward lithium aluminum hydride
but reduction with lithium in ethylamine gave, after acetylation, a mixture of the 3,12-diacetate 281, β-amyrin acetate

Scheme 21

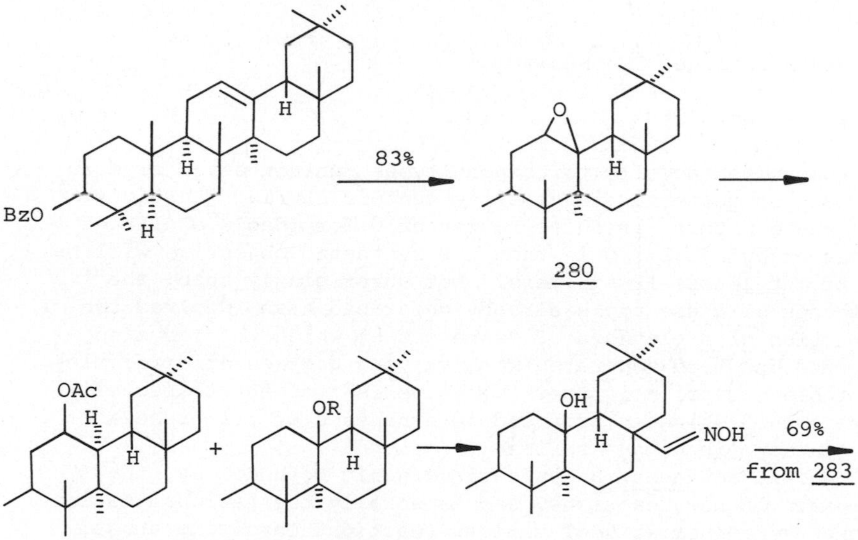

281 (45%)   282 R=H (24%)   284
             283 R=NO (83%)

285 R=H, OH   287 R=Ac
286 R=O (84%)   279 R=H

and the required 13-β-hydroxy-3-acetate 282. Photolysis of
the nitrite ester 283 resulted in oximation of the C-28 methyl
group but, interestingly, not the similarly disposed C-26
methyl group. Nitrous acid treatment of the crude oxime 284
afforded the hemiacetal 285 which was oxidized to the lactone
286. Treatment of this lactone with dry hydrogen chloride in
chloroform gave an equilibrium mixture of 286 and oleanolic
acid acetate 287. The latter was converted to oleanolic acid
279.

C. Stereorational Synthesis

General Principles

The common pentacyclic triterpene types contain seven or more
tertiary or quaternary asymmetric centers at ring junctions.
Thus these natural products represent a formidable challenge
to the organic chemist in that the synthetic objective will be
one of (at least) 128 isomers. Not surprisingly then, the
syntheses of these types already described have involved the
production of a plethora of isomers from which the required
compound has been separated with varying degrees of difficulty.
A notable exception is Stork's synthesis[41] of the tetracyclic
α-onocerin 3 which in turn led to syntheses of triterpenes of
the pentacyclic hopane series.
     Recent advances in synthetic organic methodology, the
discovery of new reactions, and especially the rationalization
of the stereochemistry of various reactions involving substitution by, or attack of anions *in rigid polycyclic systems* had
set the stage for successful stereoselective synthetic assaults
on the pentacyclic triterpenes. A brief discussion of such
reactions is undertaken here using the hypothetical "rigid"
ring system 228 as a convenient frame of reference (see Ref. 63

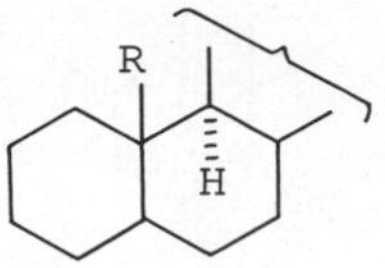

288

for a comprehensive survey). This will serve to explain the principles involved so that no further comment need be offered when such reactions are met in the synthetic work described in the sequel.

Alkylations of enolate anions of the type 289 (Scheme 22) have been much used.[64,65] If R be H then alkylation takes place predominantly from the β side to give 290 (axial attack due to stereoelectronic control in the absence of overwhelming steric hindrance factors.*[66] If R be methyl then substitution takes place predominantly from the α side to give 291 (steric approach control due to the hindrance of the 1,3-axially disposed β methyl group). The alkylation of dienolate anions of the type 292, derived from α,β-unsaturated ketones, is also highly stereoselective giving 293.[66,67] However, reactive alkylating agents must be used if reasonable yields are to be obtained.

Scheme 22

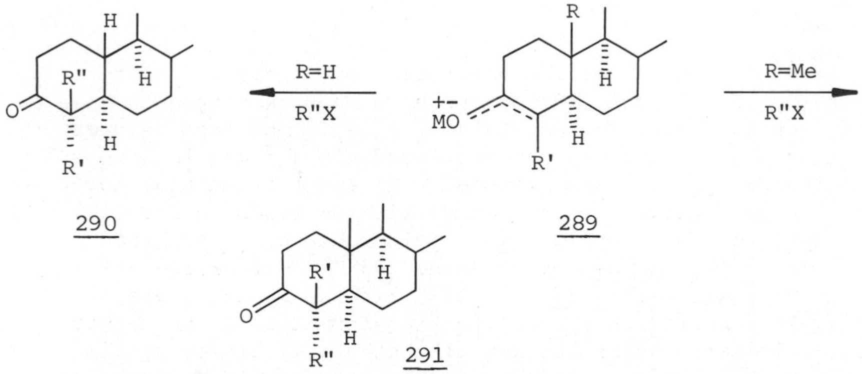

290   289

291

*G. Stork (IUPAC Conference, Boston, Mass., U.S.A., June, 1971) has suggested that steric factors alone, and especially eclipsing between a C-4 substituent and the C-6 methylene group in the transition state, may be an alternative to "stereoelectronic considerations."

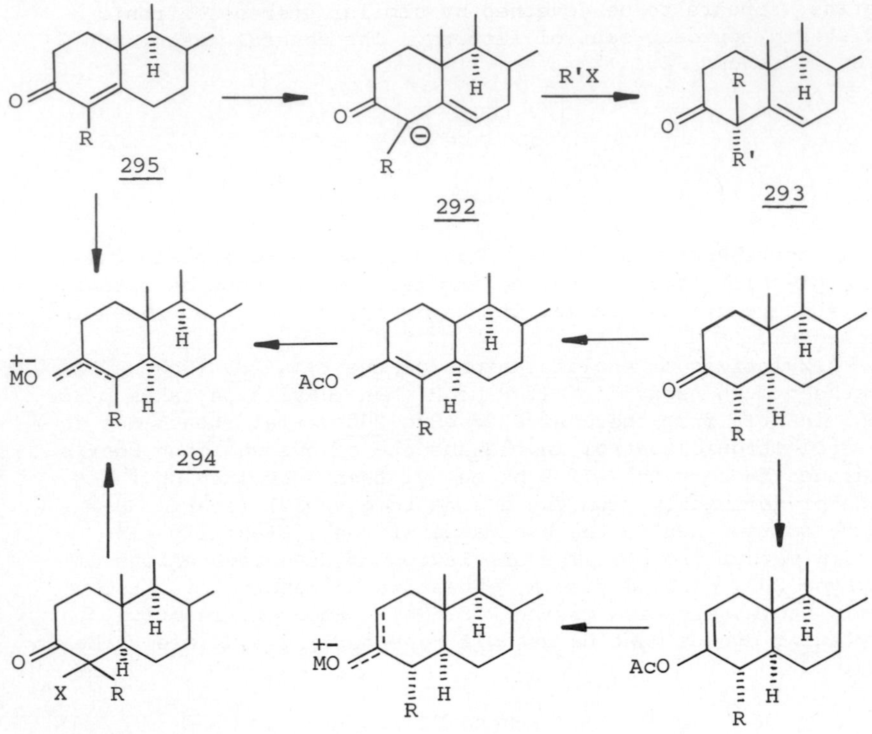

296 X=OAc or
    halogen

Several methods for the generation of specific enolate
anions for use in such alkylations have been developed in
recent years. Thus, enol acetates can be formed from ketones
and adjustment of the experimental conditions[68] can lead to a
predominance of the desired isomer. Cleavage of an enol acetate with two equivalents of methyl lithium leads to the lithium enolate, for example, 294,[68] also obtainable by Birch
reduction of the corresponding enone 295[64,65] or of the α-
acetoxy or α-haloketone 296.[69] Such lithium enolates are
relatively stable with respect to transformation to the other
possible isomeric anion but are alkylated only slowly by all
but the most reactive halides. The process of Birch reduction
followed by alkylation of the enolate anion generated, developed largely by Stork,[64a,65] will be referred to as "reductive alkylation" in the sequel.

A phenomenom related to the alkylation of ketones, the
Michael addition of enolate or dienolate anions to activated

olefins, appears to be governed by similar stereoelectronic and steric approach control factors. The general cases are given in Scheme 23.[70],[71]

Scheme 23

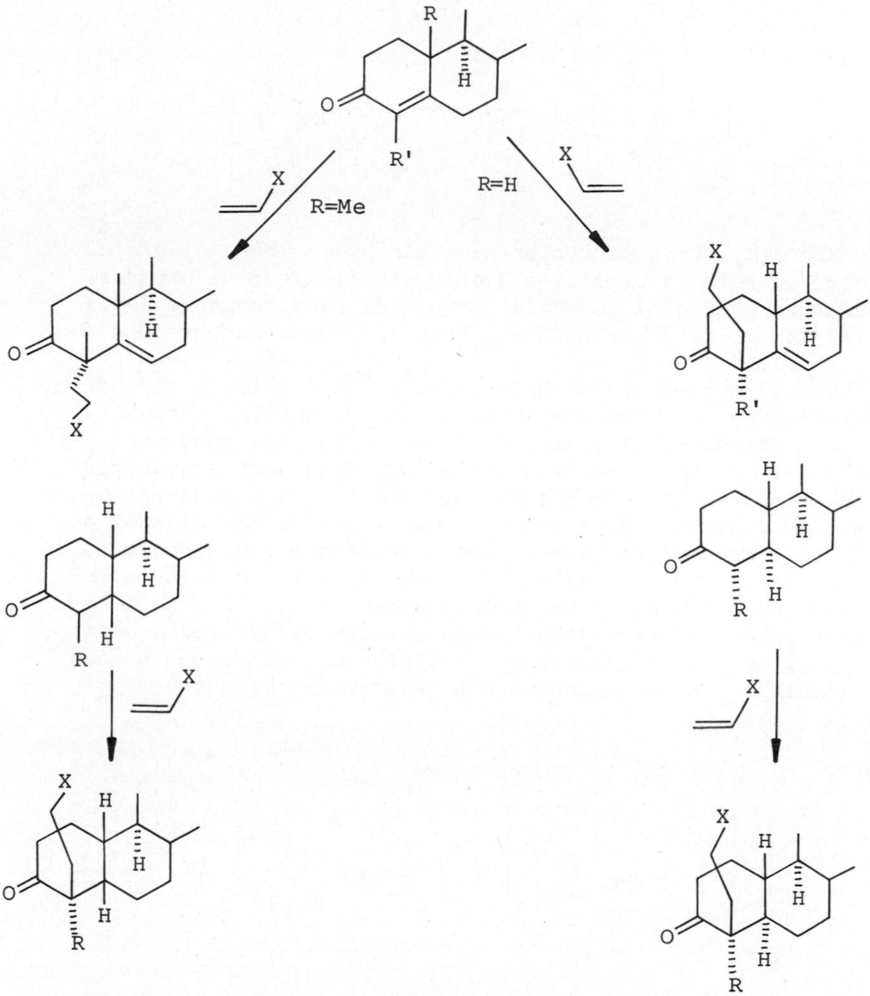

$X = CN, CO_2R, COR, NO_2$ etc.

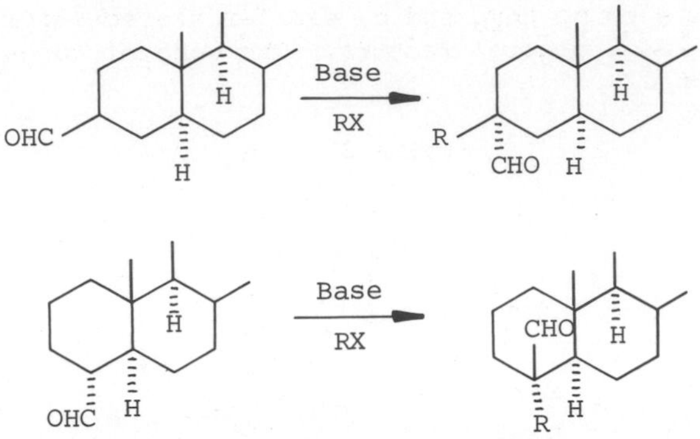

The alkylation of cyclohexane aldehydes (Scheme 23) in "rigid" systems is usually a highly stereoselective process leading to the axial aldehyde. Again stereoelectronic (equatorial attack favored) and/or steric hindrance factors are implicated.[72]

The Simmons-Smith methylenation[73,74] of allylic alcohols has been used[71,75] for the introduction of angular methyl groups. Stereospecific attack of the organozinc reagent (Scheme 24) on the same side of the molecule as the hydroxyl group (with which it complexes) gives cis-cyclopropano-alcohols, for example, 297 from 298. Oxidation of the alcohol and Birch reduction of the α,β-cyclopropanoketone 299 gives the enolate anion 300 by fission of the bond having maximum overlap with the carbonyl π orbitals (Dauben et al.[76]). Such enolate anions afford either the angularly methylated product 301 (R=H) on protonation or the additionally alkylated product 301 (R=alkyl) on quenching with a reactive alkylating agent.[65,70]

Scheme 24

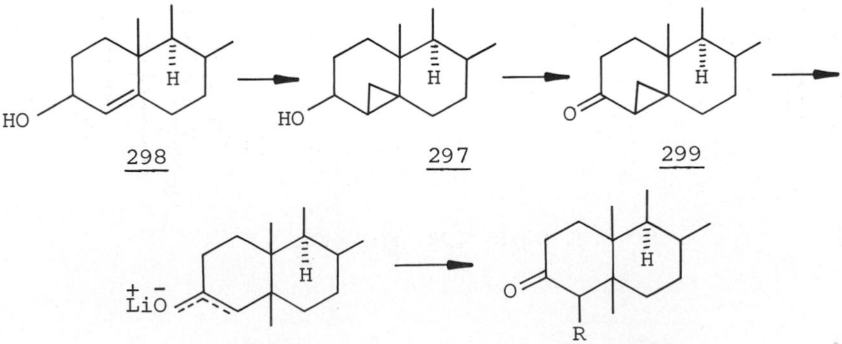

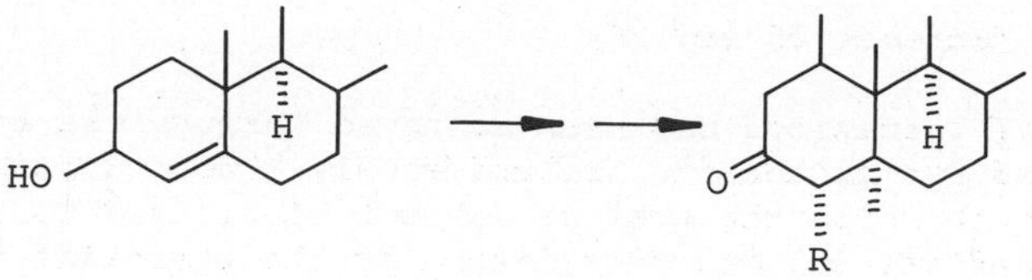

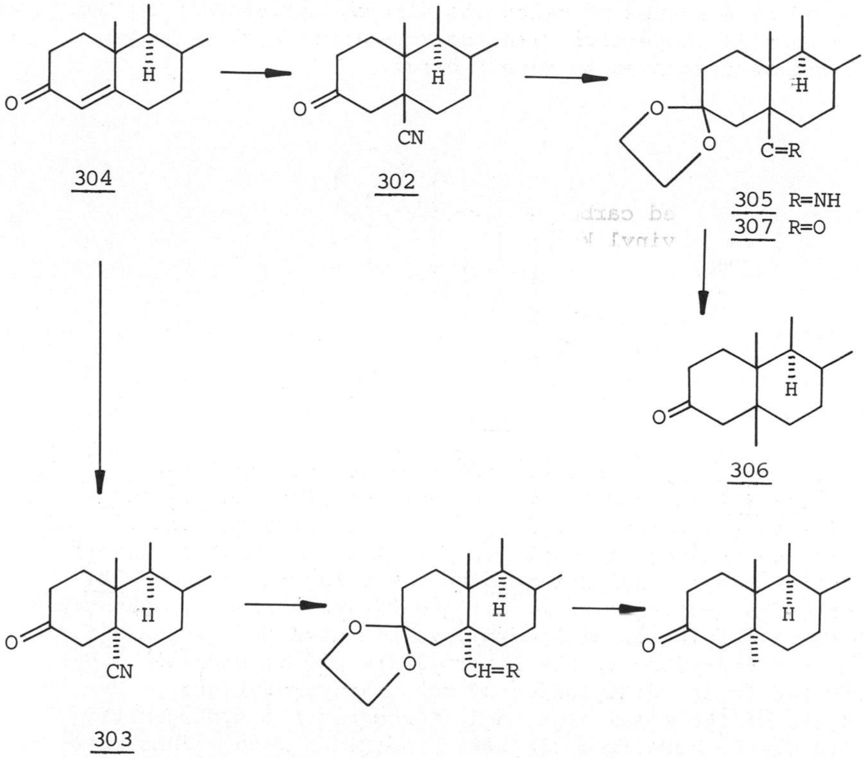

304   302

305 R=NH
307 R=O

306

303

The development of alkyl aluminum/hydrogen cyanide reagents by Nagata[77] has led to another useful method for the
introduction of angular methyl groups. These reagents add
the elements of hydrogen cyanide to α,β-unsaturated carbonyl
compounds (especially ketones) and in many cases the reagents
and conditions can be varied to give a particular isomer predominantly, for example, 302 or 303 from enone 304 (Scheme 24).
Controlled reduction of the nitrile group with lithium aluminum hydride or (best) with di-isobutylaluminum hydride gives
the imine 305 which can be directly reduced (Wolff-Kishner) to
a methyl group 306 or hydrolyzed to the aldehyde 307 for
further manipulation.

Germanicol 308 (Schemes 25 and 26)

The synthesis of germanicol has been achieved through the combined efforts of two groups.[71] Ireland et al.[78] used a total
synthetic approach while the work of Johnson et al. led to the
same key intermediate 309 via degradation of the steroidal triterpene euphol 310 to the lactone 311. The latter process,
achieved by a series of reactions already described[79] in the
literature in connection with the structural work on euphol,
will not be discussed in detail here.

308

The total synthesis started with 2-methylcyclohexane-
1,3-dione 312 (Scheme 25) which gave the diketone 313 on
annelation with ethyl vinyl ketone. Selective ketalization of
the non-conjugated carbonyl group followed by another annelation with ethyl vinyl ketone gave the tricyclic dienone 314
in which the two angular methyl groups are correctly orientated.
Reductive alkylation with methyl iodide gave the ketone 315
which was converted to the intermediate 316 by successive hydride reduction, acetylation of the C-3 hydroxyl group, hydrolysis of the ketal group and stereospecific hydrogenation
of the double bond from the less-hindered α face. This keto-
acetate was then transformed into the diketone 309 by two different routes.

In the first route the diosphenol-methyl ether 317 was
obtained via bromination of 316 and subsequent treatment of
the dibromide with alkaline methyl sulphate followed by reacetylation at C-3. This enol ether 317 was converted to a
spiro-epoxide (dimethylsulphonium methylide) which gave the
aldehyde 318 on treatment with BF_3. Methylation of the aldehyde 318 was achieved with difficulty, due to its highly con-
gested environment, but stereospecifically to give 319. Wittig reaction gave the olefin 320 which was successively hydrolyzed to the ketone and reduced to the alcohol 321. Hydroboration of this compound followed by chromium trioxide

Scheme 25

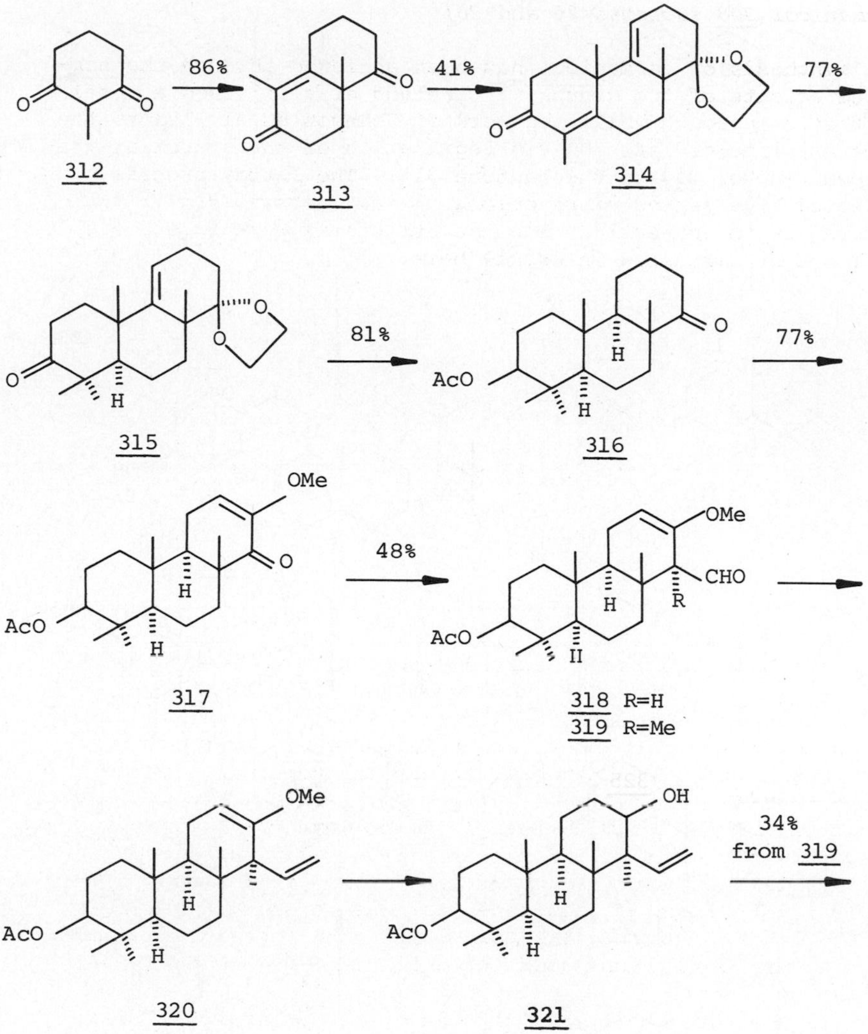

312   86%   313   41%   314   77%

315   81%   316   77%

317   48%   318 R=H   319 R=Me

320   321   34% from 319

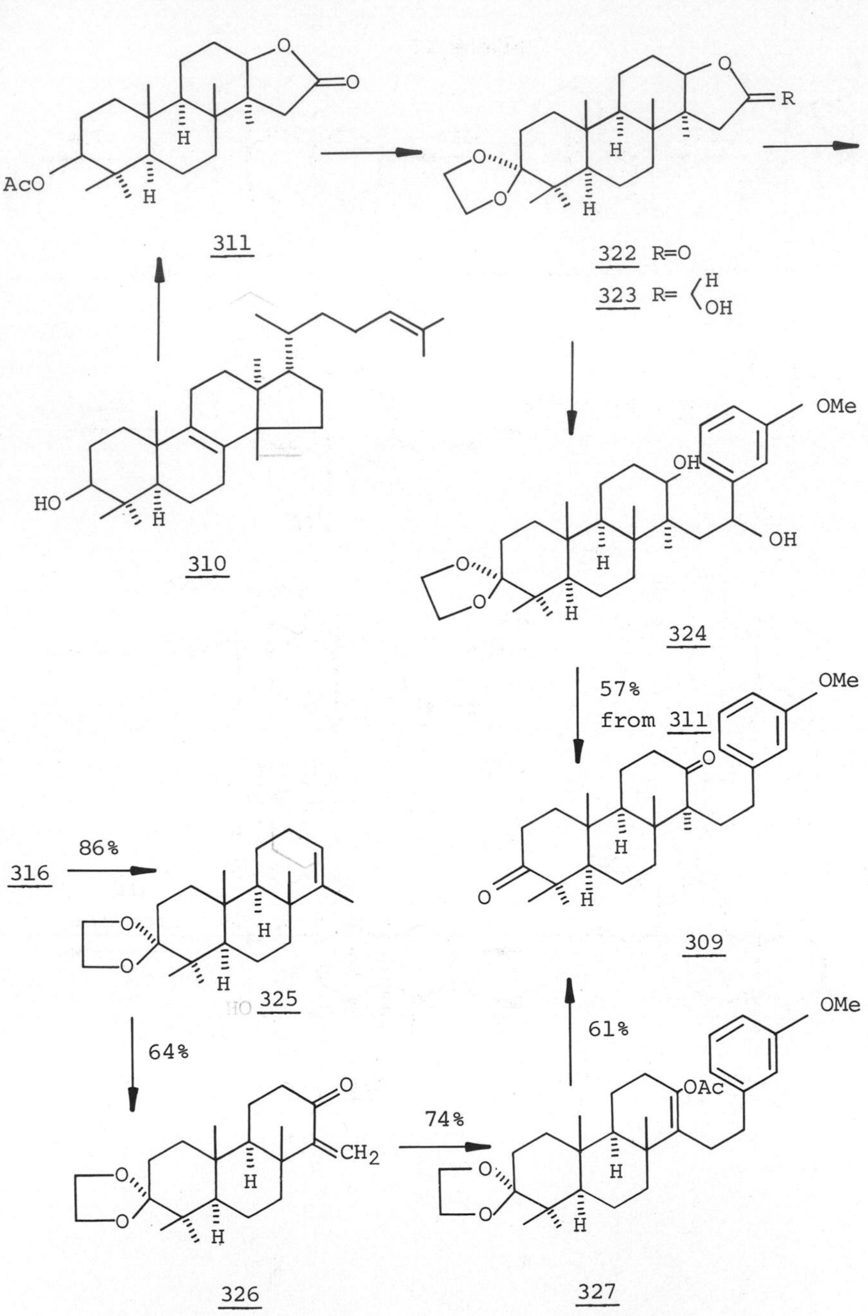

AcO
H
311
O
=O
O
R
322 R=O
323 R= H OH
HO
310
OMe
OH
OH
324
57%
from 311
OMe
O
309
316
86%
325
64%
O
CH2
326
74%
OAc
OMe
327
61%

oxidation gave the racemic lactone-acetate 311. Fortuitously, this crystallized as a racemic mixture which was resolvable by mechanical separation. Thus, the identity of the (+)-enantiomer with the same compound obtained from euphol 310 was readily established. (+)-lactone-acetate 311 (from euphol) was converted into the lactone ketal 322 via hydrolysis of the acetate function, oxidation to the ketone and ketalization. Reduction with a dialkylborane gave the hemiacetal 323 which was reacted with m-methoxyphenylmagnesium bromide. Hydrogenolysis of the benzylic hydroxyl group in the product 324, followed by hydrolysis and oxidation, then afforded the key intermediate (+)-diketone 309.

Racemic diketone 309 was obtained more conveniently from intermediate 316 through another sequence. Thus, reaction of 316 with methyl lithium, followed by dehydration of the tertiary alcohol so formed, and manipulation of the C-3 oxygen substituent, gave the olefin 325 in 86% overall yield. This was subjected to sensitized photo-oxygenation, and hydride reduction of the hydroperoxide so formed followed by oxidation of the allylic alcohol gave the enone 326. Such systems undergo conjugate addition with great facility so that treatment of 326 with m-methoxyphenylmagnesium bromide followed by trapping of the enolate anion with acetic anhydride gave the enol acetate 327. This was cleaved with methyl lithium (two equivalents) and the enolate anion so formed was stereospecifically alkylated with methyl iodide to give, after hydrolysis of the ketal function, the racemic diketone 309.

Scheme 26

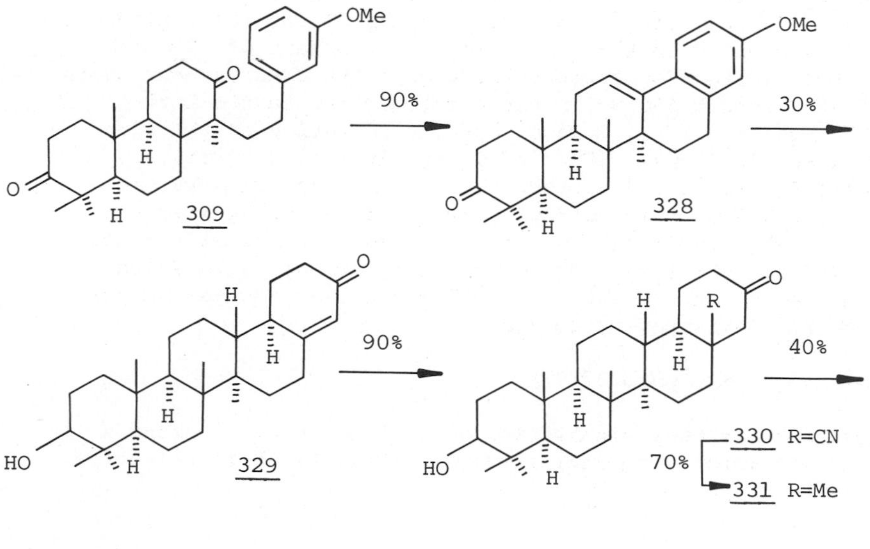

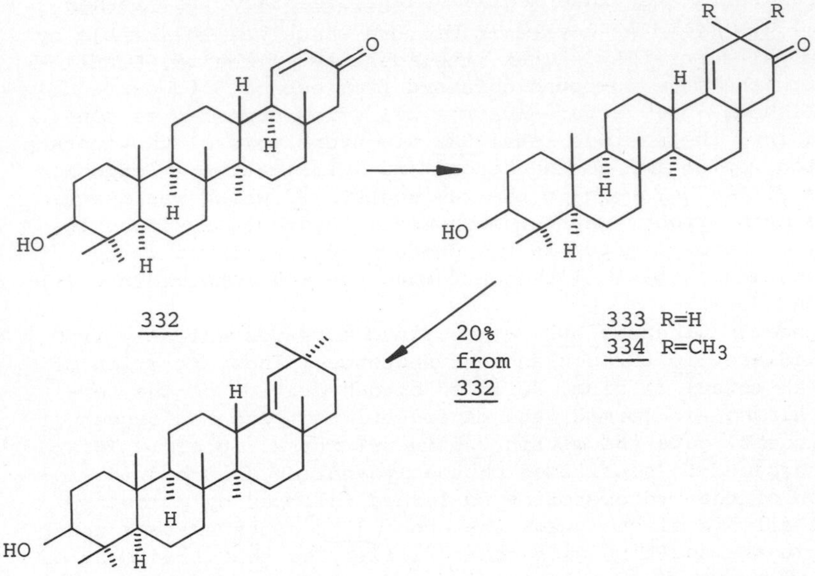

Cyclodehydration (Scheme 26) of diketone 309 using poly-phosphoric acid proceeded smoothly and the product 328 was re-duced with lithium and alcohol in liquid ammonia. This resulted in saturation of the styrenoid double bond and reduction of the ring so that acid workup gave the pentacyclic enone 329 (31% from 328). The low yield in this step may be due to the pro-duction of significant quantities of the CD *cis*-fused products, previously noted in similar systems. Treatment of 329 with triethyl aluminum/hydrogen cyanide gave the DE-*trans*-conjugate addition product 330 which was transformed into the keto-alcohol 331 via ketalization, hydride reduction of the cyano group, Wolff-Kishner reduction of the imine and hydrolysis of the ketal. Bromination/dehydrobromination of 331 gave the enone 332 which, interestingly, could not be methylated di-rectly. The double bond was first migrated by ketal formation, followed by hydrolysis, and then the β,γ-unsaturated ketone 333 was methylated. The product 334 gave racemic germanicol 308 on Wolff-Kishner reduction.

Alnusenone 6 (Schemes 27-29)

Alnusenone has been synthesized by Ireland et al.[75] using a scheme incoroporating many elegant and highly stereoselective

reactions. The key intermediate was the pentacyclic diether
335 and this was approached by two routes, the second being
developed as a result of low yields at late stages in the first.

6

In the first approach[75a] (Scheme 27) the known tricyclic
ketone 336 was stereospecifically alkylated from the α side to
give the keto-acid 337. Attempts to hydrogenate esters of
this acid led to mixtures in which the undesired cis isomer
predominated, probably due to a quasiboat conformation in the
flexible ketonic ring C of 336. Borohydride reduction of
ketone 336 gave the alcohol 338 in which the preferred equatorial configuration of the hydroxyl group assures a quasi-
chair conformation. This compound was smoothly hydrogenated
to the BC trans product which afforded trans-lactone 339 on
acid treatment. Reduction of this lactone with disiamylborane
gave not the desired lactol 340 but the diol 341, apparently
because the open chain hydroxy-aldehyde structure is preferred
over the trans-lactol 340 due to strain in the latter. However, since experiments with cis-locked lactones (less strained)
had given the corresponding lactols, cis-lactone 342 was prepared. Treatment of the trans-lactone 339 with the lithium
aluminum salt of methylamine gave the hydroxyamide 343. Reaction of this alcohol with methanesulphonyl chloride-pyridine
and then with water gave the cis-lactone 342 (R=O), presumably
via the iminolactone 342 (R=+NMe) formed by attack of the
amide oxygen on the mesylate 344 with inversion of configuration. Cis-lactone 342 was smoothly reduced to lactol 345
which, on reaction with m-ethoxyphenylmagnesium bromide followed by hydrogenolysis of the benzylic hydroxyl group and
subsequent Jones oxidation, gave the ketone 346. This was
cyclodehydrated (p-toluenesulphonic acid) to the octahydro-
picene 347, which was found (surprisingly) to be one of the
few olefins resistant to hydroboration. Epoxidation yielded
a mixture of ketones 348 and 349, readily equilibrated (95%)
to the trans-anti-cis isomer 349 and the ketol 350. The latter was transformed to give further 348 and 349 by pinacol

Scheme 27

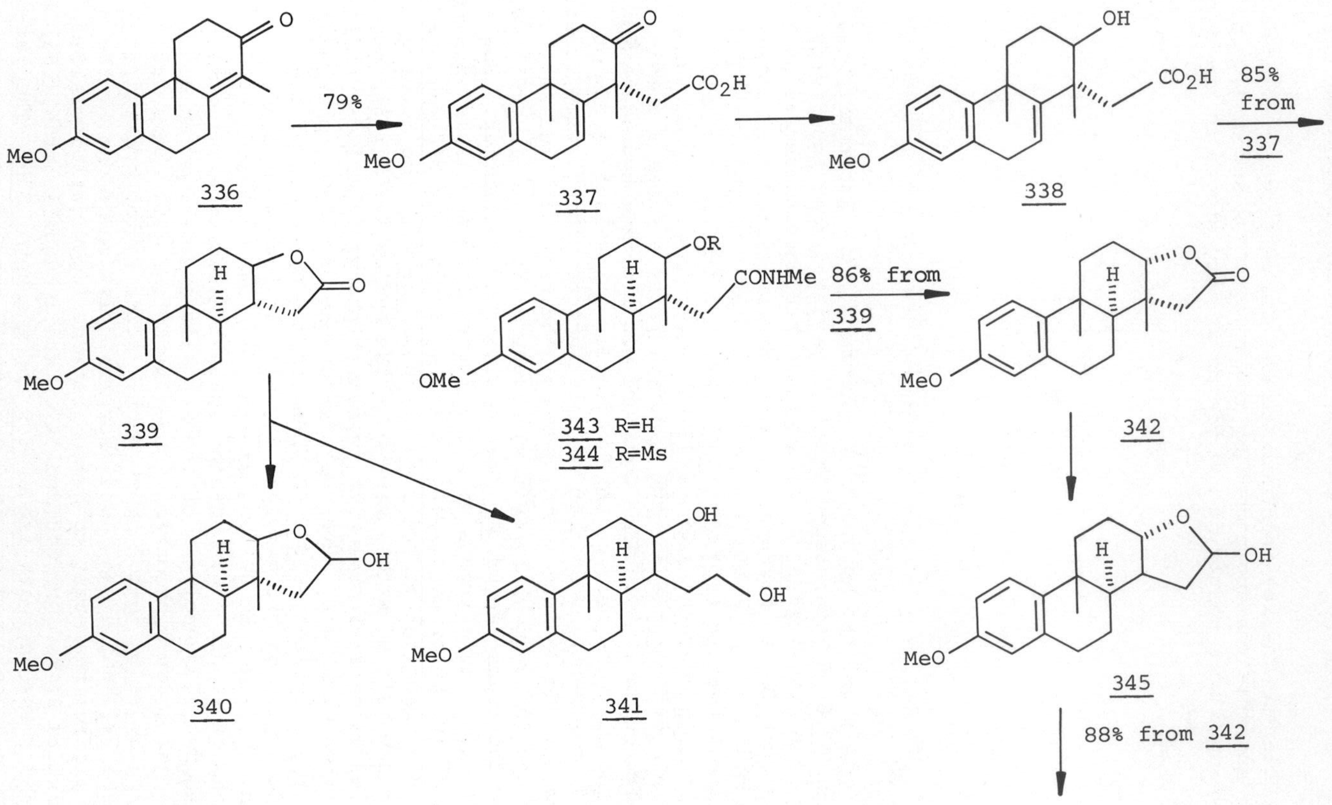
MeO
O
336
79%
CO2H
337
OH
338
85% from 337
H
339
OR
CONHMe
OMe
343 R=H
344 R=Ms
86% from 339
342
340
341
345
88% from 342

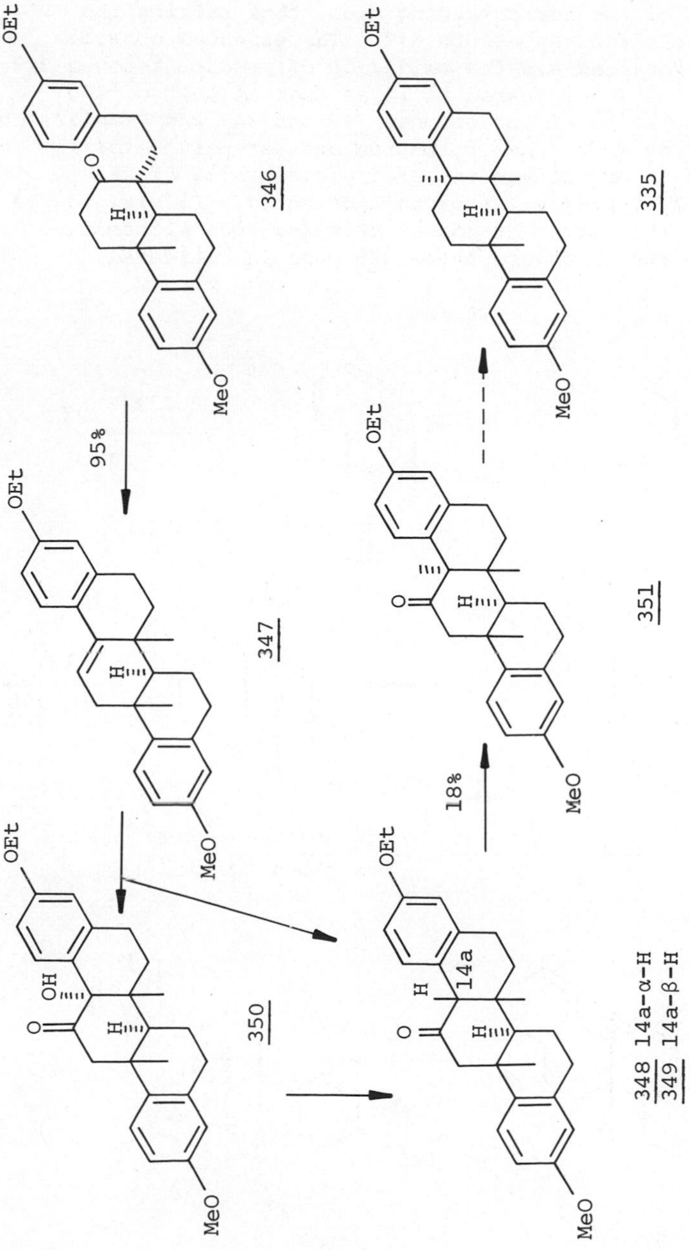
346
95%
347
350
18%
351
335
348 14a-α-H
349 14a-β-H

rearrangement of the corresponding diol, thus raising the
yield of the desired ketones to 67%. The expected epoxide
could not be detected and the oxidation of product ketones 348
and 349 to ketol 350 appeared to be as fast as that of the
olefin. Methylation of the ketones 348 and 349 was complicated
by predominating enol ether formation and extensive experi-
mentation led to an optimum isolated yield of 18% of ketone
351 as the only C-alkylation product detected. In view of the
low yields in the latter stages of this sequence alternative
approaches to the decahydropicene 335 were investigated.[78]

Scheme 28

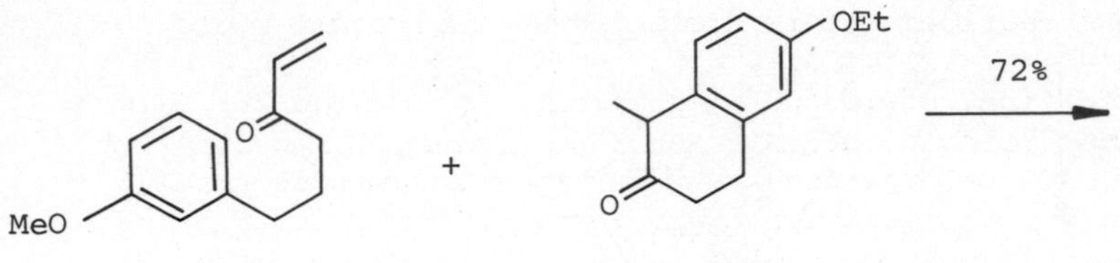

353

352

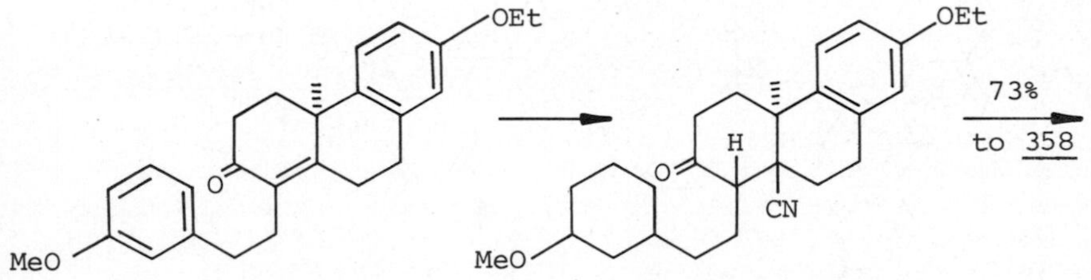

72%

354

97% 355 α-CN, β-H
85% 356 β-CN, α-H

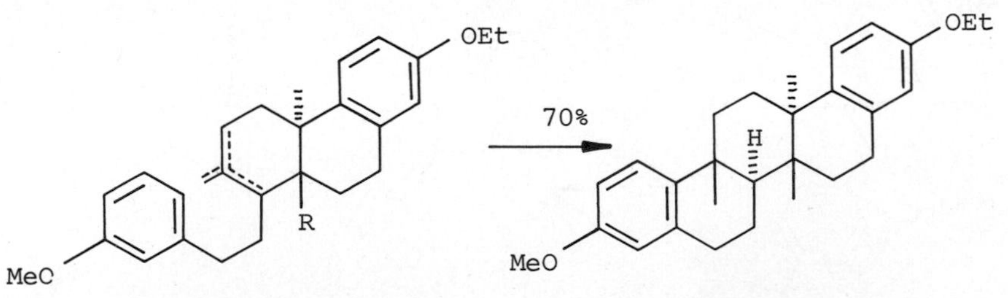

73%
to 358

357 R=CN
358 R=CH₃

335

The successful approach[75b] (Scheme 28) used a convergent type synthesis. Thus, annelation of tetralone 352 with the known vinyl ketone 353 gave the enone 354. Hydrocyanation of this compound using Nagata's alkyl aluminum cyanide reagents could be controlled to give either the *cis*-fused compound 355 (97%) [thermodynamic control-- $(C_2H_5)_2AlCN$-benzene] or the *trans*-fused isomer 356 (85%) [kinetic control-- $(C_2H_5)_3Al$-HCN-THF]. The latter failed to react with methylenetriphenylphosphorane but addition of methylmagnesium iodide followed by dehydration ($SOCl_2$-pyridine) of the tertiary alcohol gave a mixture of cyano-olefins 357, which was converted to the mixture of olefins 358 by reduction to the imine and subsequent modified Wolff-Kishner reaction. This mixture was cyclized by acid to the required intermediate 335 in an overall yield (from tetralone 352) of 31%.

In the final series of transformations (Scheme 29) selective demethylation of the diether 335, achieved with lithium diphenylphosphide, was necessary in order to protect the A ring during Birch reduction of the E ring. The latter process was complicated by the insolubility of the lithium phenolate but use of a dilute solution and excess lithium in ammonia-THF gave, after remethylation of the A ring phenolic moeity and acid treatment, the enone 359. This was reduced (di-*i*-butyl-aluminum hydride) to the *pseudo*-equatorial alcohol which gave the cyclopropylketone 360 after directed Simmons-Smith cyclo-proponation and subsequent oxidation of the hydroxyl group. *Geminal*-dialkylation of 360 gave the ketone 361, the cyclopropyl group serving as an efficient blocking agent. Lithium-ammonia cleavage of this moeity followed by Wolff-Kishner reduction of the carbonyl group (achieved with difficulty) gave the intermediate 362. This was subjected to Birch reduction followed by acid treatment to give enone 363 which was dimethylated under conditions which give the 4,5-ene, *dl*-alnus-enone 6, rather than its Δ-5,10 isomer.

Scheme 29

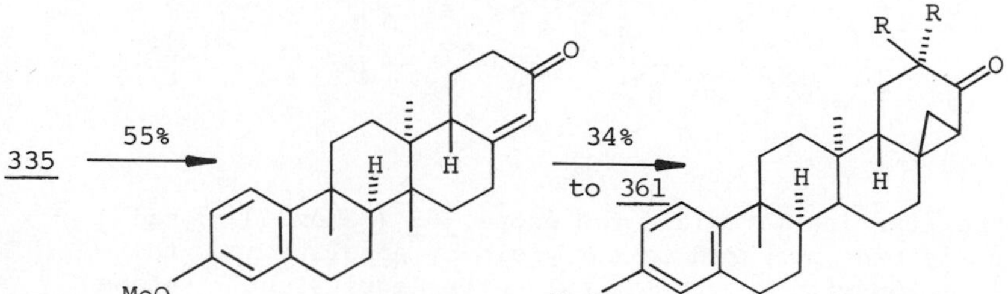

335 →(55%) 359

359 →(34% to 361)

360 R=H
361 R=Me

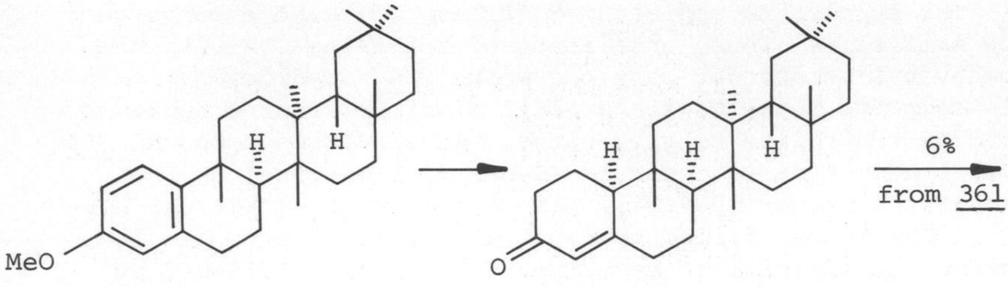

MeO

362

363

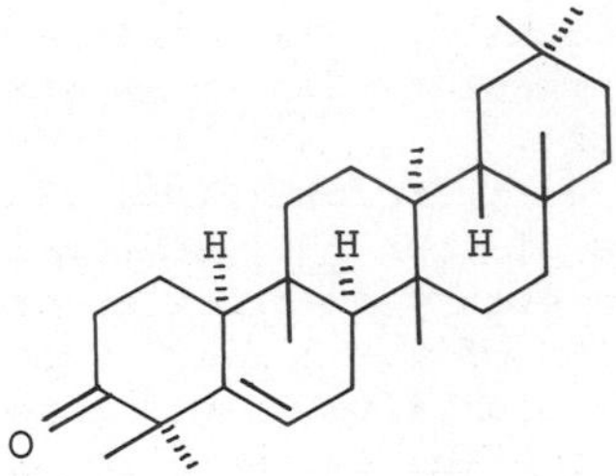

6%

from 361

O

6

Lupeol 364

Lupeol has been synthesized by Stork et al.[65,80] (Schemes 30-33) through a remarkable series of reactions which resulted in the creation of the asymmetric centers one at a time and each in a highly stereoselective fashion.

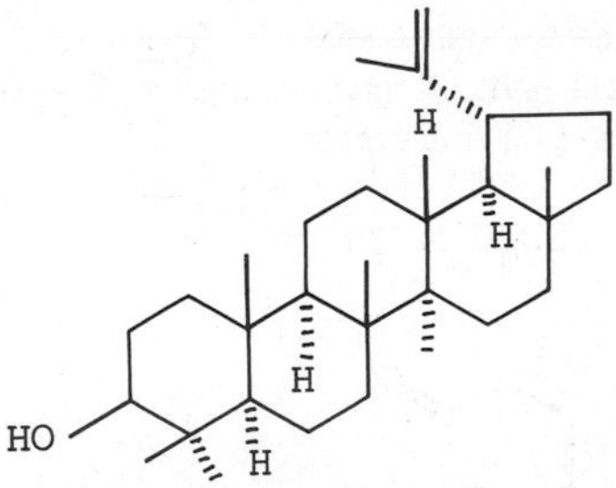

HO

364

The starting material, the enone 365 (Scheme 30), had previously been prepared in low yield by annelation of the tetralone 366 with a methyl vinyl ketone equivalent. Ketone 365 was obtained[80] more efficiently via the route shown in which selective ketalization of diketone 367 had to be performed under carefully controlled conditions. The transformation of ketone 368 to enone 365 was achieved by successive acid and base treatments.

Scheme 30

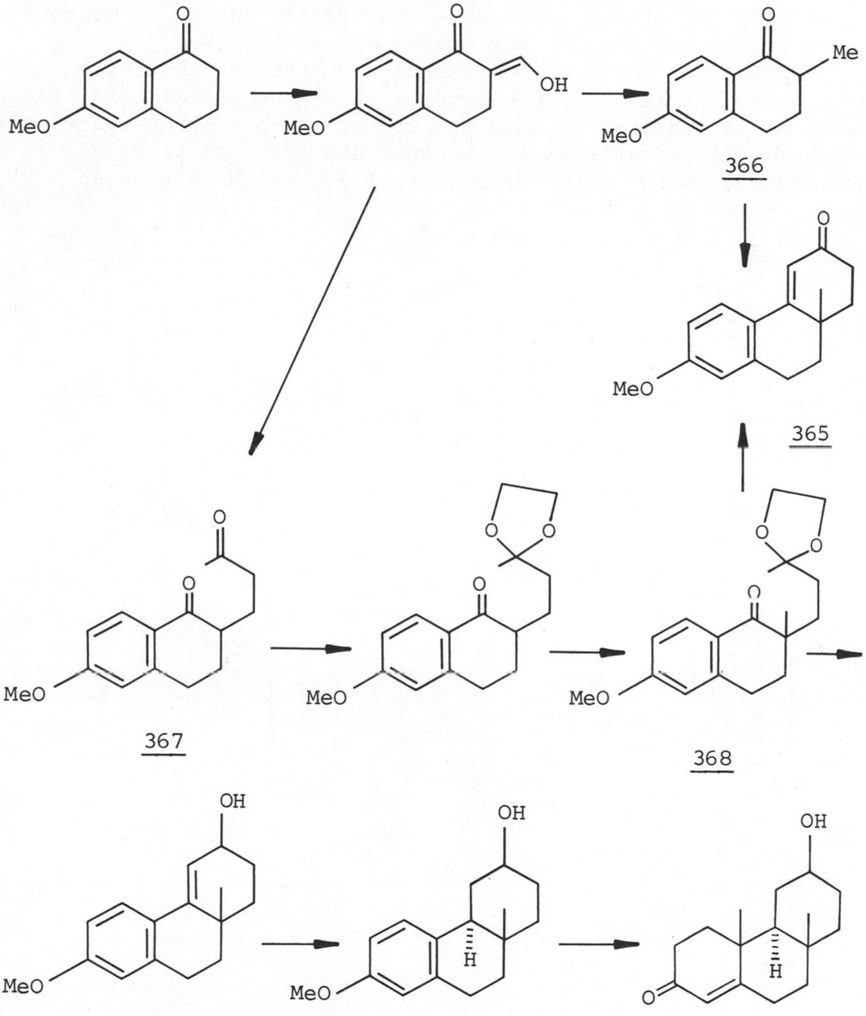

366, 365, 367, 368, 369, 370, 371

Catalytic hydrogenation of enone 365 resulted in much
hydrogenolysis of the ketonic group, presumably due to its
vinylogous benzylic nature. Both hydrogenation in the presence of base (to suppress hydrogenolysis) and Birch reduction
gave mixtures of *cis*- and *trans*-ring fused isomers. However,
carefully controlled borohydride reduction of 365 gave the
allylic alcohol 369 which could be hydrogenated in good yield
to the *trans*-fused product 370. Birch reduction of this and
hydrolysis of the product then gave the hydroxy enone 371 which
was used for both steroid[81] and triterpene syntheses.

The principal problem in triterpene synthesis, the introduction of two vicinal methyl groups in a *trans*-diaxial relationship at ring junctions, was now solved in an elegant fashion (Scheme 31). Treatment of the benzoate 372 with triallyl
orthoformate gave the allyl dienol ether 373 which rearranged

Scheme 31

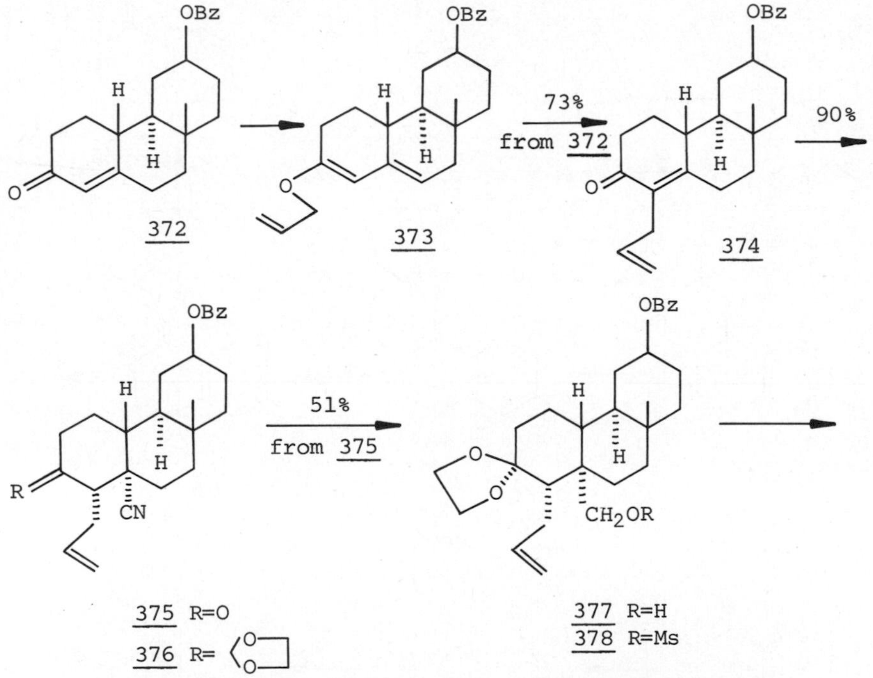

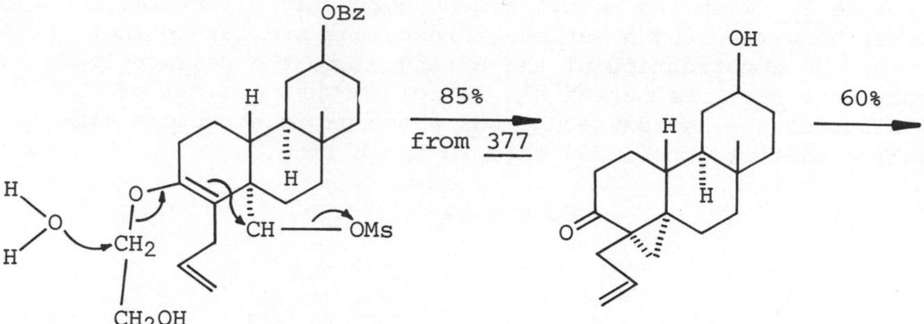

380

379

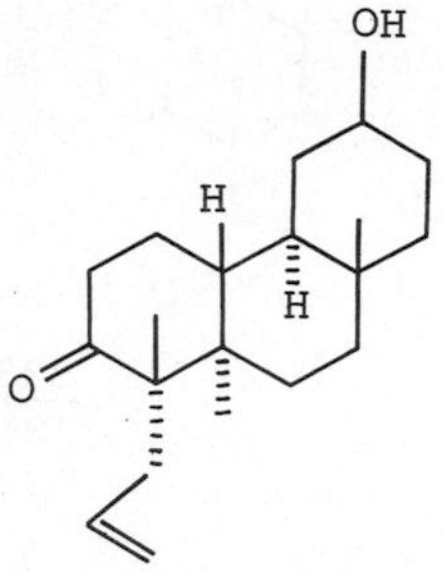

381

on heating to the allyl ketone 374. Direct alkylation of enone
372 with allyl halides results in polyalkylation, avoided here
by using the Claisen rearrangement. The allyl ketone 374
formed the *trans-anti-trans*-cyanoketone 375 on treatment with
diethylaluminum cyanide and the corresponding ketal 376 was
converted to the primary alcohol 377 via the imine and aldehyde. Acid hydrolysis of the corresponding mesylate 378 followed by removal of the benzoate protecting group gave the
α,β-cyclopropanoketone 379 directly. This unusual "solvolytic
cyclization" may have proceeded through the enol ether 380 as
shown. Reductive alkylation (methyl iodide) of cyclopropano-
ketone 379 gave the ketone 381 with the two newly introduced
methyl groups in the desired *trans*-orientation.

The construction of the A and B rings of lupeol is shown
in Scheme 32. Hydroboration (disiamylborane) of the benzoate
382 followed by oxidation (chromic acid) gave the acid 383
which was converted to the enol lactone 384. This was treated
with ethyl Grignard reagent and base catalyzed retroaldol/aldol
reaction of the resulting mixture of ketols 385 afforded the
tetracyclic enone 386. Reductive alkylation (allyl bromide)
of this and subsequent protection of the hydroxyl group gave

the ketone 387 with the methyl groups correctly orientated. This was converted, by a series of reactions similar to that used for the construction of the B ring, into the pentacyclic hydroxyenone 388. Reductive alkylation (methyl iodide) of this afforded the hydroxyketone 389 whose gross structure differs from that of lupeol 364 only in the E ring.

Scheme 32

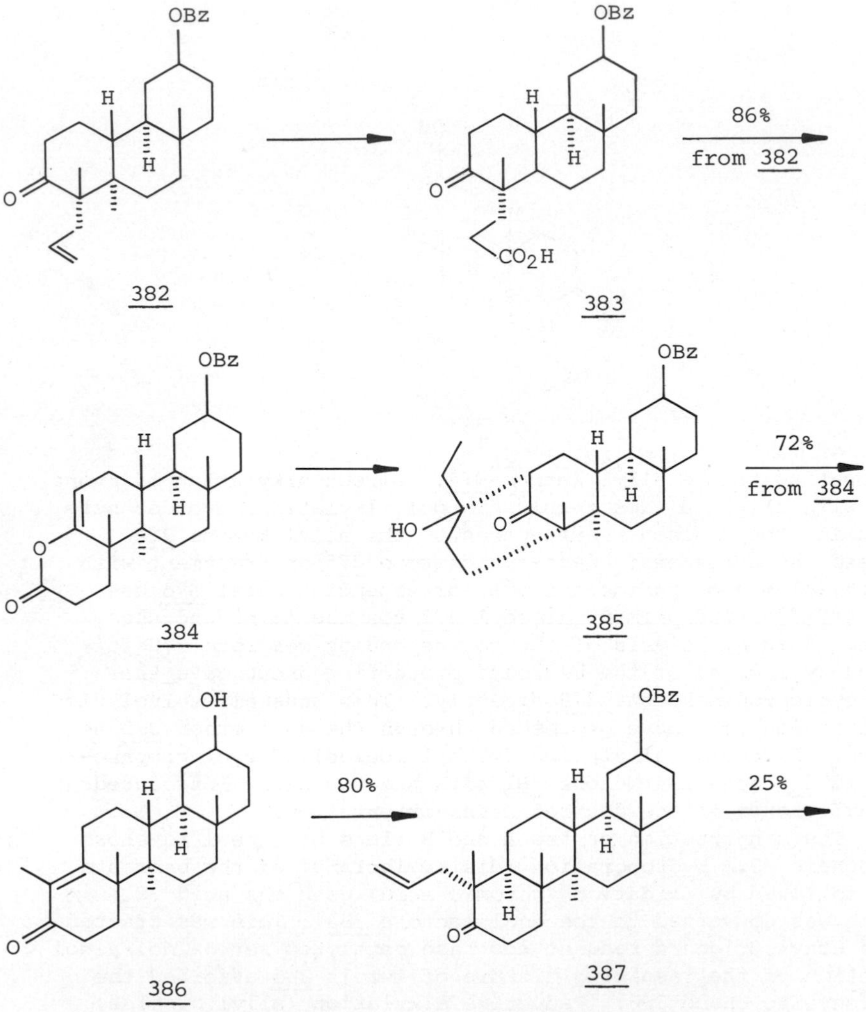

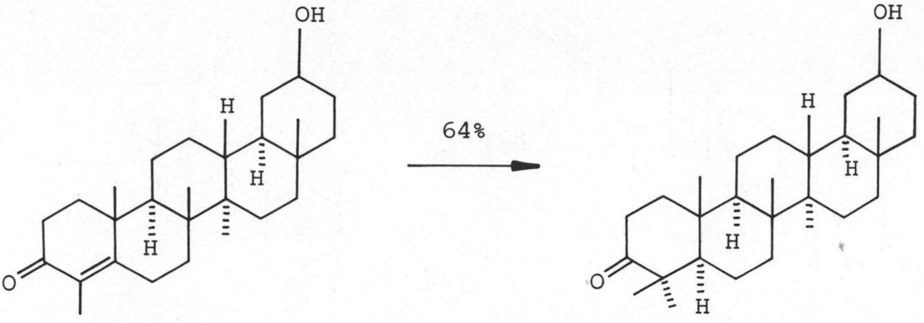

388   64% →   389

In connection with the extensive use of reductive alkylation (Birch reduction followed by trapping of the enolate
anion with an alkylating agent), it should be noted that the
presence of the E ring hydroxyl function was essential for the
success of these reactions, probably due to solubility difficulties with the hydroxyl protected compounds.[80] These were
apparently less serious with the (lithium salt of the) alcohol.

The final phase of the synthesis, the modification of the
E ring of 389, is shown in Scheme 33. Thus, formation of the
C-3 ketal followed by oxidation gave the ketone 390 which afforded the enol acetate 391 on treatment with the strong base,
sodium hexamethyldisilazane, and then acetic anhydride.

Scheme 33

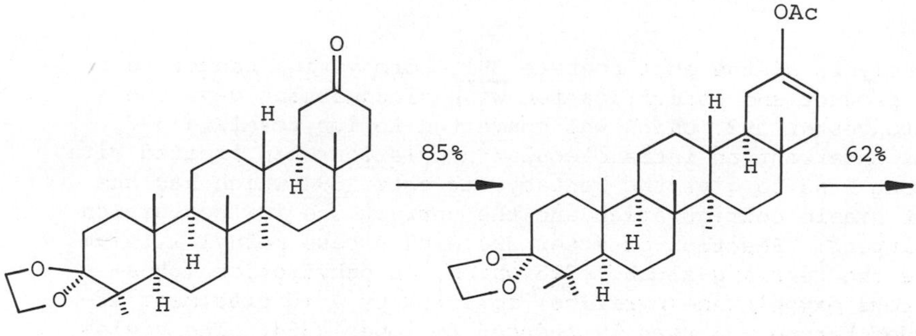

390   85% →   391   62% →

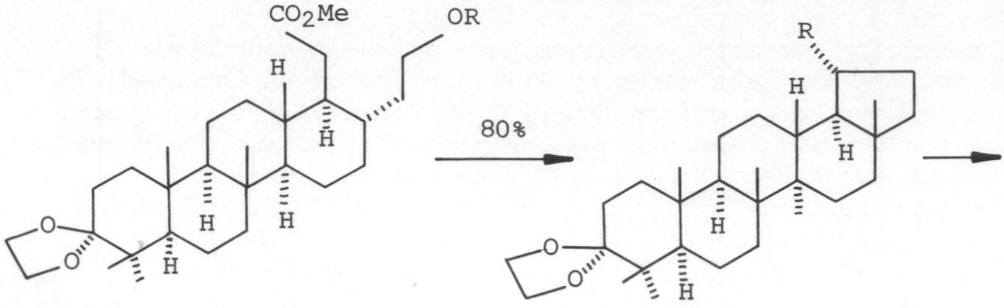

392 R=H
393 R=Ts

394 R=CO$_2$Me
395 R=-C(OH)(CH$_3$)$_2$

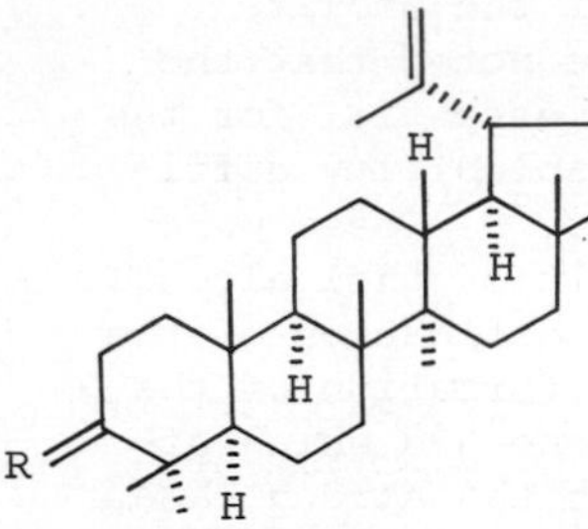

396 R=O
364 R= OH / H

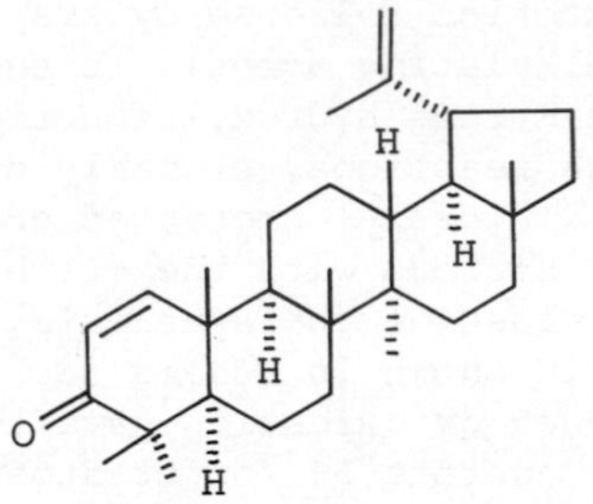

397

Ozonolysis of the enol acetate 391, borohydride reduction of
the product and esterification with diazomethane gave the
hydroxyester 392, which was converted to its tosylate 393.
This underwent an intramolecular alkylation when treated with
strong base to give the pentacyclic ester 394 which has the
most stable configuration and the correct one for conversion
to lupeol. Reaction of ester 394 with excess methyl lithium
gave the *tertiary*-alcohol 395 which, on dehydration (phosphorous oxychloride/pyridine) followed by acid treatment afforded ketone 396 readily reduced to lupeol 364. The yields
of the later stages of the synthesis were not known or opti-
mized at the time of writing.

     Glochidone 397, another natural product of the lupane
series, has been synthesized from lupeol 364 through a simple
series of reactions.[82]

5.   PRESQUALENE ALCOHOL 398

The naturally occurring pyrophosphate of the cyclopropyl-carbinol 398 has been shown by Rilling[83] to be an intermediate in the biosynthesis of squalene 1 from farnesyl pyrophosphate, hence the trivial name presqualene alcohol. Recent syntheses of the racemate of 398 have confirmed the structure.

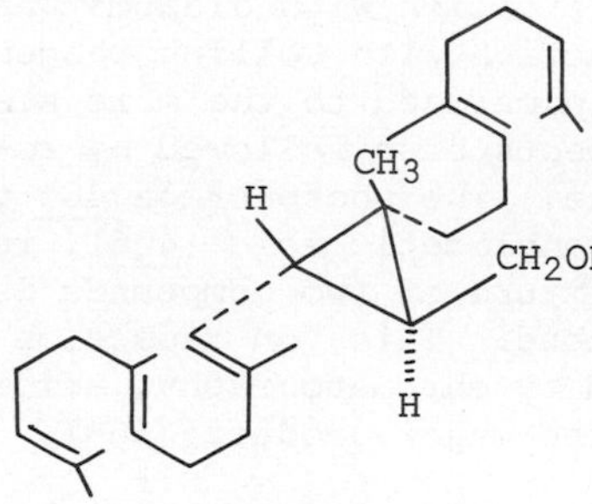

398

The simplest synthesis[84] was achieved through addition of the carbene derived from diazo compound 399 to trans,trans-farnesol 400. The product was a 70:30 mixture of racemic pre-squalene alcohol 398 and the compound isomeric at the carbon marked. The cis-relationship of the methyl and hydroxymethyl groups, expected from a cis-addition to the trans-double bond of 400, was confirmed by NMR spectroscopy. Diazo compound 399 was obtained from trans,trans-farnesaldehyde via oxidation of its hydrazone.[85]

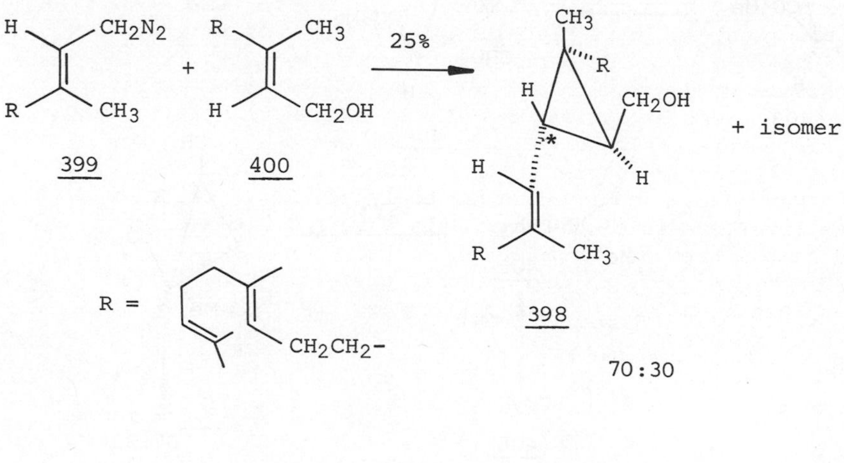

399   400

398

70:30

In the second synthesis[86] reaction of *trans,trans*-farnesol with glyoxalyl chloride tosylhydrazone and triethylamine afforded the diazoacetate 401 which gave the lactone 402 on copper-catalyzed decomposition, via internal addition of the carbenoid species to the adjacent double bond. The *cis*-stereospecificity of this reaction and the stereochemistry of the starting olefin allowed the definition of the relative stereochemistry of lactone 402 as that depicted. Successive hydrolysis of 402, esterification with diazomethane and oxidation of the alcoholic function with Collins reagent, gave the aldehyde 403. This was epimerized to the more stable isomer 404 on treatment with aqueous base followed by re-esterification of the acidic residue. The phosphorane 405 prepared by standard methods from geranylacetic acid (406), reacted with aldehyde 404 to give a mixture of two compounds isomeric about the newly formed double bond. This, on reduction with lithium aluminum hydride followed by chromatography, afforded pre-squalene alcohol 398 as the major product (66%).

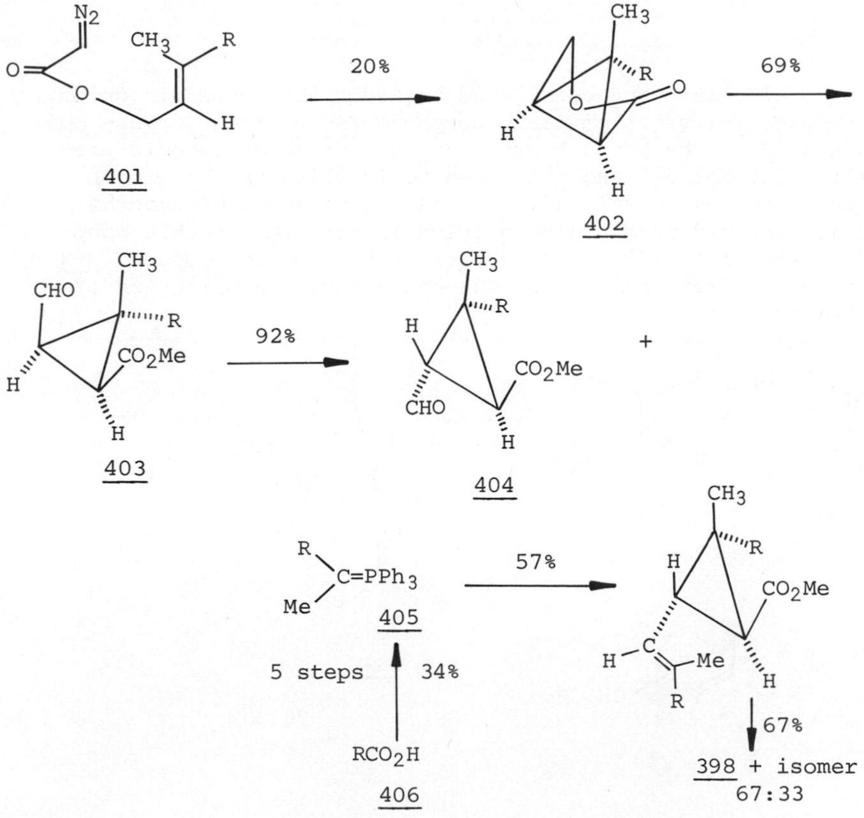

The pyrophosphates of samples of presqualene alcohol obtained by both synthetic routes were converted into squalene by yeast subcellular particles in 66% yield (assuming only one optical isomer is utilized).[84,86]

ACKNOWLEDGMENTS

We thank the National Research Council of Canada and the C. D. Howe Foundation for financial support, and Mrs. Jill Hooper for secretarial assistance.

REFERENCES

1. J. Simonsen and W. C. J. Ross, *The Terpenes* (Cambridge University Press, Cambridge, England, 1957), Vols. IV and V; P. de Mayo, *The Higher Terpenoids* (Interscience Publishers, New York, 1959); G. Ourisson, P. Crabbé, and O. Rodig, *Tetracyclic Triterpenes* (Holden Day, Inc., San Francisco, 1964).

2. J. H. Richards and J. B. Hendrickson, *The Biosynthesis of Steroids, Terpenes and Acetogenins* (W. A. Benjamin Inc., New York, 1964); R. B. Clayton, *Q. Revs. (London), 19*, 168 (1965); *19,*201 (1965).

3. P. Karrer and A. Helfenstein, *Helv. Chim. Acta, 14*, 78 (1931).

4. O. Isler, P. Rüegg, L. H. Chopard-dit-Jean, H. Wagner, and K. Bernhard, *Helv. Chim. Acta, 39*, 897 (1956).

5. E. J. Corey, M. F. Semmelhack, and L. S. Hegedus, *J. Amer. Chem. Soc., 90*, 2416 (1968); E. J. Corey, P. R. Ortiz de Montellano, and H. Yamamoto, *J. Amer. Chem. Soc., 90*, 6254 (1968).

6. (a) K. B. Sharpless, R. P. Hanzlik, and E. E. van Tamelen, *J. Amer. Chem. Soc., 90*, 209 (1968); E. E. van Tamelen, R. P. Hanzlik, K. B. Sharpless, R. B. Clayton, W. J. Richter, and A. L. Burlingame; *J. Amer. Chem. Soc., 90*, 3284 (1968); (b) E. E. van Tamelen, R. P. Hanzlik, R. B. Clayton, and A. L. Burlingame, *J. Amer. Chem. Soc., 92*, 2137 (1970).

7. E. J. Corey, K. Lin, and H. Yamamoto, *J. Amer. Chem. Soc., 91*, 2132 (1969).

8. E. E. van Tamelen, R. B. Clayton, and R. G. Nadeau, *J. Amer. Chem. Soc., 90*, 820 (1968).

9. E. E. van Tamelen, *Acc. Chem. Res., 1*, 111 (1968).

10. D. W. Dicker and M. C. Whiting, *Chem. and Ind.,* 351 (1956); *J. Chem. Soc.,* 1994 (1958); S. Trippett, *Chem. and Ind.,* 80 (1956); A. Mondon, *Annalen, 603*, 115 (1957);

M. Jayme, P. C. Schaefer, and J. H. Richards, *J. Amer. Chem. Soc.*, *92*, 2059 (1970).

11. W. P. Schneider, U. Axen, F. H. Lincoln, J. E. Pike, and J. L. Thompson, *J. Amer. Chem. Soc.*, *90*, 5895 (1968); W. P. Schneider, *Chem. Commun.*, 785 (1969), c.f. U. T. Bhalerao and H. Rapoport, *J. Amer. Chem. Soc.*, *93*, 5311 (1971).

12. J. F. Biellmann and J. P. Ducep, *Tet. Letters*, 3707 (1969); *ibid.*, *Tetrahedron 27*, 5861 (1972); see also, K. Hirai, H. Matsuda, and Y. Kishida, *Tet. Letters*, 4359 (1971).

13. G. M. Blackburn, W. D. Ollis, C. Smith, and I. O. Sutherland, *Chem. Commun.*, 99 (1969).

14. E. H. Axelrod, G. M. Milne, and E. E. van Tamelen, *J. Amer. Chem. Soc.*, *92*, 2139 (1970).

15. R. A. Heathcock, This series, "Sesquiterpene Synthesis."

16. J. W. Cornforth, R. H. Cornforth, and K. K. Mathew, *J. Chem. Soc.*, 112, 2539 (1959).

17. D. R. Boyd and M. A. McKerney, *Q. Revs. (London)*, *122*, 95 (1968).

18. S. F. Brady, M. A. Ilton, and W. S. Johnson, *J. Amer. Chem. Soc.*, *90*, 2882 (1968).

19. W. S. Johnson, L. Wethermann, W. R. Bartlett, T. J. Brocksom, T. Li, D. J. Faulkner, and M. R. Peterson, *J. Amer. Chem. Soc.*, *92*, 741 (1970).

20. (a) D. J. Faulkner and M. R. Petersen, *Tet. Letters*, 3243 (1969); (b) N. Wakabayashi, R. M. Waters and J. B. Church, *Tet. Letters*, 3253 (1969); (c) W. S. Johnson, T. J. Brockson, P. Loew, D. H. Rich, L. Wethermann, R. A. Arnold, T. Li, and D. J. Faulkner, *J. Amer. Chem. Soc.*, *92*, 4463 (1970).

21. I. M. Heilbron, E. D. Kamm and W. M. Owens, *J. Chem. Soc.*, *129*, 1630 (1926).

22. W. S. Johnson, *Acc. Chem. Res.*, *1*, 1 (1968).

23. K. B. Sharpless, *Chem. Commun.*, 1450 (1970); *J. Amer. Chem. Soc.*, *92*, 6999 (1970).

24. E. E. van Tamelen, J. Willett, M. Schwartz, and R. Nadeau, *J. Amer. Chem. Soc.*, *88*, 5937 (1966); E. E. van Tamelen, K. B. Sharpless, R. Hanzlik, R. B. Clayton, A. L. Burlingame, and P. Wszolek, *J. Amer. Chem. Soc.*, *89*, 7150 (1967).

25. O. Dürst, O. Jëger, and L. Ruzicka, *Helv. Chim. Acta*, *32*, 46 (1949).

26. P. Diedtrich and E. Lederer, *Compt. Rend.*, *234*, 637 (1952); R. E. Wolff, *Compt. Rend.*, *238*, 1041 (1954); J. A. Barltrop, D. B. Bigley, and N. A. J. Rogers, *Chem. and Ind.*, 558 (1958); *J. Chem. Soc.*, 4613 (1960); L. Ruzicka, G. Büchi, and O. Jëger, *Helv. Chim. Acta, 31,*

293 (1948).

27. R. B. Woodward, F. Sondheimer, and D. Taub, *J. Amer. Chem. Soc.*, *73*, 3548 (1951); R. B. Woodward, F. Sondheimer, D. Taub, K. Heusler, and W. M. Mclamore, *J. Amer. Chem. Soc.*, *74*, 4223 (1952); H. M. E. Cardwell, J. W. Cornforth, S. R. Duff, H. Holtermann, and R. Robinson, *J. Chem. Soc.*, 361 (1953).

28. R. B. Woodward, A. A. Patchett, D. H. R. Barton, D. A. J. Ives, and R. B. Kelly, *J. Amer. Chem. Soc.*, *76*, 2852 (1954); *Chem. and Ind.*, 605 (1954); *J. Chem. Soc.*, 1131 (1957).

29. L. F. Fieser and M. Fieser, *Steroids* (Reinhold Publishing Corp., New York, 1959), p. 378.

30. R. E. Marker, E. L. Whittle, and L. W. Mixon, *J. Amer. Chem. Soc.*, *59*, 1368 (1937).

31. U. Wrzeciono, C. F. Murphy, G. Ourisson, S. Corsano, J.-D. Ehrhardt, M.-F. Lhomme, and G. Teller, *Bull. Soc. Chim. Fr.*, 966 (1970).

32. J. Fried, J. W. Brown, and M. Applebaum, *Tet. Letters*, 849 (1965).

33. D. H. R. Barton, R. P. Budhiraja, and J. F. McGhie, *Proc. Chem. Soc.*, 170 (1963).

34. E. Altenburger, H. Wehrli, and K. Schaffner, *Helv. Chim. Acta*, *48*, 704 (1965); R. Imhof, W. Graf, H. Wehrli, and K. Schaffner, *Chem. Commun.*, 852 (1969).

35. D. H. R. Barton, D. Kumari, P. Welzel, L. J. Danks, and J. F. McGhie, *J. Chem. Soc.*, 332 (1969).

36. D. H. R. Barton and K. H. Overton, *J. Chem. Soc.*, 2639 (1955).

37. F. Sondheimer and D. Elad, *J. Amer. Chem. Soc.*, *81*, 4429 (1959).

38. E. Romann, A. J. Frey, P. A. Stadler, and A. Eschenmoser, *Helv. Chim. Acta*, *40*, 1900 (1957).

39. E. J. Corey and R. R. Sauers, *J. Amer. Chem. Soc.*, *81*, 1739 (1959).

40. D. B. Bigley, N. A. J. Rogers, and J. A. Barltrop, *J. Chem. Soc.*, 4613 (1960).

41. G. Stork, A. Meisels, and J. E. Davies, *J. Amer. Chem. Soc.*, *85*, 3419 (1963).

42. N. Danieli, Y. Mazur, and F. Sondheimer, *Tetrahedron*, *23*, 509 (1967).

43. R. F. Church, R. E. Ireland, and J. A. Marshall, *J. Org. Chem.*, *27*, 1118 (1962).

44. M. R. Johnson and B. Rickborn, *J. Org. Chem.*, *35*, 1041 (1970).

45. E. E. van Tamelen, M. A. Schwartz, E. J. Hessler, and A. Storni, *Chem. Commun.*, 409 (1966).

46. Y. Tsuda, T. Sano, K. Kawaguchi, and Y. Inubushi, *Tet. Letters*, 1279 (1964).

47. K. Schaffner, L. Caglioti, D. Arigoni, and O. Jëger, *Helv. Chim. Acta*, *41*, 152 (1958).

48. Y. Tsuda, A. Morimoto, T. Sano, and Y. Inubushi, *Tet. Letters*, 1427 (1965).

49. M. Kishi, T. Kato, and Y. Kitahara, *Chem. Pharm. Bull.*, *15*, 1073 (1967).

50. Y. Tsuda, K. Isobe, and S. Fukushima, *Tet. Letters*, 23 (1967).

51. G. V. Baddely, T. G. Halsall, and E. R. H. Jones, *J. Chem. Soc.*, 3891 (1961).

52. H. Ageta, K. Iwata, and Y. Arai, *Tet. Letters*, 5679 (1966).

53. K. Iguchi and H. Kakisawa, *Chem. Commun.*, 1486 (1970).

54. D. H. R. Barton, E. F. Lier, and J. F. McGhie, *J. Chem. Soc. (C)*, 1031 (1968).

55. E. J. Corey, H. J. Hess, and S. Proskow, *J. Amer. Chem. Soc.*, *85*, 3979 (1963).

56. J. A. Barltrop, J. D. Littlehailes, J. D. Rushton, and N. A. J. Rogers, *Tet. Letters*, 429 (1962).

57. E. Ghera and F. Sondheimer, *Tet. Letters*, 3887 (1964).

58. T. G. Halsall and D. B. Thomas, *J. Chem. Soc.*, 2431 (1956).

59. J. M. Beaton, F. S. Spring, R. Stevenson, and J. L. Stewart, *J. Chem. Soc.*, 2131 (1955).

60. F. S. Spring, R. Stevenson, and W. Laird, *J. Chem. Soc.*, 2638 (1961); O. Jëger and L. Ruzicka, *Helv. Chim. Acta*, *28*, 209 (1945); L. Ruzicka, Rüegg, E. Volli, and O. Jëger, *Helv. Chim. Acta*, *30*, 140 (1947); G. G. Allen, J. D. Johnston, and F. S. Spring, *J. Chem. Soc.*, 1546 (1954).

61. I. Agata, E. J. Corey, A. G. Hortmann, J. Klein, S. Proskow, and J. J. Ursprung, *J. Org. Chem.*, *30*, 1698 (1965).

62. R. B. Boar, D. C. Knight, J. F. McGhie, and D. H. R. Barton, *J. Chem. Soc. (C)*, 678 (1970).

63. J. W. ApSimon, in *Elucidation of Structures by Physical and Chemical Methods*, Vol. 4, part 3, edited by K. W. Bentley and G. W. Kirby (Interscience Publishers, New York), 2nd ed., pp. 251-408.

64. (a) G. Stork, P. Rosen, and N. Goldman, *J. Amer. Chem. Soc.*, *87*, 275 (1965); (b) R. S. Matthews, S. J. Girgent, and E. A. Folkers, *Chem. Commun.*, 708 (1970); R. S. Matthews, P. K. Hyer, and E. A. Folkers, *ibid.*, 38 (1970), and references quoted therein; M. E. Kuehne and J. A. Nelson, *J. Org. Chem.*, *35*, 161 (1970); M. J. Green, N. A. Abraham, E. B. Fleischer, J. Case, and J. Fried, *Chem.*

Commun., 234 (1970); G. Stork and J. E. McMurry, *J. Amer. Chem. Soc.*, *89*, 5464 (1967).

65. G. Stork (personal communication to J. W. ApSimon).

66. E. Wenkert, A. Alfonso, J. Brendenberg, C. Kaneko, and A. Tahara, *J. Amer. Chem. Soc.*, *86*, 2038 (1964).

67. V. Permutti and Y. Mazur, *J. Org. Chem.*, *31*, 705 (1966) and references cited therein; G. J. Just and K. St. C. Richardson, *Can. J. Chem.*, *42*, 464 (1964).

68. H. O. House and B. M. Trost, *J. Org. Chem.*, *30*, 2502 (1965), and references cited therein.

69. M. J. Weiss, R. E. Schaub, G. R. Allen, J. F. Poletto, C. Pidacks, R. B. Conrow, and C. J. Coscia, *Tetrahedron*, *20*, 357 (1964).

70. Y. Kitahara, A. Yoshikoshi, and S. Oida, *Tet. Letters*, 1763 (1964); G. I. Poos, G. E. Arth, R. E. Beyler, and L. H. Sarret, *J. Amer. Chem. Soc.*, *75*, 422 (1953); M. Uskokovic, J. Iacobelli, R. Phillion, and T. Williams, *J. Amer. Chem. Soc.*, *88*, 4538 (1966).

71. R. E. Ireland, S. W. Baldwin, D. J. Dawson, M. I. Dawson, J. E. Dolfini, J. Newbould, W. S. Johnson, M. Brown, R. J. Crawford, P. F. Hudrlik, G. H. Rusmussen, and K. K. Schmiesel, *J. Amer. Chem. Soc.*, *92*, 5743 (1970).

72. R. E. Ireland and L. N. Mander, *J. Org. Chem.*, *32*, 689 (1967); *34*, 142 (1969).

73. H. E. Simmons and R. D. Smith, *J. Amer. Chem. Soc.*, *80*, 5323 (1958).

74. C. D. Poulter, E. C. Friedrich, and S. Winstein, *J. Amer. Chem. Soc.*, *91*, 6892 (1969); J. H. Chan and B. Rickborn, *ibid.*, *90*, 6406 (1968), and references cited therein.

75. (a) R. E. Ireland, D. A. Evans, D. Glover, G. M. Rubottom, and H. Young, *J. Org. Chem.*, *34*, 3717 (1969); R. E. Ireland, D. A. Evans, P. Lolinger, J. Bordner, R. H. Stanford, and R. E. Dickerson, *ibid.*, *34*, 3729 (1969); (b) R. E. Ireland and S. C. Welch, *J. Amer. Chem. Soc.*, *92*, 7232 (1970).

76. W. G. Dauben and E. J. Deviny, *J. Org. Chem.*, *31*, 3794 (1966); T. Norin, *Acta Chem. Scand.*, *19*, 1289 (1965); W. G. Dauben and R. E. Wolf, *J. Org. Chem.*, *35*, 374 (1970); *35*, 2361 (1970).

77. W. Nagata and M. Yoshioka, Proceedings of the Second International Congress on Hormonal Steroids, Milan, 1966, p. 327 (unpublished), and references cited therein.

78. R. E. Ireland, M. I. Dawson, J. Bordner, and R. E. Dickerson, *J. Amer. Chem. Soc.*, *92*, 2568 (1970); R. E. Ireland, D. R. Marshall, and J. W. Tilley, *J. Amer. Chem. Soc.*, *92*, 4754 (1970).

79. D. Arigoni, R. Viterlo, M. Dunnenberger, O. Jěger, and

L. Ruzicka, *Helv. Chim. Acta, 37,* 2306 (1954); D. Arigoni, O. Jëger, and L. Ruzicka, *ibid., 38,* 222 (1955).

80. G. Stork, S. Uyeo, T. Wakamatsu, P. Grieco, and J. Laboritz, *J. Amer. Chem. Soc., 93,* 4945 (1971).

81. G. Stork, J. H. E. Loewenthal, and P. C. Mukharji, *J. Amer. Chem. Soc., 78,* 501 (1956).

82. A. S. Samson, S. J. Stevenson, and R. Stevenson, *Chem. and Ind.,* 1142 (1969).

83. H. C. Rilling, *J. Biol. Chem., 241,* 3233 (1966); H. C. Rilling and W. W. Epstein, *J. Amer. Chem. Soc., 91,* 1041 (1969); *J. Biol. Chem., 18,* 4597 (1970).

84. L. J. Altman, R. C. Kowerski, and H. C. Rilling, *J. Amer. Chem. Soc., 93,* 1782 (1971).

85. E. J. Corey and K. Achiwa, *Tet. Letters,* 3257 (1969); R. M. Coates and R. M. Freidinger, *Tetrahedron, 26,* 3487 (1970).

86. R. M. Coates and W. H. Robinson, *J. Amer. Chem. Soc., 93,* 1785 (1971).

87. E. E. van Tamelen and R. J. Anderson, *J. Amer. Chem. Soc., 94,* 8225 (1972).

88. E. E. van Tamelen, R. A. Holton, R. E. Hopla, and W. E. Konz, *J. Amer. Chem. Soc., 94,* 8228 (1972).

89. E. E. van Tamelen, M. P. Seiler, and W. Wierenga, *J. Amer. Chem. Soc., 94,* 8229 (1972).

NATURALLY OCCURRING AROMATIC STEROIDS

D. Taub

Merck, Sharp & Dohme Research Laboratories
Rahway, New Jersey

1.	Introduction	642
2.	Equilenin	642
	A. Bachmann Synthesis	642
	B. Johnson Syntheses	649
	C. Bachmann and Holmen Syntheses	654
	D. Robinson-Birch Synthesis	656
	E. Horeau Synthesis	661
	F. Syntheses from Estrone Intermediates	662
3.	Equilin	664
	A. Zderic et al. Synthesis	664
	B. Bagli et al. Synthesis	665
	C. Stein et al. Synthesis	666
	D. Bailey et al. Synthesis	668
	E. Marshall and Deghenghi Synthesis	669
4.	Estrone	670
	A. Anner and Miescher Synthesis	673
	B. Johnson Syntheses	679
	C. Johnson's First Synthesis	679
	D. Johnson's Second Synthesis	683
	E. Johnson-Walker Synthesis	687
	F. 2-Methylcyclopentane-1-3-dione	690
	G. Syntheses of Smith et al.	693
	H. Torgov Synthesis	698
	I. Resolution	703
	J. Velluz Syntheses	706
	K. Miscellaneous	711

1. INTRODUCTION[1]

The era of steroid total synthesis originated in the early
1930s following the establishment of the structure of the
steroid ring system and the isolation and characterization of
the steroidal sex hormones. The naturally occurring aromatic
estrogens were among the earliest synthetic goals. The presence of aromatic rings simplified the synthetic difficulties
and limited the number of possible racemates to figures compatible with the early nonstereoselective procedures. Initially the biological importance of the steroidal estrogens,
and later their increasing medical and industrial utility,
lent growing impetus to the development of effective syntheses.
The synthetic effort which has now progressed some three and
one-half decades has resulted not only in many ingenious and
practical routes, but also in significant enrichment of the
methodology of organic chemistry.
   Naturally occurring aromatic steroids, with few excep-
tions--such as the ring D-aromatic solanum alkaloid, veratra-
mine--are limited to the steroidal estrogens to which this review is devoted.

2. EQUILENIN

A. Bachmann Synthesis

The estrogenic steroid (+)equilenin $\underline{1}$ was isolated from mare
pregnancy urine in 1932.[2] Because of its stereochemical simplicity (two asymmetric centers) it was an early synthetic

$$\underline{1} \qquad\qquad\qquad \underline{2}$$

objective. Its synthesis in 1939 by Bachmann, Cole, and
Wilds,[3] the first of a naturally occurring steroid, was a
milestone in preparative organic chemistry. The successful
route, one of a number of approaches under general investigation,[4] utilized the addition of the elements of ring D to 1-
oxo-7-methoxy-1,2,3,4-tetrahydrophenanthrene $\underline{2}$.

The tricyclic ketone $\underline{2}$ and the corresponding phenol had been prepared in 1935 by Butenandt and Schramm[5] for estrogenic assay and as potential synthetic intermediates. Their synthesis of $\underline{2}$ as well as later syntheses are outlined in Scheme 1. By standard procedures the dye intermediate, 1-amino-naphthalene-6-sulfonic acid (Cleve's acid) $\underline{3}$ was converted into

Scheme 1.   1-Oxo-7-Methoxy-1,2,3,4-Tetrahydrophenanthrene

Butenandt:

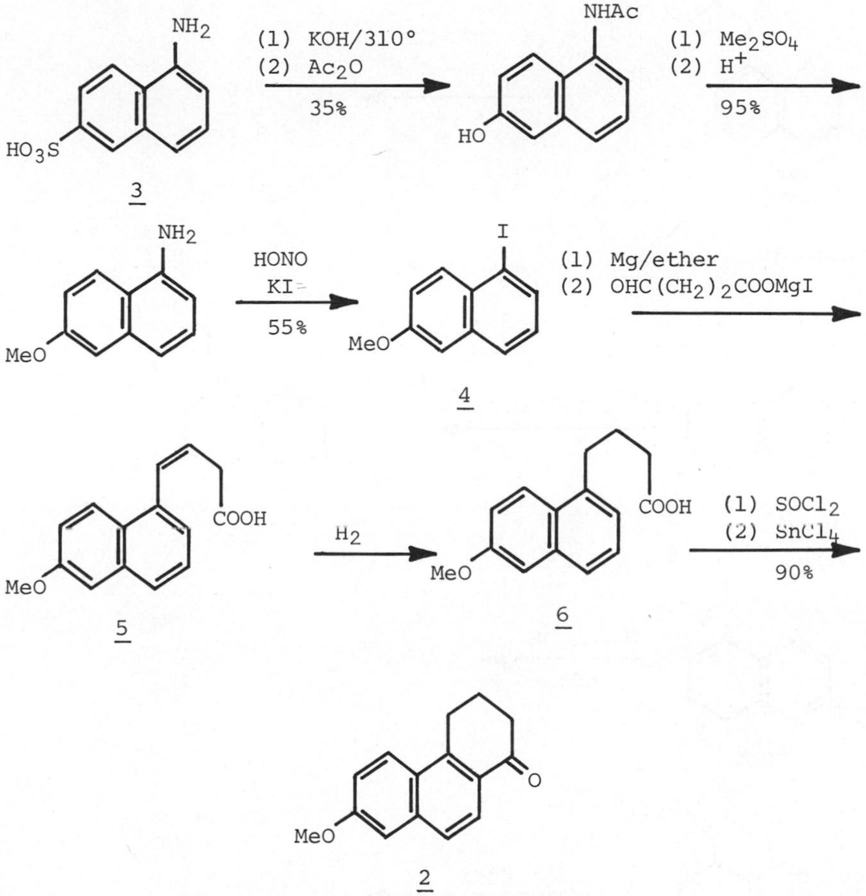

Cook:

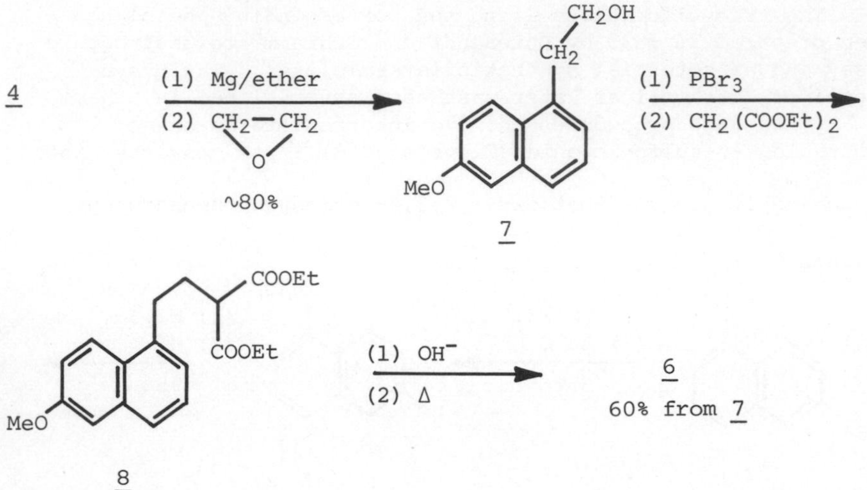

4 →
(1) Mg/ether
(2) CH₂—CH₂ (epoxide)
~80%

(structure **7**: 6-methoxynaphthalene with CH₂CH₂OH side chain)

→
(1) PBr₃
(2) CH₂(COOEt)₂

(structure **8**: 6-methoxynaphthalene with CH₂CH₂CH(COOEt)₂ side chain)

→
(1) OH⁻
(2) Δ

6
60% from **7**

Haberland:

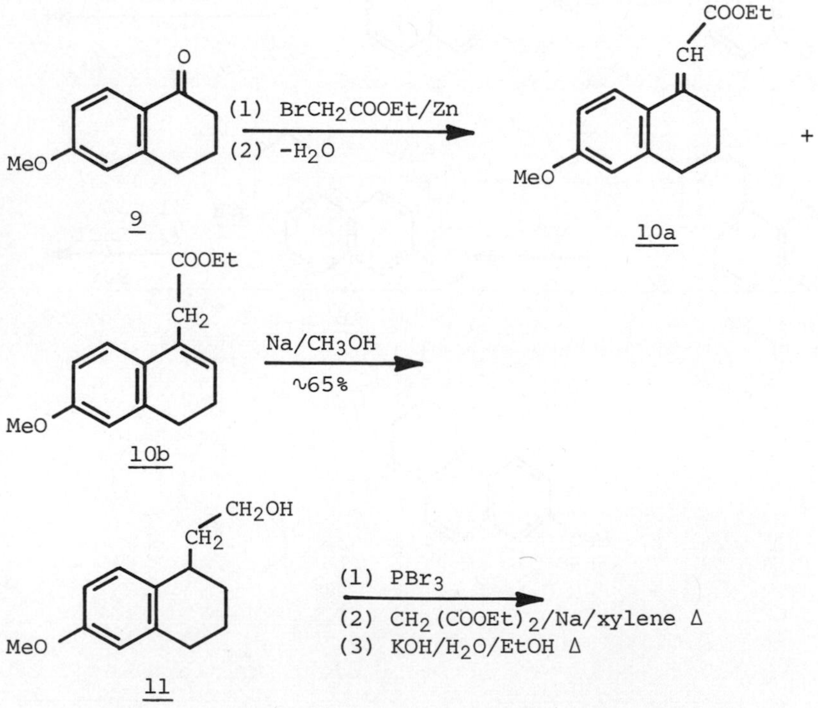

(structure **9**: 6-methoxy-1-tetralone)

→
(1) BrCH₂COOEt/Zn
(2) −H₂O

(structure **10a**: exocyclic =CH−COOEt) +

(structure **10b**: CH₂COOEt with endocyclic double bond)

→
Na/CH₃OH
~65%

(structure **11**: 1-(2-hydroxyethyl)-6-methoxytetralin, CH₂CH₂OH side chain)

→
(1) PBr₃
(2) CH₂(COOEt)₂/Na/xylene Δ
(3) KOH/H₂O/EtOH Δ

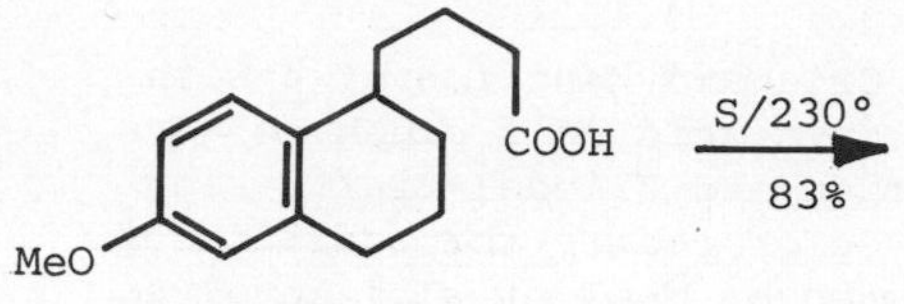

Structure **12**, S/230°, 83%

Structure **6**, H₂SO₄/25°, ~80%, Structure **2**

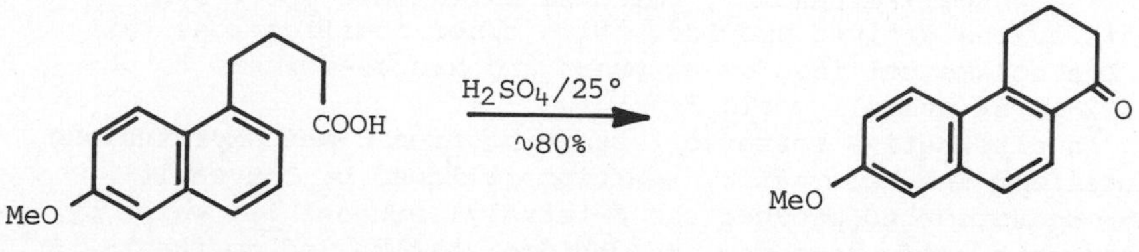

Stork:

Structure **13**, (1) MeOH/H⁺ (2) H₂/Ni/EtOH,AcOH, Structure **14**, CrO₃

Structure **9**, BrCH₂CH=CHCOOMe/Zn, 48% Direct (93% conversion)

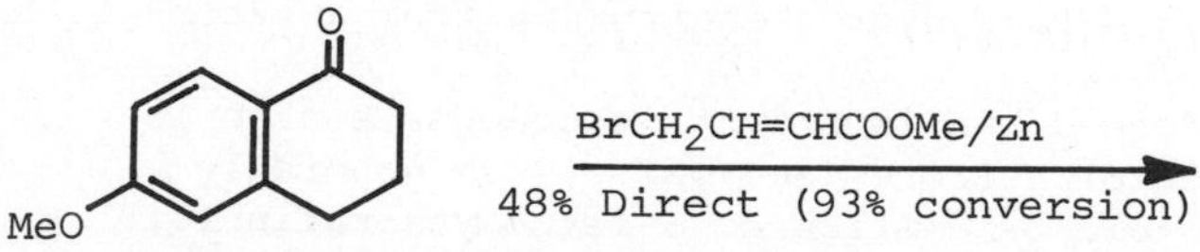

Structure **15**, 30% Pd/C 280-290°, 75%

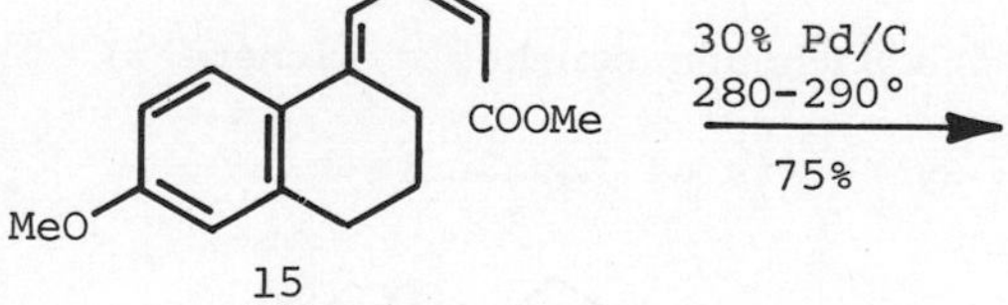

Structure **6**, HF/0°, 85%, Structure **2**

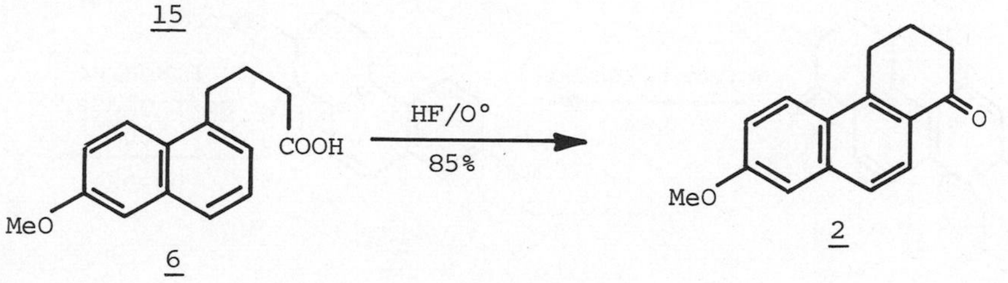

1-iodo-6-methoxy-naphthalene 4. Grignard reaction of 4 with the magnesium iodide salt of succinic acid half aldehyde gave acid 5 in poor yield. Hydrogenation and Friedel-Crafts ring closure then led to 2. An improved procedure for addition of the four carbon side chain developed by Cook et al.[6] involved Grignard reaction of 4 with ethylene oxide to give β (6-methoxynaphthyl)-ethanol 7, followed by malonic ester synthesis on the derived bromide. With minor modifications this was the scheme utilized by Bachmann and his co-workers to obtain 2 in 8% overall yield from 3.

An alternative route to 2 starting from 6-methoxytetralone 9[7] utilized the Reformatsky reaction followed by Bouveault-Blanc reduction to produce the β-tetralyl ethanol 11, which by the malonic ester sequence and dehydrogenation led to the γ-naphthylbutyric acid 6 the common precurser to 2.

Stork[8] introduced major improvements which made 2 available from β-naphthol in 30-35° overall yield. These included an improved procedure for the selective hydrogenation of the unsubstituted ring of β-naphthyl methyl ether and utilization of the Reformatsky reaction of 9 with methyl γ-bromocrotonate to introduce the four-carbon side chain in one step, followed by palladium catalyzed isomerization of the side chain double bonds into ring B. The vinylogous Reformatsky reaction had been used independently by Bachmann and Wendler in the synthesis of 1-oxo-1,2,3,4-tetrahydrophenanthrene from α-tetralone.[9]

6-Methoxytetralone 9, an important intermediate also in a number of syntheses of estrone (see below), is generally prepared by chromic acid oxidation of 6-methoxytetralin 14,[8] although tristriphenylphosphinerhodium chloride[10] and 2,3-dichloro-5,6-dicyanobenzoquinone[11] have recently been utilized as novel oxidants.

In the initial steps of the Bachmann synthesis (Scheme 2)

Scheme 2. Bachmann Synthesis of Equilenin

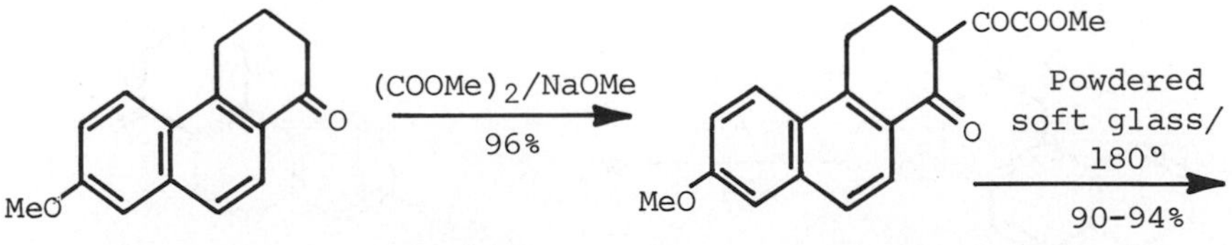

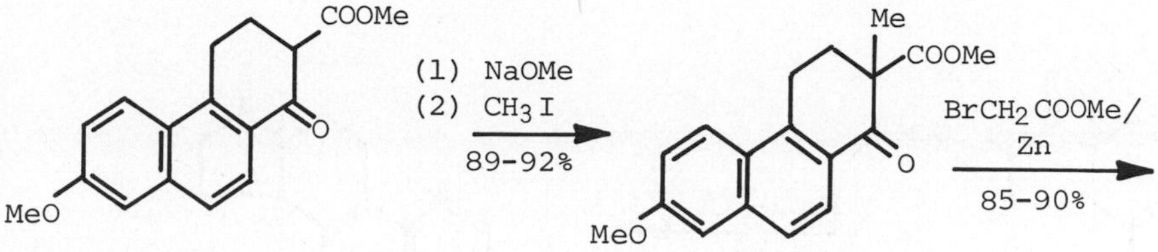

17 → (1) NaOMe (2) CH₃I 89-92% → **18** → BrCH₂COOMe/Zn 85-90% →

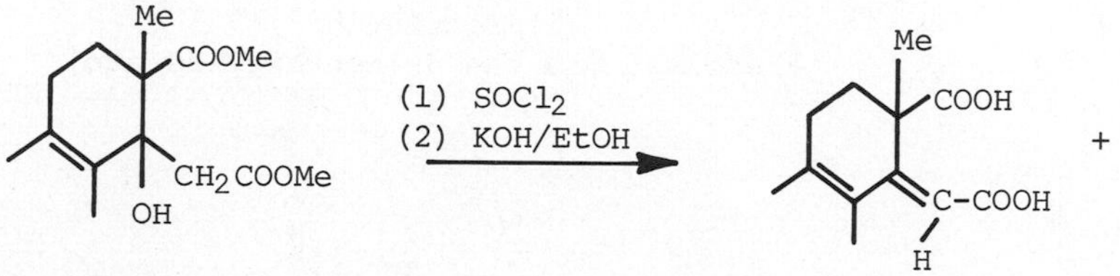

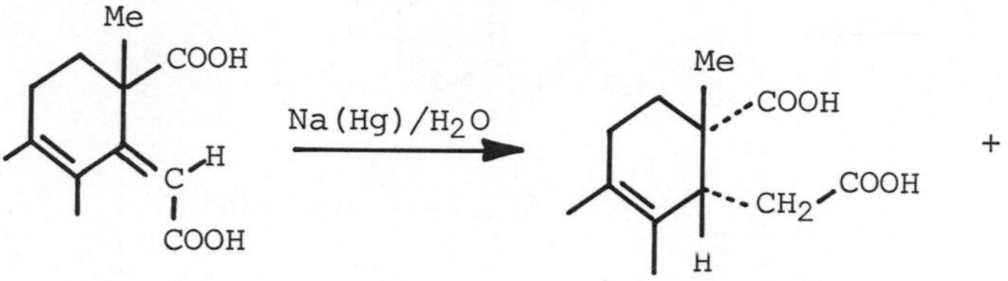

19 → (1) SOCl₂ (2) KOH/EtOH → **20** +

21 → Na(Hg)/H₂O → 45% α-isomer **22a** +

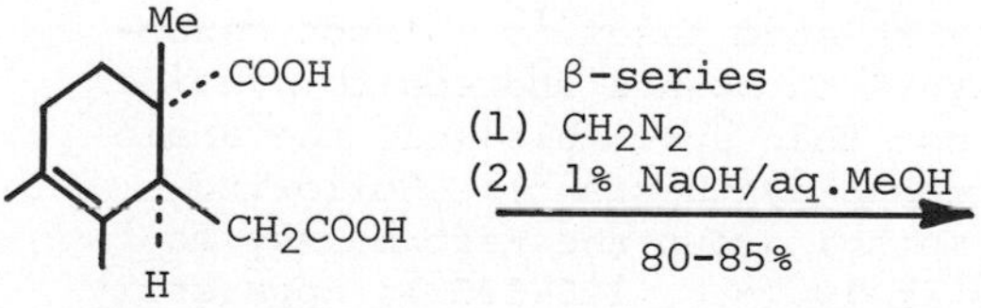

42% β-isomer **22b** → β-series (1) CH₂N₂ (2) 1% NaOH/aq.MeOH 80-85% →

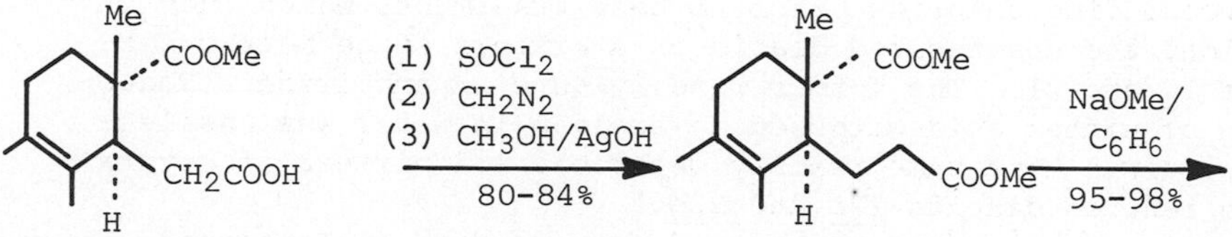

23 → (1) SOCl₂ (2) CH₂N₂ (3) CH₃OH/AgOH 80-84% → **24** → NaOMe/C₆H₆ 95-98% →

(±)-Equilenin
[Resolved via the (-)menthoxyacetate]
1

26
(±) 14-isoequilenin

successful introduction of the carbomethoxy group (cf. 17) was achieved by pyrolysis of the corresponding glyoxylate 16 in the presence of powdered soft glass. In the absence of powdered soft glass pyrolysis was erratic and, in fact, had previously failed in two closely related cases.[12] Direct carbomethoxylation employing dimethyl carbonate and sodium hydride is an alternative procedure not then available (cf. the analogous step in the CIBA aldosterone synthesis[13]). Following introduction of the angular methyl group the Reformatsky reaction gave the hydroxy diester 19 (m.p. 125-126°), apparently a single compound. Treatment of the latter with hot alkali in order to produce the corresponding unsaturated diacid 20 resulted instead in degradation to the 2-methyl derivative of 2. The desired change was accomplished by conversion of 19 to the corresponding chloride prior to base treatment, which then yielded the unsaturated diacid as a mixture of geometric isomers 20 and 21. The former readily gave an anhydride. Reduction of either acid with sodium amalgam in water was unselective, giving the same readily separable 1:1 mixture of crystalline reduced diacids 22a and 22b.

It was not known which acid corresponded to equilenin until completion of the synthesis; 22b led to (±) equilenin 1 and 22a to (±)-isoequilenin 26. Optically active 22a and 22b,

the methyl ethers of *cis* and *trans* bisdehydromarrianolic
acid, were obtained later by hypoiodite oxidation of the methyl
ethers of (+) isoequilenin and (+) equilenin, respectively.[14]

Conversion of 22b to the dimethyl ester followed by selective saponification with one equivalent of dilute alkali gave
the primary acid 23. Arndt-Eistert homologation, Dieckmann
cyclization, and simultaneous ether cleavage and decarbometh-
oxylation led to (±) equilenin 1. Resolution was accomplished
by fractional crystallization of the diastereomeric (-) menth-
oxyacetates.

Despite the nonstereospecific reduction leading to acids
22a and 22b, yields in the other steps were sufficiently high
such that about 2.5 g of (±) equilenin and an equal amount of
(±) 14-isoequilenin were obtained from 10 g of 2.

B.   Johnson Syntheses

Two syntheses of equilenin were developed by Johnson and his
associates.[15] In hopes of developing a simpler alternative to
the Bachmann procedure for constructing the propionic acid side
chain at C-1 (Scheme 2) they investigated the Stobbe reaction.[16] Previous attempts to add a propionic ester unit in
one step to ketones analogous to 18 using Grignard reagents
derived from β-halopropionic esters had given poor yields.[17]
More recently, moderate yields of β,γ-unsaturated propionic
acids have been attained via the Wittig reaction by generation
of ylids of β-carboxytriphenylphosphonium halides in the presence of the carbonyl component.[18]

In initial studies[15a] Stobbe condensation of the β-keto
ester 27 with diethyl succinate failed, possibly for steric
reasons, and the only isolable product was 28 resulting from
ketonic cleavage during work-up.

The reaction was then tried with the corresponding β-ketonitrile 29 in which the carbonyl group is sterically more accessible than in 27. Condensation occurred and in fact proceeded beyond the expected stage to yield directly the tetracyclic product 30 in 60% yield. The latter may be formed
inter al via Thorpe condensation of the intermediate Stobbe
product followed by lactonization, elimination and decarboxylation as formulated.[15a]

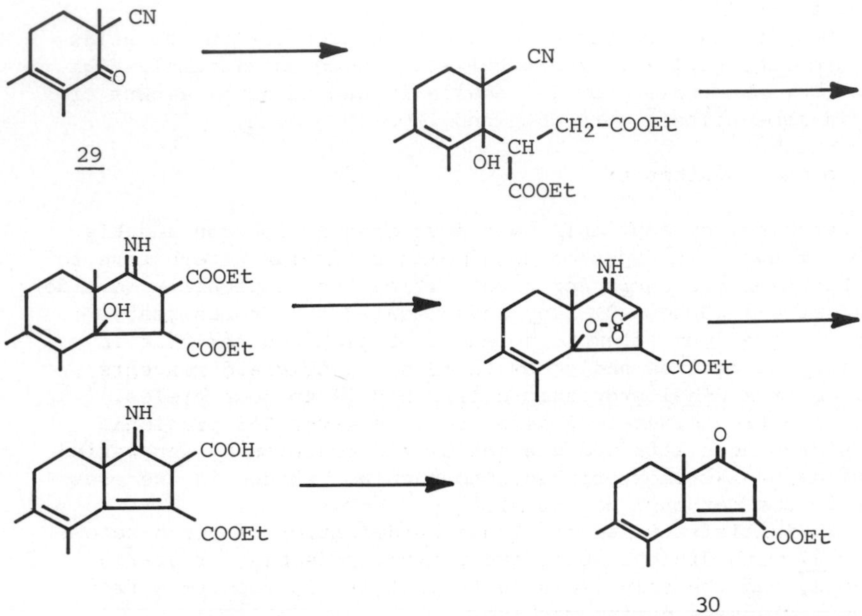

It should be noted that analogous β ketonitriles do not
undergo Stobbe condensation in the absence of an aromatic system conjugated to the carbonyl group because of competing ring
cleavage, for example, 31 → 32.[16,64]

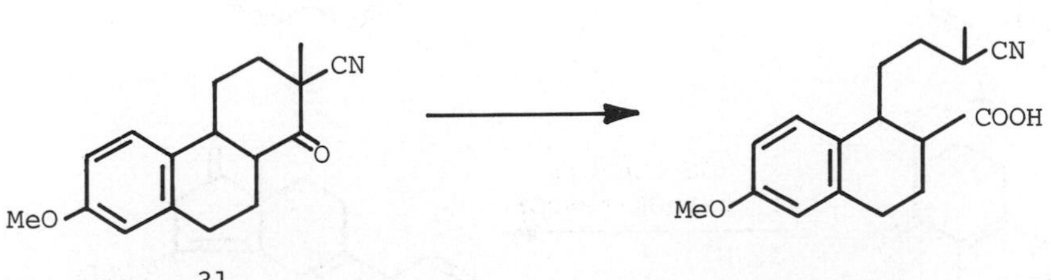

The β-ketonitrile 36 (Scheme 3) was obtained from tricyclic ketone 2 via the hydroxymethylene ketone 33 and isoxazole

Scheme 3. Johnson's First Synthesis of Equilenin

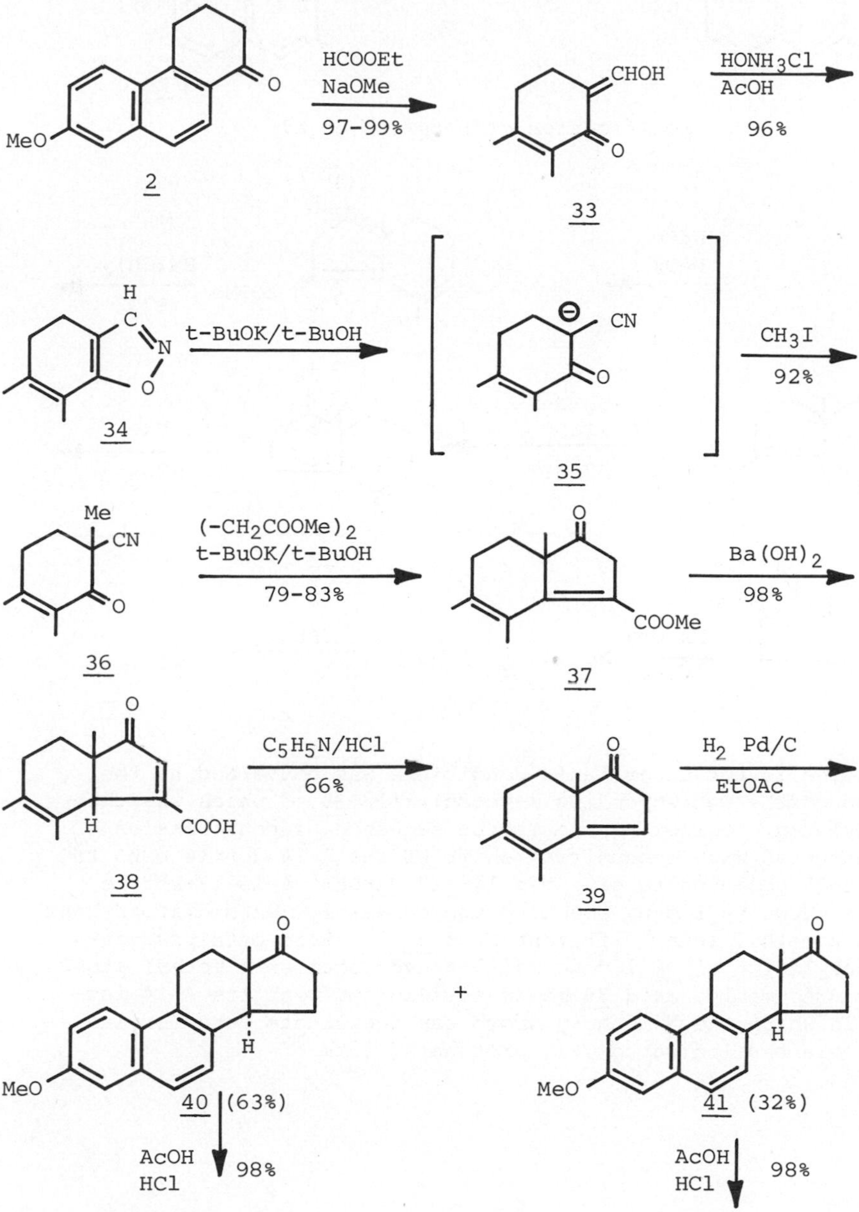

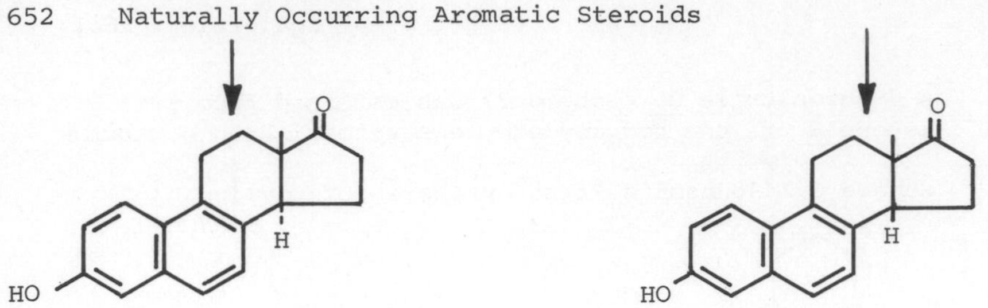

Modification of Banerjee et al.

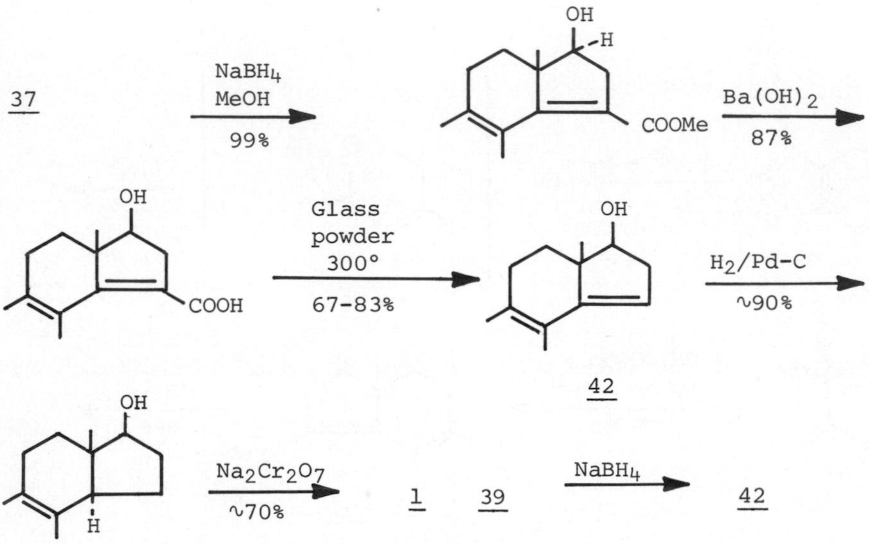

34. The latter under basic conditions was converted to the anion of the corresponding β-ketonitrile 35,[19] which was then methylated. Following the Stobbe sequence, saponification of 37 occurred with concomitant shift of the Δ-14 double bond to the Δ-15 position to give the 14β-15-carboxy-Δ-15-17-ketone 38 as shown by its uv spectrum and conversion with diazomethane into a methyl ester different from 37.[20] Acid catalyzed decarboxylation of 38 led to 14,15-dehydroequilenin methyl ether 39. Presumably, acid 38 is in equilibrium with its Δ-14 isomer in which the 3-methoxy group can facilitate decarboxylation via the indicated C-15 protonated form.[20]

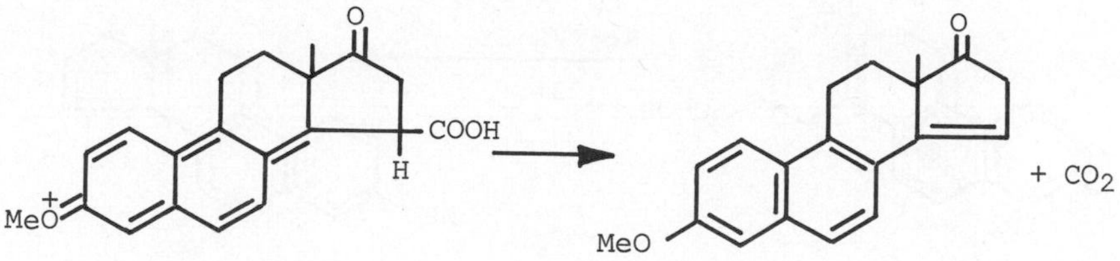

39

Catalytic hydrogenation of 39 occurred with partial stereoselectivity to give a 2:1 mixture of (±) equilenin methyl
ether 40 and (±) isoequilenin methyl ether 41, readily separated by fractional crystallization and hydrolyzed to the
respective phenols 1 and 26 by Bachmann's procedure. By this
synthesis 3.3 g of (±) equilenin 1 was obtained from 10 g of
tricyclic ketone 2.   This was further improved by Banerjee
et al.[21] who showed that the hydrogenation step could be made
stereospecific if carried out on the 17β-ol 42 corresponding
to 39.   The unsaturated 17β-ol 42 was initially obtained from
37 as indicated in Scheme 3,[21a] or better, directly from
39.[21b]

An additional synthesis of equilenin via 14,15-dehydro-
equilenin 39 based on the Stobbe reaction of 1-oxo-2-methyl-
7-methoxy-1,2,3,4-tetrahydrophenanthrene 43 with dimethyl succinate was described by Johnson and Stromberg[15b] (Scheme 4).

Scheme 4.   Johnson's Second Synthesis of Equilenin

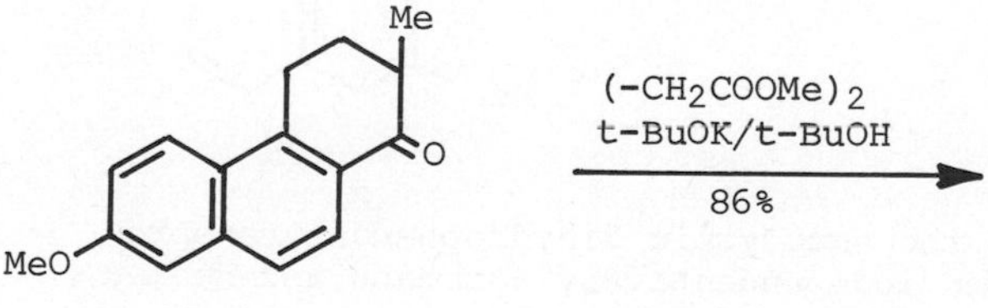

43

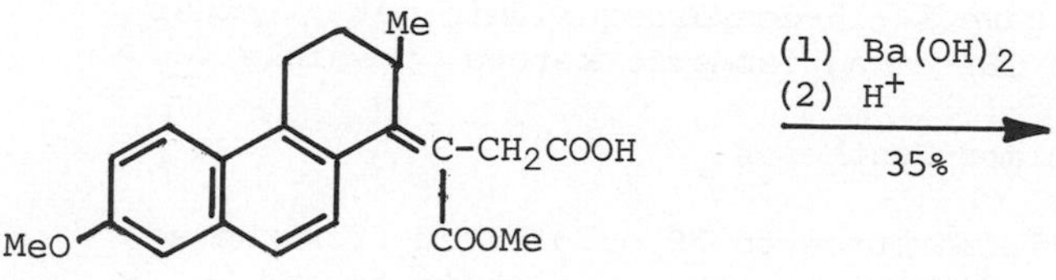

44

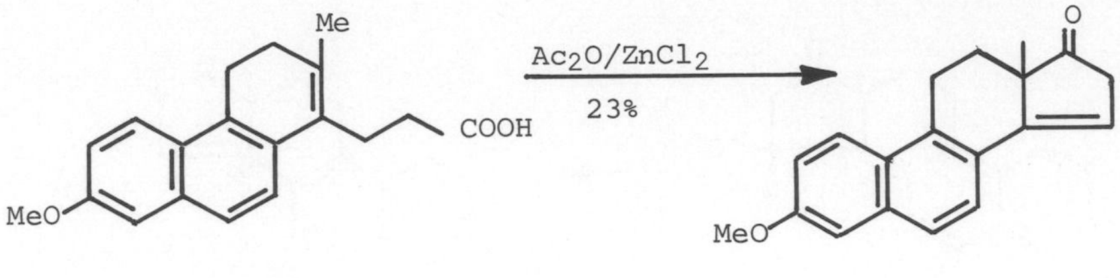

45

39

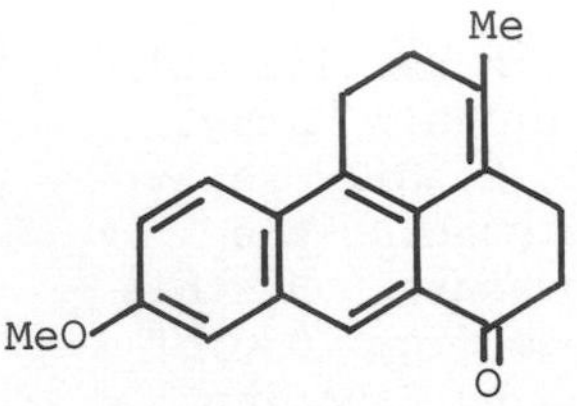

46

1

The methyl ketone 43 had been obtained by methylation of the substituted malonic ester intermediate 8 in the synthesis of 2 (Scheme 1), followed by saponification, decarboxylation, and cyclization.[6b]   The Stobbe product 44 was saponified and

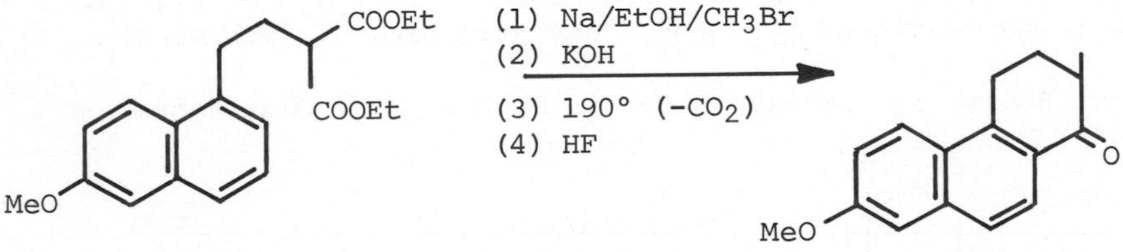

8

(1) Na/EtOH/CH$_3$Br
(2) KOH
(3) 190° (-CO$_2$)
(4) HF

43

decarboxylated to give the endocyclic dihydrophenanthrenepropionic acid 45, obtained independently by Bachmann and Holmen from 43 by Arndt-Eistert homologation of the corresponding Reformatsky product.[22] Cyclization of 45 in refluxing acetic anhydride containing a little zinc chloride under carefully defined conditions gave 14-15-dehydroequilenin methyl ether 39 (23%) as well as 18% of an isomeric ketone, possibly 46.

C. Bachmann and Holmen Syntheses

Two short but low yield routes to 39 originated from the adduct 47 derived from diethyl α-methyl-β-oxoadipate and 6-methoxy-1-naphthylethyl iodide (Scheme 5).[22] Cyclodehydration of 47 gave a mixture of two isomeric lactone esters 48, which

Scheme 5

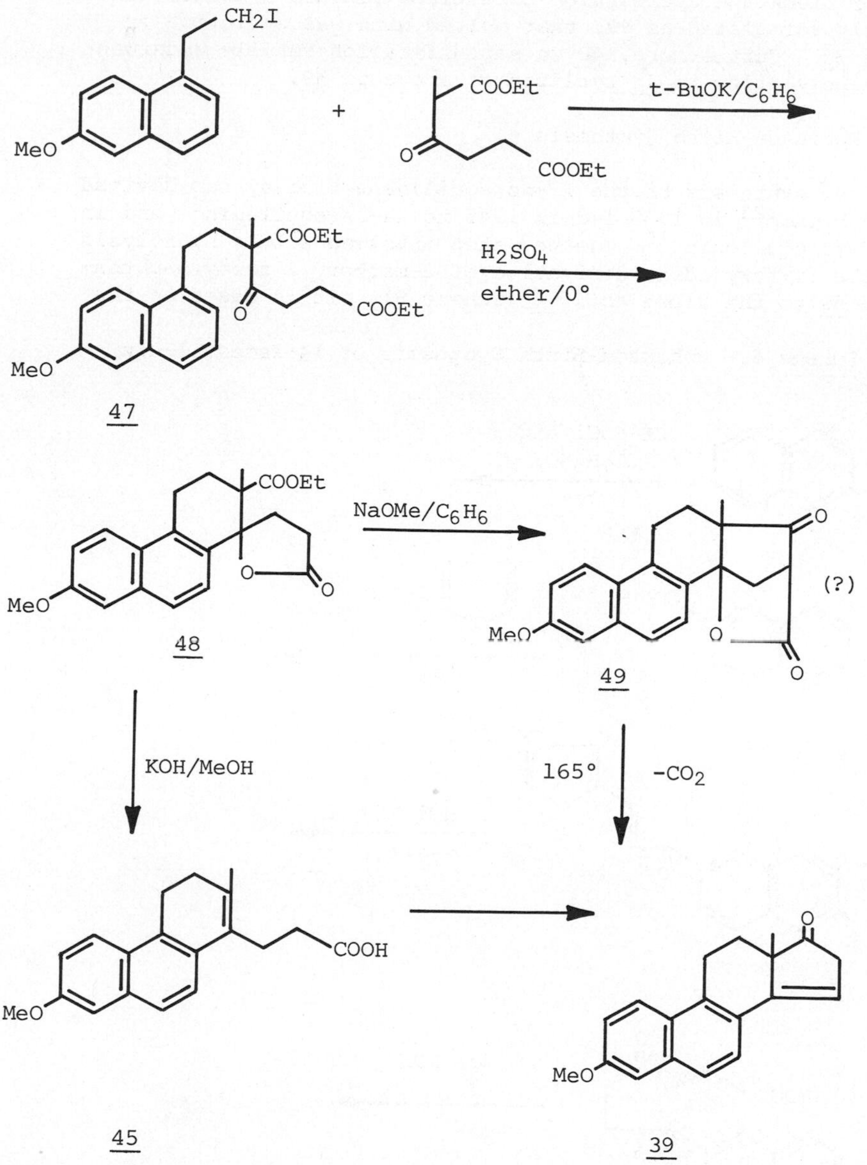
CH2I
MeO
+
COOEt
O
COOEt
t-BuOK/C6H6
COOEt
O
COOEt
MeO
47
H2SO4
ether/0°
COOEt
O
O
MeO
48
NaOMe/C6H6
O
(?)
MeO
O
O
49
KOH/MeOH
165°
-CO2
COOH
MeO
45
O
MeO
39

under Dieckmann cyclization conditions yielded a solid, tenta-
tively formulated as 49, that melted with gas evolution to
give 39. Furthermore, 48 on saponification readily underwent
decarboxylation to 45 cyclized as above to 39.

D.   Robinson-Birch Synthesis

A novel synthesis of the 13-norequilinane ring system devised
by Robinson[23] in 1938 led in 1945 to 14-isoequilenin[24] and in
1967 to equilenin.[25] The key step involved acidic hydrolysis
of the furfurylidene derivative of 6-methoxy-2-acetylnaphtha-
lene 52 to the dioxo acid 53 (Scheme 6). It is based on the

Scheme 6. Robinson-Birch Synthesis of 14-Isoequilenin

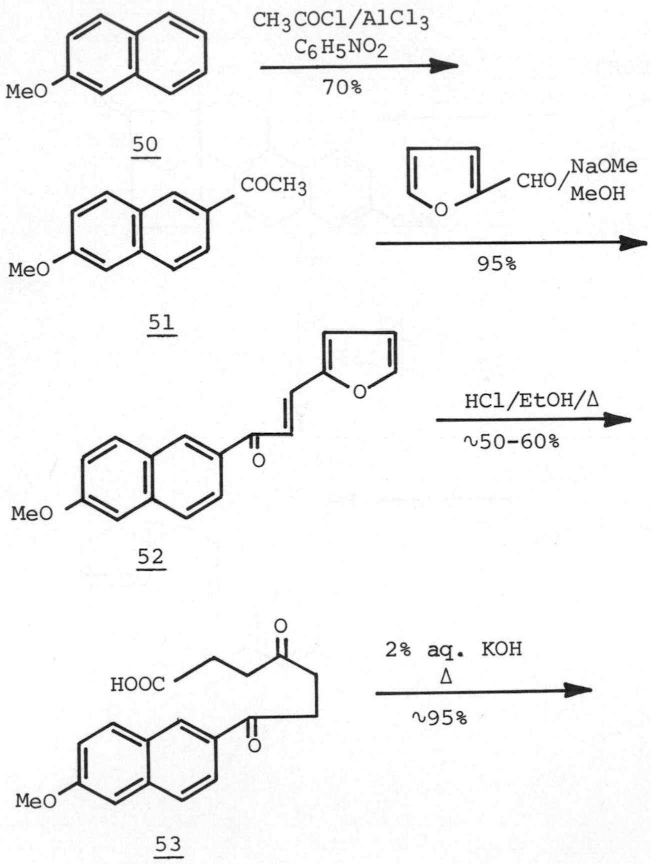

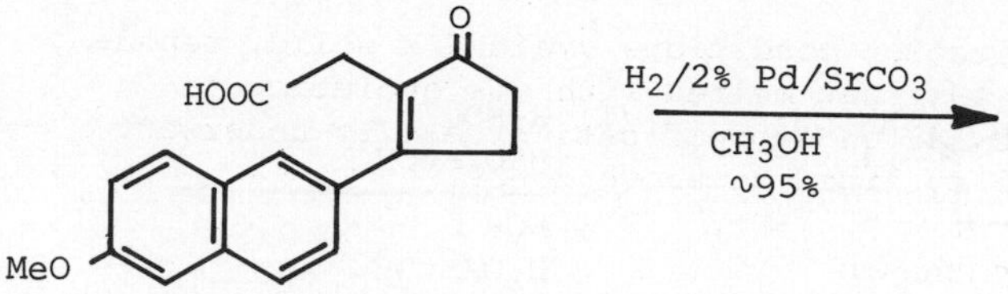

HOOC — (structure 54)

H2/2% Pd/SrCO3
CH3OH
∿95%

54

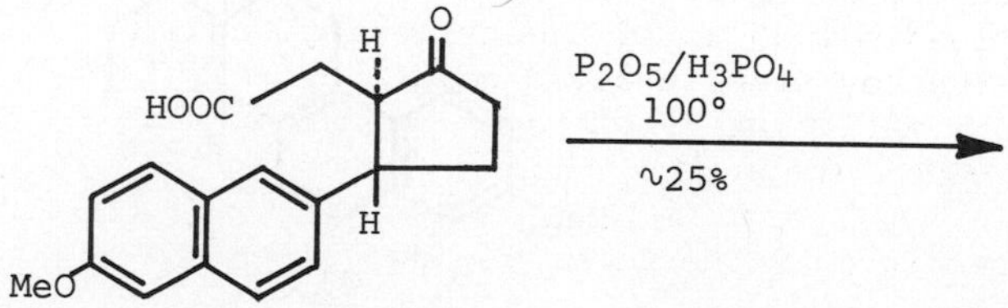

H
HOOC — (structure 55)

P2O5/H3PO4
100°
∿25%

55

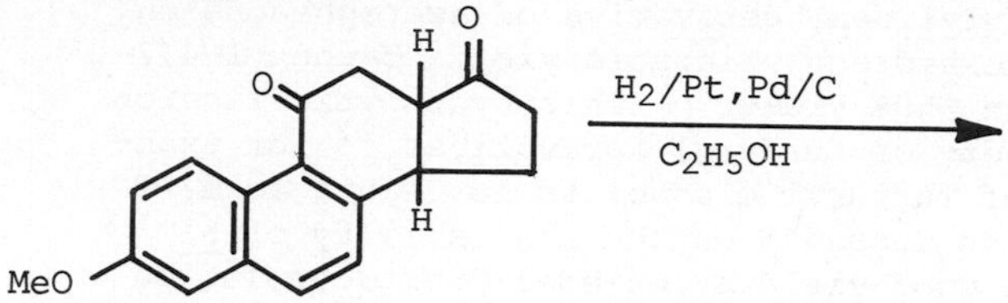

(structure 56)

H2/Pt,Pd/C
C2H5OH

56

(structure 57)

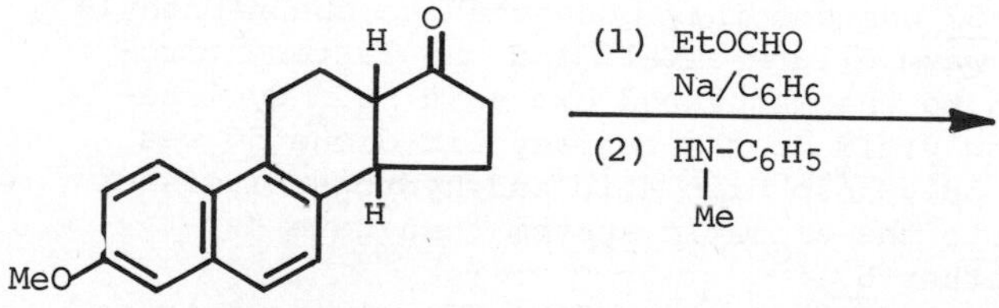

(1) EtOCHO
Na/C6H6

(2) HN—C6H5
|
Me

57

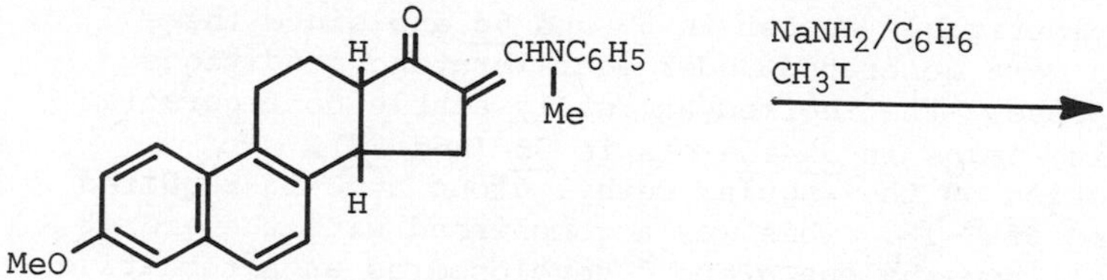

CHNC6H5
|
Me
(structure 58)

NaNH2/C6H6
CH3I

58

finding that the furfurylidene derivative of acetophenone on refluxing in ethanolic hydrochloric acid yields 7-phenyl-4,7-dioxoheptanoic acid in ~50% yield.[26] This hydrolytic fission is general for a variety of furfuryl derivatives,[27] for example, the conversion of furfuryl alcohol to levulinic acid.

The synthesis also depended on the availability of 51 which was obtained in good yield by Friedel-Crafts acylation of 50 at C-2 in nitrobenzene,[28] although in less polar solvents such as benzene or carbon disulfide acylation occurs at C-1. The dioxo acid 53 was smoothly converted to the naphthylcyclopentenone 54 in warm dilute alkali and the latter hydrogenated in high yield to the saturated oxo acid 55. Cyclization in relatively low yield to the tetracyclic dione 56 was best accomplished in polyphosphoric acid and hydrogenolysis of the carbonyl group α to the aromatic system then gave "x"-norequilenin methyl ether 57.[23]

The assumed course of cyclization 55 → 56 was shown to be correct in the 3-deoxy series by dehydrogenation of the 3-deoxy analog of 56 to 1,2-cyclopentenophenanthrene. Since C-13 is an epimerizable center in 55 and 56 and since these intermediates were generated under equilibrating conditions, they should possess the thermodynamically stable configuration shown, that is, trans in 55 and cis in 56 (and 57).

Introduction of the angular methyl group at C-13 required prior blocking of C-16. This was accomplished with the aid of the methyl-anilinomethylene group[29] developed as an alternative to the more difficultly removable benzylidene type blocking group.[30] Other derivatives of 2-hydroxymethylene ketones have since been utilized as blocking groups of which the n-butylthiomethylene is one of the most satisfactory.[31]

Methylation of the C-16 blocked ketone 58 occurred entirely in the cis sense, and following removal of the blocking group, the product was (±) 14-isoequilenin methyl ether <u>41</u>, a not unexpected result since a similar sequence on hydrindan-1-one led exclusively to *cis* 8-methyl-hydrindan-1-one.[24]

By contrast analogous methylation of α-decalone gave *cis-trans* mixtures[29,30b] in which the *trans* component, although the minor product, could be separated and transformed to *trans* 8-methylhydrindan-1-one.[32] This result led to a number of steroid syntheses based on introduction of the C-13 methyl group into a C,D decalone system followed by separation of the *trans* isomer and contraction of ring D (see below). In a large number of substituted decalones *cis* methylation pre-dominated over *trans* with the striking exception of the Δ 6-analog.[33,66,71] An important factor in determining the pre-ferred direction of reaction between enolate anion and methyl

$$\frac{cis}{trans} \sim 3$$

$$\frac{cis}{trans} > 10$$

$$\frac{cis}{trans} \sim 0.3\text{-}0.5$$

iodide in these systems appears to be the relative thermodyna-mic stabilities of the methylated products. Bucourt has ra-tionalized the results in terms of minimization of torsional strain of the dihedral angles involving the ring junctions as well as of the relative steric interference between the angu-lar methyl group and the pertinent axial hydrogens.[34]

For a number of years there was no further development of the Robinson approach. However, application of modern syn-thetic methods has led to completion of two routes to equil-enin.

In 1967 Birch and Subba Rao[25] applied Stork's[35] reductive

methylation procedure to the unsaturated oxo acid 54 and obtained the 13-methyl oxo acid 60 as a 1:3 *cis:trans* mixture,
quite different from the 3:1 *cis:trans* ratio observed with
Δ^9-octal-1-one.[35] In the least satisfactory step of the synthesis 60 was cyclized (polyphosphoric acid) in ~40% yield and

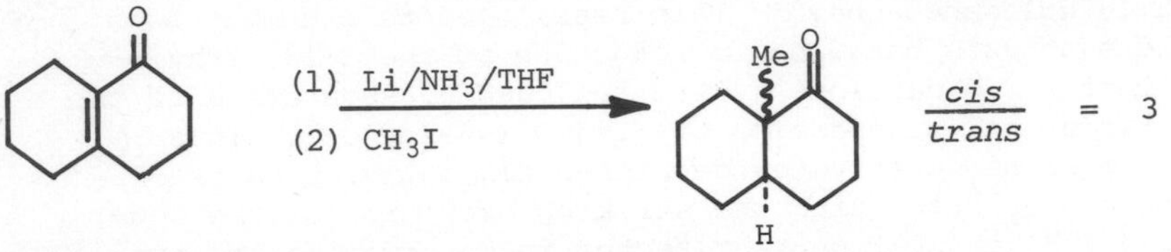

$$\frac{cis}{trans} = 3$$

the diones 61 and 62 separated chromatographically. Hydrogenolysis of each then gave (±) equilenin methyl ether 40 and
(±) isoequilenin methyl ether 41, respectively.

Scheme 7.   Birch Synthesis of Equilenin

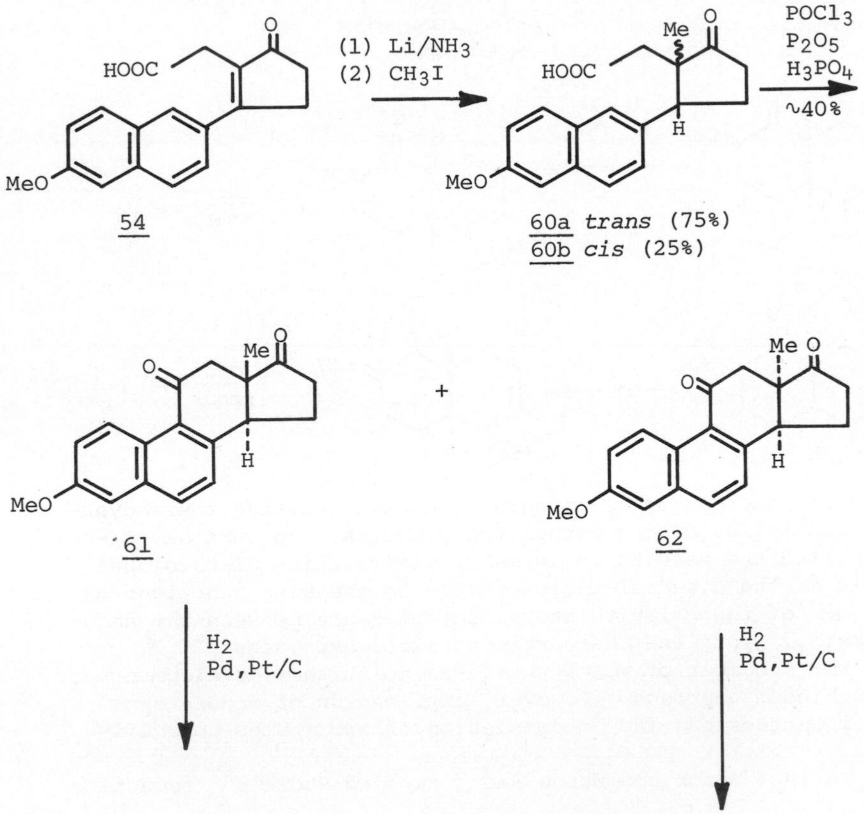

54

60a *trans* (75%)
60b *cis* (25%)

61   62

H_2
Pd,Pt/C H_2
                                                Pd,Pt/C

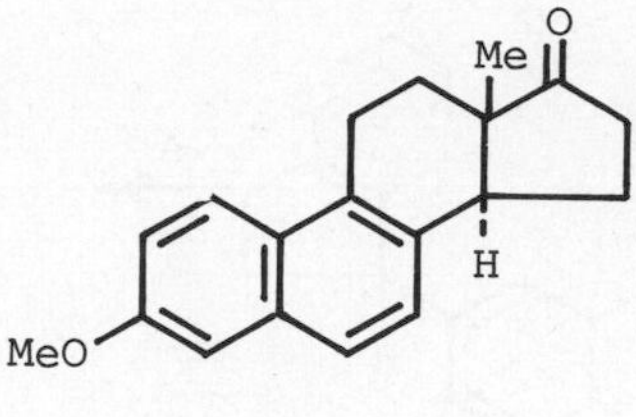

40

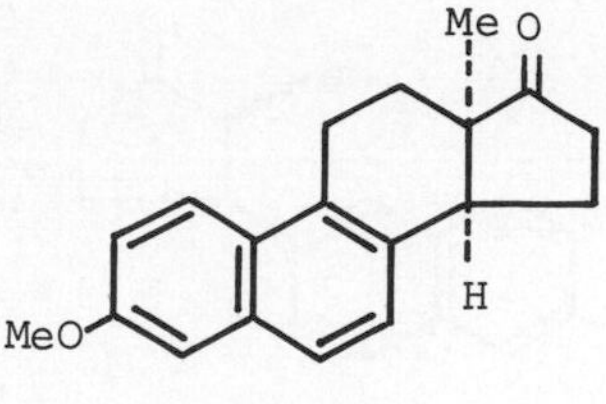

41

E. Horeau Synthesis

A stereospecific synthesis of the *trans* 13-methyl oxo acid
60a was described in 1969 by Horeau et al.[36]   2-Bromo-6-
methoxynaphthalene (available from β-naphthol in good yield)
was converted to the corresponding Grignard reagent 61 in
tetrahydrofuran and the latter condensed with the isobutyl
enol ether of 2-methylcyclopentanedione-1,3 62 to give the
tricyclic ketone 63. Hydrogenation led to a mixture of methyl
ketones 64 which gave the allyl enol ether 65 by an acid catalyzed exchange procedure involving allyl alcohol and 2,2-di-
methoxypropane. Claisen rearrangement occurred stereospecifically on warming to give a single allyl methyl ketone 66 in
good yield. Permanganate oxidation then led to the oxo acid
60a converted to equilenin methyl ether 40 by the previously
described procedures. The stereospecificity of the Claisen
rearrangement may be rationalized sterically--the large naph-
thyl residue impeding bond formation from its side of the
cyclopentane ring.

Scheme 8.   Horeau Synthesis of Equilenin

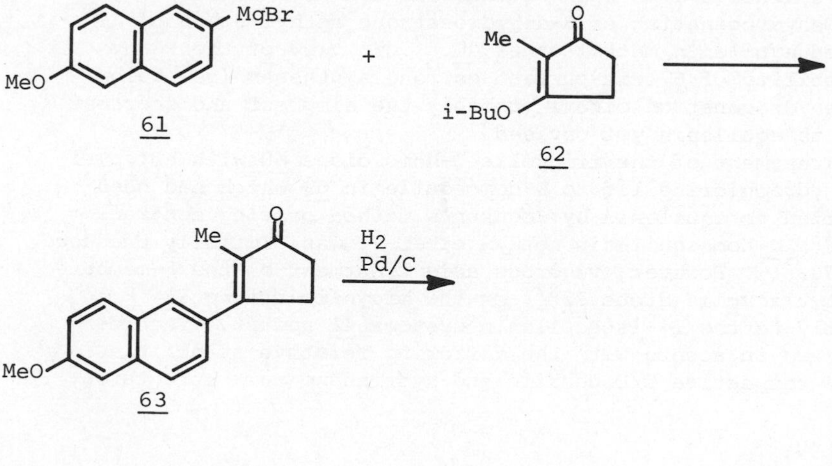

61

+

62

63

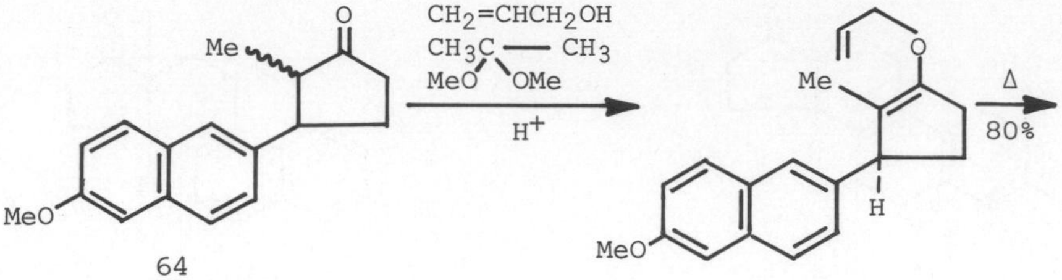

F. Synthesis from Estrone Intermediates

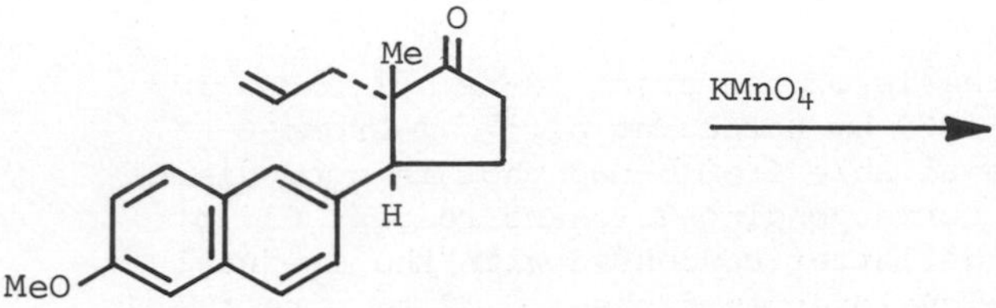

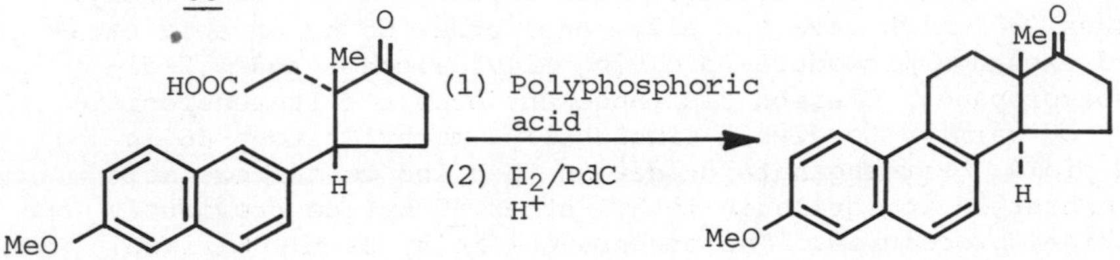

A number of routes to equilenin from intermediates in estrone
total synthesis have been devised and are formulated below.

   Dehydrogenation of 8-dehydroestrone methyl ether 67
yielded equilenin methyl ether 40.[37] In view of the ready
availability of 67 via current estrone syntheses (see below)
its dehydrogenation offers probably the simplest and shortest
route to equilenin yet devised.

   Treatment of the tricyclic D-homo dione 68 with hot pyridine hydrochloride led to D-homoequilenin 69 which had been
converted to equilenin by Johnson's method of ring contraction.[38] D-Homoequilenin methyl ether 71 was similarly produced
from 70.[37b] However, vigorous acid treatment of the 5-membered
ring D-tricyclic dione 72[37c] or the bicyclic trione 73[37b]
led only to the 14-isoequilenin systems 41 and 26. The results are in accord with the differing relative stabilities
of the respective C/D decalin and hydrindane part structures.

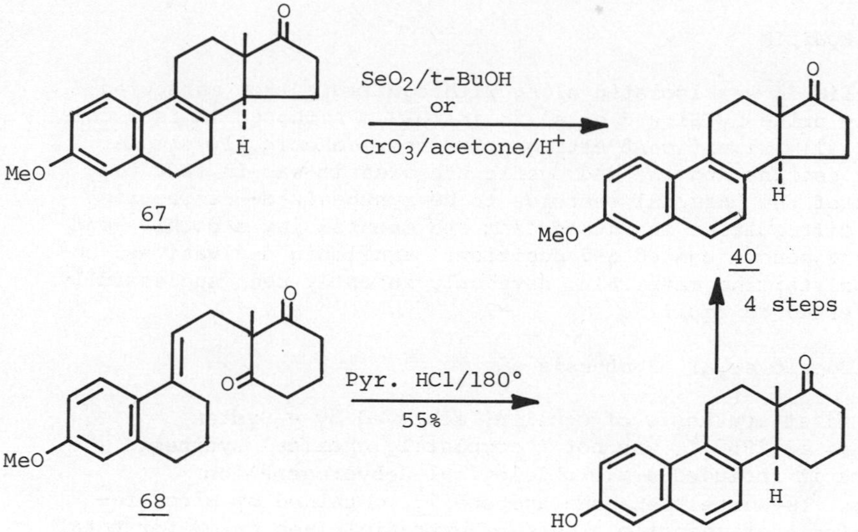

67 → SeO₂/t-BuOH or CrO₃/acetone/H⁺ → 40

68 → Pyr. HCl/180° 55% → 69 → 4 steps → 40

70 → Polyphosphoric acid/70° → 71

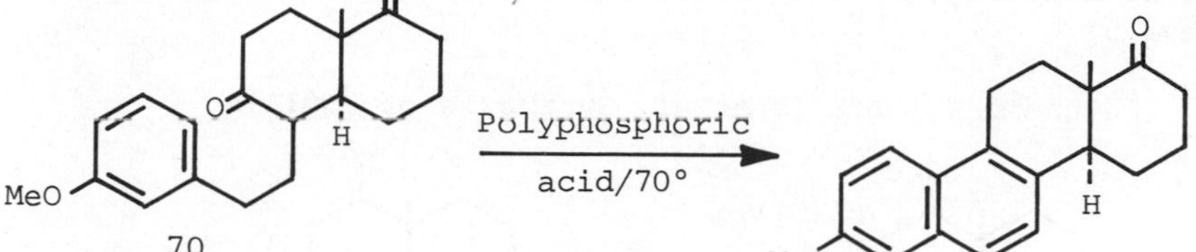

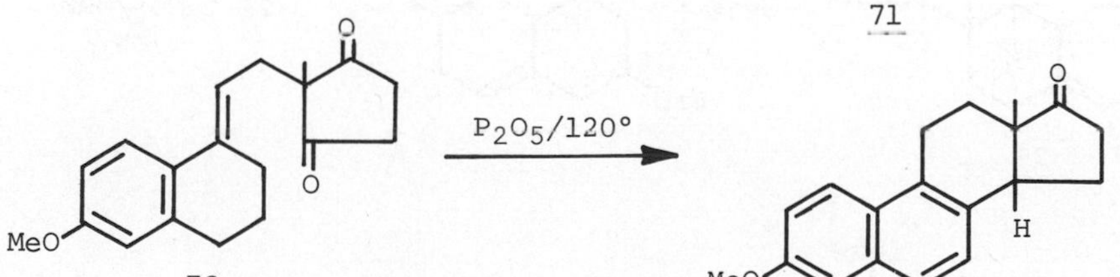

72 → P₂O₅/120° → 41

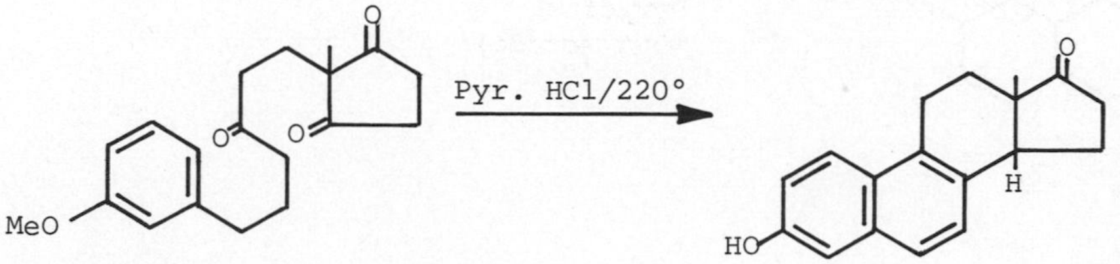

73 → Pyr. HCl/220° → 26

3. EQUILIN

Equilin 74 was isolated along with equilenin from mare pregnancy urine by Girard et al.[39] in 1932.   Although it is a commercially significant estrogen and stereochemically simpler
than estrone and the androgenic steroids it was in fact the
last of the hormonal steroids to be synthesized--reflecting
the difficulties in introducing and maintaining a double bond
in the nonconjugated Δ-7 position.   Equilenin derivatives, obvious starting materials, have only recently been successfully
converted to equilin.

A.   Zderic et al. Synthesis

The first synthesis of equilin, achieved by a Syntex
group, in 1958,[40] was not a completely chemical synthesis
since it included a microbiological dehydrogenation
step.   19-Nortestosterone acetate 75, obtained by Birch reduction[41] of phenolic ethers of estradiol (see below for total
synthesis), was converted to the corresponding 4,6-dien-3-one
77 by bromination and dehydrobromination of the enol ether 76
(Scheme 9).

Scheme 9.   Zderic et al. Synthesis of Equilin

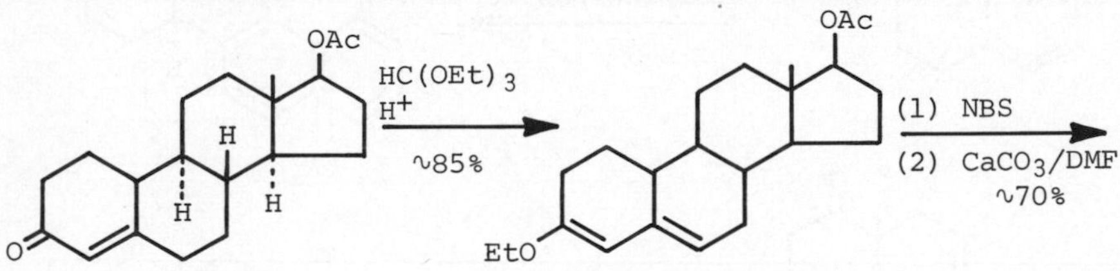

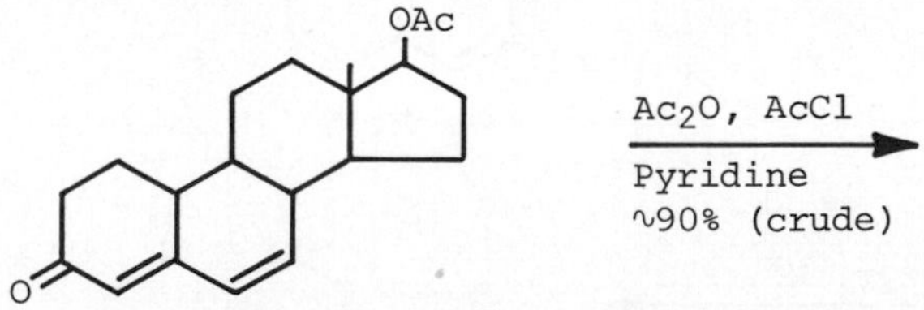

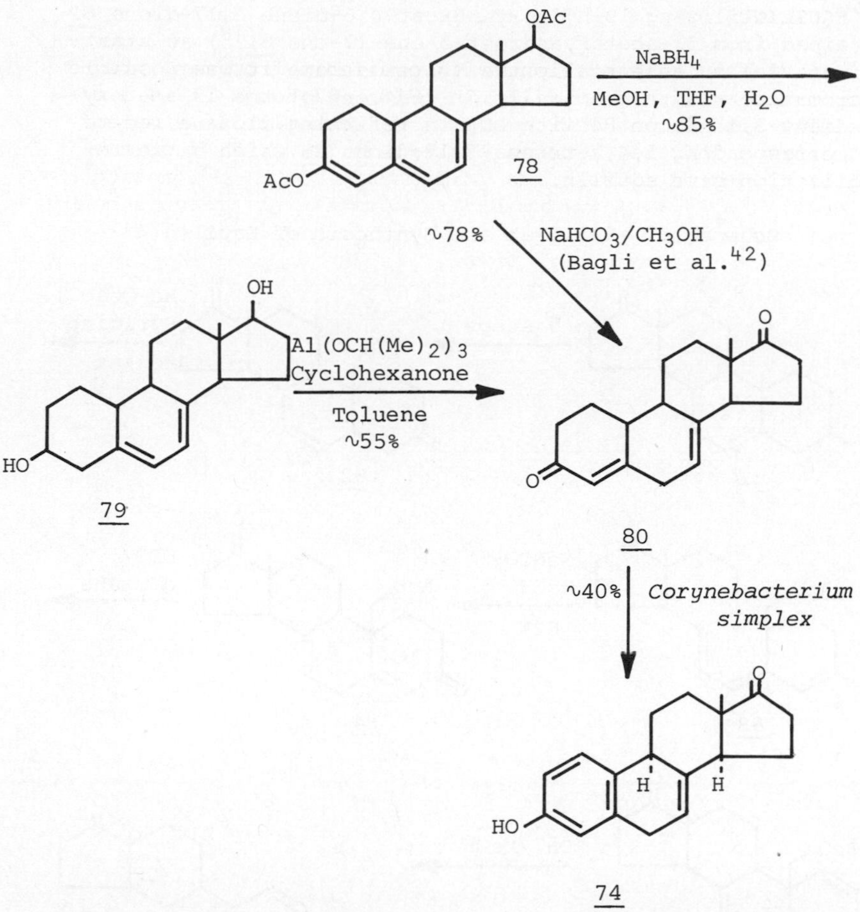

Enol acetylation of 77 led to the 3,17β-diacetoxy-3,5,7-triene
78 which on sodium borohydride reduction followed by Oppenauer
oxidation gave androsta-4,7-diene-3,17-dione 80. This two step
sequence was simplified by Bagli et al.[42] who converted enol
acetate 78 directly to 80 by brief treatment with sodium bicarbonate in dry methanol. Under these mild conditions the
Δ-7 double bond did not shift into conjugation. Attempted
ring A dehydrogenation utilizing dichlorodicyanoquinone (DDQ)
by procedures which were successful with 10-substituted-4-en-
3-ones failed. However, microbiological dehydrogenation with
Corynebacterium simplex gave equilin 74 in 40% yield.

B. Bagli et al. Synthesis

     A second synthesis of equilin was completed by Bagli et

al.[42] By utilizing 19-hydroxyandrost-4,6-diene-3,17-dione 82 (obtained from 3β-acetoxyandrost-5-ene-17-one 81[43]) as starting material in essentially the Syntex scheme it was possible to aromatize ring A chemically. Treatment of the 19-acetoxy-4,7-diene-3,17-dione 84 with DDQ in refluxing dioxane led to the corresponding 1,4,7-triene-3,17-dione 85 which on retro-aldolization gave equilin.

Scheme 10. Bagli et al. Synthesis of Equilin

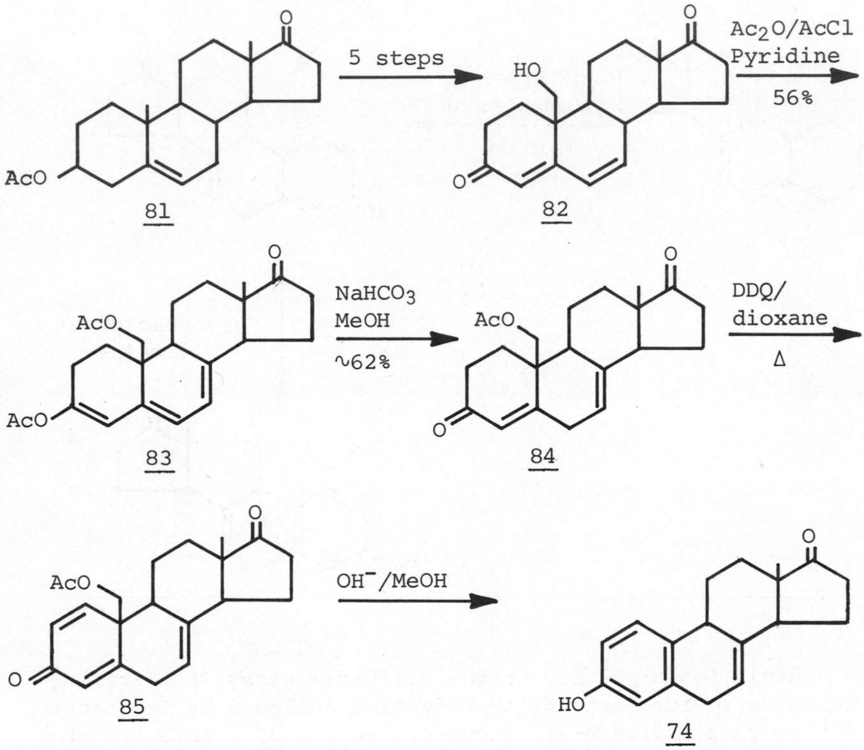

C. Stein et al. Synthesis

The third synthesis of equilin[44a,44b] proceeded from 3-methoxy-17β-hydroxyestra-1,3,5(10),8-tetraene 86, an intermediate in estrone total synthesis and consisted essentially in transfer of the Δ-8 double bond to the Δ-7 position. Synthetic resolved 86 on treatment with m-chloroperbenzoic acid in benzene-hexane gave a mixture of α-oxide 87 and allylic alcohol 88, which was converted entirely to 88 by treatment with benzoic acid in chloroform. Catalytic

hydrogenation then yielded the ring C saturated diol 89 which
with methanesulfonyl chloride in refluxing pyridine gave the
Δ-7 olefin 90a as the expected product of bimolecular *trans*
elimination, contaminated with ∿10% of its Δ-8 isomer.   Lithium
aluminum hydride converted 90a to the 17β-carbinol 91 and the
synthesis was completed by ether cleavage at C-3 and Oppenauer
oxidation at C-17.

The original 16% overall yield[44a] was later improved to
∿38%, principally by utilizing phosphorus oxychloride in dimethylformamide at 0-25° in the elimination step to give the
17-formate 90b in 81% yield (converted to 91 by aqueous methanolic alkali) and only a trace of the Δ-8 isomer.[44b]

Scheme 11.   Stein et al. Synthesis of Equilin

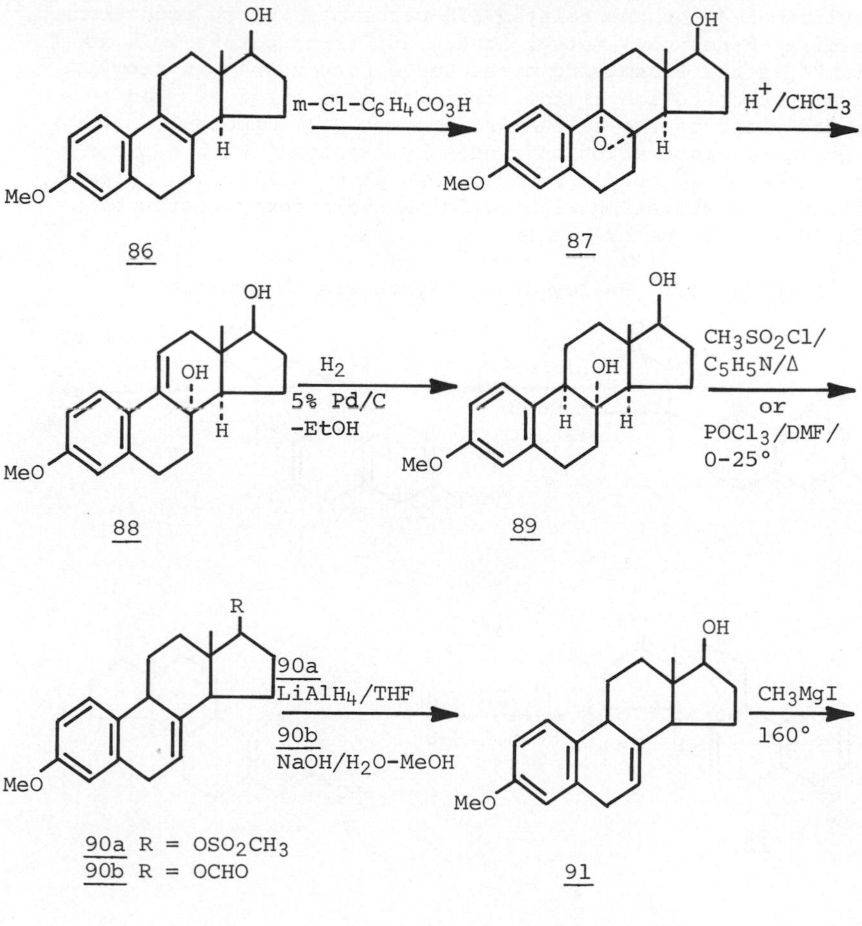

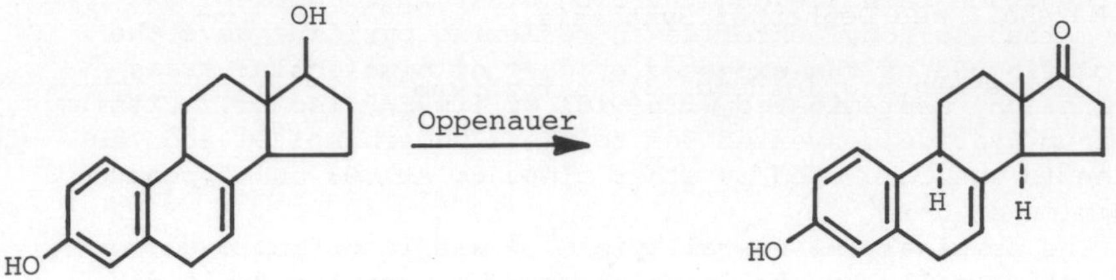

92   →(Oppenauer)→   74

D. Bailey et al. Synthesis

The fourth synthesis of equilin[45] proceeded from equilenin methyl ether 40 or the related 17β-carbinol. Birch reduction of similar β-naphthyl methyl ethers utilizing potassium[46] or sodium[47] as the dissolving metal had occurred only in ring A. However, substitution of the more efficient lithium[48] led to a moderate yield of the tetrahydro product 93. Oppenauer oxidation and aromatization of ring A by treatment with N-bromo-succinimide in aqueous t-butanol then gave equilin 74. Alternatively, aromatization with pyridine hydrobromide perbromide produced equilin methyl ether.

Scheme 12.   Bailey et al. Synthesis of Equilin

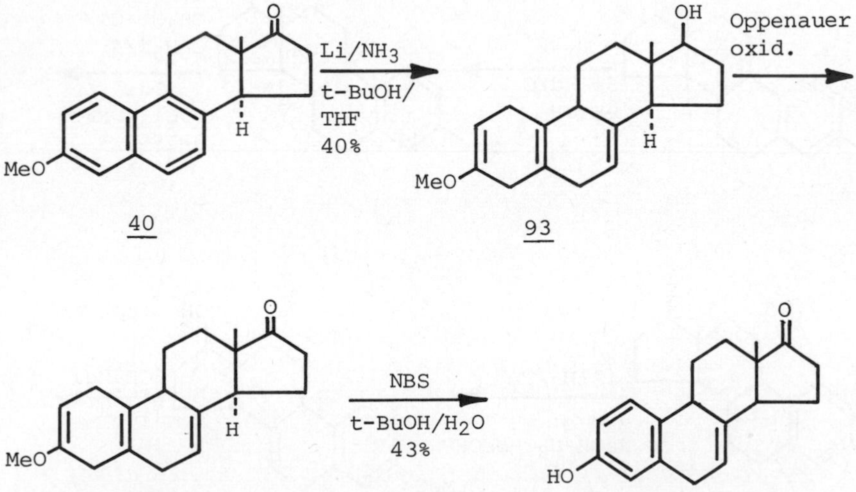

40    $\xrightarrow[\substack{t\text{-BuOH}/ \\ THF \\ 40\%}]{Li/NH_3}$ 93

94    $\xrightarrow[\substack{t\text{-BuOH}/H_2O \\ 43\%}]{NBS}$ 74

E.   Marshall and Deghenghi Synthesis

The fifth synthesis (Scheme 13), a remarkably efficient conversion of equilenin to equilin, involved low temperature

Scheme 13.   Marshall and Deghenghi Synthesis of Equilin

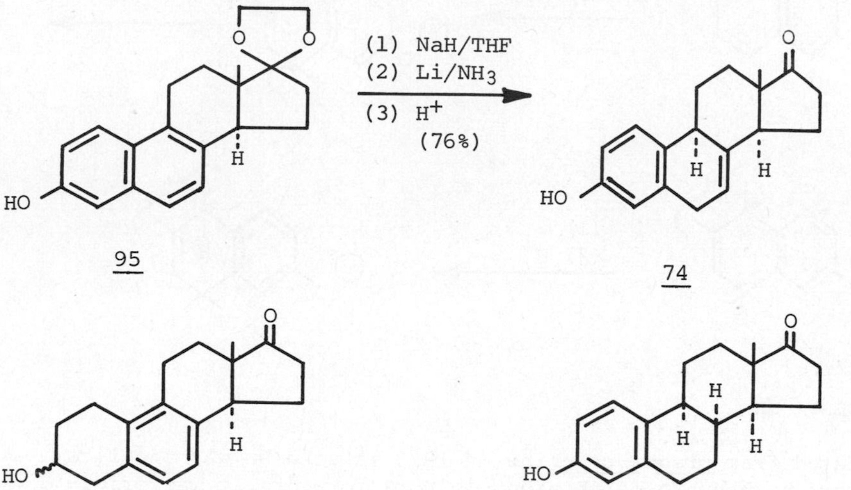

(1) NaH/THF
(2) Li/NH$_3$
(3) H$^+$
      (76%)

95

74

96a, 3α-OH
96b, 3β-OH

97

alkali metal reduction of equilenin ethylene ketal 95.[49]   The
naphthoxide anion, generated in situ or preferably by sodium
hydride in tetrahydrofuran before addition to a solution of
lithium in liquid ammonia at -70°, was preferentially reduced
in ring B to give, following deketalization, a 55% isolated
yield (76% by g.l.c. analysis) of equilin 74. By-products included the ring A reduced epimeric alcohols 96 and in some
runs the corresponding ketone. The absence of 9-isoequilin is
noteworthy. Compound 96b was identical with a ring B aromatic
isomer of estrone isolated from mare pregnancy urine[50] and
synthesized via catalytic hydrogenation of equilenin derivatives.[51] At -33° equilin was the major product with lithium
or sodium, but potassium led to 8α-estrone 97 by a novel reduction of the unconjugated Δ-7 double bond.
     Although Birch reduction of phenols should be more dif-
ficult than reduction of the corresponding phenolic ethers because of the presumed intermediacy of radical dianions rather
than of radical monoanions, it has been accomplished in good
yield by utilizing high concentrations of lithium in liquid

ammonia.[52] In the present case electrostatic repulsion of like charges would favor reduction in ring B rather than in ring A.

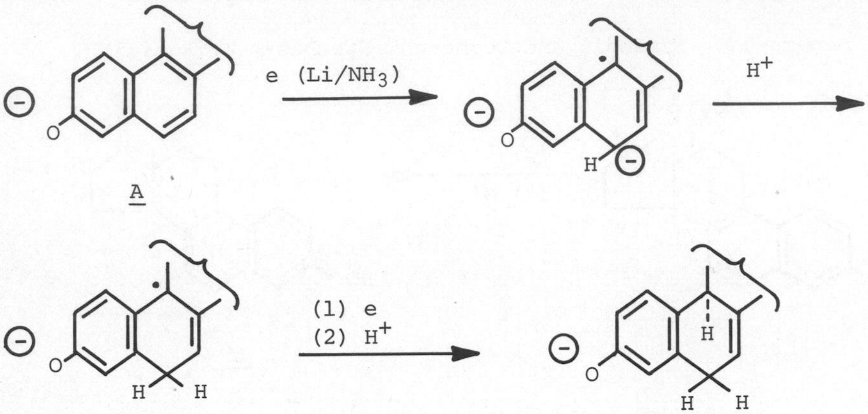

4. ESTRONE

Isolated from pregnancy urine in 1929 estrone 98 was the first steroid hormone to be obtained in pure form.[53] It was originally believed to be the main estrogenic hormone but it has since been recognized that estradiol-17β 99 is the primary estrogen secreted by the ovary.

Both compounds have been interrelated enzymatically and chemically with each other and with a number of structurally similar metabolites, for example, estriol 16α,17β-100. Estrone has been isolated from all major classes of vertebrates as well as from higher plants and recently from water beetle species[54] along with other C-18, C-19, and C-21 hormonal steroids.

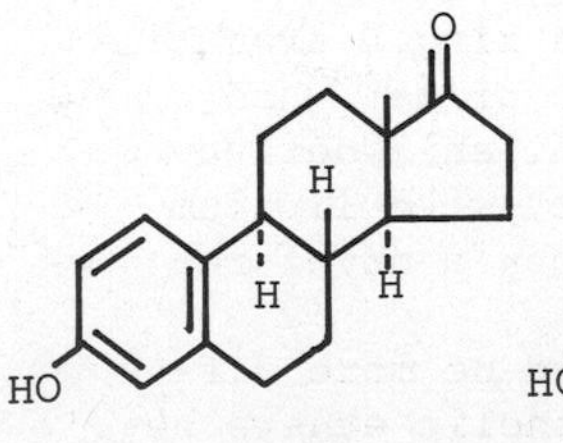

98 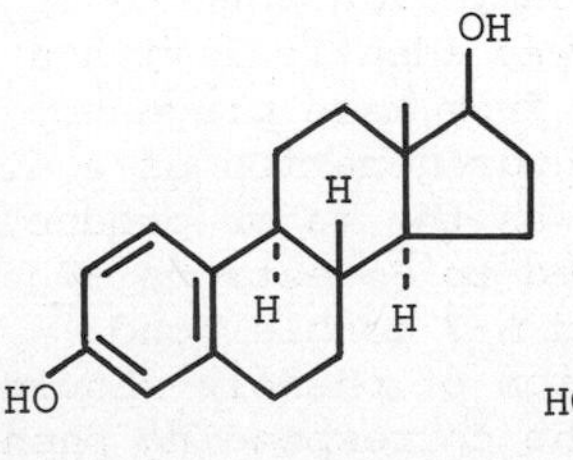   99 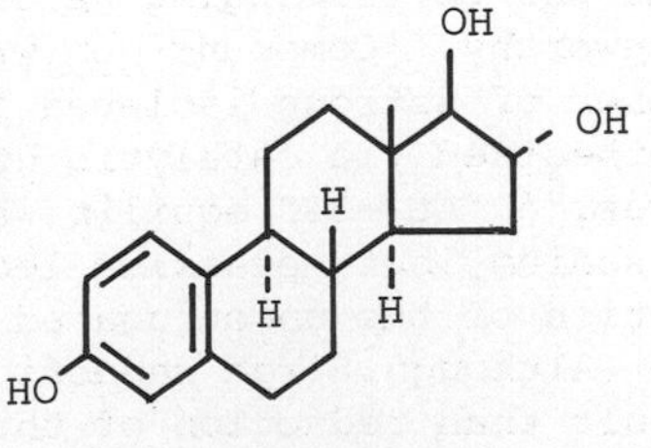   100

The medical importance of estrone engendered early interest in its synthesis. More recently its position as a precurser to commercially important 19-nor-steroids has stimulated
development of industrially feasible totally synthetic routes,
as well as partial syntheses from steroidal raw materials such
as diosgenin, stigmasterol, or cholesterol.[55] Eight racemates
of the estrone structure are possible and stereochemical control in the generation of the four asymmetric centers is man-
datory in any practical synthetic scheme.

One of the first attempts to prepare the estrone structure
involved the Diels-Alder reaction between 1-vinyl-6-methoxy-
3,4-dihydronaphthalene 101 and methylcyclopent-1-en-4,5-dione
102.[56] The crystalline adduct, isolated in low yield, a priori
103 or 104 was converted to an estrogenically inactive mono-
ketone m.p. 210° isomeric with estrone methyl ether.

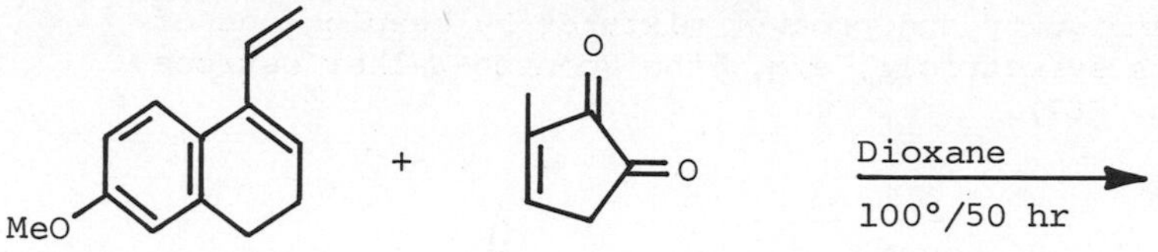

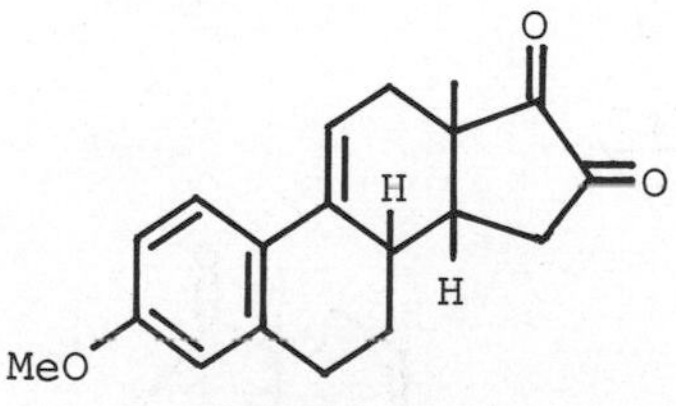

It was shown later that the crystalline adduct (23%) has
in fact the double bond isomerized Δ-8 14-methyl structure
105 and that an equivalent quantity of the non-crystalline
Δ-8 13-methyl isomer 106 is also formed.[57] In any event the
Diels-Alder reaction generates C/D *cis* rather than the required
C/D *trans* stereochemistry.

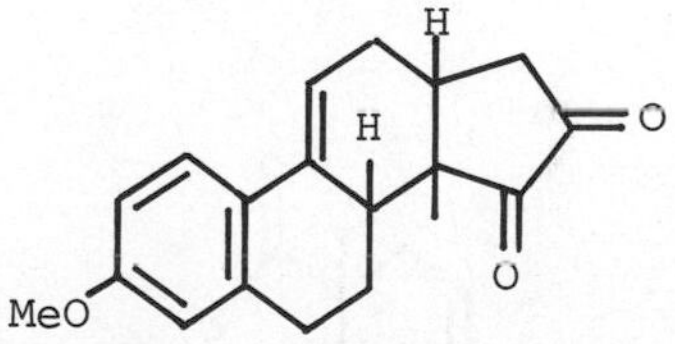

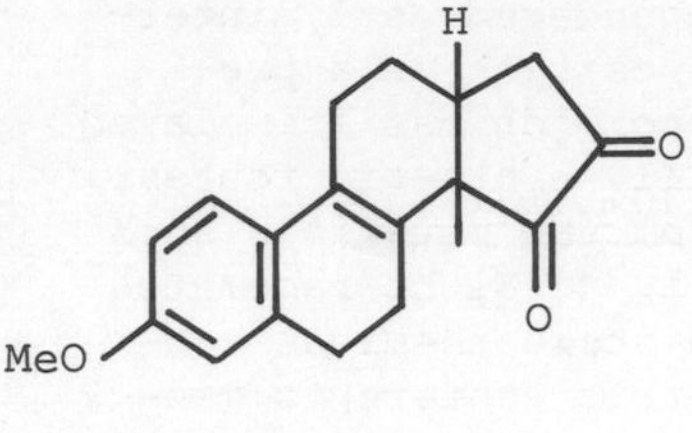

105

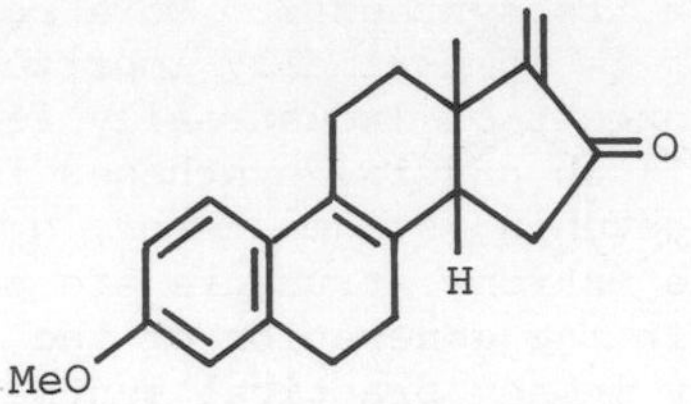

106

In the analogous condensation of diene 101 with citraconic anhydride 107 the 13-methyl adduct 108 predominated over the 14-methyl adduct 109 by about 2:1.[58] Successful examples of the Diels-Alder reaction in steroid synthesis have avoided structural ambiguity and product mixtures by keeping one of the reactants symmetrical, e.g., the Johnson-Walker estrone synthesis (p. 687).

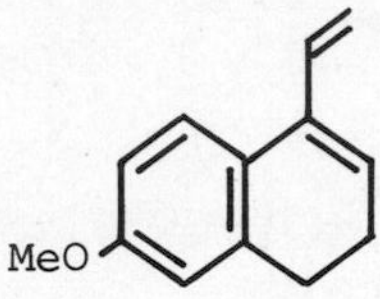

101

+

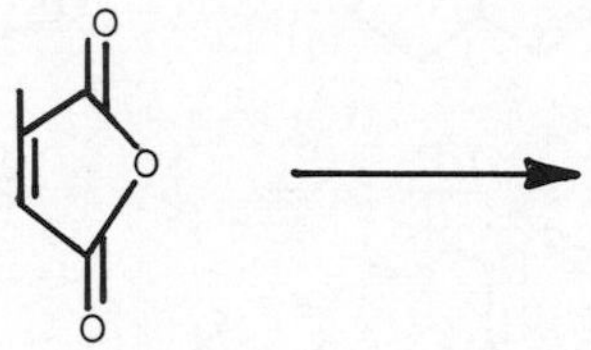

107

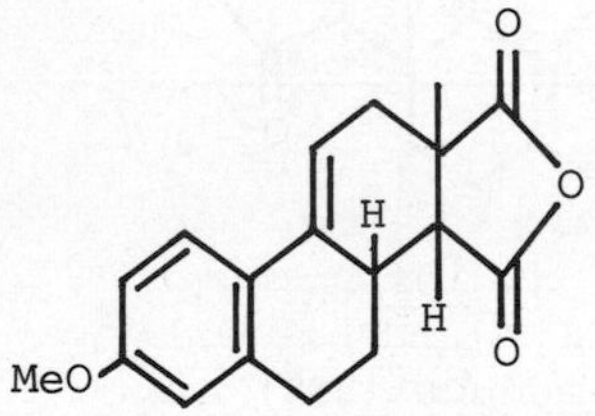

108

+

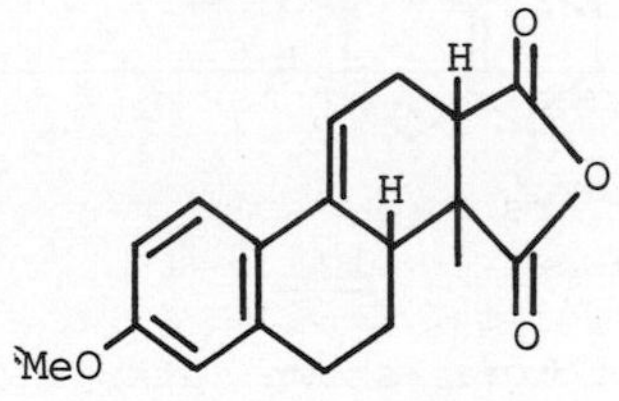

109

The exclusive formation of the 14-methyl adduct observed in the normal Diels-Alder condensation of 101 and 2,6-xylo-quinone is altered in favor of the desired 13-methyl adduct by boron trifluoride catalysis. This recent finding led Valenta and his colleagues to an effective synthesis of (±) estrone methyl ether 119.[118]

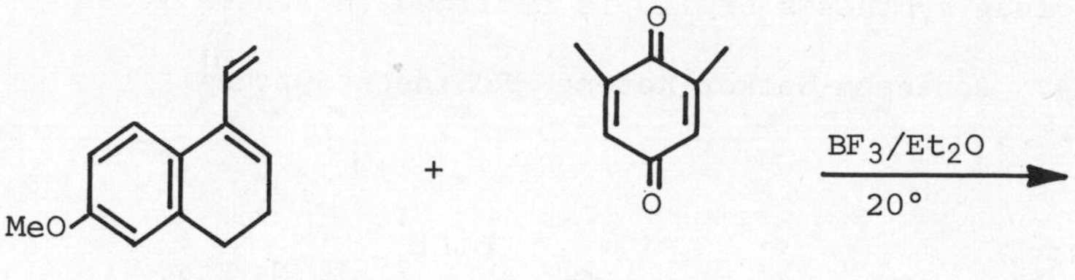

101

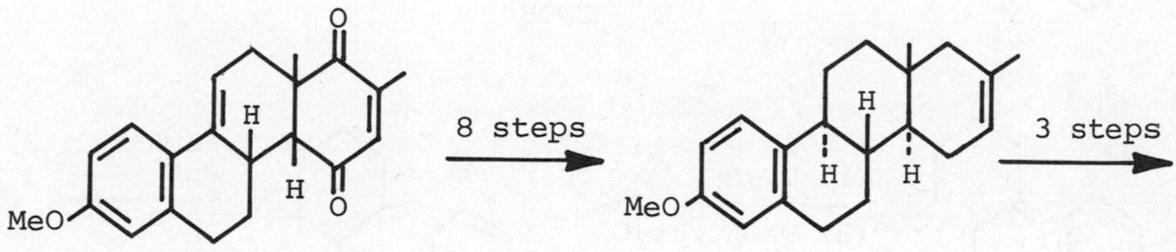

69% (+ 14% 14-methyl adduct)

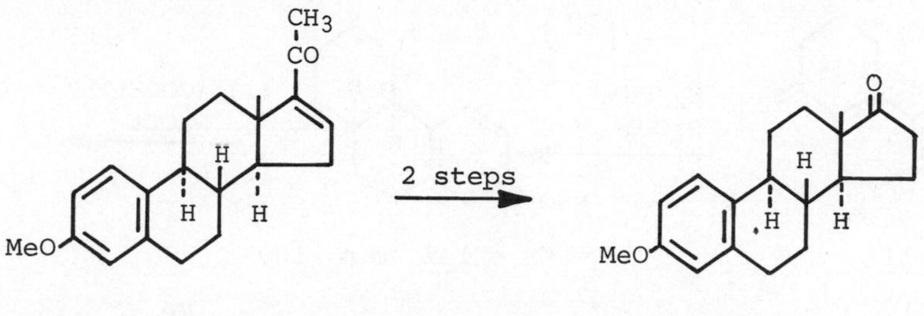

8 steps

3 steps

22% from 101

2 steps

119

A. Anner and Miescher Synthesis

The first synthesis of estrone was accomplished by Anner and Miescher in 1948.[59] Their approach followed that envisioned by Robinson during the nineteen thirties when he and his collaborators synthesized the tricyclic ketone 110 (R=Et) [60] and hoped to add the elements of ring D by reactions analogous to those developed later by Bachmann in the first equilenin synthesis.

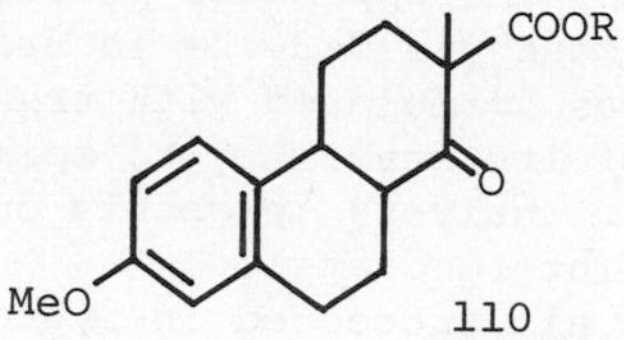

110

The original synthesis of <u>110</u> is outlined in Scheme 14.

Scheme 14.   Robinson-Walker Ketone--Robinson Synthesis

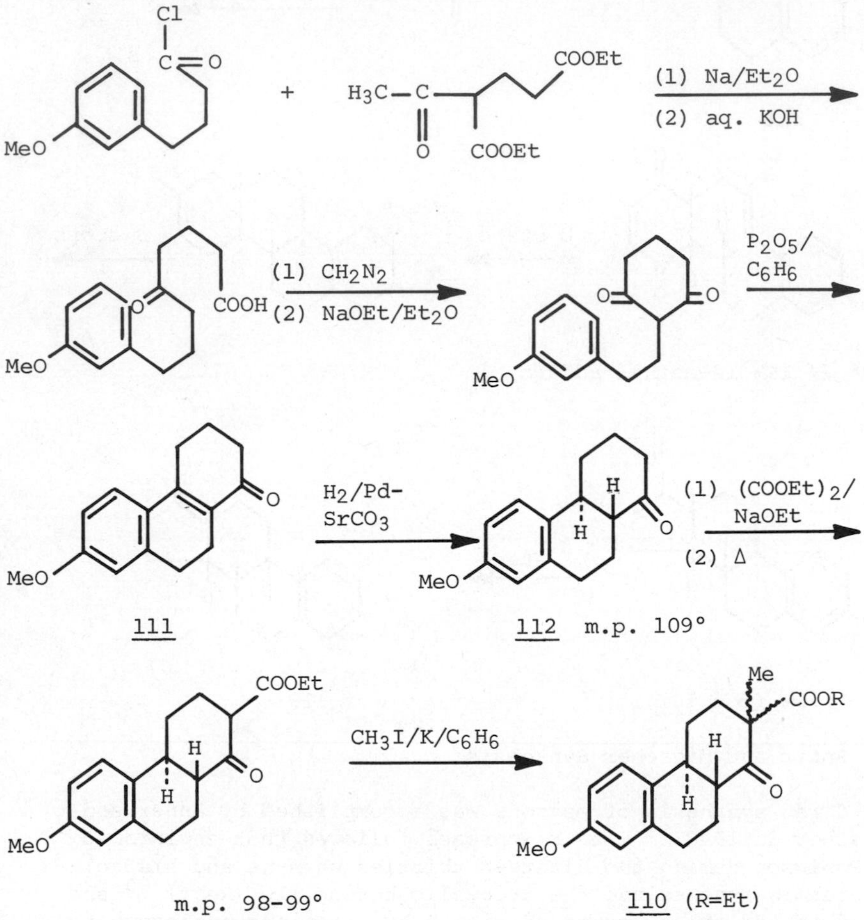

<u>111</u>

<u>112</u>   m.p. 109°

m.p. 98-99°

<u>110</u> (R=Et)

Hydrogenation of the hexahydrophenanthenone <u>111</u> gave a mixture of dihydro (ketone) and tetrahydro (alcohol) products. Back oxidation of the latter by the Oppenauer procedure and purification via the semicarbazone yielded the saturated ketone <u>112</u>, m.p. 109°. The latter was formulated with *trans* B/C stereochemistry on the basis of its stability to epimerization and an x-ray crystallographic analysis (probably one of the first of an organic synthetic intermediate).

In 1942 Bachmann et al. succeeded in synthesizing an

isomer of estrone, 8α,13α-estrone (estrone-A) 113c from the
hexahydrophenanthrene ketoester 113.[61]   Hydrogenation of the
Reformatsky product 113a yielded a mixture of stereoisomers 113b.

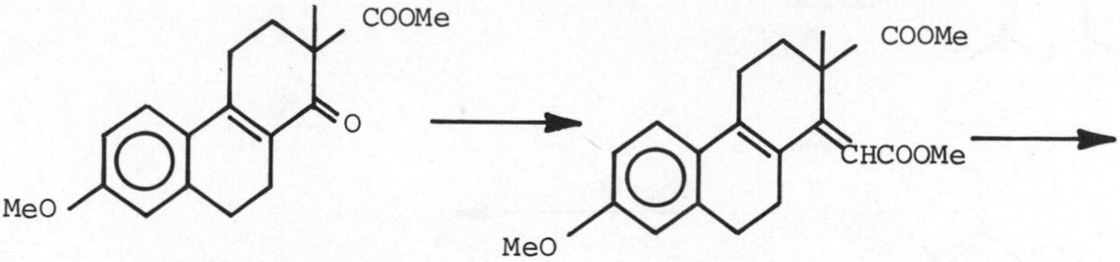

113   113a

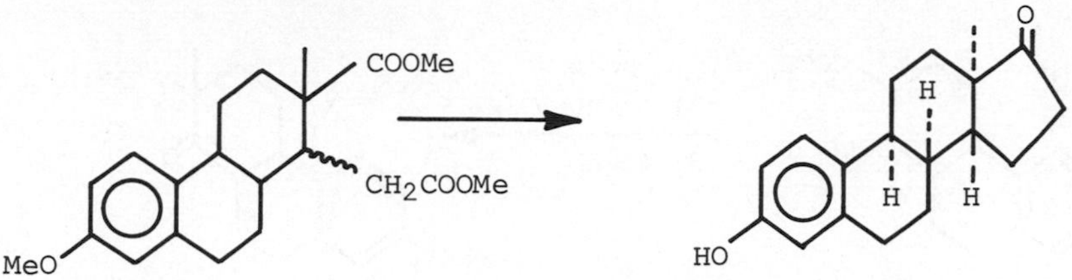

113b   113c

from which at the final stage 113c was obtained by fractional
crystallization.   In connection with this work they developed
an improved route (Scheme 15) to the Robinson-Walker ketone

Scheme 15.   Robinson-Walker Ketone--Bachmann Synthesis

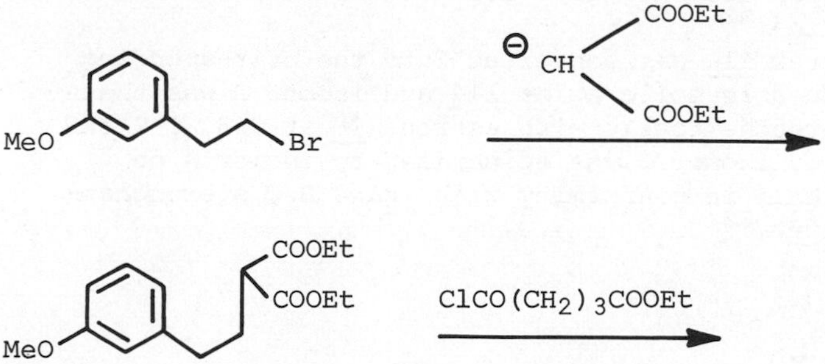

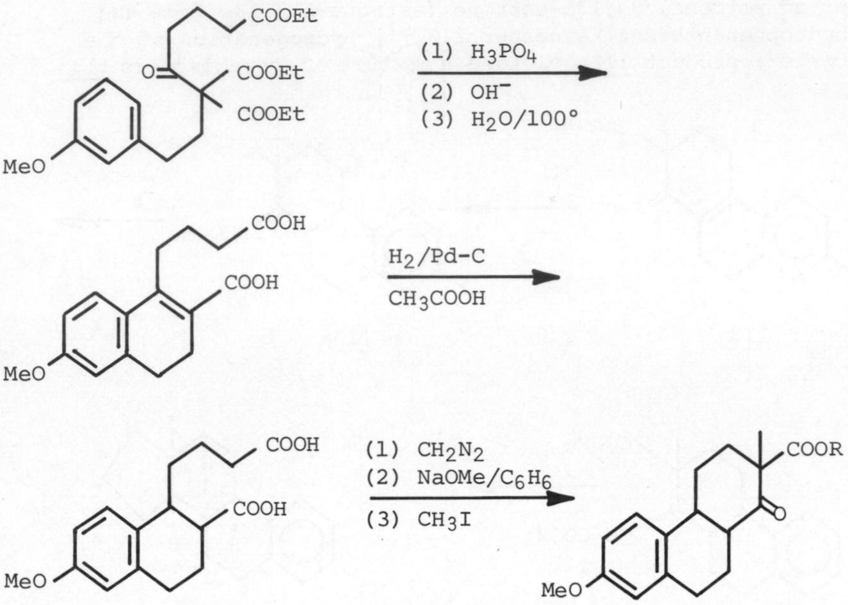
COOEt
O
COOEt
COOEt
MeO
(1) H_3PO_4
(2) OH^-
(3) $H_2O/100°$
COOH
COOH
MeO
$H_2/Pd-C$
CH_3COOH
COOH
COOH
MeO
(1) CH_2N_2
(2) $NaOMe/C_6H_6$
(3) CH_3I
COOR
O
MeO

110 (R=Me)

A - m.p. 113-135°
B - m.p. 127-128°
C - m.p. 87- 89°

110 (as the methyl ester) which was utilized by Anner and Miescher. Since structure 110 possesses three asymmetric centers, four racemates are possible. By simple fractional crystallization from acetone the Swiss workers obtained three of these crystalline.[62]

Each keto-ester 110 was converted into the corresponding pair of estrogenic doisynolic acids 114 and isomer A was thereby correlated stereochemically with estrone 98 at C-8, C-9 and C-13. Furthermore, isomer C was epimerized to isomer A on treatment with alkali in conformity with *trans* B/C stereochemistry for the latter.[62]

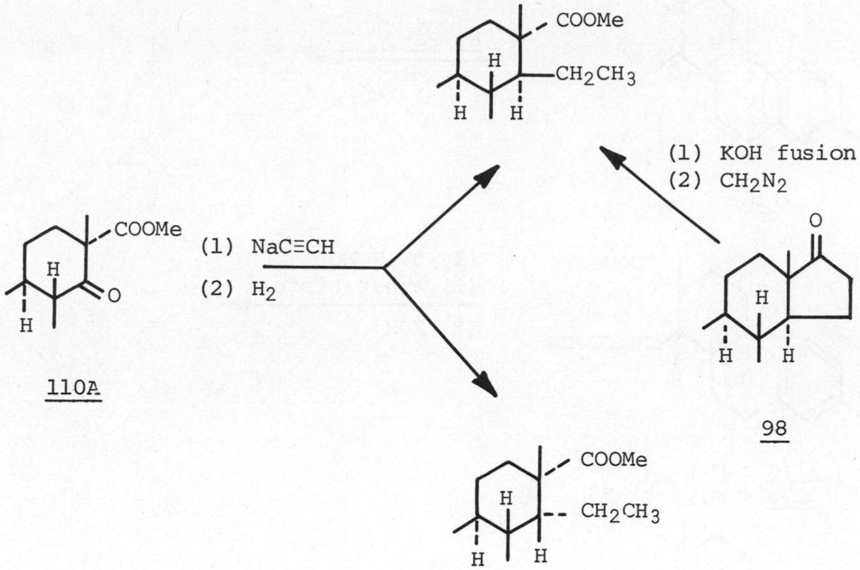
COOMe
H
CH2CH3
H H
(1) KOH fusion
(2) CH2N2
COOMe
(1) NaC≡CH
H
O
(2) H2
H
O
H
H H
COOMe
H
CH2CH3
H H

110A

98

114

Isomer A was therefore chosen as substrate for the Bachmann ring D sequence (Scheme 16), which was carried out with

Scheme 16. Estrone--Anner-Miescher Synthesis

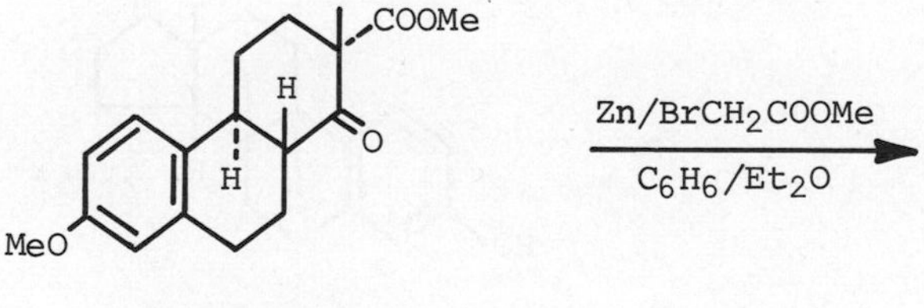
COOMe
H
O
H
MeO
Zn/BrCH2COOMe
C6H6/Et2O

110 (isomer A)

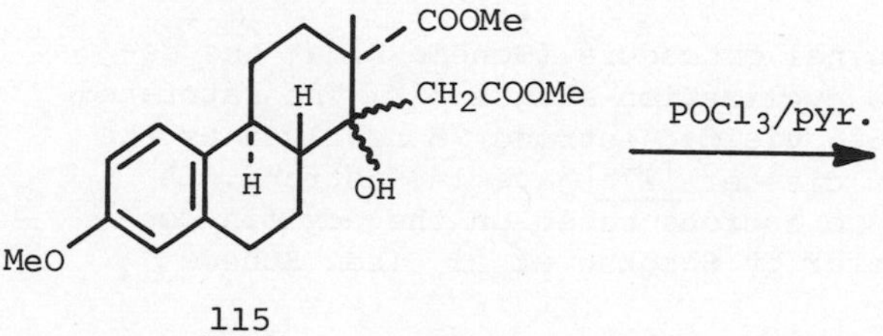
COOMe
H
CH2COOMe
H
OH
MeO
POCl3/pyr.

115

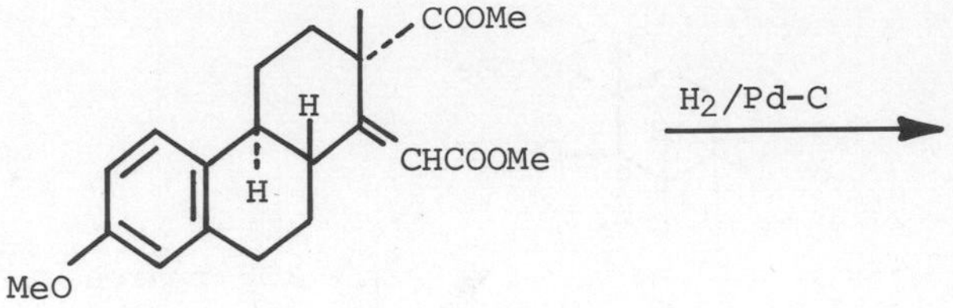
COOMe
H
H
CHCOOMe
MeO
H_2/Pd-C

116

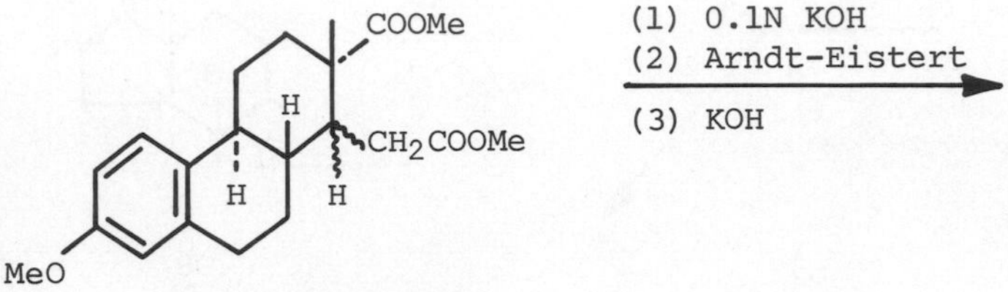
COOMe
H
H
H
CH_2COOMe
MeO
(1) 0.1N KOH
(2) Arndt-Eistert
(3) KOH

117a - m.p. 95-96°
117b - m.p. 91-93°

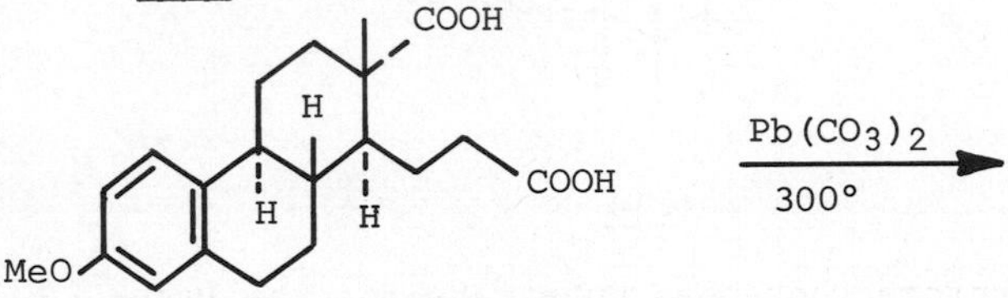
COOH
H
H
H
COOH
MeO
$Pb(CO_3)_2$
300°

118

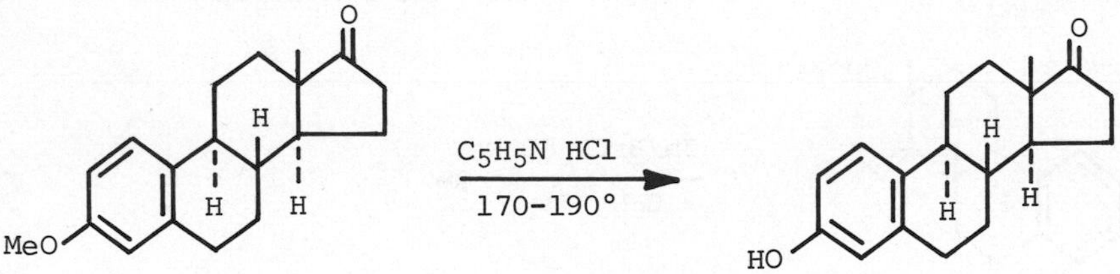
O
H
H
H
MeO
C_5H_5N HCl
170-190°
O
H
H
H
HO

119 98

modification of the original procedure (Scheme 2) at the dehydration, reduction and cyclization stages.[59a] The saturated diester 117a, m.p. 95-96°, yielded estrone 98 resolved as the (-) menthoxyacetate, and diester 117b gave 14β-estrone. A shorter route from 117a to estrone based on the acyloin condensation was devised later by Sheehan et al. (see Scheme 18).[69]

Three additional racemates of the estrone structure were
also synthesized.[59b] An endocyclic isomer m.p. 95-97° of 116
produced at the dehydration stage gave a mixture on hydrogenation which led to 9β-estrone (estrone d) and 8α,9β-estrone
(estrone e). Estrone d and e were also obtained from tricyclic keto-ester C (110C). A fifth racemate, 8α,13α-estrone
(estrone f, Bachmann's estrone a) was obtained from tricyclic
keto-ester B (110B). The configurational assignments were
deduced later by Johnson et al. following their preparation of
seven of the eight possible racemates of the estrone structure.[63] The exception was estrone e, the structure of which
was assigned by difference.

Despite the lack of stereochemical control gram quantities
of (+) estrone were produced by the above route.

B. Johnson Syntheses

Following the initial success of Anner and Miescher, Johnson
and his collaborators devised three different approaches to
estrone.[63-66] The first[63,64] and third[66] applied Johnson's
method of angular methylation, which had been developed earlier
in the preparation of *cis* and *trans* 8-methylhydrindan-1-one
from α-decalone[32] to appropriate tetracyclic intermediates.
The second synthesis[65] utilized a novel A C → B → D approach.

C. Johnson's First Synthesis

In preliminary work it was established that a route analogous
to Johnson's first equilenin synthesis, that is, Stobbe condensation on the cyanoketones 31 or 31a, was inoperative.[64]
The hydrochrysene synthesis shown in Scheme 17 was then developed.

31
31a - Δ-4a(10a)

Scheme 17. Estrone--Johnson's First Synthesis

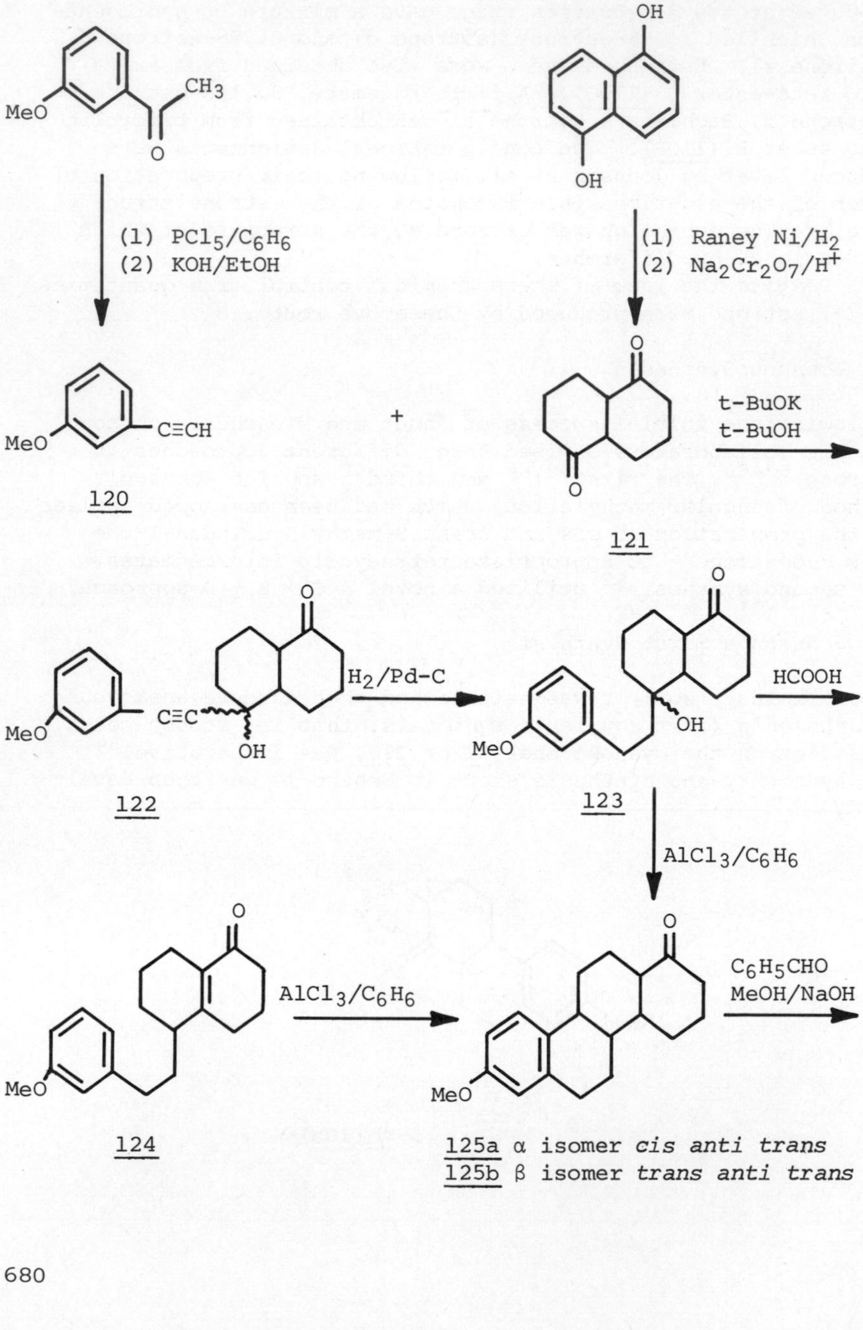

(1) PCl$_5$/C$_6$H$_6$
(2) KOH/EtOH

(1) Raney Ni/H$_2$
(2) Na$_2$Cr$_2$O$_7$/H$^+$

MeO—C≡CH

120

+

121

t-BuOK
t-BuOH

122

H$_2$/Pd-C

123

HCOOH

AlCl$_3$/C$_6$H$_6$

124

AlCl$_3$/C$_6$H$_6$

125a α isomer *cis anti trans*
125b β isomer *trans anti trans*

C$_6$H$_5$CHO
MeOH/NaOH

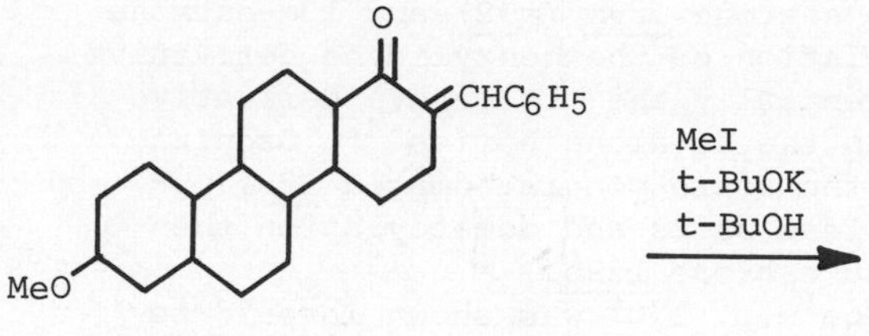
CHC6H5
MeI
t-BuOK
t-BuOH
MeO

126

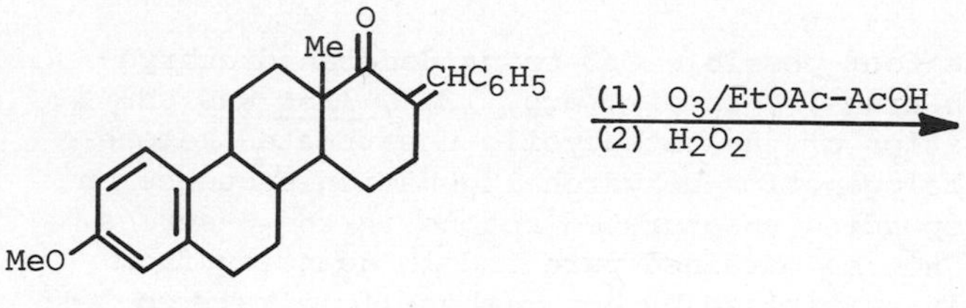
Me
CHC6H5
(1) O3/EtOAc-AcOH
(2) H2O2
MeO

127

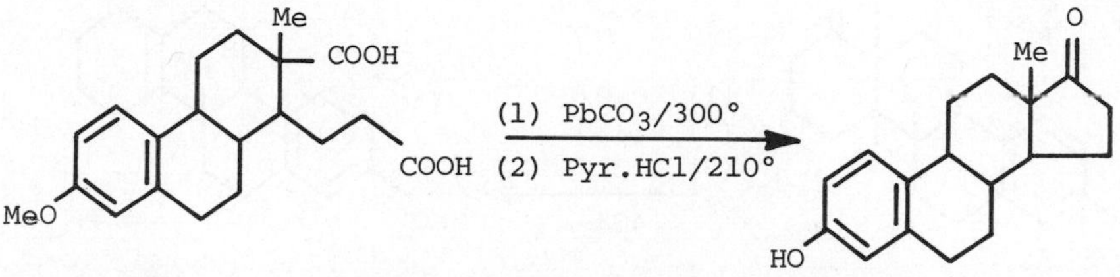
Me
COOH
COOH
(1) PbCO3/300°
(2) Pyr.HCl/210°
MeO
Me
HO

128

129a *trans anti cis*...13α-estrone
129b *trans anti trans*...estrone 98
129c *cis anti cis*...9β,13α-estrone
129d *cis anti trans*...9β-estrone

Condensation of the potassium derivative of m-methoxyphenylacetylene 120 with *cis* and/or *trans* decalin-1,5-dione 121[67] gave two acetylenic carbinols 122, hydrogenated to the corresponding saturated carbinols 123. Cyclodehydration of the mixture or the individual components led to a dodecahydrochrysenone m.p. 170° 125a. However, dehydration of the saturated carbinols 123 to the unsaturated ketone 124 (70% from 121) followed by cyclization gave in ca. 25% yield two dodecahydrochrysenones, largely the isomer 125a and a lesser amount of a β isomer m.p. 155° 125b. On subsequent repetition of this step using very active aluminum chloride only the latter could be isolated (ca. 10%).[68] The β isomer was shown to have the *trans anti trans* configuration at the B/C and C/D ring

junctions by conversion to estrone 129b (=98) and 13α-estrone
(lumiestrone) 129a. Methylation of the benzylidene derivative
126 (β series) gave preferentially the *cis* methyl derivative
127 (*cis:trans* = 3:1) as in the decalone series.[32] Ozonization of each component to the corresponding diacid 128, pyrolysis of the respective lead salts and demethylation then
led to 13α-estrone 129a and estrone 129b.

The dodecahydrochrysene m.p. 170° was shown later[63] to
possess the *cis anti trans* configuration and the corresponding
estrones are therefore the 9β, 13α 125c and the 9β 125d racemates. The methylation *cis:trans* ratio in this series was
4:1.

A third of the four possible C/D *trans* dodecahydrochrysenones, the *cis syn trans*, γ-isomer, m.p. 138°, 125c was obtained by hydrogenation of the tetracyclic unsaturated ketone
130a derived by a halogenation-dehydrohalogenation sequence on
125a.[63] The corresponding unsaturated ketone in the β-series
(B/C *trans*) could not be obtained pure and in quantity from
125b. It was partly epimerized by hot methanolic alkali to

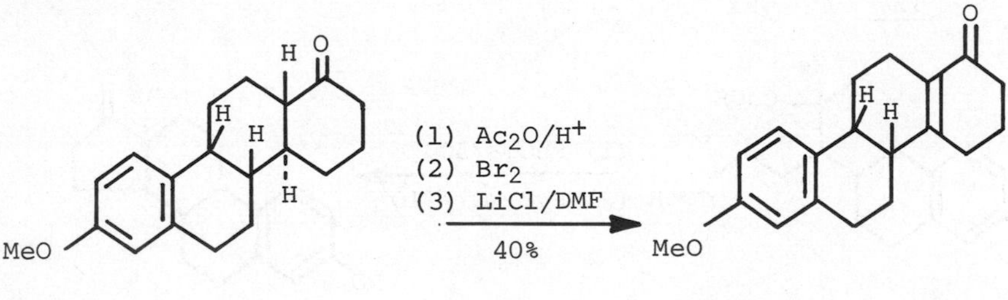

125a

(1) Ac₂O/H⁺
(2) Br₂
(3) LiCl/DMF
40%

130a B/C *cis*

130a

H₂/Pd-C
NaOH/EtOH
60%

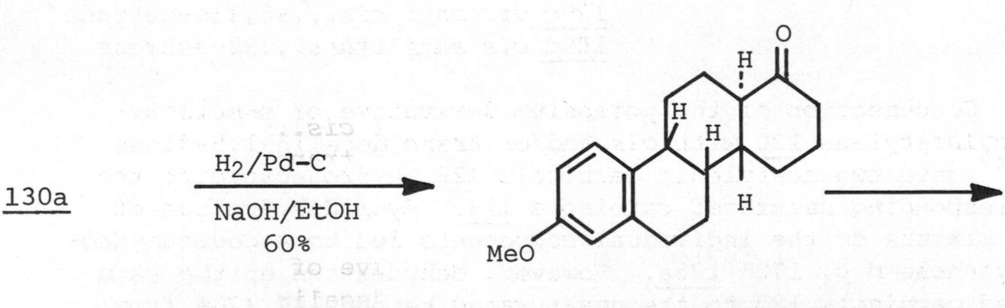

125c -isomer *cis syn trans*

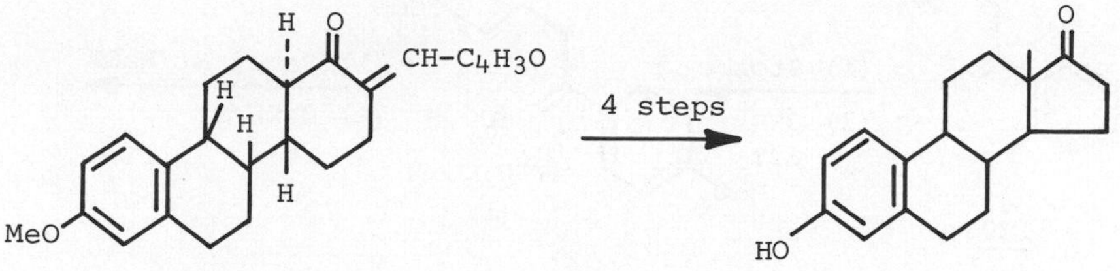

131

129e *cis syn cis* 8α, 13α estrone
129f *cis syn trans* 8α estrone

the B/C *cis* isomer 130a, which in turn was stable to base.
Assuming hydrogenation of 130a from the *exo* face and concomitant isomerization in the basic medium leads to the *cis syn
trans* formulation for 125c. The latter was methylated as the
furfurylidene derivative 131, the mixture of methyl isomers
(*cis:trans* = 11:1) separated and each converted to the corresponding estrone racemate, 8α, 13α-estrone 129e and 8α-
estrone 129f, respectively. The 8α-estrone structural assignment was confirmed by an independent preparation of the methyl
ether via hydrogenation of the B ring of (±) equilenin.[63]

This synthesis therefore led to six racemates of the
estrone structure with the desired isomer a minor product at
the cyclization (124 → 125) and methylation (126 → 127) stages.

The structures of the three dodecahydrochrysenones 125a,
b and c were confirmed later by preparation from intermediates
of the Johnson-Walker estrone synthesis,[66] which led also to
a fourth isomer, the *trans syn cis* from which 14β-estrone was
obtained.

D. Johnson's Second Synthesis

Johnson's second synthesis[65] (Scheme 18) was considerably more

Scheme 18.   Estrone--Johnson's Second Synthesis

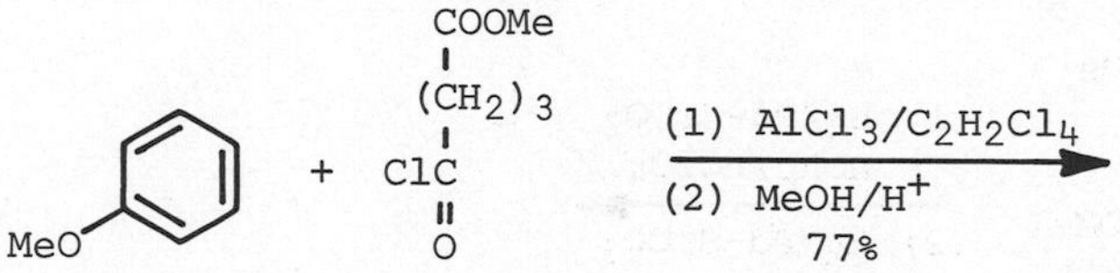

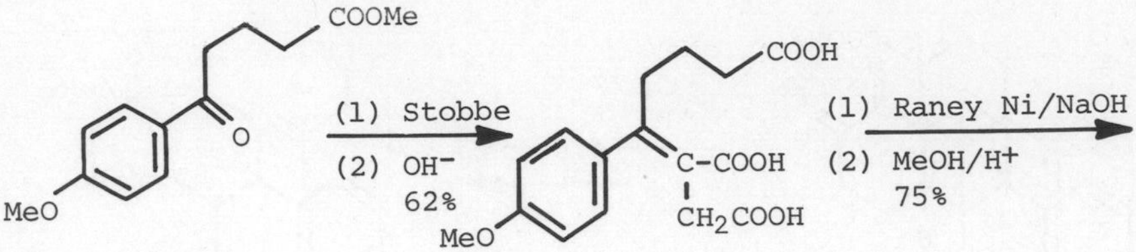
COOMe
O
MeO
(1) Stobbe
(2) OH⁻
62%
COOH
C-COOH
CH2COOH
MeO
(1) Raney Ni/NaOH
(2) MeOH/H+
75%

132

133

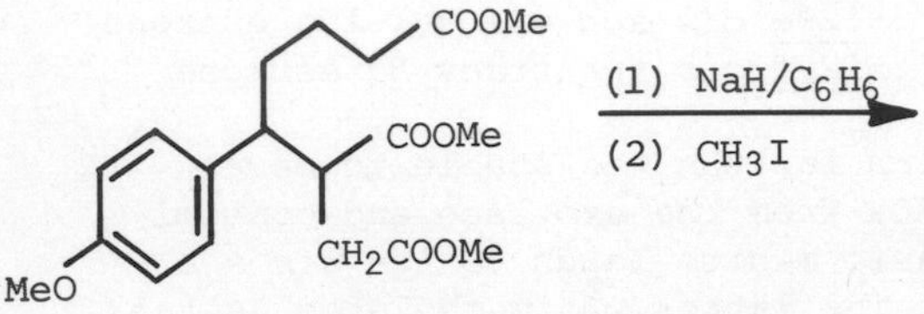
COOMe
COOMe
CH2COOMe
MeO
(1) NaH/C6H6
(2) CH3I

134

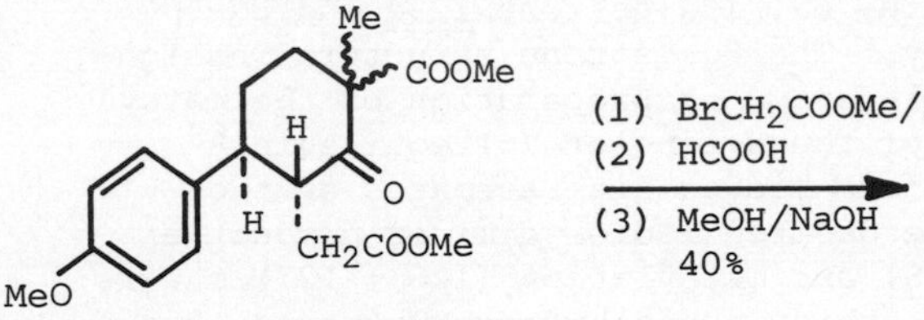
Me
COOMe
H
O
H
CH2COOMe
MeO
(1) BrCH2COOMe/Zn
(2) HCOOH
(3) MeOH/NaOH
40%

135a Ar/Me *cis* 43%
135b Ar/Me *trans* 29%

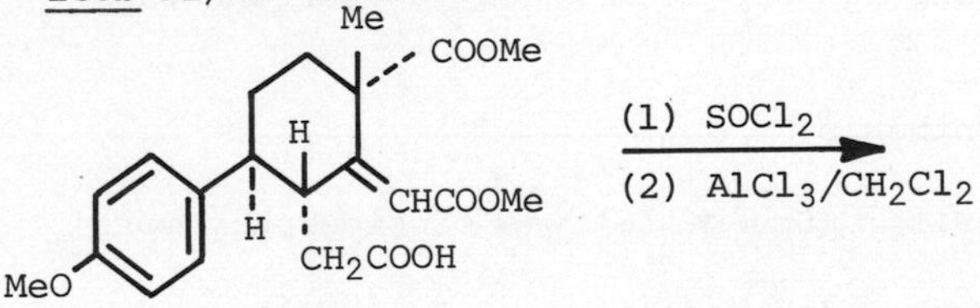
Me
COOMe
H
CHCOOMe
H
CH2COOH
MeO
(1) SOCl2
(2) AlCl3/CH2Cl2

136

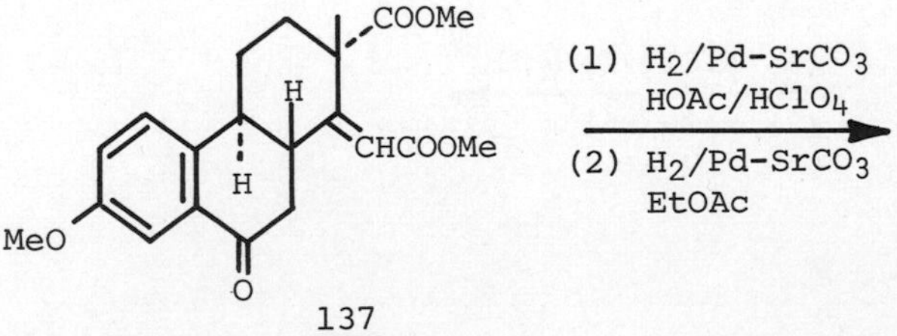
COOMe
H
CHCOOMe
H
MeO
O
(1) H2/Pd-SrCO3
HOAc/HClO4
(2) H2/Pd-SrCO3
EtOAc

137

117a

138

139

98

stereoselective than the first and by a modification led also
to a seventh racemate of the estrone structure--the *trans syn
cis* (14β-estrone). In contrast to the previous estrone syntheses which used *meta* substituted anisole starting materials
the present route utilized a more easily available *para* substituted derivative, the keto-ester 132, readily prepared by
Friedel Crafts condensation of anisole with either glutaric
anhydride or the corresponding half ester acid chloride.
Stobbe condensation with ethyl succinate and potassium t-
butoxide in t-butanol at room temperature followed by saponification led to the unsaturated tricarboxylic acid 133. The
latter was reduced and converted to the trimethyl ester 134
which on Dieckmann cyclization and methylation gave the keto
diesters 135a and 135b. As a partly equilibrated cyclohexanone
the major component 135a would be expected to have its three
bulky groups disposed equatorially, and this was shown to be
so by its conversion to estrone. [The isomer 135b presumably
would have led to 13α-estrone derivatives including 13α, 14β-
estrone (8α, 9β-estrone, Anner and Miescher's estrone e) the
one racemate Johnson did not synthesize.] Reformatsky reaction on 135a followed by lactonization and elimination gave
the diester acid 136 which was cyclized as the acid chloride
to 137. Hydrogenolysis of the keto group in the latter and
hydrogenation of the double bond then led to the methyl ether
of (±) dimethyl marrianolate 117a converted to estrone by

Arndt-Eistert homologation, cyclization, and demethylation.[59a]
A superior route to (+) estrone from naturally derived 117a
based on the acyloin condensation was devised by Sheehan et
al.[69] 16-Keto-estradiol methyl ether 138 produced in high
yield from 117a was reduced to a mixture of triols which on
fusion with pyridine hydrochloride led directly to estrone.

   Hydrogenation of the double bond of 136 *prior* to cyclization occurred from the side opposite the acetic acid side
chain leading to 140 and finally to 14β-estrone.

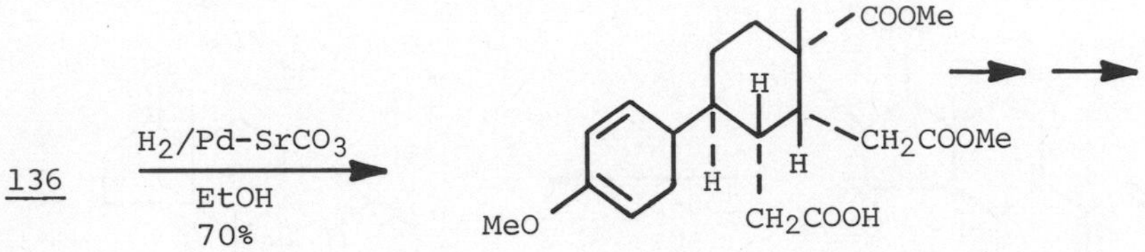

140

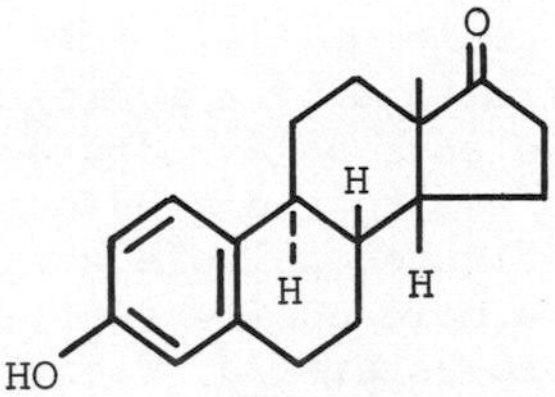

14β-estrone

   Improved routes to the keto diester 135a via the cyano-
keto ester 141 have been developed by Banerjee[70a] and by
Johnson.[70b]

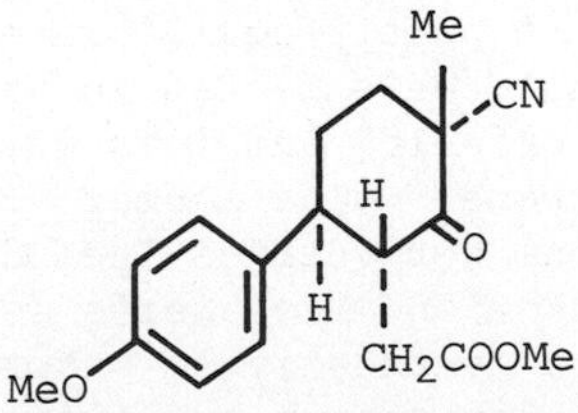

141

E.   Johnson-Walker Synthesis

Johnson's method of introducing the angular methyl group, for example, 126 → 127, although a high yield reaction, unfortunately produced mainly *cis* rather than the desired *trans* methyldecalone systems. However, in 1957 Johnson and Allen[71] succeeded in obtaining predominantly *trans* products on methylating the Δ-6 decalone system 142 or the equivalent tetracyclic system 143. It was reasoned that introduction of the double bond, by eliminating a β-axial hydrogen might facilitate β-face approach of the methylating agent. The estrone synthesis of Johnson, Walker et al. was designed with this point in mind

142   143

and the decahydrochrysenones 144 and 145 were prepared.[66] In the latter the effect of eliminating the axial 8-β hydrogen could be tested.

144   145

Compounds 144 and 145 were obtained via the Diels-Alder adduct 147 (Scheme 19) of the diene 101 and benzoquinone. This adduct had been prepared originally by Dane et al. in 1937,[72] but its chemistry was first studied in detail by Robins and Walker considerably later.[73] An improved route to diene 101 involving direct preparation of the vinyl carbinol 146 by reaction of 6-methoxytetralone 9 with vinyl magnesium bromide in tetrahydrofuran developed by Nazarov et al.[74] as well as other process improvements, led to adduct 147 in 75%

Scheme 19. Estrone—Johnson-Walker Synthesis

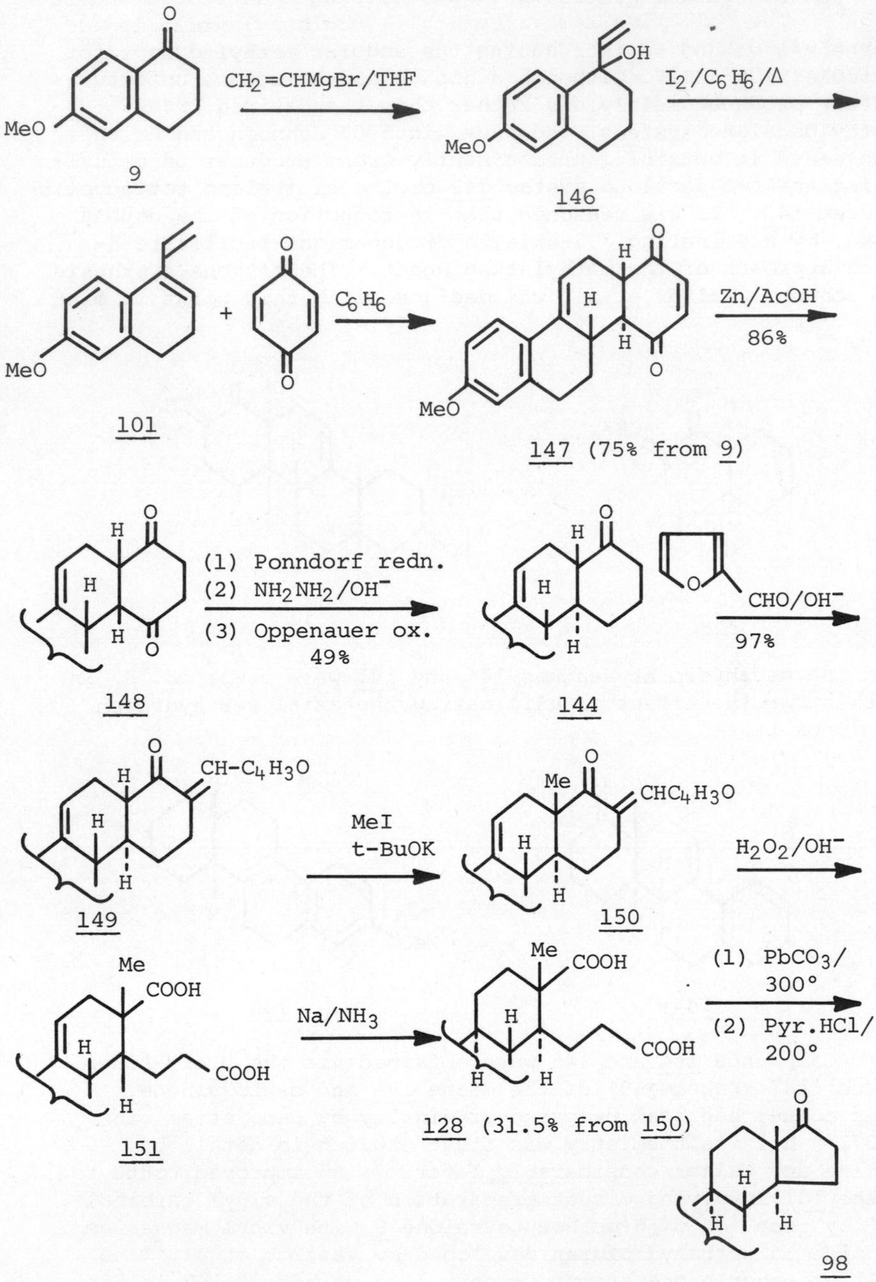

688

overall yield from 9 and thence to the ring D saturated adduct
148.[66] The selective removal of the 15-carbonyl group in 148
was studied in detail. The best procedure developed involved
selective Meerwein-Ponndorf reduction of the 17α-carbonyl
group, Huang Minlon-Wolff-Kishner reduction at C-15 with concomitant isomerization at C-14 and Oppenauer oxidation to give
the C/D trans ketone 144 in 49% yield from 148. Protection of
the 17α-carbonyl group as the dimethyl ketal led to 144 in only
22% yield.

Methylation of the furfurylidene derivative 149 produced,
as anticipated, mainly (56%) the trans isomer 150 and 33% of
its cis counterpart, which were separated by fractional crystallization (the trans:cis ratio as determined by nmr was
60:40). Conversion of 150 to the Δ-9(11) homomarrianolic acid
151 by hot alkaline hydrogen peroxide [ozonization as in 127 →
128 could not be used because of the 9(11) double bond] followed by chemical reduction led to the methyl ether of homomarrianolic acid 128, previously converted to estrone. Despite the stereoselective introduction of all asymmetric centers the low yield in the ring D cleavage sequence (150 → 128)
limited the overall yield of estrone from 9 to 3.7%.

The preparation of 145 and its C/D cis isomer 145a also
proceeded from 148. Conversion to the monoethylene ketal 152
with double bond shift followed by Huang Minlon reduction and
acid hydrolysis gave a mixture of cis and trans ketones 145a
and 145. Alkali treatment of 152 prior to hydrazone formation
led exclusively to 145 whereas hydrazone formation before
addition of alkali led to reduction without epimerization.
Either 145 or 145a produced the same 75-80% cis:5-7% trans mixture of furfurylidene derivatives 154, pointing up the sensitivity of the C/D cis:trans ratio in these systems to structural detail.

Methylation of the C/D *cis* furfurylidene ketone 154 led exclusively to the *cis* methyl compound 155, in contrast to the predominantly *trans* methylation of 149. As discussed earlier (p. 659) these results have been rationalized in terms of product stability--angular methylated *trans* and *cis* octalin systems preferring the double bond in the Δ-2 and Δ-1 positions, respectively, in order to minimize axial steric interactions and torsional strain.

F. 2-Methylcyclopentane-1,3-dione

In the syntheses of equilenin and estrone considered so far the C-13 methyl group was introduced either prior to construction of ring D or following assemblage of the tetracyclic system. Utilization of a preformed methylated ring D component in the form of 2-methylcyclohexane-1,3-dione 156, or better 2-methylcyclopentane-1,3-dione 157, has been the key factor in the development of a number of relatively short industrially feasible syntheses of estrone exemplified by the approaches of Smith, Torgov, and Velluz.

156   157

2-methylcyclohexane-1,3-dione (available in 60% yield by reduction and methylation of resorcinol[75]) was first utilized in steroid synthesis in 1953 by Miescher et al. as a ring D component in their preparation of D-homoandrostane-3,17α-dione.[76] Since use of 156 in estrone synthesis requires subsequent D-ring contraction, 2-methylcyclopentane-1,3-dione 157 a priori would be the reagent of choice. However, the latter was a relatively rare substance until 1955 when it became available in 20% yield from butanone-2 by a four-stage synthesis, the key step of which was the selective removal of the C-4 carbonyl group of the intermediate trione 159 by Wolff-Kishner reduction of the corresponding monosemicarbazone[77] (Scheme 20). Since then several superior methods have been developed and 157 is now a readily accessible intermediate.

Scheme 20. 2-Methylcyclopentane-1,3-dione 157

Panouse and Sannié[77]

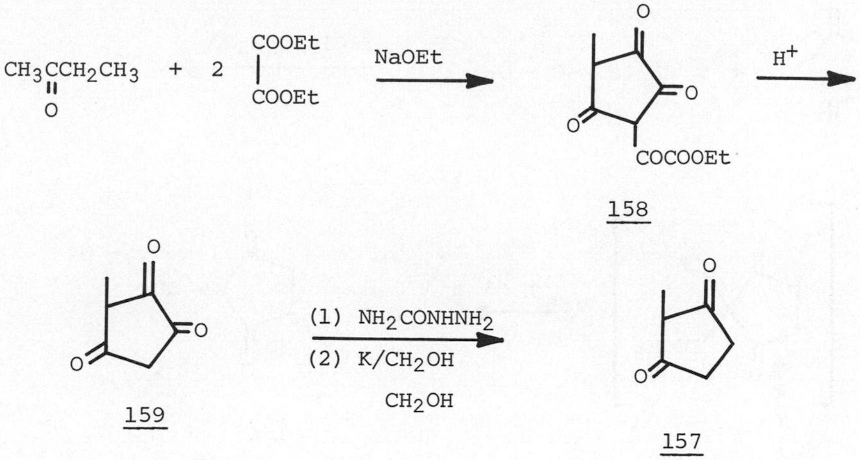

$CH_3CCH_2CH_3$ + 2 $\overset{COOEt}{\underset{COOEt}{|}}$ $\xrightarrow{\text{NaOEt}}$ $\xrightarrow{\text{H}^+}$
$\overset{||}{O}$

158

159

$\xrightarrow[\substack{(2)\ K/CH_2OH \\ CH_2OH}]{(1)\ NH_2CONHNH_2}$

157

Bucourt et al.[78]

160

$\xrightarrow[\text{xylene } \Delta]{\text{Excess KOBu}^t}$

161

$\xrightarrow[(2)\ \Delta]{(1)\ \text{Aq. KOH}}$

157

691

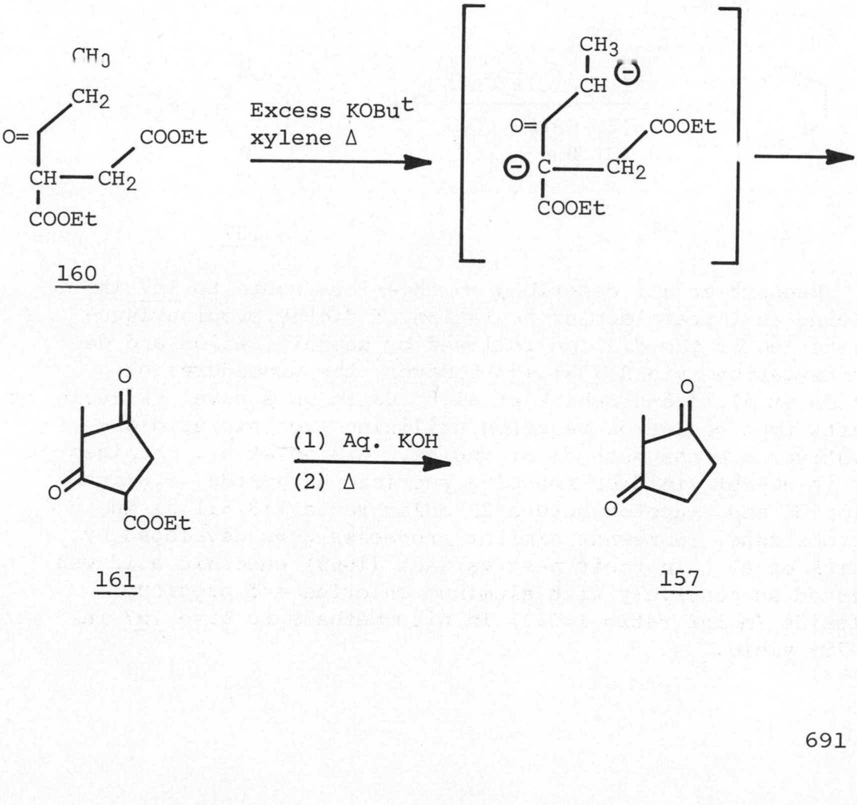

Grenda et al.[79]

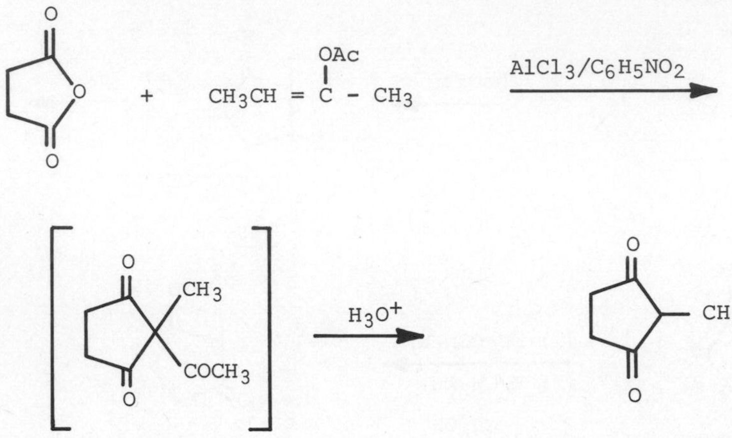

$$+ \quad CH_3CH = \underset{\underset{OAc}{|}}{C} - CH_3 \quad \xrightarrow{AlCl_3/C_6H_5NO_2}$$

$$\xrightarrow{H_3O^+}$$

157

Schick et al.[80]

(1) $AlCl_3/CH_3NO_2$

(2) CH_3CH_2COCl

(3) H_3O^+

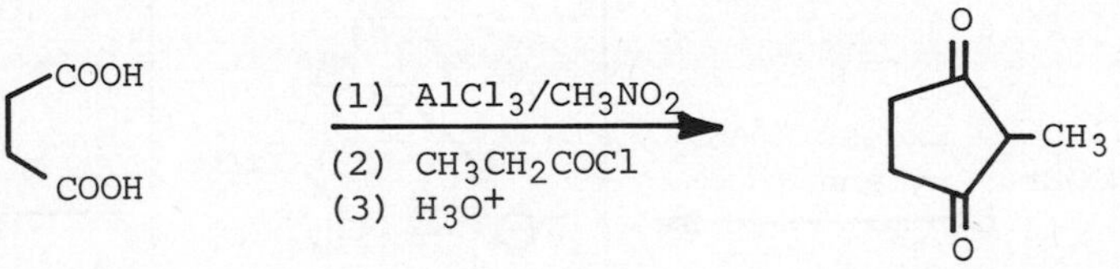

157

Bucourt et al. described an ingenious route to 157 involving an intramolecular acylation of diethylpropionylsuccinate 160 as the *dianion* followed by saponification and decarboxylation (yield 73%).[78] However, the procedures of Grenda et al.[79] and Schick et al.[80] based on a novel Friedel-Crafts type acylation reaction utilizing succinic acid derivatives are the methods of choice. Grenda et al. obtained 157 in 80-85% yield by reacting succinic anhydride, aluminum chloride and 2-acetoxybutene-2 (molar ratio 1:3.5:1.5) in nitrobenzene. Somewhat similar processes were developed by Schick et al. In their best variant (1969) succinic acid was treated successively with aluminum chloride and propionyl chloride (molar ratio 1:3:3) in nitromethane to give 157 in 72-78% yield.

G. Syntheses of Smith et al.

Preliminary accounts of two estrone syntheses based on an A → D → ABCD approach and utilizing 156[81] and 157[37a], respectively, as D components were published in 1960 by Hughes and Smith (Schemes 21 and 22). This work was described in detail in 1963[37b] along with a number of related routes.

Scheme 21

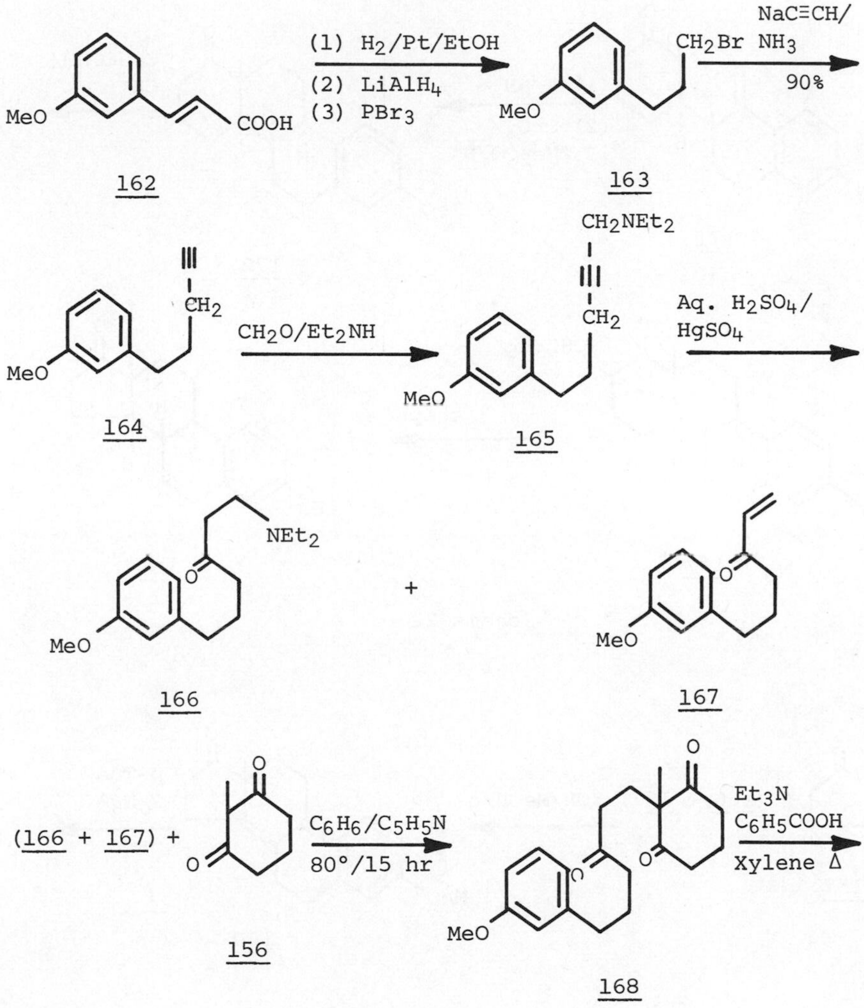
MeO
COOH
162
(1) H_2/Pt/EtOH
(2) $LiAlH_4$
(3) PBr_3
MeO
CH_2Br
163
NaC≡CH/
NH_3
90%
MeO
CH_2
164
CH_2O/Et_2NH
CH_2NEt_2
CH_2
MeO
165
Aq. H_2SO_4/
$HgSO_4$
NEt_2
O
MeO
166
+
O
MeO
167
(166 + 167) +
O
O
156
C_6H_6/C_5H_5N
80°/15 hr
O
O
O
MeO
168
Et_3N
C_6H_5COOH
Xylene Δ

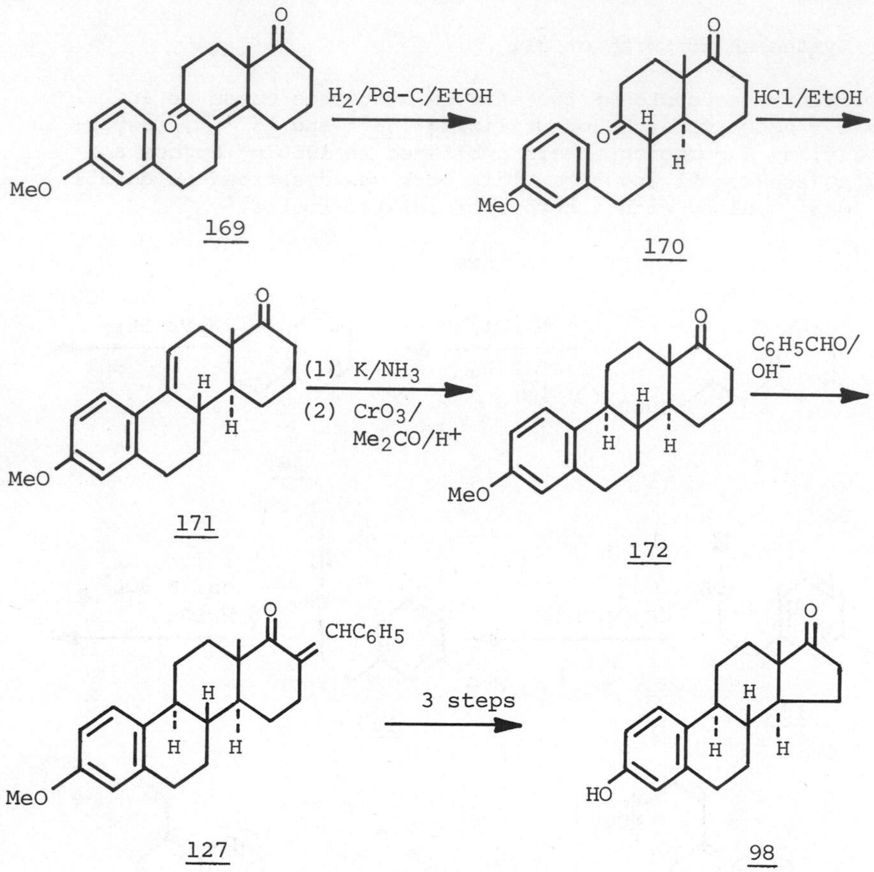

MeO
169
H2/Pd-C/EtOH
MeO
170
HCl/EtOH
MeO
171
(1) K/NH3
(2) CrO3/
Me2CO/H+
MeO
172
C6H5CHO/
OH−
CHC6H5
MeO
127
3 steps
HO
98

Scheme 22

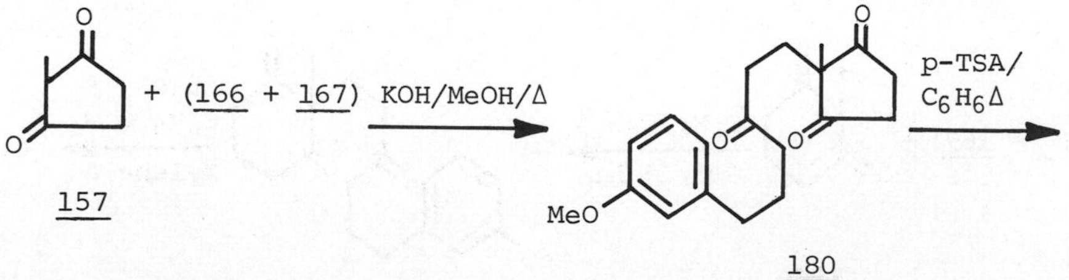

157
+ (166 + 167) KOH/MeOH/Δ
MeO
180
p-TSA/
C6H6Δ

181

H$_2$/Ni/
dioxane

or

H$_2$/Pd-CaCO$_3$/
C$_6$H$_6$

182

NaBH$_4$/
EtOH

(1) K/NH$_3$
(2) CrO$_3$/
Me$_2$CO/H$^+$

183

(1) Li/NH$_3$/C$_6$H$_5$NH$_2$
(2) CrO$_3$/Me$_2$CO/H$^+$

119

3-m-methoxyphenylpropyl bromide 163, prepared efficiently from m-methoxycinnamic acid 162, was condensed with sodium acetylide to give 5-m-methoxyphenylpentyne 164. The latter, on Mannich condensation with formaldehyde and diethylamine followed by hydration of the triple bond and distillation, gave a mixture of Mannich base 166 and vinyl ketone 167 which, in the first step of a Robinson annelation reaction,[82] was condensed with 2-methylcyclohexane-1,3-dione 156 to give the trione 168. Aldolization to close the C ring was accomplished with triethylammonium benzoate in refluxing xylene. Hydrogenation of the double bond in the unsaturated dione 169 occurred stereoselectively from the side opposite the angular methyl group, and subsequent epimerization of the alkyl substituent α to the carbonyl group led to 170 with the thermodynamically preferred anti-trans stereochemistry. Alternatively, 169 was converted to 170 by lithium-ammonia reduction followed by back oxidation at C-17a. Cyclodehydration in ethanolic hydrochloric acid then gave Δ9(11)-dehydro-D-homoestrone methyl ether 171 reduced to D-homoestrone methyl ether 172 by potassium-liquid ammonia reduction and back oxidation

at C-17a.   The overall yield of 172 from 3-m-methoxyphenyl-
propyl bromide 163 was 7.5%. The synthesis was completed by
preparation of the benzylidene derivative 127, an intermediate
in Johnson's D ring contraction procedure[64] (Scheme 17).

    2-Methylcyclopentane-1,3-dione 157 was utilized success-
fully in the above sequence with some modifications to yield
estrone methyl ether 119 in 14% yield from 163.[37b] The principal modification involved hydrogenation of the ketol 174
rather than the dione 173 which increased the yield of 175
from 173 from 40% to 60%. (See Ref. 105b for formation of
(+)-173 by closure of the C ring under asymmetric conditions.)
It is striking that the C-4 substituted tetrahydroindanes 173

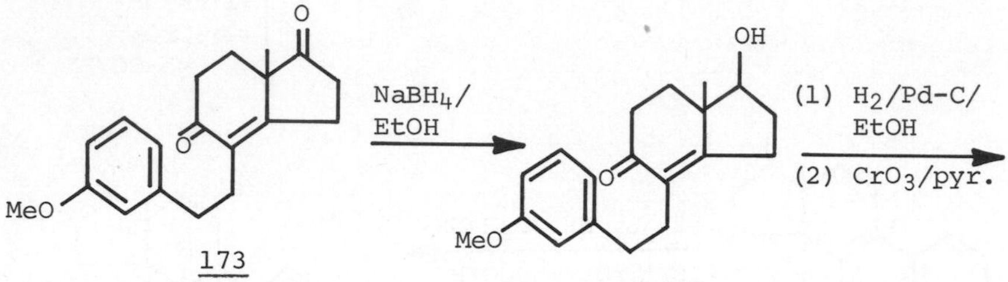

$$NaBH_4/EtOH$$

173

174

(1) H_2/Pd-C/EtOH

(2) CrO_3/pyr.

(1) HCl/EtOH

(2) H_2/Pd-C/EtOH

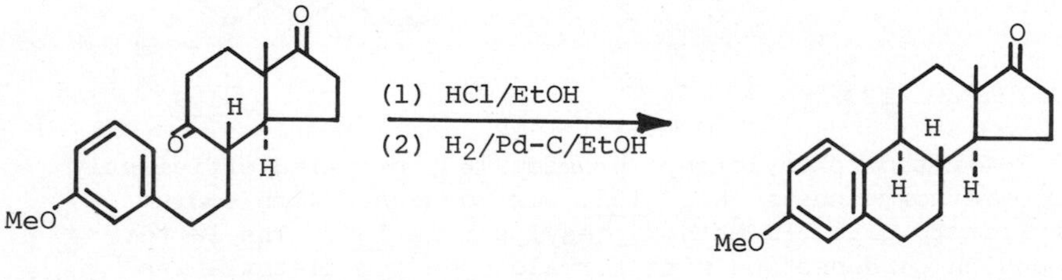

175

119

and 174 are hydrogenated stereoselectively *trans*, whereas the
unsubstituted tetrahydroindanes 176 and 177 are hydrogenated
stereoselectively *cis*[82a,82b], in accord with general experience
in this area[83] [see also Velluz second estrone synthesis
(Scheme 26)].

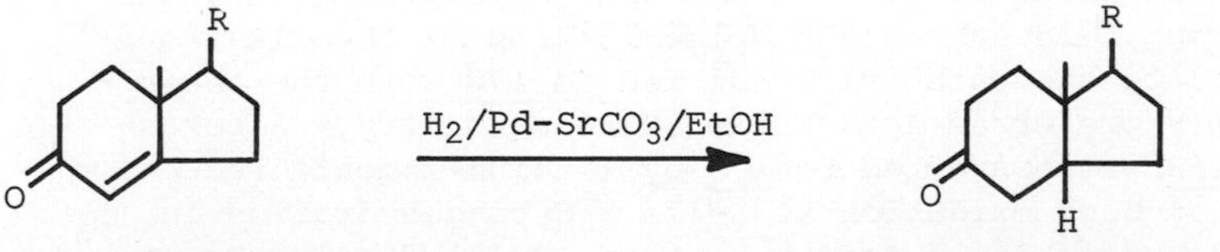

$$H_2/Pd-SrCO_3/EtOH$$

176, R=β-OH
177, R=O

178, R=β-OH
179, R=O

Further yield improvement to 18% of 119 from 163 was achieved as shown in Scheme 22.[37a,37b] Cyclodehydration of 180 under more strongly acidic conditions than had been used earlier led to the estrapentaene 181, a key intermediate also of the Torgov route to estrone (see below). The requisite *trans anti trans* stereochemistry was introduced into 181, which has one chiral center, by an essentially two step reductive sequence. Hydrogenation of the 14,15 double bond occurred mainly from the α side (cf. 14-dehydro-equilenin, Scheme 3) to give 8-dehydroestrone methyl ether 182, converted by alkali metal-ammonia reduction and back oxidation at C-17 to estrone methyl ether 119. Yields were improved by carrying out the chemical reduction of the Δ-8 double bond on the 17β-ol 183 and later by carrying out both steps on the 17β-ol derived from 181 (see Torgov synthesis).

A novel route to 167 utilizing the inexpensive essential oil, eugenol 184, as starting material was described by Horeau et al. in 1969.[84] The key intermediate, m-methoxyallylbenzene 185, was obtained from eugenol phosphate. Conversion of 185 to the Grignard reagent 186 was accomplished directly by titanium tetrachloride catalyzed exchange with n-propyl magnesium bromide in refluxing ether, a general reaction of terminal olefins.[84a] Addition of acrolein and oxidation of the intermediate vinyl carbinol completed the synthesis.

The important intermediate, 6-methoxytetralone 9 also was obtained from 186 by reaction with carbon dioxide followed by Friedel-Craft ring closure.

H.   Torgov Synthesis

In the Smith syntheses just discussed the carbon system was
assembled by condensation of the anion of the β-dicarbonyl
compound 156 or 157 with the vinyl ketone 167.   The Torgov
approach had its genesis in studies on analogous base-catalyzed
alkylations of 156 or 157 with allylic bromides 188.   The latter, prepared from vinyl carbinols 187 by anionotropic rearrangement with hydrogen bromide, were condensed with the anion
of the β-diketone and the adducts 189 cyclized to heteroannular
dienes by heating with phosphorus pentoxide.[85]

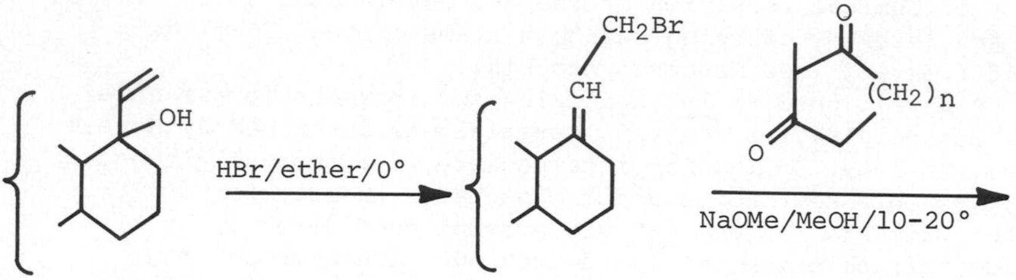

    187   188

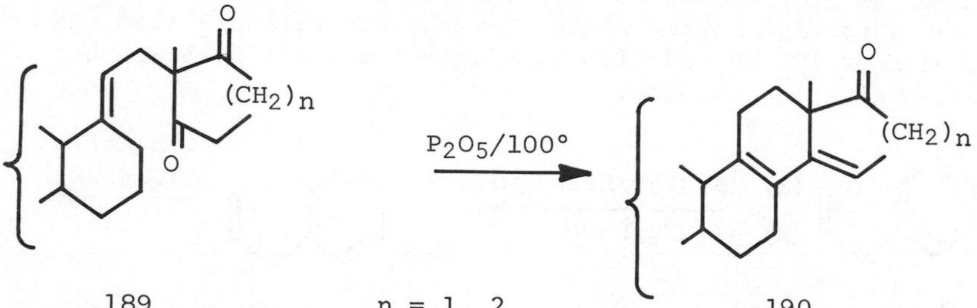

    189   n = 1, 2   190

For estrone synthesis by this route the required vinyl
carbinol 146 was readily available from 6-methoxytetralone[74]
as described earlier, but the corresponding primary bromide
191 was too unstable to be obtained by the hydrogen bromide
procedure.[86]   It was, however, prepared later by Whitehurst
and co-workers using milder conditions.[87]   Reaction of 191
with the anion of 175 [methanol, one equivalent of Triton B
(benzyltrimethylammonium hydroxide) 70°/24 hr] gave the tricyclic dione 192 in poor yield and cyclization of the latter
gave the estrapentaene 181 in only 6% yield over the condensation and cyclization steps.[87]

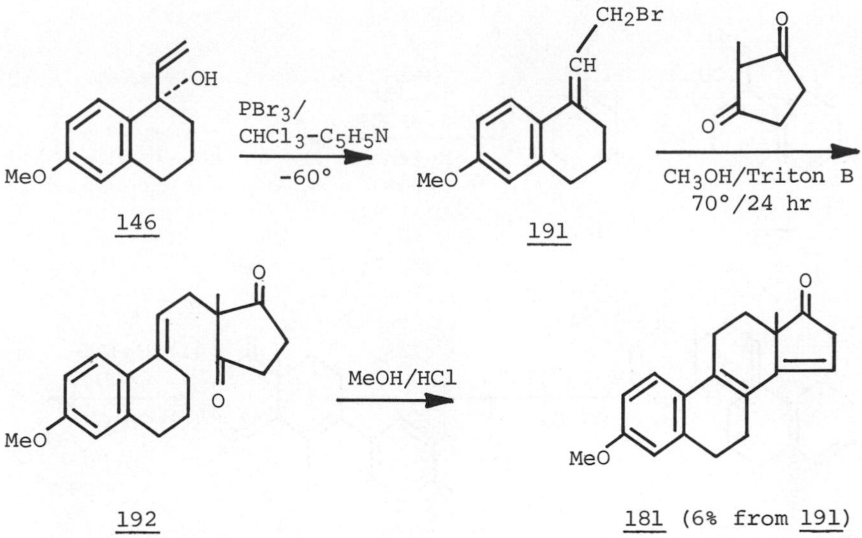

The chemical schemes depicting compounds 146, 191, 192, 181, 193, 156, and 194 are shown with reaction conditions:

146 → (PBr₃/CHCl₃–C₅H₅N, –60°) → 191 → (CH₃OH/Triton B, 70°/24 hr)

192 → (MeOH/HCl) → 181 (6% from 191)

193 (R=H, Me) + 156 → (C₆H₅CH₂NMe₃]OH⁻/140°, ~50%) → 194

Following their unsuccessful attempts to prepare 191 and other analogous multiunsaturated primary bromides the Russian workers in 1959 condensed the precurser bis-vinyl carbinols 193 with 156 in the presence of ~10% Triton B and obtained the adducts 194 in 50% yield.[86] This novel reaction which later became known as the Torgov reaction[88] failed when applied to

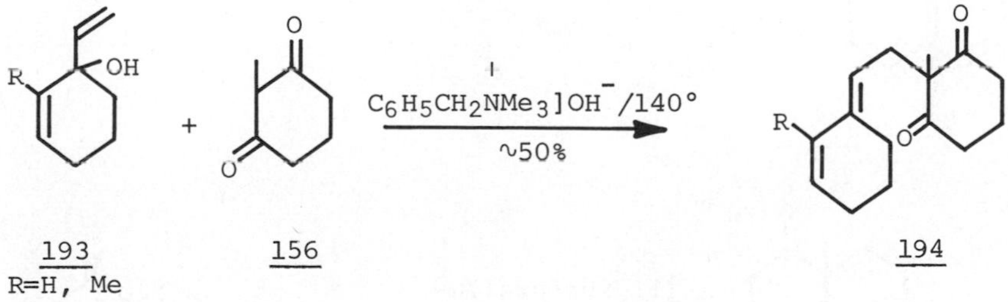

singly vinylic carbinols. Application to the bis-vinyl carbinol 146 then led to the tricyclic adduct 195 (Scheme 23) in 60% yield (based on 156 consumed) which was cyclized to the D-homoestropentaene 196 in high yield.[89] Catalytic hydrogenation of the Δ-14 double bond followed by potassium-liquid ammonia reduction of the Δ-8 17α-ethylene ketal 197 produced, after hydrolysis, D-homoestrone methyl ether 172 in 16% overall yield from 6-methoxytetralone. Estrone methyl ether was obtained from 172 in ~50% yield by oxidation of the furfurylidene derivative and pyrolysis of the lead salt of the

Scheme 23

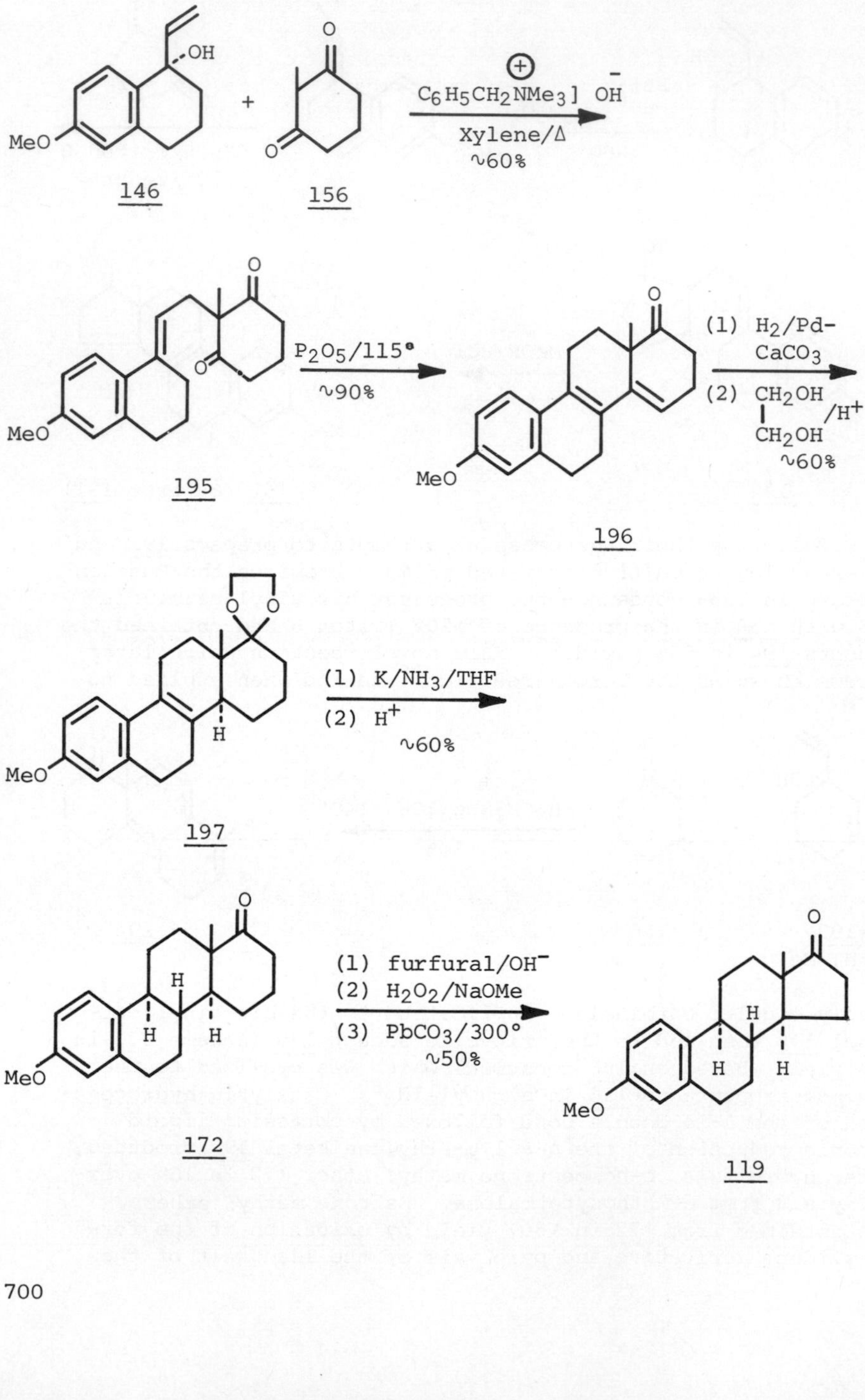

146

156

$C_6H_5CH_2NMe_3]^{\oplus}$ OH^-

Xylene/Δ

~60%

195

$P_2O_5/115°$

~90%

196

(1) H_2/Pd–CaCO$_3$

(2) $\begin{matrix}CH_2OH \\ | \\ CH_2OH\end{matrix}$ /H$^+$

~60%

197

(1) K/NH$_3$/THF

(2) H$^+$

~60%

172

(1) furfural/OH$^-$

(2) H$_2$O$_2$/NaOMe

(3) PbCO$_3$/300°

~50%

119

derived diacid [procedure of Johnson et al. (Scheme 19)].[90]

Utilization of 2-methylcyclopentane-1,3-dione 157 in the above sequence would obviate the necessity for ring D contraction, and five independent syntheses based on this approach were published in 1963.[37b,37c,87,91] Yields were similar to those in the D-homo series and estrone methyl ether 119 was produced from 6-methoxytetralone in essentially five steps via the key intermediates 192 and 181 in about 20% yield (Scheme 24). Estrapentaene 181 had already been converted to

Scheme 24

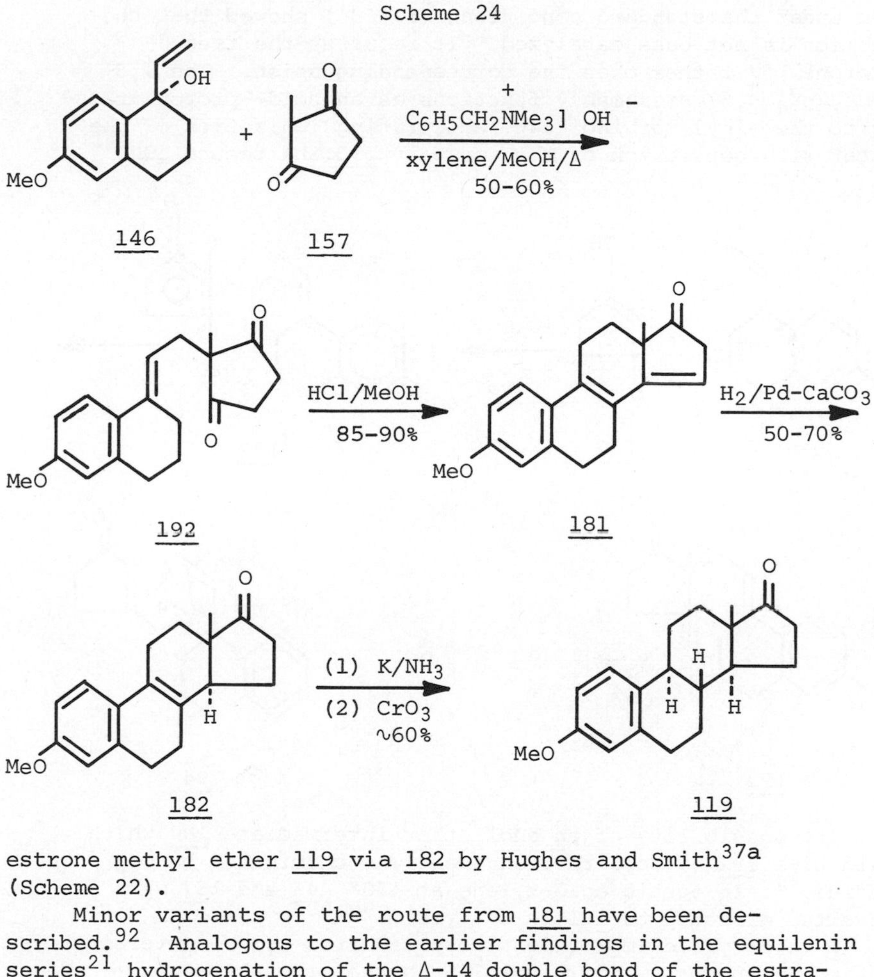

$$C_6H_5CH_2NMe_3]\ OH^-$$

xylene/MeOH/Δ
50-60%

146   157

HCl/MeOH
85-90%

H_2/Pd-CaCO$_3$
50-70%

192   181

(1) K/NH$_3$
(2) CrO$_3$
∿60%

182   119

estrone methyl ether 119 via 182 by Hughes and Smith[37a] (Scheme 22).

Minor variants of the route from 181 have been described.[92] Analogous to the earlier findings in the equilenin series[21] hydrogenation of the Δ-14 double bond of the estrapentaene-17β-ol[37c] or 17β-acetate[93] derived from 181 gave a greater proportion (80-90%) of the desired 14α product than

from the 17-ketone (∿50-70%). However, contrary to expecta-
tions based on simple steric considerations, hydrogenation of
the corresponding 17α-ol (but not the 17α-acetate) over Raney
nickel in dioxane also gave ∿90% of 14α-product.[94]

The Torgov reaction was considered originally to be base
catalyzed and was generally carried out in the presence of
about 10% of the quaternary ammonium hydroxide, Triton B.[37b,]
[87,91] However, the observations that condensation did not
occur when a full molar equivalent of base was used and actually occurred in higher (70%) yield in the absence of base
then under the standard conditions (50-60%) showed that the
reaction is not base catalyzed. It requires the free β-
diketone 157 rather than the corresponding anion. The 1,3-
dione (pK$_a$ 4.5) presumably functions as an acid--proton transfer to the vinyl carbinol 146 facilitating ionization of the
latter with generation of the ion pair 193 and thence 192.[94,95]

146   157   193

192   194

The possibility of an enol ether intermediate 194 which
could give 192 by Cope rearrangement was eliminated by O$_{18}$
studies.[94] In acetic acid-xylene at 120° 146 and 157 were
converted directly to 181 in 60% yield.[94,95a]

A major yield improvement resulted from prior conversion
of the unstable vinyl carbinol 146 to the crystalline isothiuronium acetate 195. The latter coupled with 157 under

mild conditions (water-ether/25°) to give tricyclic dione 192
in 90% yield.[94,95a]

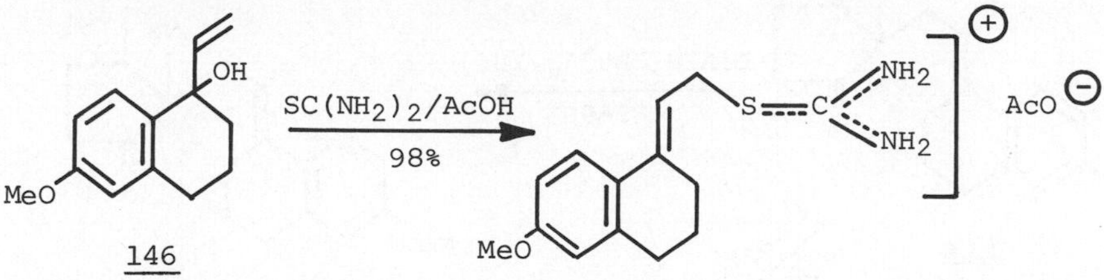

146

195

The above modifications raise the overall yield of 119
from 6 to about 35-40%. A variant of the Torgov reaction involving reaction of the pyrollidyl analog 196 of vinyl carbinol 146 with 157 in refluxing methanol led to crude 192 in
80% yield.[96]

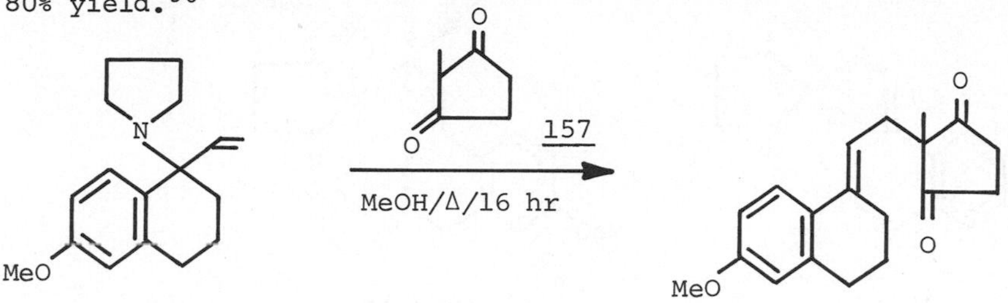

196   192

I. Resolution

Conventional resolution of estrone itself[59a] and the estra-
pentaen-17β-ol derived from 181[37c] has been accomplished via
the corresponding l-menthoxyacetates. It was subsequently
recognized that the prochiral tricyclic dione 192 offered the
unique possibility of resolution without enantiomer loss,
either by reconversion of the "wrong" enantiomer to 192 or by
asymmetric operations on 192.

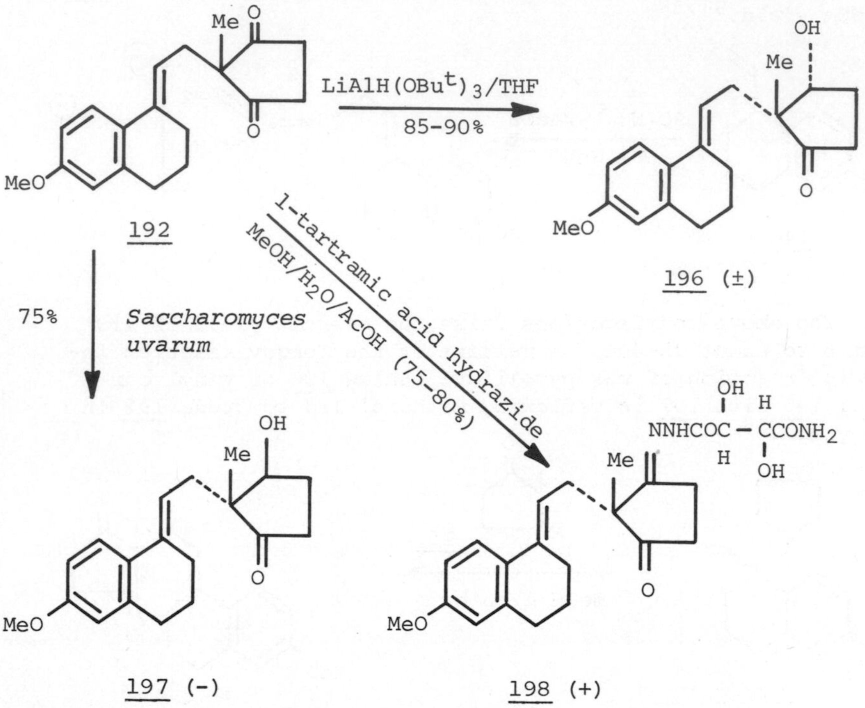

Reduction of 192 with lithium tri-t-butoxy aluminum
hydride occurred under reagent approach control to give the
(±) ketol 196 (Me/OH *trans*) in high yield. The latter was
resolved via the quinine salt of its hemisuccinate, and the
(−) ketol 196 converted to (+) estrone by way of 17α-hydroxy
intermediates. The (+) ketol 196 was recycled by oxidation
to 192.[94]

Microbiological reduction of 192 with the yeast, *Sacchar-
omyces uvarum,* gave the (−) ketol 197 (Me/OH *cis*) in 75% yield,
which was converted to (+) estradiol methyl ether via 17β-
acetoxy intermediates in 55% yield.[93]

Derivatization of dione 192 with l-tartramic acid hydra-
zide under equilibrating conditions led to the (+) derivative
198 in 75-80% yield. Reversible reaction occurs at either
carbonyl group, but equilibrium is displaced in favor of diastereomer 198 because of its greater insolubility. Cyclization and hydrolysis gave (−) estrapentaen-17-one 181.[97]

An additional synthesis of (−) 181 from a prochiral de-
rivative of 2-methylcyclopentan-1,3-dione 157, namely, the
Michael adduct 199 of 157 and ethyl acrylate, is outlined in
Scheme 25.[98] Conversion of 199 to the monoethylene ketal and

Scheme 25

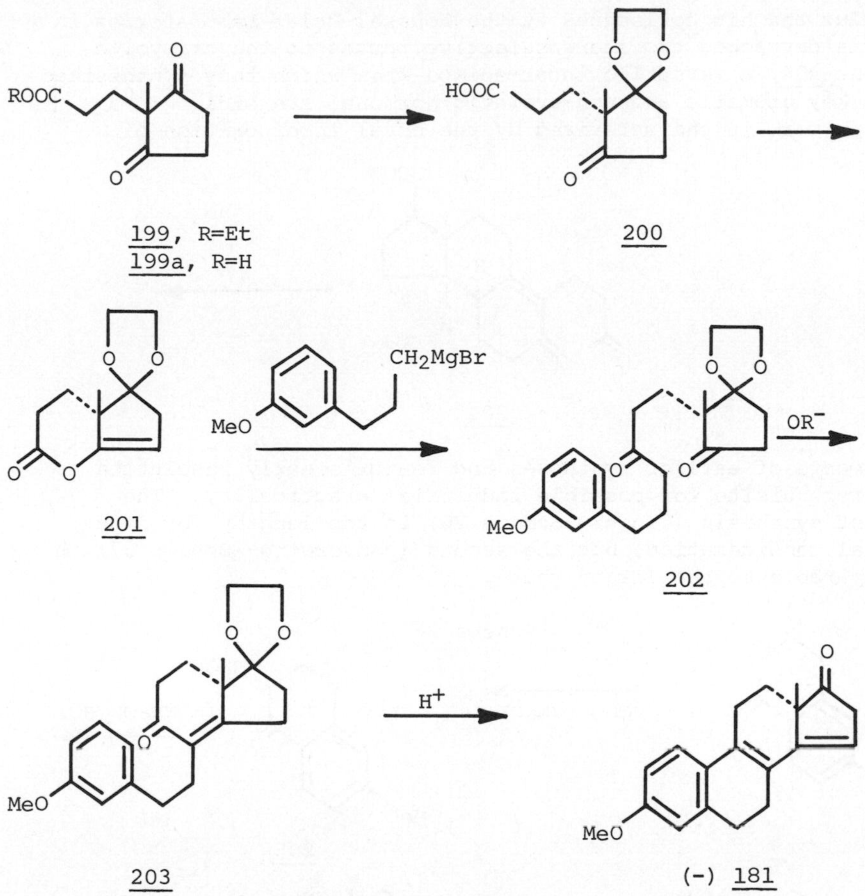
ROOC
HOOC
CH2MgBr
MeO
OR⁻
H⁺
199, R=Et
199a, R=H
200
201
202
203
(-) 181

saponification led to the acid 200 which was resolved as the cinchonine salt. The "wrong" enantiomer was recycled by hydrolysis to 199a. Preparation of the enol lactone 201 and Grignard reaction with 3-m-methoxyphenylpropyl magnesium bromide (cf. 163, Scheme 21) under conditions compatible with ketal functionality led to the dioxoketal 202, which was cyclized under alkaline conditions to give 203. Acid catalyzed cyclization and ketal cleavage then afforded (-) 181.

J. Velluz Syntheses

Velluz and his colleagues at the Roussel-Uclaf Laboratories in Paris developed two stereoselective routes to the tricyclic dione 204, a versatile intermediate from which they synthesized the key aromatic and non-aromatic hormonal steroids.[99,100] Their work is characterized by the novel incorporation of

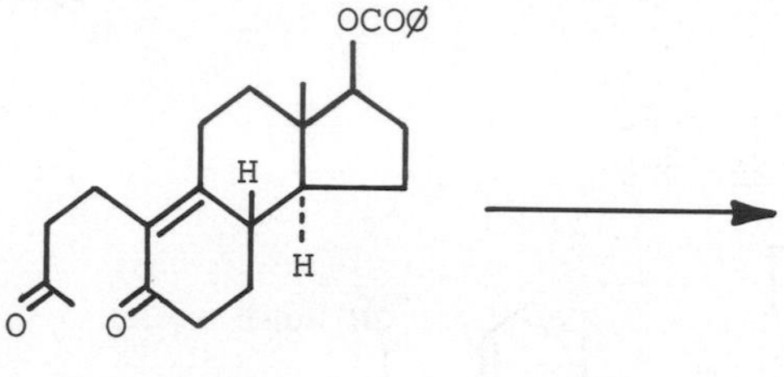

204

elements of earlier syntheses and features early resolution as a prerequisite for possible industrial practicality. The first synthesis (linear--Scheme 26) is too lengthy for industrial consideration, but the second (converging--Scheme 27) is comparable to the Torgov route.

Scheme 26

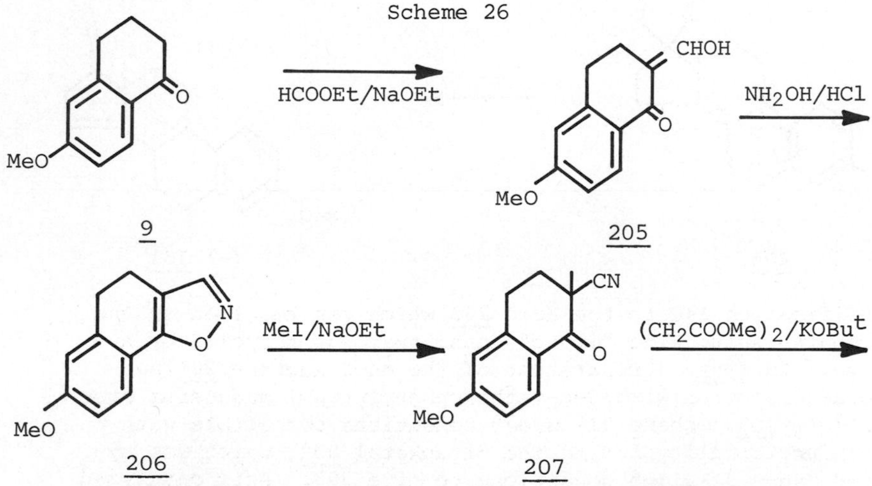

9   205

206   207

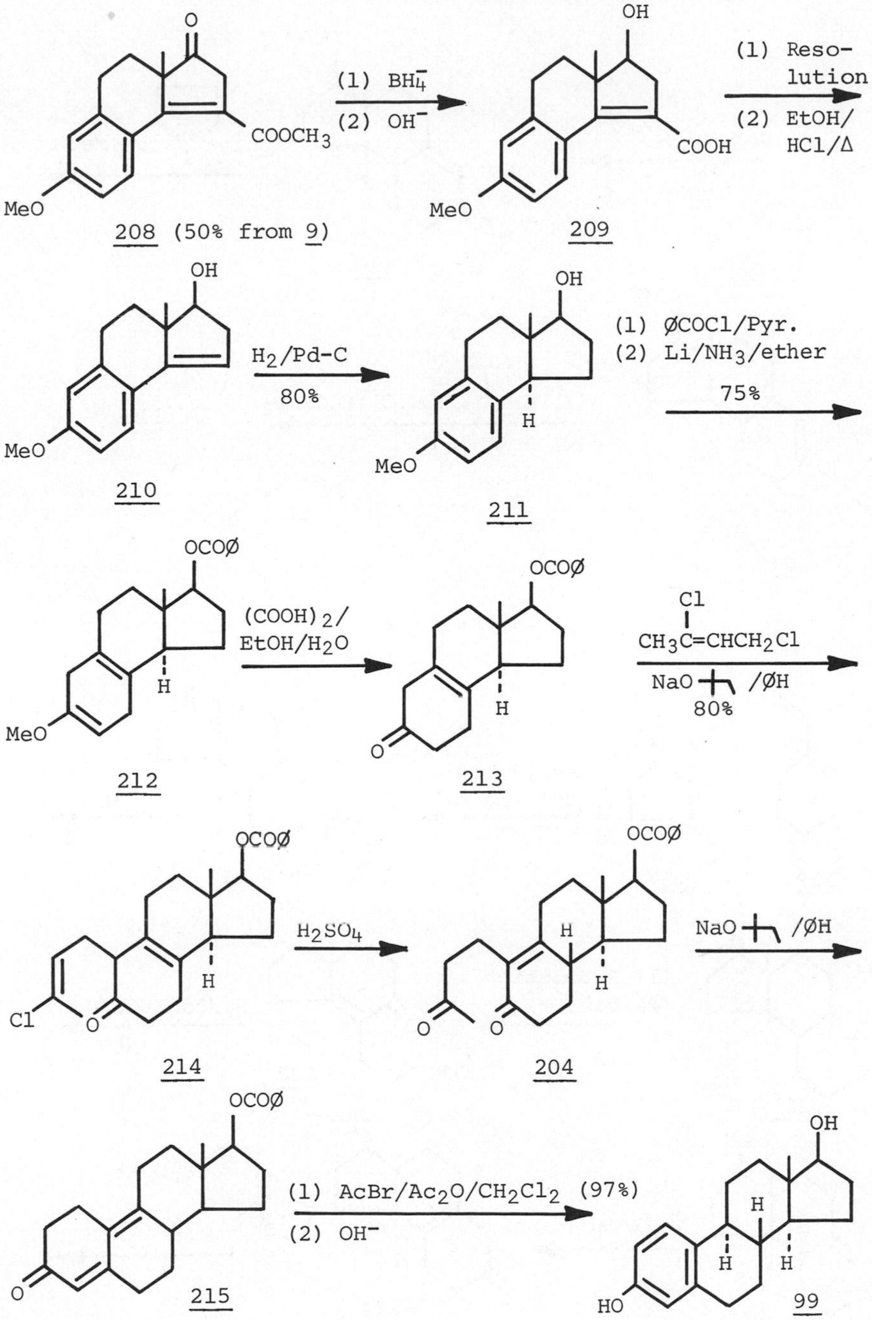
208 (50% from 9)
(1) BH4−
(2) OH−
209
(1) Resolution
(2) EtOH/HCl/Δ
210
H2/Pd-C
80%
211
(1) ØCOCl/Pyr.
(2) Li/NH3/ether
75%
212
(COOH)2/EtOH/H2O
213
CH3C(Cl)=CHCH2Cl
NaO-t-Am/ØH
80%
214
H2SO4
204
NaO-t-Am/ØH
215
(1) AcBr/Ac2O/CH2Cl2 (97%)
(2) OH−
99
MeO
OCOØ
OH
COOCH3
COOH
Cl
HO

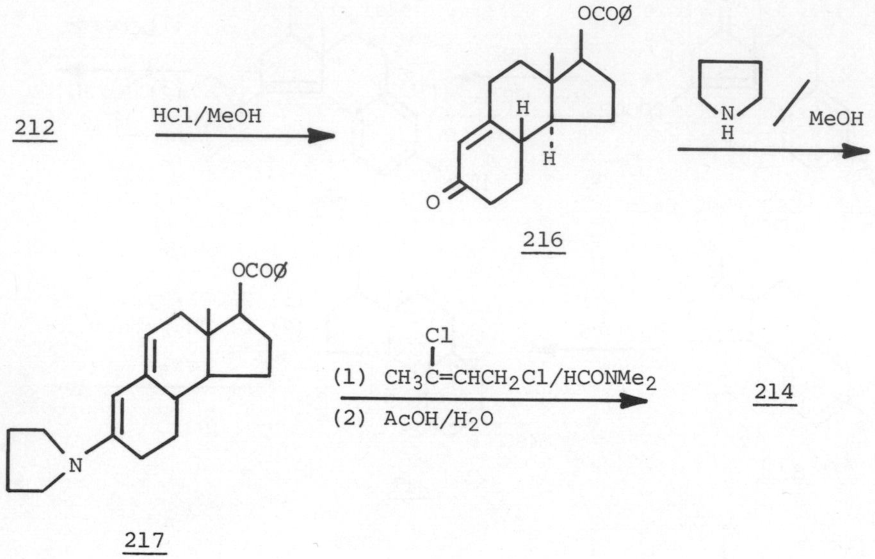
212
HCl/MeOH
OCOØ
H
H
O
216
N
H
MeOH
OCOØ
N
217
Cl
(1) CH3C=CHCH2Cl/HCONMe2
(2) AcOH/H2O
214

Scheme 27

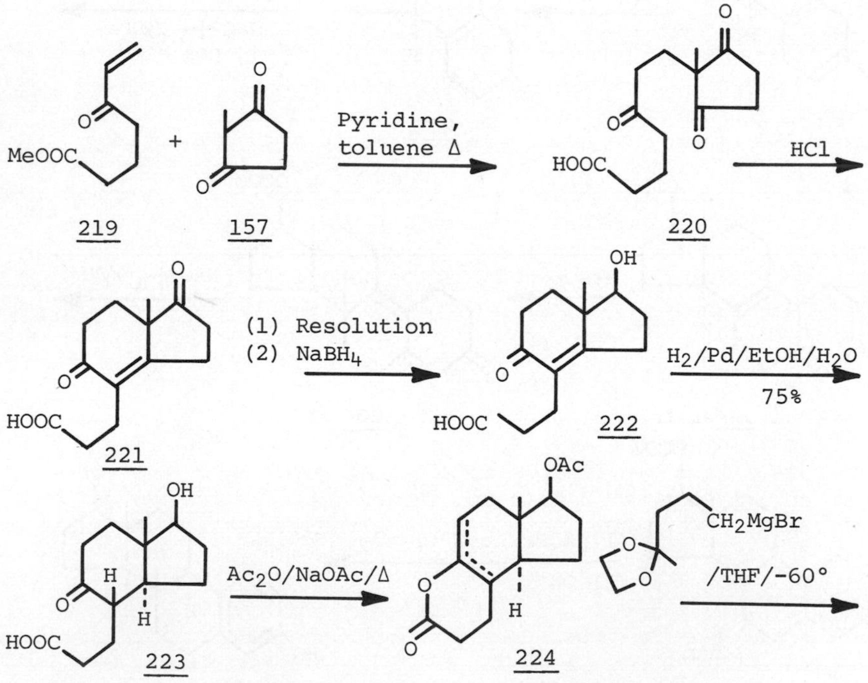
O
MeOOC
219
O
O
157
Pyridine,
toluene Δ
O
O
O
HOOC
220
HCl
O
O
HOOC
221
(1) Resolution
(2) NaBH4
OH
O
HOOC
222
H2/Pd/EtOH/H2O
75%
OH
H
O
H
HOOC
223
Ac2O/NaOAc/Δ
OAc
O
O
H
224
CH2MgBr
O
O
/THF/-60°

225

(1) KOH/MeOH

(2) H$^+$

(3) C$_6$H$_5$COCl/pyr.

204

220

Rhizopus arrhizus

>70%

226

The 1960 synthesis[99] followed a BC → D → A scheme. An efficient route to the racemic tricyclic intermediate 211 from 6-methoxytetralone 9 described by Banerjee in 1956[21a] was employed and modified to include resolution of 209. Construction of ring D followed the sequence developed by Johnson in his first equilenin synthesis (Scheme 3) with Banerjee's modification for increased stereoselectivity in the hydrogenation of the ring D double bond (208 → 211). Intermediate 209 was resolved as the chloramphenicol salt. The alkylation reaction leading to ring A was studied using methyl vinyl ketone, 3-benzyloxybutyl bromide and 1,3-dichlorobutene-2 (preferred) with the β,γ-unsaturated ketone 213 as well as with the corresponding α,β-unsaturated ketone 216. In ketone 213 the doubly activated hydrogens on the carbon α to both the carbonyl group and the double bond are clearly the most acidic, and alkylation was directed to this site. However, it was discovered later that alkylation of the dienamine 217 derived from 216 was a more efficient route to 214.[101] Sulfuric acid hydrolysis of the vinyl chloride 214 then gave the enedione 204. The 9,10-double bond in the latter inhibits further acid catalyzed condensation to a 3,3,1-bicyclononene system observed in analogous saturated cases.[102] In the present example the resultant system 218 would violate Bredt's rule. Closure of ring A (204 → 215) was best accomplished with sodium t-amylate in benzene. Aromatization of the A ring by isomerization of the 9,10-double bond was carried out catalytically in 40% yield by palladium on carbon in refluxing

218

ethanol. An alternative isomerization procedure involving reaction of 215 with acetyl bromide and acetic anhydride in methylene chloride at 20° followed by addition to aqueous ammonia led to estradiol 17-benzoate in 97% yield.[103] Saponification of the latter then gave estradiol 99. The use of benzoate as a protecting group for hydroxyl over a considerable variety of conditions is noteworthy.

The 1963 Velluz synthesis[100] utilized a D → C → B → A scheme analogous to a sequence that had been studied earlier with D-homo intermediates.[76,104] Michael condensation under weakly basic conditions of 2-methylcyclopentane-1,3-dione 157 with methyl 5-oxo-6-heptenoate 219 led to the trione acid 220, which as a base unstable 2,2-disubstituted-1,3-diketone was cyclized to the enedione acid 221 under acidic conditions. The latter was resolved conventionally via its ephedrine salt. However, its prochiral precurser 220 was later reduced microbiologically in >70% yield to the correct optically active enantiomer 226 (methyl/OH-*cis*),[105a] which was cyclized with hydrochloric acid to 222 identical with material obtained from 221. Alternatively asymmetric cyclization of 220 methyl ester to (+)-221 methyl ester of 64% optical purity was accomplished in the presence of L-phenylalanine.[105b] Hydrogenation of the latter over palladium in aqueous ethanol led to the requisite C/D *trans* product 223 with equilibrated propionic acid side chain in 75% yield.[83b] Earlier experience with the C-4 unsubstituted tetrahydroindanes 176 and 177 had suggested that hydrogenation would lead to a C/D *cis* fused product.[82a] However, in most cases with alkyl substitution at C-4 *trans* fused products were found to predominate (cf. 174 → 175[37b]).[83b,111] Conversion of 223 into a mixture of enol lactones 224 followed by Grignard reaction with 4,4-ethylenedioxypentylmagnesium bromide led to 225. Aldol condensation to close ring B and concomitant cleavage of the 17-acetoxy group followed by ketal hydrolysis and benzoylation then gave intermediate 204 which had been converted into estradiol in the earlier synthesis. A similar sequence with methyl magnesium bromide led to the

17-acetate of the tricyclic ketone 216 of the earlier synthesis[106] (cf. Ref. 104).

Reaction of Grignard reagents with enol lactones followed
by cyclization to α,β-unsaturated ketones has been a frequently
used annelation procedure in the steroid field[107] (see also
201 → 202). A useful variant involves condensation of enol
lactones with phosphonium or phosphonate ylids. Thus, the
17-t-butyl ether of 224 on reaction with the anion of diethyl
4,4-ethylenedioxypentyl phosphonate followed by treatment with
aqueous sodium hydroxide (loss of diethyl phosphate) and aqueous acetic acid yielded the t-butyl ether of 204.[108]

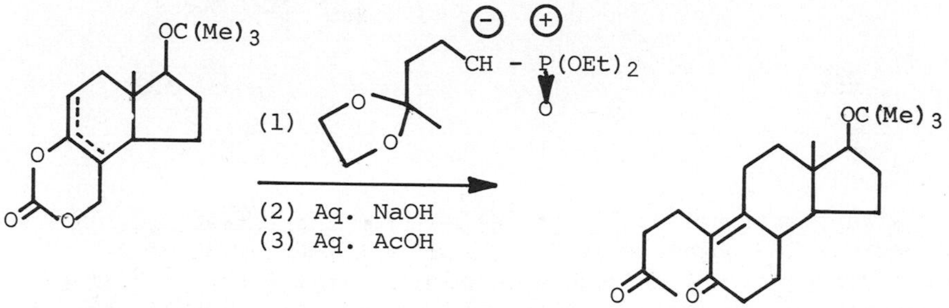

K. Miscellaneous

A novel AD → ABCD route to estrapentaene 181 involving triple
bond participation in a cyclization sequence was described by
Hiraoka and Iwai in 1966.[109] 5-m-Methoxyphenylpentyne 164
(Scheme 21) was converted into the propargyl bromide 227 and
the latter condensed with the sodium salt of 2-methylcyclo-
pentane-1,3-dione in dimethylsulfoxide. The alkylation reaction failed when attempted in methanol, dioxane, or acetone.
Treatment of the adduct 228 with polyphosphoric acid at room
temperature then led to 181 in 63% yield, possibly via the
tricyclic dione 192 (Scheme 24).

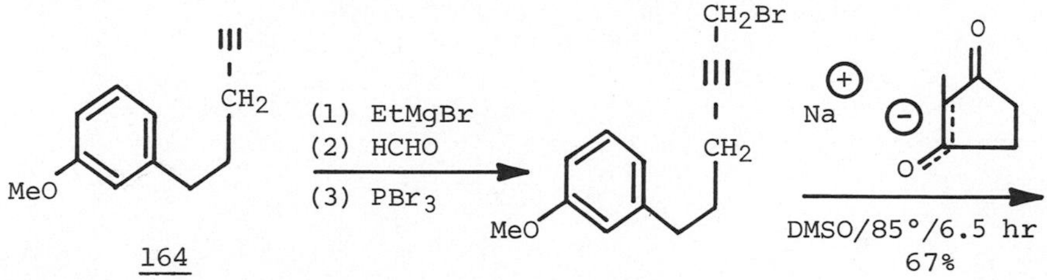

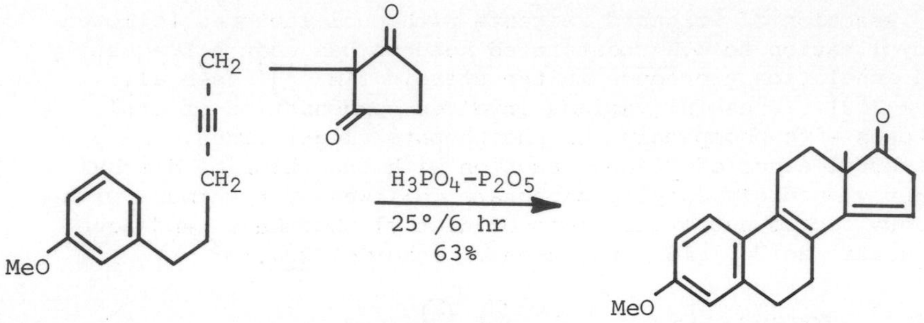

H₃PO₄–P₂O₅
25°/6 hr
63%

228   181

A further route to 181 proceeded from the tetrahydro-
indanedione 177 obtained in 70% yield by condensation of 2-
methylcyclopentane-1,3-dione with methyl vinyl ketone.[110] (For
an asymmetric synthesis yielding (+)-177 see Ref. 105b.)   Alkylation of 177 with m-methoxyphenethyl bromide (t-BuOK/C₆H₆ -
25%)[37b,111] or the derived tetrahydropyranyl ether 229 with
m-methoxyphenethyl-p-toluenesulfonate (t-BuOK/t-BuOH - 32%)[87b]
followed by hydrolysis and oxidation gave tricyclic dione
173.   A generally more efficient procedure for the preparation
of 4-substituted 5,6,7,7α-tetrahydroindan-5-ones is by a
Robinson annelation sequence in which the substituent is in-
corporated in the vinyl ketone (e.g., 167 + 157 → 173; 219 +
157 → 221).   Cyclodehydration of 173 in warm polyphosphoric
acid led to 181 in ∿50% yield.[87a-87c,37b,111]

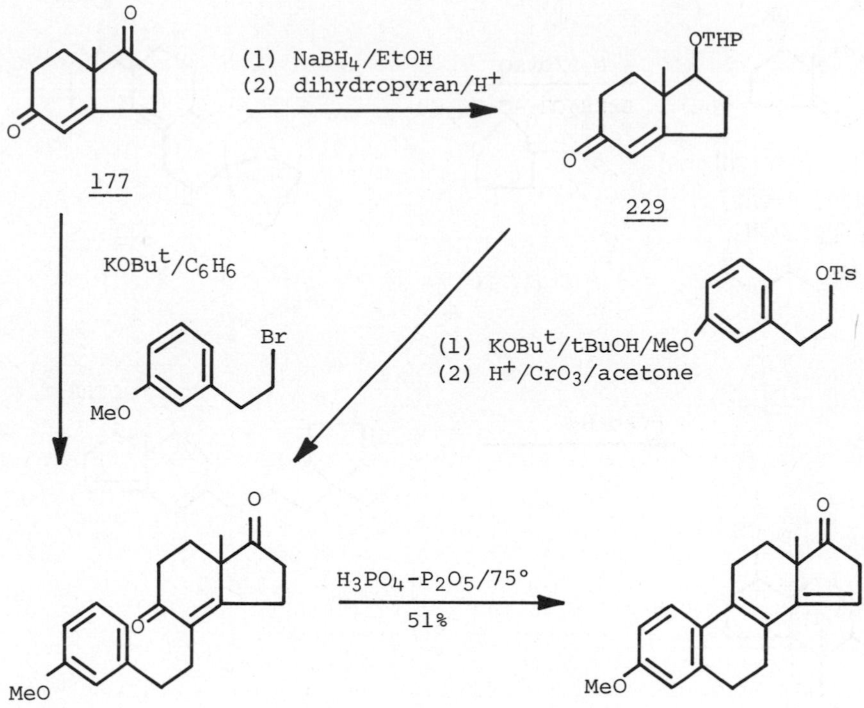

(1) NaBH$_4$/EtOH
(2) dihydropyran/H$^+$

177

229

KOBu^t/C$_6$H$_6$

(1) KOBu^t/tBuOH/MeO
(2) H$^+$/CrO$_3$/acetone

MeO

Br

OTs

O

O

MeO

173

H$_3$PO$_4$-P$_2$O$_5$/75°

51%

O

MeO

181

Improved yields resulted when alkylation of 229 was carried out in dimethylsulfoxide. Thus, in a synthesis of the tricyclic enone 232 reaction of 229 with 1-bromo-3-pentanone ethylene ketal (DMSO/NaH) led to 58% of C-alkylated 230 and 32% of o-alkylated product 231 (VPC yields). Separation and acid hydrolysis of the latter gave reusable ketal 229a.[112a] Similar results were obtained with the t-butyl ether of 229a.[112b]

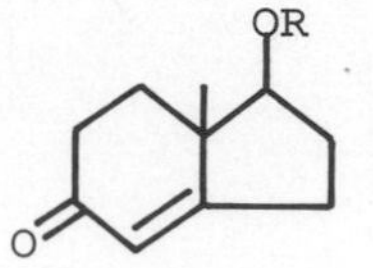
OR
O

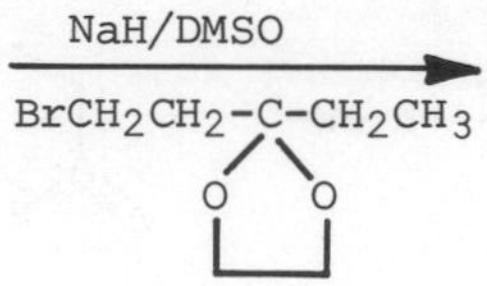
NaH/DMSO
BrCH2CH2-C-CH2CH3
O O

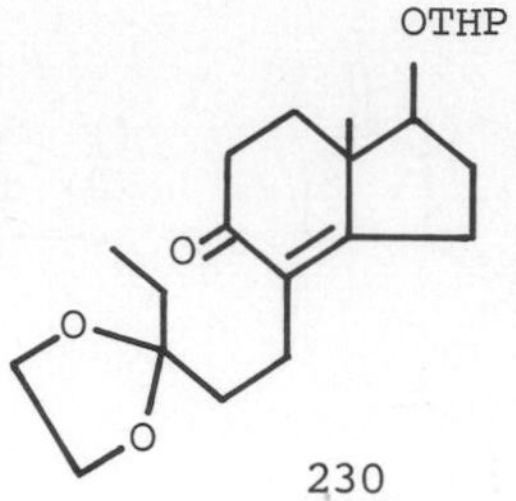
OTHP
O
O
O

229, R=THP
229a, R=H

230

+

229a

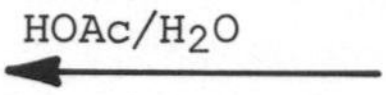
HOAc/H2O

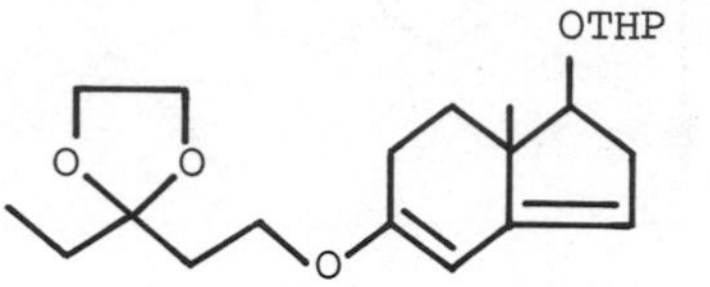
OTHP
O
O
O

231

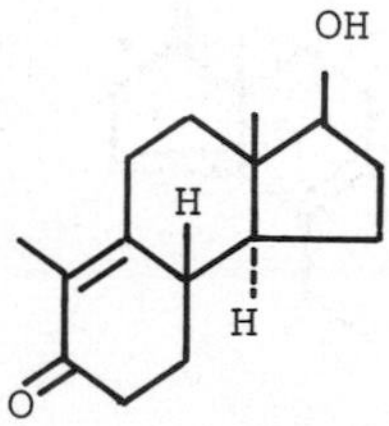
OH
H
H
O

232

A route to (±) estrone 98 from the tetrahydroindanone 177 utilized alkylation with 1,3-dichlorobutene-2 for generation of the B and A rings.[112c] It led to (±) estra-4,9-diene 3,17-dione which was aromatized by the Roussel procedure[101,103] (cf. Scheme 26: 215 → 99).

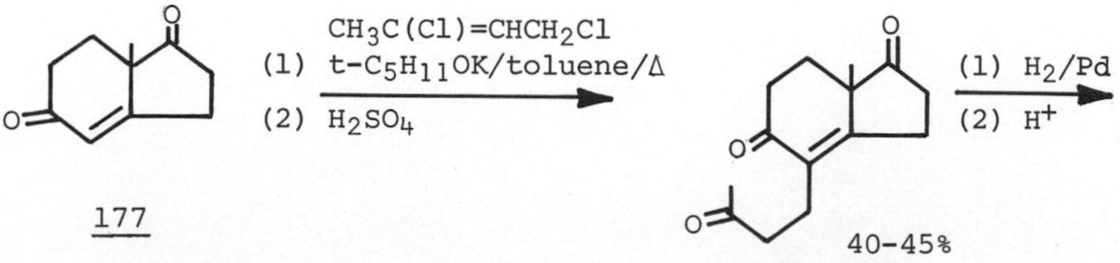
O
O
CH3C(Cl)=CHCH2Cl
(1) t-C5H11OK/toluene/Δ
(2) H2SO4
O
O
O
(1) H2/Pd
(2) H+

177

40-45%

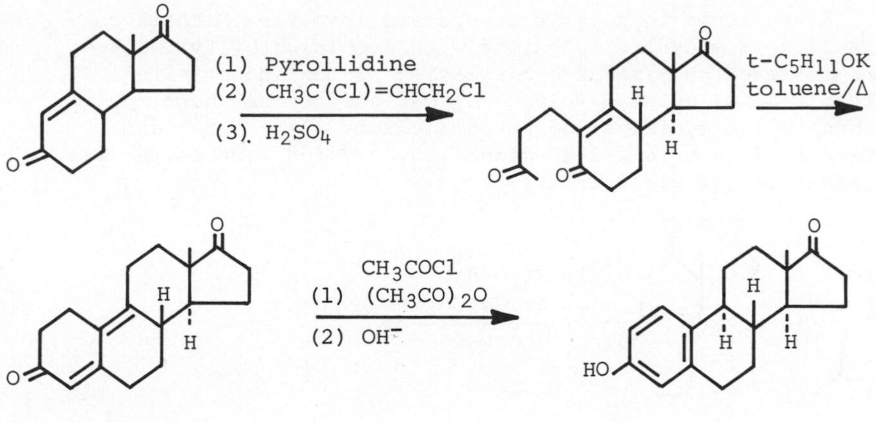

~5% from 177

98

Annelation of 2-methylcyclopentane-1,3-dione with ethoxy-carbonylmethyl vinyl ketone led to the 4-carbethoxytetrahydro-indane-dione 241 which was transformed efficiently to (±) estrone methyl ether 119 in eight steps.[122]

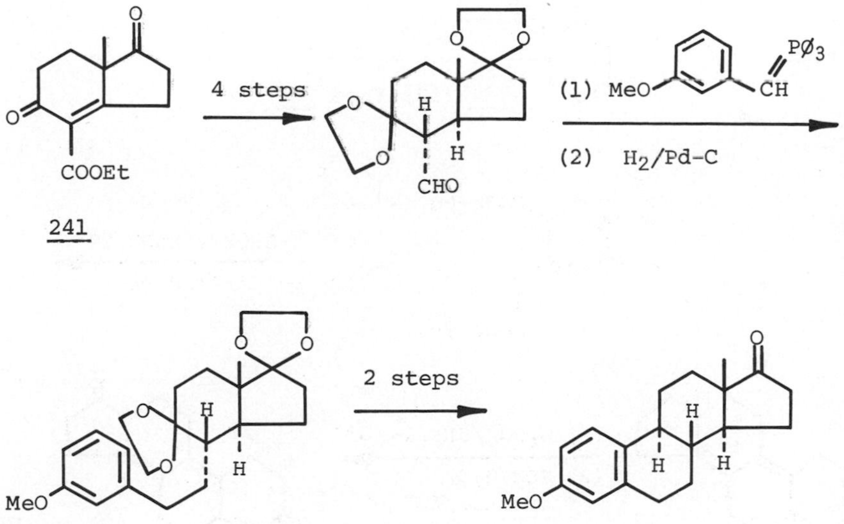

241

119

A new route to polycyclic systems involving Michael condensation of a cyclic 1,3-dione with β-chloroethyl vinyl ketone followed by similar condensation of the adduct with a
β-keto t-butyl ester developed by Danishefsky has been applied
by him to the syntheses of the dienedione 234[113a,114] and
estr-4,8(14),9-triene-3,17-dione 235.[113b,115] The latter was
obtained in 34% yield from 157.

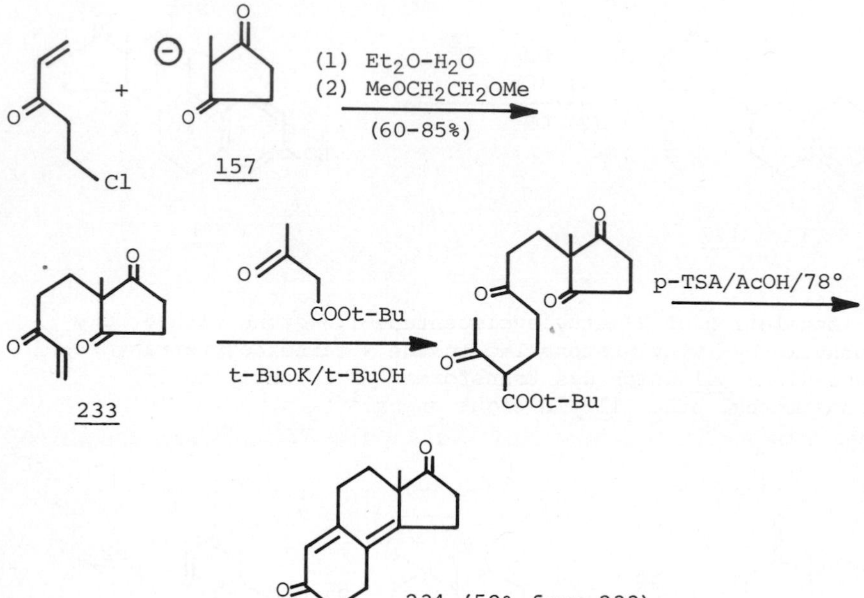

234 (52% from 233)

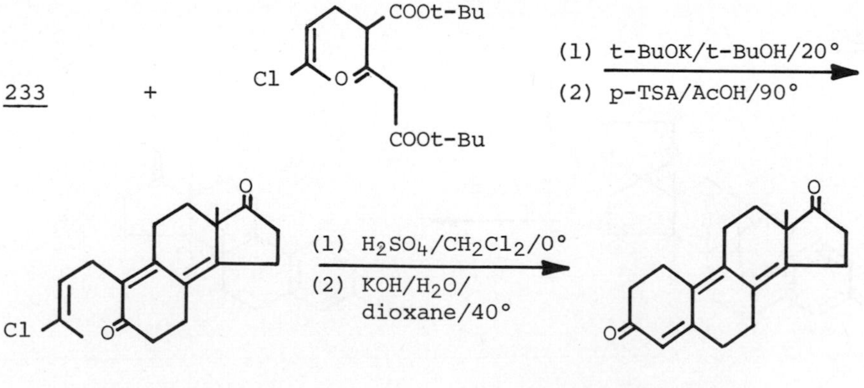

235

  G. Saucy and co-workers at the Hoffmann-LaRoche laboratories have developed several variants of a novel and efficient route to (±) and (+) 19-nor steroid-4-ene-3-ones (e.g.
238),[119] which is adaptable to the synthesis of estrone. The
following scheme[119] is illustrative.

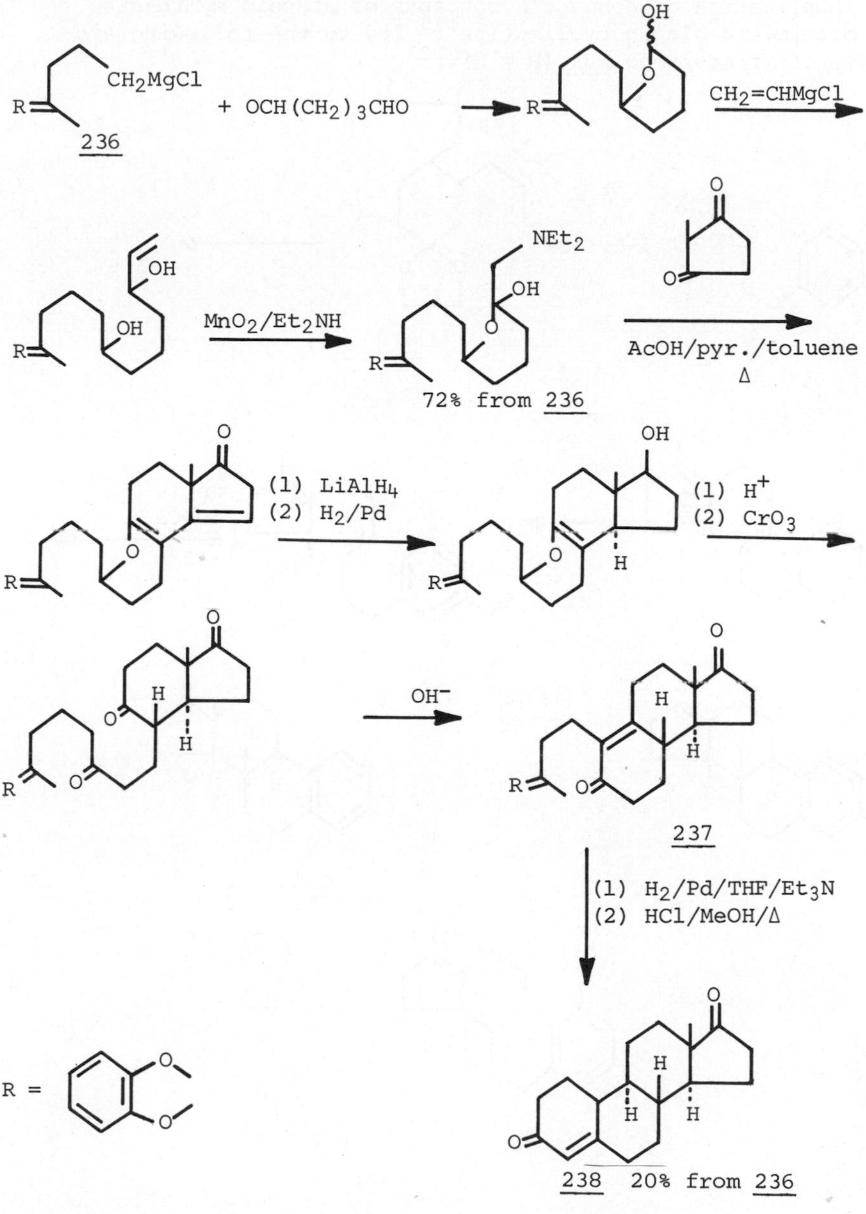

CH_2MgCl

236 + $OCH(CH_2)_3CHO$ $\longrightarrow$ $CH_2=CHMgCl$ $\longrightarrow$

MnO_2/Et_2NH $\longrightarrow$ 72% from 236

$\xrightarrow{\text{AcOH/pyr./toluene} \atop \Delta}$

(1) $LiAlH_4$
(2) H_2/Pd $\longrightarrow$

(1) H^+
(2) CrO_3 $\longrightarrow$

OH^- $\longrightarrow$ 237

237 $\downarrow$
(1) $H_2/Pd/THF/Et_3N$
(2) $HCl/MeOH/\Delta$

$R =$

238 20% from 236

Intermediate 237 is analogous to 204. Removal of the
protecting group, closure of the A ring, and double bond isomerization should yield estrone.

Application of Johnson's concepts of steroid synthesis
via biomimetic olefin cyclization[120] led to the following synthesis of (±) estrone 239 (R = H).[122]

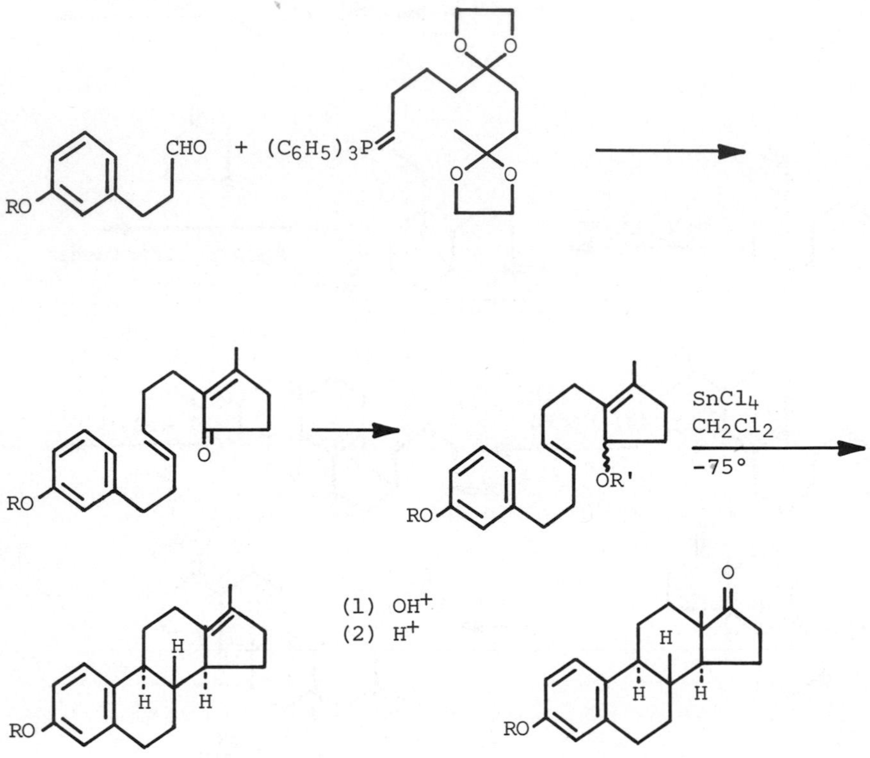

239

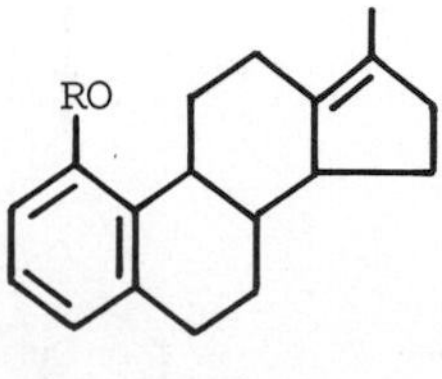

240

In the key cyclization step some ortho cyclization product 240 was also produced. The para/ortho ratio was dependent on the nature of R and R'.

R	R'	P/O
Me$_3$Si	H	8.4
Me$_3$Si	Me$_3$Si	2.6
Me	H	3.0
C$_6$H$_5$CO	C$_6$H$_5$CO	1.4

Additional (somewhat circuitous) syntheses of estrone have been described by Birch[116] and Miki.[117]

REFERENCES

1. Previous general reviews. (a) J. W. Cornforth, in *Progress in Organic Chemistry,* Vol. 3 (Academic Press, New York, 1955), pp. 1-43; (b) I. V. Torgov, *Pure Appl. Chem.*, *6*, 525 (1963); (c) I. V. Torgov, in *Recent Developments in the Chemistry of Natural Carbon Compounds,* Vol. 1 (Akadémiai Kiadó, Budapest, 1965), pp. 235-319; (d) L. Velluz, J. Valls, and G. Nominé, *Angew. Chem. Int. Ed.*, *4*, 181 (1965); (e) A. A. Akhrem and Y. A. Titov, *Total Steroid Synthesis* (Plenum Press, New York, 1970) (English translation by B. J. Hazzard of the 1967 Russian edition with an Appendix added in 1968).
2. A. Girard, G. Sandulescu, A. Fridenson, and J. J. Rutgers, *Compt. Rend.*, *195*, 981 (1932).
3. W. E. Bachmann, W. Cole, and A. L. Wilds, *J. Am. Chem. Soc.*, *61*, 977 (1939); *62*, 824 (1940).
4. R. P. Linstead, *Ann. Reports Chem. Soc.*, *33*, 312 (1936); H. D. Springall, *ibid.*, *36*, 286 (1939).
5. A. Butenandt and S. Schramm, *Chem. Ber.*, *68*, 2083 (1935).
6. (a) A. Cohen, J. W. Cook, and C. L. Hewett, *J. Chem. Soc.*, 445 (1935); (b) A. L. Wilds and W. J. Close, *J. Am. Chem. Soc.*, *69*, 3079 (1947).
7. G. Haberland, *Chem. Ber.*, *69*, 1380 (1936).
8. G. Stork, *J. Am. Chem. Soc.*, *69*, 576, 2936 (1947).
9. W. E. Bachmann and N. L. Wendler, *J. Am. Chem. Soc.*, *68*,

2580 (1946).

10. A. J. Birch and G. S. R. Subba Rao, *Tet. Letters*, 2917 (1968).

11. J. W. A. Findlay and A. B. Turner, *Chem. and Ind.*, 158 (1970).

12. R. D. Haworth, *J. Chem. Soc.*, 1130 (1932); R. Robinson and G. Walker, *ibid.*, 61 (1937).

13. J. Schmidlin, G. Anner, J. R. Billeter, K. Heusler, H. Ueberwasser, P. Wieland, and A. Wettstein, *Helv. Chim. Acta, 40*, 1034 (1957).

14. J. Heer and K. Miescher, *Helv. Chim. Acta, 29*, 1895 (1946); *30*, 550 (1947).

15. (a) W. S. Johnson, J. W. Petersen, and C. D. Gutsche, *J. Am. Chem. Soc., 67*, 2274 (1945); *69*, 2942 (1947); (b) W. S. Johnson and V. L. Stromberg, *J. Am. Chem. Soc., 72*, 505 (1950).

16. W. S. Johnson and G. H. Daub, *Org. Reactions, 6*, 1 (1951).

17. (a) R. Robinson and J. Walker, *J. Chem. Soc.*, 60 (1937); (b) G. Haberland and E. Heinrich, *Chem. Ber., 72*, 1222 (1939).

18. H. S. Corey, Jr., J. R. P. McCormick, and W. E. Swensen, *J. Am. Chem. Soc., 86*, 1884 (1964).

19. Cf. W. S. Johnson and W. E. Shelberg, *J. Am. Chem Soc., 67*, 1745 (1945); K. von Auwers, T. Bohr, and E. Frese, *Ann., 441*, 54 (1925).

20. W. S. Johnson, C. D. Gutsche, R. Hirschmann, and V. L. Stromberg, *J. Am. Chem. Soc., 73*, 322 (1951).

21. (a) D. K. Banerjee, S. Chatterjee, C. N. Pillai, and M. V. Bhatt, *J. Am. Chem. Soc., 78*, 3769 (1956); (b) D. K. Banerjee, *J. Ind. Chem. Soc., 47* (1970).

22. W. E. Bachmann and R. E. Holmen, *J. Am. Chem. Soc., 73*, 3660 (1951).

23. R. Robinson, *J. Chem. Soc.*, 1390 (1938); A. Koebner and R. Robinson, *ibid.*, 1944 (1938).

24. A. J. Birch, R. Jaeger, and R. Robinson, *J. Chem. Soc.*, 582 (1945).

25. (a) A. J. Birch and G. S. R. Subba Rao, *Tet. Letters*, 2763 (1967); (b) *Australian J. Chem., 23*, 547 (1970).

26. E. A. Kehrer and P. Igler, *Chem. Ber., 32*, 1176 (1899); E. A. Kehrer, *ibid., 34*, 1263 (1901).

27. A. P. Dunlap and F. N. Peters, *The Furans* (Reinhold Publishing Corp., New York, 1953), pp. 652-657.

28. R. D. Haworth and G. S. Sheldrick, *J. Chem. Soc.*, 865 (1934).

29. A. J. Birch and R. Robinson, *J. Chem. Soc.*, 501 (1944).

30. (a) A. Koebner and R. Robinson, *J. Chem. Soc.*, 566 (1941); (b) W. S. Johnson, *J. Am. Chem. Soc., 65*, 1317 (1943); (c) A. J. Birch, *J. Chem. Soc.*, 661 (1943).

31. R. E. Ireland and J. A. Marshall, *J. Org. Chem.*, *27*, 1615 (1962).

32. W. S. Johnson, *J. Am. Chem. Soc.*, *66*, 215 (1944).

33. W. S. Johnson, D. S. Allen, Jr., R. R. Hendersinn, G. N. Sausen, and R. Pappo, *J. Am. Chem. Soc.*, *84*, 2181 (1962).

34. R. Bucourt, *Bull. Soc. Chim. France*, 1983 (1962); 1262 (1963); 2080 (1964); L. Velluz, J. Valls, and G. Nominé, *Angew. Chem. Int. Ed.*, 181 (1965) (see p. 191). See also, E. J. Corey and R. A. Sneen, *J. Am. Chem. Soc.*, *77*, 2505 (1955).

35. G. Stork, P. Rosen, N. Goldman, R. V. Coombs, and J. Tsuji, *J. Am. Chem. Soc.*, *87*, 275 (1965).

36. A. Horeau, E. Lorthioy, and J. P. Suette, *Compt. Rend.*, *269C*, 558 (1969).

37. (a) G. A. Hughes and H. S. Smith, *Chem. Ind. (London)*, 1022 (1960); (b) G. H. Douglas, J. M. H. Graves, D. Hartley, G. A. Hughes, B. J. McLoughlin, J. Siddall, and H. S. Smith, *J. Chem. Soc.*, 5072 (1963); (c) T. Miki, K. Hiraga, and T. Asako, *Proc. Chem. Soc.*, 139 (1963); *Chem. Pharm. Bull. (Tokyo)*, *13*, 1285 (1965).

38. S. N. Ananchenko, V. H. Limanov, V. N. Leonov, V. N. Rzheznikov, and I. V. Torgov, *Tetrahedron*, *18*, 1355 (1962); S. N. Ananchenko, V. N. Leonov, A. V. Platanova, and I. V. Torgov, *Dokl. Akad. Nauk. S.S.S.R.*, *135*, 73 (1960).

39. A. Girard, G. Sandulesco, A. Friedenson, and J. J. Rutgers, *Compt. Rend.*, *194*, 909 (1932).

40. J. A. Zderic, A. Bowers, H. Carpio, and C. Djeressi, *J. Am. Chem. Soc.*, *80*, 2596 (1958); *Steroids*, *1*, 223 (1963).

41. A. J. Birch, *J. Chem. Soc.*, *367* (1950).

42. J. F. Bagli, P. F. Morand, K. Wiesner, and R. Guadry, *Tet. Letters*, 387 (1964).

43. K. Heusler, J. Kalvoda, Ch. Meystre, H. Ueberwasser, P. Wieland, G. Anner, and A. Wettstein, *Experientia*, *18*, 464 (1962).

44. (a) R. P. Stein, G. C. Buzby, Jr., and H. Smith, *Tet. Letters*, 5015 (1966); (b) *Tetrahedron*, *26*, 1917 (1970).

45. E. J. Bailey, A. Gale, G. H. Phillipps, P. T. Siddons, and G. Smith, *Chem. Commun.*, 1253 (1967).

46. A. J. Birch and H. Smith, *J. Chem. Soc.*, 1882 (1951).

47. J. Hannah and J. H. Fried, *J. Med. Chem.*, *8*, 536 (1965).

48. A. L. Wilds and N. A. Nelson, *J. Am. Chem. Soc.*, *75*, 5360 (1953); H. L. Dryden, G. M. Webber, R. R. Burtner, and J. A. Cella, *J. Org. Chem.*, *26*, 3237 (1961).

49. D. J. Marshall and R. Deghenghi, *Can. J. Chem.*, *47*, 3127 (1969).

50. R. D. H. Heard and M. M. Hoffmann, *J. Biol. Chem.*, *135*, 801 (1940); *138*, 651 (1941).

51. D. K. Banerjee and G. Nadamuni, *Indian J. Chem.*, 529 (1969).

52. J. Fried, N. A. Abraham, and T. S. Santharakrishnan, *J. Am. Chem. Soc.*, *89*, 1044 (1967).

53. (a) A. Butenandt, *Naturwiss.*, *17*, 879 (1929); A. Butenandt and E. von Ziegner, *Z. Physiol.*, *188*, 1 (1930); (b) E. A. Doisy, C. D. Veler, and S. A. Thayer, *Am. J. Physiol.*, *90*, 329 (1929); *J. Biol. Chem.*, *86*, 499 (1930); *87*, 357 (1930).

54. H. Schildknecht and H. Birringer, *Z. Naturforsch.*, *24b*, 1529 (1969); H. Schildknecht, *Angew. Chem. Int. Ed.*, *9*, 1 (1970).

55. For previous reviews see (a) T. B. Windholz and M. Windholz, *Angew. Chem. Int. Ed.*, *3*, 353 (1964); (b) P. Morand and J. Lyall, *Chem. Rev.*, *85* (1968); (c) N. Appelzweig, *Steroid Drugs* (McGraw-Hill, New York, 1962).

56. E. Dane and J. Schmitt, *Ann.*, *536*, 196 (1938); *537*, 246 (1939).

57. G. Singh, *J. Am. Chem. Soc.*, *78*, 6109 (1956).

58. J. Heer and K. Miescher, *Helv. Chim. Acta*, *31*, 219 (1948); W. E. Bachmann and J. M. Chemerda, *J. Am. Chem. Soc.*, *70*, 1468 (1948).

59. (a) G. Anner and K. Miescher, *Helv. Chim. Acta*, *31*, 2173 (1948); (b) *32*, 1957 (1949); (c) *33*, 1379 (1950).

60. R. Robinson and E. Schlittler, *J. Chem. Soc.*, 1288 (1935); R. Robinson and J. Walker, *ibid.*, 747 (1936); 183 (1938).

61. W. E. Bachmann, S. Kushner and A. C. Stevenson, *J. Am. Chem. Soc.*, *64*, 974 (1942).

62. G. Anner and K. Miescher, *Helv. Chim. Acta*, *30*, 1422 (1947).

63. W. S. Johnson, I. A. David, H. C. Dehan, R. J. Highet, E. W. Warnhoff, W. D. Wood, and E. T. Jones, *J. Am. Chem. Soc.*, *80*, 661 (1958).

64. W. S. Johnson, D. K. Banerjee, W. P. Schneider, and C. D. Gutsche, *J. Am. Chem. Soc.*, *72*, 1426 (1950); W. S. Johnson, D. K. Banerjee, W. P. Schneider, C. D. Gutsche, W. E. Shelberg, and L. J. Chinn, *ibid.*, *74*, 2832 (1952).

65. W. S. Johnson and R. G. Christiansen, *ibid.*, *73*, 5511 (1951); W. S. Johnson, R. G. Christiansen and R. E. Ireland, *ibid.*, *79*, 1995 (1957).

66. J. E. Cole, W. S. Johnson, P. A. Robins, and J. Walker, *Proc. Chem. Soc.*, 144 (1958); *J. Chem. Soc.*, 244 (1962).

67. B. J. F. Hudson and R. Robinson, *J. Chem. Soc.*, 691 (1942); W. S. Johnson, C. D. Gutsche, and D. K. Banerjee, *J. Am. Chem. Soc.*, *73*, 5464 (1951).

68. W. L. Meyer, D. D. Cameron, and W. S. Johnson, *J. Org. Chem.* *27*, 1130 (1962).

69. J. C. Sheehan, R. A. Coderre, and P. A. Cruikshank, *J.*

Am. Chem. Soc., 75, 6231 (1953); J. C. Sheehan, W. F. Erman, and P. A. Cruikshank, *ibid., 79,* 147 (1957).

70. (a) D. K. Banerjee and K. Sivahandaiah, *Tet. Letters,* 20 (1960); *J. Ind. Chem. Soc.,* 652 (1961); (b) W. S. Johnson, R. E. Ireland, and R. E. Tarney as described in L. F. Fieser and M. Fieser, *Steroids* (Reinhold Publishing Co., New York, 1959), p. 500.

71. W. S. Johnson and D. S. Allen, *J. Am. Chem. Soc., 79,* 1261 (1957).

72. E. Dane, O. Höss, A. W. Bindseil, and J. Schmitt, *Ann., 532,* 39 (1937).

73. P. A. Robins and J. Walker, *J. Chem. Soc.,* 3249 (1956).

74. I. N. Nazarov, I. V. Torgov, and G. P. Verkholetova, *Doklady Akad. Nauk. S.S.S.R., 112,* 1067 (1957).

75. For example, M. S. Newman and A. B. Mekler, *J. Am. Chem. Soc., 82,* 4039 (1960).

76. P. Wieland, H. Ueberwasser, G. Anner, and K. Miescher, *Helv. Chim. Acta, 36,* 376, 646, 1231, 1803 (1953).

77. J. J. Panouse and C. Sannié, *Bull. Soc. Chim. France,* 1036 (1955); see also J. P. John, S. Swaminathan, and P. S. Venkataramani, *Org. Synth., 47,* 83 (1967).

78. R. Bucourt, A. Pierdet, G. Costerousse, and E. Toromanoff, *Bull. Soc. Chim. France,* 645 (1965).

79. V. J. Grenda, G. W. Lindberg, N. L. Wendler, and S. H. Pines, *J. Org. Chem., 32,* 1236 (1967).

80. H. Schick, G. Lehman and G. Hilgetag, *Angew. Chem., 79,* 97, 378 (1967); *Chem. Ber., 100,* 2973 (1967); *102,* 3238 (1969).

81. G. A. Hughes and H. Smith, *Proc. Chem. Soc.,* 74 (1960).

82. E. C. DuFeu, F. J. McQuillin and R. Robinson, *J. Chem. Soc.,* 53 (1937); review: E. D. Bergmann, D. Ginsburg, and R. Pappo, *Organic Reactions, 10,* 179 (1959).

82a. C. B. C. Boyce and J. S. Whitehurst, *J. Chem. Soc.,* 4547 (1960); (b) K. H. Baggaley, S. G. Brooks, J. Green, and B. T. Redman, *J. Chem. Soc.* (C) 2671 (1971).

83. (a) R. L. Augustine, *Catalytic Hydrogenation* (M. Dekker Inc., New York, 1965), p. 61; (b) see also G. Nominé, G. Amiard, and V. Torelli, *Bull. Soc. Chim. France,* 3664 (1968).

84. A. Horeau, L. Ménager, and H. Kagan, *Compt. Rend., 269C,* 602 (1969); *Bull. Soc. Chim. France,* 3571 (1971).

84a. G. D. Cooper and H. L. Finkbeiner, *J. Org. Chem., 27,* 1493 (1962).

85. I. N. Nazarov, S. N. Ananchenko, and I. V. Torgov, *Zh. Obshch. Khim., 26,* 819 (1956); I. N. Nazarov, G. P. Verkholetova, S. N. Ananchenko, I. V. Torgov, and G. V. Aleksandrova, *ibid., 26,* 1482 (1956).

86. I. N. Nazarov, S. N. Ananchenko, and I. V. Torgov, *Isv.*

Akad. Nauk. S.S.S.R., Ser Khim, 103 (1959).

87. (a) D. J. Crispin and J. S. Whitehurst, *Proc. Chem. Soc.,* 22 (1963); (b) D. J. Crispin, A. E. Vanstone, and J. S. Whitehurst, *J. Chem. Soc.* (C) 10 (1970).

88. Review: J. Weill-Raynal, *Bull. Soc. Chim. France,* 4561 (1969).

89. S. N. Ananchenko and I. V. Torgov, *Dokl. Akad. Nauk. S.S.S.R., 127,* 553 (1959).

90. S. N. Ananchenko, V. N. Leonov, A. V. Platonova, and I. V. Torgov, *Dokl. Akad. Nauk. S.S.S.R., 135,* 73 (1960); S. N. Ananchenko, V. Y. Limanov, V. N. Leonov, V. N. Rzheznikov, and I. V. Torgov, *Tetrahedron, 18,* 1355 (1962).

91. (a) S. N. Ananchenko and I. V. Torgov, *Tet. Letters,* 1553 (1963); (b) T. B. Windholz, J. H. Fried, and A. A. Patchett, *J. Org. Chem., 28,* 1092 (1963).

92. K. K. Koshoev, S. N. Ananchenko, and I. V. Torgov, *Khim. Prir. Soedin., 1,* 172 (1965).

93. (a) H. Gibian, K. Kieslich, H. J. Koch, H. Kosmol, C. Rufer, E. Schröder, and R. Vössing, *Tet. Letters,* 2321 (1966); (b) C. Rufer, E. Schröder, and H. Gibian, *Ann., 701,* 206 (1967).

94. C. H. Kuo, D. Taub, and N. L. Wendler, *J. Org. Chem., 33,* 3126 (1968).

95. (a) C. H. Kuo, D. Taub, and N. L. Wendler, *Angew. Chem., 77,* 1142 (1965); (b) D. P. Strike, T. Y. Jen, G. A. Hughes, G. H. Douglas, and H. Smith, *Steroids, 8,* 309 (1966); (c) A. V. Zakharychev, D. R. Lagidze, and S. N. Ananchenko, *Tet. Letters,* 803 (1967).

96. U. K. Pandit, F. A. van der Vlugt, and A. C. van Dalen, *Tet. Letters,* 3697 (1969).

97. R. Bucourt, L. Nédélec, J. C. Gasc, and J. Weill-Raynal, *Bull. Soc. Chim. France,* 561 (1967).

98. R. Bucourt, M. Vignau and J. Weill-Raynal, *Compt. Rend., 265C,* 834 (1967).

99. (a) L. Velluz, G. Nominé, and J. Mathieu, *Angew. Chem., 72,* 725 (1960); (b) L. Velluz, G. Nominé, J. Mathieu, E. Toromonoff, D. Bertin, M. Vignau, and J. Tessier, *Compt. Rend., 250,* 1510 (1960); (c) See also L. L. Chinn and H. L. Dryden, *J. Org. Chem., 26,* 3904 (1961).

100. L. Velluz, G. Nominé, G. Amiard, V. Torelli, and J. Cérède, *Compt. Rend., 257,* 3086 (1963).

101. (a) L. Velluz, G. Nominé, R. Bucourt, A. Pierdet, and P. Dufay, *Tet. Letters,* 127 (1961); (b) French Patent No. 1,305,992 (1962).

102. See, for example, S. Julia, *Bull. Soc. Chim. France,* 780 (1954).

103. Roussel-Uclaf, Belgian Patent No. 634,308 (1963).

104. N. K. Chaudhuri and P. C. Mukharji, *J. Indian Chem. Soc.*,
      33, 81 (1956).
105. (a) P. Bellet, G. Nominé, and J. Mathieu, *Compt. Rend.*,
      263C, 88 (1966); (b) U. Eder, G. Sauer, and R. Wiechert,
      Angew. Chem., *83*, 492 (1971).
106. L. Velluz, J. Mathieu and G. Nominé, *Tet. Suppl.*, *8*,
      495 (1966).
107. Review: J. Weill-Raynal, *Synthesis*, *2*, 49 (1969).
108. C. A. Henrick, E. Böhme, J. A. Edwards, and J. H. Fried,
      J. Am. Chem. Soc., *90*, 5926 (1968).
109. T. Hiraoka and I. Iwai, *Chem. Pharm. Bull. (Tokyo)*, *14*,
      262 (1966).
110. (a) P. Wieland and K. Miescher, *Helv. Chim. Acta*, *33*,
      2215 (1950); (b) W. Acklen, V. Prelog, and A. P. Prieto,
      Helv. Chim. Acta, *41*, 1416 (1958); (c) C. B. C. Boyce
      and J. S. Whitehurst, *J. Chem. Soc.*, 2022 (1959).
111. H. Smith, G. A. Hughes, and B. J. McLoughton, *Experi-
      entia*, *19*, 177 (1963).
112. (a) Z. G. Hajos, D. R. Parrish, and E. P. Oliveto, *Tet.
      Letters*, 6495 (1966); *Tetrahedron*, *24*, 2039 (1968); (b)
      Z. G. Hajos, R. A. Micheli, D. R. Parrish, and E. P.
      Oliveto, *J. Org. Chem.*, *32*, 3008 (1967); (c) O. I.
      Fedorova, C. S. Grinenko, and V. I. Maksimov, *Dokl. Akad.
      Nauk. S.S.S.R.*, *171*, 880 (1966).
113. (a) S. Danishefsky and B. H. Migdalof, *J. Am. Chem. Soc.*,
      91, 2806 (1969); (b) S. Danishefsky, L. S. Crawley,
      D. M. Solomon, and P. Heggs, *J. Am. Chem. Soc.*, *93*,
      2356 (1971).
114. G. Saucy, W. Koch, M. Müller, and A. Fürst, *Helv. Chim.
      Acta*, *53*, 964 (1970).
115. T. B. Windholz, J. H. Fried, H. Schwam, and A. A.
      Patchett, *J. Am. Chem. Soc.*, *85*, 1707 (1963).
116. A. J. Birch and G. S. R. Subba Rao, *Australian J. Chem.*,
      23, 547 (1970).
117. K. Hiraga, T. Asako, and T. Miki, *Chem. Comm.*, 1013
      (1969).
118. R. A. Dickinson, R. Kubela, G. A. MacAlpine, Z. Stojanac,
      and Z. Valenta, *Can. J. Chem.*, *50*, 2377 (1972).
119. M. Rosenberger, A. J. Duggan, and G. Saucy, *Helv. Chim.
      Acta*, *55*, 1313 (1972) and earlier papers.
120. W. S. Johnson, *Accts. Chem. Res.*, *1*, 1 (1968).
121. Work of P. A. Bartlett, and W. S. Johnson, described by
      the latter at the International Symposium on Organic
      Synthesis, Vancouver, August 1972.
122. G. S. Grinenko, E. V. Popova, and V. J. Maksimov,
      J. Org. Chem. USSR, *7*, 950 (1971).

COMPOUND INDEX

3β-Acetoxyandrost-5-ene-17 one, 666
2-Acetoxybutene-2, 692
Acetyl bromide, 710
0-Acetyldehydroisophotsantonic lactone, 413
Acetylene dicarboxylic acid, 353
4-Acetyl-1-ethoxycyclohexene, 365
1-Acetyl-4-isopropenyl-1-cyclopentene, 87
Acetylmethylenetriphenylphosphorane, 463
0-Acetyl photosantonic lactone, 419
Achillene, 12, 13
Achillenol, 13
Achillin, 412–417
Acrolein, 697
Actinidiolide, 170
Actinidol, 170
Adiantone, 592
α-Agarofuran, 300–303
β-Agarofuran, 301
Agarospiron (Epihinesol), 466–474
Agnosterol, 571
(±)-Alantolactone, 326–328
Aldosterone, 646
Alloocimene, 9
Alnusenone, 560, 620–626
Ambertriene, 569
Ambrein, 569, 595
Ambreinolide, 569, 595
1-Amino-naphthalene-6-sulfonic acid
    (Cleve's acid), 643
Amyrins, 594–610
β-Amyrin, 560
β-Amyrin acetate, 609
β-Amyrin-3-benzoate, 609
δ-Amyrin, 602
Androsta-4,7-diene-3,17-dione, 665
Aplysin, 453–459
Apoaromadendrene, 418
Arborescin, 412–417
Aristolene, 380–395
Aristolone, 380–395
Aromadendrene, 417–422
Artemisia alcohol, 42
Artemisia Ketone, 40
(±)-Artemisin, 324, 412, 414
Atractylon, 310–313

Batatic acid, 165
Benzyl α-bromopropionate, 505
3-Benzyloxybutyl bromide, 709
α-cis-Bergamotene, 446–449
β-cis-Bergamotene, 446–449
α-trans-Bergamotene, 520
Bicyclofarnesic acid, see Farnesic acid
Bicyclofarnesol, see Drimenol
Bilobanone, 264
    Bisabolenes, 233–238
(±)-Bisabolol, 241
cis and trans Bis-dehydromanianolic acid, 649
Bisnoronoceradione, 579
Boll Weevil pheromone, 58
Bornyl acetate, 150
Boschnia lactone, 77
α-Bourbonene, 530–534

β-Bourbonene, 530–534
trans-π-Bromocamphor, 481
2-Bromo-6-methoxynophthalene, 661
1-Bromo-3-pentanone ethylene ketal, 715
trans-1-Bromo-3-pentene, 374
π-Bromotricyclene, 481, 485
1-Bromo-1,2,2-trimethylcyclopentane, 454
Bulneool, 395–412
α-Bulnesene, 395–412
Butenolide, 312

ε-Cadinene, 330, 336
(±)-Cadinene dihydrochloride, 330–336
(+)-Calamenene, 331
Camphene, 151, 153
(±)-Camphene-1-carboxylic acid, 543
Campholenic aldehyde, 88
Camphonanic acid, 454
Camphor, 150–153
Carabrone, 459–463
trans-Carane, 157
(-)-trans-Caran-2-one, 391
4-Carbethoxycyclohexanone, 397
(±)-Δ2-Carene, 157
Carissone, 288, 289
7-epi-Carissone, 302, 303
Car-2-one, 157
Carvacrol, 104, 105
Carvacryl trifluoroacetate, 104
Carvenone, 114, 152
Carveols, 106, 109
Carveyl acetate, 110
Carvomenthols, 106
Carvomenthones, 106, 360, 464
Carvone, 112
Carvotanacetols, 106, 109
Carvotanacetone, 106, 111
Caryophyllene, 474–481
α-Caryophyllene alcohol, 540–545
Cedrene, 504–514
Ccdrol, 504–514
Chamaecynone, 305
Chamene, 59
Chamigrene, 449–453
β-Chloroethyl vinyl ketone, 717
4-Chloro-methyl-3,5-dimethylesoxazole, 3–7
Cholesterol, 671
Chretien-Bessiere monoterpene, 36, 39
Chrysanthemic acid, 37, 49, 50, 54–56
Chrysanthemic esters, 53
Chrysanthemol, 47
Chrysanthemum dicarboxylic acid, 54
Chrysanthenone, 156
1,4-Cineol, 133
1,8-Cineol, 134
1,4-Cineol-2,3-diol, 134
Citraconic anhydride, 672
Citral, 28
Citral epoxide, 21
Citronellal, 27
(+)-Citronellol, 15
(+)-α-Citronellol, 15
Clovene, 540–545

(±)-Conterifolin, 347, 348
Copaene, 520—525
Cosmene, 14
Costal, 299
Costic acid, 299
Costol, 299
(±)-Cryptomerion, 251
Cryptone, 97, 98
β-Cubebene, 380—395
Cubebol, 380—395
Culmorin, 528—530
Cuminaldehyde, 128
Cuparene, 453—459
β-Cuparenone, 453—459
α-Curcumene, 241—246
(−)-α-Curcumene, see Iso-α-curcumene
β-Curcumene, 241
γ-Curcumene, 242—243
(±)-Curcumone, 247—248
Cycloartenol, 572—574
β-Cyclocitral, 393
β-Cyclocitrylidineacetic acid, 353
Cyclocolorenone, 417—422
β-Cyclogeraniol, 393
β-Cyclohomogeranic acid, 171
β-Cyclolavandulal, 48
β-Cyclolavandulic acid, 48
Cyclolavandulols, 48
1,2-cyclononadiene, 480
1,2-cyclopentenophenanthrene, 658
Cyclosativene, 525—528
Cyperolone, 422—424
Cyperolone acetate, 424
(+)-α-Cyperone, 422
α-Cyperone, 285—288
β-Cyperone, 285, 286
7-epi-Cyperone, 269, 270, 284, 381

Debromoaplysin, 453—459
cis and trans Decalin-1,5-dione, 681
α-Decalone, 357, 679
Dehydroagnosterol, 571
14,15-Dehydroequilenin, 653
14,15-Dehydroequilenin methyl ether, 652, 654
8-Dehydroestrone methyl ether, 662, 697
Δ9(11)-dehydro-D-homo-estrone methyl ether, 695
Dehydrofukinone, 362—380
(±)-Dehydroilleiden M, 536
Dehydro-γ-ionone, 569
Dehydrojuvabione, 253, 261, 262
24,25-Dehydrolanosterol, 571
Dehydrolinalool, 28
1,2-Dehydronerlidol, 202
Dehydroparkeol, 571
20,21-dehydrosqualene, 561
dl-Δ12-Dehydrotetrahymanol, 590
Dendrolasin, 228—231
Desacetoxymatricarin, 412—417
(±)-Desoxotodomatuic acid, 253, 254
Desoxyvaleranone, 360
2-Diazo-propane, 384
1,3-Dichlorobutene-2, 709, 716

2,3-Dichloro-5,6-dicyanobenzoquinone, 646
Diethyl p-benzyloxyphenylmalonate, 512
Diethyl cyanomethylphosphonate, 386
Diethyl 4,4-ethylenedioxypentyl phosphonate,711
Diethyl isopropylidene-malonate, 363
Diethyl α-methyl-β-oxoadipate, 654
Diethylpropionylsuccinate, 692
Dihydroactinidiolide, 170
Dihydroagarofuran, 4-hydroxy, 302
Dihydroaristolone, 384
Dihydrocarveols, 106, 113
Dihydrocarvones, 106, 108, 113, 114, 353
Dihydrocostal, 300
Dihydrocostunolide, 277, 279
Dihydro-β-ionine, 452
5,6-Dihydronorcaryophillene, 481
24,25-Dihydro-Δ13(17)-protosterol, 571
Dihydrotagetone, 31—34
(±)-Dihydro-ar-turmerone, 248
3β,11α-Dihydroxycycloartane, 572
1,4-Dimethoxy-2-butanone, 359
γ,γ-Dimethylallyl mesitoate, 481
2,3-Dimethyl-6-n-butylthiomethylene-cyclohexanone, 371
2,3-Dimethylcyclohexanone, 384
3,3-Dimethylcyclopentene, 542
Dimethyl-γ-Ketopimelate, 365
3,7-Dimethylocta-1,3-diene-5-one, see Tagetone
2,6-Dimethylocta-1,3,5,7-tetraene, 13
cis-2,6-Dimethylocta-1,4,7-triene, see Achillene
2,6-Dimethylocta-2,4,7-triene, 11
3,7-Dimethyloct-6-en-1-ol, 15
3,7-Dimethyloct-7-en-1-ol, 15
Dimethylstyrene, 92—93
Dimethyl succinate, 653
2,5-Dimethyl-3-Vinylpent-4-en-2-ol, see Chrétien-Bessiére monoterpene
Dinoronocerane, 575
Diosgenin, 671
Dipentene, 8, 23
Diplopterol, 592
Diprenyl ether, 41
Dolichoidial, 78
Drimanic acid, 343
(±)-Drimenin, 343—347
(±)-Drimenol, 339—342, 347, 348
(±)-epi-Drimenol, 339—342

Elemane, 266, 267
β-Elemene, 269—272
Elemol, 269—271, 273, 274
Elsholtzione, 160
Epicamphor, 150
7-Epicyperone, 353
11-Epideoxygeigerin, 414
Epihinesol (Agarospirol), 466—474
4-Epinootkatone, 367
Epiizanoic acid, 514—516
Equilenin, 642—663
Equilenin ethylene ketal, 669
Equilenin methyl ether, 653, 660, 662, 668
Equilin, 664—670

Eremoligenol, 361–380
Eremophil-3,11-diene, 362–380
Eremophilene, 361–380
(±)Estra-4,9-diene 3, 17-dione, 716
Estradiol, 664
Estradiol-17β, 670
Estrapentaene, 701, 713
Estrial 16α,17β, 670
Estrone, 646, 662, 670–718
8α-Estrone, 683
8α,13α-Estrone (estrone-A), 675, 683
13α-Estrone (lumiestrone), 682
8α,9β-Estrone (estrone-e), 679
13α,14β-Estrone, 685
9β-Estrone (estrone-d), 679
14β-Estrone, 685
Estrone methyl ether, 696, 699
Estr-4,8(14),9-triene-3,17-dione, 717
Ethyl β-acetoxyacrylate, 462
Ethyl 1-acetylcyclopropanecarboxylate
    ethylene ketal, 536
Ethyl γ-bromocrotonate, 509
Ethyl dimethylacetoacetate ethylene ketal,
    534
Ethyl p-tolyacetate, 457
Eucarvone, 140, 141
α-Eudesmol, 295–296
β-Eudesmol, 289–294
γ-Eudesmol, 289, 295
Eugenol, 697
Euphol, 616
Evodone, 116, 123
Exo-3-methylnorbornanone, 491

trans, trans-Farnesaldehyde, 633
(±)-Farnesiferol A, 342–343
Farnesol, 200–206, 353
cis, trans-Farnesol, 435, 444
trans, trans-Farnesol, 435, 633, 634
Farnesyl bromide, 561, 562
Farnesyl nerolidyl sulphide, 562
Farnesyl phenyl thioether, 562
Farnesyl pyrophosphate, 633
Fern-8-ene, 593
Filifolone, 144, 145
Fukinone, 361–380
3-Furoic acid, 162
Furopelargone, 70, 72, 274–276
Furoventalene, 264

Geigerin, 412–417
Geigerin mesylate, 414
Genepin, 84
Genipic acid, 85
Genipinic acid, 85
Geraniol, 17, 23, 201, 484
Geranyl acetate, 440
Geranylacetic acid, 634
Geranylacetone, 201–205, 562
Geranyl bromide, 429
Geranyl chloride, 202
Germacrane, 277, 278
Germanicol, 604, 616–620

Glochidone, 632
Glyoxalyl chloride tosylhydrazone, 634

Helminthosporal, 463–466
trans-α-Himachalene, 427
β-Himachalene, 424–428
trans-γ-Himachalene, 427
Hinesol, 466–474
Hinokitiol, see Thujaplicins
D-Homoandrostane-3,17α-dione, 690
Homo-camphor, 496
cis-Homocaronic acid, 51
D-Homoequilenin, 662
D-Homoequilenin methyl ether, 662
D-Homoestrone methyl ether, 695, 699
Homosafranic acid, 171
Homoterpenyl methyl ketone, 420
Hop-21(22)-ene, 592
Hop-17(21)-enone, 590
Hop-21(22)-enone, 592
Hotrienol, 25
Hotrienol acetate, 25
Humulene, 280, 479
Hydroxyadientone, 592
19-Hydroxyandrost-4,6-diene-3,17-dione, 666
Hydrobuyenolide, 123
Hydroxycitronellal, 27
8-Hydroxy-P-Cymene, 103
Hydroxyhopane, 560
Hydroxyhopanone, 590, 592
4α-Hydroxyisochamaecynone, 306
11β hydroxylanostenyl acetate, 572
1-Hydroxymenth-2-ene, 90
2-Hydroxy-3-methylene-6-methylbenzofuran,
    123
Hymentherenes, 12
Hyposantonin, 308
Hyposantonous acid, 308

Illudin M, 534–540
Illudin S, 534
Illudol, 534
1-Iodo-6-methoxy-naphthalene, 646
Ionones, 5
Ipomeanarone, 231
Iresin, 348
Iridane, 59
Iridodial, 72, 74, 78
Iridomyrmecins, 74, 75
(±)-Isolantolactone, 329
Isobisabolene, 239–241
Isobutylene, 475
Isocaryophyllene, 474–481
Iso-α-curcumene, 241, 244–246
Isodihydrocarveol, 113
(±)-Isodrimenin, 344, 345, 347, 348
Isoegomaketone, 164
(±)-Isoequilenin, 648
14-Isoequilenin, 656
(±)-Isoequilenin methyl ether, 653, 660
9-Isoequilin. 669
Isoeuphenol, 571
γ-Isogeigerin acetate, 414

(±)-Isoiresin diacetate, 348–352
Isoiridomyrmecin, 74, 75
Isolavandulol, 43
Isolongifolene, 540–545
Isomenthol, 124, 125
(±)-Isomenthone, 389
Isonepetalactone, 71
Isonootkatone (a-Vetivone), 361–380
Isophotosantonic lactone, 412
Isopiperitenone epoxide, 137
3-Isopropenylcyclohexanone, 376
4-Isopropenylcyclohexanone, 366
(+)-4-Isopropylcyclohex-2-en-1-one, see
    Cryptone
4-Isopropylcyclohex-3-enone, 96–99
4-Isopropylidenecyclohexanone, 96, 97
Isopropylidenetriphenylphosphorane, 566
4-Isopropyl-6-methoxy-1-tetralone, 331–
    333
Isopulegone, 121
cis-Isopyrethric acid, 51
(±)-Isosirenin, 435
Isoterpinolene, 90
Isothujone, 147
(-)-Isotirucallol, 571
Isovaleraldehyde, 534
Isoxylitone, 118

Juvabione, 253, 255–261
Juvenile Hormone, 207–222

Karahana Ether, 139
Karahanaenone, 142
Kessane, 395–412
Ketohakonanol, 592
Khusimol, 514

(±)-Lanceol, 238, 239
Lanosterol, 560, 569, 571
Lavandulol, 34, 43–48
Lavandulyl acetate, 46
Lavandulyl bromide, 45
Lavandulylic acid, ethyl ester of, 45
Levulinic acid, 658
Limonene, 89
Linalool, 5, 17, 20, 21, 23
Linalool acetate, 22
Linalyl phosphate, 24
Lindestrene, 313–315
Lippione, 119
Loganin, 81
Loganin acetate, 82
Loliolide, 170
Longiborneol, 528
(±)-Longicamphenylol, 520
(±)-Longicamphenylone, 520
Longifolene, 528, 517–520
Lumisantonin, 467
Lupeol, 626–632
Lyratol, 36, 39

Maaliol, 380
Matatabiether, 80

p-Mentha, 1(7), 8-diene, 9
Mentha-1,3-dien-7al, 126, 128
Menthadienes, 24, 89
Mentha-1,3-dienoic acid, 126
trans-Mentha-1(7),8-dien-2-ol, 110
Mentha-1,8-dien-4-ol, 100, 101, 109
Mentha-1,8-dien-10-ol, 130
Mentha-4,8-dien-3-one, 122
Menthane-1,3-diol, 137
Menthan-8-ol, cis and trans, 93, 94
1,3,8-Menthatriene, 91
Menth-1-en-9-al, 129
Menth-1-en-4-ol, 100
Menth-1(7)-en-4-ol, 100
m-Menth-1-en-8-ol, 138
Menth-1-en-9-ol, 131
trans-Menth-2-en-1-ol, 94
Menth-2-en-8-ol, 103
Menth-3-en-1-ol, 96
Menth-4(8)-en-2-ol, 106
m-Menth-5-en-8-ol, 138
Menth-3-en-1,2-trans-8-triol, 132
Menthofuran, 122
Menthols, 115, 124
Menthones, 124, 125, 389
Methacrolein, 487
p-Methoxyacetophenone, 508
6-Methoxy-2-acetylnaphthalene, 656
m-Methoxyallylbenzene, 697
2-Methoxyborane, 154
m-Methoxycinnamic acid, 695
3-Methoxy-17β-hydroxyestra-1,3,5(10),8-
    tetraene, 666
3-Methoxyisoprene, 566
β(-6-Methoxy-naphthyl)-ethanol, 646
6-Methoxy-1-naphthylethyl iodide, 654
m-Methoxy-phenyl-acetylene, 681
m-Methoxyphenethyl bromide, 714
5-m-Methoxyphenylpentyne, 695, 713
3-m-Methoxyphenylpropyl-bromide, 695,
    696
3-m-Methoxyphenylpropyl magnesium
    bromide, 705
6-Methoxytetralin, 646
6-Methoxy-1-tetralone, 471, 581, 646, 687,
    701, 709
Methyl γ-bromocrotonate, 646
3-Methylbut-2-enol, 47
Methyl (+)-camphenecarboxylate, 516
3-Methylcyclohex-2-enone, 454
2 Methyl-cyclohexan-1,3-dione, 363, 616,
    690–717
2-Methylcyclopentane-1,3-dione, 661, 690–
    693
Methylcyclopent-1-en-4,5-dione, 671
trans-Methyl-1-decalyl p-toluenesulphonate,
    402
Methyl trans, trans, cis-10-epoxy-7-ethyl-3,
    11,-dimethyl-2,6-tridecadienoate, see
    Juvenile Hormone
Methyl-trans-trans-farnesate, 588
Methyl geranate, 22, 23
6-Methylhept-5-en-2-one, 4

cis 8-Methyl-hydrindan-1-one, 659
trans 8-Methylhydrindan-1-one, 659
2-Methyl-5-Isopropenylanisole, 103
1-Methyl-3-isopropylcyclopentene, 531
2-Methyl-3-methylenehept-5-ene, *see* Salvan
2-Methyl-6-methyleneocta-2,7-diene-4-ol, 26
2-Methyl-6-methylene-oct-7-en-4-ol, 26
7-Methyl-3-methyleneoct-6-enylacetate, 29
4-Methyl-1-naphthoic acid, 379
Methyl nerolate, 22, 23
Methyl 5-oxo-6-heptenoate, 710
3-Methylpent-4-enal, 31
trans-Monocyclofarnesic acid, 453
Mullilam Diol, *see* 1,4-Cineol-2,3-diol
Multifluorenol, 594, 606−609
є-Muurolene, 335
Myrcene, 8, 10, 11
Myrtenal, 126, 127, 156
Myrtenol, 126, 127

Naginate ketone, 160
β-Naphthol, 646
γ-Naphthybutyric acid, 646
Neodihydrocarveol, 113
Neoisodihydrocarveol, 113
Neoisomenthol, 124, 125
Neomatatabiols, 78
Neomenthol, 124, 125
Neonepetalactone, 73
Neotorreyol, 228
Nepetalactone, 69, 71, 72
Nepetalic acid, 70, 72
(±)-Nepetalinic acids, 76
cis, trans-Nepetonic acid, 71
trans, cis-Nepetonic acid, 70
Neral, *see* Citral *b*
Nerol, 17, 23
Nerolidol, 200−206
Neryl didphenyl phosphate, 24
Nezukone, 143
Nickel tetracarbonyl, 561
Nootkatin, 276
Nootkatone, 361−380
Nopinic acid, 126, 128
Norbornanone, 482
Norcedrenedicarboxylic acid, 504, 507
"X"-Norequilenin methyl ether, 658
Nor-γ-Eusdesmol, 292−294
Nor-Keto-agarofuran, 303
1-Norsqualene, 561
10-Norsqualene, 561
10′-Norsqualene, 561
15-Norsqualene, 561
10,15-bis-Norsqualene *(cis and trans)*, 561
(+) 19-Nor steroid-4-ene-3-ones, 711
(±) 19-Nor steroid-4-ene-3-ones, 711
19-Nortestosterone acetate, 664
Norvaleranone, 356
Nuciferal, 250, 251
β-Nymenthrene, 12

Occidentalol, 309
Occidol, 282−284, 308

Ocimene, *cis* and *trans*, 8, 10
Olean-11, 12, 13, 18-diene, 595
Olean-12-ene, 602
18-Olean-12-ene, 598
(-)-18-Olean-12-ene, 599
18*a*-Olean-12-ene, 602
Olean-13(18)-ene, 594, 595, 598, 599, 602
Olean-12-ene-3-one, 604
Oleanolic acid, 609
Oleanolic acid acetate, 610
Oleuropic acid, 132, 135
Onoceradienes, 577−579
a-Onoceradiene, 579
β-Onoceradiene, 579
γ-Onocerene, 579
Onocerins, 574−594
a-Onocerin, 560, 574
β-Onocerin, 574
γ-Onocerin, 574
(+)-β-Onocerin diacetate, 589
γ-Onocerin diacetate, 589, 590, 593
a-Onocerindienedione, 590
Osmane, 59
11-Oxocycloartenyl benzoate, 573
11-Oxolanostenol, 572
1-Oxo-7-methoxy-1,2,3,4-tetra-hydro-
    phenanthrene, 642
1-Oxo-2-methyl-7-methoxy-1,2,3,4-tetra-
    hydrophenanthrene, 653
1-Oxo-1,2,3,4-tetrahydrophenanthrene, 646

Parkeol, 571
a-Patchoulene, 492−499
β-Patchoulene, 492−499
γ-Patchoulene, 492
Patchouli alcohol, 492−499
(±)-epi-Patchouli alcohol, 499
Patchouli alcohol acetate, 493
trans-3-Pentene-2-one, 363, 365, 368
Peracetic acid, 566
Perezone, 262, 263
Perilla alcohol, 125−127
    acetate of, 126, 127
Perilla aldehyde, 125−127, 417
    oxime of, 126
Perilla Ketone, 163
Perillene, 163
Perillic acid, 126, 128
Phellandral, 125, 126, 128
a-Phellandrene, 89, 90
β-Phellandrene, 89, 90
7-Phenyl-4,7-dioxoheptanoic acid, 658
12-Phenylthiosqualene, 562
Photocitrals *A* and *B*, 60
Picric acid, 568
trans-Pinecarveol, 156
Pinocampheols, 156
Pinocamphones, 156
Pinol, 131, 132
Pinonic acid, 51
Piperitenone, 107, 118
Piperitol, *cis* and *trans*, 119
Piperitone, 11

Plinols, 60, 61
Prenyl alcohol, *see* 3-Methylbut-2-enol
Presqualene alcohol, 633–635
Procerin, 276
Propargaldehyde dimethyl acetal, 477
Pseudoionone, 6
Pulegene, 145
Pulegenic acid, 72
Pulegone, 120, 121
Pyrethric acid, 49, 57, 58
Pyrethrum, 49
Pyrocine, 58

Rhodinal, 15
Rhodinol, 15
Rose furan, 161
Rose Oxides, *cis* and *trans,* 168

Sabinaketone, 146, 147
Sabinene, 146–148
Sabinene hydrates, *cis* and *trans,* 147, 148
cis-Sabinol, 147
Safranal, 138
Salvan, 147, 148
α-Santalene, 481–491
β-Santalene, 481–491
(±)-epi-β-Santalene, 482, 485
α-Santalol, 481–491
β-Santalol, 481–491
Santolinatriene, 34, 36, 37
Santonin, 410, 412
(+)-Santonins, A,B,C,D, 316–322
(±)-α-Santonin, 322–324, 413
(±)-β-Santonin, 322
Santonic acid, 517
Sativene, 525–528
Saussaurea Lactone, 267, 269
Sclareol, 579, 595, 599
α-Selinene, 298
β-Selinene, 296–298
Epi-γ-Selinene, 304
Serrantenediol, 589–590
Sesquicarene, 428–446
Seychellene, 499–503
Shonanic acid, 140, 141
α-Sinesal, 222, 227
β-Sinesal, 222–227
Sirenin, 428–446
Sobrerol, 131, 132
Squalene, 560
trans-Squalene, 561
Squalene-2,3-diol, 566
Stigmasterol, 671
Strychnine, 582
Succinaldehyde, 566
Sylvestrene dihydrochloride, 138
Sylveterpineol, 138

Tagetone, *cis* and *trans,* 30–34
Taraxerol, 594, 605
(±)-Telekin, 328
cis-1, 4-Terpin, 133
β-Terpinene, 89, 90

γ-Terpinene, 88
α-Terpineol, 23, 102, 420
β-Terpineol, *cis* and *trans,* 94, 96
γ-Terpineol, 96, 97
δ-Terpineol, 101, 102
Terpin -1-en-ol, 102
Terpinolene, 23
γ-Terpinyl acetate, 97
Tetrahydroelemol, 269, 270
Tetrahydroeremophilone, 362–380
Tetrahydrosaussaurea Lactone, 267, 268
Tetrahymanol, 590–594
β-Tetralone, 582
Thujane, 145, 146
α-Thujaplicin, 143, 144
2-Thujene, 147
3-Thujene, 147
Thujone, 147
Thujopsene, 380–395
Thujyl alcohol, 147
Thujyl toluenesulfonate, 60
Thymol, 117
Todomatuic acid, 255–258
Torreyal, 228
Tricycloekasantalal, 486, 489
Triethyl phosphonoacetate, 386, 453
2,2,5-Trimethylcyclohept-4-enone, 142
2,2,3-Trimethylcyclopent-3-ene acetaldehyde,
  see Camphonelic aldehyde
2,6,6-Trimethyl-2-Vinyltetrahydropyran, 168
(±)-ar-Turmerone, 247–250
Umbelliferone, Sodium salt, 343
cis, trans-Umbelliprenin, 342
Umbellulone, 147
Uroterpinol, 129
  glycoside of, 129

Valencene, 361–380
Valeranone, 353–361
Valerianol, 361–380
Veratramine, 642
Verbenalin, 85
Verbenalol, 85–87
Verbenol, *cis* and *trans,* 156
Verbenone, 156
(±)-Veticadinene, 338
(±)-Veticadinol, 336, 337
β-Vetivone, 466–474
Vitamin A, 5
1-Vinyl-6-methoscy-3,4-dihydronaphthalene,
  671
1-Vinyl-2,6,6-trimethylcyclohexane, 353

Widdrol, 424–428
(±)-Winterin, 352, 353

2,6-Xyloquinone, 672
Ylangene, 520–525
Yomogi alcohol, 42, 43

Zingiberene, 246–247
Zizaene, 514
Zizanoic acid, 514

REACTION INDEX

Acetals, acid elimination of acetic acid, 413
  claisen rearrangement, 393
  elimination of ethanol, 393
  oxidation, 477
  reaction with vinylethers, 28, 29, 52
Acetates, elimination of acetic acid, 409
  hydrolysis of, 416, 424, 568
  with LAH, 496
  pyrolysis, 48, 96, 101, 109, 122, 130,
    485, 490
  reduction with lithium-ethylamine, 474
  transesterification, 440
Acetonylation of 1,2-diols, 86, 87
Acetoxy dienones, retroaldolization, 666
Acetoxy epoxides, thermal rearrangement,
  502
α-Acetoxyketones, pyrolysis of, 122
Acetylation, 536
  selective of, diols, 400, 423, 473, 496
    triols, 536
α-Acetyl-γ-butyrolactone, 21
  hydrochlorination of, 22
Acetylenes, 205, 306
  alkylation, 206, 210
  carboxylation, 445
  condensation with acetone, 5
  cyclization, 713
  formylation, 205
  homogenization, 689
  hydration, 695
  hydroboration, 449, 489, 491
  hydrogenation, 30, 687
  iodination, 206, 440
  lithium salts, reaction with, bromides, 489
    iodides, 449
    ketones, 477
    paraformaldehyde, 445
  reaction with Raney Nickel, 431
  reduction of, 5, 30, 39
Acetylenic alcohol, acetate, pyrolysis of, 29
Acetylenic ethers, *33*
  claisen rearrangement, 33, 34
  reaction with Acid, 44
Acetylide, potassium, addition to ketones,
  135
Acetyls, *epi*merization of, 499
Achilla Filipendulina, 12
Acid catalysed rearrangement, of caryophyllene, 543
  of cyclohexanols, 542
  of diacetylenic diols, 587
  of dienes, 593
  of humulene, 540
  of hydroperoxides, 606
  of longifolene, 528
  of 18α-olean-12-ene, 602
  of (+)-β-onocerin diacetate, 589
  of (+)-γ-onocerin diacetate, 590
  of (+)-α-onocerin dienedione, 590
  of spiro[4,5] decalenes, 514
  of spiro-decane diols, 511
  of trienes, 595-596

  of tricyclic diols, 474
Acid chlorides, *57*
  cyclizations of, 386
  in dehydrations, 648
  reaction with, diazomethane, 386
    olefins, 41
    organocadmium compounds, 32, 164
Acrylonitrile, addition to substituted ethyl-
  acetoacetate, 99
Actinidia polygama, 63, 64, 65, 67, 68, 78,
  80, 170
Acylation, Friedel-Crafts, 658
  intramolecular, 530, 692
Acyloin condensation, 277, 278, 678, 686
Acyloins, reduction of, 383
α-Agarofuran, *300-303*
  double bond isomer of, *301*
  photochemical irradiation of, 302
Alcohols, bromination, 599
  dehydration, 265, 267, 287, 414, 619,
    625
  hydrogenolysis, 356
  inversion of configuration, 414
  Moffatt oxidation, 365, 516
  Oppenauer oxidation, 665, 667, 689
  oxidation, 374, 400, 442, 496, 586, 593,
    600, 614, 616, 619, 631, 634, 695
  promoting effect in reductive alkylations,
    631
  protection as and regeneration from, ace-
    tate esters, 404, 410, 414, 471, 496,
    582, 593, 616, 619
    benzoate esters, 298, 413, 431, 629,
    709, 710
    benzyl ethers, 404, 406, 514, 588, 589
    brosylate esters, 595
    esters, pyrolysis of, 151
    mesylate esters, 359, 577, 629, 667
    phenoxy ethers, 398
    silyl ethers, 568
    tetrahydropyranyl ethers, 407, 440,
    503, 509, 714
    tosylate esters, 595, 632
  removal of, 495
  replacement with bromine, 487
  replacement with chlorine, 221
  selective acetylation, 324, 325, 400
    dehydration, 667
    oxidation, 525
  with hydrobromic acid, 420
  with mercuric oxide, 446
  with phosphorous tribromide, 363
p-Alcohols, oxidation of, 431
  protection of, 221, 270, 273
  replacement with bromine, 499
  tosylation, 503
s-Alcohols, dehydration, 410, 414
  oxidation, 359, 404, 423, 424, 463, 468,
    469, 471, 503, 507, 525
  protection, 313
t-Alcohols, dehydration, 13, 22, 27, 100,
    353, 360, 372, 374, 378, 387. 392,

733

411, 440, 451, 458, 459, 471, 483,
    491, 496, 503, 509, 525, 527, 530,
    595, 632
protection as phenyl urethane, 90
Aldehydes, 27
    alkylation, 566, 616
    conversion to nitriles, 160, 595
    condensation with, acetone, 516
        methylacetoacetate, 367
        triethylphosphonoacetate, 442
    decarbonylation, 361
    dehomologation of, 55
    epimerization, 464, 634
    formylation, 393
    from pyrolysis of homoallyl alcohol, 27
    Jones oxidation, 409
    ketalization, 75
    oxidation with, oxygen, 55
        silver oxide, 74, 372, 431
    reduction, 85, 112, 125, 163, 365, 442,
        629
    Wittig reaction, 11, 43, 440, 463, 491,
        634
δ-Aldehydoacid, cyclization of, 70, 71, 73
Aldehydoesters, lactone from, 57
Aldehydoketal, Wittig reaction on, 32
Alder-Stein rule, 260
Aldols, dehydration of, 507
Aldol cyclizations, 372, 376, 377, 420, 507,
    695, 709
Alkyl aluminium-hydrogen cyanide reagents,
    615
Alylation of, aldehydes, 616
    allyl alcohols, 206
    angular position in ketones, 357, 379
    benzylic positions, 458
    benzylidene blocked ketones, 682
    blocked ketones, 683
    carbethoxymethylenetriphenylphospho-
        rane, 389
    conjugate addition, 376
    cycloheptanes, stereospecificity in, 398
    cyclohexals, 614
    cyclohexanones, 471
    cyclohexylimines, 532
    cyclopropyl, as blocking group in, 625
    α-decalones, 659
    dienamines, 709
    dienolate anions, 569, 611
    dienones, 319
    diethyl p-benzyloxyphenylnalonate, 512
    diethyl malonate, 363
    β-diketones, 698, 702
    enamines, 534
    enol acetates, 619
    enolate anions, 611
    enolic β-Keto esters, 477
    enol lactones, 372
    enones, 625, 715, stereospecificity in, 502
    esters, 457, 632, 685
    Fridel-Crafts reaction, 454
    furfurylidene blocked ketones, 454, 683,
        689
    gem diesters, 158
    geminal triketones, 625

internal during solvolysis, 399
internal Michael, 517
isoxazoles, 377, 652
ketones, 141, 374, 377, 520, 530, 624,
    625, 648, 690
β-keto esters, 208, 369, 397, 538
lithio salt of 4-bromo-3-methyl anisole,
    457
(+)-nopinone, 446
norbornanone, 483
olefins, 206, 210
reductive, 612
reductive, ketones, 616
sodio ethylmalonate, 442
tetrahydroeucarvone, 529
α,β-unsaturated ketones, 211, 326, 575,
    599, 621
using Triton B, 698, 699
via butylthiomethylene derivative, 82
Alkylations with, allyl bromides, 491, 698
    benzyl α-bromopropionate, 505
    3-benzyloxybutyl bromide, 709
    bis-vinyl carbinols, 699
    1-bromo-2-butanone, 538
    1-bromo-3-pentanone, 420
    trans 1-bromo-3-pentene, 374
    1-bromo-1,2,2-trimethyl cyclopentane,
        454
    1-chloro-methyl-3,5-dimethylisoxazole,
        377
    1,3-dichlorobut-2-ene, 709, 716
    2,3 dichloropropene, 457
    dimethylsulfoxide as solvent, 715
    ethyl α-bromopropionate, 371, 529
    isopropyl bromide, 532
    lithium dimethyl copper, 366, 502
    methallyl chloride, 371
    6-methoxy-1-napthylethyl iodide, 654
    methyl iodide, 427, 582
    4-methyl-3-pentenyl chloride, 483
    methyl vinyl ketone, 384, 514, 534, 709
    propargyl bromide, 613
    propargyltetrahydropyranyl ether, 429
    2-propenyllithium, 566
    Vilsmeiers reagent, 406
    hydroxymethylenes, 442
    β-keto esters, 369
Alkynes, see Acetylenes
Allene acetate, 29
    alkaline hydrolysis of, 29
Allenone, 33
    isomerization with base, 33, 34
Allyl acetates, 22
    elimination of acetic acid, 471
    homologation of, 220
    hydrogenation of, 503
    hydrolysis of, 130
    pyrolysis of, 10
    reaction with lithium dimethyl copper,
        220
    reduction of, 40, 348
Allyl alcohol, 12, 21, 42, 43, 48, 128, 163,
    207, 221, 258, 327, 344, 499, 561,
    666
    acetonylation, 203

acetylation, 201
alkylation, 206
carbene addition, 147
cleavage with osmic acid, periodate, 59
conversion to, acetate, 22
  bromide, 208, 210, 224, 281, 589
  chloride, 201, 221, 301
  α,β-unsaturated ester, 210, 211, 221
cyclization during oxidation, 328, 329, 347
dehydration with alkali, 10
exchange reaction with ketone, 661
hydrogenation, 327, 499, 628, 667
hydrogenolysis, 431, 604
isomerization, 12, 128, 264
methylenation, 614
p-nitrobenzoate, pyrolysis of, 12
oxidation, 12, 31, 79, 81, 107, 109, 127, 128, 147, 150, 156, 210, 287, 299, 328, 329, 334, 347, 435, 446, 577, 619
phenyl urethane derivative, pyrolysis of, 90
rearrangement, during acetylation, 201
  during thermal dehydration, 46
  to allylic chloride, 499, 532
reaction with, diketene, 204
  enamine, 39, 40
  hydrobromic acid, 589
  phosphorous trilbromide, 389
  sulphur trioxide-pyridine, 431
  thionyl chloride, 532
  vinyl ether, 6, 206
reduction, 14, 147, 431
Simmons-Smith reaction, 392
Allyl bromides, addition to β-ketoesters, 45
  in alkylations, 698
  base treatment, 487
  oxidation to α,β-unsaturated aldehyde, 28
Allyl chlorides, 499
  reduction, 532
  solvolysis, 499
Allyl dienol ether, thermal rearrangement, 628
Allyl halides, coupling with phosphorous ylids, 562
Allylic oxidation, 381, 414, 417, 469
Allylic sulfonium ylids, rearrangement of, 41
  sigmatropic rearrangement of, 562
Allyl iodides, conversion to allyl alcohol, epoxidation of, 21
Allyl magnesium bromide, in Grignard reaction, 391
Allyl sulfones, reaction with, α,β-unsaturated ester, 54
Allyl vinyl ether, thermal rearrangement of, 6, 7, 32, 36, 142, 163, 206, 222, 228, 238, 252, 272, 277, 661
Allylic carbonium ions, attack by peroxides, 606
Ambergris, 569
Amides, 36
  from ketones, 151
  γ-hydroxy, cyclization of, 39

reduction of, 36, 39, 40, 361
Amines, oxidation of, 39, 40, 419
Amine salts, hydrogenolysis of, 406
Amyrin group, 594
Angular methylation, directional effects of double bonds, 687
  in estrone synthesis, 679
Anhydrides, 56
  β-diketones from, 345
  Grignard reaction on, 56
  reaction with dimethylcadmium, 345
Anisomropha buprestoides, 65
Annelations, 581, 616
  via enol-lactones, 711
Anthonomus grandis Boheman, 58
Arens' method, for removal of trimethyl-silyl group, 209, 210
Arndt-Eistert reaction, 507, 649, 654, 685
Aromatization, 668, 709, 710
Artemisia alcohol, 41, 42
  acetate of, 41
Artemisia filifolia, 144
Artemisyl skeleton, 40
  conversion to santolinyl skeleton, 37
  from trans-chrysanthemic acid, 34
Arylmethyl ethers, cleavage of, 459
Ascaridole, 132, 133
  reduction of, 133, 134
Aspidosperma in dole alkaloids, 81

Baeyer Villiger oxidation, 308
Bamford-Stevens reaction, 376, 433
Barton reaction, 604
Base catalysed rearrangement, of santonin, 517
Beckman reaction, on oxime of bicyclo[3,1, 0]-hexan-2-one, 50
Benzaldehyde, in the protection of 1,2-diols as acetal, 38
Benzaldehyde acetals, reaction wtih n-butyl lithium, 38
Benzoates, elimination, 431
  pyrolysis, 414
Benzofuran system, 264
  Grignard reaction, 123
  reduction in, 265
Benzyl alcohols, dehydration, 263
  hydrogenolysis, 619, 621
  with methyl lithium and titanium trichlo-ride, 561
Benzyl ethers, hydrogenolysis of, 55, 507
Benzylic alkylation, 458
Benzylidenes, 540, 696
  blocking group in ketone alkylation, 682
  intermediate in ring contraction, 696
  ozonolysis, 69, 540, 682
  reduction, 540
Betaines, 490
  with n-butyl lithium, 490
  with paraformaldehyde, 490
Bicyclic enones, 442
Bicyclization, of diazoketones, 431
Bicyclo [4.3.1] decanes, rearrangement of hydroazulenes, 396
Bicyclo [3.1.0]-hexan-2-one, oxime,

Beckman rearrangement, 50
Bicyclo [3.3.1] nonene, oxidative cleavage
    of, 464
Biogenesis, α-cedrene, 507
    cedrol, 507
    squalene, 633
Biogenetic rearrangement, patchouli alcohol,
    499
Birch reduction, 379, 593, 599, 604, 605,
    612, 620, 625, 628, 668
    allyl alcohol, 147
    aromatic rings, 582
    carissone, 295, 296
    o-cresol, methyl ether, 251
    α,β-cyclopropanoketones, 614
    α-cyperone, 298
    3,5-dimethoxytoluene, 210
    6-methoxytetralone, 311
    ρ-methoxytoluene, 210
    phenolic ethers, 664
    phenols, 669
    protection of aromatic ring during, 625
    sabinol, 147
    p-substituted anisole, 255
    thymol methyl ether, 120
Bis vinyl carbinals, in alkylation of β dike-
    tones, 699
Boric acid, in the dehydration of t-alcohol,
    13
Bornylene, from α-pinene hydration, 103
Boron trifluoride, catalysed rearrangements,
    452
    cleavage of epoxides, 365
    cyclization of, diketones, 464
        epoxydienes, 588
    elimination of acetic acid, 469
    in the addition of vinylether to acetal, 28,
        29, 52
    rearrangement of, enol acetates, 511
        α-onocerindienedione, 590
    with spiro epoxides, 616
Boschniaka rossica, 64
Boschnialactone, 74, 77
    stereochemistry of, 76
Bouveault-Blanc reduction, 646
Bredt's rule, 709
Bromides, alkylation of, p-bromo toluene,
    454
    lithium salts of acetylenes, 429
    elimination of hydrogen bromide, 384
    internal cyclizations, 509, 516
    with lithio-1-trimethylsilylpropyne, 445
    with lithium acetylides, 489
    with phenyl lithium, 457
Bromination, alcohols, 363, 499
    allyl alcohols, 389, 440
    allylic, 25, 141
    with N-bromosuccimide, 25, 142, 480,
        568
    cyano ketones, 604
    facilitation by oxalylation, 509
    ketones, 98, 146, 384, 616, 620
    olean-12-ene-1-one, 604
    sodio oxalylates, 509
    terminal olefins, 381, 417, 491

Bromoacetals, in condensation reactions,
    123
N-Bromoacetamide, bromination of ketones
    with, 98
p-Bromobenzenesulphonates, with dimethyl-
    amine, 419
α-Bromoketones, 98
    dehydrobromination of, 58, 98
γ-Bromoketones, in the synthesis of cyclo-
    propanes, 158
ε-Bromoketones, internal reductive alkyla-
    tion of, 499
Bromomethylbutadiene, coupling with thio-
    ketal anion, 26
N-Bromosuccinimide, allylic bromination,
    25
    of α,β-unsaturated acid, 213
    of α,β-unsaturated ketone, 141, 142
    aromatizations, 668
    cyclization of, dienol, 142
        methyl farnesate, 342
    formation of bromohydrin, 568
    with humulene, 480
    oxidation of, β-cyclocitral, 138
        linalool, 142
        linalyl acetate, 25
    selective reaction with olefin, 340
π-Bromotricyclene, with lithium acetylides,
    489
Building Block Principle, in synthesis, 11,
    112
η-Butyl lithium, reaction with, allylether, 42
    benzaldehyde acetals, 38
η-Butyl thioesters, with Raney nickel, 463
η-Butylthiomethylenes, as blocking group,
    82, 357, 385
    borohydride reduction, 468
    cleavage, 83, 385
    hydrolysis, 357

Camphene, from α-pinene hydration, 103
Camphor, 149, 150, 152, 153
    photolysis of, 88
    reaction with sulfuric acid, 114
Camphor oil, 61
Carbanionic reactants, aldol condensations,
    595
    lithium ethoxyacetylide, 584
    methyl lithium, 595, 599, 619
    stabilization of, 562
Carbenes, addition to, allyl alcohol, 147
    farnesol, 633
    pulegene, 145
    reduction product of p-isopropylanisole,
        143
    intramolecular insertion, 533
    orientation of addition, 633
    stereochemistry of addition, 462
Carbomethoxylation, 648
Carbonate esters, pyrolysis of, 411
Carbonyl compounds, reduction of, 616
Carboxylic acids, Arndt-Eistert homologation,
    649, 654
    converstion to, acid chloride, 57
    diazoketone, 50, 51, 390, 431

decarboxylation, 363, 652, 653
degradation of, 449, 454
esterification with, diazomethane, 55, 56, 74, 79, 87, 135, 379
    methanol, 84
Grignard reaction on, 94
magnesium iodide salts of, 646
pyrolysis of lead salts of, 699
reaction with, η-butyl thiol, 463
    diethylcarbonate, 431
    methyl lithium, 543
thermal decomposition of, 353
Δ³ Carene, 54, 157
    acid, catalysed ring opening in, 138
    epoxide, 54
    ozonolysis, 55
Carissone, 288, 289
    Birch reduction, 296
Carrol reaction, 5
Carvacrol, 104, 105, 106, 141
    methyl ether, 104, 106
    oxidation of methyl ether, 103
    synthesis from p-cymene, 104
    trifluroacetate, 104
Carvenone, 106, 114, 286
    from dihydrocarvone, 152
Carvomenthone, 266
    alkylation, 267
    formylation, 267
Carvone, 105, 107, 109, 112
    conversion to eucarvone, 141
    epoxidation, 107
    interconversion, 107
    oxidation to carvacrol, 105
    reduction products, 105, 113
    selective oxidation, 263, 264
Cationic rearrangement, of endesmanoid
        precursors, 361
Catnip oil, 62, 70
Cecropia moth, 207
Chamaecyparis obtusa (Hinoki), 94, 103
Chinese star anise oil, 136
Chlorination, with sulfuryl chloride, 458
γ-Chloroesters, cyclopropyl esters from, 53
α-Chloroketones, with Grignard reagents,
        564
β-Chloroketones, in the condensation reac-
        tions, 119
γ-Chloroketone, cyclopropane synthesis
        from, 51
    Grignard reaction on, 22
2-Chloro-2-methylbut-3-ene, 7
    isomerization of, 7
1-Chloro-4-methylpent-3-ene, 22
    in the synthesis of linalool, 22
Chromatography of, diols, 568
    presqualene alcohol, 634
Chromous chloride, in lactone cleavage, 414
    on 0-acetyl photosantonic acid, 419
Chrysanthemic acid, 49, 50, 54, 55
    conversion to pyrethric acid, 57
    cyclopropane ring fission in, 35
    esters of, 54
    formation by biogenetic pathway, 36
    irradiation of reduction products, 47

cis-trans isomerization in the synthesis, 51, 56
C¹⁴-labelled methyl ester, 58
    nitrile, 50
    pyrolysis product, 58
Chrysanthemol, 47
    irradiation of, 48
Chrysanthemum cinerariifolium, 49
Chrysopidae, 62
1,4-Cineole, 132-134
1,8-Cineole, 132-134
Circuitous syntheses of ertrone, 718
Citral a and b, 27
    conversion to geraniol, 201
    cyclization to dimethyl styrene, 42
    diepoxide, 165, 166
    epoxidation, 21
    photocyclization, 60, 275
    thermal isomerization, 27
Citronellal α and β, 27
    conversion to, curcumenes, 246
        dolichodiol, 78
        isoiridomyrmecin, 75
        zingiberene, 246
    hydration of, 26
Citronellic acid, cyclization to Pulegone,
        116
Citronellol, 15
    cyclization, 117
    monoacetate, 168
    photoxidation, 169
    triacetate, pyrolysis, 168
Citrus junes oil, 100
Citrus medica L., bar. acida, 133
Citrus natsudaidi, 131
Citrus unishu, 131
Claisen rearrangement, 5, 31, 34, 36, 44,
        112, 163, 206, 222, 228, 238, 251,
        271, 393, 565, 566, 586, 629, 661
    Eschenmoser's variation, 36, 361
Condensation with, bromoacetals, 123
    β-chloroketones, 119
    ethyl cyano acetate, 420
    Mannich bases, 114
    sodium acetylide, 695
Condensing agents, sodium acetylide, 695
Conjugate addition, to enones, 353
Cope elimination, on N-oxides, 39, 40, 419,
        420
Cope rearrangement, 36, 40, 44, 163, 222,
        228, 265, 274, 702
Copper powder, in the cyclization of diazo-
        ketones, 50, 148, 473
Corynanthe indold alkaloids, 81
Conjnebacterium simplex, 665
Cosmos bipinnatus, Cav, 13
Coupling reaction, allylic bromides, 589
Cram's rule, 564
(+)-Cryptone, 90, 97, 98
    as precursor of, β-phellandrene, 90
        cis and trans-β-terpineols, 94
    resolution, 100
    semicarbazone, 97
(−)-Cryptone, Diels-Alder reaction on, 335
Cryptophenol see Thymol

Cuminaldehyde, 125, 128, 129
Cumin seeds oil, 128
Cupric acetate, in cyclizations of terminal
    olefins, 528
Cuprous iodide, in decomposition of diazo
    compounds, 435
Cyanoesters, selective reduction of, 78
Cyanohydrins, dehydration of, 171
    from α,β-unsaturated aldehyde, 171
Cyanoketones, Stobbe condensation, 679
Cyclic ethers, fragmentation with sodium,
    449
    from 1.4-diols, 134
    oxidation, 446
Cyclizations, 678, 686, 712
    of acetylenes, 713
    of acid chlorides, 685
    of δ-aldehydo acids, 70, 71, 73
    acid catalysed, 397, 595, 625, 705, 710
    aldol condensations, 709
    asymetric, 710
    base catalysed, 627, 705
    with boron trifluoride, 588
    on cleavage of aryl ethers, 359
    of diacids, 56, 97
    of 1,5-dialdehyde, 75
    of diazoketones, 50, 51, 148, 158, 386,
        389
    Dieckman, 649
    Diels-Alder, 451
    of 1,5-dienes, 49
    of 1,6-dienes, by pyrolysis, 61
    of dienoic acid, 144
    of dienol with NBS, 142
    of diesters, 530
    of diketones, 69, 114, 146, 464, 507, 516,
        534, 536, 540
    of diols, 579
    of dioxocarboxylic acids, 658, 660
    of epoxides, 571, 604
        abiological, 590
    of epoxydienes, biogenetic like, 317
    of epoxydiols, with picric acid, 568
    Friedel-Crafts, 646
    of hydrazones, 446
    of δ-hydroxyacids, 76
    of γ-hydroxyamides, 39
    inhibition of, by olefins, 709
    internal Michael reaction, 520
    intramolecular, 397
    of ketoacids, 55, 99
    of ketoaldehydes, 55, 74, 87, 113, 536
    of δ-ketoesters, 507
    of ketones, 662
    of β-ketosulfoxides, 534
    of lactone-esters, 656
    of cis and trans monocyclofarnesols, 453
    of olefins, biomimetic, 712, 713
        terminal, 528
    of β-onocerin, 574
    of phenolic bromoesters, 509
    with phosphorous pentoxide, 698
    of spiro-[4,5]-decanes, 508
    with sodium t-amylate in benzene, 709
    solvolytic, 629

    of p-substituted phenols, 514
    of tosylates, 428, 503
    of triene carboxylic acids, 577
    with triethyl ammonium benzoate, 695
    of unsaturated acids, 496, 543
π-Cyclization, 353
    stereochemistry of, 374
Cycloaddition, of dimethylketene to meth-
    ylcyclopentadiene, 144
    of ethylene to α,β-unsaturated ketone,
        58
Cyclobutanol, fragmentation of, 425
Cyclobutanones, elimination of ethanol
    from, 540
Cyclobutene, synthesis from 1,3-diene, 155
Cyclobutenones, hydride reduction of, 540
Cyclobutylhexenols, acid catalysed rear-
    rangement of, 542
β-Cyclocitral, oxidation of, 138
Cyclodehydration, 595, 697
    with aluminum chloride, 599, 681
    with concomittant dechlorination, 457
    diketones, 620
    t-diols, 601
    with ethanolic hydrochloric acid, 695
    with polyphosphoric acid, 714
    with pyridine hydrochloride, 661
    in synthesis of lactone esters, 654
    with p-toluene sulphuric acid, 621, 697
Cyclodehydrobromination, 420
Cyclofarnesyl cation, rearrangement of, 265
Cyclohexanes, oxidation of, 646
Cyclohexanes, with methyl lithium, 511
Cyclopentano pyrans, table of, 63-68
Cyclopentenanes, 457
    hydrogenation, 658
    with lithio salts, 458
    photoaddition, 532
Cyclopropanes, 49-54, 156, 417, 420, 467,
    487
    cleavage of, 380, 427, 625
    formation from, γ-bromoketones, 156,
        158
        conjugated dienes, 49, 50
        diazoketones, 148, 158
        α,β-unsaturated ketones, 156, 158
    mode of ring fission in trans-chrysanthemic
        acid, 35
    reductive rearrangement, 480
    specific electrophilic cleavage, 360
Cyclopropyl alcohol, 205
    conversion to bromide, 214
    opening of, 205, 214, 217
    photolysis of, 48
Cyclopropyl carbinyl bromide, from alcohol,
    214, 230
    rearrangement of, 214
Cyclopropyl esters, from γ-chloroesters, 53
Cyclopropyl ketones, 51, 572, 625
    alkylation, 213, 230, 629
    carbonation, 213
    1,4-dienes from, 147
    Grignard reaction, 205, 217
    reductive alkylation, 629
Cyclopropyl nitrile, hydrolysis of, 50

Cyclopropyl olefins, hydroboration of, 147
  ozonolysis of, 57
Cymbopogon, 4
Cymbopogon densiflorus, 111
p-Cymene, 92
  8-hydroxy, 102, *103*
  microbial oxidation of, 128
  photochemical oxidation of, 103, 128
  in the synthesis of, carvacrol, 104, 105
    menthatriene, 91
    thallation of, 104
α-Cyperone, *285-288*
  Birch reduction of, 298
  epoxidation, selective, 289
7-*epi*-α-cyperone, *269, 270, 284, 304*
  agarafurans synthesis from, 300, 301
  deoxygenation of, 304
  peracid oxidation of, 301

Dammar resin, 592
Deacetonylation, 87
Deacetylation, 446, 595
Deamination, 572
Debenzylation, by hydrobromic acid, 525
Debromination, with alkaline methylsulfate,
  616
  by hydrogenation, 459
α-Decalones, *cis* and *trans* methylation of,
  659
cis Decalones, *601*
Decarbonylation, by photochemical reaction, 156
  by reduction of, dithioketal, 516
    thioketal, 543
Decarboxylation, 363, 368, 374, 377, 380,
  420, 446, 456, 530, 540, 587, 649,
  650, 692
  acid-catalysed, 652
  β-cyclocitrylideneacetic acid, 353
  β-Keto acids, 477
  malonic acids, 473
  with trisbiphenyl phosphinechlororhodium
    chloride, 361
Deesterification, 453, 473, 530
  with rearrangement, 652
Deformylation, 376, 464
Dehomologation, of aldehydes, 55
Dehydration, 451, 458, 678
  acid catalysed, 353, 527
  of acid chlorides, 648
  of allyl alcohol, 10, 46
  of cyanohydrins, 171
  of diols, 55, 576, 601
  of diplopterol, 592
  of homoallylic alcohol, 13
  of β-hydroxy ketones, 161, 171
  of hydroxy olefins, 593
  with mesylchloride/pyridine, 667
  of patchouli alcohol, 492, 493
  with phosphorous oxychloride, 427, 446,
    579, 667
  of primary alcohols, 414
  of secondary alcohols, 414
  spontaneous, 392, 581
  of terminal alcohols, 463

  of tertiary alcohols, 22, 100, 360, 372,
    374, 378, 387, 411, 459, 471, 480,
    483, 491, 496, 503, 507, 509, 525,
    530, 595, 619, 632
  two-phase medium, 458
Dehydrobromination, 46, 98, 141, 142,
    420, 491
  base catalysed, 381
  bromocyanoketones, 604
  bromoketones, 58, 604, 620
  with HMPT, 384
  methyl shift, during, 605
  in the synthesis of cyclopropanes, 417
Dehydrochlorination, 47, 53, 514
Dehydrogenation, 8-dehydroestrone methyl
    ether, 662
  with DDQ, 367, 665, 666
  hydroxymethylenes, 372
  microbiological, 664
Dehydroiodination, 86, 171
Dehydrolinalool, *28*
  acetate, pyrolysis of, 29
Dehydrotosylation, pyridine catalysed, 522
Deketalization, 11, 32, 75, 79, 365, 404,
    420, 427, 431, 473, 507, 522, 536,
    689, 705
de Mayo reaction, 83
Deoxalylation, 509
Desulphurization, lithium in ethylamine,
    562
Dethioketalization, by Georgian's method,
    520
  with mercuric chloride-cadmium carbonate,
    26
  with silver nitrate, 26
Diacetates, hydrolysis, selective, 436, 538
  from olefins, *168*
  pyrolysis, 130
  selective, 109
1,5-Dialdehydes, cyclization of, 75
Diazoketones, *391, 441, 473*
  decomposition with, copper powder, 391,
    473
    cuprous iodide, 435
    mercuric iodide, 435
  reaction with, copper catalyst, 391
    cupric sulfate, 386, 389, 431
  ring closure of, 148, 158
Diazomethane, in esterification, 55, 56, 74,
    79, 87, 104, 135, 634
  ring expansion with, 141
1,2-Dibromides, reduction to olefins, 97
Dicarboxylic acids, cyclization of, 56, 97
Dichlorides, in alkylation of ethyl *p*-tolyl-
    acetate, 457
Dichlorocarbene, addition to enol ether, 43
Dieckman cyclization, 241, 363, 365, 420,
    454, 477, 649
  dimethyl γ-ketopimelate, 365
  lactone esters, 656
Diels-Alder reaction, 89, 150, 151, 162,
    259, 335, 579, 671
  of acetylene dicarboxylic acid, 353
  with benzoquinone, 687
  with borontrifluoride catalyst, 672

of citraconic acid, 672
of cyclopentadiene, 485
of 2-ethoxy-1,3-butadiene, 451
of ethyl-β-acetoxyacrylate, 462
of geraniol, 485
with methyl vinyl ketone, 499
of trimethyl cyclohexadienone, 499
with 2,6-xyloquinone, 672
Dienamines, alkylation of, 709
Diendiones, hydride reduction of, 536
Dienes, hydroboration, selective, 15
  hydroxylation, selective, 130, 520
  partial hydrogenation, 419
  photooxygenation, 133
  reduction, selective, 485, 511
    with Raney nickel, 496
1,3-Dienes, addition of ethyldiazoacetate,
    49
  cycloaddition to ketenes, 144, 145
  cyclobutenes from, 155
  Diels-Alder addition to, vinyl acetate, 150
  α,β-unsaturated ketone, 151
  epoxidation, selective, 171
  hydroboration, selective, 171
  hydrochlorination, 18
  ozonolysis of, 146
  reduction, selective, 84
1,4-Dienes, from cyclopropyl ketones, 147
1,5-Dienes, 43, 562
  cyclization of, 49
  photooxidation of, 43
  synthesis by Wurtz coupling, 43
1,6-Dienes, pyrolysis of, 61
Dienoic acids, addition of ethyldiazoacetate,
    50
  cyclization of, 144
Dienols, selective epoxidation, 166
Dienones, 1,6-addition of, sodio diethyl-
    malonate, 324, 325, 473
    sodio methyl diethyl malonate, 319
  Birch reduction, 605
  cyanide addition, 604
  enol acetylation, 665
  hydrogenation, 471, 509, 511
  irradiation, 410
  photochemical rearrangement of, 467
  photolysis of, 414
  reaction with lithium dimethyl copper,
    471
Dienyne, 576
  hydrogenation, 576
Diesters, alkylation of geminal, 158
  cyclization, 363, 454, 530
  hydride reduction, 462
  partial saponification, 11, 507
Dihydrocarvone, 108, 113, 269, 285, 291
  alkylation, 309
  annelation with ethyl vinyl ketone, 284
  conversion to camphor, 152
  enol acetates, 152
  hydrobromide, reaction with alkali, 157
  hydroxylation, 309
Dihydrolavandulic acid, cyclization to
    piperitone, 116
Dihydronepetalactones, 78

reduction, 78
Diketones, cyclization, 464, 536, 540, 627
  fragmentation, 397
  hydride reduction, 423
  reaction with Grignard, selective, 536, 579
  reduction with sodium-alcohol, 528
  selective ketalization, 627
  selective reaction with methyl lithium,
    530
  selective and stereospecific, 536
  reaction with Grignard, 536
  vigorous acid treatment, 662
α-Diketones, mono enol acetylation of, 540
β-Diketones, 99, 345
  alkylation, 698
  chlorination, 214, 216
  deacylation, 214, 216
  Grignard reaction with, 160, 161
  Michael reaction, 710, 717
  reaction with, methyl vinyl ketone, 714
    propargyl bromides, 713
  reduction, 517
γ-Diketones, 146, 534
  cyclizations of, 69, 146, 507, 534
δ-Diketones, 83
  cyclizations, 516, 540
Dimethyl acrolein, Grignard reaction with
    prenyl halide, 42
Dimethyl butynedioate, Diels-Alder reaction
    with, 162
Dimethylstyrene, 92, 93
  in the synthesis of menthatriene, 91
2,5-Dimethyl-3-vinylpent-4-en-2-ol, 39
  formation from chrysanthemic acid, 34
2,4-Dinitrophenylhydrazones, as carbonyl
    protecting groups, 428
  ozonolytic cleavage of, 428
Diols, acetylation, selective, 423, 448, 474,
    496
  cleavage, 84, 87, 540
  dehydration, 55, 427
  oxidation, 383, 391, 538, 540
  protection as acetonyls, 86
    benzaldehyde acetate, 38
  rearrangement in acid, 510, 582
  selective monomesylation, 471, 516
  selective monotosylation, 417, 462, 478,
    479, 520
1,3-Diols, monoacetate, pyrolysis of, 48
1,4-Diols, cyclic ethers from, 134
Diones, aldol cyclization of, 420
  selective reaction with vinyl lithium, 499
Dioxo carboxylic acids, cyclization, 658
Dipentene, 8, 88
  irradiation, 9
  oxidation with selenium dioxide, 100
1,3-Dipolar addition, of 2-diazo propane,
    384
Disiamylborane, in hydroboration, 440,
    449, 489
Disproportionation, during gas chromatog-
    raphy, 128
Dithio ketals, with Raney nickel, 507
Dolichoderus, 65
Dolichodial, 78

correlation with iridodial, 79
isomers, 80
Doronicum austriacum, 116
Drimic acid, 345, 347
  anhydride from, 345, 347

Electrocyclic reaction, 4, 44
Electrolysis of acetate, 579
Elemol, 269, *270*
  hydroboration of, 270
  tetrahydro, *269, 270*
Elimination, of tosylate in *trans*-decalin,
  314, 315
Elsholtzia cristata, 159
Elsholzia densa, 159
Emmons reaction, 209
Enamines, alkylation of, 534
Enediones, reduction, 575
Enol acetates, *55, 73, 152, 612, 631*
  alkylative cleavage, 619
  cleavage with methyl lithium, 612
  epoxidation, 502
  hydrogenation, 73
  irradiation in presence of 1,1,-diethoxy
    ethylene, 540
  oxidation of, 592
  ozonolysis of, 55, 632
  rearrangement with boron trifluoride, 511
  with bicarbonate, 665
Enolates, with acetic anhydride, 619
  with acetyl chloride, 502
  alkylation, 614
  generation, 612
  intramolecular alkylation, 526
  protonation, 614
Enol ethers, *624*
  addition of dichlorocarbene, 143
  bromination-dehydrobromination, 664
  cleavage of, 120
  condensation with methyl acetoacetate,
    367
  consecutive acid-base treatment, 366
  with dimethyl sulphonium methylide, 616
  with ethylene glycol, 465
  with Grignard reagents, 661
  hydrolysis, 366, 378, 409, 451, 616
  mechanism of reductions, 369
  as protecting group, 378, 442
  reduction with lithium-ammonia, 369
  with Vilsmeier's reagent, 406
Enolic β-keto esters, *477*
  with methyliodide, 417
Enol lactones, with Grignard reagents, 629,
  710
  hydrogenation of, 595
  as intermediates in annelations, 711
  with methyl lithium, 372
  with phosphonium or phosphonate ylids,
    711
Enols, *665*
  borohydride reduction, 665
  ketonization, 584
Enones, *451, 682*
  acetylation, 406
  alkylation, 620, 714, 715, 716

0-alkylation, 715
borohydride reduction, 525, 628
conjugate addition, 353, 376
dimethylation, 625
dissolving metal reduction, 469
  with ethyl cyanoacetate, 543
formylation, 373
with Grignard reagents, 705, 619
with hydrazine, 387
hydride reduction, 422, 471, 474
hydroboration, 146
hydrocyanation, 516, 625, 629
hydrogenation, 377, 507, 540, 628, 682,
  710
hydrolysis, 407
with lithium dimethyl copper, 366, 376
with methyl lithium, 511
reduction, 383, 384, 463
reductive alkylation, 629, 630
stereospecific angular methylation, 502
stereospecific hydrogenation, 372, 527,
  695, 696
specific reduction, 361
with triallyl orthoformate, 628
with tri-*t*-butoxy lithium alluminum hydride, 471
with triethyl aluminum-hydrogen cyanide,
  620
Wolf-Kishner reduction, 381
β-γ-enones, *569*
trans-Enyal, Grignard addition to, 30, 31
Epimerization, acetyls, 499
  during allylic oxidation, 416
  esters, 400
  hydroazulenes, 397
  ketones, 592, 695
  lactone carbonyls, 414
Epoxidation, of β-amyrin-3-benzoate, 609
  of benzylic olefins, 581
  of 1,3-dienes, selective, 171
  of dienols, selective, 166
  with dimethyl sulfonium methylide, 451
  of enolacetates, 306, 313, 502
  of homoallylic alcohols, 211
  of octalones, 522
  of olefines, 54, 101, 127, 365, 413, 422,
    471
  of β-patchoulene, 496
  of squalene, 566
  of α,β-unsaturated ketone, 107, 137
  mono and di, 208
  selective, 100, 210, 216, 219, 220, 271,
    289, 306, 588
  Van Tamelen method, 219
  via aerial oxidation, 606
Epoxides, *208, 210, 211, 214, 216, 217,
  219, 221, 564*
  cleavage, 88, 216, 365
    with boron trifluoride, 496
    and rearrangement, 422
    with sodium benzylate, 522
  controlled hydrolysis, 566
  conversion to, allyl alcohols, 221, 666
    enol acetate, 73
  formation of thiourea clathrates, 568

with Grignard reagents, 646
intermediates in cyclization, 571
rearrangement, to ketones, 609
    with chlorine, 609
reduction of, 101, 289, 451, 471, 609
ring opening, to α,β-unsaturated, ketone,
    108
    with benzyl alcohol-acid, 54
    with dimethylamine, 100
separation from diols, 568
Epoxyacetate, basic hydrolysis of, 307
    thermal rearrangement of, 306, 307, 313
α,β-Epoxy aldehyde, *21*
    from epoxidation of citral, 21
    in the synthesis of allyl alcohol, 21
    Wharton reaction on, 21, 165
α,β-Epoxy Ketones, cleavage of, 95, 107
    reduction of, 136
Equilibration, of Ketones, 385, 478
    ring junctions, 481
Erythro-diols, *568*
Esterification, 400, 407, 409, 419, 445,
    473, 514, 516
    with diazomethane, 379, 634, 652
    with methyl chlorocarbonate, 411
    with silver oxide-methyl iodide, 373
Esters, alkylation, 457, 632, 685
    axial-equatorial equilibration, 369
    cleavage, 649, 682
    epimerization, 400
    with Grignard reagents, 135, 582
    hydrolysis, 363, 368, 371, 374, 377, 397,
        442, 463, 503, 514, 534, 538, 540,
        604, 654, 685, 692
    with methyl lithium, 372, 387, 401, 409
    with methyl magnesium iodide, 473
    with methyl sulfinyl carbanion, 534
    reduction, 59, 363, 365, 374, 397, 411,
        419, 489, 649
Ethers, cleavage, 667
    exchange, 393
    selective cleavage, 625
Ethoxyacetals, with ethyl propenylether,
    393
Ethoxyacetylide, addition to ketone, 44
Ethylene acetals, hydrolysis of, 466
Ethyl diazo-acetate addition to, 2,5-dimeth-
    ylhexa-2,4-diene, 49
    2,5-dimethylhexa-2,4-dienoic acid, 50
Ethynylation of, Ketone, 201-203
Ethynyl carbinols, (Acelylenic Alcohols),
    reduction of, 477
Eucalyptus dives, 136
Eucalyptus globulus, 87, 97
Eudesmane group, members of, 282
Evodia hortensis, 123
Exocyclic olefins, hydroboration-oxidation
    of, 503

Farnesic acid, bicyclo, *339,* 340
    monocyclo, cyclization of, 343, 344
Farnesol, *200, 201, 203, 204, 206*
    acetate of, cyclization, 340
        epoxidation, 221
    bromide, 343

cyclization to drimenol, 339
    pyrophosphate, cyclization of, 233
Favorskii rearrangement of, (+)-Pulegone,
    72, 145
Felidae, 62
α-Fenchol, from α-pinene hydration, 103
    from reduction of fenchone, 154
Fenchone, reaction with sulfuric acid, 114
    reduction with lithium aluminum hydride,
    154
Fennel oil, 133
Foeniculum vulgare var. dulce M., 133
Formylation of, 373
    alkynes, 205
    hexones, 442
    ketones, 376
    lactones, 329
    menthones, 389
    3 methyl furan, 160
    γ,δ-unsaturated ketones, 29
1-Formyl-2,6,6-trimethylcyclohexa-1,3-
        diene, *see* Safranal
Fractional crystallization, 676
Fractional distillation, 368
Fragmentation, bicyclic diol monotosylate,
    475
    Grob type, of 1,3-diol monotosylate, 211,
        212, 471
    hydroxyenone, 315
    β-hydroxy-β-ketoester, 51
    monotosylates, 470
Friedel-Crafts reaction, 117, 241, 253, 454
    acylation of methoxy napthalere, 658
    of anisole, 685
    cyclization, 646
    intramolecular in furans, 232
    in ring closure, 697
    in synthesis of 2 methyl cyclopentan-1,3
        dione, 692
Fullers earth, in rearrangements, 590, 593
Furan ring system, *122, 264, 275, 312, 314*
    oxidative cleavage, 454
Furfurylidenes, as blocking group, 454, 683,
        689, 690
    cleavage of, 689
    hydrolysis, 656, 658
    oxidation, 699
Fused ring systems, by hydrogenation, 583,
    586
    stereoselectivity of hydrogenation, 696

Genipa americana, 66, 84, 85
Geranial, *see* Citral *a*
Geranic acid, cyclization to piperitenone,
    116
Geraniol, *200*
    acetate, 139
    acetate oxide, 165
    acetonylation, 202
    in the biogenesis of, indole alkaloids, 24
        Karahana ether, 139
    oxidation, 200
    phosphate ester in the biosynthesis, 24
Geranyl bromide, *28*
    oxidation of, 28

Geranyl diphenyl phosphate, 24
  allylic rearrangement of, 24
Gingergrass oil, 111
Glyoxalyl chloride tosylhydrazone, with
  trans, trans-farnesol, 634
Glyoxylates, pyrolysis of, 648
Grignard reactions, 579, 600, 601, 621, 625
  1,4-addition to unsaturated ketones, 32,
    86, 114, 619
  of allyl magnesium bromide, 391, 442
  of 2-bromo-6-methoxy naphthalene, 661
  in carbonation, 697
  of cycloheptalones, 427
  of enol lactone, 595, 629, 705, 710
  of π-halotricyclenes, 487
  of homocamphor, 496
  of 1-iodo-6-methoxynaphthalene, 646
  of isobutylbromide, 30, 31
  on acid chloride, 236
  on acids, 94
  on anhydrides, 56
  on benzofuran system, 123
  on bicyclic heptanone, 151
  on α-chloroketones, 214, 564
  on γ-chloroketones, 22
  on 2-chlorotropone, 143
  on 2-Cyano-3-methyl furan, 160
  on cyclopropyl ketone, 205, 391
  on β-diketones, 160, 161
  on esters, 135, 250, 474, 514, 582
  on hemiacetals, 619
  on β-hydroxy ester, 261
  on ketal aldehyde, 206
  on β-ketoacetal, 29
  on ketones, 47, 92, 96, 97, 151, 168, 540
  on nitriles, 160, 217, 241
  on prenyl halides, 42
  on protected aldehydo ketone, 29
  on 4-substituted methyl cyclopentenes,
    103
  on tropolone system, 143
  on α,β-unsaturated acid chloride, 112
  on α,β-unsaturated aldehyde, 30, 31
  on α,β-unsaturated ketones, 86, 95, 114
  selective and stereospecific with diketones,
    436
  with acrolein, 697
  with allylmagnesium bromide, 391, 496
  with trans-caran-2-one, 391
  with γ,γ-dimethylallyl mesitoate, 481
  with ethoxy acetylene magnesium bro-
    mide, 587
  with ethylene oxide, 646
  with β-halopropionic esters, 649
  with methacrolein, 487
  with spontaneous dehydration, 581
Grob fragmentation, see Fragmentation

Halogenation of, β-diketone, 214, 216
β-Halopropionic esters, as Grignard reagents,
  649
π-Halotricyclenes, homologation of, 486
Helinium species, 116, 123
Hemiacetals, 572
  in Grignard reaction, 619

with methane sulphonyl chloride, 572
  oxidation, 610
Heptenes, hydroboration, 496
Hexanones, formylation, 442
  reduction, 442
Ho leaf oil, 25
Homoallylic acetates, allylic oxidation of,
  471
Homoallylic alcohols, dehydration, 13
  pyrolysis, 27, 101
  synthesis by Prins reaction, 101
Homoallylic chloride, condensation with
  methyl vinyl ketone, 22
Homogenization, of acetylenic isomers, 689
  of ring junctions, 475
Homologation, of π-halotricyclenes, 486
Hop oil, 167
Huang-Minlon reduction, 456, 689
Hunsdieker reaction, 454
Hydration, of acetylenes, 69, 695
  of alkenes, 27
Hydrazine, with enones, 387
Hydrazones, 435
  decomposition with cuprous iodide, 446
  cis, trans-farnesal, 435
  of ketones, 689
  oxidation, 435, 633
Hydrindones, internal bond scission of, 474
Hydride transfer, during pyrolysis of 1,6-
  diene, 61
Hydroazulenes, 395, 402, 404
Hydroboration, of acetylenes, 449
  of cyclopropyl olefin, 147
  of diene, selective, 15
  of elemol, 270
  of exocyclic olefin, 503
  of heptenes, 496
  of homodiene, 171
  of hydroxyolefins, cyclization during, 134
  of methyl heptenone, 146
  of olefinic ketal, 290
  of olefins, 74, 77, 356, 404, 420, 528
  of olefins, resistant to, 621
  of terminal olefins, 391, 530
  of α,β unsaturated ketone, 136
  ring opening of ketals, 291
  selective, 80
  with disiamylborane, 629
  with tri-isobutylaluminum, 15
Hydrochlorination, of α-acetyl-γ-butyrolactone, 21
  of 1,3-dienes, 18
  of Isoprene, 7
  of Myrcene, 12
  of olefinic ketones, 51
Hydrocyanation, 311, 312, 346, 347, 350
  of enones, 625, 629
  stereochemical control, 625
Hydrogenation, 353, 392, 586, 601, 604,
  674, 679
  control by neighbouring group, 621, 686
  in presence of perchloric acid, 496
  of acetylenes, 30, 681
  of allyl acetates, 503
  of allyl alcohols, 628, 667

of aromatic nitro group, 104
of bicyclo olefins, 150, 151
of cyclopentenones, 656
of dienes, selective, 419
of dienones, 471
    selective, 511
of dienynes, 576
of enol acetate, 73
of enol lactones, 595
of enones, 377, 515, 540, 628, 682, 710
of hexenes, 499
of hydroazulenes, 410
of olefins, 353, 459, 525, 575, 621, 646
of spiro 4.5 decadienones, 514
of spiro 4.5 decanes, 468
of α,β-unsaturated esters, 372
of α,β-γ,δ-unsaturated esters, 509
of α,β-unsaturated ketones, 69, 118, 661
over nickel, 646, 697, 702
over palladium-calcium carbonate, 697,
    699
over palladium-strontium carbonate, 407
of Phenol, 98
over platinum oxide, 379, 583, 601
over tris (triphenylphosphine) rhodium,
    411
replacement of bromine, 459
selective, with Wilkinson's catalyst, 287
stereoselective, 400, 696
stereospecific, 414, 507, 522, 525, 616,
    653, 695, 696
Hydrogenolysis, of alcohols via their ace-
    tates, 356
of amine salts, 406
of benzyl ether, 55
of cyclopropanes, 427
of ketones, 685
    selective, of ketones, 658
1,2-Hydrogen shifts, 594
Hydrolysis, of allyl acetate, 130
of diesters, 507, 590
of cyclopropyl nitrile, 50
of enol ethers, 366, 378, 616
of epoxides, 566
of esters, 353, 363, 368, 371, 374, 377,
    397, 442, 463, 473, 503, 514, 534,
    538, 540, 568, 584, 586, 588, 604,
    685, 692
of ethylene acetals, 466
of β-hydroxynitrile, 53
of β-keto esters, 69, 456, 477
of ketals, 699
of lactones, 463, 579, 634
of malonates, 473
of methyl ethers, 653
of nitriles, 78, 385, 462, 516, 543, 595,
    615
of trifluoroacetate, 105
of trimethylsilyl ethers, 568
of vinyl chlorides, 709
in synthesis of oleon-12-ene-3-one, 604
partial, of diacetates, 536
Hydrolytic cleavage, of tetrahydrofurans,
    536
Hydroperoxides, by aerial oxidation, 606

hydride reduction, 619
γ-Hydroxy acetates, oxidation of, 536
δ-Hydroxy acids, cyclization of, 76
γ-Hydroxy acids, lactonization, 463, 621
γ-Hydroxy amides, cyclization, 39
    with methane sulphonyl chloride-pyridine,
    621
Hydroxy epoxides, cleavage of, 422
2-Hydroxyisopinocamphone, dehydration
    of, 141
Hydroxy ketal, oxidation of, 32
β-Hydroxy-β-keto ester, fragmentation of,
    51
α-Hydroxy ketones, with calcium-ammonia,
    374
    oxidation, 540
β-Hydroxy ketones, dehydration, 161, 171
Hydroxylation, of dienes, 520, 582
of olefins, 84, 86
    selective, 130
    with osmium tetroxide, 417
t-Hydroxyl group, replacement by chloride,
    47
1-Hydroxymenth-2-ene, 90
    pyrolysis of phenyl urethane derivative of,
    90
Hydroxymethylenes, 532
    as blocking group, 82, 83, 357
    with n-butyl mercaptan, 357, 389
    0-alkylation, 442
    deformylation, 376
    with dichlorodicyanoquinone, 372
    Grignard reaction on, 29
    with hydroxylamine hydrochloride, 651
    of lactones, 463
    with methyl iodide, 367
    with methyl vinyl ketone, 464
    reduction, 463, 532
β-Hydroxynitrile, lactone during hydrolysis,
    53
Hymentherene, 11
    isomerization of trans to cis, 13

Iboga indole alkeloids, 81
Illicum verum, 133
Imidazolines, 158
    in the formation of cyclopropanes, 158
Imines, reduction, 620
    in synthesis of aldehydes, 629
Iminolactones, 621
Immonium salts, reduction of, 406
Intramolecular alkylation, of tosylates, 632
Iodides, with lithium acetylides, 449
Ips confusus, 26
Iresin celosiodes, 348
Iridodial, 72, 74
    correlation with dolichodials, 78
Iridoids, table of, 63-68
Iridolactones, stereochemistry of, 76
Iridomyrmecin, 74, 77
    permanganate oxidation of, 76
    stereochemistry of, 76
Iridomyrmex, 62, 65
Iridomyrmex conifer, 65
Iridomyrmex defectus, 65

Iridomyrmex humilis, 64
Iridomyrmex nitridus, 64
Isobutene, reaction with acid chloride, 41
cis-Isochrysanthemic acid, double bond iso-
      merization in, 51
   nitrile of, *50*
Isolavandulol, dehydration of, 46
Isomenthone, 124
   separation from menthone, 125
Isomerization, in alkaline medium, 676
   during hydrolysis of nitriles, 50
   of allenones, 33, 34
   of *trans*-lactones, 579
   of α-onocerin, 574
   of terminal olefins, 532
Isonepetalactone, *71, 72*
   conversion to *cis, trans*-nepetonic acid, 71
   isomerization to nepetalactone, 72
Isopiperitenone epoxide, 137
   diosphenolene from, 137
Isoprene, dimerization, 89
   with formic acid-perchloric acid, 23
   with halo-acids, 7
   telomerization, 23
Isopropoxymethylenes, as blocking groups,
      377
   cleavage, 377
Isopropylols, *397*
Isopulegol, pyrolysis of, 27
Isopulegone, 116
   4-acetoxy, 122
   5-acetoxy, 122
   conversion to menthofuran, 122
   interconversion to pulegone, 121
cis-Isopyrethric acid, *51*
   cis-trans isomerization, 52
Isoxazoles, alkylation, 377, 652
   annelation, 377
   cleavage, 377

Japanese hops, 139
Japanese pepper, 100
Japanese peppermint oil, 136
Jones oxidation, 121, 621
Juniperus communis, 88, 100
Juniperus sabina, 148

Kentranthus ruber, 67
Ketals, hydrolysis, 699, 710
   rearrangement, 536
Ketalization, 11, 32, 33, 75, 112, 121, 135,
      366, 397, 404, 420, 463, 473, 518,
      522, 586, 616, 619, 627, 629, 631,
      689, 704
Ketenes, cycloaddition to 1,3-dienes, 144,
      145
Keto acids, *127*
   conversion to enol lactones, 710
   cyclization of, 99
   decarboxylation, 477
   lactonization, 372
   reduction, 399
Keto aldehydes, aldolization, 465
   cyclizations, 55, 74, 87, 113, 536
   enol acylation, selective, 55

   oxidation, 586
   in ring annelation, 366
α-Keto carbenes, with olefins, 388
11-Keto compounds, reduction, 573
Keto epoxides, cleavage, 423
β-Keto esters, addition to allylbromides, 45
   alkylation, 369
   hydrolysis, 69, 456, 477
   ketalization of, 32, 121
   reaction with methyl vinyl ketone, 397
   reduction of, 32, 577
   in Stobbe condensation, 649
   Wittig reaction, 45
γ-Keto esters, intramolecular acylation, 530
δ-Keto esters, cyclization of, 507
Ketols, chromatography on alumina, 592
   oxidation of, 73
   pinacol rearrangement, 621
Ketones, addition of, ethoxyacetylide, 44
      methyl lithium, 48
   alkylation, 82, 141, 371, 385, 491, 520
      stereospecific, 621, 658
   angular methylation, 357
   Bamford-Stevens reaction, modified, 309
   bromination, 98, 141, 146, 384, 620
      selective, 252, 253
   bromination-dehydrobromination, 385,
      471, 481
   α-carbethoxylation, 369
   carbonation, 295
   carboxymethylation, 473
   condensation with, aldehydes, 516
      methoxyacetylene, 23
   conversion to, alkynes, 205
      amides, 150
      enol acetates, 152
      olefins, 296, 390
   cyclodehydration, 620
   dehydrogenation, 367
   cpimcrization, 695
   equilibration, 621
   ethynylation, 201-203
   formylation, 29, 267, 376
   geminal alkylation, 625
   Grignard reaction on, 47, 92, 96, 97, 151,
      168
   hydroxymethylation, 357
   hydrogenolysis, 685
   in Wittig reaction, 59, 76, 89, 104, 164,
      478
      modified 207, 208, 216, 217, 218
   methylation, 427, 483, 624
   α-oxidation, 414
   preferential thioketalization, 292
   protected as 2.4-dinitrophenylhydrazone,
      528
      enol ether, 378
   reduction, 69, 86, 356, 360, 389, 407,
      410, 413, 427, 431, 463, 478, 514,
      536, 621
      selective, 211
      by Meerwein-Ponndorf, 689
   Reformatsky reaction on, 204
   removal *via* dithio ketal, 507
   selective hydrogenolysis, 658

separation as semicarbazones, 465
with acetylenic alcohols, 575
with dimethylsulfonium methylide, 451
with diethylcyanomethylphosphorate, 386
with ethyl magnesium bromide, 540
with isopropyllithium, 525, 527
with lithio acetylides, 477
with methylenetriphenylphosphorane,
    451, 503
with methyl lithium, 374, 404, 428, 469,
    471, 483, 491, 503, 527, 530, 542,
    619
with methyl magnesium bromide, 392
with methyl magnesium iodide, 391, 451,
    458, 459
with Raney nickel, 479
with sodium acetylide, 575
with sodium and isopropyl alcohol, 427
with sodium hexamethyldisilazane, 631
with triethylphosphonoacetate, 453
Ketonic cleavage, during work up, 649
β-Keto nitriles, in tobbe condensation, 650
β-Keto sulfoxides, cyclization with iodine,
    534
Kochi-Hunsdiecker reaction, 449
Kolbe electrolysis, 579, 586

Lactols, with m-ethoxyphenylmagnesium
    bromide, 621
Lactone esters, cyclization of, 656
Lactones, 53, 56, 57, 75, 76, 312, 314, 317,
    320, 322, 323, 325, 327-329, 344,
    346, 514, 634
  cleavage, 53, 56, 87, 308, 414, 514
  conversion to furan, 314
  elimination, 650
  epimerization, 414
  formylation, 329
  hydride reduction, 77, 399, 446
  hydrolysis, 420, 463, 579, 595, 634
  hydroxy, dehydration of, 351
  hydroxymethylenation, 461
  inversion of C-O linkage, 463
  α-methylene, 326, 327-330
  reduction, 77, 621
  reductive acetylation, 77
  self-condensation, 167
  with ethyldiazoacetate, 463
  with hydrogen chloride, 610
  with lithium aluminum salt of methyl-
    amine, 621
Lactonization, 399, 650, 685
  of hydroxyamides, 621
  of γ-hydroxy acids, 463
  of keto acids, 372
  specific of γ-hydroxy acids, 463
  with aqueous acid, 414
Lanosterol transformations, 571
Lavandulybromide, 45
  dehydrobromination of, 46
  Wurtz Coupling of, 45
Lavandulylic acid, cyclization of, 49, 116
  ethyl ester of, 45
Lavandulyl skeleton, 43
  formation from, trans-chrysanthemic

acid, 34
  prenylbromide, 45
Lead tetraacetate, cleavage of glycols, 592
  ketone oxidations, 414
  oxidation of α-ketols, 73
  selective allylic oxidation, 92
  selective cleavage of 1.2-diols, 84
Leukart reaction, 311, 312
Libocedrus formosana, 140
Limonene, 88, 89
  addition of aluminum alkyls to, 131
  conversion to menthatriene, 92
  dihydrobromination, 97
  hydroboration, 131, 257
  oxidation, 92, 94, 108, 109, 110, 128,
    130, 131
  photooxidation, 94, 111
  tribromide, 97
Limonene epoxide, 73
  chromatography over alumina of, 108
  diacetate from, 109
  hydration of, 137
  isomerization with Lewis acids, 108
  ring opening and contraction in, 74, 87,
    96, 108
Linalool, conversion to, geraniol, 201
    hotrienol, 25
  dehydration, 60
  from α-acetyl-γ-butyrolactone, 21
  from citral, 21
  from epoxide of geranyl iodide, 21
  from α-pinene, 20
  oxides of, 165
  phosphate ester in biosynthetic pathway,
    24
  photooxidation, 25
  pyrolysis, 61
  reaction with, acidic reagents, 22, 165
    NBS, 142
Linalool acetate, 22
  allylic rearrangement in the preparation,
    22
  chromatography of, 10
  conversion to myrcene, 10
  pyrolysis of, 10
  reaction with N-bromosuccinimide, 25
Lindlar catalyst, for selective reduction of
    triple bond, 30, 31
Lithium aluminum tri-t-butoxide
  in reductions, 704
Lithium dimethylcopper, alkylation of,
    alkynes, 206, 210
  in 1,4 additions, 376, 543
  reaction with, allylic acetate, 220
    allylic chloride, 221
Lithium diphenylphosphide, in selective
    ether cleavage, 625
Lyratol, 39
  formation from chrysanthemic acid, 34

Malonic esters, addition to bromides, 646
  hydrolysis of, 473
Manganese dioxide, in the oxidation of
    allylic alcohol, 28, 31, 79, 81, 127,
    156, 575

Mannich bases, *695*
in condensation, 114
Mannich condensation, 695
Mare pregnancy urine, isolation of (+) equilenin, 642
Medium ring sesquiterpenes, synthesis, 277-
282
Melaleucia species, 133
Mentha arvensis, var piperasceus, 136
Mentha-1,4-dien-7-al, 126
disproportionation of, 128
Mentha-1(7),8-dien-2-ol, 106, *111*
acetate of, 110
ethyl ether solvolysis, 114
Mentha-1,8-dien-10-ol, 129, *130*
derivatives of, 130
glycoside of, 130
Mentha spp., 120
Menth-1-ene, oxidation of, 108, 109, 110,
111
photooxidation of, 94, 128
stereochemistry of 9-oxygenated, 131
Menth-1-en-9-ol, 131
3,5-dinitrobenzoate of, 131
Menthofuran, 115, 116, 120, *122*
photooxidation of, 123
Menthones, 116, 126
dibromination of, 146
formylation of, 331
reduction of, 125
separation of, 125
Mercuric chloride, in cleavage of thio ethers,
26, 536
Mercuric iodide, in decomposition of diazo
compounds, 435
Mercuric oxide, in ether formation, 446,
449
Meerwein-Ponndorf reduction, selective, of
ketones, 689
Mesitoylation, 440
Mesityl oxide, condensation with methyl
vinyl ketone, 118
Mesylates, solvolytic rearrangement, 410
solvolytic ring cleavage, 425
Methanesulfonates, with sodium β-chloro-
phenoxide, 397
solvolysis of, 400
Methanesulphonation, selective, of diols,
471
Methane sulphonyl chloride, rearrangement
of hemi acetals, 572
Methoxy acetylene, condensation with
ketone, 23
Methyl-anilinomethylenes, as blocking group
α to a carbonyl, 658
2-Methylbut-3-en-2-ol, in the synthesis of
methylheptenone, 5, 6
transesterification with ethylacetvacetate,
4
Methyl cyclopropyl ketones, 204, 213, 214
α-Methylenebutyrolactone moiety, 326,
327-330
Methyl ethers, cleavage of, 685
hydrolysis of, 653
selective saponification, 649

Methyl group functionalization, 572
6-Methylhept-5-en-2-one, *5*
acetal, 29
as precursor of, citral, *28*
α-curcume, *242*
ionones, *5*
isobisabolene, *239*
lavandulol, *47*
linalool, *5*
methyl geranate, *23*
methyl nerolate, *23*
pseudoionone, *6*
vitamin A, *5*
hydroboration-oxidation, 146
reaction with acetylene, 28
Reformatsky reaction, 23
Methyl lithium, addition to, ketones, 48
α,β-unsaturated esters, 43
α,β-unsaturated ketones, 59
Methyllithium-titanium trichloride, as coupling agent, 561
Methylphenyl ether, cleavage of, 509
1,2-Methyl shifts, 594
Methylsulfinyl carbanion, in internal cyclization of tosylates, 522
Methyl vinylketone, condensation with,
homoallylic chloride, 22
mesityloxide, 118
Michael reaction, 397, 595, 612
1,6-addition, 319, 324, 325
of β-chloroethyl vinyl ketone, 717
of β-ketosulfoxides, 536
of methyl 5-oxo-6-heptenoate, 710
of sodiumborohydride, 367
with cyclopentenones, 536
with ethylacrylate, 704
Microbiological dehydrogenation, 664, 665
Mitcham peppermint oil, 136, 137
Moffatt oxidation, 351, 352, 365
Monotosylation, of diols, 520
Monotropa hypopithys, 67
Myrcene, *10*
as precursor in the synthesis of, geraniol,
17
linalool, *17*
nerol, *17*
β-pinene, *154*
β-sinesal, *224, 225*
α-terpineol, *17*
oxidation, allylic, 224
ozonolysis, 225

Nagata method of, hydrocyanation, 311,
312, 346, 347, 350
Neonepetalactone, *73*
conversion from matatabiether, 80
Nepta cataria, 62, 63
Nepetalactones, 69
optically active stereoisomers of, *70*
Nickel carbonyl, carboxylation of acetylenes,
445
cyclization with, 272-274, 561
Nickel peroxide, as oxidizing agent, 28
Nitrile groups, *407*
epimerization, 311

from aldehydes, 160
Grignard reaction, 160
hydrolysis, 78, 386, 407, 420, 462, 516,
    543, 615
introduction by Nagata method, 311, 312,
    346, 347, 350
isomerization during hydrolysis, 50
reduction, 620, 625, 629
replacement with methoxyl, 604
selective reduction, 311
Nitrite esters, photolysis of, 610
Noriridomyrmecin, *see* Boschina lactone

N-Oxides, *39, 40*
Cope reaction, 39, 40, 419
trans-Occidentalol, photolysis of, 309
Octalones, epoxidation, 522
    with *N*-bromosuccinimide, 524
Olea europaea, 135
Olefins, *97, 147, 385, 433, 571, 664*
addition of carbenes, 145, 462
biomimetic cyclizations, 712, 713
cleavage, 465
cyclizations with iodine, 453
1,2-diacetates from, 168
Diels-Alder reaction, 462
directional effect in angular methylation,
    687
epoxidation, 54, 101, 127, 365, 413, 422,
    471
equilibrium in Δ^{14} dehydroequilenin ester,
    652
from ketones, 296, 390, 392
hydration, 27, 309
hydroboration, 77, 80, 356, 404, 481,
    496, 528, 600, 616
hydrobromination, 420
hydrochlorination, 51, 514
hydrogenation, 353, 459, 525, 616, 621,
    646
isomerization, 227, 646, 709, 710, 712
    photolytic, 491
migration, 404, 584, 620
osmylation, 516
oxidation, 51, 55, 86, 208, 358, 383, 458
ozonolysis, 516
photochemical addition, 530
isomerization, 281, 327
oxidation, 258, 344, 374, 619
rearrangement, 361
rearrangement during hydrolysis, 652
reduction, 367, 648
resistant to hydroboration, 621
role in angular methylation, 690
selective oxidation, 210
    reduction, 496
Simmons-Smith reaction, 360
stereochemistry of photo addition, 532
terminal, by Wittig reaction, 404
    bromination, 381, 417, 491
    hydroboration, 391, 530, 629
    hydrogenation, 414
    isomerization, 532
    oxidation, 381, 661
    ozonolysis, 374

trisubstituted, stereospecific synthesis,
    206, 210, 214, 222
with α keto-carbenes, 388
with osmium tetroxide, 417
Olive tree, 135
Oppenauer oxidation, 665, 667, 668, 674,
    689
Organocadmium Compounds, reaction with
    acid chloride, 32, 164
Origanum Vulgare, 96
Osmiumtetroxide-periodate, in the cleavage
    of allyl alcohol, 59
Oxalic acid, as dehydrating agent, 100
    in enol ether cleavage, 120
Oxalylation, 509
Oxidation of, aerial, of β-amyrin, 606
    alcohols, 121, 431, 586, 593, 600, 606,
        614, 619, 631, 695
    secondary, 359, 374, 404, 423, 463,
        468, 469, 471, 503, 507, 525
    aldehydes, 372, 431
    allylic alcohols, 12, 79, 81, 107-109, 127,
        128, 147, 150, 156, 210, 577, 619
    allylic position, 75, 92, 108, 110, 127,
        222, 264, 322, 414, 417
    amines, 39, 40, 312, 419
    benzylic position, 103, 333
    cyclic ethers, 446
    cyclopropyl alcohol, 147
    diols, 206, 216, 391, 540, 579
        selective, 84, 538, 590
    enone to dienone, 322
    cis, trans farnesol, 435
    furans, 454
    furfurylidenes, 689, 699
    hemiacetals, 610
    homo allylic acetates, 471
    hydrazones, 435, 633
    hydroxyacetates, 536, 596
    hydroxy ketal, 32
    ketoaldehyde, 55
    α-ketols, 73
    ketones, 609
    lactones, 87
    longicamphenylol, 520
    olean-12-ene, 604
    olefins, 54, 55, 86, 206
    β-patchoulene, 496
    sclareol, 579, 599
    terminal olefins, 400, 496, 661
    tetrahydrofurans, 268
    tropolones, 144
    α,β-unsaturated aldehydes, 74, 139
    α,β-unsaturated ketones, 135, 141
Oxidation with, bismuth oxide, 540
    N-bromosuccinimide and lead tetraacetate,
        604
    chromic acid, 629, 646
    chromium trioxide, 399, 583, 616
    Collin's reagent, 634
    2,3-dichloro-5,6-dicyanobenzoquinone,
        646
    hydrogen peroxide, 582, 605
    hypoiodite, 649
    Jones reagent, 360, 409, 442, 463, 525, 621

lead tetraacetate, 572
manganese dioxide, 28, 435, 446, 575
Moffat reagent, 516
nickel peroxide, 28
Osmium tetroxide, 381, 383, 417, 465, 592, 601
performic acid, 458
periodic acid, 582
ruthenium tetroxide, 530, 582
Sarrett's reagent, 107, 383
selenium dioxide, 381, 435, 584, 586, 602
tristriphenylphosphine rhodium chloride, 646
Oxidative degradation, of (±)-α-terpineol, 420
Oximes, with nitrous acid, 604, 610
reduction, 572
Oxo-carboxylic acids, cyclization, 658, 660
Ozonolysis, 579, 582, 586
of benzylidenes, 69, 540
of cyclopropyl olefins, 57
of 1,3-dienes, 146
of 2,4-dinitrophenylhydrazones, 528
of enol acetates, 55, 270, 632
of enol ethers, 599
of geranyl acetate, 440
of ketobenzylidenes, 682
of terminal olefins, 374, 516
of triene, selective, 225
of α,β-unsaturated ketone, 56
of vinyl ether, selective, 209

Palladium, in isomerization of olefins, 646
Peppermint oil, 120
Perilla frutescens, 159, 160
Petroselenium sativum, 91
α-Phellandrene, 89, 90
peracid oxidation of, 136
β-Phellandrene, 89, 90
Phenolic ethers, Birch reduction of, 664
Phenylurethane derivative, of allyl alcohol, 90
pyrolysis of, 90
Phosphonium salts, vigorous hydrolysis of, 389
Phosphorus tribromide, with alcohols, 363
Phosphorous ylids, coupling with allylic halides, 562
Photochemical, addition of, 3,3-dimethyl-cyclopentene, 542
isobutylene, 475
1-methyl-3-isopropyl cyclopentene, 532
olefins, stereochemistry of, 532
bromination of dibromide, 97
cleavage of cyclopropyl ketone, 147
cycloaddition, of acetylacetone, 425
of ethylene to α,β-unsaturated ketone, 58
of 2-methyl-1,3-cyclopentane dione enol acetate, 426
in the synthesis of loganin acetate, 83
cyclization of, citral, 60, 275
di-α,β-unsaturated esters, 534
decarbonylation of thujone, 148
isomerization, induced, 491
of trans to cis decalin, 309

of cis to trans double bond, 281
of trans to cis double bond, 12
of diene to triene, 279
of endocyclic to exocyclic double bond, 327
oxidation of, β-carotene, 170
citronellol, 169
p-cymene, 103
1,5-diene, 43
homodiene, 300
trans-α-hymenthrene, 12
β-ionol, 170
limonene, 94, 111
menth-1-ene, 128
methyl farnesate, 220
olefins, 374
α-pinene, 156
α-terpinene, 133
terpinolene, 100
2-thujene, 147
verbenone, 156
reaction, in the synthesis of β-amyrin, 602
Photocitrals, A and B, 60, 275, 276
Reformatsky reaction on, 275
Photolysis of, β-amyrin, 606
cyclopropyl alcohol, 48
dienones, 410, 414
1,1-diethoxyethylene, 540
enol acetates, 540
enones, 467
nitrite esters, 572, 573, 610
olefins, 361
11-oxolanosterol, 572
α-santonin, 413
tosylhydrazones, 393
Photolytic functionalization, of C-19 methyl, 572
Photoperoxides, reduction with sodium sulfite, 43
Pinacol rearrangement, 417, 516, 621
Pinenes, summary of various reactions on, 156
Pinene oxide, 60
isomerization of, 156
α-Pinene, conversion to linalool, 20
epoxide, reaction with acid, 60, 88
hydration of, 102
lead tetraacetate oxidation, 156
oxidation, 51, 126, 127, 156
Prins reaction on, 102
photochemical conversion to cyclofenchene, 154
photoisomerization, 9
photo-oxidation, 156
pyrolysis, 8
γ-radiolytic conversion to ocimene, 9
reaction with NBS, 156
β-Pinene, 155
acid treatment of, 89
conversion to oleuropic acid, 135
epoxide ring opening of, 126, 127
hydrochlorination of pyrosylate, 17
pyrolysis of, 8
reaction with NBS, 156
Pinus longifolia, 157

Pinus longifolla, 54
Pinus silvestris, 93
Piperitenone, *107,* 115, 116, *118,* 137, 158
  diosphenolene from oxide of, 137
  epoxide of, *119*
Piperitols, 115
  cis and *trans, 119*
  conversion to dl-menthone, 125
  dehydration of, 120
Piperitone, *111,* 116, *119, 120*
  as precursor of, menth-2-en-1-ol, 95
    menthane-1,3-diol, 136
  diosphenol from, 136
  hydroboration of, 136, 137
  oxide of, 95, 136
  reduction of, 119
Piqueria trinervia, Cav., 139
Polyisoprenoid chains, construction of, 204
Powdered soft glass, in pyrolysis, 648
Prenyl halides, Grignard reactions of, 42,
  161
  in the synthesis of artemisia alcohol, *42*
  Wurtz coupling of, 42, 45
Prins reaction, in the synthesis of, citral, *30*
  β-cyclolavandulal, *48*
  cyclolavandulols, *48*
  iridomyrmecin, *75*
  lavandulol, *47*
  lavandulyl acetate, *47*
  δ-terpineol, *102*
Propargyl bromides, in alkylation, 713
Propionic acids, cyclization, 654
Pseudomonas spp., 128
Pulegenic acid, *72*
  conversion to iridodial, 73
  pyrolysis of, 145
Pulegol, 115, 120
  dehydration of, 122
Pulegone, 115, *116, 119-121*
  as precursor in the synthesis of, *trans*-
    carane, 157
    iridodial, 72
    isopulegone, 121
    menthofuran, 122
  epoxide, pyrolysis of, 142
  Favorskii rearrangement, 72, 145
  imidazoline derivative, 157
  lead tetraacetate oxidation, 122
  mercuric acetate oxidation, 122
  reduction of, 122
  sultone of, 122
Pummerer's rearrangement, 536
Pyrazolines, cleavage of, 387
  irradiation of, 387
  ring contraction in, 384
Pyridine, as dehydrochlorinating agent, 47
  as dehydroiodinating agent, 86
Pyridine hydrobromide perbromide, 668
Pyrollidenes, in Torgov reaction, 703
Pyrolysis, of acetates, 10, 48, 96, 101, 127,
  383, 485, 496
  of acetoxy ketones, 122
  of γ-aldehydo acids, 70, 71, 73
  of allyl acetate, 10
  of allyl alcohol nitrobenzoate, 12

of β-amyrin, 609
of benzoates, 414
of carbonate esters, 411
of carboxylic lead salts, 682
of diacetates, 109, 130
of 1,6-diene, 61
of 1,3-diol monacetate, 48
of ester of alcohols, 151
of homoallylic alcohol, 27, 101
of patchouli alcohol acetate, 492
of phenylurethane derivative, 90
of α-pinene, 8
of β-pinene, 8
of spirolactones, 391
of sultones, 122
of tricyclic formates, 511
of α,β-unsaturated ketone, 121
of vinyl ethers, 393
ring contraction during, 96
Pyrolytic rearrangement, 566
Pyrophosphates, conversion by yeast subcellular particles, 635

Racemates, of estrone, 679
Raney Nickel, for desulfurization, 82-84
  reduction of, cresolic acid methyl ester,
    121
    dienol, selective, 100
    with *n*-butyl thioesters, 463
Raspberries, oil of, 94
Rearrangement, during oxidation of allyl
    alcohol, 147
  of allyl esters through enolates, 44
  of allyl sulfonium ylids, 41
  with lithium-ethylenediamine, 105
Reduction, Bouveault-Blanc, 646
  Huang-Minlon reaction, 456
  microbiological, 704, 710
  novel, of a conjugated olefin, 669
  of acetylenes, 5, 30, 39, 477
  of acid chlorides, 350
  of aldehydes, 442, 629
  of aldehydo ester, 209
  of allylic alcohol, 14
  of allylic chlorides, 532
  of amides, 36, 39, 40
  of benzofuran system, 265
  of benzylic alcohol, 251, 265
  of benzylidines, 540
  of bicycloketones, 154, 156
  of cresolic acid methyl ether, 121
  of cyano esters, 78
  of cyclobutenones, 540
  of β-cyclocitral, 393
  of cyclopropyl ketone, 214
  of dienes, 511
  of dienol, selective, 100
  of diesters, 462
  of enediones, 575
  of enol ethers, 369
  of enones, 361, 422, 463, 474
  of epoxides, 101, 211, 359, 451, 471
  of α,β-epoxy ketones, 136
  of esters, 32, 59, 363, 365, 397, 489
  of geminal diesters, 273

of hydroperoxides, 619
of α-hydroxyketone, 278
of hydroxymethylene, 532
of imines, 620
of immonium salts, 406
of β-keto acids, 399
of 11-keto compounds, 573
of β-keto esters, 473
of ketones, 69, 356, 360, 407, 410, 419, 427, 442, 514, 536, 569, 616
of lactones, 308, 446, 621
of methyl (+)-camphene carboxylate, 516
of nitriles, 615, 620, 625, 629
of serratenedione, 590
of tetrahydrofuranones, 536
of thioethers, 536
of thioketals, 482
of tosylates, 381, 399
of α,β-unsaturated aldehydes, 80, 112, 125, 163, 389
of α,β-unsaturated cyanoesters, 420
of α,β-unsaturated epoxides, 13
of α,β-unsaturated esters, 79, 204, 207, 208, 473
of α,β-unsaturated ketones, 26, 119, 269, 270, 582
with amalgamated aluminum, 104, 536
with amalgamated sodium, 140, 648
with di-i-butyl aluminum hydride, 615, 625
with hydride-aluminum chloride, 431, 435, 440
with isopropyl lithium, 525, 527
with lithium aluminum hydride, 13, 32, 36, 39, 40, 80, 154, 156, 573, 577, 586, 595, 599, 604, 615, 616, 634
with lithium aluminum tri-t-butoxide, 39, 704
with lithium diethoxy aluminohydride, 361
with metal-amine, 105, 474, 609
with metal-ammonia, 84, 98, 120, 123, 147, 374, 383, 384, 469, 582, 593, 599, 605, 689, 695, 697, 699
with Raney nickel, 82-84, 100, 121, 479, 686
with sodium borohydride, 69, 80, 367, 586, 621, 628, 665
with sodium-alcohol, 113, 528, 579
Reductive alkylation, definition, 612
of cyclopropyl ketones, 629
of enones, 629, 630
internal, of ε-bromoketones, 499
of unsaturated oxoacids, 659
promoting effect of alcohols, 631
Reductive dechlorination, with lithium in ethylamine, 609
Reductive elimination, with zinc and acetic acid, 568
Reformatsky reaction, 23, 247, 275, 276, 313, 314, 449, 508, 509, 646, 648, 654, 675, 685
Relay synthesis, of α-bulnesene, 410
Resolution of optical isomers, 507, 579, 582, 589, 649, 666, 703, 704, 705, 709, 710

Retroaldol reaction, 629
of acetoxy dienones, 666
vinylogous, 315
Rigid ring system, 610
Ring annelation, with acrolein, 417
effect of potassium hydroxide-pyrolidine on, 363
of keto aldehydes, 366
Ring closure, by intramolecular alkylation, 521
Ring contraction, benzylidene intermediate in, 696
during pyrolysis, 96
of pyrazolines, 384, 387
of six-membered carbonium ions, 41
to cyclopentanone, 662
Ring expansion, of bicyclo [4,1,0] system, 143
by solvolysis of tosylates, 520
with diazomethane, 141, 428
with ethyl diazoacetate, 397
Ring, fusion, on to camphene, 542
Ring-opening of, lactones, 56
Robinson annelation, 521, 714
of 2,3-dimethyl cyclohexanone, 369
side products in, 464
with (−) dihydrocarvone, 353
with 1,4-dimethoxy-2-butanone, 359
with Mannich bases, 376
with 2-methyl cyclohexane-1,3-dione, 695
with trans-3-pent-2-one, 363, 365, 368
with vinylketones, 371, 615
Rose oil, 129, 159
Rupe rearrangement, 350, 351
Ruthenium tetroxide, in the cleavage of lactones, 87

Sabinene, peracid oxidation of, 134
pyrolysis of, 89
Sabinol, acetate of, 148
Birch reduction of, 147
trans-cis conversion, 147
Sacchanmyces uvarum, 704
Safranal, 138
ethoxy, 139
reduction of, 139
Salvia species, 148
Santolinyl skeleton, 36
formation from, artemesyl skeleton, 37
trans chrysanthemic acid, 34
yomogi alcohol epoxide, 38
members of, 36
Santonin, 265, 267, 268, 277, 315
aromatization of, 308
epimerization of α to epi-α, 287, 305
hydroxyanalog, see Artemisin
oxime of, 308
reduction of, 271, 279
in the synthesis of sesquiterpenes, 316
tetrahydro, 267
Sarett reagent, in allylic oxidation, 108
Selenium dioxide, 75, 435, 584, 602
Semicarbazone formation, 674
Serini reaction, 601
Seseli indicum, 48

Sesquiterpene quinone, *see* Perezone
Sigmatropic rearrangement, 41
  of allylic sulphonium ylids, 562
Silver nitrate, use in, chromatography, 568
  dethioketalization, 26
Silver oxide, in the oxidation of α,β-unsaturated aldehyde, 74
Simmons-Smith methylenation, 360, 379, 392, 427, 614, 625
Sodium acetylide, 575
Sodium *t*-amylate, in cyclopropyl ester formation, 53
Sodium hydride, in the rearrangement of allyl esters, 45
Sodium sulfite, in the reduction of photoperoxides, 43
Solvolytic rearrangement, of tosylates, 402, 592
Sommelet reaction, 151
Spiro [4,5] decadienones, *514*
  hydrogenation of, 514
Spiro [4,5] decalenes, acid catalysed rearrangement, 514
Spiro [4,5] decanes, cyclization, 508
  hydrogenation, 468
Spiro-epoxides, with boron trifluoride, 616
Spiro lactones, *477*
  pyrolysis, 391
Squalenes, *561*
  cyclization, 566
  degradation, 561
  epoxidation, 566
  stepwise synthesis, 562
Stannic chloride-nitromethane, in cyclizations, 604
Steric interferance of axial hydrogens, 659
Stereochemistry, of pentacylic triterpenes, 610
Stobbe reaction of, 655
  dimethyl succinate, 653
  β-keto esters, 649
  β-keto nitriles, 650
Structure elucidation, of patchouli alcohol, 492-494
  of taraxerol, 605
Strychnos nux Vomica, 65, 81
Sulphur trioxide-pyridine complex, in hydrogenolysis of allylic alcohols, 431
Sulfur ylid, attachment of enzyme to, 36
  rearrangement of, 37

Tagetes glandulifera, 30
Tagetes spp (compositae), 30
Tagetone, *30*
  cis-trans conversion, 31
Tea trees, 133
Telomerization, of isoprene, 23
Terminal alcohols, dehydration of, 464
Terminal olefin, isomerization of, 373
1,4 Terpin, 100, 131, 133
  cis, 133
  hydrate method, 93
  solubility in water, 133
α-Terpinene, 88
  photooxidation of, 133

α-Terpineol, *17, 102*
  esters of, 102
  hydroboration of, 134
trans-β-Terpineol, *94, 96*
  ozonolysis-dehydration of, 236
γ-Terpineol, 94, *96,* 97
  acetate of, *97*
  separation from α-Terpineol, 96
Terpinolene, autooxidation of, 132
  epoxide of, 100
  photooxidation of, 100
  from pyrolysis of terpinyl acetate, 101
γ-Terpinyl acetate, *97*
  pyrolysis of, 101
Tetrahydrofurans, *300, 302, 303*
  hydrolytic cleavage, 536
  oxidation, 268, 269
  reduction, 536
Tetralones, annelation with vinyl ketones, 625, 626
Tetramethylbicyclo [3,1,0]-hexan-2-one, *50*
  Beckman reaction on oxime of, 50
Thermal rearrangement, of acetoxy epoxides, 502
Thioethers, reduction, 536
Thioketal anion, coupling with bromomethylbutadiene, 26
Thioketals, *368, 481*
  desulphurization, 600
  reduction, 368, 481
Thiols, acid hydrolysis, 468
Thiourea clathrates, in separation of epoxides, 568
Thorpe condensation, 650
Thujaplicins, α and β, occurrence, 143
Thuja species, 140
Thuja standishii, 143
Thujene, acid treatment of, 89
  hydroboration of, 147
  photochemical oxidation of, 147
Thujic acid, 140
  p-bromophenacyl ester of, 140
  dihydro, 140
  n.m.r. spectrum of, 140
  reduction of, 140
Thujone, *147*
  photochemical decarbonylation of, 148
  reaction with acid, 60
  reduction of, 147
Thymol, 104, *117*
  catalytic reduction of, 124, 125
  methyl ether, Birch reduction of, 120
  reduced products from, 115
Todomatuic acid, 253, *255, 256*
  diasteromer of, *256*
  methyl ester, 253, *256*
  stereochemistry of, 256
P-Toluenesulfonylhydrazine, reaction with iodoepoxide, 21
p-Toluenesulfonylhydrazones, with *n*-butyl lithium, 433
  with copper-sodium hydride, 434
  photochemical decomposition, 393, 434
  pyrolysis, 434
Torgov reaction, mechanism of, 702

pyrollidenes, 703
Torsional strain in ring junctions, 659
Tosylates, cyclization of, 503
  Grob fragmentation, 478
  intramolecular alkylation, 500, 632
  internal cyclization, 528
  pinacol rearrangement, 417
  replacement with cyanide, 407
  replacement with iodide, 449
  solvolytic rearrangement, 404, 406, 407
  reduction, 381, 399
Tosylation, 446, 528
  of secondary alcohols, 503, 522
  selective of diols, 417, 462, 478
Transesterification, with ethanol, 440
Triallyl orthoformate, with enones, 628
Tricyclic enones, solvolytic rearrangement
  of, 467
Tricyclic formates, pyrolysis of, 511
Trienes, equilibrium in acid, 595
  selective hydroboration, 442
Triethylammonium benzoate, in aldolization, 695
Triols, oxidation, 530
  selective acetylation, 536
Trimethylsilylethers, in preparative gas
  liquid chromatography, 568
Tristriphenyl phosphinechlororhodium, in
  the oxidation of menthene, 111
  in selective hydrogenation
Triton B, in alkylations, 698, 699
Trityl chloride, as alcoholic protecting
  group, 221
Troponoid system, *276, 277*
Turpentine, hydration of, 102

δ,ε-Unsaturated acids, cyclization, 543
α,β-Unsaturated acid chlorides, Grignard reaction on, 112
α,β-Unsaturated aldehydes, *28, 29*
  addition to vinyl ether, 75
  Grignard addition with prenylhalides, 42
  methyl lithium addition to, 468
  oxidation with, selenium dioxide, 139
  silver oxide, 74, 300
  reaction with p-toluenesulfonylhydrazine,
    373
  reduction of, 80, 112, 125, 163, 389
  reduction, selective, 85
  Wittig reaction on, 11, 43
α,-β,γ,δ-Unsaturated aldehydes, *139*
  Grignard addition, 30
  reduction of, 40
α,β-Unsaturated cyano esters, reduction of,
  420
α,β-Unsaturated epoxides, reduction of, 13
α,β-Unsaturated esters, *11, 204, 221, 262*
  hydrogenation, 372
  hydrolysis, 57
  methyllithium addition to, 43
  photocyclization of, 534
  reaction with allyl sulfone, 54
  rearrangement through enolates, 45
  reduction, 79, 204, 207, 208, 431, 435,
    440, 453, 473

separation from β,γ-by chromatography,
  588
α,β-γ,δ-Unsaturated esters, hydrogenation,
  509
α,β-Unsaturated ketones, *306, 350, 605*
  addition to vinyl magnesium bromide, 32
  alkylation, 575, 582, 599, 709
  bromination, 141, 317, 320
  cyclopropane ring formation from, 158
  deconjugation of, 293, 294
  Diels-Alder addition to 1,3-diene, 151
  Diels-Alder reaction with, 89, 335
  dienes from, 300
  dienones from, 322, 325
  epoxidation of, 95, 107, 137
  Grignard reaction, 95
  Grignard reaction, 1,4-addition, 86, 114
  hydroboration, 136
  hydrocyanation, 295
  hydrogenation, 69, 295, 661
  ketalization of, 112, 135, 290
  metal-ammonia reduction, 269, 296, 298,
    334
  methyl lithium addition to, 59
  oxidation of, 135
  ozonolysis of, 56
  photo addition to ethylene, 58
  pyrolysis of, 121
  reduction, 26, 113, 118, 119, 146, 156
  sultones from, 122
  synthesis by Rupe rearrangement, 350
  thioketalization, 289, 292
  Wittig reaction, 90
β,γ-Unsaturated ketones, alkylation, 620,
  709
  reduction, 575, 582
Unsaturated lactones, in strong base, 584
α,β-Unsaturated oxo acids, reductive alkylation, 660
α,β-Unsaturated nitriles, conjugate addition
  to, 543

Valencia orange oil, 129, 131, 136
Verbena officinalis, 67
Verbenone, 137
  conversion to *cis*-verbenol, 156
  photochemical oxidation of, 156
  from pinenes, 156
Vilsmeiers reagent, with enol ethers, 406
Vinyl acetate, Diels-Alder addition to, 1,3-
  dienes, 150
Vinyl carbinols, anionotropic rearrangement,
  698
  oxidation, 697
  stabilization as isothiuronium acetates,
    702
Vinyl chlorides, hydrolysis of, 709
Vinyl ethers, addition to, acetals, 28, 29, 52
  α,β-unsaturated aldehyde, 75
  Claisen rearrangement, 565
  pyrolytic rearrangement, 393
  reaction with allyl alcohol, 112
  rearrangement, 361
Vinyl ketones, in Robinson annelations,
  695

Vinyl magnesium bromide, addition to
    ketone, 92
  1,4-addition to α,β-unsaturated ketone, 32

Wagner-Meerwein rearrangement, 403, 492
  in camphene, 153
  in α-pinene, 153
Wharton reaction, 21, 165
Wittig reaction, 503, 586, 616, 649
  epimerization during, 336
  methylenation, 366, 476
    selective, 365
  modified, 207, 208
  of 3-acyl furan, 164
  of aldehydes, 417, 440, 449, 491, 634
  of aldehydo ketal, 32
  of aromatic ketones, 104
  of cyclopentenones, 496
  of α-cyclopropyl ketones, 147, 148
  of diene aldehyde, 11
  of enones, 520
  of geranyl acetate, 562
  of hindered ketones, 451
  of β-keto esters, 45
  of ketones, 59, 76, 89, 104, 418, 453, 514
  of methyl heptenones, 435
  of sodium salt of keto acids, 516
  of 3-substituted furans, 164
  of tricycloekasantal, 489
  of α,β-unsaturated aldehydes, 11, 43
  of γ,δ-unsaturated aldehydes, 206, 231,
    272
  of α,β-unsaturated ketones, 90
  stereochemical control, 562
  to give terminal olefins, 392, 404

  with acetylmethylenetriphenylphospho-
    rane, 463
  with (carbethoxyethylidine) triphenyl-
    phosphorane, 489, 491
  with ethylidinetriphenylphosphorane,
    404, 490, 520
  with isopropylidenetriphenylphosphorane,
    566
  with methylenetriphenylphosphorane,
    478, 514, 532
  with methoxymethylenetriphenylphospho-
    rane, 409, 465
  with triethylphosphonoacetate, 534
Wolf-Kishner reduction, 312, 360, 374, 569,
    575, 590, 592, 593, 609, 620, 625, 689
  forcing conditions, 605
  of aldehydes, 365
  of acyloins, 383
  of enones, 381
  of imines, 615
  of ketones, selective, 690
  of todomatuic acid, 254
Wurtz coupling, in the synthesis of 1,5-
    dienes, 43
  of lavandulyl bromide, 45
  of prenyl halides, 42

Yomogi alcohol, 34, 42
  conversion to santolinatriene, 38
  epoxides, 37, 43
  formation from chrysanthemic acid, 35
  oxidation of, 42

Zanthoxlium rhetsa, 134
Zieria Smithii, 144